LINEAR STATISTICAL MODELS

LINEAR STATISTICAL MODELS
AN APPLIED APPROACH

SECOND EDITION

Bruce L. Bowerman
Richard T. O'Connell

Miami University, Ohio

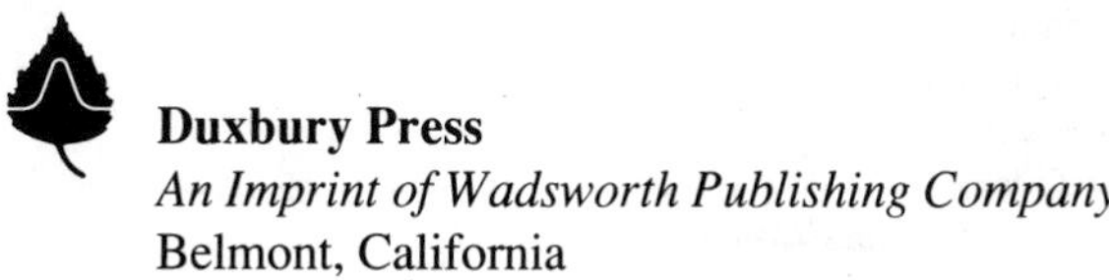

Duxbury Press
An Imprint of Wadsworth Publishing Company
Belmont, California

Duxbury Press
An Imprint of Wadsworth Publishing Company
A division of Wadsworth, Inc.

To our wives and children:
Drena and Jean
Michael, Christopher, Bradley, Asa, and Nicole

Library of Congress Cataloging-in-Publication Data
Bowerman, Bruce L.
Linear statistical models: an applied approach/by Bruce L. Bowerman, Richard T. O'Connell.—2nd ed.
p. cm.
Includes bibliographical references.

1. Regression analysis. 2. Linear models (Statistics)
I. O'Connell, Richard T. II. Title.
QA278.2.B687 1990
519.5′36—dc20 89-16367
CIP

Printed in the United States of America
3 4 5 6 7 8 9 10—97

Sponsoring editor: *Michael Payne*
Production editor: *Pamela Rockwell*
Production services: Technical Texts/*Sylvia Dovner*
Interior design: *Catherine Johnson*
Cover design: *Catherine Johnson*
Typesetting: *The Alden Press*
Printing and binding: *Arcata Graphics/Halliday*

PREFACE

Linear Statistical Models: An Applied Approach, Second Edition, is designed for applied courses in regression analysis, analysis of variance, and experimental design. The book will appeal to advanced undergraduates and graduate students in the fields of business, engineering, and the sciences (including mathematics, statistics, operations research, and computer science). The book is also written for the practitioner who wants to learn the subject through self-study.

FOCUS

The focus of *Linear Statistical Models* is **model building**, **data analysis**, and **interpretation**. Stress is on conceptual, concrete, and applied aspects rather than the abstract and theoretical. The text is written for accessibility to students and practitioners with a limited statistical background, yet it does not sacrifice the depth and breadth of topic coverage. Topics are covered in enough detail to give the reader a thorough understanding of the concepts involved, along with the ability to carry out complete, correct analysis of real data. We have endeavored to provide clear and concise explanations. Wherever possible, we supply intuitive motivation as well as technical details, which we feel helps build an understanding of the conceptual ideas behind linear models.

LEVEL

While the book is intended for those who have had college-level algebra and a course in basic statistics, Chapters 2 and 3 provide a comprehensive review of many of the most important basic statistical methods. Chapter 7 presents the needed background in matrix algebra.

ORGANIZATION

The text can be divided into several segments. Chapter 1 is an introduction to regression analysis, while Chapters 2 and 3 give the previously mentioned review of basic statistics. Chapters 4 through 6 give a thorough discussion of simple linear regression. This includes a careful explanation of the simple linear regression model, as well as the presentation of inference procedures, the regression assumptions, and residual analysis. Chapters 8 and 9 cover the basic concepts of multiple regression—the multiple regression model, inference, assumptions and residual analysis, higher-order terms, and interaction. Chapters 10 through 13 discuss advanced topics in regression analysis. These include multicollinearity, diagnostics for identifying outlying and influential observations, various model-building techniques, dummy variables, inferences for linear combinations of regression parameters, simultaneous inference, and remedies for violations of the regression assumptions. Finally, Chapters 14 through 17 deal with the analysis of designed experiments. These chapters cover one- and two-factor analysis, randomized block designs, and Latin squares. Both the ANOVA and regression approaches to these problems are fully explained. (See the section entitled "To the Instructor" for suggestions on using the text for different courses.)

EXAMPLES AND EXERCISES

To give an overall picture of how linear models are used, the book contains a wide variety of realistic examples. Numerous (ranging from simple to quite complex) real data sets are used in both the text material and end-of-chapter exercises. Furthermore, heavy emphasis is placed on fully analyzing data, from tentative model specification to using the final model for problem solving. Appendix F describes three large macroeconomic data bases that are available on diskette and can be used to provide the student with a large number of data analysis exercises.

COMPUTER IMPLEMENTATION

Implementation of most of the techniques presented in this book requires use of a computer. Therefore, we make extensive use of SAS System outputs. However, it is important to point out that *Linear Statistical Models* can be easily used even if an instructor does not have access to SAS. The SAS outputs in the book are used as a *convenient* way to present regression results, but knowledge of or exposure to SAS programming is not a prerequisite for use of this text. The exercises contain a multitude of SAS outputs that can be fully interpreted without having access to SAS, or a computer for that matter. Of course, many instructors will want to make SAS programming an integral part of the linear models course. Therefore, this book contains optional sections that show how to write SAS programs that will implement virtually every technique presented in the book.

In addition to the usual SAS programs, we include information on how to use PROC AUTOREG and PROC ARIMA to analyze autocorrelated error structures. We also show how to use PROC GLM to analyze data obtained from designed experiments. In addition, optional exercises allow the students to become proficient in using SAS.

While we believe it is important to integrate the use of a computer into regression and linear models courses, we do not feel that a "black box approach" should be taken. Therefore, we show how regression calculations are done using matrix algebra. Also, for those instructors who wish to derive regression results, the most crucial matrix algebra proofs are given in Appendices B, C, and D.

SPECIAL FEATURES

Linear Statistical Models places special emphasis on several topics that are slighted in other texts:

- Geometric interpretations are used to motivate regression models.
- Data plots are used to tentatively specify reasonable models.
- The meanings of terms such as "level of confidence" and "prediction interval" are explained in the context of repeated sampling.
- Residual analysis is carefully integrated into the text and exercises to reinforce the importance of checking the aptness of regression models.
- Model building and the motivation behind an interpretation of interaction terms are recurring themes throughout the book.
- In addition to the usual hypothesis tests, emphasis is placed on confidence interval estimation and the use of prediction intervals.
- Diagnostics for identifying outlying and influential observations are given thorough treatment in Chapter 10.
- Weighted least squares and the use of models with autocorrelated error structures receive careful attention in Chapter 13.
- Both the ANOVA (sums of squares) approach and the linear models (regression) approach to analysis of data from designed experiments are stressed. These approaches can be taught independently or in an integrated fashion.

- Balanced data (equal sample sizes), unbalanced data (unequal sample sizes), and "missing data" experimental design situations are carefully discussed.
- Confidence intervals for comparing treatment means and simultaneous inferences using Tukey, Scheffé, and Bonferroni techniques are given thorough coverage.

Acknowledgments

A great many people have provided invaluable assistance in the writing of this text. First, we would like to thank David Dickey of North Carolina State University for supplying the examples in Chapter 13. We also wish to thank the many reviewers of this book for their advice and constructive criticism. We would especially like to thank the following people for their extremely useful comments and suggestions:

Herbert Moskowitz Purdue University	Stergios Fotopoulos Washington State University
Dean Wichern Texas A&M University	James Holstein University of Missouri—Columbia
Burt Holland Temple University	Mark McNulty Kansas State University
Wayne Winston Indiana University	James Daly California Polytechnic Institute

We also thank John Skillings of Miami University for his helpful discussions of experimental design. In addition, we want to thank one of our students, Patricia Amend, for running much of the computer output that appears in the book. Of course, we also want to thank all of the fine people at PWS-KENT. In particular, we thank our editor Michael Payne for his patience and continuing commitment to making this book a success. We also thank Sylvia Dovner and the staff at Technical Texts for their excellent work. Finally, we thank our wives and children and ask their forgiveness for our sometimes misplaced priorities during the writing process.

Bruce L. Bowerman
Richard T. O'Connell

TO THE INSTRUCTOR

Linear Statistical Models can be used as the primary text for a variety of courses.

1. A one-quarter or one-semester course in regression analysis might be based on:
   - Chapter 1 (introduction)
   - Chapters 2 and 3 (review of basic statistics)
   - Chapters 4, 5, and 6 (simple linear regression)
   - Chapter 7 (matrix algebra)
   - Chapters 8 and 9 (multiple regression)
   - Chapter 10 (multicollinearity and other selected topics)
   - Chapter 11 (model building)
   - Chapters 12 and 13 (selected topics)
2. A two-quarter or two-semester course in linear statistical models might be based on:
   - Chapter 1 (introduction)
   - Chapters 2 and 3 (review of basic statistics)
   - Chapters 4, 5, and 6 (simple linear regression)

- Chapter 7 (matrix algebra)
- Chapters 8 and 9 (multiple regression)
- Chapter 10 (multicollinearity and other selected topics)
- Chapter 11 (model building)
- Chapters 12 and 13 (selected topics)
- Chapter 14 (one-factor analysis)
- Chapters 15 and 16 (two-factor analysis)
- Chapter 17 (selected topics)

3. A one-quarter or one-semester course in analysis of variance and experimental design might be based on (here we assume prior background in basic statistics and simple linear regression):
   - Chapters 7, 8, and 9 (matrix algebra and multiple regression)
   - Chapter 12 (dummy variables and advanced inference)
   - Chapter 14 (one-factor analysis)
   - Chapters 15 and 16 (two-factor analysis)
   - Chapter 17 (randomized blocks and latin squares)
4. A one-quarter or one-semester graduate level survey course in statistical methods might be based on:
   - Chapter 1 (introduction)
   - Review of Chapters 2, 3, 4, 5, and 6 (basic statistics and simple linear regression)
   - Chapters 7, 8, and 9 (matrix algebra and multiple regression)
   - Chapter 10 (multicollinearity and other selected topics)
   - Chapter 11 (model building)
   - Chapters 12 and 13 (selected topics)
   - Sections 14.1–14.4 (one-way ANOVA)
   - Sections 15.1 and 15.2 (two-way ANOVA)
5. A one-quarter or one-semester undergraduate or graduate level basic statistics course might be based on:
   - Chapter 1 (introduction)
   - Chapters 2 and 3 (populations, samples, probability distributions, and basic statistical inference)
   - Chapters 4, 5, and 6 (simple linear regression)
   - Chapters 7, 8, and 9 (matrix algebra and multiple regression)
   - Chapter 10 (multicollinearity and other selected topics)
   - Chapter 11 (model building)
   - Section 12.1 (introduction to dummy variables)
   - Sections 14.1–14.4 (one-way ANOVA)
   - Sections 15.1 and 15.2 (two-way ANOVA)

CONTENTS

CHAPTER 3

BASIC STATISTICAL INFERENCE 44

CHAPTER 4

THE SIMPLE LINEAR REGRESSION MODEL 106

CHAPTER 5

INFERENCE IN SIMPLE LINEAR REGRESSION 140

CHAPTER 6

THE ASSUMPTIONS BEHIND REGRESSION ANALYSIS 215

CHAPTER 7

MATRIX ALGEBRA 276

CHAPTER 8

MULTIPLE REGRESSION: I 295

CHAPTER 12

DUMMY VARIABLES AND ADVANCED STATISTICAL INFERENCES 554

CHAPTER 13

REMEDIES FOR VIOLATIONS OF THE REGRESSION ASSUMPTIONS 631

CHAPTER 14

ONE-FACTOR ANALYSIS 729

CHAPTER 15

TWO-FACTOR ANALYSIS: I 800

CHAPTER 16

TWO-FACTOR ANALYSIS: II 854

CHAPTER 17

THE RANDOMIZED BLOCK AND LATIN SQUARE DESIGNS 923

CHAPTER 1

INTRODUCTION TO REGRESSION ANALYSIS

This chapter introduces regression analysis. We begin in Section 1.1, which explains the general nature of *statistical studies*. In particular, we discuss using a *sample* to draw *inferences* about a *population*. We also describe the idea of point estimation and emphasize that a statistical study assesses the reliability of the point estimate. Section 1.2 explains the basic ideas involved in a *regression analysis*. We emphasize that the goal of a regression study is to build an equation that allows us to describe, predict, and control a *dependent variable* on the basis of one or more *independent variables*. Section 1.3 presents several of the many useful applications of regression analysis. Section 1.4 explains and illustrates several types of data employed in regression studies—*observational data*, *experimental data*, *time series data*, and *cross-sectional data*. Section 1.5 briefly discusses *cause-and-effect relationships* in the context of a regression study. Finally, Section 1.6 briefly previews subsequent chapters in this book.

1.1

STATISTICAL STUDIES

This book is about *regression analysis*, which is one of the most important techniques in the field of *statistics*. In this chapter we will briefly introduce the basic idea of regression analysis. We will also give a few examples of how regression analysis can be used to make decisions.

Before we introduce regression analysis, we need to explain how a statistical study is carried out. In a statistical study we wish to describe a *population*. A population is simply a large set of measurements. Because many populations of interest are very large, it is usually impossible to examine the entire population. Therefore to describe the population, we must select a *sample* of measurements from the population. The science of statistics then involves making statements about the population based on the information contained in the sample. This is called making *statistical inferences*. Because we have not examined the entire population, the inferences that we make may be somewhat in error. Therefore in any statistical study we must assess how reliable our inferences are. This allows us to make informed decisions based on our inferences.

As an example, suppose that a government agency wishes to decide whether or not the mean price of a "market basket" of 12 grocery items exceeds \$20. Here the population consists of all "market basket prices" in the United States. The only way to find the "*population mean* price" would be to find and average the prices of this market basket at each and every grocery store in the country. Because of cost and time considerations, it is not possible to do this. Therefore the agency randomly selects a sample of, say, 100 grocery stores around the country. The price of the market basket is then ascertained at each store in the sample. The obvious best guess, or *point estimate*, of the population mean is the average of the 100 market basket prices obtained at the grocery stores in the sample. Suppose, for instance, that this *sample mean* is \$23. This result provides some evidence that the population mean market basket price is higher than \$20. However, since the point estimate of \$23 is based on a sample, we cannot expect it to be exactly equal to the population mean. Thus it is necessary to assess the reliability of the point estimate. We do this by finding an *error bound* on the estimate. The error bound is determined so that we can be very confident that the population mean and the point estimate differ by no more than the bound. This bound depends on the size of the sample that has been taken, the amount of variation in the sample market basket prices, and the *level of confidence* we require. For example, if we obtain a "95% error bound" of \$2.10, we are 95% confident that the population mean market basket price and the point estimate \$23 differ by no more than \$2.10. In other words, we are 95% confident that the population mean is contained in the interval

$$[\$23 - \$2.10, \$23 + \$2.10] = [\$20.90, \$25.10]$$

Such an interval is called a *confidence interval.* It is a convenient way to express the reliability of the point estimate.

In our example, we used statistics to make an inference about a population based on a sample selected from the population. In addition, the statistics have enabled us to assess how reliable the inference is. This information allows us to make intelligent decisions. For example, if the government agency is 95% confident that the population mean is in the interval [$20.90, $25.10], then it has very strong evidence that the mean market basket price exceeds $20. It can then make policy decisions accordingly.

The kind of statistical study described above is one of the most basic of such studies. We will explain how to compute "95% error bounds" and "confidence intervals" for a population mean in Chapter 3. That chapter also reviews several other *basic statistical inference procedures.* Besides making statistical inferences, the science of statistics also helps us in deciding how to select samples and in deciding how large our samples must be. These issues are also discussed in Chapter 3 as well as other chapters throughout this book.

1.2 REGRESSION ANALYSIS

Regression analysis is a statistical methodology that is used to relate variables. Here we wish to relate a variable of interest, which is called the *dependent variable* or *response variable*, to one or more *predictor*, or *independent*, *variables.* The dependent variable is denoted by y, and the independent variables are denoted by $x_1, x_2, \ldots, x_p$. The objective is to build a *regression model* or *prediction equation* —an equation relating y to $x_1, x_2, \ldots, x_p$. We use the model to *describe*, *predict*, and *control* y on the basis of the independent variables. When we predict y for a particular set of values of $x_1, x_2, \ldots, x_p$, we will wish to place a bound on the *error of prediction*. The goal is to build a regression model that produces an error bound that will be small enough to meet our needs.

A regression model can employ *quantitative independent variables* and/or *qualitative independent variables.* A *quantitative independent variable* assumes numerical values corresponding to points on the real line. A *qualitative independent variable* is nonnumerical. The levels of such a variable are defined by describing them. As an example, suppose that we wish to build a regression model relating the dependent variable

y = demand for a consumer product

to the independent variables

x_1 = the price of the product,

x_2 = the average industry price of competitors' similar products,

x_3 = advertising expenditures made to promote the product, and

x_4 = the type of advertising campaign (television, radio, print media, etc.) used to promote the product.

Here x_1, x_2, and x_3 are quantitative independent variables. In contrast, x_4 is a qualitative independent variable, since we would define the levels of x_4 by describing the different advertising campaigns. After constructing an appropriate regression model relating y to x_1, x_2, x_3, and x_4, we would use the model

1. To *describe* the relationships between y and x_1, x_2, x_3, and x_4. For instance, we might wish to describe the effect that increasing advertising expenditure has on the demand for the product. We might also wish to determine whether this effect depends upon the price of the product.
2. To *predict* future demands for the product on the basis of future values of x_1, x_2, x_3, and x_4.
3. To *control* future demands for the product by controlling the price of the product, advertising expenditures, and the types of advertising campaigns used.

Note that we cannot control the price of competitors' products, nor can we control competitors' advertising expenditures or other factors that affect demand. Therefore we cannot perfectly control or predict future demands.

1.3 SOME EXAMPLES OF REGRESSION ANALYSIS

Regression analysis has many useful applications. We now illustrate a few of them.

Example 1.1 American Manufacturing Company produces a line of commercial dishwashers and sells this product to restaurants, hotels, hospitals, and other such institutions. The company wished to predict dollar sales volume (y) for the dishwasher line on the basis of advertising expenditure (x). To do this, the company selected ten sales regions of equal sales potential. Then advertising expenditure was varied from region to region during the third quarter of 1988, and dollar sales volume was observed for each sales region. The SAS* output of a plot of y (in units of \$10,000) versus x (also in units of \$10,000) is shown in Figure 1.1. The plot suggests that there is a straight-line relationship between y and x. We will see (in Chapter 4) that this implies that a *simple linear regression model* may appropriately relate y to x. Such a model is often used to relate a dependent variable to a single independent variable. Using methods to be presented in Chapter 4, we can use the data in

*SAS is a computer package that can be used to analyze data. Among other things (such as data storage and retrieval, report writing, and file handling), SAS can be used to perform a wide variety of statistical analyses, including regression analysis.

FIGURE 1.1 **Plot of y (dollar sales volume) versus x (advertising expenditure) for the American Manufacturing Company**

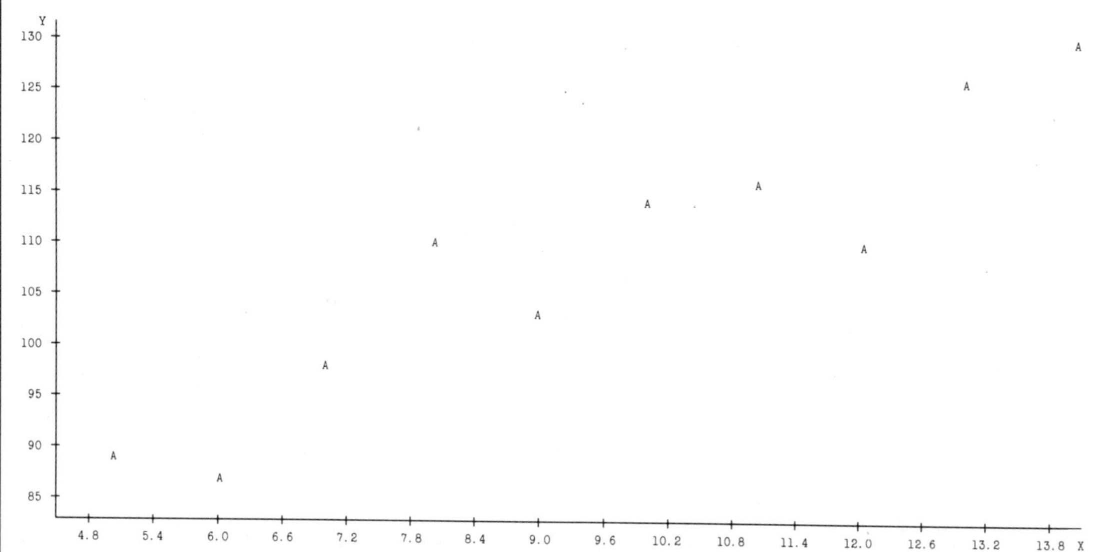

Figure 1.1 to obtain a prediction equation relating y to x. If we let $\hat{y}$ denote predicted sales, the prediction equation is

$$\hat{y} = 66.2121 + 4.4303x$$

For instance, suppose that the company wishes to predict dollar sales when advertising expenditures are \$100,000. The needed prediction is

$$\begin{aligned}\hat{y} &= 66.2121 + 4.4303(10)\\ &= 110.5 \quad \text{(or \$1,105,000)}\end{aligned}$$

Furthermore, we can use regression (see Chapter 5) to place an error bound on this prediction. We will find that we can be 95% confident that dollar sales volume will be no less than \$977,000 and no more than \$1,233,000. The company can use information such as this to study its cash flow situation.

Example 1.2 American Manufacturing Company owns a nine-building industrial complex. The company wishes to predict the weekly amount of fuel (y) that will be used to heat the complex in future weeks. It will base its prediction on two independent variables, x_1 (the average hourly temperature during the week) and x_2 (a variable called the "weekly chill index"). Here the chill index expresses the combined effects on fuel consumption of all major weather-related factors other than temperature. The company observes these variables for eight consecutive weeks. The SAS

FIGURE 1.2 Plot of y (fuel consumption) versus x_1 (average hourly temperature) for the American Manufacturing Company

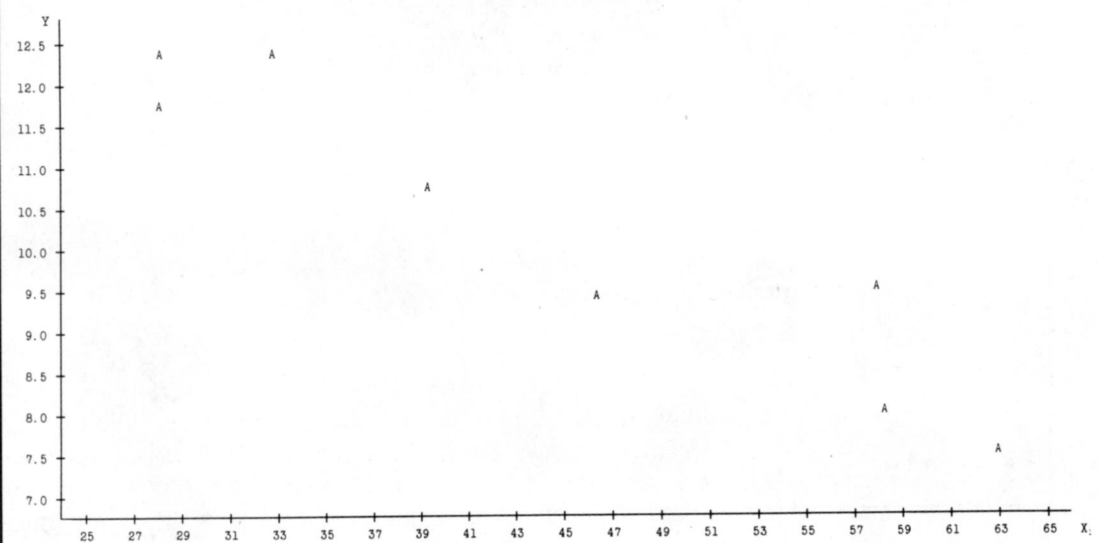

FIGURE 1.3 Plot of y (fuel consumption) versus x_2 (chill index) for the American Manufacturing Company

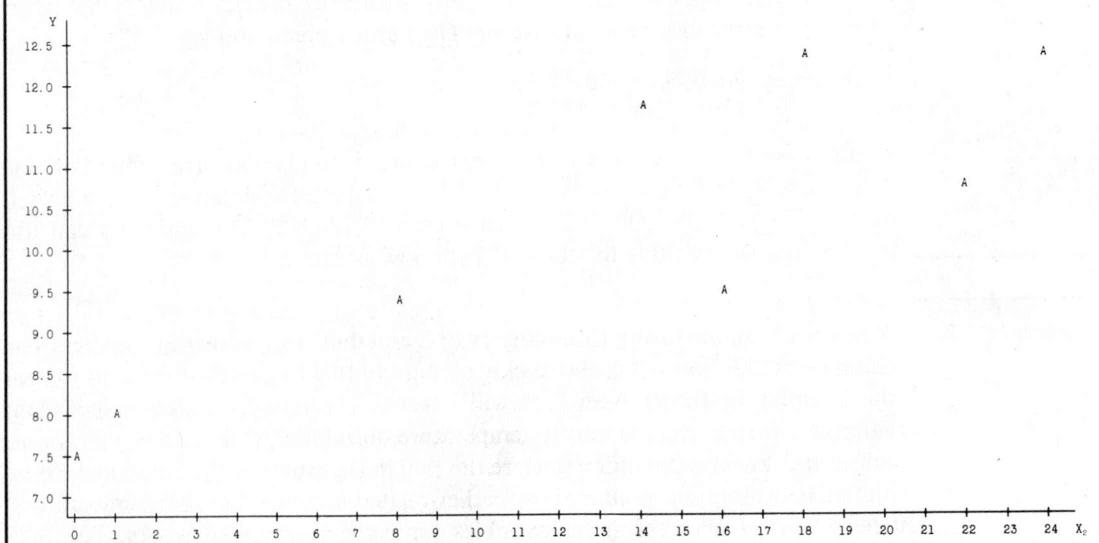

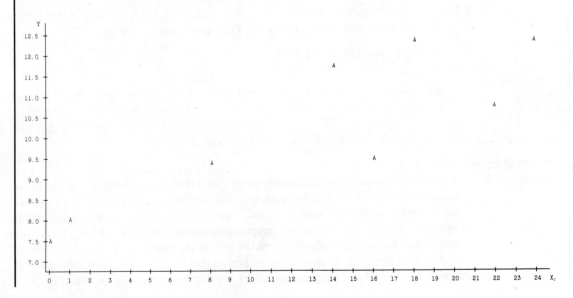

output of a plot of y (in tons of coal) versus x_1 (in °F) is shown in Figure 1.2, and the SAS output of a plot of y versus x_2 is shown in Figure 1.3. Both of these plots appear to be linear. In Chapter 7 we will use this data to construct a *multiple regression model* relating y to x_1 and x_2. Such a model is employed to relate a dependent variable to more than one independent variable. Using methods to be presented in Chapter 7, we can use the data in Figures 1.2 and 1.3 to obtain a prediction equation relating y to x_1 and x_2. If we let $\hat{y}$ denote predicted weekly fuel consumption, the prediction equation is

$$\hat{y} = 13.109 - .09x_1 + .0825x_2$$

For instance, suppose that the company wishes to predict weekly fuel consumption when the average hourly temperature during the week is 40°F and the weekly chill index is 10. The needed prediction is

$$\begin{aligned}\hat{y} &= 13.109 - .09(40) + .0825(10) \\ &= 10.333 \quad \text{(tons of coal)}\end{aligned}$$

Moreover, we can use regression (see Chapter 8) to place an error bound on this prediction. We will find that we can be 95% confident that weekly fuel consumption will be no less than 9.293 tons and no more than 11.373 tons. The company can use this information in planning its coal acquisitions for future weeks.

Example 1.3

An investor wishes to decide whether she should purchase, hold, or sell a particular stock. To do this, she will build a regression model relating the dependent variable

$$y = \text{the rate of return on the particular stock}$$

to the independent variable

$$x = \text{the rate of return on the overall stock market}$$

Using historical data, the investor can employ regression analysis to find a prediction equation that enables her to predict y on the basis of x. She can use this equation to see whether the mean rate of return on the particular stock changes more quickly or more slowly than the rate of return on the overall stock market. Suppose that the rate of return on the particular stock changes more quickly than the rate of return on the overall stock market (and the changes are in the same direction). Then the stock should be purchased if the investor expects the market to rise and sold if she expects the market to fall. On the other hand, suppose that the rate of return on the particular stock changes more slowly than the rate of return on the overall stock market (and the changes are in the same direction). Then the stock might be held if the investor expects the market to fall and sold off in favor of other stocks if the investor expects the overall market to rise.

Example 1.4 A farming cooperative wishes to predict its quarterly propane gas bills. It is believed that these bills are related to trends in natural gas prices, trends in gas consumption, and the season of the year. The cooperative can use regression analysis to find a prediction equation relating gas bills to these trend and seasonal effects. The analysis can provide point predictions of gas bills and can place bounds on the errors of prediction. Thus we can obtain reasonable estimates of the maximum and minimum gas bills that might be received in future quarters.

Example 1.5 A chemical product is produced by using a batch process. The viscosity of the product is felt to depend on the type of catalyst used in the reaction (A, B, or C) and on the chemical composition (D, E, F, or G) of a raw material used in the reaction. Five batches are produced, using each possible combination of catalyst type and raw material composition. Regression analysis can be used

1. To determine whether the catalyst type has a significant effect on the viscosity of the product. In addition, if catalyst type has a significant effect on viscosity, regression can be used to estimate the effects of changing catalyst types.
2. To determine whether the raw material composition has a significant effect on viscosity. If this composition has a significant effect on viscosity, regression can be used to estimate the effects of changing chemical compositions.
3. To determine the combination of catalyst type and chemical composition that will produce the most desirable product viscosity.
4. To predict the viscosity that will be obtained by a given combination of catalyst type and chemical composition and place an error bound on this prediction.

1.4 DATA EMPLOYED IN REGRESSION STUDIES

The data employed in a regression analysis can be *observational* or *experimental*. In addition, the data can be *time series data* or can be *cross-sectional data*. We say that data is *observational* when the values of the independent variables $x_1, x_2, \ldots, x_p$ cannot be controlled but are measured without error. If the values of the independent variables are set before the values of the dependent variable are observed, we say that the data is *experimental*. Often, such data are obtained by using a *designed experiment*. When data is observed in time order, we call it *time series data*. Time series data often consist of yearly, quarterly, or monthly observations, but any other time period may be used. When data is observed at one point in time, the data is called *cross-sectional*.

In the sales volume problem of Example 1.1, American Manufacturing Company was able to set the value of advertising expenditure in each of the ten sales regions. Therefore the data in this study is experimental. Furthermore, since all of the sales volumes were observed in the third quarter of 1988, this data is cross-sectional. In contrast, in the fuel consumption problem of Example 1.2, American Manufacturing Company cannot control x_1 (average hourly temperature) and x_2 (the chill index). Therefore the fuel consumption data is observational. Moreover, since this data was observed over eight consecutive weeks, it is time series data.

In general, the amount of information obtained in a regression study is affected by the *appropriateness of the regression model*, the *amount of data* collected, and the *values of the independent variables* that are employed. Therefore if the values of $x_1, x_2, \ldots, x_p$ can be controlled, it is advantageous to employ a designed experiment. By employing such an experiment, we can increase the amount of information obtained by the study.

1.5 CAUSE-AND-EFFECT RELATIONSHIPS

In the study of regression analysis, we sometimes speak of the effect of an independent variable upon a dependent variable. However, we *cannot prove that a change in an independent variable causes a change in the dependent variable*. Rather, regression can be used only to establish that the two variables move together and that the independent variable contributes information for predicting the dependent variable. For instance, regression analysis might be used to establish that as liquor sales have increased over the years, college professors' salaries have also increased. However, this does not prove that increases in liquor sales cause increases in college professors' salaries. Rather, both variables are influenced by a third variable—long-run growth in the national economy.

1.6 UPCOMING CHAPTERS

Most readers of this book will have already completed a basic statistics course. However, for readers who have not had such a course (or who need a refresher), the needed background in basic statistics is reviewed in optional Chapters 2 and 3. Readers who do not wish to review this material may proceed to Chapter 4, in which we begin our presentation of regression analysis with the simple linear regression model. After completion of our discussion of simple regression in Chapter 6, Chapters 7 through 13 cover multiple regression models in detail. Included in these chapters are presentations of the multiple regression model, basic and advanced inference, model assumptions, model building, and remedies for violations of the model assumptions. Then Chapters 14 through 17 cover basic analysis of designed experiments. Included are discussions of one-way and two-way analysis of variance, randomized block designs, nested designs, and Latin squares.

CHAPTER 2

POPULATIONS, SAMPLES, AND PROBABILITY DISTRIBUTIONS

In this chapter we discuss populations, samples, and probability distributions. We begin in Section 2.1 by considering *populations*. We see that populations are described by *population parameters*. We define four such parameters—the population *mean*, *range*, *variance*, and *standard deviation*. Next, in Section 2.2 we present some elementary concepts concerning probability. In Section 2.3 we find that since the true values of population parameters are unknown, we must compute estimates of these parameters. To compute such estimates, we must randomly select a sample from the population of interest. Section 2.3 describes how this is done. We also show how sample statistics, which are descriptive measures of samples, can be used as point estimates of population parameters. To compute probabilities, we use probability distributions. We introduce *continuous probability distributions* in Section 2.4. In Sections 2.5 and 2.7 we discuss several useful continuous probability distributions—the *normal*, *t*-, and *F-distributions*. Section 2.6 presents some additional ways to describe populations and samples using *medians*, *quartiles*, *box plots*, and *trimmed means*.

2.1

POPULATIONS

Frequently, we seek information about a collection of objects, or *elements*.

We define a *population* to be the entire collection of elements about which information is desired.

A population may contain a finite number of elements. In this case we call the population *finite*. We denote the size of a finite population by the symbol N. If there is no finite limit to the number of elements that could potentially exist in a population, we say that the population is infinite. We are often interested in studying properties of some numerical characteristic of the population elements. Here each element in the population possesses a particular value of the numerical characteristic under study. In this book a value of a numerical characteristic will always be a number on the real line.

Example 2.1 Table 2.1 lists a numerical characteristic that might be of interest for each of the indicated finite populations.

TABLE 2.1

Finite population of elements	Numerical characteristic of an element
The countries on the planet Earth	Number of inhabitants below poverty level in a country
The states in the U.S.A.	Number of family farms in a state
The cities with a population of at least 5000 in the state of Nebraska	Percentage of workers unemployed in a city
The inhabitants of Cleveland, Ohio	Yearly income (in dollars) of an inhabitant
The members of the United States Senate	Number of votes missed during 1990 by a Senator
The teams in the National Football League	Number of games won last year by a team

We now consider a population of numerical values. We define a *parameter* to be a description measure of the population. We define four parameters—the *population mean, range, variance,* and *standard deviation*—in the following box.

Some Population Parameters

1. The *population mean*, denoted μ, is the average of the values in the population.
2. The *population range*, denoted RNG, is the difference between the largest value and the smallest value in the population.

3. The *population variance*, denoted σ^2, is the average of the squared deviations of the values in the population from the population mean μ.
4. The *population standard deviation*, denoted σ, is the positive square root of the population variance.

Example 2.2

National Motors Company, Inc. is an automobile manufacturer that produces a model called the Hawk. The company wishes to study the gasoline mileage (measured in miles per gallon, or mpg) obtained by the Hawk when it is driven 50,000 miles under normal conditions.

Consider the population of the first $N = 3$ Hawks produced by National Motors. Assume that each automobile is driven under the above conditions and that the mileages obtained are 32.4, 30.6, and 31.8. For this population,

1. The mean is

$$\mu = \frac{32.4 + 30.6 + 31.8}{3} = \frac{94.8}{3} = 31.6$$

2. The range is

$$RNG = 32.4 - 30.6 = 1.8$$

3. The variance is

$$\sigma^2 = \frac{(32.4 - 31.6)^2 + (30.6 - 31.6)^2 + (31.8 - 31.6)^2}{3} = \frac{(.8)^2 + (-1)^2 + (.2)^2}{3} = \frac{.64 + 1 + .04}{3} = \frac{1.68}{3} = .56$$

4. The standard deviation is

$$\sigma = \sqrt{\sigma^2} = \sqrt{.56} = .7483$$

The population mean is a measure of the central tendency of the values in the population, while the population range, variance, and standard deviation are measures of the spread, or variation, of the values. To see that the population variance measures spread, first suppose that the values are spread far apart. Then, many values will be far away from the mean μ. This means that many of the squared deviations will be large. Thus the sum of the squared deviations will be large, and the average of the squared deviations—the population variance—will be relatively large. On the other hand, if the values are clustered close together, many values will be close to μ. This means that many of the squared deviations will be small. Therefore the average of the squared deviations—the population variance—will be small. We conclude that the greater the spread of the values, the larger is the population variance. Note that one might be tempted at this point to look for some "hidden meaning" behind (or practical interpretation of) the population variance. However, for now, all that should be understood is that (1) the population variance is as defined above and (2) the reason we are studying the population variance is that the population standard deviation will be found to be important in later sections.

In Example 2.2 we can calculate μ, RNG, σ^2, and σ because the population is small. In many situations the population is so large that the parameters cannot be computed. It would simply be too time consuming and/or expensive to do so. For instance, it would be impossible to calculate the mean μ of the population of mileages that would be obtained by the (theoretically) infinite population of all Hawks that could potentially be produced by the manufacturing process for this automobile. Suppose that, in general, the value of a population parameter is unknown. Then if information concerning the parameter is desired, our only recourse is to take a *sample*, or a subset of elements, from the population of interest. *Statistical inference* is the science of using the information contained in a sample to make a generalization about a population. One type of statistical inference is *statistical estimation.* This is the science of using the information contained in a sample (1) to find an estimate of an unknown population parameter and (2) to place a reasonable bound on how far the estimate might deviate from the unknown population parameter. That is, we wish to place a reasonable bound on how wrong the estimate might be. We continue our discussion of estimation in Section 2.3.

2.2 PROBABILITY

The concept of probability is employed in describing populations and in using sample information to make statistical inferences. To begin our discussion of probability, we first define an *experiment* to be any process of observation that has an uncertain outcome. An *event* is an experimental outcome that may or may not occur. The *probability* of an event is a number that measures the chance, or likelihood, that the event will occur when the experiment is performed. Let the symbol A denote an event that may or may not occur when an experiment, denoted by EXP, is performed. We denote the probability that the event A will occur by $P(A)$. Suppose that the experiment is performed n_{EXP} times, and the event A occurs

n_A of these n_{EXP} times. Then the proportion of the time event A has occurred is

$$\frac{n_A}{n_{EXP}}$$

To interpret $P(A)$, consider repeating the experiment a number of times approaching infinity, and consider the sequence of numbers obtained by calculating the ratio n_A/n_{EXP} after each repetition. The limit of this sequence is interpreted to be the probability of the event A. Stated mathematically,

$$P(A) = \lim_{n_{EXP}\to\infty} \frac{n_A}{n_{EXP}}$$

For example, the probability of a head appearing when we toss a fair coin is .5. This means that if we tossed the coin a number of times approaching infinity, the proportion of heads obtained would approach one-half. For instance, suppose we toss a fair coin a number of times and obtain the results in Table 2.2 (where H denotes a head and T denotes a tail). If we define the event A as "a head appears," Table 2.2 shows n_A (the number of repetitions on which a head has appeared), n_{EXP} (the number of times the coin has been tossed), and the ratio n_A/n_{EXP} for each repetition. Thus we obtain the following sequence of ratios n_A/n_{EXP}:

$$\frac{1}{1} \quad \frac{1}{2} \quad \frac{2}{3} \quad \frac{3}{4} \quad \frac{3}{5} \quad \frac{4}{6} \quad \frac{4}{7} \quad \frac{4}{8} \quad \frac{5}{9} \quad \frac{6}{10} \quad \cdots$$

When we say that the probability of a head is .5, we are saying that the limit of this sequence as n_{EXP} approaches infinity is .5.

Of course, in practice we cannot perform an experiment a number of times approaching infinity. So from a practical standpoint the probability of an event is roughly equal to the proportion of the time the event would occur if the experiment

TABLE 2.2 **Calculation of the ratio n_A/n_{EXP} for repeated coin tosses**

Repetition	*Outcome*	*Number of heads, n_A*	*Number of repetitions, n_{EXP}*	$\frac{n_A}{n_{EXP}}$
1	H	1	1	1/1
2	T	1	2	1/2
3	H	2	3	2/3
4	H	3	4	3/4
5	T	3	5	3/5
6	H	4	6	4/6
7	T	4	7	4/7
8	T	4	8	4/8
9	H	5	9	5/9
10	H	6	10	6/10
⋮	⋮	⋮	⋮	⋮

were performed a very large number of times. Consequently, one way to estimate the probability of an event is to perform the related experiment a great many times. Then we estimate the probability to be the proportion of times the event occurs during the repetitions of the experiment.

Since $P(A)$ is a long-run proportion, it follows that $P(A)$ is greater than or equal to zero and less than or equal to 1. If $P(A) = 0$, this means that the event A cannot occur. If $P(A) = 1$, this means that the event A is sure to occur. Sometimes we wish to estimate the probability of an event when we cannot perform the related experiment a very large number of times. Then we might estimate this probability using previous experience with similar situations and intuitive judgment. For example, a company president might estimate the probability of success for a one-time business venture to be .7. Here, on the basis of his knowledge of the success of previous similar ventures, the opinions of company personnel, and other pertinent information, he believes that there is a 70 percent chance the venture will be successful. If we can neither perform the related experiment a very large number of times nor use subjective judgment, many other methods are available to estimate the probability of an event. One such method—the use of continuous probability distributions—is discussed in Section 2.4.

2.3 RANDOM SAMPLES AND SAMPLE STATISTICS

The calculation of many population parameters—such as a mean or standard deviation—requires knowledge of all the values in the (finite or infinite) population. If we do not know all the values, we must randomly select a sample of n values from the population. Then the information contained in the sample can be used to make statistical inferences concerning the population parameter. For example, the parameter can be estimated.

We can randomly select a sample of n values by first randomly selecting a sample of n elements from the population of elements. This is done in such a way that on any particular selection *each element remaining in this population on that selection is given the same probability, or chance, of being selected.* Here we can randomly select the sample with or without replacement. If we *sample with replacement*, we place the element selected on a particular selection back into the population. Thus we give this element a chance to be selected on any succeeding selection. In such a case, all the elements in the population remain for each and every selection. If we *sample without replacement*, we do not place the element chosen on a particular selection back into the population. Thus we do not give this element a chance to be selected on any succeeding selection. In this case the elements remaining in the population for a particular selection are all the elements except for the elements that have previously been selected. *It is best to sample without replacement.* To see why, suppose that on a particular selection we randomly select an element from the population that is unrepresentative of the population. That is, the element is considerably different from the other elements in the population. This might throw off our estimates of population parameters. Then, if we sample with replacement, we might select the unrepresentative element again. This would result in our estimates of population parameters being thrown off even more drastically. On the other hand, if we sample without replacement, the unrepresentative element cannot be selected again. We assume in this book that all sampling is done *without* replacement.

If we have randomly selected a sample of n elements, the values of the numerical characteristic of interest possessed by these elements make up a randomly selected sample of n numerical values. For $i = 1, 2, \ldots, n$ we let y_i denote the value of the numerical characteristic under study possessed by the ith randomly selected element. Therefore the set

$$\{y_1, y_2, \ldots, y_n\}$$

denotes the randomly selected sample of n numerical values.

We now define a *sample statistic* to be a descriptive measure of the randomly selected sample of numerical values. Very often we use sample statistics as point estimates of population parameters.

A *point estimate* of a population parameter is a single number used as an estimate, or guess, of the population parameter.

Following are definitions of three sample statistics that are used as point estimates of the population mean, variance, and standard deviation.

Point Estimates of the Population Mean, Variance, and Standard Deviation

Suppose that the sample

$$\{y_1, y_2, \ldots, y_n\}$$

has been randomly selected from a finite or infinite population.

1. The *sample mean* is defined to be

$$\bar{y} = \frac{\sum_{i=1}^{n} y_i}{n}$$

and is a point estimate of the population mean μ.

2. The *sample variance* is defined to be

$$s^2 = \frac{\sum_{i=1}^{n} (y_i - \bar{y})^2}{n - 1}$$

and is a point estimate of the population variance σ^2.

3. The *sample standard deviation* is defined to be

$$s = \sqrt{s^2} = \sqrt{\frac{\sum_{i=1}^{n} (y_i - \bar{y})^2}{n - 1}}$$

and is a point estimate of the population standard deviation σ.

The rationale behind the use of these sample statistics as point estimates is that each sample statistic is exactly, or nearly, the sample counterpart of the corresponding population parameter. Here we define the *sample counterpart* of a population parameter to be the same function of the n values in the sample that the population parameter is of the values in the population. For example, the sample mean

$$\bar{y} = \frac{\sum_{i=1}^{n} y_i}{n}$$

is the sample counterpart of the population mean. However, although

$$\frac{\sum_{i=1}^{n} (y_i - \bar{y})^2}{n}$$

is the sample counterpart of the population variance, we use the sample variance (with $n - 1$ as divisor) as the point estimate of σ^2. This is because it can be shown that dividing by $n - 1$ rather than n makes s^2 a better estimate (in some senses) of σ^2.

Example 2.3 Suppose that we wish to estimate the mean μ and standard deviation σ of the (theoretically) infinite population of all Hawk mileages. To do this, we will randomly select a sample of $n = 5$ mileages from this population. We first randomly select a sample of five Hawks from a subpopulation of 1000 Hawks that National Motors has produced and that we assume is representative of the population of all Hawks. Then we will test drive each randomly selected automobile under the previously described conditions and record the mileage obtained.

Suppose that we obtain the following sample of mileages:

$$\{y_1, y_2, y_3, y_4, y_5\} = \{30.7, 31.8, 30.2, 32.0, 31.3\}$$

1. The sample mean is

$$\bar{y} = \frac{\sum_{i=1}^{5} y_i}{5} = \frac{30.7 + 31.8 + 30.2 + 32.0 + 31.3}{5}$$
$$= \frac{156}{5} = 31.2$$

and is the point estimate of the population mean μ.

2. The sample variance is

$$s^2 = \frac{\sum_{i=1}^{5} (y_i - \bar{y})^2}{5 - 1}$$
$$= \frac{(y_1 - \bar{y})^2 + (y_2 - \bar{y})^2 + (y_3 - \bar{y})^2 + (y_4 - \bar{y})^2 + (y_5 - \bar{y})^2}{4}$$
$$= [(30.7 - 31.2)^2 + (31.8 - 31.2)^2 + (30.2 - 31.2)^2 + (32.0 - 31.2)^2 + (31.3 - 31.2)^2] \div 4$$
$$= \frac{(-.5)^2 + (.6)^2 + (-1)^2 + (.8)^2 + (.1)^2}{4}$$
$$= \frac{2.26}{4}$$
$$= .565$$

and is the point estimate of the population variance σ^2.

3. The sample standard deviation is

$$s = \sqrt{s^2} = \sqrt{.565} = .7517$$

and is the point estimate of the population standard deviation σ.

Note that $\bar{y}$, s^2, and s have been calculated from a sample of $n = 5$ mileages. Therefore unless we are very lucky, these estimates will not be equal to the respective population parameters μ, σ^2, and σ. For example, suppose (a supernatural power knows) that μ, the true mean Hawk mileage, is 31.5 mpg. Then the sample mean $\bar{y} = 31.2$ is .3 mpg smaller than μ. We call the difference

$$\bar{y} - \mu = 31.2 - 31.5 = -.3$$

the *error of estimation* obtained when estimating μ by $\bar{y}$. We do not know the size of this error of estimation because we do not know the true value of μ. However, we will see in Chapter 3 that we can utilize the sample standard deviation, s (along with other quantities), to provide a bound on the error of estimation. This bound will tell us the farthest that $\bar{y}$ might reasonably be from μ.

2.4

CONTINUOUS PROBABILITY DISTRIBUTIONS

Before we randomly select a value, y, from a population, y can potentially be any of the values in the population. Thus, we can consider calculating probabilities concerning the value y might attain when the random selection is actually made. Specifically, consider the closed interval from a to b ($a < b$) on the real line. Denoting this interval as $[a, b]$, we often wish to find

$$P(\,y \text{ will be in the inverval } [a, b])$$

which can be written more simply as

$$P(a \leq y \leq b)$$

This probability can be interpreted as the proportion of values in the population that are greater than or equal to a and less than or equal to b.

We can often use a *continuous probability distribution* to calculate probabilities concerning y. Specifically, such distributions assign probabilities to intervals of numbers on the real line. To understand this idea, suppose that $f(v)$ is a continuous function of the numbers on the real line. Consider the continuous curve that results when $f(v)$ is graphed. A hypothetical curve $f(v)$ is illustrated in Figure 2.1.

The curve $f(v)$ is the *continuous probability distribution* of y if the probability that y will be in the interval $[a, b]$ is the area under the curve $f(v)$ corresponding to the interval $[a, b]$.

As an example, suppose that the curve $f(v)$ illustrated in Figure 2.1 is the continuous probability distribution of y = the mileage of a randomly selected automobile. Assume that we wish to find the probability that y will be between

FIGURE 2.1 **A hypothetical continuous probability curve $f(v)$**

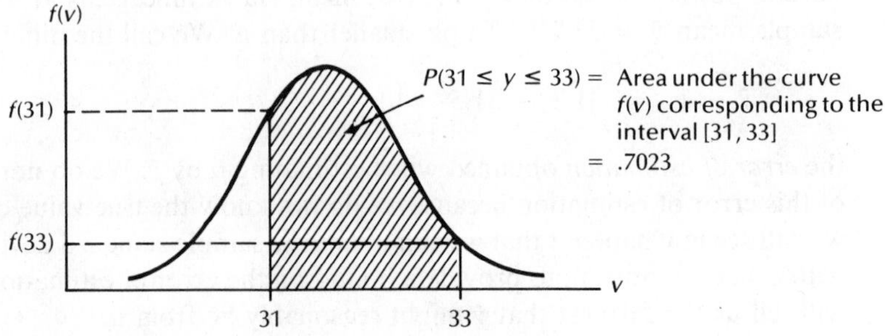

31 mpg and 33 mpg. Then, we would find the area under the curve $f(v)$ corresponding to the interval [31, 33]. If, for instance, this area was found to equal .7023, then as shown in Figure 2.1,

$$P(31 \leq y \leq 33) = .7023$$

This says that 70.23 percent of all mileages are between 31 mpg and 33 mpg.

Suppose that the curve $f(v)$ is the continuous probability distribution of y. Then we say that the population of values from which y will be randomly selected *is distributed according to the continuous probability curve* $f(v)$. Alternatively, we say that the population has the continuous probability distribution defined by $f(v)$. In Sections 2.5 and 2.6 we discuss several methods for determining the specific curve that describes a given population. We also show how to use statistical tables to find areas under probability curves.

The height of the curve $f(v)$ at a given point on the real line represents the relative probability, or chance, that y will be in a small interval of numbers around the given point. For example, suppose that the continuous probability distribution of y is as shown in Figure 2.1. From this figure we see that $f(31)$, the height of the curve at the point 31, is greater than $f(33)$, the height of the curve at the point 33. Thus it is more probable that y will be in a small interval of numbers around 31 than it is probable that y will be in a small interval of numbers around 33. Said another way, the height of the curve at a given point represents the relative proportion of values in the population that are in a small interval of numbers around the given point.

We next consider two general properties that are satisfied by a continuous probability distribution.

1. For any number v, $f(v) \geq 0$. Intuitively, this property must be satisfied because the height of the probability curve $f(v)$ at the point v represents a relative probability and because any probability must be greater than or equal to zero.
2. The total area under a continuous probability curve equals 1. This property holds because the total area under $f(v)$ equals the probability that y will fall between $-\infty$ and ∞, and y is sure to fall between $-\infty$ and ∞.

Two common shapes displayed by continuous probability curves describing actual populations are shown in Figure 2.2. The curve illustrated in Figure 2.2(a) is symmetrical. Thus in this figure, $f(\mu + \varepsilon)$ equals $f(\mu - \varepsilon)$. On the other hand, the curve depicted in Figure 2.2(b) is skewed to the right. Notice that whereas the population mean μ is not located under the highest point of the skewed curve, μ is located under the highest point of the symmetrical curve. We further discuss the location of μ for probability distributions with various shapes in Section 2.6.

Again consider Figure 2.2(a). By the symmetry of the curve $f(v)$ the area under this curve to the right of μ equals the area under the curve to the left of μ. Since

FIGURE 2.2 **Some common shapes of continuous probability distributions**

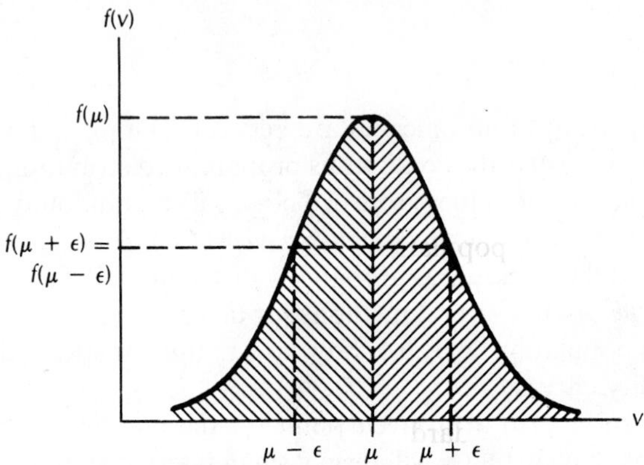

(a) A symmetrical continous probability distribution

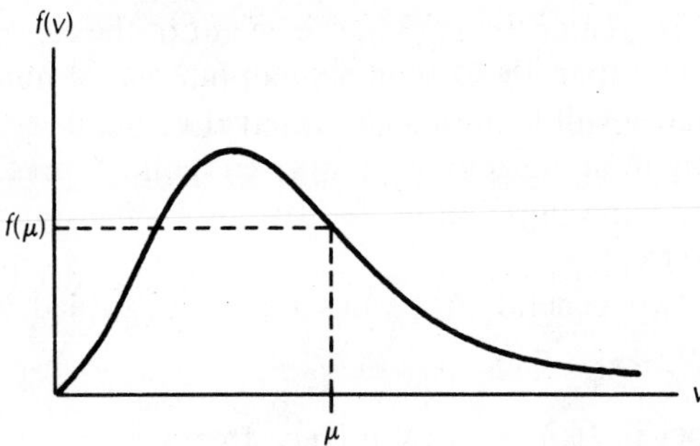

(b) A continous probability distribution that is skewed to the right

the total area under the curve equals 1, these equal areas must each equal .5. Thus if a population is described by a symmetrical continuous probability curve, then 50 percent of the values are less than or equal to the population mean μ, and 50 percent of the values are greater than or equal to μ.

2.5

THE NORMAL PROBABILITY DISTRIBUTION

THE NORMAL CURVE

Consider a population with mean μ and standard deviation σ. Sometimes such a population is distributed according to a *normal probability distribution.*

The Normal Probability Curve

The *normal probability distribution* is defined by the probability curve

$$f(v) = \frac{1}{\sigma\sqrt{2\pi}} \exp\left(-\frac{(v-\mu)^2}{2\sigma^2}\right) \quad \text{for} \quad -\infty < v < \infty$$

Here $\pi = 3.14159\ldots$ is the ratio of the circumference to the diameter of a circle, and exp denotes taking $e = 2.71828\ldots$, the base of Naperian logarithms, to the power in parentheses. If a population is distributed according to a normal distribution, we say that y is *normally distributed* with mean μ and standard deviation σ.

We denote the normal probability distribution by $N(\mu, \sigma)$. This means that the shape of the *normal (probability) curve* that results when $f(v)$ above is graphed depends on the mean and standard deviation of the population. The normal probability curve is illustrated in Figure 2.3. It is often described as a bell-shaped curve. The following are several important properties of this curve.

1. The normal curve is centered at the population mean μ.
2. The mean μ corresponds to the highest point on the normal curve.
3. The normal curve is symmetrical around the population mean.
4. Since the normal curve is a probability distribution, the total area under the curve is equal to 1.
5. Since the normal curve is symmetrical, the area under the normal curve above the mean equals the area under the curve below the mean, and each of these areas equals .5 (see Figure 2.3).

FIGURE 2.3 **The normal probability curve**

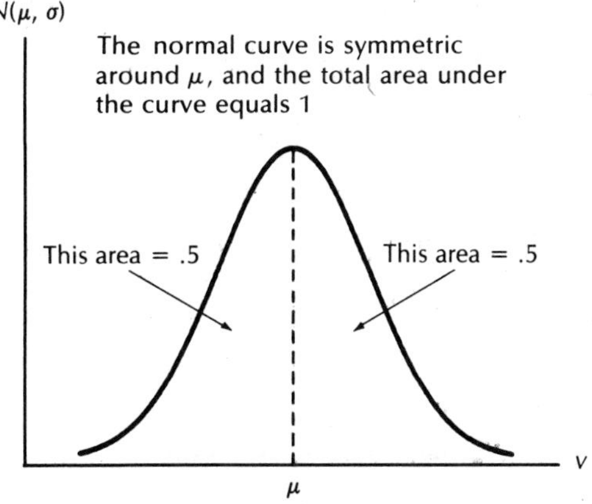

FIGURE 2.4 **Two normal curves with different means and equal standard deviations**

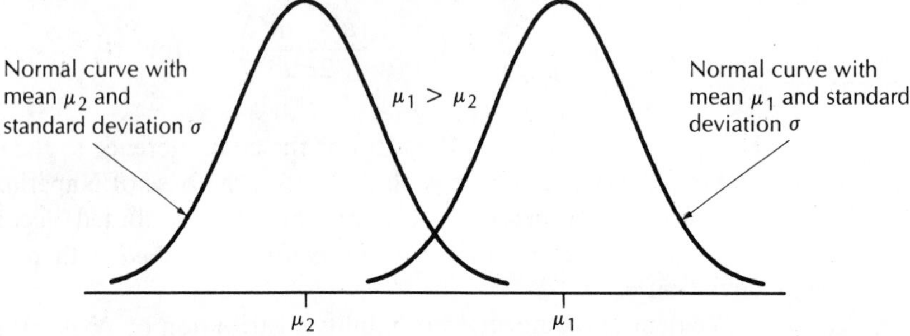

Intuitively, the population mean μ positions the normal curve on the real line. This is illustrated in Figure 2.4. This figure shows two normal curves with different means μ_1 and μ_2 (where $\mu_1 > \mu_2$) and with the same standard deviation σ. The variance σ^2 (or the standard deviation σ) measures the spread of the normal curve. This is illustrated in Figure 2.5, which shows two normal curves with the same mean μ but different standard deviations σ_1 and σ_2. Since $\sigma_1 > \sigma_2$, the normal curve with standard deviation σ_1 is more spread out than the normal curve with standard deviation σ_2.

If y is normally distributed with mean μ and standard deviation σ, then

$$P(a \leq y \leq b)$$

equals the area under the normal curve with mean μ and standard deviation σ corresponding to the interval $[a, b]$. Such an area is illustrated in Figure 2.6. To find areas under a normal curve, we can use a statistical table called a *normal table*. Such a table is presented as Table E.1 in Appendix E. In the next subsection we briefly discuss how to find areas under a normal curve.

FIGURE 2.5 **Two normal curves with the same mean and different standard deviations**

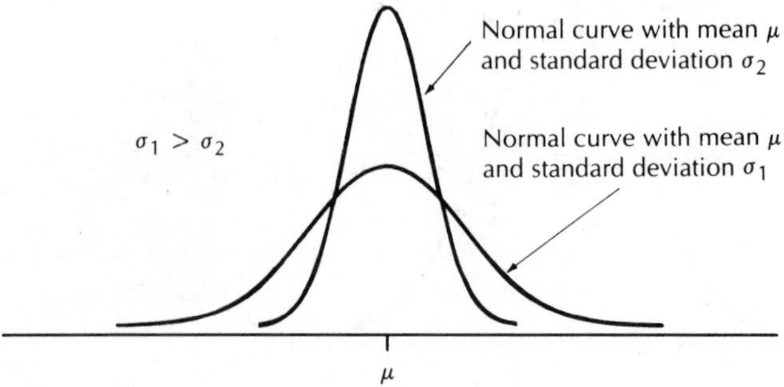

FIGURE 2.6 An area under a normal curve corresponding to the interval $[a, b]$

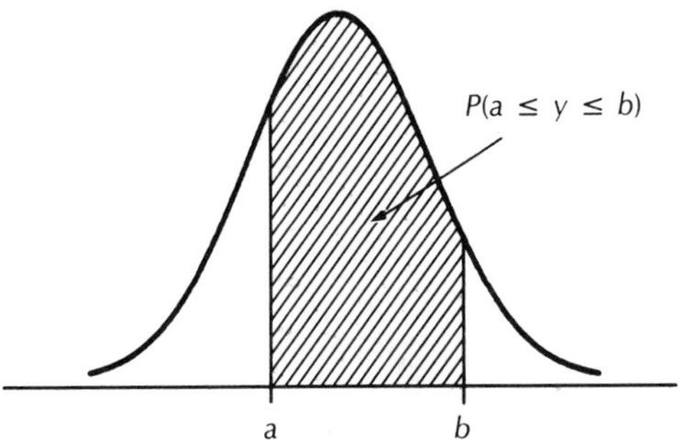

There are three important areas under the normal curve that we wish to emphasize (see Figure 2.7). If y is normally distributed with mean μ and standard deviation σ, it can be shown (using a normal table) that

1. $P(\mu - \sigma \leq y \leq \mu + \sigma) = .6826$

   This means that 68.26 percent of the values in the population are within (plus or minus) 1 standard deviation of the population mean.

2. $P(\mu - 2\sigma \leq y \leq \mu + 2\sigma) = .9544$

   This means that 95.44 percent of the values in the population are within (plus or minus) 2 standard deviations of the population mean.

FIGURE 2.7 Three important percentages concerning a normally distributed population with mean μ and standard deviation σ

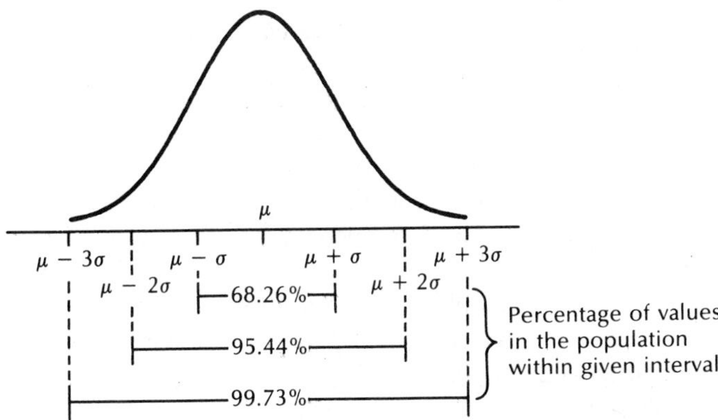

3. $P(\mu - 3\sigma \leq y \leq \mu + 3\sigma) = .9973$

   This means that 99.73 percent of the values in the population are within (plus or minus) 3 standard deviations of the population mean.

Example 2.4 In the gasoline mileage problem, suppose that the population of all mileages is normally distributed with mean $\mu = 31.5$ and standard deviation $\sigma = .8$. Then,

1. 68.26 percent of all mileages lie in the interval

$$\begin{aligned}[\mu - \sigma, \mu + \sigma] &= [31.5 - .8, 31.5 + .8] \\ &= [30.7, 32.3]\end{aligned}$$

2. 95.44 percent of all mileages lie in the interval

$$\begin{aligned}[\mu - 2\sigma, \mu + 2\sigma] &= [31.5 - 2(.8), 31.5 + 2(.8)] \\ &= [31.5 - 1.6, 31.5 + 1.6] \\ &= [29.9, 33.1]\end{aligned}$$

3. 99.73 percent of all mileages lie in the interval

$$\begin{aligned}[\mu - 3\sigma, \mu + 3\sigma] &= [31.5 - 3(.8), 31.5 + 3(.8)] \\ &= [31.5 - 2.4, 31.5 + 2.4] \\ &= [29.1, 33.9]\end{aligned}$$

Since National Motors does not know the true values of μ and σ, it cannot calculate these intervals. However, we can use the sample mean $\bar{y} = 31.2$ as the point estimate of μ and the sample standard deviation $s = .7517$ as the point estimate of σ (see Example 2.3). Thus we can *estimate* that

1. 68.26 percent of the mileages in this population lie in the interval

$$[31.2 - 1(.7517), 31.2 + 1(.7517)] = [30.4483, 31.9517]$$

2. 95.44 percent of the mileages lie in the interval

$$[31.2 - 2(.7517), 31.2 + 2(.7517)] = [29.6966, 32.7034]$$

3. 99.73 percent of the mileages lie in the interval

$$[31.2 - 3(.7517), 31.2 + 3(.7517)] = [28.9449, 33.4551]$$

The results of Example 2.4 depend on the assumption that the population of mileages is normally distributed. If we wish to verify the validity of this assumption, a sample of only $n = 5$ mileages is not sufficient. We would need to select a larger sample of mileages.

Table 2.3 lists a sample of $n = 49$ randomly selected mileages. Table 2.4 groups the 49 mileages into a *frequency distribution* having six intervals. We have chosen the number of intervals by using a general rule that says that the number of intervals should be the smallest integer K such that $2^K > n$. Here n is the number

TABLE 2.3 A sample of $n = 49$ mileages

$y_1 = 30.8$	$y_{11} = 30.9$	$y_{21} = 32.0$	$y_{31} = 32.3$	$y_{41} = 32.6$
$y_2 = 31.7$	$y_{12} = 30.4$	$y_{22} = 31.4$	$y_{32} = 32.7$	$y_{42} = 31.4$
$y_3 = 30.1$	$y_{13} = 32.5$	$y_{23} = 30.8$	$y_{33} = 31.2$	$y_{43} = 31.8$
$y_4 = 31.6$	$y_{14} = 30.3$	$y_{24} = 32.8$	$y_{34} = 30.6$	$y_{44} = 31.9$
$y_5 = 32.1$	$y_{15} = 31.3$	$y_{25} = 30.6$	$y_{35} = 31.7$	$y_{45} = 32.8$
$y_6 = 33.3$	$y_{16} = 32.1$	$y_{26} = 31.5$	$y_{36} = 31.4$	$y_{46} = 31.5$
$y_7 = 31.3$	$y_{17} = 32.5$	$y_{27} = 32.4$	$y_{37} = 32.2$	$y_{47} = 31.6$
$y_8 = 31.0$	$y_{18} = 31.8$	$y_{28} = 31.0$	$y_{38} = 31.5$	$y_{48} = 32.2$
$y_9 = 32.0$	$y_{19} = 30.4$	$y_{29} = 29.8$	$y_{39} = 31.7$	$y_{49} = 32.0$
$y_{10} = 32.4$	$y_{20} = 30.5$	$y_{30} = 31.1$	$y_{40} = 30.6$	

$$\bar{y} = \frac{\sum_{i=1}^{49} y_i}{49} = \frac{1546.1}{49} = 31.553061 \approx 31.6$$

$$s^2 = \frac{\sum_{i=1}^{49} (y_i - \bar{y})^2}{48} = \frac{30.666}{48} = .638875 \approx .64$$

$$s = \sqrt{s^2} = \sqrt{.638875} = .799 \approx .8$$

of observations that we wish to group into a frequency distribution. Since $n = 49$, and since $2^5 = 32 < 49$ and $2^6 = 64 > 49$, it follows that we should use $K = 6$ intervals. The first interval, [29.8, 30.3], is then formed by adding .5 to 29.8, the smallest mileage in Table 2.3. This yields an interval containing six measurement values—29.8, 29.9, 30.0, 30.1, 30.2, and 30.3. The decision to include six measurement values in the first interval (and in the other intervals) is based on calculating

$$\frac{[\text{Largest mileage} - \text{Smallest mileage}]}{K} = \frac{33.3 - 29.8}{6} = \frac{3.5}{6} \approx .6$$

This means that to include the smallest measurement and the largest measurement in the $K = 6$ classes, each class should contain six measurement values. The five

TABLE 2.4 A frequency distribution of the $n = 49$ mileages

Interval	*Frequency*
[29.8, 30.3]	3
[30.4, 30.9]	9
[31.0, 31.5]	12
[31.6, 32.1]	13
[32.2, 32.7]	9
[32.8, 33.3]	3

FIGURE 2.8 **A histogram of the $n = 49$ mileages**

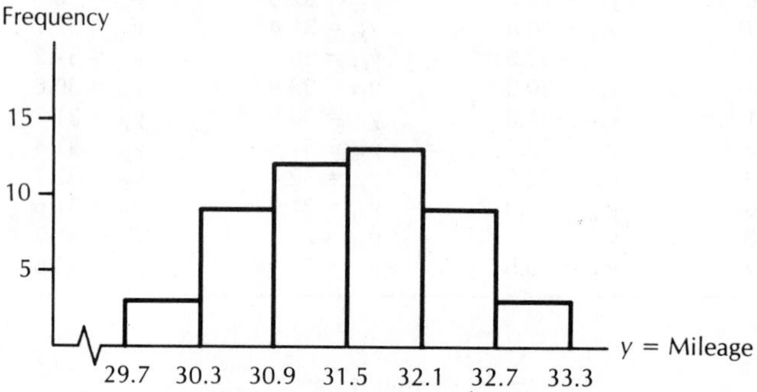

other intervals are formed in exactly the same way. The frequency distribution of Table 2.4 is depicted graphically in the form of a histogram in Figure 2.8. We see that this histogram looks reasonably bell-shaped and symmetrical. Since it is customary to look for pronounced rather than subtle departures from the normality assumption, the histogram in Figure 2.8 does not seriously contradict this assumption.

Finally, note that an alternative to constructing a frequency distribution and histogram is to make a *stem-and-leaf diagram*. The procedure for doing this is best illustrated by an example.

Consider the $n = 49$ mileages in Table 2.3. These mileages range from 29.8 to 33.3. To construct a stem-and-leaf diagram, the first two digits of the mileages —29, 30, 31, 32, and 33—are placed in a column on the left of the diagram. The respective third digits are recorded in the appropriate row. Therefore the first three mileages—30.8, 31.7, and 30.1—would be represented as

29	
30	8 1
31	7
32	
33	

We continue this procedure, and we order the data within each row to be neater (although this is not required). We obtain the following stem-and-leaf diagram:

29	8
30	1 3 4 4 5 6 6 6 8 8 9
31	0 0 1 2 3 3 4 4 4 5 5 5 6 6 7 7 7 8 8 9
32	0 0 0 1 1 2 2 3 4 4 5 5 6 7 8 8
33	3

Alternatively, we can obtain more classes by placing 29, 30*, 30, 31*, 31, 32*, 32, and 33 in the column on the left side of the diagram. Here, for instance, in the row headed by 30* we place the mileages from 30.0 to 30.4, and in the row headed by 30 we place the mileages from 30.5 to 30.9. Doing this, we obtain the following stem-and-leaf diagram:

29	8
30*	1 3 4 4
30	5 6 6 6 8 8 9
31*	0 0 1 2 3 3 4 4 4
31	5 5 5 6 6 7 7 7 8 8 9
32*	0 0 0 1 1 2 2 3 4 4
32	5 5 6 7 8 8
33	3

As we see, a stem-and-leaf diagram looks like a histogram turned sideways. The advantage of the stem-and-leaf diagram is that it not only reflects frequencies but also contains all of the observations. Note that the above stem-and-leaf display looks reasonably bell-shaped and symmetrical. Therefore it does not provide much evidence to contradict the normality assumption.

z VALUES AND FINDING NORMAL PROBABILITIES

If y is randomly selected from a normally distributed population with mean μ and standard deviation σ, then we define the z value corresponding to y to be

$$z = \frac{y - \mu}{\sigma} = \text{the number of standard deviations that } y \text{ is from the mean } \mu$$

Clearly, there is a z value corresponding to each value in the population. Therefore there is a population of z values corresponding to the population of y values. Thus we can assume that z is randomly selected from the population of z values. Moreover, it can be proved that if y is normally distributed, with mean μ and standard deviation σ, then z (or the population of z values) is normally distributed with mean zero and standard deviation 1. A normal distribution with mean zero and standard deviation 1 is referred to as a *standard normal distribution.*

If we subtract μ from the inequality

$$a \leq y \leq b$$

and divide by σ, we obtain the following inequalities:

$$\frac{a - \mu}{\sigma} \leq \frac{y - \mu}{\sigma} \leq \frac{b - \mu}{\sigma}$$

FIGURE 2.9 $P(a \leq y \leq b) = P(z_a \leq z \leq z_b)$

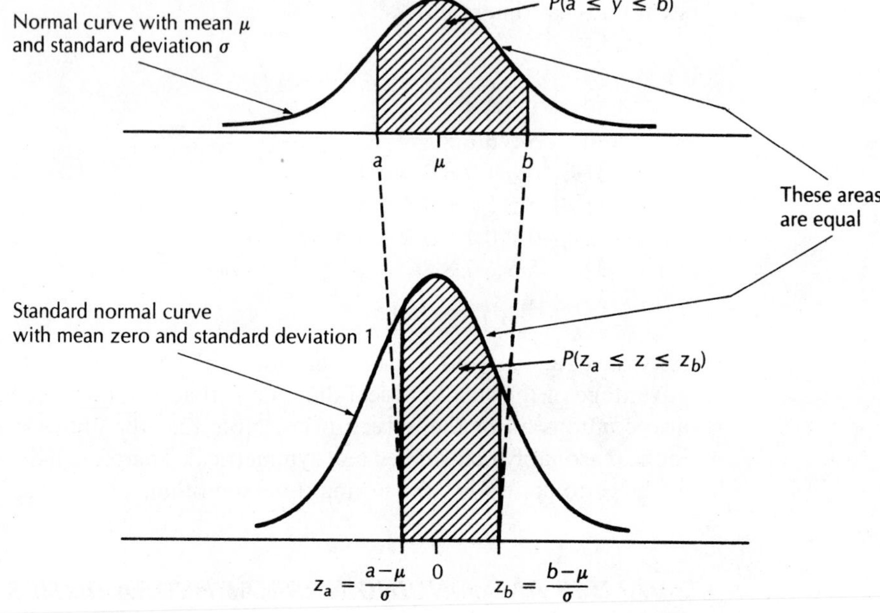

$$\frac{a - \mu}{\sigma} \leq z \leq \frac{b - \mu}{\sigma}$$

$$z_a \leq z \leq z_b$$

Here $z_a = (a - \mu)/\sigma$ is the z value corresponding to a, and $z_b = (b - \mu)/\sigma$ is the z value corresponding to b. We see that

$$P(a \leq y \leq b) = P(z_a \leq z \leq z_b)$$

Thus to find the probability

$$P(a \leq y \leq b)$$

we can calculate the z values corresponding to a and b and then find

$$P(z_a \leq z \leq z_b)$$

This is the area under the standard normal curve corresponding to the interval $[z_a, z_b]$. We illustrate this procedure in Figure 2.9.

Example 2.5 In the gasoline mileage problem the population of all mileages is normally distributed with mean $\mu = 31.5$ and standard deviation $\sigma = .8$. We now show that

$$P(29.9 \leq y \leq 33.1) = .9544$$

The z value corresponding to the mileage 29.9 mpg is

$$z_{29.9} = \frac{29.9 - \mu}{\sigma} = \frac{29.9 - 31.5}{.8} = \frac{-1.6}{.8} = -2$$

This says that the mileage 29.9 is 2 standard deviations below $\mu = 31.5$. The z value corresponding to the mileage 33.1 mpg is

$$z_{33.1} = \frac{33.1 - \mu}{\sigma} = \frac{33.1 - 31.5}{.8} = \frac{1.6}{.8} = 2$$

This says that the mileage 33.1 is 2 standard deviations above $\mu = 31.5$. It follows that

$$\begin{aligned} P(29.9 \leq y \leq 33.1) &= P(z_{29.9} \leq z \leq z_{33.1}) \\ &= P(-2 \leq z \leq 2) \end{aligned}$$

This probability can be found using the normal table (Table E.1) in Appendix E. This table gives

$$P(0 \leq z \leq z_c)$$

for values of z_c ranging from 0.00 to 3.09. Looking at this table, we see that

$$P(0 \leq z \leq 2) = .4772$$

Since the curve of the standard normal distribution is symmetrical about its mean, it follows that

$$\begin{aligned} P(-2 \leq z \leq 2) &= P(-2 \leq z \leq 0) + P(0 \leq z \leq 2) \\ &= .4772 + .4772 = .9544 \end{aligned}$$

Thus

$$P(29.9 \leq y \leq 33.1) = P(-2 \leq z \leq 2) = .9544$$

As another example, consider finding

$$P(y \leq 31)$$

The z value corresponding to 31 is

$$z_{31} = \frac{31 - \mu}{\sigma} = \frac{31 - 31.5}{.8} = \frac{-.5}{.8} = -.62$$

It follows that

$$\begin{aligned} P(y \leq 31) &= P(z \leq z_{31}) \\ &= P(z \leq -.62) \end{aligned}$$

By the symmetry of the normal curve, Table E.1 tells us that the area under the standard normal curve between $-.62$ and 0 is .2324. Since the area under the normal curve to the left of the mean is .5, it follows by examining Figure 2.10 that

$$\begin{aligned} P(y \leq 31) &= P(z \leq -.62) \\ &= .5 - .2324 \\ &= .2676 \end{aligned}$$

FIGURE 2.10 **Finding $P(z \leq -.62)$**

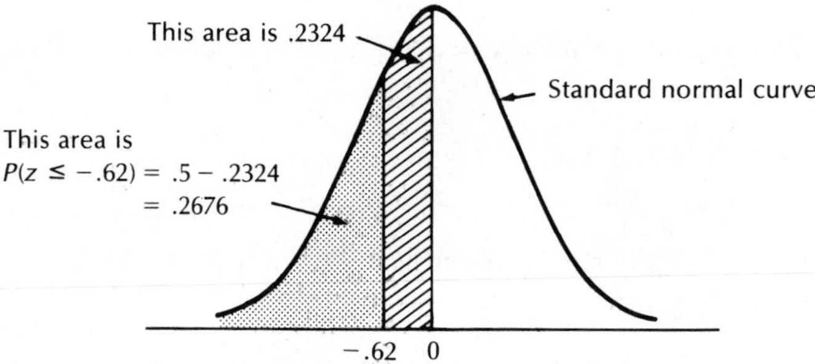

Next, recall that the sample of $n = 49$ mileages in Table 2.3 yields $\bar{y} = 31.5531$ and $s = .799$. Moreover, to be conservative, suppose that National Motors rounds the sample mean down to 31.5 and rounds the sample standard deviation up to .8. It is conservative to overestimate variation. Therefore National Motors claims that $\mu = 31.5$ and $\sigma = .8$. Now suppose that we purchase (randomly select) a Hawk and obtain a mileage of 31 mpg in 50,000 miles of driving. Should we regard the fact that our mileage is less than 31.5 as substantial evidence that the claim is incorrect? The previously calculated probability

$$P(y \leq 31) = .2676$$

says that if $\mu = 31.5$ and $\sigma = .8$, then there is a 26.76 percent chance that a randomly selected Hawk would get a mileage of 31 mpg or less. Since our car obtained such a mileage, if we are to believe the claim, we have to believe that a 26.76 percent chance has occurred. It is not particularly difficult to believe that such a chance would occur. Therefore we do not have much evidence that the claim is false. On the other hand, if National Motors had claimed that $\mu = 34$ and $\sigma = 1$, then the fact that our car got 31 mpg would cast great doubt on the claim. This is because if $\mu = 34$ and $\sigma = 1$, then

$$\begin{aligned} P(y \leq 31) &= P\left(z \leq \frac{31 - \mu}{\sigma}\right) \\ &= P\left(z \leq \frac{31 - 34}{1}\right) \\ &= P(z \leq -3) \\ &= .5 - .4987 \\ &= .0013 \end{aligned}$$

This would say that if we are to believe the claim, then we would have to believe that the mileage (31 mpg) obtained by our car is a 13 in 10,000 chance. It is extremely difficult to believe that such a small chance would occur. Therefore we would have substantial evidence that the claim is false.

To perform the calculations involved in some statistical inference procedures, we need to find the z value so that the area under the standard normal curve to the right of this z value is γ. We denote this z value as $z_{[\gamma]}$. We refer to $z_{[\gamma]}$ as *the point on the scale of the standard normal curve so that the area under this curve to the right of this point is* γ. The point $z_{[\gamma]}$ is illustrated in Figure 2.11. This point can be easily found. For example, suppose we wish to find $z_{[.025]}$, which is the point on the scale of the standard normal curve so that the area under this curve to the right of this point is .025. Then we note that the area under the standard normal curve between zero and $z_{[.025]}$ must equal $.5 - .025 = .475$. Looking at Table E.1, we see that the z value corresponding to an area of .4750 is 1.96. Thus $z_{[.025]}$ equals 1.96. This says that we must be 1.96 standard deviations above the mean of a standard normal distribution to obtain a "*right-hand tail area*" of .025.

FIGURE 2.11 $z_{[\gamma]}$ **= the point on the scale of the standard normal curve so that the area under this curve to the right of this point is** γ

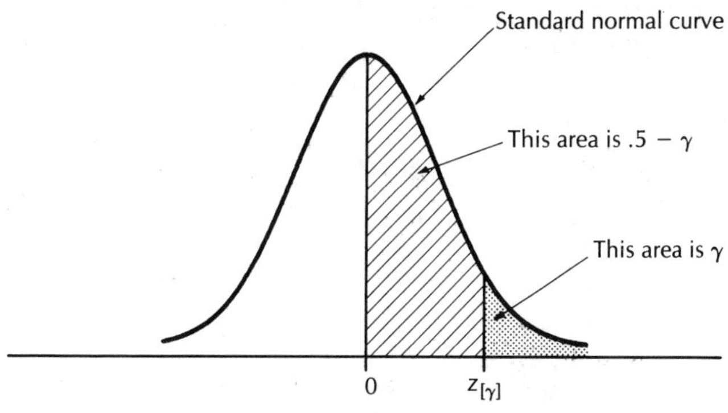

2.6

MEDIANS, QUARTILES, BOX PLOTS, AND TRIMMED MEANS

The Median and Quartiles

Consider a data set containing n observations and arrange the n observations in increasing order. Then,

1. The *median*, also called the *second quartile* and denoted Q_2, is the value below which and above which 50 percent of the observations lie. If n is odd, Q_2 is the middlemost observation. If n is even, Q_2 is the mean of the two middlemost observations.
2. The *first quartile*, denoted Q_1, is the value below which approximately 25 percent of the observations lie and above which approximately 75 percent of the observations lie. If n is even, Q_1 is the median of the smallest $n/2$ observations. If n is odd, we add Q_2 to the data set and define Q_1 to be the median of the smallest $(n + 1)/2$ observations.
3. The *third quartile*, denoted Q_3, is the value below which approximately 75 percent of the observations lie and above which approximately 25 percent of the observations lie. If n is even, Q_3 is the median of the largest $n/2$ observations. If n is odd, we add Q_2 to the data set and define Q_3 to be the median of the largest $(n + 1)/2$ observations.

We demonstrate in the following example how to use Q_1, Q_2, and Q_3 to construct a *box-and-whiskers display* (sometimes called a *box plot*). Such a display can be useful in estimating the shape of a probability distribution.

Example 2.6 Unified Medicine is a large clinic being planned in a midwestern city. The clinic plans to pay surgeons a yearly salary of $175,000 and wishes to compare this salary with the "average yearly income" of surgeons in private practice in the Midwest. To this purpose, the clinic has randomly selected a sample of $n = 12$ surgeons from the thousands of surgeons in private practice in the Midwest and has recorded the previous year's income for each surgeon in the sample. These incomes are arranged below in increasing order and are expressed in units of $1000.

127 132 138 144 146 152 154 162 171 177 192 241

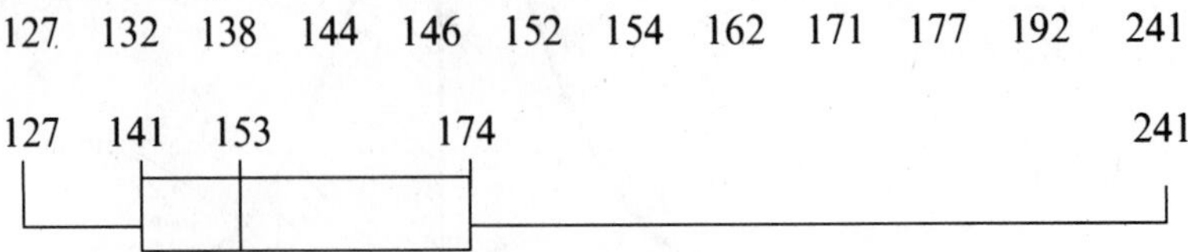

The five-number summary below the sample is called a *box-and-whiskers display*.

In addition to using the smallest sample value (127) and largest sample value (241), this display utilizes Q_1, Q_2, and Q_3. These quantities divide the data into four approximately equal parts. To compute Q_1, Q_2, and Q_3, we first arrange the n observations in increasing order and then compute Q_2, the *second quartile*, or *median*. Since $n = 12$ is even, Q_2 is the mean of the two middlemost observations, 152 and 154. Therefore $Q_2 = 153$. Furthermore, the *first quartile* is the median of the smallest $n/2 = 6$ observations. Thus $Q_1 = (138 + 144)/2 = 141$. The *third quartile* is the median of the largest $n/2 = 6$ observations. Thus $Q_3 = (171 + 177)/2 = 174$.

The relative positions in the box-and-whiskers display of the smallest observation (127), Q_1, Q_2, Q_3, and the largest observation (241) indicate that the largest 25 percent of the data is considerably more spread out than the smallest 25 percent of the data. Also, the second largest 25 percent of the data is somewhat more spread out than the second smallest 25 percent of the data. This indicates that the probability distribution of the population of all incomes is not symmetrical. Rather, the distribution is skewed to the right with a long tail.

Figure 2.12(a) depicts a population described by a symmetrical probability curve. For such a curve, the population mean equals the population median. In addition, both of these quantities equal the *population mode* (denoted M_o). The population mode is the most frequently occurring population value. It is located under the highest point of the probability curve. Figure 2.12(b) depicts a population described

FIGURE 2.12 **Relationships among the mean μ, the median Q_2, and the mode M_o**

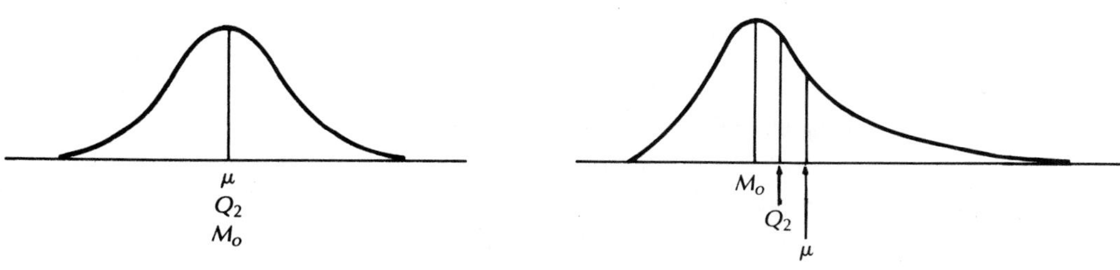

(a) A symmetrical curve

(b) A curve skewed to the right

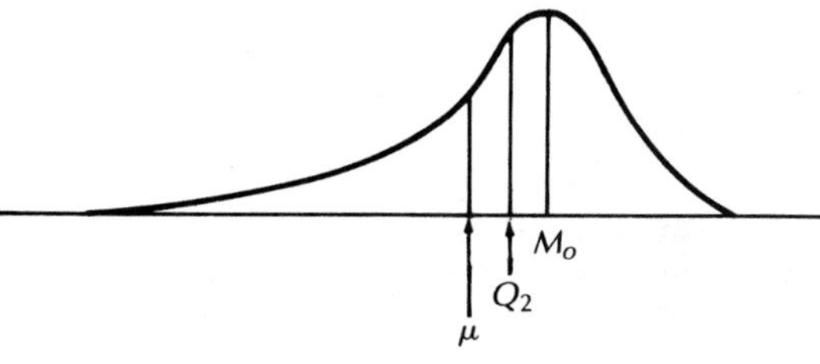

(c) A curve skewed to the left

by a probability curve that is skewed to the right. Here the population median is larger than the population mode, and the population mean is larger than the population median. In this case the population mean "averages in" the large values in the upper tail of the distribution. Thus the population mean is more affected by these large values than is the population median. For instance, for the income data in Example 2.6 the sample mean

$$\bar{y} = \frac{\sum_{i=1}^{12} y_i}{12} = 161.333$$

has been calculated by "averaging in" the large incomes 192 and 241. Thus this mean is larger than the sample median $Q_2 = 153$. Figure 2.12(c) depicts a population described by a probability curve that is skewed to the left. Here the population median is smaller than the population mode, and the population mean is smaller than the population median.

When a distribution is skewed to the right or left with a very long tail, the population mean can be greatly affected by the "extreme" population values in the tail of the distribution. In such a case the population median is a better measure of the population central tendency than is the population mean. However, for small samples the theory of the sample median is not well enough developed to use some of the methods to be presented in Chapter 3. Therefore it is sometimes useful to estimate the central tendency of a skewed population by computing the *trimmed population mean* μ_T. This is the mean of all of the population values except for the smallest 10 percent of the values and the largest 10 percent of the values. Actually, any percentage can be trimmed off both ends of the population, but 10 percent trimming seems to work well in practice.

Example 2.7 Consider the income data in Example 2.6. We wish to estimate the trimmed mean, μ_T, of the population of incomes. To do this, we consider the following two samples:

RS:	127	132	138	144	146	152	154	162	171	177	192	241
TS:			138	144	146	152	154	162	171	177		

Here *RS* denotes the random sample of $n = 12$ incomes given in Example 2.6, and *TS* denotes the *trimmed sample*. The trimmed sample is obtained by eliminating the smallest 10 percent and the largest 10 percent of the incomes from the sample of 12 incomes. Here, since $.10(12) = 1.2$, we round up and eliminate the two lowest incomes and the two highest incomes. The trimmed sample has a mean equal to $\bar{y}_T = 155.5$. This value is the estimate of μ_T, the trimmed population mean. Therefore we have evidence that the trimmed mean of all yearly incomes is smaller than Unified Medicine's yearly salary of $175,000.

2.7

THE t-DISTRIBUTION AND THE F-DISTRIBUTION

THE t-DISTRIBUTION

Sometimes a population has what is called a *t-distribution*. The probability curve of the t-distribution has the following properties:

1. The curve is symmetrical and bell-shaped.
2. The curve is symmetrical about zero, which is the mean of the t-distribution.
3. The standard deviation σ of the t-distribution is always greater than 1.
4. The exact spread, or standard deviation σ, of the t-distribution depends on a parameter that is called *the number of degrees of freedom.*
5. As the number of degrees of freedom approaches infinity, the standard deviation σ of the t-distribution approaches 1.
6. As the number of degrees of freedom approaches infinity, the curve approaches (that is, becomes shaped more and more like) the probability curve of a standard normal distribution.

To carry out the calculations in later chapters, we must find the point on the scale of the t-distribution having a given number of degrees of freedom so that the area under this curve to the right of this point is γ. Such a point is illustrated in Figure 2.13 and is denoted $t_{[\gamma]}^{(df)}$. Here the superscript (df) refers to the number of degrees of freedom and the subscript $[\gamma]$ refers to the size of the area under the curve to the right of the point $t_{[\gamma]}^{(df)}$. In general, we will refer to the point $t_{[\gamma]}^{(df)}$ as *the point on the scale of the t-distribution having df degrees of freedom so that the area under this curve to the right of this point is* γ.

FIGURE 2.13 $t_{[\gamma]}^{(df)}$ = **the point on the scale of the t-distribution having *df* degrees of freedom so that the area under this curve to the right of this point is** γ

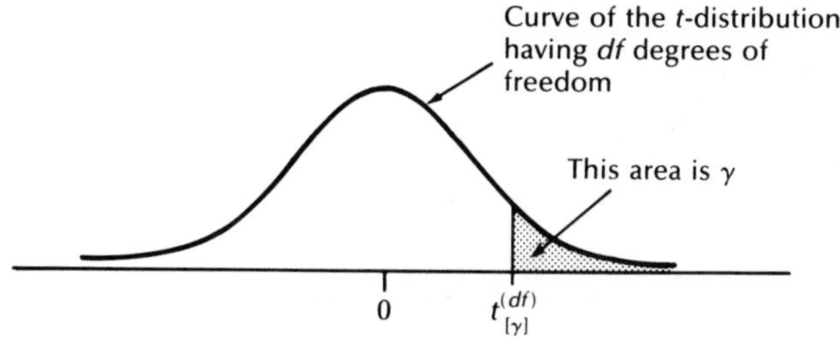

The t-point $t_{[\gamma]}^{(df)}$ can be found by using a *t-table* (see Table E.2 of Appendix E). This table lists values of $t_{[\gamma]}^{(df)}$ for values of γ ranging from .40 to .0005. The values of $t_{[\gamma]}^{(df)}$ are tabulated according to the number of degrees of freedom, *df*. For example, $t_{[.025]}^{(11)}$ is 2.201. Notice that the t-table lists values of $t_{[\gamma]}^{(df)}$ for degrees of freedom from 1 to 29 and ∞. Values of *df* greater than 29 are not listed because when the number of degrees of freedom is large, the value $t_{[\gamma]}^{(df)}$ is very close to the value $z_{[\gamma]}$. Values of $z_{[\gamma]}$ for values of γ ranging from .40 to .0005 are given in the t-table in the row corresponding to ∞. Generally, if the number of degrees of freedom is 30 or more, it is sufficient to use the value $z_{[\gamma]}$ for $t_{[\gamma]}^{(df)}$.

THE F-DISTRIBUTION

Sometimes a population has what is called an *F-distribution*. The probability curve of the F-distribution has the following properties:

1. The curve $f(v)$ of the F-distribution is positive for all values of $v > 0$.
2. The curve is skewed to the right.
3. The exact form of the curve depends on two parameters, called the numerator degrees of freedom (denoted r_1) and the denominator degrees of freedom (denoted r_2).

To perform manipulations in later chapters, we must know how to find the point on the scale of the F-distribution having r_1 and r_2 degrees of freedom so that the area under this curve to the right of this point is γ. Such a point is illustrated in Figure 2.14 and is denoted as $F_{[\gamma]}^{(r_1,r_2)}$. This point can be found using Table E.3. This table lists values of $F_{[.05]}^{(r_1,r_2)}$ tabulated according to values of the parameters r_1 and r_2. To find the point $F_{[.05]}^{(r_1,r_2)}$, we scan across the top of the F-table to find the column corresponding to the parameter r_1, and we scan down the side to find the row

FIGURE 2.14 $F_{[\gamma]}^{(r_1,r_2)}$ **= the point on the scale of the F-distribution having r_1 and r_2 degrees of freedom so that the area under this curve to the right of this point is γ**

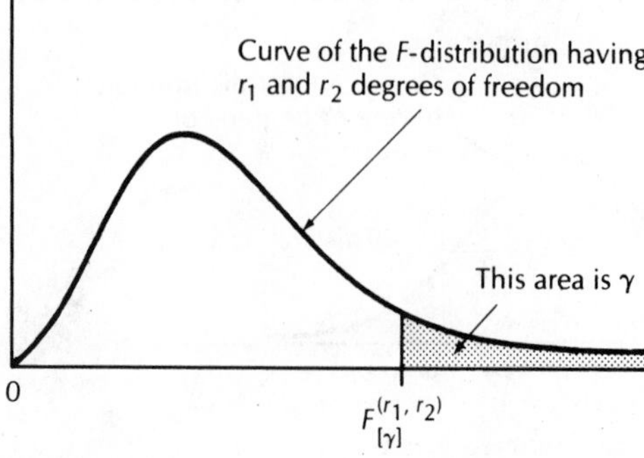

corresponding to the parameter r_2. The value in the column corresponding to r_1 and in the row corresponding to r_2 is $F_{[.05]}^{(r_1,r_2)}$. Thus, for example, $F_{[.05]}^{(8,10)}$ is 3.07. Although Table E.3 tabulates values of $F_{[\gamma]}^{(r_1,r_2)}$ for the value $\gamma = .05$, tables for other values of γ are also available. For example, see Table E.4 in Appendix E, which tabulates $F_{[.01]}^{(r_1,r_2)}$.

EXERCISES

2.1 Consider the following sample of six gasoline mileages:

32.3 30.5 31.7 31.6 31.4 32.6

a. Calculate the sample mean $\bar{y}$.
b. Calculate the sample variance s^2.
c. Calculate the sample standard deviation s.

2.2 Consider the following sample of seven dollar amounts owed by customers with delinquent charge accounts:

$99 $123 $75 $138 $105 $65 $116

a. Calculate the sample mean $\bar{y}$.
b. Calculate the sample variance s^2.
c. Calculate the sample standard deviation s.

2.3 Consider the following sample of "percent berry content" values for eight randomly selected jars of strawberry preserves:

65.4 64.9 65.2 65.7 65.0 65.7 65.3 64.7

a. Calculate the sample mean $\bar{y}$.
b. Calculate the sample variance s^2.
c. Calculate the sample standard deviation s.

2.4 An engine manufacturer tests nine engines for pollution (measured in milligrams of particulate matter per cubic yard) with the following results:

72 74 75 75 79 81 70 77 85

a. Calculate the sample mean $\bar{y}$.
b. Calculate the sample variance s^2.
c. Calculate the sample standard deviation s.

2.5 Consider the first 30 observations in Table 2.3.

a. Calculate the sample mean $\bar{y}$.
b. Calculate the sample variance s^2.
c. Calculate the sample standard deviation s.
d. Calculate Q_1, Q_2, and Q_3.
e. Construct a histogram for the data.
f. Construct a stem-and-leaf display of the data.
g. Draw a box-and-whiskers display for the data.
h. Calculate the 10 percent trimmed mean.

2.6 The following is a sample of "viscosity" readings for a chemical process:

1.93	1.96	1.98	1.94	1.99	1.98	2.01	2.01
1.94	1.97	1.98	2.02	2.00	1.98	2.01	1.96
1.97	1.99	2.00	2.02	1.99	1.95	2.02	1.99
2.00	2.01	2.00	2.02	1.99	1.96	2.01	1.99
1.95	1.99	2.02	2.02	1.98	1.98	2.01	2.04

a. Calculate the sample mean $\bar{y}$.
b. Calculate the sample variance s^2.
c. Calculate the sample standard deviation s.
d. Calculate Q_1, Q_2, and Q_3.
e. Construct a histogram for the data.
f. Construct a stem-and-leaf display of the data.
g. Draw a box-and-whiskers display for the data.
h. Calculate the 10 percent trimmed mean.

2.7 The following is a sample of "bag weights" (in pounds) for an industrial bagging operation:

50.6	49.8	50.8	50.5	50.2	50.4
50.6	51.4	50.4	50.3	49.9	50.1
50.8	50.8	50.6	50.8	52.2	50.7
50.8	50.6	50.7	50.6	50.3	49.8
50.8	50.6	49.1	51.2	50.2	52.0
49.8	50.8	49.0	51.1	46.8	50.5

a. Calculate the sample mean $\bar{y}$.
b. Calculate the sample variance s^2.
c. Calculate the sample standard deviation s.
d. Calculate Q_1, Q_2, and Q_3.
e. Construct a histogram for the data.
f. Construct a stem-and-leaf display of the data.
g. Draw a box-and-whiskers display for the data.
h. Calculate the 5 percent trimmed mean.
i. Calculate the 10 percent trimmed mean.

2.8 The following is a sample of pH values for jars of cherry preserves produced by a major manufacturer of jams and jellies:

3.15	3.12	3.06	3.09	3.19	3.08	3.14
3.10	3.11	3.14	3.02	3.09	3.12	3.15
3.17	3.15	3.10	3.06	3.07	3.07	3.12
3.12	3.12	3.11	3.11	3.08	3.08	3.11
3.11	3.10	3.06	3.08	3.11	3.15	3.07
3.11	3.12	3.12	3.06	3.07	3.15	3.13
3.13	3.14	2.97	3.12	3.27	3.09	3.10
3.15	3.16	3.06	3.14	3.11	3.11	3.08
3.09	3.06	3.09	3.09	3.10	3.12	3.10
3.12	3.08	3.10	3.14	3.13	3.11	3.11
3.07	3.12	3.14	3.14	3.10	3.06	
3.16	3.08	3.08	3.19	3.20	3.06	
3.09	3.13	3.04	3.11	3.06	3.07	
3.10	3.06	3.11	3.10	3.10	3.05	
3.12	3.12	3.14	3.11	3.05	3.12	

a. Calculate the sample mean $\bar{y}$.
b. Calculate the sample variance s^2.
c. Calculate the sample standard deviation s.
d. Calculate Q_1, Q_2, and Q_3.
e. Construct a histogram for the data.
f. Construct a stem-and-leaf display of the data.
g. Draw a box-and-whiskers display for the data.
h. Calculate the 10 percent trimmed mean.
i. Calculate the 20 percent trimmed mean.

2.9 Consider the following sample of $n = 15$ physicians' salaries (in thousands of dollars):

127	134	140	144	146	149	153	155
161	165	171	179	182	196	245	

a. Calculate the sample mean $\bar{y}$.
b. Calculate the sample standard deviation s.
c. Calculate Q_1, Q_2, and Q_3.
d. Draw a box-and-whiskers display for the data.
e. Calculate the 10 percent trimmed mean.

2.10 Suppose that the population of all gasoline mileages for the GSX-50 is normally distributed with mean $\mu = 31.5$ mpg and standard deviation $\sigma = .8$ mpg. Let y denote a mileage randomly selected from this population. Find the following probabilities:

a. $P(30.7 \le y \le 32.3)$ b. $P(29.1 \le y \le 33.9)$ c. $P(29.5 \le y \le 32.3)$
d. $P(31.0 \le y \le 31.3)$ e. $P(y \le 29.5)$ f. $P(y \ge 29.5)$
g. $P(y \ge 33.4)$ h. $P(y \le 33.4)$

2.11 Using Table E.1 in Appendix E, find

a. $z_{[.05]}$ b. $z_{[.02]}$ c. $z_{[.01]}$ d. $z_{[.005]}$

2.12 Using Table E.2 in Appendix E, find

a. $t_{[.05]}^{(7)}$ b. $t_{[.01]}^{(7)}$ c. $t_{[.005]}^{(7)}$

2.13 Using Table E.3 in Appendix E, find

a. $F_{[.05]}^{(2,5)}$ b. $F_{[.05]}^{(5,2)}$

2.14 The daily water consumption for an Ohio community is normally distributed with a mean consumption of 300,000 gallons and a standard deviation of 20,000 gallons.

a. Find the probability that daily water consumption will be less than 250,000 gallons.
b. Find the probability that daily water consumption will be between 260,000 gallons and 330,000 gallons.
c. The community water system will experience a noticeable drop in water pressure when daily water consumption exceeds 346,000 gallons. What is the probability of experiencing such a drop in water pressure?

2.15 In the paper industry the moisture content of paper is very important (for example, paper with excessive moisture content tends to curl). For a certain type of paperboard, moisture content is normally distributed with a mean of 6 percent and a standard deviation of .2 percent.

a. What percentage of this paperboard will have a moisture content exceeding 6.4 percent?
b. What percentage of this paperboard will have a moisture content between 6.2 and 6.6 percent?
c. What percentage of this paperboard will have a moisture content less than 6.13 percent?
d. Moisture content specifications for paperboard are 6 percent $\pm$ .5 percent (that is, paperboard with moisture contents below 5.5 percent or above 6.5 percent is unacceptable and must be recycled). What percentage of this paperboard is out of specifications?

2.16 Weekly demand for VHS video tapes at a major discount store is normally distributed with mean $\mu = 300$ tapes and standard deviation $\sigma = 30$ tapes. The store policy is to stock enough tapes so that there is at most a 2.5 percent chance of running short of tapes during the week. How many VHS video tapes should be ordered each week?

2.17 A filling machine is being used to fill 50-lb bags with a chemical product. The machine can be adjusted so that it dispenses an average of μ pounds per bag. If the fills are normally distributed with a standard deviation of $\sigma = .1$ lb, find the setting for μ so that 50-lb bags are underfilled only 2.5 percent of the time.

2.18 A company sells bags of Chemical XL-500. The amount of chemical in each bag is supposed to weigh 50 lb. However, because of variability in the bagging operation, not all bags of Chemical XL-500 weigh exactly 50 lb. While the mean bag fill is 50 lb, the standard deviation of the bag fills is .2 lb. Furthermore, the bag fills are normally distributed.

a. A customer purchases a bag of Chemical XL-500 and weighs the contents. Find a reasonable estimate of the maximum weight that might be obtained for the bag contents.
b. A customer purchases a bag of Chemical XL-500 and weighs the contents. Find a reasonable estimate of the minimum weight that might be obtained for the bag contents.
c. Would a bag content of 50.8 lb be considered to be unusually high in this situation? Explain your answer.
d. Would a bag content of 49.7 lb be considered to be unusually low in this situation? Explain your answer.
e. The company is considering adjusting the bag filler so that substantially all of the bag fills (more than 99% of the fills) will be at least 50 lb. How high must the mean bag fill be set in order to meet this goal?

2.19 A mechanical jar filler is being used to fill jars with grape jelly. The filler has been set so that the mean jar fill is 515 grams. The standard deviation of the jar fills is 2.5 grams. Furthermore, the population of all jar fills is normally distributed.

a. What percentage of the jar fills will be greater than 512.5 grams?
b. What percentage of the jar fills will be between 511.25 grams and 518.75 grams?
c. What percentage of the jar fills will be greater than 522.5 grams?
d. What percentage of 510-gram-capacity jars will be underfilled by the jar filler?
e. If we wish to reset the jar filler (the filler can be reset so that the mean jar fill will be any number of grams desired), where must the mean jar fill μ be set so that only .4 percent of the jars will contain less than the label weight (510 grams)?

2.20 The present value of the cash flow from an investment is normally distributed with mean \$10,000 and standard deviation \$4000.

a. What is the probability that the present value of the cash flow from this investment will be greater than \$18,000?
b. What is the probability that the present value of the cash flow from this investment will be between \$5000 and \$15,000?
c. What is the probability that the investment will not pay? That is, what is the probability that the present value of the cash flow from this investment will be less than zero?
d. What is the probability that the investment will lose the present value equivalent of \$2000 or more?

2.21 The annual winter snowfall in Oxford, Ohio, is normally distributed with a mean of 14 inches and a standard deviation of 4 inches.

a. Find a reasonable estimate of the minimum snowfall in Oxford during the winter season.
b. Find a reasonable estimate of the maximum snowfall in Oxford during the winter season.
c. If Oxford receives 8 inches of snow this winter, would this snowfall total be considered to be unusually low? Explain your answer.
d. If Oxford receives 27 inches of snow this winter, would this snowfall total be considered to be unusually high? Explain your answer.
e. The Oxford Highway Department has purchased enough road salt to treat 18 inches of snow this winter. What are the chances that the Oxford Highway Department will run out of road salt during this winter season?

2.22 United Motors claims that one of its cars, the Starbird 300, gets gasoline mileages that are normally distributed with a mean of 30 mpg and a standard deviation of .9 mpg. Suppose that you purchase (randomly select) a Starbird 300 and that this car gets 28 mpg.

a. Determine how much probabilistic doubt this result casts on the United Motors claim by computing $P(y \leq 28)$ where y = gasoline mileage for the Starbird 300.
b. Interpret your results of part (a).

CHAPTER 3

BASIC STATISTICAL INFERENCE

When we draw conclusions about a population based on the information contained in a sample drawn from the population, we say that we are making *statistical inferences*. In this chapter we will see how probability distributions—for example, the normal and *t*-distributions—can be used to make statistical inferences. Specifically, we will study *confidence interval estimation* of population parameters. We will also study how to carry out *hypothesis tests* concerning population parameters.

We begin in Section 3.1 by discussing *confidence intervals for a population mean*. Then in Section 3.2 we study *hypothesis tests about a population mean*. We explain both the *rejection point* and *probability-value methods* of hypothesis testing. In Sections 3.3 and 3.4 we discuss *comparing two population means*. In all cases, both large and small sample methods are presented.

3.1

CONFIDENCE INTERVALS FOR A POPULATION MEAN

A CONFIDENCE INTERVAL BASED ON THE t-DISTRIBUTION

Suppose that we randomly select a sample of n values

$$\{y_1, y_2, \ldots, y_n\}$$

from a population. Then we use the sample mean and the sample standard deviation

$$\bar{y} = \frac{\sum_{i=1}^{n} y_i}{n} \quad \text{and} \quad s = \sqrt{\frac{\sum_{i=1}^{n} (y_i - \bar{y})^2}{n-1}}$$

as the point estimates of the population mean μ and population standard deviation σ. In this section we consider using a *confidence interval* to estimate the population mean. The reason for doing this is that the point estimate $\bar{y}$ does not provide any indication of how close it is to the unknown μ. A *confidence interval* for the population mean μ is an interval constructed around the sample mean $\bar{y}$ so that we are reasonably sure, or confident, that this interval contains μ.

We can construct a confidence interval so that we are 99%, 95%, 90%, and so on, confident that this interval contains μ. In general, we let $100(1 - \alpha)\%$ denote our *level of confidence*. For example, if $\alpha = .05$, then we are $100(1 - \alpha)\% = 100(1 - .05)\% = 95\%$ confident. Below we present the formula for a $100(1 - \alpha)\%$ confidence interval for μ based on the t-distribution. Then, after illustrating the use of this interval, we discuss the logic behind deriving the interval.

A Confidence Interval for μ Based on the t-Distribution

Suppose that $\bar{y}$ is the mean and s is the standard deviation of a sample of n values that has been randomly selected from a normally distributed population having mean μ and standard deviation σ. Then a *100(1 − α)% confidence interval for μ* is

$$\left[\bar{y} \pm t_{[\alpha/2]}^{(n-1)}\left(\frac{s}{\sqrt{n}}\right)\right] = \left[\bar{y} - t_{[\alpha/2]}^{(n-1)}\left(\frac{s}{\sqrt{n}}\right), \bar{y} + t_{[\alpha/2]}^{(n-1)}\left(\frac{s}{\sqrt{n}}\right)\right]$$

Here, $t_{[\alpha/2]}^{(n-1)}$ is the point on the scale of the t-distribution having $n - 1$ degrees of freedom so that the area under this curve to the right of this point is $\alpha/2$.

The $100(1 - \alpha)\%$ confidence interval for μ says that we are $100(1 - \alpha)\%$

confident that μ is greater than or equal to the lower bound

$$\bar{y} - t_{[\alpha/2]}^{(n-1)}\left(\frac{s}{\sqrt{n}}\right)$$

and less than or equal to the upper bound

$$\bar{y} + t_{[\alpha/2]}^{(n-1)}\left(\frac{s}{\sqrt{n}}\right)$$

Since we do not know the true value of μ, we are not absolutely certain (not 100% confident) that μ is contained in the $100(1 - \alpha)\%$ confidence interval. Exactly what we mean by $100(1 - \alpha)\%$ confidence will be discussed later. For now, suffice it to say that if we choose a high level of confidence, we are very sure that μ is in the interval.

Before presenting an example we make three comments. First, if the number of degrees of freedom for the t-point $t_{[\alpha/2]}^{(n-1)}$ is 30 or more, it is sufficient to use the normal point $z_{[\alpha/2]}$. Second, it has been shown that the confidence interval formula approximately holds for many populations that are not normally distributed. In particular, the formula approximately holds for a population described by a probability curve that is mound-shaped (even if this curve is somewhat skewed to the right or left). This assumption can be checked by using sample data to construct a histogram. Third, although the formula applies to infinite populations, it also approximately applies to many finite (mound-shaped) populations. However, better formulas, which yield shorter confidence intervals, often exist for finite populations.

Example 3.1 Federal gasoline mileage standards state that μ, the mean gasoline mileage obtained by the fleet of all Hawks, must be at least 30 mpg. If this standard is not met, a heavy fine will be imposed. To demonstrate that this standard is being met, National Motors randomly selects a sample of $n = 5$ Hawks and tests them for gasoline mileage. The sample

$$\{y_1, y_2, y_3, y_4, y_5\} = \{30.7, 31.8, 30.2, 32.0, 31.3\}$$

is obtained. In Chapter 2 we saw that for this sample the sample mean is $\bar{y} = 31.2$ mpg and the sample standard deviation is $s = .7517$ mpg. This sample of mileages has been randomly selected from the (theoretically) infinite population of all Hawk mileages. Therefore $\bar{y} = 31.2$ is the point estimate of μ.

Suppose that we wish to calculate a 95% confidence interval for μ. Then, since $100(1 - \alpha)\% = 95\%$ implies that $\alpha = .05$, we use $t_{[\alpha/2]}^{(n-1)} = t_{[.05/2]}^{(5-1)} = t_{[.025]}^{(4)} = 2.776$. (See Table E.2 in Appendix E.) It follows that a 95% confidence interval for μ is

$$\begin{aligned}\left[\bar{y} \pm t_{[\alpha/2]}^{(n-1)}\left(\frac{s}{\sqrt{n}}\right)\right] &= \left[31.2 \pm 2.776\left(\frac{.7517}{\sqrt{5}}\right)\right] \\ &= [31.2 \pm .9333] \\ &= [30.3, 32.1]\end{aligned}$$

This interval says that we can be 95% confident that μ is between 30.3 and 32.1 mpg. Therefore since the lower bound of this interval is above 30 mpg, we can be at least 95% confident that μ is greater than 30 mpg. We conclude that we have strong evidence that the Hawk not only meets, but exceeds, the federal mileage standard.

Next, suppose that we wish to calculate a 99% confidence interval for μ. Then, since $100(1 - \alpha)\% = 99\%$ implies that $\alpha = .01$, we use $t_{[\alpha/2]}^{(n-1)} = t_{[.01/2]}^{(5-1)} = t_{[.005]}^{(4)} = 4.604$. It follows that a 99% confidence interval for μ is

$$\left[\bar{y} \pm t_{[\alpha/2]}^{(n-1)}\left(\frac{s}{\sqrt{n}}\right)\right] = \left[31.2 \pm 4.604\left(\frac{.7517}{\sqrt{5}}\right)\right]$$
$$= [31.2 \pm 1.5479]$$
$$= [29.7, 32.7]$$

This interval is longer than the 95% confidence interval for μ. Hence increasing the level of confidence from 95% to 99% (1) has the advantage of making us more confident that μ is contained in our interval for μ but (2) has the disadvantage of increasing the length of our confidence interval. This results in a less precise guess of the true value of μ. Note that the lower bound of the 99% confidence interval for μ is not greater than or equal to 30. Therefore on the basis of this confidence interval we cannot be 99% confident that μ is at least 30. However, since National Motors is at least 95% confident that μ is greater than 30, there is substantial evidence that current gasoline mileage standards are being met.

FIGURE 3.1 **An illustration of the meaning of a 95% confidence interval for μ**

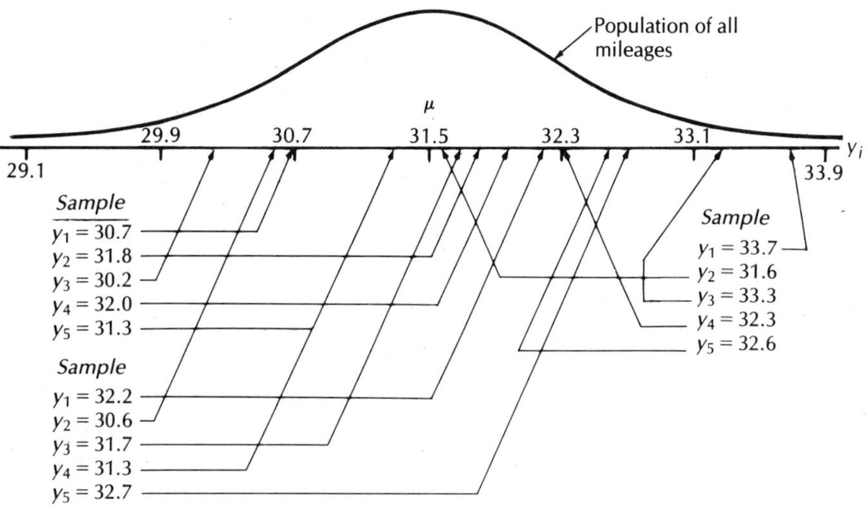

(a) Three possible samples

(continues)

FIGURE 3.1 **Continued**

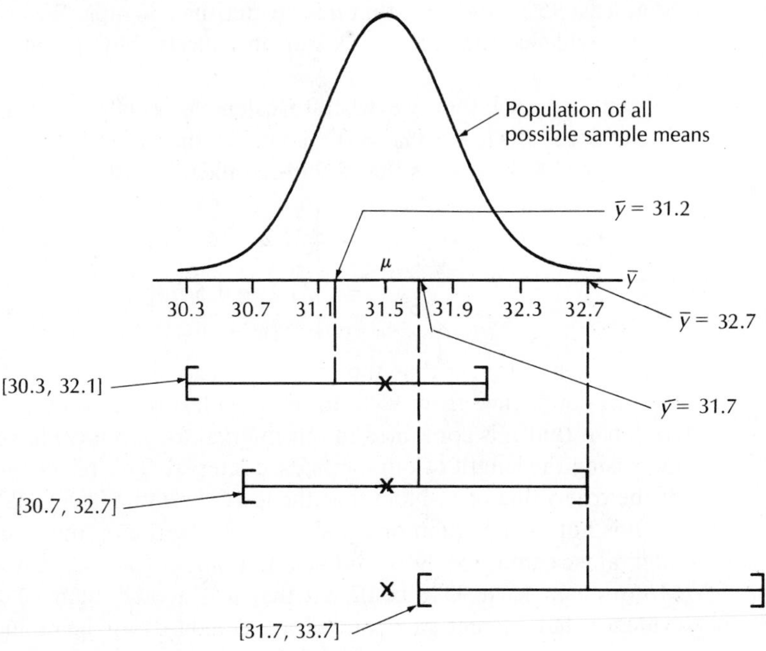

(b) The sample mean $\bar{y}$, sample standard deviation s, and 95% confidence interval for μ, $[\bar{y} \pm 2.776 \frac{s}{\sqrt{5}}]$, given by each of the three possible samples

Sample	*Sample*	*Sample*
$y_1 = 30.7$	$y_1 = 32.2$	$y_1 = 33.7$
$y_2 = 31.8$	$y_2 = 30.6$	$y_2 = 31.6$
$y_3 = 30.2$	$y_3 = 31.7$	$y_3 = 33.3$
$y_4 = 32.0$	$y_4 = 31.3$	$y_4 = 32.3$
$y_5 = 31.3$	$y_5 = 32.7$	$y_5 = 32.6$
$\bar{y} = 31.2$	$\bar{y} = 31.7$	$\bar{y} = 32.7$
$s = .7517$	$s = .8093$	$s = .8276$
$[31.2 \pm 2.776 \frac{.7517}{\sqrt{5}}]$	$[31.7 \pm 2.776 \frac{.8093}{\sqrt{5}}]$	$[32.7 \pm 2.776 \frac{.8276}{\sqrt{5}}]$
$= [30.3, 32.1]$	$= [30.7, 32.7]$	$= [31.7, 33.7]$

Suppose that federal mileage standards also state that three years from now, μ must be at least 33 mpg. The 95% confidence interval for μ makes us at least 95% confident that μ is less than 33 mpg. Therefore the company is very confident that the current model of the Hawk will not meet gasoline mileage standards three years from now. The company should probably begin a research and development project to increase Hawk gasoline mileage.

To complete this example, we will discuss the meaning of 95% confidence as it pertains to the above 95% confidence interval. Figure 3.1(a) depicts three possible samples from the *population of all possible samples of five mileages* that could have been randomly selected from the infinite population of all Hawk mileages. Note that the population of all mileages is illustrated as normally distributed with mean μ and standard deviation σ equal to 31.5 and .8, respectively. In Figure 3.1(b) we summarize and depict the sample mean $\bar{y}$, sample standard deviation s, and 95% confidence interval for μ—$[\bar{y} \pm 2.776(s/\sqrt{5})]$—given by each of the three possible samples. Note that two of the three 95% confidence intervals for $\mu = 31.5$ in Figure 3.1 contain μ. The interpretation of 95% confidence here is that 95 percent of the 95% confidence intervals for μ *in the population of all such intervals* contain $\mu = 31.5$, while 5 percent of the confidence intervals in this population do not contain μ. Thus when we compute a 95% confidence interval for μ, we can be 95% confident that μ is contained in our interval. This is because 95 percent of the intervals in the population of all possible 95% confidence intervals for μ contain μ and because we have obtained one of the confidence intervals in this population.

THE DERIVATION OF THE INTERVAL

To derive the $100(1 - \alpha)\%$ confidence interval for μ, we state several important results.

Properties of the Population of All Possible Sample Means

The *population of all possible sample means* (that is, point estimates of μ)

1. Has mean $\mu_{\bar{y}} = \mu$.
2. Has variance $\sigma_{\bar{y}}^2 = \sigma^2/n$ (if the population sampled is infinite).
3. Has *standard deviation* $\sigma_{\bar{y}} = \sigma/\sqrt{n}$ (if the population sampled is infinite).
4. Has a *normal distribution* (if the population sampled has a normal distribution).

See Appendix A for a proof of (1), (2), and (3).

Result 1,

$$\mu_{\bar{y}} = \mu$$

says that $\mu_{\bar{y}}$, the mean of all possible sample means, equals μ, the population mean. For this reason, when we use the sample mean $\bar{y}$ as the point estimate of μ, we are using an *unbiased* estimation procedure. This property of unbiasedness tells us that although the sample mean $\bar{y}$ that we calculate probably does not equal μ, the average of all the different sample means that we could have calculated (from all the different possible samples) *is* equal to μ.

We note that $\sigma_{\bar{y}}^2$ and $\sigma_{\bar{y}}$, the variance and standard deviation of the population of all possible sample means, measure the variation, or spread, of the different possible sample means. Here a large variance $\sigma_{\bar{y}}^2$ indicates that the sample means are widely dispersed around μ. A small variance $\sigma_{\bar{y}}^2$ indicates that the sample means are clustered closely around μ. Now consider result 2:

$$\sigma_{\bar{y}}^2 = \frac{\sigma^2}{n}$$

Note that since each possible sample mean is an average of n sample values, the sample mean "averages out" high and low sample values. Therefore we would expect that the sample means are more closely clustered around μ than are the individual population values. That is, intuitively, $\sigma_{\bar{y}}^2$ should be smaller than the population variance σ^2. Result 2 says that this is the case—that the division by n makes $\sigma_{\bar{y}}^2$ smaller than σ^2. Furthermore, this result says that the larger the sample size is, the smaller is $\sigma_{\bar{y}}^2$. That is, when n is larger, more sample values are used to compute each possible sample mean. This results in the sample means being clustered even more closely around μ.

Example 3.2

In the gasoline mileage problem, recall that the true values of μ, σ^2, and σ, the mean, variance, and standard deviation of the infinite population of all mileages, are 31.5, .64, and .8, respectively. Therefore the mean, variance, and standard deviation of the infinite population of all possible sample means (that would be calculated by using all possible samples of size $n = 5$) are, respectively,

$$\mu_{\bar{y}} = \mu = 31.5$$

$$\sigma_{\bar{y}}^2 = \frac{\sigma^2}{n} = \frac{.64}{5} = .128$$

$$\sigma_{\bar{y}} = \frac{\sigma}{\sqrt{n}} = \frac{.8}{\sqrt{5}} = .358$$

Since $\sigma = .8$ and $\sigma_{\bar{y}} = .358$, we see that the sample means are more closely clustered around μ than are the individual mileages. Moreover, assume that the population of all mileages is normally distributed. Then the population of all possible sample means is also normally distributed (see Figure 3.1). Thus as discussed in Section 2.5, 68.26%, 95.44%, and 99.73% of all possible sample means lie in, respectively, the intervals

$$[\mu_{\bar{y}} \pm \sigma_{\bar{y}}] = [31.5 \pm .358] = [31.142, 31.858]$$

$$[\mu_{\bar{y}} \pm 2\sigma_{\bar{y}}] = [31.5 \pm 2(.358)] = [31.5 \pm .716] = [30.784, 32.216]$$

$$[\mu_{\bar{y}} \pm 3\sigma_{\bar{y}}] = [31.5 \pm 3(.358)] = [31.5 \pm 1.074] = [30.426, 32.574]$$

These intervals are narrower than the intervals containing 68.26%, 95.44%, and 99.73% of the individual mileages. (See Example 2.4.)

Results 1, 2, and 3 imply that if the population that is sampled is normally distributed, then the population of all possible values of

$$\frac{\bar{y} - \mu_{\bar{y}}}{\sigma_{\bar{y}}} = \frac{\bar{y} - \mu}{\sigma/\sqrt{n}}$$

has a standard normal distribution. We estimate $\sigma_{\bar{y}} = \sigma/\sqrt{n}$ by $s_{\bar{y}} = s/\sqrt{n}$, which is called the *standard error of the estimate* $\bar{y}$. Then it can be proven that if the population that is sampled is normally distributed, the population of all possible values of

$$\frac{\bar{y} - \mu}{s_{\bar{y}}} = \frac{\bar{y} - \mu}{s/\sqrt{n}}$$

has a t-distribution with $n - 1$ degrees of freedom. This implies that

$$P\left(-t_{[\alpha/2]}^{(n-1)} \leq \frac{\bar{y} - \mu}{s/\sqrt{n}} \leq t_{[\alpha/2]}^{(n-1)}\right)$$

is the area under the curve of the t-distribution having $n - 1$ degrees of freedom between $-t_{[\alpha/2]}^{(n-1)}$ and $t_{[\alpha/2]}^{(n-1)}$. As illustrated in Figure 3.2, this area equals $1 - \alpha$. Multiplying the inequality in the probability statement

$$P\left(-t_{[\alpha/2]}^{(n-1)} \leq \frac{\bar{y} - \mu}{s/\sqrt{n}} \leq t_{[\alpha/2]}^{(n-1)}\right) = 1 - \alpha$$

by $s/\sqrt{n}$ (which is positive), we obtain

$$P\left[-t_{[\alpha/2]}^{(n-1)}\left(\frac{s}{\sqrt{n}}\right) \leq \bar{y} - \mu \leq t_{[\alpha/2]}^{(n-1)}\left(\frac{s}{\sqrt{n}}\right)\right] = 1 - \alpha$$

This implies (subtracting $\bar{y}$ through the above inequality) that

$$P\left[-\bar{y} - t_{[\alpha/2]}^{(n-1)}\left(\frac{s}{\sqrt{n}}\right) \leq -\mu \leq -\bar{y} + t_{[\alpha/2]}^{(n-1)}\left(\frac{s}{\sqrt{n}}\right)\right] = 1 - \alpha$$

FIGURE 3.2 $\mathbf{P\left(-t_{[\alpha/2]}^{(n-1)} \leq \frac{\bar{y} - \mu}{s/\sqrt{n}} \leq t_{[\alpha/2]}^{(n-1)}\right) = 1 - \alpha}$

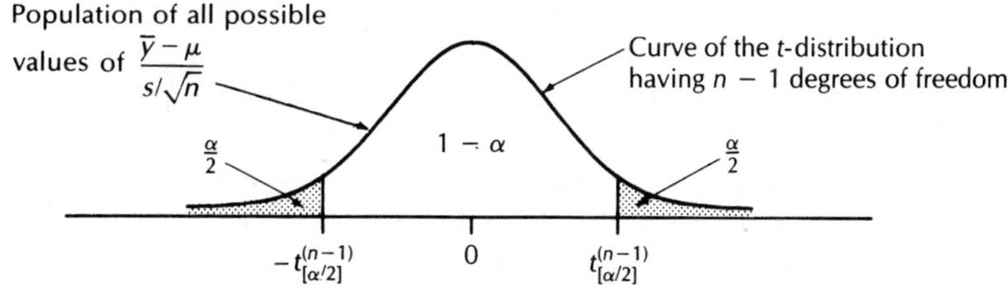

which in turn implies (multiplying the above inequality by -1) that

$$P\left[\bar{y} + t_{[\alpha/2]}^{(n-1)}\left(\frac{s}{\sqrt{n}}\right) \geq \mu \geq \bar{y} - t_{[\alpha/2]}^{(n-1)}\left(\frac{s}{\sqrt{n}}\right)\right] = 1 - \alpha$$

This probability statement is equivalent to

$$P\left[\bar{y} - t_{[\alpha/2]}^{(n-1)}\left(\frac{s}{\sqrt{n}}\right) \leq \mu \leq \bar{y} + t_{[\alpha/2]}^{(n-1)}\left(\frac{s}{\sqrt{n}}\right)\right] = 1 - \alpha$$

This says that the proportion of confidence intervals containing the population mean μ in the population of all possible $100(1 - \alpha)\%$ confidence intervals for μ is equal to $1 - \alpha$. That is, suppose that we compute a $100(1 - \alpha)\%$ confidence interval for μ by using the formula

$$\left[\bar{y} \pm t_{[\alpha/2]}^{(n-1)}\left(\frac{s}{\sqrt{n}}\right)\right]$$

Then $100(1 - \alpha)\%$ (for example, 95%) of the confidence intervals in the population of all possible $100(1 - \alpha)\%$ confidence intervals for μ contain μ, and $100(\alpha)\%$ (for example, 5%) of the confidence intervals in this population do not contain μ.

CONFIDENCE INTERVALS BASED ON THE NORMAL DISTRIBUTION

The preceding confidence interval is based on the t-distribution. It assumes that the population sampled is normally distributed (or at least is mound-shaped). We now consider a confidence interval that is valid for any population.

A Large Sample Confidence Interval for μ Based on the Normal Distribution

The *Central Limit Theorem* states that if the sample size n is large (say, at least 30), then the population of all possible sample means approximately has a normal distribution (with mean $\mu_{\bar{y}} = \mu$ and standard deviation $\sigma_{\bar{y}} = \sigma/\sqrt{n}$), no matter what probability distribution describes the population sampled (see Figure 3.3). Therefore *if n is large*, the population of all possible values of

$$\frac{\bar{y} - \mu_{\bar{y}}}{\sigma_{\bar{y}}} = \frac{\bar{y} - \mu}{\sigma/\sqrt{n}}$$

approximately has a *standard normal distribution*. This implies that

$$\left[\bar{y} \pm z_{[\alpha/2]}\left(\frac{\sigma}{\sqrt{n}}\right)\right] \quad \text{and} \quad \left[\bar{y} \pm z_{[\alpha/2]}\left(\frac{s}{\sqrt{n}}\right)\right]$$

are *approximately correct $100(1 - \alpha)\%$ confidence intervals for μ, no matter what probability distribution describes the population sampled.* Here the second interval follows from the first by approximating σ by s.

FIGURE 3.3 **The Central Limit Theorem**

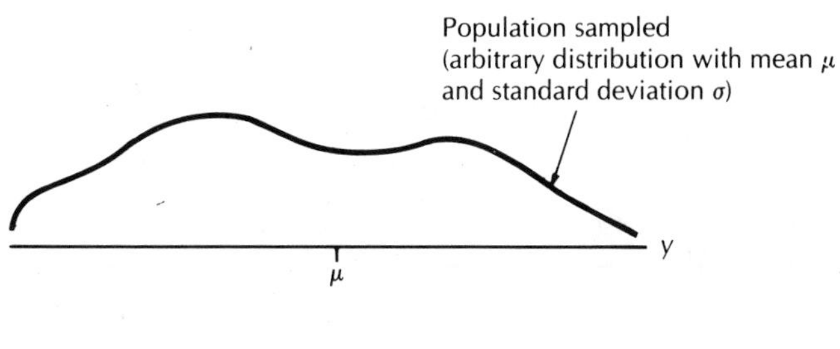

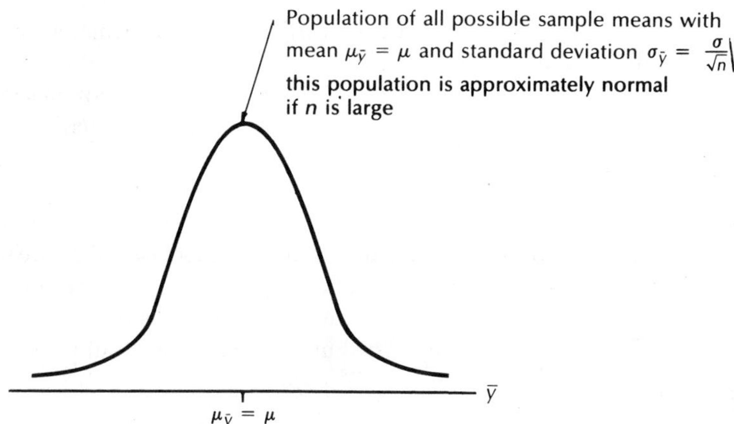

To derive the first of these intervals, consider the z value $z_{[\alpha/2]}$ and Figure 3.4. The fact that the population of all possible values of

$$\frac{\bar{y} - \mu}{\sigma/\sqrt{n}}$$

approximately has a standard normal distribution implies that

$$P\left(-z_{[\alpha/2]} \leq \frac{\bar{y} - \mu}{\sigma/\sqrt{n}} \leq z_{[\alpha/2]}\right) \approx 1 - \alpha$$

This probability is the area under the curve of the standard normal distribution between $-z_{[\alpha/2]}$ and $z_{[\alpha/2]}$. Using algebraic manipulations analogous to those carried out in the preceding subsections, we find that

$$P\left[\bar{y} - z_{[\alpha/2]}\left(\frac{\sigma}{\sqrt{n}}\right) \leq \mu \leq \bar{y} + z_{[\alpha/2]}\left(\frac{\sigma}{\sqrt{n}}\right)\right] \approx 1 - \alpha$$

FIGURE 3.4

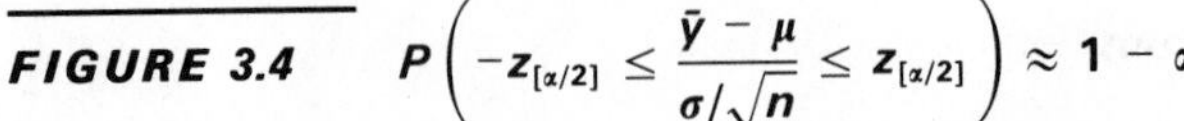

$$P\left(-z_{[\alpha/2]} \leq \frac{\bar{y} - \mu}{\sigma/\sqrt{n}} \leq z_{[\alpha/2]}\right) \approx 1 - \alpha$$

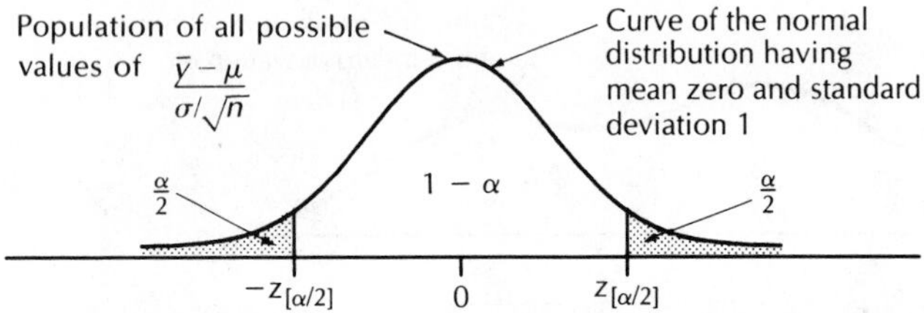

This implies that $[\bar{y} \pm z_{[\alpha/2]}(\sigma/\sqrt{n})]$ is an approximately correct $100(1 - \alpha)\%$ confidence interval for μ.

A more precise statement of the Central Limit Theorem says that the larger the sample size n is, the more nearly normally distributed is the population of all possible sample means. This is illustrated in Figure 3.5 for several sampled population shapes. This figure also illustrates that the larger n is, then the smaller is $\sigma_{\bar{y}} = \sigma/\sqrt{n}$. Recall that $\sigma_{\bar{y}}$ measures the spread of the population of all possible sample means. Figure 3.5 shows that as n increases, the spread of this population decreases. There are some indications that, for many populations, computing sample means for samples of size $n = 10$, or possibly even $n = 5$, will produce an approximately normally distributed population of all possible sample means. In such cases,

$$[\bar{y} \pm z_{[\alpha/2]}(\sigma/\sqrt{n})]$$

is an approximately correct $100(1 - \alpha)\%$ confidence interval for μ.

However, in practice we rarely know the true value of the population standard deviation σ. Moreover, practice indicates that n should be at least 30 for the sample standard deviation s to be an accurate point estimate of σ. Therefore n should probably be at least 30 for $[\bar{y} \pm z_{[\alpha/2]}(s/\sqrt{n})]$ to be an approximately correct $100(1 - \alpha)\%$ confidence interval for μ.

To summarize, when we do not know the true value of the population standard deviation σ, we should use the $100(1 - \alpha)\%$ confidence interval for μ based on the normal distribution

$$\left[\bar{y} \pm z_{[\alpha/2]}\left(\frac{s}{\sqrt{n}}\right)\right]$$

if the sample size n is large (say, at least 30). If the sample size n is small and the population sampled is normally distributed (or at least mound-shaped), we should use the $100(1 - \alpha)\%$ confidence interval for μ based on the t-distribution

$$\left[\bar{y} \pm t_{[\alpha/2]}^{(n-1)}\left(\frac{s}{\sqrt{n}}\right)\right]$$

FIGURE 3.5 **An illustration of the Central Limit Theorem**

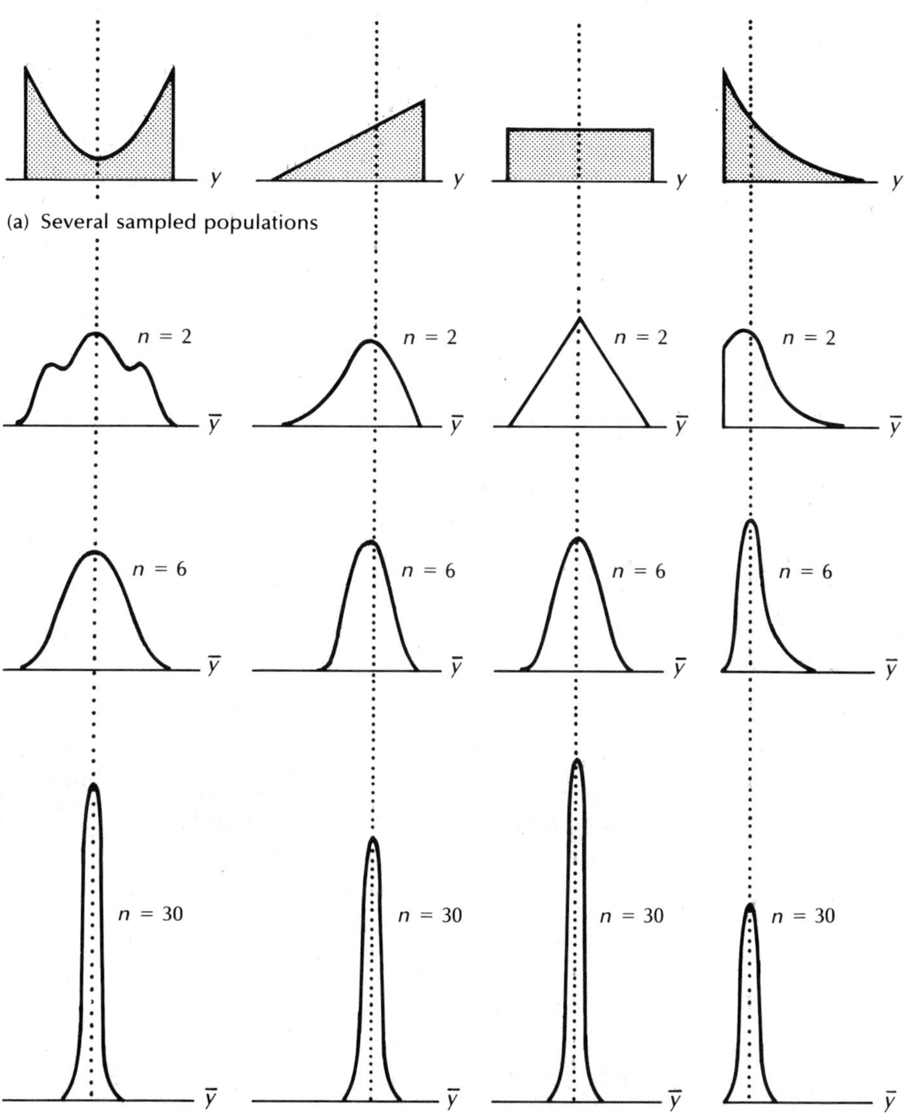

(b) Corresponding populations of all possible sample means for different sample sizes

Example 3.3 Suppose that National Motors wishes to claim that the mean of all Hawk mileages is greater than the 31 mpg average claimed by a competitor. The company randomly selects the sample of $n = 49$ mileages in Table 2.3. This sample has mean $\bar{y} = 31.5531$ and standard deviation $s = .799$. It follows that $\bar{y} = 31.5531$ is the point estimate of the population mean, μ. Moreover, since $n = 49$ is large, a 95% confidence interval for μ is

$$\left[\bar{y} \pm z_{[.025]}\left(\frac{s}{\sqrt{n}}\right)\right] = \left[31.5531 \pm 1.96\left(\frac{.799}{\sqrt{49}}\right)\right]$$
$$= [31.5531 \pm .2237]$$
$$= [31.3, 31.8]$$

Here $z_{[.025]}$ is obtained from Table E.1 in Appendix E. Since this 95% interval for μ has a lower bound of 31.3, we can be very confident that μ is greater than 31 mpg. This convinces the company that it can legitimately claim in a new advertising campaign that μ is greater than 31 mpg.

SAMPLE SIZE DETERMINATION

Consider the large sample $100(1 - \alpha)\%$ confidence interval for μ. The quantity

$$z_{[\alpha/2]}\left(\frac{\sigma}{\sqrt{n}}\right)$$

is the *large sample $100(1 - \alpha)\%$ bound on the error of estimation obtained when estimating μ by $\bar{y}$*. Consequently, we are $100(1 - \alpha)\%$ confident that the sample mean $\bar{y}$ is within

$$z_{[\alpha/2]}\left(\frac{\sigma}{\sqrt{n}}\right)$$

units of the population mean, μ. Suppose we wish to determine the sample size n so that we are $100(1 - \alpha)\%$ confident that $\bar{y}$ is within B units of μ. By setting the above bound equal to B and solving for n we can determine the necessary sample size. We do this as follows

$$z_{[\alpha/2]}\left(\frac{\sigma}{\sqrt{n}}\right) = B$$

This implies that

$$z_{[\alpha/2]}\sigma = \sqrt{n}B$$

and that

$$n = \left[\frac{z_{[\alpha/2]}\sigma}{B}\right]^2$$

To summarize,

Sample Size Determination When Estimating a Population Mean

A sample of size

$$n = \left[\frac{z_{[\alpha/2]}\sigma}{B}\right]^2$$

makes us $100(1 - \alpha)\%$ confident that the sample mean, $\bar{y}$, is within B units of the population mean μ.

This formula involves the population standard deviation σ, which is probably unknown. Thus we must often find an estimate of σ. The following are three ways to obtain this estimate:

1. A theorem called Chebyshev's Theorem implies that

   $$\sigma \approx \frac{R}{4}$$

   Here R is the range of the values in the population to be sampled. Therefore if we know R, we can obtain a rough estimate of σ by dividing R by 4.
2. Sometimes theory or knowledge concerning a population similar to the population to be sampled allows us to obtain a rough estimate of σ.
3. We can calculate the standard deviation, denoted by s_p, of a preliminary sample. This preliminary sample consists of n_p values randomly selected from the population. If $n_p - 1$ is at least 30, we calculate n by the formula

   $$n = \left[\frac{z_{[\alpha/2]}s_p}{B}\right]^2$$

   If $n_p - 1$ is less than 30, we calculate n by the formula

   $$n = \left[\frac{t_{[\alpha/2]}^{(n_p-1)}s_p}{B}\right]^2$$

Example 3.4 In Example 3.3, National Motors calculated a 95% confidence interval for the mean mileage μ by using a sample of $n = 49$ mileages. The sample size $n = 49$ was obtained by using the above sample size formula. Specifically, recall from Example 3.1 that we used a sample of $n = 5$ mileages to calculate a 95% confidence interval for μ. We obtained the interval [30.3, 32.1]. Here $\bar{y} = 31.2$ is greater than 31, the mean mileage claimed by a competitor. However, the bound

$$t_{[\alpha/2]}^{(n-1)}\left(\frac{s}{\sqrt{n}}\right) = 2.776\left(\frac{.7517}{\sqrt{5}}\right) = .9333$$

is greater than .2. Therefore the lower bound of the above interval is less than 31. It follows that National Motors cannot use this interval to conclude that μ is at least 31. Since we wish to show that μ is at least 31, it might seem reasonable to determine the sample size so that we are 95% confident that $\bar{y}$ is within $B = .2$ of μ. Or to prove that μ is greater than 31, it might seem reasonable to choose B equal to .19. However, a large sample will probably give a sample mean that is different from the previously obtained 31.2 mpg. Suppose that we believe that the true value of μ is greater than 31.2. Therefore we believe that a large sample will give a sample mean greater than 31.2. On the basis of this belief, National Motors feels (somewhat arbitrarily) that a confidence interval with a bound of .3 will have a lower bound that is greater than 31. Therefore we wish to determine the sample size n so that we are 95% confident that $\bar{y}$ is within $B = .3$ of μ. The company can regard the previously observed sample of five mileages, which has standard deviation .7517, as a preliminary sample. Then, since

$$t_{[.025]}^{(n_p-1)} = t_{[.025]}^{(5-1)} = 2.776$$

we should randomly select a sample of size

$$n = \left[\frac{t_{[\alpha/2]}^{(n_p-1)} s_p}{B}\right]^2 = \left[\frac{(2.776)(.7517)}{.3}\right]^2 = 48.38, \text{ or } 49 \quad \text{(rounding up)}$$

Note that 48.38 has been rounded up to 49 to guarantee that our confidence interval is at least as precise as specified by setting $B = .3$ mpg. Rounding down would result in a less precise interval.

Next, recall that the true value of μ is 31.5. Also recall that the mean of the sample of $n = 49$ mileages in Table 2.3 is $\bar{y} = 31.5531$. Thus for this sample, $\bar{y}$ is within $B = .3$ of μ. Another sample of 49 mileages would yield a different sample mean $\bar{y}$, which would be a different distance from μ. When we say that a sample of 49 mileages makes us 95% confident that $\bar{y}$ is within .3 of μ, we mean that 95 percent of all possible sample means based on $n = 49$ mileages are within .3 of μ and 5 percent of such sample means are not. Thus when we compute the point estimate of μ by using the observed sample and obtain the sample mean $\bar{y} = 31.5531$, we can be 95 percent confident that this sample mean is within .3 of μ. This is because 95 percent of the sample means in the population of all possible sample means are within .3 of μ and because we have obtained one of the sample means in this population.

3.2

HYPOTHESIS TESTING FOR A POPULATION MEAN

TESTING $H_0: \mu = c$ VERSUS $H_1: \mu \neq c$

Example 3.5 The G & B Corporation produces a 16-ounce bottle of Gem Shampoo. The bottles are filled by an automated bottle-filling process. If, for a particular adjustment of the bottle-filling process, this process is substantially overfilling bottles or underfilling bottles, then this process must be shut down and readjusted. Overfilling results in lost profits for G & B, while underfilling is unfair to consumers. For a given adjustment of the bottle-filling process we consider the infinite population of all bottles that could potentially be produced. For each bottle there is a corresponding bottle fill (measured in ounces). We let μ denote the mean of the infinite population of all the bottle fills that could potentially be produced by (the particular adjustment of) the bottle-filling process. G & B has decided that it will shut down and readjust the process if it can be very certain that it should reject the null hypothesis H_0: $\mu = 16$ in favor of the alternative hypothesis H_1: $\mu \neq 16$. Here H_0 says that the mean bottle fill is at the appropriate level. H_1 says that the mean fill is above or below the desired 16 ounces.

As illustrated in Example 3.5, we sometimes wish to test the null hypothesis $H_0: \mu = c$ versus the alternative hypothesis H_1: $\mu \neq c$. Here μ is a population mean, and c is an arbitrary constant. The classical approach to testing these hypotheses utilizes the *test statistic*

$$t = \frac{\bar{y} - c}{s/\sqrt{n}}$$

Here $\bar{y}$ and s are the mean and standard deviation of a sample of size n that has been randomly selected from the population having mean μ. The test statistic t measures the distance between $\bar{y}$ and c (the value that makes H_0 true). A test statistic nearly (or exactly) equal to zero results when $\bar{y}$ is nearly (or exactly) equal to c. Such a test statistic provides little or no evidence to support rejecting H_0 in favor of H_1. This is because the point estimate $\bar{y}$ indicates that μ is nearly or exactly equal to c. However, a positive test statistic substantially greater than zero results when $\bar{y}$ is substantially greater than c. This provides evidence to support rejecting H_0 in favor of H_1 because the point estimate $\bar{y}$ indicates that μ is greater than c. Similarly, a negative test statistic substantially less than zero results when $\bar{y}$ is substantially smaller than c. This also provides evidence to support rejecting H_0 in favor of H_1 because the point estimate $\bar{y}$ indicates that μ is smaller than c.

To decide how large in absolute value the test statistic must be before we reject H_0, we consider the errors that can be made in hypothesis testing. These errors are summarized in Table 3.1 in the context of the Gem Shampoo bottle-filling example.

TABLE 3.1 **Type I and Type II errors in the Gem bottle fill example**

	State of nature	
Decisions made in hypothesis test	***Null hypothesis $H_0: \mu = 16$ is true: Process mean is at the correct level***	***Null hypothesis $H_0: \mu = 16$ is false: Process mean is not correct***
Reject null hypothesis $H_0: \mu = 16$. Readjust the process	Type I error: Readjust the process when the process mean is at the correct level	Correct action: Readjust the process when the process mean is not correct
Do not reject null hypothesis $H_0: \mu = 16$. Do not readjust the process	Correct action: Do not readjust the process when the process mean is at the correct level	Type II error: Do not readjust the process when the process mean is not correct

A *Type I error* is committed if we reject $H_0: \mu = 16$ when H_0 is true. This means that we would readjust the process when the process mean is at the correct level. A Type II error is committed if we do not reject $H_0: \mu = 16$ when H_0 is false. This means that we would not readjust the process when the process mean is not correct.

We obviously desire that both *the probability of a Type I error, denoted by α, and the probability of a Type II error be small.* It is common procedure to base an hypothesis test on taking a sample of a fixed size and on setting α equal to a specified value. Here we sometimes choose α to be as high as .1, but we usually choose α to be between .05 and .01. The most frequent choice for α is .05. We usually do not set α lower than .01 because setting α extremely small often leads to a probability of a Type II error that is unacceptably large. Generally, for a fixed sample size the lower the probability of a Type I error, the higher the probability of a Type II error. We consider how to determine the precise value of α that should be used after discussing the procedure for testing H_0.

Testing $H_0: \mu = c$ Versus $H_1: \mu \neq c$ by Using Rejection Points

Define the test statistic

$$t = \frac{\bar{y} - c}{s/\sqrt{n}}$$

If the population sampled is normally distributed with mean μ, we can reject $H_0: \mu = c$ in favor of $H_1: \mu \neq c$ by setting the probability of a Type I error equal to α if and only if

$$|t| > t_{[\alpha/2]}^{(n-1)} \quad \text{that is, if} \quad t > t_{[\alpha/2]}^{(n-1)} \quad \text{or} \quad t < -t_{[\alpha/2]}^{(n-1)}$$

FIGURE 3.6 **The rejection points for testing $H_0: \mu = c$ versus $H_1: \mu \neq c$. The probability that we reject $H_0: \mu = c$ when H_0 is true equals $\alpha/2 + \alpha/2 = \alpha$.**

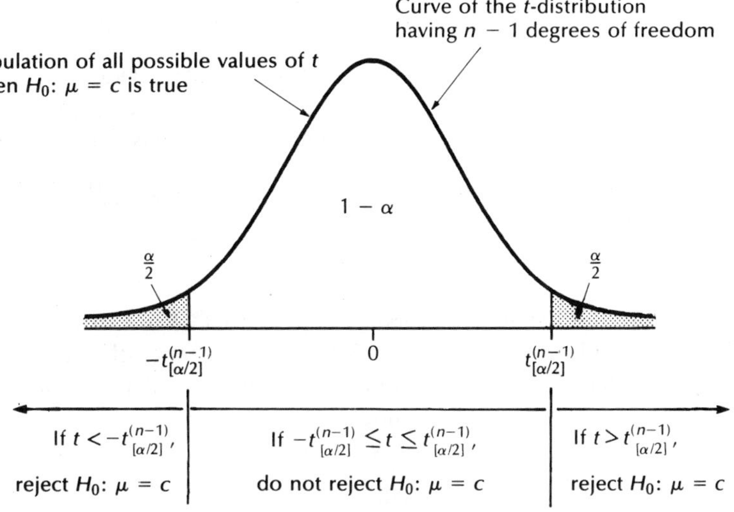

In the above procedure we call the points $-t_{[\alpha/2]}^{(n-1)}$ and $t_{[\alpha/2]}^{(n-1)}$ *rejection points*. This is because these points tell us how different from zero t must be for us to be able to reject H_0 by setting the probability of a Type I error equal to α.

We now consider why using this rejection point procedure ensures that the probability of a Type I error equals α. Recall that if the population sampled is normally distributed with mean μ, then the population of all possible values of

$$\frac{\bar{y} - \mu}{s/\sqrt{n}}$$

has a t-distribution with $n - 1$ degrees of freedom. It follows that if the null hypothesis $H_0: \mu = c$ is true, then the population of all possible values of the test statistic

$$t = \frac{\bar{y} - c}{s/\sqrt{n}}$$

has a t-distribution with $n - 1$ degrees of freedom. Therefore using the above rejection points says that if $H_0: \mu = c$ is true, then as illustrated in Figure 3.6:

1. The probability that

   $$-t_{[\alpha/2]}^{(n-1)} \leq t \leq t_{[\alpha/2]}^{(n-1)}$$

   is $1 - \alpha$. That is, $100(1 - \alpha)\%$ (for example, $100(1 - .05)\% = 95\%$)

of all possible values of t are between $-t_{[\alpha/2]}^{(n-1)}$ and $t_{[\alpha/2]}^{(n-1)}$ and thus lead us to fail to reject $H_0: \mu = c$ (when $H_0: \mu = c$ is true), a correct decision.

2. The probability that

$$t < -t_{[\alpha/2]}^{(n-1)} \quad \text{or} \quad t > t_{[\alpha/2]}^{(n-1)}$$

is α. That is, $100(\alpha)\%$ (for example, $100(.05)\% = 5\%$) of all possible values of t are either less than $-t_{[\alpha/2]}^{(n-1)}$ or greater than $t_{[\alpha/2]}^{(n-1)}$ and lead us to reject $H_0: \mu = c$ (when $H_0: \mu = c$ is true), a Type I error.

Therefore using these rejection points implies that the probability of a Type I error equals α.

Example 3.6

To test $H_0: \mu = 16$ versus $H_1: \mu \neq 16$, G & B Corporation will randomly select a sample of $n = 6$ bottle fills from the current adjustment of the Gem bottle-filling process. We assume that the population of all possible bottle fills is normally distributed or at least mound-shaped. The rejection point condition tells us that we can reject $H_0: \mu = 16$ by setting α, the probability of a Type I error, equal to .05 if and only if the absolute value of

$$t = \frac{\bar{y} - 16}{s/\sqrt{n}} = \frac{\bar{y} - 16}{s/\sqrt{6}}$$

is greater than $t_{[\alpha/2]}^{(n-1)} = t_{[.05/2]}^{(6-1)} = t_{[.025]}^{(5)} = 2.571$. That is, we can reject H_0 if t is greater than $t_{[.025]}^{(5)} = 2.571$ or less than $-t_{[.025]}^{(5)} = -2.571$.

Now suppose that G & B Corporation observes the following sample of $n = 6$ bottle fills:

$$\{y_1, y_2, y_3, y_4, y_5, y_6\} = \{15.68, 16.00, 15.61, 15.93, 15.86, 15.72\}$$

It can be verified that this sample has mean $\bar{y} = 15.8$ and standard deviation $s = .1532$. It follows that

$$t = \frac{\bar{y} - 16}{s/\sqrt{n}} = \frac{15.8 - 16}{.1532/\sqrt{6}} = -3.2$$

Since $|t| = 3.2 > 2.571 = t_{[.025]}^{(5)}$, we can reject $H_0: \mu = 16$ by setting α equal to .05.

As another example, we can reject $H_0: \mu = 16$ in favor of $H_1: \mu \neq 16$ by setting α equal to .01 if and only if the absolute value of t is greater than $t_{[\alpha/2]}^{(n-1)} = t_{[.01/2]}^{(6-1)} = t_{[.005]}^{(5)} = 4.032$. That is, we can reject H_0 if $t > 4.032$ or $t < -4.032$. Specifically, since $t = -3.2$ is between -4.032 and 4.032, we cannot reject $H_0: \mu = 16$ by setting α equal to .01.

We next discuss a link between hypothesis testing and confidence intervals.

Testing $H_0: \mu = c$ Versus $H_1: \mu \neq c$ by Using a Confidence Interval

If the population sampled is normally distributed with mean μ, we can reject $H_0: \mu = c$ in favor of $H_1: \mu \neq c$ by setting the probability of a Type I error equal to α if and only if the $100(1 - \alpha)\%$ confidence interval for μ,

$$\left[\bar{y} \pm t_{[\alpha/2]}^{(n-1)}\left(\frac{s}{\sqrt{n}}\right)\right]$$

does not contain c.

This condition is intuitive and can easily be shown to be mathematically equivalent to the previously discussed rejection point condition.

Example 3.7 Consider the Gem Shampoo bottle fill problem. We can reject $H_0: \mu = 16$ in favor of $H_1: \mu \neq 16$ by setting the probability of a Type I error equal to α if and only if the $100(1 - \alpha)\%$ confidence interval for μ does not contain 16. From the information in Example 3.6 the $100(1 - \alpha)\%$ confidence interval for μ is

$$\begin{aligned}\left[\bar{y} \pm t_{[\alpha/2]}^{(n-1)}\left(\frac{s}{\sqrt{n}}\right)\right] &= \left[15.8 \pm t_{[\alpha/2]}^{(6-1)}\left(\frac{.1532}{\sqrt{6}}\right)\right] \\ &= [15.8 \pm t_{[\alpha/2]}^{(5)}(.0625)]\end{aligned}$$

For example, the 95% confidence interval for μ,

$$\begin{aligned}[15.8 \pm t_{[.025]}^{(5)}(.0625)] &= [15.8 \pm 2.571(.0625)] \\ &= [15.64, 15.96]\end{aligned}$$

does not contain 16. Thus we can reject $H_0: \mu = 16$ in favor of $H_1: \mu \neq 16$ by setting α equal to .05. As another example, the 98% confidence interval for μ,

$$\begin{aligned}[15.8 \pm t_{[.01]}^{(5)}(.0625)] &= [15.8 \pm 3.365(.0625)] \\ &= [15.59, 16.01]\end{aligned}$$

does contain 16. Thus we cannot reject $H_0: \mu = 16$ in favor of $H_1: \mu \neq 16$ by setting α equal to .02.

CONSIDERATIONS IN SETTING α AND THE DISTINCTION BETWEEN STATISTICAL AND PRACTICAL SIGNIFICANCE

Note that the lower we set α, the smaller the probability that we will reject $H_0: \mu = c$ when H_0 is true. Therefore the lower the value of α at which H_0 can be rejected by using the test statistic t, the stronger the evidence that $H_0: \mu = c$ is false and $H_1: \mu \neq c$ is true. Practice has shown that

1. If we can reject $H_0: \mu = c$ in favor of $H_1: \mu \neq c$ by setting α *equal to .05*, we have *strong evidence* that H_0 is false and H_1 is true.

2. If we can reject $H_0 : \mu = c$ in favor of $H_1 : \mu \neq c$ by setting α *equal to .01*, we have *very strong evidence* that H_0 is false and H_1 is true.

However, we usually do not require that H_0 be rejected in favor of H_1 at a value of α less than .01 before deciding that H_0 is false and H_1 is true. This is because, as was previously stated, setting α extremely small leads to an unacceptably large probability of a Type II error (not rejecting H_0 when H_0 is false). For example, consider the Gem Shampoo bottle fill problem. We saw in Example 3.6 that we can reject $H_0 : \mu = 16$ in favor of $H_1 : \mu \neq 16$ by setting α equal to .05 but not by setting α equal to .01. Therefore we have strong evidence that $H_0 : \mu = 16$ is false and $H_1 : \mu \neq 16$ is true.

If we have decided to take a particular action if and only if we can reject $H_0 : \mu = c$ in favor of $H_1 : \mu \neq c$ by setting the probability of a Type I error equal to α, then we must decide how to set α. We stated previously that it is reasonable to set α between .05 and .01. In setting a specific value of α (say, between .05 and .01) we should take into account the relative costs of making Type I and Type II errors. For example, suppose that G & B Corporation has decided that it will shut down and readjust the bottle-filling process if it can reject $H_0 : \mu = 16$ in favor of $H_1 : \mu \neq 16$ by setting α equal to .05. From Table 3.1 it is clear that setting α equal to .05 implies that this procedure will cause the process to be readjusted 5 percent of the time when the process mean is at the correct level. Here G & B has decided to set α at .05 rather than .01 because the company feels that making a Type II error (not readjusting when the process mean is not correct) is very serious. Setting α at .05 makes the probability of a Type II error smaller than would setting α at .01.

We have discussed two equivalent methods of testing hypotheses, the rejection point and confidence interval methods. One advantage of using confidence intervals is that in addition to determining whether we can reject $H_0 : \mu = c$ by setting α equal to a fixed value, we obtain an interval of numbers that we are quite sure contains μ. This can help us to distinguish between *statistical significance* and *practical significance*. For example, in the bottle-filling problem we can reject $H_0 : \mu = 16$ in favor of $H_1 : \mu \neq 16$ by setting $\alpha = .05$. Thus we say that the hypothesis test has *statistical significance* at a .05 *significance level* (or, equivalently, at a 95% *confidence level*). Whether this result has practical significance, however, is somewhat questionable. The 95% confidence interval for μ was calculated in Example 3.7 to be [15.64, 15.96]. Therefore we can be 95% confident that μ is between 15.64 and 15.96 ounces. So while the mean bottle fill is probably somewhat below the desired 16 ounces, consumers probably (on the average) are not being seriously cheated. However, since the company wishes to maintain good customer relations, it will nevertheless readjust the process if it can reject $H_0 : \mu = 16$ with $\alpha = .05$.

It is also important to point out that keeping the mean bottle fill at 16 ounces does not guarantee that most *individual* bottle fills are 16 ounces. To evaluate individual bottle fills, we can use the concept of a *prediction interval*. We discuss these intervals in detail in Chapters 5 and 7.

TESTING $H_0: \mu = c$ VERSUS $H_1: \mu > c$

Example 3.8 In the gasoline mileage problem, recall that mileage standards state that the mean mileage μ must be at least 30 mpg. Here we might be tempted to say that National Motors can "prove" that $\mu \geq 30$ if it can accept the null hypothesis $H_0: \mu \geq 30$ instead of the alternative hypothesis $H_1: \mu < 30$. However, *hypothesis testing seeks to find how confident we can be that the null hypothesis should be rejected in favor of the alternative hypothesis.* It does not seek to find how confident we can be that the null hypothesis should be accepted. Therefore we cannot use hypothesis testing to "prove" that a null hypothesis is true. Thus we cannot make the statement that National Motors wishes to prove—that $\mu \geq 30$—the null hypothesis. It might therefore be tempting to make $\mu \geq 30$ the alternative hypothesis. But we cannot do this either because in hypothesis testing the alternative hypothesis always involves a strict inequality ($>$ or $<$), and the null hypothesis involves an equality ($=$) or a nonstrict inequality ($\geq$ or $\leq$). The only way out of this predicament is to use the following procedure:

1. State what you wish to justify written as a *strict inequality* ($<$, $>$, or $\neq$), and make it the *alternative hypothesis* H_1.
2. State what is reasonably possible if the alternative hypothesis if false, and make this statement the *null hypothesis* H_0.

National Motors can use this procedure by rewriting what it wishes to prove ($\mu \geq 30$) as the inequality $\mu > 30$. This statement becomes the alternative hypothesis. We then make $\mu \leq 30$ (which is reasonably possible if $\mu > 30$ is false) the null hypothesis. We then attempt to justify that the mileage standard is being met by attempting to reject $H_0: \mu \leq 30$ in favor of $H_1: \mu > 30$ by setting α, the probability of a Type I error, equal to .05. Here H_1 says that the mileage standard is in fact being exceeded.

As illustrated in Example 3.8, we sometimes need to test $H_0: \mu \leq c$ versus $H_1: \mu > c$. We note that it can be shown that testing these hypotheses is equivalent to testing $H_0: \mu = c$ versus $H_1: \mu > c$, and we consider the following result.

Testing $H_0: \mu = c$ Versus $H_1: \mu > c$ by Using a Rejection Point

Define the test statistic

$$t = \frac{\bar{y} - c}{s/\sqrt{n}}$$

If the population sampled has a normal distribution with mean μ, we can reject $H_0: \mu = c$ in favor of $H_1: \mu > c$ by setting the probability of a Type I error equal to α if and only if

$$t > t_{[\alpha]}^{(n-1)}$$

Example 3.9 Consider Example 3.8 and note that testing $H_0: \mu \leq 30$ versus $H_1: \mu > 30$ is equivalent to testing $H_0: \mu = 30$ versus $H_1: \mu > 30$. Suppose that we use a sample of $n = 5$ mileages. Then we can reject H_0 with $\alpha = .05$ if and only if

$$t = \frac{\bar{y} - 30}{s/\sqrt{n}}$$

is greater than

$$t_{[\alpha]}^{(n-1)} = t_{[.05]}^{(5-1)} = t_{[.05]}^{(4)} = 2.132$$

This condition is intuitively reasonable because a value of t substantially greater than zero results when $\bar{y}$ is substantially greater than 30. This provides substantial evidence for rejecting $H_0: \mu = 30$ in favor of $H_1: \mu > 30$. Setting α equal to .05 means that if $H_0: \mu = 30$ is true, then

1. $100(1 - \alpha)\% = 95\%$ of all possible values of t are less than or equal to the rejection point $t_{[.05]}^{(4)} = 2.132$ and thus lead us to fail to reject H_0: $\mu = 30$ (when H_0 is true), a correct decision.
2. $100(\alpha)\% = 5\%$ of all possible values of t are greater than the rejection point $t_{[.05]}^{(4)} = 2.132$ and lead us to reject $H_0: \mu = 30$ (when H_0 is true), a Type I error.

Now, from Example 3.1, $\bar{y} = 31.2$ is the mean and $s = .7517$ is the standard deviation of the sample of $n = 5$ mileages that National Motors has randomly selected from the population of all mileages. It follows that

$$t = \frac{\bar{y} - 30}{s/\sqrt{n}} = \frac{31.2 - 30}{.7517/\sqrt{5}} = 3.569$$

Since $t = 3.569 > 2.132 = t_{[.05]}^{(4)}$, we can reject $H_0: \mu = 30$ in favor of $H_1: \mu > 30$ by setting α equal to .05.

The confidence intervals previously discussed are *two-sided confidence intervals.* Such intervals provide reasonable minimum and maximum values for μ. A *lower one-sided confidence interval,* for μ provides a reasonable minimum value for μ. In addition, such an interval can be used to test $H_0: \mu = c$ versus $H_1: \mu > c$.

A Lower One-Sided Confidence Interval for μ and Testing $H_0: \mu = c$ Versus $H_1: \mu > c$

If the population sampled is normally distributed with mean μ,

1. A *lower one-sided $100(1 - \alpha)\%$ confidence interval for μ* is

$$\left[\bar{y} - t_{[\alpha]}^{(n-1)}\left(\frac{s}{\sqrt{n}}\right), \infty\right)$$

2. We can reject $H_0: \mu = c$ in favor of $H_1: \mu > c$ by setting the probability of a Type I error equal to α if and only if the interval in (1) does not contain c.

Example 3.10 Consider the sample of $n = 5$ mileages in Example 3.1. A lower one-sided 95% confidence interval for the mean mileage μ based on this sample is

$$\begin{aligned}\left[\bar{y} - t_{[\alpha]}^{(n-1)}\left(\frac{s}{\sqrt{n}}\right), \infty\right) &= \left[31.2 - t_{[.05]}^{(5-1)}\left(\frac{.7517}{\sqrt{5}}\right), \infty\right) \\ &= [31.2 - 2.132(.3362), \infty) \\ &= [31.2 - .72, \infty) \\ &= [30.5, \infty)\end{aligned}$$

Therefore we are 95% confident that μ is at least 30.5 mpg. Furthermore, this interval does not contain 30. Thus we can reject $H_0: \mu = 30$ in favor of $H_1: \mu > 30$ with $\alpha = .05$. Also note that using the one-sided 95% interval is "more powerful" than using the two-sided 95% interval that was calculated in Example 3.1. For example, the two-sided interval [30.3, 32.1] does not allow us to reject $H_0: \mu = 30.4$ in favor of $H_1: \mu > 30.4$ with 95% confidence. However, the one-sided 95% interval $[30.5, \infty)$ does allow us to reject $H_0: \mu = 30.4$ in favor of $H_1: \mu > 30.4$ with 95% confidence.

TESTING $H_0: \mu = c$ VERSUS $H_1: \mu < c$

Example 3.11 National Motors equips the Hawk with a newly designed disc brake system. A major competitor claims that its automobile achieves a mean stopping distance of 60 feet. Here we define the stopping distance to be the distance (in feet) required to bring an automobile to a complete stop from a speed of 35 mph under normal driving conditions. National Motors would like to conduct a new television advertising campaign claiming that the mean stopping distance achieved by the Hawk is less than the 60-foot distance claimed by its competitor.

We define μ to be the mean of the (theoretically) infinite population of all Hawk stopping distances. The television networks will allow National Motors to make its claim only if the company can statistically justify that μ is less than 60 ft. To do this, the company establishes an alternative hypothesis of $H_1: \mu < 60$; this is the claim we wish to make. Therefore the null hypothesis becomes $H_0: \mu \geq 60$. If we can reject H_0 in favor of H_1 by setting the probability of a Type I error equal to .05, National Motors will be allowed to make the claim in the television ad campaign.

As illustrated in Example 3.11, we sometimes need to test $H_0: \mu \geq c$ versus $H_1: \mu < c$. We note that testing these hypotheses can be shown to be equivalent to testing $H_0: \mu = c$ versus $H_1: \mu < c$, and we consider the following result.

Testing $H_0: \mu = c$ Versus $H_1: \mu < c$ by Using a Rejection Point

Define the test statistic

$$t = \frac{\bar{y} - c}{s/\sqrt{n}}$$

If the population sampled has a normal distribution with mean μ, we can reject $H_0: \mu = c$ in favor of $H_1: \mu < c$ by setting the probability of a Type I error equal to α if and only if

$$t < -t_{[\alpha]}^{(n-1)}$$

Example 3.12 Consider Example 3.11 and note that testing $H_0: \mu \geq 60$ versus $H_1: \mu < 60$ is equivalent to testing $H_0: \mu = 60$ versus $H_1: \mu < 60$. Suppose that we employ a sample of $n = 64$ stopping distances. Then we can reject H_0 in favor of H_1 by setting α equal to .05 if and only if

$$t = \frac{\bar{y} - 60}{s/\sqrt{n}}$$

is less than

$$-t_{[\alpha]}^{(n-1)} = -t_{[.05]}^{(64-1)} = -t_{[.05]}^{(63)} \approx -z_{[.05]} = -1.645$$

Here the point $-t_{[.05]}^{(63)}$ employs more than 30 degrees of freedom. Therefore we approximate this t-point by using the normal point $-z_{[.05]} = -1.645$. The use of this rejection point is intuitively reasonable because a value of t substantially less than zero results when $\bar{y}$ is substantially less than 60. This provides substantial evidence to support rejecting $H_0: \mu = 60$ in favor of $H_1: \mu < 60$. Setting α equal to .05 means that if $H_0: \mu = 60$ is true, then,

1. $100(1 - \alpha)\% = 95\%$ of all possible values of t are greater than or equal to the rejection point $-t_{[.05]}^{(63)} \approx -1.645$ and thus lead us to not reject $H_0: \mu = 60$ (when $H_0: \mu = 60$ is true), a correct decision.
2. $100(\alpha)\% = 5\%$ of all possible values of t are less than the rejection point $-t_{[.05]}^{(63)} \approx -1.645$ and lead us to reject $H_0: \mu = 60$ (when $H_0: \mu = 60$ is true), a Type I error.

National Motors randomly selects a sample of $n = 64$ Hawks and obtains a stopping distance for each automobile. It calculates the mean and standard deviation of the sample to be $\bar{y} = 58.12$ ft and $s = 6.13$ ft. It follows that

$$t = \frac{\bar{y} - 60}{s/\sqrt{n}} = \frac{58.12 - 60}{6.13/\sqrt{64}} = -2.45$$

Since $t = -2.45 < -1.645 \approx -t_{[.05]}^{(63)}$, we can reject $H_0: \mu = 60$ in favor of $H_1: \mu < 60$ by setting α equal to .05. On the basis of this result, National Motors will be allowed to make the television claim that μ is less than 60.

We now consider an *upper one-sided confidence interval* for μ. Such an interval provides a reasonable maximum value for μ. In addition, such an interval can be used to test $H_0: \mu = c$ versus $H_1: \mu < c$.

An Upper One-Sided Confidence Interval for μ and Testing $H_0: \mu = c$ Versus $H_1: \mu < c$

If the population sampled is normally distributed with mean μ,

1. An *upper one-sided 100(1 − α)% confidence interval for* μ is

$$\left(-\infty, \bar{y} + t_{[\alpha]}^{(n-1)}\left(\frac{s}{\sqrt{n}}\right)\right]$$

2. We can reject $H_0: \mu = c$ in favor of $H_1: \mu < c$ by setting the probability of a Type I error equal to α if and only if the interval in (1) does not contain c.

Example 3.13 Consider the sample of $n = 64$ stopping distances in Example 3.12. An upper one-sided 95% confidence interval for μ, the mean stopping distance, is

$$\begin{aligned}\left(-\infty, \bar{y} + t_{[\alpha]}^{(n-1)}\left(\frac{s}{\sqrt{n}}\right)\right] &= \left(-\infty, 58.12 + t_{[.05]}^{(64-1)}\left(\frac{6.13}{\sqrt{64}}\right)\right] \\ &\approx (-\infty, 58.12 + z_{[.05]}(.76625)] \\ &= (-\infty, 58.12 + 1.64(.76625)] \\ &= (-\infty, 59.38]\end{aligned}$$

Therefore we are 95% confident that μ is less than or equal to 59.38 ft. Furthermore, this interval does not contain 60. Thus we can reject $H_0: \mu = 60$ in favor of $H_1: \mu < 60$ by setting α equal to .05.

A SUMMARY OF TESTING $H_0: \mu = c$

We now summarize our results concerning testing $H_0: \mu = c$.

Testing $H_0: \mu = c$ by Using the t-Distribution

Suppose that the population sampled is normally distributed with mean μ, and define the test statistic

$$t = \frac{\bar{y} - c}{s/\sqrt{n}}$$

Then we can reject $H_0: \mu = c$ in favor of a particular alternative hypothesis by setting the probability of a Type I error equal to α by using the following rejection point conditions.

Alternative hypothesis	Rejection point condition: Reject H_0 in favor of H_1 if and only if
$H_1 : \mu \neq c$	$\lvert t \rvert > t_{[\alpha/2]}^{(n-1)}$
$H_1 : \mu > c$	$t > t_{[\alpha]}^{(n-1)}$
$H_1 : \mu < c$	$t < -t_{[\alpha]}^{(n-1)}$

As was demonstrated in the preceding examples, for 30 or more degrees of freedom we approximate the t-points $t_{[\alpha/2]}^{(n-1)}$ and $t_{[\alpha]}^{(n-1)}$ by the normal points $z_{[\alpha/2]}$ and $z_{[\alpha]}$. Furthermore, when n is at least 30, the Central Limit Theorem tells us that the population of all possible values of

$$\frac{\bar{y} - \mu}{s/\sqrt{n}}$$

approximately has a standard normal distribution, no matter what probability distribution describes the population being sampled. Therefore we obtain the following large sample tests of $H_0 : \mu = c$.

Testing $H_0 : \mu = c$ by Using the Normal Distribution

Define the test statistic

$$z = \frac{\bar{y} - c}{s/\sqrt{n}}$$

and *suppose that the sample size n is large* (say, at least 30). Then we can reject $H_0 : \mu = c$ in favor of a particular alternative hypothesis by setting the probability of a Type I error equal to α by using the following rejection point conditions.

Alternative hypothesis	Rejection point condition: Reject H_0 in favor of H_1 if and only if
$H_1 : \mu \neq c$	$\lvert z \rvert > z_{[\alpha/2]}$
$H_1 : \mu > c$	$z > z_{[\alpha]}$
$H_1 : \mu < c$	$z < -z_{[\alpha]}$

USING PROB-VALUES

We next introduce a *probability-value* (or *prob-value*) condition for testing $H_0 : \mu = c$ versus $H_1 : \mu \neq c$.

Testing $H_0: \mu = c$ Versus $H_1: \mu \neq c$ by Using Prob-Values

Define the *prob-value* to be twice the area under the curve of the t-distribution having $n - 1$ degrees of freedom to the right of the absolute value of

$$t = \frac{\bar{y} - c}{s/\sqrt{n}}$$

Then if the population sampled is normally distributed with mean μ, we can reject $H_0: \mu = c$ in favor of $H_1: \mu \neq c$ by setting the probability of a Type I error equal to α if and only if

$$\text{prob-value} < \alpha$$

Although we have initially defined the prob-value in terms of an area under a curve, we will soon show that the prob-value is a *probability* that has an important interpretation. First, however, we note that the above prob-value condition for rejecting $H_0: \mu = c$ in favor of $H_1: \mu \neq c$ is equivalent to the previously discussed rejection point condition. To see this, suppose that the prob-value is less than a specified value of α. Then,

$$\begin{bmatrix}\text{The area under the curve of} \\ \text{the } t\text{-distribution having } n-1 \\ \text{degrees of freedom to the right} \\ \text{of } |t|\end{bmatrix} < \alpha/2$$

This implies (as can be seen from Figure 3.7) that $|t| > t_{[\alpha/2]}^{(n-1)}$. Therefore we can reject $H_0: \mu = c$ in favor of $H_1: \mu \neq c$ by setting the probability of a Type I error equal to α.

Comparing the rejection point and prob-value conditions, first note that the rejection point condition is simpler from a computational standpoint. This is because it requires only that we calculate the test statistic t and look up the value $t_{[\alpha/2]}^{(n-1)}$ in a t-table. The prob-value condition is more complicated; it requires that we calculate an area under the t-curve. However, this problem is not serious, since most regression computer packages (and some hand-held calculators) compute areas under t-curves and prob-values. Furthermore, the prob-value condition has an advantage. Suppose that several hypothesis testers all wish to use different values of α. Then, using the rejection point condition, each hypothesis tester would have to look up a different rejection point $t_{[\alpha/2]}^{(n-1)}$ to decide whether to reject H_0. However, when the prob-value condition is used only the prob-value needs to be calculated. Then each hypothesis tester knows that if the prob-value is less than his or her particular chosen value of α, H_0 should be rejected. Another advantage of the prob-value is that it may be interpreted as a measure of probabilistic doubt about the validity of H_0. This is illustrated in the following example.

FIGURE 3.7 **The equivalence of the rejection point and prob-value conditions for rejecting $H_0: \mu = c$ in favor of $H_1: \mu \neq c$**

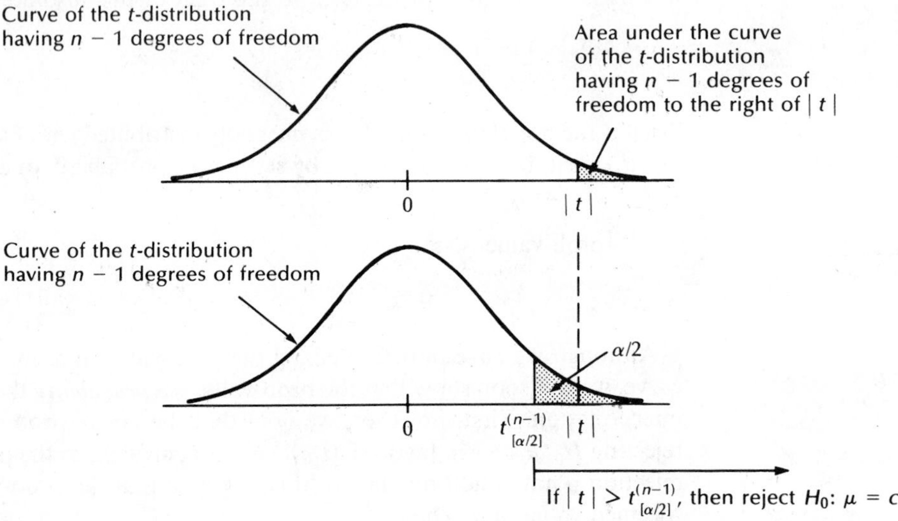

Example 3.14 We saw in Example 3.6 that the t statistic for testing $H_0: \mu = 16$ versus $H_1: \mu \neq 16$ in the Gem shampoo bottle fill problem was

$$t = \frac{\bar{y} - 16}{s/\sqrt{n}} = \frac{15.8 - 16}{.1532/\sqrt{6}} = -3.2$$

It follows that the prob-value for testing these hypotheses is twice the area under the curve of the t-distribution having $n - 1 = 6 - 1 = 5$ degrees of freedom to the right of $|t| = |-3.2| = 3.2$. This prob-value can be computer calculated to be .026. We see that the prob-value is less than .05 but, for example, not less than .02. It follows that we can reject $H_0: \mu = 16$ in favor of $H_1: \mu \neq 16$ by setting α equal to .05 but not by setting α equal to .02. We would regard this as strong evidence that $H_0: \mu = 16$ is false and $H_1: \mu \neq 16$ is true.

In addition to its use as a decision rule, the prob-value can be interpreted as a probability. If $H_0: \mu = 16$ is true, then the population of all possible values of the test statistic

$$t = \frac{\bar{y} - 16}{s/\sqrt{n}}$$

has a t-distribution with $n - 1 = 5$ degrees of freedom. Furthermore, twice the area under the curve of the t-distribution having $n - 1$ degrees of freedom to the right of $|t|$ is (by the symmetry of the t-distribution) equal to the area to the right of $|t|$ plus the area to the left of $-|t|$. It follows that the prob-value equals the sum of these latter two areas. Therefore the prob-value equals the proportion of all possible values of t that, if $H_0: \mu = 16$ is true, are at least as far away from zero as the value of t that we have actually observed ($t = -3.2$). Notice here that the larger (in absolute value) t is, the more this test statistic contradicts H_0. It follows that the prob-value equals the proportion of all possible values of t that are at least as contradictory to H_0 as the value of t that we have actually observed. The prob-value of .026 leads us to reach one of two possible conclusions. The first conclusion is that $H_0: \mu = 16$ is true and only .026 (that is, 26 in 1000) of all possible values of t contradict H_0 at least as much as the value of t that we have observed. That is, if we believe that H_0 is true, we must also believe that we have observed a value of t so rare that it can be described as a 26 in 1000 chance. The second conclusion is that $H_0: \mu = 16$ is false and $H_1: \mu \neq 16$ is true. A reasonable person would probably believe the second conclusion. That is, the fact that the prob-value .026 is small casts strong doubt on the validity of the null hypothesis $H_0: \mu = 16$ and thus lends strong support to the validity of the alternative hypothesis $H_1: \mu \neq 16$.

We now consider testing one-sided hypotheses by using prob-values.

Testing $H_0: \mu = c$ Versus $H_1: \mu > c$ by Using Prob-Values

Define the prob-value to be the area under the curve of the t-distribution having $n - 1$ degrees of freedom to the right of

$$t = \frac{\bar{y} - c}{s/\sqrt{n}}$$

Then, if the population sampled is normally distributed with mean μ, we can reject $H_0: \mu = c$ in favor of $H_1: \mu > c$ by setting the probability of a Type I error equal to α if and only if

$$\text{prob-value} < \alpha$$

The prob-value condition for rejecting $H_0: \mu = c$ in favor of $H_1: \mu > c$ is equivalent to the previously discussed rejection point condition. To see this, suppose that for a specified value of α,

$$\text{prob-value} = \begin{bmatrix} \text{The area under the curve of} \\ \text{the } t\text{-distribution having } n-1 \\ \text{degrees of freedom to the} \\ \text{right of } t \end{bmatrix} < \alpha$$

This implies (as can be seen from Figure 3.8) that $t > t_{[\alpha]}^{(n-1)}$. Therefore we can reject $H_0: \mu = c$ in favor of $H_1: \mu > c$ by setting the probability of a Type I error equal to α.

FIGURE 3.8 **The equivalence of the rejection point and prob-value conditions for rejecting $H_0: \mu = c$ in favor of $H_1: \mu > c$**

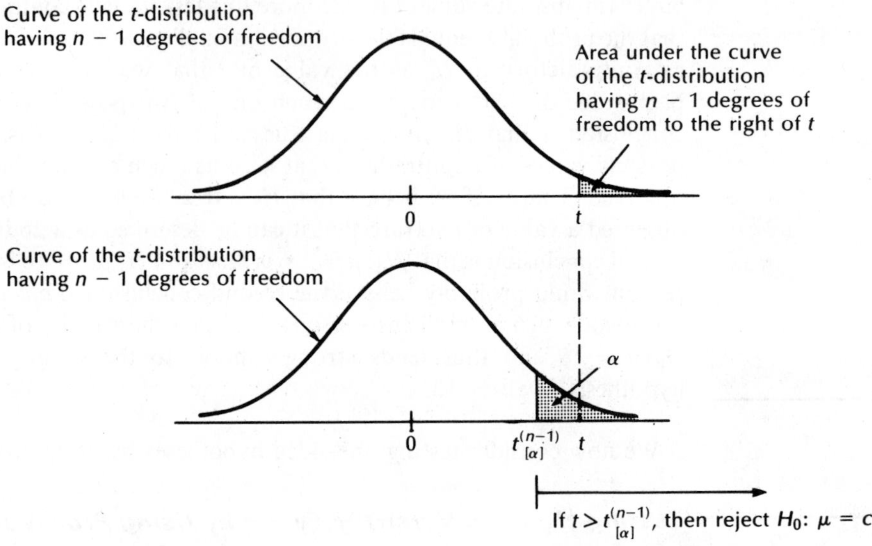

Example 3.15 We saw in Example 3.9 that the t statistic for testing $H_0: \mu = 30$ versus $H_1: \mu > 30$ is

$$t = \frac{\bar{y} - 30}{s/\sqrt{n}} = \frac{31.2 - 30}{.7517/\sqrt{5}} = 3.569$$

The corresponding prob-value is the area under the curve of the t-distribution having $n - 1 = 4$ degrees of freedom to the right of $t = 3.569$. This prob-value can be computer calculated to be .013. We see that the prob-value is less than .05 but not less than .01. It follows that we can reject $H_0: \mu = 30$ in favor of $H_1: \mu > 30$ by setting α equal to .05 but not by setting α equal to .01. As a probability, the prob-value says that if we are to believe that $H_0: \mu = 30$ is true, then we must believe that we have observed a value of t that is so rare that only .013 (that is, 13 in 1000) of all possible values of t are at least as contradictory to $H_0: \mu = 30$.

Testing $H_0: \mu = c$ Versus $H_1: \mu < c$ by Using Prob-Values

Define the prob-value to be the area under the curve of the t-distribution having $n - 1$ degrees of freedom to the left of

$$t = \frac{\bar{y} - c}{s/\sqrt{n}}$$

Then, if the population sampled is normally distributed with mean μ, we can reject $H_0: \mu = c$ in favor of $H_1: \mu < c$ by setting the probability of a Type I error equal to α if and only if

$$\text{prob-value} < \alpha$$

The prob-value condition for rejecting $H_0: \mu = c$ in favor of $H_1: \mu < c$ is equivalent to the previously discussed rejection point condition. To see this, suppose that for a specified value of α,

$$\text{prob-value} = \left[\begin{array}{c}\text{The area under the curve of} \\ \text{the } t\text{-distribution having } n - 1 \\ \text{degrees of freedom to the} \\ \text{left of } t\end{array}\right] < \alpha$$

This implies (as can be seen from Figure 3.9) that $t < -t_{[\alpha]}^{(n-1)}$. Therefore we can reject $H_0: \mu = c$ in favor of $H_1: \mu < c$ by setting the probability of a Type I error equal to α.

FIGURE 3.9 **The equivalence of the rejection point and prob-value conditions for rejecting $H_0: \mu = c$ in favor of $H_1: \mu < c$**

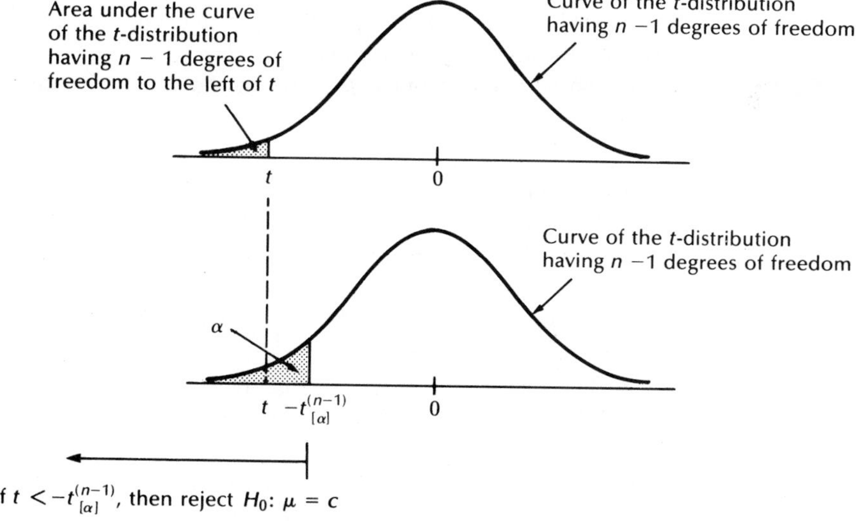

Example 3.16 We saw in Example 3.12 that the t statistic for testing $H_0: \mu = 60$ versus $H_1: \mu < 60$ is

$$t = \frac{\bar{y} - 60}{s/\sqrt{n}} = \frac{58.12 - 60}{6.13/\sqrt{64}} = -2.45$$

The corresponding prob-value is the area under the curve of the t-distribution having $n - 1 = 63$ degrees of freedom to the left of -2.45. This is approximately the area under the standard normal curve to the left of -2.45. This area equals $(.5 - .4929) = .0071$. (See Table E.1.) Since this prob-value is less than .01, we can reject $H_0 : \mu = 60$ in favor of $H_1 : \mu < 60$ by setting α equal to .01. We would regard this as very strong evidence that H_0 is false and H_1 is true. As a probability, the prob-value says that if we are to believe that $H_0 : \mu = 60$ is true, then we must believe that we have observed a value of t that is so rare that only .0071 (that is, 7 in 1000) of all possible values of t are at least as contradictory to $H_0 : \mu = 60$.

In general, recall that if the sample size n is large (say, at least 30), we use the test statistic

$$z = \frac{\bar{y} - c}{s/\sqrt{n}}$$

In this case we redefine the prob-values in the above results to be the corresponding areas under the standard normal curve. It follows by the Central Limit Theorem that if n is large, the previous prob-value results approximately hold for populations that are not normally distributed.

We can summarize our discussion of the prob-value for testing $H_0 : \mu = c$ as follows:

1. If the *prob-value* is *less than .05*, we can reject $H_0 : \mu = c$ by setting α equal to .05. This means that we have *strong evidence* that $H_0 : \mu = c$ is false.
2. If the *prob-value* is *less than .01*, we can reject $H_0 : \mu = c$ by setting α equal to .01. This means that we have *very strong evidence* that $H_0 : \mu = c$ is false.
3. The smaller the prob-value is, the stronger is the evidence that $H_0 : \mu = c$ is false.

One way to report the result of a statistical test is to state whether we "reject" or "do not reject" H_0 on the basis of setting the probability of a Type I error equal to α. As an alternative to this, reporting the prob-value is becoming increasingly popular. This is because the result of the statistical test may be used by several people to make different decisions at different times. Thus by reporting the prob-value the individual performing the test permits any other decision makers to reach their own conclusions.

3.3

COMPARING TWO POPULATION MEANS BY USING INDEPENDENT SAMPLES

INTRODUCTION

In this section we discuss comparing two population means, μ_1 and μ_2. Here, for $l = 1$ and 2, μ_l is the mean of the *lth population of values*. In order to compare μ_1 and μ_2, we assume that we have randomly selected the two samples listed in Table 3.2. Specifically, we assume that for $l = 1$ and 2,

$$y_{l,1}, y_{l,2}, \ldots, y_{l,n_l}$$

is a sample of n_l values that we have randomly selected from the lth population. The mean, variance, and standard deviation of this sample

$$\bar{y}_l = \frac{\sum_{k=1}^{n_l} y_{l,k}}{n_l} \qquad s_l^2 = \frac{\sum_{k=1}^{n_l} (y_{l,k} - \bar{y}_l)^2}{n_l - 1} \qquad s_l = \sqrt{s_l^2}$$

are point estimates of, respectively, μ_l, σ_l^2, and σ_l, the mean, variance, and standard deviation of the lth population. It follows that the point estimate of the difference $\mu_1 - \mu_2$ is $\bar{y}_1 - \bar{y}_2$.

TABLE 3.2 **Two samples randomly selected from two populations**

Sample of n_1 values from population 1	***Sample of n_2 values from population 2***
$y_{1,1}, y_{1,2}, \ldots, y_{1,n_1}$	$y_{2,1}, y_{2,2}, \ldots, y_{2,n_2}$
Sample mean $= \bar{y}_1$	Sample mean $= \bar{y}_2$
Sample variance $= s_1^2$	Sample variance $= s_2^2$

Example 3.17 Bargain Town is a large discount chain. Management wishes to compare the performance of its credit managers in Ohio and Illinois. It will do this by comparing the mean dollar amounts owed by customers with delinquent charge accounts in these two states. Here, a small mean dollar amount owed is desirable. This is because such a mean would say that bad credit risks are not being extended large amounts of credit. Let μ_O, μ_I, σ_O^2, σ_I^2, σ_O, and σ_I denote the means, variances, and standard deviations of the populations of dollar amounts in Ohio and Illinois. Suppose that Bargain Town randomly selects two samples from these populations in order to make the required comparison. The sample results obtained are as

follows.

Sample of Ohio accounts	***Sample of Illinois accounts***
$n_O = 103$	$n_I = 103$
$\bar{y}_O = \$122$	$\bar{y}_I = \$69$
$s_O^2 = 1521$	$s_I^2 = 529$

It follows that the point estimate of $\mu_O - \mu_I$ is

$$\begin{aligned}\bar{y}_O - \bar{y}_I &= 122 - 69\\ &= 53\end{aligned}$$

This says that we estimate that μ_O, the mean dollar amount of delinquent charge accounts in Ohio, is \$53 more than μ_I, the mean dollar amount of delinquent charge accounts in Illinois.

The various confidence interval and hypothesis testing formulas for comparing μ_1 and μ_2 that we present in this section are valid under different sets of assumptions. There are three assumptions to be considered. The first assumption says that the two *populations* have equal variances (that is, $\sigma_1^2 = \sigma_2^2$). One way to check the validity of this assumption is to compare the sample variances s_1^2 and s_2^2. If these sample variances are nearly equal, then it is reasonable to believe that this "equal variances" assumption approximately holds. The second assumption is that there is no relationship between the values of one sample and the values of the other sample. In this case we say that we have performed an *independent samples experiment*. The last assumption states that each of the two populations is normally distributed. *All of the formulas in this section require that the independence assumption holds.* However, only some of the formulas require that the equal variance assumption holds, and only some require that the normality assumption holds. We discuss these assumptions in more detail as we present the formulas. Also, note that in Example 3.23 we show that it is not always best to make the samples independent of each other.

CONFIDENCE INTERVALS FOR $\mu_1 - \mu_2$ BASED ON LARGE SAMPLES

Confidence Intervals for $\mu_1 - \mu_2$ Based on the Normal Distribution

1. The *population of all possible values of* $\bar{y}_1 - \bar{y}_2$ (that is, point estimates of $\mu_1 - \mu_2$)
   a. Has *mean* $\mu_{\bar{y}_1 - \bar{y}_2} = \mu_1 - \mu_2$.
   b. Has *variance* $\sigma^2_{\bar{y}_1 - \bar{y}_2} = \dfrac{\sigma_1^2}{n_1} + \dfrac{\sigma_2^2}{n_2}$ (if each of the populations sampled is infinite, or very large).

c. Has *standard deviation* $\sigma_{\bar{y}_1 - \bar{y}_2} = \sqrt{(\sigma_1^2/n_1) + (\sigma_2^2/n_2)}$ (if each of the populations sampled is infinite or very large).
d. Has a *normal distribution* (if the normality assumption holds).
e. *Approximately* has a *normal distribution* no matter what probability distributions describe the sampled populations (if each of the sample sizes n_1 and n_2 is large).

2. If each of n_1 and n_2 is *large*, the population of all possible values of

$$\frac{(\bar{y}_1 - \bar{y}_2) - (\mu_1 - \mu_2)}{\sqrt{\dfrac{\sigma_1^2}{n_1} + \dfrac{\sigma_2^2}{n_2}}}$$

approximately has a *standard normal distribution.* This implies that

$$\left[(\bar{y}_1 - \bar{y}_2) \pm z_{[\alpha/2]} \sqrt{\frac{\sigma_1^2}{n_1} + \frac{\sigma_2^2}{n_2}}\right]$$

and

$$\left[(\bar{y}_1 - \bar{y}_2) \pm z_{[\alpha/2]} \sqrt{\frac{s_1^2}{n_1} + \frac{s_2^2}{n_2}}\right]$$

are *approximately correct 100*$(1 - \alpha)$% *confidence intervals for* $\mu_1 - \mu_2$, no matter what probability distributions describe the sampled populations. To use the second interval, each sample size should be at least 30.

Example 3.18 It is reasonable to assume that the samples in Example 3.17 are independent. Also each of the sample sizes n_O and n_I is large. Therefore a 95% confidence interval for $\mu_O - \mu_I$ is

$$\begin{aligned}\left[(\bar{y}_O - \bar{y}_I) \pm z_{[.025]} \sqrt{\frac{s_O^2}{n_O} + \frac{s_I^2}{n_I}}\right] &= \left[(122 - 69) \pm 1.96 \sqrt{\frac{1521}{103} + \frac{529}{103}}\right] \\ &= [53 \pm 8.74] \\ &= [44.26, 61.74]\end{aligned}$$

This interval says that management can be 95% confident that μ_O, the mean dollar amount of delinquent charge accounts in Ohio, is between \$44.26 and \$61.74 more than μ_I, the mean dollar amount of delinquent charge accounts in Illinois. This might convince management that new credit policies should be instituted to reduce the mean dollar amount of delinquent charge accounts in Ohio.

TESTING $H_0: \mu_1 - \mu_2 = c$ BY USING LARGE SAMPLES

Testing $H_0: \mu_1 - \mu_2 = c$ by Using the Normal Distribution

Define the test statistic

$$z = \frac{(\bar{y}_1 - \bar{y}_2) - c}{s_{\bar{y}_1 - \bar{y}_2}}$$

where

$$s_{\bar{y}_1 - \bar{y}_2} = \sqrt{\frac{s_1^2}{n_1} + \frac{s_2^2}{n_2}}$$

If each of n_1 and n_2 is large (say, at least 30), we can reject $H_0: \mu_1 - \mu_2 = c$ in favor of a particular alternative hypothesis by setting the probability of a Type I error equal to α by using the following rejection point conditions.

Alternative hypothesis	Rejection point condition: Reject H_0 in favor of H_1 if and only if
$H_1: \mu_1 - \mu_2 \neq c$	$\lvert z \rvert > z_{[\alpha/2]}$
$H_1: \mu_1 - \mu_2 > c$	$z > z_{[\alpha]}$
$H_1: \mu_1 - \mu_2 < c$	$z < -z_{[\alpha]}$

It should be noted that in the (unlikely) case that we know the population variances σ_1^2 and σ_2^2, we should replace s_1^2 and s_2^2 in the above result by σ_1^2 and σ_2^2. Furthermore, if σ_1^2 and σ_2^2 are known and if each of the populations sampled is normally distributed, the above result (with σ_1^2 and σ_2^2 replacing s_1^2 and s_2^2) is valid for any sample sizes.

Example 3.19 Reconsider the Bargain Town problem of Example 3.17. Suppose that before the samples were taken, management felt that if it could be shown that μ_O is more than \$40 greater than μ_I, then new credit policies should be instituted in Ohio to reduce μ_O. Therefore such new policies will be implemented if we can reject

$$H_0: \mu_O - \mu_I \leqslant 40$$

in favor of

$$H_1: \mu_O - \mu_I > 40$$

by setting α equal to .05. Testing these hypotheses is equivalent to testing $H_0: \mu_O - \mu_I = 40$ versus $H_1: \mu_O - \mu_I > 40$. To do this, we compute the test statistic

$$\begin{aligned} z &= \frac{(\bar{y}_O - \bar{y}_I) - 40}{\sqrt{(s_O^2/n_O) + (s_I^2/n_I)}} \\ &= \frac{(122 - 69) - 40}{\sqrt{(1521/103) + (529/103)}} \\ &= \frac{53 - 40}{4.4613} \\ &= 2.91 \end{aligned}$$

and use the rejection point $z_{[\alpha]} = z_{[.05]} = 1.645$. Since

$$z = 2.91 > 1.645 = z_{[.05]}$$

we can reject H_0 in favor of H_1 by setting α equal to .05.

Furthermore, a lower one-sided 95% confidence interval for $\mu_O - \mu_I$ is

$$\begin{aligned} \left[(\bar{y}_O - \bar{y}_I) - z_{[\alpha]}\sqrt{\frac{s_O^2}{n_O} + \frac{s_I^2}{n_I}}, \infty\right) &= \left[(122 - 69) - z_{[.05]}\sqrt{\frac{1521}{103} + \frac{529}{103}}, \infty\right) \\ &= [53 - 1.64(4.4613), \infty) \\ &= [53 - 7.32, \infty) \\ &= [45.68, \infty) \end{aligned}$$

Therefore management is 95% confident that μ_O is at least \$45.68 greater than μ_I. Finally, the prob-value for the test is the area under the standard normal curve to the right of $z = 2.91$. This prob-value equals .0018 and provides very strong evidence against $H_0: \mu_O - \mu_I = 40$.

CONFIDENCE INTERVALS FOR $\mu_1 - \mu_2$ BASED ON SMALL SAMPLES

If we cannot select large samples, we usually cannot utilize confidence intervals based on the normal distribution. In this case we must use confidence intervals based on the t-distribution. Consider the statistic

$$\frac{(\bar{y}_1 - \bar{y}_2) - (\mu_1 - \mu_2)}{\sqrt{\dfrac{\sigma_1^2}{n_1} + \dfrac{\sigma_2^2}{n_2}}}$$

It might seem reasonable to replace σ_1^2 and σ_2^2 by s_1^2 and s_2^2 and to form the statistic

$$\frac{(\bar{y}_1 - \bar{y}_2) - (\mu_1 - \mu_2)}{\sqrt{\dfrac{s_1^2}{n_1} + \dfrac{s_2^2}{n_2}}}$$

However, the population of all such statistics does not always have a t-distribution.

Nevertheless, if the equal variances assumption holds ($\sigma_1^2 = \sigma_2^2 = \sigma^2$), then

$$\sigma_{\bar{y}_1 - \bar{y}_2} = \sqrt{\frac{\sigma_1^2}{n_1} + \frac{\sigma_2^2}{n_2}} = \sqrt{\frac{\sigma^2}{n_1} + \frac{\sigma^2}{n_2}} = \sqrt{\sigma^2\left(\frac{1}{n_1} + \frac{1}{n_2}\right)}$$

To find a point estimate of the common variance σ^2, we "pool" the observations in both samples. We obtain the "pooled estimate"

$$\begin{aligned} s_p^2 &= \frac{\sum_{k=1}^{n_1} (y_{1,k} - \bar{y}_1)^2 + \sum_{k=1}^{n_2} (y_{2,k} - \bar{y}_2)^2}{(n_1 - 1) + (n_2 - 1)} \\ &= \frac{(n_1 - 1)\left(\frac{\sum_{k=1}^{n_1} (y_{1,k} - \bar{y}_1)^2}{n_1 - 1}\right) + (n_2 - 1)\left(\frac{\sum_{k=1}^{n_2} (y_{2,k} - \bar{y}_2)^2}{n_2 - 1}\right)}{n_1 + n_2 - 2} \\ &= \frac{(n_1 - 1)s_1^2 + (n_2 - 1)s_2^2}{n_1 + n_2 - 2} \end{aligned}$$

This motivates us to define the statistic

$$\frac{(\bar{y}_1 - \bar{y}_2) - (\mu_1 - \mu_2)}{\sqrt{s_p^2\left(\frac{1}{n_1} + \frac{1}{n_2}\right)}} = \frac{(\bar{y}_1 - \bar{y}_2) - (\mu_1 - \mu_2)}{\sqrt{\frac{(n_1 - 1)s_1^2 + (n_2 - 1)s_2^2}{n_1 + n_2 - 2}\left(\frac{1}{n_1} + \frac{1}{n_2}\right)}}$$

Below we present a confidence interval for $\mu_1 - \mu_2$ based on this statistic. We also present an approximate confidence interval for $\mu_1 - \mu_2$ that can be used if the equal variance assumption does not hold.

Confidence Intervals for $\mu_1 - \mu_2$ Based on the t-Distribution

1. Assume that the populations sampled are *normally distributed* with *equal variances*. Then the application of all possible values of

$$\frac{(\bar{y}_1 - \bar{y}_2) - (\mu_1 - \mu_2)}{\sqrt{\frac{(n_1 - 1)s_1^2 + (n_2 - 1)s_2^2}{n_1 + n_2 - 2}\left(\frac{1}{n_1} + \frac{1}{n_2}\right)}}$$

has a *t-distribution with $n_1 + n_2 - 2$ degrees if freedom*. This implies that a *100(1 − α)% confidence interval for* $\mu_1 - \mu_2$ is

$$\left[(\bar{y}_1 - \bar{y}_2) \pm t_{[\alpha/2]}^{(n_1+n_2-2)} \sqrt{\frac{(n_1 - 1)s_1^2 + (n_2 - 1)s_2^2}{n_1 + n_2 - 2}\left(\frac{1}{n_1} + \frac{1}{n_2}\right)}\right]$$

We refer to this as formula 1.

2. Assume that the populations sampled are *normally distributed.* Then the population of all possible values of

$$\frac{(\bar{y}_1 - \bar{y}_2) - (\mu_1 - \mu_2)}{\sqrt{\frac{s_1^2}{n_1} + \frac{s_2^2}{n_2}}}$$

approximately has a *t-distribution with df degrees of freedom.* Here

$$df = \frac{(n_1 - 1)(n_2 - 1)}{(n_2 - 1)g^2 + (1 - g)^2 (n_1 - 1)}$$

where

$$g = \frac{s_1^2/n_1}{s_1^2/n_1 + s_2^2/n_2}$$

Note that if df is not an integer, we round down to the nearest integer. This implies that an *approximately correct 100(1 − α)% confidence interval for* $\mu_1 - \mu_2$ is

$$\left[(\bar{y}_1 - \bar{y}_2) \pm t_{[\alpha/2]}^{(df)} \sqrt{\frac{s_1^2}{n_1} + \frac{s_2^2}{n_2}}\right]$$

We refer to this as formula 2.

Examining these results, we see that formula 1 holds *exactly* under the equal variances, independence, and normality assumptions. Formula 2 holds *approximately* under the last two of these assumptions. When the sample sizes n_1 and n_2 are equal, studies have shown that even when the equal variances assumption does not hold, it is best to use formula 1. When the sample sizes are not equal, it is common practice to use formula 1 if the sample variances indicate that the population variances are approximately equal. This is because formula 1 is exactly valid, while formula 2 is only approximately valid, under the appropriate assumptions. Finally, it can be shown that the normality assumption is not extremely crucial to the validity of either formula. In particular, it has been shown that both formulas are approximately valid for mound-shaped populations.

In the following example we will illustrate the use of formula 1, and in the exercises the reader will have an opportunity to use formula 2.

Example 3.20 National Motors wishes to compare the gasoline mileage obtained by the Hawk with the gasoline mileage obtained by a major competitor. Let μ_1, σ_1^2, and σ_1 denote the mean, variance, and standard deviation of the population of Hawk mileages, and let μ_2, σ_2^2, and σ_2 denote the mean, variance, and standard deviation of the population of the competitor's mileages. Then suppose that we obtain the sample results summarized in Table 3.3. We might conclude (somewhat arbitrarily) that

TABLE 3.3 **Two independent random samples of gasoline mileages**

Sample of $n_1 = 5$ Hawk mileages	*Sample of $n_2 = 3$ competitor's mileages*
$y_{1,1} = 30.3$	$y_{2,1} = 28.3$
$y_{1,2} = 31.5$	$y_{2,2} = 29.0$
$y_{1,3} = 31.4$	$y_{2,3} = 29.7$
$y_{1,4} = 32.0$	
$y_{1,5} = 32.6$	
$\bar{y}_1 = 31.56$	$\bar{y}_2 = 29.0$
$s_1^2 = .723$	$s_2^2 = .49$
$s_1 = .8503$	$s_2 = .7$

the sample variances $s_1^2 = .723$ and $s_2^2 = .49$ do not differ substantially. Note that the sample standard deviations $s_1 = .8503$ and $s_2 = .7$ are more nearly equal. Thus we conclude that the equal variances assumption approximately holds. Therefore assuming that the samples are independent and that the populations of mileages are approximately normal, it follows that a 95% confidence interval for $\mu_1 - \mu_2$ is

$$\left[(\bar{y}_1 - \bar{y}_2) \pm t_{[\alpha/2]}^{(n_1+n_2-2)} \sqrt{\frac{(n_1 - 1)s_1^2 + (n_2 - 1)s_2^2}{n_1 + n_2 - 2}\left(\frac{1}{n_1} + \frac{1}{n_2}\right)}\right]$$

$$= \left[31.56 - 29 \pm t_{[.025]}^{(5+3-2)} \sqrt{\frac{(5 - 1)(.723) + (3 - 1)(.49)}{5 + 3 - 2}\left(\frac{1}{5} + \frac{1}{3}\right)}\right]$$

$$= [2.56 \pm t_{[.025]}^{(6)}(.5866)]$$

$$= [2.56 \pm 2.447(.5866)]$$

$$= [1.1245, 3.9955]$$

This interval says that National Motors can be 95% confident that μ_1 is between 1.1245 mpg and 3.9955 mpg greater than μ_2.

TESTING $H_0: \mu_1 - \mu_2 = c$ BY USING SMALL SAMPLES

Testing $H_0: \mu_1 - \mu_2 = c$ by Using the t-Distribution

Define the test statistic

$$t = \frac{(\bar{y}_1 - \bar{y}_2) - c}{s_{\text{equal}}}$$

where

$$s_{\text{equal}} = \sqrt{\frac{(n_1 - 1)s_1^2 + (n_2 - 1)s_2^2}{n_1 + n_2 - 2}\left(\frac{1}{n_1} + \frac{1}{n_2}\right)}$$

If the populations sampled are normally distributed with equal variances, we can reject $H_0: \mu_1 - \mu_2 = c$ in favor of a particular alternative hypothesis by setting the probability of a Type I error equal to α by using the following rejection point conditions.

Alternative hypothesis	*Rejection point condition: Reject H_0 in favor of H_1 if and only if*
$H_1: \mu_1 - \mu_2 \neq c$	$\lvert t \rvert > t_{[\alpha/2]}^{(n_1+n_2-2)}$
$H_1: \mu_1 - \mu_2 > c$	$t > t_{[\alpha]}^{(n_1+n_2-2)}$
$H_1: \mu_1 - \mu_2 < c$	$t < -t_{[\alpha]}^{(n_1+n_2-2)}$

If the equal variances assumption does not approximately hold, we can use the above result if we replace s_{equal} by

$$s_{\text{unequal}} = \sqrt{\frac{s_1^2}{n_1} + \frac{s_2^2}{n_2}}$$

and the degrees of freedom $n_1 + n_2 - 2$ by

$$df = \frac{(n_1 - 1)(n_2 - 1)}{(n_2 - 1)g^2 + (1 - g)^2 (n_1 - 1)} \quad \text{where} \quad g = \frac{s_1^2/n_1}{s_1^2/n_1 + s_2^2/n_2}$$

Here, if df is not an integer, we round down to the nearest integer.

Example 3.21 Consider the gasoline mileage comparison problem of Example 3.20. Suppose that we wish to test

$$H_0: \mu_1 - \mu_2 = 0$$

versus

$$H_1: \mu_1 - \mu_2 \neq 0$$

by setting α equal to .05. Here the null hypothesis says that the mean gasoline mileages μ_1 and μ_2 do not differ. The alternative hypothesis says that μ_1 and μ_2 differ. If we can reject H_0 in favor of H_1, we conclude that there is a statistically significant difference between μ_1 and μ_2. To test these hypotheses, we compute

$$\begin{aligned} t = \frac{(\bar{y}_1 - \bar{y}_2) - 0}{s_{\text{equal}}} &= \frac{(\bar{y}_1 - \bar{y}_2)}{\sqrt{\dfrac{(n_1 - 1)s_1^2 + (n_2 - 1)s_2^2}{n_1 + n_2 - 2}\left(\dfrac{1}{n_1} + \dfrac{1}{n_2}\right)}} \\ &= \frac{(31.56 - 29)}{\sqrt{\dfrac{(5 - 1)(.723) + (3 - 1)(.49)}{5 + 3 - 2}\left(\dfrac{1}{5} + \dfrac{1}{3}\right)}} \\ &= \frac{2.56}{.5866} \\ &= 4.364 \end{aligned}$$

We use the rejection points

$$t_{[\alpha/2]}^{(n_1+n_2-2)} = t_{[.05/2]}^{(n_1+n_2-2)} = t_{[.025]}^{(5+3-2)} = t_{[.025]}^{(6)} = 2.447$$

and

$$-t_{[\alpha/2]}^{(n_1+n_2-2)} = -2.447$$

Since

$$t = 4.364 > 2.447 = t_{[.025]}^{(6)}$$

we can reject H_0 in favor of H_1 with α equal to .05. We conclude that μ_1 and μ_2 differ. Note here that we must use the confidence interval for $\mu_1 - \mu_2$ computed in Example 3.20 to judge the practical importance of this difference.

In addition, the prob-value for the test is twice the area under the curve of the t-distribution having $n_1 + n_2 - 2 = 6$ degrees of freedom to the right of $|t| = 4.364$. This prob-value can be computer calculated to be .0048. Since this prob-value is very small, it provides very strong evidence against $H_0: \mu_1 - \mu_2 = 0$.

AN F-TEST FOR THE EQUALITY OF TWO POPULATION VARIANCES

Testing $H_0: \sigma_1^2 = \sigma_2^2$ by Using the F-Distribution

Define the test statistic

$$F = \frac{\max(s_1^2, s_2^2)}{\min(s_1^2, s_2^2)}$$

where

$$\max(s_1^2, s_2^2) = \text{the larger of } s_1^2 \text{ and } s_2^2$$

$$\min(s_1^2, s_2^2) = \text{the smaller of } s_1^2 \text{ and } s_2^2$$

Furthermore, let

$$r_1 = \{\text{the size of the sample having the largest variance}\} - 1$$

and

$$r_2 = \{\text{the size of the sample having the smallest variance}\} - 1$$

Then if the independence assumption holds and if the sampled populations are normally distributed, we can reject $H_0: \sigma_1^2 = \sigma_2^2$ in favor of $H_1: \sigma_1^2 \neq \sigma_2^2$ by setting the probability of a Type I error equal to α if and only if either of the following equivalent conditions holds:

$$F > F_{[\alpha]}^{(r_1, r_2)} \qquad \text{or} \qquad \text{prob-value} < \alpha$$

Here, $F_{[\alpha]}^{(r_1, r_2)}$ is the point on the scale of the F-distribution having r_1 and r_2 degrees of freedom so that the area under this curve to the right of this point is α, and the prob-value is defined to be the area under the curve of the F-distribution having r_1 and r_2 degrees of freedom to the right of F.

The first condition is intuitive. This is because a large value of

$$F = \frac{\max(s_1^2, s_2^2)}{\min(s_1^2, s_2^2)}$$

results when $\max(s_1^2, s_2^2)$ and $\min(s_1^2, s_2^2)$ differ substantially. This would provide strong evidence that σ_1^2 and σ_2^2 are not equal and that $H_0: \sigma_1^2 = \sigma_2^2$ should be rejected.

Example 3.22 Again consider the gasoline mileage comparison problem of Example 3.20. To use the data in Table 3.3 to test $H_0: \sigma_1^2 = \sigma_2^2$ versus $H_1: \sigma_1^2 \neq \sigma_2^2$ by setting α equal to .05, we calculate

$$F = \frac{\max(s_1^2, s_2^2)}{\min(s_1^2, s_2^2)} = \frac{s_1^2}{s_2^2} = \frac{.723}{.49} = 1.4755$$

Note that

$$\begin{aligned} r_1 &= \{\text{the size of the sample having the largest variance}\} - 1 \\ &= n_1 - 1 = 5 - 1 = 4 \end{aligned}$$

and

$$\begin{aligned} r_2 &= \{\text{the size of the sample having the smallest variance}\} - 1 \\ &= n_2 - 1 = 3 - 1 = 2 \end{aligned}$$

We see from Appendix Table E.3 that $F_{[.05]}^{(r_1, r_2)} = F_{[.05]}^{(4,2)} = 19.25$. Since

$$F = 1.4755 < 19.25 = F_{[.05]}^{(4,2)}$$

we cannot reject $H_0: \sigma_1^2 = \sigma_2^2$ by setting α equal to .05.

It has been suggested that the result of the F-test be employed to choose between statistics for making inferences concerning differences between two population means. Specifically, it has been suggested that if we cannot reject $H_0: \sigma_1^2 = \sigma_2^2$ by setting α equal to .05, then we should make inferences based on the *equal variances statistic*

$$t = \frac{(\bar{y}_1 - \bar{y}_2) - (\mu_1 - \mu_2)}{s_{\text{equal}}}$$

with $n_1 + n_2 - 2$ degrees of freedom. If we can reject $H_0: \sigma_1^2 = \sigma_2^2$ with $\alpha = .05$, it has been suggested that we should make inferences based on the *unequal variances statistic*

$$t = \frac{(\bar{y}_1 - \bar{y}_2) - (\mu_1 - \mu_2)}{s_{\text{unequal}}}$$

with df (as previously defined) degrees of freedom. The F-test is one approach to making this choice. However, studies have shown that the validity of this F-test is more sensitive to violations of the normality assumption than the validity of the "equal variances statistic" is sensitive to violations of the equal variances assumption! This fact calls the appropriateness of the F-test into question. However, it probably does not hurt to perform the test as long as its result is wisely used. Furthermore, if the choice between statistics is in question, it would be reasonable to use both. If both give essentially the same results, the choice is unimportant.

3.4

PAIRED DIFFERENCE EXPERIMENTS

INTRODUCTION

Example 3.23 Home State Life and Casualty, an insurance company, wishes to compare the repair costs of moderately damaged automobiles at two garages—garage 1 and garage 2. An automobile is moderately damaged if its repair cost is between $700 and $1400. One experimental procedure that could be used to carry out this study would be to take a sample of $n = 7$ cars that have been recently involved in accidents to garage 1 and obtain repair cost estimates for these cars and then to take a different sample of $n = 7$ cars that have been recently involved in accidents to garage 2 and obtain repair cost estimates for these cars. This procedure would yield two samples that are independent of each other. However, there are substantial differences in damages to moderately damaged cars. These differences would tend to conceal any real differences between the repair costs at the two garages. For example, if the repair cost estimates for the cars taken to garage 1 were higher than those for the cars taken to garage 2, we might not be able to tell whether the higher estimates at garage 1 were due to the way garage 1 charges customers for repair work or to the severity of the damage to the cars taken to garage 1. To overcome this difficulty, we can perform a *paired difference experiment*. Here we would take each of $n = 7$ cars to both garages and obtain repair cost estimates for each car at the two garages. The advantage of this paired difference experiment is that the repair cost estimates at the two garages are obtained by using the same cars. Thus any true differences in these estimtes would not be concealed by differences in the damages to the cars.

If, as illustrated in Example 3.23, we obtain two samples by taking two different measurements on the same n elements, then the two samples are related to (or depend upon) each other. In such a case we say that we have performed a *paired difference experiment*. The needed formulas in this kind of situation employ what we call the sample of *n paired differences* $d_1, d_2, \ldots, d_n$ summarized in Table 3.4.

TABLE 3.4 **Sample of *n* paired differences**

First sample	*Second sample*	*Sample of n paired differences (assuming that $n_1 = n_2 = n$)*
$y_{1,1}$	$y_{2,1}$	$d_1 = y_{1,1} - y_{2,1}$
$y_{1,2}$	$y_{2,2}$	$d_2 = y_{1,2} - y_{2,2}$
⋮	⋮	⋮
$y_{1,n}$	$y_{2,n}$	$d_n = y_{1,n} - y_{2,n}$
Sample mean = $\bar{y}_1$	Sample mean = $\bar{y}_2$	$\bar{d} = \bar{y}_1 - \bar{y}_2$ = mean of the sample of paired differences

The mean, variance, and standard deviation of these paired differences are

$$\bar{d} = \frac{\sum_{k=1}^{n} d_k}{n} \qquad s_d^2 = \frac{\sum_{k=1}^{n} (d_k - \bar{d})^2}{n - 1} \qquad s_d = \sqrt{s_d^2}$$

These are the point estimates of μ_d, σ_d^2, and σ_d, the mean, variance, and standard deviation of the population of all possible paired differences. Here it is useful to note that

$$\begin{aligned} \bar{d} &= \frac{\sum_{k=1}^{n} d_k}{n} = \frac{\sum_{k=1}^{n} (y_{1,k} - y_{2,k})}{n} \\ &= \frac{\sum_{k=1}^{n} y_{1,k}}{n} - \frac{\sum_{k=1}^{n} y_{2,k}}{n} \\ &= \bar{y}_1 - \bar{y}_2 \end{aligned}$$

That is, $\bar{d}$, the mean of the sample of n paired differences, equals $\bar{y}_1 - \bar{y}_2$, the difference between the sample means. Similarly, it can be shown that μ_d, the mean of the population of all possible paired differences, equals $\mu_1 - \mu_2$, the difference between the population means. Thus, to summarize, $\bar{d} = \bar{y}_1 - \bar{y}_2$ is the point estimate of $\mu_d = \mu_1 - \mu_2$.

Example 3.24 Suppose that Home State Life and Casualty obtains the samples, means, variance, and standard deviation in Table 3.5. Let

$$\mu_d = \mu_1 - \mu_2$$

denote the mean of all possible paired differences of the repair cost estimates at garages 1 and 2, or, the difference between the mean repair cost estimates at garages

TABLE 3.5 **Sample of $n = 7$ paired differences of the repair cost estimates at garages 1 and 2 (cost estimates in hundreds of dollars)**

Sample of $n = 7$ damaged cars	Sample of $n = 7$ repair cost estimates at garage 1	Sample of $n = 7$ repair cost estimates at garage 2	Sample of $n = 7$ paired differences
Car 1	$y_{1,1} = 7.1$	$y_{2,1} = 7.9$	$d_1 = -.8$
Car 2	$y_{1,2} = 9.0$	$y_{2,2} = 10.1$	$d_2 = -1.1$
Car 3	$y_{1,3} = 11.0$	$y_{2,3} = 12.2$	$d_3 = -1.2$
Car 4	$y_{1,4} = 8.9$	$y_{2,4} = 8.8$	$d_4 = .1$
Car 5	$y_{1,5} = 9.9$	$y_{2,5} = 10.4$	$d_5 = -.5$
Car 6	$y_{1,6} = 9.1$	$y_{2,6} = 9.8$	$d_6 = -.7$
Car 7	$y_{1,7} = 10.3$	$y_{2,7} = 11.7$	$d_7 = -1.4$
	$\bar{y}_1 = 9.329$	$\bar{y}_2 = 10.129$	$\bar{d} = \bar{y}_1 - \bar{y}_2 = -.8$ $s_d^2 = .2533$ $s_d = .5033$

1 and 2. Then,

$$\bar{d} = \bar{y}_1 - \bar{y}_2$$
$$= -.8 \qquad (-\$80)$$

is the point estimate of μ_d.

CONFIDENCE INTERVALS FOR $\mu_d = \mu_1 - \mu_2$

Confidence Intervals for $\mu_d = \mu_1 - \mu_2$

1. The *population of all possible values of $\bar{d} = \bar{y}_1 - \bar{y}_2$* (that is, point estimates of $\mu_d = \mu_1 - \mu_2$)
   a. Has *mean* $\mu_{\bar{d}} = \mu_d$.
   b. Has *variance* $\sigma_{\bar{d}}^2 = (\sigma_d^2/n)$ (if the population of all possible paired differences is infinite or very large).
   c. Has *standard deviation* $\sigma_{\bar{d}} = (\sigma_d/\sqrt{n})$ (if the population of all possible paired differences is infinite or very large).
   d. Has a *normal distribution* (if the population of all possible paired differences has a normal distribution).
   e. *Approximately* has a *normal distribution* no matter what probability distribution describes the population of all possible paired differences (if the sample size n is large).
2. A point estimate of $\sigma_{\bar{d}} = \sigma_d/\sqrt{n}$ is $s_{\bar{d}} = s_d/\sqrt{n}$, which is called the *standard error of the estimate $\bar{d}$.*

3. Assume that the population of all possible paired differences has a *normal distribution.* Then the population of all possible values of

$$\frac{\bar{d} - \mu_d}{s_{\bar{d}}} = \frac{\bar{d} - \mu_d}{s_d/\sqrt{n}}$$

has a *t-distribution with n − 1 degrees of freedom.* This implies that a *100(1 − α)% confidence interval for* $\mu_d = \mu_1 - \mu_2$ is

$$\left[\bar{d} \pm t_{[\alpha/2]}^{(n-1)}\left(\frac{s_d}{\sqrt{n}}\right)\right]$$

4. If n is *large,* the population of all possible values of

$$\frac{\bar{d} - \mu_d}{\sigma_d/\sqrt{n}}$$

approximately has a *standard normal distribution.* This implies that

$$\left[\bar{d} \pm z_{[\alpha/2]}\left(\frac{\sigma_d}{\sqrt{n}}\right)\right] \quad \text{and} \quad \left[\bar{d} \pm z_{[\alpha/2]}\left(\frac{s_d}{\sqrt{n}}\right)\right]$$

are *approximately correct 100(1 − α)% confidence intervals for* $\mu_d = \mu_1 - \mu_2$, no matter what probability distribution describes the population of all possible paired differences. To use the second interval, the sample size n should be at least 30.

Example 3.25 Consider Example 3.24. Assuming that the population of all possible paired differences is normally distributed (or at least mound-shaped), it follows that a 95% confidence interval for $\mu_d = \mu_1 - \mu_2$ is

$$\begin{aligned}\left[\bar{d} \pm t_{[.025]}^{(n-1)}\left(\frac{s_d}{\sqrt{n}}\right)\right] &= \left[(\bar{y}_1 - \bar{y}_2) \pm t_{[.025]}^{(7-1)}\left(\frac{s_d}{\sqrt{n}}\right)\right] \\ &= \left[-.8 \pm t_{[.025]}^{(6)}\left(\frac{.5033}{\sqrt{7}}\right)\right] \\ &= [-.8 \pm 2.447(.1902)] \\ &= [-.8 \pm .4654] \\ &= [-1.2654, -.3346]\end{aligned}$$

This interval says that Home State Life and Casualty can be 95% confident that μ_d, the mean of all possible paired differences of the repair cost estimates at garages 1 and 2, is between −\$126.54 and −\$33.46. That is, we are 95% confident that μ_1, the mean of all possible repair cost estimates at garage 1, is between \$126.54 and \$33.46 less than μ_2, the mean of all possible repair cost estimates at garage 2.

TESTING $H_0: \mu_d = c$

Testing $H_0: \mu_d = c$

Define the test statistic

$$t = \frac{\bar{d} - c}{s_d/\sqrt{n}}$$

If the population of all possible paired differences has a normal distribution, we can reject $H_0: \mu_d = c$ in favor of a particular alternative hypothesis by setting the probability of a Type I error equal to α by using the following rejection point conditions.

Alternative hypothesis	Rejection point condition: Reject H_0 in favor of H_1 if and only if
$H_1: \mu_d \neq c$	$\lvert t \rvert > t_{[\alpha/2]}^{(n-1)}$
$H_1: \mu_d > c$	$t > t_{[\alpha]}^{(n-1)}$
$H_1: \mu_d < c$	$t < -t_{[\alpha]}^{(n-1)}$

Note that if n is large (say, at least 30), then we can replace the above t-points by the corresponding z-points. In this case the Central Limit Theorem says that the result is approximately valid no matter what probability distribution describes the population of all possible paired differences.

Example 3.26 Suppose that although Home State Life and Casualty has been having its moderately damaged automobiles repaired at garage 2, the insurance company has heard that garage 1 provides less expensive repair service of equal quality. However, because of its past experience with garage 2, Home State has decided that it will have its moderately damaged automobiles repaired at garage 1 only if it is very confident that $\mu_d = \mu_1 - \mu_2$ is less than zero. Specifically, Home State has decided to have its moderately damaged automobiles repaired at garage 1 only if it can use the sample of $n = 7$ paired differences of the repair cost estimates at garages 1 and 2 to reject

$$H_0: \mu_d \geq 0 \quad (\mu_1 - \mu_2 \geq 0)$$

in favor of

$$H_1: \mu_d < 0 \quad (\mu_1 - \mu_2 < 0)$$

by setting α equal to .01. Note that setting α so small means that there is a very small probability that Home State will conclude that garage 1 is less expensive than garage 2 when garage 1 is not less expensive than garage 2. Testing H_0 versus H_1

is equivalent to testing $H_0: \mu_d = 0$ versus $H_1: \mu_d < 0$. To do this, we compute

$$\begin{aligned} t &= \frac{\bar{d} - 0}{s_d/\sqrt{n}} \\ &= \frac{\bar{y}_1 - \bar{y}_2 - 0}{s_d/\sqrt{n}} \\ &= \frac{-.8 - 0}{.5033/\sqrt{7}} \\ &= -4.206 \end{aligned}$$

and use the rejection point

$$-t_{[\alpha]}^{(n-1)} = -t_{[.01]}^{(n-1)} = -t_{[.01]}^{(7-1)} = -t_{[.01]}^{(6)} = -3.143$$

Since

$$t = -4.206 < -3.143 = -t_{[.01]}^{(6)}$$

we can reject H_0 in favor of H_1 by setting α equal to .01.

Furthermore, an upper one-sided 99% confidence interval for $\mu_d = \mu_1 - \mu_2$ is

$$\begin{aligned} \left(-\infty, \bar{d} + t_{[\alpha]}^{(n-1)}\left(\frac{s_d}{\sqrt{n}}\right)\right] &= \left(-\infty, -.8 + t_{[.01]}^{(7-1)}\left(\frac{.5033}{\sqrt{7}}\right)\right] \\ &= (-\infty, -.8 + 3.143(.1902)] \\ &= (-\infty, -.2022] \end{aligned}$$

Therefore Home State Life and Casualty is 99% confident that μ_1 is at least \$20.22 less than μ_2. Finally, the prob-value for the test is the area under the curve of the t-distribution having $n - 1 = 6$ degrees of freedom to the left of $t = -4.206$. This prob-value equals .0014 and provides very strong evidence against $H_0: \mu_d = 0$.

EXERCISES

3.1 Assume that the population of all Hawk mileages is normally distributed. Use the following sample of $n = 5$ mileages

$$\{32.3, 30.5, 31.7, 31.4, 32.6\}$$

to find 90%, 95%, 98%, and 99% confidence intervals for μ, the population mean Hawk mileage.

3.2 a. Using the three samples in Figure 3.1, calculate three 99% confidence intervals for μ.
b. What percentage of these three confidence intervals contain μ (=31.5)?
c. Explain the meaning of 99% confidence in this situation.

3.3 Zenex Radio Corporation has developed a new way to assemble an electrical component used in the manufacture of radios. The company wishes to determine whether μ, the mean assembly time of this component using the new method, is less than 20 minutes, which is known to be the mean assembly time of the component using the current method.

Suppose that Zenex Radio randomly selects a sample of $n = 6$ employees. The company thoroughly trains each employee to use the new assembly method and has each employee assemble one component using the new method. The company records the assembly times and calculates the mean and standard deviation of the sample of $n = 6$ assembly times to be $\bar{y} = 14.29$ minutes and $s = 2.19$ minutes.

a. Assuming that the population of all assembly times has a normal distribution, calculate a 99% confidence interval for μ.
b. Using the confidence interval that you calculated in part (a), can Zenex Radio be at least 99% confident that μ is less than 20 minutes? Justify your answer.

3.4 Referring to Exercise 3.3, determine the sample size n so that Zenex Radio is 99% confident that $\bar{y}$ is within $B = 2$ minutes of μ. Regard the sample in Exercise 3.3 as a preliminary sample.

3.5 National Motors has equipped the ZX-900 with a new disc brake system. We define the stopping distance for a ZX-900 to be the distance (in feet) required to bring the automobile to a complete stop from a speed of 35 mph under normal driving conditions using this new brake system. In addition, we define μ to be the mean stopping distance of all ZX-900s. One of the ZX-900's major competitors is advertised to achieve a mean stopping distance of 60 ft. National Motors would like to claim in a new advertising campaign that the ZX-900 achieves a shorter mean stopping distance.

Suppose that National Motors randomly selects a sample of $n = 81$ ZX-900s. The company records the stopping distances of each of these automobiles and calculates the mean and standard deviation of the sample of $n = 81$ stopping distances to be $\bar{y} = 57.8$ ft and $s = 6.02$ ft.

a. Calculate 90%, 95%, 98%, and 99% confidence intervals for μ.
b. Using the 95% confidence interval, can National Motors be at least 95% confident that μ is less than 60 ft? Explain.
c. Using the 98% confidence interval, can National Motors be at least 98% confident that μ is less than 60 ft? Explain.

3.6 Referring to Exercise 3.5, determine the sample size n so that National Motors is 98% confident that $\bar{y}$ is within 1 ft of μ. Regard the sample in Exercise 3.5 as a preliminary sample.

3.7 Consider the Gem Shampoo problem. Suppose that a sample of $n = 6$ bottle fills is randomly selected from a particular adjustment of the Gem bottle-filling process. A sample mean of $\bar{y} = 15.7665$ and a sample standard deviation of $s = .1524$ are obtained.

a. Test $H_0: \mu = 16$ versus $H_1: \mu \neq 16$ by setting $\alpha = .05$. On the basis of this test, should the process be readjusted?
b. Test $H_0: \mu = 16$ versus $H_1: \mu \neq 16$ by setting $\alpha = .01$. On the basis of this test, should the process be reajusted?

3.8 Use the sample information in Exercise 3.3 to test $H_0: \mu = 20$ versus $H_1: \mu < 20$. Set $\alpha = .05$.

3.9 Use the sample information in Exercise 3.5 to test $H_0: \mu = 60$ versus $H_1: \mu < 60$. Set $\alpha = .05$. Also, calculate the prob-value.

3.10 An engine manufacturer is testing the amount of air pollution (in milligrams per cubic yard of exhaust) emitted by a new rotary engine. The manufacturer randomly selects four rotary engines and subjects them to a pollution test. It finds that these four engines emit 72, 74, 75, and 75 milligrams per cubic yard.

a. Assuming normality, calculate a 99% confidence interval for the mean emission of air pollution by the rotary engine.
b. A competing piston engine has a mean air pollution emission of 80 milligrams per cubic yard of exhaust. On the basis of your answer to part (a), would it be reasonable to claim that the mean emission of air pollution for the rotary engine is less than that of the competing piston engine? Explain your answer.
c. Letting μ denote the mean emission of air pollution by the rotary engine, test $H_0: \mu \geq 80$ versus $H_1: \mu < 80$ by setting $\alpha = .01$.
d. Calculate an upper one-sided 99% confidence interval for μ. Using this interval, can $H_0: \mu \geq 80$ be rejected in favor of $H_1: \mu < 80$?
e. What do the results of parts (c) and (d) say about whether the mean emission for the rotary engine is less than that of the competing piston engine? Explain.

3.11 Consider a population that has a standard deviation of 25. We wish to estimate the mean of this population.

a. How large a random sample is needed if we wish to construct a 95.45% confidence interval for the mean of this population with a total length that will not exceed 5?
b. Suppose that we now take a sample of the size we determined in part (a). If we obtain a sample mean of 100, calculate the 95.45% confidence interval for the population mean.

3.12 The State Department of Agriculture (SDA) requires that the average "berry solids" content for strawberry preserves be at least 65 percent (if the berry solids content is less than 65 percent, the product cannot be labeled as a "preserve"). The SDA enforces this requirement by randomly sampling eight jars from each production run for all manufacturers of strawberry (and other) preserves. The SDA samples eight jars of strawberry preserves from a production run and measures the "berry solids" content for each jar. The following results are obtained.

65.4 64.9 65.2 65.7 65.0 65.7 65.3 64.7

a. Use this data to compute a lower one-sided 95% confidence interval for the mean "berry solids" content of all jars of strawberry preserves in the production run. Assume that "berry solids" contents are normally distributed.
b. On the basis of this interval, can the SDA conclude that the mean "berry solids" content of all jars of strawberry preserves is at least 65 percent? Explain.

3.13 The *bad debt ratio* for a financial institution is defined to be the dollar value of loans defaulted divided by the total dollar value of all loans made. A random sample of seven Ohio banks is selected. The bad debt ratios (written as percentages) for these banks are 7, 4, 6, 7, 5, 4, and 9 percent.

a. The mean bad debt ratio for all federally insured banks is 3.5%. Federal banking officials claim that the mean bad debt ratio for Ohio banks is higher than the mean for all federally insured banks. Set up the null and alternative hypotheses that should be used to statistically justify this claim.

b. Assuming that bad debt ratios for Ohio banks are normally distributed, use the sample results given above to test the hypotheses you set up in part (a) with $\alpha = .01$.
c. Use the above sample to calculate a lower one-sided 95% confidence interval for the mean bad debt ratio for Ohio banks. Use this interval to test the hypotheses you set up in part (a) with $\alpha = .01$.
d. What do the results of parts (b) and (c) say about whether or not the mean bad debt ratio for Ohio banks is greater than the average for all federally insured banks? Explain.

3.14 The State Environmental Protection Agency (SEPA) is responsible for monitoring the air pollution level for a large western metropolis. The air pollution level is considered to be acceptable (or safe) if the mean pollution level is at or below a reading of 100 milligrams of pollution per cubic yard of air. Air pollution levels substantially above the 100 milligrams per cubic yard level are considered to be dangerous. To monitor air pollution levels, the SEPA will take a pollution reading 100 times a day. If the mean pollution reading for this sample of 100 readings casts considerable doubt on the hypothesis that the air pollution level is acceptable, then the SEPA must declare an air pollution emergency and must impose emergency measures to reduce pollution levels.

a. The SEPA wishes to set up a hypothesis test so that an air pollution emergency will be declared when the null hypothesis is rejected. Set up the null and alternative hypotheses that should be used in this situation.
b. Environmentalist groups have been applying great political pressure to force the SEPA to enforce air pollution standards in a vigorous fashion. Therefore the SEPA feels that it will suffer very serious political consequences if no air pollution emergency has been declared when air pollution levels are in fact dangerous. Other things being equal, should the SEPA use a level of significance of $\alpha = .01$ or $\alpha = .05$ in testing the hypotheses set up in part (a). Explain.
c. Suppose that it is extremely expensive to issue an air pollution emergency and to impose emergency measures to reduce pollution levels when pollution levels are in fact acceptable. In this case, other things being equal, should the SEPA use a level of significance of $\alpha = .01$ or $\alpha = .05$ in testing the hypotheses set up in part (a). Explain.
d. Suppose that the SEPA decides to use $\alpha = .01$. Determine the rejection point that should be used to test the hypotheses set up in part (a).
e. If the SEPA obtains a sample mean of $\bar{y} = 100.327$ milligrams per cubic yard and a sample standard deviation of $s = 2$ milligrams per cubic yard, should the SEPA declare an air pollution emergency? Explain.
f. If the SEPA obtains $\bar{y} = 100.503$ and $s = 2$, should the SEPA declare an air pollution emergency? Explain.
g. Suppose that the SEPA obtains $\bar{y} = 100.1$ and $s = 2$. Calculate a lower one-sided 99% confidence interval for the mean air-pollution level.
h. Based on the confidence interval calculated in part (g), should the SEPA declare an air pollution emergency? Explain your answer.

3.15 Consolidated Power, a large electric power utility, has just built a modern nuclear power plant. This plant discharges waste water, which is allowed to flow into the Atlantic

Ocean. The State Environmental Protection Agency (SEPA) has issued an order stating that the waste water discharged into the ocean may not be excessively warm so that thermal pollution of the marine environment near the plant can be avoided. Because of this order, the waste water is allowed to cool in specially constructed ponds and is then released into the ocean. This cooling system is working properly if the mean temperature of waste water discharged is 50°F or cooler. Consolidated Power is required to monitor the temperature of the waste water discharged into the ocean. A temperature reading will be made 100 times a day. If the mean temperature for this sample of 100 readings casts substantial doubt on the hypothesis that the mean temperature of waste water discharged is 50°F or cooler, the power plant must be shut down.

a. Consolidated Power wishes to set up a hypothesis test so that the power plant will be shut down when the null hypothesis is rejected. Set up the null and alternative hypotheses that should be used in this situation.
b. The SEPA periodically conducts spot checks to determine whether or not the waste water being discharged is too warm. Suppose the SEPA has the power to impose very severe penalties (for example, very heavy fines) when the waste water is excessively warm. Other things being equal, should Consolidated Power use a level of significance of $\alpha = .01$ or $\alpha = .05$ in testing the hypotheses set up in part (a)? Explain.
c. Suppose that Consolidated Power has been experiencing technical problems with the cooling system. Because the system is unreliable, the company feels that it must guard against Type II errors. Other things being equal, should Consolidated Power use a level of significance of $\alpha = .01$ or $\alpha = .05$ in testing the hypotheses set up in part (a)? Explain.
d. Suppose that Consolidated Power decides to use $\alpha = .01$. Determine the rejection point that should be used to test the hypotheses in part (a).
e. If Consolidated Power obtains a sample mean of $\bar{y} = 50.481$°F and a sample standard deviation of $s = 2$°F, should the power plant be shut down? Explain.
f. If Consolidated Power obtains $\bar{y} = 50.263$°F and $s = 2$°F, should the power plant be shut down? Explain.
g. Suppose that Consolidated Power obtains $\bar{y} = 50.2$°F and $s = 2$°F. Calculate a lower one-sided 99% confidence interval for the mean waste water temperature.
h. Based on the confidence interval calculated in part (g), should the power plant be shut down? Explain.

3.16 The Rola-Cola Bottling Company has recently installed a new bottling process that will be used to fill 16-ounce bottles of the Rola-Cola Classic soft drink. Both overfilling and underfilling bottles are undesirable, since underfilling will result in consumer complaints and overfilling costs the company substantial money. To monitor the bottling process, 36 bottles of Rola-Cola Classic will be sampled during each shift, and the bottle fill for each of the bottles will be recorded. If the mean bottle fill for this sample casts a substantial amount of doubt on the hypothesis that the process mean bottle fill is the desired 16 ounces, the bottling process will be shut down and adjusted.

a. Rola-Cola wishes to set up a hypothesis test so that the bottling process will be shut down and adjusted when the null hypothesis is rejected. Set up the null and alternative hypotheses that should be used in this situation.
b. If the bottling process is very *reliable* (which says that the process very seldom needs

to be shut down and adjusted), other things being equal, should Rola-Cola use a level of significance of $\alpha = .01$ or $\alpha = .05$ for the above hypothesis test? Explain.

c. If the bottling process is very *unreliable* (which says that the process needs to be shut down and adjusted very often), other things being equal, should Rola-Cola use a level of significance of $\alpha = .01$ or $\alpha = .05$ for the above hypothesis test? Explain.

d. Suppose that Rola-Cola decides to use $\alpha = .05$. Determine the rejection point that should be used to test the hypotheses in part (a).

e. If Rola-Cola obtains a sample mean of $\bar{y} = 16.05$ ounces and a sample standard deviation of $s = .1$ ounce, should the bottling process be shut down and readjusted? Explain.

f. If Rola-Cola obtains $\bar{y} = 15.96$ ounces and $s = .1$ ounce, should the bottling process be shut down and readjusted? Explain.

g. Suppose that Rola-Cola obtains $\bar{y} = 16.2$ ounces and $s = .1$ ounce. Calculate a 95% confidence interval for the process mean bottle fill.

h. Based on the confidence interval calculated in part (g), should the bottling process be shut down and adjusted? Explain.

3.17 General Nails, Inc. owns a machine that is set to produce nails with a mean length of 2 inches. Nails that are too long or too short do not meet the customer's specifications and must be rejected. To avoid producing too many defective nails, 100 nails produced by the machine are sampled each half-hour and measured as a check to see whether the machine is still producing nails with a mean length of 2 inches. If the mean nail length for the sample casts a substantial amount of doubt on the hypothesis that the machine is producing nails with a mean length of 2 inches, the machine will be shut down and reset.

a. General Nails, Inc. wishes to set up a hypothesis test so that the machine will be shut down and reset when the null hypothesis is rejected. Set up the null and alternative hypotheses that should be used in this situation.

b. Suppose that General Nails, Inc. believes that it is very important to maintain its image as a consumer-conscious firm. Therefore the company wishes to avoid selling out-of-specification nails. Other things being equal, should the company use a level of significance of $\alpha = .01$ or $\alpha = .05$ for the above hypothesis test? Explain.

c. If the machine is very *reliable* (which says that the machine very seldom needs to be shut down and reset), other things being equal, should General Nails, Inc. use a level of significance of $\alpha = .01$ or $\alpha = .05$ for the above hypothesis test? Explain.

d. Suppose that General Nails, Inc. decides to use $\alpha = .05$. Find the rejection point that should be used to test the hypotheses in part (a).

e. If General Nails, Inc. obtains a sample mean of $\bar{y} = 0.99510$ inches and a sample standard deviation of $s = .02$ inch, should the machine be shut down and reset? Explain.

f. If General Nails, Inc. obtains $\bar{y} = 1.00284$ inches and $s = .02$ inch, should the machine be shut down and reset? Explain.

g. If General Nails, Inc. obtains $\bar{y} = 1.00546$ inches and $s = .02$ inch, should the machine be shut down and reset? Explain.

h. Suppose that General Nails, Inc. obtains $\bar{y} = 1.002$ inches and $s = .02$ inch. Calculate a 95% confidence interval for the population mean nail length.

i. Based on the confidence interval calculated in part (h), should the machine be shut down and reset? Explain.

3.18 Consolidated Chemical Corporation receives a large weekly shipment of bags of chemical product H132. This chemical product is supposed to have an average "acid value" of 200, while the standard deviation of the acid values is known to be 2.

a. Suppose that the mean acid value for this week's entire shipment of bags of product H132 is 200. If we draw a random sample of 100 bags, what is the probability that the *average* acid value for the 100 bags is greater than 200.35?
b. The management of Consolidated Chemical Corporation is worried that the mean acid value of the bags of chemical product H132 being received from its supplier may not be at the desired value of 200. Because of this, a random sample of 100 bags is drawn from each shipment, and the shipment is rejected (sent back to the supplier) if the average acid value for the 100 sample bags is either greater than 200.6 or less than 199.4. Suppose that the average acid value for this week's entire shipment is actually the desired 200 value. What is the probability that this shipment will be rejected and sent back to the supplier?

3.19 A sample of 81 60-watt lightbulbs produced by Dynamics, Inc. obtained a mean lifetime of 1347 hours with a variance of 729 hours. A sample of 50 60-watt lightbulbs produced by National Electronics Corporation obtained a mean lifetime of 1282 hours with a variance of 800 hours.

a. Calculate a 99% confidence interval for $\mu_1 - \mu_2$, the difference between the true mean lifetimes of the two types of lightbulbs. Interpret the results.
b. Test $H_0: \mu_1 - \mu_2 = 50$ versus $H_1: \mu_1 - \mu_2 > 50$ by setting $\alpha = .01$. Interpret the results.

3.20 Reconsider Example 3.17. Suppose that Bargain Town randomly selected the following small samples.

Sample of Ohio accounts	Sample of Illinois accounts
$n_O = 10$	$n_I = 5$
$\bar{y}_O = \$124$	$\bar{y}_I = \$68$
$s_O^2 = 1681$	$s_I^2 = 484$

a. Assuming that the equal variances, independent samples, and normality assumptions hold, compute a 95% confidence interval for $\mu_O - \mu_I$.
b. Assuming that only the independent samples and normality assumptions hold, calculate a 95% confidence interval for $\mu_O - \mu_I$.
c. Carry out the F-test for equality of population variances. What does F say about the approximate equality of variances?
d. Carry out an appropriate test of $H_0: \mu_O - \mu_I = 0$ versus $H_1: \mu_O - \mu_I \neq 0$. Interpret the results.

3.21 American International Paper Corporation must purchase a new papermaking machine. The company must choose between two machines, machine 1 and machine 2. Since the machines produce paper of equal quality, the company will choose the machine that produces the most paper in a one-hour period (measured in rolls of paper per hour). To compare the two machines, five randomly selected machine operators produce paper for one hour using machine 1, and another five randomly selected machine operators produce paper for one hour using machine 2. The operators are of approximately equal skill. The following results are obtained.

	Hourly production of paper (no. of rolls)				
Machine 1 (Sample 1)	50	47	50	49	48
Machine 2 (Sample 2)	42	44	43	41	43

a. Calculate the means and variances of these samples.
b. Discuss why it is reasonable to believe that these samples are independent of each other.
c. Assuming that the normal populations assumption holds, select an appropriate formula and calculate a 95% confidence interval for $\mu_1 - \mu_2$, the difference between the true mean numbers of rolls of paper produced per hour by the two machines. Interpret the results.
d. Test $H_0: \mu_1 - \mu_2 = 0$ versus $H_1: \mu_1 - \mu_2 \neq 0$ by setting $\alpha = .05$. Interpret the results.

3.22 A company wishes to evaluate the merits of a new sales training program. In a study done by the company, 100 sales agents who were given traditional sales training had an average transaction time (the time until a sale is closed) of 17 minutes with a standard deviation of 4 minutes. In addition, 100 sales agents who were given the new sales training program had an average transaction time of 14 minutes with a standard deviation of 3 minutes.

a. Assuming that the above groups of sales agents are independently and randomly selected, compute a 99.73% confidence interval for the difference between the mean transaction time for sales agents given the traditional sales training program and the mean transaction time for sales agents given the new sales training program. That is, if μ_1 is the mean transaction time for sales agents given the traditional sales training program and μ_2 is the mean transaction time for sales agents given the new sales training program, find a 99.73% confidence interval for $\mu_1 - \mu_2$.
b. On the basis of your confidence interval, do you believe that the new sales training program is effective in reducing the mean transaction time for sales agents? Why or why not?
c. Set up the null and alternative hypotheses that should be used to justify that the new sales training program reduces the mean transaction time for sales agents.
d. Test the hypotheses that you set up in part (c) with $\alpha = .01$. Interpret the results of this test.

3.23 The president of a labor union claims that the mean daily wage rate for union workers in the plastics industry is higher than the mean daily wage rate for nonunion workers. A random sample of 200 union workers yielded an average daily wage rate of \$68 with a variance of 9, while a random sample of 400 nonunion workers yielded an average daily wage rate of \$63.50 with a variance of 7.

a. Assuming that the independence assumption holds, compute a 95% confidence interval for the difference between the mean daily wage rate for union workers and the mean daily wage rate for nonunion workers. That is, if μ_1 is the mean daily wage rate for union workers and μ_2 is the mean daily wage rate for nonunion workers, find a 95% confidence interval for $\mu_1 - \mu_2$.
b. On the basis of your confidence interval, do you believe that the president of the labor union is justified in claiming that the mean daily wage rate is higher for union workers? Why or why not?
c. Set up the null and alternative hypotheses that should be used to justify that the mean daily wage rate is higher for union workers than for nonunion workers.
d. Test the hypotheses that you set up in part (c) with $\alpha = .05$. Interpret the results of this test.

3.24 An Ohio university wishes to attempt to prove that car ownership is detrimental to academic achievement. A random sample of 100 students who do not own cars had an average GPA of 2.68 with a sample standard deviation of .7, while a random sample of 100 students who own cars had an average GPA of 2.55 with a sample standard deviation of .6.

a. Assuming that the independence assumption holds and letting

μ_1 = the mean GPA for all students who are not car owners

μ_2 = the mean GPA for all students who are car owners

use the above data to compute a 95% confidence interval for $\mu_1 - \mu_2$.
b. On the basis of the interval calculated in part (a), can the university statistically justify that car ownership is detrimental to academic achievement? That is, can the university justify that μ_1 is greater than μ_2? Explain.
c. Set up the null and alternative hypotheses that should be used to justify that the mean GPA for non-car-owners is higher than the mean GPA for car owners.
d. Test the hypotheses that you set up in part (c) with $\alpha = .05$. Interpret the results of this test.

3.25 National Paper Company must purchase a new machine for producing cardboard boxes. The company must choose between two machines, machine 1 and machine 2. Since the machines produce boxes of equal quality, the company will choose the machine that produces the most boxes in a one-hour period. It is known that there are substantial differences in the abilities of the company's machine operators. Therefore National Paper has decided to compare the machines using a paired difference experiment. Suppose that eight randomly selected machine operators produce boxes for one hour using machine 1 and for one hour using machine 2, with the following results.

	Machine operator							
	1	*2*	*3*	*4*	*5*	*6*	*7*	*8*
Machine 1	53	60	58	48	46	54	62	49
Machine 2	50	55	56	44	45	50	57	47

a. Assuming that the population of all possible paired differences has a normal distribution, calculate a 95% confidence interval for $\mu_d = \mu_1 - \mu_2$. Interpret the results.
b. Test $H_0: \mu_d = 0$ versus $H_1: \mu_d \neq 0$ by setting α equal to .05. Interpret the results.

Exercises 3.26 through 3.28 require the following background information. It can be shown that if μ and σ^2 are the mean and variance of a finite population of N values and if n is the size of a sample randomly selected from this population, then the population of all possible sample means (that is, point estimates of μ) has mean $\mu_{\bar{y}} = \mu$, variance $\sigma_{\bar{y}}^2 = (\sigma^2/n)((N - n)/(N - 1))$, and standard deviation $\sigma_{\bar{y}} = (\sigma/\sqrt{n})\sqrt{(N - n)/(N - 1)}$. Moreover, if the sample size, n, is large, the population of all possible sample means approximately has a normal distribution.

3.26 If we define

$$z_{[\bar{y},\mu]} = \frac{\bar{y} - \mu}{(\sigma/\sqrt{n})\sqrt{(N - n)/(N - 1)}}$$

then the above background information implies that if n is large, the population of all possible $z_{[\bar{y},\mu]}$ statistics approximately has a standard normal distribution.

a. Use the preceding information to show that if n is large, then

$$\left[\bar{y} - z_{[\alpha/2]}\left(\frac{\sigma}{\sqrt{n}}\right)\sqrt{\frac{N - n}{N - 1}}, \bar{y} + z_{[\alpha/2]}\left(\frac{\sigma}{\sqrt{n}}\right)\sqrt{\frac{N - n}{N - 1}}\right]$$

is an approximately correct $100(1 - \alpha)\%$ confidence interval for μ.

b. It can be shown that

$$s\sqrt{\frac{N - 1}{N}} = \sqrt{\frac{\sum_{i=1}^{n}(y_i - \bar{y})^2}{n - 1}}\sqrt{\frac{N - 1}{N}}$$

is the appropriate point estimate of σ when N is finite. Use this fact and the confidence interval in part (a) to show that

$$\left[\bar{y} - z_{[\alpha/2]}\left(\frac{s}{\sqrt{n}}\right)\sqrt{\frac{N - n}{N}}, \bar{y} + z_{[\alpha/2]}\left(\frac{s}{\sqrt{n}}\right)\sqrt{\frac{N - n}{N}}\right]$$

is an approximately correct $100(1 - \alpha)\%$ confidence interval for μ.

3.27 Dynamics, Inc. employs 900 scientists. Because of a lack of secretaries and laboratory assistants, the company is concerned that these scientists must spend too much time performing "trivial tasks." A sample of $n = 70$ scientists is randomly selected, and each scientist is asked to record the number of hours he or she spends doing "trivial tasks" during the upcoming week. When the results are summarized, a sample mean of $\bar{y} = 9.75$ hours and a sample standard deviation of $s = 2.14$ hours are obtained. Use the formula in part (b) of Exercise 3.26 to calculate a 95% confidence interval for μ, the *mean* number of hours spent last week doing "trivial tasks" per scientist by all 900 scientists. Interpret the results.

3.28 Again consider Exercise 3.27. If we let τ denote the *total* number of hours spent last week doing "trivial tasks" by all 900 scientists, then $\mu = \tau/N$, and thus $\tau = N\mu$. It follows that a point estimate of $\tau = N\mu$ is $\hat{\tau} = N\bar{y}$ and that an approximately correct $100(1 - \alpha)\%$ confidence interval for τ is

$$\left[N\bar{y} - z_{[\alpha/2]}N\left(\frac{s}{\sqrt{n}}\right)\sqrt{\frac{N-n}{N}},\ N\bar{y} + z_{[\alpha/2]}N\left(\frac{s}{\sqrt{n}}\right)\sqrt{\frac{N-n}{N}}\right]$$

This has been obtained by multiplying through the formula in part (b) of Exercise 3.26 by N. Use the sample information in Exercise 3.27 to calculate a point estimate of, and a 95% confidence interval for, τ. Interpret the results.

3.29 Refer to Example 2.7. To estimate the 10% trimmed mean, μ_T, of the population of physicians' salaries, we consider the following samples:

RS:	127	132	138	144	146	152	154	162	171	177	192	241
TS:			138	144	146	152	154	162	171	177		
WS:	138	138	138	144	146	152	154	162	171	177	177	177

The sample denoted *RS* is the previously discussed random sample of $n = 12$ physicians' salaries. The sample denoted *TS*, which has mean $\bar{y}_T = 155.5$, is the *trimmed sample* obtained by eliminating the smallest 10% and the largest 10% of the 12 physicians' salaries from the random sample. Note that since $.10(12) = 1.2$, we round up and eliminate the two smallest and two largest salaries. Thus the trimmed sample contains $h = 8$ observations. The point estimate of μ_T is $\bar{y}_T = 155.5$. The quantity $s_W^2 = 256.333$ is the variance of the sample denoted *WS*. This sample is called the Winsorized sample and is found by replacing the trimmed observation(s) in each tail of the trimmed sample by the most extreme observation remaining in that tail. We can calculate a 95% confidence interval for μ_T by first calculating the *h-Winsorized standard deviation* $s_{Wh} = ((n - 1)s_w^2/(h - 1))^{1/2} = ((12 - 1)256.333/(8 - 1))^{1/2} = 20.07$. Then the interval is

$$\left[\bar{y}_T \pm t_{[.025]}^{(h-1)} \frac{s_{Wh}}{\sqrt{h}}\right] = \left[155.5 \pm 2.365\,\frac{(20.07)}{\sqrt{8}}\right]$$
$$= [138.718,\ 172.282]$$

Using the above procedure, calculate a 95% confidence interval for the 10% trimmed mean μ_T of the population of physicians' salaries by using the sample of $n = 15$ physicians' salaries in Exercise 2.9.

3.30 Consider testing $H_0: \mu = 80$ versus $H_1: \mu > 80$ by using a sample of $n = 100$ observations. Moreover, assume that the population standard deviation σ is known to equal 5 (that is, $\sigma = 5$). *If we wish to set α equal to .05*, then we should reject $H_0: \mu = 80$ in favor of $H_1: \mu > 80$ if

$$\frac{\bar{y} - 80}{\sigma/\sqrt{n}} > z_{[.05]}$$

Thus we obtain the rejection condition

$$\frac{\bar{y} - 80}{5/\sqrt{100}} > 1.64$$

or

$$\bar{y} - 80 > 1.64(5/\sqrt{100})$$

FIGURE 3.10 **Probability of committing a Type II error when $\mu = 82$**

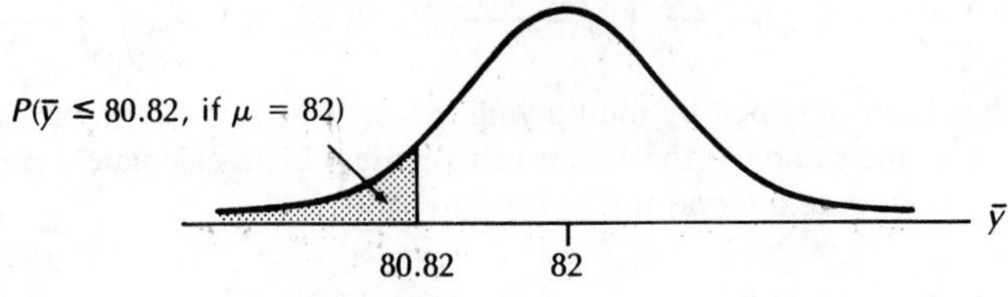

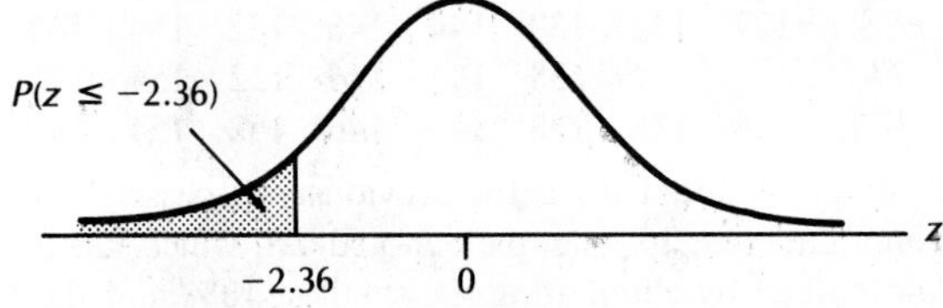

or

$$\bar{y} > 80 + 1.64(.5)$$

or

$$\bar{y} > 80.82$$

Therefore we should *not* reject $H_0: \mu = 80$ in favor of $H_1: \mu > 80$ if

$$\bar{y} \leq 80.82$$

It follows that the probability that we will commit a Type II error if μ really equals 82 is (as illustrated in Figure 3.10)

$$\begin{aligned} &P(\text{We will not reject } H_0: \mu = 80 \text{ if } \mu \text{ really equals } 82) \\ &\quad = P(\bar{y} \leq 80.82 \text{ if } \mu \text{ really equals } 82) \\ &\quad = P\left\{\frac{\bar{y} - \mu}{\sigma/\sqrt{n}} \leq \frac{80.82 - 82}{5/\sqrt{100}}\right\} \\ &\quad = P(z \leq -2.36) \\ &\quad = .0091 \end{aligned}$$

Given this background, consider testing $H_0: \mu = 60$ versus $H_1: \mu < 60$ by using a sample of $n = 64$ observations. Moreover, assume that the population standard deviation σ is known to equal 4 (that is, $\sigma = 4$). *If we wish to set α equal to .01*, then calculate the probability that we will commit a Type II error if μ really equals 58. That is, compute

$$P(\text{We will not reject } H_0: \mu = 60 \text{ if } \mu \text{ really equals } 58)$$

Hint: Determine the rejection point condition for rejecting $H_0: \mu = 60$ in favor of $H_1: \mu < 60$ when setting α equal to .01. Remember that the alternative hypothesis $H_1: \mu < 60$ is a *less than* the alternative hypothesis.

3.31 Consider estimating the difference between two population means μ_1 and μ_2. Suppose that from past experience we know that the population variances are approximately $\sigma_1^2 = 5$ and $\sigma_2^2 = 4$. If we wish to compute a 95.45% confidence interval for $\mu_1 - \mu_2$ by drawing equally sized independent random samples ($n_1 = n_2 = n$) from the two populations, find the sample size needed so that this 95.45% confidence interval will have a length equal to 1.

CHAPTER 4

THE SIMPLE LINEAR REGRESSION MODEL

In this chapter we begin our study of regression analysis by discussing the *simple linear regression model*. In Section 4.1 we describe this model. We will see that the model employs two parameters: *the slope* and *the y-intercept*. Sections 4.2 and 4.3 show how to compute "least squares point estimates" of these parameters and how to use these estimates to compute point predictions of the dependent variable. We conclude this chapter with Section 4.4, which explains the assumptions behind the simple linear regression model.

4.1 DESCRIPTION OF THE MODEL

We introduce the simple linear regression model by using an example. This example is motivated by an application of regression analysis recently carried out by the management services division of a large accounting firm. (For purposes of confidentiality we will use a fictitious name and will not discuss all the details of this application.)

Example 4.1

Suppose that we are analysts in the management services division of the accounting firm of Johnson and Lilly. One of our firm's clients is American Manufacturing Company, a major manufacturer of a wide variety of commercial and industrial products. American Manufacturing owns a large nine-building complex in Central City and heats this complex by using a modern coal-fueled heating system. In the past, American Manufacturing has encountered problems in determining the proper amount of coal to order each week to heat the complex adequately. Because of this, the firm has requested that we develop an accurate way to predict the amount of fuel (in tons of coal) that will be used to heat the nine-building complex in future weeks. Our experience indicates that (1) weekly fuel consumption substantially depends on the average hourly temperature (in degrees Fahrenheit) during the week and (2) weekly fuel consumption also depends on factors other than average hourly temperature that contribute to an overall "chill factor." Some of these factors are:

Wind velocity (in miles per hour) during the week

"Cloud cover" during the week

Variations in temperature, wind velocity, and cloud cover during the week (perhaps caused by the movement of weather fronts)

In this chapter we use regression analysis to predict the *dependent variable* weekly fuel consumption (y) on the basis of the *independent variable* average hourly temperature (x). In subsequent chapters we will use additional independent variables, which measure the effects of factors such as wind velocity and cloud cover, to help us predict weekly fuel consumption. Suppose that we have gathered data concerning y and x for the $n = 8$ weeks prior to the current week. This data is given in Table 4.1. Here the letter i denotes the time order of a previously observed week, where x_i denotes the average hourly temperature and y_i denotes the fuel consumption that has been observed in week i. It should be noted that it would, of course, be better to have more than eight weeks of data. However, sometimes data availability is initially limited. Furthermore, we have purposely

TABLE 4.1 **The fuel consumption data**

Week, i	Average hourly temperature, x_i (°F)	Weekly fuel consumption, y_i (tons)
1	$x_1 = 28.0$	$y_1 = 12.4$
2	$x_2 = 28.0$	$y_2 = 11.7$
3	$x_3 = 32.5$	$y_3 = 12.4$
4	$x_4 = 39.0$	$y_4 = 10.8$
5	$x_5 = 45.9$	$y_5 = 9.4$
6	$x_6 = 57.8$	$y_6 = 9.5$
7	$x_7 = 58.1$	$y_7 = 8.0$
8	$x_8 = 62.5$	$y_8 = 7.5$

limited the amount of data to simplify subsequent discussions in this, our first regression example.

To develop a regression model describing the fuel consumption data, we first consider the fifth week in Table 4.1 (for the purposes of our discussion we could consider any particular week). In the fifth week the average hourly temperature was $x_5 = 45.9$, and the fuel consumption was $y_5 = 9.4$. If we were to observe another week having the same average hourly temperature of 45.9, we might well observe a fuel consumption that is different from 9.4. This is because factors other than average hourly temperature—factors such as average hourly wind velocity and average hourly thermostat setting—affect weekly fuel consumption. Therefore although two weeks might have the same average hourly temperature of $x_5 = 45.9$, there could be a lower average hourly wind velocity and thus a smaller fuel consumption in one such week than in the other week. It follows that there is an infinite population of potential weekly fuel consumptions that could be observed when the average hourly temperature is $x_5 = 45.9$. Letting μ_5 denote the mean of this population, we may express y_5 in the form

$$y_5 = \mu_5 + \varepsilon_5$$

Here, the mean μ_5 represents the effect on y_5 of the average hourly temperature $x_5 = 45.9$. Furthermore, the *error term* ε_5 describes the effects on y_5 of all other factors that have occurred in the fifth week.

To generalize the preceding discussion, consider all eight fuel consumptions in Table 4.1. For $i = 1, 2, \ldots, 8$ we may express y_i in the form

$$y_i = \mu_i + \varepsilon_i$$

Here, μ_i is the mean of the population of potential weekly fuel consumptions that could be observed when the average hourly temperature is x_i. Furthermore, the error term ε_i describes the effect on y_i of all factors that have occurred in the ith week other than the average hourly temperature x_i.

In Figure 4.1(a) we plot the eight fuel consumptions against the eight average hourly temperatures. Note that the fuel consumptions tend to decrease in a straight-line fashion as the temperatures increase. We can define a regression model by assuming that the eight mean fuel consumptions $\mu_1, \mu_2, \ldots, \mu_8$ are related to the eight average hourly temperatures $x_1, x_2, \ldots, x_8$ by a straight line. We describe this straight line by using the equation (see Figure 4.1(b)).

$$\mu_i = \beta_0 + \beta_1 x_i$$

Here, β_0 is the *y-intercept* of the straight line, and β_1 is the *slope* of the straight line. To interpret the meaning of the y-intercept, β_0, assume that $x_i = 0$. Then

$$\begin{aligned} \mu_i &= \beta_0 + \beta_1 x_i \\ &= \beta_0 + \beta_1(0) = \beta_0 \end{aligned}$$

So β_0 is the mean weekly fuel consumption for all potential weeks having an average hourly temperature of 0°F. We later discuss the fact that since we have not

FIGURE 4.1 **The data plot and simple linear regression model relating y (weekly fuel consumption) to x (average hourly temperature)**

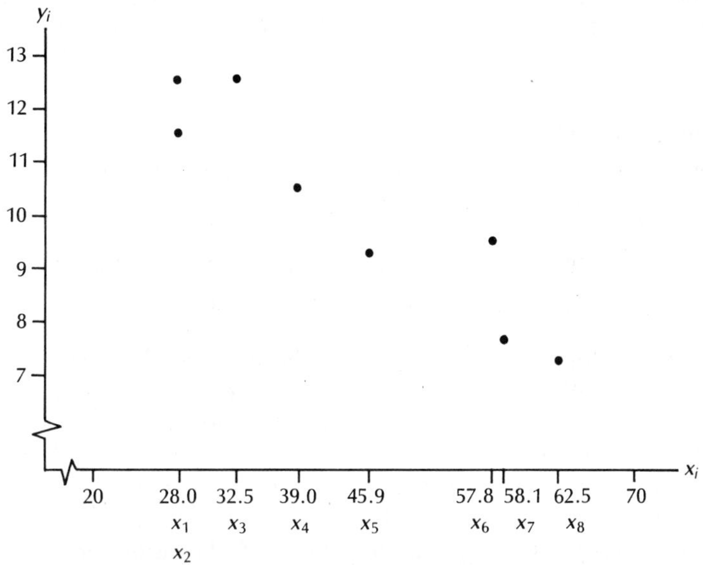

(a) The plot of y_i versus x_i

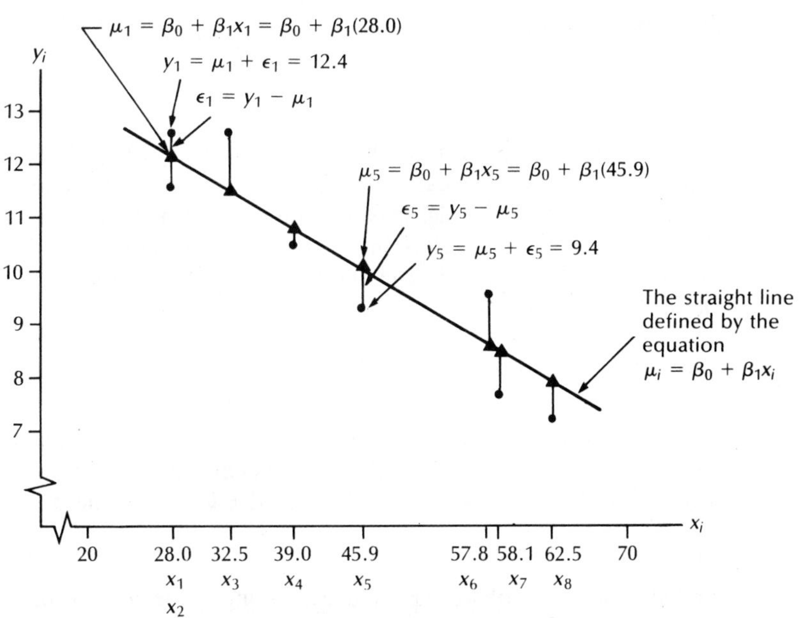

(b) The simple linear regression model relating y to x

observed any average hourly temperatures near 0°F (see Table 4.1), this interpretation of β_0 is of dubious practical value. To interpret the meaning of the slope, β_1, consider two different weeks. Suppose that for the first week average hourly temperature is c. The mean weekly fuel consumption for all such potential weeks is

$$\beta_0 + \beta_1(c)$$

For the second week, suppose that the average hourly temperature is $c + 1$. The mean weekly fuel consumption for all such potential weeks is

$$\beta_0 + \beta_1(c + 1)$$

The difference between these mean weekly fuel consumptions is

$$[\beta_0 + \beta_1(c + 1)] - [\beta_0 + \beta_1(c)] = \beta_1$$

Thus the slope β_1 is the change in the mean weekly fuel consumption that is associated with a 1-degree increase in average hourly temperature. Note that since we do not know the true values of β_0 and β_1, the position of the straight line depicted in Figure 4.1b is only hypothetical. However, we do know that β_1 must be a negative number because this would say that the mean weekly fuel consumption decreases as the average hourly temperature increases.

Since we are assuming that the ith mean fuel consumption, μ_i equals $\beta_0 + \beta_1 x_i$, we can describe the ith observed fuel consumption

$$y_i = \mu_i + \varepsilon_i$$

by using the equation

$$\begin{aligned} y_i &= \mu_i + \varepsilon_i \\ &= \beta_0 + \beta_1 x_i + \varepsilon_i \end{aligned}$$

We refer to this equation as the *simple linear* (or *straight-line*) *regression model* relating y_i to x_i, and we call β_0 and β_1 the *parameters* of the model. As illustrated in Figure 4.1(b), this model says that the eight error terms cause the eight observed fuel consumptions (the dots in the figure) to deviate from the eight mean fuel consumptions (the triangles in the figure), which exactly lie on the straight line defined by the equation

$$\mu_i = \beta_0 + \beta_1 x_i$$

As is illustrated in the preceding example, we sometimes need to use the simple linear regression model to relate a dependent variable y to a single independent variable x. Assume that we have observed n values of the dependent variable

$$y_1, y_2, \ldots, y_n$$

and that we have observed n corresponding values of the independent variable

$$x_1, x_2, \ldots, x_n$$

The Simple Linear Regression Model

The *simple linear* (or *straight line*) *regression model* is

$$\begin{aligned} y_i &= \mu_i + \varepsilon_i \\ &= \beta_0 + \beta_1 x_i + \varepsilon_i \end{aligned}$$

Here,

1. $\mu_i = \beta_0 + \beta_1 x_i$ is the mean value of the dependent variable when the value of the independent variable x is x_i.
2. ε_i is an error term that describes the effects on y_i of all factors other than the value x_i of the independent variable x.
3. β_0 (the y-intercept) is the mean value of the dependent variable when the value of the independent variable x is zero.
4. β_1 (the slope) is the change in the mean value of the dependent variable that is associated with a one-unit increase in the value of the independent variable. If β_1 is positive, the mean value of the dependent variable increases as the value of the independent variable increases. See Figure 4.2(a). If β_1 is negative, the mean value of the dependent variable decreases as the value of the independent variable increases. See Figure 4.2(b).

FIGURE 4.2 **The simple linear regression model**

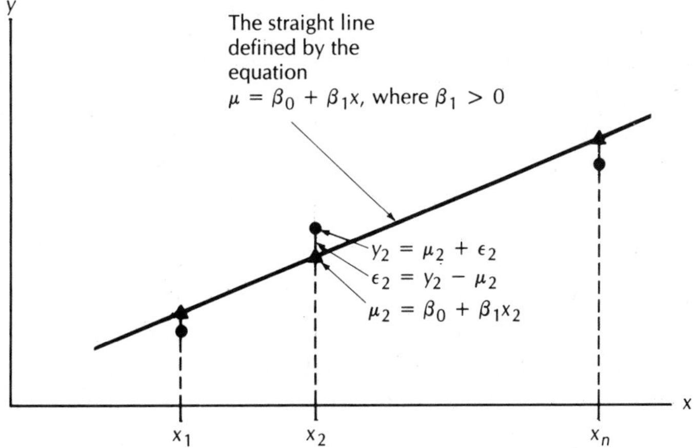

(a) The simple linear regression model having a positive slope

(continues)

FIGURE 4.2 **Continued**

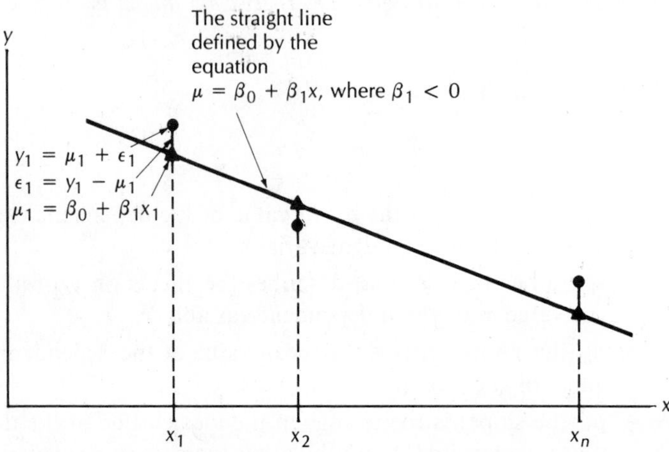

(b) The simple linear regression model having a negative slope

Note that y_i is assumed to be randomly selected from the infinite population of potential values of the dependent variable that could be observed when the value of the independent variable x is x_i. As defined above, μ_i is the mean of this population.

Since the fuel consumption data in Table 4.1 was observed sequentially over time (in eight consecutive weeks), this data is called *time series data*. Many applications of regression analysis utilize time series data. For example, in the actual application that served as the motivation for our fuel consumption problem, the accounting firm observed the natural gas consumptions and the associated weather conditions in a small city over time. Using this data, the firm developed an equation for predicting future monthly natural gas consumption in the city. The predictions obtained have enabled the city to order appropriate amounts of natural gas.

Another type of data that is frequently used in regression analysis is *cross-sectional data*, which is data observed at one point in time.

Example 4.2 American Manufacturing Company produces a line of commercial dishwashers and sells this product to restaurants, hotels, hospitals, and other such institutions. To develop a regression model that can be used to study the relationship between quarterly advertising expenditure and quarterly sales volume for the dishwashers in its sales regions for the third quarter of 1988, American Manufacturing has gathered the data given in Table 4.2. The data represent the advertising expenditures and associated sales volumes for the dishwashers in ten sales regions of equal sales potential for the third quarter of 1988. Here, for $i = 1, 2, \ldots, 10$,

x_i = the advertising expenditure (in units of \$10,000) in sales region i

y_i = the sales volume (in units of \$10,000) in sales region i

Since this data has been observed at one point (that is, quarter) in time in ten different sales regions, it is cross-sectional. In contrast, if, for example, the advertising expenditures and sales volumes had been observed in one sales region over ten consecutive quarters, the data would be time series data.

TABLE 4.2 American Manufacturing Company sales volume data

Sales region, i	Advertising expenditure, x_i	Sales volume, y_i
1	5	89
2	6	87
3	7	98
4	8	110
5	9	103
6	10	114
7	11	116
8	12	110
9	13	126
10	14	130

FIGURE 4.3 Plot of American Manufacturing Company sales volume data

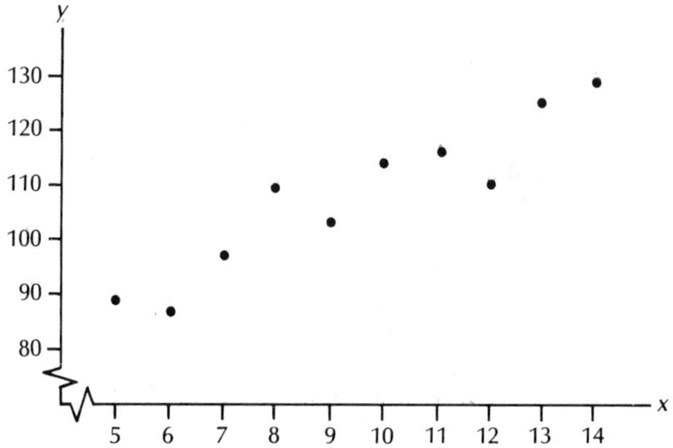

In Figure 4.3 the values of y are plotted against the values of x. We see that as the values of x increase, the values of y tend to increase in a straight-line fashion. Thus the simple linear regression model having a positive slope ($\beta_1 > 0$)

$$\begin{aligned} y_i &= \mu_i + \varepsilon_i \\ &= \beta_0 + \beta_1 x_i + \varepsilon_i \end{aligned}$$

might be an appropriate regression model relating y_i to x_i. Note that

$$\mu_i = \beta_0 + \beta_1 x_i$$

is the mean of all of the sales volumes that could have been observed in a sales region for the third quarter of 1988 when x_i was spent on advertising. It follows that β_1, the slope of the regression model, is the change in mean third quarter sales volume (in units of \$10,000) in a sales region associated with each additional \$10,000 spent on third quarter advertising in the sales region.

It is sometimes more complicated to use time series data to relate an independent variable to a dependent variable than it is to use cross-sectional data to relate an independent variable to a dependent variable. This is because, sometimes, after values of an independent variable are related to values of a dependent variable observed sequentially over time, the time order in which the values of the dependent variable have been observed is found to have an "additional" effect on these values that is not explained by the values of the independent variable. For example, sometimes, after advertising expenditures are related to sales volumes observed over consecutive time periods, the time order in which the sales volumes have been observed is found to have an additional effect on the sales volumes that is not explained by the advertising expenditures. Specifically, we might find that increasing sales volumes over time are related to changing consumer tastes that are not completely explained by the advertising expenditures. One approach to analyzing time series data is to initially ignore the time order of the data and use the techniques of this chapter to relate the independent variable to the dependent variable. We used this approach in Example 4.1 to relate the weekly fuel consumptions to the average hourly temperatures by using the simple linear regression model. Then we can use the techniques of later chapters (specifically, Chapters 6 and 13) to determine whether time does have an additional effect and to modify our results accordingly.

The data in Table 4.1 and 4.2 have been artificially constructed so that we may simply illustrate some of the basic ideas of regression analysis. In practice, regression analysis is performed with real data. Two real data sets are presented in Tables 4.3 and 4.4. Table 4.3 presents the sales price y and square footage x for each of 63 single-family residences sold during 1988 in Oxford, Ohio. The reader can use a data plot to verify that it is reasonable to describe this data by using a simple linear regression model having a positive slope. The slope expresses how much higher the mean sales price is for a residence that is one square foot larger than another residence. The slope multiplied by 100 expresses how much higher the mean sales

TABLE 4.3 Measurements taken on 63 single-family residences

Residence, i	Sales price, y (×$1000)	Square feet, x	Residence, i	Sales price, y (×$1000)	Square feet, x
1	53.5	1008	33	63.0	1053
2	49.0	1290	34	60.0	1728
3	50.5	860	35	34.0	416
4	49.9	912	36	52.0	1040
5	52.0	1204	37	75.0	1496
6	55.0	1204	38	93.0	1936
7	80.5	1764	39	60.0	1904
8	86.0	1600	40	73.0	1080
9	69.0	1255	41	71.0	1768
10	149.0	3600	42	83.0	1503
11	46.0	864	43	90.0	1736
12	38.0	720	44	83.0	1695
13	49.5	1008	45	115.0	2186
14	105.0	1950	46	50.0	888
15	152.5	2086	47	55.2	1120
16	85.0	2011	48	61.0	1400
17	60.0	1465	49	147.0	2165
18	58.5	1232	50	210.0	2353
19	101.0	1736	51	60.0	1536
20	79.4	1296	52	100.0	1972
21	125.0	1996	53	44.5	1120
22	87.9	1874	54	55.0	1664
23	80.0	1580	55	53.4	925
24	94.0	1920	56	65.0	1288
25	74.0	1430	57	73.0	1400
26	69.0	1486	58	40.0	1376
27	63.0	1008	59	141.0	2038
28	67.5	1282	60	68.0	1572
29	35.0	1134	61	139.0	1545
30	142.5	2400	62	140.0	1993
31	92.2	1701	63	55.0	1130
32	56.0	1020			

Source: Reprinted with permission from Alpha, Inc., Oxford, Ohio.

price is for a residence that is 100 square feet larger than another residence*. Table 4.4 presents the sales territory performance y (measured by aggregate sales in units) and the number of accounts x of 25 randomly selected sales representatives of a company. Here, each representative had the sole responsibility for one sales territory. This data is analyzed in "An Analytical Approach for Evaluating Sales Territory Performance" by David W. Cravens, Robert B. Woodruff, and Joe C. Stomper, in the *Journal of Marketing* (Vol. 36, January 1972, pp. 31–37). Furthermore, the data were coded for purposes of respondent confidentiality. The reader can use a data plot to verify that it is reasonable to describe this data by using a

*If certain points are removed from the plot of the data in Table 4.3, this plot has more of a curved appearance than a straight-line appearance. This situation will be investigated in the exercises at the end of Chapter 6.

TABLE 4.4 **Data for sales territory performance study**

Sales, y	Accounts, x	Sales, y	Accounts, x
3669.88	74.86	2337.38	84.55
3473.95	107.32	4586.95	119.51
2295.10	96.75	2729.24	80.49
4675.56	195.12	3289.40	136.58
6125.96	180.44	2800.78	78.86
2134.94	104.88	3264.20	136.58
5031.66	256.10	3453.62	138.21
3367.45	126.83	1741.45	75.61
6519.45	203.25	2035.75	102.44
4876.37	119.51	1578.00	76.42
2468.27	116.26	4167.44	136.58
2533.31	142.28	2799.97	88.62
2408.11	89.43		

Source: David W. Cravens, Department of Marketing, Texas Christian University.

simple linear regression model having a positive slope. The slope may be interpreted to be the increase in the mean unit sales of a representative associated with each additional account unit handled by the representative.

4.2 THE LEAST SQUARES POINT ESTIMATES

Although we do not know the true values of the parameters β_0 and β_1 in the simple linear regression model

$$\begin{aligned} y_i &= \mu_i + \varepsilon_i \\ &= \beta_0 + \beta_1 x_i + \varepsilon_i \end{aligned}$$

we can use the n observed values of the independent variable x,

$$x_1, x_2, \ldots, x_n$$

and the n observed values of the dependent variable y,

$$y_1, y_2, \ldots, y_n$$

to calculate point estimates b_0 and b_1 of β_0 and β_1. These estimates are called the *least squares point estimates* of the parameters β_0 and β_1. We first present the formulas for computing b_0 and b_1. We then present an example in which we use these formulas and discuss the rationale for the formulas.

The Least Squares Point Estimates of β_0 and β_1

The *least squares point estimates b_1 and b_0* of the parameters β_1 and β_0 in the *simple linear regression model* are calculated by using the formulas

$$b_1 = \frac{n \sum_{i=1}^{n} x_i y_i - \left(\sum_{i=1}^{n} x_i\right)\left(\sum_{i=1}^{n} y_i\right)}{n \sum_{i=1}^{n} x_i^2 - \left(\sum_{i=1}^{n} x_i\right)^2}$$

and

$$b_0 = \bar{y} - b_1 \bar{x}$$

where $\bar{y} = \frac{\sum_{i=1}^{n} y_i}{n}$ and $\bar{x} = \frac{\sum_{i=1}^{n} x_i}{n}$

Example 4.3 In Table 4.5 we use the above formulas and the fuel consumption data from Table 4.1 to calculate least squares point estimates of the parameters β_1 and β_0 in the fuel consumption model

$$\begin{aligned} y_i &= \mu_i + \varepsilon_i \\ &= \beta_0 + \beta_1 x_i + \varepsilon_i \end{aligned}$$

These estimates are found to be $b_1 = -.1279$ and $b_0 = 15.84$. Since $b_1 = -.1279$ is the point estimate of β_1, we estimate that mean weekly fuel consumption decreases (since b_1 is negative) by .1279 tons of coal when average hourly temperature increases by 1 degree.

To explain the meaning of the term *least squares*, note that we can use the point estimates b_0 and b_1 of the parameters β_0 and β_1 to calculate the *point prediction*

$$\begin{aligned} \hat{y}_i &= b_0 + b_1 x_i \\ &= 15.84 - .1279 x_i \end{aligned}$$

of the fuel consumption observed in the ith week

$$y_i = \beta_0 + \beta_1 x_i + \varepsilon_i$$

Here, we have predicted ε_i to be zero because of several assumptions to be discussed in Section 4.4. One implication of these assumptions is that ε_i has a 50 percent chance of being positive and a 50 percent chance of being negative. For example, since the average hourly temperature in the fifth week was $x_5 = 45.9$, it follows that the point prediction of y_5 is

$$\begin{aligned} \hat{y}_5 &= b_0 + b_1 x_5 \\ &= 15.84 - .1279(45.9) \\ &= 9.9663 \end{aligned}$$

TABLE 4.5 **The calculation of point estimates b_0 and b_1 of the parameters in the fuel consumption model $y_i = \mu_i + \varepsilon_i = \beta_0 + \beta_1 x_i + \varepsilon_i$**

y_i	x_i	x_i^2	$y_i x_i = x_i y_i$
12.4	28.0	$(28.0)^2 = 784$	$(12.4)(28.0) = 347.2$
11.7	28.0	$(28.0)^2 = 784$	$(11.7)(28.0) = 327.6$
12.4	32.5	$(32.5)^2 = 1056.25$	$(12.4)(32.5) = 403$
10.8	39.0	$(39.0)^2 = 1521$	$(10.8)(39.0) = 421.2$
9.4	45.9	$(45.9)^2 = 2106.81$	$(9.4)(45.9) = 431.46$
9.5	57.8	$(57.8)^2 = 3340.84$	$(9.5)(57.8) = 549.1$
8.0	58.1	$(58.1)^2 = 3375.61$	$(8.0)(58.1) = 464.8$
7.5	62.5	$(62.5)^2 = 3906.25$	$(7.5)(62.5) = 468.75$
$\sum_{i=1}^{8} y_i = 81.7$	$\sum_{i=1}^{8} x_i = 351.8$	$\sum_{i=1}^{8} x_i^2 = 16{,}874.76$	$\sum_{i=1}^{8} x_i y_i = 3413.11$

$$\bar{y} = \frac{81.7}{8} = 10.21 \qquad \bar{x} = \frac{351.8}{8} = 43.98$$

$$b_1 = \frac{8\sum_{i=1}^{8} x_i y_i - \left(\sum_{i=1}^{8} x_i\right)\left(\sum_{i=1}^{8} y_i\right)}{8\sum_{i=1}^{8} x_i^2 - \left(\sum_{i=1}^{8} x_i\right)^2}$$

$$= \frac{8(3413.11) - (351.8)(81.7)}{8(16{,}874.76) - (351.8)^2} = -.1279$$

$$b_0 = \bar{y} - b_1\bar{x} = 10.21 - (-.1279)(43.98) = 15.84$$

In general, if any particular values of b_0 and b_1 are good point estimates of β_0 and β_1, they will, for $i = 1, 2, \ldots, 8$, make $\hat{y}_i$ fairly close to y_i. Therefore the *ith residual*

$$e_i = y_i - \hat{y}_i = y_i - (b_0 + b_1 x_i)$$

will be fairly small. We use the previously given formulas to calculate the point estimates b_0 and b_1 of the parameters β_0 and β_1 because it can be proved (see Appendix B) that these point estimates give a value of the *sum of squared residuals*

$$SSE = \sum_{i=1}^{8} e_i^2 = \sum_{i=1}^{8} (y_i - \hat{y}_i)^2$$

$$= \sum_{i=1}^{8} [y_i - (b_0 + b_1 x_i)]^2$$

that is smaller than would be given by any other values of b_0 and b_1. Since these point estimates minimize *SSE*, we call them the *least squares point estimates.*

To illustrate the fact that the least squares point estimates minimize *SSE*, consider Table 4.6. Comparing the upper and lower parts of the table, we see that $SSE = 2.57$, the value given by the least squares point estimates $b_0 = 15.84$ and

TABLE 4.6 The values of SSE given by the least squares point estimates $b_0 = 15.84$ and $b_1 = -.1279$ and by the point estimates $b_0 = 16.22$ and $b_1 = -.1152$

Predictions using the least squares point estimates $b_0 = 15.84$ and $b_1 = -.1279$

Week, i	Average hourly temperature, x_i (°F)	Observed fuel consumption, y_i (tons)	Predicted fuel consumption, $\hat{y}_i = b_0 + b_1x_i = 15.84 - .1279x_i$	Residual, $e_i = y_i - \hat{y}_i$
1	28.0	12.4	12.2560	.1440
2	28.0	11.7	12.2560	−.5560
3	32.5	12.4	11.6804	.7196
4	39.0	10.8	10.8489	−.0489
5	45.9	9.4	9.9663	−.5663
6	57.8	9.5	8.4440	1.0560
7	58.1	8.0	8.4056	−.4056
8	62.5	7.5	7.8428	−.3428

$$SSE = \sum_{i=1}^{8} e_i^2 = 2.57$$

Predictions using the point estimates $b_0 = 16.22$ and $b_1 = -.1152$

Week, i	Average hourly temperature, x_i (°F)	Observed fuel consumption, y_i (tons)	Predicted fuel consumption, $\hat{y}_i = b_0 + b_1x_i = 16.22 - .1152x_i$	Residual, $e_i = y_i - \hat{y}_i$
1	28.0	12.4	12.9944	−.5944
2	28.0	11.7	12.9944	−1.2944
3	32.5	12.4	12.4760	−.0760
4	39.0	10.8	11.7272	−.9272
5	45.9	9.4	10.9323	−1.5323
6	57.8	9.5	9.5614	−.0614
7	58.1	8.0	9.5269	−1.5269
8	62.5	7.5	9.0200	−1.5200

$$SSE = \sum_{i=1}^{8} e_i^2 = 9.8877$$

$b_1 = -.1279$, is less than $SSE = 9.8877$, the value given by $b_0 = 16.22$ and $b_1 = -.1152$, which are not the least squares point estimates. There is nothing special about the values 16.22 and $-.1152$. The point is that if we choose any values of b_0 and b_1 that are different from the least squares values, we will obtain a larger SSE.

In Figure 4.4 we plot the eight observed fuel consumptions (the dots in the figure) and the eight predicted fuel consumptions (the squares in the figure). Note that the predicted fuel consumptions lie on the straight line defined by the prediction equation

$$\begin{aligned}\hat{y}_i &= b_0 + b_1x_i \\ &= 15.84 - .1279x_i\end{aligned}$$

FIGURE 4.4 **The eight residuals given by the prediction equation $\hat{y} = b_0 + b_1 x = 15.84 - .1279x$**

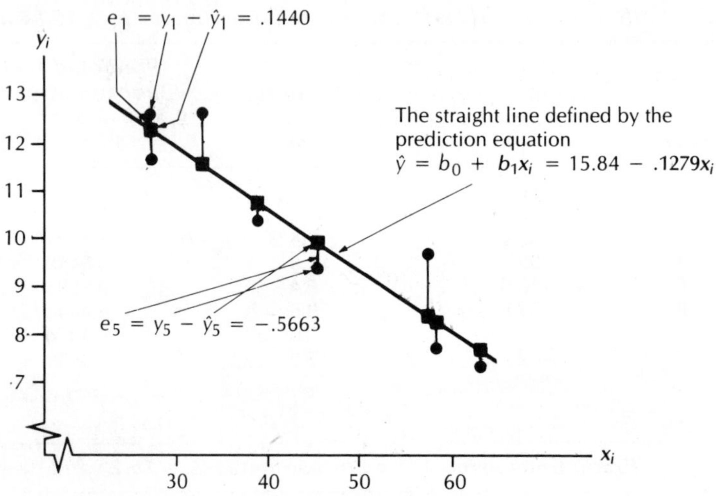

Furthermore, the distances between the observed and predicted fuel consumptions are the residuals. Since the least squares point estimates minimize *SSE*, we can interpret them as positioning the straight line prediction equation so as to minimize the sum of squared distances between the observed and predicted fuel consumptions. In this sense we can say that the straight line defined by the above prediction equation is the best straight line that can be fit to the eight observed fuel consumptions.

Example 4.4 Consider the sales volume data in Table 4.2 and the model

$$\begin{aligned} y_i &= \mu_i + \varepsilon_i \\ &= \beta_0 + \beta_1 x_i + \varepsilon_i \end{aligned}$$

We find (omitting the detailed calculations) that the least squares point estimates of the parameters in this model are

$$b_1 = \frac{10 \sum_{i=1}^{10} x_i y_i - \left(\sum_{i=1}^{10} x_i\right)\left(\sum_{i=1}^{10} y_i\right)}{10 \sum_{i=1}^{10} x_i^2 - \left(\sum_{i=1}^{10} x_i\right)^2} = 4.4303$$

and

$$b_0 = \bar{y} - b_1 \bar{x} = 66.2121$$

The point estimate $b_1 = 4.4303$ says that we estimate that mean third quarter sales volume in a sales region increased by \$44,303 for each additional \$10,000 spent on advertising during the quarter in the sales region.

As another example, if we use the simple linear regression model to describe the residential sales data in Table 4.3, we find that the least squares point estimates of β_0 and β_1 are $b_0 = -3.72103$ and $b_1 = .05477$. The point estimate $b_1 = .05477$ says that we estimate that the mean sales price is \$5477 higher for each increase of 100 square feet in residence size.

4.3

POINT ESTIMATES AND POINT PREDICTIONS

We now consider using the simple linear regression model to estimate and predict.

Example 4.5 Consider the fuel consumption problem. Suppose that we have employed a weather forecasting service, which has told us that the average hourly temperature in a future week will be 40°F. Moreover, suppose that we wish to predict y_0, the amount of fuel (in tons of coal) that will be used to heat the nine-building complex during the future week. We can express this future (individual) fuel consumption by the equation

$$y_0 = \mu_0 + \varepsilon_0$$

Here, μ_0 is the mean of the population of potential weekly fuel consumptions that could be observed when the average hourly temperature is $x_0 = 40$, and ε_0 is the error term for the future week.

Recall that we concluded that the straight line defined by the equation

$$\mu = \beta_0 + \beta_1 x$$

relates the eight mean fuel consumptions $\mu_1, \mu_2, \ldots, \mu_8$ to the eight observed average hourly temperatures $x_1, x_2, \ldots, x_8$. Note from Figure 4.5 that $x_0 = 40.0$, the future average hourly temperature, is in the *experimental region* (the range of the eight observed average hourly temperatures—28.0°F to 62.5°F). It follows that this straight line (or equation) also relates the future mean fuel consumption μ_0 to $x_0 = 40.0$. That is, it seems reasonable to conclude that

$$\begin{aligned} \mu_0 &= \beta_0 + \beta_1 x_0 \\ &= \beta_0 + \beta_1(40.0) \end{aligned}$$

Therefore

$$\begin{aligned} \hat{y}_0 &= b_0 + b_1 x_0 \\ &= 15.84 - .1279(40.0) \\ &= 10.72 \quad \text{tons of coal} \end{aligned}$$

FIGURE 4.5 **A hypothetical illustration of $\hat{y}_0$, y_0, and μ_0 in the fuel consumption problem and the danger of extrapolation**

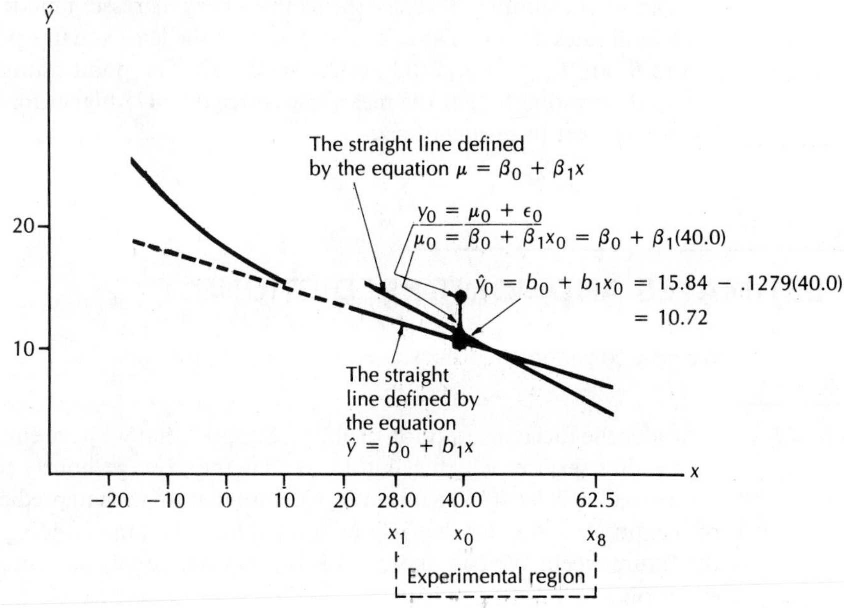

is the point estimate of the mean μ_0 and the point prediction of the individual value y_0. Here, the error term ε_0 is predicted to be zero.

Examining Figure 4.5, note that the relative positions of $\hat{y}_0$, μ_0, and y_0 are only hypothetical, since we cannot know the true values of β_0, β_1, and ε_0. However, Figure 4.5 does illustrate the fact that (unless we are extremely fortunate) $\hat{y}_0$ will differ from both μ_0 and y_0.

In Chapter 5 we will explain how we can place bounds (defined with a high level of confidence) on how far $\hat{y}_0$ might be from μ_0 and on how far $\hat{y}_0$ might be from y_0.

Next, we note that it would be dangerous to use the straight-line prediction equation

$$\begin{aligned}\hat{y} &= b_0 + b_1 x \\ &= 15.84 - .1279x\end{aligned}$$

to predict a future value of y on the basis of a future value of x that is far outside the experimental region. For example, -10.0 (in Figure 4.5) is a future value of x that is far outside the experimental region. Figure 4.1(a) indicates that for values of x in the experimental region the observed values of y tend to decrease in a straight-line fashion as the values of x increase. However, there are no observed data to tell us whether the same relationship between y and x holds for temperatures between -20.0 and 20.0. If, for example, we observe values of x between -20.0 and 20.0, and if the corresponding observed values of y tend to fluctuate around the *curve* illustrated in Figure 4.5, then instead of using a straight-line prediction equation, we should use a "curved prediction equation." Some curved prediction equations are discussed in Chapter 8.

Finally, recall that we have interpreted the y-intercept β_0 of the regression model

$$\begin{aligned} y &= \mu + \varepsilon \\ &= \beta_0 + \beta_1 x + \varepsilon \end{aligned}$$

to be the mean weekly fuel consumption for all potential weeks having an average hourly temperature of 0°F. However, since zero degrees is not in the experimental region, we do not know whether this regression model appropriately describes weeks having an average hourly temperature of 0°F; thus this interpretation of β_0 is of dubious practical value.

Generalizing the above example, we obtain the following results.

Point Estimation and Prediction in Simple Linear Regression

Let b_0 and b_1 be the *least squares point estimates* of the parameters β_0 and β_1 in the *simple linear regression model* and suppose that x_0, a specified value of the independent variable x, is inside the *experimental region* (the range of the n observed values $x_1, x_2, \ldots, x_n$). Then,

1. The *point estimate* of

$$\mu_0 = \beta_0 + \beta_1 x_0$$

the *mean value of the dependent variable* when the value of the independent variable is x_0, is

$$\hat{y}_0 = b_0 + b_1 x_0$$

2. The *point prediction* of

$$y_0 = \beta_0 + \beta_1 x_0 + \varepsilon_0$$

the *individual value of the dependent variable* when the value of the independent variable is x_0, is also

$$\hat{y}_0 = b_0 + b_1 x_0$$

since we predict the error term ε_0 to be zero.

Example 4.6 Recall from Example 4.4 that $b_0 = 66.2121$ and $b_1 = 4.4303$ are the least squares point estimates of the parameters β_0 and β_1 in the simple linear regression model describing the sales volume data in Table 4.2. Consider a sales region in which the third quarter advertising expenditure was $x_0 = 10$ (that is, \$100,000). It follows that

$$\begin{aligned} \hat{y}_0 &= b_0 + b_1 x_0 \\ &= 66.2121 + 4.4303(10) \\ &= 110.5 \qquad \text{(that is, \$1,105,000)} \end{aligned}$$

is the point estimate of

$$\begin{aligned} \mu_0 &= \beta_0 + \beta_1 x_0 \\ &= \beta_0 + \beta_1(10) \end{aligned}$$

the mean of all of the third quarter sales volumes that could have been observed in a sales region when \$100,000 was spent on advertising, and is the point prediction of

$$\begin{aligned} y_0 &= \mu_0 + \varepsilon_0 \\ &= \beta_0 + \beta_1(10) + \varepsilon_0 \end{aligned}$$

the actual third quarter sales volume in a sales region when \$100,000 was spent on advertising.

There are two interesting ways in which the point prediction $\hat{y}_0 = 110.5$ (\$1,105,000) might be used. First, suppose that American Manufacturing is confident that the sales forces in the ten sales regions represented in Table 4.2 use their advertising expenditures effectively. However, the firm is concerned that the sales forces in certain other sales regions do not make effective use of their advertising expenditures. All regions are assumed to be of equal sales potential. Also suppose that the sales volume for the third quarter of 1988 in a sales region in which the advertising expenditure was \$100,000 and in which the sales force is questionable was \$890,800. Since this sales figure is substantially smaller than the point prediction \$1,105,000, we have evidence that the questionable sales force was not, in fact, performing effectively. An even more convincing proof of this is provided in Chapter 5, where we study *prediction intervals*. A second way in which $\hat{y}_0 = 110.5$ might be used is as a point prediction of the actual sales volume in a *future quarter* in a sales region in which \$100,000 will be spent on advertising and in which there is an effective sales force. It is important to realize, however, that to predict sales volumes in future quarters, the regression relationships between advertising expenditure and sales volume that we have developed by using data from the third quarter of 1988 must apply to future quarters. Therefore there must not be any trend, seasonal, or other time-related influences affecting quarterly sales volumes.

4.4

MODEL ASSUMPTIONS, THE MEAN SQUARE ERROR, AND THE STANDARD ERROR

MODEL ASSUMPTIONS

The validity of all of the regression results in this chapter and in Chapter 5 depends upon the fundamental assumption that the straight-line equation

$$\mu = \beta_0 + \beta_1 x$$

appropriately relates μ, the mean value of the dependent variable, to x, the independent variable being utilized. For example, suppose that the true relationship between μ and x in the fuel consumption problem was described by a curve instead of a straight line. Then, the results obtained by using the regression model

$$\begin{aligned} y &= \mu + \varepsilon \\ &= \beta_0 + \beta_1 x + \varepsilon \end{aligned}$$

would not be valid. However, we have used the data plot in Figure 4.1 to reason that a linear equation relates μ to x. Furthermore, in subsequent chapters we present statistical evidence showing that this linear equation is appropriate in the fuel consumption problem.

In addition to the above assumption the validity of the formulas for the confidence intervals, prediction intervals, and statistical hypothesis tests to be presented in Chapter 5 depends on three assumptions that we refer to as the *inference assumptions*. These assumptions can be stated as follows.

Inference Assumption 1 (Constant Variance)

For any value x_i of the independent variable x the corresponding population of potential values of the dependent variable has a variance σ^2 that does not depend on the value x_i of x. That is, the different populations of potential values of the dependent variable corresponding to different values of x have *equal variances*.

Inference Assumption 2 (Independence)

Any one value of the dependent variable y is *statistically independent* of any other value of y.

Inference Assumption 3 (Normality)

For any value x_i of the independent variable x the corresponding population of potential values of the dependent variable has a *normal distribution*.

FIGURE 4.6 An illustration of the constant variance and normality assumptions

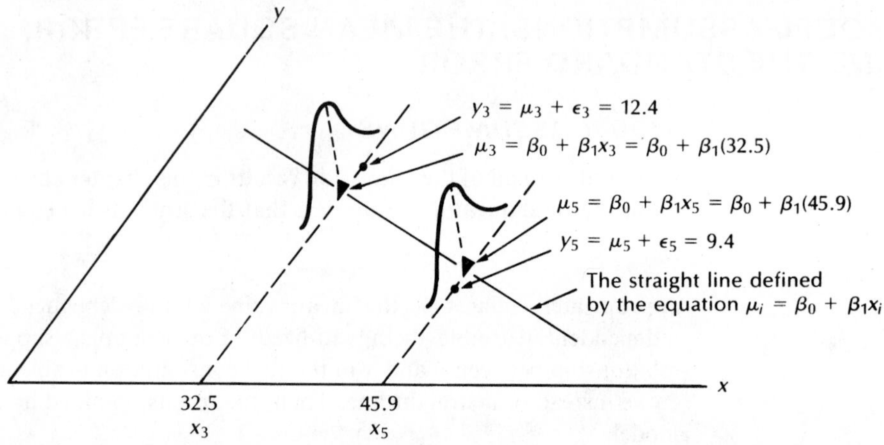

Figure 4.6 illustrates the constant variance and normality assumptions for two populations in the fuel consumption problem. Specifically, this figure illustrates the population of potential weekly fuel consumptions that could be observed when the average hourly temperature is $x_3 = 32.5$ and the population of potential weekly fuel consumptions that could be observed when the average hourly temperature is $x_5 = 45.9$. Note that these populations are depicted to be normally distributed with equal variances (that is, the normal curves have the same spread).

For future reference we denote the constant variance and constant standard deviation of each of the different populations by the symbols σ^2 and σ.

Intuitively, the independence assumption is most likely to be violated for data observed over time. For such data this assumption intuitively says that there is *no pattern* of greater than average values of the dependent variable being followed (in time) by greater than average values of the dependent variable and that there is *no pattern* of greater than average values of the dependent variable being followed by less than average values of the dependent variable. A more complete discussion of the meaning of the inference assumptions, along with various techniques that can be used to check their validity, is given in Chapter 6. We assume that the inference assumptions are satisfied for the examples prior to Chapter 6. Finally, it is important to point out that although most of the formulas for prediction intervals, confidence intervals, and hypothesis tests in this book are strictly valid only when the inference assumptions hold, these formulas are still approximately correct when mild departures from the inference assumptions can be detected. In fact, these assumptions very seldom, if ever, hold exactly in any practical regression problem. Thus in practice, only pronounced departures from the inference assumptions require attention.

In Section 4.2 we stated that when predicting y_i, it is customary to predict the error term ε_i to be zero. The rationale for this comes from the independence and normality assumptions. The normality assumption says that each population of potential values of the dependent variable is normally distributed. The independence assumption says that any two values of the dependent variable are statistically independent. It follows that y_i has a 50 percent chance of being greater than, and a 50 percent chance of being less than, the mean μ_i. Hence,

$$\varepsilon_i = y_i - \mu_i$$

has a 50 percent chance of being positive and a 50 percent chance of being negative. Therefore we predict this error term to be zero.

THE MEAN SQUARE ERROR AND THE STANDARD ERROR

To present the statistical inference formulas in Chapter 5, we need to see how to compute point estimates of σ^2 and σ (the constant variance and standard deviation of each of the populations of potential values of the dependent variable). We now summarize how to compute these point estimates.

The Mean Square Error and the Standard Error

Assume that b_0 and b_1 are the least squares point estimates of β_0 and β_1 in the simple linear regression model

$$\begin{aligned} y_i &= \mu_i + \varepsilon_i \\ &= \beta_0 + \beta_1 x_i + \varepsilon_i \end{aligned}$$

If the inference assumptions are satisfied:

1. A point estimate of σ^2 is the *mean square error*

$$s^2 = \frac{SSE}{n - 2}$$

2. A point estimate of σ is the *standard error*

$$s = \sqrt{s^2} = \sqrt{\frac{SSE}{n - 2}}$$

Here, the sum of squared residuals

$$SSE = \sum_{i=1}^{n} (y_i - \hat{y}_i)^2 \qquad \text{where} \qquad \hat{y}_i = b_0 + b_1 x_i$$

can be calculated by the alternative formula

$$SSE = \sum_{i=1}^{n} y_i^2 - \left[b_0 \sum_{i=1}^{n} y_i + b_1 \sum_{i=1}^{n} x_i y_i \right]$$

Example 4.7 Recall that $b_0 = 15.84$ and $b_1 = -.1279$ are the least squares point estimates of β_0 and β_1 in the fuel consumption regression model

$$\begin{aligned} y_i &= \mu_i + \varepsilon_i \\ &= \beta_0 + \beta_1 x_i + \varepsilon_i \end{aligned}$$

It follows, as illustrated in Table 4.6, that

$$\hat{y}_i = b_0 + b_1 x_i = 15.84 - .1279x_i$$

is the point prediction of y_i and thus that the sum of squared residuals is

$$\begin{aligned} SSE &= \sum_{i=1}^{8} (y_i - \hat{y}_i)^2 \\ &= (.1440)^2 + (-.5560)^2 + \cdots + (-.3428)^2 = 2.57 \end{aligned}$$

Also, recall from Table 4.5 that

$$\sum_{i=1}^{8} y_i = 81.7 \quad \text{and} \quad \sum_{i=1}^{8} x_i y_i = 3413.11$$

and note that

$$\begin{aligned} \sum_{i=1}^{8} y_i^2 &= y_1^2 + y_2^2 + \cdots + y_8^2 \\ &= (12.4)^2 + (11.7)^2 + \cdots + (7.5)^2 = 859.91 \end{aligned}$$

It follows that SSE can be calculated in the following alternative fashion:

$$\begin{aligned} SSE &= \sum_{i=1}^{8} y_i^2 - \left[b_0 \sum_{i=1}^{8} y_i + b_1 \sum_{i=1}^{8} x_i y_i \right] \\ &= 859.91 - [15.84(81.7) + (-.1279)(3413.11)] \\ &= 859.91 - 857.34 \\ &= 2.57 \end{aligned}$$

Therefore a point estimate of σ^2 is the mean square error

$$s^2 = \frac{SSE}{n-2} = \frac{2.57}{8-2} = \frac{2.57}{6} = .428$$

and a point estimate of σ is the standard error

$$s = \sqrt{s^2} = \sqrt{.428} = .6542$$

EXERCISES

4.1 Suppose that the simple linear regression model $y = \beta_0 + \beta_1 x + \varepsilon$ describes the following data.

x	10	8	6	10	4	12
y	26	25	20	31	13	34

a. Compute the least squares point estimates of β_0 and β_1.
b. Write the least squares prediction equation relating y to x.
c. Plot y versus x and graph the least squares prediction equation on your data plot.
d. Calculate SSE, s^2, and s.

4.2 Suppose that the simple linear regression model $y = \beta_0 + \beta_1 x + \varepsilon$ describes the following data.

x	6	2	4	1	5	6	3
y	33	10	22	6	29	30	18

a. Compute the least squares point estimates of β_0 and β_1.
b. Write the least squares prediction equation relating y to x.
c. Plot y versus x and graph the least squares prediction equation on your data plot.
d. Calculate SSE, s^2, and s.

4.3 The chairman of the accountancy department at a large state university undertakes a study to relate starting salary (y) after graduation for accounting majors to grade point average (GPA) in major courses. To do this, records of seven recent accounting graduates are randomly selected.

Accounting graduate, i	GPA, x_i	Starting salary, y_i (thousands of dollars)
1	3.26	28.2
2	2.60	24.8
3	3.35	27.9
4	2.86	25.3
5	3.82	30.3
6	2.21	23.0
7	3.47	29.4

a. Plot y versus x. Explain why the plot suggests that the simple linear regression model having a positive slope

$$\begin{aligned} y_i &= \mu_i + \varepsilon_i \\ &= \beta_0 + \beta_1 x_i + \varepsilon_i \end{aligned}$$

might appropriately relate y to x.
b. Discuss the meaning of the third historical population of potential starting salaries.
c. Discuss the meaning of μ_3, the third mean starting salary.
d. Discuss the conceptual difference between μ_3 and $y_3 = \mu_3 + \varepsilon_3$. What does ε_3 measure in this situation?

e. Discuss the meanings of β_0 and β_1 in this model. Why does the interpretation of β_0 fail to make practical sense?
f. Calculate the least squares point estimates b_0 and b_1 of β_0 and β_1.
g. Using the prediction equation $\hat{y}_i = b_0 + b_1 x_i$, calculate a point estimate of μ_3 and a point prediction of $y_3 = \mu_3 + \varepsilon_3$.
h. Suppose that an accounting major will graduate with a GPA of $x_0 = 3.25$. The starting salary of this graduate may be expressed in the form $y_0 = \mu_0 + \varepsilon_0$.

   (1) Discuss the conceptual difference between μ_0 and $y_0 = \mu_0 + \varepsilon_0$.
   (2) Is the grade point average $x_0 = 3.25$ in the experimental region?
   (3) Using an appropriate prediction equation, calculate a point estimate of μ_0 and a point prediction of y_0.

i. Calculate SSE, s^2, and s.

4.4 Consider the sales volume data in Table 4.2 and the model

$$\begin{aligned} y_i &= \mu_i + \varepsilon_i \\ &= \beta_0 + \beta_1 x_i + \varepsilon_i \end{aligned}$$

a. Carry out the detailed calculations and verify that the least squares point estimates are $b_0 = 66.2121$ and $b_1 = 4.4303$.
b. Calculate SSE, s^2, and s.
c. Use the least squares point estimates to compute a point estimate of mean third quarter sales volume when \$125,000 is spent on advertising.
d. Use the least squares point estimates to compute a point prediction of an individual third quarter sales volume in a sales region when \$135,000 is spent on advertising.
e. Suppose that the third quarter sales volume in a sales region in which \$80,000 was spent on advertising was \$780,000. Do you believe that the sales force in this sales region effectively utilized the \$80,000 that was spent on advertising?

4.5 Consider the following data concerning the demand (y) and price (x) of a consumer product.

Demand, y	252	244	241	234	230	223
Price, x	\$2.00	\$2.20	\$2.40	\$2.60	\$2.80	\$3.00

a. Plot y versus x. Does it seem reasonable to use the simple linear regression model to relate y to x?
b. Calculate the least squares point estimates of the parameters in the simple linear model.
c. Write the least squares prediction equation. Graph this equation on the plot of y versus x.
d. Calculate SSE, s^2, and s.
e. Use the least squares prediction equation to predict demand for the following prices: \$2.10, \$2.75, and \$3.10.

4.6 The following data concerns weekly natural gas consumption (y) and production (x) at an industrial plant.

Week, i	Natural gas consumption, y_i (hundreds of cubic feet)	Production, x_i (units)
1	1611	9084
2	1497	8359
3	1505	9001
4	1551	11154
5	1681	14371
6	2357	16607
7	2075	14074
8	2213	15473
9	1687	14775
10	1749	16061
11	2017	18518
12	1856	17571

a. Plot y versus x.
b. If we use the simple linear regression model $y = \beta_0 + \beta_1 x + \varepsilon$ to relate y to x, calculate the least squares point estimates of β_0 and β_1.
c. Calculate SSE, s^2, and s.
d. Use the least squares prediction equation to predict natural gas consumption when weekly production is 17,000 units.

4.7 In an article in the *Journal of Accounting Research*, Benzion Barlev and Haim Levy consider relating accounting rates on stocks and market returns. Fifty-four companies were selected. For each company the authors recorded values of

x = mean yearly accounting rate for the period 1959 to 1974

y = mean yearly market return rate for the period 1959 to 1974

The data in Table 4.7 was obtained. Here the accounting rate can be interpreted to represent input into investment and therefore is a logical predictor of market return.

TABLE 4.7 **Accounting rates on stocks and market returns for 54 companies**

Company	Market rate	Accounting rate
McDonnell Douglas	17.73	17.96
NCR	4.54	8.11
Honeywell	3.96	12.46
TRW	8.12	14.70
Raython	6.78	11.90
W.R. Grace	9.69	9.67

(*continues*)

TABLE 4.7 Continued

Company	Market rate	Accounting rate
Ford Motors	12.37	13.35
Textron	15.88	16.11
Lockheed Aircraft	−1.34	6.78
Getty Oil	18.09	9.41
Atlantic Richfield	17.17	8.96
Radio Corporation of America	6.78	14.17
Westinghouse Electric	4.74	9.12
Johnson & Johnson	23.02	14.23
Champion International	7.68	10.43
R.J. Reynolds	14.32	19.74
General Dynamics	−1.63	6.42
Colgate-Palmolive	16.51	12.16
Coca-Cola	17.53	23.19
International Business Machines	12.69	19.20
Allied Chemical	4.66	10.76
Uniroyal	3.67	8.49
Greyhound	10.49	17.70
Cities Service	10.00	9.10
Philip Morris	21.90	17.47
General Motors	5.86	18.45
Phillips Petroleum	10.81	10.06
FMC	5.71	13.30
Caterpillar Tractor	13.38	17.66
Georgia Pacific	13.43	14.59
Minnesota Mining & Manufacturing	10.00	20.94
Standard Oil (Ohio)	16.66	9.62
American Brands	9.40	16.32
Aluminum Company of America	.24	8.19
General Electric	4.37	15.74
General Tire	3.11	12.02
Borden	6.63	11.44
American Home Products	14.73	32.58
Standard Oil (California)	6.15	11.89
International Paper	5.96	10.06
National Steel	6.30	9.60
Republic Steel	.68	7.41
Warner Lambert	12.22	19.88
U.S. Steel	.90	6.97
Bethelehem Steel	2.35	7.90
Armco Steel	5.03	9.34
Texaco	6.13	15.40
Shell Oil	6.58	11.95
Standard Oil (Indiana)	14.26	9.56
Owens Illinois	2.60	10.05
Gulf Oil	4.97	12.11
Tenneco	6.65	11.53
Inland Steel	4.25	9.92
Kraft	7.30	12.27

Source: Reprinted by permission from Benzion Barlev and Haim Levy, "On the Variability of Accounting Income Numbers," *Journal of Accounting Research* (Autumn 1979): 305–315.

a. If we relate y to x by using a simple linear regression model, the least squares point estimates of the model parameters can be calculated to be

$$b_0 = .84801 \quad \text{and} \quad b_1 = .61033$$

Verify that these values are correct.

b. Estimate the mean market return for all stocks having an accounting rate of 15.00.

c. Predict the individual market rate of a stock having an accounting rate of 21.00.

4.8 Enterprise Industries produces Fresh, a liquid laundry detergent. The company wishes to study the relationships between price and demand for the large size bottle of Fresh in its sales regions. The company has gathered data (see Table 4.8) concerning first quarter 1989 demand for Fresh in 30 sales regions of equal sales potential. Here, for $i = 1, 2, \ldots, 30$,

y_i = the demand for the large size bottle of Fresh (in hundreds of thousands of bottles) in sales region i

TABLE 4.8 **Fresh Detergent demand data**

Sales region, i	Price for Fresh, x_{i1}	Average industry price, x_{i2} (dollars)	Price difference, $x_{i4} = x_{i2} - x_{i1}$ (dollars)	Demand for Fresh, y_i (hundreds of thousands of bottles)
1	3.85	3.80	−.05	7.38
2	3.75	4.00	.25	8.51
3	3.70	4.30	.60	9.52
4	3.70	3.70	0	7.50
5	3.60	3.85	.25	9.33
6	3.60	3.80	.20	8.28
7	3.60	3.75	.15	8.75
8	3.80	3.85	.05	7.87
9	3.80	3.65	−.15	7.10
10	3.85	4.00	.15	8.00
11	3.90	4.10	.20	7.89
12	3.90	4.00	.10	8.15
13	3.70	4.10	.40	9.10
14	3.75	4.20	.45	8.86
15	3.75	4.10	.35	8.90
16	3.80	4.10	.30	8.87
17	3.70	4.20	.50	9.26
18	3.80	4.30	.50	9.00
19	3.70	4.10	.40	8.75
20	3.80	3.75	−.05	7.95
21	3.80	3.75	−.05	7.65
22	3.75	3.65	−.10	7.27
23	3.70	3.90	.20	8.00
24	3.55	3.65	.10	8.50
25	3.60	4.10	.50	8.75
26	3.65	4.25	.60	9.21
27	3.70	3.65	−.05	8.27
28	3.75	3.75	0	7.67
29	3.80	3.85	.05	7.93
30	3.70	4.25	.55	9.26

x_{i1} = the price (in dollars) of Fresh as offered by Enterprise Industries in sales region i

x_{i2} = the average industry price (in dollars) of competitors' similar detergents in sales region i

$x_{i4} = x_{i2} - x_{i1}$ = the "price difference" in sales region i

a. Plot y versus x_4. Is it reasonable to use the simple linear model to relate y to x_4?
b. Suppose that we wish to use the model $y = \beta_0 + \beta_1 x_4 + \varepsilon$ to predict demand. Calculate the least squares point estimates b_0 and b_1 of the model parameters.
c. Compute SSE, s^2, and s.
d. Calculate a point estimate of μ_0, the mean demand for Fresh when x_{01} = \$3.80 and x_{02} = \$3.90.
e. Calculate a point prediction of y_0, an individual demand for Fresh when x_{01} = \$3.75 and x_{02} = \$4.00.

4.9 Below we present data concerning hospital labor needs at 13 U.S. Navy Hospitals. Here

y = monthly labor hours required

x = monthly occupied bed days

Hospital	*y*	*x*
1	566.52	472.92
2	696.82	1339.75
3	1033.15	620.25
4	1603.62	568.33
5	1611.37	1497.60
6	1613.27	1365.83
7	1854.17	1687.00
8	2160.55	1639.92
9	2305.58	2872.33
10	3503.93	3655.08
11	3571.89	2912.00
12	3741.40	3921.00
13	4026.52	3865.67

Source: Procedures and Analysis for Staffing Standards Development: Regression Analysis Handbook (San Diego, Calif.: Navy Manpower and Material Analysis Center, 1979).

a. Plot y versus x.
b. Consider the simple linear regression model relating y to x. Compute the least squares point estimates of the model parameters.
c. Interpret the meanings of the least squares point estimates b_0 and b_1.
d. Write the least squares prediction equation. Graph this equation on the plot of y versus x.
e. Calculate a point estimate of μ_0, the mean monthly labor hours requirement for all hospitals, when $x_0 = 2500$.
f. Calculate a point prediction of y_0, an individual monthly labor hours requirement for a hospital when $x_0 = 3000$.

4.10 Ott (1987) presents a study of the amount of heat loss for a certain brand of thermal pane windows. Three different windows were randomly assigned to each of three different outdoor temperatures. For each trial the indoor window temperature was controlled at 68°F and 50 percent relative humidity.

Outdoor temperature, x (°F)	Heat loss, y	Outdoor temperature, x (°F)	Heat loss, y
20	86	40	75
20	80	60	33
20	77	60	38
40	78	60	43
40	84		

a. Use the model $y = \beta_0 + \beta_1 x + \varepsilon$ to predict heat loss for outdoor temperatures of 25°F, 30°F, and 50°F.
b. Calculate SSE, s^2, and s.

4.11 The following data concerns quarterly sales y of a personal computer (in thousands of units) over a four-year period of time.

Time period, t	Quarterly sales, y	Time period, t	Quarterly sales, y
1	21.74	9	28.26
2	25.41	10	27.87
3	25.44	11	28.40
4	25.40	12	30.16
5	23.91	13	32.61
6	27.05	14	30.33
7	26.63	15	30.18
8	26.98	16	33.33

a. Plot sales (y) versus time (t). Is it reasonable to use the simple linear regression model to relate y to t?
b. When we use the simple linear regression model to relate a dependent variable to time, the prediction equation relating y to t is called a *trend line*. The intercept and slope of the trend line are the least squares point estimates b_0 and b_1 calculated by using the dependent variable y and the independent variable t. Find the trend line relating personal computer sales y to time t.
c. Use the trend line to predict personal computer sales in time periods 17 and 18. Note here that we are extrapolating the trend line beyond the experimental region. It is dangerous to extrapolate this line too far beyond time period 16.
d. Calculate SSE, s^2, and s for the trend line.

4.12 The following data concerns monthly sales of an electronic calculator for the last year.

Month	Calculator sales, y	Month	Calculator sales, y
January	197	July	308
February	211	August	262
March	203	September	258
April	247	October	256
May	239	November	261
June	269	December	288

a. Plot sales (y) versus time (t). Is it reasonable to use a least squares trend line to relate y to t? (Trend lines are discussed in Exercise 4.11.)
b. Use a least squares trend line to predict calculator sales for the next three months (January, February, and March of the next year).

4.13 The following data concerns the number of work stoppages (y) in the United States and labor union membership (x) as a percent of the civilian labor force for the period 1920–1970.

Year	Number of work stoppages in U.S., y	Percent of U.S. civilian labor force with union membership, x
1920	3411	12.2
1925	1301	7.9
1930	637	7.5
1935	2014	7.1
1940	2508	16.1
1945	4000*	27.5
1950	4843	24.1
1955	4320	27.3
1960	3333	26.0
1965	3963	24.9
1970	5716	25.1

*Estimated by the authors on the basis of 1946 statistics.

Source: U.S. Department of Commerce, Bureau of the Census, *Bicentennial Statistics*, Washington, D.C., 1976.

Use simple linear regression to predict the number of work stoppages when 20 percent of the labor force has union membership.

4.14 The following data concerns U.S. illiteracy rates (y) and U.S. school enrollment rates (x).

Year	Percent illiterate, y*	School enrollment rate† (per 100 population)
1870	11.5	54.4
1890	7.7	57.9
1910	5.0	61.3
1930	3.0	71.2
1950	1.8‡	79.3
1960	1.6‡	84.8
1970	.7‡	88.3

*White population 10 years old and over 1870–1930; 14 years old and over thereafter.

†White population 5–19 years old, except 5–20 years old 1910–1930.

‡1952, 1959, 1969 statistics.

Source: U.S. Department of Commerce, Bureau of the Census, *Bicentennial Statistics*, Washington, D.C., 1976.

Use simple linear regression to predict the percent illiterate when the school enrollment rate is 75.

4.15 The following data concerns the divorce rate (y) per 1000 women and the percent of female population in the labor force (x).

Year	*Divorce rate, y (number of divorces per 1000 women)*	*Percent of female population in the labor force, x**
1890	3.0	18.9
1900	4.1	20.6
1910	4.7	25.4
1920	8.0	23.7
1930	7.5	24.8
1940	8.8	27.4
1950	10.3	31.4
1960	9.2	34.8
1970	14.9	42.6

*15 years old and over 1890–1930; 14 and over 1940–1960; 16 and over thereafter.

Source: U.S. Department of Commerce, Bureau of the Census, *Bicentennial Statistics*, Washington, D.C., 1976.

Use simple linear regression to predict the divorce rate for women when 45 percent of the female population is in the labor force.

4.16 The following data concerns total U.S. energy consumption (y) and total U.S. population (x).

Year	*U.S. energy consumption, y* (in quadrillion Btu)*	*U.S. total population* (in millions)*
1900	7.6	76.0
1910	14.8	86.0
1920	19.8	105.7
1930	22.3	120.0
1940	23.9	131.7
1950	34.2	151.3
1960	44.6	179.3
1970	67.1	203.2

*Prior to 1950 excludes Alaska and Hawaii.

Source: U.S. Department of Commerce, Bureau of the Census, *Bicentennial Statistics*, Washington, D.C., 1976.

a. Use simple linear regression to predict U.S. energy consumption when the total U.S. population is 213 million.

b. In 1975 the total U.S. population was 213 million, and total energy consumption was 71.1 quadrillion Btu. What was the error of prediction for the prediction you made in part (a) based on the 1900–1970 data?

Exercises 4.17 through 4.20 refer to the data in Table 4.9. This data concerns U.S. social welfare expenditures (y) as a percent of gross national product (GNP), total U.S. GNP (in 1972 dollars), per capita U.S. GNP (in 1972 dollars), and U.S. unemployment rate (civilian labor force only).

4.17 Use simple linear regression to relate social welfare expenditures (y) to total GNP (x_1). Predict social welfare expenditures when total GNP (in billions of 1972 dollars) is 1186.

TABLE 4.9 **Social welfare expenditure data**

Year	Total GNP (in billions of 1972 dollars)	Per capita GNP (in 1972 dollars)	Unemployment rate (%)	Social welfare expenditures (% of GNP)
1930	286	2323	8.7	4.2
1935	263	2067	20.1	9.5
1940	348	2630	14.6	9.2
1945	564	4032	1.9	4.4
1950	534	3517	5.3	8.2
1955	655	3962	4.4	8.2
1960	737	4078	5.5	10.3
1965	926	4765	4.5	11.2
1970	1075	5248	4.9	14.8

Source: U.S. Department of Commerce, Bureau of the Census, *Bicentennial Statistics*, Washington, D.C., 1976.

TABLE 4.10 **Sales price data for 24 condominiums in Oxford, Ohio**

Selling price, y (in thousands of dollars)	Square footage, x_1	Age, x_2 (in years)
64.5	1200	6
54.9	1030	4
47.0	630	4
47.0	630	3
70.0	1148	3
42.5	630	3
48.0	630	2
45.5	630	4
44.0	630	3
72.0	1148	3
65.0	1200	6
53.0	966	17
71.0	1148	3
72.0	1148	3
72.0	1148	3
73.5	1148	2
73.0	1148	2
73.5	1148	3
40.3	630	3
45.0	630	4
45.0	630	4
72.5	1148	4
74.0	1148	2
73.0	1148	2

Source: Reprinted by permission from Alpha, Inc., Oxford, Ohio.

4.18 Use simple linear regression to relate social welfare expenditures (y) to per capita GNP (x_2). Predict social welfare expenditures when per capita GNP (in 1972 dollars) is 5552.

4.19 Plot social welfare expenditures (y) versus unemployment rate (x_3). Does it seem reasonable to use simple linear regression to relate y to x_3? Explain why or why not. Try using simple linear regression to predict social welfare expenditures when the unemployment rate is 8.5%.

4.20 In 1975 the total GNP was 1186, the per capita GNP was 5552, the unemployment rate was 8.5 percent, and social welfare expenditures were 19.1. Did any of the predictor variables employed in Exercises 4.17, 4.18, and 4.19 give a good prediction of 1975 social welfare expenditures? Which predictor variable gave the best prediction of 1975 social welfare expenditures?

Exercises 4.21 and 4.22 refer to the data in Table 4.10. This data concerns the selling price (y) in thousands of dollars, square footage (x_1), and age (x_2) in years for 24 condominiums sold in Oxford, Ohio, during 1988.

4.21 Use simple linear regression to predict the selling price of a condominium with 1000 square feet.

4.22 Plot selling price (y) versus age (x_2). Does it seem reasonable to use a simple linear regression model to relate y to x_2? Explain why or why not.

4.23 The following data concerns the time (y) required to perform service on a service call and the number of copiers serviced (x) on the service call.

Service call	*Time required to perform service, y (minutes)*	*Number of copiers serviced, x*
1	92	3
2	63	2
3	145	6
4	195	8
5	49	2
6	90	4
7	119	5
8	145	6
9	67	2
10	115	4
11	154	6
12	267	11
13	77	3
14	210	10
15	27	1

a. Plot y versus x. Does it seem reasonable to relate y to x by using a simple linear regression model?
b. Calculate the least squares point estimates of the regression parameters in the simple linear regression model relating y to x.
c. Write the least squares prediction equation. Graph this equation on your plot of y versus x.
d. Predict the service time required to perform service on 12 copiers.
e. Estimate the average service time required when performing service on 9 copiers.
f. Compute SSE, s^2, and s.

C H A P T E R 5

INFERENCE IN SIMPLE LINEAR REGRESSION

In this chapter we consider using the simple linear regression model to make statistical inferences. Sections 5.1 through 5.6 discuss sampling properties concerning the least squares point estimates, testing the significance of the independent variable, calculating confidence intervals for means, and calculating prediction intervals for individual values of the dependent variable. Section 5.7 presents an example of simple linear regression utilizing SAS, while Sections 5.8 and 5.9 explain several measures of the utility of the simple linear regression model. These include the simple coefficient of determination and an *F*-test for the simple linear model. We conclude this chapter with optional Sections 5.10, 5.11, and 5.12, which discuss an *F*-test of lack of fit, regression analysis through the origin, and using SAS to perform simple linear regression.

5.1

SAMPLING PROPERTIES CONCERNING b_1, b_0, s^2, AND s

The formulas presented in Sections 4.2 and 4.4 for computing b_1, b_0, s^2, and s define what we will call the least squares estimation procedure for estimating β_1, β_0, σ^2, and σ.

To understand certain properties concerning b_1, b_0, s^2, and s, recall that we have observed n values $x_1, x_2, \ldots, x_n$ of the independent variable x. For $i = 1, 2, \ldots, n$, define the *ith historical population* to be the infinite population of potential values of the dependent variable that could have been observed when the value of the independent variable x was x_i. The $n = 8$ historical populations in the fuel consumption problem are listed in Table 5.1(a). Before the sample of n observed values of the dependent variable was randomly selected, y_i could have been any of the potential values of the dependent variable in the ith historical population. Therefore there is an infinite population of possible samples of n values of the dependent variable that could have been randomly selected from the n historical populations. Three such possible samples in the fuel consumption problem are listed in Table 5.1(b). Note that the first of these samples is the sample that we have actually observed. Also note that, in general, the different possible samples correspond to the same fixed set of n observed values $x_1, x_2, \ldots, x_n$ of the independent variable x. Table 5.1 illustrates that each possible sample gives values of b_1, b_0, s^2, and s that are different from the values given by other samples. Therefore there is an infinite population of possible values of each of b_1, b_0, s^2, and s.

We first consider the population of all possible values of b_1. This population is also called the population of all possible least squares point estimates of β_1. It can be proven* that μ_{b_1}, the mean of this population, equals β_1. For this reason, when we use the observed sample to calculate the least squares point estimate b_1, we are using an *unbiased* estimation procedure. This unbiasedness property tells us that although the least squares point estimate b_1 that we calculate probably does not equal β_1, the average of all the different possible least squares point estimates of β_1 that we could have calculated is equal to β_1.

We next consider $\sigma^2_{b_1}$, the variance of the population of all possible least squares point estimates of β_1. This variance is the average of the squared deviations of the different possible values of b_1 from β_1. Letting

$$c_{11} = \frac{1}{\sum_{i=1}^{n} (x_i - \bar{x})^2}$$

it can be proven that if the constant variance and independence assumptions hold,

*All of the proofs in Appendix D are given in terms of the general linear regression model (to be presented in Chapter 8). The proofs of the results in Chapter 5 are special cases of the proofs in Appendix D.

TABLE 5.1 **Sampling properties concerning b_1, b_0, and s^2 in the fuel consumption problem**

(a) The eight historical populations

Week, i	*Average hourly temperature, x_i*	*Historical population of potential weekly fuel consumptions*
1	$x_1 = 28.0$	First historical population
2	$x_2 = 28.0$	Second historical population
3	$x_3 = 32.5$	Third historical population
4	$x_4 = 39.0$	Fourth historical population
5	$x_5 = 45.9$	Fifth historical population
6	$x_6 = 57.8$	Sixth historical population
7	$x_7 = 58.1$	Seventh historical population
8	$x_8 = 62.5$	Eighth historical population

(b) Three possible samples

Sample 1	*Sample 2*	*Sample 3*
$y_1 = 12.4$	$y_1 = 12.0$	$y_1 = 10.7$
$y_2 = 11.7$	$y_2 = 11.8$	$y_2 = 10.2$
$y_3 = 12.4$	$y_3 = 12.3$	$y_3 = 10.5$
$y_4 = 10.8$	$y_4 = 11.5$	$y_4 = 9.8$
$y_5 = 9.4$	$y_5 = 9.1$	$y_5 = 9.5$
$y_6 = 9.5$	$y_6 = 9.2$	$y_6 = 8.9$
$y_7 = 8.0$	$y_7 = 8.5$	$y_7 = 8.5$
$y_8 = 7.5$	$y_8 = 7.2$	$y_8 = 8.0$

(c) Three possible values of b_1

$b_1 = -.1279$	$b_1 = -.1285$	$b_1 = -.0666$

(d) Three possible values of b_0

$b_0 = 15.84$	$b_0 = 15.85$	$b_0 = 12.44$

(e) Three possible values of s^2

$s^2 = .428$	$s^2 = .47$	$s^2 = .0667$

(f) Three possible values of s

$s = .6542$	$s = .686$	$s = .2582$

then

$$\sigma_{b_1}^2 = \sigma^2 c_{11}$$

This result is not at all intuitive, but it plays a major role in the statistical inference procedures of Sections 5.2 and 5.3. This result is also related to an important theorem called the *Gauss–Markov Theorem*. To understand what this theorem says, note that the formula for the least squares point estimate b_1 can be shown to be equivalent to

$$b_1 = \sum_{i=1}^{n} k_i y_i \qquad \text{where} \qquad k_i = \frac{(x_i - \bar{x})}{\sum_{i=1}^{n} (x_i - \bar{x})^2}$$

It follows that b_1 is a linear function of $y_1, y_2, \ldots, y_n$, and thus we are calculating b_1 by using a *linear estimation procedure* (which, from the preceding discussion, is also an *unbiased* estimation procedure). We now state the Gauss–Markov Theorem.

The Gauss–Markov Theorem

If the constant variance and independence assumptions hold, then the *variance* of the population of all possible *least squares* point estimates of β_1 is *smaller* than the variance of the population of all possible point estimates of β_1 that could be obtained by using any other unbiased, linear estimation procedure.

Intuitively, this means that the least squares point estimates of β_1 are clustered more closely around β_1 than are the point estimates of β_1 that could be obtained by using any other unbiased, linear estimation procedure. Thus when we calculate the least squares point estimate b_1, we are likely to obtain a point estimate of β_1 that is closer to the true value of β_1 than would be a point estimate of β_1 obtained by using any other unbiased, linear estimation procedure.

Similar results apply to the population of all possible values of b_0 (least squares point estimates of β_0). Specifically, it can be proven that

$$\mu_{b_0} = \beta_0$$

This says that the average of all possible least squares point estimates of β_0 equals β_0. Furthermore, it can be proven that the Gauss–Markov Theorem also says that if the constant variance and independence assumptions hold, then $\sigma^2_{b_0}$, the variance of the population of all possible least squares point estimates of β_0, is smaller than the variance of the population of all possible point estimates of β_0 that could be obtained by using any other unbiased linear estimation procedure. For future reference it can be proven that, if the constant variance and independence assumptions hold, then

$$\sigma^2_{b_0} = \sigma^2 c_{00}$$

where

$$c_{00} = \frac{1}{n} + \frac{\bar{x}^2}{\sum_{i=1}^{n} (x_i - \bar{x})^2}$$

Finally, we note that we compute the point estimate s^2 of σ^2 by using the formula

$$s^2 = \frac{SSE}{n - 2}$$

because it can be proven that, if the constant variance and independence assumptions hold, then using this formula implies that the mean of the population of all possible values of s^2 equals σ^2. This says that the least squares procedure for estimating σ^2 is unbiased. It should be noted, however, that the average of all possible values of s does not equal σ. Nevertheless, s is used as the point estimate of σ.

5.2

A CONFIDENCE INTERVAL FOR THE SLOPE

We summarize the logic behind (and formula for) a $100(1 - \alpha)\%$ confidence interval for the slope β_1 in the following box.

Confidence Interval for β_1

1. The *population of all possible least squares point estimates of* β_1
   a. Has *mean* $\mu_{b_1} = \beta_1$.
   b. Has *variance* $\sigma^2_{b_1} = \sigma^2 c_{11}$ (if inference assumptions 1 and 2 hold).
   c. Has *standard deviation* $\sigma_{b_1} = \sigma\sqrt{c_{11}}$ (if inference assumptions 1 and 2 hold).
   d. Has a *normal distribution* (if inference assumptions 1, 2, and 3 hold).
2. A point estimate of $\sigma_{b_1} = \sigma\sqrt{c_{11}}$ is $s_{b_1} = s\sqrt{c_{11}}$, which is called the *standard error of the estimate* b_1.
3. If the inference assumptions hold, the population of all possible values of

$$\frac{b_1 - \beta_1}{s_{b_1}} = \frac{b_1 - \beta_1}{s\sqrt{c_{11}}}$$

has a *t-distribution with $n - 2$ degrees of freedom.*

4. If the inference assumptions hold, a *$100(1 - \alpha)\%$ confidence interval for* β_1 is

$$[b_1 \pm t^{(n-2)}_{[\alpha/2]} s\sqrt{c_{11}}] = \left[b_1 \pm t^{(n-2)}_{[\alpha/2]} s \sqrt{\frac{1}{\sum_{i=1}^{n} (x_i - \bar{x})^2}}\right]$$

FIGURE 5.1 $P(-t_{[\alpha/2]}^{(n-2)} \le [(b_1 - \beta_1)/s_{b_1}] \le t_{[\alpha/2]}^{(n-2)}) = 1 - \alpha$

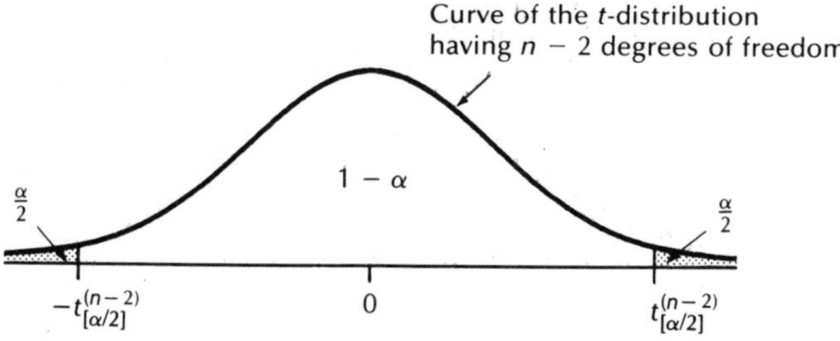

Note that the above interval for β_1 follows from (3) above. This is because, if the inference assumptions hold, then (3) implies that

$$P\left(-t_{[\alpha/2]}^{(n-2)} \le \frac{b_1 - \beta_1}{s_{b_1}} \le t_{[\alpha/2]}^{(n-2)}\right) = 1 - \alpha$$

which is the area under the curve of the t-distribution having $n - 2$ degrees of freedom between $-t_{[\alpha/2]}^{(n-2)}$ and $t_{[\alpha/2]}^{(n-2)}$ (see Figure 5.1). Multiplying the inequality in the above probability statement by s_{b_1} (which is positive), we obtain the equivalent probability statement

$$P(-t_{[\alpha/2]}^{(n-2)} s_{b_1} \le b_1 - \beta_1 \le t_{[\alpha/2]}^{(n-2)} s_{b_1}) = 1 - \alpha$$

which implies (subtracting b_1 through the inequality) that

$$P(-b_1 - t_{[\alpha/2]}^{(n-2)} s_{b_1} \le -\beta_1 \le -b_1 + t_{[\alpha/2]}^{(n-2)} s_{b_1}) = 1 - \alpha$$

which in turn implies (multiplying the inequality by -1) that

$$P(b_1 + t_{[\alpha/2]}^{(n-2)} s_{b_1} \ge \beta_1 \ge b_1 - t_{[\alpha/2]}^{(n-2)} s_{b_1}) = 1 - \alpha$$

This probability statement is equivalent to

$$P(b_1 - t_{[\alpha/2]}^{(n-2)} s_{b_1} \le \beta_1 \le b_1 + t_{[\alpha/2]}^{(n-2)} s_{b_1}) = 1 - \alpha$$

which, since $s_{b_1} = s\sqrt{c_{11}}$, says that

$$P(b_1 - t_{[\alpha/2]}^{(n-2)} s\sqrt{c_{11}} \le \beta_1 \le b_1 + t_{[\alpha/2]}^{(n-2)} s\sqrt{c_{11}}) = 1 - \alpha$$

This last probability statement says that the proportion of confidence intervals that contain β_1 in the population of all possible $100(1 - \alpha)\%$ confidence intervals for β_1 *is equal to* $1 - \alpha$. That is, if we compute a $100(1 - \alpha)\%$ confidence interval for β_1 using the formula

$$[b_1 \pm t_{[\alpha/2]}^{(n-2)} s\sqrt{c_{11}}]$$

then $100(1-\alpha)\%$ (for example, 95%) of all possible $100(1-\alpha)\%$ confidence intervals for β_1 *contain* β_1 and $100(\alpha)\%$ (for example, 5%) of all possible $100(1-\alpha)\%$ confidence intervals for β_1 do not contain β_1. This interpretation will be discussed further in the following example.

Example 5.1

Part 1: Calculating a Confidence Interval for β_1

Recall from Table 4.5 that $b_1 = -.1279$ is the least squares point estimate of β_1 in the fuel consumption regression model

$$\begin{aligned} y_i &= \mu_i + \varepsilon_i \\ &= \beta_0 + \beta_1 x_i + \varepsilon_i \end{aligned}$$

and recall from Example 4.7 that the standard error s is .6542. Also, noting that it can be proven that

$$\sum_{i=1}^{n} (x_i - \bar{x})^2 = \sum_{i=1}^{n} x_i^2 - n\bar{x}^2$$

we can use the information in Table 4.5 to calculate c_{11} to be

$$\begin{aligned} c_{11} &= \frac{1}{\sum_{i=1}^{n} (x_i - \bar{x})^2} = \frac{1}{\sum_{i=1}^{8} x_i^2 - 8\bar{x}^2} \\ &= \frac{1}{16{,}874.76 - 8(43.98)^2} = .000712 \end{aligned}$$

If we wish to calculate a $100(1-\alpha)\% = 95\%$ confidence interval for β_1, then, since $\alpha = .05$, we would use $t_{[\alpha/2]}^{(n-2)} = t_{[.05/2]}^{(8-2)} = t_{[.025]}^{(6)} = 2.447$. It follows that a 95% confidence interval for β_1 is

$$\begin{aligned} [b_1 \pm t_{[\alpha/2]}^{(n-2)} s\sqrt{c_{11}}] &= [-.1279 \pm 2.447(.6542)\sqrt{.000712}] \\ &= [-.1279 \pm .0427] \\ &= [-.1706, -.0852] \end{aligned}$$

Recall (from Example 4.1) that

β_1 = the change in mean weekly fuel consumption associated with a 1-degree increase in the average hourly temperature

Thus, the 95% confidence interval for β_1 states that we are 95% confident that if the average hourly temperature increases by 1-degree; the mean weekly fuel consumption will decrease (because both the lower bound and the upper bound of the confidence interval are negative) by at least .0852 tons of coal and by at most .1706 tons of coal. Also note that, since the 95% confidence interval for β_1 does not contain zero, we have strong evidence that there is in fact a change (a decrease) in mean weekly fuel consumption associated with a 1-degree increase in average hourly temperature. That is, we have strong evidence that x, average hourly temperature, is significantly related to y, weekly fuel consumption.

Part 2: The Meaning of 95% Confidence

To understand the meaning of 95% confidence, we assume that (a supernatural power knows that) the true value of β_1 is $-.1281$. In Table 5.2 we calculate the 95% confidence interval for β_1, $[b_1 \pm 2.447s\sqrt{.000712}]$, given by each of the samples in Table 5.1. Note that two of the three 95% confidence intervals for $\beta_1 = -.1281$ in Table 5.2 contain β_1 (see Figure 5.2). When we compute the 95% confidence interval for β_1 by using the observed sample (which is the first sample in Tables 5.1 and 5.2) and obtain the interval $[-.1706, -.0852]$, we are 95% confident that this interval contains β_1. This is because 95 percent of the confidence intervals in the population of all possible 95% confidence intervals for β_1 contain β_1 and because we have obtained one of the confidence intervals in this population.

TABLE 5.2 The 95% confidence interval for β_1, $[b_1 \pm 2.447s\sqrt{.000712}]$, given by each of the three samples in Table 5.1

Sample 1	Sample 2	Sample 3
$[-.1279 \pm 2.447(.6542)\sqrt{.000712}]$ $= [-.1706, -.0852]$	$[-.1285 \pm 2.447(.686)\sqrt{.000712}]$ $= [-.1733, -.0837]$	$[-.0666 \pm 2.447(.2582)\sqrt{.000712}]$ $= [-.0835, -.0497]$

FIGURE 5.2 95% confidence intervals for β_1 obtained by using three samples

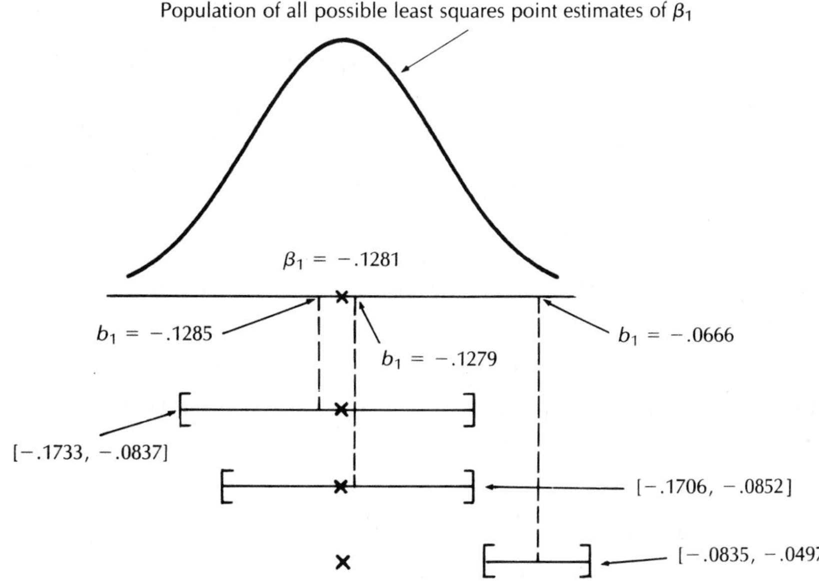

Example 5.2 Recall from Example 4.2 that $b_0 = 66.2121$ and $b_1 = 4.4303$ are the least squares point estimates of β_0 and β_1 in the American Manufacturing sales volume regression model

$$\begin{aligned} y_i &= \mu_i + \varepsilon_i \\ &= \beta_0 + \beta_1 x_i + \varepsilon_i \end{aligned}$$

Also note that we can use the data in Table 4.2 to obtain s and c_{11} by making the following calculations:

$$SSE = \sum_{i=1}^{n} (y_i - \hat{y}_i)^2 = \sum_{i=1}^{10} [y_i - (66.2121 + 4.4303x_i)]^2 = 222.8242$$

$$s = \sqrt{\frac{SSE}{n-2}} = \sqrt{\frac{222.8242}{10-2}} = 5.2776$$

$$c_{11} = \frac{1}{\sum_{i=1}^{10} (x_i - \bar{x})^2} = .01212$$

It follows that a 95% confidence interval for β_1 is

$$\begin{aligned} [b_1 \pm t_{[\alpha/2]}^{(n-2)} s\sqrt{c_{11}}] &= [4.4303 \pm t_{[.025]}^{(10-2)}(5.2776)\sqrt{.01212}] \\ &= [4.4303 \pm 2.306(5.2776)\sqrt{.01212}] \\ &= [3.0905, 5.7701] \end{aligned}$$

This interval says that American Manufacturing Company is 95% confident that mean third quarter sales volume in a sales region increased by \$30,905 to \$57,701 for each additional \$10,000 spent on third quarter advertising in the sales region.

As another example, a 95% confidence interval for the slope β_1 in the simple linear regression model describing the residential sales data in Table 4.3 can be calculated to be

$$\begin{aligned} [b_1 \pm t_{[\alpha/2]}^{(n-2)} s\sqrt{c_{11}}] &= [.05477 \pm t_{[.025]}^{(68-2)}(.005337)] \\ &= [.05477 \pm 1.96(.005337)] \\ &= [.04401, .06493] \end{aligned}$$

Therefore we are 95% confident that the mean sales price of a residence is between \$4401 and \$6493 higher for each increase of 100 square feet in residence size.

Example 5.3 In analysis of the stock market we sometimes use the model

$$\begin{aligned} y &= \mu + \varepsilon \\ &= \beta_0 + \beta_1 x + \varepsilon \end{aligned}$$

to relate

$$y = \text{the rate of return on a particular stock}$$

to

$$x = \text{the rate of return on the overall stock market}$$

When using the preceding model, we can interpret β_1 to be the percentage point change in the mean (or expected) rate of return on the particular stock that is associated with an increase of one percentage point in the rate of return on the overall stock market.

1. If regression analysis can be used to conclude (at a high level of confidence) that β_1 is greater than 1 (for example, if the 95% confidence interval for β_1 were [1.1826, 1.4723]), this would indicate that the mean rate of return on the particular stock changes more quickly than the rate of return on the overall stock market. Such a stock is called an *aggressive stock*, since gains for such a stock tend to be greater than overall market gains (which occur when the market is bullish). However, losses for such a stock tend to be greater than overall market losses (which occur when the market is bearish). Aggressive stocks should be purchased if you expect the market to rise and avoided if you expect the market to fall.
2. If regression analysis can be used to conclude (at a high level of confidence) that β_1 is less than 1 (for example, if the 95% confidence interval for β_1 were [.4729, .7861]), this would indicate that the mean rate of return on the particular stock changes more slowly than the rate of return on the overall stock market. Such a stock is called a *defensive stock*. Losses for such a stock tend to be less than overall market losses, whereas gains for such a stock tend to be less than overall market gains. Defensive stocks should be held if you expect the market to fall and sold off if you expect the market to rise.
3. If the least squares point estimate b_1 of β_1 is nearly equal to 1, and if the 95% confidence interval for β_1 contains 1, this might indicate that the mean rate of return on the particular stock changes at roughly the same rate as the rate of return on the overall stock market. Such a stock is called a *neutral stock*.

For most stocks, β_1 is positive. This is because general economic conditions that affect the overall market tend to affect most individual stocks in the same direction. Although stocks having zero or negative β_1 values are rare, they do exist. For example, gold stocks have negative values of β_1 because when the overall market falls, the value of gold rises, and when the overall market rises, the value of gold falls. In the exercises at the end of this chapter we consider an interesting journal article concerning how β_1 is affected by the length of time for which the rate of return is calculated.

Since the parameter β_0 equals the mean (or "expected") rate of return on the particular stock when the rate of return on the overall stock market is zero, the value of β_0 is determined by factors specific to the firm associated with the stock. For instance, if regression analysis can be used to conclude (at a high level of confidence) that β_0 is greater than zero (say, if the 95% confidence interval for β_0 were [.1381, .3742]), this would indicate that there is a positive mean rate of return on the particular stock when the rate of return on the overall market is zero. In Section 5.4 we show how to calculate a confidence interval for the intercept β_0.

5.3

TESTING THE SIGNIFICANCE OF THE INDEPENDENT VARIABLE

TESTING $H_0: \beta_1 = 0$ VERSUS $H_1: \beta_1 \neq 0$ USING REJECTION POINTS

In this section we discuss how to test the significance of the independent variable x in the simple linear regression model. It would be reasonable to conclude that x is significantly related to y if we can be quite certain that we should reject the null hypothesis

$$H_0: \beta_1 = 0$$

which says that there is no change in the mean value of the dependent variable associated with an increase in x, in favor of the alternative hypothesis

$$H_1: \beta_1 \neq 0$$

which says that there is a (positive or negative) change in the mean value of the dependent variable associated with an increase in x.

Notice that we use the *two-sided* alternative $H_1: \beta_1 \neq 0$ for this test of significance. However, in some situations we might wish to employ a *one-sided* alternative ($H_1: \beta_1 > 0$ or $H_1: \beta_1 < 0$). For example, in the fuel consumption problem we have previously reasoned that if β_1 does not equal zero, then β_1 must be a negative number. This is because a negative β_1 would express the fact that the larger x (the average hourly temperature) is, the smaller is the mean weekly fuel consumption. Hence it would really be more appropriate to decide that x is significantly related to y if we can reject $H_0: \beta_1 = 0$ in favor of the *one-sided* alternative $H_1: \beta_1 < 0$. Although this is slightly more effective than attempting to reject $H_0: \beta_1 = 0$ in favor of the *two-sided* alternative $H_1: \beta_1 \neq 0$, it makes little practical difference whether we use a one-sided or a two-sided alternative hypothesis. Consequently, to simplify our discussion of hypothesis testing in regression analysis, we will attempt to reject a null hypothesis in favor of a two-sided alternative hypothesis, even if we can determine an appropriate one-sided alternative hypothesis. We will explain how to test the significance of an independent variable by using a one-sided alternative hypothesis later in this section.

The classical approach to testing H_0 versus H_1 uses the test statistic

$$t = \frac{b_1}{s_{b_1}} = \frac{b_1}{s\sqrt{c_{11}}}$$

where b_1 is the least squares point estimate of β_1, s is the standard error, and

$$c_{11} = \frac{1}{\sum_{i=1}^{n} (x_i - \bar{x})^2}$$

Since we can write the test statistic in the form

$$t = \frac{b_1 - 0}{s\sqrt{c_{11}}}$$

we see that it measures the distance between b_1 and zero (the value that makes the null hypothesis $H_0 : \beta_1 = 0$ true). If the absolute value of t is "large," this implies that the distance between b_1 and zero is "large" and provides evidence that we should reject $H_0 : \beta_1 = 0$. Rejecting H_0 would lead us to conclude that x is significantly related to y.

To decide how large (in absolute value) t must be before we reject $H_0 : \beta_1 = 0$, we consider the errors that can be made in hypothesis testing. These errors are summarized in Table 5.3. A *Type I error* is committed if we reject $H_0 : \beta_1 = 0$ when $H_0 : \beta_1 = 0$ is true. This error means that we would conclude that x is significantly related to y when in fact x is not significantly related to y. A *Type II error* is committed if we do not reject $H_0 : \beta_1 = 0$ when $H_0 : \beta_1 = 0$ is false. This error

TABLE 5.3 **Type I and Type II errors when testing $H_0 : \beta_1 = 0$**

	State of nature	
Decisions made in hypothesis test	***Null hypothesis $H_0 : \beta_1 = 0$ is true: x is not significantly related to y***	***Null hypothesis $H_0 : \beta_1 = 0$ is false: x is significantly related to y***
Reject null hypothesis $H_0 : \beta_1 = 0$. Decide that x is significantly related to y	Type I error: Conclude that x is significantly related to y when x is not significantly related to y	Correct decision: Conclude that x is significantly related to y when x is significantly related to y
Do not reject null hypothesis $H_0 : \beta_1 = 0$. Decide that x is not significantly related to y	Correct decision: Conclude that x is not significantly related to y when x is not significantly related to y	Type II error: Conclude that x is not significantly related to y when x is significantly related to y

means that we would conclude that x is not significantly related to y when in fact x is significantly related to y. We obviously desire that both *the probability of a Type I error, denoted by* α, *and the probability of a Type II error be small.*

It is common procedure to base an hypothesis test on taking a sample of a fixed size and on setting α equal to a specified value. Here, we sometimes choose α to be as high as .1, but we usually choose α to be between .05 and .01, .05 being the most frequent choice. We usually do not set α lower than .01 because setting α extremely small often leads to a probability of a Type II error that is unacceptably large. (Generally, for a fixed sample size, the lower the probability of a Type I error, the higher the probability of a Type II error.) We consider how to determine the precise value of α that should be used after discussing the procedure for testing $H_0: \beta_1 = 0$ versus $H_1: \beta_1 \neq 0$.

Testing $H_0: \beta_1 = 0$ Versus $H_1: \beta_1 \neq 0$ Using Rejection Points

Define the test statistic

$$t = \frac{b_1}{s\sqrt{c_{11}}}$$

If the inference assumptions are satisfied, we can reject $H_0: \beta_1 = 0$ in favor of $H_1: \beta_1 \neq 0$ by setting the probability of a Type I error equal to α if and only if

$$|t| > t_{[\alpha/2]}^{(n-2)}$$

that is, if $t > t_{[\alpha/2]}^{(n-2)}$ or $t < -t_{[\alpha/2]}^{(n-2)}$.

In the above procedure we call the points $-t_{[\alpha/2]}^{(n-2)}$ and $t_{[\alpha/2]}^{(n-2)}$ *rejection points.* This is because these points tell us how different from zero (positive or negative) t must be in order for us to be able to reject $H_0: \beta_1 = 0$ by setting the probability of a Type I error equal to α. To see why using this rejection point procedure ensures that the probability of a Type I error equals α (for example, .05), recall that, if the inference assumptions are satisfied, then the population of all possible values of

$$\frac{b_1 - \beta_1}{s\sqrt{c_{11}}}$$

has a t-distribution with $n - 2$ degrees of freedom. It follows that, if the null hypothesis $H_0: \beta_1 = 0$ is true, then the population of all possible values of the test statistic

$$t = \frac{b_1}{s\sqrt{c_{11}}}$$

has a t-distribution with $n - 2$ degrees of freedom. Therefore using the rejection points $-t_{[\alpha/2]}^{(n-2)}$ and $t_{[\alpha/2]}^{(n-2)}$ says that, if $H_0: \beta_1 = 0$ is true, then as illustrated in

FIGURE 5.3 **The rejection points for testing $H_0: \beta_1 = 0$ versus $H_1: \beta_1 \neq 0$. The probability that we reject $H_0: \beta_1 = 0$ when $H_0: \beta_1 = 0$ is true equals $\alpha/2 + \alpha/2 = \alpha$.**

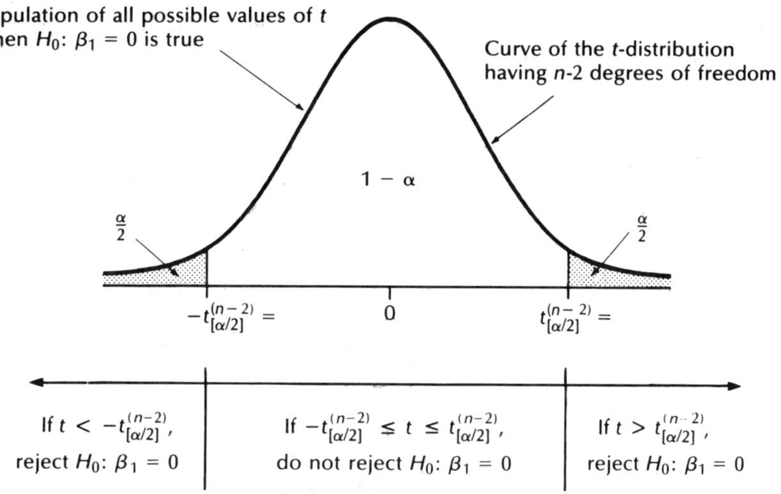

Figure 5.3:

1. The probability that

   $$-t_{[\alpha/2]}^{(n-2)} \leq t \leq t_{[\alpha/2]}^{(n-2)}$$

   is $1 - \alpha$. That is, $100(1 - \alpha)\%$—for example, $100(1 - .05)\% = 95\%$—of all possible values of t are between $-t_{[\alpha/2]}^{(n-2)}$ and $t_{[\alpha/2]}^{(n-2)}$ and thus lead us to not reject $H_0: \beta_1 = 0$ (when $H_0: \beta_1 = 0$ is true), a correct decision.
2. The probability that

   $$t < -t_{[\alpha/2]}^{(n-2)} \qquad \text{or} \qquad t > t_{[\alpha/2]}^{(n-2)}$$

   is α. That is, $100(\alpha)\%$—for example, $100(.05)\% = 5\%$—of all possible values of t are either less than $-t_{[\alpha/2]}^{(n-2)}$ or greater than $t_{[\alpha/2]}^{(n-2)}$ and lead us to reject $H_0: \beta_1 = 0$ (when $H_0: \beta_1 = 0$ is true), a Type I error.

Therefore using the rejection points $-t_{[\alpha/2]}^{(n-2)}$ and $t_{[\alpha/2]}^{(n-2)}$ implies that the probability of a Type I error equals α.

Example 5.4 Consider the fuel consumption model

$$\begin{aligned} y &= \mu + \varepsilon \\ &= \beta_0 + \beta_1 x + \varepsilon \end{aligned}$$

It follows from the rejection point procedure that we can reject $H_0: \beta_1 = 0$ in favor of $H_1: \beta_1 \neq 0$ by setting α, the probability a Type I error, equal to .05 if and only if the absolute value of

$$t = \frac{b_1}{s\sqrt{c_{11}}}$$

is greater than $t_{[\alpha/2]}^{(n-2)} = t_{[.05/2]}^{(8-2)} = t_{[.025]}^{(6)} = 2.447$. That is, we reject H_0 if t is greater than $t_{[.025]}^{(6)} = 2.447$ or less than $-t_{[.025]}^{(6)} = -2.447$.

Noting from Table 4.5 and Example 4.7 that $b_1 = -.1279$ and $s = .6542$, we can compute

$$c_{11} = \frac{1}{\sum_{i=1}^{8}(x_i - \bar{x})^2} = .000712$$

and

$$t = \frac{b_1}{s\sqrt{c_{11}}} = \frac{-.1279}{.6542\sqrt{.000712}} = -7.328$$

Since $t = -7.328 < -2.447 = -t_{[.025]}^{(6)}$ (that is, $|t| = 7.328 > 2.447 = t_{[.025]}^{(6)}$), we can reject $H_0: \beta_1 = 0$ by setting α equal to .05.

The rejection point procedure also says that we can reject $H_0: \beta_1 = 0$ in favor of $H_1: \beta_1 \neq 0$ by setting $\alpha = .01$ if and only if the absolute value of t is greater than $t_{[\alpha/2]}^{(n-2)} = t_{[.01/2]}^{(6)} = t_{[.005]}^{(6)} = 3.707$, that is, if $t > 3.707$ or $t < -3.707$. Specifically, since $t = -7.328 < -3.707$, we can reject $H_0: \beta_1 = 0$ by setting α equal to .01.

Note that the lower we set α, the smaller is the probability that we will reject $H_0: \beta_1 = 0$ when $H_0: \beta_1 = 0$ is true. That is, the lower we set α, the smaller is the probability that we will conclude that the independent variable x is significantly related to the dependent variable y when x is not significantly related to y. Therefore the lower the value of α at which $H_0: \beta_1 = 0$ can be rejected, the stronger is the evidence that x really is significantly related to y. Practice has shown that

1. If we can reject $H_0: \beta_1 = 0$ in favor of $H_1: \beta_1 \neq 0$ by setting α *equal to .05*, we have *strong evidence* that x is significantly related to y.
2. If we can reject $H_0: \beta_1 = 0$ in favor of $H_1: \beta_1 \neq 0$ by setting α *equal to .01*, we have *very strong evidence* that x is significantly related to y.

For example, for the fuel consumption model,

$$y = \beta_0 + \beta_1 x + \varepsilon$$

we have seen in Example 5.4 that we can reject $H_0: \beta_1 = 0$ in favor of $H_1: \beta_1 \neq 0$ by setting α equal to .01. Therefore we have very strong evidence that x is significantly related to y.

In addition to using the above hypothesis test to decide whether an independent variable x is significantly related to a dependent variable y, we should also use *prior belief* about the significance of the independent variable x. In particular, suppose that we believe on theoretical grounds that x is significantly related to y. For instance, in the fuel consumption problem we would believe that average hourly temperature during the week is significantly related to weekly fuel consumption. We therefore believe that $H_0: \beta_1 = 0$ is false, which implies that we do not believe that we will make a Type I error. Thus we might allow a higher probability of a Type I error (say an α of .05 or even higher) to reduce the probability of a Type II error.

As another example, suppose that we are firmly convinced on theoretical grounds that x is significantly related to y. Suppose, moreover, that we can reject $H_0: \beta_1 = 0$ in favor of $H_1: \beta_1 \neq 0$ by setting α equal to .10 but not by setting α equal to .05. Then, deciding that x is not significantly related to y would contradict both our prior belief and the statistical evidence (which at least mildly supports our prior belief). Thus we should seriously consider concluding that x is significantly related to y. On the other hand, suppose that there are no strong theoretical grounds for believing that x is significantly related to y—x is just being tried out to see whether it might be significantly related to y. Then the fact that we can reject $H_0: \beta_1 = 0$ by setting α equal to .10 would probably not be enough statistical evidence to lead us to conclude that x is significantly related to y. In general, if prior belief does not indicate that the independent variable x is significantly related to the dependent variable y, which implies that we believe that $H_0: \beta_1 = 0$ may be true, then we wish to guard against making a Type I error. Therefore we might set α equal to .01 or even a smaller value.

USING PROB-VALUES

We next introduce a *probability-value* (or *prob-value*) condition for deciding whether to reject $H_0: \beta_1 = 0$ in favor of $H_1: \beta_1 \neq 0$ by setting the probability of a Type I error equal to α.

Testing $H_0: \beta_1 = 0$ Versus $H_1: \beta_1 \neq 0$ by Using Prob-Values

Define the *prob-value* to be twice the area under the curve of the t-distribution having $n - 2$ degrees of freedom to the right of the absolute value of

$$t = \frac{b_1}{s\sqrt{c_{11}}}$$

If the inference assumptions are satisfied, we can reject $H_0: \beta_1 = 0$ in favor of $H_1: \beta_1 \neq 0$ by setting the probability of a Type I error equal to α if and only if

$$\text{prob-value} < \alpha$$

Although we have initially defined the prob-value in terms of an area under a curve, we will soon show that the prob-value is a *probability* that has an important interpretation. First, however, we note that the above prob-value condition for rejecting $H_0: \beta_1 = 0$ is equivalent to the previously discussed rejection point condition. To see this, suppose that the prob-value is less than a specified value of α. Then,

$$\begin{bmatrix} \text{the area under the curve of the} \\ t\text{-distribution having } n - 2 \text{ degrees} \\ \text{of freedom to the right of } |t| \end{bmatrix} < \alpha/2$$

This implies (as can be seen from Figure 5.4) that $|t| > t_{[\alpha/2]}^{(n-2)}$. Therefore we can reject $H_0: \beta_1 = 0$ by setting the probability of a Type I error equal to α.

Comparing the rejection point condition and the prob-value condition, first note that (if we are doing hand calculations) the rejection point condition is simpler from a computational standpoint. This is because it requires only that we calculate the test statistic t and look up the value $t_{[\alpha/2]}^{(n-2)}$ in a t-table. The prob-value condition is more complicated from a computational standpoint—it requires that we calculate an area under the t-curve. However, this problem is not serious, since most regression computer packages (and some hand-held calculators) can be used to calculate areas under t-curves and prob-values. Furthermore, the prob-value condition has an advantage over the rejection point method. Suppose that several hypothesis testers all wish to use different values of α in testing $H_0: \beta_1 = 0$. Then, if the rejection point condition were used, each hypothesis tester would have to look up a different rejection point $t_{[\alpha/2]}^{(n-2)}$. However, if the prob-value condition is

FIGURE 5.4 **The equivalence of the rejection point and prob-value conditions for rejecting $H_0: \beta_1 = 0$**

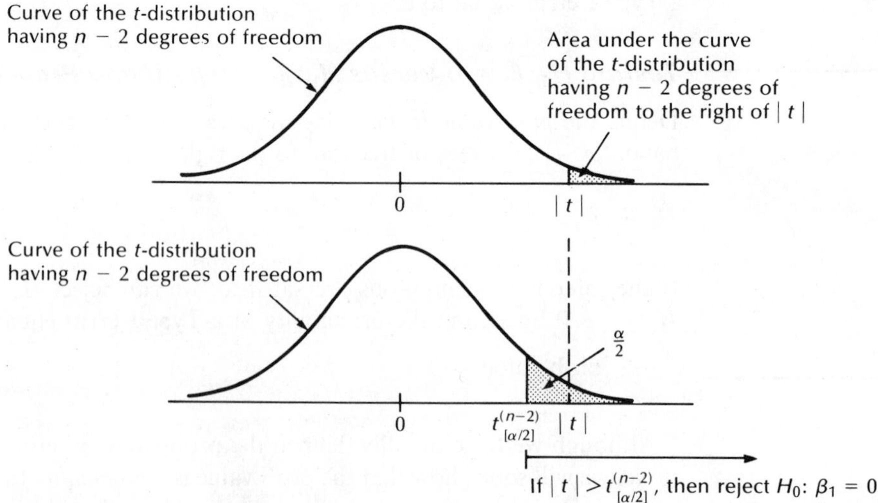

used, only the prob-value needs to be calculated, and each hypothesis tester knows that if the prob-value is less than his or her particular chosen value of α, then H_0 should be rejected. Another advantage of the prob-value is that it may be interpreted as a measure of probabilistic doubt about the validity of H_0. This is illustrated in the following example.

Example 5.5 Consider the fuel consumption regression model

$$y = \beta_0 + \beta_1 x + \varepsilon$$

Recall from Example 5.4 that the t statistic for testing $H_0: \beta_1 = 0$ versus $H_1: \beta_1 \neq 0$ is

$$t = \frac{b_1}{s\sqrt{c_{11}}} = \frac{-.1279}{.6542\sqrt{.000712}} = -7.328$$

It follows that the prob-value for testing $H_0: \beta_1 = 0$ versus $H_1: \beta_1 \neq 0$ is twice the area under the curve of the t-distribution having $n - 2 = 8 - 2 = 6$ degrees of freedom to the right of $|t| = |-7.328| = 7.328$. This prob-value can be computer calculated to be .00033. Since this prob-value is less than .001, and not less than .0001, it follows by the prob-value condition that we can reject $H_0: \beta_1 = 0$ in favor of $H_1: \beta_1 \neq 0$ by setting α equal to .05, .01, or .001 but not by setting α equal to .0001. The fact that we can reject $H_0: \beta_1 = 0$ by setting α equal to .001 should be regarded as extremely strong evidence that x is significantly related to y.

In addition to its use as a decision rule the prob-value can be interpreted as a probability. If $H_0: \beta_1 = 0$ is true, then the population of all possible values of the test statistic

$$t = \frac{b_1}{s\sqrt{c_{11}}}$$

has a t-distribution with $n - 2 = 6$ degrees of freedom. Furthermore, *twice* the area under the curve of the t-distribution having $n - 2$ degrees of freedom to the right of $|t|$ is (by the symmetry of the t-distribution) equal to the area to the right of $|t|$ plus the area to the left of $-|t|$. Therefore the prob-value equals the sum of these latter two areas. This says that the prob-value equals the proportion of all possible values of t that, if $H_0: \beta_1 = 0$ is true, are at least as far away from zero and thus at least as contradictory to $H_0: \beta_1 = 0$ as the value of t that we have actually observed ($t = -7.328$). The prob-value of .00033 leads us to reach one of two possible conclusions. The first conclusion is that $H_0: \beta_1 = 0$ is true, and we have observed a value of t that is so rare that only .00033 (that is, 33 in 100,000) of all possible values of t are at least at contradictory to $H_0: \beta_1 = 0$. The second conclusion is that $H_0: \beta_1 = 0$ is false and $H_1: \beta_1 \neq 0$ is true. A reasonable person would probably make the second conclusion. That is, the fact that the prob-value of .00033 is so small casts great doubt on the validity of the null hypothesis $H_0: \beta_1 = 0$ and thus lends great support to the validity of the alternative hypothesis $H_1: \beta_1 \neq 0$.

We can summarize our discussion of the prob-value for testing $H_0 : \beta_1 = 0$ as follows:

1. If the *prob-value* is *less than .05*, we can reject $H_0 : \beta_1 = 0$ in favor of $H_1 : \beta_1 \neq 0$ by setting α equal to .05, which means that we have *strong evidence* that x is significantly related to y.
2. If the *prob-value* is *less than .01*, we can reject $H_0 : \beta_1 = 0$ in favor of $H_1 : \beta_1 \neq 0$ by setting α equal to .01, which means that we have *very strong evidence* that x is significantly related to y.
3. The smaller the prob-value is, the stronger is the evidence that x is significantly related to y.

Reporting the prob-value as the result of a statistical test (rather than merely whether we "reject H_0" or "do not reject H_0" on the basis of setting the probability of a Type I error equal to α) is becoming increasingly popular. This is because the result of the statistical test may be used by several people to make different decisions at different times. Thus by reporting the prob-value the individual performing the statistical test permits any other potential decision makers to reach their own conclusions.

Finally, we note that the $100(1 - \alpha)\%$ confidence interval for β_1 (presented in Section 5.2) can be used to test $H_0 : \beta_1 = 0$. Specifically, we can reject $H_0 : \beta_1 = 0$ by setting the probability of a Type I error equal to α if and only if the $100(1 - \alpha)\%$ confidence interval for β_1 does not contain zero. This condition is intuitive and can easily be shown to be mathematically equivalent to the rejection point and prob-value conditions for rejecting $H_0 : \beta_1 = 0$.

TESTING ONE-SIDED ALTERNATIVE HYPOTHESES

We conclude this section by recalling that we sometimes wish to test a null hypothesis versus a one-sided alternative hypothesis. For example, considering the fuel consumption model

$$y = \beta_0 + \beta_1 x + \varepsilon$$

we have previously reasoned that we might be interested in testing $H_0 : \beta_1 = 0$ versus $H_1 : \beta_1 < 0$.

Testing $H_0 : \beta_1 = 0$ Versus One-Sided Alternatives

To test $H_0 : \beta_1 = 0$, define the test statistic

$$t = \frac{b_1}{s\sqrt{c_{11}}}$$

Then, if the inference assumptions are satisfied,

1. We can reject $H_0: \beta_1 = 0$ in favor of $H_1: \beta_1 > 0$ by setting the probability of a Type I error equal to α if and only if $t > t_{[\alpha]}^{(n-2)}$ or, equivalently, if prob-value $< \alpha$. Here, the prob-value is defined to be the area under the curve of the t-distribution having $n - 2$ degrees of freedom to the right of t.
2. We can reject $H_0: \beta_1 = 0$ in favor of $H_1: \beta_1 < 0$ by setting the probability of a Type I error equal to α if and only if $t < -t_{[\alpha]}^{(n-2)}$ or, equivalently, if prob-value $< \alpha$. Here, the prob-value is defined to be the area under the curve of the t-distribution having $n - 2$ degrees of freedom to the left of t.

5.4

STATISTICAL INFERENCE FOR THE INTERCEPT

In this section we explain how to carry out statistical inferences for the intercept β_0 in the simple linear regression model.

A Confidence Interval for β_0

1. The *population of all possible least squares point estimates of* β_0
   a. Has *mean* $\mu_{b_0} = \beta_0$.
   b. Has *variance* $\sigma_{b_0}^2 = \sigma^2 c_{00}$ (if inference assumptions 1 and 2 hold).
   c. Has *standard deviation* $\sigma_{b_0} = \sigma\sqrt{c_{00}}$ (if inference assumptions 1 and 2 hold).
   d. Has a *normal distribution* (if inference assumptions 1, 2, and 3 hold).
2. A point estimate of $\sigma_{b_0} = \sigma\sqrt{c_{00}}$ is $s_{b_0} = s\sqrt{c_{00}}$, which is called the *standard error of the estimate* b_0.
3. If the inference assumptions hold, the population of all possible values of

   $$\frac{b_0 - \beta_0}{s_{b_0}} = \frac{b_0 - \beta_0}{s\sqrt{c_{00}}}$$

   has a *t-distribution with $n - 2$ degrees of freedom.*
4. If the inference assumptions hold, a *100(1 − α)% confidence interval for* β_0 is

   $$\left[b_0 \pm t_{[\alpha/2]}^{(n-2)} s\sqrt{c_{00}}\right] = \left[b_0 \pm t_{[\alpha/2]}^{(n-2)} s \sqrt{\frac{1}{n} + \frac{\bar{x}^2}{\sum_{i=1}^{n}(x_i - \bar{x})^2}}\right]$$

Testing $H_0: \beta_0 = 0$ Versus $H_1: \beta_0 \neq 0$

Define the test statistic

$$t = \frac{b_0}{s\sqrt{c_{00}}}$$

Also, define the prob-value to be twice the area under the curve of the t-distribution having $n - 2$ degrees of freedom to the right of $|t|$. If the inference assumptions are satisfied, we can reject $H_0: \beta_0 = 0$ in favor of $H_1: \beta_0 \neq 0$ by setting the probability of a Type I error equal to α if and only if either of the following equivalent conditions holds:

1. $|t| > t_{[\alpha/2]}^{(n-2)}$, that is, if $t > t_{[\alpha/2]}^{(n-2)}$ or $t < -t_{[\alpha/2]}^{(n-2)}$.
2. Prob-value $< \alpha$.

Example 5.6 Consider the fuel consumption model

$$\begin{aligned} y &= \mu + \varepsilon \\ &= \beta_0 + \beta_1 x + \varepsilon \end{aligned}$$

We saw in Table 4.5 and Example 4.7 that $b_0 = 15.84$ and $s = .6542$, and we can use the information in Table 4.5 to calculate c_{00} to be

$$\begin{aligned} c_{00} &= \frac{1}{n} + \frac{\bar{x}^2}{\sum_{i=1}^{n}(x_i - \bar{x})^2} = \frac{1}{8} + \frac{\bar{x}^2}{\sum_{i=1}^{8} x_i^2 - 8\bar{x}^2} \\ &= \frac{1}{8} + \frac{(43.98)^2}{16{,}874.76 - 8(43.98)^2} = 1.5020 \end{aligned}$$

It follows that

$$t = \frac{b_0}{s\sqrt{c_{00}}} = \frac{15.84}{.6542\sqrt{1.5020}} = 19.7535$$

Furthermore, the prob-value, which is twice the area under the curve of the t-distribution having $n - 2 = 8 - 2 = 6$ degrees of freedom to the right of $|t| = 19.7535$, can be computer calculated to be .000001. Since the prob-value is less than .01, we can reject $H_0: \beta_0 = 0$ by setting α equal to .05 or .01. Therefore we have very strong evidence that the intercept β_0 does not equal zero and thus should be included in the above model.

In general, if we cannot reject $H_0 : \beta_0 = 0$ in favor of $H_1 : \beta_0 \neq 0$ by setting α equal to a low value (say, .05), it might seem reasonable to conclude that β_0 should not be included in the simple linear regression model. However, since

$$\mu = \beta_0 + \beta_1 x$$

we see that β_0 equals the mean value of the dependent variable when the independent variable x equals zero. If, logically speaking, μ would not equal zero when x equals zero (for example, in the fuel consumption problem, mean fuel consumption would not equal zero when the average hourly temperature equals zero), it follows that β_0 would not equal zero. In such a situation it is common practice to include the intercept β_0 even if we cannot reject $H_0 : \beta_0 = 0$ by setting α equal to a low value. In fact, experience indicates that it is definitely safest, when in doubt, to include the intercept β_0. However, if logic and statistical evidence (for example, a data plot or the t statistic) indicate that μ would equal zero when x equals zero, it might well be reasonable to exclude β_0 and thus use the model

$$\begin{aligned} y &= \mu + \varepsilon \\ &= \beta_1 x + \varepsilon \end{aligned}$$

Since this model assumes that μ equals zero when x equals zero, when we use this model, we say that we are performing a "regression analysis through the origin" (see Section 5.11).

5.5

A CONFIDENCE INTERVAL FOR A MEAN VALUE OF THE DEPENDENT VARIABLE

Since the point estimate $\hat{y}_0$ of μ_0, the mean value of the dependent variable when the value of the independent variable x is x_0, is not likely to be equal to μ_0, we are motivated to compute a confidence interval for μ_0. Recall that each possible sample of n values of the dependent variable gives values of b_0 and b_1 that are different from the values given by other samples. It follows that, if x_0 is a specified value of the independent variable x, then each possible sample gives a point estimate

$$\hat{y}_0 = b_0 + b_1 x_0$$

of the mean value of the dependent variable

$$\mu_0 = \beta_0 + \beta_1 x_0$$

that is different from the point estimates given by other samples. We now state the logic behind (and formula for) a $100(1 - \alpha)\%$ confidence interval for μ_0.

A Confidence Interval for μ_0

1. The *population of all possible point estimates of* μ_0
   a. Has *mean* $\mu_{\hat{y}_0} = \mu_0$.
   b. Has *variance* $\sigma^2_{\hat{y}_0} = \sigma^2 h_{00}$ where

$$h_{00} = \frac{1}{n} + \frac{(x_0 - \bar{x})^2}{\sum_{i=1}^{n} (x_i - \bar{x})^2}$$

   (if inference assumptions 1 and 2 hold).
   c. Has *standard deviation* $\sigma_{\hat{y}_0} = \sigma\sqrt{h_{00}}$ (if inference assumptions 1 and 2 hold).
   d. Has a *normal distribution* (if inference assumptions 1, 2, and 3 hold).
2. A point estimate of $\sigma_{\hat{y}_0} = \sigma\sqrt{h_{00}}$ is $s_{\hat{y}_0} = s\sqrt{h_{00}}$, which is called the *standard error of the estimate* $\hat{y}_0$.
3. If the inference assumptions hold, the population of all possible values of

$$\frac{\hat{y}_0 - \mu_0}{s_{\hat{y}_0}} = \frac{\hat{y}_0 - \mu_0}{s\sqrt{h_{00}}}$$

   has a *t-distribution with n − 2 degrees of freedom.*
4. If the inference assumptions hold, a *100(1 − α)% confidence interval for* μ_0 is

$$[\hat{y}_0 \pm t^{(n-2)}_{[\alpha/2]} s\sqrt{h_{00}}] = \left[\hat{y}_0 \pm t^{(n-2)}_{[\alpha/2]} s \sqrt{\frac{1}{n} + \frac{(x_0 - \bar{x})^2}{\sum_{i=1}^{n} (x_i - \bar{x})^2}}\right]$$

The derivation of the above interval (and of the prediction interval for y_0 to be presented in the next section) is analogous to the derivation of the confidence interval for β_1 given in Section 5.2.

Before considering an example, note that

$$h_{00} = \frac{1}{n} + \frac{(x_0 - \bar{x})^2}{\sum_{i=1}^{n} (x_i - \bar{x})^2}$$

is called a *distance value*. It is so named because this quantity is a measure of the distance between the value x_0 of the independent variable and $\bar{x}$, the average of the previously observed values of the independent variable. The farther x_0 is from $\bar{x}$, which represents the center of the experimental region, the greater is the distance value h_{00}, and thus the longer is the $100(1 - \alpha)\%$ confidence interval for μ_0.

Example 5.7 In the fuel consumption problem, recall that a weather forecasting service has told us that the average hourly temperature in the future week will be $x_0 = 40$. We saw in Example 4.5 that

$$\begin{aligned}\hat{y}_0 &= b_0 + b_1 x_0 \\ &= 15.84 - .1279(40) \\ &= 10.72 \quad \text{tons of coal}\end{aligned}$$

is the point estimate of the future mean fuel consumption

$$\begin{aligned}\mu_0 &= \beta_0 + \beta_1 x_0 \\ &= \beta_0 + \beta_1(40)\end{aligned}$$

Moreover, we can use the information in Table 4.5 to calculate the distance value h_{00} to be

$$\begin{aligned}h_{00} &= \frac{1}{n} + \frac{(x_0 - \bar{x})^2}{\sum_{i=1}^{n}(x_i - \bar{x})^2} = \frac{1}{8} + \frac{(40 - 43.98)^2}{\sum_{i=1}^{8} x_i^2 - 8\bar{x}^2} \\ &= \frac{1}{8} + \frac{(-3.98)^2}{16{,}874.76 - 8(43.98)^2} = .1362\end{aligned}$$

Therefore since we recall from Example 4.7 that the standard error s is .6542, it follows that a 95% confidence interval for μ_0 is

$$\begin{aligned}[\hat{y}_0 \pm t_{[\alpha/2]}^{(n-2)} s\sqrt{h_{00}}] &= [10.72 \pm t_{[.025]}^{(8-2)}(.6542)\sqrt{.1362}] \\ &= [10.72 \pm 2.447(.6542)\sqrt{.1362}] \\ &= [10.72 \pm .59] \\ &= [10.13, 11.31]\end{aligned}$$

Thus we are 95% confident that μ_0, the future mean fuel consumption, is between 10.13 tons of coal and 11.31 tons of coal.

Here it is important to note that this 95% confidence interval for μ_0 is not meant to contain and thus cannot be expected to contain $y_0 = \mu_0 + \varepsilon_0$, the actual amount of fuel that will be used to heat the nine-building complex during the future week, when the average hourly temperature will be $x_0 = 40$. An interval meant to contain y_0 is called a *prediction* interval and is discussed in the next section.

To understand the meaning of 95% confidence, we assume that (a supernatural power knows that) the true values of β_0 and β_1 are 15.77 and $-.1281$, respectively. This implies that the true value of μ_0 is

$$\begin{aligned}\mu_0 &= \beta_0 + \beta_1 x_0 \\ &= 15.77 - .1281(40) \\ &= 10.65\end{aligned}$$

In Table 5.4 we calculate the point estimate of μ_0 and the 95% confidence interval for μ_0 given by each of the samples in Table 5.1. Note that two of the three 95% confidence intervals for $\mu_0 = 10.65$ in Table 5.4 contain μ_0. When we compute the

TABLE 5.4 The point estimate of μ_0, $\hat{y}_0 = b_0 + b_1(40)$, and the 95% confidence interval for μ_0, $[\hat{y}_0 \pm 2.447s\sqrt{.1362}]$, given by each of the three samples in Table 5.1

Sample 1	Sample 2	Sample 3
$\hat{y}_0 = 15.84 - .1279(40) = 10.72$ $[10.72 \pm 2.447(.6542)\sqrt{.1362}]$ $= [10.13, 11.31]$	$\hat{y}_0 = 15.85 - .1285(40) = 10.71$ $[10.71 \pm 2.447(.686)\sqrt{.1362}]$ $= [10.09, 11.33]$	$\hat{y}_0 = 12.44 - .0666(40) = 9.78$ $[9.78 \pm 2.447(.2582)\sqrt{.1362}]$ $= [9.54, 10.01]$

95% confidence interval for μ_0 by using the observed sample (which is the first sample in Tables 5.1 and 5.4) and obtain the interval [10.13, 11.31], we can be 95% confident that this interval contains μ_0. This is because 95 percent of the confidence intervals in the population of all possible 95% confidence intervals for μ_0 contain μ_0 and because we have obtained one of the confidence intervals in this population.

5.6

A PREDICTION INTERVAL FOR AN INDIVIDUAL VALUE OF THE DEPENDENT VARIABLE

THE INTERVAL AND ITS MEANING

Since the point prediction $\hat{y}_0$ of y_0, an individual value of the dependent variable when the value of the independent variable x is x_0, is not likely to be equal to y_0, we are motivated to compute an interval for y_0. Such an interval is called a *prediction interval.* Recall that each possible sample of n values of the dependent variable gives values of b_0 and b_1 that are different from the values given by other samples. It follows that, if x_0 is a specified value of the independent variable x, then each possible sample gives a point prediction

$$\hat{y}_0 = b_0 + b_1 x_0$$

of the individual value of the dependent variable

$$\begin{aligned} y_0 &= \mu_0 + \varepsilon_0 \\ &= \beta_0 + \beta_1 x_0 + \varepsilon_0 \end{aligned}$$

that is different from the point predictions given by other samples. Now consider the *prediction error*

$$y_0 - \hat{y}_0$$

After observing each possible sample and calculating the point prediction based on that sample, we could observe any one of an infinite number of different possible individual values of the dependent variable (because of different possible error terms). Therefore there are an infinite number of different prediction errors that could be observed.

Below we summarize the logic behind (and formula for) a $100(1 - \alpha)\%$ prediction interval for y_0.

A Prediction Interval for y_0

1. The *population of all possible prediction errors*
   a. Has *mean* $\mu_{(y_0 - \hat{y}_0)} = 0$.
   b. Has *variance* $\sigma^2_{(y_0 - \hat{y}_0)} = \sigma^2(1 + h_{00})$ where

$$h_{00} = \frac{1}{n} + \frac{(x_0 - \bar{x})^2}{\sum_{i=1}^{n} (x_i - \bar{x})^2}$$

   (if inference assumptions 1 and 2 hold).
   c. Has *standard deviation* $\sigma_{(y_0 - \hat{y}_0)} = \sigma\sqrt{1 + h_{00}}$ (if inference assumptions 1 and 2 hold).
   d. Has a *normal distribution* (if inference assumptions 1, 2, and 3 hold).
2. A point estimate of $\sigma_{(y_0 - \hat{y}_0)} = \sigma\sqrt{1 + h_{00}}$ is $s_{(y_0 - \hat{y}_0)} = s\sqrt{1 + h_{00}}$, which is called the *standard error of the prediction error* $(y_0 - \hat{y}_0)$.
3. If the inference assumptions hold, the population of all possible values of

$$\frac{y_0 - \hat{y}_0}{s_{(y_0 - \hat{y}_0)}} = \frac{y_0 - \hat{y}_0}{s\sqrt{1 + h_{00}}}$$

   has a *t-distribution with $n - 2$ degrees of freedom.*
4. If the inference assumptions hold, a *100(1 − α)% prediction interval for* y_0 is

$$[\hat{y}_0 \pm t^{(n-2)}_{[\alpha/2]} s\sqrt{1 + h_{00}}] = \left[\hat{y}_0 \pm t^{(n-2)}_{[\alpha/2]} s \sqrt{1 + \frac{1}{n} + \frac{(x_0 - \bar{x})^2}{\sum_{i=1}^{n} (x_i - \bar{x})^2}}\right]$$

Example 5.8 In the fuel consumption problem we saw in Example 4.5 that when $x_0 = 40$,

$$\begin{aligned} \hat{y}_0 &= b_0 + b_1 x_0 \\ &= 15.84 - .1279(40) \\ &= 10.72 \quad \text{tons of coal} \end{aligned}$$

is the point prediction of the future individual fuel consumption

$$\begin{aligned} y_0 &= \mu_0 + \varepsilon_0 \\ &= \beta_0 + \beta_1 x_0 + \varepsilon_0 \\ &= \beta_0 + \beta_1(40) + \varepsilon_0 \end{aligned}$$

Therefore since we recall from Example 5.7 that $h_{00} = .1362$ and $s = .6542$, it follows that a 95% prediction interval for y_0 is

$$\begin{aligned} [\hat{y}_0 \pm t_{[\alpha/2]}^{(n-2)} s\sqrt{1 + h_{00}}] &= [10.72 \pm t_{[.025]}^{(8-2)}(.6542)\sqrt{1.1362}] \\ &= [10.72 \pm 2.447(.6542)\sqrt{1.1362}] \\ &= [10.72 \pm 1.71] \\ &= [9.01, 12.43] \end{aligned}$$

This interval says that we are 95% confident that y_0, the future individual fuel consumption, will be between 9.01 tons of coal and 12.43 tons of coal. The upper bound of this interval is important because it says that if we plan to acquire 12.43 tons of coal in the future week, then we can be very confident that we will have enough fuel to adequately heat the nine-building complex (since we are very confident that y_0 will be no greater than 12.43 tons of coal). As will be illustrated in subsequent examples, there are many regression problems in which both the upper and lower bounds of a prediction interval for y_0 are important. However, in this problem the lower bound is probably not very important.

As illustrated in Figure 5.5, both the 95% confidence interval for μ_0 (calculated in Example 5.7) and the 95% prediction interval for $y_0 = \mu_0 + \varepsilon_0$ are centered at $\hat{y}_0$. However, the prediction interval is longer than the confidence interval. This is because the prediction interval utilizes an "extra 1 under the radical." This extra 1 accounts for the added uncertainty with respect to our not knowing the value of the future error term ε_0. Suppose that (a supernatural power knows that) the future mean fuel consumption μ_0 equals 10.65 tons of coal and the future individual fuel consumption will be $y_0 = \mu_0 + \varepsilon_0 = 10.65 + 1.05 = 11.7$ tons of coal. As illustrated in Figure 5.5, the 95% confidence interval for μ_0 contains μ_0. However, this interval is not long enough to contain $y_0 = \mu_0 + \varepsilon_0$; remember, it is not meant to contain y_0. However, the 95% prediction interval for y_0 is long enough and does contain y_0. Of course, the relative positions of μ_0, y_0, and $\hat{y}_0$ will be different in different situations. However, Figure 5.5 emphasizes that we must include the extra 1 under the radical when calculating a prediction interval for y_0.

To understand the meaning of 95% confidence, we calculate in Table 5.5 the point prediction of $y_0 = \mu_0 + \varepsilon_0$ and the 95% prediction interval for y_0 given by each of the samples in Table 5.1. After using a sample to calculate a prediction interval for y_0 we will observe a future fuel consumption. In general, we call a prediction interval–future fuel consumption combination successful if the future fuel consumption that we observe (after having observed our sample) falls in the prediction interval that we have calculated by using our sample. We call this combination unsuccessful if the future fuel consumption does not fall in the prediction interval. Suppose that after calculating the prediction intervals in

FIGURE 5.5 A comparison of a confidence interval for μ_0 and a prediction interval for $y_0 = \mu_0 + \varepsilon_0$

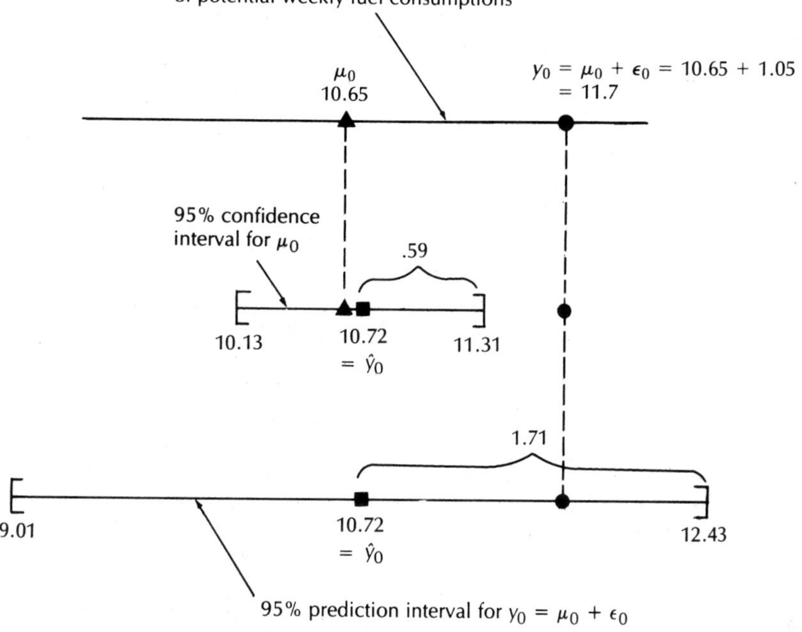

TABLE 5.5 The point prediction of y_0, $\hat{y}_0 = b_0 + b_1(40)$, and the 95% prediction interval for y_0, $[\hat{y}_0 \pm 2.447s\sqrt{1.1362}]$, given by each of the samples in Table 5.1

Sample 1	*Sample 2*	*Sample 3*
$\hat{y}_0 = 15.84 - .1279(40) = 10.72$	$\hat{y}_0 = 15.85 - .1285(40) = 10.71$	$\hat{y}_0 = 12.44 - .0666(40) = 9.78$
$[10.72 \pm 2.447(.6542)\sqrt{1.1362}]$	$[10.71 \pm 2.447(.686)\sqrt{1.1362}]$	$[9.78 \pm 2.447(.2582)\sqrt{1.1362}]$
$= [9.01, 12.43]$	$= [8.92, 12.50]$	$= [9.10, 10.45]$

Table 5.5 we observe the respective future fuel consumptions $y_0 = 11.7$, $y_0 = 9.9$, and $y_0 = 11.1$. We can see that the prediction interval–future fuel consumption combinations

$$\{[9.01, 12.43]; y_0 = 11.7\}$$

$$\{[8.92, 12.50]; y_0 = 9.9\}$$

are successful, while the prediction interval–future fuel consumption combination

$$\{[9.10, 10.45]; y_0 = 11.1\}$$

is unsuccessful. Thus two of the three prediction interval–future fuel consumption combinations are successful. Before we compute the 95% prediction interval for y_0 by using the observed sample (which is the first sample in Tables 5.1 and 5.5) and obtain the interval [9.01, 12.43], and before we observe y_0, we can be 95% confident that we will be successful (that is, obtain a 95% prediction interval such that y_0 falls in our interval). This is because 95% of the prediction interval–future fuel consumption combinations in the population of all such combinations are successful and because we know that we will obtain one prediction interval–future fuel consumption combination in this population.

Example 5.9 Consider the American Manufacturing sales volume problem. Recall from Example 4.6 that

$$\begin{aligned}\hat{y}_0 &= b_0 + b_1 x_0 \\ &= 66.2121 + 4.4303(10) \\ &= 110.5 \qquad \text{(that is, \$1,105,000)}\end{aligned}$$

is the point estimate of

$$\begin{aligned}\mu_0 &= \beta_0 + \beta_1 x_0 \\ &= \beta_0 + \beta_1(10)\end{aligned}$$

the mean of all of the third quarter sales volumes that could have been observed in a sales region when $100,000 was spent on advertising. Also remember that $\hat{y}_0$ is the point prediction of

$$\begin{aligned}y_0 &= \mu_0 + \varepsilon_0 \\ &= \beta_0 + \beta_1(10) + \varepsilon_0\end{aligned}$$

the individual third quarter sales volume in a sales region when $100,000 was spent on advertising. Furthermore, we can use the information in Table 4.2 to calculate the distance value h_{00} to be

$$h_{00} = \frac{1}{n} + \frac{(x_0 - \bar{x})^2}{\sum_{i=1}^{n}(x_i - \bar{x})^2} = \frac{1}{10} + \frac{(10 - 9.5)^2}{82.50825} = .10303$$

Therefore since we recall from Example 5.2 that $s = 5.2776$, it follows that a 95% confidence interval for μ_0 is

$$\begin{aligned}[\hat{y}_0 \pm t_{[.025]}^{(10-2)} s\sqrt{h_{00}}] &= [110.5 \pm 2.306(5.2776)\sqrt{.10303}] \\ &= [106.6, 114.4]\end{aligned}$$

that is, [\$1,066,000, \$1,144,000]. Furthermore, a 95% prediction interval for y_0 is

$$[\hat{y}_0 \pm t_{[.025]}^{(10-2)} s\sqrt{1 + h_{00}}] = [110.5 \pm 2.306(5.2776)\sqrt{1.10303}]$$
$$= [97.7, 123.3]$$

that is, [\$977,000, \$1,233,000].

Recall that the data in Table 4.2, upon which this 95% prediction interval is based, comes from sales regions in which the sales forces use their advertising expenditure effectively. It follows that we are quite sure that the third quarter sales volume in a sales region in which \$100,000 was spent on advertising and in which there is an effective sales force should have been at least \$977,000. Therefore if the third quarter sales volume was \$890,800 in a sales region in which \$100,000 was spent on advertising and in which there is a questionable sales force, we have substantial evidence that the questionable sales force was not, in fact, performing effectively.

The 95% confidence interval for μ_0 and the 95% prediction interval for y_0 (when $x_0 = 10$) are illustrated in Figure 5.6. Also shown in this figure are the 95% confidence intervals for μ_0 and the 95% prediction intervals for y_0 corresponding to values of x_0 in the experimental region. Note that the farther x_0 is from $\bar{x} = 9.5$ (which represents the center of the experimental region), the greater is the distance value h_{00} and thus the longer are the 95% confidence interval for μ_0 and the 95% prediction interval for y_0.

FIGURE 5.6 **Sales volume 95% confidence and prediction intervals**

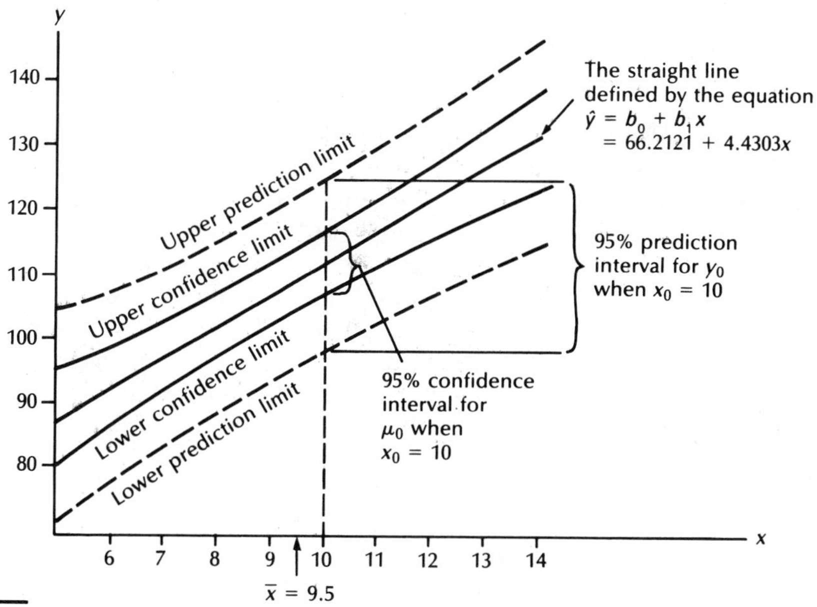

THE CASE WHEN x IS RANDOM

Before continuing, we note that all of our interpretations of "100(1 − α)% confidence" for confidence and prediction intervals refer to repeated sampling of y_i values when the x_i values remain the same from sample to sample. In relating sales to advertising expenditure it might be reasonable to keep the advertising expenditures (the x_i values) constant from sample to sample because we can control advertising expenditure. However, in relating fuel consumption to average hourly temperature the average hourly temperatures would not remain constant from sample to sample. This is because we cannot control the weather, and it is extremely unlikely that the eight observed temperatures would exactly repeat themselves. In this kind of situation, in which the x_i values are not controllable, we say that x is "random." In such a case it can be proven that all of the results in this chapter (and book) still hold if inference assumptions 1, 2, and 3 hold for $x_1, x_2, \ldots, x_n$ (the values of the independent variable that we have actually observed) and x_0. Furthermore, it is required that

1. The probability distribution of the potential x_i values that could be obtained for the ith observation is independent of the parameters β_0, β_1, and σ^2 (for $i = 1, 2, \ldots, n$, and 0).
2. The values $x_1, x_2, \ldots, x_n$, and x_0 are statistically independent.

In this case the previously discussed interpretation of "100(1 − α)% confidence" for confidence and prediction intervals refers to repeated sampling of pairs of (x_i, y_i) values, where the x_i values (for example, the temperatures) as well as the y_i values (for example, the fuel consumptions) vary from sample to sample.

5.7

AN EXAMPLE USING SAS

Example 5.10 Comp-U-Systems, a computer manufacturer, sells and services the Comp-U-Systems Microcomputer. As part of its standard purchase contract, Comp-U-Systems agrees to perform regular service on its microcomputer. To better schedule its service calls, Comp-U-Systems wishes to obtain information concerning the time it takes to perform the required service. To this end, Comp-U-Systems has collected data for 11 service calls. These data are presented in Table 5.6, which lists the number of microcomputers serviced (x) and the time (in minutes) required to perform the needed service (y) for each of these service calls. Here, for $i = 1, 2, \ldots, 11$:

x_i = the number of microcomputers serviced on the ith service call

y_i = the number of minutes required to perform service on the ith service call

TABLE 5.6 Comp-U-Systems service call data

Service call, i	Number of microcomputers serviced, x_i	Number of minutes required on call, y_i
1	4	109
2	2	58
3	5	138
4	7	189
5	1	37
6	3	82
7	4	103
8	5	134
9	2	68
10	4	112
11	6	154

FIGURE 5.7 Plot of Comp-U-Systems service call data

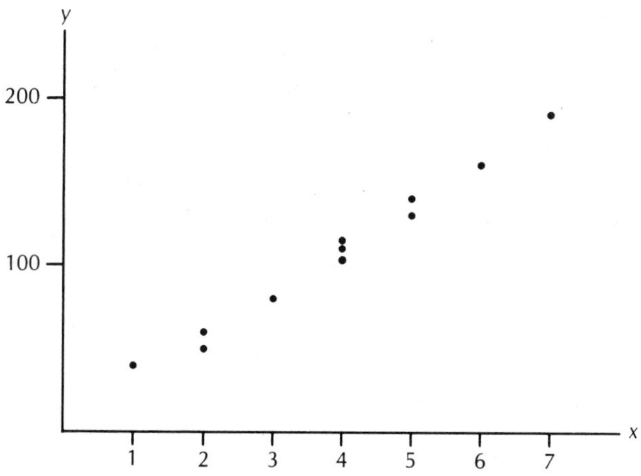

In Figure 5.7 we plot y_i versus x_i. We see that as the values of x increase, the observed values of y tend to increase in a straight line fashion. This implies that the simple linear regression model having a positive slope ($\beta_1 > 0$)

$$\begin{aligned} y &= \mu + \varepsilon \\ &= \beta_0 + \beta_1 x + \varepsilon \end{aligned}$$

might be an appropriate regression model relating y to x.

In Figure 5.8 we present the SAS output from using this model to perform a regression analysis of the $n = 11$ service times in Table 5.6. Note that some of the

FIGURE 5.8 **SAS output for a regression analysis of the Comp-U-Systems data**

DEP VARIABLE: Y

SOURCE	DF	SUM OF SQUARES	MEAN SQUARE	F VALUE	PROB>F
MODEL	1	19918.844[a]	19918.844[q]	935.149[r]	0.0001[s]
ERROR	9	191.702[b]	21.300184[f]		
C TOTAL	10	20110.545[c]			

Analysis of variance table

ROOT MSE	4.615212[d]	R-SQUARE	0.9905[h]
DEP MEAN	107.636	ADJ R-SQ	0.9894[i]
C.V.	4.287782		

VARIABLE	DF	PARAMETER ESTIMATE[e]	STANDARD ERROR[g]	T FOR HO: PARAMETER=0[j]	PROB > |T|[k]
INTERCEP	1	11.464088	3.439026	3.334	0.0087
X	1	24.602210	0.804514	30.580	0.0001

OBS	ACTUAL[l]	PREDICT VALUE	STD ERR PREDICT	LOWER95% MEAN	UPPER95% MEAN	LOWER95% PREDICT	UPPER95% PREDICT	RESIDUAL
1	37.000	36.066	2.723	29.907	42.226	23.944	40.188	0.933702
2	58.000	60.669	2.073	55.980	65.357	49.224	72.113	-2.669
3	68.000	60.669	2.073	55.980	65.357	49.224	72.113	7.331
4	82.000	85.271	1.572	81.715	88.827	74.241	96.300	-3.271
5	103.000	109.873	1.393	106.721	113.025	98.967	120.779	-6.873
6	109.000	109.873	1.393	106.721	113.025	98.967	120.779	-.872928
7	112.000	109.873	1.393	106.721	113.025	98.967	120.779	2.127
8	134.000	134.475	1.645	130.753	138.197	123.391	145.559	-.475138
9	138.000	134.475	1.645	130.753	138.197	123.391	145.559	3.525
10	154.000	159.077	2.183	154.139	164.016	147.528	170.627	-5.077
11	189.000	183.680	2.850	177.233	190.126	171.409	195.950	5.320
12	.	109.873[m]	1.393[n]	106.721	113.025	98.967	120.779	

95% confidence interval (LOWER95% MEAN, UPPER95% MEAN); *95% prediction interval* (LOWER95% PREDICT, UPPER95% PREDICT)

SUM OF RESIDUALS	-1.06581E-14
SUM OF SQUARED RESIDUALS	191.7017[p]

[a] *Explained variation* [b] *SSE* [c] *Total variation* [d] *Standard error s*
[e] b_0, b_1 [f] *Mean square error* s^2 [g] $s\sqrt{c_{00}}$, $s\sqrt{c_{11}}$ [h] r^2
[i] $\bar{r}^2$ [j] t [k] *Prob-value for t* [l] *Observed values = y*
[m] *Point prediction* $\hat{y}_0$ [n] $s\sqrt{h_{00}}$ [p] *SSE* [q] MS_{model}
[r] *F(model)* [s] *Prob-value for F(model)*

quantities on the SAS output—explained variation, total variation, analysis of variance table, r^2, and $\bar{r}^2$—will be explained in later sections. To carry out this analysis, we (arbitrarily) rearranged the data in Table 5.6 according to increasing observed service times. Noting that the important quantities are labeled on the output, we can summarize the quantities from the output that are used to test

$H_0: \beta_0 = 0$ and $H_0: \beta_1 = 0$:

$$b_0 = 11.4641 \quad \text{and} \quad b_1 = 24.6022$$

$$SSE = \sum_{i=1}^{11} (y_i - \hat{y}_i)^2 = 191.702$$

$$s^2 = \frac{SSE}{11-2} = 21.3002 \quad \text{and} \quad s = \sqrt{s^2} = 4.6152$$

$$s\sqrt{c_{00}} = 3.4390 \quad \text{and} \quad s\sqrt{c_{11}} = .8045$$

$$t = \frac{b_0}{s\sqrt{c_{00}}} = 3.334 \quad \text{and} \quad \text{prob-value} = .0087$$

$$t = \frac{b_1}{s\sqrt{c_{11}}} = 30.580 \quad \text{and} \quad \text{prob-value} = .0001$$

Since the prob-value for testing $H_0: \beta_0 = 0$ is .0087, we can reject $H_0: \beta_0 = 0$ in favor of $H_1: \beta_0 \neq 0$ by setting α equal to .01. Since the prob-value for testing $H_0: \beta_1 = 0$ is .0001, we can reject $H_0: \beta_1 = 0$ in favor of $H_1: \beta_1 \neq 0$ by setting α equal to .01. Note that when a SAS output indicates that a prob-value is equal to .0001, this means that the prob-value is .0001 or less. Furthermore, although the SAS output does not include the $100(1 - \alpha)\%$ confidence interval for β_1, it does give b_1 and $s\sqrt{c_{11}}$. Thus, for example, we can use these quantities to calculate the 95% confidence interval for β_1 to be

$$\begin{aligned}[b_1 \pm t_{[.025]}^{(11-2)} s\sqrt{c_{11}}] &= [24.6022 \pm 2.262(.8045)] \\ &= [22.7824, 26.422]\end{aligned}$$

This interval says that Comp-U-Systems is 95% confident that mean service time increases by 22.7824 to 26.422 minutes for each additional microcomputer serviced on a call.

Suppose that we will service four microcomputers on a future service call, and consider the SAS output under the observation (OBS) numbered 12 (which, since there were $n = 11$ historical service calls, represents the future service call on which $x = 4$ microcomputers will be serviced). It follows that

$$\begin{aligned}\hat{y}_0 &= b_0 + b_1 x_0 \\ &= 11.4641 + 24.6022(4) \\ &= 109.873 \quad \text{minutes}\end{aligned}$$

is the point estimate of

$$\begin{aligned}\mu_0 &= \beta_0 + \beta_1 x_0 \\ &= \beta_0 + \beta_1(4)\end{aligned}$$

the mean time required to service four microcomputers. In addition, $\hat{y}_0 = 109.873$

is the point prediction of

$$\begin{aligned} y_0 &= \mu_0 + \varepsilon_0 \\ &= \beta_0 + \beta_1(4) + \varepsilon_0 \end{aligned}$$

an individual time required to service four microcomputers. Moreover, the 95% confidence interval for μ_0, [106.721, 113.025], says that Comp-U-Systems is 95% confident that the mean service time for four microcomputers is between 106.721 and 113.025 minutes. The 95% prediction interval for $y_0 = \mu_0 + \varepsilon_0$, [98.967, 120.779], says that Comp-U-Systems is 95% confident that an individual service time for four microcomputers will be between 98.967 and 120.779 minutes. Furthermore, in addition to giving these intervals, the SAS output tells us that

$$s\sqrt{h_{00}} = 1.393$$

This implies that

$$h_{00} = \left(\frac{1.393}{s}\right)^2 = \left(\frac{1.393}{4.6152}\right)^2 = .0911$$

It follows, for example, that a 99% prediction interval for y_0 is

$$\begin{aligned} [\hat{y}_0 \pm t_{[.005]}^{(9)} s\sqrt{1 + h_{00}}] &= [109.873 \pm 3.250(4.6152)\sqrt{1.0911}] \\ &= [94.205, 125.541] \end{aligned}$$

5.8

SIMPLE COEFFICIENTS OF DETERMINATION AND CORRELATION

THE SIMPLE COEFFICIENT OF DETERMINATION

To introduce the *simple coefficient of determination*, denoted r^2, assume that we have the n observed values of the dependent variable y, but we do not have the n observed values of the independent variable x with which to predict y_i. In such a case the only reasonable prediction of y_i would be

$$\bar{y} = \frac{\sum_{i=1}^{n} y_i}{n}$$

which is the average of the n observed values $y_1, y_2, \ldots, y_n$. The error of prediction would then be $y_i - \bar{y}$. However, in reality we do have the observed values of the independent variable x to use in predicting y_i. The prediction of y_i is

$$\hat{y}_i = b_0 + b_1 x_i$$

which means that the error of prediction is $y_i - \hat{y}_i$. Therefore using the independent variable has decreased the error of prediction from $y_i - \bar{y}$ to $y_i - \hat{y}_i$, or by an amount equal to

$$(y_i - \bar{y}) - (y_i - \hat{y}_i) = (\hat{y}_i - \bar{y})$$

It can be shown (Searle, 1971) that in general

$$\sum_{i=1}^{n} (y_i - \bar{y})^2 - \sum_{i=1}^{n} (y_i - \hat{y}_i)^2 = \sum_{i=1}^{n} (\hat{y}_i - \bar{y})^2$$

The quantity

$$\sum_{i=1}^{n} (y_i - \bar{y})^2$$

is called the *total variation.* It equals the sum of squared prediction errors that would be obtained if we did not use the observed values of the independent variable to predict the observed values of the dependent variable. The quantity

$$\sum_{i=1}^{n} (y_i - \hat{y}_i)^2$$

is called the *unexplained variation* (note that this is another name for SSE). It equals the sum of squared prediction errors that is obtained when we use the observed values of the independent variable to predict the observed values of the dependent variable. The quantity

$$\sum_{i=1}^{n} (\hat{y}_i - \bar{y})^2$$

is called the *explained variation.* Since

$$\sum_{i=1}^{n} (\hat{y}_i - \bar{y})^2 = \sum_{i=1}^{n} (y_i - \bar{y})^2 - \sum_{i=1}^{n} (y_i - \hat{y}_i)^2$$

this explained variation represents the reduction in the sum of squared prediction errors that has been accomplished by using the independent variable in predicting the observed values of the dependent variable. That is, the explained variation measures the amount of the total variation that is explained by the simple linear regression model. It follows that

$$\sum_{i=1}^{n} (y_i - \bar{y})^2 = \sum_{i=1}^{n} (\hat{y}_i - \bar{y})^2 + \sum_{i=1}^{n} (y_i - \hat{y}_i)^2$$

$$\text{Total variation} = \text{Explained variation} + \text{Unexplained variation}$$

We define the *simple coefficient of determination* r^2 by the equation

$$r^2 = \frac{\sum_{i=1}^{n} (\hat{y}_i - \bar{y})^2}{\sum_{i=1}^{n} (y_i - \bar{y})^2} = \frac{\text{Explained variation}}{\text{Total variation}}$$

That is, r^2 is the proportion of the total variation in the n observed values of the dependent variable that is explained by the simple linear regression model. Since neither the explained variation nor the total variation can be negative (both of these quantities are sums of squares), r^2 cannot be negative. Since the explained variation cannot be more than the total variation, it follows that r^2 cannot be more than 1.

Another way to express r^2 is

$$\begin{aligned} r^2 = \frac{\text{Explained variation}}{\text{Total variation}} &= \frac{\text{Total variation} - \text{Unexplained variation}}{\text{Total variation}} \\ &= \frac{\text{Total variation}}{\text{Total variation}} - \frac{\text{Unexplained variation}}{\text{Total variation}} \\ &= 1 - \frac{\text{Unexplained variation}}{\text{Total variation}} \end{aligned}$$

Thus we see that the nearer r^2 is to 1, the smaller is the proportion of the total variation that is the unexplained variation. This says that the nearer r^2 is to 1, the smaller are the prediction errors (compared to the total variation) and the greater is the utility of the simple linear regression model in describing the dependent variable.

Recall that a formula that often provides a simpler way to calculate the unexplained variation is

$$SSE = \sum_{i=1}^{n} (y_i - \hat{y}_i)^2 = \sum_{i=1}^{n} y_i^2 - \left[b_0 \sum_{i=1}^{n} y_i + b_1 \sum_{i=1}^{n} x_i y_i \right]$$

It can also be shown that there are formulas that often provide simpler ways to calculate the total variation and explained variation. To summarize:

The Simple Coefficient of Determination, r^2

For the simple linear regression model

$$y = \beta_0 + \beta_1 x + \varepsilon$$

1. Total variation $= \sum_{i=1}^{n} (y_i - \bar{y})^2 = \sum_{i=1}^{n} y_i^2 - n\bar{y}^2$

2. Explained variation $= \sum_{i=1}^{n} (\hat{y}_i - \bar{y})^2 = \left[b_0 \sum_{i=1}^{n} y_i + b_1 \sum_{i=1}^{n} x_i y_i \right] - n\bar{y}^2$

3. Unexplained variation $= \sum_{i=1}^{n} (y_i - \hat{y}_i)^2 = \sum_{i=1}^{n} y_i^2 - \left[b_0 \sum_{i=1}^{n} y_i + b_1 \sum_{i=1}^{n} x_i y_i \right]$

4. Total variation $=$ Explained variation $+$ Unexplained variation

5. Simple coefficient of determination $= r^2 = \dfrac{\text{Explained variation}}{\text{Total variation}}$
$= 1 - \dfrac{\text{Unexplained variation}}{\text{Total variation}}$

6. r^2 is the proportion of the total variation in the n observed values of the dependent variable that is explained by the simple linear regression model.

Example 5.11 Consider the fuel consumption model

$$y = \beta_0 + \beta_1 x + \varepsilon$$

Using the observed fuel consumption data, we made the following calculations in Examples 4.3 and 4.7:

$$\sum_{i=1}^{8} y_i^2 = 859.91 \qquad \bar{y} = \frac{\sum_{i=1}^{8} y_i}{8} = 10.21$$

$$b_0 \sum_{i=1}^{8} y_i + b_1 \sum_{i=1}^{8} x_i y_i = 857.34$$

$$\begin{aligned} \text{Unexplained variation} = SSE &= \sum_{i=1}^{8} (y_i - \hat{y}_i)^2 \\ &= \sum_{i=1}^{8} y_i^2 - \left[b_0 \sum_{i=1}^{8} y_i + b_1 \sum_{i=1}^{8} x_i y_i \right] \\ &= 859.91 - 857.34 = 2.57 \end{aligned}$$

We can calculate the total variation to be

$$\begin{aligned} \text{Total variation} = \sum_{i=1}^{8} (y_i - \bar{y})^2 &= \sum_{i=1}^{8} y_i^2 - 8\bar{y}^2 \\ &= 859.91 - 8(10.21)^2 = 25.55 \end{aligned}$$

Moreover, we can calculate the explained variation by either of the following two methods:

$$\begin{aligned} \text{Explained variation} &= \text{Total variation} - \text{Unexplained variation} \\ &= 25.55 - 2.57 = 22.98 \end{aligned}$$

or

$$\text{Explained variation} = \sum_{i=1}^{8} (\hat{y}_i - \bar{y})^2 = \left[b_0 \sum_{i=1}^{8} y_i + b_1 \sum_{i=1}^{8} x_i y_i \right] - n\bar{y}^2$$

$$= 857.34 - 8(10.21)^2 = 22.98$$

Next, we can calculate the simple coefficient of determination by either of the following two methods:

$$r^2 = \frac{\text{Explained variation}}{\text{Total variation}} = \frac{22.98}{25.55} = .90$$

or

$$r^2 = 1 - \frac{\text{Unexplained variation}}{\text{Total variation}} = 1 - \frac{2.57}{25.55} = .90$$

This value of r^2 says that the regression model explains 90% of the total variation in the eight observed fuel consumptions.

Example 5.12 Consider the Comp-U-Systems model

$$y = \beta_0 + \beta_1 x + \varepsilon$$

and the SAS output in Figure 5.8. Examining this output, we see that

Total variation = 20,110.545

Explained variation = 19,918.844

Unexplained variation = 191.702

$r^2 = .9905$

This value of r^2 says that the regression model explains 99.05% of the total variation in the eleven observed service times.

As two other examples, r^2 can be calculated to be .6332 for the residential sales data in Table 4.3, and r^2 can be calculated to be .5685 for the sales performance data in Table 4.4.

THE SIMPLE CORRELATION COEFFICIENT

We next define the *simple correlation coefficient*, denoted by r, to be $r = \pm\sqrt{r^2}$, where we define $r = +\sqrt{r^2}$ if b_1 is positive and $r = -\sqrt{r^2}$ if b_1 is negative.

Since for the simple linear regression model the prediction of y_i is

$$\hat{y}_i = b_0 + b_1 x_i$$

it can be shown (by utilizing the algebraic formulas for the least squares point estimates b_0 and b_1) that

$$r^2 = \frac{\sum_{i=1}^{n} (\hat{y}_i - \bar{y})^2}{\sum_{i=1}^{n} (y_i - \bar{y})^2} = \frac{\left[\sum_{i=1}^{n} (x_i - \bar{x})(y_i - \bar{y})\right]^2}{\sum_{i=1}^{n} (x_i - \bar{x})^2 \sum_{i=1}^{n} (y_i - \bar{y})^2}$$

where

$$\bar{x} = \frac{\sum_{i=1}^{n} x_i}{n} \quad \text{and} \quad \bar{y} = \frac{\sum_{i=1}^{n} y_i}{n}$$

Furthermore, taking square roots implies the following:

The Simple Correlation Coefficient, r

1. The *simple correlation coefficient* is

$$r = \frac{\sum_{i=1}^{n} (x_i - \bar{x})(y_i - \bar{y})}{\left[\sum_{i=1}^{n} (x_i - \bar{x})^2 \sum_{i=1}^{n} (y_i - \bar{y})^2\right]^{1/2}}$$

2. It can be shown that the above formula for r is equivalent to the formula

$$r = b_1 \frac{s_x}{s_y}$$

where

$$s_x = \sqrt{\frac{\sum_{i=1}^{n} (x_i - \bar{x})^2}{n - 1}} \quad \text{and} \quad s_y = \sqrt{\frac{\sum_{i=1}^{n} (y_i - \bar{y})^2}{n - 1}}$$

and b_1 is the least squares point estimate of the slope β_1 in the simple linear regression model.

The formula in (2) above automatically yields a value of the simple correlation coefficient that is positive if b_1 is positive or negative if b_1 is negative. Since r^2 is always between zero and 1, it follows that r is always between -1 and 1. A value of r close to 1 indicates that the independent variable x and the dependent variable y have a strong tendency to move together in a linear fashion with a positive slope and therefore that x and y are highly related and *positively correlated.* A value of r close to -1 indicates that x and y have a strong tendency to move together in a linear fashion with a negative slope and therefore that x and y are highly related and *negatively correlated.*

Example 5.13 Using the fuel consumption data in Table 4.1, we make the following calculations:

$$\bar{x} = \frac{\sum_{i=1}^{8} x_i}{8} = \frac{351.8}{8} = 43.98$$

$$\bar{y} = \frac{\sum_{i=1}^{8} y_i}{8} = \frac{81.7}{8} = 10.21$$

$$r = \frac{\sum_{i=1}^{8} (x_i - \bar{x})(y_i - \bar{y})}{\left[\sum_{i=1}^{8} (x_i - \bar{x})^2 \sum_{i=1}^{8} (y_i - \bar{y})^2\right]^{1/2}}$$

$$= \frac{(x_1 - \bar{x})(y_1 - \bar{y}) + \cdots + (x_8 - \bar{x})(y_8 - \bar{y})}{[[(x_1 - \bar{x})^2 + \cdots + (x_8 - \bar{x})^2][(y_1 - \bar{y})^2 + \cdots + (y_8 - \bar{y})^2]]^{1/2}}$$

$$= \frac{(28.0 - 43.98)(12.4 - 10.21) + \cdots + (62.5 - 43.98)(7.5 - 10.21)}{[[(28.0 - 43.98)^2 + \cdots + (62.5 - 43.98)^2][(12.4 - 10.21)^2 + \cdots + (7.5 - 10.21)^2]]^{1/2}}$$

$$= -.9487$$

The simple correlation coefficient of $r = -.9487$ indicates that (as illustrated in Figure 4.1), x and y have a strong tendency to move together in a linear fashion with a negative slope. An r of $-.9487$ also means that the simple coefficient of determination equals

$$r^2 = (-.9487)^2 = .90$$

INFERENCES FOR THE POPULATION CORRELATION COEFFICIENT

We have seen that the simple correlation coefficient, r, measures the linear relationship, or correlation, between the n observed values of the independent variable x and the n observed values of the dependent variable y that constitute the sample. A similar coefficient of linear correlation can be defined for the population of all possible combinations of observed values of x and y. We call this coefficient the *population correlation coefficient* and denote it by ρ. We use r as the point estimate of ρ. Moreover, if we assume that the population of all possible combinations of observed values of x and y has a bivariate normal probability distribution (see Wonnacott and Wonnacott (1981) for a discussion of this distribution), we can use r to carry out an hypothesis test. Here we test the null hypothesis $H_0 : \rho = 0$, which says that there is no linear relationship between x and y, versus the alternative hypothesis $H_1 : \rho \neq 0$, which says that there is a (positive or negative) linear relationship between x and y. However, it can be shown that this hypothesis test is equivalent to using the t statistic (and the related prob-value) to test $H_0 : \beta_1 = 0$ versus $H_1 : \beta_1 \neq 0$, where β_1 is the slope—that is, the parameter describing the *linear relationship* between x and y—in the simple linear regression model. Keep in mind, however, that although the mechanics involved in testing $H_0 : \rho = 0$ and $H_0 : \beta_1 = 0$ are the same, the assumptions on which these tests are based are different. Testing $H_0 : \rho = 0$ is based on the bivariate normality assumption, and testing $H_0 : \beta_1 = 0$ is based on the inference assumptions.

Example 5.14 Recall from Example 5.5 that we calculated the prob-value for testing $H_0 : \beta_1 = 0$ in the simple linear regression model describing the fuel consumption data to be .00033. Thus under the appropriate assumptions, since the prob-value is very small, we have very strong evidence that $H_0 : \beta_1 = 0$ (or $H_0 : \rho = 0$) is false and $H_1 : \beta_1 \neq 0$ (or $H_1 : \rho \neq 0$) is true. Therefore we have very strong evidence that there is a negative linear relationship, or negative correlation, between x and y.

We next note that the above test of $H_0 : \rho = 0$ versus $H_1 : \rho \neq 0$ can also be carried out by expressing the t statistic in terms of r. It can be proven that

$$t = \frac{r\sqrt{n-2}}{\sqrt{1-r^2}}$$

Also, if it is desired to test the more general hypothesis

$$H_0 : \rho = c \qquad \text{versus} \qquad H_1 : \rho \neq c$$

we can make use of the z' *transformation*, which is due to R. A. Fisher. It can be proven that for large samples ($n \geq 25$) the population of all possible values of

$$z' = \tfrac{1}{2} \ln \left(\frac{1+r}{1-r} \right)$$

is approximately normally distributed with mean

$$\tfrac{1}{2} \ln \left(\frac{1+\rho}{1-\rho} \right)$$

and variance

$$\frac{1}{n-3}$$

This implies the following large sample results.

Testing $H_0: \rho = c$ Versus $H_1: \rho \neq c$ and a Confidence Interval for ρ

1. We can reject $H_0: \rho = c$ in favor of $H_1: \rho \neq c$ by setting the probability of a Type I error equal to α if the absolute value of

$$\frac{\frac{1}{2}\ln\left(\frac{1+r}{1-r}\right) - \frac{1}{2}\ln\left(\frac{1+c}{1-c}\right)}{\sqrt{\frac{1}{n-3}}}$$

is greater than $z_{[\alpha/2]}$.

2. An approximate $100(1-\alpha)\%$ confidence interval for

$$\frac{1}{2}\ln\left(\frac{1+\rho}{1-\rho}\right)$$

is

$$\left[\frac{1}{2}\ln\left(\frac{1+r}{1-r}\right) \pm z_{[\alpha/2]}\sqrt{\frac{1}{n-3}}\right]$$

3. If the interval in (2) above is calculated to be $[a, b]$, then an approximate $100(1-\alpha)\%$ confidence interval for ρ is

$$\left[\frac{e^{2a}-1}{e^{2a}+1}, \frac{e^{2b}-1}{e^{2b}+1}\right]$$

Example 5.15 Suppose that the sample correlation coefficient between the productivities and aptitude test scores of $n = 250$ word processing specialists is .84. Then a 95% confidence interval for

$$\frac{1}{2}\ln\left(\frac{1+\rho}{1-\rho}\right)$$

is

$$\left[\frac{1}{2}\ln\left(\frac{1+r}{1-r}\right) \pm z_{[.025]}\sqrt{\frac{1}{n-3}}\right]$$

$$= \left[\frac{1}{2}\ln\left(\frac{1+.84}{1-.84}\right) \pm 1.96\sqrt{\frac{1}{250-3}}\right] = [1.0965, 1.3459]$$

It follows that a 95% confidence interval for ρ is

$$\left[\frac{e^{2(1.0965)}-1}{e^{2(1.0965)}+1}, \frac{e^{2(1.3459)}-1}{e^{2(1.3459)}+1}\right] = [.80, .87]$$

A COMPARISON OF CORRELATION ANALYSIS AND REGRESSION ANALYSIS

We now briefly compare correlation analysis and regression analysis. Suppose, for example, that we were to use the simple linear regression model to relate

y = the productivity of a word processing specialist

to

x = the score on a word processing aptitude test

Then, while $r = .84$ might indicate a fairly strong positive linear relationship, or correlation, between x and y, regression analysis could be used to predict the productivity of prospective word processing specialists on the basis of their aptitude test scores. As another example, suppose we were to use the simple linear regression model to relate

y = the rate of return on a particular stock

to

x = the rate of return on the overall stock market

Then, while $r = .74$ might indicate a moderate positive linear relationship, or correlation, between x and y, we could use regression analysis to learn about the nature of this relationship.

That is, regression analysis is more informative than (and thus preferred to) correlation analysis. Correlation analysis is usually used to better understand certain aspects of regression analysis (for example, multicollinearity, which we discuss in Chapter 10).

5.9

AN F-TEST FOR THE SIMPLE LINEAR REGRESSION MODEL

In this section we present an *F-test* of $H_0: \beta_1 = 0$ versus $H_1: \beta_1 \neq 0$. Note that in presenting this test we denote the *explained variation* (also called the *sum of squares due to the overall model*) by SS_{model} and we denote the *mean square error* s^2 by MSE (MS denotes "mean square," and E denotes "error").

An F-Test of $H_0: \beta_1 = 0$ Versus $H_1: \beta_1 \neq 0$

Assume that the inference assumptions are satisfied, and define the *overall F-statistic* to be

$$F(\text{model}) = \frac{MS_{\text{model}}}{MSE} = \frac{SS_{\text{model}}/1}{SSE/(n-2)}$$

where

$$MS_{\text{model}} = \frac{SS_{\text{model}}}{1} = \frac{\text{Explained variation}}{1}$$

$$MSE = \frac{SSE}{n - 2} = s^2$$

Also, define the prob-value to be the area to the right of $F(\text{model})$ under the curve of the F-distribution having 1 and $n - 2$ degrees of freedom (see Figure 5.9b).

FIGURE 5.9 **An F-test for the simple linear regression model**

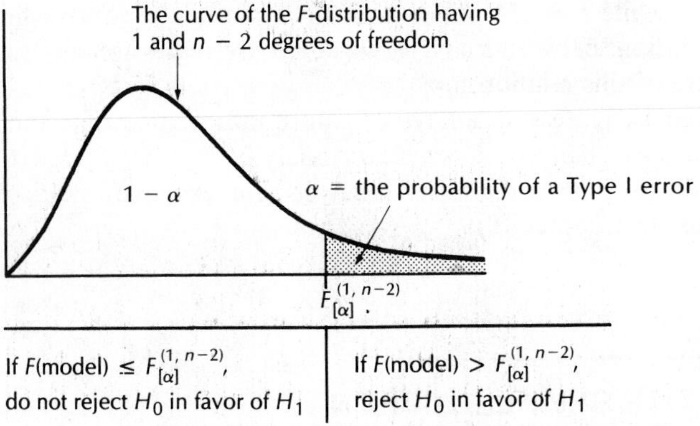

(a) The rejection point $F_{[\alpha]}^{(1, n-2)}$ based on setting the probability of a Type I error equal to α

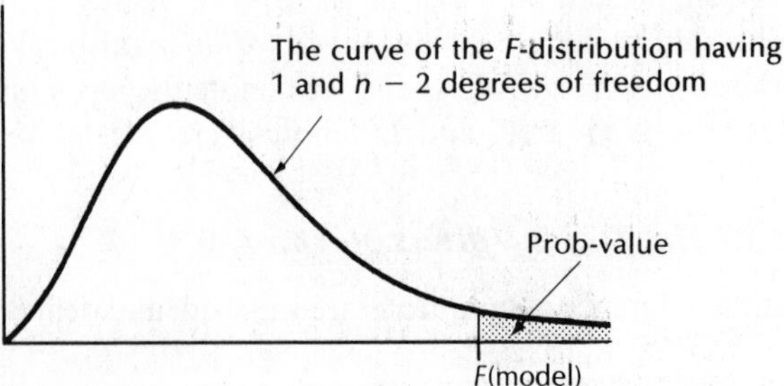

(b) If prob-value is smaller than α, then $F(\text{model}) > F_{[\alpha]}^{(1, n-2)}$. Reject H_0

Then we can reject H_0 in favor of H_1 by setting the probability of a Type I error equal to α if and only if either of the following equivalent conditions hold:

1. $F(\text{model}) > F_{[\alpha]}^{(1,n-2)}$, where $F_{[\alpha]}^{(1,n-2)}$ is the point on the scale of the F-distribution having 1 and $n - 2$ degrees of freedom so that the area under this curve to the right of $F_{[\alpha]}^{(1,n-2)}$ is α. See Figure 5.9(a).
2. Prob-value $< \alpha$. See Figure 5.9(b).

First consider condition (1), which says that we should reject $H_0: \beta_1 = 0$ when $F(\text{model})$ is large. This is intuitively reasonable because a large value of $F(\text{model})$ would be caused by an explained variation that is large in relation to the unexplained variation (SSE). This would occur if the independent variable x significantly affects the dependent variable, which would imply that $H_0: \beta_1 = 0$ is false and $H_1: \beta_1 \neq 0$ is true. A more precise rationalization for condition (1) comes from using the technique of *expected mean squares*. Recall from Section 5.1 that, if the inference assumptions hold, then the average of all possible values of $MSE(=s^2)$ is equal to σ^2. It can also be proven that, if the inference assumptions hold, then the average of all possible values of MS_{model} is

$$\sigma^2 + \beta_1^2 \sum_{i=1}^{n} (x_i - \bar{x})^2$$

Therefore if $H_0: \beta_1 = 0$ is true, the expected value of MS_{model} is equal to the expected value of MSE, which implies that we would expect

$$F(\text{model}) = \frac{MS_{\text{model}}}{MSE}$$

to be roughly equal to 1. On the other hand, if $H_1: \beta_1 \neq 0$ is true, the expected value of MS_{model} is greater than the expected value of MSE. This implies that we would expect $F(\text{model})$ to be greater than 1. Therefore a large value of $F(\text{model})$ indicates that we should reject $H_0: \beta_1 = 0$ in favor of $H_1: \beta_1 \neq 0$. It should be noted that all F-tests in this book are based on analogous uses of expected mean squares.

To understand the meaning of the probability of a Type I error for the F-test, consider the population of all possible samples. Since different samples yield different overall F-statistics, it is clear that an infinite population of overall F-statistics exists. We can make the following statement about this population.

The Population of Overall F-Statistics

If the inference assumptions hold, and if the null hypothesis $H_0: \beta_1 = 0$ is true, the *population of all possible overall F statistics* has an *F-distribution* with 1 and $n - 2$ degrees of freedom.

The probability of a Type I error for the F-test is depicted in Figure 5.9(a). Setting the probability of a Type I error equal to α (for example, equal to .05) implies that, if $H_0: \beta_1 = 0$ is true:

1. The probability that

   $$F(\text{model}) \leq F_{[\alpha]}^{(1,n-2)}$$

   is $1 - \alpha$. That is, $100(1 - \alpha)\%$—for example, $100(1 - .05)\% = 95\%$—of all possible overall F statistics are less than $F_{[\alpha]}^{(1,n-2)}$ and thus are small enough to cause us to not reject $H_0: \beta_1 = 0$ (when $H_0: \beta_1 = 0$ is true), a correct decision, and
2. The probability that

   $$F(\text{model}) > F_{[\alpha]}^{(1,n-2)}$$

   is α. That is, $100(\alpha)\%$—for example, $100(.05)\% = 5\%$—of all possible overall F statistics are greater than $F_{[\alpha]}^{(1,n-2)}$ and thus are large enough to cause us to reject $H_0: \beta_1 = 0$ (when $H_0: \beta_1 = 0$ is true), a Type I error.

Furthermore, the rationale for the prob-value condition can be understood by comparing Figures 5.9(a) and 5.9(b).

Example 5.16 Consider the SAS output for the Comp-U-Systems model

$$y = \beta_0 + \beta_1 x + \varepsilon$$

in Figure 5.8. We see that

$$SS_{\text{model}} = \text{Explained variation} = 19918.844$$

$$MS_{\text{model}} = \frac{SS_{\text{model}}}{1} = \frac{19918.844}{1} = 19918.844$$

$$SSE = \text{Unexplained variation} = 191.702$$

$$s^2 = MSE = \frac{SSE}{n-2} = \frac{191.702}{11-2} = \frac{191.702}{9} = 21.300184$$

$$F(\text{model}) = \frac{MS_{\text{model}}}{MSE} = \frac{19918.844}{21.300184} = 935.149$$

Furthermore, the prob-value is the area to the right of 935.149 under the curve of the F-distribution having 1 and 9 degrees of freedom. This prob-value can be computer calculated to be less than .0001 (for the purposes of subsequent discussion we suppose that the prob-value is equal to .0001). If we wish to use condition (1) with α set equal to .05, then we use the rejection point

$$F_{[\alpha]}^{(1,n-2)} = F_{[.05]}^{(1,9)} = 5.12$$

Since $F(\text{model}) = 935.149 > 5.12 = F_{[.05]}^{(1,9)}$, we can reject $H_0: \beta_1 = 0$ in favor of

$H_1 : \beta_1 \neq 0$ by setting α equal to .05. Alternatively, since prob-value $=$ .0001 is less than .05 and .01, it follows by condition (2) that we can reject H_0 in favor of H_1 by setting α equal to .05 or .01.

To conclude this section, note that testing $H_0 : \beta_1 = 0$ versus $H_1 : \beta_1 \neq 0$ by using the overall F statistic and its prob-value is equivalent to testing these hypotheses by using the t statistic and its prob-value. This is because it can be proved that

$$(t)^2 = F(\text{model})$$

and that

$$(t_{[\alpha/2]}^{(n-2)})^2 = F_{[\alpha]}^{(1,n-2)}$$

Hence the rejection point conditions

$$|t| > t_{[\alpha/2]}^{(n-2)} \qquad \text{and} \qquad F(\text{model}) > F_{[\alpha]}^{(1,n-2)}$$

are equivalent. Furthermore, it can be shown that the prob-values for testing $H_0 : \beta_1 = 0$ computed by using the t statistic and the overall F statistic are equal to each other.

Since the F-test and t-test of $H_0 : \beta_1 = 0$ versus $H_1 : \beta_1 \neq 0$ are equivalent, the reader might ask why we have presented the F-test. There are two reasons. First, most standard regression computer packages include the results of the F-test as a part of the regression output (see Figure 5.8). Second, the F-test has a useful generalization in the context of multiple regression analysis. This generalization is not equivalent to a t-test (see Chapter 8).

*5.10

AN *F*-TEST OF LACK OF FIT

When more than one value of the dependent variable is observed at some of the values of the independent variables, we can perform what is called a *lack-of-fit test* of the *functional form* of the simple linear regression model. Here, when we test the functional form of this model, we are attempting to determine whether the linear equation

$$\mu = \beta_0 + \beta_1 x$$

adequately represents the relationship between μ and x. If the linear equation is not adequate, an equation having a different functional form, such as a curve, is needed to represent the relationship between μ and x. Performing a lack-of-fit test requires calculating a quantity called the *sum of squares due to pure error*. This

*This section is optional.

quantity is denoted by SS_{PE} and is given by the equation

$$SS_{PE} = \sum_{l=1}^{m} \left[\sum_{k=1}^{n_l} (y_{l,k} - \bar{y}_l)^2 \right]$$

In this equation, m is the total number of values of the independent variable at which at least one corresponding value of the dependent variable has been observed, and

$$\bar{y}_l = \frac{\sum_{k=1}^{n_l} y_{l,k}}{n_l}$$

is the average of the n_l values $y_{l,1}, y_{l,2}, \ldots, y_{l,n_l}$ of the dependent variable that have been observed at the lth value of the independent variable. The calculation of this quantity is illustrated in Example 5.17. We now show how SS_{PE} can be used to perform a lack-of-fit test.

An F-Test of Lack of Fit

Assume that the inference assumptions are satisfied, and consider testing

H_0: The functional form of the simple linear regression model is correct

versus

H_1: The functional form of the simple linear regression model is not correct

Define the *lack-of-fit F statistic* to be

$$F(LF) = \frac{MS_{LF}}{MS_{PE}} = \frac{SS_{LF}/(m-2)}{SS_{PE}/(n-m)} = \frac{(SSE - SS_{PE})/(m-2)}{SS_{PE}/(n-m)}$$

where

$$SSE = \text{the unexplained variation}$$

and

$$SS_{PE} = \sum_{l=1}^{m} \left[\sum_{k=1}^{n_l} (y_{l,k} - \bar{y}_l)^2 \right] = \text{the sum of squares due to pure error}$$

Also, define the prob-value to be the area to the right of $F(LF)$ under the curve of the F-distribution having $m - 2$ and $n - m$ degrees of freedom. Then we can reject H_0 in favor of H_1 by setting the probability of a Type I error equal to α if and only if either of the following two equivalent conditions holds:

1. $F(LF) > F_{[\alpha]}^{(m-2,n-m)}$
2. Prob-value $< \alpha$

Condition (1), which says that we should reject H_0 in favor of H_1 if $F(LF)$ is large, is reasonable because a large value of $F(LF)$ would be caused by a large value of MS_{LF}. This would be caused by a large value of

$$SS_{LF} = SSE - SS_{PE}$$

which would occur if the regression model yields an unexplained variation that is much larger than the sum of squares due to pure error. This would imply that the simple linear regression model *lacks the appropriate terms needed to fit the observed data.*

Example 5.17 Consider the Comp-U-Systems problem. In Table 5.7 we rearrange the Comp-U-Systems data originally given in Table 5.6. In this rearrangement, x_l denotes the lth smallest number of microcomputers serviced, and $y_{l,k}$ denotes the number of minutes required to perform service on the kth service call involving x_l microcomputers. For this data there are $n = 11$ values of the dependent variable, and there are $m = 7$ values of the independent variable at which at least one corresponding value of the dependent variable has been observed. Therefore the sum of squares due to pure error is

$$\begin{aligned} SS_{PE} &= \sum_{l=1}^{7}\left[\sum_{k=1}^{n_l}(y_{l,k} - \bar{y}_l)^2\right] \\ &= [(y_{2,1} - \bar{y}_2)^2 + (y_{2,2} - \bar{y}_2)^2] \\ &\quad + [(y_{4,1} - \bar{y}_4)^2 + (y_{4,2} - \bar{y}_4)^2 + (y_{4,3} - \bar{y}_4)^2] \\ &\quad + [(y_{5,1} - \bar{y}_5)^2 + (y_{5,2} - \bar{y}_5)^2] \\ &= [(58 - 63)^2 + (68 - 63)^2] \\ &\quad + [(103 - 108)^2 + (109 - 108)^2 + (112 - 108)^2] \\ &\quad + [(134 - 136)^2 + (138 - 136)^2] \\ &= 100 \end{aligned}$$

TABLE 5.7 **Comp-U-Systems service call data**

The lth smallest number of microcomputers serviced, x_l	*Number of minutes required to perform service, $y_{l,k}$*			*Mean, $\bar{y}_l$*
$x_1 = 1$	$y_{1,1} = 37$			$\bar{y}_1 = 37$
$x_2 = 2$	$y_{2,1} = 58$	$y_{2,2} = 68$		$\bar{y}_2 = 63$
$x_3 = 3$	$y_{3,1} = 82$			$\bar{y}_3 = 82$
$x_4 = 4$	$y_{4,1} = 103$	$y_{4,2} = 109$	$y_{4,3} = 112$	$\bar{y}_4 = 108$
$x_5 = 5$	$y_{5,1} = 134$	$y_{5,2} = 138$		$\bar{y}_5 = 136$
$x_6 = 6$	$y_{6,1} = 154$			$\bar{y}_6 = 154$
$x_7 = 7$	$y_{7,1} = 189$			$\bar{y}_7 = 189$

and

$$\begin{aligned} MS_{PE} &= \frac{SS_{PE}}{n - m} \\ &= \frac{100}{11 - 7} \\ &= \frac{100}{4} \\ &= 25 \end{aligned}$$

Consider the simple linear regression model

$$y_{l,k} = \beta_0 + \beta_1 x_l + \varepsilon_{l,k}$$

and note that the unexplained variation resulting from using this model to perform a regression analysis of the data in Table 5.7 is $SSE = 191.702$. Then we can make the following calculations:

$$SS_{LF} = SSE - SS_{PE} = 191.702 - 100 = 91.702$$

$$MS_{LF} = \frac{SS_{LF}}{m - 2} = \frac{91.702}{7 - 2} = \frac{91.702}{5} = 18.3404$$

$$F(LF) = \frac{MS_{LF}}{MS_{PE}} = \frac{18.3404}{25} = .7336$$

Employing condition (1) to test

H_0: The functional form of the regression model $y_{l,k} = \beta_0 + \beta_1 x_l + \varepsilon_{l,k}$ is correct

versus

H_1: The functional form of the regression model $y_{l,k} = \beta_0 + \beta_1 x_l + \varepsilon_{lk}$ is not correct

with α equal to .05, we use the rejection point

$$F_{[\alpha]}^{(m-2,n-m)} = F_{[.05]}^{(5,4)} = 6.26$$

Since $F(LF) = .7336 < 6.26 = F_{[.05]}^{(5,4)}$, we cannot reject H_0 in favor of H_1 by setting α equal to .05. Thus there is not much evidence that the functional form of the simple linear model is incorrect. Intuitively, this conclusion should be expected, because the plot of the Comp-U-Systems data in Figure 5.7 indicates that as the values of x increase, the values of y tend to increase in a straight line fashion. If the plot of the Comp-U-Systems data had indicated that y and x are related in a curved fashion, the F-test of lack of fit would probably have indicated that the functional form of the simple linear model is not correct.

*5.11

REGRESSION ANALYSIS THROUGH THE ORIGIN

If logic and statistical evidence indicate that μ equals zero when x equals zero, it might well be reasonable to exclude the intercept from the simple linear regression model and use the model

$$\begin{aligned} y &= \mu + \varepsilon \\ &= \beta_1 x + \varepsilon \end{aligned}$$

We now consider using this model, which is called "*regression analysis through the origin.*"

Statistical Inference for Regression Analysis Through the Origin

Assume that the inference assumptions hold for the model

$$\begin{aligned} y &= \mu + \varepsilon \\ &= \beta_1 x + \varepsilon \end{aligned}$$

1. The *least squares point estimate* b_1 of the parameter β_1 is the value of b_1 that minimizes the sum of squared residuals

$$\begin{aligned} SSE &= \sum_{i=1}^{n} e_i^2 = \sum_{i=1}^{n} (y_i - \hat{y}_i)^2 \\ &= \sum_{i=1}^{n} (y_i - b_1 x_i)^2 \end{aligned}$$

and is calculated by the equation

$$b_1 = \frac{\sum_{i=1}^{n} x_i y_i}{\sum_{i=1}^{n} x_i^2}$$

2. The *point estimates of σ^2 and σ*, the variance and standard deviation of the populations of potential values of the dependent variable, are

$$s^2 = \sqrt{\frac{SSE}{n-1}} \quad \text{and} \quad s = \sqrt{s^2}$$

respectively. Here,

$$SSE = \sum_{i=1}^{n} (y_i - b_1 x_i)^2 = \sum_{i=1}^{n} y_i^2 - b_1 \sum_{i=1}^{n} x_i y_i$$

*This section is optional.

3. Define

$$c_{11} = \frac{1}{\sum_{i=1}^{n} x_i^2}$$

and define the *test statistic*

$$t = \frac{b_1}{s\sqrt{c_{11}}}$$

Also, define the prob-value to be twice the area under the curve of the t-distribution having $n - 1$ degrees of freedom to the right of $|t|$. Then we can reject $H_0 : \beta_1 = 0$ in favor of $H_1 : \beta_1 \neq 0$ by setting the probability of a Type I error equal to α if and only if either of the following equivalent conditions holds:

a. $|t| > t_{[\alpha/2]}^{(n-1)}$, that is, if $t > t_{[\alpha/2]}^{(n-1)}$ or $t < -t_{[\alpha/2]}^{(n-1)}$.
b. Prob-value $< \alpha$.

4. A *100(1 − α)% confidence interval for* β_1 is

$$[b_1 \pm t_{[\alpha/2]}^{(n-1)} s\sqrt{c_{11}}] \qquad \text{where} \qquad c_{11} = \frac{1}{\sum_{i=1}^{n} x_i^2}$$

5. If x_0, a specified value of the independent variable x, is in the experimental region, then

$$\hat{y}_0 = b_1 x_0$$

is the *point estimate of the mean value of the dependent variable*

$$\mu_0 = \beta_1 x_0$$

and is the *point prediction of the individual value of the dependent variable*

$$\begin{aligned} y_0 &= \mu_0 + \varepsilon_0 \\ &= \beta_1 x_0 + \varepsilon_0 \end{aligned}$$

6. Define the *distance value* h_{00} to be

$$h_{00} = \frac{x_0^2}{\sum_{i=1}^{n} x_i^2}$$

Then

a. A *100(1 − α)% confidence interval for* μ_0 is

$$[\hat{y}_0 \pm t_{[\alpha/2]}^{(n-1)} s\sqrt{h_{00}}] = \left[\hat{y}_0 \pm t_{[\alpha/2]}^{(n-1)} s \sqrt{\frac{x_0^2}{\sum_{i=1}^{n} x_i^2}}\right]$$

b. A *100(1 − α)% prediction interval for* $y_0 = \mu_0 + \varepsilon_0$ is

$$[\hat{y}_0 \pm t_{[\alpha/2]}^{(n-1)} s\sqrt{1 + h_{00}}] = \left[\hat{y}_0 \pm t_{[\alpha/2]}^{(n-1)} s \sqrt{1 + \frac{x_0^2}{\sum_{i=1}^{n} x_i^2}}\right]$$

Example 5.18 In Table 5.8 we present data sampled from 22 naval installations. Here, for $i = 1, 2, \ldots, 22$,

$$x_i = \text{the number of items processed in a month at installation } i$$

and

$$y_i = \text{the number of labor hours used in the month at installation } i$$

Theory suggests that, as the number of items processed approaches zero, the number of required labor hours approaches zero. Furthermore, the plot in Figure 5.10 of the data in Table 5.8 confirms that this is true. This suggests that

TABLE 5.8 **Monthly labor hours as a function of items processed**

Installation, i	*Items processed, x_i*	*Monthly labor hours, y_i*
1	15	85
2	25	125
3	57	203
4	67	293
5	197	763
6	166	639
7	162	673
8	131	499
9	158	657
10	241	939
11	399	1546
12	527	2158
13	533	2182
14	563	2302
15	563	2202
16	932	3678
17	986	3894
18	1021	4034
19	1643	6622
20	1985	7890
21	1640	6610
22	2143	8522

Source: *Procedures and Analyses for Staffing Standards Development: Regression Analysis Handbook* (San Diego, Calif.: Navy Manpower and Material Analysis Center, 1979).

FIGURE 5.10 **Plot of monthly labor hours versus items processed**

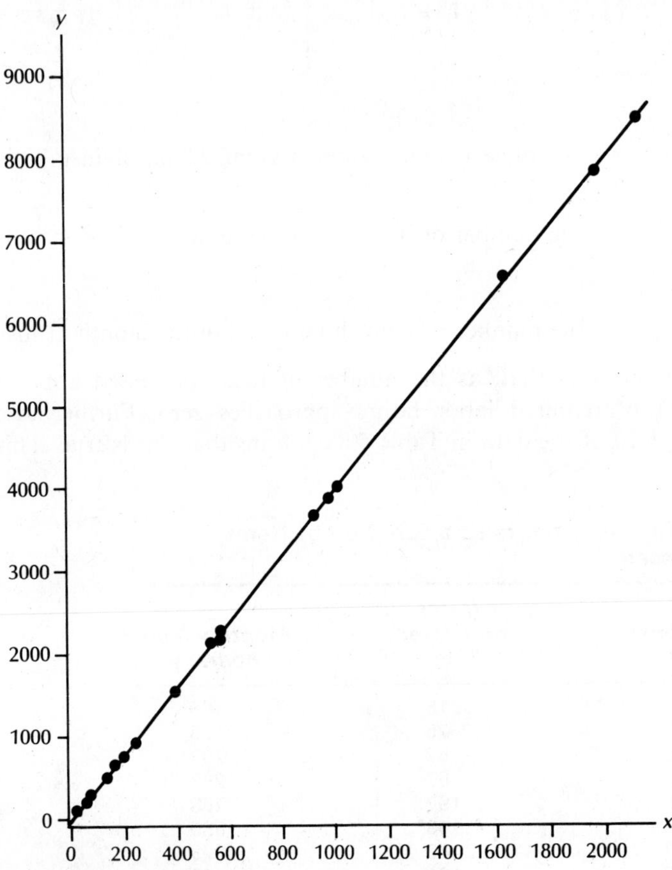

we use the model

$$\begin{aligned} y_i &= \mu_i + \varepsilon_i \\ &= \beta_1 x_i + \varepsilon_i \end{aligned}$$

to relate y_i to x_i. Since the least squares point estimate of β_1 is

$$b_1 = \frac{\sum_{i=1}^{22} x_i y_i}{\sum_{i=1}^{22} x_i^2} = 3.9910$$

it follows that

$$\begin{aligned}\hat{y}_0 &= b_1 x_0 \\ &= 3.9910(500) \\ &= 1995.5\end{aligned}$$

is the point prediction of

$$\begin{aligned}y_0 &= \mu_0 + \varepsilon_0 \\ &= \beta_1 x_0 + \varepsilon_0 \\ &= \beta_1(500) + \varepsilon_0\end{aligned}$$

the number of labor hours that will be needed to process 500 items in a month at a naval installation. Other statistical inferences can be made by using the formulas summarized above.

*5.12

USING SAS

In Figure 5.11 we present the SAS statements that generate the Comp-U-Systems service call output of Figure 5.8. In Figure 5.12 we present the SAS statements

FIGURE 5.11 **SAS program for a simple linear regression analysis of the Comp-U-Systems service call data**

```
DATA COMP; } ——→ Assigns name COMP to file
INPUT Y X; } ——→ Assigns variable names: Y = service time; X = number of microcomputers
CARDS;
 37   1
 58   2
 68   2
 82   3
103   4
109   4  }——→ Comp-U-Systems data: see Table 5.6
112   4
134   5
138   5
154   6
189   7
  .   4} ——→ Decimal point denotes missing value
             Used to obtain point prediction when x0 = 4
PROC PRINT; } ——→ Print procedure: prints the data
PROC REG DATA = COMP; } ——→ Specifies regression procedure
MODEL Y = X/P CLM CLI; } Specifies model y_i = β0 + β1 x_i + ε_i
```

Note: P = *Point predictions desired*
CLM = *95% confidence intervals desired*
CLI = *95% prediction intervals desired*

*This section is optional.

FIGURE 5.12 **SAS program for a regression through the origin of the naval data in Table 5.8**

```
DATA HOURS; }  ——→ Assigns name HOURS to data file
INPUT Y X; }   ——→ Assigns variable names:
                   Y = labor hours, X = items processed
CARDS;
  85   15  )
 125   25  |
   .       |
   .       }——→ Naval data: see Table 5.8
   .       |
 8522  2143)
   .    500} → Decimal point denotes missing value. Used
               to obtain point prediction when x0 = 500
PROC PRINT; }  ——→ Print procedure: prints the data
PROC REG DATA = HOURS; } ——→ Specifies regression procedure
MODEL Y = X/NOINT P CLM CLI; }→ NOINT omits the intercept in the usual
                                model and thus specifies the model
                                     y_i = β_1 x_i + ε_i

                                Note: P = Point predictions desired
                                      CLM = 95% confidence intervals desired
                                      CLI = 95% prediction intervals desired
```

needed to carry out a regression analysis through the origin for the naval data in Table 5.8.

EXERCISES

5.1 Consider Exercise 4.3 in which we use simple linear regression to relate starting salary to grade point average (GPA).

a. Use a t-test to test the significance of the regression relationship. That is, decide whether or not we can reject $H_0: \beta_1 = 0$ in favor of $H_1: \beta_1 \neq 0$ by setting the probability of a Type I error equal to .05.

b. Use a t-test to decide whether or not we can reject $H_0: \beta_0 = 0$ in favor of $H_1: \beta_0 \neq 0$ by setting $\alpha = .05$. Can we reject $H_0: \beta_0 = 0$ with $\alpha = .01$?

c. Calculate the total variation, explained variation, unexplained variation (SSE), r^2, and r. Interpret r^2.

d. Calculate s^2 and s.

e. Calculate a 95% confidence interval for β_1. Interpret this interval.

f. Calculate a 95% confidence interval for β_0.

g. Calculate a 95% confidence interval for μ_0, the mean starting salary for all accounting majors who graduate with a 3.25 GPA.

h. Calculate a 95% prediction interval for y_0, an individual starting salary for an accounting major who graduates with a 3.25 GPA.
i. Calculate F(model). Use this statistic to test the significance of the regression relationship. That is, determine whether we can reject $H_0 : \beta_1 = 0$ in favor of $H_1 : \beta_1 \neq 0$ by setting α equal to .05.

5.2 Consider the market return rate data in Table 4.7. If we use the simple linear regression model to relate mean yearly market return rate (y) to mean yearly accounting rate (x), we obtain the following results:

$$b_0 = .84801 \qquad b_1 = .61033$$

$$s\sqrt{c_{00}} = 1.9765 \qquad s\sqrt{c_{11}} = .14316$$

a. Test to see whether there is a significant regression relationship between x and y.
b. Predict the market return rate for a stock having an accounting rate of 15.00.

5.3 Consider the Fresh Detergent demand data in Table 4.8. Suppose that we employ the model $y = \beta_0 + \beta_1 x_4 + \varepsilon$ to relate demand (y) to price difference (x_4). Using this model:

a. Test $H_0 : \beta_1 = 0$ versus $H_1 : \beta_1 \neq 0$ with $\alpha = .05$. Can we reject $H_0 : \beta_1 = 0$ with $\alpha = .05$?
b. Compute a 95% confidence interval for the slope β_1. Interpret this interval.
c. Test $H_0 : \beta_0 = 0$ versus $H_1 : \beta_0 \neq 0$ with $\alpha = .05$. Can we reject $H_0 : \beta_0 = 0$ with $\alpha = .05$?
d. Compute a 95% confidence interval for the intercept β_0.
e. Calculate a point estimate of μ_0, the mean demand for Fresh, when $x_{01} = \$3.80$ and $x_{02} = \$3.90$ (which implies that $x_{04} = x_{02} - x_{01} = \$3.90 - \$3.80 = \$.10$).
f. Calculate a 95% confidence interval for μ_0 when $x_{01} = \$3.80$ and $x_{02} = \$3.90$. Interpret this interval.
g. Calculate a point prediction of y_0, an individual demand for Fresh, when $x_{01} = \$3.80$ and $x_{02} = \$3.90$.
h. Calculate a 95% prediction interval for y_0 when $x_{01} = \$3.80$ and $x_{02} = \$3.90$. Interpret this interval.
i. Calculate F(model).
j. Use F(model) to test $H_0 : \beta_1 = 0$ versus $H_1 : \beta_1 \neq 0$ with $\alpha = .05$. Can we reject $H_0 : \beta_1 = 0$ with $\alpha = .05$?
k. Calculate the total variation, explained variation, unexplained variation, and r^2. Interpret r^2.

5.4 Consider the hospital labor requirements data given in Exercise 4.9. Suppose that we employ the simple linear regression model to relate monthly labor hours (y) to monthly occupied bed days (x). Using this model:

a. Employ a t statistic to test $H_0 : \beta_1 = 0$ versus $H_1 : \beta_1 \neq 0$ with $\alpha = .05$. Can we reject $H_0 : \beta_1 = 0$ with $\alpha = .05$?
b. Compute a 99% confidence interval for the slope β_1. Interpret this interval.
c. Calculate F(model).
d. Using F(model), test $H_0 : \beta_1 = 0$ versus $H_1 : \beta_1 \neq 0$ with $\alpha = .05$. Can we reject $H_0 : \beta_1 = 0$ with $\alpha = .05$?
e. Show that $F(\text{model}) = (t)^2$ and that $F_{[.05]}^{(1,13-2)} = (t_{[.025]}^{(13-2)})^2$.

f. What does the result of part (e) say about the tests carried out in parts (a) and (d)?
g. Using a t statistic, test $H_0 : \beta_0 = 0$ versus $H_1 : \beta_0 \neq 0$ with $\alpha = .05$? Can we reject $H_0 : \beta_0 = 0$ with $\alpha = .05$?
h. Compute a 99% confidence interval for the intercept β_0.
i. Calculate a point estimate of μ_0, the mean monthly labor hour requirement when $x_0 = 2500$.
j. Calculate a 95% confidence interval for μ_0 when $x_0 = 2500$. Interpret this interval.
k. Calculate a point prediction for y_0, an individual monthly labor hours requirement for a hospital, when $x_0 = 3000$.
l. Calculate a 95% prediction interval for y_0 when $x_0 = 3000$.
m. Calculate the total variation, explained variation, unexplained variation, and r^2. Interpret r^2.

5.5 Consider the Fresh Detergent demand data in Table 4.8.

a. Plot y versus each of x_1, x_2, and x_4.
b. For each of the models

$$y = \beta_0 + \beta_1 x_1 + \varepsilon$$
$$y = \beta_0 + \beta_1 x_2 + \varepsilon$$
$$y = \beta_0 + \beta_1 x_4 + \varepsilon$$

calculate r^2, the simple coefficient of determination. Which model describes the highest proportion of variation in the observed demands?
c. Test the significance of each of the above regression models using (1) an appropriate t statistic and (2) F(model), the overall F statistic.

5.6 Table 5.9 presents data concerning the need for hospital labor in $n = 13$ U.S. Navy hospitals. Here, x_1 denotes average daily patient load, x_2 denotes monthly X-ray exposures, x_3 denotes monthly occupied bed days, x_4 denotes eligible population in the area (divided by 1000), x_5 denotes average length of patients' stay in days, and y denotes monthly labor hours.

TABLE 5.9 **Hospital labor requirements data**

Hospital	x_1	x_2	x_3	x_4	x_5	y
1	15.57	2463	472.92	18.0	4.45	566.52
2	44.02	2048	1339.75	9.5	6.92	696.82
3	20.42	3940	620.25	12.8	4.28	1033.15
4	18.74	6505	568.33	36.7	3.90	1603.62
5	49.20	5723	1497.60	35.7	5.50	1611.37
6	44.92	11520	1365.83	24.0	4.60	1613.27
7	55.48	5779	1687.00	43.3	5.62	1854.17
8	59.28	5969	1639.92	46.7	5.15	2160.55
9	94.39	8461	2872.33	78.7	6.18	2305.58
10	128.02	20106	3655.08	180.5	6.15	3503.93
11	96.00	13313	2912.00	60.9	5.88	3571.89
12	131.42	10771	3921.00	103.7	4.88	3741.40
13	127.21	15543	3865.67	126.8	5.50	4026.52

Source: Procedures and Analysis for Staffing Standards Development: Regression Analysis Handbook (San Diego, Calif.: Navy Manpower and Material Analysis Center, 1979).

a. Plot y versus each of x_1, x_2, x_3, x_4, and x_5.
b. For each of the models

$$\begin{aligned} y &= \beta_0 + \beta_1 x_1 + \varepsilon \\ y &= \beta_0 + \beta_1 x_2 + \varepsilon \\ y &= \beta_0 + \beta_1 x_3 + \varepsilon \\ y &= \beta_0 + \beta_1 x_4 + \varepsilon \\ y &= \beta_0 + \beta_1 x_5 + \varepsilon \end{aligned}$$

calculate r^2, the simple coefficient of determination. Which model describes the highest proportion of variation in the observed labor requirements?
c. Test the significance of the regression relationship for each of the above models.

In Exercises 5.7 through 5.19, test the significance of the indicated regression relationship using the appropriate data. In each exercise, test the significance of the model (that is, test $H_0: \beta_1 = 0$) by employing (1) a t statistic and (2) F(model), the overall F statistic. Also, in each case compute and interpret a 95% confidence interval for the slope.

5.7 Using the demand data in Exercise 4.5, test the significance of the simple linear regression model relating demand (y) to price (x).

5.8 Using the natural gas consumption data in Exercise 4.6, test the significance of the simple linear regression model relating natural gas consumption (y) to production (x).

5.9 Using the heat loss data in Exercise 4.10, test the significance of the simple linear regression model relating heat loss (y) to outdoor temperature (x).

5.10 Using the personal computer sales data in Exercise 4.11, test the significance of the trend line relating sales (y) to time (t).

5.11 Using the electronic calculator sales data in Exercise 4.12, test the significance of the trend line relating sales (y) to time (t).

5.12 Using the work stoppage data in Exercise 4.13, test the significance of the simple linear regression model relating number of work stoppages (y) to the percent of the labor force having union membership (x).

5.13 Using the illiteracy rate data in Exercise 4.14, test the significance of the simple linear regression model relating the percent illiterate (y) to the school enrollment rate (x).

5.14 Using the divorce rate data in Exercise 4.15, test the significance of the simple linear regression model relating the divorce rate (y) to the percent of female population in the labor force (x).

5.15 Using the energy consumption data in Exercise 4.16, test the significance of the simple linear regression model relating U.S. energy consumption (y) to U.S. total population (x).

5.16 Using the social welfare expenditure data in Table 4.9, test the significance of the simple linear regression model relating social welfare expenditures (y) to total GNP (x_1).

5.17 Using the social welfare expenditure data in Table 4.9, test the significance of the simple linear regression model relating social welfare expenditures (y) to per capita GNP (x_2).

5.18 Using the social welfare expenditure data in Table 4.9, test the significance of the simple linear regression model relating social welfare expenditures (y) to unemployment rate (x_3).

5.19 Using the condominium selling price data in Table 4.10, test the significance of the simple linear regression model relating selling price (y) to square footage (x).

5.20 Using the residential sales data in Table 4.3, calculate a 95% prediction interval for the (individual) sales price of a house having 2000 square feet.

5.21 Consider the sales territory performance data in Table 4.4. For this data:

a. Calculate the least squares point estimates of the parameters in the simple linear regression model relating sales (y) to accounts (x).
b. Interpret the least squares point estimates computed in part (a).
c. Calculate a 95% confidence interval for mean unit sales for all representatives handling 130 accounts.
d. Calculate a 95% prediction interval for the individual unit sales of a representative handling 120 accounts.

5.22 Consider Example 5.3. In an article in *Financial Analysts Journal*, Haim Levy considers how a stock's value of β_1 depends on the length of time for which the rate of return is calculated. Levy calculated estimated values of β_1 for return length times varying from 1 to 30 months for each of 38 aggressive stocks, 38 defensive stocks, and 68 neutral stocks. Each estimated value was based on data from 1946 to 1975. Table 5.10 gives the average estimate of β_1 for each stock type for different return length times.

Let

$$y = \text{average estimate of } \beta_1$$
$$x = \text{return length time}$$

TABLE 5.10 **Estimates of β_1 for different return length times**

Return length time	Average estimate of β_1		
	Aggressive stocks	Defensive stocks	Neutral stocks
1	1.37	.50	.98
3	1.42	.44	.95
6	1.53	.41	.94
9	1.69	.39	1.00
12	1.83	.40	.98
15	1.67	.38	1.00
18	1.78	.39	1.02
24	1.86	.35	1.14
30	1.83	.33	1.22

Source: Reprinted by permission from Haim Levy, "Measuring Risk and Performance over Alternative Investment Horizons," *Financial Analysts Journal* (March–April 1984): 61–68.

Consider relating y to x for each stock type by using the simple linear regression model

$$y = \beta_0^* + \beta_1^* x + \varepsilon$$

Calculate a 95% confidence interval for β_1^* for each stock type. Carefully interpret the meaning of each interval.

5.23 Use the demand data given in Exercise 4.5 to calculate:

a. A 95% confidence interval for the mean demand when the product price is \$2.75.
b. A 99% prediction interval for an individual demand when the product price is \$3.10.

5.24 Use the natural gas consumption data in Exercise 4.6 to compute a 99% confidence interval for the mean natural gas consumption when weekly production is 17,000 units.

5.25 Use the heat loss data in Exercise 4.10 to compute a 95% confidence interval for the average heat loss when the outdoor temperature is 30°F.

5.26 Use the personal computer sales data in Exercise 4.11 to compute a 95% prediction interval for sales in time period 17.

5.27 Use the work stoppage data in Exercise 4.13 to compute a 98% prediction interval for the number of work stoppages when 20 percent of the labor force has union membership.

5.28 Use the divorce rate data in Exercise 4.15 to compute a 95% prediction interval for the divorce rate when 45% of the female population is in the labor force.

5.29 Use the U.S. energy consumption data in Exercise 4.16 to compute a 95% prediction interval for energy consumption when the U.S. total population is 213 million.

5.30 Use the social welfare expenditure data in Table 4.9 to compute:

a. A 95% prediction interval for the social welfare expenditure when total GNP is 1186.
b. A 95% prediction interval for the social welfare expenditure when per capita GNP is 5552.

5.31 Use the condominium selling price data in Table 4.10 to compute a 99% confidence interval for the mean selling price of condominiums having 1000 square feet.

5.32 Use the service call data in Exercise 4.23 to compute a 99% confidence interval for the average time required to perform service on nine copiers.

5.33 Using the data in Exercise 4.3, compute the simple correlation coefficient between starting salary (y) and GPA (x). Interpret this correlation coefficient.

5.34 Using the data in Exercise 4.6, compute the simple correlation coefficient between natural gas consumption (y) and production (x). Interpret this correlation coefficient.

5.35 Use the data in Table 4.7 to do the following:

a. Calculate the simple correlation coefficient between mean market rate (y) and mean accounting rate (x). Interpret the correlation coefficient.
b. If ρ is the population correlation coefficient between y and x, test the null hypothesis $H_0: \rho = 0$ versus $H_1: \rho \neq 0$ with $\alpha = .05$.
c. Calculate an approximate 95% confidence interval for ρ.

5.36 Using the data in Exercise 4.13, compute the simple correlation coefficient between the number of work stoppages (y) and the percent of the civilian labor force with union membership (x). Interpret this correlation coefficient.

5.37 Using the data in Exercise 4.14, compute and interpret the simple correlation coefficient between the illiteracy rate (y) and school enrollment rate (x).

5.38 Using the data in Exercise 4.15, compute and interpret the simple correlation coefficient between divorce rate (y) and the percent of female population in the labor force (x).

5.39 Using the data in Exercise 4.16, compute and interpret the simple correlation coefficient between U.S. energy consumption (y) and U.S. total population (x).

5.40 Using the data in Table 4.9, calculate and interpret the simple correlation coefficients between:

a. Social welfare expenditures and total GNP.
b. Social welfare expenditures and per capita GNP.
c. Social welfare expenditures and unemployment rate.

5.41 The data in Table 5.11 concerns hotel room rates. The variables are defined as follows:

y = daily rate of the room

x_1 = population of the city in which the hotel is located

x_2 = rating of the hotel (1, 2, or 3)

x_3 = number of rooms (units) in the hotel

TABLE 5.11 Hotel room rate data

Population, x_1 (hundreds)	Rating, x_2	Number of rooms, x_3	Daily rate, y (dollars)
385	2	59	56.0
385	1	62	38.0
1256	2	139	56.0
1256	2	51	49.0
1256	3	150	67.0
1256	3	151	60.5
350	3	89	50.5
292	2	44	49.0
185	3	101	55.0
290	2	101	65.0
3645	3	317	94.0
3645	3	159	57.0
3645	3	257	67.0
820	2	98	66.0
360	2	116	54.0
200	2	130	63.0
131	2	32	46.5
515	3	128	59.0
550	3	118	65.0
260	3	117	57.0
650	2	53	53.0

TABLE 5.11 Continued

Population, x_1 (hundreds)	Rating, x_2	Number of rooms, x_3	Daily rate, y (dollars)
580	2	147	44.0
363	2	61	42.0
250	2	170	47.0
462	3	117	66.0
2140	2	190	43.0
2140	1	49	51.0
5750	3	175	65.0
5750	2	169	41.0
5840	3	160	78.0
5840	3	201	83.5
5840	3	117	71.0
5840	2	148	46.0
3954	3	250	94.0
3954	3	246	78.0

Using this data, do the following:

a. Plot y versus x_1, x_2, x_3. Then consider the models

$$y = \beta_0 + \beta_1 x_1 + \varepsilon$$
$$y = \beta_0 + \beta_1 x_2 + \varepsilon$$
$$y = \beta_0 + \beta_1 x_3 + \varepsilon$$

Calculate r^2, the simple coefficient of determination, for each of these models. Which model explains the highest proportion of the total variation in the observed room rates?

b. Test the significance of the regression relationship for each of the above models.
c. Which of the above models seems to be best? Explain your answer.
d. Using the model that you think is best, compute a 95% prediction interval for an individual room rate for a hotel having 100 units, given a rating of 3, and located in a city with a population of 150,000.
e. Using the model that you think is best, compute a 95% confidence interval for the average room rate for hotels having 100 units, given a rating of 3, and located in a city with a population of 150,000.

Exercises 5.42–5.44 refer to the data in Table 5.12. This table contains economic data for selected U.S. cities. The variables are defined as follows.

CITY = city name

HCOST = indexed housing cost (mean U.S.A. cost = 100)

FCOST = indexed food cost

OCOST = indexed cost of other items

INC = mean personal income

TABLE 5.12 Economic data for selected U.S. cities

CITY	HCOST	FCOST	OCOST	INC	TAXES	BITE	BCI	IG	JG
Athens	80	96	99	23251	808	3.82	106	39.00	15.54
Baltimore	110	99	106	32368	1267	3.91	98	44.52	6.86
Battle Creek	76	95	106	29868	1387	4.64	116	32.53	−14.97
Bremerton	109	103	112	32631	506	1.55	107	38.87	15.60
Bridgeport	161	110	103	47258	546	1.16	104	53.37	17.60
Chattanooga	74	98	97	25782	314	1.22	114	32.68	−0.82
Dayton	92	105	97	30232	865	2.86	103	34.69	−4.60
Fall River	108	94	97	27591	1198	4.34	125	43.46	1.96
Gainesville	89	98	99	22756	265	1.16	85	44.38	23.05
Harrisburg	95	98	106	31447	945	3.01	103	42.24	3.21
Jackson	87	107	95	27499	683	2.48	95	38.01	8.66
Kansas City	85	102	105	30783	789	2.56	102	37.41	−2.60
Lake Charles	83	104	101	29288	296	1.01	119	45.44	10.31
Lake County	159	101	106	38912	1228	3.16	75	38.47	17.29
Lexington	100	97	101	30570	924	3.02	95	45.94	19.68
Macon	75	101	96	26944	972	3.61	106	46.69	7.46
Medford	108	109	103	24699	1046	4.23	103	29.08	0.12
Milwaukee	122	94	94	35272	1703	4.83	106	40.33	−4.03
New London	123	100	103	32922	438	1.29	109	58.43	16.48
Ocala	81	97	101	23777	265	1.11	108	50.70	39.38
Portland	97	94	97	30794	999	3.24	99	46.99	14.66
Reno	149	102	111	38472	374	0.97	91	32.32	20.26
San Antonio	83	99	96	28367	214	0.75	100	52.94	22.65
San Diego	173	97	101	32586	997	3.06	93	44.37	16.82
San Francisco	206	108	120	47966	2030	4.23	81	48.69	12.53
Shreveport	78	104	105	30817	319	1.04	109	54.97	11.84
Tallahassee	96	99	96	24144	280	1.16	77	45.36	25.82
Washington	165	107	107	41888	2371	5.66	75	47.67	13.44
Youngstown	84	100	93	28960	814	2.81	122	30.30	−14.27

Source: From the book *Places Rated Almanac* by Richard Boyer and David Savageau. © 1985. Used by permission of the publisher, Prentice Hall Press, New York, NY.

TAXES = mean local and state taxes

BITE = TAXES/INC

BCI = blue-collar population index (U.S.A. mean level = 100)

IG = income growth (personal) over last five years

JG = jobs growth over last five years

5.42 Find a good simple linear regression model for predicting HCOST (indexed housing cost). Justify your model choice and use the model to make a relevant prediction.

5.43 Find a good simple linear regression model for predicting JG (jobs growth over the last five years). Justify your model choice and use the model to make a relevant prediction.

5.44 Find a good simple linear regression model for predicting IG (personal income growth over the last five years). Justify your model choice and use the model to make a relevant prediction.

5.45 Consider the Fresh Detergent demand data in Table 4.8. Figure 5.13 gives the SAS output of a simple linear regression relating demand (y) to price difference (x_4). Using this SAS output:

a. Find and report the least squares point estimates b_0 and b_1. Use them to write the least squares prediction equation.
b. Find and report the total variation, explained variation, *SSE*, and r^2. Interpret the value of r^2.
c. Find and report the t statistic for testing $H_0: \beta_1 = 0$ and the associated prob-value. Use the prob-value to test the significance of the regression relationship.
d. Find and report F(model) and the associated prob-value. Use the prob-value to test the significance of the regression relationship.
e. Find and report the t statistic for testing $H_0: \beta_0 = 0$ and the associated prob-value. Use the prob-value to test $H_0: \beta_0 = 0$.
f. Find and report s^2 and s.
g. Calculate a 95% confidence interval for β_1. Interpret this interval.
h. Find and report the 95% confidence interval for mean demand when the price difference is \$.25. *Hint*: Note that sales region 2 employed a price difference of \$.25 (see Table 4.8).
i. Find and report the 95% prediction interval for an individual demand when the price difference is \$.60. Interpret this interval.

FIGURE 5.13 **SAS output of the simple linear regression relating demand (y) to price difference (x_4)**

SAS

ANALYSIS OF VARIANCE

SOURCE	DF	SUM OF SQUARES	MEAN SQUARE	F VALUE	PROB>F
MODEL	1	10.65268464	10.65268464	106.303	0.0001
ERROR	28	2.80590202	0.10021079		
C TOTAL	29	13.45858667			

ROOT MSE	0.3165609	R-SQUARE	0.7915
DEP MEAN	8.382667	ADJ R-SQ	0.7841
C.V.	3.776374		

PARAMETER ESTIMATES

VARIABLE	DF	PARAMETER ESTIMATE	STANDARD ERROR	T FOR H0: PARAMETER=0	PROB > |T|
INTERCEP	1	7.81408758	0.07988432	97.818	0.0001
X4	1	2.66521449	0.25849959	10.310	0.0001

OBS	ACTUAL	PREDICT VALUE	STD ERR PREDICT	LOWER95% MEAN	UPPER95% MEAN	LOWER95% PREDICT	UPPER95% PREDICT	RESIDUAL
1	7.3800	7.6808	0.0893	7.4979	7.8637	7.0071	8.3546	-0.3008
2	8.5100	8.4804	0.0586	8.3604	8.6004	7.8209	9.1398	0.0296
3	9.5200	9.4132	0.1155	9.1767	9.6497	8.7230	10.1034	0.1068

(continues)

FIGURE 5.13 **Continued**

OBS	ACTUAL	PREDICT VALUE	STD ERR PREDICT	LOWER95% MEAN	UPPER95% MEAN	LOWER95% PREDICT	UPPER95% PREDICT	RESIDUAL
4	7.5000	7.8141	0.0799	7.6505	7.9777	7.1453	8.4829	-0.3141
5	9.3300	8.4804	0.0586	8.3604	8.6004	7.8209	9.1398	0.8496
6	8.2800	8.3471	0.0579	8.2285	8.4657	7.6879	9.0063	-0.0671
7	8.7500	8.2139	0.0601	8.0908	8.3369	7.5539	8.8739	0.5361
8	7.8700	7.9473	0.0716	7.8007	8.0940	7.2825	8.6122	-0.0773
9	7.1000	7.4143	0.1103	7.1884	7.6402	6.7276	8.1010	-0.3143
10	8.0000	8.2139	0.0601	8.0908	8.3369	7.5539	8.8739	-0.2139
11	7.8900	8.3471	0.0579	8.2285	8.4657	7.6879	9.0063	-0.4571
12	8.1500	8.0806	0.0648	7.9479	8.2133	7.4187	8.7425	0.0694
13	9.1000	8.8802	0.0753	8.7259	9.0344	8.2136	9.5467	0.2198
14	8.8600	9.0134	0.0842	8.8410	9.1858	8.3425	9.6844	-0.1534
15	8.9000	8.7469	0.0677	8.6082	8.8857	8.0838	9.4100	0.1531
16	8.8700	8.6137	0.0620	8.4867	8.7406	7.9529	9.2744	0.2563
17	9.2600	9.1467	0.0940	8.9542	9.3392	8.4703	9.8231	0.1133
18	9.0000	9.1467	0.0940	8.9542	9.3392	8.4703	9.8231	-0.1467
19	8.7500	8.8802	0.0753	8.7259	9.0344	8.2136	9.5467	-0.1302
20	7.9500	7.6808	0.0893	7.4979	7.8637	7.0071	8.3546	0.2692
21	7.6500	7.6808	0.0893	7.4979	7.8637	7.0071	8.3546	-0.0308
22	7.2700	7.5476	0.0995	7.3437	7.7514	6.8678	8.2273	-0.2776
23	8.0000	8.3471	0.0579	8.2285	8.4657	7.6879	9.0063	-0.3471
24	8.5000	8.0806	0.0648	7.9479	8.2133	7.4187	8.7425	0.4194
25	8.7500	9.1467	0.0940	8.9542	9.3392	8.4703	9.8231	-0.3967
26	9.2100	9.4132	0.1155	9.1767	9.6497	8.7230	10.1034	-0.2032
27	8.2700	7.6808	0.0893	7.4979	7.8637	7.0071	8.3546	0.5892
28	7.6700	7.8141	0.0799	7.6505	7.9777	7.1453	8.4829	-0.1441
29	7.9300	7.9473	0.0716	7.8007	8.0940	7.2825	8.6122	-0.0173
30	9.2600	9.2800	0.1045	9.0660	9.4940	8.5971	9.9628	-0.0200
31	.	8.3471	0.0579	8.2285	8.4657	7.6879	9.0063	.

```
SUM OF RESIDUALS                 3.73035E-14
SUM OF SQUARED RESIDUALS            2.805902
PREDICTED RESID SS (PRESS)          3.155519

DURBIN-WATSON D                2.414
(FOR NUMBER OF OBS.)              30
1ST ORDER AUTOCORRELATION     -0.223
```

5.46 Consider the residential sales price data in Table 4.3. Figure 5.14 gives the SAS output of the simple linear regression relating sales price (y) to square footage (x). Using this SAS output:

a. Find and report the least squares point estimates b_0 and b_1. Use them to write the least squares prediction equation.

b. Find and report the total variation, explained variation, SSE, and r^2. Interpret the value of r^2.

c. Find and report the t statistic for testing $H_0: \beta_1 = 0$ and the associated prob-value. Use the prob-value to test the significance of the regression relationship.

FIGURE 5.14 SAS output of the simple linear regression relating sales price (y) to square footage (x)

SAS

ANALYSIS OF VARIANCE

SOURCE	DF	SUM OF SQUARES	MEAN SQUARE	F VALUE	PROB>F
MODEL	1	48147.30258	48147.30258	105.307	0.0001
ERROR	61	27889.68726	457.20799		
C TOTAL	62	76036.98984			

ROOT MSE	21.38242	R-SQUARE	0.6332
DEP MEAN	78.80159	ADJ R-SQ	0.6272
C.V.	27.13451		

PARAMETER ESTIMATES

VARIABLE	DF	PARAMETER ESTIMATE	STANDARD ERROR	T FOR H0: PARAMETER=0	PROB > |T|
INTERCEP	1	-3.72102928	8.48086264	-0.439	0.6624
SQFT	1	0.05477049	0.005337250	10.262	0.0001

OBS	ACTUAL	PREDICT VALUE	STD ERR PREDICT	LOWER95% MEAN	UPPER95% MEAN	LOWER95% PREDICT	UPPER95% PREDICT	RESIDUAL
1	53.5000	51.4876	3.7871	43.9149	59.0603	8.0654	94.9099	2.0124
2	49.0000	66.9329	2.9317	61.0706	72.7952	23.7761	110.1	-17.9329
3	50.5000	43.3816	4.3784	34.6264	52.1368	-0.2624	87.0256	7.1184
4	49.9000	46.2297	4.1632	37.9049	54.5544	2.6700	89.7894	3.6703
5	52.0000	62.2226	3.1412	55.9414	68.5039	19.0069	105.4	-10.2226
6	55.0000	62.2226	3.1412	55.9414	68.5039	19.0069	105.4	-7.2226
7	80.5000	92.8941	3.0238	86.8477	98.9405	49.7119	136.1	-12.3941
8	86.0000	83.9118	2.7396	78.4336	89.3899	40.8054	127.0	2.0882
9	69.0000	65.0159	3.0103	58.9965	71.0354	21.8375	108.2	3.9841
10	149.0	193.5	11.4927	170.5	216.4	144.9	242.0	-44.4527
11	46.0000	43.6007	4.3616	34.8791	52.3223	-0.0366	87.2380	2.3993
12	38.0000	35.7137	4.9887	25.7382	45.6893	-8.1914	79.6188	2.2863
13	49.5000	51.4876	3.7871	43.9149	59.0603	8.0654	94.9099	-1.9876
14	105.0	103.1	3.5854	95.9119	110.3	59.7277	146.4	1.9186
15	152.5	110.5	4.1009	102.3	118.7	66.9942	154.1	41.9698
16	85.0000	106.4	3.8081	98.8076	114.0	62.9928	149.9	-21.4224
17	60.0000	76.5177	2.7031	71.1125	81.9229	33.4206	119.6	-16.5177
18	58.5000	63.7562	3.0671	57.6233	69.8892	20.5618	107.0	-5.2562
19	101.0	91.3605	2.9589	85.4439	97.2772	48.1963	134.5	9.6395
20	79.4000	67.2615	2.9192	61.4242	73.0989	24.1081	110.4	12.1385
21	125.0	105.6	3.7520	98.0983	113.1	62.1908	149.0	19.3991
22	87.9000	98.9189	3.3317	92.2567	105.6	55.6461	142.2	-11.0189
23	80.0000	82.8164	2.7222	77.3730	88.2597	39.7144	125.9	-2.8164
24	94.0000	101.4	3.4818	94.4759	108.4	58.1183	144.8	-7.4383
25	74.0000	74.6008	2.7249	69.1521	80.0495	31.4982	117.7	-0.6008
26	69.0000	77.6679	2.6962	72.2765	83.0593	34.5725	120.8	-8.6679
27	63.0000	51.4876	3.7871	43.9149	59.0603	8.0654	94.9099	11.5124
28	67.5000	66.4947	2.9488	60.5982	72.3913	23.3333	109.7	1.0053
29	35.0000	58.3887	3.3487	51.6925	65.0850	15.1107	101.7	-23.3887

(continues)

FIGURE 5.14 Continued

SAS

OBS	ACTUAL	PREDICT VALUE	STD ERR PREDICT	LOWER95% MEAN	UPPER95% MEAN	LOWER95% PREDICT	UPPER95% PREDICT	RESIDUAL
30	142.5	127.7	5.4762	116.8	138.7	83.5914	171.9	14.7718
31	92.2000	89.4436	2.8866	83.6714	95.2158	46.2989	132.6	2.7564
32	56.0000	52.1449	3.7423	44.6616	59.6281	8.7381	95.5516	3.8551
33	63.0000	53.9523	3.6223	46.7091	61.1955	10.5863	97.3183	9.0477
34	60.0000	90.9224	2.9415	85.0405	96.8043	47.7629	134.1	-30.9224
35	34.0000	19.0635	6.4144	6.2370	31.8900	-25.5758	63.7028	14.9365
36	52.0000	53.2403	3.6690	45.9036	60.5770	9.8586	96.6220	-1.2403
37	75.0000	78.2156	2.6945	72.8276	83.6037	35.1207	121.3	-3.2156
38	93.0000	102.3	3.5366	95.2428	109.4	58.9770	145.7	-9.3146
39	60.0000	100.6	3.4284	93.7065	107.4	57.2591	143.9	-40.5620
40	73.0000	55.4311	3.5276	48.3773	62.4849	12.0963	98.7659	17.5689
41	71.0000	93.1132	3.0335	87.0473	99.1791	49.9282	136.3	-22.1132
42	83.0000	78.5990	2.6940	73.2120	83.9860	35.5042	121.7	4.4010
43	90.0000	91.3605	2.9589	85.4439	97.2772	48.1963	134.5	-1.3605
44	83.0000	89.1150	2.8753	83.3654	94.8645	45.9733	132.3	-6.1150
45	115.0	116.0	4.5169	107.0	125.0	72.3069	159.7	-1.0073
46	50.0000	44.9152	4.2616	36.3935	53.4368	1.3174	88.5129	5.0848
47	55.2000	57.6219	3.3937	50.8359	64.4080	14.3299	100.9	-2.4219
48	61.0000	72.9577	2.7535	67.4518	78.4636	29.8478	116.1	-11.9577
49	147.0	114.9	4.4274	106.0	123.7	71.1933	158.5	32.1429
50	210.0	125.2	5.2593	114.6	135.7	81.1228	169.2	84.8461
51	60.0000	80.4064	2.6985	75.0105	85.8024	37.3105	123.5	-20.4064
52	100.0	104.3	3.6640	96.9598	111.6	60.9064	147.7	-4.2864
53	44.5000	57.6219	3.3937	50.8359	64.4080	14.3299	100.9	-13.1219
54	55.0000	87.4171	2.8217	81.7747	93.0595	44.2896	130.5	-32.4171
55	53.4000	46.9417	4.1105	38.7222	55.1611	3.4020	90.4814	6.4583
56	65.0000	66.8234	2.9359	60.9526	72.6941	23.6654	110.0	-1.8234
57	73.0000	72.9577	2.7535	67.4518	78.4636	29.8478	116.1	0.0423
58	40.0000	71.6432	2.7828	66.0787	77.2077	28.5258	114.8	-31.6432
59	141.0	107.9	3.9113	100.1	115.7	64.4350	151.4	33.0988
60	68.0000	82.3782	2.7164	76.9464	87.8099	39.2777	125.5	-14.3782
61	139.0	80.8994	2.7017	75.4970	86.3017	37.8026	124.0	58.1006
62	140.0	105.4	3.7409	97.9563	112.9	62.0303	148.8	34.5634
63	55.0000	58.1696	3.3615	51.4479	64.8913	14.8877	101.5	-3.1696

```
SUM OF RESIDUALS                 7.60281E-13
SUM OF SQUARED RESIDUALS         27889.69
PREDICTED RESID SS (PRESS)       31800.02
```

d. Find and report F(model) and the associated prob-value. Use the prob-value to test the significance of the regression relationship.

e. Find and report the t statistic for testing $H_0: \beta_0 = 0$ and the associated prob-value. Use the prob-value to test $H_0: \beta_0 = 0$.

f. Find and report s^2 and s.
g. Calculate a 95% confidence interval for β_0. What does this interval say about whether or not we should include the intercept in the model?
h. Find and report the 95% confidence interval for the average sales price of all residences having 1290 square feet. *Hint*: Note that residence 2 has 1290 square feet (see Table 4.3).
i. Calculate the 99% confidence interval for the average sales price of all residences having 1290 square feet. Interpret this interval.
j. Find and report the 95% prediction interval for an individual sales price of a residence having 1400 square feet.
k. Calculate the 99% prediction interval for an individual sales price of a residence having 1400 square feet. Interpret this interval.

5.47 Consider the service time data in Exercise 4.23. Figure 5.15 gives the SAS output of a simple linear regression relating the time required to perform service (y) to the number of copiers serviced (x). Using this SAS output:

a. Find and report the least squares point estimates b_0 and b_1. Use them to write the least squares prediction equation.
b. Find and report the total variation, explained variation, *SSE*, and r^2. Interpret the value of r^2.
c. Find and report the t statistic for testing $H_0: \beta_1 = 0$ and the associated prob-value. Use the prob-value to test the significance of the regression relationship.
d. Find and report F(model) and the associated prob-value. Use the prob-value to test the significance of the regression relationship.

FIGURE 5.15 **SAS output of a simple linear regression relating service time (y) to number of copiers serviced (x)**

```
                                        SAS
DEP VARIABLE:  SERVTIME
                                ANALYSIS OF VARIANCE

                                SUM OF              MEAN
              SOURCE     DF     SQUARES             SQUARE    F VALUE    PROB>F

              MODEL       1   59492.77572     59492.77572    502.465    0.0001
              ERROR      13    1539.22428       118.40187
              C TOTAL    14   61032.00000

                  ROOT MSE      10.88126        R-SQUARE     0.9748
                  DEP MEAN           121        ADJ R-SQ     0.9728
                  C.V.          8.992779

                                PARAMETER ESTIMATES

                         PARAMETER          STANDARD      T FOR HO:
          VARIABLE DF     ESTIMATE             ERROR    PARAMETER=0      PROB > |T|

          INTERCEP  1    15.13838812      5.49516796          2.755          0.0164

          COPIERS   1    21.75238600      0.97040732         22.416          0.0001
```

(continues)

FIGURE 5.15 Continued

OBS	ACTUAL	PREDICT VALUE	STD ERR PREDICT	LOWER95% MEAN	UPPER95% MEAN	LOWER95% PREDICT	UPPER95% PREDICT	RESIDUAL
1	92.0000	80.3955	3.3429	73.1737	87.6174	55.8037	105.0	11.6045
2	63.0000	58.6432	3.9537	50.1016	67.1847	33.6319	83.6544	4.3568
3	145.0	145.7	3.0171	139.1	152.2	121.3	170.0	-0.6527
4	195.0	189.2	4.1399	180.2	198.1	164.0	214.3	5.8425
5	49.0000	58.6432	3.9537	50.1016	67.1847	33.6319	83.6544	-9.6432
6	90.0000	102.1	2.9327	95.8122	108.5	77.8016	126.5	-12.1479
7	119.0	123.9	2.8125	117.8	130.0	99.6202	148.2	-4.9003
8	145.0	145.7	3.0171	139.1	152.2	121.3	170.0	-0.6527
9	67.0000	58.6432	3.9537	50.1016	67.1847	33.6319	83.6544	8.3568
10	115.0	102.1	2.9327	95.8122	108.5	77.8016	126.5	12.8521
11	154.0	145.7	3.0171	139.1	152.2	121.3	170.0	8.3473
12	267.0	254.4	6.5816	240.2	268.6	226.9	281.9	12.5854
13	77.0000	80.3955	3.3429	73.1737	87.6174	55.8037	105.0	-3.3955
14	210.0	232.7	5.7191	220.3	245.0	206.1	259.2	-22.6622
15	27.0000	36.8908	4.6875	26.7640	47.0175	11.2948	62.4868	-9.8908

SUM OF RESIDUALS	-3.19744E-14
SUM OF SQUARED RESIDUALS	1539.224
PREDICTED RESID SS (PRESS)	2462.054

e. Find and report the t statistic for testing $H_0: \beta_0 = 0$ and the associated prob-value. Use the prob-value to test $H_0: \beta_0 = 0$.
f. Find and report s^2 and s.
g. Find and report the 95% confidence interval for the average service time required to service eight copiers. *Hint*: Note that eight copiers were serviced on call 4 (see Exercise 4.23).
h. Calculate the 98% confidence interval for the average service time required to service two copiers. Interpret this interval.
i. Find and report the 95% prediction interval for an individual service time when six copiers are serviced.
j. Calculate the 99% prediction interval for an individual service time when five copiers are serviced. Interpret this interval.

5.48 Consider Example 4.2 and the American Manufacturing sales volume data in Table 4.2. The prob-value related to testing $H_0: \beta_1 = 0$ in the simple linear model relating sales volume (y) to advertising expenditure (x) can be computed to be .0009.

a. Draw a figure that illustrates how this prob-value is computed.
b. Using the prob-value, determine whether we can reject $H_0: \beta_1 = 0$ in favor of $H_1: \beta_1 \neq 0$ by setting α equal to .05, .01, .001, and .0001.
c. Interpret the prob-value as a probability.

5.49 Consider the fuel consumption problem and the three samples in Table 5.1.

a. Using each of the three samples in Table 5.1, calculate a 99% confidence interval for β_1 in the fuel consumption model

$$y = \beta_0 + \beta_1 x + \varepsilon$$

b. What percentage of the three 99% confidence intervals computed in part (a) contains the slope β_1 (= $-.1281$)? Explain the meaning of "99% confidence" in the context of this situation.

To answer parts (c) through (g), suppose that in a future week the average hourly temperature will be $x_0 = 37.0$°F.

c. Using each of the three samples in Table 5.1, calculate a 95% confidence interval for μ_0, the future mean fuel consumption.
d. By using the true values of β_0 and β_1 ($\beta_0 = 15.77$ and $\beta_1 = -.1281$), calculate the true value of μ_0.
e. What percentage of the three 95% confidence intervals computed in part (c) contains the true value of μ_0? Explain the meaning of "95% confidence" in the context of this situation.
f. Using each of the three samples in Table 5.1, calculate a 95% prediction interval for $y_0 = \mu_0 + \varepsilon_0$, the future individual fuel consumption.
g. Suppose that
   (1) after using the first sample to calculate a prediction interval, we observe the future fuel consumption $y_0 = 11.9$,
   (2) after using the second sample to calculate a prediction interval, we observe the future fuel consumption $y_0 = 10.5$, and
   (3) after using the third sample to calculate a prediction interval, we observe the future fuel consumption $y_0 = 11.4$.

What percentage of the three prediction interval–future fuel consumption combinations are successful? Explain the meaning of "95% confidence" in the context of this situation.

5.50 Reconsider the heat loss data of Exercise 4.10. The data is repeated below.

Outdoor temperature, (°F)	***Heat loss, y***
20	86, 80, 77
40	78, 84, 75
60	33, 38, 43

If we perform a regression analysis of this data by using the model

$$y = \beta_0 + \beta_1 x + \varepsilon$$

we obtain an unexplained variation of $SSE = 894.5$.

a. Perform a lack-of-fit test of the functional form of this model by setting $\alpha = .05$.
b. Does a data plot of y versus x confirm the result of the lack-of-fit test? Explain.

5.51 Consider the lack-of-fit test performed in Example 5.17.

a. If the next service call requires 84 minutes to service three microcomputers, add this observation to the data set and perform a new test of lack of fit by setting α equal to .05.

b. What does the lack-of-fit test done in part (a) say about the adequacy of the simple linear regression model

$$y = \beta_0 + \beta_1 x + \varepsilon$$

That is, is the linear functional form appropriate? Why?

5.52 Consider the regression through the origin in Example 5.18, which employs the data in Table 5.8. Using the model

$$\begin{aligned} y_i &= \mu_i + \varepsilon_i \\ &= \beta_1 x_i + \varepsilon_i \end{aligned}$$

do the following:

a. Calculate SSE, s^2, and s.
b. Using a t statistic, test $H_0 : \beta_1 = 0$ versus $H_1 : \beta_1 \neq 0$ with $\alpha = .05$. Can we reject $H_0 : \beta_1 = 0$ with $\alpha = .05$?
c. Calculate a 95% confidence interval for β_1. Interpret this interval.
d. Calculate a point estimate of and a 95% confidence interval for

$$\mu_0 = \beta_1 x_0 = \beta_1(500)$$

Interpret the 95% confidence interval.
e. Calculate a point prediction of and a 95% prediction interval for

$$y_0 = \mu_0 + \varepsilon_0 = \beta_1(500) + \varepsilon_0$$

Interpret the 95% prediction interval.

5.53 Consider the residential sales price data in Table 4.3. Carry out a regression through the origin for this data using the model

$$\begin{aligned} y_i &= \mu_i + \varepsilon_i \\ &= \beta_1 x_i + \varepsilon_i \end{aligned}$$

a. Calculate b_1, SSE, s^2, and s.
b. Using a t statistic, test the significance of the regression relationship.
c. Calculate and interpret a 99% confidence interval for β_1.
d. Calculate a point estimate of and a 95% confidence interval for

$$\mu_0 = \beta_1 x_0 = \beta_1(1290)$$

Interpret this 95% confidence interval.
e. Calculate a point prediction of and a 95% prediction interval for

$$y_0 = \mu_0 + \varepsilon_0 = \beta_1(1290) + \varepsilon_0$$

Interpret this 95% prediction interval.

5.54 Use fact (3) in the first box in Section 5.4 to prove that a $100(1 - \alpha)\%$ confidence interval for β_0 in the simple linear regression model is

$$[b_0 \pm t_{[\alpha/2]}^{(n-2)} s\sqrt{c_{00}}]$$

5.55 Use fact (3) in the box in Section 5.5 to prove that a $100(1 - \alpha)\%$ confidence interval for μ_0, the mean value of the dependent variable when the value of the independent variable x is x_0, is

$$[\hat{y}_0 \pm t_{[\alpha/2]}^{(n-2)} s\sqrt{h_{00}}]$$

when we use the simple linear regression model to relate y to x.

5.56 Use fact (3) in the box in Section 5.6 to prove that a $100(1 - \alpha)\%$ prediction interval for y_0, the individual value of the dependent variable when the value of the independent variable x is x_0, is

$$[\hat{y}_0 \pm t_{[\alpha/2]}^{(n-2)} s\sqrt{1 + h_{00}}]$$

when we use the simple linear regression model to relate y to x.

5.57 Consider the simple linear regression model

$$y = \beta_0 + \beta_1 x + \varepsilon$$

and consider testing $H_0: \beta_1 = 0$ versus $H_1: \beta_1 \neq 0$ by using the $100(1 - \alpha)\%$ confidence interval for β_1. Show that, if the $100(1 - \alpha)\%$ confidence interval for β_1 does not contain zero, then the rejection point condition for rejecting $H_0: \beta_1 = 0$ (by setting the probability of a Type I error equal to α) holds.

5.58 (Optional) Write the SAS program needed to carry out a regression analysis of the hospital labor requirement data (given in Table 5.9) using the model

$$y = \beta_0 + \beta_1 x_3 + \varepsilon$$

Make point and interval predictions of y_0 when $x_3 = 2500$ and when $x_3 = 3000$.

5.59 (Optional) Write the SAS program needed to carry out a regression analysis of the Fresh Detergent data given in Table 4.8 using the model

$$y = \beta_0 + \beta_1 x_4 + \varepsilon$$

Make point and interval predictions of y_0 when $x_1 = \$3.80$ and $x_2 = \$3.90$.

5.60 (Optional) Write the SAS program needed to carry out a simple linear regression analysis relating market rate (y) to accounting rate (x). Use the data given in Table 4.7 and make point and interval predictions for a stock having an accounting rate of 21.00.

5.61 In Appendix F, consider Data Base 1: Manufacturing Data. Use the simple linear regression model to relate each of the following dependent variables to the indicated independent variable. Note that the variables are defined in Appendix F.

a. PAY to NEMP
b. PRODPAY to NPRODH
c. NCAPEX to VALSHIP

In each case, calculate r^2 and test $H_0: \beta_1 = 0$ by setting α equal to .05.

5.62 In Appendix F, consider Data Base 2: Ohio Local Government and Payroll Data. Use the simple linear regression model to relate each of the following dependent variables to the indicated independent variable. Note that the variables are defined in Appendix F.

a. POLEMP to POP
b. FIREMP to POP
c. OCTPAY to EMP
d. AVPAYIN to POP

In each case, calculate r^2 and test $H_0: \beta_1 = 0$ by setting α equal to .05.

5.63 In Appendix F, consider Data Base 3: Population Data. Use the simple linear regression model to relate each of the following dependent variables to the indicated independent variable. Note that the variables are defined in Appendix F.

a. PCTUNEM to OV25CG
b. PCTUNEM to OV25HSG
c. PCTUNEM to PCTPOPCH
d. PCTMARML to PCTMARFM

In each case, calculate r^2 and test $H_0: \beta_1 = 0$ by setting α equal to .05.

CHAPTER 6

THE ASSUMPTIONS BEHIND REGRESSION ANALYSIS

What are the assumptions behind regression analysis? What do these assumptions mean? How can we tell whether or not these assumptions are met? The goal of this chapter is to discuss the answers to these questions.

We begin this chapter by discussing *residual plots* in Section 6.1. We use these plots in later sections to investigate the validity of the assumptions behind regression analysis. In Section 6.2 we see how these plots can help us to discover when the use of the straight-line functional form may be inappropriate. Next, in Section 6.3, in order to discuss the inference assumptions more fully, we introduce the infinite population of potential error terms and present the properties of this population. Inference assumption 1, the *constant variance assumption*, is explained in Section 6.4, where we also explain how to use residual plots to decide whether or not this assumption is valid. The second inference assumption, the *independence assumption*, is the topic of Section 6.5. When this assumption is violated, the error terms for the regression model are said to be *autocorrelated*. We can detect violations of the independence assumption by using residual plots and a statistical test called the *Durbin-Watson test for autocorrelation*. Inference assumption 3, the *normality assumption*, is covered in Section 6.6. Here we look at several ways in which residuals can be used to detect violations of this assumption (including use of a *normal plot*). In Section 6.7 we present an introductory discussion of outlying and influential observations. We conclude this chapter with optional Section 6.8, which shows how to use SAS to implement the techniques of this chapter.

It is important to reiterate that although the formulas for the confidence intervals, prediction intervals, and hypothesis tests in this book are strictly valid only when the inference assumptions hold, these formulas are still approximately correct even when mild departures from the inference assumptions can be detected. In fact, these assumptions very seldom, if ever, exactly hold in any practical regression problem. Therefore, in practice, only pronounced depatures from the inference assumptions are considered to be serious enough to need remedial action. However, if these assumptions are seriously violated, then remedies can and should be employed. A detailed discussion of the available remedies is given in Chapter 13.

6.1 RESIDUAL PLOTS

Residual plots can be used to help decide whether or not the regression assumptions are valid. Recall that the *ith residual*, e_i, is defined to be

$$e_i = y_i - \hat{y}_i = y_i - (b_0 + b_1 x_i)$$

where b_0 and b_1 are the least squares point estimates of the parameters β_0 and β_1 in the simple linear regression model. This is the difference between the ith observed value of the dependent variable and the ith predicted value of the dependent variable. To construct a residual plot, we compute the residual for each of the observations $y_1, y_2, \ldots, y_n$. The calculated residuals, denoted $e_1, e_2, \ldots, e_n$, are then plotted against some criterion. The resulting plot is called a *residual plot*. To validate the regression assumptions, we should make residual plots against

1. Values of the independent variable x.
2. Values of $\hat{y}$, the predicted value of the dependent variable.
3. The time order in which the historical data have been observed.

In the following example we illustrate these residual plots.

Example 6.1 In Figure 6.1 we repeat the SAS output resulting from using the simple linear regression model

$$\begin{aligned} y_i &= \mu_i + \varepsilon_i \\ &= \beta_0 + \beta_1 x_i + \varepsilon_i \end{aligned}$$

to analyze the Comp-U-Systems service call data of Table 5.6. Here, we recall that x_i denotes the number of microcomputers serviced on the ith service call, and y_i

FIGURE 6.1 **SAS output for a regression analysis of the Comp-U-Systems data**

DEP VARIABLE: Y

SOURCE	DF	SUM OF SQUARES	MEAN SQUARE	F VALUE	PROB>F
MODEL	1	19918.844[a]	19918.844[q]	935.149[r]	0.0001[s]
ERROR	9	191.702[b]	21.300184[f]		
C TOTAL	10	20110.545[c]			

Analysis of variance table

ROOT MSE	4.615212[d]	R-SQUARE	0.9905[h]
DEP MEAN	107.636	ADJ R-SQ	0.9894[i]
C.V.	4.287782		

VARIABLE	DF	PARAMETER ESTIMATE[e]	STANDARD ERROR[g]	T FOR H0: PARAMETER=0[j]	PROB > |T|[k]
INTERCEP	1	11.464088	3.439026	3.334	0.0087
X	1	24.602210	0.804514	30.580	0.0001

OBS	ACTUAL[l]	PREDICT[t] VALUE	STD ERR PREDICT	LOWER95% MEAN	UPPER95% MEAN	LOWER95% PREDICT	UPPER95% PREDICT	RESIDUAL[u]
1	37.000	36.066	2.723	29.907	42.226	23.944	40.188	0.933702
2	58.000	60.669	2.073	55.980	65.357	49.224	72.113	-2.669
3	68.000	60.669	2.073	55.980	65.357	49.224	72.113	7.331
4	82.000	85.271	1.572	81.715	88.827	74.241	96.300	-3.271
5	103.000	109.873	1.393	106.721	113.025	98.967	120.779	-6.873
6	109.000	109.873	1.393	106.721	113.025	98.967	120.779	-.872928
7	112.000	109.873	1.393	106.721	113.025	98.967	120.779	2.127
8	134.000	134.475	1.645	130.753	138.197	123.391	145.559	-.475138
9	138.000	134.475	1.645	130.753	138.197	123.391	145.559	3.525
10	154.000	159.077	2.183	154.139	164.016	147.528	170.627	-5.077
11	189.000	183.680	2.850	177.233	190.126	171.409	195.950	5.320
12	.	109.873[m]	1.393[n]	106.721	113.025	98.967	120.779	

95% confidence interval (LOWER95% MEAN – UPPER95% MEAN); *95% prediction interval* (LOWER95% PREDICT – UPPER95% PREDICT)

SUM OF RESIDUALS -1.06581E-14

SUM OF SQUARED RESIDUALS 191.7017[p]

[a]*Explained variation*	[b]*SSE*	[c]*Total variation*	[d]*Standard error, s*
[e]b_0, b_1	[f]s^2	[g]$s\sqrt{c_{00}}$, $s\sqrt{c_{11}}$	[h]r^2
[i]$\bar{r}^2$	[j]t	[k]*Prob-value for t*	[l]*Observed values* $= y$
[m]*Point prediction* $\hat{y}_0$	[n]$s\sqrt{h_{00}}$	[p]*SSE*	[q]MS_{model}
[r]*F(model)*	[s]*Prob-value for F(model)*	[t]*Predicted values* $= \hat{y}$	
[u]*Residuals* $e_i = y_i - \hat{y}_i$			

denotes the number of minutes required to perform service on the ith service call. Note that the residuals in Figure 6.1 have been calculated, for $i = 1, 2, \ldots, 11$, as

$$\begin{aligned} e_i &= y_i - \hat{y}_i \\ &= y_i - (b_0 + b_1 x_i) \\ &= y_i - (11.4641 + 24.6022x_i) \end{aligned}$$

FIGURE 6.2 **SAS output of the residuals for the Comp-U-Systems model plotted versus *x***

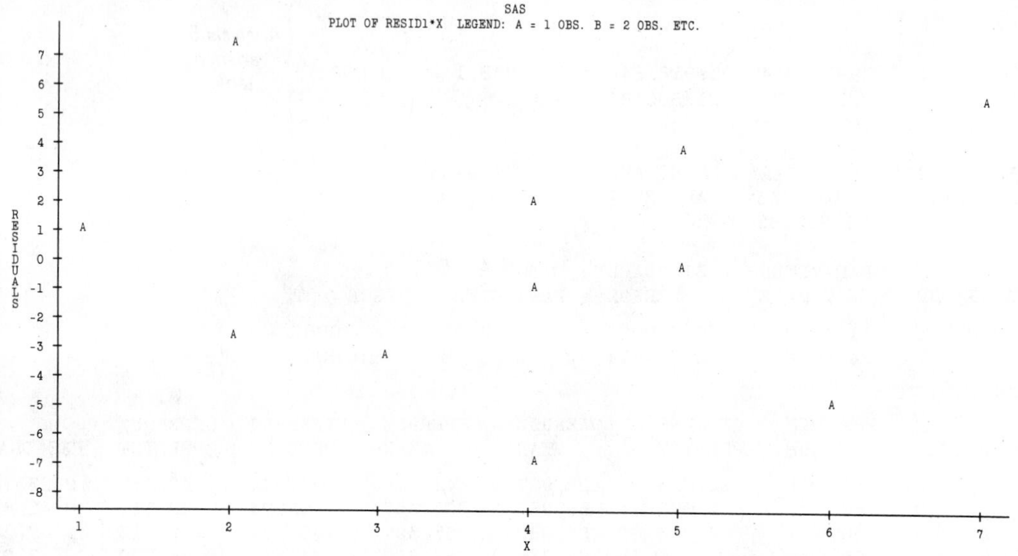

```
SAS
PLOT OF RESID1*X  LEGEND: A = 1 OBS. B = 2 OBS. ETC.
```

RESIDUALS

X

FIGURE 6.3 **SAS output of the residuals for the Comp-U-Systems model plotted versus $\hat{y}$**

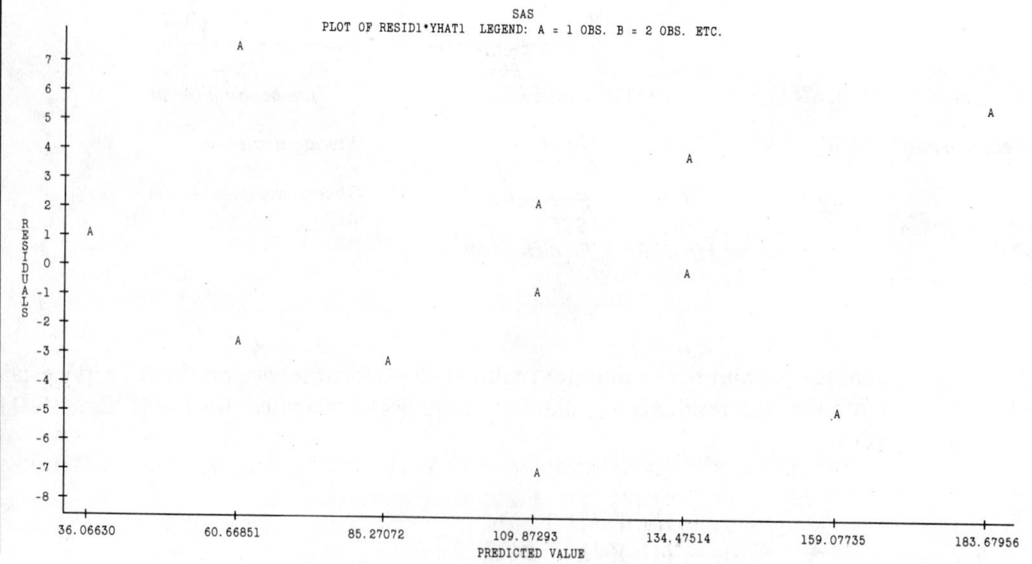

FIGURE 6.4 **SAS output of the residuals for the Comp-U-Systems model plotted in time order**

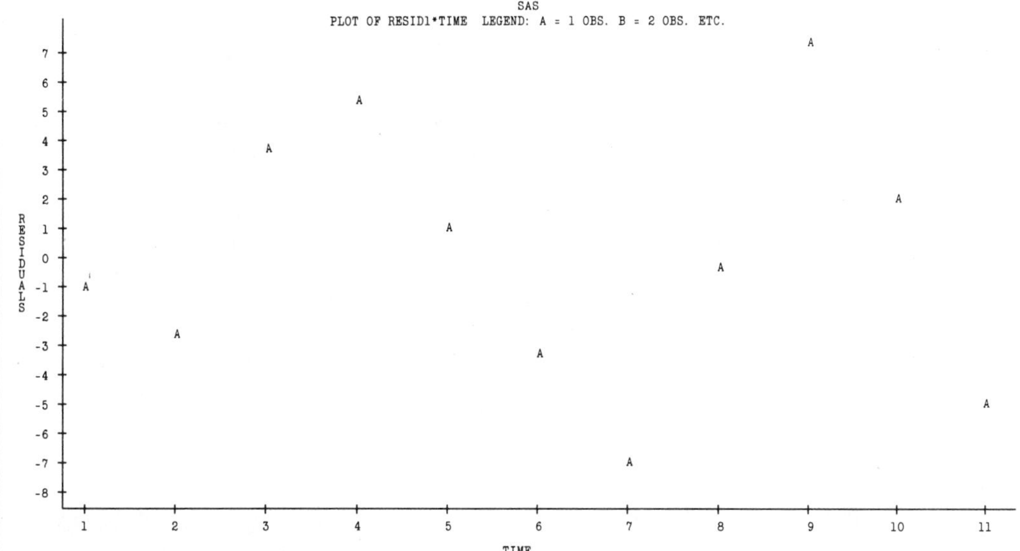

We present the SAS output of a plot of the residuals against

1. Values of x in Figure 6.2,
2. Values of $\hat{y}$ in Figure 6.3, and
3. The time order in which the data has been observed in Figure 6.4. We assume that the 11 service calls were made on 11 consecutive days as listed in Table 5.6.

6.2

THE ASSUMPTION OF CORRECT FUNCTIONAL FORM

If the functional form of a regression model is incorrect, the residual plots constructed by using the model will often display a pattern. This pattern can then be used to determine a more appropriate model. As an illustration of this idea, consider Figure 6.5. In this figure we plot the values of a dependent variable y against the values of an independent variable x. These plotted data points (the dots in the figure) indicate that there is a linear relationship between y and x. Suppose that, instead of using the simple linear regression model

$$y = \beta_0 + \beta_1 x + \varepsilon$$

FIGURE 6.5 **When the regression model does not account for the linear relationship between y and x, the residual plot displays a linear trend**

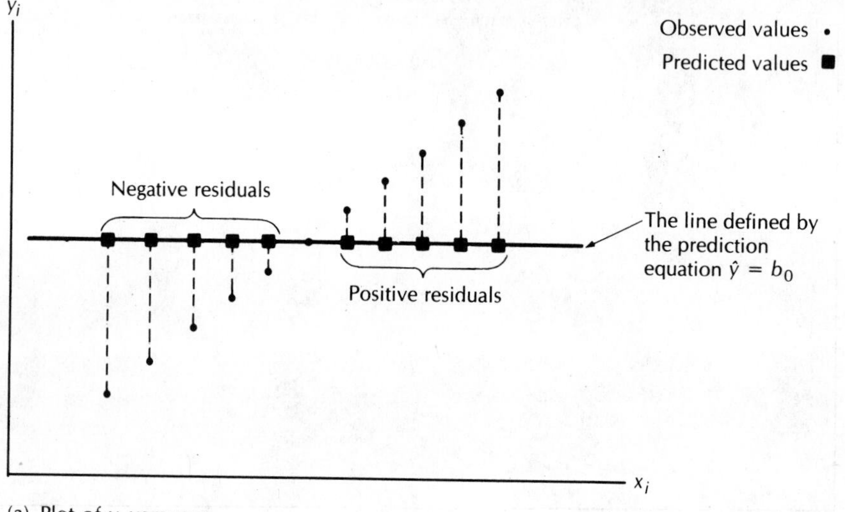

(a) Plot of y_i versus x_i

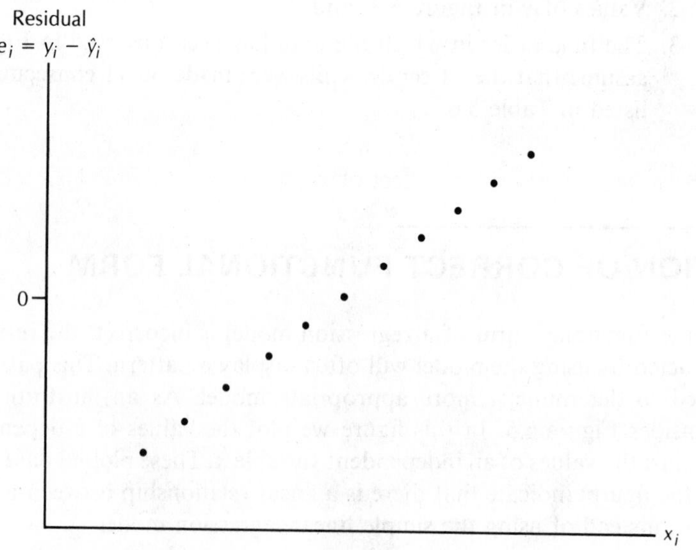

(b) Residual plot ($e_i = y_i - \hat{y}_i$ versus x_i)

we mistakenly used the model

$$y = \beta_0 + \varepsilon$$

which does not properly account for the linear relationship between y and x. Then the least squares point estimate of β_0 is $b_0 = \bar{y}$, the mean of the n observed values $y_1, y_2, \ldots, y_n$ of the dependent variable. This implies that the predicted values of the dependent variable would be calculated by the equation

$$\hat{y} = b_0 = \bar{y}$$

It follows that these predicted values would be the squares that are plotted along the horizontal line in Figure 6.5(a). The resulting residuals, whose magnitudes are denoted by the dashed lines in Figure 6.5(a), are the observed values of y minus the corresponding predicted values of y. Notice that because the above prediction equation does not account for the linear relationship between y and x:

1. The residuals for low values of x are negative, and the residuals for high values of x are positive.
2. When these residuals are plotted against x (see Figure 6.5(b)), they display a straight-line pattern.

This linear pattern in the residuals suggests that a more appropriate model for predicting y is

$$y = \beta_0 + \beta_1 x + \varepsilon$$

Generalizing the above discussion, suppose that we have used the simple linear regression model

$$y = \beta_0 + \beta_1 x_1 + \varepsilon$$

to relate the dependent variable y to an independent variable x_1, and suppose there is another independent variable x_2 that we think might be related to y. If a plot of the residuals from this model against the values of the independent variable x_2 has a straight-line appearance, this indicates that y is linearly related to x_2. Thus we must somehow include the effect of x_2 in the model. How this is done is discussed in Chapter 8. Note that if the values of the dependent variable y and independent variable x_1 have been observed in consecutive time periods, then the independent variable x_2 might be t $(= 1, 2, 3, \ldots)$, the time order in which the data has been observed.

Next, in Figure 6.6(a) we plot the values of a dependent variable y against the values of an independent variable x. This plot (the dots in the figure) indicates that there is a curved relationship between y and x. Suppose that we mistakenly use the regression model

$$y = \beta_0 + \beta_1 x + \varepsilon$$

and the resulting prediction equation

$$\hat{y} = b_0 + b_1 x$$

FIGURE 6.6 **When the regression model does not account for the curved relationship between y and x, the residual plot displays a curved pattern**

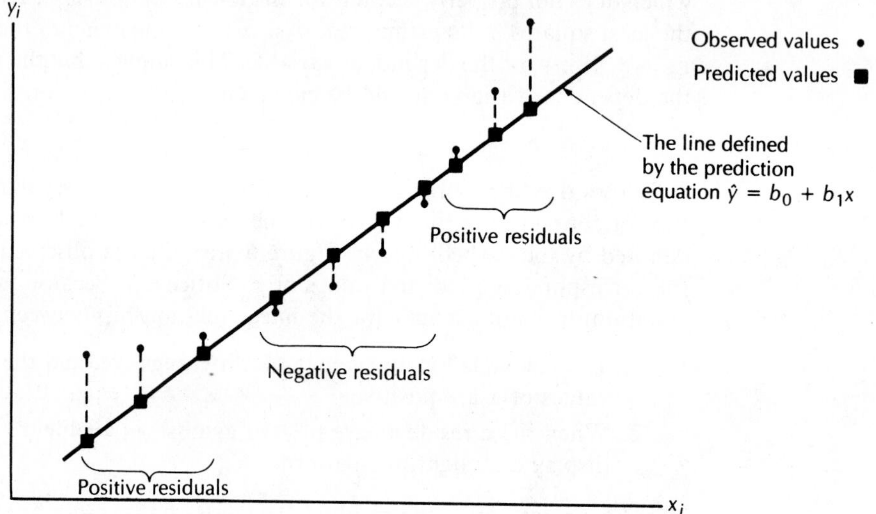

(a) Plot of y_i versus x_i

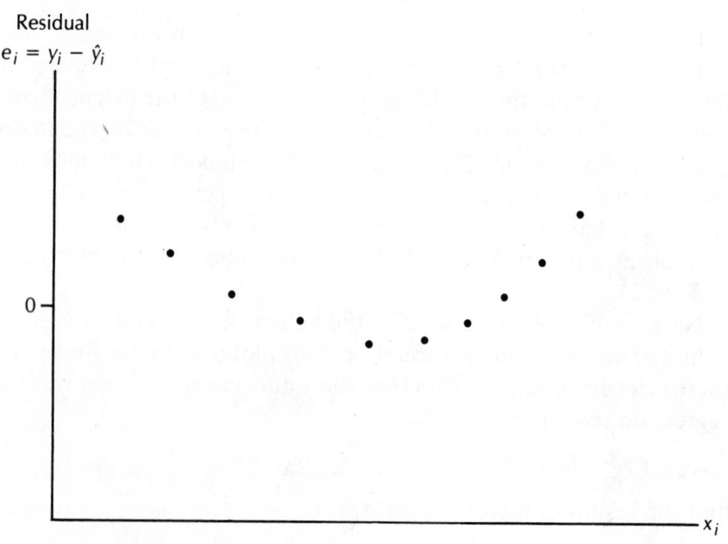

(b) Residual plot ($e_i = y_i - \hat{y}_i$ versus x_i)

which does not properly account for the curved relationship between y and x. Then the predicted values of y would be the squares that are plotted along the line defined by the prediction equation. The resulting residuals, whose magnitudes are denoted by the dashed lines in Figure 6.6(a), are the observed values of y minus the corresponding predicted values of y. Notice that because the prediction equation does not account for the curved relationship between y and x:

1. The residuals for very low values of x and for very high values of x are positive, while the residuals for intermediate values of x are negative.
2. When these residuals are plotted against x [see Figure 6.6(b)], they display a curved pattern.

This pattern in the residuals suggests that we must modify the simple linear model to account for the curvature. How this might be done is explained in Chapter 8.

Of course, it is possible for residual plots to display other curved patterns. For example, if the residual plot from a straight-line model has the curved appearance of Figure 6.7, this also indicates that there is a curved relationship between y and x. Again, this says that we must modify the straight-line model to account for the curvature.

If little or no pattern exists in the appropriate residual plots, the relationship between the dependent variable and the independent variable is (probably) properly accounted for. For example, Figure 6.2 is a plot of the residuals obtained from the Comp-U-Systems model

$$y = \beta_0 + \beta_1 x + \varepsilon$$

against x. Since there is little or no pattern in this figure, the model probably properly accounts for the relationship between y and x.

We conclude this section with a more detailed example of the use of residual plots.

FIGURE 6.7 **Another curved residual plot**

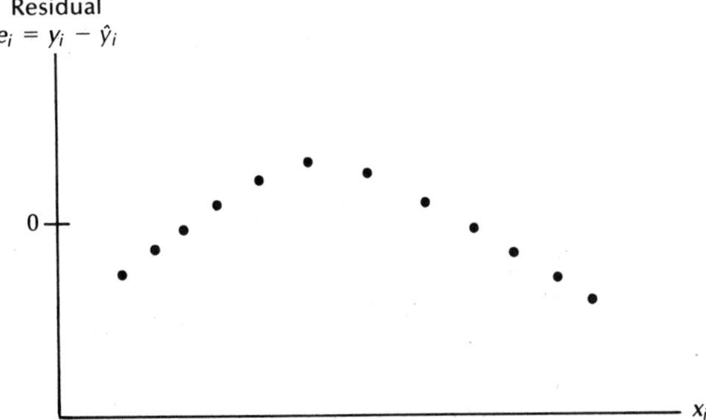

Example 6.2 Enterprise Industries produces Fresh, a brand of liquid laundry detergent. The company wishes to study the factors affecting the quarterly demand for the large size bottle of Fresh in its sales regions. To carry out this study, the company has gathered data concerning demand for Fresh in $n = 30$ sales regions of equal sales potential for the first quarter of 1989. This data is presented in Table 6.1. Here, for $i = 1, 2, \ldots, 30$:

y_i = the demand for the large size bottle of Fresh (in hundreds of thousands of bottles) in sales region i

x_{i1} = the price (in dollars) of Fresh as offered by Enterprise Industries in sales region i

TABLE 6.1 **Historical data, including price differences, concerning demand for Fresh Detergent**

Sales region, i	Price for Fresh, x_{i1} (dollars)	Average industry price, x_{i2} (dollars)	Price difference, $x_{i4} = x_{i2} - x_{i1}$ (dollars)	Advertising expenditure for Fresh, x_{i3} (hundreds of thousands of dollars)	Demand for Fresh, y_i (hundreds of thousands of bottles)
1	3.85	3.80	−.05	5.50	7.38
2	3.75	4.00	.25	6.75	8.51
3	3.70	4.30	.60	7.25	9.52
4	3.70	3.70	0	5.50	7.50
5	3.60	3.85	.25	7.00	9.33
6	3.60	3.80	.20	6.50	8.28
7	3.60	3.75	.15	6.75	8.75
8	3.80	3.85	.05	5.25	7.87
9	3.80	3.65	−.15	5.25	7.10
10	3.85	4.00	.15	6.00	8.00
11	3.90	4.10	.20	6.50	7.89
12	3.90	4.00	.10	6.25	8.15
13	3.70	4.10	.40	7.00	9.10
14	3.75	4.20	.45	6.90	8.86
15	3.75	4.10	.35	6.80	8.90
16	3.80	4.10	.30	6.80	8.87
17	3.70	4.20	.50	7.10	9.26
18	3.80	4.30	.50	7.00	9.00
19	3.70	4.10	.40	6.80	8.75
20	3.80	3.75	−.05	6.50	7.95
21	3.80	3.75	−.05	6.25	7.65
22	3.75	3.65	−.10	6.00	7.27
23	3.70	3.90	.20	6.50	8.00
24	3.55	3.65	.10	7.00	8.50
25	3.60	4.10	.50	6.80	8.75
26	3.65	4.25	.60	6.80	9.21
27	3.70	3.65	−.05	6.50	8.27
28	3.75	3.75	0	5.75	7.67
29	3.80	3.85	.05	5.80	7.93
30	3.70	4.25	.55	6.80	9.26

FIGURE 6.8 **Plot of y (demand for Fresh Detergent) against x_3 (advertising expenditure for Fresh)**

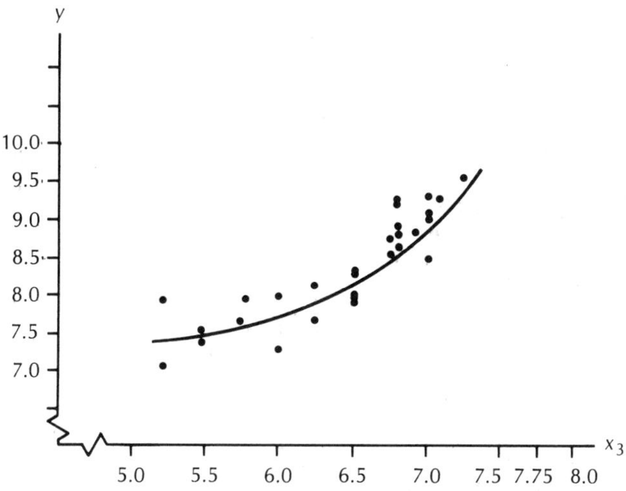

$$x_{i2} = \text{the average industry price (in dollars) of competitors' similar detergents in sales region } i$$

$$x_{i3} = \text{Enterprise Industries' advertising expenditure (in hundreds of thousands of dollars) to promote Fresh in sales region } i$$

$$x_{i4} = x_{i2} - x_{i1} = \text{the "price difference" in sales region } i$$

To begin our analysis, suppose that Enterprise Industries believes on theoretical grounds that the single independent variable x_4 adequately describes the effects of x_1 and x_2 on y. We will assess the validity of this assumption in later chapters, but for now we continue the analysis by studying the relationship between the dependent variable y and the independent variables x_3 and x_4. In Figure 6.8 we plot y against x_3, and in Figure 6.9 we plot y against x_4. Examining these plots, we see that there appears to be a curved relationship between y and x_3 and a straight-line relationship between y and x_4.

Next, consider the simple linear model

$$y = \beta_0 + \beta_1 x_3 + \varepsilon$$

Using this model and the Fresh Detergent data of Table 6.1, we obtain the prediction equation

$$\begin{aligned}\hat{y} &= b_0 + b_1 x_3 \\ &= 1.6490 + 1.0434 x_3\end{aligned}$$

FIGURE 6.9 **Plot of y (demand for Fresh Detergent) against x_4 (price difference)**

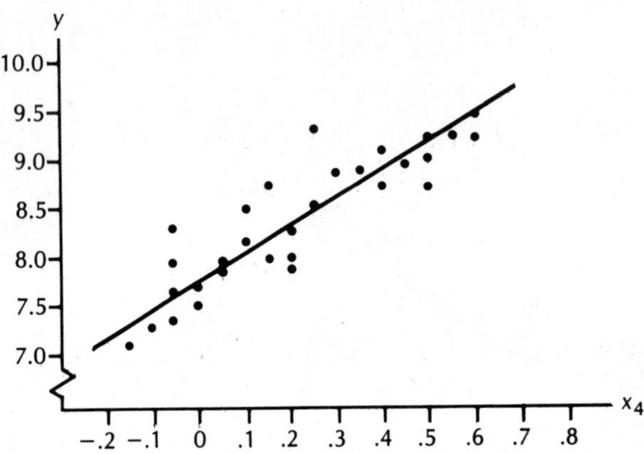

The residuals are calculated, for $i = 1, 2, \ldots, 30$, as

$$e_i = y_i - \hat{y}_i = y_i - 1.6490 - 1.0434x_{i3}$$

A plot of these residuals against x_3 is shown in Figure 6.10 and has the general

FIGURE 6.10 **Residuals for Fresh Detergent model $y = \beta_0 + \beta_1 x_3 + \varepsilon$ plotted against x_3**

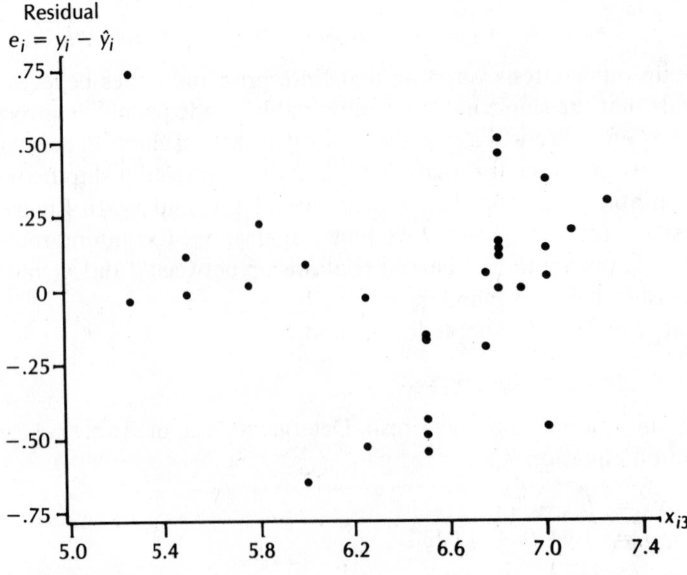

FIGURE 6.11 **Residuals for Fresh Detergent model $y = \beta_0 + \beta_1 x_3 + \varepsilon$ plotted against x_4**

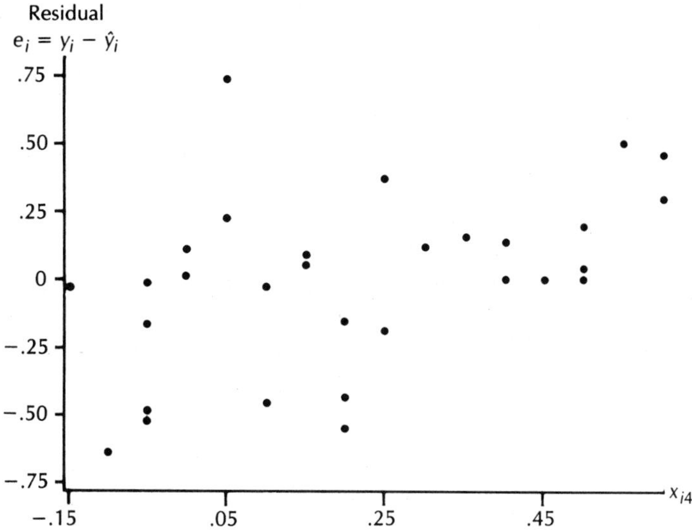

appearance of Figure 6.6(b). Notice the pattern of positive residuals associated with high or low values of x_3 and negative residuals associated with intermediate values of x_3. This indicates a curved relationship between y and x_3 and is consistent with the curved data plot of y versus x_3 in Figure 6.8. In Figure 6.11 a plot of the residuals against x_4 gives a slight indication of a linear relationship between y and x_4. Here we see a predominance of positive residuals on the right side of the plot. This is consistent with the straight line data plot of y versus x_4 in Figure 6.9. In summary, the residual plots in Figures 6.10 and 6.11 from the model

$$y = \beta_0 + \beta_1 x_3 + \varepsilon$$

indicate that we must modify this model to account for the curvature between y and x_3 and to account for the linear effect of the omitted variable x_4. How this is done is described in Chapter 8.

The reader might inquire as to why we need residual plots to tell us that there is a curved relationship between y and x_3 and a straight-line relationship between y and x_4 when the data plots in Figures 6.8 and 6.9 already tell us these facts. The answer is that, whereas in this case the residual plots do not tell us any more than the data plots tell us, in some situations, residual plots do tell us more. It all depends upon the nature of the data, and therefore it is good practice to make both data plots and residual plots.

6.3

THE POPULATION OF ERROR TERMS

For the regression techniques we have presented in Chapters 4 and 5 to be valid, the simple linear regression model

$$y = \mu + \varepsilon = \beta_0 + \beta_1 x + \varepsilon$$

must, in addition to having the correct functional form, at least approximately satisfy the inference assumptions. To fully discuss these assumptions, we need to introduce what we call the *ith population of potential error terms*. Recall that y_i is assumed to be randomly selected from the ith population of potential values of the dependent variable (the population of values of the dependent variable that could be observed when the value of the independent variable x is x_i). The ith error term

$$\varepsilon_i = y_i - \mu_i$$

is the difference between y_i and μ_i, the ith mean value of the dependent variable. Therefore since (before selection) y_i could potentially be any one of an infinite number of values, and since an error term could be defined for each of these potential y_i values, we see that ε_i could potentially be any one of an infinite number of values. This population of potential ε_i values is the *ith population of potential error terms*.

The population of potential error terms has two important properties that are stated below.

The Mean and Variance of the ith Population of Potential Error Terms

Define the *ith population of potential error terms* to be the population of error terms that could be observed when the value of the independent variable x is x_i. Then

1. The ith population of potential error terms has a mean equal to zero.
2. The variance of the ith population of potential error terms is equal to the variance of the ith population of potential values of the dependent variable.

The first property follows because the mean of the y_i values in the ith population of potential values of the dependent variable is μ_i and because the ε_i values are obtained by simply subtracting μ_i from the potential y_i values. The second property follows because it is clear that the amount of variability in the ε_i values is the same as the amount of variability in the y_i values in the ith population of potential values of the dependent variable. This is because the ε_i values are simply the y_i values minus the constant μ_i.

6.4

INFERENCE ASSUMPTION 1: CONSTANT VARIANCE

We are now ready to further discuss the first inference assumption, which can be stated as follows.

Inference Assumption 1 (Constant Variance)

For any value x_i of the independent variable x the corresponding population of potential values of the dependent variable has a variance σ^2 that does not depend on the value x_i of x. That is, the different populations of potential values of the dependent variable corresponding to different values of x have *equal variances.* Said equivalently, the different populations of potential error terms corresponding to different values of x have *equal variances.*

When this assumption holds for a regression model, we say that the model displays a *constant error variance.* We now present an example of a regression problem in which the constant variance assumption does not hold.

Example 6.3 The National Association of Retail Hardware Stores (NARHS), a nationally known trade association, wishes to investigate the relationship between home value x (in thousands of dollars) and yearly expenditure y (in dollars) on upkeep (such as lawn care, painting, repairs). A random sample of 40 homeowners is taken, and the results are given in Table 6.2. Figure 6.12 gives a plot of y versus x. From this

TABLE 6.2 **NARHS upkeep expenditure data**

House	Value of house, x (thousands of dollars)	Expenditure on upkeep, y (dollars)
1	118.50	706.04
2	76.54	398.60
3	92.43	436.24
4	111.03	501.71
5	80.34	426.45
6	49.84	144.24
7	114.52	644.23
8	50.89	211.54
9	128.93	675.87
10	48.14	189.02
11	85.50	459.04
12	115.51	813.62
13	114.16	602.39
14	102.95	428.52
15	92.86	387.50

(continues)

TABLE 6.2 Continued

House	Value of house, x (thousands of dollars)	Expenditure on upkeep, y (dollars)
16	84.39	434.63
17	123.53	698.00
18	77.77	355.75
19	112.10	737.59
20	101.02	706.66
21	76.52	424.57
22	116.09	656.92
23	62.72	301.03
24	84.91	321.07
25	88.64	519.40
26	81.41	348.50
27	60.22	162.17
28	95.55	482.55
29	79.39	460.07
30	89.25	475.45
31	136.10	835.16
32	24.45	62.70
33	52.28	239.89
34	143.09	1005.32
35	41.86	184.18
36	43.10	212.80
37	66.79	313.45
38	106.43	658.47
39	61.01	195.08
40	99.01	545.42

FIGURE 6.12 Upkeep expenditure versus value of house for NARHS data

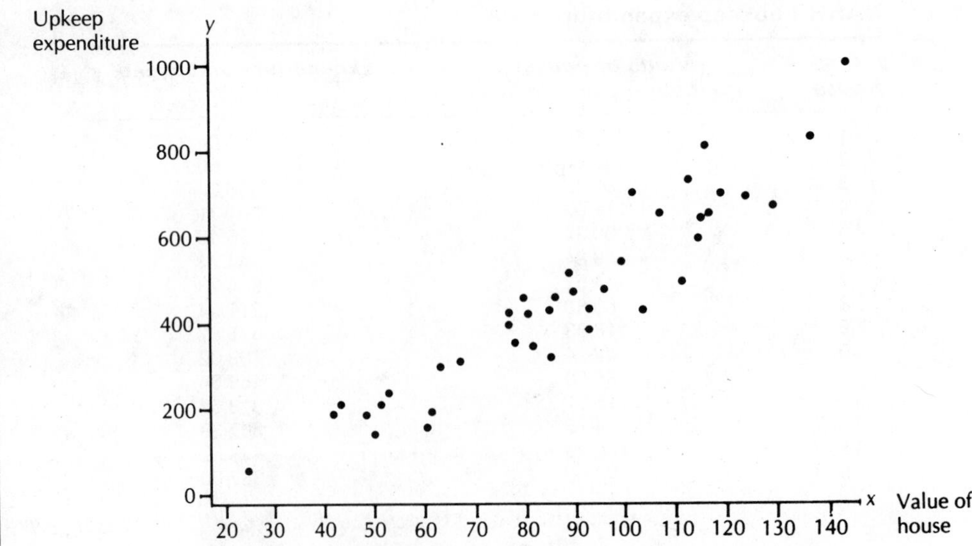

plot it appears that y is increasing (in either a straight line or a somewhat curved fashion) as x increases. Because the y values seem to fan out as x increases, the variance of the population of potential yearly upkeep expenditures for houses worth x (thousand dollars) appears to increase as x increases. For example, the variance of the population of potential yearly upkeep expenditures for houses worth \$100,000 would be larger than the variance of the population of potential yearly upkeep expenditures for houses worth \$50,000. Increasing variance makes some intuitive sense here. This is because people with more expensive homes generally have higher incomes and can afford to pay to have upkeep done by lawn services, painters, and so on if they wish or can perform upkeep chores themselves if they wish, thus causing a relatively large variation in upkeep expenses.

Residual plots can be used to assess the validity of the constant variance assumption. These plots should be made against the following criteria:

1. Values of the independent variable.
2. Values of $\hat{y}$, the predicted value of the dependent variable.
3. The time order in which the historical data have been observed.

To see how the residuals relate to the constant variance assumption, consider the model

$$y_i = \mu_i + \varepsilon_i = \beta_0 + \beta_1 x_i + \varepsilon_i$$

which implies that for $i = 1, 2, \ldots, n$,

$$\varepsilon_i = y_i - \mu_i = y_i - (\beta_0 + \beta_1 x_i)$$

It follows that the point estimate of the ith error term ε_i is the ith residual

$$e_i = y_i - \hat{y}_i = y_i - (b_0 + b_1 x_i)$$

Therefore the residuals $e_1, e_2, \ldots, e_n$ provide point estimates of the error terms $\varepsilon_1, \varepsilon_2, \ldots, \varepsilon_n$. Since the ith population of potential error terms has mean zero, we would expect the residuals $e_1, e_2, \ldots, e_n$ to fluctuate around zero.

The pattern in which the residuals fluctuate around zero indicates whether or not the constant variance assumption holds. Since the residuals $e_1, e_2, \ldots, e_n$ are estimates of the error terms $\varepsilon_1, \varepsilon_2, \ldots, \varepsilon_n$, we can interpret patterns in the residual plots as follows. A residual plot that "fans out" [for example, see Figure 6.13(a)] indicates that the error terms $\varepsilon_1, \varepsilon_2, \ldots, \varepsilon_n$ are increasing in absolute value as the horizontal plot criterion increases. This would suggest that σ_i^2, the variance of the ith population of potential error terms, is increasing with increasing values of the criterion. Thus such a plot suggests that the constant variance assumption is violated, and we would say that a *nonconstant error variance* exists; in this case we have an *increasing error variance*. A residual plot that "funnels in" [for example, see Figure 6.13(b)] suggests that the error terms $\varepsilon_1, \varepsilon_2, \ldots, \varepsilon_n$ are decreasing in absolute value as the horizontal plot criterion increases and that σ_i^2 is decreasing with increasing values of the criterion. In this case we would have a *decreasing error*

FIGURE 6.13 Residual plots and the constant variance assumption

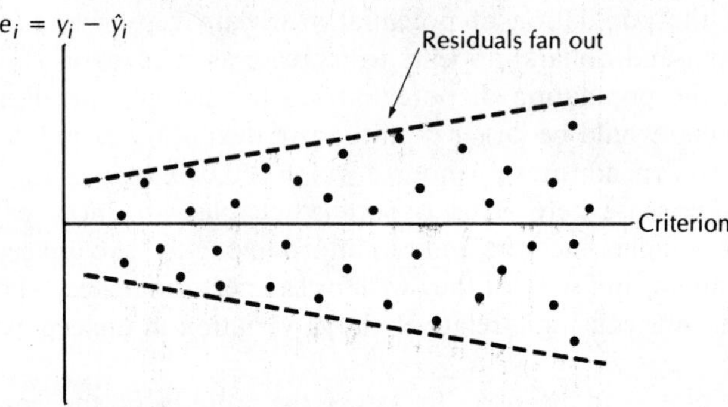

(a) Nonconstant error variance: Error variance increases with increasing values of the criterion

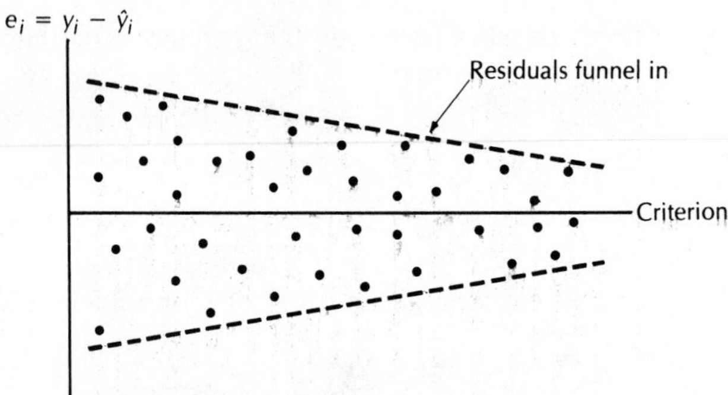

(b) Nonconstant error variance: Error variance decreases with increasing values of the criterion

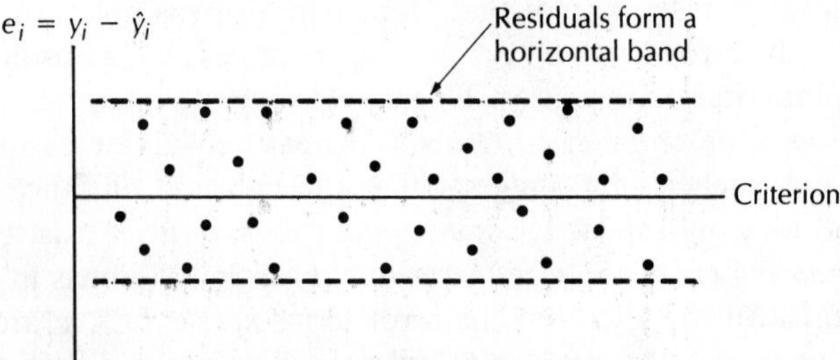

(c) Constant error variance with increasing values of the criterion

variance and a violation of the constant variance assumption. A residual plot with a "horizontal band appearance" [for example, see Figure 6.13(c)] indicates that the error terms are remaining relatively constant in absolute value and that σ_i^2 is not changing much with increasing values of the horizontal plot criterion. Such a plot suggests that the constant variance assumption (approximately) holds.

Example 6.4 Consider the plot of the NARHS upkeep expenditure data in Figure 6.12. Since y appears to increase in either a straight-line or somewhat curved fashion as x increases, we consider the simple linear model

$$y_i = \beta_0 + \beta_1 x_i + \varepsilon_i$$

When we estimate β_0 and β_1 by using the data in Table 6.2, we obtain the prediction equation

$$\hat{y}_i = -174.1965 + 7.2583x_i$$

Residuals (for $i = 1, 2, \ldots, 40$) for this model are calculated as

$$e_i = y_i - \hat{y}_i = y_i - (-174.1965 + 7.2583x_i)$$

Figure 6.14 is the SAS output of a plot of these residuals against x. We see that the residuals appear to fan out as x increases, indicating that the variance

FIGURE 6.14 **SAS output of the residuals for the NARHS simple linear model plotted against x**

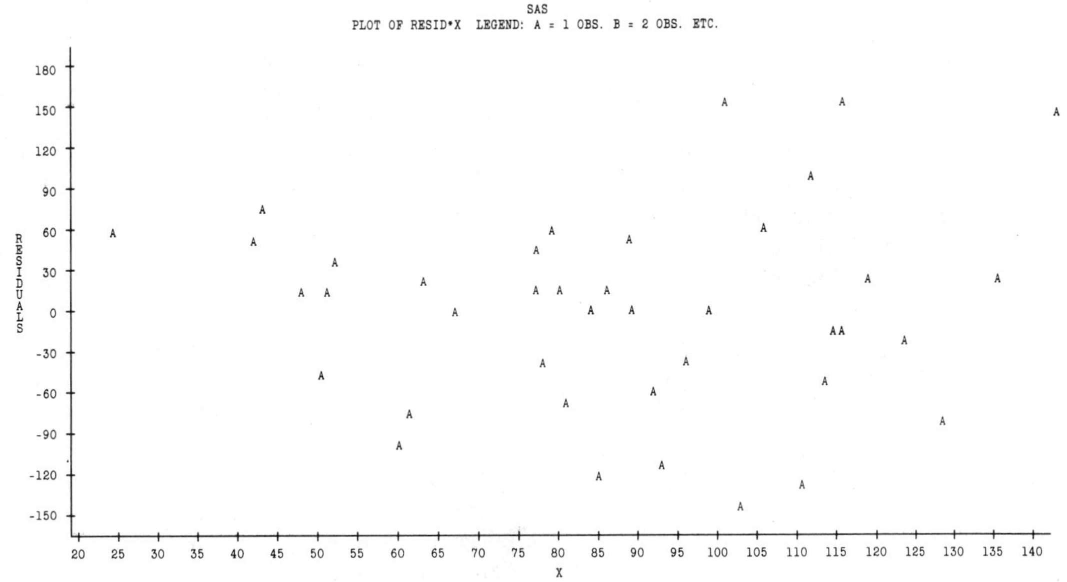

FIGURE 6.15 **SAS output of the residuals for the NARHS simple linear model plotted against $\hat{y}$**

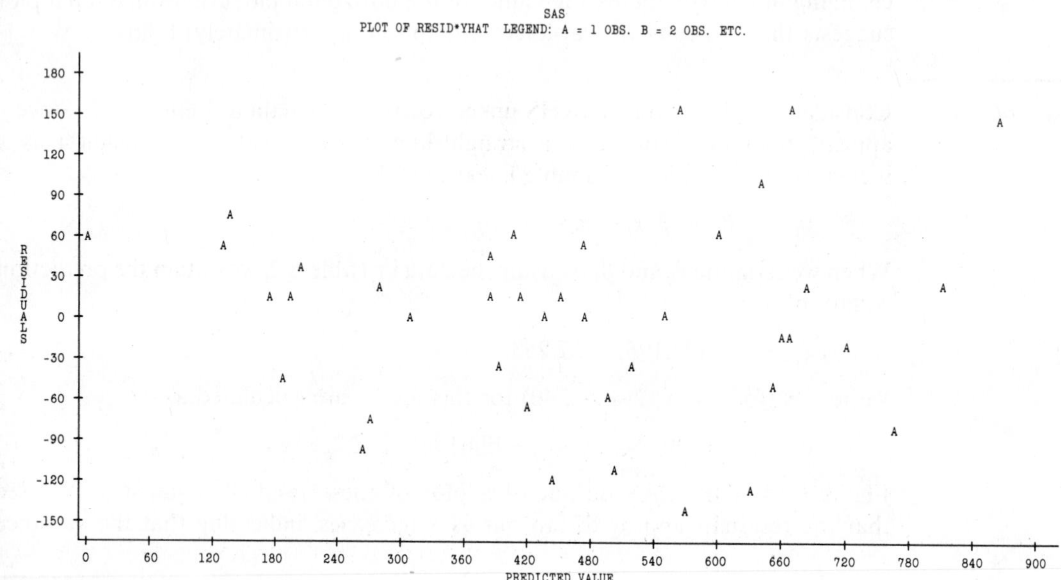

of the population of potential yearly upkeep expenditures (in dollars) for houses worth x (thousand dollars) increases as x increases. Another indication of a nonconstant error variance is the SAS output in Figure 6.15. This plot indicates that the residuals appear to fan out as $\hat{y}$ increases (this is logical, since $\hat{y} = -174.1965 + 7.2583x$ is an increasing function of x). Since a nonconstant error variance seems to exist, we cannot use the formulas of Chapters 4 and 5 to make statistical inferences. In Chapter 13 we discuss how we can make statistical inferences when a nonconstant error variance exists. Also note that there is a slight "dip" in the residual plot of Figure 6.14. This implies (as is also indicated by Figure 6.12) that there might be a somewhat curved relationship between y and x. In Chapter 13 we will investigate this curved relationship further, and we will completely analyze the data in Table 6.2.

Finally, note that each of the residual plots in Figures 6.2, 6.3, and 6.4 probably has a horizontal band appearance. We conclude that the constant variance assumption approximately holds for the Comp-U-Systems model

$$y = \beta_0 + \beta_1 x + \varepsilon$$

When a nonconstant error variance seems to exist, we can often use a *data transformation* to equalize the variances and thus remedy the situation. Several such transformations are discussed in Chapter 13. When the variances increase as $\hat{y}$ increases (as in Figure 6.15), we can sometimes equalize the variances by performing a regression analysis using either the square roots or the natural logarithms of the y values.

6.5

INFERENCE ASSUMPTION 2: INDEPENDENCE

The second inference assumption can be stated as follows.

Inference Assumption 2 (Independence)

Any one value of the dependent variable y is *statistically independent* of any other value of y. Said equivalently, any one value of the error term ε is *statistically independent* of any other value of ε.

The independence assumption is most likely to be violated when the data being used in a regression problem are *time series data,* that is, data that have been collected in a time sequence. For such data the time-ordered error terms can be *autocorrelated.* Intuitively, we say that error terms occurring over time have *positive autocorrelation* if a positive error term in time period t tends to produce, or be followed by, another positive error term in time period $t + k$ (a later time period) and if a negative error term in time period t tends to produce, or be

FIGURE 6.16 **Positive and negative autocorrelation**

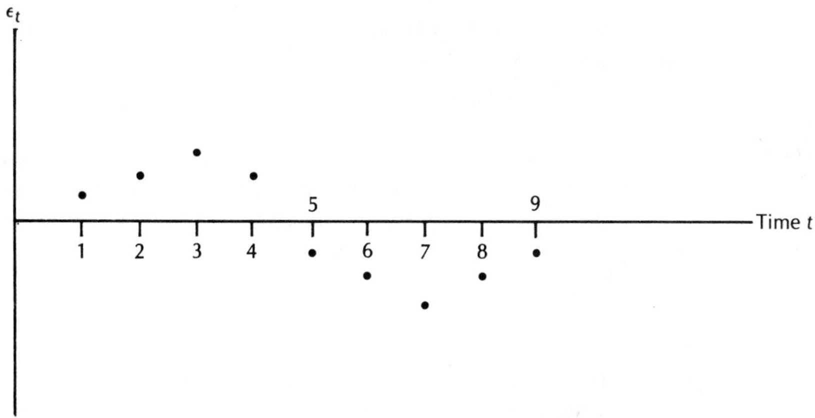

(a) Positive autocorrelation in the error terms: Cyclical pattern

(continues)

FIGURE 6.16 Continued

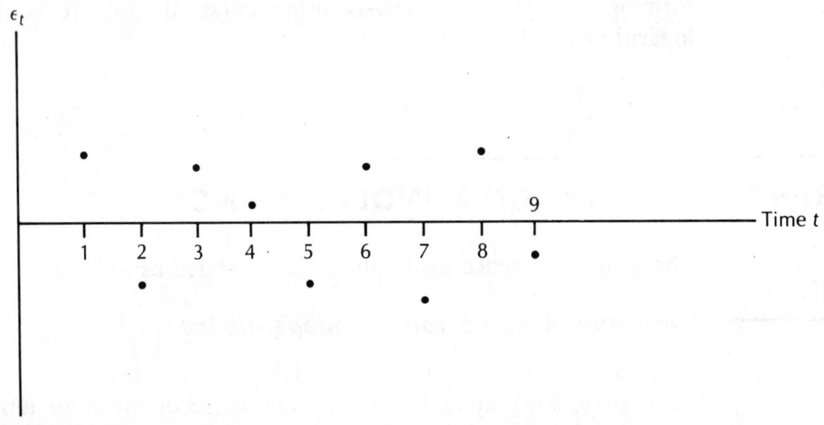

(b) Negative autocorrelation in the error terms: Alternating pattern

followed by, another negative error term in time period $t + k$. In other words, positive autocorrelation exists when positive error terms tend to be followed over time by positive error terms, and negative error terms tend to be followed over time by negative error terms. An example of positive autocorrelation in the error terms is depicted in Figure 6.16(a). This figure illustrates that positive autocorrelation in the error terms can produce a cyclical pattern over time. If we consider μ_i, the mean of the ith historical population of potential values of the dependent variable, we have seen that a positive error term produces a value of the dependent variable y that is greater than μ_i, and a negative error term produces a value of y_i that is smaller than μ_i. This says that positive autocorrelation in the error terms means that greater than average values of y tend to be followed by greater than average values of y, and smaller than average values of y tend to be followed by smaller than average values of y.

Example 6.5 Suppose that we have observed sales volume (y) and advertising expenditure (x) for a product in one sales region over 20 consecutive months. Also suppose that a data plot and the t statistic indicate that a reasonable model relating y to x is

$$\begin{aligned} y &= \mu + \varepsilon \\ &= \beta_0 + \beta_1 x + \varepsilon \end{aligned}$$

In this model the effect of competitors' advertising expenditure is included in ε, the error term. Assume for the moment that competitors' advertising expenditure significantly affects sales volume (y). Then a higher than average competitors' advertising expenditure probably causes sales volume to be lower than average and hence a negative error term to occur. On the other hand, a lower than average competitors' advertising expenditure probably causes sales volume to be higher than average and hence a positive error term to occur. Suppose that competitors tend to spend money on advertising in a cyclical fashion—spending large amounts for several consecutive months (during an advertising campaign) and then spending lesser amounts for several consecutive months. It follows that a negative error term in one month will tend to be followed by a negative error in the next month. Similarly, a positive error term in one month will tend to be followed by a positive error term in the next month. In this case the error terms

$$\varepsilon_1, \varepsilon_2, \ldots, \varepsilon_{20}, \quad \text{and} \quad \varepsilon_0$$

would display positive autocorrelation. Therefore, these error terms and the demand values

$$y_1, y_2, \ldots, y_{20}, \quad \text{and} \quad y_0$$

would be dependent.

Intuitively, error terms occurring over time have *negative autocorrelation* if a positive error term in time period t tends to produce, or be followed by, a negative error term in time period $t + k$ and if a negative error term in time period t tends to produce, or be followed by, a positive error term in time period $t + k$. In other words, negative autocorrelation exists when positive error terms tend to be followed over time by negative error terms, and negative error terms tend to be followed over time by positive error terms. An example of negative autocorrelation in the error terms is depicted in Figure 6.16(b). This figure illustrates that negative autocorrelation in the error terms can produce an alternating pattern over time. It follows that negative autocorrelation in the error terms means that greater than average values of y tend to be followed by smaller than average values of y and smaller than average values of y tend to be followed by greater than average values of y. An example of negative autocorrelation might be provided by a retailer's weekly stock orders. Here, a larger than average stock order one week might result in an oversupply and hence a smaller than average order the next week.

These ideas allow us to give a relatively simple interpretation of the independence assumption. In essence, this assumption says that the randomly selected error terms (listed here in time order)

$$\varepsilon_1, \varepsilon_2, \ldots, \varepsilon_n, \quad \text{and} \quad \varepsilon_0$$

display no positive autocorrelation and display no negative autocorrelation. This would say that the error terms occur in a random pattern over time, as illustrated in Figure 6.17. Such a pattern would imply that these error terms are statistically independent. This would in turn imply that the randomly selected values of the dependent variable (also listed in time order)

$$y_1, y_2, \ldots, y_n, \quad \text{and} \quad y_0$$

are statistically independent.

FIGURE 6.17 **Little or no autocorrelation in the error terms: random pattern**

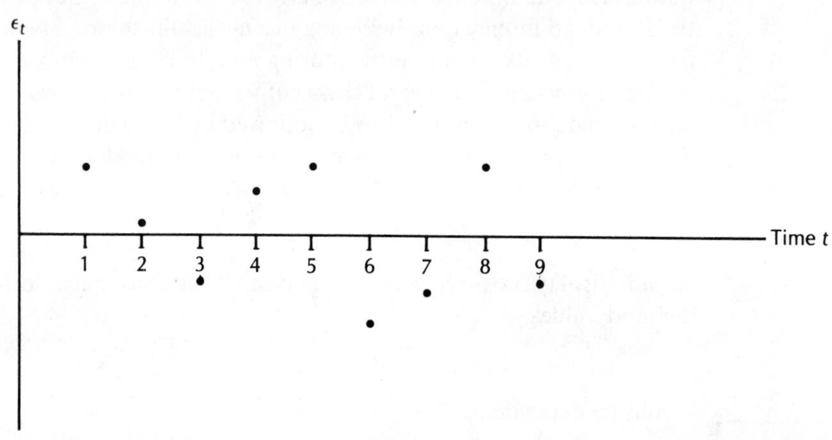

Since the residuals $e_1, e_2, \ldots, e_n$ are point estimates of the error terms $\varepsilon_1, \varepsilon_2, \ldots, \varepsilon_n$, a residual plot against time can be used to detect violations of the independence assumption. If a residual plot against the time sequence in which the data have been collected has a cyclical appearance (for example, see Figure 6.16(a)), the error terms $\varepsilon_1, \varepsilon_2, \ldots, \varepsilon_n$ are positively autocorrelated, and the independence assumption does not hold. Another way to detect positive autocorrelation is to look at the signs of the time-ordered residuals. Letting $+$ denote a positive residual and $-$ denote a negative residual, we call a sequence of residuals with the same sign (for instance, $+++$) a *run*. If positive autocorrelation exists, the signs of the residuals should display relatively few runs of fairly long duration. For instance, the pattern $++++---+++-----++++---$ in the time-ordered residuals would indicate that positive autocorrelation exists and that the independence assumption is violated.

If a plot of the time-ordered residuals has an alternating pattern (for example, see Figure 6.16(b)), the error terms $\varepsilon_1, \varepsilon_2, \ldots, \varepsilon_n$ are negatively autocorrelated, and the independence assumption does not hold. If we look at the signs of the time-ordered residuals, negative autocorrelation is characterized by many runs of relatively short duration. For example, the pattern $+-+-++-+-+--+-$ in the time-ordered residuals would indicate that negative autocorrelation exists and that the independence assumption is violated.

However, if a plot of the time-ordered residuals displays a random pattern, as illustrated in Figure 6.17, the error terms $\varepsilon_1, \varepsilon_2, \ldots, \varepsilon_n$ have little or no autocorrelation. This suggests that these error terms are independent and that the independence assumption holds. For example, Figure 6.4 indicates that the time-ordered residuals from the Comp-U-Systems model display a random pattern. Thus it is reasonable to assume that the independence assumption holds for this model. Similarly, it can be verified that a residual plot of the time-ordered residuals from the fuel consumption model of Example 4.1 displays a random pattern. It is therefore reasonable to assume that the independence assumption also holds for this model.

One type of positive or negative autocorrelation in the error terms is called *first-order autocorrelation*. It says that ε_t, the error term in time period t, is related to ε_{t-1}, the error term in time period $t - 1$, by the equation

$$\varepsilon_t = \rho\varepsilon_{t-1} + u_t$$

Here, we assume that ρ is the correlation coefficient between error terms separated by one time period and $u_1, u_2, \ldots$ are values randomly and independently selected from a normal distribution having mean zero and a variance independent of time. We discuss first-order autocorrelation in more detail in Chapter 13. For now, we present the *Durbin-Watson test*, which is a formal test for first-order (positive or negative) autocorrelation.

The Durbin-Watson Test for (First-Order) Positive Autocorrelation

The *Durbin-Watson statistic* is

$$d = \frac{\sum_{t=2}^{n} (e_t - e_{t-1})^2}{\sum_{t=1}^{n} e_t^2}$$

where $e_1, e_2, \ldots, e_n$ are the time-ordered residuals.

Consider testing the null hypothesis

H_0: The error terms are not autocorrelated

versus the alternative hypothesis

H_1: The error terms are positively autocorrelated

Durbin and Watson have shown that there are points (denoted $d_{L,\alpha}$ and $d_{U,\alpha}$) such that, if α is the probability of a Type I error, then

1. If $d < d_{L,\alpha}$, we reject H_0.
2. If $d > d_{U,\alpha}$, we do not reject H_0.
3. If $d_{L,\alpha} \leq d \leq d_{U,\alpha}$, the test is inconclusive.

Here, small values of d lead to the conclusion of positive autocorrelation, because if d is small, the differences $(e_t - e_{t-1})$ are small. This indicates that the adjacent residuals e_t and e_{t-1} are of the same magnitude, which in turn says that the adjacent error terms ε_t and ε_{t-1} are positively correlated.

So that the Durbin-Watson test may be easily done, tables containing the points $d_{L,\alpha}$ and $d_{U,\alpha}$ have been constructed. These tables give the appropriate $d_{L,\alpha}$ and $d_{U,\alpha}$

points for various values of α; $k - 1$, the number of independent variables in the regression model; and n, the number of observations. Tables E.5 and E.6 in Appendix E give values for $\alpha = .05$ and $\alpha = .01$. Note that $k - 1$ equals 1 for the simple linear regression model, but the Durbin-Watson statistic can be used for models containing more than one independent variable.

Example 6.6 Suppose that the demand and price difference (average industry price minus product price) data in Table 6.3 had been observed over 30 consecutive sales periods. A plot of y_t (demand in sales period t) versus x_t (price difference in sales period t) suggests that the simple linear model

$$\begin{aligned} y_t &= \mu_t + \varepsilon_t \\ &= \beta_0 + \beta_1 x_t + \varepsilon_t \end{aligned}$$

TABLE 6.3 **Demand and price difference data observed over 30 consecutive sales periods**

Sales period, t	*Demand, y_t*	*Price difference, x_t*
1	7.38	−0.05
2	8.51	0.25
3	9.52	0.60
4	7.50	0.00
5	9.33	0.25
6	8.28	0.20
7	8.75	0.15
8	7.87	0.05
9	7.10	−0.15
10	8.00	0.15
11	7.89	0.20
12	8.15	0.10
13	9.10	0.40
14	8.86	0.45
15	8.90	0.35
16	8.87	0.30
17	9.26	0.50
18	9.00	0.50
19	8.75	0.40
20	7.95	−0.05
21	7.65	−0.05
22	7.27	−0.10
23	8.00	0.20
24	8.50	0.10
25	8.75	0.50
26	9.21	0.60
27	8.27	−0.05
28	7.67	0.00
29	7.93	0.05
30	9.26	0.55

FIGURE 6.18 **SAS output of residuals and the Durbin-Watson statistic for a regression analysis of the data in Table 6.3 using the model $y_t = \beta_0 + \beta_1 x_t + \varepsilon_t$**

PARAMETER ESTIMATES

VARIABLE	DF	PARAMETER ESTIMATE	STANDARD ERROR	T FOR HO: PARAMETER=0	PROB > |T|
INTERCEP	1	7.81408758	0.07988432	97.818	0.0001
X4	1	2.66521449	0.25849959	10.310	0.0001

OBS	ACTUAL	PREDICT VALUE	STD ERR PREDICT	LOWER95% MEAN	UPPER95% MEAN	LOWER95% PREDICT	UPPER95% PREDICT	RESIDUAL[a]
1	7.3800	7.6808	0.0893	7.4979	7.8637	7.0071	8.3546	-0.3008
2	8.5100	8.4804	0.0586	8.3604	8.6004	7.8209	9.1398	0.0296
3	9.5200	9.4132	0.1155	9.1767	9.6497	8.7230	10.1034	0.1068
4	7.5000	7.8141	0.0799	7.6505	7.9777	7.1453	8.4829	-0.3141
5	9.3300	8.4804	0.0586	8.3604	8.6004	7.8209	9.1398	0.8496
6	8.2800	8.3471	0.0579	8.2285	8.4657	7.6879	9.0063	-0.0671
7	8.7500	8.2139	0.0601	8.0908	8.3369	7.5539	8.8739	0.5361
8	7.8700	7.9473	0.0716	7.8007	8.0940	7.2825	8.6122	-0.0773
9	7.1000	7.4143	0.1103	7.1884	7.6402	6.7276	8.1010	-0.3143
10	8.0000	8.2139	0.0601	8.0908	8.3369	7.5539	8.8739	-0.2139
11	7.8900	8.3471	0.0579	8.2285	8.4657	7.6879	9.0063	-0.4571
12	8.1500	8.0806	0.0648	7.9479	8.2133	7.4187	8.7425	0.0694
13	9.1000	8.8802	0.0753	8.7259	9.0344	8.2136	9.5467	0.2198
14	8.8600	9.0134	0.0842	8.8410	9.1858	8.3425	9.6844	-0.1534
15	8.9000	8.7469	0.0677	8.6082	8.8857	8.0838	9.4100	0.1531
16	8.8700	8.6137	0.0620	8.4867	8.7406	7.9529	9.2744	0.2563
17	9.2600	9.1467	0.0940	8.9542	9.3392	8.4703	9.8231	0.1133
18	9.0000	9.1467	0.0940	8.9542	9.3392	8.4703	9.8231	-0.1467
19	8.7500	8.8802	0.0753	8.7259	9.0344	8.2136	9.5467	-0.1302
20	7.9500	7.6808	0.0893	7.4979	7.8637	7.0071	8.3546	0.2692
21	7.6500	7.6808	0.0893	7.4979	7.8637	7.0071	8.3546	-0.0308
22	7.2700	7.5476	0.0995	7.3437	7.7514	6.8678	8.2273	-0.2776
23	8.0000	8.3471	0.0579	8.2285	8.4657	7.6879	9.0063	-0.3471
24	8.5000	8.0806	0.0648	7.9479	8.2133	7.4187	8.7425	0.4194
25	8.7500	9.1467	0.0940	8.9542	9.3392	8.4703	9.8231	-0.3967
26	9.2100	9.4132	0.1155	9.1767	9.6497	8.7230	10.1034	-0.2032
27	8.2700	7.6808	0.0893	7.4979	7.8637	7.0071	8.3546	0.5892
28	7.6700	7.8141	0.0799	7.6505	7.9777	7.1453	8.4829	-0.1441
29	7.9300	7.9473	0.0716	7.8007	8.0940	7.2825	8.6122	-0.0173
30	9.2600	9.2800	0.1045	9.0660	9.4940	8.5971	9.9628	-0.0200

SUM OF RESIDUALS	3.73035E-14
SUM OF SQUARED RESIDUALS	2.805902
PREDICTED RESID SS (PRESS)	3.155519

DURBIN-WATSON D	2.414[b]
(FOR NUMBER OF OBS.)	30
1ST ORDER AUTOCORRELATION	-0.223

[a]*Residuals* [b]*Durbin-Watson statistic*

may be an appropriate model relating y_t to x_t. When we estimate β_0 and β_1 using the data in Table 6.3, we obtain the prediction equation

$$\hat{y}_t = 7.8141 + 2.6652x_t$$

A portion of the SAS output for this regression analysis is shown in Figure 6.18. Note that the residuals shown on this output have been computed using the equation

$$e_t = y_t - \hat{y}_t = y_t - (7.8141 + 2.6652x_t)$$

If we wish to test

H_0: The error terms are not autocorrelated

versus the alternative

H_1: The error terms are positively autocorrelated

we compute the Durbin-Watson statistic to be

$$d = \frac{\sum_{t=2}^{30}(e_t - e_{t-1})^2}{\sum_{t=1}^{30} e_t^2}$$

$$= \frac{(.0296 - (-.3008))^2 + (.1068 - .0296)^2 + \cdots + (-.0200 - (-.0173))^2}{(-.3008)^2 + (.0296)^2 + \cdots + (-.0200)^2}$$

$$= 2.414$$

Notice that the Durbin-Watson statistic is given on the SAS output of Figure 6.18. To test for positive autocorrelation at $\alpha = .05$, we use (see Table E.5) $d_{L,.05} = 1.35$ and $d_{U,.05} = 1.49$. Since $d = 2.414 > d_{U,.05} = 1.49$, we do not reject the null hypothesis of no positive autocorrelation. That is, there is no evidence of positive (first-order) autocorrelation.

The Durbin-Watson test can also be used to test for negative autocorrelation as follows.

The Durbin-Watson Test for (First-Order) Negative Autocorrelation

Consider testing the null hypothesis

H_0: The error terms are not autocorrelated

versus the alternative hypothesis

H_1: The error terms are negatively autocorrelated

Durbin and Watson have shown that based on setting the probability of a Type I error equal to α, the points $d_{L,\alpha}$ and $d_{U,\alpha}$ are such that

1. If $(4 - d) < d_{L,\alpha}$, we reject H_0.
2. If $(4 - d) > d_{U,\alpha}$, we do not reject H_0.
3. If $d_{L,\alpha} \leq (4 - d) \leq d_{U,\alpha}$, the test is inconclusive.

Here, large values of d (and hence small values of $4 - d$) lead to the conclusion of negative autocorrelation because if d is large, this indicates that the differences $(e_t - e_{t-1})$ are large. This says that the adjacent error terms ε_t and e_{t-1} are negatively autocorrelated. As an example, for the data and model in Example 6.6 we see that

$$(4 - d) = (4 - 2.414) = 1.586 > d_{U,.05} = 1.49$$

Therefore on the basis of setting α equal to .05, we do not reject the null hypothesis of no negative autocorrelation. That is, there is no evidence of negative (first-order) autocorrelation.

Finally, we can also use the Durbin-Watson statistic to test for positive or negative autocorrelation.

The Durbin-Watson Test for (First-Order) Positive or Negative Autocorrelation

Consider testing the null hypothesis

H_0: The error terms are not autocorrelated

versus the alternative hypothesis

H_1: The error terms are positively or negatively autocorrelated

Durbin and Watson have shown that, based on setting the probability of a Type I error equal to α,

1. If $d < d_{L,\alpha/2}$ or if $(4 - d) < d_{L,\alpha/2}$, we reject H_0.
2. If $d > d_{U,\alpha/2}$ and if $(4 - d) > d_{U,\alpha/2}$, we do not reject H_0.
3. If $d_{L,\alpha/2} \leq d \leq d_{U,\alpha/2}$ and $d_{L,\alpha/2} \leq (4 - d) \leq d_{U,\alpha/2}$, the test is inconclusive.

Before we conclude our presentation of the Durbin-Watson test, several comments are relevant. First, the validity of the Durbin-Watson test depends upon the assumption that the population of all possible residuals at any time t has a normal distribution. We will see how to tell whether or not this assumption is reasonable in the next section. Second, positive autocorrelation is found in practice more commonly than negative autocorrelation. Therefore the first test we have presented (the test for positive autocorrelation) is used more often than the others. Third, most regression computer packages print the Durbin-Watson d statistic.

When residual plots or statistical tests indicate that the error terms in a regression problem are autocorrelated, remedies can be employed. We discuss these remedies in Chapter 13.

6.6 INFERENCE ASSUMPTION 3: NORMAL POPULATIONS

The third inference assumption is as follows.

Inference Assumption 3 (Normal Populations)

For any value x_i of the independent variable x the corresponding population of potential values of the dependent variable has a *normal distribution*. Said equivalently, for any value x_i of the independent variable x the corresponding population of potential error terms has a *normal distribution*.

Validation of the normality assumption can be accomplished by using several methods. First, we can construct a bar chart or histogram of the residuals. If the normality assumption holds, then each error term ε_i (for $i = 1, 2, \ldots, n$) has been randomly selected from a normal distribution with mean zero and (if the constant variance assumption holds) variance σ^2. Thus if the normality assumption holds, the histogram of the residuals should look reasonably bell-shaped and reasonably symmetric about zero.

It is important to point out again that mild departures from the inference assumptions do not seriously hinder our ability to make statistical inferences. Hence we are looking for pronounced, rather than subtle, departures from the normality assumption. We require only that the histogram of the residuals look fairly normal. Also, if one wishes to employ a formal statistical test for normality, two goodness-of-fit tests can be used to analyze the residuals—the chi-square test and the Kolmogorov-Smirnov test. See Pfaffenberger and Patterson (1987) for descriptions of these tests. One other point worth making here is that violations of inference assumptions 1 or 2, as well as an incorrect functional form, can often cause a histogram of the residuals to look non-normal. Because of this, it is usually a good idea to use residual plots to check for incorrect functional form, non-constant error variance, and positive or negative autocorrelation before attempting to validate the normality assumption.

Another way to validate the normality assumption is to use *standardized residuals*. Standardized residuals are computed by dividing the residuals by the standard error s. Recall that 68.26 percent of the values in a normally distributed population are within 1 standard deviation of the mean, while 95.44 percent of such values are within 2 standard deviations of the mean. It follows that the normality assumption implies (if the constant variance assumption holds) that

$$P\left(-1 \leq \frac{\varepsilon_i}{\sigma} \leq 1\right) = .6826$$

$$P\left(-2 \leq \frac{\varepsilon_i}{\sigma} \leq 2\right) = .9544$$

Now, since the standardized residual e_i/s is a point estimate of ε_i/σ, these expressions intuitively say that if the normality assumption holds, then about 68 percent of the standardized residuals will be between -1 and 1, while about 95 percent of the standardized residuals will be between -2 and 2. For small samples the corresponding values from the t-distribution with $n - 2$ degrees of freedom should be used. For example, about 95 percent of the standardized residuals should be between $-t_{[.025]}^{(n-2)}$ and $t_{[.025]}^{(n-2)}$.

Example 6.7 Consider the demand (y_t) and price difference (x_t) data given in Table 6.3. The $n = 30$ residuals for the model

$$y_t = \beta_0 + \beta_1 x_t + \varepsilon_t$$

are given in the SAS output of Figure 6.18. A histogram of these residuals is shown in Figure 6.19. We see that the histogram is reasonably bell-shaped and symmetric about zero. These conclusions are obviously subjective, but we are looking only for serious violations of the normality assumption. The standardized residuals are computed by dividing the residuals in Figure 6.18 by the standard error $s = .3165$

FIGURE 6.19 Histogram of the residuals for the demand and price difference model $y_t = \beta_0 + \beta_1 x_t + \varepsilon_t$

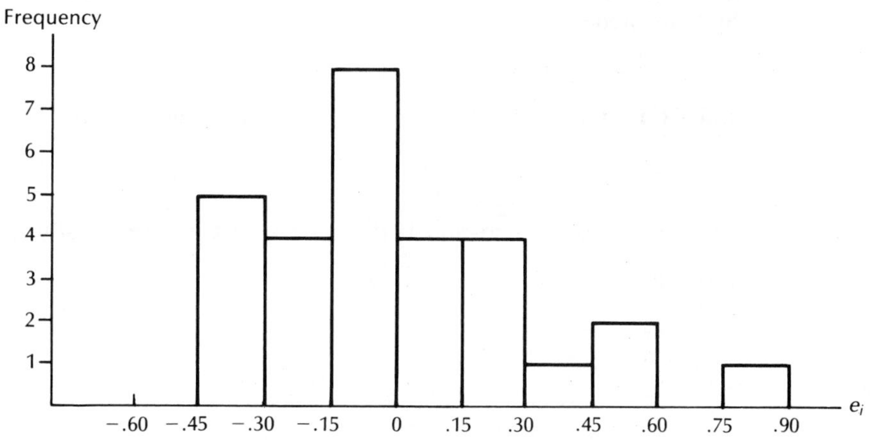

TABLE 6.4 **Standardized residuals for the demand and price difference model $y_t = \beta_0 + \beta_1 x_t + \varepsilon_t$**

Sales period	*Standardized residual*	*Sales period*	*Standardized residual*
1	−.9504	16	.8098
2	.0935	17	.3580
3	.3374	18	−.4635
4	−.9924	19	−.4114
5	2.6844	20	.8506
6	−.2120	21	−.0973
7	1.6938	22	−.8771
8	−.2442	23	−1.0967
9	−.9930	24	1.3251
10	−.6758	25	−1.2534
11	−1.4442	26	−.6420
12	.2193	27	1.8616
13	.6945	28	−.4553
14	−.4847	29	−.0547
15	.4837	30	−.0632

(which is not shown on the computer output). These standardized residuals are given in Table 6.4. We find that 23 of the 30 standardized residuals (76.7 percent) are between −1 and 1, while 29 of the 30 standardized residuals (96.7 percent) are between −2 and 2. We conclude that the normality assumption is not seriously violated.

In the following example we illustrate that if there are not enough residuals to construct a histogram, we can construct a stem-and-leaf diagram to check the normality assumption.

Example 6.8 In Table 6.5 we list (in increasing order) the 11 residuals from the Comp-U-Systems model

$$y = \beta_0 + \beta_1 x + \varepsilon$$

and we round these residuals to the nearest integer values. In Figure 6.20 we

FIGURE 6.20 **SAS output of a stem-and-leaf diagram of the Comp-U-Systems residuals**

```
STEM LEAF                  #
   0 57                    2
   0 124                   3
  -0 3310                  4
  -0 75                    2
     ---+----+----+----+
  MULTIPLY STEM.LEAF BY 10**+01
```

TABLE 6.5 **The ordered residuals and rounded ordered residuals for the Comp-U-Systems model**

Ordered residual, e_i	*Rounded ordered residual*
−6.873	−7
−5.077	−5
−3.271	−3
−2.669	−3
−.872928	−1
−.475138	0
.933702	1
2.127	2
3.525	4
5.320	5
7.331	7

present the SAS output of a stem-and-leaf diagram of these residuals. Since this diagram looks reasonably bell-shaped, we might conclude that the normality assumption approximately holds for the Comp-U-Systems model.

Another graphical technique for examining the validity of the normality assumption is the *normal plot*. This technique (along with several other techniques for examining residuals) is discussed by Anscombe and Tukey (1963). The normal plot technique first requires that the residuals $e_1, e_2, \ldots, e_n$ be arranged in order from smallest to largest. Letting $e_{(i)}$ denote the ith residual in the ordered listing, we plot $e_{(i)}$ on the vertical axis against the point $z_{(i)}$ on the horizontal axis. Here, $z_{(i)}$ is defined to be the point on the scale of the standard normal curve so that the area under this curve to the left of $z_{(i)}$ is $(3i - 1)/(3n + 1)$. If the normality assumption holds, and if the model has the correct functional form, then this plot should have a straight-line appearance. Substantial departures from a straight-line appearance (admittedly a subjective decision) indicate a violation of the normality assumption.

Example 6.9 In this example we construct a normal plot of the residuals for the Comp-U-Systems model

$$y = \beta_0 + \beta_1 x + \varepsilon$$

To construct this plot, we must first arrange the residuals in order from smallest to largest. These ordered residuals are given in Table 6.5. Denoting the ith ordered residual as $e_{(i)}$ $(i = 1, 2, \ldots, 11)$, we next compute for each value of i the point $z_{(i)}$. These computations are summarized in Table 6.6. We now plot the ordered residuals from Table 6.5 on the vertical axis against the values of $z_{(i)}$ from Table 6.6 on the horizontal axis. The resulting normal plot is depicted in Figure 6.21. Looking at this figure, we see that while this normal plot is certainly not exactly

TABLE 6.6 **Normal plot calculations for the Comp-U-Systems model**

i	$\frac{3i-1}{3n+1}$	$z_{(i)}$
1	$\frac{3(1)-1}{3(11)+1} = \frac{2}{34} = .0588$	−1.565
2	$\frac{3(2)-1}{3(11)+1} = \frac{5}{34} = .1470$	−1.05
3	$\frac{3(3)-1}{3(11)+1} = \frac{8}{34} = .2353$	−.72
4	.3235	−.46
5	.4118	−.22
6	.5000	0
7	.5882	.22
8	.6765	.46
9	.7647	.72
10	.8529	1.05
11	.9412	1.565

FIGURE 6.21 **Normal plot of the residuals for the Comp-U-Systems model**

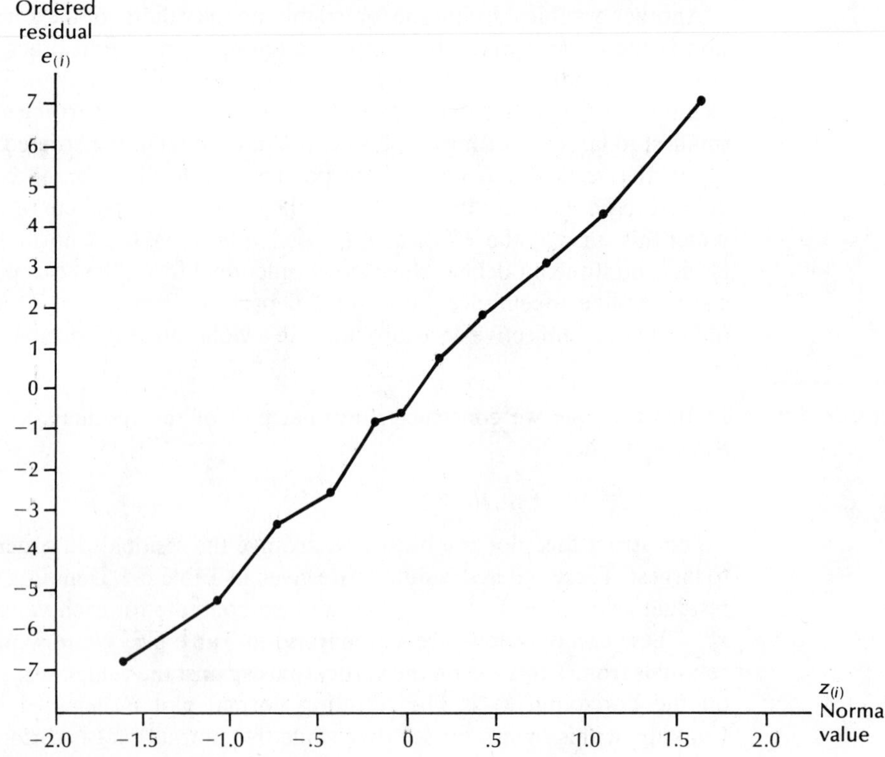

a straight line, the plot does have a distinct straight-line appearance. This indicates that the normality assumption approximately holds for the Comp-U-Systems model.

If analysis of the residuals indicates that the normality assumption is seriously violated, then remedies should be employed. We discuss what can be done to remedy violations of this assumption in Chapter 13.

6.7 OUTLYING AND INFLUENTIAL OBSERVATIONS

An observation that is well separated from the rest of the data is called an *outlier*, and an observation that causes the least squares point estimates to be substantially different from what they would be if the observaton were removed from the data set is called *influential*. An observation may be an outlier with respect to its y value and/or its x value, but an outlier may or may not be influential. We illustrate these ideas by considering Figure 6.22, which is a hypothetical plot of the values of a dependent variable y against an independent variable x. Observation 1 in Figure 6.22 is outlying with respect to its y value. However, it is not outlying with

FIGURE 6.22 Data plot illustrating outlying and influential observations

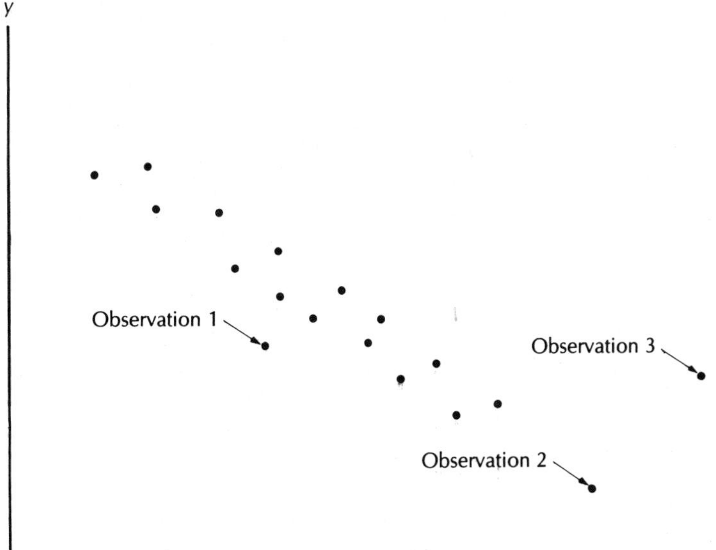

FIGURE 6.23 **SAS output of the residuals for the fuel consumption model plotted against *x***

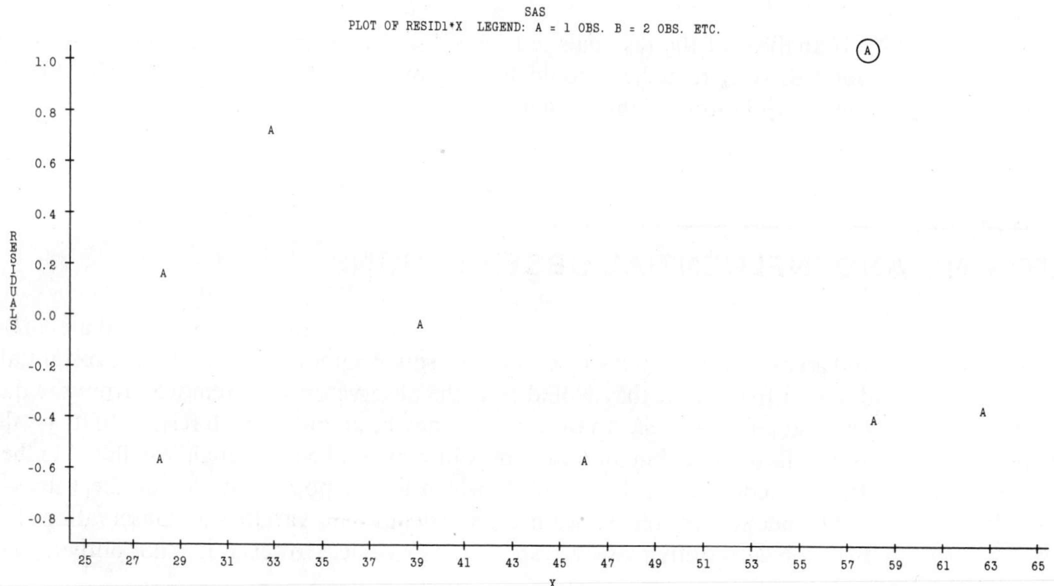

respect to its x value, since its x value is near the middle of the other x values. Moreover, observation 1 may not be influential because there are several observations with similar x values and nonoutlying y values, which will keep the least squares point estimates from being excessively influenced by observation 1. Observation 2 in Figure 6.22 is outlying with respect to its x value, but since its y value is consistent with the regression relationship displayed by the nonoutlying observations, it is probably not influential. Observation 3, however, is probably influential, because it is outlying with respect to its x value and because its y value is not consistent with the regression relationship displayed by the other observations.

In addition to using data plots (such as Figure 6.22) we can use residual plots against values of the independent variable, predicted values of the dependent variable, or time to indicate the presence of outliers with respect to their y values. Any residual that is substantially different from the others is suspect. As a rule of thumb, if the absolute value of the standardized residual is greater than 3 or 4, an outlier with respect to its y value is suspected. For example, Table 4.6 tells us that the largest residual for the fuel consumption model

$$y = \beta_0 + \beta_1 x + \varepsilon$$

is $e_6 = y_6 - \hat{y}_6 = 1.0560$. This largest residual is circled in Figure 6.23, which is the SAS output of a plot of the fuel consumption model residuals against x.

However, since the standard error for this model is $s = .6542$, it follows that the largest standardized residual is

$$\frac{e_6}{s} = \frac{1.0560}{.6542} = 1.6142$$

Since the absolute value of this largest standardized residual is less than 3, we do not have evidence that there are outliers with respect to their y values in the fuel consumption problem.

We give a much more complete discussion of outlying and influential observations in Chapter 10.

*6.8 Using SAS

In Figure 6.24 we present a SAS program that produces residual plots for the Comp-U-Systems problem (these plots are shown in Figures 6.2, 6.3, and 6.4).

FIGURE 6.24 **SAS program to obtain residual plots and other diagnostics for the Comp-U-Systems model $y = \beta_0 + \beta_1 x + \varepsilon$**

```
DATA COMP;      } ——→ Assigns name COMP to data file
INPUT Y X;      } ——→ Assigns variable names:
                        Y = service time; X = number of microcomputers
TIME=_N_;       } ——→ Defines variable TIME to be observation number

CARDS;
109 4
 58 2
138 5
189 7
 37 1
 82 3   ——→ Comp-U-Systems data (see Table 5.6)
103 4
134 5
 68 2
112 4
154 6

PROC PRINT; } ——→ Print procedure: prints the data
PROC REG DATA=COMP; } ——→ Specifies regression procedure

MODEL Y=X/P DW;
OUTPUT OUT=ONE PREDICTED=YHAT RESIDUAL=RESID;
PROC PLOT DATA=ONE;
PLOT RESID*(X YHAT TIME);
    } ——→ Plot residuals from model y = β0 + β1x + ε against
          x, ŷ, and time. DW requests Durbin-Watson statistic

PROC UNIVARIATE PLOT DATA=ONE;
VAR RESID;
    } ——→ Requests stem-and-leaf diagram, box plot,
          and normal plot of residuals for model y = β0 + β1x + ε
```

*This section is optional.

EXERCISES

Note: To employ residual analysis most profitably, it is desirable to have collected a reasonably large number of observations. Some of the exercises in this chapter employ only a small number of residuals. This has been done to illustrate residual analysis while minimizing calculations. Of course, in some situations a small number of residuals can be very informative.

6.1 Consider the simple linear regression model in Exercise 4.3. This model relates starting salary (y) to grade point average (x).

a. Compute the residuals.
b. Plot the residuals versus x and $\hat{y}$ (predicted starting salary).
c. Interpret the plots. Do they suggest a violation of the regression assumptions?
d. Compute and interpret the standardized residuals.
e. Construct and interpret a stem-and-leaf diagram of the residuals.

6.2 Consider the sales volume data in Table 4.2 and the simple linear regression model relating sales volume (y) to advertising expenditure (x).

a. Compute the residuals; see Exercise 4.4 for b_0 and b_1.
b. Plot the residuals versus x and $\hat{y}$ (predicted sales volume).
c. Interpret the plots. Do they suggest a violation of the regression assumptions?
d. Compute and interpret the standardized residuals.
e. Construct and interpret a stem-and-leaf diagram of the residuals.

6.3 Consider the simple linear regression model in Exercise 4.5. This model relates demand (y) to price (x).

a. Compute the residuals.
b. Plot the residuals versus x and $\hat{y}$ (predicted demand).
c. Interpret the plots. Do they suggest a violation of the regression assumptions?
d. Compute and interpret a stem-and-leaf diagram of the residuals.
e. Compute and interpret the standardized residuals.

6.4 Consider the simple linear regression model in Exercise 4.6. This model relates natural gas consumption (y) to production (x).

a. Compute the residuals.
b. Construct and interpret a histogram of the residuals.
c. Construct and interpret a normal plot of the residuals.
d. Compute and interpret the standardized residuals.

6.5 Consider the data in Table 6.3 and the simple linear regression model relating demand (y) to price difference (x) in Example 6.6. The residuals for this model are listed on the SAS output of Figure 6.18.

a. Plot these residuals versus x. Interpret the plot.
b. Plot these residuals versus $\hat{y}$ (predicted sales). Interpret the plot.
c. Plot these residuals versus time. Interpret the plot.

6.6 Consider the simple linear regression model in Exercise 4.9. This model relates monthly labor hours required (y) to monthly occupied bed days (x).

a. Compute the residuals.
b. Construct and interpret residual plots versus x and $\hat{y}$.
c. Compute the standardized residuals and check for any outliers with respect to their y-values.
d. Construct a stem-and-leaf diagram of the residuals. Interpret.
e. Construct and interpret a normal plot of the residuals.

6.7 Consider the simple linear trend model in Exercise 4.11. This model relates quarterly sales (y) to time (t).

a. Compute the residuals and plot them versus time.
b. Does the plot suggest that the error terms are autocorrelated? Explain.
c. Calculate the Durbin-Watson statistic.
d. Test for positive (first-order) autocorrelation with $\alpha = .05$.

6.8 Consider the simple linear trend model in Exercise 4.12. This model relates monthly calculator sales (y) to time (t).

a. Compute the residuals and plot them versus time.
b. Does the plot suggest that the error terms are autocorrelated? Explain.
c. Calculate the Durbin-Watson statistic.

6.9 Figure 6.25 presents the SAS output of the simple linear regression analysis of the heat loss data in Exercise 4.10. The data plot of y (heat loss) versus x (outdoor temperature) and a plot of the model's residuals versus $\hat{y}$ are also given.

FIGURE 6.25 SAS output of a simple linear regression analysis of the heat loss data

```
                                  GENERAL LINEAR MODELS PROCEDURE

DEPENDENT VARIABLE: H_LOSS     HEAT LOSS

SOURCE               DF   SUM OF SQUARES     MEAN SQUARE    F VALUE        PR > F    R-SQUARE          C.V.
MODEL                 1    2773.50000000   2773.50000000      21.70        0.0023    0.756134       17.1276
ERROR                 7     894.50000000    127.78571429                  ROOT MSE                H LOSS MEAN
CORRECTED TOTAL       8    3668.00000000                               11.30423435                66.00000000

SOURCE         DF          TYPE I SS     F VALUE      PR > F      DF        TYPE III SS      F VALUE    PR > F
TEMP            1      2773.50000000       21.70      0.0023       1      2773.50000000        21.70    0.0023

                                         T FOR HO:        PR > |T|       STD ERROR
PARAMETER              ESTIMATE       PARAMETER=0                        OF ESTIMATE

INTERCEPT          109.00000000             10.93          0.0001        9.96939762
TEMP                -1.07500000             -4.66          0.0023        0.23074672
```

(continues)

FIGURE 6.25 Continued

OBSERVATION	OBSERVED VALUE	PREDICTED VALUE	RESIDUAL
1	86.00000000	87.50000000	-1.50000000
2	80.00000000	87.50000000	-7.50000000
3	77.00000000	87.50000000	-10.50000000
4	78.00000000	66.00000000	12.00000000
5	84.00000000	66.00000000	18.00000000
6	75.00000000	66.00000000	9.00000000
7	33.00000000	44.50000000	-11.50000000
8	38.00000000	44.50000000	-6.50000000
9	43.00000000	44.50000000	-1.50000000

SUM OF RESIDUALS	0.00000000
SUM OF SQUARED RESIDUALS	894.50000000
SUM OF SQUARED RESIDUALS - ERROR SS	0.00000000
FIRST ORDER AUTOCORRELATION	0.36109558
DURBIN-WATSON D	1.27277809

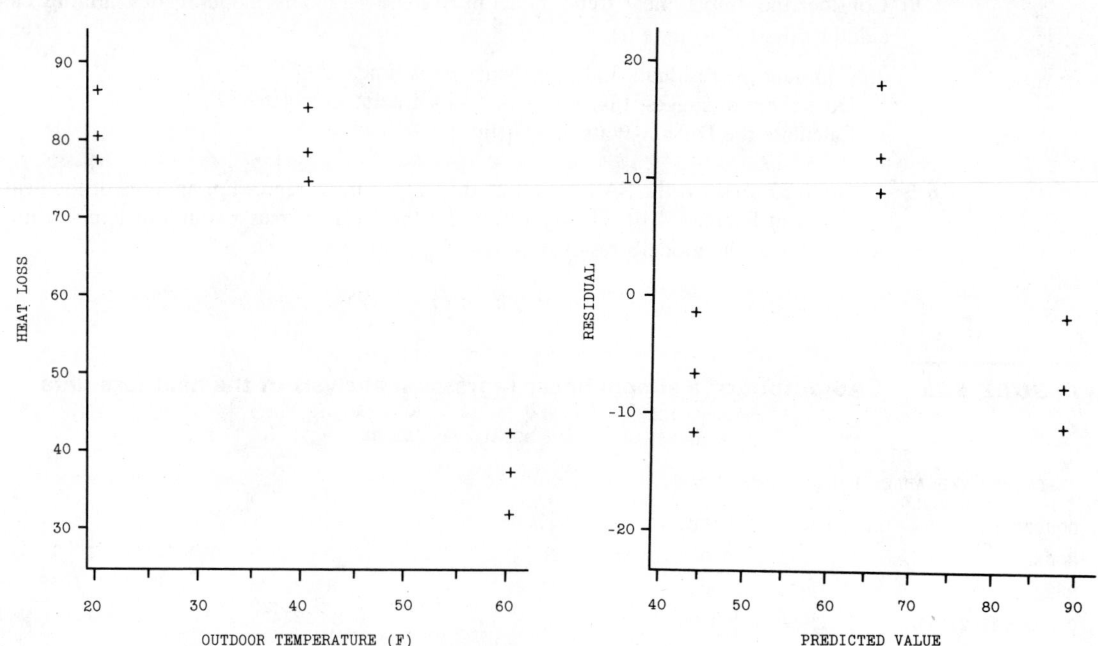

a. Is the regression relationship significant? That is, can we reject $H_0: \beta_1 = 0$ at a reasonable setting of α?
b. What do the data plot and residual plot say about the functional form of the simple linear model?
c. Are the data plots consistent with the result of the lack-of-fit test performed in Exercise 5.50?

6.10 Consider the simple linear regression model in Exercise 4.13. This model relates the number of work stoppages (y) to the percent of the civilian labor force with union membership (x).

a. Construct and interpret residual plots versus x, $\hat{y}$, and time.
b. Calculate the standardized residuals. Use them to check for outliers.
c. Construct and interpret a normal plot of the residuals.

6.11 Consider the simple linear regression model in Exercise 4.14. This model relates illiteracy rate (y) to school enrollment rate (x).

a. Construct and interpret residual plots versus x, $\hat{y}$, and time.
b. Calculate and interpret the standardized residuals.
c. Construct and interpret a stem-and-leaf diagram of the residuals.

6.12 Consider the simple linear regression model in Exercise 4.15. This model relates the divorce rate (y) per 1000 women to the percent of female population in the labor force (x).

a. Plot y versus x. Do there appear to be any outliers? If so, do they appear to be influential?
b. Construct and interpret residual plots versus x, $\hat{y}$, and time.
c. Compute and interpret the standardized residuals.
d. Construct and interpret a normal plot of the residuals.
e. Construct a stem-and-leaf diagram of the residuals. Interpret.

6.13 Consider the simple linear regression model in Exercise 4.16. This model relates U.S. energy consumption (y) to U.S. total population (x).

a. Plot y versus x. Do there appear to be any outliers? If so, do they appear to be influential?
b. Compute and interpret the standardized residuals.
c. Construct and interpret residual plots versus x, $\hat{y}$, and time.
d. Construct a stem-and-leaf diagram of the residuals. Interpret.

6.14 Consider the simple linear regression model in Exercise 4.17. This model relates social welfare expenditures (y) to total GNP (x_1).

a. Plot y versus x_1. Do there appear to be any outliers? If so, do they appear to be influential?
b. Construct and interpret residual plots versus x_1, $\hat{y}$, and time.
c. Compute and interpret the standardized residuals.

6.15 Consider the simple linear regression model in Exercise 4.23. This model relates the time (y) required to perform service on a service call to the number of copiers serviced (x). Figure 6.26 gives the SAS output of residual plots for this model versus x and $\hat{y}$.

a. Interpret the residual plots. Do they suggest a violation of any of the regression assumptions?
b. Construct and interpret a histogram of the residuals.
c. Construct and interpret a normal plot of the residuals.

FIGURE 6.26 SAS output of residual plots for a simple linear regression relating service time (y) to number of copiers serviced (x)

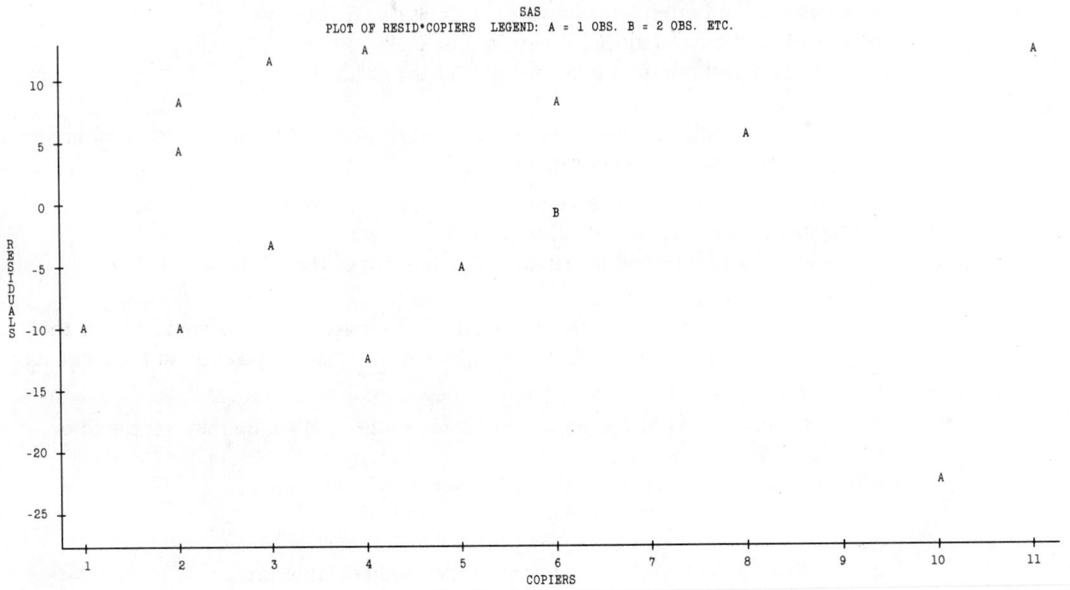

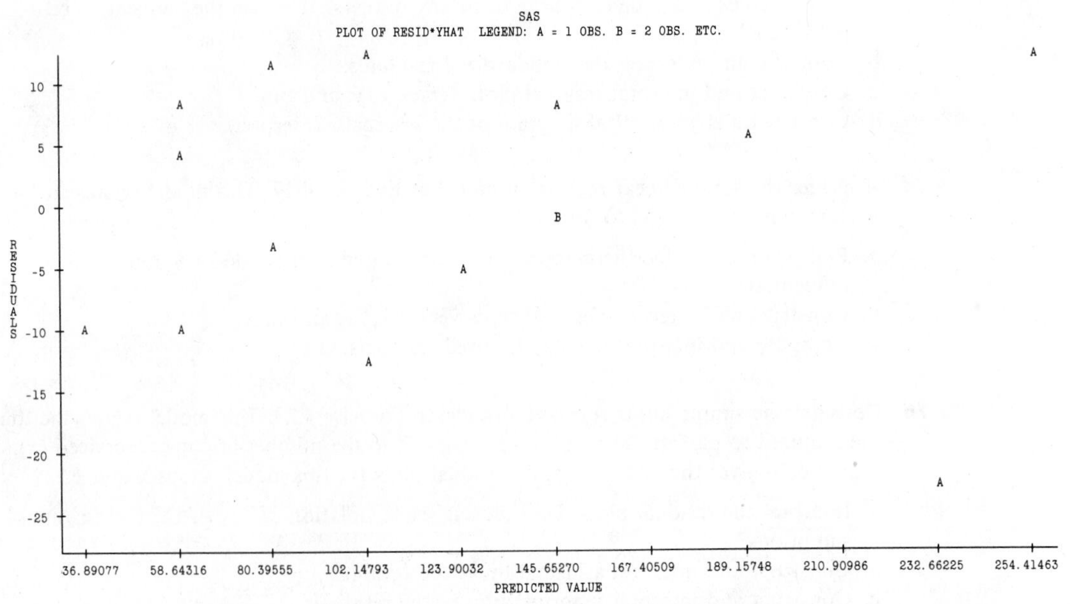

FIGURE 6.27 **Residual plot versus x_3 (advertising expenditure) for the Fresh Detergent demand model $y = \beta_0 + \beta_1 x_4 + \varepsilon$**

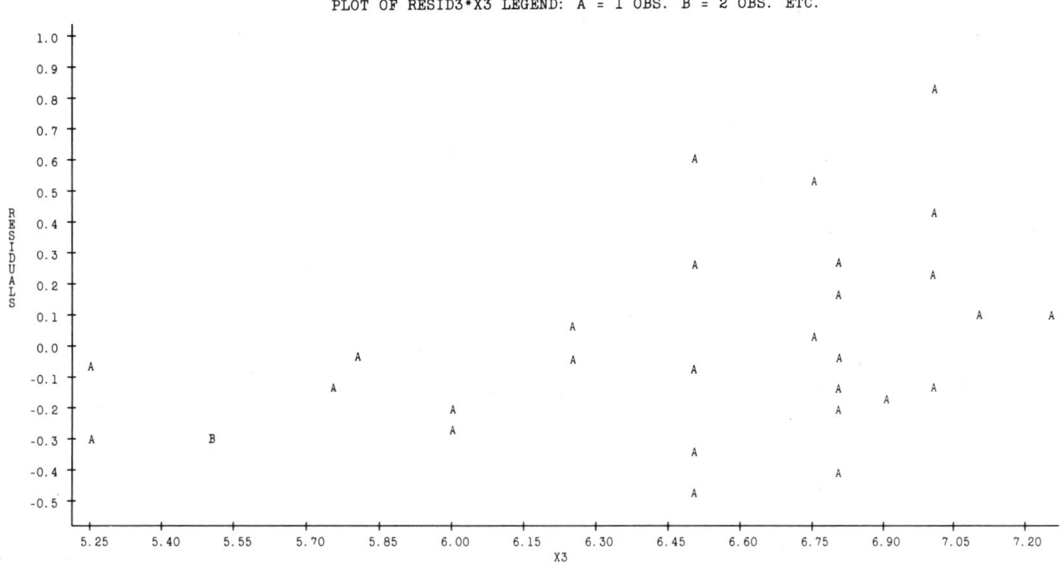

6.16 Consider the Fresh demand data in Table 6.1 and the simple linear regression model relating demand (y) to price difference (x_4). Figure 6.27 gives the SAS output of a residual plot for this model versus x_3. Interpret this residual plot.

6.17 Ott (1987) discusses a test of the abrasive effect of a wear-testing machine. Forty-eight identical 5-inch-square pieces of a particular fabric were cut. Eight squares were randomly assigned to each of six machine speeds, and the wear over a 3-minute test period was measured for each square. The data is given in Table 6.7.

TABLE 6.7 **Wear-testing data**

Machine speed (rpm)	*Wear*
100	23.0, 23.5, 24.4, 25.2, 25.6, 26.1, 24.8, 25.6
120	26.7, 26.1, 25.8, 26.3, 27.2, 27.9, 28.3, 27.4
140	28.0, 28.4, 27.0, 28.8, 29.8, 29.4, 28.7, 29.3
160	32.7, 32.1, 31.9, 33.0, 33.5, 33.7, 34.0, 32.5
180	43.1, 41.7, 42.4, 42.1, 43.5, 43.8, 44.2, 43.6
200	54.2, 43.7, 53.1, 53.8, 55.6, 55.9, 54.7, 54.5

Source: Lyman Ott, *An Introduction to Statistical Methods and Data Analysis*, 3rd ed., PWS-KENT Publishing Company, Boston. © 1987. Used with permission.

FIGURE 6.28 **SAS output of a simple linear regression analysis of the wear-testing data**

GENERAL LINEAR MODELS PROCEDURE

DEPENDENT VARIABLE: WEAR

SOURCE	DF	SUM OF SQUARES	MEAN SQUARE	F VALUE	PR > F	R-SQUARE	C.V.
MODEL	1	4326.79207143	4326.79207143	291.47	0.0001	0.863693	11.0305
ERROR	46	682.84709524	14.84450207		ROOT MSE		WEAR MEAN
CORRECTED TOTAL	47	5009.63916667			3.85285635		34.92916667

SOURCE	DF	TYPE I SS	F VALUE	PR > F	DF	TYPE III SS	F VALUE	PR > F
SPEED	1	4326.79207143	291.47	0.0001	1	4326.79207143	291.47	0.0001

PARAMETER	ESTIMATE	T FOR HO: PARAMETER=0	PR > |T|	STD ERROR OF ESTIMATE
INTERCEPT	-6.76547619	-2.70	0.0096	2.50470943
SPEED	0.27796429	17.07	0.0001	0.01628129

OBSERVATION	OBSERVED VALUE	PREDICTED VALUE	RESIDUAL
1	23.00000000	21.03095238	1.96904762
2	23.50000000	21.03095238	2.46904762
3	24.40000000	21.03095238	3.36904762
4	25.20000000	21.03095238	4.16904762
5	25.60000000	21.03095238	4.56904762
6	26.10000000	21.03095238	5.06904762
7	24.80000000	21.03095238	3.76904762
8	25.60000000	21.03095238	4.56904762
9	26.70000000	26.59023810	0.10976190
10	26.10000000	26.59023810	-0.49023810
11	25.80000000	26.59023810	-0.79023810
12	26.30000000	26.59023810	-0.29023810
13	27.20000000	26.59023810	0.60976190
14	27.90000000	26.59023810	1.30976190
15	28.30000000	26.59023810	1.70976190
16	27.40000000	26.59023810	0.80976190
17	28.00000000	32.14952381	-4.14952381
18	28.40000000	32.14952381	-3.74952381
19	27.00000000	32.14952381	-5.14952381
20	28.80000000	32.14952381	-3.34952381
21	29.80000000	32.14952381	-2.34952381
22	29.40000000	32.14952381	-2.74952381
23	28.70000000	32.14952381	-3.44952381
24	29.30000000	32.14952381	-2.84952381
25	32.70000000	37.70880952	-5.00880952
26	32.10000000	37.70880952	-5.60880952
27	31.90000000	37.70880952	-5.80880952
28	33.00000000	37.70880952	-4.70880952
29	33.50000000	37.70880952	-4.20880952
30	33.70000000	37.70880952	-4.00880952
31	34.00000000	37.70880952	-3.70880952
32	32.50000000	37.70880952	-5.20880952
33	43.10000000	43.26809524	-0.16809524
34	41.70000000	43.26809524	-1.56809524
35	42.40000000	43.26809524	-0.86809524
36	42.10000000	43.26809524	-1.16809524
37	43.50000000	43.26809524	0.23190476
38	43.80000000	43.26809524	0.53190476
39	44.20000000	43.26809524	0.93190476
40	43.60000000	43.26809524	0.33190476
41	54.20000000	48.82738095	5.37261905

OBSERVATION	OBSERVED VALUE	PREDICTED VALUE	RESIDUAL
42	43.70000000	48.82738095	-5.12738095
43	53.10000000	48.82738095	4.27261905
44	53.80000000	48.82738095	4.97261905
45	55.60000000	48.82738095	6.77261905
46	55.90000000	48.82738095	7.07261905
47	54.70000000	48.82738095	5.87261905
48	54.50000000	48.82738095	5.67261905

SUM OF RESIDUALS	0.00000000
SUM OF SQUARED RESIDUALS	682.84709524
SUM OF SQUARED RESIDUALS - ERROR SS	-0.00000000
FIRST ORDER AUTOCORRELATION	0.73342816
DURBIN-WATSON D	0.48034159

LINEAR REGRESSION ANALYSIS
PLOT OF RESIDUALS VS. PREDICTED VALUES

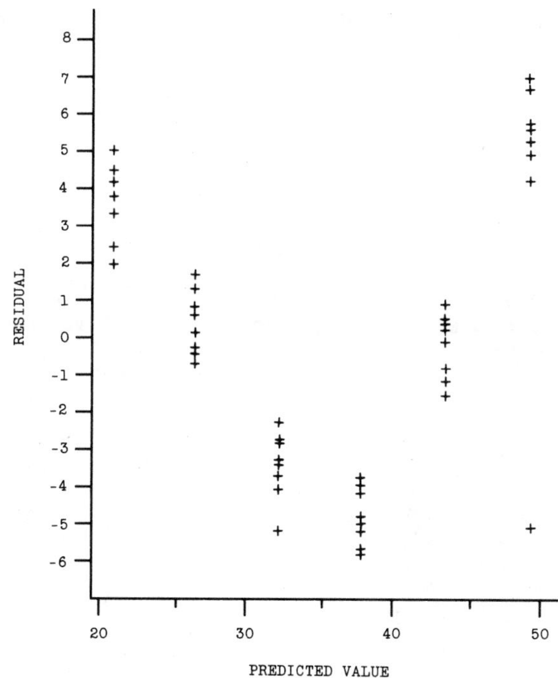

Figure 6.28 gives the SAS output of a simple linear regression analysis relating wear (y) to machine speed (x). The figure also gives the residual plot for this model versus $\hat{y}$.

a. Test the significance of the regression relationship.
b. Interpret the value of r^2, the simple coefficient of determination.
c. What does the residual plot say about the functional form of the model that has been employed? Explain.
d. Compute and interpret the standardized residuals.
e. Construct and interpret a histogram of the residuals.

6.18 Consider the simple linear regression model in Exercise 4.18. This model relates social welfare expenditures (y) to per capita GNP (x_2).

a. Plot y versus x_2. Do there appear to be any outliers? If so, do they appear to be influential?
b. Construct and interpret residual plots versus x_2, $\hat{y}$, and time.
c. Compute and interpret the standardized residuals.
d. Construct and interpret a stem-and-leaf diagram of the residuals.

6.19 Consider the hospital labor needs data in Table 5.9, and consider using simple linear regression to relate monthly labor requirements (y) to monthly occupied bed days (x). You have computed the residuals for this model in Exercise 6.6.

a. Plot the residuals for the model $y = \beta_0 + \beta_1 x_3 + \varepsilon$ versus x_1, x_2, x_4, and x_5. Interpret these plots. What do they say about variables that may need to be included in a hospital labor model?
b. Plot the residuals for the model $y = \beta_0 + \beta_1 x_5 + \varepsilon$ versus x_1, x_2, x_3, and x_4. Interpret the plots.

6.20 A company carries out an experiment that is designed to relate

$$y = \text{Sales volume (in units of \$10,000)}$$

TABLE 6.8 Sales volume–advertising expenditure data

Sales region, i	Media expenditures, x_{i1}	Point of sale expenditures, x_{i2}	Sales volume, y_i
1	1	1	3.27
2	1	2	5.73
3	1	3	7.92
4	1	4	8.71
5	1	5	10.21
6	2	1	8.30
7	2	2	10.54
8	2	3	12.46
9	2	4	12.87
10	2	5	13.75
11	3	1	12.91
12	3	2	14.63
13	3	3	14.75
14	3	4	16.67
15	3	5	17.21
16	4	1	16.82
17	4	2	18.11
18	4	3	18.24
19	4	4	18.37
20	4	5	19.57
21	5	1	20.51
22	5	2	21.14
23	5	3	21.47
24	5	4	21.73
25	5	5	22.26

to

$$x_1 = \text{Media advertising expenditures (in units of \$1000)}$$
$$x_2 = \text{Point of sale advertising expenditures (in units of \$1000)}$$

The company observes the data in Table 6.8.

a. Plot y versus x_1 and versus x_2. Interpret the plots.
b. Consider the model $y = \beta_0 + \beta_1 x_1 + \varepsilon$. Plot the residuals for this model versus x_1 and x_2. Interpret the results.
c. Consider the model $y = \beta_0 + \beta_1 x_2 + \varepsilon$. Plot the residuals for this model versus x_1 and x_2. Interpret the results.

6.21 An economist studied the relationship between x, 1970 yearly income (in thousands of dollars) for a family of four, and y, 1970 yearly clothing expenditure (in hundreds of dollars) for the family. The following data were observed:

x_i:	8	10	12	14	16	18	20
y_i:	6.47	6.17	7.4	10.57	11.93	10.3	14.67

When we use simple linear regression to relate y to x, we find that the least squares point estimates of β_0 and β_1 are $b_0 = .2968$ and $b_1 = .6677$ and that the standard error s is equal to 1.3434.

a. Calculate the residuals and the standardized residuals.
b. Construct residual plots versus x and $\hat{y}$.

FIGURE 6.29 A residual plot versus time for 40 monthly observations

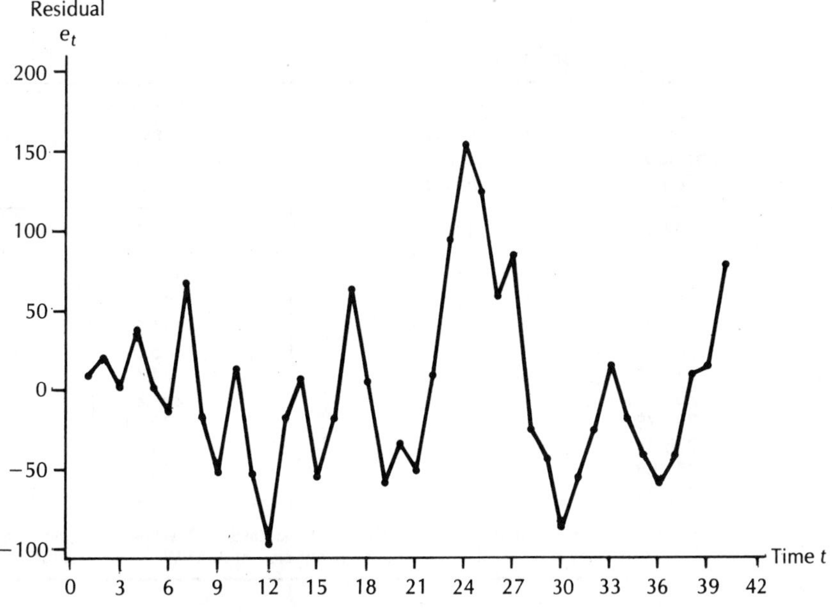

c. Construct plots of the standardized residuals versus x and $\hat{y}$.
d. Do these residual plots suggest any violations of the regression assumptions? If so, which assumption is violated and why?
e. If one of the assumptions is violated, why might the violation be logical?

6.22 A simple linear regression model is employed to analyze 40 monthly observations. Residuals are computed and are plotted versus time. The resulting residual plot is shown in Figure 6.29.

a. Discuss why this residual plot suggests the existence of positive autocorrelation.
b. Assuming that the Durbin-Watson statistic is equal to .8397, test for positive autocorrelation at $\alpha = .05$, and test for positive autocorrelation at $\alpha = .01$.

6.23 Consider the simple linear regression model in Exercise 5.22. This model relates average estimate of β_1 (y) to return length time (x). For each stock type (aggressive, defensive, and neutral), do the following:

a. Calculate and plot the residuals versus x and $\hat{y}$. Interpret the plots.
b. Compute and interpret the standardized residuals.
c. Construct and interpret a stem-and-leaf display of the residuals.

6.24 Makridakis, Wheelwright, and McGee present data concerning monthly total sales (y) and advertising expenditures (x). The first 24 observations are given in Table 6.9. The SAS output of the simple linear regression analysis relating y to x is given in Figure 6.30. Residual plots versus x, $\hat{y}$, and time are included.

a. Test the significance of the regression relationship.
b. Examine the residual plots for the model. Do they suggest a violation of any of the regression assumptions? Explain.
c. Test for positive (first-order) autocorrelation. Are your results consistent with the way you interpreted the residual plots?

TABLE 6.9 **Sales and advertising data**

Monthly total sales, y (thousands of cases)	*Advertising expenditures, x (thousands of dollars)*	*Monthly total sales, y (thousands of cases)*	*Advertising expenditures, x (thousands of dollars)*
202.66	116.44	260.51	129.85
232.91	119.58	266.34	122.65
272.07	125.74	281.24	121.64
290.97	124.55	286.19	127.24
299.09	122.35	271.97	132.35
296.95	120.44	265.01	130.86
279.49	123.24	274.44	122.90
255.75	127.55	291.81	117.15
242.78	121.19	290.91	109.47
255.34	118.00	264.95	114.34
271.58	121.81	228.40	123.72
268.27	126.54	209.33	130.33

Source: Forecasting: Methods and Applications, S. Makridakis, S.C. Wheelwright, and V.E. McGee.

FIGURE 6.30 **SAS output of a simple linear regression relating total sales (y) to advertising expenditure (x)**

SAS

DEP VARIABLE: SALES

ANALYSIS OF VARIANCE

SOURCE	DF	SUM OF SQUARES	MEAN SQUARE	F VALUE	PROB>F
MODEL	1	140.19790	140.19790	0.201	0.6580
ERROR	22	15320.93263	696.40603		
C TOTAL	23	15461.13053			

ROOT MSE	26.38951	R-SQUARE	0.0091
DEP MEAN	264.9567	ADJ R-SQ	-0.0360
C.V.	9.959933		

PARAMETER ESTIMATES

VARIABLE	DF	PARAMETER ESTIMATE	STANDARD ERROR	T FOR HO: PARAMETER=0	PROB > |T|
INTERCEP	1	320.08119	122.97657	2.603	0.0162
ADVER	1	-0.44848131	0.99955076	-0.449	0.6580

OBS	ACTUAL	PREDICT VALUE	STD ERR PREDICT	LOWER95% MEAN	UPPER95% MEAN	LOWER95% PREDICT	UPPER95% PREDICT	RESIDUAL
1	202.7	267.9	8.4195	250.4	285.3	210.4	325.3	-65.2000
2	232.9	266.5	6.3341	253.3	279.6	210.2	322.7	-33.5418
3	272.1	263.7	6.0826	251.1	276.3	207.5	319.9	8.3809
4	291.0	264.2	5.6295	252.5	275.9	208.3	320.2	26.7472
5	299.1	265.2	5.4161	254.0	276.4	209.3	321.1	33.8805
6	296.9	266.1	5.9271	253.8	278.4	210.0	322.2	30.8839
7	279.5	264.8	5.3966	253.6	276.0	208.9	320.7	14.6797
8	255.8	262.9	7.1058	248.1	277.6	206.2	319.6	-7.1274
9	242.8	265.7	5.6556	254.0	277.5	209.8	321.7	-22.9497
10	255.3	267.2	7.2897	252.0	282.3	210.4	323.9	-11.8204
11	271.6	265.5	5.4986	254.0	276.9	209.5	321.4	6.1283
12	268.3	263.3	6.4927	249.9	276.8	207.0	319.7	4.9396
13	260.5	261.8	8.7798	243.6	280.1	204.2	319.5	-1.3359
14	266.3	265.1	5.3932	253.9	276.3	209.2	320.9	1.2650
15	281.2	265.5	5.5352	254.0	277.0	209.6	321.4	15.7121
16	286.2	263.0	6.9077	248.7	277.3	206.4	319.6	23.1736
17	272.0	260.7	10.8618	238.2	283.3	201.5	319.9	11.2453
18	265.0	261.4	9.5970	241.5	281.3	203.2	319.6	3.6171
19	274.4	265.0	5.3868	253.8	276.1	209.1	320.8	9.4772
20	291.8	267.5	7.8872	251.2	283.9	210.4	324.7	24.2684
21	290.9	271.0	14.4772	241.0	301.0	208.6	333.4	19.9241
22	264.9	268.8	10.1223	247.8	289.8	210.2	327.4	-3.8518
23	228.4	264.6	5.4467	253.3	275.9	208.7	320.5	-36.1951
24	209.3	261.6	9.1634	242.6	280.6	203.7	319.6	-52.3006

SUM OF RESIDUALS	5.20117E-12
SUM OF SQUARED RESIDUALS	15320.93
PREDICTED RESID SS (PRESS)	18513.86
DURBIN-WATSON D	0.473
(FOR NUMBER OF OBS.)	24
1ST ORDER AUTOCORRELATION	0.536

(continues)

FIGURE 6.30 **Continued**

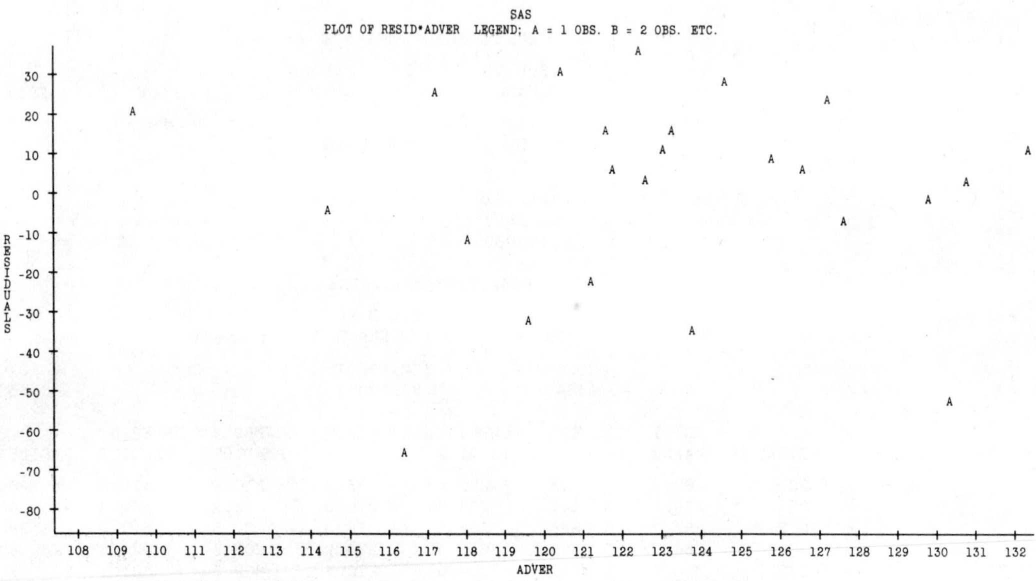

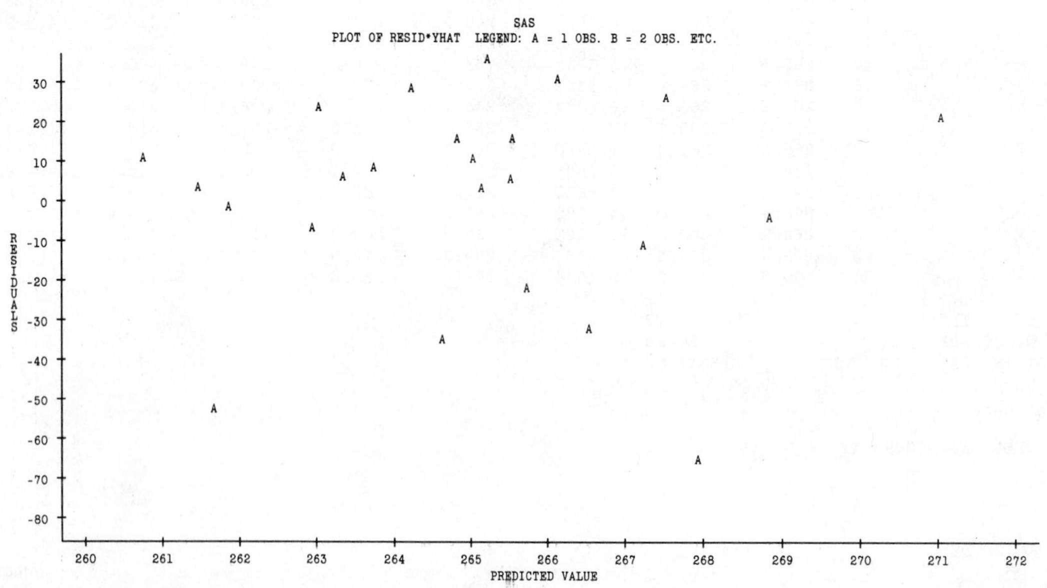

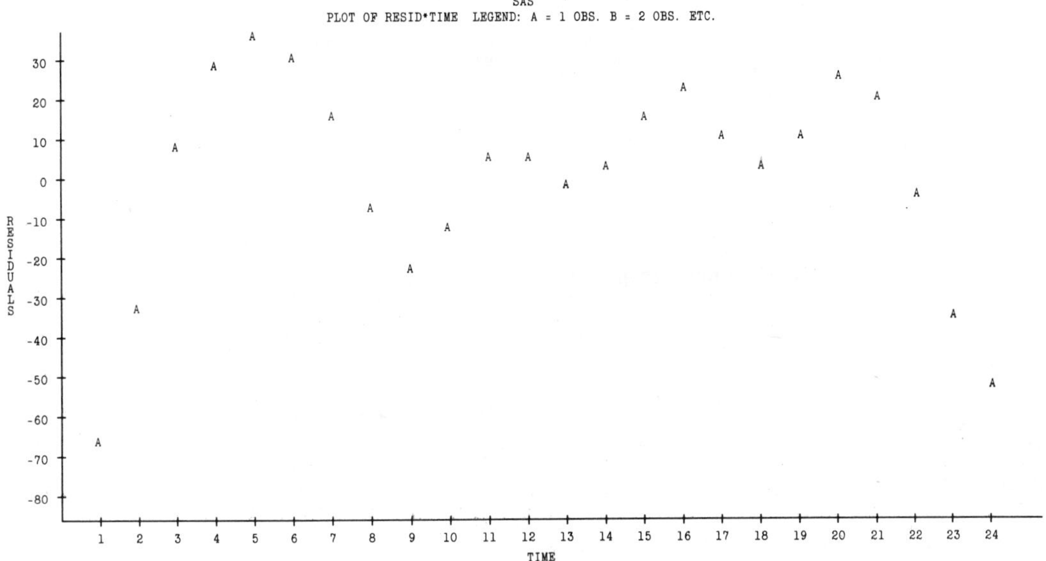

6.25 Table 6.10 presents data concerning the time (y) required to perform service and the number of microcomputers serviced (x) for 15 service calls. The SAS output of the simple linear regression relating y to x is given in Figure 6.31. Residual plots versus x and $\hat{y}$ are included.

TABLE 6.10 **Service time data for 15 service calls**

Service time, y (minutes)	*Number of microcomputers serviced, x*
92	3
63	2
126	6
247	8
49	2
90	4
119	5
114	6
67	2
115	4
188	6
298	11
77	3
151	10
27	1

a. Test the significance of the regression relationship.
b. Examine the residual plots for the model. Do they suggest a violation of any of the regression assumptions? Explain.
c. Plot y versus x. Is the data plot consistent with your conclusion of part (b)?

FIGURE 6.31 SAS output of a simple linear regression relating service time (y) to number of microcomputers serviced (x)

SAS

DEP VARIABLE: SERVTIME

ANALYSIS OF VARIANCE

SOURCE	DF	SUM OF SQUARES	MEAN SQUARE	F VALUE	PROB>F
MODEL	1	60896.00693	60896.00693	49.960	0.0001
ERROR	13	15845.72641	1218.90203		
C TOTAL	14	76741.73333			

ROOT MSE	34.91278	R-SQUARE	0.7935
DEP MEAN	121.5333	ADJ R-SQ	0.7776
C.V.	28.72692		

PARAMETER ESTIMATES

VARIABLE	DF	PARAMETER ESTIMATE	STANDARD ERROR	T FOR HO: PARAMETER=0	PROB > |T|
INTERCEP	1	14.43054083	17.63137142	0.818	0.4278
MICROS	1	22.00742312	3.11357395	7.068	0.0001

OBS	ACTUAL	PREDICT VALUE	STD ERR PREDICT	LOWER95% MEAN	UPPER95% MEAN	LOWER95% PREDICT	UPPER95% PREDICT	RESIDUAL
1	92.0000	80.4528	10.7256	57.2815	103.6	1.5494	159.4	11.5472
2	63.0000	58.4454	12.6857	31.0397	85.8511	-21.8037	138.7	4.5546
3	126.0	146.5	9.6805	125.6	167.4	68.2049	224.7	-20.4751
4	247.0	190.5	13.2830	161.8	219.2	109.8	271.2	56.5101
5	49.0000	58.4454	12.6857	31.0397	85.8511	-21.8037	138.7	-9.4454
6	90.0000	102.5	9.4097	82.1319	122.8	24.3444	180.6	-12.4602
7	119.0	124.5	9.0240	105.0	144.0	46.5645	202.4	-5.4677
8	114.0	146.5	9.6805	125.6	167.4	68.2049	224.7	-32.4751
9	67.0000	58.4454	12.6857	31.0397	85.8511	-21.8037	138.7	8.5546
10	115.0	102.5	9.4097	82.1319	122.8	24.3444	180.6	12.5398
11	188.0	146.5	9.6805	125.6	167.4	68.2049	224.7	41.5249
12	298.0	256.5	21.1173	210.9	302.1	168.4	344.7	41.4878
13	77.0000	80.4528	10.7256	57.2815	103.6	1.5494	159.4	-3.4528
14	151.0	234.5	18.3498	194.9	274.1	149.3	319.7	-83.5048
15	27.0000	36.4380	15.0400	3.9461	68.9299	-45.6874	118.6	-9.4380

SUM OF RESIDUALS	-2.13163E-14
SUM OF SQUARED RESIDUALS	15845.73
PREDICTED RESID SS (PRESS)	26663.22

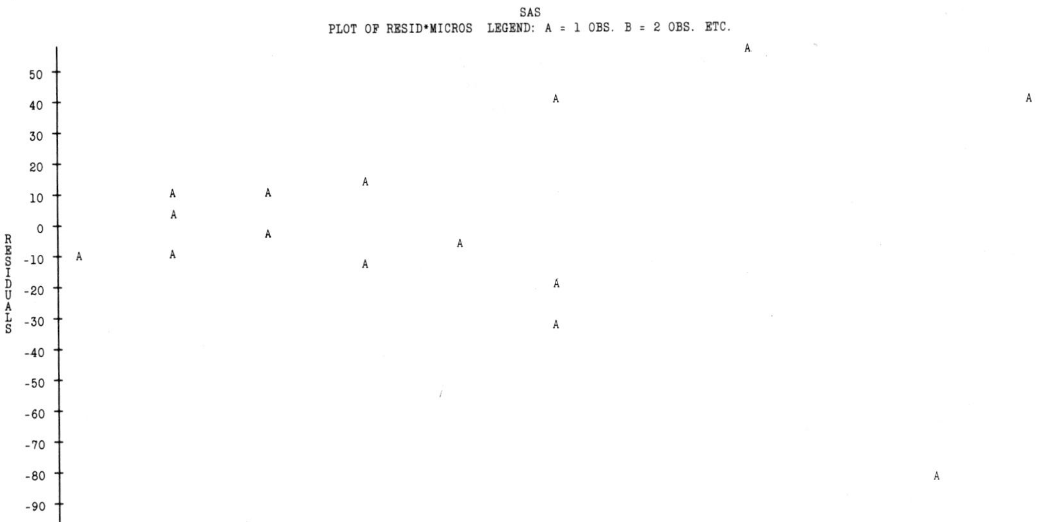
SAS
PLOT OF RESID*MICROS LEGEND: A = 1 OBS. B = 2 OBS. ETC.
RESIDUALS
50
40
30
20
10
0
-10
-20
-30
-40
-50
-60
-70
-80
-90
-100
1 2 3 4 5 6 7 8 9 10 11
MICROS

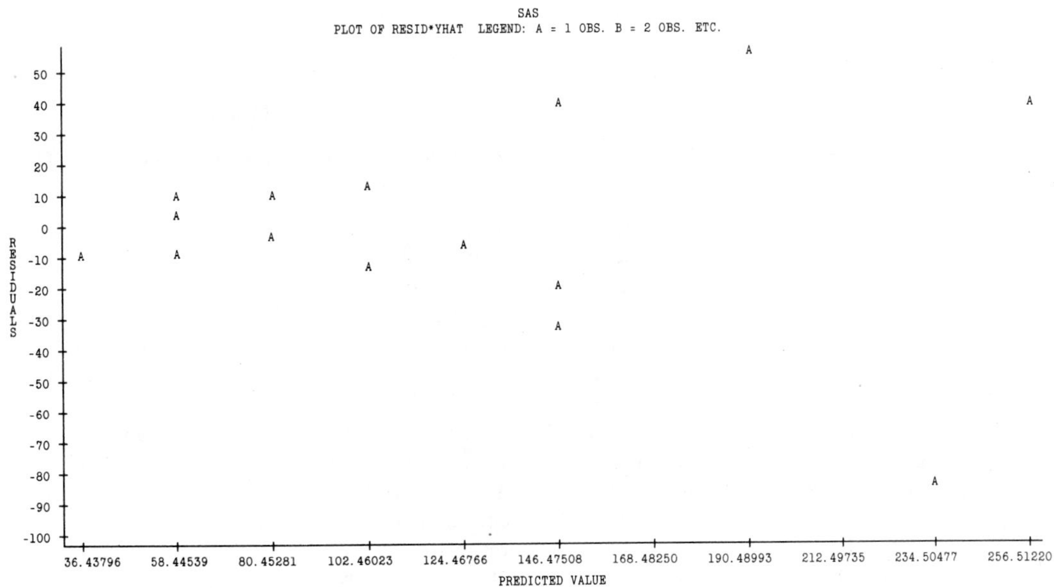
SAS
PLOT OF RESID*YHAT LEGEND: A = 1 OBS. B = 2 OBS. ETC.
RESIDUALS
50
40
30
20
10
0
-10
-20
-30
-40
-50
-60
-70
-80
-90
-100
36.43796 58.44539 80.45281 102.46023 124.46766 146.47508 168.48250 190.48993 212.49735 234.50477 256.51220
PREDICTED VALUE

TABLE 6.11 Service time data for 17 service calls

Service time, y (minutes)	*Number of copiers serviced, x*
92	3
401	13
101	2
145	6
229	8
87	2
428	14
90	4
119	5
145	6
123	2
115	4
154	6
366	11
134	3
321	10
71	1

6.26 Table 6.11 presents data concerning the time (y) required to perform service and the number of copiers serviced (x) for 17 service calls. The SAS output of the simple linear regression relating y to x is given in Figure 6.32. Residual plots versus x and $\hat{y}$ are included.

a. Test the significance of the regression relationship.
b. Examine the residual plots for the model. Do they suggest a violation of any of the regression assumptions? Explain.
c. Plot y versus x. Is the data plot consistent with your conclusion of part (b)?

6.27 Consider the measurements taken on 63 single-family residences in Oxford, Ohio. This data, which is presented in Table 4.3, concerns sales price (y) and square footage (x).

a. Figure 6.33 presents the SAS output of a plot of y versus x. Is there a clear regression relationship between y and x? Do you think the regression relationship is more nearly linear or more nearly curved? Explain.
b. Suppose that we use simple linear regression to relate y to x. The SAS output of this regression is given in Figure 6.34. Residual plots versus x and $\hat{y}$ for this model are shown in Figure 6.35. Use Figure 6.34 to find the three residuals that are largest in absolute value.
c. Examine the residual plots in Figure 6.35. Are they easily interpreted? Do they say anything about the functional form of the simple linear model relating y to x?
d. Circle the plot points on the plot of y versus x that correspond to the three largest residuals. Do the points you have circled appear to be outliers? If you ignore these points, is the nature of the regression relationship clearer? If you ignore the circled points, does the regression relationship appear to be more nearly linear or more nearly curved?

e. If we drop the observations corresponding to the three largest (in absolute value) residuals from the data set and rerun the simple linear regression relating y to x, we obtain the residual plots in Figure 6.36. What do these residual plots say about the functional form of the regression relating y and x? Has removing the three outliers clarified this relationship and made the residual plots easier to interpret? Explain.

FIGURE 6.32 SAS output of a simple linear regression analysis relating service time (y) to number of copiers serviced (x)

SAS

ANALYSIS OF VARIANCE

SOURCE	DF	SUM OF SQUARES	MEAN SQUARE	F VALUE	PROB>F
MODEL	1	210389.92	210389.92	201.443	0.0001
ERROR	15	15666.20059	1044.41337		
C TOTAL	16	226056.12			

ROOT MSE	32.31738	R-SQUARE	0.9307
DEP MEAN	183.5882	ADJ R-SQ	0.9261
C.V.	17.60319		

PARAMETER ESTIMATES

VARIABLE	DF	PARAMETER ESTIMATE	STANDARD ERROR	T FOR HO: PARAMETER=0	PROB > |T|
INTERCEP	1	15.53308991	14.19989741	1.094	0.2913
COPIERS	1	28.56937471	2.01291060	14.193	0.0001

OBS	ACTUAL	PREDICT VALUE	STD ERR PREDICT	LOWER95% MEAN	UPPER95% MEAN	LOWER95% PREDICT	UPPER95% PREDICT	RESIDUAL
1	92.0000	101.2	9.7518	80.4557	122.0	29.2909	173.2	-9.2412
2	401.0	386.9	16.3311	352.1	421.7	309.8	464.1	14.0650
3	101.0	72.6718	11.0683	49.0804	96.2633	-0.1387	145.5	28.3282
4	145.0	186.9	7.8417	170.2	203.7	116.1	257.8	-41.9493
5	229.0	244.1	8.9222	225.1	263.1	172.6	315.5	-15.0881
6	87.0000	72.6718	11.0683	49.0804	96.2633	-0.1387	145.5	14.3282
7	428.0	415.5	18.1228	376.9	454.1	336.5	494.5	12.4957
8	90.0000	129.8	8.7059	111.3	148.4	58.4724	201.1	-39.8106
9	119.0	158.4	8.0368	141.2	175.5	87.3993	229.4	-39.3800
10	145.0	186.9	7.8417	170.2	203.7	116.1	257.8	-41.9493
11	123.0	72.6718	11.0683	49.0804	96.2633	-0.1387	145.5	50.3282
12	115.0	129.8	8.7059	111.3	148.4	58.4724	201.1	-14.8106
13	154.0	186.9	7.8417	170.2	203.7	116.1	257.8	-32.9493
14	366.0	329.8	12.9443	302.2	357.4	255.6	404.0	36.2038
15	134.0	101.2	9.7518	80.4557	122.0	29.2909	173.2	32.7588
16	321.0	301.2	11.4077	276.9	325.5	228.2	374.3	19.7732
17	71.0000	44.1025	12.5706	17.3089	70.8960	-29.8077	118.0	26.8975

SUM OF RESIDUALS	1.70530E-13
SUM OF SQUARED RESIDUALS	15666.2
PREDICTED RESID SS (PRESS)	19335.42

DURBIN-WATSON D	1.795
(FOR NUMBER OF OBS.)	17
1ST ORDER AUTOCORRELATION	0.077

(continues)

FIGURE 6.32 Continued

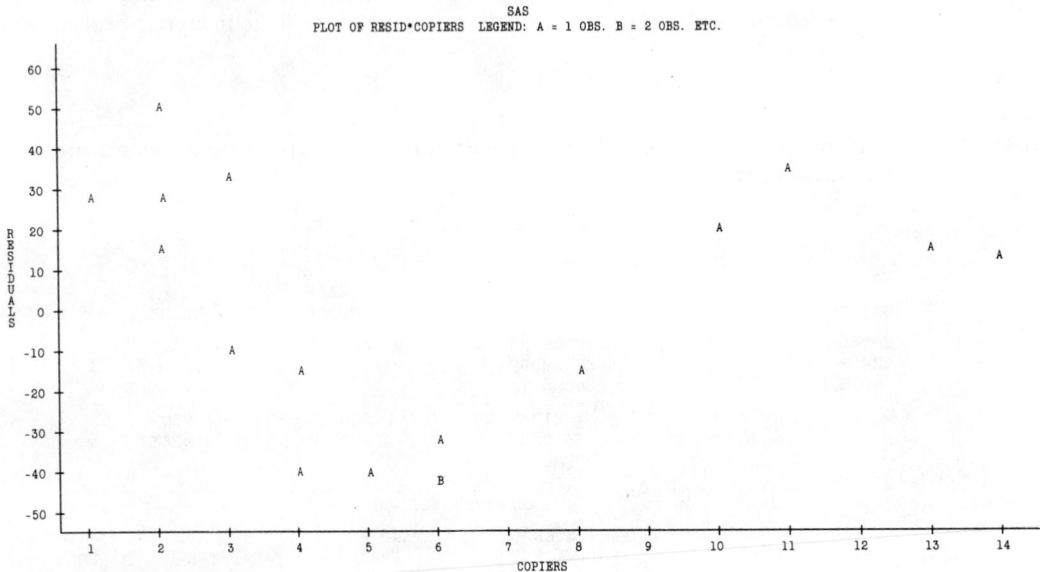

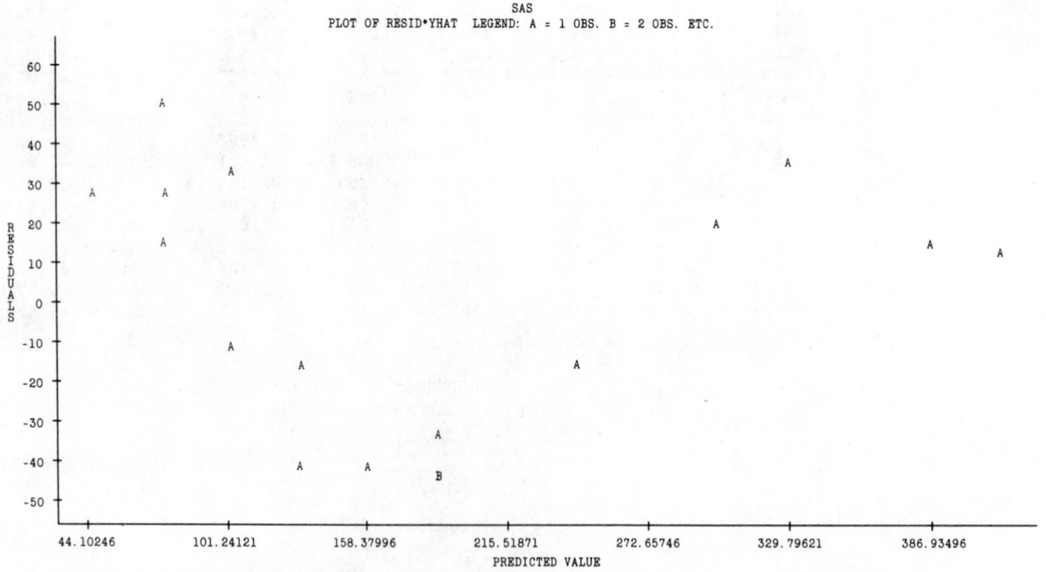

FIGURE 6.33 SAS output of a plot of sales price (y) versus square footage (x)

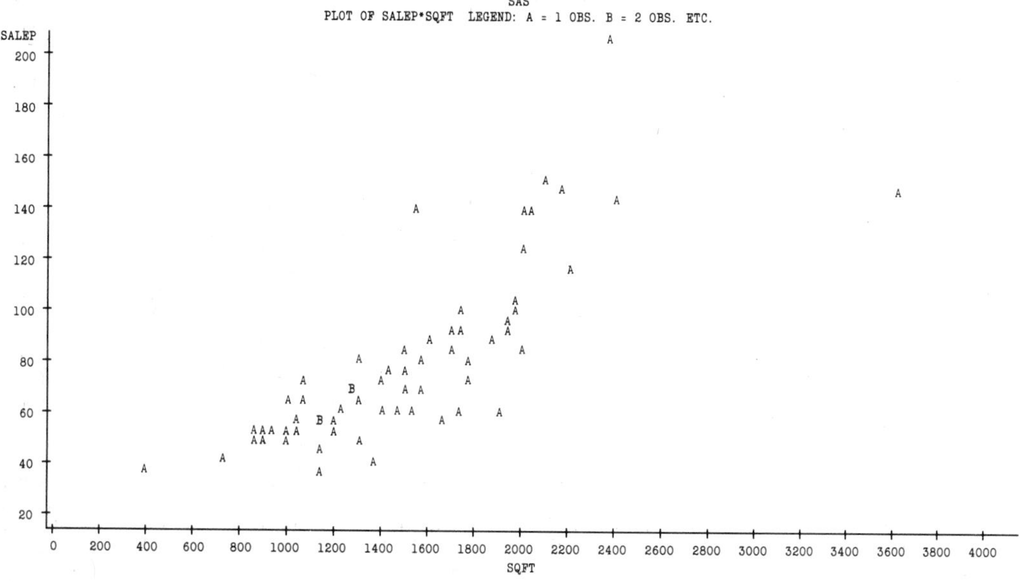

FIGURE 6.34 SAS output of the simple linear regression relating sales price (y) to square footage (x)

SAS

ANALYSIS OF VARIANCE

SOURCE	DF	SUM OF SQUARES	MEAN SQUARE	F VALUE	PROB>F
MODEL	1	48147.30258	48147.30258	105.307	0.0001
ERROR	61	27889.68726	457.20799		
C TOTAL	62	76036.98984			

ROOT MSE	21.38242	R-SQUARE	0.6332
DEP MEAN	78.80159	ADJ R-SQ	0.6272
C.V.	27.13451		

PARAMETER ESTIMATES

VARIABLE	DF	PARAMETER ESTIMATE	STANDARD ERROR	T FOR H0: PARAMETER=0	PROB > |T|
INTERCEP	1	-3.72102928	8.48086264	-0.439	0.6624
SQFT	1	0.05477049	0.005337250	10.262	0.0001

OBS	ACTUAL	PREDICT VALUE	STD ERR PREDICT	LOWER95% MEAN	UPPER95% MEAN	LOWER95% PREDICT	UPPER95% PREDICT	RESIDUAL
1	53.5000	51.4876	3.7871	43.9149	59.0603	8.0654	94.9099	2.0124
2	49.0000	66.9329	2.9317	61.0706	72.7952	23.7761	110.1	-17.9329
3	50.5000	43.3816	4.3784	34.6264	52.1368	-0.2624	87.0256	7.1184
4	49.9000	46.2297	4.1632	37.9049	54.5544	2.6700	89.7894	3.6703
5	52.0000	62.2226	3.1412	55.9414	68.5039	19.0069	105.4	-10.2226

(continues)

FIGURE 6.34 Continued

SAS

OBS	ACTUAL	PREDICT VALUE	STD ERR PREDICT	LOWER95% MEAN	UPPER95% MEAN	LOWER95% PREDICT	UPPER95% PREDICT	RESIDUAL
6	55.000	62.2226	3.1412	55.9414	68.5039	19.0069	105.4	-7.2226
7	80.5000	92.8941	3.0238	86.8477	98.9405	49.7119	136.1	-12.3941
8	86.0000	83.9118	2.7396	78.4336	89.3899	40.8054	127.0	2.0882
9	69.0000	65.0159	3.0103	58.9965	71.0354	21.8375	108.2	3.9841
10	149.0	193.5	11.4927	170.5	216.4	144.9	242.0	-44.4527
11	46.0000	43.6007	4.3616	34.8791	52.3223	-0.0366	87.2380	2.3993
12	38.0000	35.7137	4.9887	25.7382	45.6893	-8.1914	79.6188	2.2863
13	49.5000	51.4876	3.7871	43.9149	59.0603	8.0654	94.9099	-1.9876
14	105.0	103.1	3.5854	95.9119	110.3	59.7277	146.4	1.9186
15	152.5	110.5	4.1009	102.3	118.7	66.9942	154.1	41.9698
16	85.0000	106.4	3.8081	98.8076	114.0	62.9928	149.9	-21.4224
17	60.0000	76.5177	2.7031	71.1125	81.9229	33.4206	119.6	-16.5177
18	58.5000	63.7562	3.0671	57.6233	69.8892	20.5618	107.0	-5.2562
19	101.0	91.3605	2.9589	85.4439	97.2772	48.1963	134.5	9.6395
20	79.4000	67.2615	2.9192	61.4242	73.0989	24.1081	110.4	12.1385
21	125.0	105.6	3.7520	98.0983	113.1	62.1908	149.0	19.3991
22	87.9000	98.9189	3.3317	92.2567	105.6	55.6461	142.2	-11.0189
23	80.0000	82.8164	2.7222	77.3730	68.2597	39.7144	125.9	-2.8164
24	94.0000	101.4	3.4818	94.4759	108.4	58.1183	144.8	-7.4383
25	74.0000	74.6008	2.7249	69.1521	80.0495	31.4982	117.7	-0.6008
26	69.0000	77.6679	2.6962	72.2765	83.0593	34.5725	120.8	-8.6679
27	63.0000	51.4876	3.7871	43.9149	59.0603	8.0654	94.9099	11.5124
28	67.5000	66.4947	2.9488	60.5982	72.3913	23.3333	109.7	1.0053
29	35.0000	58.3887	3.3487	51.6925	65.0850	15.1107	101.7	-23.3887
30	142.5	127.7	5.4762	116.8	138.7	83.5914	171.9	14.7718
31	92.2000	89.4436	2.8866	83.6714	95.2158	46.2989	132.6	2.7564
32	56.0000	52.1449	3.7423	44.6616	59.6281	8.7381	95.5516	3.8551
33	63.0000	53.9523	3.6223	46.7091	61.1955	10.5863	97.3183	9.0477
34	60.0000	90.9224	2.9415	85.0405	96.8043	47.7629	134.1	-30.9224
35	34.0000	19.0635	6.4144	6.2370	31.8900	-25.5758	63.7028	14.9365
36	52.0000	53.2403	3.6690	45.9036	60.5770	9.8586	96.6220	-1.2403
37	75.0000	78.2156	2.6945	72.8276	83.6037	35.1207	121.3	-3.2156
38	93.0000	102.3	3.5366	95.2428	109.4	58.9770	145.7	-9.3146
39	60.0000	100.6	3.4284	93.7065	107.4	57.2591	143.9	-40.5620
40	73.0000	55.4311	3.5276	48.3773	62.4849	12.0963	98.7659	17.5689
41	71.0000	93.1132	3.0335	87.0473	99.1791	49.9282	136.3	-22.1132
42	83.0000	78.5990	2.6940	73.2120	83.9860	35.5042	121.7	4.4010
43	90.0000	91.3605	2.9589	85.4439	97.2772	48.1963	134.5	-1.3605
44	83.0000	89.1150	2.8753	83.3654	94.8645	45.9733	132.3	-6.1150
45	115.0	116.0	4.5169	107.0	125.0	72.3069	159.7	-1.0073
46	50.0000	44.9152	4.2616	36.3935	53.4368	1.3174	88.5129	5.0848
47	55.2000	57.6219	3.3937	50.8359	64.4080	14.3299	100.9	-2.4219
48	61.0000	72.9577	2.7535	67.4518	78.4636	29.8478	116.1	-11.9577
49	147.0	114.9	4.4274	106.0	123.7	71.1933	158.5	32.1429
50	210.0	125.2	5.2593	114.6	135.7	81.1228	169.2	84.8461
51	60.0000	80.4064	2.6985	75.0105	85.8024	37.3105	123.5	-20.4064
52	100.0	104.3	3.6640	96.9598	111.6	60.9064	147.7	-4.2864
53	44.5000	57.6219	3.3937	50.8359	64.4080	14.3299	100.9	-13.1219
54	55.0000	87.4171	2.8217	81.7747	93.0595	44.2896	130.5	-32.4171
55	53.4000	46.9417	4.1105	38.7222	55.1611	3.4020	90.4814	6.4583
56	65.0000	66.8234	2.9359	60.9526	72.6941	23.6654	110.0	-1.8234
57	73.0000	72.9577	2.7535	67.4518	78.4636	29.8478	116.1	0.0423
58	40.0000	71.6432	2.7828	66.0787	77.2077	28.5258	114.8	-31.6432
59	141.0	107.9	3.9113	100.1	115.7	64.4350	151.4	33.0988
60	68.0000	82.3782	2.7164	76.9464	87.8099	39.2777	125.5	-14.3782
61	139.0	80.8994	2.7017	75.4970	86.3017	37.8026	124.0	58.1006
62	140.0	105.4	3.7409	97.9563	112.9	62.0303	148.8	34.5634
63	55.0000	58.1696	3.3615	51.4479	64.8913	14.8877	101.5	-3.1696

SUM OF RESIDUALS	7.60281E-13
SUM OF SQUARED RESIDUALS	27889.69
PREDICTED RESID SS (PRESS)	31800.02

FIGURE 6.35 SAS output of residual plots for the simple linear regression relating sales price (y) to square footage (x)

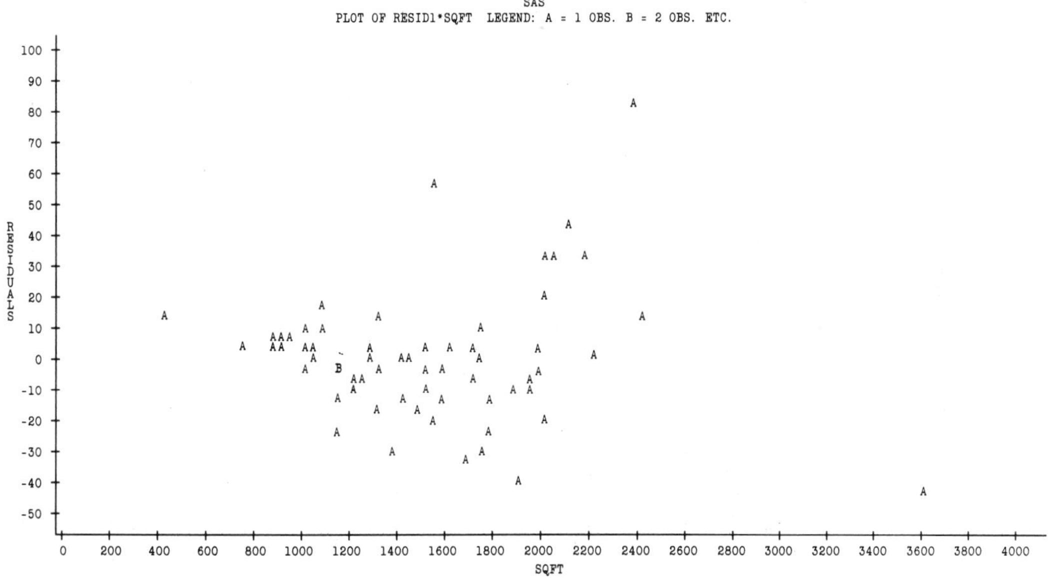

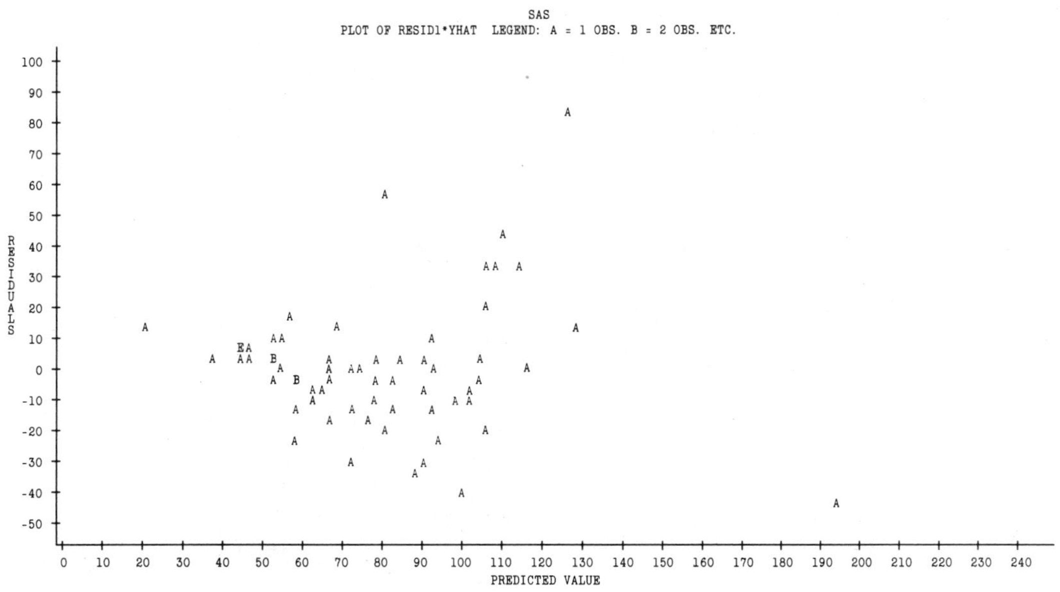

FIGURE 6.36 SAS output of residual plots for the simple linear regression relating selling price (y) to square footage (x) with outliers removed from the data set

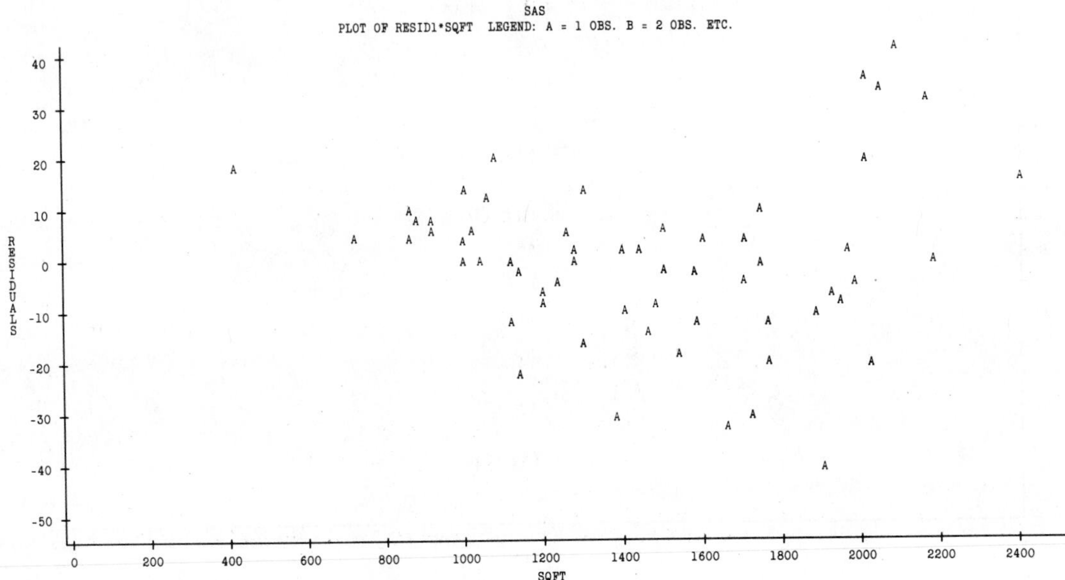

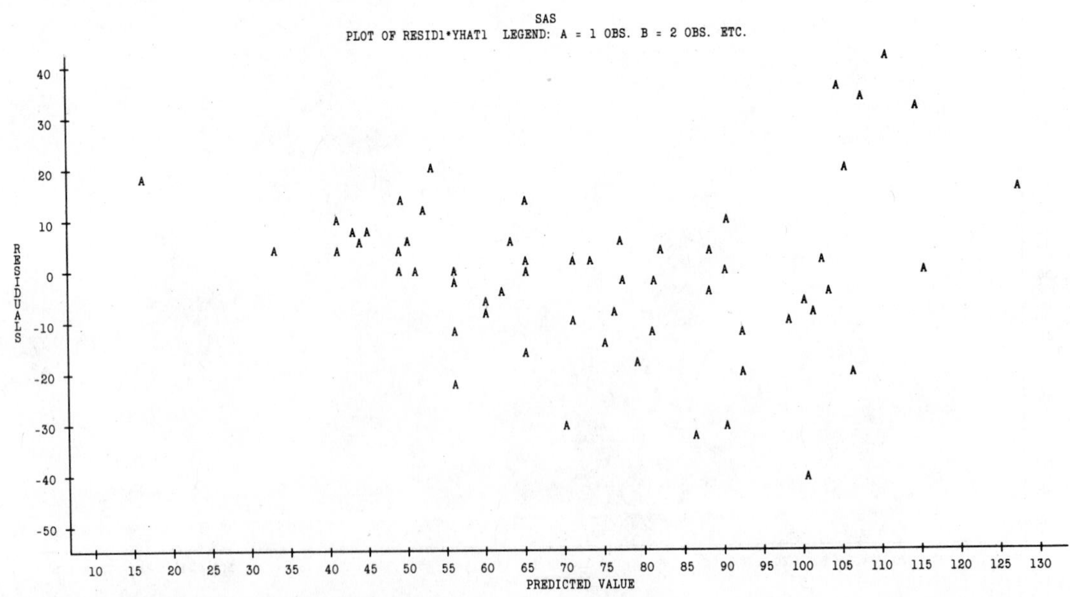

6.28 Consider the hotel room rate data in Table 5.11 and the simple linear regression model that you feel best describes hotel room rate (y). Completely analyze the residuals for the model that you feel is best. Interpret the residual analysis. Do any of the regression assumptions seem to be violated? Explain.

6.29 Consider the economic data in Table 5.12 and the simple linear regression model that you feel best describes HCOST (indexed housing cost). Completely analyze the residuals for this model. Interpret the residual analysis. Do any of the regression assumptions seem to be violated? Explain.

6.30 Consider the economic data in Table 5.12 and the simple linear regression model that you feel best describes JG (jobs growth). Completely analyze the residuals for this model. Interpret the residual analysis. Do any of the regression assumptions seem to be violated? Explain.

6.31 Consider the economic data in Table 5.12 and the simple linear regression model that you feel best describes IG (personal income growth). Completely analyze the residuals for this model. Interpret the residual analysis. Do any of the regression assumptions seem to be violated? Explain.

6.32 (Optional) Consider the simple linear regression model in Exercise 4.7. This model relates market rate (y) to accounting rate (x). Write a SAS program that will construct residual plots versus x and $\hat{y}$ for this model. Interpret the plots.

6.33 (Optional) Consider the simple linear regression model in Exercise 4.21. This model relates condominium selling price (y) to square footage (x_1). Write a SAS program that will construct residual plots versus x_1, $\hat{y}$, and x_2 (age) for this model. Interpret the plots.

6.34 (Optional) Consider the sales price data in Table 4.3 and the simple linear regression model relating sales price (y) to square footage (x). Write a SAS program that will construct the residual plots given in Figure 6.35. Also write the program so that it will produce a stem-and-leaf diagram, box plot, and normal plot of the residuals. Interpret the results.

CHAPTER 7

MATRIX ALGEBRA

Before we begin our discussion of multiple regression analysis, it is necessary to present some introductory concepts in a branch of mathematics known as matrix algebra.

7.1 MATRICES AND VECTORS

A *matrix* is a rectangular array of numbers (called elements) that is composed of rows and columns. An example of a matrix is

$$\mathbf{A} = \begin{bmatrix} 1 & 5 & 3 & 10 \\ 12 & 6 & 7 & 4 \\ 9 & 2 & 11 & 8 \end{bmatrix}$$

The notation **A** is used to indicate that we are referring to a matrix rather than a number.

The *dimension* of a matrix is determined by the number of rows and columns in the matrix. Since the matrix **A** has 3 rows and 4 columns, this matrix is said to have dimension 3 by 4 (commonly written 3 × 4). In general, a matrix with m rows and n columns is said to have dimension $m \times n$. As another example, the matrix

$$\mathbf{X} = \begin{bmatrix} 1 & 0 & 0 \\ 1 & 1 & 0 \\ 1 & 2 & 0 \\ 1 & 0 & 1 \\ 1 & 1 & 1 \\ 1 & 2 & 1 \\ 1 & 0 & 2 \\ 1 & 1 & 2 \\ 1 & 2 & 2 \end{bmatrix}$$

has dimension 9 × 3, since it has 9 rows and 3 columns.

In general, a matrix with dimension $m \times n$ can be represented as

$$\mathbf{A}_{m \times n} = \begin{bmatrix} a_{11} & a_{12} & \cdots & a_{1j} & \cdot & a_{1n} \\ a_{21} & a_{22} & \cdots & a_{2j} & \cdot & a_{2n} \\ \vdots & \vdots & & \vdots & & \vdots \\ a_{i1} & a_{i2} & \cdots & a_{ij} & \cdot & a_{in} \\ \vdots & \vdots & & \vdots & & \vdots \\ a_{m1} & a_{m2} & \cdots & a_{mj} & \cdot & a_{mn} \end{bmatrix}$$

where a_{ij} is the number in the matrix in row i and column j, and the subscript $m \times n$ indicates the dimension of **A**.

A matrix that consists of one column is a *column vector*, for example,

$$\mathbf{B}_{4 \times 1} = \begin{bmatrix} 5 \\ 3 \\ 4 \\ 1 \end{bmatrix} \quad \text{and} \quad \mathbf{C}_{3 \times 1} = \begin{bmatrix} 101 \\ 73 \\ 51 \end{bmatrix}$$

A matrix that consists of one row is a *row vector*, for example,

$$\mathbf{E}'_{1 \times 4} = [10 \quad 7 \quad 6 \quad 12]$$

$$\mathbf{F}'_{1 \times 6} = [1 \quad 2 \quad 7 \quad 11 \quad 5 \quad 8]$$

Note that the prime mark (′) is used to distinguish a row vector from a column vector.

7.2

THE TRANSPOSE OF A MATRIX

The *transpose* of a matrix is formed by interchanging the rows and columns of the matrix.

For example, consider the matrix

$$\mathbf{A}_{2\times 3} = \begin{bmatrix} 5 & 6 & 7 \\ 3 & 2 & 1 \end{bmatrix}$$

The transpose of **A**, which is denoted **A**′, is

$$\mathbf{A}'_{3\times 2} = \begin{bmatrix} 5 & 3 \\ 6 & 2 \\ 7 & 1 \end{bmatrix}$$

Thus the first row of **A** is the first column of **A**′, and the second row of **A** is the second column of **A**′.

Notice that the transpose of the column vector

$$\mathbf{E}_{4\times 1} = \begin{bmatrix} 10 \\ 7 \\ 6 \\ 12 \end{bmatrix}$$

is the row vector

$$\mathbf{E}'_{1\times 4} = [10 \quad 7 \quad 6 \quad 12]$$

As a last example, consider the matrix

$$\mathbf{X}_{9\times 3} = \begin{bmatrix} 1 & 0 & 0 \\ 1 & 1 & 0 \\ 1 & 2 & 0 \\ 1 & 0 & 1 \\ 1 & 1 & 1 \\ 1 & 2 & 1 \\ 1 & 0 & 2 \\ 1 & 1 & 2 \\ 1 & 2 & 2 \end{bmatrix}$$

The transpose of **X** is

$$\mathbf{X}'_{3\times 9} = \begin{bmatrix} 1 & 1 & 1 & 1 & 1 & 1 & 1 & 1 & 1 \\ 0 & 1 & 2 & 0 & 1 & 2 & 0 & 1 & 2 \\ 0 & 0 & 0 & 1 & 1 & 1 & 2 & 2 & 2 \end{bmatrix}$$

7.3

SUMS AND DIFFERENCES OF MATRICES

Consider two matrices, **A** and **B**, that have the same dimensions.

The *sum* of **A** and **B** is a matrix obtained by adding the corresponding elements of **A** and **B**.

For example, for the matrices

$$\mathbf{A}_{2\times 3} = \begin{bmatrix} 1 & 4 & 2 \\ 5 & 3 & 2 \end{bmatrix} \quad \text{and} \quad \mathbf{B}_{2\times 3} = \begin{bmatrix} 7 & 0 & 4 \\ 3 & 1 & 5 \end{bmatrix}$$

the sum is

$$\mathbf{C}_{2\times 3} = \mathbf{A}_{2\times 3} + \mathbf{B}_{2\times 3} = \begin{bmatrix} 1+7 & 4+0 & 2+4 \\ 5+3 & 3+1 & 2+5 \end{bmatrix} = \begin{bmatrix} 8 & 4 & 6 \\ 8 & 4 & 7 \end{bmatrix}$$

In general, if **A** and **B** have the same dimensions and $\mathbf{C} = \mathbf{A} + \mathbf{B}$,

$$c_{ij} = a_{ij} + b_{ij}$$

where

c_{ij} = the number in **C** in row i and column j

a_{ij} = the number in **A** in row i and column j

b_{ij} = the number in **B** in row i and column j

Again consider two matrices, **A** and **B**, that have the same dimensions.

The *difference* of **A** and **B** is a matrix obtained by subtracting the corresponding elements of **A** and **B**.

For example, for the matrices

$$\mathbf{A}_{2\times 3} = \begin{bmatrix} 1 & 4 & 2 \\ 5 & 3 & 2 \end{bmatrix} \quad \text{and} \quad \mathbf{B}_{2\times 3} = \begin{bmatrix} 7 & 0 & 4 \\ 3 & 1 & 5 \end{bmatrix}$$

the difference is

$$\mathbf{D}_{2\times 3} = \mathbf{A}_{2\times 3} - \mathbf{B}_{2\times 3} = \begin{bmatrix} 1-7 & 4-0 & 2-4 \\ 5-3 & 3-1 & 2-5 \end{bmatrix} = \begin{bmatrix} -6 & 4 & -2 \\ 2 & 2 & -3 \end{bmatrix}$$

In general, if **A** and **B** have the same dimensions and $\mathbf{D} = \mathbf{A} - \mathbf{B}$,

$$d_{ij} = a_{ij} - b_{ij}$$

where

d_{ij} = the number in **D** in row i and column j

a_{ij} = the number in **A** in row i and column j

b_{ij} = the number in **B** in row i and column j

7.4 MATRIX MULTIPLICATION

We now consider *multiplication of a matrix by a number.*

The *product of a number λ and a matrix* **A** is a matrix obtained by multiplying each element of **A** by the number λ.

For example, multiplying the following matrix

$$\mathbf{Z}_{2\times 3} = \begin{bmatrix} 1 & 4 & 7 \\ 3 & 2 & 3 \end{bmatrix}$$

by $\lambda = 5$, we get

$$5\mathbf{Z}_{2\times 3} = 5\begin{bmatrix} 1 & 4 & 7 \\ 3 & 2 & 3 \end{bmatrix} = \begin{bmatrix} 5(1) & 5(4) & 5(7) \\ 5(3) & 5(2) & 5(3) \end{bmatrix} = \begin{bmatrix} 5 & 20 & 35 \\ 15 & 10 & 15 \end{bmatrix}$$

In general, if λ is a number, **A** is a matrix, and $\mathbf{E} = \lambda\mathbf{A}$,

$$e_{ij} = \lambda a_{ij}$$

where

e_{ij} = the number in **E** in row i and column j

a_{ij} = the number in **A** in row i and column j

We next consider *multiplication of a matrix by a matrix.*

Consider two matrices **A** and **B** where the number of *columns* in **A** is equal to the number of *rows* in **B**. Then the *product of the two matrices* **A** and **B** is a matrix calculated so that the element in row i and column j of the product is obtained by multiplying the elements in row i of matrix **A** by the corresponding elements in column j of matrix **B** and adding the resulting products.

For example, consider the following matrices.

$$\mathbf{A}_{2\times 2} = \begin{bmatrix} 4 & 3 \\ 2 & 2 \end{bmatrix} \quad \text{and} \quad \mathbf{B}_{2\times 2} = \begin{bmatrix} 2 & 1 \\ 3 & 5 \end{bmatrix}$$

Suppose we wish to find the product **AB**. The number in row 1 and column 1 of the product is obtained by multiplying the elements in row 1 of **A** by the corresponding elements in column 1 of **B** and adding these products. We obtain

$$4(2) + 3(3) = 8 + 9 = 17$$

The number in row 1 and column 2 of the product is obtained by multiplying the elements in row 1 of **A** by the corresponding elements in column 2 of **B** and adding these products. We obtain

$$4(1) + 3(5) = 4 + 15 = 19$$

The number in row 2 and column 1 of the product is obtained by multiplying the elements in row 2 of **A** by the corresponding elements in column 1 of **B** and adding these products. We obtain

$$2(2) + 2(3) = 4 + 6 = 10$$

The number in row 2 and column 2 of the product is obtained by multiplying the elements in row 2 of **A** by the corresponding elements in column 2 of **B** and adding these products. We obtain

$$2(1) + 2(5) = 2 + 10 = 12$$

Thus the product **AB** is as follows:

$$\mathbf{A}_{2\times 2}\mathbf{B}_{2\times 2} = \begin{bmatrix} 4 & 3 \\ 2 & 2 \end{bmatrix}\begin{bmatrix} 2 & 1 \\ 3 & 5 \end{bmatrix} = \begin{bmatrix} 4(2) + 3(3) & 4(1) + 3(5) \\ 2(2) + 2(3) & 2(1) + 2(5) \end{bmatrix}$$

$$= \begin{bmatrix} 17 & 19 \\ 10 & 12 \end{bmatrix}$$

In general, we can multiply a matrix **A** with m rows and r columns by a matrix **B** with r rows and n columns and obtain a matrix **C** with m rows and n columns. Moreover, c_{ij}, the number in the product in row i and column j, is obtained by multiplying the elements in row i of **A** by the corresponding elements in column j

of **B** and adding the resulting products. Note that the number of columns in **A** must equal the number of rows in **B** in order for this multiplication procedure to be defined.

The multiplication procedure is illustrated in Figure 7.1.

We now present several more examples. Consider the following matrices:

$$\mathbf{W}_{3\times 2} = \begin{bmatrix} 1 & 6 \\ 2 & 5 \\ 3 & 4 \end{bmatrix} \qquad \mathbf{U}_{2\times 2} = \begin{bmatrix} 2 & 2 \\ 1 & 3 \end{bmatrix}$$

$$\mathbf{X}_{9\times 3} = \begin{bmatrix} 1 & 0 & 0 \\ 1 & 1 & 0 \\ 1 & 2 & 0 \\ 1 & 0 & 1 \\ 1 & 1 & 1 \\ 1 & 2 & 1 \\ 1 & 0 & 2 \\ 1 & 1 & 2 \\ 1 & 2 & 2 \end{bmatrix} \qquad \mathbf{y}_{9\times 1} = \begin{bmatrix} 18 \\ 21 \\ 20 \\ 20 \\ 22 \\ 20 \\ 19 \\ 21 \\ 20 \end{bmatrix}$$

FIGURE 7.1 An illustration of matrix multiplication

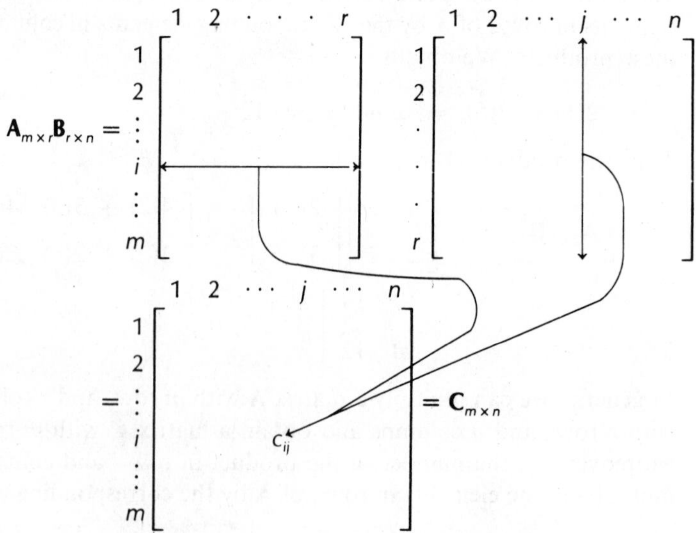

Then we find that

$$\mathbf{W}_{3\times 2}\mathbf{U}_{2\times 2} = \begin{bmatrix} 1 & 6 \\ 2 & 5 \\ 3 & 4 \end{bmatrix}\begin{bmatrix} 2 & 2 \\ 1 & 3 \end{bmatrix} = \begin{bmatrix} 8 & 20 \\ 9 & 19 \\ 10 & 18 \end{bmatrix}$$

but that

$$\mathbf{U}_{2\times 2}\mathbf{W}_{3\times 2} = \begin{bmatrix} 2 & 2 \\ 1 & 3 \end{bmatrix}\begin{bmatrix} 1 & 6 \\ 2 & 5 \\ 3 & 4 \end{bmatrix}$$

does not exist because the number of columns in $\mathbf{U}_{2\times 2}$ does not equal the number of rows in $\mathbf{W}_{3\times 2}$. Also, we find that

$$\mathbf{X}'_{3\times 9}\mathbf{X}_{9\times 3} = \begin{bmatrix} 1 & 1 & 1 & 1 & 1 & 1 & 1 & 1 & 1 \\ 0 & 1 & 2 & 0 & 1 & 2 & 0 & 1 & 2 \\ 0 & 0 & 0 & 1 & 1 & 1 & 2 & 2 & 2 \end{bmatrix}\begin{bmatrix} 1 & 0 & 0 \\ 1 & 1 & 0 \\ 1 & 2 & 0 \\ 1 & 0 & 1 \\ 1 & 1 & 1 \\ 1 & 2 & 1 \\ 1 & 0 & 2 \\ 1 & 1 & 2 \\ 1 & 2 & 2 \end{bmatrix}$$

$$= \begin{bmatrix} 9 & 9 & 9 \\ 9 & 15 & 9 \\ 9 & 9 & 15 \end{bmatrix}$$

and that

$$\mathbf{X}'_{3\times 9}\mathbf{y}_{9\times 1} = \begin{bmatrix} 1 & 1 & 1 & 1 & 1 & 1 & 1 & 1 & 1 \\ 0 & 1 & 2 & 0 & 1 & 2 & 0 & 1 & 2 \\ 0 & 0 & 0 & 1 & 1 & 1 & 2 & 2 & 2 \end{bmatrix}\begin{bmatrix} 18 \\ 21 \\ 20 \\ 20 \\ 22 \\ 20 \\ 19 \\ 21 \\ 20 \end{bmatrix} = \begin{bmatrix} 181 \\ 184 \\ 182 \end{bmatrix}$$

As a last example, consider the matrices

$$\mathbf{A}_{2\times 2} = \begin{bmatrix} 1 & 1 \\ 2 & 2 \end{bmatrix} \quad \text{and} \quad \mathbf{B}_{2\times 2} = \begin{bmatrix} 0 & 1 \\ 1 & 0 \end{bmatrix}$$

Then we find that

$$\mathbf{A}_{2\times 2}\mathbf{B}_{2\times 2} = \begin{bmatrix} 1 & 1 \\ 2 & 2 \end{bmatrix}\begin{bmatrix} 0 & 1 \\ 1 & 0 \end{bmatrix} = \begin{bmatrix} 1 & 1 \\ 2 & 2 \end{bmatrix}$$

In this case, **B** is said to be premultiplied by **A**, or **A** is said to be postmultiplied by **B**. Now consider

$$\mathbf{B}_{2\times 2}\mathbf{A}_{2\times 2} = \begin{bmatrix} 0 & 1 \\ 1 & 0 \end{bmatrix}\begin{bmatrix} 1 & 1 \\ 2 & 2 \end{bmatrix} = \begin{bmatrix} 2 & 2 \\ 1 & 1 \end{bmatrix}$$

Here **A** is said to be premultiplied by **B**, or **B** is said to be postmultiplied by **A**. Note that in this case **AB** is not equal to **BA**. In general, if **A** and **B** are matrices, then $\mathbf{AB} \neq \mathbf{BA}$.

7.5 THE IDENTITY MATRIX

A matrix in which the number of rows is equal to the number of columns is called a *square matrix*.

For example, the following matrices are square matrices:

$$\mathbf{A}_{3\times 3} = \begin{bmatrix} 3 & 4 & 1 \\ 6 & 10 & 2 \\ 3 & 1 & 5 \end{bmatrix} \qquad \mathbf{B}_{2\times 2} = \begin{bmatrix} 1 & 6 \\ 2 & 3 \end{bmatrix}$$

A square matrix in which the numbers on the main diagonal (the diagonal that runs from upper left to lower right) are 1's and in which all numbers off this diagonal are zeros is called an *identity matrix* and is denoted **I**.

The following are examples of identity matrices:

$$\mathbf{I}_{2\times 2} = \begin{bmatrix} 1 & 0 \\ 0 & 1 \end{bmatrix} \qquad \mathbf{I}_{3\times 3} = \begin{bmatrix} 1 & 0 & 0 \\ 0 & 1 & 0 \\ 0 & 0 & 1 \end{bmatrix}$$

Such a matrix is called an identity matrix because premultiplication or postmultiplication of a square $n \times n$ matrix $\mathbf{A}$ by the $n \times n$ identity matrix $\mathbf{I}$ leaves the matrix $\mathbf{A}$ unchanged. For example, if

$$\mathbf{A}_{3\times 3} = \begin{bmatrix} 2 & 1 & 3 \\ 4 & 1 & 2 \\ 2 & 2 & 1 \end{bmatrix}$$

then we see that

$$\mathbf{I}_{3\times 3}\mathbf{A}_{3\times 3} = \begin{bmatrix} 1 & 0 & 0 \\ 0 & 1 & 0 \\ 0 & 0 & 1 \end{bmatrix}\begin{bmatrix} 2 & 1 & 3 \\ 4 & 1 & 2 \\ 2 & 2 & 1 \end{bmatrix} = \begin{bmatrix} 2 & 1 & 3 \\ 4 & 1 & 2 \\ 2 & 2 & 1 \end{bmatrix}$$

and also that

$$\mathbf{A}_{3\times 3}\mathbf{I}_{3\times 3} = \begin{bmatrix} 2 & 1 & 3 \\ 4 & 1 & 2 \\ 2 & 2 & 1 \end{bmatrix}\begin{bmatrix} 1 & 0 & 0 \\ 0 & 1 & 0 \\ 0 & 0 & 1 \end{bmatrix} = \begin{bmatrix} 2 & 1 & 3 \\ 4 & 1 & 2 \\ 2 & 2 & 1 \end{bmatrix}$$

In general, if $\mathbf{A}_{m\times n}$ is not a square matrix, then

$$\mathbf{A}_{m\times n}\mathbf{I}_{n\times n} = \mathbf{A}_{m\times n} \quad \text{and} \quad \mathbf{I}_{m\times m}\mathbf{A}_{m\times n} = \mathbf{A}_{m\times n}$$

7.6

LINEAR DEPENDENCE AND LINEAR INDEPENDENCE

We now discuss two concepts known as linear independence and linear dependence. Consider the following matrix $\mathbf{A}$:

$$\mathbf{A}_{3\times 3} = \begin{bmatrix} 1 & 4 & 2 \\ 3 & 2 & 6 \\ 2 & 1 & 4 \end{bmatrix}$$

Notice that the third column in this matrix is a multiple of the first column in the matrix. In particular, the third column is simply the first column multiplied by 2. That is,

$$\begin{bmatrix} 2 \\ 6 \\ 4 \end{bmatrix} = 2\begin{bmatrix} 1 \\ 3 \\ 2 \end{bmatrix}$$

In a situation like this, when one column in a matrix **A** is a multiple of another column in matrix **A**, the columns of **A** are said to be linearly dependent. More generally,

If one of the columns of a matrix **A** can be written as a linear combination of some of the other columns in **A**, then the columns of **A** are said to be *linearly dependent*.

As an example, consider the matrix

$$\mathbf{A}_{3\times 4} = \begin{bmatrix} 1 & 3 & 10 & 4 \\ 5 & 2 & 5 & 1 \\ 2 & 2 & 7 & 3 \end{bmatrix}$$

In this case, column 3 is the sum of column 4 plus 2 times column 2. That is,

$$\begin{bmatrix} 4 \\ 1 \\ 3 \end{bmatrix} + 2\begin{bmatrix} 3 \\ 2 \\ 2 \end{bmatrix} = \begin{bmatrix} 4 \\ 1 \\ 3 \end{bmatrix} + \begin{bmatrix} 6 \\ 4 \\ 4 \end{bmatrix} = \begin{bmatrix} 10 \\ 5 \\ 7 \end{bmatrix}$$

Thus the columns of **A** are linearly dependent because column 3 can be expressed as a linear combination of columns 2 and 4.

If none of the columns in a matrix **A** can be written as a linear combination of other columns in **A**, then the columns of **A** are *linearly independent*. The maximum number of linearly independent columns in a matrix **A** is called the *rank* of the matrix. When the rank of a matrix **A** is equal to the number of columns in **A**, the matrix **A** is said to be of *full rank*.

7.7 THE INVERSE OF A MATRIX

Now consider a square matrix $\mathbf{A}_{n\times n}$, which is of full rank.

The *inverse* of the matrix **A** is another matrix, denoted $\mathbf{A}^{-1}$, which satisfies the condition

$$\mathbf{A}\mathbf{A}^{-1} = \mathbf{A}^{-1}\mathbf{A} = \mathbf{I}_{n\times n}$$

where $\mathbf{I}_{n\times n}$ is the identity matrix with dimension $n \times n$.

It should be emphasized that $\mathbf{A}^{-1}$ exists if and only if $\mathbf{A}$ is a square matrix of full rank. As an example, consider the matrix

$$\mathbf{A}_{3\times 3} = \begin{bmatrix} 9 & 9 & 9 \\ 9 & 15 & 9 \\ 9 & 9 & 15 \end{bmatrix}$$

The inverse of $\mathbf{A}$ is

$$\mathbf{A}_{3\times 3}^{-1} = \begin{bmatrix} \frac{4}{9} & -\frac{1}{6} & -\frac{1}{6} \\ -\frac{1}{6} & \frac{1}{6} & 0 \\ -\frac{1}{6} & 0 & \frac{1}{6} \end{bmatrix}$$

since

$$\begin{aligned} \mathbf{A}_{3\times 3}\mathbf{A}_{3\times 3}^{-1} &= \begin{bmatrix} 9 & 9 & 9 \\ 9 & 15 & 9 \\ 9 & 9 & 15 \end{bmatrix} \begin{bmatrix} \frac{4}{9} & -\frac{1}{6} & -\frac{1}{6} \\ -\frac{1}{6} & \frac{1}{6} & 0 \\ -\frac{1}{6} & 0 & \frac{1}{6} \end{bmatrix} \\ &= \begin{bmatrix} 1 & 0 & 0 \\ 0 & 1 & 0 \\ 0 & 0 & 1 \end{bmatrix} \end{aligned}$$

and

$$\begin{aligned} \mathbf{A}_{3\times 3}^{-1}\mathbf{A}_{3\times 3} &= \begin{bmatrix} \frac{4}{9} & -\frac{1}{6} & -\frac{1}{6} \\ -\frac{1}{6} & \frac{1}{6} & 0 \\ -\frac{1}{6} & 0 & \frac{1}{6} \end{bmatrix} \begin{bmatrix} 9 & 9 & 9 \\ 9 & 15 & 9 \\ 9 & 9 & 15 \end{bmatrix} \\ &= \begin{bmatrix} 1 & 0 & 0 \\ 0 & 1 & 0 \\ 0 & 0 & 1 \end{bmatrix} \end{aligned}$$

It can be shown that $\mathbf{A}^{-1}\mathbf{A} = \mathbf{I}_{n\times n}$ if and only if $\mathbf{A}\mathbf{A}^{-1} = \mathbf{I}_{n\times n}$.

Although we will not discuss them here, general formulas exist that allow the calculation of matrix inverses. Also, computer programs are often used to calculate matrix inverses.

7.8

SOME REGRESSION-RELATED MATRIX CALCULATIONS

We now demonstrate how to perform two types of matrix calculations that we will encounter in later chapters of this book. The first calculation is of the form $(\mathbf{X}'\mathbf{X})^{-1}\mathbf{X}'\mathbf{y}$ where $\mathbf{X}$ is a matrix and $\mathbf{y}$ is a column vector. Suppose that

$$\mathbf{y} = \begin{bmatrix} 18 \\ 21 \\ 20 \\ 20 \\ 22 \\ 20 \\ 19 \\ 21 \\ 20 \end{bmatrix} \quad \text{and} \quad \mathbf{X} = \begin{bmatrix} 1 & 0 & 0 \\ 1 & 1 & 0 \\ 1 & 2 & 0 \\ 1 & 0 & 1 \\ 1 & 1 & 1 \\ 1 & 2 & 1 \\ 1 & 0 & 2 \\ 1 & 1 & 2 \\ 1 & 2 & 2 \end{bmatrix}$$

To compute $(\mathbf{X}'\mathbf{X})^{-1}\mathbf{X}'\mathbf{y}$, we make the following calculations:

$$\mathbf{X}' = \begin{bmatrix} 1 & 1 & 1 & 1 & 1 & 1 & 1 & 1 & 1 \\ 0 & 1 & 2 & 0 & 1 & 2 & 0 & 1 & 2 \\ 0 & 0 & 0 & 1 & 1 & 1 & 2 & 2 & 2 \end{bmatrix}$$

$$\mathbf{X}'\mathbf{X} = \begin{bmatrix} 1 & 1 & 1 & 1 & 1 & 1 & 1 & 1 & 1 \\ 0 & 1 & 2 & 0 & 1 & 2 & 0 & 1 & 2 \\ 0 & 0 & 0 & 1 & 1 & 1 & 2 & 2 & 2 \end{bmatrix} \begin{bmatrix} 1 & 0 & 0 \\ 1 & 1 & 0 \\ 1 & 2 & 0 \\ 1 & 0 & 1 \\ 1 & 1 & 1 \\ 1 & 2 & 1 \\ 1 & 0 & 2 \\ 1 & 1 & 2 \\ 1 & 2 & 2 \end{bmatrix}$$

$$= \begin{bmatrix} 9 & 9 & 9 \\ 9 & 15 & 9 \\ 9 & 9 & 15 \end{bmatrix}$$

$$(\mathbf{X}'\mathbf{X})^{-1} = \begin{bmatrix} \frac{4}{9} & -\frac{1}{6} & -\frac{1}{6} \\ -\frac{1}{6} & \frac{1}{6} & 0 \\ -\frac{1}{6} & 0 & \frac{1}{6} \end{bmatrix}$$

since

$$(\mathbf{X}'\mathbf{X})(\mathbf{X}'\mathbf{X})^{-1} = \begin{bmatrix} 9 & 9 & 9 \\ 9 & 15 & 9 \\ 9 & 9 & 15 \end{bmatrix} \begin{bmatrix} \frac{4}{9} & -\frac{1}{6} & -\frac{1}{6} \\ -\frac{1}{6} & \frac{1}{6} & 0 \\ -\frac{1}{6} & 0 & \frac{1}{6} \end{bmatrix}$$

$$= \begin{bmatrix} 1 & 0 & 0 \\ 0 & 1 & 0 \\ 0 & 0 & 1 \end{bmatrix}$$

Here we assume that $(\mathbf{X}'\mathbf{X})^{-1}$ has been calculated by computer. Next,

$$\mathbf{X}'\mathbf{y} = \begin{bmatrix} 1 & 1 & 1 & 1 & 1 & 1 & 1 & 1 & 1 \\ 0 & 1 & 2 & 0 & 1 & 2 & 0 & 1 & 2 \\ 0 & 0 & 0 & 1 & 1 & 1 & 2 & 2 & 2 \end{bmatrix} \begin{bmatrix} 18 \\ 21 \\ 20 \\ 20 \\ 22 \\ 20 \\ 19 \\ 21 \\ 20 \end{bmatrix} = \begin{bmatrix} 181 \\ 184 \\ 182 \end{bmatrix}$$

Finally, we find that

$$(\mathbf{X}'\mathbf{X})^{-1}\mathbf{X}'\mathbf{y} = \begin{bmatrix} \frac{4}{9} & -\frac{1}{6} & -\frac{1}{6} \\ -\frac{1}{6} & \frac{1}{6} & 0 \\ -\frac{1}{6} & 0 & \frac{1}{6} \end{bmatrix} \begin{bmatrix} 181 \\ 184 \\ 182 \end{bmatrix}$$

$$= \begin{bmatrix} 19.44 \\ .50 \\ .167 \end{bmatrix}$$

The second type of matrix calculation that we will encounter is of the form $\mathbf{x}_0'(\mathbf{X}'\mathbf{X})^{-1}\mathbf{x}_0$, where $\mathbf{x}_0$ is a column vector and $\mathbf{X}$ is a matrix.

We illustrate this type of calculation using

$$\mathbf{x}_0 = \begin{bmatrix} 1 \\ 2 \\ 1 \end{bmatrix} \quad \text{and} \quad \mathbf{X} = \begin{bmatrix} 1 & 0 & 0 \\ 1 & 1 & 0 \\ 1 & 2 & 0 \\ 1 & 0 & 1 \\ 1 & 1 & 1 \\ 1 & 2 & 1 \\ 1 & 0 & 2 \\ 1 & 1 & 2 \\ 1 & 2 & 2 \end{bmatrix}$$

As was shown previously,

$$(\mathbf{X}'\mathbf{X})^{-1} = \begin{bmatrix} \frac{4}{9} & -\frac{1}{6} & -\frac{1}{6} \\ -\frac{1}{6} & \frac{1}{6} & 0 \\ -\frac{1}{6} & 0 & \frac{1}{6} \end{bmatrix}$$

Then we find that

$$\mathbf{x}_0'(\mathbf{X}'\mathbf{X})^{-1}\mathbf{x}_0 = [1 \quad 2 \quad 1] \begin{bmatrix} \frac{4}{9} & -\frac{1}{6} & -\frac{1}{6} \\ -\frac{1}{6} & \frac{1}{6} & 0 \\ -\frac{1}{6} & 0 & \frac{1}{6} \end{bmatrix} \begin{bmatrix} 1 \\ 2 \\ 1 \end{bmatrix}$$

$$= [-\tfrac{1}{18} \quad \tfrac{1}{6} \quad 0] \begin{bmatrix} 1 \\ 2 \\ 1 \end{bmatrix} = \tfrac{5}{18}$$

EXERCISES

7.1 Let

$$\mathbf{A} = \begin{bmatrix} 1 & 2 \\ 3 & 1 \\ 2 & 2 \end{bmatrix}$$

a. Calculate $\mathbf{A}'$. b. Calculate $\mathbf{A}'\mathbf{A}$.

7.2 Let

$$\mathbf{A} = \begin{bmatrix} 1 & 3 & 1 \\ 2 & 1 & 1 \\ 1 & 3 & 3 \end{bmatrix} \quad \text{and} \quad \mathbf{B} = \begin{bmatrix} 0 & .6 & -.2 \\ .5 & -.2 & -.1 \\ -.5 & 0 & .5 \end{bmatrix}$$

a. Calculate $\mathbf{A} + \mathbf{B}$. b. Calculate $\mathbf{AB}$.
c. Calculate $\mathbf{BA}$. d. How are $\mathbf{A}$ and $\mathbf{B}$ related?

7.3 Let

$$\mathbf{A} = \begin{bmatrix} .02 & 0 & 0 & 0 & 0 \\ 0 & .01 & 0 & 0 & 0 \\ 0 & 0 & .004 & 0 & 0 \\ 0 & 0 & 0 & .005 & 0 \\ 0 & 0 & 0 & 0 & .002 \end{bmatrix} \quad \mathbf{c} = \begin{bmatrix} 1000 \\ 300 \\ 500 \\ 80 \\ 250 \end{bmatrix} \quad \mathbf{x} = \begin{bmatrix} 1 \\ 2 \\ 5 \\ 4 \\ 10 \end{bmatrix}$$

a. Calculate $\mathbf{c} + \mathbf{x}$. b. Calculate $\mathbf{Ac}$. c. Calculate $(\mathbf{Ac})'$.
d. Calculate $(\mathbf{Ac})'\mathbf{c}$. e. Calculate $\mathbf{x}'$. f. Calculate $\mathbf{x}'\mathbf{Ax}$.

7.4 Suppose that $\mathbf{x}_0' = [1 \quad 2 \quad 5 \quad 25]$ and

$$(\mathbf{X}'\mathbf{X})^{-1} = \begin{bmatrix} .05 & 0 & 0 & 0 \\ 0 & .01 & 0 & 0 \\ 0 & 0 & .10 & 0 \\ 0 & 0 & 0 & .02 \end{bmatrix}$$

a. Calculate $\mathbf{x}_0'(\mathbf{X}'\mathbf{X})^{-1}$.
b. Calculate $(\mathbf{X}'\mathbf{X})^{-1}\mathbf{x}_0$.
c. Calculate $\mathbf{x}_0'(\mathbf{X}'\mathbf{X})^{-1}\mathbf{x}_0$.

7.5 Suppose that $\mathbf{x}_0' = [1 \quad 10 \quad 20 \quad 100 \quad 200]$ and

$$(\mathbf{X}'\mathbf{X})^{-1} = \begin{bmatrix} .001 & 0 & 0 & 0 & 0 \\ 0 & .010 & 0 & 0 & 0 \\ 0 & 0 & .008 & 0 & 0 \\ 0 & 0 & 0 & .050 & 0 \\ 0 & 0 & 0 & 0 & .200 \end{bmatrix}$$

a. Calculate $\mathbf{x}_0'(\mathbf{X}'\mathbf{X})^{-1}$.
b. Calculate $(\mathbf{X}'\mathbf{X})^{-1}\mathbf{x}_0$.
c. Calculate $\mathbf{x}_0'(\mathbf{X}'\mathbf{X})^{-1}\mathbf{x}_0$.

7.6 Suppose that $\mathbf{x}_0' = [1 \quad 5 \quad 10 \quad 15 \quad 20]$

$$\mathbf{X}'\mathbf{y} = \begin{bmatrix} 5000 \\ 1000 \\ 750 \\ 2000 \\ 500 \end{bmatrix} \quad \text{and} \quad (\mathbf{X}'\mathbf{X})^{-1} = \begin{bmatrix} .10 & 0 & 0 & 0 & 0 \\ 0 & .20 & 0 & 0 & 0 \\ 0 & 0 & .30 & 0 & 0 \\ 0 & 0 & 0 & .40 & 0 \\ 0 & 0 & 0 & 0 & .50 \end{bmatrix}$$

a. Calculate $(\mathbf{X}'\mathbf{X})^{-1}\mathbf{X}'\mathbf{y}$.
b. Calculate $\mathbf{x}_0'(\mathbf{X}'\mathbf{X})^{-1}$.
c. Calculate $\mathbf{x}_0'(\mathbf{X}'\mathbf{X})^{-1}\mathbf{x}_0$.

To work Exercises 7.7 through 7.14, it is useful to know that if the matrix

$$\mathbf{A} = \begin{bmatrix} a & b \\ c & d \end{bmatrix}$$

is a 2 × 2 matrix, then the inverse of **A** is

$$\mathbf{A}^{-1} = \begin{bmatrix} d/D & -b/D \\ -c/D & a/D \end{bmatrix} \quad \text{where} \quad D = ad - bc$$

7.7 Suppose that

$$\mathbf{A} = \begin{bmatrix} 5 & 10 \\ 5 & 20 \end{bmatrix}$$

a. Calculate $\mathbf{A}^{-1}$.
b. Verify your result of part (a) by computing $\mathbf{A}^{-1}\mathbf{A}$ and $\mathbf{A}\mathbf{A}^{-1}$.

7.8 Suppose that

$$\mathbf{A} = \begin{bmatrix} 100 & .5 \\ 200 & .10 \end{bmatrix}$$

a. Calculate $\mathbf{A}^{-1}$.
b. Verify your result of part (a) by computing $\mathbf{A}^{-1}\mathbf{A}$ and $\mathbf{A}\mathbf{A}^{-1}$.

7.9 Suppose that

$$\mathbf{y} = \begin{bmatrix} 4 \\ 5 \\ 3 \\ 7 \\ 1 \\ 8 \\ 10 \end{bmatrix} \quad \text{and} \quad \mathbf{X} = \begin{bmatrix} 1 & 1.5 \\ 1 & 2.0 \\ 1 & 1.0 \\ 1 & 0.5 \\ 1 & 2.0 \\ 1 & 3.0 \\ 1 & 2.5 \end{bmatrix}$$

a. Calculate $\mathbf{X}'\mathbf{X}$.
b. Calculate $(\mathbf{X}'\mathbf{X})^{-1}$.
c. Calculate $\mathbf{X}'\mathbf{y}$.
d. Calculate $(\mathbf{X}'\mathbf{X})^{-1}\mathbf{X}'\mathbf{y}$.

7.10 Suppose that **X** is as defined in Exercise 7.9 and that

$$\mathbf{x}_0' = [1 \quad 4.0]$$

a. Calculate $\mathbf{x}_0'(\mathbf{X}'\mathbf{X})^{-1}$.
b. Calculate $\mathbf{x}_0'(\mathbf{X}'\mathbf{X})^{-1}\mathbf{x}_0$.

7.11 Suppose that

$$\mathbf{y} = \begin{bmatrix} 10 \\ 12 \\ 8 \\ 14 \\ 6 \\ 15 \end{bmatrix} \quad \text{and} \quad \mathbf{X} = \begin{bmatrix} 1 & 1 \\ 1 & 2 \\ 1 & 3 \\ 1 & 2 \\ 1 & 3 \\ 1 & 1 \end{bmatrix}$$

a. Calculate $\mathbf{X}'\mathbf{X}$.
b. Calculate $(\mathbf{X}'\mathbf{X})^{-1}$.
c. Calculate $\mathbf{X}'\mathbf{y}$.
d. Calculate $(\mathbf{X}'\mathbf{X})^{-1}\mathbf{X}'\mathbf{y}$.

7.12 Suppose that $\mathbf{X}$ is as defined in Exercise 7.11 and that

$$\mathbf{x}_0' = [1 \quad 2.5]$$

a. Calculate $\mathbf{x}_0'(\mathbf{X}'\mathbf{X})^{-1}$.
b. Calculate $(\mathbf{X}'\mathbf{X})^{-1}\mathbf{x}_0$.
c. Calculate $\mathbf{x}_0'(\mathbf{X}'\mathbf{X})^{-1}\mathbf{x}_0$.

7.13 Suppose that

$$\boldsymbol{\lambda} = \begin{bmatrix} 1 \\ -1 \end{bmatrix} \quad \text{and} \quad \mathbf{X} = \begin{bmatrix} 1 & 5 \\ 1 & 6 \\ 1 & 4 \\ 1 & 5 \\ 1 & 8 \\ 1 & 10 \end{bmatrix}$$

a. Calculate $(\mathbf{X}'\mathbf{X})^{-1}$.
b. Calculate $\boldsymbol{\lambda}'(\mathbf{X}'\mathbf{X})^{-1}\boldsymbol{\lambda}$.

7.14 Suppose that $\mathbf{X}$ is as defined in Exercise 7.13 and that $\boldsymbol{\lambda}' = [-1 \quad 1]$. Compute $\boldsymbol{\lambda}'(\mathbf{X}'\mathbf{X})^{-1}\boldsymbol{\lambda}$.

7.15 Define the row vector $\boldsymbol{\beta}' = [\beta_0 \ \ \beta_1 \ \ \beta_2 \ \ \beta_3 \ \ \beta_4 \ \ \beta_5]$. For each expression below, determine the row vector $\boldsymbol{\lambda}' = [\lambda_0 \ \ \lambda_1 \ \ \lambda_2 \ \ \lambda_3 \ \ \lambda_4 \ \ \lambda_5]$ so that the expression can be written as the product $\boldsymbol{\lambda}'\boldsymbol{\beta}$. For example, if we wish to write $\beta_0 + \beta_5$ as a product $\boldsymbol{\lambda}'\boldsymbol{\beta}$, then

$$\boldsymbol{\lambda}' = [1 \quad 0 \quad 0 \quad 0 \quad 0 \quad 1]$$

since $\boldsymbol{\lambda}'\boldsymbol{\beta}$ equals

$$(1)(\beta_0) + (0)(\beta_1) + (0)(\beta_2) + (0)(\beta_3) + (0)(\beta_4) + (1)(\beta_5) = \beta_0 + \beta_5$$

a. $\beta_1 + \beta_2 + \beta_3$
b. $\beta_5 - \beta_3$
c. $\beta_2 + \beta_4 - \beta_5$
d. $2\beta_1 + 3\beta_2 + \beta_3$
e. $\beta_4 - (\beta_1 + \beta_2 + \beta_3)/3$
f. $\beta_0 + 2\beta_1 + 5\beta_2 + 3\beta_3 + \beta_4 + 6\beta_5$

7.16 Define the row vector

$$\boldsymbol{\beta}' = [\beta_0 \quad \beta_1 \quad \beta_2 \quad \beta_3 \quad \beta_4 \quad \beta_5 \quad \beta_6 \quad \beta_7 \quad \beta_8]$$

For each expression below, determine the row vector

$$\boldsymbol{\lambda}' = [\lambda_0 \quad \lambda_1 \quad \lambda_2 \quad \lambda_3 \quad \lambda_4 \quad \lambda_5 \quad \lambda_6 \quad \lambda_7 \quad \lambda_8]$$

so that the expression can be written as the product $\boldsymbol{\lambda}'\boldsymbol{\beta}$ (see Exercise 7.15).

a. $\beta_6 - \beta_3 + \beta_1$
b. $(\beta_1 + \beta_2 + \beta_3) - (\beta_7 + \beta_8)/2$
c. $3\beta_4 + 2\beta_6 + \beta_8$
d. $\beta_7 - \beta_6$
e. $\beta_5 - (\beta_3 + \beta_4)/2$
f. $\beta_0 + 2\beta_2 - 3\beta_5 + \beta_8$

CHAPTER 8

MULTIPLE REGRESSION: I

In this chapter we extend our discussion of regression analysis by studying regression models that employ more than one independent variable. Sometimes these models are called *multiple regression models*. We begin in Section 8.1 by introducing the (multiple) linear regression model. Here, we include the use of polynomial terms. In Section 8.2 we show how to compute the least squares point estimates of the model parameters, and in Section 8.3 we use the model to compute point estimates and point predictions. Section 8.4 presents the regression assumptions: model appropriateness, constant variance, independence, and normality. Here, we also discuss validation of these assumptions. Sections 8.5 and 8.6 consider some ways to judge overall model utility. Included are the overall F-test and the multiple coefficients of determination and correlation. In Section 8.7 we study statistical inference for a single regression parameter β_j (including the test of H_0: $\beta_j = 0$ versus $H_1: \beta_j \neq 0$), and in Section 8.8 we study how to compute confidence intervals for means and prediction intervals for individual values. We conclude this chapter with optional Section 8.9, which discusses using SAS to perform (multiple) linear regression analysis.

8.1

THE LINEAR REGRESSION MODEL

THE GENERAL LINEAR MODEL

Example 8.1 ***Part 1: The Data and a Regression Model***

In Chapters 4 and 5 we considered using the independent variable x, average hourly temperature, to predict y, weekly fuel consumption, at American Manufacturing Company's nine-building complex. We now consider predicting y on the basis of average hourly temperature and a second independent variable—the chill index. Here, the chill index for a given average hourly temperature expresses the combined effects of all other major weather-related factors that influence weekly fuel consumption at the nine-building complex. Among these factors would be, for example, wind velocity, cloud cover, and the passage of weather fronts. The chill index is expressed as an integer between 0 and 30. A weekly chill index near zero indicates that, given the average hourly temperature during the week, all other major weather-related factors will only minimally increase weekly fuel consumption. A weekly chill index near 30 indicates that, given the average hourly temperature during the week, all other major weather-related factors will maximally increase weekly fuel consumption.

Data concerning weekly fuel consumption (y), average hourly temperature (x_1), and chill index (x_2) has been observed for the last $n = 8$ weeks. This data is presented in Table 8.1. In this table the subscript i refers to the week for which the data has been observed. Thus y_i, x_{i1}, and x_{i2} denote the weekly fuel consumption, average hourly temperature, and chill index observed in week i.

In Figure 8.1 we plot y versus x_1. This plot indicates that values of y tend to decrease in a straight-line fashion as x_1 increases. This suggests that the simple

TABLE 8.1 **Fuel consumption data**

Week, i	*Average hourly temperature, x_{i1} (°F)*	*Chill index, x_{i2}*	*Fuel consumption, y_i (tons)*
1	$x_{11} = 28.0$	$x_{12} = 18$	$y_1 = 12.4$
2	$x_{21} = 28.0$	$x_{22} = 14$	$y_2 = 11.7$
3	$x_{31} = 32.5$	$x_{32} = 24$	$y_3 = 12.4$
4	$x_{41} = 39.0$	$x_{42} = 22$	$y_4 = 10.8$
5	$x_{51} = 45.9$	$x_{52} = 8$	$y_5 = 9.4$
6	$x_{61} = 57.8$	$x_{62} = 16$	$y_6 = 9.5$
7	$x_{71} = 58.1$	$x_{72} = 1$	$y_7 = 8.0$
8	$x_{81} = 62.5$	$x_{82} = 0$	$y_8 = 7.5$

FIGURE 8.1 **Plot of y (weekly fuel consumption) versus x_1 (average hourly temperature)**

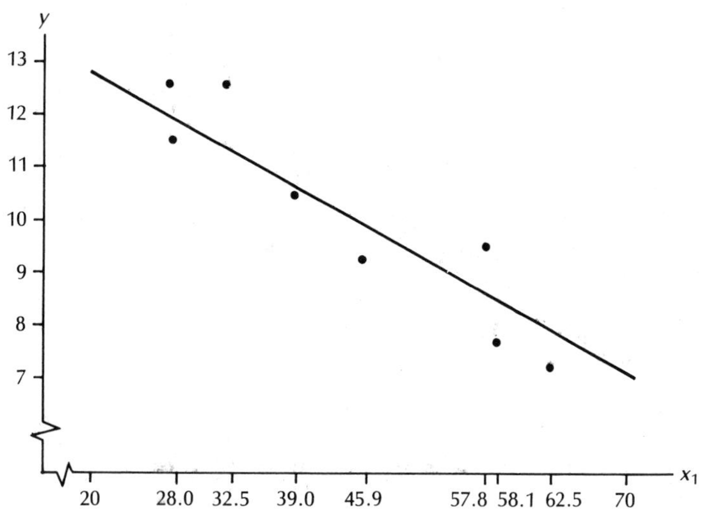

linear model having a negative slope

$$\begin{aligned} y &= \mu + \varepsilon \\ &= \beta_0 + \beta_1 x_1 + \varepsilon \end{aligned}$$

relates y to x_1. In Figure 8.2 we plot y versus x_2. This plot indicates that values of y tend to increase in a straight-line fashion as x_2 increases. This suggests that the simple linear model having a positive slope

$$\begin{aligned} y &= \mu + \varepsilon \\ &= \beta_0 + \beta_1 x_2 + \varepsilon \end{aligned}$$

relates y to x_2. Therefore it seems reasonable to combine these two models to form the model

$$\begin{aligned} y_i &= \mu_i + \varepsilon_i \\ &= \beta_0 + \beta_1 x_{i1} + \beta_2 x_{i2} + \varepsilon_i \end{aligned}$$

which relates y to x_1 and x_2. This regression model says that

1. $\mu_i = \beta_0 + \beta_1 x_{i1} + \beta_2 x_{i2}$ is the mean value of y when the values of the independent variables x_1 and x_2 are x_{i1} and x_{i2}. For example,

$$\begin{aligned} \mu_5 &= \beta_0 + \beta_1 x_{51} + \beta_2 x_{52} \\ &= \beta_0 + \beta_1(45.9) + \beta_2(8) \end{aligned}$$

FIGURE 8.2 **Plot of y (weekly fuel consumption) versus x_2 (the chill index)**

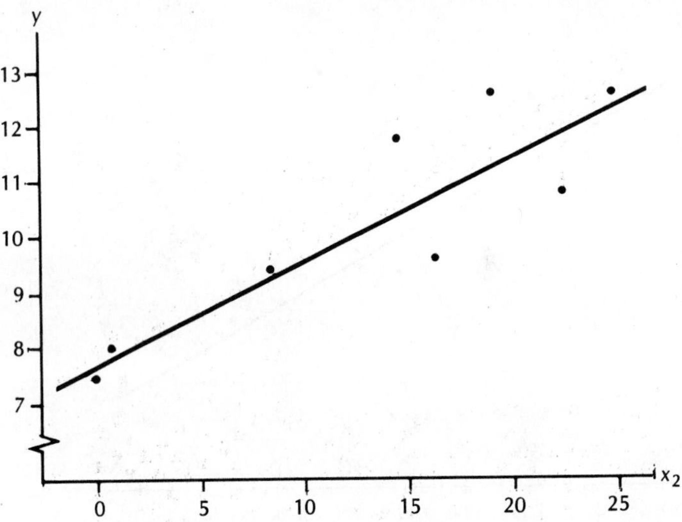

is intuitively the average fuel consumption for all weeks having an average hourly temperature equal to 45.9 and a chill index equal to 8.

2. β_0, β_1, and β_2 are parameters relating μ_i to x_{i1} and x_{i2}. We will soon interpret the exact meaning of these parameters.
3. ε_i is an error term that describes the effects on y_i of all factors other than x_{i1} and x_{i2}.

Part 2: Interpreting the Regression Parameters β_0, β_1, and β_2

The exact interpretations of the parameters β_0, β_1, and β_2 are quite simple. First, suppose that $x_{i1} = 0$ and $x_{i2} = 0$. Then

$$\begin{aligned} \mu_i &= \beta_0 + \beta_1 x_{i1} + \beta_2 x_{i2} \\ &= \beta_0 + \beta_1(0) + \beta_2(0) = \beta_0 \end{aligned}$$

So β_0 is the mean weekly fuel consumption for all weeks having an average hourly temperature of 0°F and a chill index of zero. The parameter β_0 is called the *intercept* in the regression model. One might wonder whether β_0 has any practical interpretation, since it is unlikely that a week having an average hourly temperature of 0°F would also have a chill index of zero. Indeed, sometimes the parameter β_0 and other parameters in a regression analysis do not have practical interpretations because the situations related to the interpretations would not be likely to occur in practice. In fact, sometimes each parameter does not, by itself, have much practical importance. Rather, the real importance of the parameters is that they relate the mean of the dependent variable to the independent variables in an overall sense.

We next interpret the individual meanings of β_1 and β_2. To examine the interpretation of β_1, consider two different weeks. Suppose that for the first week the average hourly temperature is c and the chill index is d. The mean weekly fuel consumption for all such weeks is

$$\beta_0 + \beta_1(c) + \beta_2(d)$$

For the second week, suppose that the average hourly temperature is $c + 1$ and the chill index is d. The mean weekly fuel consumption for all such weeks is

$$\beta_0 + \beta_1(c + 1) + \beta_2(d)$$

Now consider the difference between these mean fuel consumptions. This difference is

$$\begin{aligned}&[\beta_0 + \beta_1(c + 1) + \beta_2(d)] - [\beta_0 + \beta_1(c) + \beta_2(d)]\\ &= (\beta_0 - \beta_0) + [\beta_1(c + 1) - \beta_1(c)] + [\beta_2(d) - \beta_2(d)] = \beta_1\end{aligned}$$

Since weeks 1 and 2 differ only in that the average hourly temperature during week 2 is one degree higher than the average hourly temperature during week 1, we can interpret the parameter β_1 as the change in mean weekly fuel consumption that is associated with a one-degree increase in average hourly temperature when the chill index does not change.

The interpretation of β_2 can be established in a similar fashion. It is easy to see that we can interpret β_2 as the change in mean weekly fuel consumption that is associated with a one-unit increase in the chill index when the average hourly temperature does not change.

Part 3: A Geometric Interpretation of the Regression Model

We now interpret our fuel consumption model geometrically. We begin by defining the *experimental region* to be the range of the combinations of the observed values of the independent variables x_1, average hourly temperature, and x_2, chill index. From the data in Table 8.1, it is reasonable to depict the experimental region as the shaded region in Figure 8.3. Here, the combinations of x_1 and x_2 values are the ordered pairs in the figure.

Next, consider the equation

$$\mu = \beta_0 + \beta_1 x_1 + \beta_2 x_2$$

which relates mean fuel consumption to x_1 and x_2. This equation is the equation of a plane in three-dimensional space. We illustrate the portion of this plane corresponding to the (x_1, x_2) combinations in the experimental region in Figure 8.4. As illustrated in this figure, the model

$$\begin{aligned}y_i &= \mu_i + \varepsilon_i\\ &= \beta_0 + \beta_1 x_{i1} + \beta_2 x_{i2} + \varepsilon_i\end{aligned}$$

says that the eight error terms cause the eight observed fuel consumptions (the

FIGURE 8.3 **The experimental region**

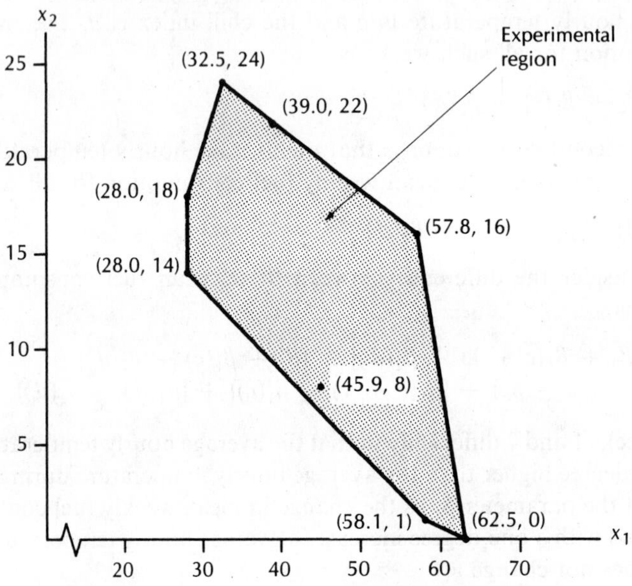

FIGURE 8.4 **A geometrical interpretation of the regression model relating *y* to x_1 and x_2**

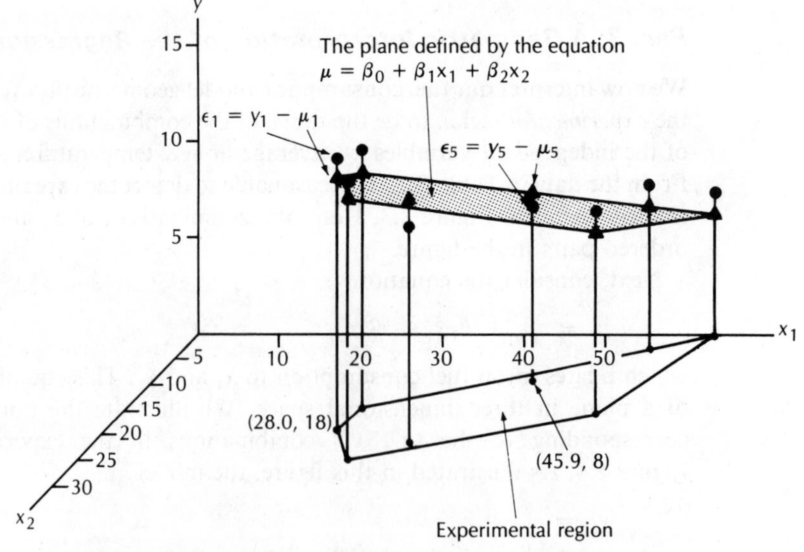

When looking at this figure, it is best to pretend that you are sitting high in a football stadium and that you are looking down at the playing field, which is the (x_1, x_2) plane

heavy dots in the figure) to deviate from the eight mean fuel consumptions (the triangles in the figure), which exactly lie on the plane defined by the equation

$$\mu = \beta_0 + \beta_1 x_1 + \beta_2 x_2$$

For example, the error term ε_1, which is depicted to be positive in Figure 8.4, causes y_1 to be higher than μ_1 (mean fuel consumption when $x_1 = 28$ and $x_2 = 18$). As another example, ε_5, which is depicted to be negative, causes y_5 to be lower than μ_5 (mean fuel consumption when $x_1 = 45.9$ and $x_2 = 8$).

We now define the linear regression model relating a dependent variable y to several independent variables $x_1, x_2, \ldots, x_p$. Here, we assume that we have obtained n observations, with each observation consisting of an observed value of y and corresponding observed values of $x_1, x_2, \ldots, x_p$. For $i = 1, 2, \ldots, n$ we denote the values of $y, x_1, x_2, \ldots, x_p$ that comprise the ith observation as $y_i, x_{i1}, x_{i2}, \ldots, x_{ip}$. We now state the model.

The Linear Regression Model

The *linear regression model* relating y to $x_1, x_2, \ldots, x_p$ is

$$\begin{aligned} y_i &= \mu_i + \varepsilon_i \\ &= \beta_0 + \beta_1 x_{i1} + \beta_2 x_{i2} + \cdots + \beta_p x_{ip} + \varepsilon_i \end{aligned}$$

Here,

1. $\mu_i = \beta_0 + \beta_1 x_{i1} + \beta_2 x_{i2} + \cdots + \beta_p x_{ip}$ is the *mean value of the dependent variable* when the values of the independent variables $x_1, x_2, \ldots, x_p$ are $x_{i1}, x_{i2}, \ldots, x_{ip}$.
2. $\beta_0, \beta_1, \beta_2, \ldots, \beta_p$ are (unknown) *parameters* relating μ_i to $x_{i1}, x_{i2}, \ldots, x_{ip}$.
3. ε_i is an *error term* that describes the effects on y_i of all factors other than the values $x_{i1}, x_{i2}, \ldots, x_{ip}$ of the independent variables $x_1, x_2, \ldots, x_p$.

Note that we assume that y_i has been randomly selected from the infinite population of potential values of the dependent variable that could be observed when the values of the independent variables $x_1, x_2, \ldots, x_p$ are $x_{i1}, x_{i2}, \ldots, x_{ip}$. As defined in the preceding box, μ_i is the mean of this population.

The reason that we call this model a *linear* regression model is that the equation

$$\mu_i = \beta_0 + \beta_1 x_{i1} + \beta_2 x_{i2} + \cdots + \beta_p x_{ip}$$

expresses μ_i as a linear function of the parameters $\beta_0, \beta_1, \beta_2, \ldots, \beta_p$. For example, the equation

$$\mu_i = \beta_0 + \beta_1 x_{i1} + \beta_2 x_{i2}$$

is a linear equation because it expresses μ_i as a linear function of the parameters

β_0, β_1, and β_2. Furthermore, although the equation

$$\mu_i = \beta_0 + \beta_1 x_{i1} + \beta_2 x_{i2} + \beta_3 x_{i1}^2 + \beta_4 x_{i1} x_{i2}$$

utilizes x_{i1}^2 and $x_{i1}x_{i2}$, this equation is also linear because it expresses μ_i as a linear function of the parameters β_0, β_1, β_2, β_3, and β_4. The need for terms such as x_{i1}^2 will be discussed in the next subsection, and the need for terms such as $x_{i1}x_{i2}$ will be discussed in Chapter 9. The equation

$$\mu_i = \beta_0 + \beta_1 x_{i1}^{\beta_2}$$

is not a linear equation, because it does not express μ_i as a linear function of the parameters β_0, β_1, and β_2. We emphasize the concept of linearity because the techniques of regression analysis are easiest to use and best developed when we assume that a linear equation relates μ_i to $x_{i1}, x_{i2}, \ldots, x_{ip}$.

Example 8.2 Table 8.2 presents the sales price y, square footage x_1, number of rooms x_2, number of bedrooms x_3, and age x_4 for each of 63 single-family residences sold during 1988 in Oxford, Ohio. Suppose that we relate y to x_1, x_2, x_3, and x_4 by the regression model

$$\begin{aligned} y_i &= \mu_i + \varepsilon_i \\ &= \beta_0 + \beta_1 x_{i1} + \beta_2 x_{i2} + \beta_3 x_{i3} + \beta_4 x_{i4} + \varepsilon_i \end{aligned}$$

Here, $\mu_i = \beta_0 + \beta_1 x_{i1} + \beta_2 x_{i2} + \beta_3 x_{i3} + \beta_4 x_{i4}$ is, intuitively, the mean selling price of all single-family residences that have x_{i1} square feet, x_{i2} rooms, and x_{i3} bedrooms and are x_{i4} years old. Furthermore, the parameter β_1 tells us how much higher the mean selling price is for a residence that is one square foot larger than another residence when both residences are of the same age and have the same number of rooms and the same number of bedrooms. The parameter β_2 tells us how much higher (or possibly lower) the mean selling price is for a residence that has one more room than another residence when both residences are the same age and have the same square footage and the same number of bedrooms. The parameter β_3 tells us how much higher (or possibly lower) the mean selling price is for a residence that has one more bedroom than another residence when both residences are the same age and have the same square footage and the same number of rooms. The parameter β_4 tells us how much lower (or possibly higher) the mean selling price is for a residence that is one year older than another residence when both residences have the same square footage, the same number of rooms, and the same number of bedrooms. The error term ε_i describes the effect on selling price of all factors other than x_{i1}, x_{i2}, x_{i3}, and x_{i4}.

TABLE 8.2 Measurements taken on 63 single-family residences

Residence, i	Sales price, y ($\times$\$1000)	Square feet, x_1	Rooms, x_2	Bedrooms, x_3	Age, x_4
1	53.5	1008	5	2	35
2	49.0	1290	6	3	36
3	50.5	860	8	2	36
4	49.9	912	5	3	41
5	52.0	1204	6	3	40
6	55.0	1204	5	3	10
7	80.5	1764	8	4	64
8	86.0	1600	7	3	19
9	69.0	1255	5	3	16
10	149.0	3600	10	5	17
11	46.0	864	5	3	37
12	38.0	720	4	2	41
13	49.5	1008	6	3	35
14	105.0	1950	8	3	52
15	152.5	2086	7	3	12
16	85.0	2011	9	4	76
17	60.0	1465	6	3	102
18	58.5	1232	5	2	69
19	101.0	1736	7	3	67
20	79.4	1296	6	3	11
21	125.0	1996	7	3	9
22	87.9	1874	5	2	14
23	80.0	1580	5	3	11
24	94.0	1920	5	3	14
25	74.0	1430	9	3	16
26	69.0	1486	6	3	27
27	63.0	1008	5	2	35
28	67.5	1282	5	3	20
29	35.0	1134	5	2	74
30	142.5	2400	9	4	15
31	92.2	1701	5	3	15
32	56.0	1020	6	3	16
33	63.0	1053	5	2	24
34	60.0	1728	6	3	26
35	34.0	416	3	1	42
36	52.0	1040	5	2	9
37	75.0	1496	6	3	30
38	93.0	1936	8	4	39
39	60.0	1904	7	4	32
40	73.0	1080	5	2	24
41	71.0	1768	8	4	74
42	83.0	1503	6	3	14
43	90.0	1736	7	3	16
44	83.0	1695	6	3	12
45	115.0	2186	8	4	12
46	50.0	888	5	2	34
47	55.2	1120	6	3	29
48	61.0	1400	5	3	33
49	147.0	2165	7	3	2
50	210.0	2353	8	4	15
51	60.0	1536	6	3	36
52	100.0	1972	8	3	37
53	44.5	1120	5	3	27
54	55.0	1664	7	3	79
55	53.4	925	5	3	20
56	65.0	1288	5	3	2
57	73.0	1400	5	3	2
58	40.0	1376	6	3	103
59	141.0	2038	12	4	62
60	68.0	1572	6	3	29
61	139.0	1545	6	3	9
62	140.0	1993	6	3	4
63	55.0	1130	5	2	21

Reprinted by permission from Alpha, Inc., Oxford, Ohio.

SOME SPECIAL CASES: QUADRATIC AND HIGHER-ORDER MODELS

We have already studied one special case of the general linear regression model: the simple linear regression model

$$y = \beta_0 + \beta_1 x + \varepsilon$$

Another useful regression model is called the *quadratic regression model*. Suppose that we have observed n values of the dependent variable

$$y_1, y_2, \ldots, y_n$$

and that we have observed n corresponding values of the independent variable

$$x_1, x_2, \ldots, x_n$$

The Quadratic Regression Model

The *quadratic regression model* relating y_i to x_i is expressed by

$$\begin{aligned} y_i &= \mu_i + \varepsilon_i \\ &= \beta_0 + \beta_1 x_i + \beta_2 x_i^2 + \varepsilon_i \end{aligned}$$

This model employs a quadratic equation to relate μ_i to x_i. Therefore as illustrated in Figure 8.5, the mean value of the dependent variable is either

1. Increasing at an increasing rate or increasing at a decreasing rate as x increases

or

2. Decreasing at an increasing rate or decreasing at a decreasing rate as x increases.

The numerical values of β_0, β_1, and β_2 determine exactly how the mean value of the dependent variable changes as x increases.

As shown in Figure 8.6, the quadratic model says that the n error terms, $\varepsilon_1, \varepsilon_2, \ldots, \varepsilon_n$, cause the n observed values of the dependent variable (the dots in the figure) to deviate from the n mean values of the dependent variable (the triangles in the figure), which exactly lie on the quadratic curve

$$\mu = \beta_0 + \beta_1 x + \beta_2 x^2$$

We can determine whether this model might be appropriate by plotting y versus x. If, as the values of x increase, the values of y tend to change according to any of the quadratic curves illustrated in Figure 8.5, then the quadratic regression model might appropriately relate y to x.

We now consider an example of a regression model that employs a quadratic term.

FIGURE 8.5 **The mean value of the dependent variable changing in a quadratic fashion as *x* increases**

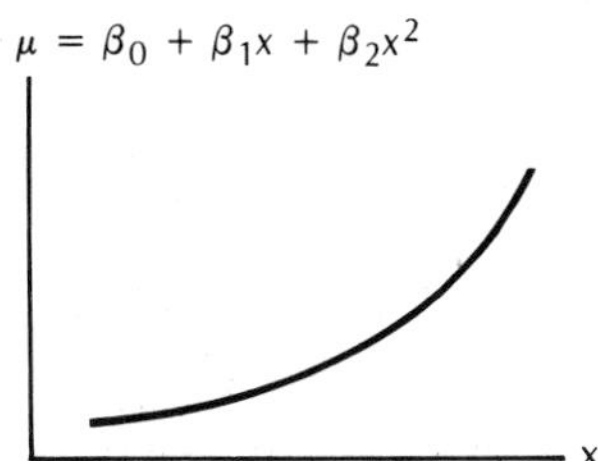

(a) The mean value increasing at an increasing rate as x increases

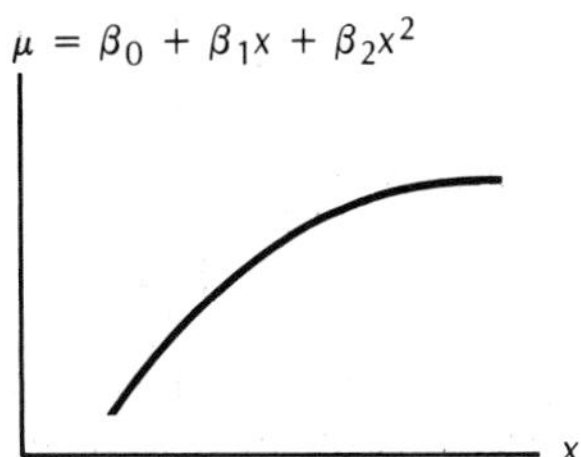

(b) The mean value increasing at a decreasing rate as x increases

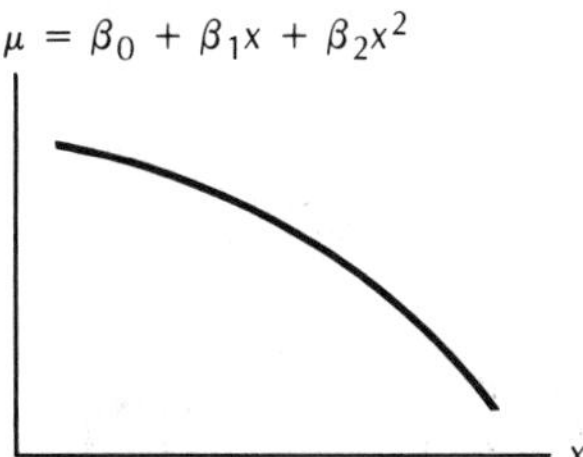

(c) The mean value decreasing at an increasing rate as x increases

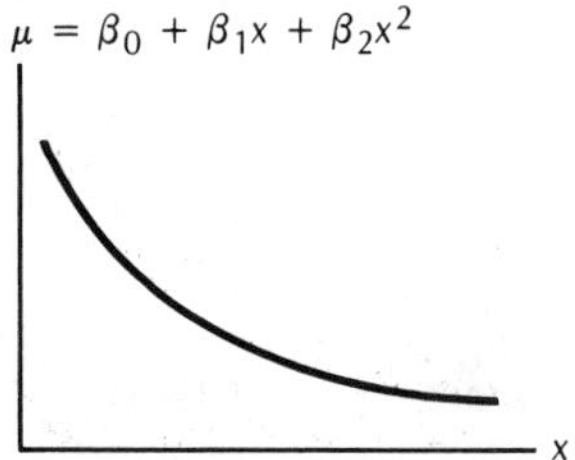

(d) The mean value decreasing at a decreasing rate as x increases

FIGURE 8.6 **The quadratic regression model**

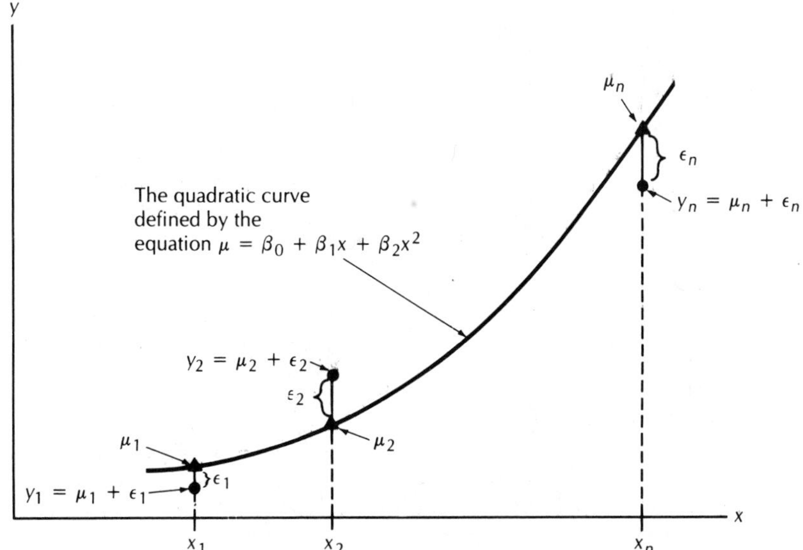

Example 8.3 Enterprise Industries produces Fresh, a brand of liquid laundry detergent. The company wishes to study the factors affecting the quarterly demand for the large size bottle of Fresh in its sales regions. To do this, the company has gathered data concerning demand for Fresh in $n = 30$ sales regions of equal sales potential for the first quarter of 1989. This data is presented in Table 8.3. Here, for $i = 1, 2, \ldots, 30$,

y_i = the demand for the large size bottle of Fresh (in hundreds of thousands of bottles) in sales region i

x_{i1} = the price (in dollars) of Fresh as offered by Enterprise Industries in sales region i

TABLE 8.3 **Historical data, including price differences, concerning demand for Fresh Detergent**

Sales region, i	Price for Fresh, x_{i1} (dollars)	Average industry price, x_{i2} (dollars)	Price difference, $x_{i4} = x_{i2} - x_{i1}$ (dollars)	Advertising expenditure for Fresh, x_{i3} (hundreds of thousands of dollars)	Demand for Fresh, y_i (hundreds of thousands of bottles)
1	3.85	3.80	−.05	5.50	7.38
2	3.75	4.00	.25	6.75	8.51
3	3.70	4.30	.60	7.25	9.52
4	3.70	3.70	0	5.50	7.50
5	3.60	3.85	.25	7.00	9.33
6	3.60	3.80	.20	6.50	8.28
7	3.60	3.75	.15	6.75	8.75
8	3.80	3.85	.05	5.25	7.87
9	3.80	3.65	−.15	5.25	7.10
10	3.85	4.00	.15	6.00	8.00
11	3.90	4.10	.20	6.50	7.89
12	3.90	4.00	.10	6.25	8.15
13	3.70	4.10	.40	7.00	9.10
14	3.75	4.20	.45	6.90	8.86
15	3.75	4.10	.35	6.80	8.90
16	3.80	4.10	.30	6.80	8.87
17	3.70	4.20	.50	7.10	9.26
18	3.80	4.30	.50	7.00	9.00
19	3.70	4.10	.40	6.80	8.75
20	3.80	3.75	−.05	6.50	7.95
21	3.80	3.75	−.05	6.25	7.65
22	3.75	3.65	−.10	6.00	7.27
23	3.70	3.90	.20	6.50	8.00
24	3.55	3.65	.10	7.00	8.50
25	3.60	4.10	.50	6.80	8.75
26	3.65	4.25	.60	6.80	9.21
27	3.70	3.65	−.05	6.50	8.27
28	3.75	3.75	0	5.75	7.67
29	3.80	3.85	.05	5.80	7.93
30	3.70	4.25	.55	6.80	9.26

x_{i2} = the average industry price (in dollars) of competitors' similar detergents in sales region i

x_{i3} = Enterprise Industries' advertising expenditure (in hundreds of thousands of dollars) to promote Fresh in sales region i

$x_{i4} = x_{i2} - x_{i1}$ = the "price difference" in sales region i

To begin our analysis, suppose that Enterprise Industries believes on theoretical grounds that the single independent variable x_4 adequately describes the effects of x_1 and x_2 on y. We determine whether this assumption is valid in later chapters, but for now we continue the analysis by studying the relationship between the dependent variable y and the independent variables x_3 and x_4.

In Figure 8.7 we plot y versus x_4, and in Figure 8.8 we plot y versus x_3. Figure 8.7 indicates that y tends to increase in a straight-line fashion as x_4 increases. This suggests that the model

$$\begin{aligned} y_i &= \mu_i + \varepsilon_i \\ &= \beta_0 + \beta_1 x_{i4} + \varepsilon_i \end{aligned}$$

might relate y_i to x_{i4}. Figure 8.8 indicates that y tends to increase in a curved fashion as x_3 increases. Since this curve appears to be quadratic [it has the shape of Figure 8.5(a)], Figure 8.8 suggests that the quadratic model

$$\begin{aligned} y_i &= \mu_i + \varepsilon_i \\ &= \beta_0 + \beta_1 x_{i3} + \beta_2 x_{i3}^2 + \varepsilon_i \end{aligned}$$

might appropriately relate y_i to x_{i3}. To construct a model that relates y_i to x_{i4} and x_{i3}, it seems reasonable to combine the two models above to form the regression

FIGURE 8.7 **Plot of y (demand for Fresh Detergent) against x_4 (price difference)**

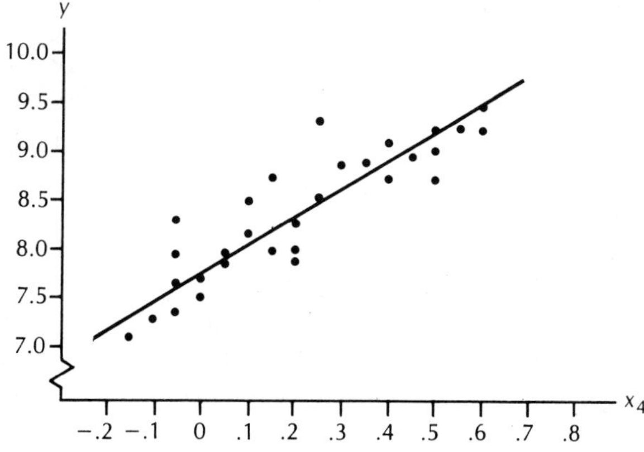

FIGURE 8.8 **Plot of y (demand for Fresh Detergent) against x_3 (advertising expenditure for Fresh)**

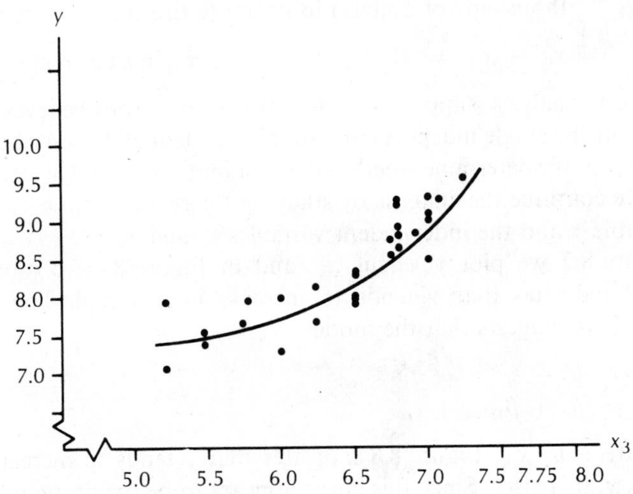

model

$$
\begin{aligned}
y_i &= \mu_i + \varepsilon_i \\
&= \beta_0 + \beta_1 x_{i4} + \beta_2 x_{i3} + \beta_3 x_{i3}^2 + \varepsilon_i
\end{aligned}
$$

Here, $\mu_i = \beta_0 + \beta_1 x_{i4} + \beta_2 x_{i3} + \beta_3 x_{i3}^2$ is the mean demand for Fresh when the price difference is x_{i4} and the advertising expenditure is x_{i3}. The error term ε_i describes the effects on demand of all factors other than x_{i4} and x_{i3}.

In addition to the quadratic model there are other regression models that can be used to model situations in which y tends to change in a curved fashion. One such model is:

The pth-Order Polynomial Model

The *pth-order polynomial regression model* is

$$
\begin{aligned}
y_i &= \mu_i + \varepsilon_i \\
&= \beta_0 + \beta_1 x_i + \beta_2 x_i^2 + \beta_3 x_i^3 + \cdots + \beta_p x_i^p + \varepsilon_i
\end{aligned}
$$

Clearly the quadratic model is a special case of this more general polynomial model, where $p = 2$. Third-order and higher-order polynomials are used to model situations in which y changes according to a curve that displays one or more reversals in curvature. A third-order polynomial model

$$\begin{aligned} y_i &= \mu_i + \varepsilon_i \\ &= \beta_0 + \beta_1 x_i + \beta_2 x_i^2 + \beta_3 x_i^3 + \varepsilon_i \end{aligned}$$

describes a curve with one reversal in curvature (a curve with one peak and one trough). A pth-order polynomial model describes a curve that displays a total of $(p - 1)$ peaks and troughs.

Although third- and higher-order polynomial models are occasionally found to be useful, the quadratic model seems to appropriately describe most curved regression relationships. In addition to polynomial models, however, there are still other models that can be used to describe such relationships. We study some of these models in Chapter 13.

8.2

THE LEAST SQUARES POINT ESTIMATES

Let $b_0, b_1, b_2, \ldots, b_p$ denote point estimates of the parameters $\beta_0, \beta_1, \beta_2, \ldots, \beta_p$ in the linear regression model. Then the point prediction of

$$y_i = \beta_0 + \beta_1 x_{i1} + \beta_2 x_{i2} + \cdots + \beta_p x_{ip} + \varepsilon_i$$

is

$$\hat{y}_i = b_0 + b_1 x_{i1} + b_2 x_{i2} + \cdots + b_p x_{ip}$$

(Here we predict ε_i to be zero.) Next, we define, for $i = 1, 2, \ldots, n$, the *residual*

$$e_i = y_i - \hat{y}_i$$

and we also define the *sum of squared residuals* to be

$$\begin{aligned} SSE &= \sum_{i=1}^{n} e_i^2 = \sum_{i=1}^{n} (y_i - \hat{y}_i)^2 \\ &= \sum_{i=1}^{n} (y_i - (b_0 + b_1 x_{i1} + b_2 x_{i2} + \cdots + b_p x_{ip}))^2 \end{aligned}$$

Intuitively, if any particular values of $b_0, b_1, b_2, \ldots, b_p$ are good point estimates, they will make (for $i = 1, 2, \ldots, n$) the predicted value $\hat{y}_i$ fairly close to the observed value y_i and thus will make SSE fairly small. We define the *least squares point estimates* to be the values of $b_0, b_1, b_2, \ldots, b_p$ that minimize SSE. It can be proved (see Appendix B) that these estimates can be calculated by the following matrix algebra formula.

The Least Squares Point Estimates

The *least squares point estimates* $b_0, b_1, b_2, \ldots, b_p$ are calculated by using the formula

$$\begin{bmatrix} b_0 \\ b_1 \\ b_2 \\ \vdots \\ b_p \end{bmatrix} = \mathbf{b} = (\mathbf{X}'\mathbf{X})^{-1}\mathbf{X}'\mathbf{y}$$

where $\mathbf{y}$ and $\mathbf{X}$ are the following column vector and matrix:

$$\mathbf{y} = \begin{bmatrix} y_1 \\ y_2 \\ \vdots \\ y_n \end{bmatrix} \quad \text{and} \quad \mathbf{X} = \begin{array}{c} \begin{array}{cccccc} 0 & 1 & 2 & \cdot & \cdot & \cdot & p \\ & x_1 & x_2 & \cdot & \cdot & \cdot & x_p \end{array} \\ \begin{bmatrix} 1 & x_{11} & x_{12} & \cdot & \cdot & \cdot & x_{1p} \\ 1 & x_{21} & x_{22} & \cdot & \cdot & \cdot & x_{2p} \\ \vdots & \vdots & \vdots & & & & \vdots \\ 1 & x_{n1} & x_{n2} & \cdot & \cdot & \cdot & x_{np} \end{bmatrix} \end{array}$$

Here $\mathbf{y}$ is a column vector of the n observed values of the dependent variable, $y_1, y_2, \ldots, y_n$. To define the matrix $\mathbf{X}$, consider the regression model

$$y = \beta_0 + \beta_1 x_1 + \beta_2 x_2 + \cdots + \beta_p x_p + \varepsilon$$

If k is the number of parameters in this model, then the matrix $\mathbf{X}$ will consist of k columns. The columns in the matrix $\mathbf{X}$ contain the observed values of the independent variables corresponding to (that is, multiplied by) the k parameters $\beta_0, \beta_1, \beta_2, \ldots, \beta_p$. The columns of this matrix are numbered in the same manner as the parameters are numbered (see the preceding $\mathbf{X}$ matrix). In the following examples we demonstrate how to calculate the least squares point estimates.

Example 8.4 The least squares point estimates b_0, b_1, and b_2 of the parameters β_0, β_1, and β_2 in the fuel consumption model

$$y = \beta_0 + \beta_1 x_1 + \beta_2 x_2 + \varepsilon$$

are calculated by using the formula

$$\begin{bmatrix} b_0 \\ b_1 \\ b_2 \end{bmatrix} = \mathbf{b} = (\mathbf{X}'\mathbf{X})^{-1}\mathbf{X}'\mathbf{y}$$

where

$$\mathbf{y} = \begin{bmatrix} y_1 \\ y_2 \\ y_3 \\ y_4 \\ y_5 \\ y_6 \\ y_7 \\ y_8 \end{bmatrix} = \begin{bmatrix} 12.4 \\ 11.7 \\ 12.4 \\ 10.8 \\ 9.4 \\ 9.5 \\ 8.0 \\ 7.5 \end{bmatrix}$$

and

$$\mathbf{X} = \overset{\begin{matrix} 0 & 1 & 2 \\ & x_1 & x_2 \end{matrix}}{\begin{bmatrix} 1 & x_{11} & x_{12} \\ 1 & x_{21} & x_{22} \\ 1 & x_{31} & x_{32} \\ 1 & x_{41} & x_{42} \\ 1 & x_{51} & x_{52} \\ 1 & x_{61} & x_{62} \\ 1 & x_{71} & x_{72} \\ 1 & x_{81} & x_{82} \end{bmatrix}} = \overset{\begin{matrix} 0 & 1 & 2 \\ & x_1 & x_2 \end{matrix}}{\begin{bmatrix} 1 & 28.0 & 18 \\ 1 & 28.0 & 14 \\ 1 & 32.5 & 24 \\ 1 & 39.0 & 22 \\ 1 & 45.9 & 8 \\ 1 & 57.8 & 16 \\ 1 & 58.1 & 1 \\ 1 & 62.5 & 0 \end{bmatrix}}$$

Here, the column vector **y** is simply a vector of the observed weekly fuel consumptions, and the three columns of the **X** matrix contain the observed values of the independent variables corresponding to (that is, multiplied by) the three parameters in the model. Therefore since the number 1 is multiplied by β_0, the column of the **X** matrix corresponding to β_0 is a column of 1's. Since the independent variable x_1 is multiplied by β_1, the column of the **X** matrix corresponding to β_1 is a column containing the observed average hourly temperatures (see Table 8.1). The independent variable x_2 is multiplied by β_2, and thus the column of the **X** matrix corresponding to β_2 is a column containing the observed chill indices (see Table 8.1).

To calculate $\mathbf{b} = (\mathbf{X}'\mathbf{X})^{-1}\mathbf{X}'\mathbf{y}$, we first find

$$\mathbf{X}'\mathbf{X} = \begin{bmatrix} 1 & 1 & 1 & 1 & 1 & 1 & 1 & 1 \\ 28.0 & 28.0 & 32.5 & 39.0 & 45.9 & 57.8 & 58.1 & 62.5 \\ 18 & 14 & 24 & 22 & 8 & 16 & 1 & 0 \end{bmatrix} \begin{bmatrix} 1 & 28.0 & 18 \\ 1 & 28.0 & 14 \\ 1 & 32.5 & 24 \\ 1 & 39.0 & 22 \\ 1 & 45.9 & 8 \\ 1 & 57.8 & 16 \\ 1 & 58.1 & 1 \\ 1 & 62.5 & 0 \end{bmatrix}$$

$$= \begin{bmatrix} 8.0 & 351.8 & 103.0 \\ 351.8 & 16874.76 & 3884.1 \\ 103.0 & 3884.1 & 1901.0 \end{bmatrix}$$

Since the columns of the matrix $\mathbf{X}'\mathbf{X}$ can be verified to be linearly independent of each other, the matrix $\mathbf{X}'\mathbf{X}$ possesses an inverse matrix $(\mathbf{X}'\mathbf{X})^{-1}$. This matrix can be computer calculated to be

$$(\mathbf{X}'\mathbf{X})^{-1} = \begin{bmatrix} 8.0 & 351.8 & 103.0 \\ 351.8 & 16874.76 & 3884.1 \\ 103.0 & 3884.1 & 1901.0 \end{bmatrix}^{-1}$$

$$= \begin{bmatrix} 5.43405 & -.085930 & -.118856 \\ -.085930 & .00147070 & .00165094 \\ -.118856 & .00165094 & .00359276 \end{bmatrix}$$

We next calculate

$$\mathbf{X}'\mathbf{y} = \begin{bmatrix} 1 & 1 & 1 & 1 & 1 & 1 & 1 & 1 \\ 28.0 & 28.0 & 32.5 & 39.0 & 45.9 & 57.8 & 58.1 & 62.5 \\ 18 & 14 & 24 & 22 & 8 & 16 & 1 & 0 \end{bmatrix} \begin{bmatrix} 12.4 \\ 11.7 \\ 12.4 \\ 10.8 \\ 9.4 \\ 9.5 \\ 8.0 \\ 7.5 \end{bmatrix}$$

$$= \begin{bmatrix} 81.7 \\ 3413.11 \\ 1157.4 \end{bmatrix}$$

Finally, we compute

$$\begin{bmatrix} b_0 \\ b_1 \\ b_2 \end{bmatrix} = \mathbf{b} = (\mathbf{X}'\mathbf{X})^{-1}\mathbf{X}'\mathbf{y}$$

$$= \begin{bmatrix} 5.43405 & -.085930 & -.118856 \\ -.085930 & .00147070 & .00165094 \\ -.118856 & .00165094 & .00359276 \end{bmatrix} \begin{bmatrix} 81.7 \\ 3413.11 \\ 1157.4 \end{bmatrix}$$

$$= \begin{bmatrix} 13.109 \\ -.0900 \\ .0825 \end{bmatrix}$$

The point estimate $b_1 = -.0900$ of β_1 says that we estimate that mean weekly fuel consumption decreases (since b_1 is negative) by .0900 tons of coal when average hourly temperature increases by 1 degree and the chill index does not change. The point estimate $b_2 = .0825$ of β_2 says that we estimate that the mean weekly fuel consumption increases (since b_2 is positive) by .0825 tons of coal when there is a one-unit increase in the chill index and average hourly temperature does not change.

To illustrate the fact that the least squares point estimates minimize *SSE*, compare Table 8.4(a) and Table 8.4(b). We see that the least squares point estimates $b_0 = 13.109$, $b_1 = -.0900$, and $b_2 = .0825$ give the value $SSE = .674$.

TABLE 8.4 **Values of *SSE* given by the least squares point estimates, $b_0 = 13.109$, $b_1 = -.0900$ and $b_2 = .0825$, and by the point estimates, $b_0 = 12.54$, $b_1 = -.095$ and $b_2 = .077$**

(a) Predictions using the least squares point estimates, $b_0 = 13.109$, $b_1 = -.0900$ and $b_2 = .0825$

Week, i	*Average hourly temperature, x_{i1} (°F)*	*Chill index, x_{i2}*	*Observed fuel consumption, y_i (tons)*	*Predicted fuel consumption, $\hat{y}_i = b_0 + b_1x_{i1} + b_2x_{i2} = 13.109 - .0900x_{i1} + .0825x_{i2}$*	*Residual, $e_i = y_i - \hat{y}_i$*
1	28.0	18	12.4	12.0733	.3267
2	28.0	14	11.7	11.7433	−.0433
3	32.5	24	12.4	12.1632	.2368
4	39.0	22	10.8	11.4131	−.6131
5	45.9	8	9.4	9.6371	−.2371
6	57.8	16	9.5	9.2259	.2741
7	58.1	1	8.0	7.9614	.0386
8	62.5	0	7.5	7.4829	.0171

$$SSE = \sum_{i=1}^{8} e_i^2 = .674$$

(continues)

TABLE 8.4 **Continued**

(b) Predictions using the point estimates $b_0 = 12.54$, $b_1 = -.095$, and $b_2 = .077$

Week, i	Average hourly temperature, x_{i1} (°F)	Chill index, x_{i2}	Observed fuel consumption, y_i (tons)	Predicted fuel consumption, $\hat{y}_i = b_0 + b_1x_{i1} + b_2x_{i2} = 12.54 - .095x_{i1} + .077x_{i2}$	Residual, $e_i = y_i - \hat{y}_i$
1	28.0	18	12.4	11.2660	1.1340
2	28.0	14	11.7	10.9580	.7420
3	32.5	24	12.4	11.3005	1.0995
4	39.0	22	10.8	10.5290	.2710
5	45.9	8	9.4	8.7955	.6045
6	57.8	16	9.5	8.2810	1.2190
7	58.1	1	8.0	7.0975	.9025
8	62.5	0	7.5	6.6025	.8975

$$SSE = \sum_{i=1}^{8} e_i^2 = 6.590$$

This is less than $SSE = 6.590$, which is the value given by $b_0 = 12.54$, $b_1 = -.095$, and $b_2 = .077$, which are not the least squares point estimates. There is nothing special about the values 12.54, $-.095$, and .077. The point is that if we choose any values of b_0, b_1, and b_2 that are different from the least squares values, we will obtain a larger SSE.

FIGURE 8.9 **A geometrical interpretation of the prediction equation relating $\hat{y}$ to x_1 and x_2**

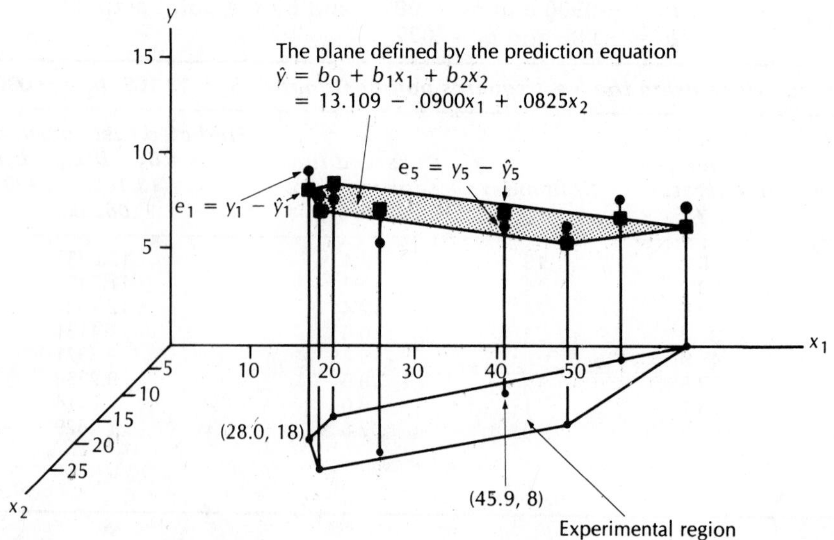

The prediction equation

$$\begin{aligned}\hat{y} &= b_0 + b_1x_1 + b_2x_2 \\ &= 13.109 - .0900x_1 + .0825x_2\end{aligned}$$

is the equation of a plane. We illustrate the portion of this plane corresponding to the (x_1, x_2) combinations in the experimental region in Figure 8.9. Looking at this figure, we see that the observed fuel consumptions in Table 8.4 (the heavy dots in the figure) differ from the predicted fuel consumptions in Table 8.4(a) (the squares in the figure), which exactly lie on the plane defined by the above prediction equation. The eight line segments drawn between the squares and the dots represent the sizes of the residuals. Since the least squares point estimates minimize *SSE*, we can interpret them as positioning the planar prediction equation in three spaces so as to minimize the sum of squared distances between the observed and predicted fuel consumptions. In this sense we can say that the plane defined by the least squares point estimates is the best plane that can be positioned between the observed fuel consumptions.

Example 8.5 Consider the residential sales model of Example 8.2:

$$\begin{aligned}y_i &= \mu_i + \varepsilon_i \\ &= \beta_0 + \beta_1x_{i1} + \beta_2x_{i2} + \beta_3x_{i3} + \beta_4x_{i4} + \varepsilon_i\end{aligned}$$

By using the data in Table 8.2 we define the column vector $\mathbf{y}$ and matrix $\mathbf{X}$ as follows:

$$\mathbf{y} = \begin{bmatrix} y_1 \\ y_2 \\ \vdots \\ y_{63} \end{bmatrix} = \begin{bmatrix} 53.5 \\ 49.0 \\ \vdots \\ 55.0 \end{bmatrix} \qquad \mathbf{X} = \begin{array}{c} \begin{array}{ccccc} 1 & x_1 & x_2 & x_3 & x_4 \end{array} \\ \begin{bmatrix} 1 & 1008 & 5 & 2 & 35 \\ 1 & 1290 & 6 & 3 & 36 \\ \vdots & \vdots & \vdots & \vdots & \vdots \\ 1 & 1130 & 5 & 2 & 21 \end{bmatrix} \end{array}$$

Therefore we can calculate the least squares point estimates of β_0, β_1, β_2, β_3, and β_4 to be

$$\begin{bmatrix} b_0 \\ b_1 \\ b_2 \\ b_3 \\ b_4 \end{bmatrix} = \mathbf{b} = (\mathbf{X}'\mathbf{X})^{-1}\mathbf{X}'\mathbf{y} = \begin{bmatrix} 10.36762 \\ .05001 \\ 6.32178 \\ -11.10316 \\ -.43186 \end{bmatrix}$$

Recalling the interpretation of β_1 in Example 8.2, the point estimate $b_1 = .05001$ says that we estimate that the mean selling price is \$5001 higher for a residence that is 100 square feet larger than another residence when both residences are the same age and have the same number of rooms and the same number of bedrooms.

Example 8.6 Consider the Fresh Detergent model of Example 8.3:

$$\begin{aligned} y_i &= \mu_i + \varepsilon_i \\ &= \beta_0 + \beta_1 x_{i4} + \beta_2 x_{i3} + \beta_3 x_{i3}^2 + \varepsilon_i \end{aligned}$$

By using the data in Table 8.3, we define the column vector

$$\mathbf{y} = \begin{bmatrix} y_1 \\ y_2 \\ \cdot \\ \cdot \\ \cdot \\ y_{30} \end{bmatrix} = \begin{bmatrix} 7.38 \\ 8.51 \\ \cdot \\ \cdot \\ \cdot \\ 9.26 \end{bmatrix}$$

and the matrix

$$\mathbf{X} = \begin{array}{c} \begin{array}{cccc} 1 & x_4 & x_3 & x_3^2 \end{array} \\ \begin{bmatrix} 1 & -.05 & 5.50 & (5.50)^2 \\ 1 & .25 & 6.75 & (6.75)^2 \\ \cdot & \cdot & \cdot & \cdot \\ \cdot & \cdot & \cdot & \cdot \\ \cdot & \cdot & \cdot & \cdot \\ 1 & .55 & 6.80 & (6.80)^2 \end{bmatrix} \end{array} = \begin{bmatrix} 1 & -.05 & 5.50 & 30.25 \\ 1 & .25 & 6.75 & 45.5625 \\ \cdot & \cdot & \cdot & \cdot \\ \cdot & \cdot & \cdot & \cdot \\ \cdot & \cdot & \cdot & \cdot \\ 1 & .55 & 6.80 & 46.24 \end{bmatrix}$$

Thus we can calculate the least squares point estimates of β_0, β_1, β_2, and β_3 to be

$$\begin{bmatrix} b_0 \\ b_1 \\ b_2 \\ b_3 \end{bmatrix} = \mathbf{b} = (\mathbf{X}'\mathbf{X})^{-1}\mathbf{X}'\mathbf{y} = \begin{bmatrix} 17.3244 \\ 1.3070 \\ -3.6956 \\ 0.3486 \end{bmatrix}$$

In concluding this section we note that the column vector $\mathbf{y}$ and the matrix $\mathbf{X}$ used to calculate the least squares point estimates b_0 and b_1 of the parameters β_0 and β_1 in the simple linear regression model are

$$\mathbf{y} = \begin{bmatrix} y_1 \\ y_2 \\ \cdot \\ \cdot \\ \cdot \\ y_n \end{bmatrix} \quad \text{and} \quad \mathbf{X} = \begin{bmatrix} 1 & x_1 \\ 1 & x_2 \\ \cdot & \cdot \\ \cdot & \cdot \\ \cdot & \cdot \\ 1 & x_n \end{bmatrix}$$

By using this **y** vector and **X** matrix it can be shown that

$$\begin{bmatrix} b_0 \\ b_1 \end{bmatrix} = \mathbf{b} = (\mathbf{X}'\mathbf{X})^{-1}\mathbf{X}'\mathbf{y} = \begin{bmatrix} \bar{y} - b_1\bar{x} \\ \dfrac{n\sum_{i=1}^{n} x_i y_i - \left(\sum_{i=1}^{n} x_i\right)\left(\sum_{i=1}^{n} y_i\right)}{n\sum_{i=1}^{n} x_i^2 - \left(\sum_{i=1}^{n} x_i\right)^2} \end{bmatrix}$$

These are the same formulas for b_0 and b_1 that we presented in Chapter 4. In fact, each general matrix algebra formula of this chapter can be shown to reduce, when considering the simple linear regression model, to the corresponding formula in Chapters 4 and 5.

8.3

POINT ESTIMATES AND POINT PREDICTIONS

Example 8.7 Consider the fuel consumption problem. Suppose that we have employed a weather forecasting service, which has told us that in a future week the average hourly temperature will be $x_{01} = 40$ and the chill index will be $x_{02} = 10$. Also, suppose that we wish to predict y_0, the amount of fuel that will be used in the future week at the nine-building complex. We assume that y_0 can be expressed as

$$y_0 = \mu_0 + \varepsilon_0$$

Here, μ_0 is the mean weekly fuel consumption for all weeks when $x_{01} = 40$ and $x_{02} = 10$, and ε_0 is the error term in the future week.

We concluded in Example 8.1 that the plane defined by the equation

$$\mu = \beta_0 + \beta_1 x_1 + \beta_2 x_2$$

relates the eight mean fuel consumptions $\mu_1, \mu_2, \ldots, \mu_8$ to the eight previously observed combinations of temperatures and chill indices. As illustrated in Figure 8.10, the future combination ($x_{01} = 40$, $x_{02} = 10$) is in the experimental region. Therefore it is reasonable to believe that the future mean fuel consumption μ_0 lies on this plane. That is, it seems reasonable to conclude that

$$\begin{aligned} \mu_0 &= \beta_0 + \beta_1 x_{01} + \beta_2 x_{02} \\ &= \beta_0 + \beta_1(40) + \beta_2(10) \end{aligned}$$

It follows (predicting ε_0 to be zero) that

$$\begin{aligned} \hat{y}_0 &= b_0 + b_1 x_{01} + b_2 x_{02} \\ &= 13.109 - .0900(40) + .0825(10) \\ &= 10.333 \quad \text{tons of coal} \end{aligned}$$

is a point estimate of μ_0 and is a point prediction of y_0.

FIGURE 8.10 **A geometrical interpretation of the mean fuel consumption μ_0 and the individual fuel consumption $y_0 = \mu_0 + \varepsilon_0$**

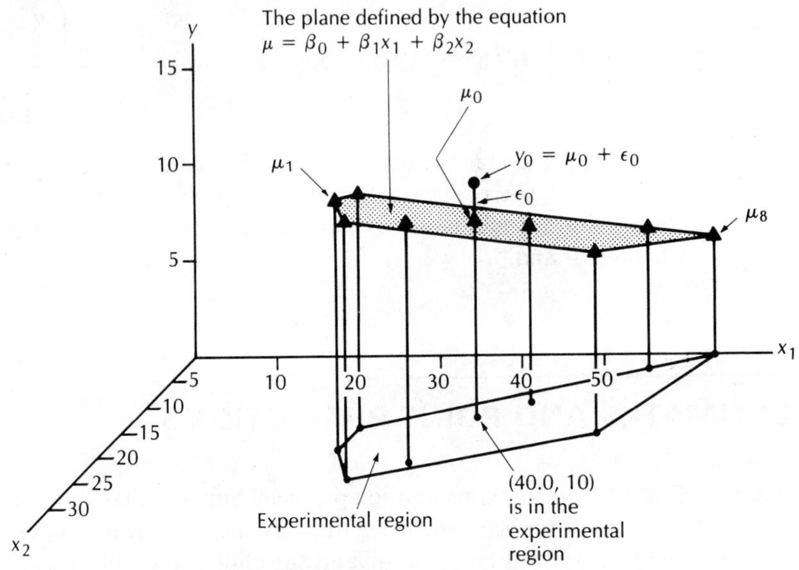

Since the combination ($x_{01} = 40$, $x_{02} = 10$) is in the experimental region, it is reasonable to use the preceding prediction equation to compute $\hat{y}_0$. However, extrapolating the plane defined by this prediction equation to predict y on the basis of a combination of values of x_1 and x_2 far outside the experimental region can lead to unacceptable prediction errors. Moreover, the farther outside the experimental region this combination is, the more likely we are to obtain large prediction errors. For example, since Figure 8.11 reveals that the combination (10.0, 5) of values of x_1 and x_2 is far outside the experimental region, extrapolating the plane defined by our prediction equation to predict y on the basis of this combination might yield an unacceptable prediction error. Intuitively, it is reasonable to use a planar prediction equation to predict y on the basis of values of x_1 and x_2 that are in the experimental region. However, it is possibly more appropriate to use a curved prediction equation to predict y on the basis of the combination (10.0, 5). The only way to tell what sort of prediction equation should be used is to observe data so that the combination (10.0, 5) is in the experimental region. As another example, although an average hourly temperature of 40.0°F is in the range of the observed average hourly temperatures—28.0°F to 62.5°F—and although a chill index of 5 is in the range of the observed chill indices—0 to 18—note from Figure 8.11 that

FIGURE 8.11 **A geometrical interpretation of the point prediction $\hat{y}_0$**

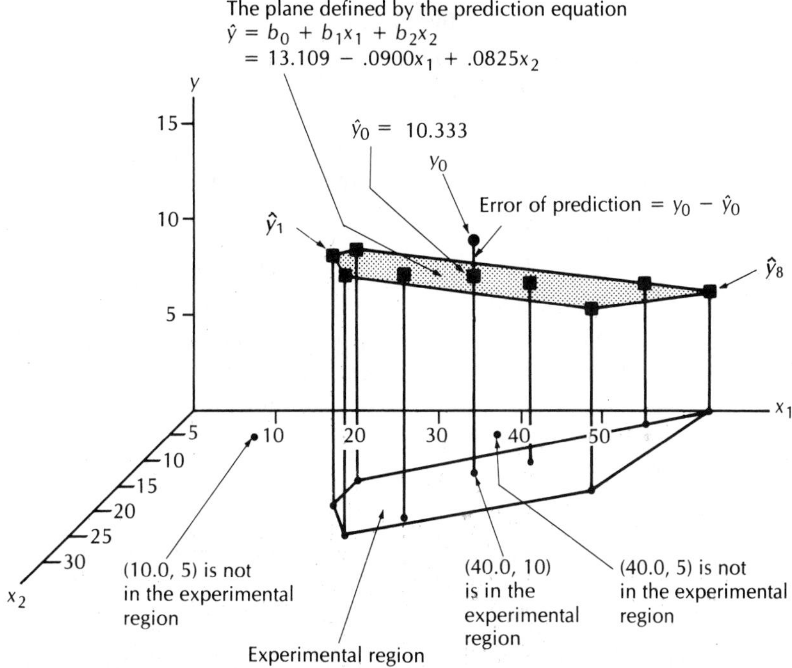

the point representing the *combination* (40.0, 5) of values of x_1 and x_2 is outside the experimental region. Hence extrapolating the plane defined by the prediction equation above to predict for this combination might also yield a large prediction error.

Generalizing from Example 8.7, we have the following result.

Point Estimation and Prediction Using the Linear Regression Model

Let $b_0, b_1, b_2, \ldots, b_p$ be the *least squares point estimates* of the parameters $\beta_0, \beta_1, \beta_2, \ldots, \beta_p$ in the *linear regression model*. Suppose that $(x_{01}, x_{02}, \ldots, x_{0p})$, a specified combination of values of the independent variables $x_1, x_2, \ldots, x_p$, is in the *experimental region* (the range of the combinations of the observed values of the p independent variables). Then,

$$\hat{y}_0 = b_0 + b_1x_{01} + b_2x_{02} + \cdots + b_px_{0p}$$

is the following:

1. The *point estimate of* μ_0, *the mean value of the dependent variable* when the values of the independent variables are $x_{01}, x_{02}, \ldots, x_{0p}$ and
2. The *point prediction of* y_0, the *individual value of the dependent variable* when the values of the independent variables are $x_{01}, x_{02}, \ldots, x_{0p}$.

Example 8.8 Suppose that we wish to use the residential sales model

$$\begin{aligned} y_i &= \mu_i + \varepsilon_i \\ &= \beta_0 + \beta_1 x_{i1} + \beta_2 x_{i2} + \beta_3 x_{i3} + \beta_4 x_{i4} + \varepsilon_i \end{aligned}$$

to predict the 1988 sales price (y_0) of a residence that has 1700 square feet ($x_{01} = 1700$), seven rooms ($x_{02} = 7$), and three bedrooms ($x_{03} = 3$) and is 15 years old ($x_{04} = 15$). It can be verified that the combination ($x_{01} = 1700$, $x_{02} = 7$, $x_{03} = 3$, and $x_{04} = 15$) is in the experimental region (the range of the combinations of the observed values of x_1, x_2, x_3, and x_4 in Table 8.2). Therefore by using the least squares point estimates of β_0, β_1, β_2, β_3, and β_4 calculated in Example 8.5, it follows that

$$\begin{aligned} \hat{y}_0 &= b_0 + b_1 x_{01} + b_2 x_{02} + b_3 x_{03} + b_4 x_{04} \\ &= 10.36762 + .05001(1700) + 6.32178(7) - 11.10316(3) - .43186(15) \\ &= 99.849 \qquad \text{(that is, \$99,849)} \end{aligned}$$

Here, $\hat{y}_0$ is a point estimate of μ_0, the mean selling price of all residences that have 1700 square feet, seven rooms, and three bedrooms and are 15 years old. Furthermore, $\hat{y}_0$ is also the point prediction of the individual selling price $y_0 = \mu_0 + \varepsilon_0$ of one such residence.

Example 8.9 Consider Figure 8.12, which presents the SAS output of a regression analysis of the Fresh demand data in Table 8.3 by using the model

$$\begin{aligned} y &= \mu + \varepsilon \\ &= \beta_0 + \beta_1 x_4 + \beta_2 x_3 + \beta_3 x_3^2 + \varepsilon \end{aligned}$$

Note that x_3^2 is denoted as X3SQ on the output. Also note that in addition to presenting the least squares point estimates of β_0, β_1, β_2, and β_3 the SAS output presents other quantities that we will discuss as we proceed through this chapter. Although the above model has been developed by using cross-sectional data, it might be appropriate to use this model to predict demand for Fresh in a sales region for a future quarter. To do this, the regression relationship that we have developed by using the data in Table 8.3 must apply to the future quarter and particular sales region. Therefore there must be no trend, seasonal, or other time-related influences affecting quarterly demand for Fresh in the sales regions. We will assume that this is the case. To predict future demands, Enterprise Industries will have to know future values of $x_4 = x_2 - x_1$ and x_3. Of course, the

FIGURE 8.12 **SAS output of a regression analysis of the Fresh demand data in Table 8.3 using the model $y = \beta_0 + \beta_1 x_4 + \beta_2 x_3 + \beta_3 x_3^2 + \varepsilon$**

SAS

ANALYSIS OF VARIANCE

SOURCE	DF	SUM OF SQUARES	MEAN SQUARE	F VALUE	PROB>F
MODEL	3	12.18531809[a]	4.06177270[p]	82.941[q]	0.0001[r]
ERROR	26	1.27326857[b]	0.04897187[f]		
C TOTAL	29	13.45858667[c]			

ROOT MSE	0.2212959[d]	R-SQUARE	0.9054[h]
DEP MEAN	8.382667	ADJ R-SQ	0.8945[i]
C.V.	2.639922		

PARAMETER ESTIMATES

VARIABLE	DF	PARAMETER[e] ESTIMATE	STANDARD[g] ERROR	T FOR H0: PARAMETER=0[j]	PROB > |T|[k]
INTERCEP	1	17.32436855	5.64145865	3.071	0.0050
X4	1	1.30698873	0.30361008	4.305	0.0002
X3	1	-3.69558671	1.85026611	-1.997	0.0564
X3SQ	1	0.34861167	0.15117669	2.306	0.0293

OBS	ACTUAL	PREDICT VALUE	STD ERR PREDICT	LOWER95% MEAN	UPPER95% MEAN	LOWER95% PREDICT	UPPER95% PREDICT	RESIDUAL
1	7.3800	7.4788	0.0829	7.3084	7.6492	6.9930	7.9645	-0.0988
2	8.5100	8.5895	0.0487	8.4895	8.6896	8.1238	9.0553	-0.0795
3	9.5200	9.6395	0.1167	9.3996	9.8793	9.1252	10.1537	-0.1195
4	7.5000	7.5441	0.0838	7.3719	7.7164	7.0578	8.0305	-0.0441
5	9.3300	8.8640	0.0833	8.6927	9.0352	8.3779	9.3500	0.4660
6	8.2800	8.2933	0.0581	8.1738	8.4128	7.8230	8.7636	-0.0133
7	8.7500	8.4588	0.0626	8.3302	8.5874	7.9861	8.9315	0.2912
8	7.8700	7.5965	0.1285	7.3323	7.8607	7.0705	8.1225	0.2735
9	7.1000	7.3351	0.1312	7.0654	7.6048	6.8063	7.8639	-0.2351
10	8.0000	7.8969	0.0792	7.7341	8.0598	7.4138	8.3801	0.1031
11	7.8900	8.2933	0.0581	8.1738	8.4128	7.8230	8.7636	-0.4033
12	8.1500	7.9753	0.0667	7.8382	8.1124	7.5002	8.4504	0.1747
13	9.1000	9.0600	0.0649	8.9267	9.1934	8.5860	9.5341	0.0400
14	8.8600	9.0104	0.0588	8.8894	9.1313	8.5397	9.4810	-0.1504
15	8.9000	8.7716	0.0493	8.6702	8.8731	8.3056	9.2377	0.1284
16	8.8700	8.7063	0.0479	8.6078	8.8047	8.2409	9.1717	0.1637
17	9.2600	9.3127	0.0820	9.1441	9.4813	8.8276	9.7978	-0.0527
18	9.0000	9.1907	0.0685	9.0499	9.3315	8.7146	9.6669	-0.1907
19	8.7500	8.8370	0.0551	8.7237	8.9502	8.3682	9.3057	-0.0870
20	7.9500	7.9665	0.0878	7.7861	8.1470	7.4772	8.4559	-0.0165
21	7.6500	7.7792	0.0770	7.6209	7.9376	7.2976	8.2609	-0.1292
22	7.2700	7.5702	0.0759	7.4142	7.7262	7.0893	8.0511	-0.3002
23	8.0000	8.2933	0.0581	8.1738	8.4128	7.8230	8.7636	-0.2933
24	8.5000	8.6679	0.1176	8.4263	8.9096	8.1528	9.1830	-0.1679
25	8.7500	8.9677	0.0750	8.8136	9.1218	8.4874	9.4479	-0.2177
26	9.2100	9.0984	0.1002	8.8923	9.3044	8.5990	9.5977	0.1116
27	8.2700	7.9665	0.0878	7.7861	8.1470	7.4772	8.4559	0.3035
28	7.6700	7.6007	0.0678	7.4613	7.7401	7.1250	8.0765	0.0693
29	7.9300	7.6826	0.0701	7.5385	7.8267	7.2055	8.1598	0.2474
30	9.2600	9.0330	0.0872	8.8538	9.2123	8.5441	9.5219	0.2270
31		8.2933[l]	0.0581[m]	8.1738	8.4128	7.8230	8.7636	

(LOWER95%/UPPER95% MEAN: 95% confidence interval; LOWER95%/UPPER95% PREDICT: 95% prediction interval)

SUM OF RESIDUALS	-4.19664E-14
SUM OF SQUARED RESIDUALS	1.273269[n]
PREDICTED RESID SS (PRESS)	1.773602

[a] *Explained variation* = SS_{model}; [b] *SSE*; [c] *Total variation*; [d] *s = standard error*; [e] b_j: b_0, b_1, b_2, b_3; [f] $MSE = s^2$; [g] $s\sqrt{c_{jj}}$; [h] R^2; [i] $\bar{R}^2$; [j] *t*; [k] *Prob-value for t*; [l] *Point prediction* $\hat{y}_0$; [m] $s\sqrt{h_{00}}$; [n] *SSE*; [p] MS_{model}; [q] *F(model)*; [r] *Prob-value for F(model)*.

company can set x_1 (its price for Fresh) and x_3 (its advertising expenditure) for future quarters. Also, it feels that by examining the prices of competitors' similar products immediately prior to a future quarter it can very accurately predict x_2 (the average industry price for competitors' similar detergents) for a future quarter. Furthermore, the company can react to any change in competitors' price to maintain any desired price difference $x_4 = x_2 - x_1$.

Suppose that Enterprise Industries will maintain a price difference of $x_{04} = .10$ and will spend $x_{03} = 6.80$ (that is, \$680,000) on advertising in a future quarter in a particular sales region. Since it can be verified that the combination ($x_{04} = .10$, $x_{03} = 6.80$) is in the experimental region defined by the data in Table 8.3, it is reasonable to compute

$$\begin{aligned}\hat{y}_0 &= b_0 + b_1 x_{04} + b_2 x_{03} + b_3 x_{03}^2 \\ &= 17.3244 + 1.3070(.10) - 3.6956(6.8) + .3486(6.8)^2 \\ &= 8.2933 \qquad \text{(that is, 829,330 bottles)}\end{aligned}$$

This quantity is a point estimate of μ_0, the mean of all possible demands when the price difference is $x_{04} = .10$ and the advertising expenditure is $x_{03} = 6.80$, and is a point prediction of $y_0 = \mu_0 + \varepsilon_0$, the actual demand in the future quarter in the sales region.

8.4

THE REGRESSION ASSUMPTIONS, THE STANDARD ERROR, AND RESIDUAL ANALYSIS

THE REGRESSION ASSUMPTIONS

The validity of all of the regression results in this chapter depends upon the fundamental assumption that the linear equation

$$\mu = \beta_0 + \beta_1 x_1 + \beta_2 x_2 + \cdots + \beta_p x_p$$

appropriately relates μ, the mean value of the dependent variable, to the independent variables $x_1, x_2, \ldots, x_p$ being utilized. For example, if the true linear equation relating μ to x_1 and x_2 in the fuel consumption problem were

$$\mu = \beta_0 + \beta_1 x_1 + \beta_2 x_2 + \beta_3 x_1^2 + \beta_4 x_1 x_2$$

but we mistakenly used the equation

$$\mu = \beta_0 + \beta_1 x_1 + \beta_2 x_2$$

then the results obtained by using the regression model

$$\begin{aligned}y &= \mu + \varepsilon \\ &= \beta_0 + \beta_1 x_1 + \beta_2 x_2 + \varepsilon\end{aligned}$$

would not be valid. Note, however, that we will subsequently present statistical evidence showing that the latter linear equation above is in fact appropriate.

In addition to the above assumption the validity of the formulas for the confidence intervals, prediction intervals, and statistical hypothesis tests to be presented in subsequent sections of this chapter depends on three assumptions that we refer to as the *inference assumptions*. These assumptions can be stated as follows.

Inference Assumption 1 (Constant Variance)

For any combination of values $(x_{i1}, x_{i2}, \ldots, x_{ip})$ of the independent variables x_1, $x_2, \ldots, x_p$ the corresponding population of potential values of the dependent variable has a variance σ^2 that does not depend on the combination of values $(x_{i1}, x_{i2}, \ldots, x_{ip})$ of $x_1, x_2, \ldots, x_p$. That is, the different populations of potential values of the dependent variable corresponding to different combinations of values of $x_1, x_2, \ldots, x_p$ have *equal variances.*

Said equivalently, the different populations of potential error terms corresponding to different combinations of values of $x_1, x_2, \ldots, x_p$ have *equal variances.*

Inference Assumption 2 (Independence)

Any one value of the dependent variable y is *statistically independent* of any other value of y.

Said equivalently, any one value of the error term ε is *statistically independent* of any other value of ε.

Inference Assumption 3 (Normal Populations)

For any combination of values $(x_{i1}, x_{i2}, \ldots, x_{ip})$ of the independent variables x_1, $x_2, \ldots, x_p$ the corresponding population of potential values of the dependent variable has a *normal distribution.*

Said equivalently, for any combination of values $(x_{i1}, x_{i2}, \ldots, x_{ip})$ of the independent variables $x_1, x_2, \ldots, x_p$ the corresponding population of potential error terms has a *normal distribution.*

THE MEAN SQUARE ERROR AND STANDARD ERROR

We summarize below how to calculate point estimates of σ^2 and σ.

The Mean Square Error and Standard Error

Suppose that the linear regression model

$$\begin{aligned} y_i &= \mu_i + \varepsilon_i \\ &= \beta_0 + \beta_1 x_{i1} + \beta_2 x_{i2} + \cdots + \beta_p x_{ip} + \varepsilon_i \end{aligned}$$

has k parameters $\beta_0, \beta_1, \beta_2, \ldots, \beta_p$ and that $b_0, b_1, b_2, \ldots, b_p$ are the least squares point estimates of these parameters.

If the inference assumptions are satisfied,

1. A point estimate of σ^2 is the *mean square error*

$$s^2 = \frac{SSE}{n - k}$$

2. A point estimate of σ is the *standard error*

$$s = \sqrt{s^2} = \sqrt{\frac{SSE}{n - k}}$$

Furthermore, the sum of squared residuals

$$SSE = \sum_{i=1}^{n} (y_i - \hat{y}_i)^2$$

where

$$\hat{y}_i = b_0 + b_1 x_{i1} + b_2 x_{i2} + \cdots + b_p x_{ip}$$

can be calculated by the alternative formula

$$SSE = \sum_{i=1}^{n} y_i^2 - \mathbf{b}'\mathbf{X}'\mathbf{y}$$

Here, $\mathbf{b}' = [b_0\ b_1\ b_2\ \ldots\ b_p]$ is a row vector (the transpose of $\mathbf{b}$) containing the least squares point estimates, and $\mathbf{X}'\mathbf{y}$ is the column vector used in calculating the least squares point estimates by the formula $\mathbf{b} = (\mathbf{X}'\mathbf{X})^{-1}\mathbf{X}'\mathbf{y}$.

Example 8.10 For the fuel consumption model

$$y = \beta_0 + \beta_1 x_1 + \beta_2 x_2 + \varepsilon$$

we have calculated *SSE* to be .674 (see Table 8.4). Also, from Example 8.4, we have computed

$$\mathbf{X}'\mathbf{y} = \begin{bmatrix} 81.7 \\ 3413.11 \\ 1157.40 \end{bmatrix}$$

It follows that

$$\begin{aligned} \mathbf{b}'\mathbf{X}'\mathbf{y} &= [13.109\ \ -.0900\ \ .0825] \begin{bmatrix} 81.7 \\ 3413.11 \\ 1157.40 \end{bmatrix} \\ &= 13.109(81.7) + (-.0900)(3413.11) + (.0825)(1157.40) \\ &= 859.236 \end{aligned}$$

Furthermore, the eight observed fuel consumptions (see Table 8.1) can be used to calculate

$$\begin{aligned}\sum_{i=1}^{8} y_i^2 &= y_1^2 + y_2^2 + \cdots + y_8^2 \\ &= (12.4)^2 + (11.7)^2 + \cdots + (7.5)^2 = 859.91\end{aligned}$$

Therefore SSE can be calculated in the following alternative fashion:

$$\begin{aligned}SSE &= \sum_{i=1}^{8} y_i^2 - \mathbf{b}'\mathbf{X}'\mathbf{y} \\ &= 859.91 - 859.236 \\ &= .674\end{aligned}$$

Since the above fuel consumption model utilizes $k = 3$ parameters (β_0, β_1, and β_2), a point estimate of σ^2 is the *mean square error*

$$s^2 = \frac{SSE}{n-k} = \frac{.674}{8-3} = \frac{.674}{5} = .1348$$

and a point estimate of σ is the *standard error*

$$s = \sqrt{s^2} = \sqrt{.1348} = .3671$$

Example 8.11 The SAS output in Figure 8.12 gives the residuals from the Fresh demand model

$$y_i = \beta_0 + \beta_1 x_{i4} + \beta_2 x_{i3} + \beta_3 x_{i3}^2 + \varepsilon_i$$

Here, the ith residual (for $i = 1, 2, \ldots, 30$) is calculated as

$$\begin{aligned}e_i &= y_i - \hat{y}_i \\ &= (y_i - (b_0 + b_1 x_{i4} + b_2 x_{i3} + b_3 x_{i3}^2)) \\ &= (y_i - (17.3244 + 1.3070x_{i4} - 3.6956x_{i3} + .3486x_{i3}^2))\end{aligned}$$

Figure 8.12 also presents the sum of squared residuals, mean square error, and standard error:

$$SSE = 1.2733$$

$$s^2 = \frac{SSE}{n-k} = \frac{1.2733}{30-4} = .0490$$

$$s = \sqrt{.0490} = .2213$$

RESIDUAL ANALYSIS

To construct a residual plot, we compute the residual for each observation. For $y_1, y_2, \ldots, y_n$ we compute

$$e_i = y_i - \hat{y}_i = y_i - (b_0 + b_1 x_{i1} + b_2 x_{i2} + \cdots + b_p x_{ip})$$

where $b_0, b_1, b_2, \ldots, b_p$ are the least squares point estimates. The calculated residuals, denoted $e_1, e_2, \ldots, e_n$, are then plotted against

1. Values of each of the independent variables in the model.
2. Values of $\hat{y}$, the predicted value of the dependent variable.
3. The time order in which the historical data have been observed.

As discussed in Chapter 6, a residual plot that fans out or funnels in suggests that the constant variance assumption might be violated, and a cyclical or alternating residual plot against time indicates that the independence assumption might be violated. On the other hand, residual plots that have a random, horizontal band appearance indicate that these assumptions probably approximately hold. In addition, the Durbin-Watson statistic can be used to test for positive or negative first-order autocorrelation, and histograms, stem-and-leaf diagrams, and normal plots of the residuals can be used to check the normality assumption.

Example 8.12 The SAS output in Figure 8.12 gives the residuals from the Fresh demand model

$$y = \beta_0 + \beta_1 x_4 + \beta_2 x_3 + \beta_3 x_3^2 + \varepsilon$$

Examining Figure 8.13, which presents the SAS output of plots of these residuals

FIGURE 8.13 **SAS output of residual plots for the Fresh demand model $y = \beta_0 + \beta_1 x_4 + \beta_2 x_3 + \beta_3 x_3^2 + \varepsilon$**

(a) Residuals versus x_4

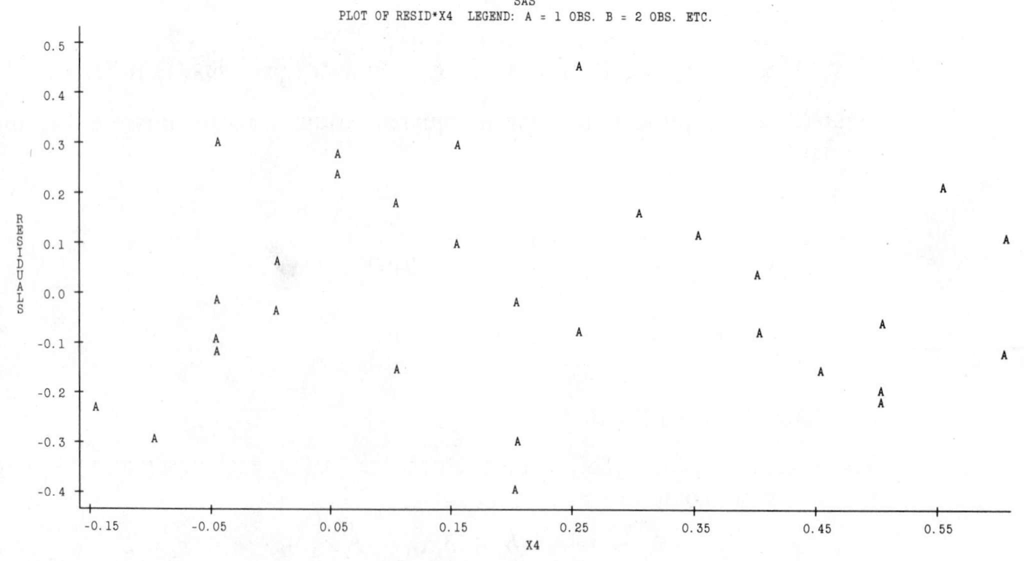

(b) Residuals versus x_3

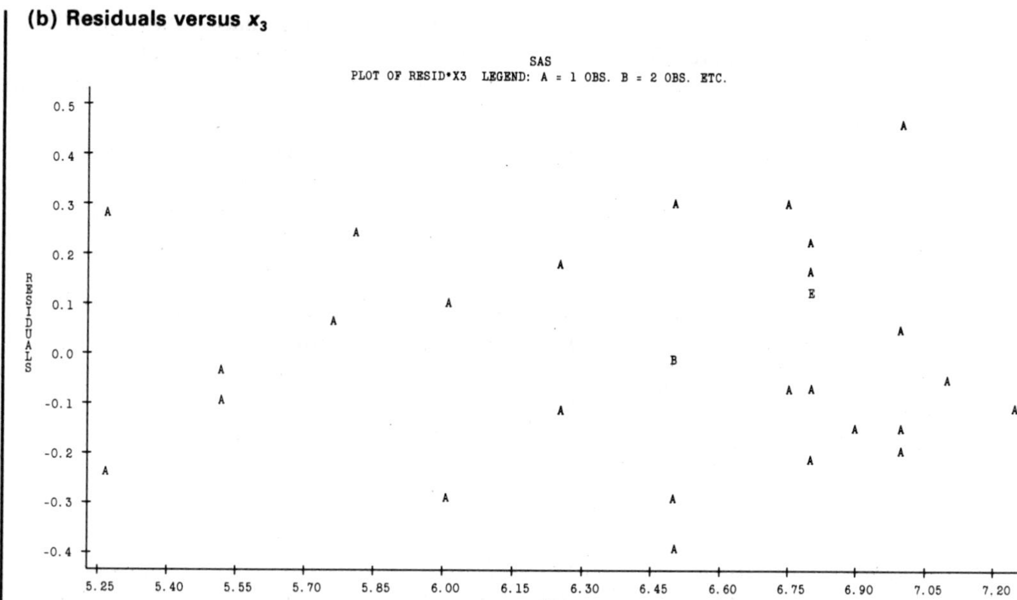

(c) Residuals versus $\hat{y}$

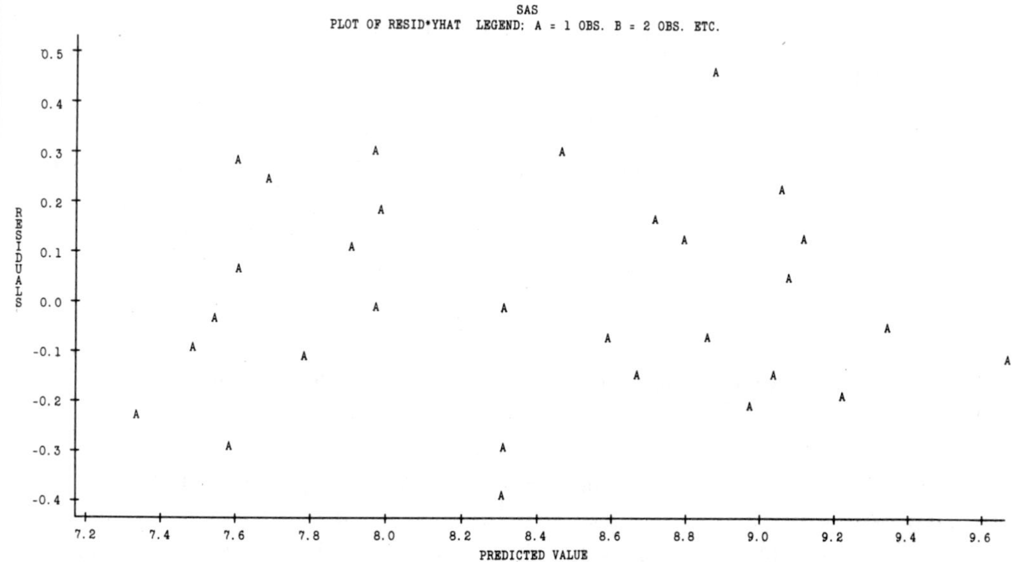

against x_4, x_3, and $\hat{y}$, we see that each of the plots has a horizontal band appearance. Therefore it is reasonable to conclude that the constant variance assumption approximately holds. In addition, a normal plot of the residuals (not shown here) indicates that the normality assumption approximately holds. Since the Fresh demand data in Table 8.4 is cross-sectional data, we do not need to test for autocorrelation.

Finally, we note that residual analysis of the fuel consumption model

$$y_i = \beta_0 + \beta_1 x_{i1} + \beta_2 x_{i2} + \varepsilon_i$$

(including a plot of the residuals against the time order in which the fuel consumption data has been observed) and residual analysis of the residential sales model

$$y_i = \beta_0 + \beta_1 x_{i1} + \beta_2 x_{i2} + \beta_3 x_{i3} + \beta_4 x_{i4} + \varepsilon_i$$

do not indicate any serious violations of the regression assumptions.

8.5 MULTIPLE COEFFICIENTS OF DETERMINATION AND CORRELATION

We now consider a measure of overall model utility.

The Multiple Coefficient of Determination, R^2

For the regression model

$$y = \beta_0 + \beta_1 x_1 + \beta_2 x_2 + \cdots + \beta_p x_p + \varepsilon$$

let

$$\hat{y}_i = b_0 + b_1 x_{i1} + b_2 x_{i2} + \cdots + b_p x_{ip}$$

be the point prediction of y_i, the ith observed value of the dependent variable. Then,

1. *Total variation* $= \sum_{i=1}^{n} (y_i - \bar{y})^2 = \sum_{i=1}^{n} y_i^2 - n\bar{y}^2$

2. *Explained variation* $= \sum_{i=1}^{n} (\hat{y}_i - \bar{y})^2 = \mathbf{b}'\mathbf{X}'\mathbf{y} - n\bar{y}^2$

3. *Unexplained variation* $= \sum_{i=1}^{n} (y_i - \hat{y}_i)^2 = \sum_{i=1}^{n} y_i^2 - \mathbf{b}'\mathbf{X}'\mathbf{y}$

4. Total variation = Explained variation + Unexplained variation

5. *Multiple coefficient of determination*

$$R^2 = \frac{\text{Explained variation}}{\text{Total variation}}$$
$$= 1 - \frac{\text{Unexplained variation}}{\text{Total variation}}$$

6. R^2 is the proportion of the total variation in the n observed values of the dependent variable that is explained by the overall regression model.
7. *Multiple correlation coefficient* $= R = \sqrt{R^2}$

Example 8.13 Consider the fuel consumption model

$$y = \beta_0 + \beta_1 x_1 + \beta_2 x_2 + \varepsilon$$

Using the fuel consumption data, we previously made the following calculations:

$$\sum_{i=1}^{8} y_i^2 = 859.91 \qquad \mathbf{b'X'y} = 859.236 \qquad \bar{y} = \frac{\sum_{i=1}^{8} y_i}{8} = 10.21$$

$$\text{Unexplained variation} = SSE = \sum_{i=1}^{8} (y_i - \hat{y}_i)^2 = \sum_{i=1}^{8} y_i^2 - \mathbf{b'X'y}$$
$$= 859.91 - 859.236 = .674$$

We can calculate the total variation to be

$$\text{Total variation} = \sum_{i=1}^{8} (y_i - \bar{y})^2 = \sum_{i=1}^{8} y_i^2 - 8\bar{y}^2$$
$$= 859.91 - 8(10.21)^2 = 25.55$$

Moreover, we can calculate the explained variation by either of the following two methods:

$$\text{Explained variation} = \text{Total variation} - \text{Unexplained variation}$$
$$= 25.55 - .674 = 24.876$$

or

$$\text{Explained variation} = \sum_{i=1}^{8} (\hat{y}_i - \bar{y})^2$$
$$= \mathbf{b'X'y} - 8\bar{y}^2$$
$$= 859.236 - 8(10.21)^2 = 24.876$$

Next, we can calculate the multiple coefficient of determination by either of the following two methods:

$$R^2 = \frac{\text{Explained variation}}{\text{Total variation}} = \frac{24.876}{25.55} = .974$$

or

$$R^2 = 1 - \frac{\text{Unexplained variation}}{\text{Total variation}} = 1 - \frac{.674}{25.55} = .974$$

This value of R^2 says that the above regression model explains 97.4% of the total variation in the eight observed fuel consumptions. Also, an R^2 of .974 implies that the multiple correlation coefficient is

$$R = \sqrt{R^2} = \sqrt{.974} = .9869$$

In Figure 8.14 we present the SAS output of the *analysis of variance table* summarizing the total variation, the explained variation, the unexplained variation (SSE), and R^2 for this model. Some of the other quantities in this figure, such as MS_{model} and F(model), will be explained later.

As two other examples, Figure 8.12 tells us that $R^2 = .9054$ for the Fresh demand model

$$y = \beta_0 + \beta_1 x_4 + \beta_2 x_3 + \beta_3 x_3^2 + \varepsilon$$

and R^2 can be calculated to be .7257 for the residential sales model

$$y = \beta_0 + \beta_1 x_1 + \beta_2 x_2 + \beta_3 x_3 + \beta_4 x_4 + \varepsilon$$

As a final example, recall that in Table 4.4 we presented the sales performance data for 25 randomly selected sales territory representatives. This data set is part of a larger data set analyzed in "An Analytical Approach for Evaluating Sales Territory Performance" by David W. Cravens, Robert B. Woodruff, and Joe C. Stomper, *Journal of Marketing* (Vol. 36, January 1972, pp. 31–37). The complete

FIGURE 8.14 **SAS output of the analysis of variance table for the two-variable fuel consumption model**

```
DEP VARIABLE:   Y
                          SUM OF            MEAN
SOURCE      DF          SQUARES          SQUARE     F VALUE          PROB>F

MODEL        2       24.875018a      12.437509e      92.303g          0.0001i
ERROR        5        0.673732b       0.134746f
C TOTAL      7       25.548750c

    ROOT MSE          0.367078d      R-SQUARE             0.9736h
    DEP MEAN         10.212500       ADJ R-SQ             0.9631
    C.V.              3.594401
```

[a] Explained variation = SS_{model} [b] SSE [c] Total variation
[d] s = standard error [e] MS_{model} [f] $MSE = s^2$
[g] F(model) [h] R^2 [i] Prob-value for F(model)

data set is presented in Table 8.5. Here,

y = performance of the sales territory representative, measured by aggregate sales, in units. Recall from Chapter 5 that each representative had the sole responsibility for one sales territory and that the sales values were coded for purposes of respondent confidentiality

x_1 = number of months the representative has been employed by the company

x_2 = unit sales of the company's product and competing products in the sales territory

x_3 = dollar advertising expenditure in the territory

x_4 = weighted average of the company's market share in the territory for the previous four years

x_5 = change in the company's market share in the territory over the previous four years

x_6 = the total number of accounts assigned to the representative

TABLE 8.5 **Data for sales territory performance study**

Sales, y	*Time with company,* x_1	*Market potential,* x_2	*Advertising,* x_3	*Market share,* x_4	*Market share change,* x_5	*Accounts,* x_6	*Workload,* x_7	*Rating,* x_8
3,669.88	43.10	74,065.11	4,582.88	2.51	0.34	74.86	15.05	4.9
3,473.95	108.13	58,117.30	5,539.78	5.51	0.15	107.32	19.97	5.1
2,295.10	13.82	21,118.49	2,950.38	10.91	−0.72	96.75	17.34	2.9
4,675.56	186.18	68,521.27	2,243.07	8.27	0.17	195.12	13.40	3.4
6,125.96	161.79	57,805.11	7,747.08	9.15	0.50	180.44	17.64	4.6
2,134.94	8.94	37,806.94	402.44	5.51	0.15	104.88	16.22	4.5
5,031.66	365.04	50,935.26	3,140.62	8.54	0.55	256.10	18.80	4.6
3,367.45	220.32	35,602.08	2,086.16	7.07	−0.49	126.83	19.86	2.3
6,519.45	127.64	46,176.77	8,846.25	12.54	1.24	203.25	17.42	4.9
4,876.37	105.69	42,053.24	5,673.11	8.85	0.31	119.51	21.41	2.8
2,468.27	57.72	36,829.71	2,761.76	5.38	0.37	116.26	16.32	3.1
2,533.31	23.58	33,612.67	1,991.85	5.43	−0.65	142.28	14.51	4.2
2,408.11	13.82	21,412.79	1,971.52	8.48	0.64	89.43	19.35	4.3
2,337.38	13.82	20,416.87	1,737.38	7.80	1.01	84.55	20.02	4.2
4,586.95	86.99	36,272.00	10,694.20	10.34	0.11	119.51	15.26	5.5
2,729.24	165.85	23,093.26	8,618.61	5.15	0.04	80.49	15.87	3.6
3,289.40	116.26	26,878.59	7,747.89	6.64	0.68	136.58	7.81	3.4
2,800.78	42.28	39,571.96	4,565.81	5.45	0.66	78.86	16.00	4.2
3,264.20	52.84	51,866.15	6,022.70	6.31	−0.10	136.58	17.44	3.6
3,453.62	165.04	58,749.82	3,721.10	6.35	−0.03	138.21	17.98	3.1
1,741.45	10.57	23,990.82	860.97	7.37	−1.63	75.61	20.99	1.6
2,035.75	13.82	25,694.86	3,571.51	8.39	−0.43	102.44	21.66	3.4
1,578.00	8.13	23,736.35	2,845.50	5.15	0.04	76.42	21.46	2.7
4,167.44	58.54	34,314.29	5,060.11	12.88	0.22	136.58	24.78	2.8
2,799.97	21.14	22,809.53	3,552.00	9.14	−0.74	88.62	24.96	3.9

x_7 = average workload per account, measured by using a weighting based on the sizes of the orders by the accounts and other workload related criteria

x_8 = an aggregate rating on eight dimensions of the representative's performance, made by a sales manager and expressed on a 1–7 scale

One point of the analysis was to develop a regression model that provided more reasonable predictions of sales territory performance than were provided by the company's sales quota system. One possible model uses all eight independent variables:

$$y = \beta_0 + \beta_1 x_1 + \beta_2 x_2 + \beta_3 x_3 + \beta_4 x_4 + \beta_5 x_5 + \beta_6 x_6 + \beta_7 x_7 + \beta_8 x_8 + \varepsilon$$

It can be shown that $R^2 = .9220$ for this model. This says that the model explains 92.2% of the total variation in the 25 sales values. However, we will find in later sections that there are better models describing the data in Table 8.5.

8.6

AN *F*-TEST FOR THE OVERALL MODEL

In this section we present an F-test related to the utility of the overall regression model

$$y = \beta_0 + \beta_1 x_1 + \cdots + \beta_p x_p + \varepsilon$$

Specifically, we test the null hypothesis

$$H_0: \beta_1 = \beta_2 = \cdots = \beta_p = 0$$

which says that none of the independent variables $x_1, x_2, \ldots, x_p$ affects y, versus the alternative hypothesis

$$H_1: \text{At least one of } \beta_1, \beta_2, \ldots, \beta_p \text{ does not equal zero}$$

which says that at least one of the independent variables $x_1, x_2, \ldots, x_p$ affects y. If we can reject H_0 in favor of H_1 with a small probability of a Type I error, then it is reasonable to conclude that at least one of $x_1, x_2, \ldots, x_p$ *significantly* affects y. In this case we should use the techniques of Section 8.7 and other procedures to determine which of these variables significantly affect y. Recalling that k denotes the number of parameters in the overall regression model, and denoting the *explained variation* (also called the *sum of squares due to the overall model*) by SS_{model}, we summarize the F-test of H_0 versus H_1.

Testing $H_0: \beta_1 = \beta_2 = \cdots = \beta_p = 0$ Versus H_1: At Least One of $\beta_1, \beta_2, \ldots, \beta_p$ Does Not Equal Zero

Assume the inference assumptions are satisfied, and define the *overall F-statistic* to be

$$F(\text{model}) = \frac{MS_{\text{model}}}{MSE} = \frac{\text{Explained variation}/(k-1)}{SSE/(n-k)}$$

where

$$MS_{\text{model}} = \frac{SS_{\text{model}}}{k-1} = \frac{\text{Explained variation}}{k-1}$$

$$MSE = \frac{SSE}{n-k} = s^2$$

Also, define the prob-value to be the area to the right of F(model) under the curve of the F-distribution having $k - 1$ and $n - k$ degrees of freedom. Then, we can reject H_0 in favor of H_1 by setting the probability of a Type I error equal to α if and only if either of the following equivalent conditions holds:

1. $F(\text{model}) > F_{[\alpha]}^{(k-1,n-k)}$, where $F_{[\alpha]}^{(k-1,n-k)}$ is the point on the scale of the F-distribution having $k - 1$ and $n - k$ degrees of freedom so that the area under this curve to the right of $F_{[\alpha]}^{(k-1,n-k)}$ is α.
2. Prob-value $< \alpha$.

Note that condition (1) is intuitively reasonable because a large value of F(model) would be caused by an explained variation that is large relative to the unexplained variation (SSE). This would occur if at least one of the independent variables in the regression model significantly affects the dependent variable, which would imply that H_0 is false and that H_1 is true.

Example 8.14 Consider the Fresh Detergent demand model

$$y = \beta_0 + \beta_1 x_4 + \beta_2 x_3 + \beta_3 x_3^2 + \varepsilon$$

which has $k = 4$ parameters. The SAS output in Figure 8.12 tells us that, when we use this model to carry out a regression analysis of the data in Table 8.3, we obtain

$$SS_{\text{model}} = \text{Explained variation} = 12.18532$$

$$MS_{\text{model}} = \frac{SS_{\text{model}}}{k-1} = \frac{12.18532}{4-1} = 4.06177$$

$$SSE = \text{Unexplained variation} = 1.27327$$

$$s^2 = MSE = \frac{SSE}{n - k} = \frac{1.27327}{30 - 4} = 0.04897$$

$$F(\text{model}) = \frac{MS_{\text{model}}}{MSE} = \frac{4.06177}{.04897} = 82.941$$

Furthermore, the prob-value, which is the area to the right of 82.941 under the curve of the F-distribution having 3 and 26 degrees of freedom, can be computer calculated to be less than .0001. For the purposes of subsequent discussion we assume that the prob-value is equal to .0001. Suppose that we wish to use condition (1) to test

$$H_0 : \beta_1 = \beta_2 = \beta_3 = 0$$

versus

$$H_1 : \text{At least one of } \beta_1, \beta_2, \text{ or } \beta_3 \text{ does not equal zero}$$

by setting α equal to .05. Then we would use the rejection point

$$F_{[\alpha]}^{(k-1, n-k)} = F_{[.05]}^{(3,26)} = 2.98$$

Since $F(\text{model}) = 82.941 > 2.98 = F_{[.05]}^{(3,26)}$, we can reject H_0 in favor of H_1 by setting α equal to .05. Alternatively, since prob-value $=$ .0001 is less than .05 and .01, it follows by condition (2) that we can reject H_0 in favor of H_1 by setting α equal to .05 or .01. Therefore we conclude that at least one of the independent variables x_4, x_3, and x_3^2 significantly affects demand for Fresh.

As another example, the prob-value for F(model) can be calculated to be less than or equal to .0001 for each of the other regression models discussed in this chapter—the fuel consumption model (see the SAS output in Figure 8.14), the residential sales model, and the sales territory performance model.

In the next section we discuss one way to help determine which independent variables are important in a given model.

8.7

STATISTICAL INFERENCE FOR β_j

Consider the regression model

$$y = \beta_0 + \beta_1 x_1 + \beta_2 x_2 + \cdots + \beta_j x_j + \cdots + \beta_p x_p + \varepsilon$$

Statistical inference procedures for a single parameter β_j in this model are similar to the inference procedures for β_0 and β_1 in the simple linear regression model. We now present a confidence interval for β_j and a hypothesis test of $H_0 : \beta_j = 0$, as well as the sampling results that provide the basis for their validity. In the following results we define (as will be illustrated in Example 8.15) c_{jj} to be the diagonal element of the matrix $(\mathbf{X}'\mathbf{X})^{-1}$ corresponding to β_j, when the rows and columns of $(\mathbf{X}'\mathbf{X})^{-1}$ are numbered as the parameters in the regression model are numbered.

A Confidence Interval for β_j

1. The *population of all possible least squares point estimates of* β_j
   a. has *mean* $\mu_{b_j} = \beta_j$
   b. has *variance* $\sigma^2_{b_j} = \sigma^2 c_{jj}$ (if inference assumptions 1 and 2 hold)
   c. has *standard deviation* $\sigma_{b_j} = \sigma\sqrt{c_{jj}}$ (if inference assumptions 1 and 2 hold)
   d. has a *normal distribution* (if inference assumptions 1, 2, and 3 hold)
2. A point estimate of $\sigma_{b_j} = \sigma\sqrt{c_{jj}}$ is $s_{b_j} = s\sqrt{c_{jj}}$, which is called the *standard error of the estimate* b_j.
3. If the inference assumptions hold, the population of all possible values of

   $$\frac{b_j - \beta_j}{s_{b_j}} = \frac{b_j - \beta_j}{s\sqrt{c_{jj}}}$$

   has a *t-distribution with* $n - k$ *degrees of freedom.*
4. If the inference assumptions hold, a *100(1 − α)% confidence interval for* β_j is

   $$[b_j \pm t^{(n-k)}_{[\alpha/2]}\, s\sqrt{c_{jj}}]$$

Testing $H_0: \beta_j = 0$ Versus $H_1: \beta_j \neq 0$

Define the test statistic

$$t = \frac{b_j}{s\sqrt{c_{jj}}}$$

Also, define the prob-value to be twice the area under the curve of the t-distribution having $n - k$ degrees of freedom to the right of $|t|$. If the inference assumptions are satisfied, we can reject $H_0: \beta_j = 0$ in favor of $H_1: \beta_j \neq 0$ by setting the probability of a Type I error equal to α if and only if either of the following equivalent conditions holds:

1. $|t| > t^{(n-k)}_{[\alpha/2]}$—that is, if $t > t^{(n-k)}_{[\alpha/2]}$ or $t < -t^{(n-k)}_{[\alpha/2]}$
2. Prob-value $< \alpha$

Note that

1. If we can reject $H_0: \beta_j = 0$ in favor of $H_1: \beta_j \neq 0$ by setting α equal to .05, we have *strong evidence* that the independent variable x_j is significantly related to the dependent variable y.

2. If we can reject $H_0: \beta_j = 0$ in favor of $H_1: \beta_j \neq 0$ by setting α equal to .01, we have *very strong evidence* that x_j is significantly related to y.
3. The smaller the value of α at which $H_0: \beta_j = 0$ can be rejected in favor of $H_1: \beta_j \neq 0$, the stronger is the evidence that x_j is significantly related to y.

Example 8.15 Consider the fuel consumption model

$$y_i = \beta_0 + \beta_1 x_{i1} + \beta_2 x_{i2} + \varepsilon_i$$

Recall that we have previously (in Example 8.4) calculated the least squares point estimates of the parameters β_0, β_1, and β_2 to be

$$\begin{bmatrix} b_0 \\ b_1 \\ b_2 \end{bmatrix} = \mathbf{b} = (\mathbf{X}'\mathbf{X})^{-1}\mathbf{X}'\mathbf{y} = \begin{bmatrix} 13.109 \\ -.0900 \\ .0825 \end{bmatrix}$$

and that we have previously (in Example 8.10) calculated the standard error, s, to be .3671. Now, numbering the rows and columns of $(\mathbf{X}'\mathbf{X})^{-1}$ (which was calculated in Example 8.4) as the parameters β_0, β_1, and β_2 are numbered, we have

$$(\mathbf{X}'\mathbf{X})^{-1} = \begin{array}{c} \text{row} \\ 0 \\ 1 \\ 2 \end{array} \begin{array}{c} \text{column} \\ \begin{array}{ccc} 0 & 1 & 2 \end{array} \\ \begin{bmatrix} 5.43405 & -.085930 & -.118856 \\ -.085930 & .00147070 & .00165094 \\ -.118856 & .00165094 & .00359276 \end{bmatrix} \end{array}$$

$$= \begin{bmatrix} c_{00} & & \\ & c_{11} & \\ & & c_{22} \end{bmatrix}$$

Thus,

1. The diagonal element of $(\mathbf{X}'\mathbf{X})^{-1}$ corresponding to β_0 is $c_{00} = 5.43405 \approx 5.434$.
2. The diagonal element of $(\mathbf{X}'\mathbf{X})^{-1}$ corresponding to β_1 is $c_{11} = .00147070 \approx .00147$.
3. The diagonal element of $(\mathbf{X}'\mathbf{X})^{-1}$ corresponding to β_2 is $c_{22} = .00359276 \approx .0036$.

Using these diagonal elements, the t statistics and related prob-values for testing $H_0: \beta_0 = 0$, $H_0: \beta_1 = 0$, and $H_0: \beta_2 = 0$ are calculated as summarized in Table 8.6.

TABLE 8.6 **The calculations and the SAS output of the t statistics and prob-values for testing $H_0: \beta_0 = 0$, $H_0: \beta_1 = 0$, and $H_0: \beta_2 = 0$ in the fuel consumption model $y = \beta_0 + \beta_1 x_1 + \beta_2 x_2 + \varepsilon$**

Independent variable	b_j	$s\sqrt{c_{jj}}$	$t = \dfrac{b_j}{s\sqrt{c_{jj}}}$	*Prob-value*
Intercept	$b_0 = 13.11$	$s\sqrt{c_{00}} = .8557$	$t = \dfrac{b_0}{s\sqrt{c_{00}}} = 15.32$	.0001
x_1	$b_1 = -.0900$	$s\sqrt{c_{11}} = .0141$	$t = \dfrac{b_1}{s\sqrt{c_{11}}} = -6.39$	.0014
x_2	$b_2 = .0825$	$s\sqrt{c_{22}} = .0220$	$t = \dfrac{b_2}{s\sqrt{c_{22}}} = 3.75$	.0133

VARIABLE	DF	PARAMETER ESTIMATE[a]	STANDARD ERROR[b]	T FOR HO: PARAMETER=0[c]	PROB > |T|[d]
INTERCEP	1	13.108737	0.855698	15.319	0.0001
X1	1	-0.090014	0.014077	-6.394	0.0014
X2	1	0.082495	0.022003	3.749	0.0133

[a] b_j: b_0, b_1, b_2 [b] $s\sqrt{c_{jj}}$ [c] t [d] Prob-value

If, for example, we wish to use condition (1) to determine whether we can reject $H_0: \beta_2 = 0$ in favor of $H_1: \beta_2 \neq 0$ by setting α equal to .05, we would use the rejection points

$$t_{[\alpha/2]}^{(n-k)} = t_{[.05/2]}^{(8-3)} = t_{[.025]}^{(5)} = 2.571 \quad \text{and} \quad -t_{[.025]}^{(5)} = -2.571$$

Table 8.6 tells us that the t statistic for testing $H_0: \beta_2 = 0$ is $t = 3.75$. Since $t = 3.75 > 2.571 = t_{[.025]}^{(5)}$, we can reject $H_0: \beta_2 = 0$ in favor of $H_1: \beta_2 \neq 0$ by setting α equal to .05. Table 8.6 also tells us that the prob-value for testing $H_0: \beta_2 = 0$ equals .0133. Since this prob-value is less than .05, less than .02, and not less than .01, it follows by condition (2) that we can reject $H_0: \beta_2 = 0$ in favor of $H_1: \beta_2 \neq 0$ by setting α equal to .05 or .02 but not by setting α equal to .01. Therefore we have strong evidence that x_2, the chill index, is significantly related to y, the weekly fuel consumption, in the above model. Furthermore, since the prob-values for testing $H_0: \beta_0 = 0$ and $H_0: \beta_1 = 0$ are less than .01, we can reject each of these hypotheses by setting α equal to .01. Thus we have very strong evidence that in the above model the intercept β_0 is important and that x_1, the average hourly temperature, is significantly related to y.

Example 8.16 Consider the residential sales model

$$y_i = \beta_0 + \beta_1 x_{i1} + \beta_2 x_{i2} + \beta_3 x_{i3} + \beta_4 x_{i4} + \varepsilon_i$$

The prob-values related to the importance of x_1, x_2, x_3, and x_4 in this model can be calculated to be .0001, .0152, .0635, and .0002, respectively. Therefore we have strong evidence that each of x_1, x_2, and x_4 is important in this model, and we have some evidence that x_3 is important. It can be further shown that $b_1 = .05001$ and $s\sqrt{c_{11}} = .008104$. This implies that a 95% confidence interval for β_1 is (since there are $n = 63$ observations in Table 8.2)

$$\begin{aligned}[b_1 \pm t_{[.025]}^{(63-5)} s\sqrt{c_{11}}] &= [.05001 \pm 1.96(.008104)] \\ &= [.03413, .06589]\end{aligned}$$

Therefore we are 95% confident that the mean selling price is between \$3413 and \$6589 higher for a residence that is 100 square feet larger than another residence when both residences are the same age and have the same number of rooms and the same number of bedrooms.

As will be further discussed in Chapter 10, in most multiple regression models the independent variables contribute somewhat overlapping information for describing the dependent variable. This phenomenon is called *multicollinearity*. In this situation the size of the t statistic measures the *additional importance* of the independent variable x_j *over and above* the combined importance of the other independent variables in the regression model. Therefore the t statistic and related prob-value must be used with caution in assessing the importance of an independent variable.

Example 8.17 Consider the Fresh demand model

$$y = \beta_0 + \beta_1 x_4 + \beta_2 x_2 + \beta_3 x_3^2 + \varepsilon$$

The SAS output in Figure 8.15(a), which is a portion of the output in Figure 8.12, tells us that the prob-values related to the importance of the intercept, x_4, and x_3^2 in this model are .005, .0002, and .0293, respectively. Therefore each of these model components seems important. The SAS output further tells us that the prob-value related to the importance of x_3 is .0564. Although this prob-value is slightly larger than .05, note that x_3 and x_3^2 contribute somewhat overlapping information in describing the effect of advertising expenditure on y. Therefore since using both x_3 and x_3^2 allows the model to describe the quadratic relationship between y and x_3, we will retain x_3 in the model.

Next, reconsider the plot of y versus x_4 in Figure 8.7. The slight curvature in this plot might suggest that we try the model

$$y = \beta_0 + \beta_1 x_4 + \beta_2 x_3 + \beta_3 x_3^2 + \beta_4 x_4^2 + \varepsilon$$

which includes the quadratic term x_4^2. Figure 8.15(b) presents the SAS output of the t statistics and related prob-values for this model. Since the prob-values related to testing $H_0: \beta_0 = 0$, $H_0: \beta_1 = 0$, $H_0: \beta_2 = 0$, and $H_0: \beta_3 = 0$ are all less than .05, we conclude that the intercept is important and that each of x_4, x_3, and x_3^2 is

FIGURE 8.15 **SAS output of the t statistics and related prob-values for two Fresh demand models**

(a) Model: $y = \beta_0 + \beta_1 x_4 + \beta_2 x_3 + \beta_3 x_3^2 + \varepsilon$

PARAMETER ESTIMATES

VARIABLE	DF	PARAMETER ESTIMATE	STANDARD ERROR	T FOR H0: PARAMETER=0	PROB > |T|
INTERCEP	1	17.32436855	5.64145865	3.071	0.0050
X4	1	1.30698873	0.30361008	4.305	0.0002
X3	1	-3.69558671	1.85026611	-1.997	0.0564
X3SQ	1	0.34861167	0.15117669	2.306	0.0293

(b) Model: $y = \beta_0 + \beta_1 x_4 + \beta_2 x_3 + \beta_3 x_3^2 + \beta_4 x_4^2 + \varepsilon$

PARAMETER ESTIMATES

VARIABLE	DF	PARAMETER ESTIMATE	STANDARD ERROR	T FOR H0: PARAMETER=0	PROB > |T|
INTERCEP	1	19.49645339	5.87103884	3.321	0.0028
X4	1	1.90607624	0.57847045	3.295	0.0030
X3	1	-4.37535740	1.91752616	-2.282	0.0313
X3SQ	1	0.40106908	0.15594774	2.572	0.0164
X4SQ	1	-1.19689638	0.98704893	-1.213	0.2366

significantly related to y. However, since the prob-value related to testing $H_0: \beta_4 = 0$ equals 0.2366, we cannot reject $H_0: \beta_4 = 0$ with α set equal to .05 (or even .10). We therefore conclude that inclusion of x_4^2 in the Fresh demand model is probably not warranted.

Example 8.18 The prob-values related to the importance of the eight independent variables in the sales territory performance model of Example 8.13 are as summarized below:

Variable	*Prob-value*
TIME	0.3134
POTEN	0.0003
ADV	0.0055
SHARE	0.0090
CHANGE	0.1390
ACCTS	0.2621
WORK	0.5649
RATE	0.9500

Since some of these independent variables may be contributing overlapping information, we must be cautious in using the prob-values to evaluate their importance.

In Chapter 11 we discuss other methods for evaluating the importance of the independent variables. We will find that we should not necessarily drop all independent variables having large prob-values from a final regression model.

If desired, the t statistic can be used to test one-sided hypotheses concerning the parameter β_j. This can be done as follows.

Testing $H_0: \beta_j = 0$ Versus $H_1: \beta_j > 0$ and Testing $H_0: \beta_j = 0$ Versus $H_1: \beta_j < 0$

To test $H_0: \beta_j = 0$, define the test statistic

$$t = \frac{b_j}{s\sqrt{c_{jj}}}$$

Then, if the inference assumptions are satisfied,

1. We can reject $H_0: \beta_j = 0$ in favor of $H_1: \beta_j > 0$ by setting the probability of a Type I error equal to α if and only if $t > t_{[\alpha]}^{(n-k)}$.
2. We can reject $H_0: \beta_j = 0$ in favor of $H_1: \beta_j < 0$ by setting the probability of a Type I error equal to α if and only if $t < -t_{[\alpha]}^{(n-k)}$.

8.8 CONFIDENCE INTERVALS AND PREDICTION INTERVALS

Recall that

$$\hat{y}_0 = b_0 + b_1 x_{01} + b_2 x_{02} + \cdots + b_p x_{0p}$$

is the point estimate of the mean value of the dependent variable

$$\mu_0 = \beta_0 + \beta_1 x_{01} + \beta_2 x_{02} + \cdots + \beta_p x_{0p}$$

and is the point prediction of the individual value of the dependent variable

$$\begin{aligned} y_0 &= \mu_0 + \varepsilon_0 \\ &= \beta_0 + \beta_1 x_{01} + \beta_2 x_{02} + \cdots + \beta_p x_{0p} + \varepsilon_0 \end{aligned}$$

To find a confidence interval for μ_0 and a prediction interval for y_0, define the *distance value* h_{00} by the equation

$$h_{00} = \mathbf{x}_0'(\mathbf{X}'\mathbf{X})^{-1}\mathbf{x}_0$$

Here, $\mathbf{x}_0'$ is a row vector containing the values of the independent variables for which we wish to estimate μ_0 or predict y_0. More specifically,

$$\mathbf{x}_0' = [1 \ x_{01} \ x_{02} \ \ldots \ x_{0p}]$$

which says that $\mathbf{x}_0'$ is a row vector containing the numbers multiplied by b_0, b_1, $b_2, \ldots, b_p$ in the above expression for $\hat{y}_0$.

We now present formulas for a confidence interval for μ_0 and a prediction interval for y_0, as well as the sampling results that provide the basis for their validity.

A Confidence Interval for μ_0

1. The *population of all possible point estimates of* μ_0
   a. has *mean* $\mu_{\hat{y}_0} = \mu_0$
   b. has *variance* $\sigma^2_{\hat{y}_0} = \sigma^2 h_{00}$ (if inference assumptions 1 and 2 hold)
   c. has *standard deviation* $\sigma_{\hat{y}_0} = \sigma\sqrt{h_{00}}$ (if inference assumptions 1 and 2 hold)
   d. has a *normal distribution* (if inference assumptions 1, 2, and 3 hold).
2. A point estimate of $\sigma_{\hat{y}_0} = \sigma\sqrt{h_{00}}$ is $s_{\hat{y}_0} = s\sqrt{h_{00}}$, which is called the *standard error of the estimate* $\hat{y}_0$.
3. If the inference assumptions hold, the population of all possible values of

$$\frac{\hat{y}_0 - \mu_0}{s_{\hat{y}_0}} = \frac{\hat{y}_0 - \mu_0}{s\sqrt{h_{00}}}$$

has a *t distribution with* $n - k$ *degrees of freedom.*

4. If the inference assumptions hold, a *100(1 − α)% confidence interval for* μ_0 is

$$[\hat{y}_0 \pm t_{[\alpha/2]}^{(n-k)} s\sqrt{h_{00}}] = [\hat{y}_0 \pm t_{[\alpha/2]}^{(n-k)} s\sqrt{\mathbf{x}_0'(\mathbf{X}'\mathbf{X})^{-1}\mathbf{x}_0}]$$

A Prediction Interval for y_0

1. The *population of all possible prediction errors*
   a. has *mean* $\mu_{(y_0 - \hat{y}_0)} = 0$
   b. has *variance* $\sigma^2_{(y_0 - \hat{y}_0)} = \sigma^2(1 + h_{00})$ (if inference assumptions 1 and 2 hold)
   c. has *standard deviation* $\sigma_{(y_0 - \hat{y}_0)} = \sigma\sqrt{1 + h_{00}}$ (if inference assumptions 1 and 2 hold)
   d. has a *normal distribution* (if inference assumptions 1, 2, and 3 hold)
2. A point estimate of $\sigma_{(y_0 - \hat{y}_0)} = \sigma\sqrt{1 + h_{00}}$ is $s_{(y_0 - \hat{y}_0)} = s\sqrt{1 + h_{00}}$, which is called the *standard error of the prediction error* $(y_0 - \hat{y}_0)$.
3. If the inference assumptions hold, the population of all possible values of

$$\frac{y_0 - \hat{y}_0}{s_{(y_0 - \hat{y}_0)}} = \frac{y_0 - \hat{y}_0}{s\sqrt{1 + h_{00}}}$$

has a *t-distribution with* $n - k$ *degrees of freedom.*

4. If the inference assumptions hold, a *100(1 − α)% prediction interval for* y_0 is

$$[\hat{y}_0 \pm t_{[\alpha/2]}^{(n-k)} s\sqrt{1 + h_{00}}] = [\hat{y}_0 \pm t_{[\alpha/2]}^{(n-k)} s\sqrt{1 + \mathbf{x}_0'(\mathbf{X}'\mathbf{X})^{-1}\mathbf{x}_0}]$$

Before considering an example, note that it can be shown that the farther the values $x_{01}, x_{02}, \ldots, x_{0p}$ of the independent variables $x_1, x_2, \ldots, x_p$ are from the center of the experimental region, that is, from the means

$$\bar{x}_1 = \frac{\sum_{i=1}^{n} x_{i1}}{n}, \quad \bar{x}_2 = \frac{\sum_{i=1}^{n} x_{i2}}{n}, \quad \ldots, \quad \bar{x}_p = \frac{\sum_{i=1}^{n} x_{ip}}{n}$$

then the larger is the *distance value*

$$h_{00} = \mathbf{x}_0'(\mathbf{X}'\mathbf{X})^{-1}\mathbf{x}_0$$

and thus the longer are the 100(1 − α)% confidence interval for μ_0 and the 100(1 − α)% prediction interval for y_0.

Example 8.19 In the fuel consumption problem, recall that a weather forecasting service has told us that the average hourly temperature in the future week will be $x_{01} = 40.0$ and the chill index in the future week will be $x_{02} = 10$. We saw in Example 8.7 that

$$\begin{aligned}\hat{y}_0 &= b_0 + b_1x_{01} + b_2x_{02} \\ &= 13.109 - .0900(40.0) + .0825(10) \\ &= 10.333 \text{ tons of coal}\end{aligned}$$

is the point estimate of the future mean fuel consumption

$$\begin{aligned}\mu_0 &= \beta_0 + \beta_1x_{01} + \beta_2x_{02} \\ &= \beta_0 + \beta_1(40.0) + \beta_2(10)\end{aligned}$$

and is the point prediction of the future individual fuel consumption

$$y_0 = \mu_0 + \varepsilon_0$$

To calculate the distance value,

$$h_{00} = \mathbf{x}_0'(\mathbf{X}'\mathbf{X})^{-1}\mathbf{x}_0$$

note that $\mathbf{x}_0'$ is a row vector containing the numbers multiplied by the least squares point estimates b_0, b_1, and b_2 in the point estimate (and prediction) $\hat{y}_0$. Since 1 is multiplied by b_0, $x_{01} = 40.0$ is multiplied by b_1, and $x_{02} = 10$ is multiplied by b_2, it follows that

$$\mathbf{x}_0' = [1 \quad x_{01} \quad x_{02}] = [1 \quad 40 \quad 10]$$

and

$$\mathbf{x}_0 = \begin{bmatrix} 1 \\ x_{01} \\ x_{02} \end{bmatrix} = \begin{bmatrix} 1 \\ 40 \\ 10 \end{bmatrix}$$

Hence, since we have previously calculated $(\mathbf{X}'\mathbf{X})^{-1}$ (see Example 8.4), it follows that

$$\begin{aligned} h_{00} &= \mathbf{x}_0'(\mathbf{X}'\mathbf{X})^{-1}\mathbf{x}_0 \\ &= [1 \quad 40 \quad 10]\begin{bmatrix} 5.43405 & -.085930 & -.118856 \\ -.085930 & .00147070 & .00165094 \\ -.118856 & .00165094 & .00359276 \end{bmatrix}\begin{bmatrix} 1 \\ 40 \\ 10 \end{bmatrix} \\ &= [.80828 \quad -.0105926 \quad -.0168908]\begin{bmatrix} 1 \\ 40 \\ 10 \end{bmatrix} = .2157 \end{aligned}$$

Therefore since we recall from Example 8.10 that the standard error, s, is .3671, it follows that a 95% confidence interval for μ_0 is

$$\begin{aligned} [\hat{y}_0 \pm t_{[\alpha/2]}^{(n-k)} s\sqrt{h_{00}}] &= [10.333 \pm t_{[.025]}^{(8-3)}(.3671)\sqrt{.2157}] \\ &= [10.333 \pm 2.571(.3671)\sqrt{.2157}] \\ &= [10.333 \pm .438] \\ &= [9.895, 10.771] \end{aligned}$$

Thus we are 95% confident that μ_0, the future mean fuel consumption, is between 9.895 and 10.771 tons of coal. It also follows that a 95% prediction interval for y_0 is

$$\begin{aligned} [\hat{y}_0 \pm t_{[\alpha/2]}^{(n-k)} s\sqrt{1 + h_{00}}] &= [10.333 \pm t_{[.025]}^{(8-3)}(.3671)\sqrt{1.2157}] \\ &= [10.333 \pm 2.571(.3671)\sqrt{1.2157}] \\ &= [10.333 \pm 1.04] \\ &= [9.293, 11.373] \end{aligned}$$

This interval says that we are 95% confident that y_0, the future individual fuel consumption, will be between 9.293 and 11.373 tons of coal. The upper bound of this interval is important. It says that if we plan to acquire 11.373 tons of coal in the future week, then we can be very confident that we will have enough fuel to adequately heat the nine-building complex.

Example 8.20 Recall from Example 8.9 that

$$\begin{aligned} \hat{y}_0 &= b_0 + b_1x_{04} + b_2x_{03} + b_3x_{03}^2 \\ &= 17.3244 + 1.3070(.10) - 3.6956(6.8) + .3486(6.8)^2 \\ &= 8.2933 \qquad \text{(that is, 829,330 bottles)} \end{aligned}$$

is a point estimate of

$$\mu_0 = \beta_0 + \beta_1(.10) + \beta_2(6.8) + \beta_3(6.8)^2$$

the mean of all possible demands for Fresh when the price difference is $x_{04} = .10$ and the advertising expenditure is $x_{03} = 6.80$. In addition, $\hat{y}_0$ is a point prediction of

$$y_0 = \mu_0 + \varepsilon_0$$

the actual demand for Fresh that will be observed in the future quarter in the particular sales region. Moreover, by using

$$\mathbf{x}_0' = [1 \quad .10 \quad 6.8 \quad (6.8)^2] = [1 \quad .10 \quad 6.8 \quad 46.24]$$

the quantity

$$h_{00} = \mathbf{x}_0'(\mathbf{X}'\mathbf{X})^{-1}\mathbf{x}_0$$

can be calculated to be .06893. Although h_{00} is not directly given on the SAS output in Figure 8.12, the SAS output tells us that the standard error of the estimate $\hat{y}_0$ is

$$s\sqrt{h_{00}} = 0.0581$$

and that the standard error is $s = 0.2213$. This implies that

$$h_{00} = \left(\frac{.0581}{s}\right)^2 = \left(\frac{.0581}{.2213}\right)^2 = .06893$$

It follows that a 95% confidence interval for μ_0 is

$$\begin{aligned}[\hat{y}_0 \pm t_{[.025]}^{(26)} s\sqrt{h_{00}}] &= [8.2933 \pm 2.056(.0581)] \\ &= [8.1738, 8.4128]\end{aligned}$$

and that a 95% prediction interval for y_0 is

$$\begin{aligned}[\hat{y}_0 \pm t_{[.025]}^{(26)} s\sqrt{1 + h_{00}}] &= [8.2933 \pm 2.056(.2213)\sqrt{1 + .06893}] \\ &= [7.8230, 8.7636]\end{aligned}$$

Note that both the 95% confidence interval and the 95% prediction interval are given on the SAS output in Figure 8.12. The prediction interval says that we are 95% confident that the actual demand for Fresh that will be observed in the future quarter in the particular sales region will be between 782,300 bottles and 876,360 bottles. Furthermore, a 99% prediction interval for y_0 is

$$\begin{aligned}[\hat{y}_0 \pm t_{[.005]}^{(26)} s\sqrt{1 + h_{00}}] &= [8.2933 \pm 2.779(.2213)\sqrt{1 + .06893}] \\ &= [8.2933 \pm .6358] \\ &= [7.6575, 8.9291]\end{aligned}$$

Example 8.21 Consider the residential sales model

$$y_i = \beta_0 + \beta_1 x_{i1} + \beta_2 x_{i2} + \beta_3 x_{i3} + \beta_4 x_{i4} + \varepsilon_i$$

The standard error, s, can be calculated to be 18.9615 for this model. Therefore this model yields a 95% prediction interval for the sales price of a residence of the form (since $n = 63$)

$$\begin{aligned}[\hat{y}_0 \pm t_{[.025]}^{(63-5)} s\sqrt{1 + \mathbf{x}_0'(\mathbf{X}'\mathbf{X})^{-1}\mathbf{x}_0}] &= [\hat{y}_0 \pm 1.96(18.9615)\sqrt{1 + \mathbf{x}_0'(\mathbf{X}'\mathbf{X})^{-1}\mathbf{x}_0}] \\ &= [\hat{y}_0 \pm 37.1645\sqrt{1 + \mathbf{x}_0'(\mathbf{X}'\mathbf{X})^{-1}\mathbf{x}_0}]\end{aligned}$$

Since $\mathbf{x}_0'(\mathbf{X}'\mathbf{X})^{-1}\mathbf{x}_0$ is greater than zero, the error bound in the prediction interval is at least 37.1645. This means that the length of the prediction interval is at least 74.3291. That is, this model provides a 95% prediction interval for the sales price of a residence that has a length of at least \$74,329! This is not very helpful because one would think that a trained real estate agent could intuitively (and accurately) provide a shorter interval.

In the exercises of Chapter 6, we saw that residences 10, 50, and 61 in Table 4.3, which are identical to residences 10, 50, and 61 in Table 8.2, are possibly outlying and influential. We will see in the exercises of this chapter that if we remove these observations from Table 8.2, then the model

$$y_i = \beta_0 + \beta_1 x_{i1} + \beta_2 x_{i1}^2 + \beta_3 x_{i2} + \beta_4 x_{i2}^2 + \beta_5 x_{i3} + \beta_6 x_{i3}^2 + \beta_7 x_{i4} + \varepsilon_i$$

seems to describe the remaining 60 residences adequately. Note that the prob-values related to the importance of the quadratic terms in the above model indicate the importance of these terms when describing the 60 residences but do not indicate such importance when describing all 63 residences. Also note that the standard error, s, for the 60-residence model is 11.1277. This is substantially less than the standard error of 18.9615 for the model describing all 63 residences. Still, a standard error of 11.1277 yields a prediction interval with a length of at least \$43,621. If we are to obtain a shorter interval, we must search for other predictor variables. An additional predictor variable that might be considered is a home condition rating. In the exercises at the end of this chapter we consider a journal article and a data set that indicate that this variable can be quite useful.

*8.9 USING SAS

In Figure 8.16 we present the SAS statements that generate the SAS output of Figure 8.12, which presents a regression analysis of the Fresh demand data using the model $y = \beta_0 + \beta_1 x_4 + \beta_2 x_3 + \beta_3 x_3^2 + \varepsilon$. We also give the SAS statements needed to perform residual analysis for the above model.

*This section is optional.

FIGURE 8.16 SAS program to carry out a regression analysis of the Fresh demand data

```
DATA DETR;                 ——→ Assigns the name DETR to the data file
INPUT Y X4 X3;             ——→ Defines variable names, Y = demand, X4 = price
                               difference, X3 = advertising expenditure
X3SQ = X3*X3;              ——→ Transformation defines the squared term x3^2
                               and assigns the name X3SQ to this variable
CARDS;
7.38  -0.05  5.50
8.51   0.25  6.75
9.52   0.60  7.25
         .                 ——→ Fresh demand data—see Table 8.3
         .
         .
9.26   0.55  6.80
 .     0.10  6.80          ——→ Decimal point represents a missing
                               value. Used to obtain point prediction
                               when x03 = 6.80 and x04 = .10
PROC PRINT;                ——→ Prints the data
PROC REG DATA = DETR;      ——→ Specifies regression procedure
MODEL Y = X4  X3  X3SQ / P CLM CLI;
```

$X4\ X3\ X3SQ$ → Specifies model $y = \beta_0 + \beta_1 x_4 + \beta_2 x_3 + \beta_3 x_3^2 + \varepsilon$ (intercept β_0 is assumed)

P CLM CLI → P = Predictions desired; CLM = 95% confidence interval desired; CLI = 95% prediction interval desired

```
PROC REG DATA = DETR;
MODEL Y = X4 X3 X3SQ/P;
OUTPUT OUT = ONE PREDICTED = YHAT RESIDUAL = RESID;
PROC PLOT DATA = ONE;
PLOT RESID * (X4 X3 YHAT);
```

→ Plots residuals from the above model against x_4, x_3, and $\hat{y}$

```
PROC UNIVARIATE PLOT DATA = ONE;
VAR RESID;
```

→ Produces stem-and-leaf display, box plot, and normal plot of residuals from the above model

EXERCISES

8.1 Consider the fuel consumption data in Table 8.1. In Example 8.4 we computed the least squares point estimates of the parameters in the two-variable model

$$\begin{aligned} y &= \mu + \varepsilon \\ &= \beta_0 + \beta_1 x_1 + \beta_2 x_2 + \varepsilon \end{aligned}$$

These estimates are $b_0 = 13.109$, $b_1 = -.09$, and $b_2 = .0825$.

a. Suppose that in a future week the average hourly temperature will be $x_{01} = 37.0$ and the chill index will be $x_{02} = 16$. Use the above model to calculate a point estimate of μ_0, the mean fuel consumption for weeks when $x_{01} = 37.0$ and $x_{02} = 16$, and a point prediction of y_0, the individual fuel consumption in a future week when $x_{01} = 37.0$ and $x_{02} = 16$.
b. Is the combination $(x_{01}, x_{02}) = (37, 16)$ in the experimental region in Figure 8.3? Why is this important?
c. Use a t statistic in Table 8.6 and an appropriate rejection point to test the importance of x_1 in the above model. That is, can we reject $H_0: \beta_1 = 0$ in favor of $H_1: \beta_1 \neq 0$ by setting α equal to .05? Can we reject H_0 with $\alpha = .01$?
d. Use an appropriate prob-value in Table 8.6 to test the importance of x_1 with $\alpha = .05$ and with $\alpha = .01$. What do you conclude about the importance of x_1?
e. Using the matrix $(\mathbf{X}'\mathbf{X})^{-1}$ given in Example 8.15, calculate a 95% confidence interval for μ_0 when $x_{01} = 37.0$ and $x_{02} = 16$. Interpret the meaning of this interval.
f. Using the matrix $(\mathbf{X}'\mathbf{X})^{-1}$ given in Example 8.15, calculate a 95% prediction interval for y_0 when $x_{01} = 37.0$ and $x_{02} = 16$. Interpret the meaning of this interval.

8.2 Consider the residential sales model of Example 8.2:

$$y_i = \beta_0 + \beta_1 x_{i1} + \beta_2 x_{i2} + \beta_3 x_{i3} + \beta_4 x_{i4} + \varepsilon_i$$

This model relates sales price (y) to square footage (x_1), number of rooms (x_2), number of bedrooms (x_3), and age (x_4). The SAS output of the regression analysis using the above model is given in Figure 8.17.

a. Identify the least squares point estimates b_0, b_1, b_2, b_3, and b_4.
b. Use the t statistics given on the SAS output to test the importance of each of the independent variables x_1, x_2, x_3, and x_4 with $\alpha = .05$. Which variables are important?

FIGURE 8.17 SAS output of a regression analysis using the residential sales model $y = \beta_0 + \beta_1 x_1 + \beta_2 x_2 + \beta_3 x_3 + \beta_4 x_4 + \varepsilon$

SAS

DEP VARIABLE: SALEP

ANALYSIS OF VARIANCE

SOURCE	DF	SUM OF SQUARES	MEAN SQUARE	F VALUE	PROB>F
MODEL	4	55183.68778	13795.92195	38.371	0.0001
ERROR	58	20853.30206	359.53969		
C TOTAL	62	76036.98984			

ROOT MSE	18.96153	R-SQUARE	0.7257
DEP MEAN	78.80159	ADJ R-SQ	0.7068
C.V.	24.06237		

PARAMETER ESTIMATES

VARIABLE	DF	PARAMETER ESTIMATE	STANDARD ERROR	T FOR H0: PARAMETER=0	PROB > |T|
INTERCEP	1	10.36761556	11.49845952	0.902	0.3710
SQFT	1	0.05001119	0.008104134	6.171	0.0001
ROOMS	1	6.32177893	2.52798952	2.501	0.0152
BED	1	-11.10316277	5.86838064	-1.892	0.0635
AGE	1	-0.43186496	0.10970614	-3.937	0.0002

c. Use the prob-values given on the SAS output to test the importance of each of the independent variables x_1, x_2, x_3, and x_4 with $\alpha = .05$. Which variables are important?
d. Use F(model) to test the adequacy of the model with $\alpha = .05$.* Interpret the result of this test.
e. Identify the standard error (s), the mean square error (s^2), the explained variation, unexplained variation (SSE), total variation, and R^2. Interpret the value of R^2.

8.3 Use the SAS output of Figure 8.17 and the residential sales model

$$y_i = \beta_0 + \beta_1 x_{i1} + \beta_2 x_{i2} + \beta_3 x_{i3} + \beta_4 x_{i4} + \varepsilon_i$$

to compute 95% confidence intervals for β_2, β_3, and β_4. Carefully interpret each interval.

8.4 Consider the fuel consumption data in Table 8.7 and the model

$$y = \beta_0 + \beta_1 x_1 + \beta_2 x_2 + \varepsilon$$

which relates fuel consumption (y) to the average hourly temperature (x_1) and the chill index (x_2).

a. Plot y versus x_1 and y versus x_2. Explain why the model

$$y = \beta_0 + \beta_1 x_1 + \beta_2 x_2 + \varepsilon$$

might be a reasonable model relating y to x_1 and x_2.

For this model it can be shown that

Explained variation $= 25.462472$

$SSE = .277528$

$b_0 = 12.917034, \quad b_1 = -.087064, \quad b_2 = .090221$

$s\sqrt{c_{00}} = 0.5492, \quad s\sqrt{c_{11}} = .009035063, \quad s\sqrt{c_{22}} = 0.014122$

Using this information, do the following:

b. Calculate F(model) and test the adequacy of the model with $\alpha = .05$. That is, can we reject $H_0: \beta_1 = \beta_2 = 0$ with $\alpha = .05$? Interpret the result of this test.
c. Calculate s, R^2, and R. Interpret the value of R^2.

TABLE 8.7 Fuel consumption data

Average hourly temperature, x_{i1}	Chill index, x_{i2}	Weekly fuel consumption, y_i
$x_{11} = 28$	$x_{12} = 18$	$y_1 = 12.4$
$x_{21} = 32.5$	$x_{22} = 24$	$y_2 = 12.3$
$x_{31} = 28$	$x_{32} = 14$	$y_3 = 11.7$
$x_{41} = 39$	$x_{42} = 22$	$y_4 = 11.2$
$x_{51} = 57.8$	$x_{52} = 16$	$y_5 = 9.5$
$x_{61} = 45.9$	$x_{62} = 8$	$y_6 = 9.4$
$x_{71} = 58.1$	$x_{72} = 1$	$y_7 = 8.0$
$x_{81} = 62.5$	$x_{82} = 0$	$y_8 = 7.5$

*Here and in other exercises where we employ the F(model), we use the term "model adequacy" to simply mean that we can reject the null hypothesis for the overall F-test.

d. Calculate an appropriate t statistic and use it to test the importance of the intercept with $\alpha = .05$ and with $\alpha = .01$.
e. Calculate appropriate t statistics and use them to test the importance of the variables x_1 and x_2 with $\alpha = .01$ and $\alpha = .05$. What do you conclude about the importance of these variables?
f. Calculate 95% confidence intervals for β_0, β_1, and β_2. Interpret each of these intervals.
g. Calculate a point prediction of y_0, an individual weekly fuel consumption when $x_{01} = 40$ and $x_{02} = 10$.
h. When $x_{01} = 40$ and $x_{02} = 10$, it can be shown that $s\sqrt{h_{00}} = 0.109411$. Compute and interpret a 99% confidence interval for μ_0, mean fuel consumption when $x_{01} = 40$ and $x_{02} = 10$.
i. Compute and interpret a 99% prediction interval for y_0, an individual fuel consumption when $x_{01} = 40$ and $x_{02} = 10$ (see part (h)).

8.5 Consider the fuel consumption data in Table 8.7 and the information given in Exercise 8.4 for the model

$$y = \beta_0 + \beta_1 x_1 + \beta_2 x_2 + \varepsilon$$

a. Compute the residuals for this model.
b. Construct residual plots versus x_1, x_2, $\hat{y}$, and time (assume that the data in Table 8.7 is in time order). Do these plots suggest a violation of any of the regression assumptions?
c. Construct a stem-and-leaf diagram of the residuals. Interpret.
d. Compute and interpret the standardized residuals.
e. Construct and interpret a normal plot of the residuals.

Exercises 8.6–8.9 refer to the following situation. Compustat, Inc., sells an electronic calculator, the CS-22. Compustat has, over the past $n = 25$ sales periods, spent different amounts of money on advertising and has had three different advertising firms handle the advertising strategy for the CS-22. The three advertising firms have different amounts of experience in the advertising field. One company has had 5 years of experience, while the others have had 10 and 15 years of experience, respectively. Compustat wishes to develop a regression model that relates the dependent variable

y_i = demand for the CS-22 during the ith sales period (measured in units of 1000 calculators sold)

to the independent variables

x_{i1} = advertising expenditure during the ith sales period (measured in units of \$100,000)

x_{i2} = the number of years of experience in the advertising field of the firm handling the advertising strategy for the CS-22 during the ith sales period

Compustat wishes to consider using the regression model

$$\begin{aligned} y_i &= \mu_i + \varepsilon_i \\ &= \beta_0 + \beta_1 x_{i1} + \beta_2 x_{i2} + \beta_3 x_{i1}^2 + \varepsilon_i \end{aligned}$$

Assume that a sample of $n = 25$ combinations of (y_i, x_{i1}, x_{i2}) values are observed and that the following calculations are made

$$(\mathbf{X}'\mathbf{X})^{-1} = \begin{bmatrix} .02 & 0 & 0 & 0 \\ 0 & .01 & 0 & 0 \\ 0 & 0 & .004 & 0 \\ 0 & 0 & 0 & .006 \end{bmatrix}$$

$$\mathbf{X}'\mathbf{y} = \begin{bmatrix} 1000 \\ 300 \\ 500 \\ 100 \end{bmatrix} \quad \text{and} \quad \sum_{i=1}^{25} y_i^2 = 22{,}162$$

8.6 In the Compustat, Inc. situation,

a. Calculate the least squares point estimates of β_0, β_1, β_2, and β_3.

b. Assume that in a future sales period Compustat will spend $200,000 on advertising and will have an advertising firm with 5 years of experience handle its advertising strategy for the CS-22. The demand for the CS-22 in this future sales period may be expressed in the form $y_0 = \mu_0 + \varepsilon_0$.

(1) Discuss the meaning of the future population of potential demands for the CS-22.

(2) Discuss the meaning of μ_0, the future mean demand for the CS-22.

(3) Discuss the conceptual difference between μ_0 and $y_0 = \mu_0 + \varepsilon_0$. What does ε_0 measure in this situation?

(4) Calculate a point estimate of μ_0 and a point prediction of y_0.

c. Show that $\mathbf{b}'\mathbf{X}'\mathbf{y} = 21{,}960$, $s^2 = 9.62$, and $s = 3.10$.

8.7 In the Compustat, Inc. situation,

a. Calculate appropriate t statistics and use them to test the importance of each of the independent variables x_1, x_2, and x_1^2 with $\alpha = .05$ and $\alpha = .01$. What do you conclude about the importance of these variables? Recall that $s = 3.10$.

b. Calculate and interpret a 95% confidence interval for each of the parameters β_0, β_1, β_2, and β_3.

8.8 In the Compustat, Inc. situation,

a. Calculate and interpret a 95% confidence interval for μ_0 when $200,000 is spent on advertising and the firm handling the advertising has 5 years of experience.

b. Calculate and interpret a 95% prediction interval for y_0 when $200,000 is spent on advertising and the firm handling the advertising has 5 years of experience.

8.9 In the Compustat, Inc. situation,

a. Using the facts that

$$\sum_{i=1}^{25} y_i^2 = 22{,}162 \quad \text{and} \quad 25\bar{y}^2 = 20{,}474$$

compute the total variation, explained variation, and R^2. Interpret the value of R^2.

b. Compute F(model) and use this statistic to test the importance of the model with $\alpha = .05$. That is, can we reject $H_0: \beta_1 = \beta_2 = \beta_3 = 0$ with $\alpha = .05$? Interpret the result of this test.

8.10 A study* was undertaken to examine the profit y per sales dollar earned by a construction company and its relationship to the size x_1 of the construction contract (in hundreds of thousands of dollars) and the number x_2 of years of experience of the construction supervisor. Data were obtained from a sample of $n = 18$ construction projects undertaken by the construction company over the past two years. This data is presented in Table 8.8, where y_i, x_{i1}, and x_{i2} denote the profit, contract size, and supervisor experience associated with construction project i. SAS plots of y versus x_1 and x_2 are presented in Figures 8.18 and 8.19. The (possibly) curved data plot in Figure 8.18 suggests that the quadratic regression model

$$y_i = \beta_0 + \beta_1 x_{i1} + \beta_2 x_{i1}^2 + \varepsilon_i$$

might appropriately relate y_i to x_{i1}. The (possibly) curved data plot in Figure 8.19 indicates that the quadratic regression model

$$y_i = \beta_0 + \beta_1 x_{i2} + \beta_2 x_{i2}^2 + \varepsilon_i$$

ppropriately relate y_i to x_{i2}. However, since there are only three values of x_{i2} in (2, 4, and 6), it is difficult to discern the relationship between y_i and x_{i2}. Thus it also consider relating y_i to x_{i2} by using the simple linear regression model

$$= \beta_0 + \beta_1 x_{i2} + \varepsilon_i$$

ıction project data

ction t, i	Contract size, x_{i1} (hundreds of thousands of dollars)	Supervisor experience, x_{i2} (years)	Profit, y_i
	5.1	4	2.0
	3.5	4	3.5
	2.4	2	8.5
	4.0	6	4.5
	1.7	2	7.0
	2.0	2	7.0
	5.0	4	2.0
	3.2	2	5.0
	5.2	6	8.0
	4.3	6	5.0
	2.9	2	6.0
	1.1	2	7.5
	2.6	4	4.0
	4.0	6	4.0
	5.3	4	1.0
	4.9	6	5.0
	5.0	6	6.5
	3.9	4	1.5

denhall and J. Reinmuth, *Statistics for Management Economics*, 4th ed., PWS-KENT Publishing n. © 1982. Used with permission.

*This exercise is based on an example in Mendenhall and Reinmuth (1982).

FIGURE 8.18 SAS plot of y versus x_1 for the construction project data

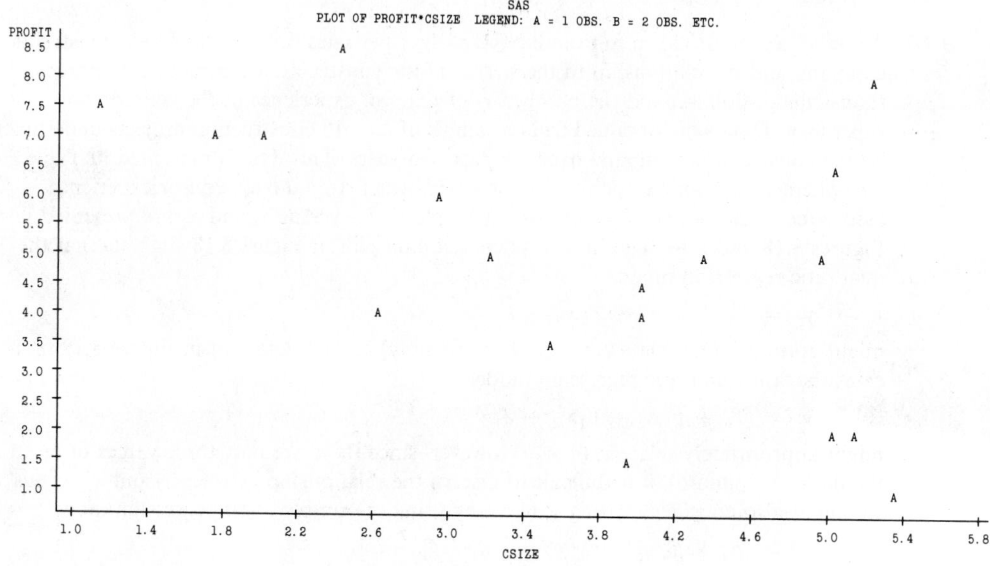

FIGURE 8.19 SAS plot of y versus x_2 for the construction project data

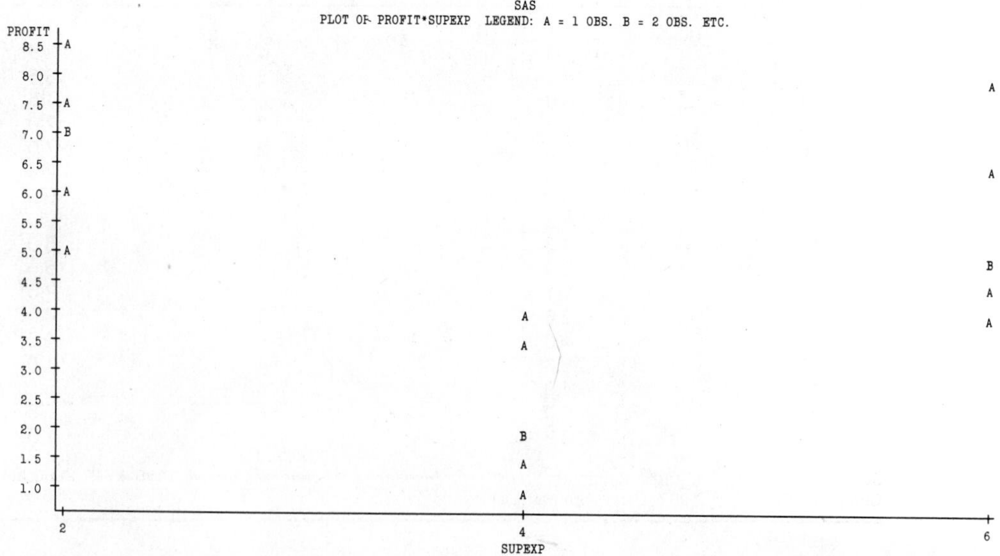

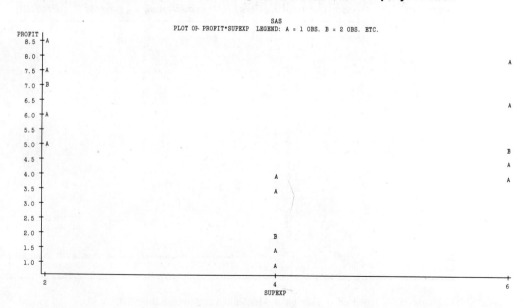

Combining the above models, we conclude that either the model

$$y_i = \beta_0 + \beta_1 x_{i1} + \beta_2 x_{i1}^2 + \beta_3 x_{i2} + \varepsilon_i$$

or the model

$$y_i = \beta_0 + \beta_1 x_{i1} + \beta_2 x_{i1}^2 + \beta_3 x_{i2} + \beta_4 x_{i2}^2 + \varepsilon_i$$

might appropriately relate y_i to x_{i1} and x_{i2}.

The SAS output for these two models is given in Figures 8.20 and 8.21. Residual plots

FIGURE 8.20 **SAS output of a regression analysis of the construction project data using the model $y = \beta_0 + \beta_1 x_1 + \beta_2 x_1^2 + \beta_3 x_2 + \varepsilon$**

SAS

DEP VARIABLE: PROFIT

ANALYSIS OF VARIANCE

SOURCE	DF	SUM OF SQUARES	MEAN SQUARE	F VALUE	PROB>F
MODEL	3	33.90007591	11.30002530	2.857	0.0749
ERROR	14	55.37770186	3.95555013		
C TOTAL	17	89.27777778			

ROOT MSE	1.988856	R-SQUARE	0.3797
DEP MEAN	4.888889	ADJ R-SQ	0.2468
C.V.	40.68116		

PARAMETER ESTIMATES

VARIABLE	DF	PARAMETER ESTIMATE	STANDARD ERROR	T FOR H0: PARAMETER=0	PROB > |T|
INTERCEP	1	11.39419114	3.45541609	3.297	0.0053
CSIZE	1	-3.77079866	2.31191279	-1.631	0.1252
SUPEXP	1	0.53152579	0.43701162	1.216	0.2440
CSIZESQ	1	0.34469469	0.32037124	1.076	0.3002

OBS	ACTUAL	PREDICT VALUE	STD ERR PREDICT	LOWER95% MEAN	UPPER95% MEAN	LOWER95% PREDICT	UPPER95% PREDICT	RESIDUAL
1	2.0000	3.2547	0.9564	1.2035	5.3059	-1.4785	7.9879	-1.2547
2	3.5000	4.5450	0.7187	3.0035	6.0865	.0093717	9.0806	-1.0450
3	8.5000	5.3928	0.7919	3.6942	7.0913	0.8014	9.9842	3.1072
4	4.5000	5.0153	0.9356	3.0086	7.0220	0.3012	9.7294	-0.5153
5	7.0000	7.0431	0.9408	5.0252	9.0609	2.3242	11.7619	-0.0431
6	7.0000	6.2944	0.7924	4.5949	7.9939	1.7027	10.8862	0.7056
7	2.0000	3.2837	0.8864	1.3825	5.1849	-1.3865	7.9538	-1.2837
8	5.0000	3.9204	1.0497	1.6689	6.1718	-0.9030	8.7437	1.0796
9	8.0000	4.2957	0.9218	2.3186	6.2728	-0.4058	8.9973	3.7043
10	5.0000	4.7423	0.8260	2.9707	6.5140	0.1234	9.3613	0.2577
11	6.0000	4.4208	0.9478	2.3881	6.4535	-0.3044	9.1460	1.5792
12	7.5000	8.7264	1.5972	5.3008	12.1521	3.2556	14.1973	-1.2264
13	4.0000	6.0464	0.8019	4.3265	7.7662	1.4470	10.6457	-2.0464
14	4.0000	5.0153	0.9356	3.0086	7.0220	0.3012	9.7294	-1.0153
15	1.0000	3.2175	1.1261	0.8023	5.6327	-1.6844	8.1195	-2.2175
16	5.0000	4.3826	0.7724	2.7258	6.0393	-0.1935	8.9586	0.6174
17	6.5000	4.3467	0.8071	2.6157	6.0777	-0.2568	8.9502	2.1533
18	1.5000	4.0570	0.6870	2.5834	5.5305	-0.4560	8.5700	-2.5570
19	.	4.4253	0.7528	2.8106	6.0400	-0.1358	8.9863	.
20	.	6.2944	0.7924	4.5949	7.9939	1.7027	10.8862	.

SUM OF RESIDUALS	3.75255E-14
SUM OF SQUARED RESIDUALS	55.3777
PREDICTED RESID SS (PRESS)	96.08057

(continues)

FIGURE 8.20 **Continued**

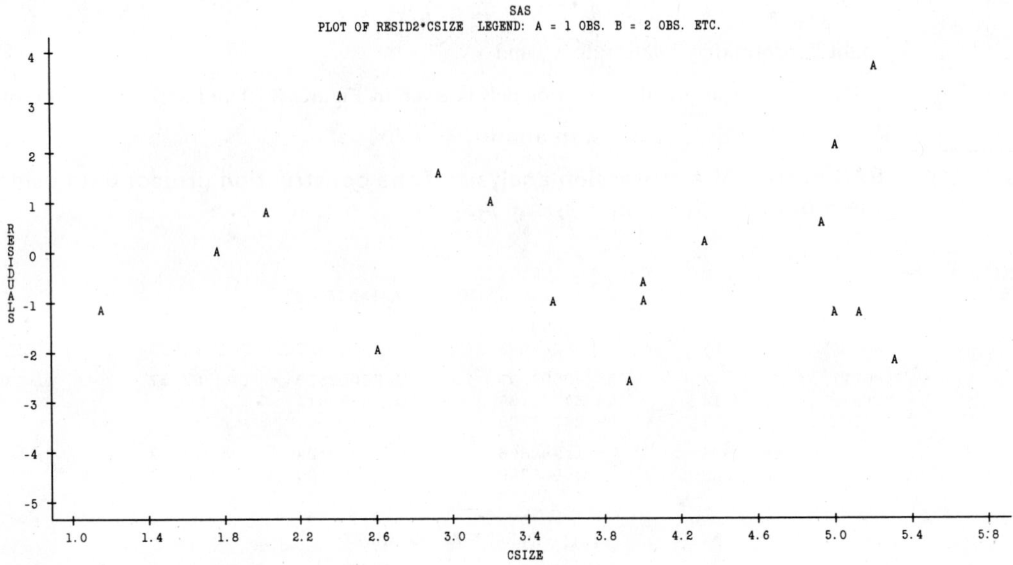

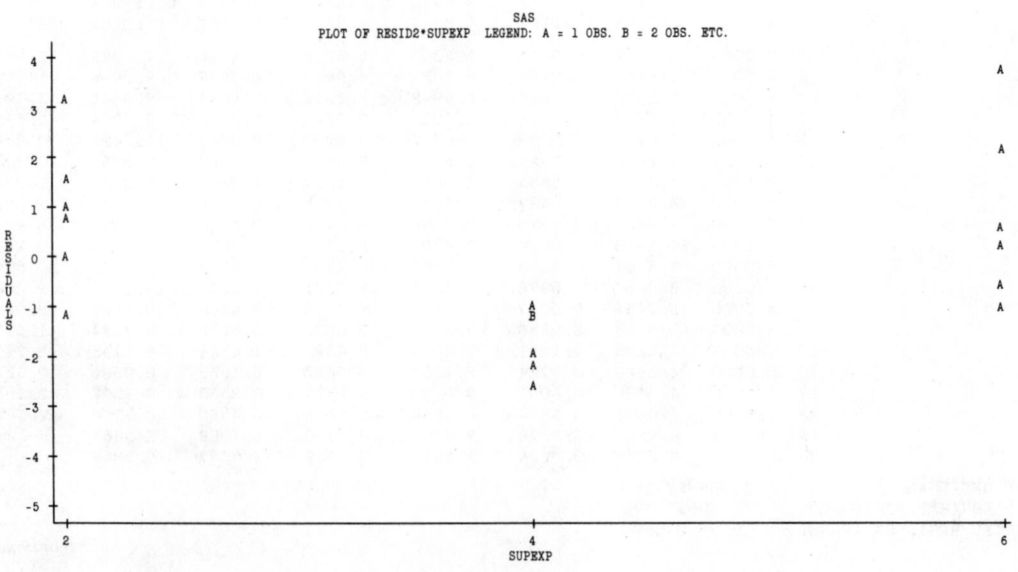

versus x_1 and x_2 are included in the output. Note that x_1, x_1^2, x_2, and x_2^2 are identified on the output as CSIZE, CSIZESQ, SUPEXP, and SUPEXPSQ, respectively.

a. Use F(model) from Figure 8.20 to test the overall adequacy of the model

$$y = \beta_0 + \beta_1 x_1 + \beta_2 x_1^2 + \beta_3 x_2 + \varepsilon$$

with $\alpha = .05$. What do you conclude about the model?

FIGURE 8.21 **SAS output of a regression analysis of the construction project data using the model $y = \beta_0 + \beta_1 x_1 + \beta_2 x_1^2 + \beta_3 x_2 + \beta_4 x_2^2 + \varepsilon$**

SAS

DEP VARIABLE: PROFIT

ANALYSIS OF VARIANCE

SOURCE	DF	SUM OF SQUARES	MEAN SQUARE	F VALUE	PROB>F
MODEL	4	68.78371736	17.19592934	10.908	0.0004
ERROR	13	20.49406041	1.57646619		
C TOTAL	17	89.27777778			

ROOT MSE	1.255574	R-SQUARE	0.7704
DEP MEAN	4.888889	ADJ R-SQ	0.6998
C.V.	25.6822		

PARAMETER ESTIMATES

VARIABLE	DF	PARAMETER ESTIMATE	STANDARD ERROR	T FOR H0: PARAMETER=0	PROB > |T|
INTERCEP	1	20.51766829	2.91895313	7.029	0.0001
CSIZE	1	-2.25634702	1.49460778	-1.510	0.1550
SUPEXP	1	-6.70010837	1.56189172	-4.290	0.0009
CSIZESQ	1	0.25185519	0.20321248	1.239	0.2371
SUPEXPSQ	1	0.83820391	0.17818909	4.704	0.0004

OBS	ACTUAL	PREDICT VALUE	STD ERR PREDICT	LOWER95% MEAN	UPPER95% MEAN	LOWER95% PREDICT	UPPER95% PREDICT	RESIDUAL
1	2.0000	2.1719	0.6462	0.7760	3.5678	-0.8787	5.2225	-0.1719
2	3.5000	2.3165	0.6560	0.8994	3.7336	-0.7439	5.3769	1.1835
3	8.5000	6.5057	0.5531	5.3108	7.7006	3.5417	9.4698	1.9943
4	4.5000	5.4967	0.5995	4.2016	6.7917	2.4909	8.5024	-0.9967
5	7.0000	7.3623	0.5978	6.0709	8.6538	4.3581	10.3666	-0.3623
6	7.0000	6.9650	0.5202	5.8413	8.0887	4.0289	9.9011	0.0350
7	2.0000	2.1431	0.6099	0.8256	3.4607	-0.8724	5.1587	-0.1431
8	5.0000	5.8290	0.7770	4.1503	7.5076	2.6390	9.0189	-0.8290
9	8.0000	5.5695	0.6419	4.1829	6.9562	2.5231	8.6159	2.4305
10	5.0000	5.4469	0.5426	4.2747	6.6190	2.4920	8.4018	-0.4469
11	6.0000	6.0450	0.6908	4.5526	7.5373	2.9490	9.1409	-0.0450
12	7.5000	8.2930	1.0125	6.1056	10.4804	4.8084	11.7776	-0.7930
13	4.0000	2.9645	0.8279	1.1759	4.7532	-0.2846	6.2137	1.0355
14	4.0000	5.4967	0.5995	4.2016	6.7917	2.4909	8.5024	-1.4967
15	1.0000	2.2445	0.7404	0.6450	3.8440	-0.9045	5.3934	-1.2445
16	5.0000	5.4833	0.5409	4.3148	6.6518	2.5298	8.4368	-0.4833
17	6.5000	5.5070	0.5661	4.2841	6.7299	2.5316	8.4824	0.9930
18	1.5000	2.1595	0.5923	0.8798	3.4391	-0.8397	5.1586	-0.6595
19	.	5.4646	0.5241	4.3323	6.5969	2.5253	8.4040	.
20	.	6.9650	0.5202	5.8413	8.0887	4.0289	9.9011	.

SUM OF RESIDUALS	9.37028E-14
SUM OF SQUARED RESIDUALS	20.49406
PREDICTED RESID SS (PRESS)	42.21029

(*continues*)

FIGURE 8.21 Continued

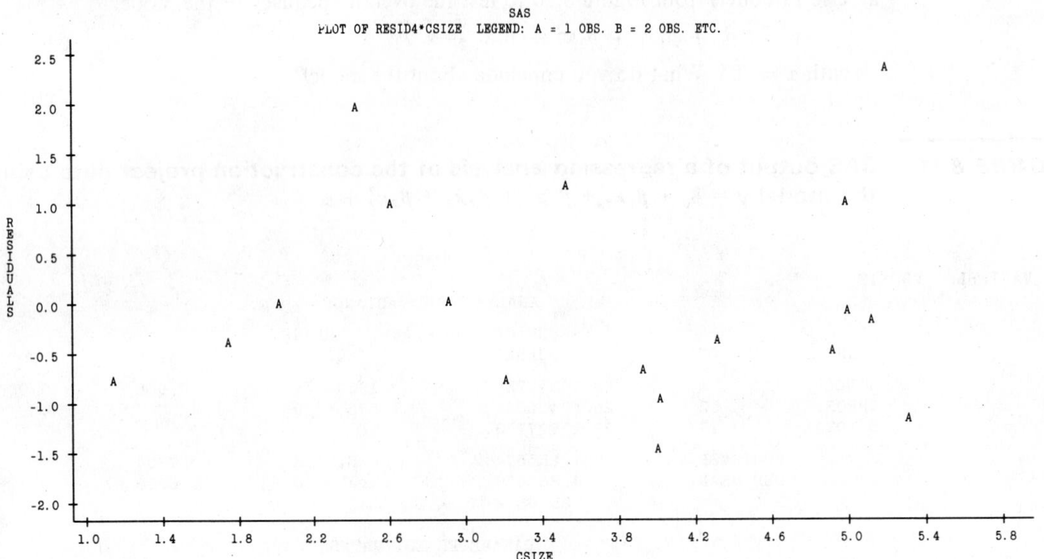

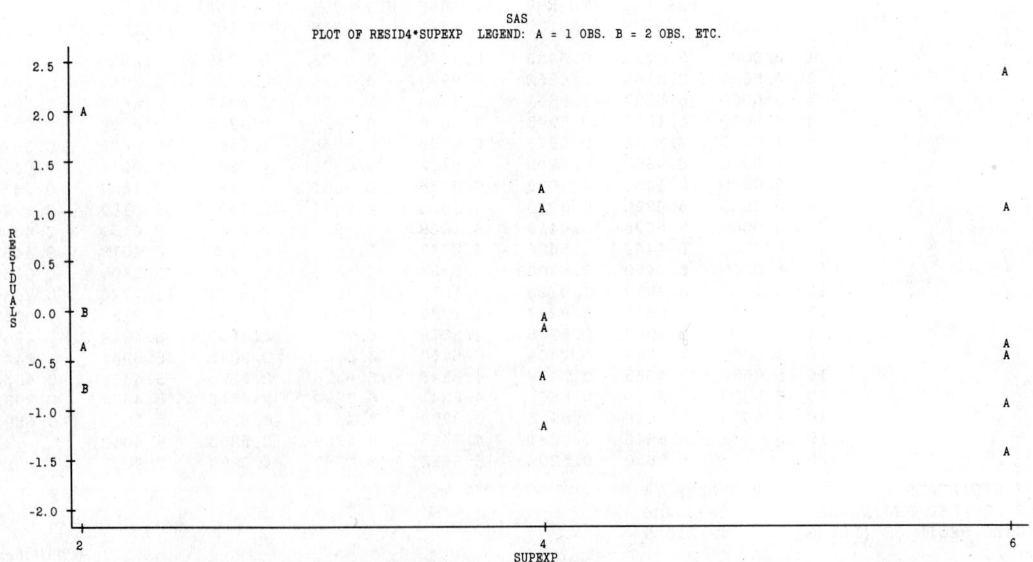

b. Use F(model) from Figure 8.21 to test the overall adequacy of the model

$$y = \beta_0 + \beta_1 x_1 + \beta_2 x_1^2 + \beta_3 x_2 + \beta_4 x_2^2 + \varepsilon$$

with $\alpha = .05$. What do you conclude about the model?

c. Use appropriate t statistics to test the importance of each of the variables x_1, x_1^2, and x_2 in the model

$$y = \beta_0 + \beta_1 x_1 + \beta_2 x_1^2 + \beta_3 x_2 + \varepsilon$$

with $\alpha = .05$. What do you conclude about the importance of these variables?

d. Use appropriate t statistics to test the importance of each of the variables x_1, x_1^2, x_2, and x_2^2 in the model

$$y = \beta_0 + \beta_1 x_1 + \beta_2 x_1^2 + \beta_3 x_2 + \beta_4 x_2^2 + \varepsilon$$

with $\alpha = .05$. What do you conclude about the importance of these variables?

e. Find R^2 for each of the above models. Which model seems best on the basis of R^2?

f. Examine the residual plots versus x_1 and x_2 for each model. What do these plots suggest about the adequacy of these models?

g. Considering all the information at your disposal, does either of these models seem to adequately relate y to x_1 and x_2? Explain. (We will consider further improvement in this model in the exercises of Chapter 9.)

8.11 Consider the construction project data in Table 8.8 of Exercise 8.10. Write out the **y**-vector and **X**-matrix that are used to compute the least squares point estimates of the parameters for each of the models

$$y = \beta_0 + \beta_1 x_1 + \beta_2 x_1^2 + \beta_3 x_2 + \varepsilon$$

and

$$y = \beta_0 + \beta_1 x_1 + \beta_2 x_1^2 + \beta_3 x_2 + \beta_4 x_2^2 + \varepsilon$$

8.12 Consider the construction project data of Exercise 8.10 and the model

$$y = \beta_0 + \beta_1 x_1 + \beta_2 x_1^2 + \beta_3 x_2 + \beta_4 x_2^2 + \varepsilon$$

Use the residuals for this model (given in Figure 8.21) to

a. Calculate and interpret the standardized residuals.
b. Construct and interpret a histogram of the residuals.
c. Construct and interpret a normal plot of the residuals.
d. Assuming that the data in Table 8.8 are listed in the time order in which they were observed, plot the residuals versus time. Interpret.
e. Calculate the Durbin–Watson statistic. Test for positive autocorrelation with $\alpha = .05$. Test for negative autocorrelation with $\alpha = .05$.

Exercises 8.13–8.16 refer to the following situation. Exron Chemical Company wishes to develop a regression model relating the dependent variable

$$y_i = \text{the yield of a chemical reaction (in pounds)}$$

to the independent variables

$$x_{i1} = \text{the amount of catalyst XST used in the reaction (in pounds)}$$

$$x_{i2} = \text{the amount of catalyst RST used in the reaction (in pounds)}$$

$$x_{i3} = \text{the amount of catalyst ZST used in the reaction (in pounds)}$$

The company wishes to use the model

$$\begin{aligned} y_i &= \mu_i + \varepsilon_i \\ &= \beta_0 + \beta_1 x_{i1} + \beta_2 x_{i1}^2 + \beta_3 x_{i2} + \beta_4 x_{i3} + \varepsilon_i \end{aligned}$$

to relate y_i to x_{i1}, x_{i2}, and x_{i3}. Suppose that a sample of $n = 28$ combinations of y_i, x_{i1}, x_{i2}, and x_{i3} values is observed. From these observations the following calculations are made.

$$(\mathbf{X}'\mathbf{X})^{-1} = \begin{bmatrix} .005 & 0 & 0 & 0 & 0 \\ 0 & .01 & 0 & 0 & 0 \\ 0 & 0 & .001 & 0 & 0 \\ 0 & 0 & 0 & .03 & 0 \\ 0 & 0 & 0 & 0 & .008 \end{bmatrix}$$

$$\mathbf{X}'\mathbf{y} = \begin{bmatrix} 2000 \\ 500 \\ 8000 \\ 200 \\ 1000 \end{bmatrix} \quad \text{and} \quad \sum_{i=1}^{28} y_i^2 = 96{,}068$$

8.13 In the Exron Chemical Company situation,

a. Calculate the least squares point estimates of the parameters β_0, β_1, β_2, β_3, and β_4.
b. Write the least squares prediction equation for predicting yield.
c. Calculate $\mathbf{b}'\mathbf{X}'\mathbf{y}$, s^2, and s.

8.14 In the Exron Chemical Company situation, suppose that

$$\text{Total variation} = \sum_{i=1}^{28} (y_i - \bar{y})^2 = 3000$$

a. Use this fact and the information given previously to compute R^2, the multiple coefficient of determination. Interpret this value of R^2.
b. Set up the null and alternative hypotheses that are needed to carry out the overall F-test for the Exron model.
c. Calculate F(model) for the Exron model.
d. Using an appropriate rejection point with $\alpha = .05$, decide whether or not the null hypothesis that you set up in part (b) can be rejected.
e. What does the result of the overall F-test say about the adequacy of the Exron model? Explain.

8.15 In the Exron Chemical Company situation,

a. Calculate appropriate t statistics and use them to test the importance of each of the independent variables x_1, x_1^2, x_2, and x_3 with $\alpha = .05$ and $\alpha = .01$. What do you conclude about the importance of these variables?
b. Calculate and interpret a 99% confidence interval for each of the parameters β_0, β_1, β_2, β_3, and β_4.

8.16 In the Exron Chemical Company situation,

a. Calculate and interpret a 99% confidence interval for the average yield of the chemical reaction when 10 pounds of catalyst XST, 20 pounds of catalyst RST, and 10 pounds of catalyst ZST are used in the reaction.
b. Calculate and interpret a 99% prediction interval for an individual yield when 20 pounds of catalyst XST, 10 pounds of catalyst RST, and 25 pounds of catalyst ZST are used in the reaction.
c. Suppose that Exron decides to run the reaction using 20 pounds of catalyst XST, 10 pounds of catalyst RST, and 25 pounds of catalyst ZST. If the reaction is run many times using these amounts of the catalysts, find a reasonable estimate of the minimum average yield that might be obtained per run.

8.17 Suppose that we wish to employ the regression model

$$y = \beta_0 + \beta_1 x_1 + \beta_2 x_2 + \beta_3 x_3 + \beta_4 x_4 + \varepsilon$$

and that

$$\sum_{i=1}^{25} y_i^2 = 22{,}102 \qquad 25\bar{y}^2 = 21{,}802 \qquad \mathbf{b}'\mathbf{X}'\mathbf{y} = 22{,}057$$

a. Calculate the total variation $= \sum_{i=1}^{25} (y_i - \bar{y})^2$.
b. Calculate the unexplained variation $= \sum_{i=1}^{25} (y_i - \hat{y}_i)^2$.
c. Calculate the explained variation $= \sum_{i=1}^{25} (\hat{y}_i - \bar{y})^2$. Show how this can be calculated in two ways.
d. Calculate the multiple coefficient of determination, R^2. Show how this can be calculated in two ways.
e. Calculate the overall F statistic. Using this statistic and the appropriate rejection point, determine whether we can, by setting α equal to .05, reject $H_0: \beta_1 = \beta_2 = \beta_3 = \beta_4 = 0$ in favor of H_1: At least one of β_1, β_2, β_3, and β_4 does not equal zero.

8.18 Consider the construction project data given in Table 8.8 (see Exercise 8.10). Suppose that we employ the model

$$y = \beta_0 + \beta_1 x_1 + \beta_2 x_2 + \varepsilon$$

to relate y (profit) to x_1 (contract size) and x_2 (supervisor experience). The residual plots for this model versus x_1 and x_2 are given Figure 8.22. Note that x_1 and x_2 are denoted as CSIZE and SUPEXP on the SAS output of these plots.

a. Does the residual plot versus x_1 indicate the need for the quadratic term x_1^2? Explain.
b. Does the residual plot versus x_2 indicate the need for the quadratic term x_2^2? Explain.

8.19 In an article in *The Real Estate Appraiser and Analyst*, Andrews and Ferguson considered predicting

$$y = \text{sale price (in thousands of dollars) of a home}$$

on the basis of

$$x_1 = \text{home size (in hundreds of square feet)}$$

FIGURE 8.22 SAS output of residual plots for a regression analysis of the construction project data using the model $y = \beta_0 + \beta_1 x_1 + \beta_2 x_2 + \varepsilon$

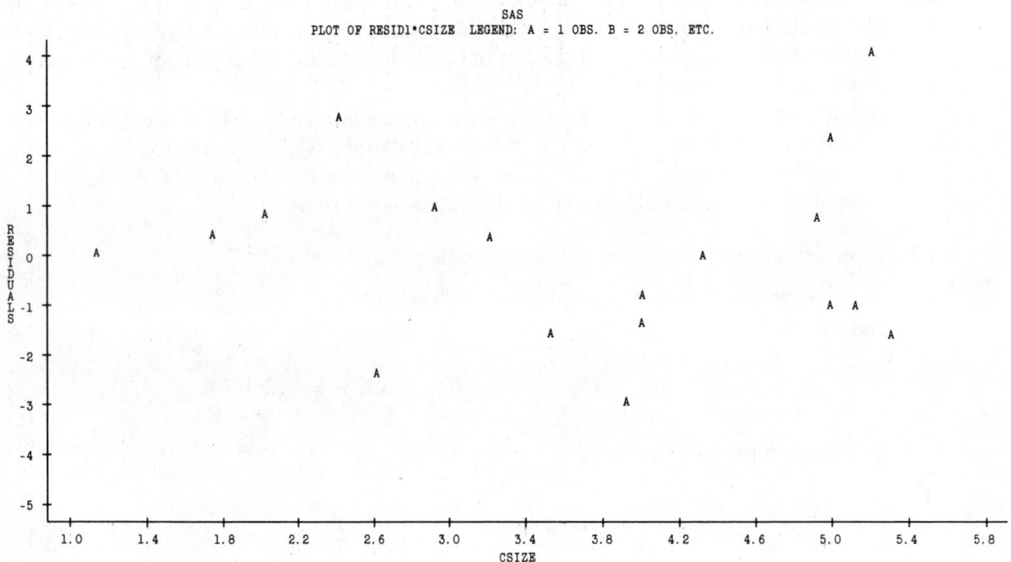

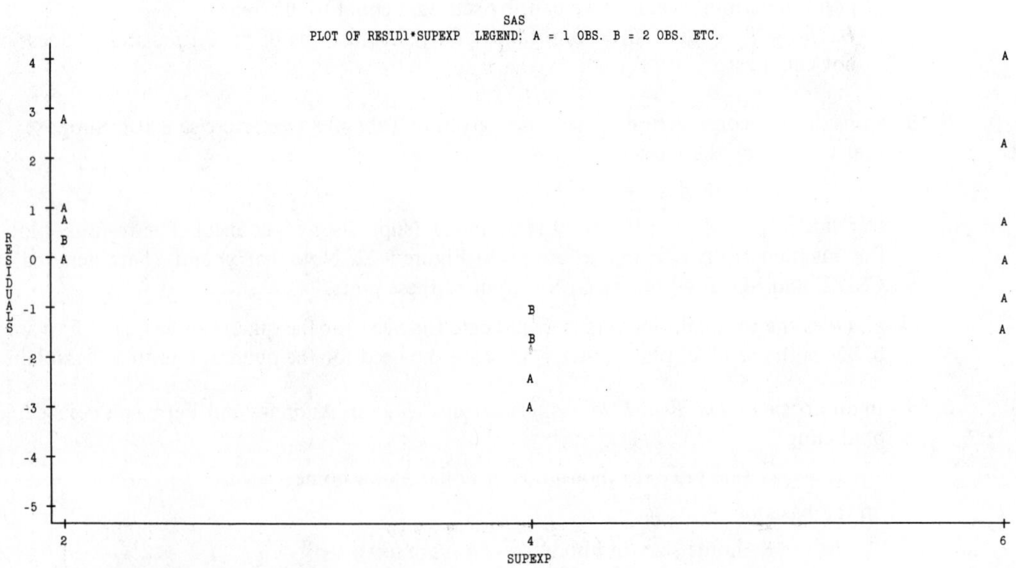

and

$$x_2 = \text{home condition rating (on a ten-point scale, 1 being worst and 10 being best)}$$

The data analyzed is given in Table 8.9. Figure 8.23 presents the SAS output resulting from analyzing this data by using the model

$$y = \beta_0 + \beta_1 x_1 + \beta_2 x_2 + \varepsilon$$

a. Use F(model) to test the significance of the overall regression model.
b. Test the importance of each of the independent variables x_1 and x_2. What do you conclude about the importance of these variables?
c. Compute and interpret 95% confidence intervals for the parameters β_1 and β_2.
d. Calculate a 95% prediction interval for the selling price of a 2000 square foot home with a condition rating of 9. *Hint*: Consider observation 3 and note that $s\sqrt{h_{00}} = .6375$.
e. Calculate a 95% confidence interval for the average selling price of all 2000 square foot homes with a condition rating of 9.

FIGURE 8.23 **SAS output of a regression analysis of the Andrews and Ferguson real estate sales data**

ANALYSIS OF VARIANCE

SOURCE	DF	SUM OF SQUARES	MEAN SQUARE	F VALUE	PROB>F
MODEL	2	819.32795	409.66398	350.866	0.0001
ERROR	7	8.17304573	1.16757796		
C TOTAL	9	827.50100			

ROOT MSE	1.080545	R-SQUARE	0.9901
DEP MEAN	51.73	ADJ R-SQ	0.9873
C.V.	2.088817		

PARAMETER ESTIMATES

VARIABLE	DF	PARAMETER ESTIMATE	STANDARD ERROR	T FOR H0: PARAMETER=0	PROB > |T|
INTERCEP	1	9.78227061	1.63048067	6.000	0.0005
X1	1	1.87093528	0.07617357	24.561	0.0001
X2	1	1.27814078	0.14440032	8.851	0.0001

OBS	ACTUAL	PREDICT VALUE	STD ERR PREDICT
1	60.0000	59.2045	0.4714
2	32.7000	32.9188	0.8200
3	57.7000	58.7042	0.6375
4	45.5000	45.4226	0.4919
5	47.0000	48.0714	0.6019
6	55.3000	54.1845	0.4276
7	64.5000	63.6317	0.5705
8	42.6000	41.7733	0.5710
9	54.5000	54.2770	0.4206
10	57.5000	59.1119	0.7657

SUM OF RESIDUALS 6.75016E-14
SUM OF SQUARED RESIDUALS 8.173046

TABLE 8.9 **Andrews and Ferguson real estate sales data**

Home size, x_1 (hundreds of square feet)	*Home condition rating, x_2*	*Sales price, y ($1000s)*
23	5	60.0
11	2	32.7
20	9	57.7
17	3	45.5
15	8	47.0
21	4	55.3
24	7	64.5
13	6	42.6
19	7	54.5
25	2	57.5

Source: R.L. Andrews and J.T. Ferguson, "Integrating Judgment with a Regression Appraisal," *The Real Estate Appraiser and Analyst*, Vol. 52, No. 2 (1986). Reprinted by permission.

8.20 Consider the Andrews and Ferguson real estate sales data and the SAS output of Figure 8.23, which presents the results from analyzing this data using the model

$$y = \beta_0 + \beta_1 x_1 + \beta_2 x_2 + \varepsilon$$

a. Compute the residuals for this model.
b. Construct residual plots versus x_1, x_2, and $\hat{y}$. Do these plots suggest any violations of the regression assumptions?
c. Compute and interpret the standardized residuals.
d. Construct and interpret a stem-and-leaf display of the residuals.
e. Construct and interpret a normal plot of the residuals.

8.21 Consider the data in Table 6.7 concerning the abrasive effect of a wear-testing machine. Recall that a plot of the residuals versus $\hat{y}$ for the model

$$y = \beta_0 + \beta_1 x + \varepsilon$$

indicates that there is a curved relationship between y and x. Figure 8.24 presents the SAS outputs resulting from using the models

$$y = \beta_0 + \beta_1 x + \beta_2 x^2 + \varepsilon$$

and

$$y = \beta_0 + \beta_1 x + \beta_2 x^2 + \beta_3 x^3 + \varepsilon$$

to relate y to x. Residual plots versus $\hat{y}$ are also presented for these models.

a. Determine which model seems most appropriate. Justify your choice.
b. For the most appropriate model, identify any observations that are outliers with respect to their x or y values.
c. Identify outliers that you think may be influential.

8.22 Consider the residential sales model

$$y = \beta_0 + \beta_1 x_1 + \beta_2 x_2 + \beta_3 x_3 + \beta_4 x_4 + \varepsilon$$

of Example 8.2, which relates sales price (y) to square footage (x_1), number of rooms

FIGURE 8.24a SAS output of a regression analysis of the wear-testing data using the model $y = \beta_0 + \beta_1 x + \beta_2 x^2 + \varepsilon$

QUADRATIC REGRESSION ANALYSIS

GENERAL LINEAR MODELS PROCEDURE

DEPENDENT VARIABLE: WEAR

SOURCE	DF	SUM OF SQUARES	MEAN SQUARE	F VALUE	PR > F	R-SQUARE	C.V.
MODEL	2	4839.89302381	2419.94651190	641.53	0.0001	0.966116	5.5604
ERROR	45	169.74614286	3.77213651		ROOT MSE		WEAR MEAN
CORRECTED TOTAL	47	5009.63916667			1.94219888		34.92916667

SOURCE	DF	TYPE I SS	F VALUE	PR > F	DF	TYPE III SS	F VALUE	PR > F
SPEED	1	4326.79207143	1147.04	0.0001	1	261.47616353	69.32	0.0001
SPEED*SPEED	1	513.10095238	136.02	0.0001	1	513.10095238	136.02	0.001

PARAMETER	ESTIMATE	T FOR HO: PARAMETER=0	PR > |T|	STD ERROR OF ESTIMATE
INTERCEPT	63.13928571	10.31	0.0001	6.12529888
SPEED	-0.70507143	-8.33	0.0001	0.08468583
SPEED*SPEED	0.00327679	11.66	0.0001	0.00028096

OBSERVATION	OBSERVED VALUE	PREDICTED VALUE	RESIDUAL
1	23.00000000	25.40000000	-2.40000000
2	23.50000000	25.40000000	-1.90000000
3	24.40000000	25.40000000	-1.00000000
4	25.20000000	25.40000000	-0.20000000
5	25.60000000	25.40000000	0.20000000
6	26.10000000	25.40000000	0.70000000
7	24.80000000	25.40000000	-0.60000000
8	25.60000000	25.40000000	0.20000000
9	26.70000000	25.71642857	0.98357143
10	26.10000000	25.71642857	0.38357143
11	25.80000000	25.71642857	0.08357143
12	26.30000000	25.71642857	0.58357143
13	27.20000000	25.71642857	1.48357143
14	27.90000000	25.71642857	2.18357143
15	28.30000000	25.71642857	2.58357143
16	27.40000000	25.71642857	1.68357143
17	28.00000000	28.65428571	-0.65428571
18	28.40000000	28.65428571	-0.25428571
19	27.00000000	28.65428571	-1.65428571
20	28.80000000	28.65428571	0.14571429
21	29.80000000	28.65428571	1.14571429
22	29.40000000	28.65428571	0.74571429
23	28.70000000	28.65428571	0.04571429
24	29.30000000	28.65428571	0.64571429
25	32.70000000	34.21357143	-1.51357143
26	32.10000000	34.21357143	-2.11357143
27	31.90000000	34.21357143	-2.31357143
28	33.00000000	34.21357143	-1.21357143
29	33.50000000	34.21357143	-0.71357143
30	33.70000000	34.21357143	-0.51357143
31	34.00000000	34.21357143	-0.21357143
32	32.50000000	34.21357143	-1.71357143
33	43.10000000	42.39428571	0.70571429
34	41.70000000	42.39428571	-0.69428571
35	42.40000000	42.39428571	0.00571429
36	42.10000000	42.39428571	-0.29428571

(continues)

FIGURE 8.24a Continued

OBSERVATION	OBSERVED VALUE	PREDICTED VALUE	RESIDUAL
37	43.50000000	42.39428571	1.10571429
38	43.80000000	42.39428571	1.40571429
39	44.20000000	42.39428571	1.80571429
40	43.60000000	42.39428571	1.20571429
41	54.20000000	53.19642857	1.00357143
42	43.70000000	53.19642857	-9.49642857
43	53.10000000	53.19642857	-0.09642857
44	53.80000000	53.19642857	0.60357143
45	55.60000000	53.19642857	2.40357143
46	55.90000000	53.19642857	2.70357143
47	54.70000000	53.19642857	1.50357143
48	54.50000000	53.19642857	1.30357143

SUM OF RESIDUALS	0.00000000
SUM OF SQUARED RESIDUALS	169.74614286
SUM OF SQUARED RESIDUALS-ERROR SS	0.00000000
FIRST ORDER AUTOCORRELATION	0.25731691
DURBIN-WATSON D	1.44142233

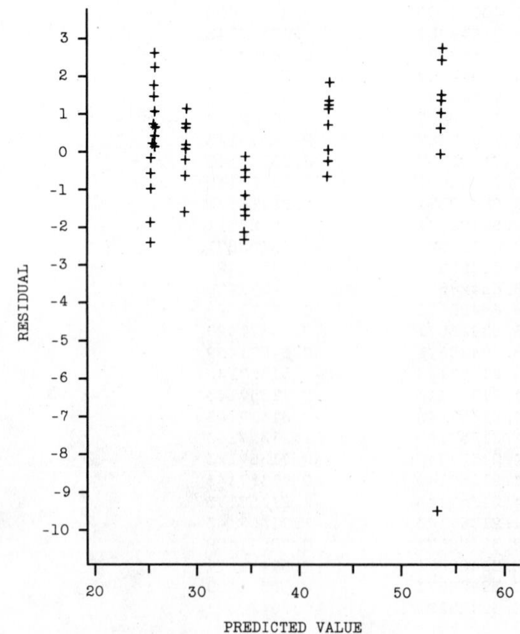

FIGURE 8.24b **SAS output of a regression analysis of the wear-testing data using the model $y = \beta_0 + \beta_1 x + \beta_2 x^2 + \beta_3 x^3 + \varepsilon$**

CUBIC REGRESSION ANALYSIS

GENERAL LINEAR MODELS PROCEDURE

DEPENDENT VARIABLE: WEAR

SOURCE	DF	SUM OF SQUARES	MEAN SQUARE	F VALUE	PR > F	R-SQUARE	C.V.
MODEL	3	4846.78202381	1615.59400794	436.49	0.0001	0.967491	5.5079
ERROR	44	162.85714286	3.70129870		ROOT MSE		WEAR MEAN
CORRECTED TOTAL	47	5009.63916667			1.92387596		34.92916667

SOURCE	DF	TYPE I SS	F VALUE	PR > F	DF	TYPE III SS	F VALUE	PR > F
SPEED	1	4326.79207143	1168.99	0.0001	1	0.43368923	0.12	0.7338
SPEED*SPEED	1	513.10095238	138.63	0.0001	1	1.67992922	0.45	0.5040
SPEED*SPEED*SPEED	1	6.88900000	1.86	0.1794	1	6.88900000	1.86	0.1794

PARAMETER	ESTIMATE	T FOR HO: PARAMETER=0	PR > |T|	STD ERROR OF ESTIMATE
INTERCEPT	18.87261905	0.57	0.5704	33.00952220
SPEED	0.23847718	0.34	0.7338	0.69668199
SPEED*SPEED	-0.00320759	-0.67	0.5040	0.00476113
SPEED*SPEED*SPEED	0.0000144	1.36	0.1794	0.00001056

OBSERVATION	OBSERVED VALUE	PREDICTED VALUE	RESIDUAL
1	23.00000000	25.05416667	-2.05416667
2	23.50000000	25.05416667	-1.55416667
3	24.40000000	25.05416667	-0.65416667
4	25.20000000	25.05416667	0.14583333
5	25.60000000	25.05416667	0.54583333
6	26.10000000	25.05416667	1.04583333
7	24.80000000	25.05416667	-0.25416667
8	25.60000000	25.05416667	0.54583333
9	26.70000000	26.20059524	0.49940476
10	26.10000000	26.20059524	-0.10059524
11	25.80000000	26.20059524	-0.40059524
12	26.30000000	26.20059524	0.09940476
13	27.20000000	26.20059524	0.99940476
14	27.90000000	26.20059524	1.69940476
15	28.30000000	26.20059524	2.09940476
16	27.40000000	26.20059524	1.19940476
17	28.00000000	28.93095238	-0.93095238
18	28.40000000	28.93095238	-0.53095238
19	27.00000000	28.93095238	-1.93095238
20	28.80000000	28.93095238	-0.13095238
21	29.80000000	28.93095238	0.86904762
22	29.40000000	28.93095238	0.46904762
23	28.70000000	28.93095238	-0.23095238
24	29.30000000	28.93095238	0.36904762
25	32.70000000	33.93690476	-1.23690476
26	32.10000000	33.93690476	-1.83690476
27	31.90000000	33.93690476	-2.03690476
28	33.00000000	33.93690476	-0.93690476
29	33.50000000	33.93690476	-0.43690476
30	33.70000000	33.93690476	-0.23690476
31	34.00000000	33.93690476	0.06309524
32	32.50000000	33.93690476	-1.43690476
33	43.10000000	41.91011905	1.18988095

(continues)

FIGURE 8.24b Continued

OBSERVATION	OBSERVED VALUE	PREDICTED VALUE	RESIDUAL
34	41.70000000	41.91011905	-0.21011905
35	42.40000000	41.91011905	0.48988095
36	42.10000000	41.91011905	0.18988095
37	43.50000000	41.91011905	1.58988095
38	43.80000000	41.91011905	1.88988095
39	44.20000000	41.91011905	2.28988095
40	43.60000000	41.91011905	1.68988095
41	54.20000000	53.54226190	0.65773810
42	43.70000000	53.54226190	-9.84226190
43	53.10000000	53.54226190	-0.44226190
44	53.80000000	53.54226190	0.25773810
45	55.60000000	53.54226190	2.05773810
46	55.90000000	53.54226190	2.35773810
47	54.70000000	53.54226190	1.15773810
48	54.50000000	53.54226190	0.95773810

SUM OF RESIDUALS	0.00000000
SUM OF SQUARED RESIDUALS	162.85714286
SUM OF SQUARED RESIDUALS-ERROR SS	-0.00000000
FIRST ORDER AUTOCORRELATION	0.23779258
DURBIN-WATSON D	1.49287270

CUBIC REGRESSION ANALYSIS
PLOT OF RESIDUALS VS. PREDICTED VALUES

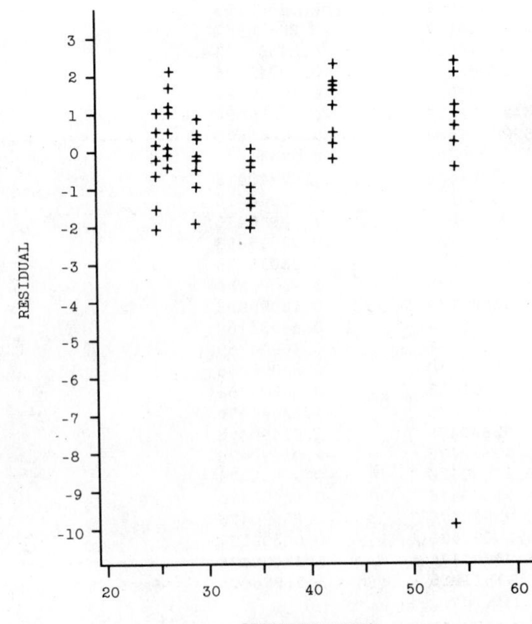

(x_2), number of bedrooms (x_3), and age (x_4). In Example 8.5 we calculated the least squares point estimates of the model parameters using the data in Table 8.2.

a. Interpret the meaning of the least squares point estimate $b_2 = 6.32178$.
b. Interpret the meaning of the least squares point estimate $b_3 = -11.10316$. Why is the negative sign here somewhat logical?
c. Interpret the meaning of the least squares point estimate $b_4 = -.43186$.

8.23 In an article in *Business Economics*, C. I. Allmon related

y = Crest toothpaste sales in a given year (in thousands of dollars)

to

x_1 = Crest advertising budget in the year (in thousands of dollars)

x_2 = ratio of Crest's advertising budget to Colgate's advertising budget in the year

and

x_3 = U.S. personal disposable income in the year (in billions of dollars)

The data analyzed is given in Table 8.10.

a. Using a computer, perform a regression analysis of the data by employing each of the models

$$y = \beta_0 + \beta_1 x_1 + \beta_2 x_2 + \beta_3 x_3 + \varepsilon$$
$$y = \beta_0 + \beta_1 x_1 + \beta_2 x_2 + \beta_3 x_3 + \beta_4 x_1^2 + \beta_5 x_2^2 + \beta_6 x_3^2 + \varepsilon$$

b. Test the overall significance of each model.
c. Which variables seem to be important in each model? Explain.
d. Write the prediction equations obtained from each model.
e. Which of the two models seems best? Explain.

TABLE 8.10 **Crest toothpaste sales data**

Year	*Crest sales, y*	*Crest budget, x_1*	*Ratio, x_2*	*U.S. personal disposable income, x_3*
1967	105,000	16,300	1.25	547.9
1968	105,000	15,800	1.34	593.4
1969	121,600	16,000	1.22	638.9
1970	113,750	14,200	1.00	695.3
1971	113,750	15,000	1.15	751.8
1972	128,925	14,000	1.13	810.3
1973	142,500	15,400	1.05	914.5
1974	126,000	18,250	1.27	998.3
1975	162,000	17,300	1.07	1,096.1
1976	191,625	23,000	1.17	1,194.4
1977	189,000	19,300	1.07	1,311.5
1978	210,000	23,056	1.54	1,462.9
1979	224,250	26,000	1.59	1,641.7
1980	245,000	28,000	1.56	1,821.7

Source: C.I. Allmon, "Advertising and Sales Relationships for Toothpaste: Another Look," *Business Economics*, September 1982, pp. 17, 58. Reprinted by permission.

8.24 Using a computer, perform a complete analysis of the residuals for each of the models given in Exercise 8.23. Interpret the results.

8.25 United Oil Company is attempting to develop a reasonably priced unleaded gasoline that will deliver higher gasoline mileages than can be achieved by its current unleaded gasolines. As part of its development process, United Oil wishes to study the effect of two independent variables—x_1, amount of gasoline additive RST (0, 1, or 2 units), and x_2, amount of gasoline additive XST (0, 1, 2, or 3 units) on the gasoline mileage y obtained by an automobile called the Empire. For testing purposes a sample of $n = 22$ Empires is randomly selected. Each Empire is test driven under normal driving conditions. The combinations of x_1 and x_2 used in the experiment, along with the corresponding values of y, are given in Table 8.11. Figure 8.25 presents SAS output of a regression analysis of the mileage data using each of the models

$$y = \beta_0 + \beta_1 x_1 + \beta_2 x_2 + \varepsilon \quad \text{(Model 1)}$$

$$y = \beta_0 + \beta_1 x_1 + \beta_2 x_1^2 + \beta_3 x_2 + \varepsilon \quad \text{(Model 2)}$$

$$y = \beta_0 + \beta_1 x_1 + \beta_2 x_2 + \beta_3 x_2^2 + \varepsilon \quad \text{(Model 3)}$$

$$y = \beta_0 + \beta_1 x_1 + \beta_2 x_1^2 + \beta_3 x_2 + \beta_4 x_2^2 + \varepsilon \quad \text{(Model 4)}$$

Residual plots versus $\hat{y}$ are also given for Models 1 and 4.

a. Test the overall significance of each model.
b. Which variables seem to be important in each model? Explain.
c. What do the residual plots say about the functional forms of models 1 and 4? Explain.
d. Which model seems best? Explain.

TABLE 8.11 **United oil company unleaded gasoline mileage data**

Gasoline mileage, y_i (mpg)	*Amount of gasoline additive RST, x_{i1}*	*Amount of gasoline additive XST, x_{i2}*
27.4	0	0
28.0	0	0
28.6	0	0
29.6	1	0
30.6	1	0
28.6	2	0
29.8	2	0
32.0	0	1
33.0	0	1
33.3	1	1
34.5	1	1
32.3	0	2
33.5	0	2
34.4	1	2
35.0	1	2
35.6	1	2
33.3	2	2
34.0	2	2
34.7	2	2
33.4	1	3
32.0	2	3
33.0	2	3

FIGURE 8.25 **SAS output of regression analysis of the mileage data using Models 1, 2, 3, and 4**

(a) Model 1

SAS

DEP VARIABLE: MILEAGE

ANALYSIS OF VARIANCE

SOURCE	DF	SUM OF SQUARES	MEAN SQUARE	F VALUE	PROB>F
MODEL	2	71.80608790	35.90304395	12.254	0.0004
ERROR	19	55.66663937	2.92982312		
C TOTAL	21	127.47273			

ROOT MSE	1.711673	R-SQUARE	0.5633
DEP MEAN	32.11818	ADJ R-SQ	0.5173
C.V.	5.329295		

PARAMETER ESTIMATES

VARIABLE	DF	PARAMETER ESTIMATE	STANDARD ERROR	T FOR HO: PARAMETER=0	PROB > |T|
INTERCEP	1	29.85320269	0.66849522	44.657	0.0001
RST	1	0.03073998	0.48323057	0.064	0.9499
XST	1	1.69494004	0.36327165	4.666	0.0002

(b) Model 2

SAS

DEP VARIABLE: MILEAGE

ANALYSIS OF VARIANCE

SOURCE	DF	SUM OF SQUARES	MEAN SQUARE	F VALUE	PROB>F
MODEL	3	86.64666304	28.88222101	12.734	0.0001
ERROR	18	40.82606423	2.26811468		
C TOTAL	21	127.47273			

ROOT MSE	1.506026	R-SQUARE	0.6797
DEP MEAN	32.11818	ADJ R-SQ	0.6263
C.V.	4.689014		

PARAMETER ESTIMATES

VARIABLE	DF	PARAMETER ESTIMATE	STANDARD ERROR	T FOR HO: PARAMETER=0	PROB > |T|
INTERCEP	1	29.26289858	0.63183159	46.314	0.0001
RST	1	3.46360078	1.40777350	2.460	0.0242
RSTSQ	1	-1.70893289	0.66808565	-2.558	0.0198
XST	1	1.65995165	0.31991944	5.189	0.0001

(continues)

FIGURE 8.25 Continued

(c) Model 3

SAS

DEP VARIABLE: MILEAGE

ANALYSIS OF VARIANCE

SOURCE	DF	SUM OF SQUARES	MEAN SQUARE	F VALUE	PROB>F
MODEL	3	110.72772	36.90924061	39.675	0.0001
ERROR	18	16.74500545	0.93027808		
C TOTAL	21	127.47273			

ROOT MSE	0.9645092	R-SQUARE	0.8686
DEP MEAN	32.11818	ADJ R-SQ	0.8467
C.V.	3.003001		

PARAMETER ESTIMATES

VARIABLE	DF	PARAMETER ESTIMATE	STANDARD ERROR	T FOR H0: PARAMETER=0	PROB > |T|
INTERCEP	1	28.57166177	0.42561667	67.130	0.0001
RST	1	0.50649308	0.28205418	1.796	0.0893
XST	1	5.52741962	0.62686683	8.818	0.0001
XSTSQ	1	-1.48277104	0.22923713	-6.468	0.0001

(d) Model 4

SAS

DEP VARIABLE: MILEAGE

ANALYSIS OF VARIANCE

SOURCE	DF	SUM OF SQUARES	MEAN SQUARE	F VALUE	PROB>F
MODEL	4	120.71374	30.17843561	75.904	0.0001
ERROR	17	6.75898482	0.39758734		
C TOTAL	21	127.47273			

ROOT MSE	0.6305453	R-SQUARE	0.9470
DEP MEAN	32.11818	ADJ R-SQ	0.9345
C.V.	1.963204		

PARAMETER ESTIMATES

VARIABLE	DF	PARAMETER ESTIMATE	STANDARD ERROR	T FOR H0: PARAMETER=0	PROB > |T|
INTERCEP	1	28.15891841	0.29017813	97.040	0.0001
RST	1	3.31330645	0.58963231	5.619	0.0001
RSTSQ	1	-1.41107685	0.28155984	-5.012	0.0001
XST	1	5.27521347	0.41289076	12.776	0.0001
XSTSQ	1	-1.39637097	0.15085150	-9.257	0.0001

(e) Residual plot versus $\hat{y}$ for the model $y = \beta_0 + \beta_1 x_1 + \beta_2 x_2 + \varepsilon$ (Model 1)

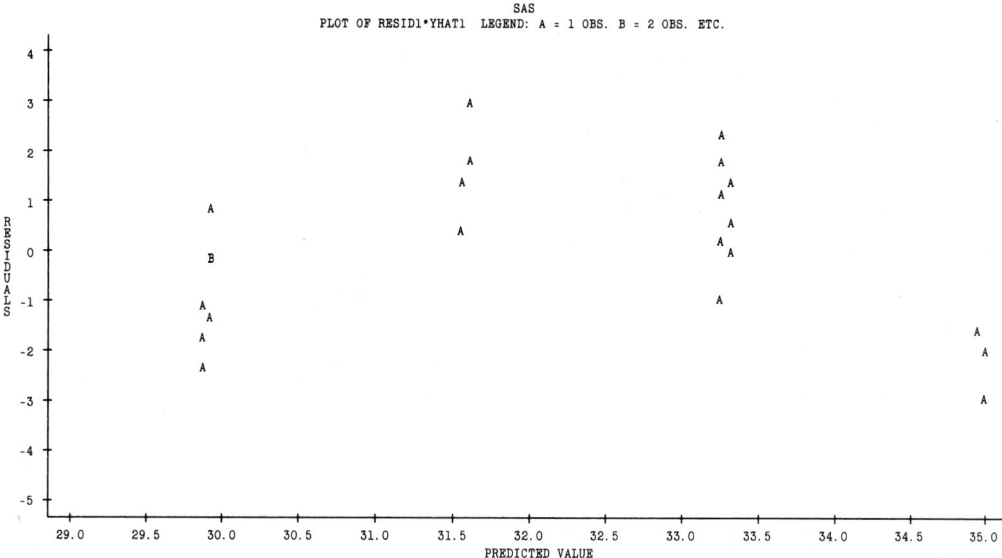

(f) Residual plot versus $\hat{y}$ for the model $y = \beta_0 + \beta_1 x_1 + \beta_2 x_1^2 + \beta_3 x_2 + \beta_4 x_2^2 + \varepsilon$ (Model 4)

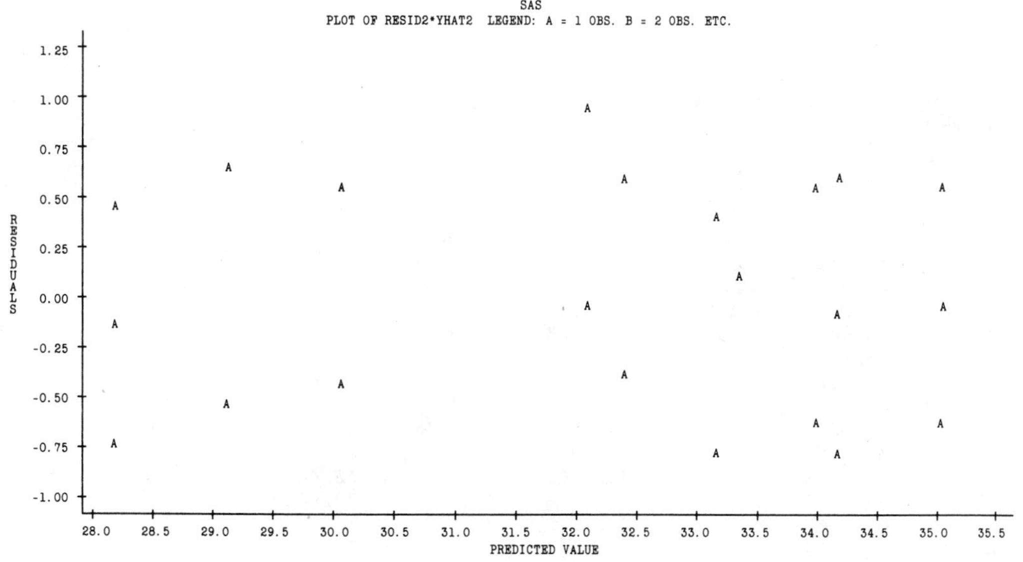

8.26 Suppose that the regression model

$$y = \beta_0 + \beta_1 x_1 + \beta_2 x_1^2 + \beta_3 x_2 + \beta_4 x_2^2 + \varepsilon$$

is an appropriate model relating y (gasoline mileage) to x_1 (amount of additive RST) and x_2 (amount of additive XST) in the United Oil Company situation of Exercise 8.25.

a. Specify the vector **y** and matrix **X** used to calculate the least squares point estimates of the parameters in this model.

b. The least squares point estimates of the model parameters are $b_0 = 28.1589$, $b_1 = 3.3133$, $b_2 = -1.4111$, $b_3 = 5.2752$, and $b_4 = -1.3964$. Noting from Table 8.11 that the combination of one unit of gasoline additive RST and two units of gasoline additive XST seems to maximize gasoline mileage, assume that United Oil Company will use this combination to make its unleaded gasoline.
  (1) Discuss the meaning of μ_0, the mean gasoline mileage (corresponding to the combination $x_{01} = 1$ and $x_{02} = 2$).
  (2) Discuss the conceptual difference between μ_0 and $y_0 = \mu_0 + \varepsilon_0$, the (individual) gasoline mileage corresponding to the above combination.
  (3) Using an appropriate prediction equation, calculate a point estimate of μ_0 and a point prediction of $y_0 = \mu_0 + \varepsilon_0$.

c. Calculate a 95% confidence interval for μ_0 and a 95% prediction interval for y_0. Use the facts that $SSE = 6.7590$ and $\mathbf{x}_0'(\mathbf{X}'\mathbf{X})^{-1}\mathbf{x}_0 = .1565$.

d. In reference to part (c), specify $\mathbf{x}_0'$.

8.27 Figure 8.26 gives the SAS output of the predictions and residuals obtained when we carry out a regression analysis of the United Oil Company gas mileage data in Table 8.11 using the model $y = \beta_0 + \beta_1 x_1 + \beta_2 x_1^2 + \beta_3 x_2 + \beta_4 x_2^2 + \varepsilon$.

FIGURE 8.26 **SAS output of the predictions and residuals obtained when using the mileage model $y = \beta_0 + \beta_1 x_1 + \beta_2 x_1^2 + \beta_3 x_2 + \beta_4 x_2^2 + \varepsilon$**

OBS	ACTUAL	PREDICT VALUE	STD ERR PREDICT	LOWER95% MEAN	UPPER95% MEAN	LOWER95% PREDICT	UPPER95% PREDICT	RESIDUAL
1	27.4000	28.1589	0.2902	27.5467	28.7711	26.6945	29.6234	-0.7589
2	28.0000	28.1589	0.2902	27.5467	28.7711	26.6945	29.6234	-0.1589
3	28.6000	28.1589	0.2902	27.5467	28.7711	26.6945	29.6234	0.4411
4	29.6000	30.0611	0.3071	29.4133	30.7090	28.5815	31.5408	-0.4611
5	30.6000	30.0611	0.3071	29.4133	30.7090	28.5815	31.5408	0.5389
6	28.6000	29.1412	0.3317	28.4413	29.8411	27.6380	30.6444	-0.5412
7	29.8000	29.1412	0.3317	28.4413	29.8411	27.6380	30.6444	0.6588
8	32.0000	32.0378	0.2677	31.4729	32.6026	30.5925	33.4830	-0.0378
9	33.0000	32.0378	0.2677	31.4729	32.6026	30.5925	33.4830	0.9622
10	33.3000	33.9400	0.2661	33.3785	34.5015	32.4960	35.3839	-0.6400
11	34.5000	33.9400	0.2661	33.3785	34.5015	32.4960	35.3839	0.5600
12	32.3000	33.1239	0.2892	32.5137	33.7341	31.6603	34.5875	-0.8239
13	33.5000	33.1239	0.2892	32.5137	33.7341	31.6603	34.5875	0.3761
14	34.4000	35.0261	0.2495	34.4997	35.5524	33.5954	36.4568	-0.6261
15	35.0000	35.0261	0.2495	34.4997	35.5524	33.5954	36.4568	-0.0261
16	35.6000	35.0261	0.2495	34.4997	35.5524	33.5954	36.4568	0.5739
17	33.3000	34.1062	0.2867	33.5013	34.7110	32.6448	35.5675	-0.8062
18	34.0000	34.1062	0.2867	33.5013	34.7110	32.6448	35.5675	-0.1062
19	34.7000	34.1062	0.2867	33.5013	34.7110	32.6448	35.5675	0.5938
20	33.4000	33.3194	0.4252	32.4224	34.2165	31.7149	34.9240	0.0806
21	32.0000	32.3995	0.3607	31.6384	33.1606	30.8669	33.9322	-0.3995

a. Identify and interpret the 95% confidence interval for the mean gasoline mileage obtained when using two units of additive RST and two units of additive XST.
b. Identify and interpret the 95% prediction interval for an individual gasoline mileage obtained when using one unit of additive RST and three units of additive XST.
c. Plot the residuals versus x_1 and x_2. What do these plots say about the equal variance assumption?
d. Carry out the residual analysis needed to check the normality assumption.

8.28 Consider the service call data given in Exercise 6.26 (see Table 6.11). Figure 8.27 presents the SAS output obtained when we analyze the data using each of the models

$$y = \beta_0 + \beta_1 x + \varepsilon \qquad \text{(Model 1)}$$
$$y = \beta_0 + \beta_1 x + \beta_2 x^2 + \varepsilon \qquad \text{(Model 2)}$$
$$y = \beta_0 + \beta_1 x + \beta_2 x^2 + \beta_3 x^3 + \varepsilon \qquad \text{(Model 3)}$$

FIGURE 8.27 SAS output of regression analysis of the service time data using Models 1, 2, and 3

(a) Model 1

SAS

DEP VARIABLE: SERVTIME

ANALYSIS OF VARIANCE

SOURCE	DF	SUM OF SQUARES	MEAN SQUARE	F VALUE	PROB>F
MODEL	1	210389.92	210389.92	201.443	0.0001
ERROR	15	15666.20059	1044.41337		
C TOTAL	16	226056.12			

ROOT MSE	32.31738	R-SQUARE	0.9307
DEP MEAN	183.5882	ADJ R-SQ	0.9261
C.V.	17.60319		

PARAMETER ESTIMATES

VARIABLE	DF	PARAMETER ESTIMATE	STANDARD ERROR	T FOR H0: PARAMETER=0	PROB > |T|
INTERCEP	1	15.53308991	14.19989741	1.094	0.2913
COPIERS	1	28.56937471	2.01291060	14.193	0.0001

(b) Model 2

SAS

DEP VARIABLE: SERVTIME

ANALYSIS OF VARIANCE

SOURCE	DF	SUM OF SQUARES	MEAN SQUARE	F VALUE	PROB>F
MODEL	2	217491.08	108745.54	177.750	0.0001
ERROR	14	8565.03314	611.78808		
C TOTAL	16	226056.12			

ROOT MSE	24.73435	R-SQUARE	0.9621
DEP MEAN	183.5882	ADJ R-SQ	0.9567
C.V.	13.47273		

(continues)

FIGURE 8.27 Continued

PARAMETER ESTIMATES

VARIABLE	DF	PARAMETER ESTIMATE	STANDARD ERROR	T FOR HO: PARAMETER=0	PROB > |T|
INTERCEP	1	70.19159148	19.37783293	3.622	0.0028
COPIERS	1	6.66705962	6.61075594	1.009	0.3303
COPIERSQ	1	1.49058745	0.43751514	3.407	0.0043

(c) Model 3

SAS

DEP VARIABLE: SERVTIME

ANALYSIS OF VARIANCE

SOURCE	DF	SUM OF SQUARES	MEAN SQUARE	F VALUE	PROB>F
MODEL	3	221045.24	73681.74704	191.157	0.0001
ERROR	13	5010.87652	385.45204		
C TOTAL	16	226056.12			

ROOT MSE	19.63293	R-SQUARE	0.9778
DEP MEAN	183.5882	ADJ R-SQ	0.9727
C.V.	10.694		

PARAMETER ESTIMATES

VARIABLE	DF	PARAMETER ESTIMATE	STANDARD ERROR	T FOR HO: PARAMETER=0	PROB > |T|
INTERCEP	1	130.85446	25.21267543	5.190	0.0002
COPIERS	1	-34.54311632	14.55040387	-2.374	0.0337
COPIERSQ	1	8.41574297	2.30687539	3.648	0.0029
COPIERCU	1	-0.31594394	0.10404637	-3.037	0.0095

OBS	ACTUAL	PREDICT VALUE	STD ERR PREDICT	LOWER95% MEAN	UPPER95% MEAN	LOWER95% PREDICT	UPPER95% PREDICT	RESIDUAL
1	92.0000	94.4363	6.6726	80.0209	108.9	49.6392	139.2	-2.4363
2	401.0	409.9	11.1561	385.8	434.0	361.1	458.7	-8.9257
3	101.0	92.9036	7.8631	75.9166	109.9	47.2141	138.6	8.0964
4	145.0	158.3	7.4141	142.3	174.3	113.0	203.7	-13.3186
5	229.0	231.4	8.9721	212.0	250.7	184.7	278.0	-2.3538
6	87.0000	92.9036	7.8631	75.9166	109.9	47.2141	138.6	-5.9036
7	428.0	429.8	16.7808	393.5	466.0	374.0	485.6	-1.7863
8	90.0000	107.1	7.2467	91.4579	122.8	61.9020	152.3	-17.1135
9	119.0	129.0	7.4571	112.9	145.1	83.6687	174.4	-10.0395
10	145.0	158.3	7.4141	142.3	174.3	113.0	203.7	-13.3186
11	123.0	92.9036	7.8631	75.9166	109.9	47.2141	138.6	30.0964
12	115.0	107.1	7.2467	91.4579	122.8	61.9020	152.3	7.8865
13	154.0	158.3	7.4141	142.3	174.3	113.0	203.7	-4.3186
14	366.0	348.7	11.4144	324.0	373.3	299.6	397.7	17.3363
15	134.0	94.4363	6.6726	80.0209	108.9	49.6392	139.2	39.5637
16	321.0	311.1	11.3687	286.5	335.6	262.0	360.1	9.9463
17	71.0000	104.4	13.9873	74.1935	134.6	52.3335	156.5	-33.4111

SUM OF RESIDUALS	-7.38964E-13
SUM OF SQUARED RESIDUALS	5010.877
PREDICTED RESID SS (PRESS)	10297.19

(d) Plot of residuals for Model 1 versus *x*

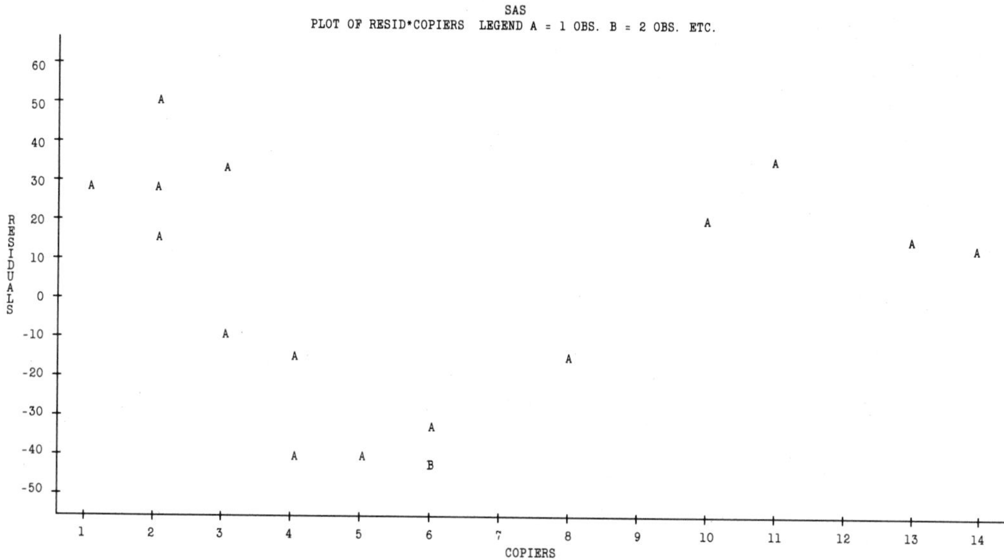

(e) Plot of residuals for Model 2 versus *x*

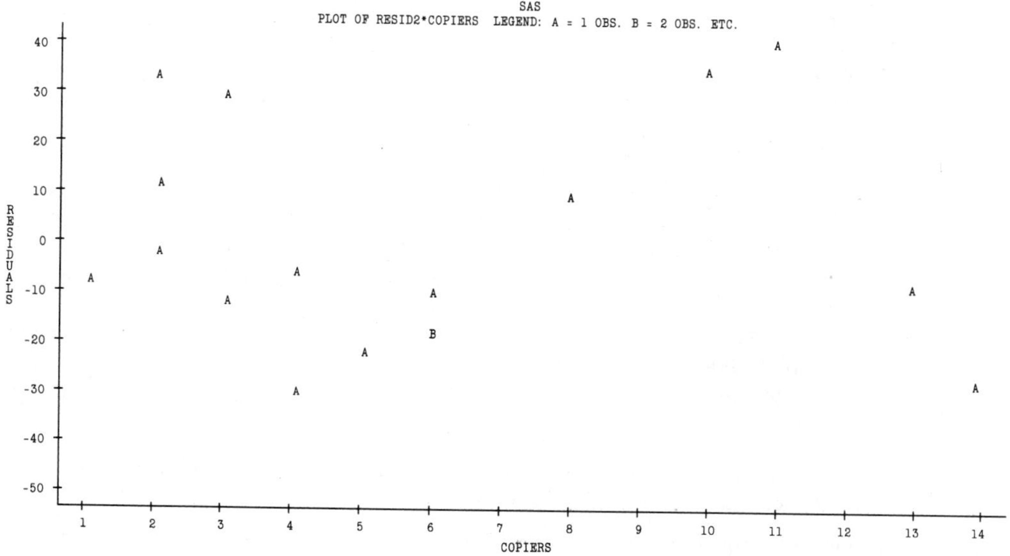

(continues)

FIGURE 8.27 Continued

(f) Plot of residuals for Model 3 versus *x*

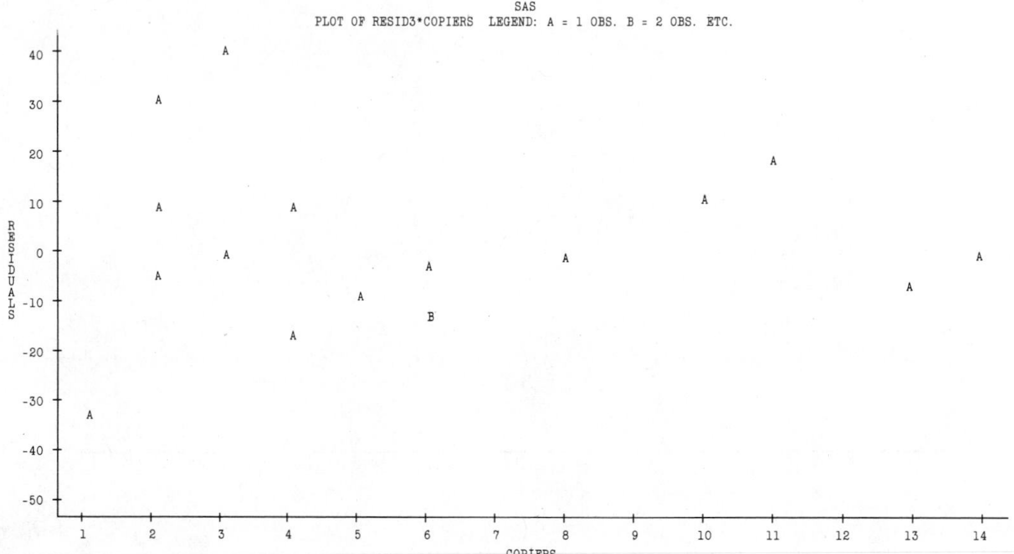

to relate y (service time) to x (number of copiers serviced). Residual plots versus x (denoted COPIERS on the SAS output) are also given for each model.

a. Use the SAS output to determine which of the above models seems best. Justify your answer.
b. What do the residual plots versus x say about the functional form of a model relating y to x? Explain.

8.29 Consider the service call situation of Exercise 8.28. Using the SAS output in Figure 8.27 and the model that seems to best relate y to x,

a. Find and report the 95% confidence interval for the mean service time when servicing eight copiers. Interpret the interval.
b. Compute a 99% confidence interval for the mean service time when servicing eleven copiers.
c. Find and report the 95% prediction interval for an individual service time when servicing six copiers. Interpret the interval.
d. Compute a 99% prediction interval for an individual service time when servicing ten copiers.

8.30 Consider the service call situation of Exercise 8.28. Using the SAS output in Figure 8.27 and the model that seems to best relate y to x:

a. Plot the residuals versus $\hat{y}$. Interpret the plot.
b. Carry out appropriate residual analysis to check the validity of the normality assumption. Fully interpret the results.

8.31 Figure 8.28 gives the SAS output of a regression analysis that relates y (daily room rate) to x_1 (population), x_2 (hotel rating), and x_3 (number of units) for 35 hotels located in the midwest United States using the model

$$y = \beta_0 + \beta_1 x_1 + \beta_2 x_2 + \beta_3 x_3 + \varepsilon$$

Here, we do not give values of x_1, x_2, and x_3 for purposes of confidentiality. Using the output,

a. Test the adequacy of the overall model by using F(model) and its associated prob-value.
b. Assess the importance of each of the independent variables by using appropriate t statistics and associated prob-values.
c. Find and report the standard error, total variation, explained variation, unexplained variation, and R^2. Interpret the value of R^2.
d. Do you think this model adequately relates y to x_1, x_2, and x_3? Explain.

8.32 Consider the hotel room rate situation of Exercise 8.31. Using the SAS output in Figure 8.28 and the model

$$y = \beta_0 + \beta_1 x_1 + \beta_2 x_2 + \beta_3 x_3 + \varepsilon$$

do the following:

a. Construct a residual plot that can be used to check the validity of the equal variances assumption. Interpret the plots.
b. Carry out and interpret the residual analysis needed to check the validity of the normality assumption.
c. Do we need to check the validity of the independence assumption? Explain why or why not.

FIGURE 8.28 **SAS output of a regression analysis of the hotel room rate data using the model $y = \beta_0 + \beta_1 x_1 + \beta_2 x_2 + \beta_3 x_3 + \varepsilon$**

SAS

DEP VARIABLE: RATE

ANALYSIS OF VARIANCE

SOURCE	DF	SUM OF SQUARES	MEAN SQUARE	F VALUE	PROB>F
MODEL	3	3825.66181	1275.22060	13.843	0.0001
ERROR	31	2855.65419	92.11787724		
C TOTAL	34	6681.31600			

ROOT MSE	9.597806	R-SQUARE	0.5726
DEP MEAN	49.08	ADJ R-SQ	0.5312
C.V.	19.55543		

PARAMETER ESTIMATES

VARIABLE	DF	PARAMETER ESTIMATE	STANDARD ERROR	T FOR H0: PARAMETER=0	PROB > |T|
INTERCEP	1	12.94848766	6.81089459	1.901	0.0666
POP	1	0.000391170	0.000946328	0.413	0.6822
RATING	1	10.06519873	3.20747265	3.138	0.0037
UNITS	1	0.08174470	0.03343919	2.445	0.0204

(continues)

FIGURE 8.28 Continued

OBS	ACTUAL	PREDICT VALUE	STD ERR PREDICT	LOWER95% MEAN	UPPER95% MEAN	LOWER95% PREDICT	UPPER95% PREDICT	RESIDUAL
1	45.0000	38.0958	2.4588	33.0811	43.1106	17.8890	58.3027	6.9042
2	29.0000	28.1124	4.2099	19.5263	36.6985	6.7374	49.4874	0.8876
3	47.0000	44.9746	2.3211	40.2408	49.7083	24.8356	65.1135	2.0254
4	38.0000	37.6175	2.7223	32.0655	43.1696	17.2707	57.9644	0.3825
5	58.0000	56.0207	2.5242	50.8726	61.1688	35.7803	76.2610	1.9793
6	50.0000	56.1024	2.5261	50.9504	61.2545	35.8611	76.3438	-6.1024
7	41.5000	50.5977	3.3080	43.8511	57.3444	29.8930	71.3025	-9.0977
8	38.0000	36.6694	2.7616	31.0370	42.3018	16.3005	57.0383	1.3306
9	46.0000	51.3514	3.2237	44.7766	57.9262	30.7020	72.0008	-5.3514
10	53.5000	41.3281	2.2497	36.7398	45.9164	21.2228	61.4333	12.1719
11	84.0000	70.6077	5.1821	60.0387	81.1767	48.3620	92.8535	13.3923
12	48.0000	57.6103	2.4910	52.5299	62.6907	37.3871	77.8336	-9.6103
13	58.0000	65.3761	3.4270	58.3867	72.3654	44.5910	86.1612	-7.3761
14	55.0000	41.4556	2.0721	37.2295	45.6817	21.4299	61.4813	13.5444
15	43.8000	42.8261	2.3657	38.0012	47.6509	22.6655	62.9866	0.9739
16	52.5000	43.9091	2.6592	38.4856	49.3326	23.5969	64.2212	8.5909
17	35.0000	35.6251	3.0129	29.4803	41.7699	15.1086	56.1416	-0.6251
18	49.0000	53.6868	2.8750	47.8232	59.5504	33.2527	74.1209	-4.6868
19	54.0000	53.1244	2.9035	47.2027	59.0461	32.6736	73.5752	0.8756
20	48.0000	52.9362	3.0261	46.7645	59.1080	32.4116	73.4609	-4.9362
21	42.0000	37.4587	2.6083	32.1391	42.7783	17.1740	57.7434	4.5413
22	35.5000	45.5260	2.8075	39.8000	51.2520	25.1310	65.9210	-10.0260
23	32.5000	38.0861	2.4614	33.0659	43.1062	17.8779	58.2942	-5.5861
24	36.0000	47.1949	3.5224	40.0109	54.3789	26.3435	68.0463	-11.1949
25	55.0000	52.6840	2.9733	46.6199	58.7481	32.1915	73.1765	2.3160
26	32.0000	49.2433	3.1347	42.8501	55.6365	28.6510	69.8356	-17.2433
27	40.0000	27.8973	4.5145	18.6900	37.1047	6.2653	49.5294	12.1027
28	54.0000	59.7409	3.5875	52.4242	67.0575	38.8434	80.6383	-5.7409
29	30.0000	49.1852	4.0224	40.9816	57.3888	27.9610	70.4095	-19.1852
30	69.5000	58.3043	3.8933	50.3638	66.2447	37.1803	79.4282	11.1957
31	72.0000	61.7376	3.4727	54.6550	68.8201	40.9209	82.5542	10.2624
32	60.0000	55.1162	4.6441	45.6446	64.5879	33.3704	76.8621	4.8838
33	35.0000	47.5851	4.1344	39.1531	56.0172	26.2715	68.8987	-12.5851
34	83.0000	65.1700	3.3068	58.4257	71.9142	44.4660	85.8739	17.8300
35	68.0000	64.8430	3.2201	58.2756	71.4104	44.1959	85.4900	3.1570

8.33 Consider the Crest toothpaste sales data of Table 8.10 and the regression model

$$y = \beta_0 + \beta_1 x_1 + \beta_2 x_2 + \beta_3 x_3 + \varepsilon$$

a. Using a computer, attempt to improve the above model by employing one or more quadratic terms. Can the model be improved? Carefully justify your answer.
b. Using a computer, attempt to improve the above model by using third-order terms (or higher-order terms of your choice). Can the model be improved? Carefully justify your answer.

8.34 Consider the residential sales data in Table 8.2. In Exercise 6.27 we found that residences 10, 50, and 61 in Table 4.3 (which are the same residences as those in Table 8.2) are possibly outlying and influential. Figure 8.29 gives the SAS output of a regression

FIGURE 8.29 **SAS output of a regression analysis of the residential sales data with residences 10, 50, and 61 removed using the model $y = \beta_0 + \beta_1 x_1 + \beta_2 x_1^2 + \beta_3 x_2 + \beta_4 x_2^2 + \beta_5 x_3 + \beta_6 x_3^2 + \beta_7 x_4 + \varepsilon$**

SAS

DEP VARIABLE: SALEP

ANALYSIS OF VARIANCE

SOURCE	DF	SUM OF SQUARES	MEAN SQUARE	F VALUE	PROB>F
MODEL	7	42692.82563	6098.97509	49.254	0.0001
ERROR	52	6438.94020	123.82577		
C TOTAL	59	49131.76583			

ROOT MSE	11.1277	R-SQUARE	0.8689
DEP MEAN	74.44167	ADJ R-SQ	0.8513
C.V.	14.94822		

PARAMETER ESTIMATES

VARIABLE	DF	PARAMETER ESTIMATE	STANDARD ERROR	T FOR HO: PARAMETER=0	PROB > |T|
INTERCEP	1	44.76848465	22.85500515	1.959	0.0555
SQFT	1	-0.05785964	0.02973061	-1.946	0.0571
SQFTSQ	1	0.000039090	.00000986478	3.963	0.0002
ROOMS	1	-11.88247653	6.99496011	-1.699	0.0953
ROOMSSQ	1	1.15033560	0.46272143	2.486	0.0162
BED	1	54.85589691	20.24428721	2.710	0.0091
BEDSQ	1	-11.31976861	3.51458392	-3.221	0.0022
AGE	1	-0.26334189	0.06943653	-3.793	0.0004

OBS	ACTUAL	PREDICT VALUE	STD ERR PREDICT	LOWER95% MEAN	UPPER95% MEAN	LOWER95% PREDICT	UPPER95% PREDICT	RESIDUAL
1	53.5000	50.7254	2.6975	45.3125	56.1382	27.7493	73.7014	2.7746
2	49.0000	58.5054	2.0781	54.3353	62.6755	35.7900	81.2208	-9.5054
3	50.5000	57.4340	6.6969	43.9957	70.8723	31.3728	83.4952	-6.9340
4	49.9000	45.7519	4.2163	37.2913	54.2125	21.8734	69.6303	4.1481
5	52.0000	54.0438	2.3243	49.3798	58.7079	31.2326	76.8551	-2.0438
6	55.0000	61.1729	2.9157	55.3221	67.0237	38.0897	84.2560	-6.1729
7	80.5000	64.3544	4.8767	54.5687	74.1401	39.9749	88.7339	16.1456
8	86.0000	83.1380	2.6886	77.7429	88.5332	60.1662	106.1	2.8620
9	69.0000	61.5442	2.7591	56.0077	67.0806	38.5387	84.5497	7.4558
10	46.0000	46.2502	4.6158	36.9880	55.5124	22.0761	70.4243	-0.2502
11	38.0000	47.8848	4.6813	38.4910	57.2786	23.6600	72.1096	-9.8848
12	49.5000	49.7536	3.4056	42.9197	56.5875	26.4019	73.1053	-0.2536
13	105.0	108.1	3.8227	100.5	115.8	84.5282	131.7	-3.1384
14	152.5	126.9	4.0022	118.9	134.9	103.2	150.6	25.6132
15	85.0000	91.0244	4.3214	82.3529	99.6958	67.0704	115.0	-6.0244
16	60.0000	49.8455	5.1365	39.5383	60.1527	25.2521	74.4389	10.1545
17	58.5000	48.4248	4.0554	40.2871	56.5625	24.6588	72.1908	10.0752
18	101.0	80.3636	3.3389	73.6636	87.0636	57.0507	103.7	20.6364

(continues)

FIGURE 8.29 Continued

OBS	ACTUAL	PREDICT VALUE	STD ERR PREDICT	LOWER95% MEAN	UPPER95% MEAN	LOWER95% PREDICT	UPPER95% PREDICT	RESIDUAL
19	79.4000	65.3483	2.6697	59.9911	70.7055	42.3853	88.3113	14.0517
20	125.0	118.5	3.3239	111.9	125.2	95.2192	141.8	6.4766
21	87.9000	103.7	5.1717	93.3319	114.1	79.0865	128.3	-15.8096
22	80.0000	80.0728	3.1987	73.6540	86.4915	56.8392	103.3	-0.0728
23	94.0000	106.1	4.1337	97.8323	114.4	82.3069	129.9	-12.1272
24	74.0000	86.6749	5.4340	75.7708	97.5790	61.8254	111.5	-12.6749
25	69.0000	70.8035	1.9689	66.8526	74.7544	48.1273	93.4797	-1.8035
26	63.0000	50.7254	2.6975	45.3125	56.1382	27.7493	73.7014	12.2746
27	67.5000	61.6062	2.7019	56.1845	67.0279	38.6281	84.5843	5.8938
28	35.0000	43.7147	3.9262	35.8362	51.5932	20.0362	67.3932	-8.7147
29	142.5	151.7	6.3235	139.0	164.3	126.0	177.3	-9.1540
30	92.2000	87.5370	3.3538	80.8071	94.2669	64.2156	110.9	4.6630
31	56.0000	55.0141	3.5213	47.9481	62.0801	31.5934	78.4348	0.9859
32	63.0000	54.6438	2.9371	48.7500	60.5376	31.5497	77.7379	8.3562
33	60.0000	87.4683	2.1282	83.1978	91.7389	64.7343	110.2	-27.4683
34	34.0000	34.6449	8.6556	17.2762	52.0136	6.3559	62.9340	-0.6449
35	52.0000	58.2825	3.3822	51.4956	65.0695	34.9445	81.6205	-6.2825
36	75.0000	70.6005	1.9340	66.7197	74.4814	47.9365	93.2646	4.3995
37	93.0000	85.8627	4.4052	77.0231	94.7023	61.8474	109.9	7.1373
38	60.0000	79.3817	4.9475	69.4539	89.3096	54.9449	103.8	-19.3817
39	73.0000	55.3328	3.0219	49.2690	61.3966	32.1948	78.4709	17.6672
40	71.0000	62.0418	4.9616	52.0855	71.9980	37.5933	86.4902	8.9582
41	83.0000	75.2296	2.3200	70.5741	79.8851	52.4201	98.0391	7.7704
42	90.0000	93.7940	2.5848	88.6071	98.9809	70.8701	116.7	-3.7940
43	83.0000	88.6489	2.3122	84.0092	93.2886	65.8427	111.5	-5.6489
44	115.0	118.8	4.8398	109.1	128.5	94.4401	143.1	-3.7900
45	50.0000	49.0382	2.8877	43.2435	54.8328	25.9692	72.1071	0.9618
46	55.2000	54.1699	2.6956	48.7608	59.5790	31.1947	77.1450	1.0301
47	61.0000	63.7263	2.8214	58.0647	69.3879	40.6904	86.7622	-2.7263
48	147.0	138.1	4.8418	128.4	147.8	113.7	162.4	8.9233
49	60.0000	71.4469	1.9509	67.5322	75.3616	48.7770	94.1168	-11.4469
50	100.0	114.2	3.6062	107.0	121.4	90.7158	137.7	-14.1884
51	44.5000	53.9253	2.7958	48.3151	59.5356	30.9020	76.9487	-9.4253
52	55.0000	71.8002	3.8278	64.1191	79.4813	48.1867	95.4137	-16.8002
53	53.4000	51.4634	3.9144	43.6086	59.3182	27.7928	75.1340	1.9366
54	65.0000	66.6020	3.1886	60.2035	73.0004	43.3740	89.8300	-1.6020
55	73.0000	71.8899	3.2630	65.3423	78.4375	48.6204	95.1594	1.1101
56	40.0000	44.8479	5.1318	34.5501	55.1456	20.2584	69.4374	-4.8479
57	141.0	134.2	10.4728	113.2	155.3	103.6	164.9	6.7539
58	68.0000	75.5810	1.9479	71.6722	79.4898	52.9121	98.2499	-7.5810
59	140.0	116.5	3.4297	109.6	123.4	93.1081	139.8	23.5260
60	55.0000	57.5493	3.2812	50.9650	64.1335	34.2694	80.8291	-2.5493

analysis of the residential sales data with residences 10, 50, and 61 removed using the model

$$y = \beta_0 + \beta_1 x_1 + \beta_2 x_1^2 + \beta_3 x_2 + \beta_4 x_2^2 + \beta_5 x_3 + \beta_6 x_3^2 + \beta_7 x_4 + \varepsilon$$

a. Test the overall adequacy of the model using F(model) and its associated prob-value.
b. Assess the importance of each of the independent variables by using appropriate t statistics and their associated prob-values.
c. Find and report the total variation, explained variation, unexplained variation, and R^2. Interpret R^2.

8.35 Consider the residential sales situation of Exercise 8.34. Usng the SAS output in Figure 8.29,

a. Find and report the 95% confidence interval for the average sales price of all 12-year-old residences having 2086 square feet, seven rooms, and three bedrooms.
b. Compute a 99% confidence interval for the average sales price of all 12-year-old residences having 2086 square feet, seven rooms, and three bedrooms.
c. Find and report the 95% prediction interval for an individual sales price of a 24-year-old residence having 1080 square feet, five rooms, and two bedrooms.
d. Compute a 99% prediction interval for an individual sales price of a 24-year-old residence having 1080 square feet, five rooms, and two bedrooms.

8.36 Consider the residential sales situation of Exercise 8.34. Figure 8.30 gives the SAS output of residual plots for the model

$$y = \beta_0 + \beta_1 x_1 + \beta_2 x_1^2 + \beta_3 x_2 + \beta_4 x_2^2 + \beta_5 x_3 + \beta_6 x_3^2 + \beta_7 x_4 + \varepsilon$$

(with residences 10, 50, and 61 removed from the data set) versus x_1, x_2, x_3, x_4, and $\hat{y}$. Interpret these plots.

FIGURE 8.30 SAS output of residual plots for a regression analysis of the residential sales data with residences 10, 50, and 61 removed using the model $y = \beta_0 + \beta_1 x_1 + \beta_2 x_1^2 + \beta_3 x_2 + \beta_4 x_2^2 + \beta_5 x_3 + \beta_6 x_3^2 + \beta_7 x_4 + \varepsilon$

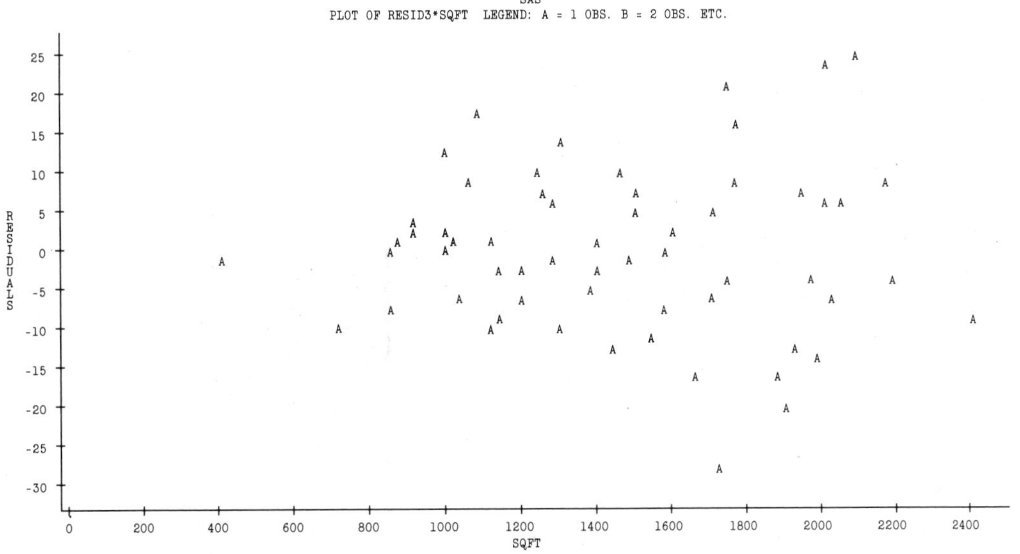

(continues)

FIGURE 8.30 Continued

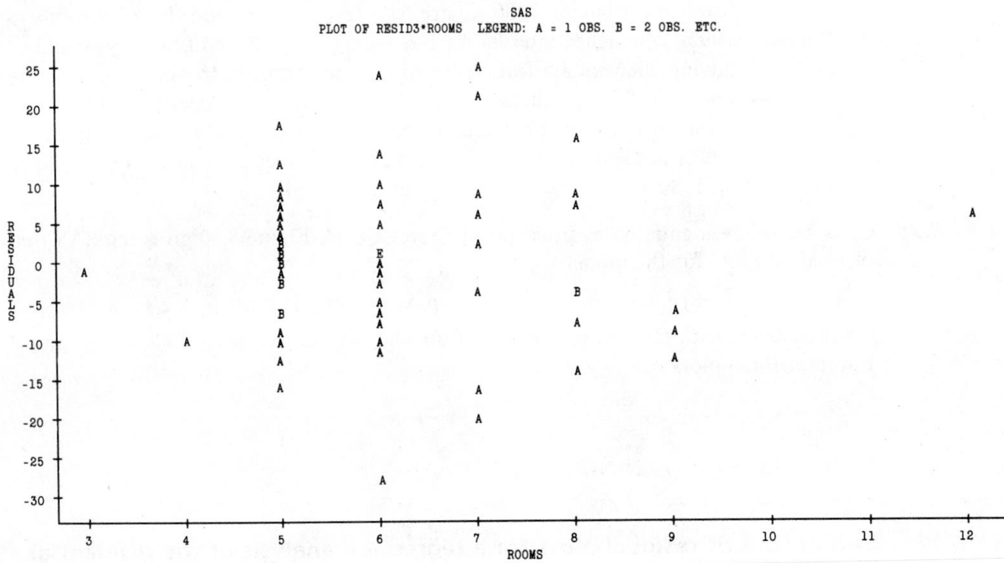

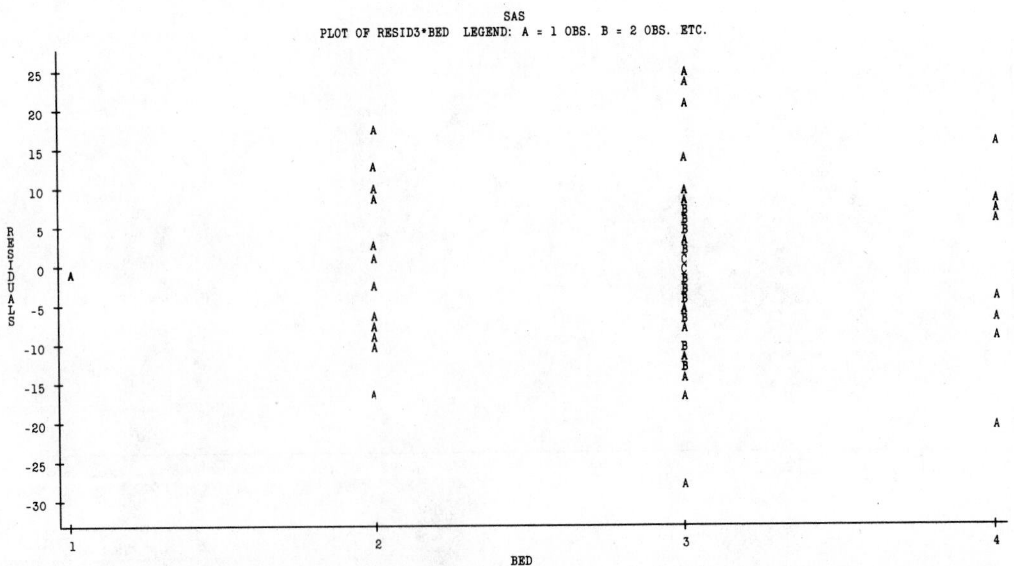

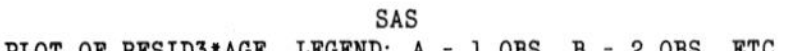

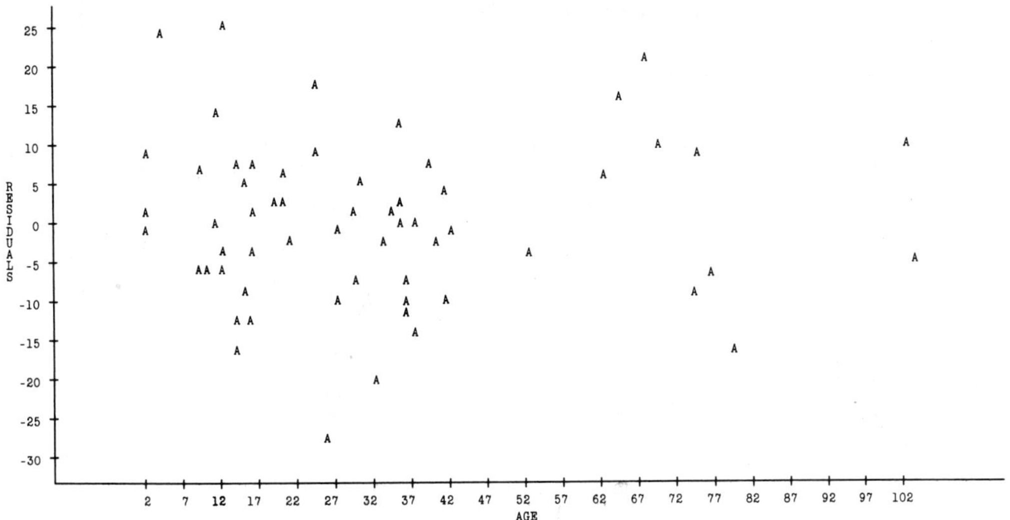
SAS
PLOT OF RESID3*AGE LEGEND: A = 1 OBS. B = 2 OBS. ETC.
RESIDUALS
AGE

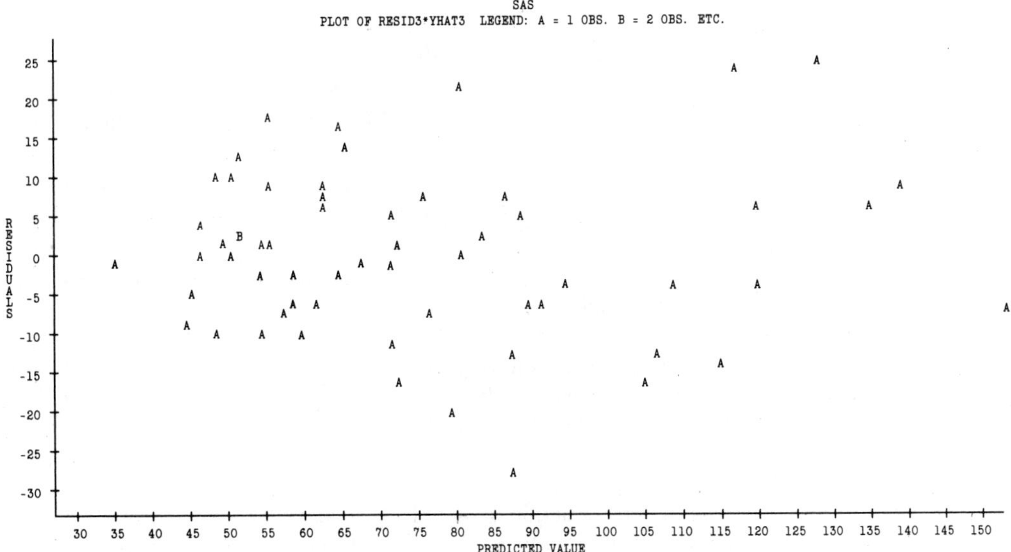
SAS
PLOT OF RESID3*YHAT3 LEGEND: A = 1 OBS. B = 2 OBS. ETC.
RESIDUALS
PREDICTED VALUE

FIGURE 8.31 SAS output of a regression analysis of all 63 residences in Table 8.2 using the model $y = \beta_0 + \beta_1 x_1 + \beta_2 x_1^2 + \beta_3 x_2 + \beta_4 x_2^2 + \beta_5 x_3 + \beta_6 x_3^2 + \beta_7 x_4 + \beta_8 x_4^2 + \varepsilon$

SAS

DEP VARIABLE: SALEP

ANALYSIS OF VARIANCE

SOURCE	DF	SUM OF SQUARES	MEAN SQUARE	F VALUE	PROB>F
MODEL	8	56625.86898	7078.23362	19.691	0.0001
ERROR	54	19411.12086	359.46520		
C TOTAL	62	76036.98984			

ROOT MSE	18.95957	R-SQUARE	0.7447
DEP MEAN	78.80159	ADJ R-SQ	0.7069
C.V.	24.05988		

PARAMETER ESTIMATES

VARIABLE	DF	PARAMETER ESTIMATE	STANDARD ERROR	T FOR H0: PARAMETER=0	PROB > |T|
INTERCEP	1	37.54494967	40.17707583	0.934	0.3542
SQFT	1	0.07223398	0.02732422	2.644	0.0107
SQFTSQ	1	-.0000065601	.00000780728	-0.840	0.4045
ROOMS	1	-6.48900758	11.90568320	-0.545	0.5880
ROOMSSQ	1	0.90739429	0.78030558	1.163	0.2500
BED	1	-8.05492290	32.63296094	-0.247	0.8060
BEDSQ	1	-0.44499985	5.72039815	-0.078	0.9383
AGE	1	-0.85542483	0.39693345	-2.155	0.0356
AGESQ	1	0.004375579	0.003913609	1.118	0.2685

8.37 Consider the residential sales data in Table 8.2. Figure 8.31 presents SAS output of a regression analysis of all 63 residences using the model

$$y = \beta_0 + \beta_1 x_1 + \beta_2 x_1^2 + \beta_3 x_2 + \beta_4 x_2^2 + \beta_5 x_3 + \beta_6 x_3^2 + \beta_7 x_4 + \beta_8 x_4^2 + \varepsilon$$

a. Using the SAS output, assess the importance of each of the quadratic terms.

b. Compare your results of part (a) to the importance of the quadratic terms in the model

$$y = \beta_0 + \beta_1 x_1 + \beta_2 x_1^2 + \beta_3 x_2 + \beta_4 x_2^2 + \beta_5 x_3 + \beta_6 x_3^2 + \beta_7 x_4 + \varepsilon$$

when the possibly outlying and influential residences 10, 50, and 61 are removed from the data set. (See Exercise 8.34 and Figure 8.29.)

c. Has removing residences 10, 50, and 61 clarified the regression relationship between y, x_1, x_2, x_3, and x_4? Explain.

8.38 Consider the social welfare expenditure data of Exercises 4.17–4.20. Using a computer, try to find an adequate multiple regression model relating y (social welfare expenditure) to x_1 (total GNP), x_2 (per capita GNP), and x_3 (unemployment rate). Justify your model choice.

8.39 Consider the social welfare situation of Exercise 8.38 and the model that you feel best relates social welfare expenditure to x_1, x_2, and x_3. Carry out a complete analysis of the residuals for your model. Check the validity of the constant variance, independence, and normality assumptions. Also check for possible outlying and influential observations. Can the model be improved by removing outliers from the data set?

8.40 Consider the hospital labor needs data of Exercise 5.6 (see Table 5.9). Using a computer, try to determine an adequate multiple regression model for predicting labor requirements. Can the model be improved by including quadratic or other higher-order terms?

8.41 Consider the situation of Exercise 8.40 and the model that you feel best describes hospital labor requirements. Carry out a complete analysis of the residuals for your model. Check the validity of the inference assumptions and also check for outliers. Can the model be improved by removing outliers from the data set?

8.42 Consider the economic data for selected cities in Table 5.12. Using a computer, try to find an adequate model for predicting each of (1) HCOST (indexed housing cost), (2) JG (jobs growth), and (3) IG (personal income growth). Consider quadratic and higher-order terms. Carry out a complete analysis of the residuals for each model. Can the models be improved by removing outliers from the data set?

8.43 Consider the sales volume–advertising expenditure data in Table 6.8. Using a computer, try to find an adequate multiple regression model for predicting sales volume. Investigate the possibility of using quadratic or other higher-order terms. Carry out a complete residual analysis of the model that you feel is best.

8.44 Consider the sales territory performance data in Table 8.5. Using a computer, try to find an adequate multiple regression model for predicting sales. Investigate the possibility of using quadratic or other higher-order terms. Carry out a complete analysis of the residuals for the model that you feel is best. Be sure to check the validity of the inference assumptions and also check for outliers.

8.45 Consider the Andrews and Ferguson real estate sales data in Table 8.9 (see Exercise 8.19). Using a computer, investigate the possibility of improving the model of Exercise 8.19 by utilizing quadratic terms. Can the model be improved? Justify your answer.

8.46 (Optional) Write a SAS program that will analyze the construction project data in Table 8.8 by using each of the following models:

$$y = \beta_0 + \beta_1 x_1 + \beta_2 x_1^2 + \beta_3 x_2 + \varepsilon$$

$$y = \beta_0 + \beta_1 x_1 + \beta_2 x_1^2 + \beta_3 x_2 + \beta_4 x_2^2 + \varepsilon$$

Write the program so that a complete analysis of the residuals will be carried out.

8.47 (Optional) Write a SAS program that will analyze the Crest toothpaste sales data in Table 8.10 by using each of the following models:

$$y = \beta_0 + \beta_1 x_1 + \beta_2 x_2 + \beta_3 x_3 + \varepsilon$$

$$y = \beta_0 + \beta_1 x_1 + \beta_2 x_2 + \beta_3 x_3 + \beta_4 x_1^2 + \beta_5 x_2^2 + \beta_6 x_3^2 + \varepsilon$$

Write the program so that a complete analysis of the residuals will be carried out. Include residual plots versus x_1, x_2, x_3, $\hat{y}$, and time as well as a stem-and-leaf diagram, normal plot, and calculated Durbin–Watson statistic.

8.48 (Optional) Write a SAS program that will analyze the United Oil Company gas mileage data in Table 8.11 by using each of the following models:

$$y = \beta_0 + \beta_1 x_1 + \beta_2 x_2 + \varepsilon \quad \text{(Model 1)}$$
$$y = \beta_0 + \beta_1 x_1 + \beta_2 x_1^2 + \beta_3 x_2 + \varepsilon \quad \text{(Model 2)}$$
$$y = \beta_0 + \beta_1 x_1 + \beta_2 x_2 + \beta_3 x_2^2 + \varepsilon \quad \text{(Model 3)}$$
$$y = \beta_0 + \beta_1 x_1 + \beta_2 x_1^2 + \beta_3 x_2 + \beta_4 x_2^2 + \varepsilon \quad \text{(Model 4)}$$

Write the program so that residual plots versus x_1, x_2, and $\hat{y}$ will be constructed for each model.

8.49 (Optional) Write a SAS program that will analyze the service call data given in Table 6.11 by using each of the following models:

$$y = \beta_0 + \beta_1 x + \varepsilon \quad \text{(Model 1)}$$
$$y = \beta_0 + \beta_1 x + \beta_2 x^2 + \varepsilon \quad \text{(Model 2)}$$
$$y = \beta_0 + \beta_1 x + \beta_2 x^2 + \beta_3 x^3 + \varepsilon \quad \text{(Model 3)}$$

Write the program to completely analyze the residuals for each model. (Include residual plots versus x and $\hat{y}$, a stem-and-leaf diagram, and a normal plot.)

8.50 (Optional) Write a SAS program that will analyze the residential sales data in Table 8.2 (with residences 10, 50, and 61 removed from the data) by using the model

$$y = \beta_0 + \beta_1 x_1 + \beta_2 x_1^2 + \beta_3 x_2 + \beta_4 x_2^2 + \beta_5 x_3 + \beta_6 x_3^2 + \beta_7 x_4 + \varepsilon$$

Write the program so that it will carry out a complete analysis of the residuals for this model.

8.51 In Appendix F, consider Data Base 1: Manufacturing Data. Relate NCAPEX to VALSHIP and NMANES by using the model

$$y = \beta_0 + \beta_1 x_1 + \beta_2 x_2 + \varepsilon$$

Calculate R^2 and test each of $H_0: \beta_1 = 0$ and $H_0: \beta_2 = 0$ by setting α equal to .05. Try to improve the model by adding squared terms.

8.52 In Appendix F, consider Data Base 2: Ohio Local Government and Payroll Data. Relate OCTPAY to EMP and INSTEMP by using the model

$$y = \beta_0 + \beta_1 x_1 + \beta_2 x_2 + \varepsilon$$

Calculate R^2 and test each of $H_0: \beta_1 = 0$ and $H_0: \beta_2 = 0$ by setting α equal to .05. Try to improve the model by adding squared terms.

8.53 In Appendix F, consider Data Base 3: Population Data. Relate PCTUNEM to OV25CG, OV25HSG, and PCTPOPCH by using the model

$$y = \beta_0 + \beta_1 x_1 + \beta_2 x_2 + \beta_3 x_3 + \varepsilon$$

Calculate R^2 and test each of $H_0: \beta_1 = 0$, $H_0: \beta_2 = 0$, and $H_0: \beta_3 = 0$ by setting α equal to .05. Try to improve the model by adding squared terms.

CHAPTER 9

MULTIPLE REGRESSION: II

In this chapter we continue our presentation of multiple regression. We begin in Section 9.1 by discussing the use of what we call interaction terms in regression models. Then in Section 9.2 we present the partial F-test, which is a test of significance for a portion of a regression model. We complete this chapter with optional Sections 9.3 and 9.4, which discuss using regression analysis to make statistical inferences when sampling from a single population and using SAS to implement the techniques of this chapter.

9.1 INTERACTION

Multiple regression models often contain *interaction variables*. We say that there is *no interaction* between two independent variables if the relationship between the mean value of the dependent variable and each one of the independent variables is independent of the value of the other independent variable. There is said to be *interaction* between two independent variables if the relationship between the mean value of the dependent variable and one of the independent variables is dependent upon the value of the other independent variable. We discuss the concept of interaction in detail in the next example.

Example 9.1

Part 1: The Data and Data Plots

Bonner Frozen Foods, Inc. has designed an experiment to study the effects of two types of advertising expenditures on sales of one of its lines of frozen foods. Twenty-five sales regions of equal sales potential were selected. Then different combinations of

$$x_1 = \text{radio and television expenditures (measured in units of \$1000)}$$

and

$$x_2 = \text{print expenditures (measured in units of \$1000)}$$

were specified and randomly assigned to the sales regions. Table 9.1 shows the

TABLE 9.1 Bonner Frozen Foods, Inc. sales volume data

Sales region, i	Radio and television expenditures, x_{i1}	Print expenditures, x_{i2}	Sales volume, y_i
1	1	1	3.27
2	1	2	8.38
3	1	3	11.28
4	1	4	14.50
5	1	5	19.63
6	2	1	5.84
7	2	2	10.01
8	2	3	12.46
9	2	4	16.67
10	2	5	19.83
11	3	1	8.51
12	3	2	10.14
13	3	3	14.75
14	3	4	17.99
15	3	5	19.85
16	4	1	9.46
17	4	2	12.61
18	4	3	15.50
19	4	4	17.68
20	4	5	21.02
21	5	1	12.23
22	5	2	13.58
23	5	3	16.77
24	5	4	20.56
25	5	5	21.05

expenditure combinations along with the associated values of

$$y = \text{sales volume (measured in units of \$10,000)}$$

for the sales regions during August 1988.

We can plot the data in Table 9.1 to help us determine whether interaction exists between x_1 and x_2. To do this, we first plot y versus x_1 for different levels of x_2 ($x_2 = 1, 2, 3, 4, 5$). This plot is shown in Figure 9.1. Examining this plot, we see that the straight line relating the five values of y to the corresponding five values of x_1 when $x_2 = 5$ appears to have a smaller slope than the straight line relating the five values of y to the corresponding five values of x_1 when $x_2 = 1$. The reader can further verify that the entire data plot in Figure 9.1 might suggest that the larger x_2 is, the smaller is the slope of the straight line relating y to x_1.

Next, we plot (in Figure 9.2) y versus x_2 for different levels of x_1 ($x_1 = 1, 2, 3, 4, 5$). Examining this plot, we see that the straight line relating the five values of y to the corresponding five values of x_2 when $x_1 = 5$ appears to have a smaller slope than the straight line relating the five values of y to the corresponding five values of x_2 when $x_1 = 1$. The reader can further verify that the entire data plot in Figure 9.2 might suggest that the larger that x_1 is, the smaller is the slope of the straight line relating y to x_2.

In summary, the data plots in Figures 9.1 and 9.2 seem to imply that the more money that is spent on one type of advertising, the smaller is the slope of the straight line relating sales volume to the amount spent on the other type of advertising. This type of interaction seems reasonable because, as Bonner Frozen

FIGURE 9.1 **Plot of y versus x_1: The larger x_2 is, the smaller is the slope of the straight line relating y to x_1**

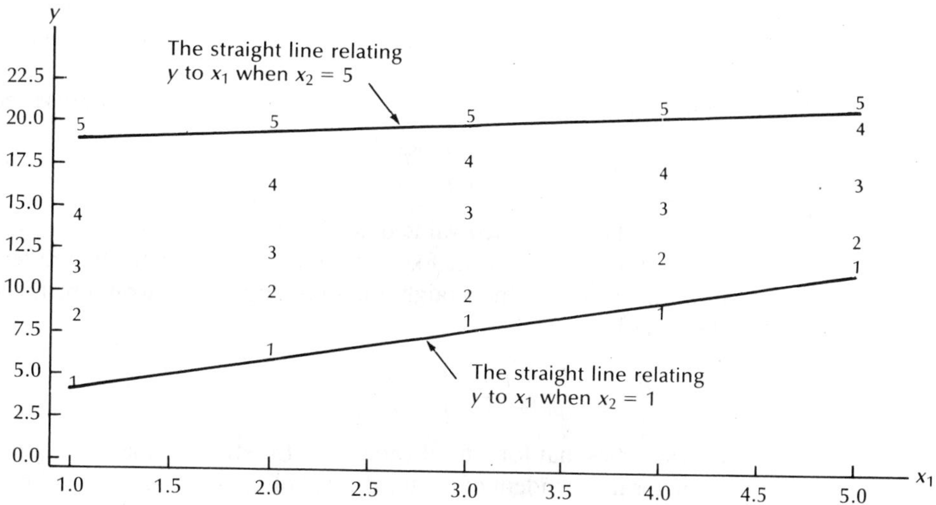

FIGURE 9.2 Plot of y versus x_2: The larger x_1 is, the smaller is the slope of the straight line relating y to x_2

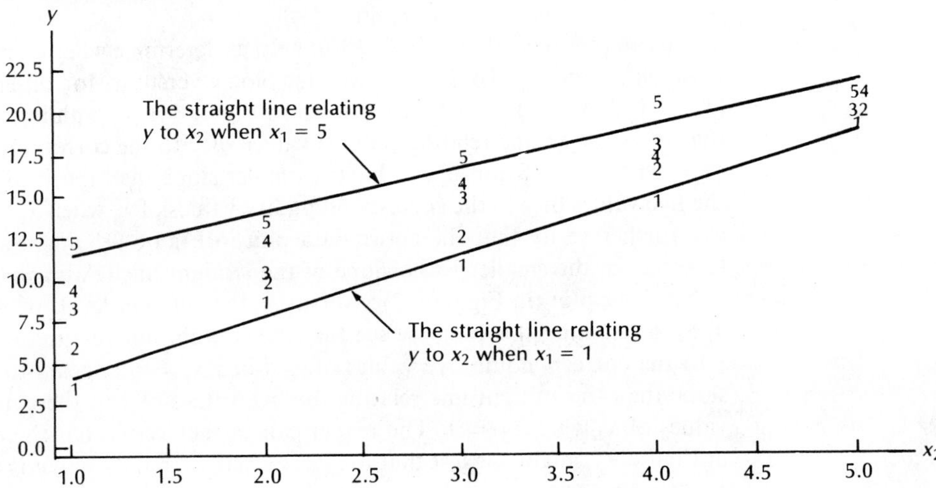

Foods spends more money on one type of advertising, increases in spending on the other type of advertising might become less effective. This implies that the slope of the straight line relating sales to the other type of advertising expenditure becomes smaller.

Part 2: Two Models

Because of the above described interaction, it would not be appropriate to use the model

$$y = \beta_0 + \beta_1 x_1 + \beta_2 x_2 + \varepsilon$$

to describe the sales volume data in Table 9.1. This is because the equation

$$\begin{aligned}\mu &= \beta_0 + \beta_1 x_1 + \beta_2 x_2 \\ &= (\beta_0 + \beta_1 x_1) + \beta_2 x_2\end{aligned}$$

assumes that for a fixed value of x_1 the straight line relating μ to x_2 has a slope β_2 that is independent of the fixed value of x_1. That is, for two different fixed values of x_1, the slopes of the straight lines relating μ to x_2 would be the same. Similarly, the equation

$$\begin{aligned}\mu &= \beta_0 + \beta_1 x_1 + \beta_2 x_2 \\ &= (\beta_0 + \beta_2 x_2) + \beta_1 x_1\end{aligned}$$

also assumes that for a fixed value of x_2 the straight line relating μ to x_1 has a slope β_1 that is independent of the fixed value of x_2. That is, for two different fixed values of x_2, the slopes of the straight lines relating μ to x_1 would be the same. Therefore the preceding model assumes that no interaction exists between x_1 and x_2. In contrast, it would be appropriate to describe the data in Table 9.1 by using the model

$$y = \beta_0 + \beta_1 x_1 + \beta_2 x_2 + \beta_3 x_1 x_2 + \varepsilon$$

which uses the *cross-product term* $x_1 x_2$. The equation

$$\begin{aligned} \mu &= \beta_0 + \beta_1 x_1 + \beta_2 x_2 + \beta_3 x_1 x_2 \\ &= (\beta_0 + \beta_1 x_1) + (\beta_2 + \beta_3 x_1) x_2 \end{aligned}$$

assumes that for a fixed value of x_1 the straight line relating μ to x_2 has a slope $(\beta_2 + \beta_3 x_1)$ that is dependent upon the fixed value of x_1. That is, for two different fixed values of x_1, the slopes of the straight lines relating μ to x_2 would be different (as in Figure 9.2). Similarly, the equation

$$\begin{aligned} \mu &= \beta_0 + \beta_1 x_1 + \beta_2 x_2 + \beta_3 x_1 x_2 \\ &= (\beta_0 + \beta_2 x_2) + (\beta_1 + \beta_3 x_2) x_1 \end{aligned}$$

assumes that for a fixed value of x_2 the straight line relating μ to x_1 has a slope $(\beta_1 + \beta_3 x_2)$ that is dependent upon the fixed value of x_2. That is, for two different fixed values of x_2, the slopes of the straight lines relating μ to x_1 would be different (as in Figure 9.1). Therefore the model employing the term $x_1 x_2$ assumes that interaction exists between x_1 and x_2.

Part 3: Statistical Inference

We now consider using the model

$$y = \beta_0 + \beta_1 x_1 + \beta_2 x_2 + \beta_3 x_1 x_2 + \varepsilon$$

to describe the data in Table 9.1. We define the column vector

$$\mathbf{y} = \begin{bmatrix} y_1 \\ y_2 \\ \vdots \\ y_{25} \end{bmatrix} = \begin{bmatrix} 3.27 \\ 8.38 \\ \vdots \\ 21.05 \end{bmatrix}$$

and the matrix

$$\begin{array}{c} \\ \mathbf{X} = \end{array} \begin{array}{c} \begin{array}{cccc} 1 & x_1 & x_2 & x_1 x_2 \end{array} \\ \begin{bmatrix} 1 & 1 & 1 & 1 \\ 1 & 1 & 2 & 2 \\ \vdots & \vdots & \vdots & \vdots \\ 1 & 5 & 5 & 25 \end{bmatrix} \end{array}$$

We can calculate the least squares point estimates of β_0, β_1, β_2, and β_3 to be

$$\begin{bmatrix} b_0 \\ b_1 \\ b_2 \\ b_3 \end{bmatrix} = \mathbf{b} = (\mathbf{X}'\mathbf{X})^{-1}\mathbf{X}'\mathbf{y} = \begin{bmatrix} -2.3497 \\ 2.3611 \\ 4.1831 \\ -.3489 \end{bmatrix}$$

Figure 9.3 presents the SAS output of these point estimates and other key quantities resulting from using the above model to perform a regression analysis of the data in Table 9.1. Note that x_1x_2 is denoted as X12 on the output. We see that the prob-values for testing the importance of the intercept and the independent variables x_1, x_2, and x_1x_2 are all less than .01. Therefore we have very strong evidence that the intercept β_0 and the independent variables x_1, x_2, and x_1x_2 are important in the above model. In particular, since the prob-value related to the importance of x_1x_2 is .0001, we have confirmed that interaction exists between x_1 and x_2. This was initially suggested by the graphical analysis in Figures 9.1 and 9.2.

Next, suppose that Bonner Frozen Foods will spend $x_{01} = 4.5$ (that is, \$4500) on radio and television advertising and $x_{02} = 3.5$ (that is, \$3500) on print advertising in a future month in a particular sales region. Also, suppose that the regression relationship between y and x_1 and x_2 that we have developed applies to the future month and particular sales region. That is, there are no trend, seasonal, or other time-related influences affecting monthly sales volume. It follows that

$$\begin{aligned} \hat{y}_0 &= b_0 + b_1x_{01} + b_2x_{02} + b_3x_{01}x_{02} \\ &= -2.3497 + 2.3611(4.5) + 4.1831(3.5) - .3489(4.5)(3.5) \\ &= 17.4209 \qquad \text{(that is, \$174,209)} \end{aligned}$$

is a point estimate of

$$\mu_0 = \beta_0 + \beta_1(4.5) + \beta_2(3.5) + \beta_3(4.5)(3.5)$$

the mean of all possible sales volumes when \$4500 is spent on radio and television advertising and \$3500 is spent on print advertising. In addition, $\hat{y}_0$ is a point prediction of

$$y_0 = \mu_0 + \varepsilon_0$$

the individual sales volume that will be observed in the future month in the particular sales region. To compute a confidence interval for μ_0 and a prediction interval for y_0, we use

$$\mathbf{x}_0' = [1 \quad 4.5 \quad 3.5 \quad (4.5)(3.5)] = [1 \quad 4.5 \quad 3.5 \quad 15.75]$$

Although $h_{00} = \mathbf{x}_0'(\mathbf{X}'\mathbf{X})^{-1}\mathbf{x}_0$ is not directly given in the SAS output of Figure 9.3, the SAS output tells us that

$$s\sqrt{h_{00}} = .1935 \qquad \text{and} \qquad s = .6257$$

FIGURE 9.3 SAS output of a regression analysis of the sales volume data in Table 9.1 by using the model $y = \beta_0 + \beta_1 x_1 + \beta_2 x_2 + \beta_3 x_1 x_2 + \varepsilon$

SAS

DEP VARABLE: Y

ANALYSIS OF VARIANCE

SOURCE	DF	SUM OF SQUARES	MEAN SQUARE	F VALUE	PROB>F
MODEL	3	590.40574[a]	196.80191[p]	502.671[q]	0.0001[r]
ERROR	21	8.22176700[b]	0.39151271[f]		
C TOTAL	24	598.62750[c]			

ROOT MSE	0.6257098[d]	R-SQUARE	0.9863[h]
DEP MEAN	14.1428	ADJ R-SQ	0.9843[i]
C.V.	4.424228		

PARAMETER ESTIMATES

VARIABLE	DF	PARAMETER ESTIMATE[e]	STANDARD ERROR[g]	T FOR HO: PARAMETER=0[j]	PROB > |T|[k]
INTERCEP	1	-2.34970000	0.68828075	-3.414	0.0026
X1	1	2.36110000	0.20752445	11.377	0.0001
X2	1	4.18310000	0.20752445	20.157	0.0001
X12	1	-0.34890000	0.06257098	-5.576	0.0001

OBS	ACTUAL	PREDICT VALUE	STD ERR PREDICT	LOWER95% MEAN	UPPER95% MEAN	LOWER95% PREDICT	UPPER95% PREDICT	RESIDUAL
1	3.2700	3.8456	0.3754	3.0649	4.6263	2.3281	5.3631	-0.5756
2	8.3800	7.6798	0.2655	7.1277	8.2319	6.2663	9.0933	0.7002
3	11.2800	11.5140	0.2168	11.0632	11.9648	10.1369	12.8911	-0.2340
4	14.5000	15.3482	0.2655	14.7961	15.9003	13.9347	16.7617	-0.8482
5	19.6300	19.1824	0.3754	18.4017	19.9631	17.6649	20.6999	0.4476
6	5.8400	5.8578	0.2655	5.3057	6.4099	4.4443	7.2713	-0.0178
7	10.0100	9.3431	0.1877	8.9527	9.7335	7.9846	10.7016	0.6669
8	12.4600	12.8284	0.1533	12.5097	13.1471	11.4887	14.1681	-0.3684
9	16.6700	16.3137	0.1877	15.9233	16.7041	14.9552	17.6722	0.3563
10	19.8300	19.7990	0.2655	19.2469	20.3511	18.3855	21.2125	0.0310
11	8.5100	7.8700	0.2168	7.4192	8.3208	6.4929	9.2471	0.6400
12	10.1400	11.0064	0.1533	10.6877	11.3251	9.6667	12.3461	-0.8664
13	14.7500	14.1428	0.1251	13.8826	14.4030	12.8158	15.4698	0.6072
14	17.9900	17.2792	0.1533	16.9605	17.5979	15.9395	18.6185	0.7108
15	19.8500	20.4156	0.2168	19.9648	20.8664	19.0385	21.7927	-0.5656
16	9.4600	9.8822	0.2655	9.3301	10.4343	8.4687	11.2957	-0.4222
17	12.6100	12.6697	0.1877	12.2793	13.0601	11.3112	14.0282	-0.0597
18	15.5000	15.4572	0.1533	15.1385	15.7759	14.1175	16.7969	0.0428
19	17.6800	18.2447	0.1877	17.8543	18.6351	16.8862	19.6032	-0.5647
20	21.0200	21.0322	0.2655	20.4801	21.5843	19.6187	22.4457	-0.0122
21	12.2300	11.8944	0.3754	11.1137	12.6751	10.3769	13.4119	0.3356
22	13.5800	14.3330	0.2655	13.7809	14.8851	12.9195	15.7465	-0.7530
23	16.7700	16.7716	0.2168	16.3208	17.2224	15.3945	18.1487	-0.0016
24	20.5600	19.2102	0.2655	18.6581	19.7623	17.7967	20.6237	1.3498
25	21.0500	21.6488	0.3754	20.8681	22.4295	20.1313	23.1663	-0.5988
26	.	17.4209[l]	0.1935[m]	17.0185	17.8233	16.0589	18.7829	.
				95% confidence interval		95% prediction interval		

SUM OF RESIDUALS 2.46692E-13

SUM OF SQUARED RESIDUALS 8.221767[n]

[a] *Explained variation* $= SS_{model}$; [b] *SSE*; [c] *Total variation*; [d] s; [e] b_j: b_0, b_1, b_2, b_3; [f] $MSE = s^2$; [g] $s\sqrt{c_{jj}}$; [h] R^2; [i] $\bar{R}^2$; [j] t; [k] *Prob-value for t*; [l] *Point prediction* $\hat{y}_0$; [m] $s\sqrt{h_{00}}$; [n] *SSE*; [p] MS_{model}; [q] *F(model)*; [r] *Prob-value for F(model)*

This implies that

$$h_{00} = \left(\frac{.1935}{s}\right)^2 = \left(\frac{.1935}{.6257}\right)^2 = .0956$$

It follows that a 95% prediction interval for y_0 is (since $n - k = 25 - 4 = 21$)

$$\begin{aligned}[\hat{y}_0 \pm t^{(21)}_{[.025]} s\sqrt{1 + h_{00}}] &= [17.4209 \pm 2.08(.6257)\sqrt{1.0965}] \\ &= [16.0589, 18.7829]\end{aligned}$$

Note that this 95% prediction interval is given in Figure 9.3 along with the 95% confidence interval for μ_0, which is [17.0185, 17.8233]. The prediction interval says that we are 95% confident that the individual sales volume in the future month in the particular sales region will be between \$160,589 and \$187,829.

Part 4: Comparing the Two Types of Advertising

The least squares point estimates of the parameters β_1 and β_2 in the Bonner Frozen Foods interaction model are $b_1 = 2.3611$ and $b_2 = 4.1831$. These estimates imply that each additional dollar spent on print advertising brings a mean sales volume increase that is substantially greater than the mean sales volume increase brought by each additional dollar spent on radio and television advertising. Therefore Bonner Frozen Foods, Inc. should probably spend a greater proportion of its

FIGURE 9.4 **Residuals for the sales volume model $y = \beta_0 + \beta_1 x_1 + \beta_2 x_2 + \beta_3 x_1 x_2 + \varepsilon$ plotted against x_1, x_2, and $\hat{y}$**

(a) Residuals versus x_1

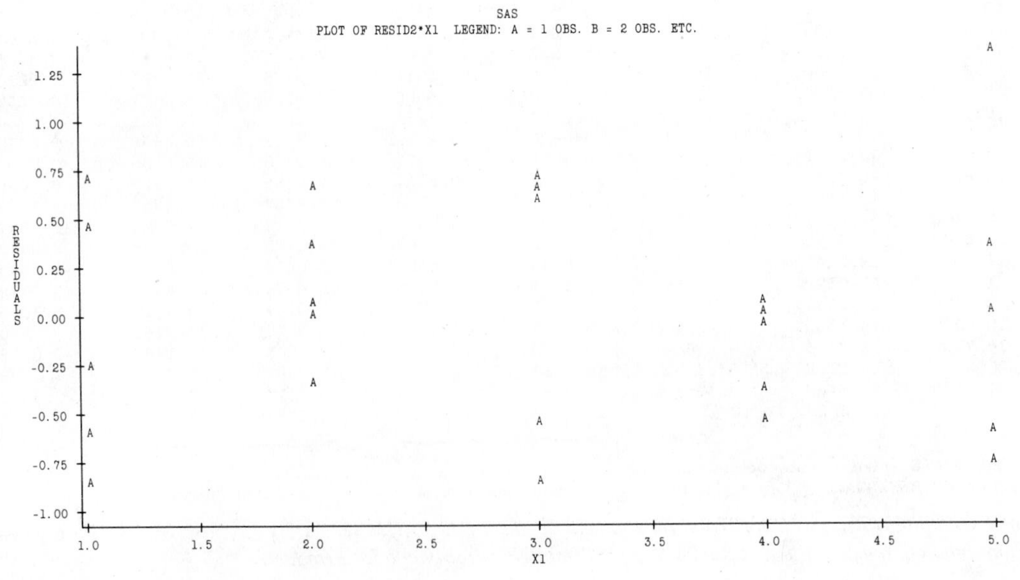

(b) Residuals versus x_2

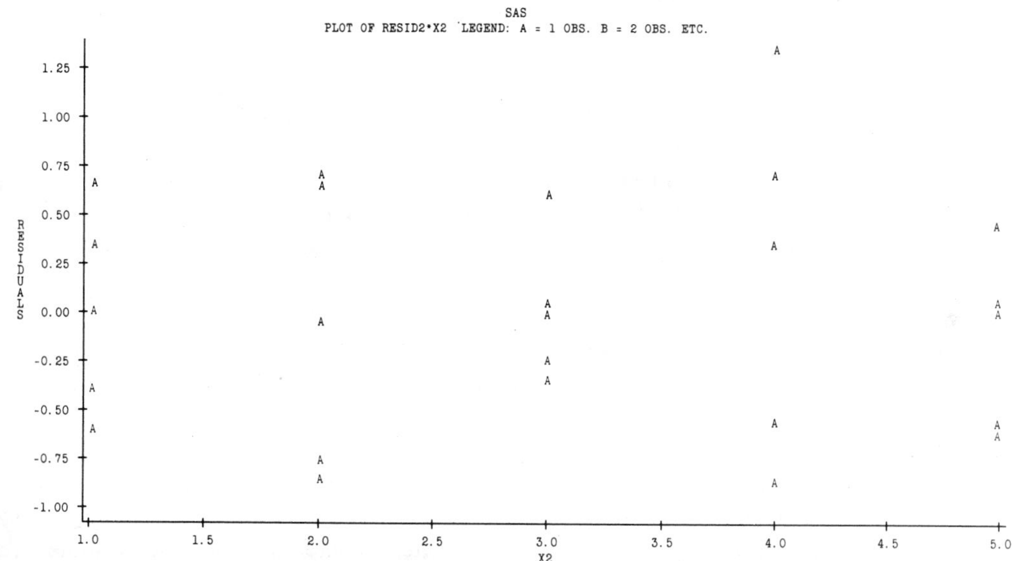

(c) Residuals versus $\hat{y}$

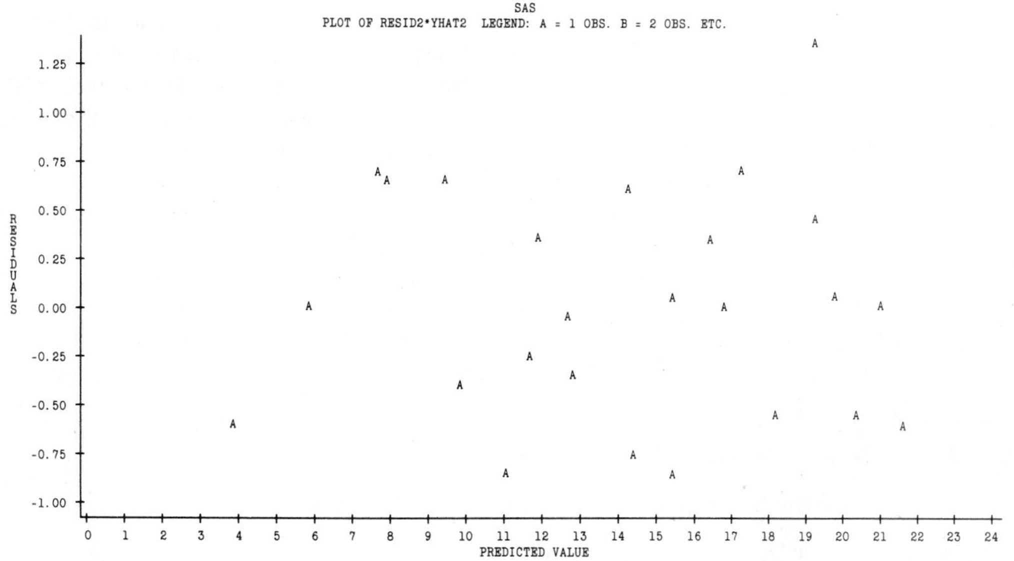

advertising budget on print advertising. As another illustration of this, suppose that the company decides to spend \$2000 ($x_{01} = 2$) on radio and television advertising and \$5000 ($x_{02} = 5$) on print advertising in the future month in the particular sales region. If we consider observation 10 (for which $x_{i1} = 2$ and $x_{i2} = 5$) in Figure 9.3, we find that the 95% prediction interval associated with this combination of expenditures is [18.3855, 21.2125]. Therefore if the company spends \$2000 on radio and television advertising and \$5000 on print advertising (a total budget that is \$1000 less than spending \$4500 on radio and television advertising and \$3500 on print advertising), we are 95% confident that the individual sales volume in the future month in the particular sales region will be between \$183,855 and \$212,125.

Part 5: Residual Analysis

We conclude this example by presenting (in Figure 9.4) residual plots for the model

$$y = \beta_0 + \beta_1 x_1 + \beta_2 x_2 + \beta_3 x_1 x_2 + \varepsilon$$

versus x_1, x_2, and $\hat{y}$. None of the plots suggest that a serious violation of the constant variance assumption exists. Further analysis (not shown here) suggests that the normality assumption approximately holds. Since the sales volume data is cross-sectional data, there is no need to test for autocorrelation.

In the Bonner Frozen Foods problem, the company has carried out a designed experiment. In many regression problems, however, we do not carry out a designed experiment, and the data is "unstructured." In such a case, it may not be possible to construct the data plots needed to detect interaction between independent variables. For example, if we consider the Fresh demand data in Table 8.3, we might suspect that there is interaction between x_3 (advertising expenditure) and x_4 (the price difference). That is, we might suspect that the relationship between mean demand for Fresh and advertising expenditure is different for different levels of the price difference. For instance, increases in advertising expenditures might be more effective at some price differences than at other price differences. To detect such interaction, we would like to construct plots of demand versus x_3 for different levels of x_4. However, examination of the Fresh demand data in Table 8.3 reveals that there are only a few observations at any one level of the price difference, and therefore the needed data plots cannot be made. However, in such a case we can use the t statistics related to potential interaction terms (as well as other model-building techniques to be discussed in Chapter 11) to try to assess the importance of interaction. We now present a detailed example in which we use this approach.

Example 9.2

Part 1: Several Models

Consider the Fresh demand model

$$y = \beta_0 + \beta_1 x_4 + \beta_2 x_3 + \beta_3 x_3^2 + \varepsilon$$

which relates y to x_3 and x_4 and which we have previously used to predict demand for Fresh. The equation

$$\mu = (\beta_0 + \beta_1 x_4) + \beta_2 x_3 + \beta_3 x_3^2$$

assumes that there is no interaction between x_4 and x_3. This is because it says that for a fixed value of x_4 the parameters β_2 and β_3, which determine the exact nature of the quadratic curve relating μ to x_3, do not depend on the fixed value of x_4. In contrast, consider the equation

$$\mu = \beta_0 + \beta_1 x_4 + \beta_2 x_3 + \beta_3 x_3^2 + \beta_4 x_4 x_3$$

which uses the cross-product term $x_4 x_3$. We can write this equation in the form

$$\mu = (\beta_0 + \beta_1 x_4) + (\beta_2 + \beta_4 x_4) x_3 + \beta_3 x_3^2$$

This says that for a fixed value of x_4 the quadratic curve relating μ to x_3 has a curvature parameter $(\beta_2 + \beta_4 x_4)$ that depends on the fixed value of x_4. We can also write this equation in the form

$$\mu = (\beta_0 + \beta_2 x_3 + \beta_3 x_3^2) + (\beta_1 + \beta_4 x_3) x_4$$

This says that for a fixed value of x_3 the straight line relating μ to x_4 has a slope $(\beta_1 + \beta_4 x_3)$ that depends on the fixed value of x_3. Therefore the equation

$$\mu = \beta_0 + \beta_1 x_4 + \beta_2 x_3 + \beta_3 x_3^2 + \beta_4 x_4 x_3$$

assumes that there is interaction between x_4 and x_3.

It should also be noted that in some situations it is appropriate to model interaction by using cross-product terms such as $x_4 x_3^2$. For example, the equation

$$\begin{aligned} \mu &= \beta_0 + \beta_1 x_4 + \beta_2 x_3 + \beta_3 x_3^2 + \beta_4 x_4 x_3 + \beta_5 x_4 x_3^2 \\ &= (\beta_0 + \beta_1 x_4) + (\beta_2 + \beta_4 x_4) x_3 + (\beta_3 + \beta_5 x_4) x_3^2 \end{aligned}$$

says that for a fixed value of x_4 both of the curvature parameters $(\beta_2 + \beta_4 x_4)$ and $(\beta_3 + \beta_5 x_4)$ in the equation relating μ to x_3 depend on the fixed value of x_4. However, we will show in Chapter 11 that it is probably not appropriate to use the term $x_4 x_3^2$ in a final Fresh demand model.

Part 2: Statistical Inference

Since we might logically suspect that there is interaction between x_4 and x_3, we consider the model

$$y = \beta_0 + \beta_1 x_4 + \beta_2 x_3 + \beta_3 x_3^2 + \beta_4 x_4 x_3 + \varepsilon$$

Figure 9.5 presents the SAS output obtained by using this model to perform a regression analysis of the Fresh demand data in Table 8.3. Here, the calculations are done by using the column vector

$$\mathbf{y} = \begin{bmatrix} y_1 \\ y_2 \\ y_3 \\ \vdots \\ y_{30} \end{bmatrix} = \begin{bmatrix} 7.38 \\ 8.51 \\ 9.52 \\ \vdots \\ 9.26 \end{bmatrix}$$

FIGURE 9.5 **SAS output of a regression analysis of the Fresh demand data by using the model $y = \beta_0 + \beta_1 x_4 + \beta_2 x_3 + \beta_3 x_3^2 + \beta_4 x_4 x_3 + \varepsilon$**

MODEL CROSSPRODUCTS X'X X'Y Y'Y

X'X		INTERCEP	X4	X3	X3SQ	X43		Y
INTERCEP		30	6.4	193.6	1258.85	44.1675		251.48
X4		6.4	2.865	44.1675	304.7814	19.71662		57.646
X3	$\mathbf{X'X} =$	193.6	44.1675	1258.85	8241.968	304.7814	$\mathbf{X'y} =$	1632.781
X3SQ		1258.85	304.7814	8241.968	54298.76	2103.453		10677.4
X43		44.1675	19.71662	304.7814	2103.453	135.8881		397.7442
Y		251.48	57.646	1632.781	10677.4	397.7442		2121.532[a]

X'X INVERSE, B, SSE

INVERSE		INTERCEP	X4	X3	X3SQ	X43		Y
INTERCEP		1315.261	543.4463	-433.586	35.50156	-83.4036		29.11329
X4		543.4463	464.2447	-179.952	14.80313	-69.5252		11.13423
X3	$(\mathbf{X'X})^{-1} =$	433.586	-179.952	143.1914	-11.7449	27.67939	$\mathbf{b} =$	-7.60801
X3SQ		35.50156	14.80313	-11.7449	0.965045	-2.28257		0.6712472
X43		-83.4036	-69.5252	27.67939	-2.28257	10.45448		-1.47772
Y		29.11329	11.13423	-7.60801	0.6712472	-1.47772		1.064397[b]

[a] $\sum_{i=1}^{n} y_i^2$; [b] SSE

DEP VARIABLE: Y

SOURCE	DF	SUM OF SQUARES	MEAN SQUARE	F VALUE	PROB>F	
MODEL	4	12.394190[a]	3.098548[p]	72.777[q]	0.0001[r]	Analysis of variance table
ERROR	25	1.064397[b]	0.042576[f]			
C TOTAL	29	13.458587[c]				

ROOT MSE	0.206339[d]	R-SQUARE	0.9209[h]
DEP MEAN	8.382667	ADJ R-SQ	0.9083[i]
C.V.	2.461498		

VARIABLE	DF	PARAMETER ESTIMATE[e]	STANDARD ERROR[g]	T FOR HO: PARAMETER=0[j]	PROB > |T|[k]
INTERCEP	1	29.113287	7.483206	3.890	0.0007
X4	1	11.134226	4.445854	2.504	0.0192
X3	1	-7.608007	2.469109	-3.081	0.0050
X3SQ	1	0.671247	0.202701	3.312	0.0028
X43	1	-1.477717	0.667165	-2.215	0.0361

OBS	ACTUAL	PREDICT VALUE	STD ERR PREDICT	LOWER95% MEAN	UPPER95% MEAN	LOWER95% PREDICT	UPPER95% PREDICT	RESIDUAL
1	7.380	7.424	0.081140	7.257	7.591	6.968	7.881	-.044139
2	8.510	8.833	0.049414	8.531	8.735	8.196	9.070	-.122850
3	9.520	9.490	0.127991	9.227	9.754	8.990	9.990	0.029866
4	7.500	7.574	0.079322	7.411	7.738	7.119	8.030	-.074478
5	9.330	8.946	0.086038	8.769	9.123	8.485	9.406	0.384097
6	8.280	8.327	0.056341	8.211	8.443	7.887	8.768	-.047250
7	8.750	8.517	0.063947	8.385	8.649	8.072	8.962	0.233113
8	7.870	7.841	0.163020	7.506	8.177	7.300	8.383	0.028687
9	7.100	7.166	0.144177	6.869	7.463	6.648	7.684	-.066071
10	8.000	7.970	0.080969	7.804	8.137	7.514	8.427	0.029666
11	7.890	8.327	0.056341	8.211	8.443	7.887	8.768	-.437250
12	8.150	7.974	0.062179	7.846	8.102	7.530	8.418	0.176312
13	9.100	9.064	0.060524	8.940	9.189	8.622	9.507	0.035566
14	8.860	8.998	0.055135	8.885	9.112	8.558	9.438	-.138209
15	8.900	8.797	0.047450	8.700	8.895	8.361	9.233	0.102677
16	8.870	8.743	0.047649	8.645	8.841	8.307	9.179	0.126964
17	9.260	9.255	0.080749	9.089	9.422	8.799	9.712	0.004773
18	9.000	9.143	0.067340	9.005	9.282	8.696	9.590	-.143455
19	8.750	8.652	0.051799	8.745	8.958	8.413	9.290	-.101611
20	7.950	7.945	0.082413	7.775	8.115	7.487	8.403	0.005016
21	7.650	7.689	0.082596	7.519	7.859	7.231	8.147	-.038914
22	7.270	7.403	0.103346	7.191	7.616	6.928	7.879	-.133353
23	8.000	8.327	0.056341	8.211	8.443	7.887	8.768	-.327250
24	8.500	8.827	0.131141	8.557	9.097	8.324	9.331	-.327373
25	8.750	8.960	0.069988	8.816	9.104	8.511	9.409	-.210185
26	9.210	9.069	0.094423	8.874	9.263	8.601	9.536	0.141240
27	8.270	7.945	0.082413	7.775	8.115	7.487	8.403	0.325016
28	7.670	7.560	0.065802	7.425	7.696	7.114	8.006	0.109641
29	7.930	7.696	0.065635	7.561	7.831	7.250	8.142	0.234223
30	9.260	9.014	0.081742	8.846	9.183	8.557	9.472	0.245527
31		8.526[l]	0.082777[m]	8.355	8.696	8.068	8.984	

SUM OF RESIDUALS 1.00586E-13
SUM OF SQUARED RESIDUALS 1.064397[n]

(LOWER95% MEAN – UPPER95% MEAN: 95% confidence interval; LOWER95% PREDICT – UPPER95% PREDICT: 95% prediction interval)

[a]Explained variation = SS_{model}; [b]SSE; [c]Total variation; [d]Standard error; [e]b_j, b_0, b_1, b_2, b_3, b_4; [f]$MSE = s^2$; [g]$s\sqrt{c_{jj}}$; [h]R^2; [i]$\bar{R}^2$; [j]t; [k]Prob-value for t; [l]Point prediction $\hat{y}_0$; [m]$s\sqrt{h_{00}}$; [n]SSE; [p]MS_{model}; [q]F(model); [r]Prob-value for F(model)

and the matrix

$$\mathbf{X} = \begin{bmatrix} 1 & -.05 & 5.50 & (5.50)^2 & (-.05)(5.50) \\ 1 & .25 & 6.75 & (6.75)^2 & (.25)(6.75) \\ 1 & .60 & 7.25 & (7.25)^2 & (.60)(7.25) \\ \vdots & \vdots & \vdots & \vdots & \vdots \\ 1 & .55 & 6.80 & (6.80)^2 & (.55)(6.80) \end{bmatrix}$$

(columns: 1, x_4, x_3, x_3^2, x_4x_3)

$$= \begin{bmatrix} 1 & -.05 & 5.50 & 30.25 & -.275 \\ 1 & .25 & 6.75 & 45.5625 & 1.6875 \\ 1 & .60 & 7.25 & 52.5625 & 4.35 \\ \vdots & \vdots & \vdots & \vdots & \vdots \\ 1 & .55 & 6.80 & 46.24 & 3.74 \end{bmatrix}$$

(columns: 1, x_4, x_3, x_3^2, x_4x_3)

Note from Figure 9.5 that the prob-values related to testing $H_0: \beta_j = 0$ ($j = 0, 1, 2, 3, 4$) are all less than .05. Therefore we have strong evidence that the intercept and all of the independent variables x_4, x_3, x_3^2, and x_4x_3 are important. In particular, since the prob-value related to testing $H_0: \beta_4 = 0$ is .0361, we have strong evidence that the interaction variable x_4x_3 is important.

Recall from Example 8.9 that Enterprise Industries desires to predict demand for Fresh in a sales region during a future quarter when the price of Fresh will be $x_{01} = \$3.80$, the average industry price will be $x_{02} = \$3.90$, the price difference will be $x_{04} = x_{02} - x_{01} = \$3.90 - \$3.80 = \$.10$, and the advertising expenditure for Fresh will be $x_{03} = \$6.80$. Using the least squares point estimates in Figure 9.5, it follows that

$$\begin{aligned}\hat{y}_0 &= b_0 + b_1x_{04} + b_2x_{03} + b_3x_{03}^2 + b_4x_{04}x_{03} \\ &= 29.1133 + 11.1342(.10) - 7.6080(6.8) + .6712(6.8)^2 \\ &\quad - 1.4777(.10)(6.8) \\ &= 8.526 \qquad (852{,}600 \text{ bottles})\end{aligned}$$

is a point estimate of μ_0, the mean demand for Fresh when $x_{04} = \$.10$ and $x_{03} = \$6.8$, and is a point prediction of $y_0 = \mu_0 + \varepsilon_0$, the individual demand for Fresh in the future quarter in the sales region. This point estimate (and prediction) is given in Figure 9.5. This figure also gives the 95% confidence interval for μ_0, which is [8.355, 8.697], and the 95% prediction interval for y_0, which is [8.068, 8.984]. Note that these intervals are calculated by using the row vector

$$\begin{aligned}\mathbf{x}_0' &= [1 \quad .10 \quad 6.80 \quad (6.80)^2 \quad (.10)(6.80)] \\ &= [1 \quad .10 \quad 6.80 \quad 46.24 \quad .68]\end{aligned}$$

Here, this vector contains the numbers multiplied by b_0, b_1, b_2, b_3, and b_4 in the preceding prediction equation.

Part 3: The Nature of Interaction Between x_3 and x_4

To understand the exact nature of the interaction between x_3 and x_4, we consider the prediction equation obtained above. This equation gives the point estimate

$$\begin{aligned}\hat{y} &= b_0 + b_1x_4 + b_2x_3 + b_3x_3^2 + b_4x_4x_3 \\ &= 29.1133 + 11.1342x_4 - 7.6080x_3 + .6712x_3^2 - 1.4777x_4x_3\end{aligned}$$

of the mean demand for Fresh

$$\mu = \beta_0 + \beta_1x_4 + \beta_2x_3 + \beta_3x_3^2 + \beta_4x_4x_3$$

We now plot $\hat{y}$ against x_3 when we fix x_4 at two different values—first at $x_4 = .10$ and then at $x_4 = .30$. If we fix x_4 at .10, the point estimate $\hat{y}$ of μ is

$$\begin{aligned}\hat{y} &= 29.1133 + 11.1342x_4 - 7.6080x_3 + .6712x_3^2 - 1.4777x_4x_3 \\ &= 29.1133 + 11.1342(.10) - 7.6080x_3 + .6712x_3^2 - 1.4777(.10)x_3 \\ &= 30.2267 - 7.7558x_3 + .6712x_3^2\end{aligned}$$

This equation may be interpreted as the point estimate of the equation relating μ to x_3 when x_4 is .10. The quadratic curve defined by this equation is plotted in Figure 9.6. Note that, for example, when $x_3 = 6.0$,

$$\hat{y} = 30.2267 - 7.7558(6.0) + .6712(6.0)^2 \approx 7.76$$

When $x_3 = 6.4$,

$$\hat{y} = 30.2267 - 7.7558(6.4) + .6712(6.4)^2 \approx 8.08$$

When $x_3 = 6.8$,

$$\hat{y} = 30.2267 - 7.7558(6.8) + .6712(6.8)^2 \approx 8.52$$

Next, if we fix x_4 at .30, the point estimate $\hat{y}$ of μ equals

$$\begin{aligned} \hat{y} &= 29.1133 + 11.1342x_4 - 7.6080x_3 + .6712x_3^2 - 1.4777x_4x_3 \\ &= 29.1133 + 11.1342(.30) - 7.6080x_3 + .6712x_3^2 - 1.4777(.30)x_3 \\ &= 32.4535 - 8.0513x_3 + .6712x_3^2 \end{aligned}$$

This equation may be interpreted as the point estimate of the equation relating μ to x_3 when x_4 is .30. The quadratic curve defined by this equation is also plotted in Figure 9.6. Note that, for example, when $x_3 = 6.0$,

$$\hat{y} = 32.4535 - 8.0513(6.0) + .6712(6.0)^2 \approx 8.31$$

When $x_3 = 6.4$,

$$\hat{y} = 32.4535 - 8.0513(6.4) + .6712(6.4)^2 \approx 8.42$$

FIGURE 9.6 **Illustration of the interaction between x_3 and x_4 in the Fresh Detergent problem**

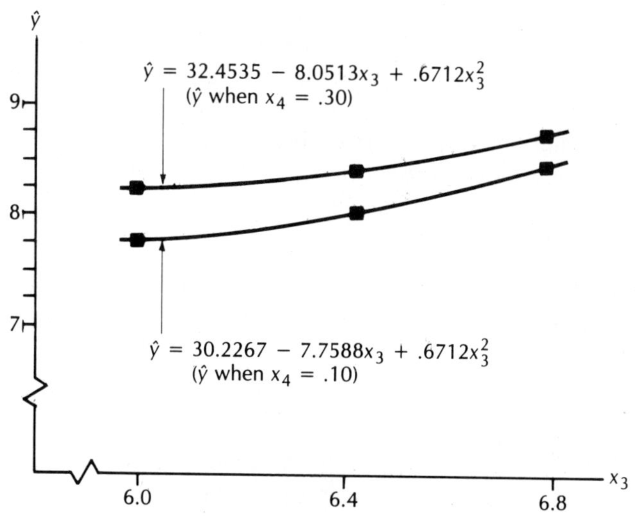

When $x_3 = 6.8$,

$$\hat{y} = 32.4535 - 8.0513(6.8) + .6712(6.8)^2 \approx 8.74$$

Examining Figure 9.6, we note that, as x_3 increases, the quadratic curve

$$\hat{y} = 32.4535 - 8.0513x_3 + .6712x_3^2$$

increases at a slower rate than the quadratic curve

$$\hat{y} = 30.2267 - 7.7558x_3 + .6712x_3^2$$

This type of interaction is logical because when the price difference is large (the price for Fresh is low relative to the average industry price), the mean demand for Fresh will be high (assuming the quality of Fresh is comparable to competing brands). Thus with mean demand already high because many consumers are buying Fresh on the basis of price, there may be little opportunity for increased advertising expenditure to increase mean demand. However, when the price difference is smaller, there may be more potential consumers who are not buying Fresh who can be convinced to do so by increased advertising. Thus when the price difference is smaller, increased advertising expenditure is more effective than it is when the price difference is larger.

It should be noted that this type of interaction between x_4 and x_3 was estimated from the observed Fresh demand data in Table 8.3. This is because we obtained the least squares point estimates using these data. We are not hypothesizing the existence of the interaction; the importance of the x_4x_3 term and the least squares point estimates tell us that this type of interaction exists. However, we can only hypothesize the reasons behind the interaction. We should also point out that this type of interaction can only be assumed to exist for values of x_4 and x_3 inside the experimental region. Examination of Table 8.3 shows that Fresh was being sold in the 30 observed sales regions at either a price advantage (when the price of Fresh is lower than the average industry price) or at a slight price disadvantage (when the price of Fresh is slightly higher than the average industry price). However, if Fresh were sometimes sold at a large price disadvantage, the type of interaction that exists between x_4 and x_3 might be different. In such a case, increases in advertising expenditure might be very ineffective because most consumers will not wish to buy a product with a much higher price.

Part 4: Residual Analysis

The residuals given by the Fresh Detergent prediction equation are calculated, for $i = 1, 2, \ldots, 30$, as

$$\begin{aligned} e_i &= y_i - \hat{y}_i \\ &= y_i - (29.1133 + 11.1342x_{i4} - 7.6080x_{i3} + .6712x_{i3}^2 \\ &\quad - 1.4777x_{i4}x_{i3}) \end{aligned}$$

and are given in Figure 9.5. Plots of the residuals against x_4, x_3, and $\hat{y}$ are given in Figures 9.7, 9.8, and 9.9. These plots probably suggest that the constant variance

FIGURE 9.7 Residuals for the Fresh Detergent model plotted against x_4

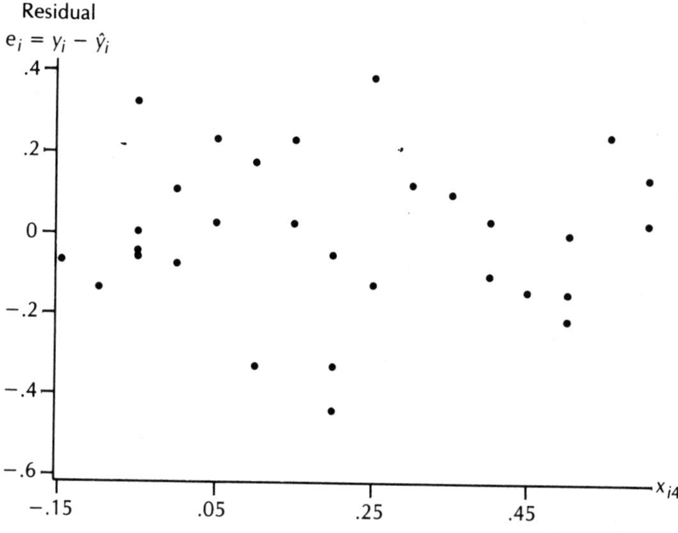

FIGURE 9.8 Residuals for the Fresh Detergent model plotted against x_3

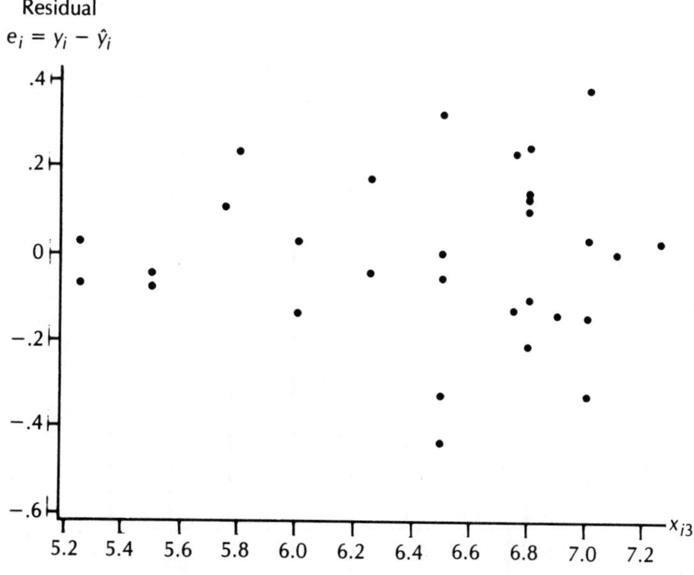

FIGURE 9.9 **Residuals for the Fresh Detergent model plotted against $\hat{y}$**

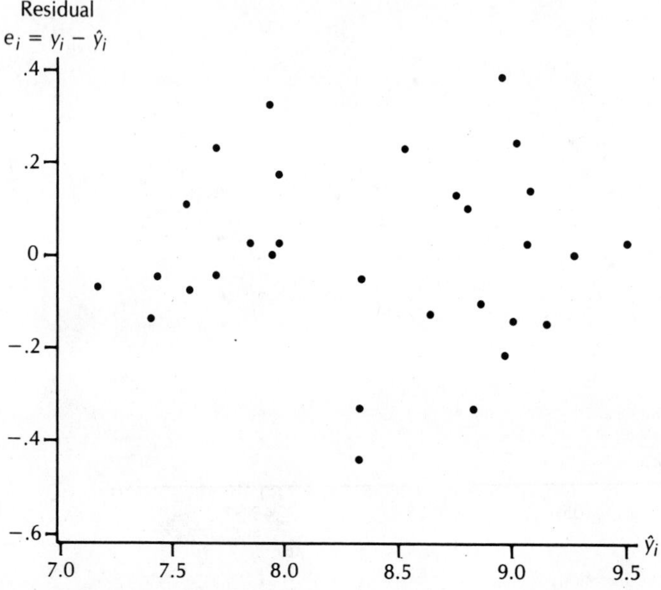

FIGURE 9.10 **Histogram of the residuals for the Fresh Detergent model**

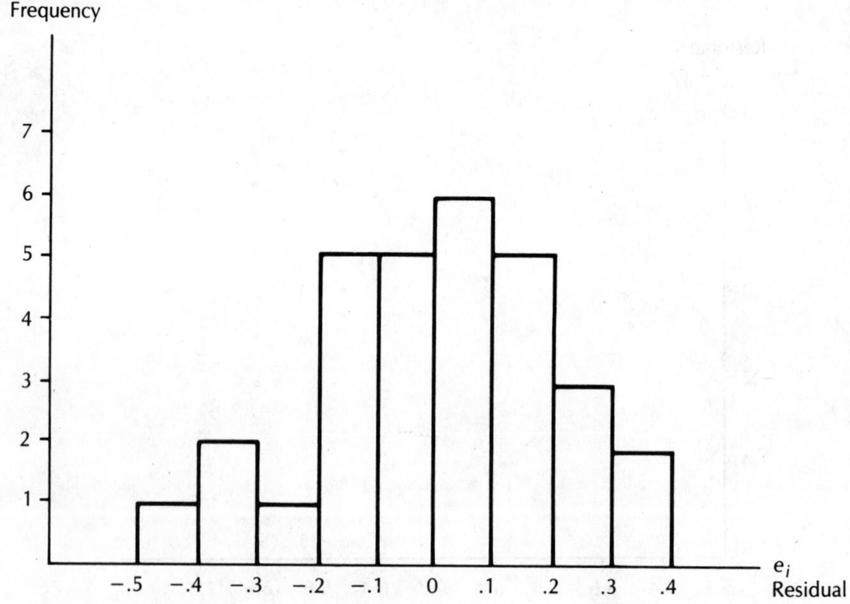

FIGURE 9.11 **Normal plot of the residuals for the Fresh Detergent model**

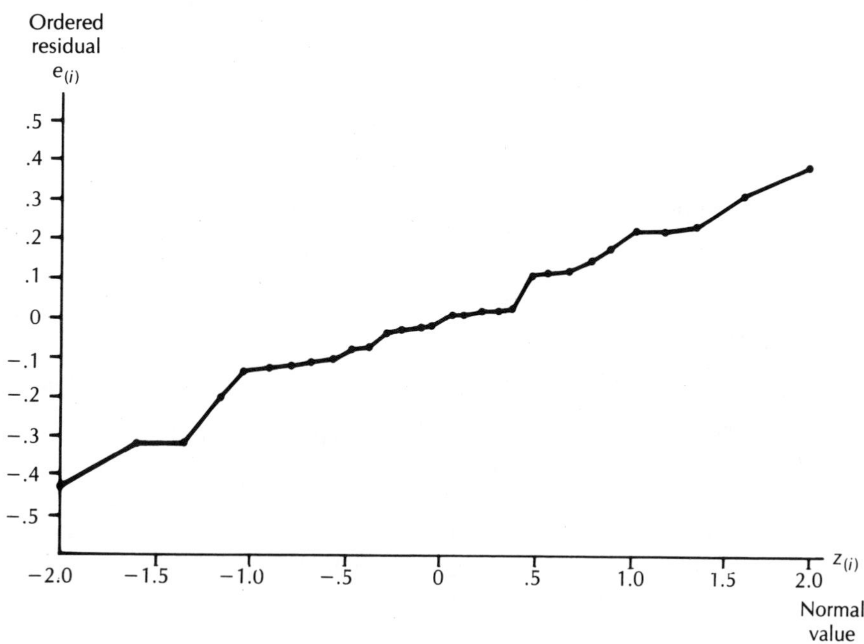

assumption approximately holds. Furthermore, a histogram of the residuals in Figure 9.10 has a bell-shaped appearance, and a normal plot of the residuals in Figure 9.11 has an approximate straight-line appearance. Therefore it is reasonable to believe that the normality assumption approximately holds. Since the Fresh demand data is cross-sectional data, we do not need to test for autocorrelation.

9.2

AN *F*-TEST FOR A PORTION OF A MODEL AND THE PARTIAL COEFFICIENT OF DETERMINATION

AN F-TEST FOR A PORTION OF A MODEL

In this section we present an *F*-test related to the utility of a portion of a regression model. For example, consider the Fresh Detergent model

$$y = \beta_0 + \beta_1 x_4 + \beta_2 x_3 + \beta_3 x_3^2 + \beta_4 x_4 x_3 + \varepsilon$$

It might be useful to test the null hypothesis

$$H_0: \beta_3 = \beta_4 = 0$$

which says that neither of the higher-order terms x_3^2 and x_4x_3 affect y, versus the alternative hypothesis

$$H_1: \text{At least one of } \beta_3 \text{ and } \beta_4 \text{ does not equal zero}$$

which says that at least one of the higher-order terms x_3^2 and x_4x_3 affects y. In general, consider the regression model

$$y = \beta_0 + \beta_1 x_1 + \cdots + \beta_g x_g + \beta_{g+1} x_{g+1} + \cdots + \beta_p x_p + \varepsilon$$

Suppose that we wish to test the null hypothesis

$$H_0: \beta_{g+1} = \beta_{g+2} = \cdots = \beta_p = 0$$

which says that none of the independent variables $x_{g+1}, x_{g+2}, \ldots, x_p$ affect y, versus the alternative hypothesis

$$H_1: \text{At least one of } \beta_{g+1}, \beta_{g+2}, \ldots, \beta_p \text{ does not equal zero}$$

which says that at least one of the independent variables $x_{g+1}, x_{g+2}, \ldots, x_p$ affects y. If we can reject H_0 in favor of H_1 by specifying a *small* probability of a Type I error, then it is reasonable to conclude that at least one of $x_{g+1}, x_{g+2}, \ldots, x_p$ *significantly* affects y. In this case we should use t statistics and other techniques to determine which of $x_{g+1}, x_{g+2}, \ldots, x_p$ significantly affect y. To test H_0 versus H_1, consider the following two models:

Complete model: $y = \beta_0 + \beta_1 x_1 + \cdots + \beta_g x_g + \beta_{g+1} x_{g+1} + \cdots + \beta_p x_p + \varepsilon$

Reduced model: $y = \beta_0 + \beta_1 x_1 + \cdots + \beta_g x_g + \varepsilon$

Here, the complete model is assumed to have k parameters, the reduced model is the complete model under the assumption that H_0 is true, and $(p - g)$ denotes the number of regression parameters we have set equal to zero in the statement of H_0. Using regression analysis, we calculate SSE_C, the unexplained variation for the complete model, and SSE_R, the unexplained variation for the reduced model. Then we consider the difference

$$SS_{\text{drop}} = SSE_R - SSE_C$$

We call this difference the *drop in the unexplained variation attributable to the independent variables* $x_{g+1}, x_{g+2}, \ldots, x_p$. It can be shown that the "extra" independent variables $x_{g+1}, x_{g+2}, \ldots, x_p$ will always make SSE_C somewhat smaller than SSE_R and hence will always make SS_{drop} positive. The question is whether this difference is large enough to conclude that at least one of the independent variables $x_{g+1}, x_{g+2}, \ldots, x_p$ significantly affects y. Since the value of SS_{drop} depends on the units in which the observed values of the dependent variable are measured, we need to modify SS_{drop} and formulate a unitless measure of the additional importance of the set $x_{g+1}, x_{g+2}, \ldots, x_p$. Such a statistic is called the *partial F statistic* and is denoted by

$$F(x_{g+1}, \ldots, x_p | x_1, x_2, \ldots, x_g)$$

Next we define this statistic and show how it is used to test H_0 versus H_1.

Testing $H_0: \beta_{g+1} = \beta_{g+2} = \cdots = \beta_p = 0$ Versus H_1: At Least One of $\beta_{g+1}, \beta_{g+2}, \ldots, \beta_p$ Does Not Equal Zero

Assume the inference assumptions are satisfied, and define the partial F statistic to be

$$F(x_{g+1}, \ldots, x_p | x_1, \ldots, x_g) = \frac{MS_{\text{drop}}}{MSE_C} = \frac{SS_{\text{drop}}/(p-g)}{SSE_C/(n-k)} = \frac{\{SSE_R - SSE_C\}/(p-g)}{SSE_C/(n-k)}$$

where

$SS_{\text{drop}} = SSE_R - SSE_C =$ the drop in the unexplained variation attributable to $x_{g+1}, x_{g+2}, \ldots, x_p$

$MSE_C = \dfrac{SSE_C}{n-k} =$ the mean square error for the complete model

Also, define the prob-value to be the area to the right of $F(x_{g+1}, \ldots, x_p | x_1, \ldots, x_g)$ under the curve of the F-distribution having $p - g$ and $n - k$ degrees of freedom.

Then we can reject H_0 in favor of H_1 by setting the probability of a Type I error equal to α if and only if either of the following equivalent conditions holds:

1. $F(x_{g+1}, \ldots, x_p | x_1, \ldots, x_g) > F_{[\alpha]}^{(p-g, n-k)}$, where $F_{[\alpha]}^{(p-g,n-k)}$ is the point on the scale of the F-distribution having $p - g$ and $n - k$ degrees of freedom so that the area under this curve to the right of $F_{[\alpha]}^{(p-g,n-k)}$ is α.
2. Prob-value $< \alpha$.

Condition 1 says we should reject H_0 in favor of H_1 if

$$F(x_{g+1}, \ldots, x_p | x_1, \ldots, x_g) = \frac{\{SSE_R - SSE_C\}/(p-g)}{SSE_C/(n-k)}$$

is large. This is reasonable because a large value of

$$F(x_{g+1}, \ldots, x_p | x_1, \ldots, x_g)$$

would be caused by a large value of $\{SSE_R - SSE_C\}$. Such a value would be obtained if at least one of the independent variables $x_{g+1}, x_{g+2}, \ldots, x_p$ makes SSE_C substantially smaller than SSE_R. This would indicate that H_0 is false and that H_1 is true.

Also, note that there is a relationship between the t statistic and the partial F statistic. It can be proved that

$$t^2 = F(x_j | x_1, \ldots, x_{j-1}, x_{j+1}, \ldots, x_p)$$

and that

$$(t_{[\alpha/2]}^{(n-k)})^2 = F_{[\alpha]}^{(1,n-k)}$$

Hence the rejection conditions

$$|t| > t_{[\alpha/2]}^{(n-k)} \quad \text{and} \quad F(x_j|x_1, \ldots, x_{j-1}, x_{j+1}, \ldots, x_p) > F_{[\alpha]}^{(1,n-k)}$$

are equivalent. Furthermore, it can be shown that the prob-values for testing H_0: $\beta_j = 0$ versus H_1: $\beta_j \neq 0$ computed by using the t statistic and the partial F statistic are equal to each other.

Example 9.3 Consider the Fresh Detergent model

$$\text{Complete model: } y_i = \beta_0 + \beta_1 x_{i4} + \beta_2 x_{i3} + \beta_3 x_{i3}^2 + \beta_4 x_{i4} x_{i3} + \varepsilon_i$$

which has $k = 5$ parameters. If we use this model to carry out a regression analysis of the data in Table 8.3, we obtain an unexplained variation of $SSE_C = 1.0644$. To test

$$H_0: \beta_3 = \beta_4 = 0$$

versus

$$H_1: \text{At least one of } \beta_3 \text{ and } \beta_4 \text{ does not equal zero}$$

note that $(p - g) = 2$, since two parameters (β_3 and β_4) are set equal to zero in the statement of H_0. Also, under the assumption that H_0 is true, the complete model becomes the following reduced model:

$$\text{Reduced model: } y_i = \beta_0 + \beta_1 x_{i4} + \beta_2 x_{i3} + \varepsilon_i$$

The unexplained variation for this reduced model is $SSE_R = 1.5337$. Thus to test H_0 versus H_1 we use the following partial F statistic and prob-value:

$$\begin{aligned} F(x_{i3}^2, x_{i4}x_{i3}|x_{i4}, x_{i3}) &= \frac{MS_{\text{drop}}}{MSE_C} \\ &= \frac{SS_{\text{drop}}/(p - g)}{SSE_C/(n - k)} \\ &= \frac{\{SSE_R - SSE_C\}/2}{SSE_C/(30 - 5)} \\ &= \frac{\{1.5337 - 1.0644\}/2}{1.0644/25} \\ &= \frac{.2347}{.0426} \\ &= 5.5094 \end{aligned}$$

and prob-value $= .0111$, which is the area to the right of 5.5094 under the curve of the F distribution having 2 and 25 degrees of freedom.

If we wish to use condition (1) to test H_0 versus H_1 by setting α equal to .05, then we would use the rejection point

$$F_{[\alpha]}^{(p-g,n-k)} = F_{[.05]}^{(2,25)} = 3.39$$

Since

$$F(x_{i3}^2, x_{i4}x_{i3} | x_{i4}, x_{i3}) = 5.5094 > 3.39 = F_{[.05]}^{(2.25)}$$

we can reject H_0 in favor of H_1 by setting α equal to .05. Alternatively, since prob-value = .0111 is less than .05, we can reject H_0 in favor of H_1 by setting α equal to .05. Thus we have substantial evidence that at least one of the higher-order terms x_3^2 and x_4x_3 significantly affects y, and we should use t statistics and other techniques to determine which of these higher-order terms significantly affects demand for Fresh (see Example 9.2).

THE PARTIAL COEFFICIENT OF DETERMINATION

Partial Coefficients of Determination and Correlation

1. The *partial coefficient of determination* is

$$R^2(x_{g+1}, \ldots, x_p | x_1, \ldots, x_g) = \frac{SSE_R - SSE_C}{SSE_R}$$

= the proportion of the unexplained variation in the reduced model that is explained by the extra independent variables in the complete model

2. The *partial coefficient of correlation* is

$$R(x_{g+1}, \ldots, x_p | x_1, \ldots, x_g) = \sqrt{R^2(x_{g+1}, \ldots, x_p | x_1, \ldots, x_g)}$$

Example 9.4 Recall from Example 9.3 that the unexplained variation for the

$$\text{Complete model: } y_i = \beta_0 + \beta_1 x_{i4} + \beta_2 x_{i3} + \beta_3 x_{i3}^2 + \beta_4 x_{i4}x_{i3} + \varepsilon_i$$

is $SSE_C = 1.0644$, and that the unexplained variation for the

$$\text{Reduced model: } y_i = \beta_0 + \beta_1 x_{i4} + \beta_2 x_{i3} + \varepsilon_i$$

is $SSE_R = 1.5337$. It follows that

$$\begin{aligned} R^2(x_{i3}^2, x_{i4}x_{i3} | x_{i4}, x_{i3}) &= \frac{SSE_R - SSE_C}{SSE_R} \\ &= \frac{1.5337 - 1.0644}{1.5337} \\ &= \frac{.4693}{1.5337} \\ &= .3060 \end{aligned}$$

This says that the extra independent variables x_{i3}^2 and $x_{i4}x_{i3}$ in the complete model explain 30.6 percent of the unexplained variation in the reduced model. Furthermore,

$$R(x_{i3}^2, x_{i4}x_{i3} \mid x_{i4}, x_{i3}) = \sqrt{.306} = .5532$$

*9.3

INFERENCES FOR A SINGLE POPULATION

Suppose that we have randomly selected a sample of size n,

$$y_1, y_2, \ldots, y_n$$

from a single population having mean μ, variance σ^2, and standard deviation σ. Then, for $i = 1, 2, \ldots, n$ we can express y_i by the regression model

$$\begin{aligned} y_i &= \mu + \varepsilon_i \\ &= \beta_0 + \varepsilon_i \end{aligned}$$

The column vector $\mathbf{y}$ and the matrix $\mathbf{X}$ used to calculate the least squares point estimate b_0 of $\beta_0 = \mu$ are

$$\mathbf{y} = \begin{bmatrix} y_1 \\ y_2 \\ \cdot \\ \cdot \\ \cdot \\ y_n \end{bmatrix} \quad \text{and} \quad \mathbf{X} = \begin{bmatrix} 1 \\ 1 \\ \cdot \\ \cdot \\ \cdot \\ 1 \end{bmatrix}$$

Thus

$$\mathbf{X}'\mathbf{X} = [1 \quad 1 \quad \cdots \quad 1] \begin{bmatrix} 1 \\ 1 \\ \cdot \\ \cdot \\ \cdot \\ 1 \end{bmatrix} = n$$

$$(\mathbf{X}'\mathbf{X})^{-1} = \frac{1}{n}$$

$$\mathbf{X}'\mathbf{y} = [1 \quad 1 \quad \cdots \quad 1] \begin{bmatrix} y_1 \\ y_2 \\ \cdot \\ \cdot \\ \cdot \\ y_n \end{bmatrix} = \sum_{i=1}^{n} y_i$$

*This section is optional.

It follows that the least squares point estimate of $\beta_0 = \mu$ is

$$b_0 = (\mathbf{X}'\mathbf{X})^{-1}\mathbf{X}'\mathbf{y} = \left(\frac{1}{n}\right)\left(\sum_{i=1}^{n} y_i\right) = \bar{y}$$

Thus the least squares point estimate of the population mean μ is the sample mean $\bar{y}$. Note that $b_0 = \bar{y}$ minimizes

$$SSE = \sum_{i=1}^{n} (y_i - \hat{y}_i)^2 = \sum_{i=1}^{n} (y_i - b_0)^2$$

where $\hat{y}_i = b_0$ is the appropriate point prediction of $y_i = \mu + \varepsilon_i = \beta_0 + \varepsilon_i$. Therefore the standard error is

$$s = \sqrt{\frac{SSE}{n-k}} = \sqrt{\frac{\sum_{i=1}^{n} (y_i - \bar{y})^2}{n-1}}$$

which is the sample standard deviation.

We can obtain point and interval estimates of μ and $y_0 = \mu + \varepsilon_0$ by applying the formulas of Section 8.8. Here:

1. The point prediction of a value that will be randomly selected from the single population, $y_0 = \mu + \varepsilon_0 = \beta_0 + \varepsilon_0$, is $\hat{y}_0 = b_0 = \bar{y}$.
2. The number 1 is multiplied by b_0 in the preceding point prediction, which implies that $\mathbf{x}_0' = [1]$ and thus that

$$\mathbf{x}_0'(\mathbf{X}'\mathbf{X})^{-1}\mathbf{x}_0 = [1]\frac{1}{n}[1] = \frac{1}{n}$$

3. When we randomly select a sample from a single population, μ_0 (the mean value of the dependent variable) equals μ (the mean of the single population).

We therefore obtain the following results.

Statistical Inferences Concerning a Single Population

Assume that $\bar{y}$ is the *mean* and s is the *standard deviation* of a sample randomly selected from a normally distributed population having mean μ, variance σ^2, and standard deviation σ. Then:

1. $\bar{y}$ is the *point estimate* of μ and is the *point prediction* of the individual value

$$y_0 = \mu + \varepsilon_0$$

2. A *100(1 − α)% confidence interval for* $\mu_0 = \mu$ is

$$[\hat{y}_0 \pm t_{[\alpha/2]}^{(n-k)} s\sqrt{\mathbf{x}_0'(\mathbf{X}'\mathbf{X})^{-1}\mathbf{x}_0}] = \left[\bar{y} \pm t_{[\alpha/2]}^{(n-1)} s\sqrt{\frac{1}{n}}\right] = \left[\bar{y} \pm t_{[\alpha/2]}^{(n-1)}\left(\frac{s}{\sqrt{n}}\right)\right]$$

3. A *100(1 − α) % prediction interval for* $y_0 = \mu + \varepsilon_0$ is

$$[\hat{y}_0 \pm t_{[\alpha/2]}^{(n-k)} s\sqrt{1 + \mathbf{x}_0'(\mathbf{X}'\mathbf{X})^{-1}\mathbf{x}_0}] = \left[\bar{y} \pm t_{[\alpha/2]}^{(n-1)} s\sqrt{1 + \frac{1}{n}}\right]$$

Example 9.5 In Chapter 2 we saw that $\bar{y} = 31.2$ mpg is the mean and $s = .7517$ mpg is the standard deviation of the sample of $n = 5$ gasoline mileages

$$y_1 = 30.7, \quad y_2 = 31.8, \quad y_3 = 30.2, \quad y_4 = 32.0, \quad y_5 = 31.3$$

This sample has been randomly selected from the infinite population of all Hawk mileages (which we assume is normally distributed). It follows that $\bar{y} = 31.2$ is the point estimate of μ, the mean of all Hawk mileages, and that a 95% confidence interval for μ is

$$\begin{aligned}\left[\bar{y} \pm t_{[.025]}^{(n-1)}\left(\frac{s}{\sqrt{n}}\right)\right] &= \left[\bar{y} \pm t_{[.025]}^{(5-1)}\left(\frac{s}{\sqrt{5}}\right)\right] \\ &= \left[31.2 \pm 2.776\left(\frac{.7517}{\sqrt{5}}\right)\right] \\ &= [31.2 \pm .9333] \\ &= [30.2667, 32.1333]\end{aligned}$$

Suppose that federal gasoline mileage standards state that μ must be greater than 30 mpg. Since the 95% confidence interval for μ makes the company 95% confident that μ is greater than or equal to the lower bound 30.2667, and since this lower bound is itself greater than 30 mpg, National Motors can be at least 95% confident that μ is greater than 30 mpg. Thus there is strong evidence that the Hawk not only meets but exceeds current gasoline mileage standards.

Next, assume that purchasing a Hawk is equivalent to randomly selecting a Hawk from the infinite population of all Hawks. It follows that $\bar{y} = 31.2$ is the point prediction of $y_0 = \mu + \varepsilon_0$, an individual gasoline mileage for a randomly selected (purchased) Hawk, and that a 95% prediction interval for y_0 is

$$\begin{aligned}\left[\bar{y} \pm t_{[.025]}^{(n-1)} s\sqrt{1 + \frac{1}{n}}\right] &= \left[\bar{y} \pm t_{[.025]}^{(5-1)} s\sqrt{1 + \frac{1}{5}}\right] \\ &= [31.2 \pm 2.776(.7517)\sqrt{1.2}] \\ &= [31.2 \pm 2.29] \\ &= [28.91, 33.49]\end{aligned}$$

This 95% prediction interval says that the purchaser of a (randomly selected) Hawk can be 95% confident that his or her car will obtain a gasoline mileage somewhere between 28.91 mpg and 33.49 mpg.

Example 9.6 Consider the sample of $n = 49$ Hawk mileages listed in Table 2.3, which has mean $\bar{y} = 31.5531$ and standard deviation $s = .799$. Using this sample, assuming that the population of all Hawk mileages is normally distributed, and replacing $t_{[\alpha/2]}^{(n-1)}$ by $z_{[\alpha/2]}$ (since $n = 49$ is large), it follows that a 95% confidence interval for μ is

$$\left[\bar{y} \pm z_{[.025]}\left(\frac{s}{\sqrt{n}}\right)\right] = \left[31.5531 \pm 1.96\left(\frac{.799}{\sqrt{49}}\right)\right]$$
$$= [31.5531 \pm .2237]$$
$$= [31.3294, 31.7768]$$

and that a 95% prediction interval for $y_0 = \mu + \varepsilon_0$ is

$$\left[\bar{y} \pm z_{[.025]}s\sqrt{1 + \frac{1}{n}}\right] = \left[31.5531 \pm 1.96(.799)\sqrt{1 + \frac{1}{49}}\right]$$
$$= [31.5531 \pm 1.5819]$$
$$= [29.9712, 33.135]$$

The Central Limit Theorem (discussed in Chapter 3) tells us that since $n = 49$ is a large sample size, this confidence interval for μ is approximately correct no matter what probability distribution describes the population of all Hawk mileages. However, even though $n = 49$ is large, the validity of the prediction interval for $y_0 = \mu + \varepsilon_0$ still depends on the assumption that the population of all Hawk mileages is normally distributed because its validity depends on the inference assumptions.

*9.4 USING SAS

In Figure 9.12 we present the SAS statements that generate the SAS output of Figure 9.5, which presents a regression analysis of the Fresh demand data using the model $y = \beta_0 + \beta_1 x_4 + \beta_2 x_3 + \beta_3 x_3^2 + \beta_4 x_4 x_3 + \varepsilon$. We also give the SAS statements needed to carry out the partial F-test of Example 9.3 and the SAS statements needed to perform residual analysis for the above model.

*This section is optional.

FIGURE 9.12 **SAS program for a regression analysis of the Fresh demand data using the model $y = \beta_0 + \beta_1 x_4 + \beta_2 x_3 + \beta_3 x_3^2 + \beta_4 x_4 x_3 + \varepsilon$**

```
DATA DETR;                 ——→ Assigns the name DETR to the data file
INPUT Y X4 X3;             ——→ Defines variable names. Y = demand, X4 = price
                               difference, X3 = advertising expenditure
X3SQ = X3 * X3;            ——→ Transformation defines the squared term x3^2
                               and assigns the name X3SQ to this variable
X43 = X4 * X3;             ——→ Transformation defines the interaction term x4x3
                               and assigns the name X43 to this variable
CARDS;
7.38  -0.05  5.50 )
8.51   0.25  6.75 |
9.52   0.60  7.25 |
        .          }——→ Fresh demand data—see Table 8.3
        .         |
        .         |
9.26   0.55  6.80 )
 .     0.10  6.80  ——→ Decimal point represents a missing value. Used to
                       obtain point prediction when x03 = 6.80 and x04 = .10
PROC PRINT;                ——→ Prints the data
PROC REG DATA = DETR;      ——→ Specifies regression procedure
MODEL Y = X4  X3  X3SQ  X43 / P  XPX  I  CLM  CLI;
```

Specifies model $y = \beta_0 + \beta_1 x_4 + \beta_2 x_3 + \beta_3 x_3^2 + \beta_4 x_4 x_3 + \varepsilon$ *(intercept β_0 is assumed)*

P = Predictions desired
*XPX = **X'X** and **X'y** desired*
I = $(\mathbf{X}'\mathbf{X})^{-1}$ desired
CLM = 95% confidence interval desired
CLI = 95% prediction interval desired

```
T1: TEST X3SQ=0, X43=0;
```

Performs partial F-test of H_0: $\beta_3 = \beta_4 = 0$ in above model

```
PROC REG DATA = DETR;
MODEL Y = X4 X3 X3SQ X43/P;
OUTPUT OUT = ONE PREDICTED = YHAT RESIDUAL = RESID;
PROC PLOT DATA = ONE;
PLOT RESID * (X4 X3 YHAT);
```

Plots residuals from the above model against x_4, x_3, and $\hat{y}$

```
PROC UNIVARIATE PLOT DATA = ONE;
VAR RESID;
```

Produces stem-and-leaf display, box plot, and normal plot of residuals from the above model

EXERCISES

Exercises 9.1–9.6 refer to the Compustat, Inc. situation of Exercises 8.6–8.9. Recall that we wish to relate y (demand for the CS-22) to x_1 (advertising expenditure) and x_2 (number of years of experience of the advertising firm). Suppose that we consider the interaction model

$$y = \beta_0 + \beta_1 x_1 + \beta_2 x_2 + \beta_3 x_1^2 + \beta_4 x_1 x_2 + \varepsilon$$

A sample of $n = 25$ combinations of (y_i, x_{i1}, x_{i2}) values is observed, and the following calculations are made:

$$(\mathbf{X}'\mathbf{X})^{-1} = \begin{bmatrix} .02 & 0 & 0 & 0 & 0 \\ 0 & .01 & 0 & 0 & 0 \\ 0 & 0 & .004 & 0 & 0 \\ 0 & 0 & 0 & .006 & 0 \\ 0 & 0 & 0 & 0 & .002 \end{bmatrix} \qquad \mathbf{X}'\mathbf{y} = \begin{bmatrix} 1000 \\ 300 \\ 500 \\ 100 \\ 250 \end{bmatrix}$$

and

$$\sum_{i=1}^{25} y_i^2 = 22{,}162$$

9.1 In the Compustat, Inc. situation,

a. Calculate the least squares point estimates of the parameters in the interaction model.
b. Write the prediction equation obtained by using the interaction model.
c. Compute a point estimate of μ_0, the mean demand when Compustat, Inc. spends \$200,000 on advertising ($x_{01} = 2$) and has a firm with 5 years of experience handle the advertising account ($x_{02} = 5$).
d. Compute a point prediction of y_0, an individual demand when Compustat, Inc. spends \$200,000 on advertising and has a firm with 5 years of experience handle the advertising account.
e. Show that $\mathbf{b}'\mathbf{X}'\mathbf{y} = 22{,}085$, $s^2 = 3.85$, and $s = 1.96$.

9.2 In the Compustat, Inc. situation,

a. Calculate appropriate t statistics and use them to test the importance of each of the independent variables x_1, x_2, x_1^2, and x_1x_2 in the interaction model with $\alpha = .05$ and with $\alpha = .01$. Is the interaction term important? Are the other independent variables important? Recall that $s = 1.96$.
b. Calculate and interpret a 95% confidence interval for each of the parameters β_0, β_1, β_2, β_3, and β_4 in the interaction model.

9.3 In the Compustat, Inc. situation,

a. Using the facts that

$$\sum_{i=1}^{25} y_i^2 = 22{,}162 \quad \text{and} \quad 25\bar{y}^2 = 20{,}474$$

compute the total variation, explained variation, unexplained variation, and R^2. Interpret the value of R^2.

b. Compute F(model) and use this statistic to test the adequacy of the interaction model with $\alpha = .05$. That is, can we reject $H_0: \beta_1 = \beta_2 = \beta_3 = \beta_4 = 0$ with $\alpha = .05$? Interpret the result of this test.

9.4 In the Compustat, Inc. situation,

a. Using the interaction model, calculate and interpret a 95% confidence interval for μ_0 when \$200,000 is spent on advertising and the firm handling the advertising has 5 years of experience.
b. Using the interaction model, calculate and interpret a 95% prediction interval for y_0 when \$200,000 is spent on advertising and the firm handling the advertising has 5 years of experience.

9.5 In the Compustat, Inc. situation,

a. Using the least squares point estimates of the parameters in the interaction model, plot

$$\hat{y} = (b_0 + b_2x_2) + (b_1 + b_4x_2)x_1 + b_3x_1^2$$

against $x_1 = 2, 3,$ and 4 when x_2 equals 5.
b. Using the least squares point estimates of the parameters in the interaction model, plot

$$\hat{y} = (b_0 + b_2x_2) + (b_1 + b_4x_2)x_1 + b_3x_1^2$$

against $x_1 = 2, 3,$ and 4 when x_2 equals 10.
c. By comparing the plots you made in parts (a) and (b), describe the exact nature of the interaction between x_1 and x_2.

9.6 In the Compustat, Inc. situation, suppose that we wish to test the importance of the higher-order terms x_1^2 and x_1x_2. That is, we wish to test

$$H_0: \beta_3 = \beta_4 = 0$$

a. For the above null hypothesis, define the appropriate complete and reduced models.
b. Assume that for the appropriate reduced model

$$(\mathbf{X'X})^{-1} = \begin{bmatrix} .02 & 0 & 0 \\ 0 & .01 & 0 \\ 0 & 0 & .004 \end{bmatrix} \quad \text{and} \quad \mathbf{X'y} = \begin{bmatrix} 1000 \\ 300 \\ 500 \end{bmatrix}$$

Calculate the partial F statistic needed to test H_0 above.
c. Use the partial F statistic of part (b) to test H_0 above by setting $\alpha = .05$.
d. Interpret the result of the hypothesis test of part (c).

Exercises 9.7–9.17 refer to the construction project study of Exercise 8.10. Recall that the construction project data is given in Table 8.8 of the exercises in Chapter 8 and that we wish to relate y (profit) to x_1 (contract size) and x_2 (supervisor experience). In Exercise 8.10 we concluded that either of the models

$$y = \beta_0 + \beta_1x_1 + \beta_2x_1^2 + \beta_3x_2 + \varepsilon$$

or

$$y = \beta_0 + \beta_1x_1 + \beta_2x_1^2 + \beta_3x_2 + \beta_4x_2^2 + \varepsilon$$

might appropriately relate y to x_1 and x_2. We now consider finding an improved model. To do this, consider Figure 9.13, which gives plots of y versus x_1 for different values of

FIGURE 9.13 **Plots of y_i versus x_{i1} for different values of x_{i2} for the construction project data**

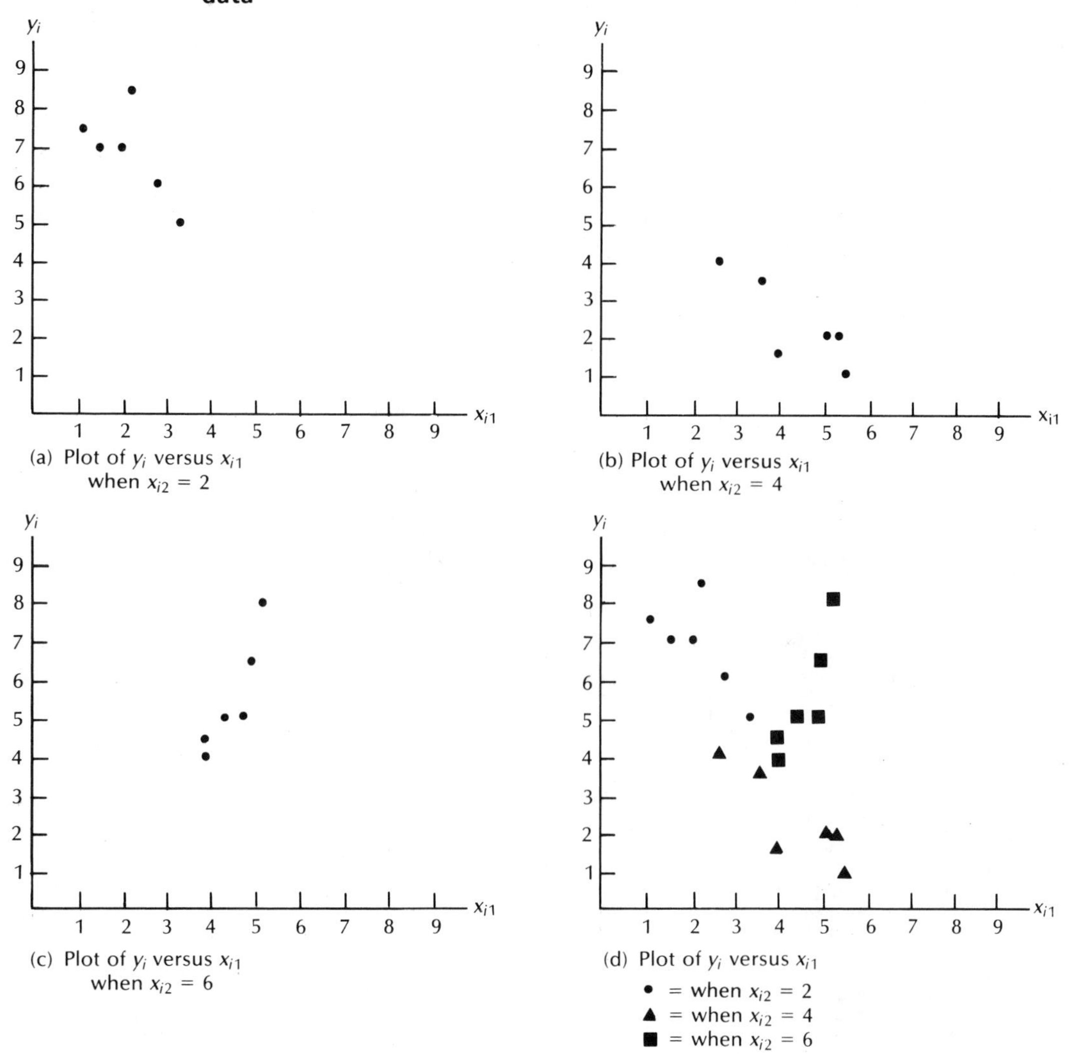

x_2. Plot (a) in Figure 9.13 shows that when x_2 (supervisor experience) equals 2, y (profit) decreases as x_1 (contract size) increases. This says that less experienced supervisors handle large contracts less effectively. Plot (b) in Figure 9.13 shows that when x_2 equals 4, y also decreases as x_1 increases. Plot (c) in Figure 9.13 shows that when x_2 equals 6, y increases as x_1 increases. This says that more experienced supervisors handle larger contracts more effectively. These plots suggest that the relationship between mean profit and contract size (x_1) depends upon supervisor experience (x_2). This says that interaction exists between x_1 and x_2 as these variables affect y. Therefore it is appropriate to include interaction terms such as x_1x_2 and $x_1^2x_2$ in a construction project model. Thus we consider models such as

$$y = \beta_0 + \beta_1x_1 + \beta_2x_2 + \beta_3x_1^2 + \beta_4x_1x_2 + \varepsilon$$
$$y = \beta_0 + \beta_1x_1 + \beta_2x_1^2 + \beta_3x_2 + \beta_4x_1x_2 + \beta_5x_1^2x_2 + \varepsilon$$
$$y = \beta_0 + \beta_1x_1 + \beta_2x_1^2 + \beta_3x_2 + \beta_4x_2^2 + \beta_5x_1x_2 + \beta_6x_1^2x_2 + \beta_7x_1x_2^2 + \varepsilon$$

FIGURE 9.14 **SAS output of a regression analysis of the construction project data using the model $y = \beta_0 + \beta_1x_1 + \beta_2x_2 + \beta_3x_1^2 + \beta_4x_1x_2 + \varepsilon$**

SAS

DEP VARIABLE: PROFIT

ANALYSIS OF VARIANCE

SOURCE	DF	SUM OF SQUARES	MEAN SQUARE	F VALUE	PROB>F
MODEL	4	77.02981991	19.25745498	20.440	0.0001
ERROR	13	12.24795787	0.94215061		
C TOTAL	17	89.27777778			

ROOT MSE	0.9706444	R-SQUARE	0.8628
DEP MEAN	4.888889	ADJ R-SQ	0.8206
C.V.	19.85409		

PARAMETER ESTIMATES

VARIABLE	DF	PARAMETER ESTIMATE	STANDARD ERROR	T FOR HO: PARAMETER=0	PROB > |T|
INTERCEP	1	19.30495716	2.05205668	9.408	0.0001
CSIZE	1	-1.48660196	1.17773388	-1.262	0.2290
SUPEXP	1	-6.37145238	1.04230807	-6.113	0.0001
CSIZESQ	1	-0.75224827	0.22523749	-3.340	0.0053
INTER	1	1.71705261	0.25377879	6.766	0.0001

OBS	ACTUAL	PREDICT VALUE	STD ERR PREDICT	LOWER95% MEAN	UPPER95% MEAN	LOWER95% PREDICT	UPPER95% PREDICT	RESIDUAL
1	2.0000	1.6994	0.5203	0.5754	2.8234	-0.6798	4.0786	0.3006
2	3.5000	3.4397	0.3869	2.6038	4.2757	1.1823	5.6972	0.0603
3	8.5000	6.9031	0.4463	5.9389	7.8674	4.5951	9.2111	1.5969
4	4.5000	4.3031	0.4686	3.2908	5.3155	1.9746	6.6316	0.1969
5	7.0000	7.6988	0.4693	6.6850	8.7126	5.3696	10.0280	-0.6988
6	7.0000	7.4481	0.4226	6.5350	8.3611	5.1610	9.7352	-0.4481
7	2.0000	1.9210	0.4772	0.8901	2.9519	-0.4157	4.2577	0.0790
8	5.0000	5.0910	0.5407	3.9228	6.2592	2.6907	7.4914	-0.0910
9	8.0000	6.5772	0.5622	5.3625	7.7918	4.1538	9.0005	1.4228
10	5.0000	5.0747	0.4061	4.1974	5.9521	2.8016	7.3478	-0.0747
11	6.0000	5.8834	0.5106	4.7804	6.9864	3.5141	8.2528	0.1166
12	7.5000	7.7941	0.7916	6.0840	9.5042	5.0882	10.4999	-0.2941
13	4.0000	2.7261	0.6277	1.3701	4.0821	0.2290	5.2233	1.2739
14	4.0000	4.3031	0.4686	3.2908	5.3155	1.9746	6.6316	-0.3031
15	1.0000	1.2110	0.6245	-0.1381	2.5601	-1.2824	3.7045	-0.2110
16	5.0000	6.2118	0.4639	5.2096	7.2140	3.8876	8.5359	-1.2118
17	6.5000	6.3486	0.4926	5.2843	7.4129	3.9970	8.7002	0.1514
18	1.5000	3.3657	0.3505	2.6085	4.1230	1.1362	5.5952	-1.8657
19	.	6.0599	0.4397	5.1099	7.0098	3.7578	8.3620	.
20	.	7.4481	0.4226	6.5350	8.3611	5.1610	9.7352	.

SUM OF RESIDUALS	-1.48992E-13
SUM OF SQUARED RESIDUALS	12.24796
PREDICTED RESID SS (PRESS)	23.09611

9.7 Figure 9.14 presents the SAS output of a regression analysis of the construction project data using the model

$$y = \beta_0 + \beta_1 x_1 + \beta_2 x_2 + \beta_3 x_1^2 + \beta_4 x_1 x_2 + \varepsilon$$

a. Use the appropriate t statistics and prob-values to test the importance of each of the independent variables x_1, x_2, x_1^2, and $x_1 x_2$. What do you conclude about the importance of the variables? Notice here that the prob-value for testing $H_0: \beta_1 = 0$ does not cast substantial doubt on this hypothesis. This might indicate that x_1 should not be included in the model. However, the prob-values for testing $H_0: \beta_3 = 0$ and $H_0: \beta_4 = 0$ indicate that x_1^2 and $x_1 x_2$ are important. In such a situation we often include the linear term x_1, even though the prob-value for testing $H_0: \beta_1 = 0$ indicates that this term may not be important. This is because of a phenomenon called *multicollinearity*, which exists when independent variables are related to each other (here, x_1, x_1^2, and $x_1 x_2$ are related). Intuitively, this causes the importance of x_1 to be spread between x_1, x_1^2, and $x_1 x_2$, making x_1 appear to have less importance than it might actually have. Therefore it might make sense to include x_1 in the regression model. We discuss multicollinearity in detail in Chapter 10.

b. Demonstrate how $t = -3.340$ has been calculated. By using the appropriate rejection point, determine whether we can reject $H_0: \beta_3 = 0$ in favor of $H_1: \beta_3 \neq 0$ by setting α equal to .05.

c. Explain how the prob-value for testing $H_0: \beta_3 = 0$ versus $H_1: \beta_3 \neq 0$ has been calculated. By using this prob-value, determine whether we can reject $H_0: \beta_3 = 0$ versus $H_1: \beta_3 \neq 0$ by setting α equal to .05; to .01; to .001.

d. Interpret the prob-value for testing $H_0: \beta_3 = 0$ as a probability.

9.8 Consider the construction project situation and the model

$$y = \beta_0 + \beta_1 x_1 + \beta_2 x_2 + \beta_3 x_1^2 + \beta_4 x_1 x_2 + \varepsilon$$

Using the SAS output in Figure 9.14,

a. Test the overall adequacy of the model by using F(model) and its associated prob-value.

b. Identify the total variation, explained variation, unexplained variation, and R^2. Interpret the value of R^2.

9.9 Consider the construction project situation. By using the least squares point estimates given in Figure 9.14 and by placing values of x_1 and x_2 into the prediction equation

$$\hat{y} = b_0 + b_1 x_1 + b_2 x_2 + b_3 x_1^2 + b_4 x_1 x_2$$

a. Plot $\hat{y}$ against x_1 (for $x_1 = 3$, 4, and 5) when $x_2 = 2$.

b. Plot $\hat{y}$ against x_1 (for $x_1 = 3$, 4, and 5) when $x_2 = 4$.

c. Plot $\hat{y}$ against x_1 (for $x_1 = 3$, 4, and 5) when $x_2 = 6$.

d. Combine these plots by graphing them on the same set of axes. Use a different color for each level of x_2 ($=2$, 4, 6). What do these plots suggest concerning the policy the company should follow in assigning supervisors to construction projects?

9.10 Consider the construction project situation and the model

$$y = \beta_0 + \beta_1 x_1 + \beta_2 x_2 + \beta_3 x_1^2 + \beta_4 x_1 x_2 + \varepsilon$$

Suppose that the construction company has been awarded a "future contract" of $x_{01} = 4.8$ (\$480,000) and has decided to assign a supervisor with $x_{02} = 6$ years of experience to this project.

a. Discuss the meaning of μ_0, the mean profit for this future contract.
b. Discuss the conceptual difference between μ_0 and $y_0 = \mu_0 + \varepsilon_0$, the individual profit for this future contract.
c. Plot the eighteen (x_1, x_2) combinations in two dimensions (with x_1 on the horizontal axis and x_2 on the vertical axis) to form the experimental region.
d. Is the point $(x_{01} = 4.8, x_{02} = 6)$ in the experimental region?
e. Compute a point estimate of μ_0 and a point prediction of y_0.

9.11 Consider the construction project situation. Using the SAS output in Figure 9.14,

a. Find and report the 95% confidence interval for mean profit when \$500,000 contracts are assigned to supervisors with 4 years of experience. *Hint*: See Table 8.8 in the exercises of Chapter 8.
b. Compute the 99% confidence interval for mean profit when \$500,000 contracts are assigned to supervisors with 4 years of experience.
c. Find and report the 95% prediction interval for an individual profit when a \$480,000 contract is assigned to a supervisor with 6 years of experience. Note here that "observation 19" on the SAS output of Figure 9.14 corresponds to the values $x_{01} = 4.8$ and $x_{02} = 6$.
d. Compute a 99% prediction interval for an individual profit when a \$480,000 contract is assigned to a supervisor with 6 years of experience.

9.12 Consider the construction project situation and the model

$$y = \beta_0 + \beta_1 x_1 + \beta_2 x_2 + \beta_3 x_1^2 + \beta_4 x_1 x_2 + \varepsilon$$

a. Using the fact that $SSE = 59.9565$ for the model

$$y = \beta_0 + \beta_1 x_1 + \beta_2 x_2 + \varepsilon$$

and using the SAS output of Figure 9.14, calculate the partial F statistic for testing $H_0: \beta_3 = \beta_4 = 0$ versus H_1: At least one of β_3 and β_4 does not equal zero.
b. Using an appropriate rejection point, decide whether or not we can reject H_0 in favor of H_1 by setting α equal to .05.

9.13 Consider the construction project data and the model

$$y = \beta_0 + \beta_1 x_1 + \beta_2 x_2 + \varepsilon$$

In Exercise 8.18 we found that plots of the residuals for this model versus x_1 and x_2 suggest the need for quadratic terms. Figure 9.15 gives SAS output of a residual plot for this model versus $x_1 x_2$. What does this plot say about the need for the interaction term $x_1 x_2$? Explain.

9.14 Consider the construction project data.

a. Figure 9.16 gives SAS output of a residual plot versus $x_1 x_2$ for the model

$$y = \beta_0 + \beta_1 x_1 + \beta_2 x_1^2 + \beta_3 x_2 + \varepsilon$$

What does this plot say about the need for the interaction term $x_1 x_2$? Explain.
b. Figure 9.17 gives SAS output of a residual plot versus $x_1 x_2$ for the model

$$y = \beta_0 + \beta_1 x_1 + \beta_2 x_1^2 + \beta_3 x_2 + \beta_4 x_2^2 + \varepsilon$$

What does this plot say about the need for the interaction term $x_1 x_2$? Explain.

FIGURE 9.15 SAS output of a residual plot versus x_1x_2 for the construction project model $y = \beta_0 + \beta_1x_1 + \beta_2x_2 + \varepsilon$

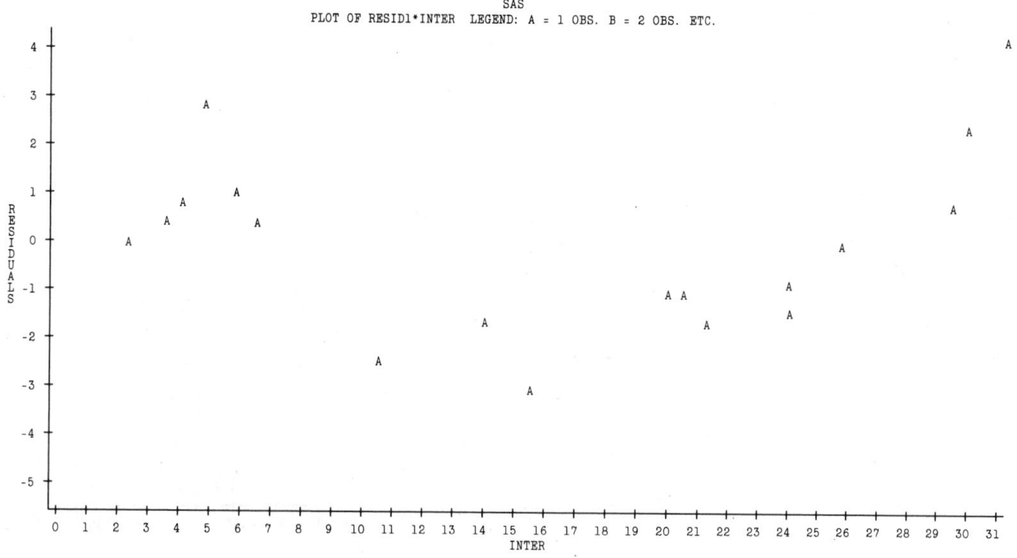

FIGURE 9.16 SAS output of a residual plot versus x_1x_2 for the construction project model $y = \beta_0 + \beta_1x_1 + \beta_2x_1^2 + \beta_3x_2 + \varepsilon$

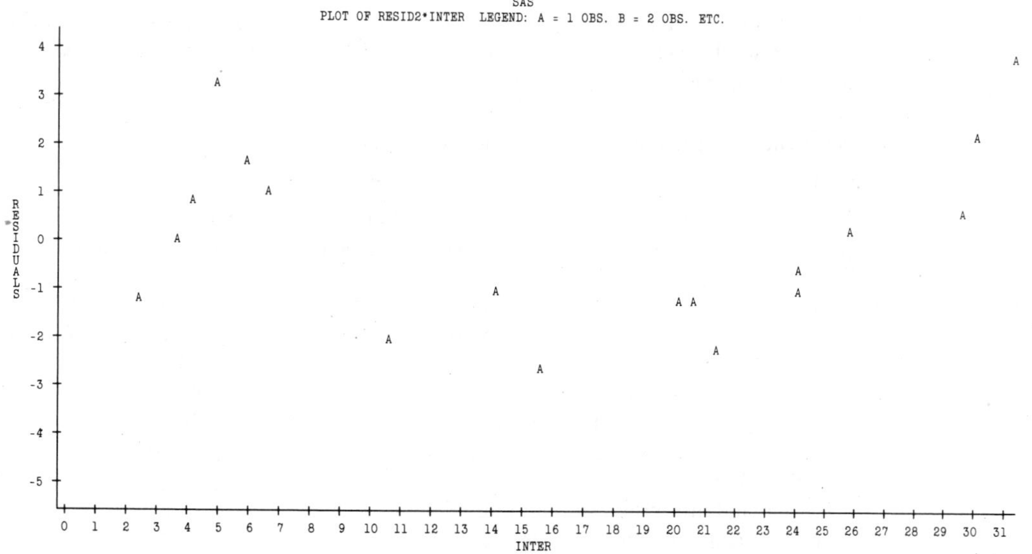

FIGURE 9.17 SAS output of a residual plot versus x_1x_2 for the construction project model $y = \beta_0 + \beta_1x_1 + \beta_2x_1^2 + \beta_3x_2 + \beta_4x_2^2 + \varepsilon$

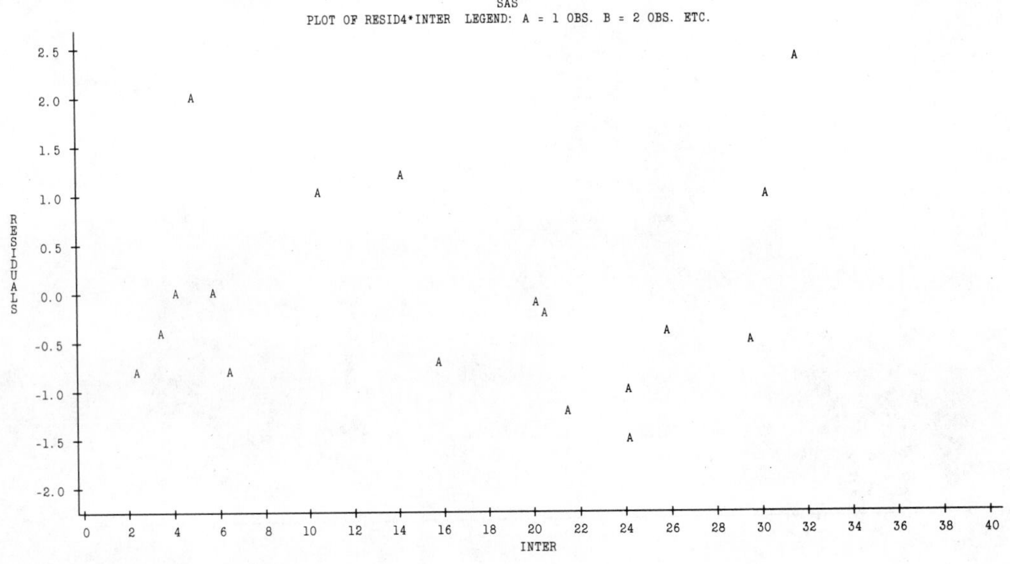

9.15 Consider the construction project data and the model

$$y = \beta_0 + \beta_1x_1 + \beta_2x_2 + \beta_3x_1^2 + \beta_4x_1x_2 + \varepsilon$$

Figure 9.18 presents the SAS output of the residual plots for this model versus x_1, x_2, $\hat{y}$, and x_1x_2. What do these plots say about the functional form of this model? Do the plots suggest any violations of the inference assumptions? Explain.

9.16 Consider the construction project data and the model

$$y = \beta_0 + \beta_1x_1 + \beta_2x_2 + \beta_3x_1^2 + \beta_4x_1x_2 + \varepsilon$$

a. Assume that the residuals given in Figure 9.14 are in time order. Plot these residuals versus time. Does the plot suggest a violation of the independence assumption? Explain.

b. Using the residuals in Figure 9.14, calculate the Durbin–Watson statistic. Test for positive autocorrelation with $\alpha = .05$. Test for negative autocorrelation with $\alpha = .05$.

c. Construct a frequency distribution of the residuals in Figure 9.14. Does the frequency distribution suggest a violation of the normality assumption? Explain.

d. Calculate the standardized residuals. What percentage of standardized residuals is between -1 and 1? Between -2 and 2? Is a violation of the normality assumption suggested? Explain.

e. Construct and interpret a normal plot of the residuals in Figure 9.14. Is a violation of the normality assumption suggested? Explain.

FIGURE 9.18 SAS output of residual plots for the construction project model $y = \beta_0 + \beta_1 x_1 + \beta_2 x_2 + \beta_3 x_1^2 + \beta_4 x_1 x_2 + \varepsilon$

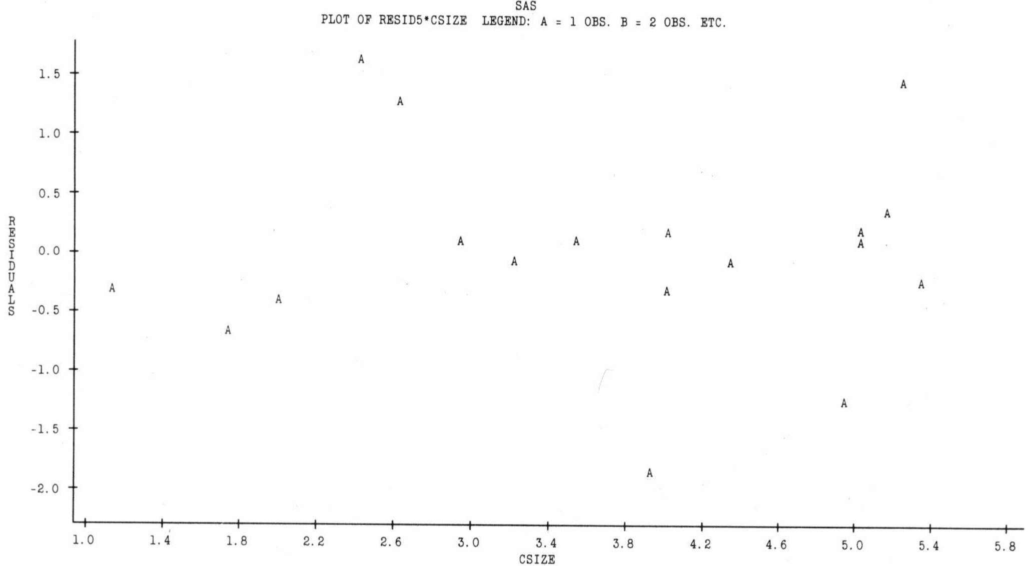

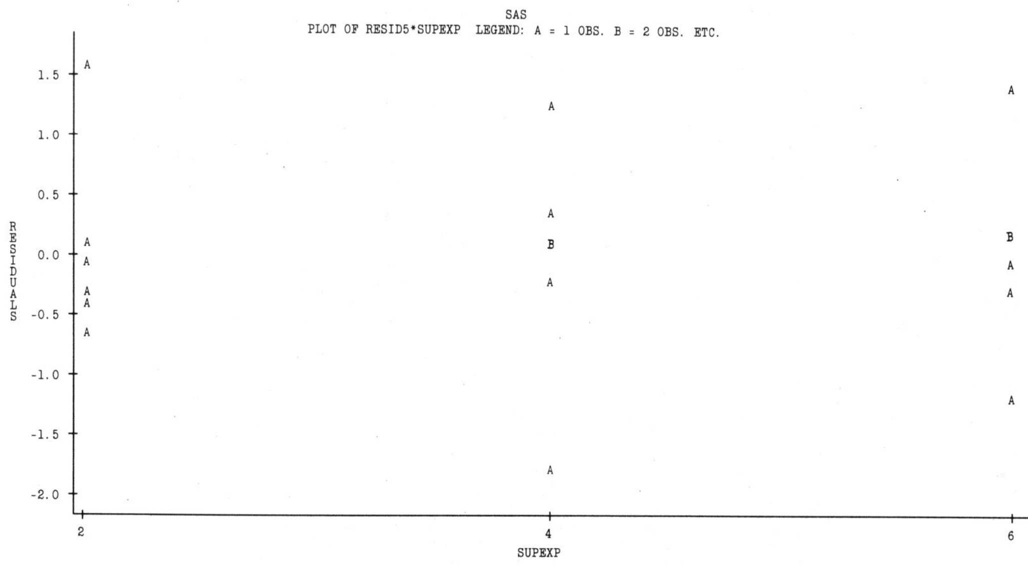

(*continues*)

FIGURE 9.18 Continued

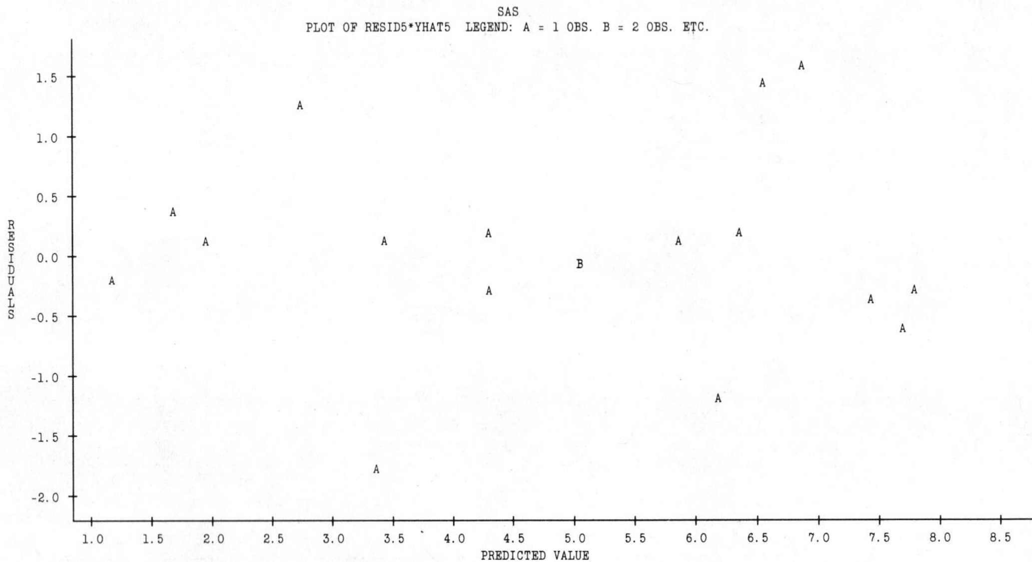

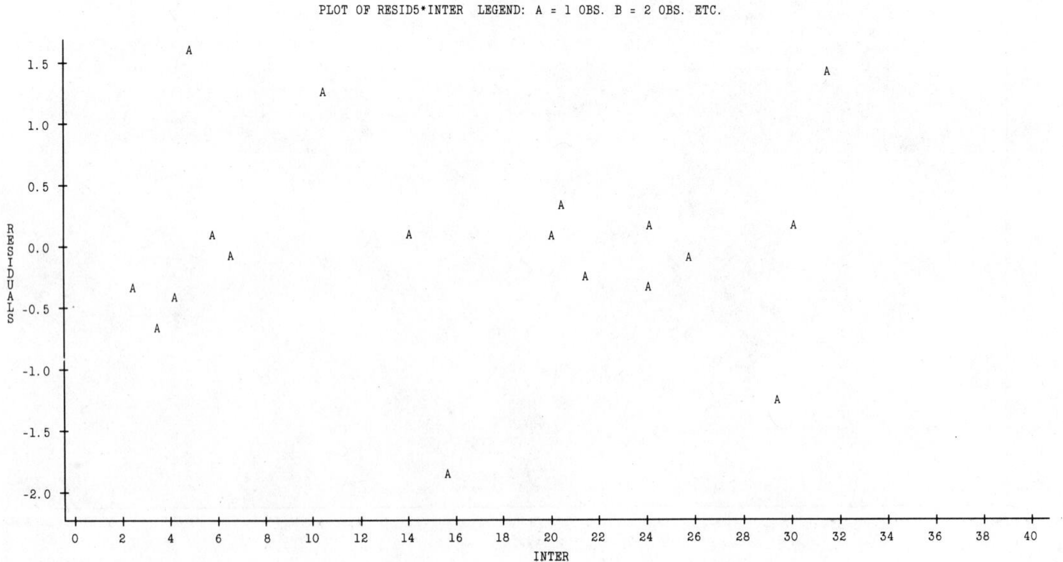

9.17 Consider the construction project data and the model

$$y = \beta_0 + \beta_1 x_1 + \beta_2 x_2 + \beta_3 x_1^2 + \beta_4 x_1 x_2 + \beta_5 x_1^2 x_2 + \varepsilon$$

When we carry out a regression analysis of the construction project data by using this model, we obtain a standard error equal to .8515 and the residuals given in Table 9.2. Using a computer, carry out a complete analysis of these residuals (for example, do all the analysis in Exercises 9.15 and 9.16). By comparing your answers with those of Exercises 9.15 and 9.16, which model

$$y = \beta_0 + \beta_1 x_1 + \beta_2 x_2 + \beta_3 x_1^2 + \beta_4 x_1 x_2 + \varepsilon$$

or

$$y = \beta_0 + \beta_1 x_1 + \beta_2 x_2 + \beta_3 x_1^2 + \beta_4 x_1 x_2 + \beta_5 x_1^2 x_2 + \varepsilon$$

seems to best satisfy the regression assumptions? Explain.

Exercises 9.18–9.22 refer to the following situation. United Oil Company is attempting to develop a reasonably priced premium gasoline that will deliver higher gasoline mileages than can be achieved by its current premium gasolines. As part of its development process, United Oil wishes to study the effect of two independent variables—x_1, amount of gasoline additive WST (0, 1, or 2 units), and x_2, amount of gasoline additive YST (0, 1, 2, or 3 units)—on the gasoline mileage y obtained by an automobile called the GT-5000. For testing purposes a sample of $n = 22$ automobiles is randomly selected. Each is tested under normal driving conditions. The combinations of x_1 and x_2 used in the experiment, along with the corresponding values of y, are given in Table 9.3.

TABLE 9.2 Time-ordered residuals for the model $y = \beta_0 + \beta_1 x_1 + \beta_2 x_2 + \beta_3 x_1^2 + \beta_4 x_1 x_2 + \beta_5 x_1^2 x_2 + \varepsilon$

Obs	y_i	x_{i1}	x_{i2}	$e_i = y_i - \hat{y}_i$	$(\hat{y}_i)$
1	2.0	5.1	4	.481948	1.518
2	3.5	3.5	4	.041071	3.459
3	8.5	2.4	2	1.533	6.967
4	4.5	4.0	6	.682848	3.817
5	7.0	1.7	2	−.593227	7.593
6	7.0	2.0	2	−.435951	7.436
7	2.0	5.0	4	.328047	1.672
8	5.0	3.2	2	−.139599	5.140
9	8.0	5.2	6	.664748	7.335
10	5.0	4.3	6	.436063	4.564
11	6.0	2.9	2	.036226	5.964
12	7.5	1.1	2	.092394	7.408
13	4.0	2.6	4	−.061724	4.062
14	4.0	4.0	6	.182848	3.817
15	1.0	5.3	4	−.197210	1.197
16	5.0	4.9	6	−1.323	6.323
17	6.5	5.0	6	−.150576	6.651
18	1.5	3.9	4	−1.578	3.078

TABLE 9.3 **United Oil Company premium gasoline mileage data**

Gasoline mileage, y_i (mpg)	Amount of gasoline additive WST, x_{i1}	Amount of gasoline additive YST, x_{i2}
28.0	0	0
28.6	0	0
27.4	0	0
33.3	1	0
34.5	1	0
33.0	0	1
32.0	0	1
35.6	1	1
34.4	1	1
35.0	1	1
34.0	2	1
33.3	2	1
34.7	2	1
33.5	0	2
32.3	0	2
33.4	1	2
33.0	2	2
32.0	2	2
29.6	1	3
30.6	1	3
28.6	2	3
29.8	2	3

9.18 Consider the United Oil Company premium gasoline mileage data in Table 9.3.

a. Plot y against x_1.
b. Plot y against x_1 when $x_2 = 0$.
c. Plot y against x_1 when $x_2 = 1$.
d. Plot y against x_1 when $x_2 = 2$.
e. Plot y against x_1 when $x_2 = 3$.
f. Combine these plots by graphing them on the same set of axes. Use a different color for each level of x_2 ($=0$, 1, 2, and 3). What do these plots say about whether interaction exists between x_1 and x_2?
g. Plot y against x_2.
h. Plot y against x_2 when $x_1 = 0$.
i. Plot y against x_2 when $x_1 = 1$.
j. Plot y against x_2 when $x_1 = 2$.
k. Combine these plots by graphing them on the same set of axes. Use a different color for each level of x_1 ($=0$, 1, and 2). What do these plots say about whether interaction exists between x_1 and x_2?
l. Discuss why the graphical analysis of parts (a)–(k) indicates that the model

$$y = \beta_0 + \beta_1 x_1 + \beta_2 x_1^2 + \beta_3 x_2 + \beta_4 x_2^2 + \beta_5 x_1 x_2 + \beta_6 x_1^2 x_2^2 + \varepsilon$$

may appropriately relate y to x_1 and x_2.

9.19 Consider the United Oil Company premium gasoline mileage data in Table 9.3 and the model

$$y = \beta_0 + \beta_1 x_1 + \beta_2 x_1^2 + \beta_3 x_2 + \beta_4 x_2^2 + \beta_5 x_1 x_2 + \beta_6 x_1^2 x_2^2 + \varepsilon$$

a. Specify the vector **y** and matrix **X** that are used to calculate the least squares point estimates of the parameters in this model.

b. Figure 9.19 gives the SAS output that is obtained when we analyze the data in Table 9.3 using the above model. We find that $b_0 = 28.1323$, $b_1 = 7.8004$, $b_2 = -2.2009$, $b_3 = 5.5718$, $b_4 = -1.5557$, $b_5 = -2.9365$, and $b_6 = .2580$. Noting from Table 9.3 that the combination of one unit of gasoline additive WST and one unit of gasoline additive YST seems to maximize gasoline mileage, assume that United Oil Company will use this combination to produce the premium gasoline.

(1) Discuss the meaning of μ_0, the mean gasoline mileage corresponding to the combination $x_{01} = 1$ and $x_{02} = 1$.

(2) Discuss the conceptual difference between μ_0 and $y_0 = \mu_0 + \varepsilon_0$, the individual gasoline mileage corresponding to $x_{01} = 1$ and $x_{02} = 1$.

(3) Using an appropriate prediction equation, calculate a point estimate of μ_0 and a point prediction of $y_0 = \mu_0 + \varepsilon_0$.

9.20 Consider the United Oil Company situation. Using the SAS output in Figure 9.19,

a. Test the significance of the overall model by using *F*(model) and its associated prob-value.

FIGURE 9.19 **SAS output of a regression analysis of the United Oil Company premium gasoline mileage data using the model $y = \beta_0 + \beta_1 x_1 + \beta_2 x_1^2 + \beta_3 x_2 + \beta_4 x_2^2 + \beta_5 x_1 x_2 + \beta_6 x_1^2 x_2^2 + \varepsilon$**

SAS

DEP VARIABLE: MILEAGE

ANALYSIS OF VARIANCE

SOURCE	DF	SUM OF SQUARES	MEAN SQUARE	F VALUE	PROB>F
MODEL	6	120.56404	20.09400691	43.628	0.0001
ERROR	15	6.90868583	0.46057906		
C TOTAL	21	127.47273			

ROOT MSE	0.6786597	R-SQUARE	0.9458
DEP MEAN	32.11818	ADJ R-SQ	0.9241
C.V.	2.113008		

PARAMETER ESTIMATES

VARIABLE	DF	PARAMETER ESTIMATE	STANDARD ERROR	T FOR H0: PARAMETER=0	PROB > |T|
INTERCEP	1	28.13229859	0.37514559	74.990	0.0001
WST	1	7.80041974	0.78706063	9.911	0.0001
WSTSQ	1	-2.20091497	0.37772724	-5.827	0.0001
YST	1	5.57177497	0.55351000	10.066	0.0001
YSTSQ	1	-1.55567399	0.21277027	-7.312	0.0001
INTER	1	-2.93645670	0.50776009	-5.783	0.0001
INTERSQ	1	0.25795688	0.05872000	4.393	0.0005

(continues)

FIGURE 9.19 **Continued**

OBS	ACTUAL	PREDICT VALUE	STD ERR PREDICT	LOWER95% MEAN	UPPER95% MEAN	LOWER95% PREDICT	UPPER95% PREDICT	RESIDUAL
1	28.0000	28.1323	0.3751	27.3327	28.9319	26.4795	29.7851	-0.1323
2	28.6000	28.1323	0.3751	27.3327	28.9319	26.4795	29.7851	0.4677
3	27.4000	28.1323	0.3751	27.3327	28.9319	26.4795	29.7851	-0.7323
4	33.3000	33.7318	0.4342	32.8062	34.6574	32.0145	35.4491	-0.4318
5	34.5000	33.7318	0.4342	32.8062	34.6574	32.0145	35.4491	0.7682
6	33.0000	32.1484	0.3041	31.5002	32.7966	30.5633	33.7335	0.8516
7	32.0000	32.1484	0.3041	31.5002	32.7966	30.5633	33.7335	-0.1484
8	35.6000	35.0694	0.3240	34.3788	35.7600	33.4665	36.6723	0.5306
9	34.4000	35.0694	0.3240	34.3788	35.7600	33.4665	36.6723	-0.6694
10	35.0000	35.0694	0.3240	34.3788	35.7600	33.4665	36.6723	-0.0694
11	34.0000	34.1045	0.3586	33.3402	34.8688	32.4684	35.7405	-0.1045
12	33.3000	34.1045	0.3586	33.3402	34.8688	32.4684	35.7405	-0.8045
13	34.7000	34.1045	0.3586	33.3402	34.8688	32.4684	35.7405	0.5955
14	33.5000	33.0532	0.4280	32.1409	33.9654	31.3430	34.7633	0.4468
15	32.3000	33.0532	0.4280	32.1409	33.9654	31.3430	34.7633	-0.7532
16	33.4000	33.8116	0.3457	33.0746	34.5485	32.1881	35.4350	-0.4116
17	33.0000	32.2318	0.3173	31.5556	32.9081	30.6350	33.8286	0.7682
18	32.0000	32.2318	0.3173	31.5556	32.9081	30.6350	33.8286	-0.2318
19	29.6000	29.9583	0.4597	28.9786	30.9380	28.2112	31.7054	-0.3583
20	30.6000	29.9583	0.4597	28.9786	30.9380	28.2112	31.7054	0.6417
21	28.6000	29.3114	0.4632	28.3241	30.2988	27.5601	31.0628	-0.7114
22	29.8000	29.3114	0.4632	28.3241	30.2988	27.5601	31.0628	0.4886

b. Use appropriate t statistics and their associated prob-values to test the importance of each of the independent variables. Which variables are important? Explain.
c. Find and interpret R^2.

9.21 Consider the United Oil Company situation. Using the SAS output in Figure 9.19,

a. Find and report the 95% confidence interval for the mean gasoline mileage obtained when using one unit of additive WST and one unit of additive YST. Interpret this interval.
b. Compute a 99% confidence interval for the mean gasoline mileage obtained when using two units of additive WST and two units of additive YST.
c. Find and report the 95% prediction interval for an individual mileage obtained when using one unit of additive WST and one unit of additive YST. Interpret this interval.
d. Compute a 99% prediction interval for an individual mileage obtained when using two units of additive WST and three units of additive YST.

9.22 Using a computer, carry out a complete analysis of the residuals obtained when the model

$$y = \beta_0 + \beta_1 x_1 + \beta_2 x_1^2 + \beta_3 x_2 + \beta_4 x_2^2 + \beta_5 x_1 x_2 + \beta_6 x_1^2 x_2^2 + \varepsilon$$

is employed to analyze the United Oil Company mileage data in Table 9.3.

9.23 Consider the United Oil Company unleaded gasoline mileage data in Table 8.11 of the exercises in Chapter 8. Recall that we wish to relate y (unleaded gasoline mileage) to x_1 (amount of additive RST) and x_2 (amount of additive XST). Using this data,

a. Plot y against x_1.
b. Plot y against x_1 when $x_2 = 0$.
c. Plot y against x_1 when $x_2 = 1$.
d. Plot y against x_1 when $x_2 = 2$.
e. Plot y against x_1 when $x_2 = 3$.
f. Combine these plots by graphing them on the same set of axes. Use a different color for each level of x_2 ($=0$, 1, 2, and 3). What do these plots say about whether interaction exists between x_1 and x_2?
g. Plot y against x_2.
h. Plot y against x_2 when $x_1 = 0$.
i. Plot y against x_2 when $x_1 = 1$.
j. Plot y against x_2 when $x_1 = 2$.
k. Combine these plots by graphing them on the same set of axes. Use a different color for each level of x_1 ($=0$, 1, and 2). What do these plots say about whether interaction exists between x_1 and x_2?
l. Discuss why the graphical analysis of parts (a)–(k) indicates that the model

$$y = \beta_0 + \beta_1 x_1 + \beta_2 x_1^2 + \beta_3 x_2 + \beta_4 x_2^2 + \varepsilon$$

may appropriately relate y to x_1 and x_2.

9.24 Bonner Frozen Foods, Inc. carries out a designed experiment to relate

$$y = \text{sales volume (in units of \$10,000) for a line of frozen foods}$$

to

$$x_1 = \text{media expenditures (in units of \$1000)}$$

$$x_2 = \text{point of sale advertising expenditures (in units of \$1000)}$$

The company observes the data in Table 9.4. Figure 9.20 presents the SAS output of a regression analysis of these data using the model

$$y = \beta_0 + \beta_1 x_1 + \beta_2 x_1^2 + \beta_3 x_2 + \beta_4 x_2^2 + \beta_5 x_1 x_2 + \varepsilon$$

Fully interpret the SAS output.

9.25 Using a computer, carry out a complete analysis of the residuals obtained when the model

$$y = \beta_0 + \beta_1 x_1 + \beta_2 x_1^2 + \beta_3 x_2 + \beta_4 x_2^2 + \beta_5 x_1 x_2 + \varepsilon$$

is used to analyze the Bonner Frozen Foods data in Table 9.4.

9.26 Consider the Crest Toothpaste sales data of Exercise 8.23 (see Table 8.10 in the exercises of Chapter 8) and the model

$$y = \beta_0 + \beta_1 x_1 + \beta_2 x_2 + \beta_3 x_3 + \beta_4 x_1^2 + \beta_5 x_2^2 + \beta_6 x_3^2 + \beta_7 x_1 x_2 + \beta_8 x_1 x_3 + \beta_9 x_2 x_3 + \varepsilon$$

Using a computer, analyze the Crest data using this model. Is the model significant? Which variables seem to be important? Carry out a complete residual analysis.

TABLE 9.4 Bonner Frozen Foods sales volume data

Sales region, i	Media expenditures, x_{i1}	Point of sale expenditures, x_{i2}	Sales volume, y_i
1	1	1	3.27
2	1	2	5.73
3	1	3	7.92
4	1	4	8.71
5	1	5	10.21
6	2	1	8.30
7	2	2	10.54
8	2	3	12.46
9	2	4	12.87
10	2	5	13.75
11	3	1	12.91
12	3	2	14.63
13	3	3	14.75
14	3	4	16.67
15	3	5	17.21
16	4	1	16.82
17	4	2	18.11
18	4	3	18.24
19	4	4	18.37
20	4	5	19.57
21	5	1	20.51
22	5	2	21.14
23	5	3	21.47
24	5	4	21.73
25	5	5	22.26

FIGURE 9.20 SAS output of a regression analysis of the Bonner Frozen Foods data in Table 9.4 using the model $y = \beta_0 + \beta_1 x_1 + \beta_2 x_1^2 + \beta_3 x_2 + \beta_4 x_2^2 + \beta_5 x_1 x_2 + \varepsilon$

```
                                SAS
                        ANALYSIS OF VARIANCE
                            SUM OF            MEAN
            SOURCE    DF   SQUARES          SQUARE    F VALUE   PROB>F

            MODEL      5   684.14962     136.82992    775.593   0.0001
            ERROR     19   3.35197729    0.17641986
            C TOTAL   24   687.50160

              ROOT MSE     0.4200236       R-SQUARE    0.9951
              DEP MEAN        14.726       ADJ R-SQ    0.9938
              C.V.          2.852259

                         PARAMETER ESTIMATES

                         PARAMETER          STANDARD      T FOR HO:
VARIABLE        DF        ESTIMATE             ERROR   PARAMETER=0       PROB > |T|

INTERCEP         1     -4.22650000        0.67856896        -6.229           0.0001
X1               1      5.83090000        0.33186804        17.570           0.0001
X1SQ             1     -0.22100000        0.05020243        -4.402           0.0003
X2               1      2.73561429        0.33186804         8.243           0.0001
X2SQ             1     -0.12228571        0.05020243        -2.436           0.0249
X12              1     -0.33010000        0.04200236        -7.859           0.0001
```

OBS	ACTUAL	PREDICT VALUE	STD ERR PREDICT	LOWER95% MEAN	UPPER95% MEAN	LOWER95% PREDICT	UPPER95% PREDICT	RESIDUAL
1	3.2700	3.6666	0.2893	3.0612	4.2721	2.5992	4.7340	-0.3966
2	5.7300	5.7053	0.2106	5.2645	6.1461	4.7218	6.6887	0.0247
3	7.9200	7.4994	0.2033	7.0739	7.9249	6.5227	8.4761	0.4206
4	8.7100	9.0489	0.2106	8.6081	9.4897	8.0654	10.0323	-0.3389
5	10.2100	10.3538	0.2893	9.7484	10.9593	9.2864	11.4212	-0.1438
6	8.3000	8.5044	0.2106	8.0636	8.9452	7.5210	9.4879	-0.2044
7	10.5400	10.2130	0.1446	9.9103	10.5157	9.2832	11.1428	0.3270
8	12.4600	11.6770	0.1523	11.3583	11.9957	10.7419	12.6121	0.7830
9	12.8700	12.8964	0.1446	12.5937	13.1991	11.9666	13.8262	-0.0264
10	13.7500	13.8712	0.2106	13.4304	14.3120	12.8878	14.8547	-0.1212
11	12.9100	12.9002	0.2033	12.4747	13.3257	11.9235	13.8769	.0097714
12	14.6300	14.2787	0.1523	13.9600	14.5974	13.3436	15.2138	0.3513
13	14.7500	15.4126	0.1650	15.0673	15.7579	14.4681	16.3571	-0.6626
14	16.6700	16.3019	0.1523	15.9832	16.6206	15.3668	17.2370	0.3681
15	17.2100	16.9466	0.2033	16.5211	17.3721	15.9699	17.9233	0.2634
16	16.8200	16.8540	0.2106	16.4132	17.2948	15.8706	17.8375	-0.0340
17	18.1100	17.9024	0.1446	17.5997	18.2051	16.9726	18.8322	0.2076
18	18.2400	18.7062	0.1523	18.3875	19.0249	17.7711	19.6413	-0.4662
19	18.3700	19.2654	0.1446	18.9627	19.5681	18.3356	20.1952	-0.8954
20	19.5700	19.5800	0.2106	19.1392	20.0208	18.5966	20.5635	-0.0100
21	20.5100	20.3658	0.2893	19.7604	20.9713	19.2984	21.4332	0.1442
22	21.1400	21.0841	0.2106	20.6433	21.5249	20.1006	22.0675	0.0559
23	21.4700	21.5578	0.2033	21.1323	21.9833	20.5811	22.5345	-0.0878
24	21.7300	21.7869	0.2106	21.3461	22.2277	20.8034	22.7703	-0.0569
25	22.2600	21.7714	0.2893	21.1660	22.3769	20.7040	22.8388	0.4886

9.27 Consider the United Oil Company premium gasoline mileage data in Table 9.3. In Table 9.5 we give the values of SSE and s^2 obtained by analyzing this data using four models.

a. Consider Model 4 and calculate the partial F statistic used to test

$$H_0: \beta_5 = \beta_6 = \beta_7 = \beta_8 = 0$$

versus

$$H_1: \text{At least one of } \beta_5, \beta_6, \beta_7, \text{ and } \beta_8 \text{ does not equal zero}$$

Using this statistic and the appropriate rejection point, determine whether we can reject H_0 in favor of H_1 by setting α equal to .05.

b. Consider Model 4 and calculate the partial F statistic used to test

$$H_0: \beta_7 = \beta_8 = 0$$

versus

$$H_1: \text{At least one of } \beta_7 \text{ and } \beta_8 \text{ does not equal zero}$$

Using this statistic and the appropriate rejection point, determine whether we can reject H_0 in favor of H_1 by setting α equal to .05.

c. Consider Model 3 and calculate the partial F statistic used to test

$$H_0: \beta_5 = \beta_6 = 0$$

versus

$$H_1: \text{At least one of } \beta_5 \text{ and } \beta_6 \text{ does not equal zero}$$

Using this statistic and the appropriate rejection point, determine whether we can reject H_0 in favor of H_1 by setting α equal to .05.

d. Which of the four models seems best? Justify your answer using your results of parts (a), (b), and (c).

TABLE 9.5 ***SSE* and s^2 for four United Oil Company models**

Model	*SSE*	s^2
1. $y = \beta_0 + \beta_1 x_1 + \beta_2 x_1^2 + \beta_3 x_2 + \beta_4 x_2^2 + \varepsilon$	22.4124	1.3184
2. $y = \beta_0 + \beta_1 x_1 + \beta_2 x_1^2 + \beta_3 x_2 + \beta_4 x_2^2 + \beta_5 x_1 x_2 + \varepsilon$	15.7971	.9873
3. $y = \beta_0 + \beta_1 x_1 + \beta_2 x_1^2 + \beta_3 x_2 + \beta_4 x_2^2 + \beta_5 x_1 x_2 + \beta_6 x_1^2 x_2^2 + \varepsilon$	6.9087	.4606
4. $y = \beta_0 + \beta_1 x_1 + \beta_2 x_1^2 + \beta_3 x_2 + \beta_4 x_2^2 + \beta_5 x_1 x_2 + \beta_6 x_1^2 x_2^2 + \beta_7 x_1 x_2^2 + \beta_8 x_1^2 x_2 + \varepsilon$	6.1569	.4736

9.28 Consider Model 4 in Table 9.5.

a. Compute and interpret the partial coefficient of determination

$$R^2(x_1x_2,\ x_1^2x_2^2,\ x_1x_2^2,\ x_1^2x_2 | x_1,\ x_1^2,\ x_2,\ x_2^2)$$

b. Compute and interpret the partial coefficient of determination

$$R^2(x_1x_2^2,\ x_1^2x_2 | x_1,\ x_1^2,\ x_2,\ x_2^2,\ x_1x_2,\ x_1^2x_2^2)$$

9.29 Consider Model 3 in Table 9.5. Compute and interpret the partial coefficient of determination

$$R^2(x_1x_2,\ x_1^2x_2^2 | x_1,\ x_1^2,\ x_2,\ x_2^2)$$

9.30 Consider the United Oil Company unleaded gasoline mileage study of Exercise 8.25 and the mileage data in Table 8.11 of the exercises in Chapter 8. Figure 8.25 in the exercises of Chapter 8 gives the SAS output of a regression analysis of this mileage data for each of the four models given in Exercise 8.25.

a. Consider the model

$$y = \beta_0 + \beta_1 x_1 + \beta_2 x_1^2 + \beta_3 x_2 + \beta_4 x_2^2 + \varepsilon$$

Using the SAS output in Figure 8.25, calculate the partial F statistic needed to test $H_0: \beta_2 = \beta_4 = 0$ versus H_1: At least one of β_2 and β_4 does not equal zero. Can we reject H_0 in favor of H_1 by setting α equal to .05? What does the result of this test say about the importance of the quadratic terms in the above model?

b. Using the SAS output of Figure 8.25, compute and interpret the partial coefficient of determination

$$R^2(x_1^2,\ x_2^2 | x_1,\ x_2)$$

9.31 Consider the hotel room rate data analyzed in Exercise 8.31. Suppose that we relate y (daily room rate) to x_1 (population), x_2 (hotel rating), and x_3 (number of units) by using the regression model

$$y = \beta_0 + \beta_1 x_1 + \beta_2 x_2 + \beta_3 x_3 + \beta_4 x_1 x_2 + \beta_5 x_1 x_3 + \beta_6 x_2 x_3 + \varepsilon$$

The SAS output of this regression analysis is given in Figure 9.21. (Note that the interaction terms x_1x_2, x_1x_3, and x_2x_3 are denoted as PR, PU, and RU, respectively, on the SAS output. Fully interpret this SAS output. Which independent variables seem important? Which interaction terms seem important?

9.32 Zenex Radio Corporation has developed a new way to assemble an electrical component used in the manufacture of radios. The company wishes to determine whether μ, the

FIGURE 9.21 **SAS output of a regression analysis of the hotel room rate data using the model $y = \beta_0 + \beta_1x_1 + \beta_2x_2 + \beta_3x_3 + \beta_4x_1x_2 + \beta_5x_1x_3 + \beta_6x_2x_3 + \varepsilon$**

SAS

DEP VARIABLE: RATE

ANALYSIS OF VARIANCE

SOURCE	DF	SUM OF SQUARES	MEAN SQUARE	F VALUE	PROB>F
MODEL	6	4968.79724	828.13287	13.540	0.0001
ERROR	28	1712.51876	61.16138423		
C TOTAL	34	6681.31600			

ROOT MSE	7.820574	R-SQUARE	0.7437
DEP MEAN	49.08	ADJ R-SQ	0.6888
C.V.	15.93434		

PARAMETER ESTIMATES

VARIABLE	DF	PARAMETER ESTIMATE	STANDARD ERROR	T FOR H0: PARAMETER=0	PROB > |T|
INTERCEP	1	56.18988442	12.21830252	4.599	0.0001
POP	1	-0.005229181	0.004755801	-1.100	0.2809
RATING	1	-8.45235468	6.00475744	-1.408	0.1703
UNITS	1	-0.25066505	0.12091334	-2.073	0.0475
PR	1	0.002631554	0.001446539	1.819	0.0796
PU	1	-0.000010266	0.000020733	-0.495	0.6243
RU	1	0.13600283	0.05377789	2.529	0.0174

mean assembly time of this component using the new method, is less than 20 minutes, which is known to be the mean assembly time for the component using the current method of assembly. Suppose that Zenex Radio randomly selects a sample of $n = 6$ employees, thoroughly trains each employee to use the new assembly method, has each employee assemble one component using the new method, records the assembly times, and calculates the mean and standard deviation of the sample of $n = 6$ assembly times to be $\bar{y} = 14.29$ (minutes) and $s = 2.19$ (minutes).

a. Assuming that the population of all assembly times is normally distributed, calculate a 95% confidence interval for μ.
b. Using the confidence interval that you computed in part (a), can Zenex Radio be at least 95% confident that μ is less than 20 minutes? Justify your answer.
c. Calculate a 95% prediction interval for y_0, an individual assembly time. Interpret this interval.

9.33 Consider the hospital labor needs data of Exercise 5.6 (see Table 5.9). Using a computer, investigate the possibility of finding an improved model for predicting labor requirements by employing one or more interaction terms. Analyze the residuals for the improved model.

9.34 Consider the economic data for selected cities in Table 5.12. Using a computer, employ one or more interaction terms to try to find improved models for predicting (1) HCOST (indexed housing cost), (2) JG (jobs growth), and (3) IG (personal income growth). Carry out a complete analysis of the residuals for each improved model.

9.35 Consider the sales territory performance data in Table 8.5. Using a computer, investigate the possibility of finding an improved model for predicting sales by utilizing one or more interaction terms. Analyze the residuals for the improved model.

9.36 Consider the hotel room rate data of Exercise 5.41 (see Table 5.11). Using a computer, investigate using one or more interaction terms to find an improved model for predicting hotel room rates. Carry out a complete analysis of the residuals for the improved model.

9.37 (Optional) Write a SAS program that will analyze the construction project data (in Table 8.8 of the exercises in Chapter 8) by using each of the following regression models:

$$y = \beta_0 + \beta_1 x_1 + \beta_2 x_2 + \beta_3 x_1^2 + \beta_4 x_1 x_2 + \varepsilon$$
$$y = \beta_0 + \beta_1 x_1 + \beta_2 x_1^2 + \beta_3 x_2 + \beta_4 x_1 x_2 + \beta_5 x_1^2 x_2 + \varepsilon$$
$$y = \beta_0 + \beta_1 x_1 + \beta_2 x_1^2 + \beta_3 x_2 + \beta_4 x_2^2 + \beta_5 x_1 x_2 + \beta_6 x_1^2 x_2 + \beta_7 x_1 x_2^2 + \varepsilon$$

Write the program so that a complete analysis of the residuals will be carried out for each model.

9.38 (Optional) Write a SAS program that will analyze the United Oil Company premium gasoline mileage data in Table 9.3 by using each of the four regression models given in Exercise 9.27. Write the program so that a complete analysis of the residuals will be carried out for each model.

9.39 (Optional) Write a SAS program that will analyze the Bonner Frozen Foods sales volume data in Table 9.4 by using the regression model

$$y = \beta_0 + \beta_1 x_1 + \beta_2 x_1^2 + \beta_3 x_2 + \beta_4 x_2^2 + \beta_5 x_1 x_2 + \varepsilon$$

Write the program so that a complete analysis of the residuals will be carried out for this model.

9.40 (Optional) Write a SAS program that will analyze the Crest Toothpaste sales data of Exercise 8.23 (see Table 8.10 in the exercises of Chapter 8) by using the regression model

$$y = \beta_0 + \beta_1 x_1 + \beta_2 x_2 + \beta_3 x_3 + \beta_4 x_1^2 + \beta_5 x_2^2 + \beta_6 x_3^2 + \beta_7 x_1 x_2 + \beta_8 x_1 x_3 + \beta_9 x_2 x_3 + \varepsilon$$

Write the program so that a complete analysis of the residuals will be carried out for this model.

9.41 Consider the premium gasoline mileage data in Table 9.3. We see that more than one value of the dependent variable has been observed at some of the combinations of values of the independent variables. Therefore letting $y_{x_1 x_2, k}$ denote the kth gasoline mileage obtained by using x_1 units of gasoline additive WST and x_2 units of gasoline additive YST, we can rearrange the data in Table 9.3 as in Table 9.6. Now the lack-of-fit test in Chapter 5 applies to testing lack of fit in multiple regression models by replacing $(m - 2)$ by $(m - k)$ and thus redefining MS_{LF} to be

$$MS_{LF} = \frac{SS_{LF}}{m - k}$$

Here, m is the total number of combinations of values of the independent variables at which at least one corresponding value of the dependent variable has been observed, and k is the number of parameters in the model under consideration.

TABLE 9.6 Rearranged data for United Oil Company premium gasoline mileage

Combination	x_1	x_2	$y_{x_1x_2,k}$			$\bar{y}_{x_1x_2}$
1	0	0	$y_{00,1} = 28.0$	$y_{00,2} = 28.6$	$y_{00,3} = 27.4$	$\bar{y}_{00} = 28.0$
2	1	0	$y_{10,1} = 33.3$	$y_{10,2} = 34.5$		$\bar{y}_{10} = 33.9$
3	0	1	$y_{01,1} = 33.0$	$y_{01,2} = 32.0$		$\bar{y}_{01} = 32.5$
4	1	1	$y_{11,1} = 35.6$	$y_{11,2} = 34.4$	$y_{11,3} = 35.0$	$\bar{y}_{11} = 35.0$
5	2	1	$y_{21,1} = 34.0$	$y_{21,2} = 33.3$	$y_{21,3} = 34.7$	$\bar{y}_{21} = 34.0$
6	0	2	$y_{02,1} = 33.5$	$y_{02,2} = 32.3$		$\bar{y}_{02} = 32.9$
7	1	2	$y_{12,1} = 33.4$			$\bar{y}_{12} = 33.4$
8	2	2	$y_{22,1} = 33.0$	$y_{22,2} = 32.0$		$\bar{y}_{22} = 32.5$
9	1	3	$y_{13,1} = 29.6$	$y_{13,2} = 30.6$		$\bar{y}_{13} = 30.1$
10	2	3	$y_{23,1} = 28.6$	$y_{23,2} = 29.8$		$\bar{y}_{23} = 29.2$

a. Verify that SS_{PE}, the sum of squares due to pure error, is 6.08.

b. Note that there are $n = 22$ values of the dependent variable and that there are $m = 10$ combinations of values of the independent variables at which at least one corresponding value of the dependent variable has been observed. Calculate MS_{PE}.

c. Using the fact that $SSE = 15.7971$ is the unexplained variation resulting from using the model

$$y = \beta_0 + \beta_1 x_1 + \beta_2 x_1^2 + \beta_3 x_2 + \beta_4 x_2^2 + \beta_5 x_1 x_2 + \varepsilon$$

to perform a regression analysis of the data in Table 9.3, calculate the lack-of-fit F statistic related to the preceding model. By using this statistic and the appropriate rejection point, determine whether we can, by setting α equal to .05, reject the null hypothesis that the functional form of this model is correct. The prob-value related to this hypothesis can be calculated to be .0153. Discuss how this prob-value has been calculated. Using this prob-value, determine whether we can reject this hypothesis by setting α equal to .05 or .01.

d. Using the fact that $SSE = 6.9087$ is the unexplained variation resulting from using the model

$$y = \beta_0 + \beta_1 x_1 + \beta_2 x_1^2 + \beta_3 x_2 + \beta_4 x_2^2 + \beta_5 x_1 x_2 + \beta_6 x_1^2 x_2^2 + \varepsilon$$

to perform a regression analysis of the data in Table 9.3, calculate the lack-of-fit F statistic related to this model. By using this statistic and the appropriate rejection point, determine whether we can, by setting α equal to .05, reject the null hypothesis that the functional form of this model is correct.

e. On the basis of the results obtained in parts (c) and (d), which of the models in parts (c) and (d) seems best?

CHAPTER 10

SOME PROBLEMS AND REMEDIES

In this chapter we study two important problems that are often encountered in regression analysis. Recall that the hypothesis test of $H_0: \beta_j = 0$ versus $H_0: \beta_j \neq 0$, where β_j is the regression parameter multiplied by x_j, is useful in assessing the importance of the single independent variable x_j. Therefore this test helps us decide whether to include x_j in a final regression model. However, we will see in Section 10.1 that *multicollinearity*, which exists when the independent variables in a regression model are related to each other, can hinder our ability to use this test to judge the importance of x_j. In Section 10.1 we discuss some simple solutions to problems caused by multicollinearity. In optional Sections 10.2 and 10.3 we discuss two more complex solutions: the *standardized regression model* and *ridge regression*. Then we present in Section 10.4 some useful diagnostic statistics for identifying another problem—outlying and influential observations. We also briefly discuss a *robust regression* procedure that can be used to estimate the regression parameters when the least squares point estimates are seriously affected by outlying observations. We complete this chapter with optional Sections 10.5 and 10.6, which discuss partial leverage residual plots and using SAS to implement the techniques of this chapter.

Finally, note that although both Sections 10.1 and 10.4 contain necessary material for a thorough understanding of regression analysis, only Section 10.1 is necessary for understanding the remaining chapters in this book. The reader may therefore proceed to Chapter 11 after studying Section 10.1 and return to the other sections of this chapter at any time.

10.1

THE PROBLEM OF MULTICOLLINEARITY

When the independent variables in a regression model are interrelated or are dependent on each other, *multicollinearity* is said to exist among the independent variables. We illustrate multicollinearity in the following example.

Example 10.1 In the fuel consumption model

$$y = \beta_0 + \beta_1 x_1 + \beta_2 x_2 + \varepsilon$$

it is logical that the independent variables x_1 (average hourly temperature) and x_2 (chill index) might be related to each other. That is, we might expect low values of the average hourly temperature to be associated with high values of the chill index. Recall that a high value of the chill index says that factors other than the average hourly temperature indicate that cold weather conditions exist. Likewise, we might expect high values of the average hourly temperature to be associated with lower values of the chill index.

There are several ways to determine the existence and extent of multicollinearity between x_1 and x_2. One simple method is to plot x_2 against x_1. The eight combinations of observed values of x_1 and x_2 are listed in Table 10.1, and the plot of x_2

TABLE 10.1 **Eight combinations of observed values of x_1 and x_2**

Week, i	*Average hourly temperature, x_{i1}*	*Chill index, x_{i2}*	*Combination of x_1 and x_2 (x_{i1}, x_{i2})*
1	28.0	18	(28.0, 18)
2	28.0	14	(28.0, 14)
3	32.5	24	(32.5, 24)
4	39.0	22	(39.0, 22)
5	45.9	8	(45.9, 8)
6	57.8	16	(57.8, 16)
7	58.1	1	(58.1, 1)
8	62.5	0	(62.5, 0)
	$\sum_{i=1}^{8} x_{i1} = 351.8$	$\sum_{i=1}^{8} x_{i2} = 103$	
	$\bar{x}_1 = \frac{351.8}{8} = 43.98$	$\bar{x}_2 = \frac{103}{8} = 12.875$	

FIGURE 10.1 **Multicollinearity between x_1 (average hourly temperature) and x_2 (chill index)**

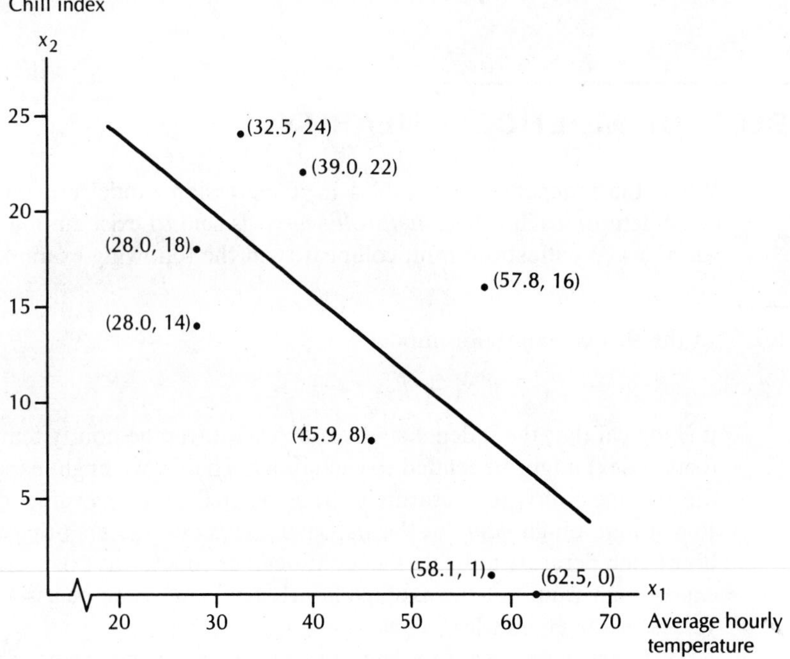

against x_1 is given in Figure 10.1. This figure indicates that (as expected) as x_1 increases, x_2 tends to decrease in a linear fashion.

One way to measure the *extent* of the multicollinearity between x_1 and x_2 is to compute the *simple correlation coefficient* between x_1 and x_2. This quantity, denoted by r_{x_1,x_2}, is computed by using the following formula

$$r_{x_1,x_2} = \frac{\sum_{i=1}^{n} (x_{i1} - \bar{x}_1)(x_{i2} - \bar{x}_2)}{\left[\sum_{i=1}^{n} (x_{i1} - \bar{x}_1)^2 \sum_{i=1}^{n} (x_{i2} - \bar{x}_2)^2\right]^{1/2}}$$

where

$$\bar{x}_1 = \frac{\sum_{i=1}^{n} x_{i1}}{n} \quad \text{and} \quad \bar{x}_2 = \frac{\sum_{i=1}^{n} x_{i2}}{n}$$

Using the data in Table 10.1, we make the following calculations:

$$\bar{x}_1 = \frac{\sum_{i=1}^{8} x_{i1}}{8} = \frac{351.8}{8} = 43.98$$

$$\bar{x}_2 = \frac{\sum_{i=1}^{8} x_{i2}}{8} = \frac{103}{8} = 12.875$$

$$r_{x_1,x_2} = \frac{\sum_{i=1}^{8} (x_{i1} - \bar{x}_1)(x_{i2} - \bar{x}_2)}{\left[\sum_{i=1}^{8} (x_{i1} - \bar{x}_1)^2 \sum_{i=1}^{8} (x_{i2} - \bar{x}_2)^2\right]^{1/2}}$$

$$= \frac{(x_{11} - \bar{x}_1)(x_{12} - \bar{x}_2) + \cdots + (x_{81} - \bar{x}_1)(x_{82} - \bar{x}_2)}{[[(x_{11} - \bar{x}_1)^2 + \cdots + (x_{81} - \bar{x}_1)^2][(x_{12} - \bar{x}_2)^2 + \cdots + (x_{82} - \bar{x}_2)^2]]^{1/2}}$$

$$= \frac{(28.0 - 43.98)(18 - 12.875) + \cdots + (62.5 - 43.98)(0 - 12.875)}{[[(28.0 - 43.98)^2 + \cdots + (62.5 - 43.98)^2][(18 - 12.875)^2 + \cdots + (0 - 12.875)^2]]^{1/2}}$$

$$= -.7182$$

We next see how to interpret what $r_{x_1,x_2} = -.7182$ says about the extent of the multicollinearity between x_1 and x_2.

In general, consider the linear regression model

$$y = \beta_0 + \beta_1 x_1 + \cdots + \beta_j x_j + \cdots + \beta_k x_k + \cdots + \beta_p x_p + \varepsilon$$

As illustrated in the preceding example, we can use the n observed values of the independent variables x_j and x_k to plot x_k against x_j or to compute the *simple correlation coefficient* between x_j and x_k.

The Simple Correlation Coefficient

The *simple correlation coefficient* between x_j and x_k is

$$r_{x_j,x_k} = \frac{\sum_{i=1}^{n} (x_{ij} - \bar{x}_j)(x_{ik} - \bar{x}_k)}{\left[\sum_{i=1}^{n} (x_{ij} - \bar{x}_j)^2 \sum_{i=1}^{n} (x_{ik} - \bar{x}_k)^2\right]^{1/2}}$$

where

$$\bar{x}_j = \frac{\sum_{i=1}^{n} x_{ij}}{n} \quad \text{and} \quad \bar{x}_k = \frac{\sum_{i=1}^{n} x_{ik}}{n}$$

The quantity r_{x_j,x_k} is a special case of the simple correlation coefficient between x and y if we consider x_j to be the independent variable x and x_k to be the dependent variable y. From our interpretation of r in Chapter 5 it follows that r_{x_j,x_k} is a measure of the linear relationship, or correlation, between x_j and x_k. For example, the simple correlation coefficient $r_{x_1,x_2} = -.7182$ in the fuel consumption problem indicates (as illustrated in Figure 10.1) that the independent variables x_1 and x_2 have some tendency to move together in a straight-line fashion with a negative slope. Therefore the independent variables x_1 and x_2 are somewhat negatively correlated. Most standard multiple regression computer packages calculate the simple correlation coefficient for each pair of independent variables in the model. However, these simple correlation coefficients are not the only measures of the multicollinearity between the independent variables. We discuss one other measure of multicollinearity later in this section.

In general, when two or more independent variables in a regression model are related to each other (or *correlated*), they contribute redundant information. That is, even though each independent variable contributes information for describing and predicting the dependent variable, some of the information overlaps. The consequences of multicollinearity in a regression analysis vary from slight to quite substantial, depending on the extent of the multicollinearity and the aspect of the regression analysis one considers. Since multicollinearity is often encountered in a regression analysis, we now summarize its effects.

EFFECTS OF MULTICOLLINEARITY ON THE LEAST SQUARES POINT ESTIMATES

If multicollinearity exists, the values of the least squares point estimates of the model parameters depend on the particular independent variables that are included in the model. That is, when an independent variable is added to a regression model, and when the independent variable being added is related to the independent variable(s) already in the model, the least squares point estimates of the regression parameters will change. As an example, consider the one- and two-variable fuel consumption models

$$y = \beta_0 + \beta_1 x_1 + \varepsilon \qquad \text{and} \qquad y = \beta_0 + \beta_1 x_1 + \beta_2 x_2 + \varepsilon$$

The least squares point estimates of the parameters β_0 and β_1 in the above simple linear model have been calculated to be $b_0 = 15.84$ and $b_1 = -.1279$. However, when the independent variable x_2, which is related to x_1, is added to the fuel consumption model, the least squares point estimates become $b_0 = 13.109$, $b_1 = -.0900$, and $b_2 = .0825$. We see that adding a correlated independent variable results in altered least squares point estimates b_0 and b_1. Thus when multicollinearity exists, the least squares point estimates are conditional—they depend upon the correlated independent variables included in the regression model. Because of this, when multicollinearity exists, a particular least squares point estimate b_j does not really measure the influence of the independent variable

x_j upon the mean value of the dependent variable. Rather, b_j measures a partial influence of x_j upon the mean value of the dependent variable, and this estimate b_j depends upon which of the correlated independent variables are included in the model.

Extreme cases of multicollinearity can cause the least squares point estimates to be far from the true values of the regression parameters. In fact, some of the least squares point estimates may have the wrong sign. That is, the sign (positive or negative) of a least squares point estimate may be different from the sign of the true value of the parameter. The reason for this is that extreme cases of multicollinearity cause the least squares point estimates to be highly dependent on the particular sample of values of the dependent variable that is observed. In other words, severe multicollinearity can cause two slightly different samples of values of the dependent variable to yield substantially different least squares point estimates. To see why this is true, consider the so-called "picket fence" display in Figure 10.2. This figure depicts two independent variables (x_1 and x_2) exhibiting strong multicollinearity. The heights of the pickets on the fence represent the y observations. If we assume that the model

$$y = \beta_0 + \beta_1 x_1 + \beta_2 x_2 + \varepsilon$$

adequately describes this data, then obtaining the least squares point estimates b_0, b_1, and b_2 of β_0, β_1, and β_2 can be likened to balancing a plane

$$\hat{y} = b_0 + b_1 x_1 + b_2 x_2$$

on top of the picket fence. Here, the least squares point estimates describe the "slant" of this plane. Note that b_1 represents the plane's slope in the x_1 direction

FIGURE 10.2 **Picket fence display of multicollinearity**

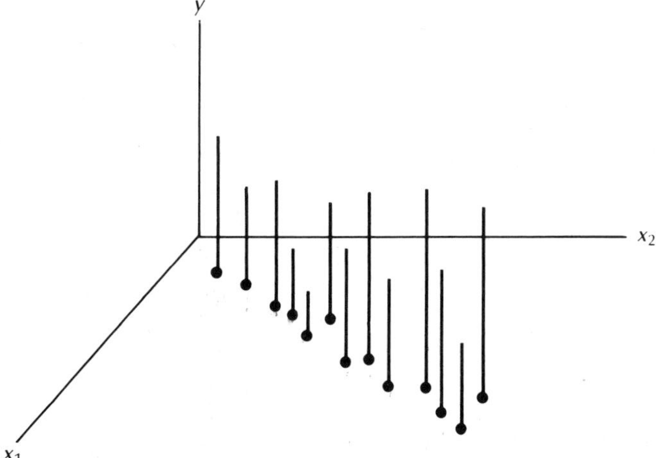

and b_2 represents the plane's slope in the x_2 direction. Clearly, this plane is quite unstable. That is, if a small change occurs in the height of one of the pickets (that is, in a y value), then the slant of the plane (that is, the least squares point estimates) might radically change.

On the other hand, if the multicollinearity did not exist, the observed combinations of values of x_1 and x_2 would not tend to fall around a line but would be more uniformly spread out in the (x_1, x_2) plane. Then the plane would be more stable. That is, a small change in the height of one of the pickets (that is, in a y value) would probably not radically change the slant of the plane and thus would probably not radically change the least squares point estimates.

It should also be mentioned that when multicollinearity exists, the practical meaning of the true regression parameters becomes clouded. For example, in the fuel consumption model

$$y = \beta_0 + \beta_1 x_1 + \beta_2 x_2 + \varepsilon$$

the parameter β_1 measures the influence on mean weekly fuel consumption of a change in the average hourly temperature *when the chill index does not change.* However, since the multicollinearity between x_1 and x_2 tells us that a change in average hourly temperature is often accompanied by a change in the chill index, the practical meaning of β_1 (and hence the meaning of the point estimate b_1) becomes uncertain.

EFFECTS OF MULTICOLLINEARITY ON t STATISTICS AND ASSOCIATED PROB-VALUES

Perhaps the most serious effect of multicollinearity is that it can hinder our ability to assess the importance of an independent variable x_j. Recall that we can, by setting the probability of a Type I error equal to α, reject $H_0: \beta_j = 0$ in favor of $H_1: \beta_j \neq 0$ if and only if

$$|t| > t_{[\alpha/2]}^{(n-k)}$$

or, equivalently, if and only if

$$\text{prob-value} < \alpha$$

Thus the larger (in absolute value) the t statistic is and the smaller the prob-value is, the stronger is the evidence that we should reject $H_0: \beta_j = 0$ in favor of H_1: $\beta_j \neq 0$. Therefore, intuitively, the size of the t statistic (how large it is) and the size of the prob-value (how small it is) measure the importance of the independent variable x_j. When multicollinearity exists, the sizes of the t statistic and of the related prob-value *measure the additional importance of the independent variable x_j over the combined importance of the other independent variables in the regression model.* Clearly, two or more correlated independent variables contribute redundant information. Therefore multicollinearity often causes the t statistics obtained by relating a dependent variable to a set of correlated independent variables to be smaller (in absolute value) than the t statistics that would be obtained if separate regression analyses were run, where each separate regression analysis relates the dependent variable to a smaller set (for example, only one) of the correlated independent variables. Thus multicollinearity can cause some of the correlated independent variables to appear to be less important than they really are. In extreme cases, the t statistics and the prob-values for a set of correlated independent variables can make each and every variable look unimportant when in fact the correlated independent variables taken together accurately describe and predict the dependent variable.

Example 10.2 Consider the fuel consumption problem and the regression model

$$y = \beta_0 + \beta_1 x_1 + \beta_2 x_2 + \varepsilon$$

Remember that although the independent variable x_2, the chill index, measures much of the effect of wind velocity on weekly fuel consumption, it probably does not measure the total effect of wind velocity. For instance, it is quite possible that two weeks with the same chill index might have somewhat different average hourly wind velocities. Thus we might consider a model that explicitly describes the effect of the average hourly wind velocity on fuel consumption. Letting x_3 denote the average hourly wind velocity, we try the model

$$y = \beta_0 + \beta_1 x_1 + \beta_2 x_2 + \beta_3 x_3 + \varepsilon$$

Suppose that we have observed the values of x_3 given in Table 10.2 in addition to the previously given values of y, x_1, and x_2 (which are repeated in Table 10.2). Table 10.3 presents the SAS outputs resulting from using the above two models to perform regression analyses of the data in Table 10.2. To interpret the results, first note that we would expect multicollinearity to exist between x_1, x_2, and x_3. For example, we would expect x_1 and x_3 to be related—high average hourly wind

TABLE 10.2 **Observed fuel consumption data, including observed average hourly wind velocities**

Week, i	Average hourly temperature, x_{i1}	Chill index, x_{i2}	Average hourly wind velocity, x_{i3}	Fuel consumption, y_i
1	$x_{11} = 28.0$	$x_{12} = 18$	$x_{13} = 10.2$	$y_1 = 12.4$
2	$x_{21} = 28.0$	$x_{22} = 14$	$x_{23} = 11.5$	$y_2 = 11.7$
3	$x_{31} = 32.5$	$x_{32} = 24$	$x_{33} = 12.1$	$y_3 = 12.4$
4	$x_{41} = 39.0$	$x_{42} = 22$	$x_{43} = 9.5$	$y_4 = 10.8$
5	$x_{51} = 45.9$	$x_{52} = 8$	$x_{53} = 8.9$	$y_5 = 9.4$
6	$x_{61} = 57.8$	$x_{62} = 16$	$x_{63} = 9.2$	$y_6 = 9.5$
7	$x_{71} = 58.1$	$x_{72} = 1$	$x_{73} = 8.0$	$y_7 = 8.0$
8	$x_{81} = 62.5$	$x_{82} = 0$	$x_{83} = 7.0$	$y_8 = 7.5$

TABLE 10.3 **The effect of multicollinearity on the t statistics and the prob-values in the fuel consumption problem**

The t statistics and prob-values for testing the importance of the independent variables in the model $y = \beta_0 + \beta_1 x_1 + \beta_2 x_2 + \varepsilon$

VARIABLE	DF	PARAMETER ESTIMATE	STANDARD ERROR	T FOR HO: PARAMETER=0	PROB > |T|
INTERCEP	1	13.108737	0.855698	15.319	0.0001
X_1	1	-0.090014	0.014077	-6.394	0.0014
X_2	1	0.082495	0.022003	3.749	0.0133

The t statistics and prob-values for testing the importance of the independent variables in the model $y = \beta_0 + \beta_1 x_1 + \beta_2 x_2 + \beta_3 x_3 + \varepsilon$

VARIABLE	DF	PARAMETER ESTIMATE	STANDARD ERROR	T FOR HO: PARAMETER=0	PROB > |T|
INTERCEP	1	10.794031	2.404032	4.490	0.0109
X_1	1	-0.076013	0.019515	-3.895	0.0176
X_2	1	0.068523	0.025741	2.662	0.0563
X_3	1	0.196742	0.191115	1.029	0.3614

velocities might be associated with low average hourly temperatures. If we use the data in Table 10.2 to calculate the simple correlation coefficient between x_1 and x_3, we find that r_{x_1,x_3} equals $-.8659$. This indicates that the independent variables x_1 and x_3 have a tendency to move together in a straight line fashion with a negative slope. Therefore x_1 and x_3 are somewhat negatively correlated. Next, recall that x_2, the chill index, is defined to measure much of the effect of wind velocity on weekly fuel consumption. Thus the independent variables x_2 and x_3 are certainly related. If we use the data in Table 10.2 to calculate the simple correlation coefficient between x_2 and x_3, we find that r_{x_2,x_3} equals .8054. This indicates that the independent variables x_2 and x_3 have a tendency to move together in a straight line fashion with a positive slope. Therefore x_2 and x_3 are somewhat positively correlated. The fact that there is multicollinearity between x_3 and x_1 and between x_3 and x_2 means that x_3 contributes information that is already partially contributed by x_1 and by x_2. Since the t statistic and prob-value for testing the hypothesis H_0: $\beta_3 = 0$ are (see Table 10.3)

$$t = 1.03 \quad \text{and} \quad \text{prob-value} = .3614$$

we cannot reject H_0 with a small probability of a Type I error. Thus we would conclude that the independent variable x_3 does not have substantial additional importance over the combined importance of the independent variables x_1 and x_2. This at least partly confirms our belief that we have defined the chill index so that it measures most of the effect of wind velocity and also implies that we probably should not include the average hourly wind velocity in a final regression model if we have included the chill index.

Next, using Table 10.3, we compare the t statistics and the corresponding prob-values for the hypotheses H_0: $\beta_0 = 0$, H_0: $\beta_1 = 0$, and H_0: $\beta_2 = 0$ as related to the two- and three-variable fuel consumption models. We find that for the two-variable model (which does not include x_3) the t statistics are larger than the corresponding t statistics for the three-variable model (which does include x_3). We also find that the corresponding prob-values are smaller for the two-variable model than for the three-variable model. This says that the t statistics and prob-values indicate that the independent variables x_1 and x_2 and the intercept β_0 have more additional importance in the model that does not include the correlated independent variable x_3 than they have in the model that does include x_3. Since we have concluded that x_3 does not have substantial additional importance over the combined importance of x_1 and x_2 in the regression model

$$y = \beta_0 + \beta_1 x_1 + \beta_2 x_2 + \beta_3 x_3 + \varepsilon$$

and since removing x_3 from the model would eliminate the multicollinearity between x_3 and x_1 and x_2, we should use the model

$$y = \beta_0 + \beta_1 x_1 + \beta_2 x_2 + \varepsilon$$

to judge the importance of x_1 and x_2. Since Table 10.3 indicates that both x_1 and x_2 are important in this model, we conclude that both x_1 and x_2 are significantly related to y and that it is reasonable to use the regression model

$$y = \beta_0 + \beta_1 x_1 + \beta_2 x_2 + \varepsilon$$

to describe and predict weekly fuel consumption.

Multicollinearity can cause the t statistic

$$t = \frac{b_j}{s\sqrt{c_{jj}}}$$

and the related prob-value to give a misleading impression of the importance of the independent variable x_j because c_{jj}, the diagonal element of $(\mathbf{X}'\mathbf{X})^{-1}$ corresponding to the parameter β_j, is used to calculate the t statistic. It can be shown that if we consider the regression model

$$y_i = \beta_0 + \beta_1 x_{i1} + \cdots + \beta_j x_{ij} + \cdots + \beta_p x_{ip} + \varepsilon_i$$

and if we let

$$\bar{x}_j = \frac{\sum_{i=1}^{n} x_{ij}}{n}$$

then

$$c_{jj} = \frac{1}{\sum_{i=1}^{n} (x_{ij} - \bar{x}_j)^2 (1 - R_j^2)}$$

Here, R_j^2 is the multiple coefficient of determination that would be calculated by running a regression analysis using the model

$$x_{ij} = \beta_0 + \beta_1 x_{i1} + \cdots + \beta_{j-1} x_{i,j-1} + \beta_{j+1} x_{i,j+1} + \cdots + \beta_p x_{ip} + \varepsilon_i$$

This model expresses the independent variable x_j as a function of the remaining independent variables $x_1, \ldots, x_{j-1}, x_{j+1}, \ldots, x_p$. Since R_j^2 is the proportion of the total variation in the n observed values of the independent variable x_j that is explained by the overall regression model, it follows that R_j^2 is a measure of the multicollinearity between x_j and $x_1, \ldots, x_{j-1}, x_{j+1}, \ldots, x_p$. The greater this multicollinearity is, the greater is R_j^2. Thus the larger is

$$c_{jj} = \frac{1}{\sum_{i=1}^{n} (x_{ij} - \bar{x}_j)^2 (1 - R_j^2)}$$

and the larger is the denominator of the t statistic

$$t = \frac{b_j}{s\sqrt{c_{jj}}}$$

If t has a large denominator, then this statistic might, depending on the size of b_j, be small. Thus strong multicollinearity between x_j and $x_1, \ldots, x_{j-1}, x_{j+1}, \ldots, x_p$ can cause the t statistic to be small (and the related prob-value to be large). This would give the impression that x_j is not important (even if it really is important).

Recalling that the variance of the population of all possible least squares point estimates of the parameter β_j is

$$\sigma_{b_j}^2 = \sigma^2 c_{jj}$$

it follows that

$$\begin{aligned} \sigma_{b_j}^2 &= \sigma^2 \frac{1}{\sum_{i=1}^{n} (x_{ij} - \bar{x}_j)^2 (1 - R_j^2)} \\ &= \frac{\sigma^2}{\sum_{i=1}^{n} (x_{ij} - \bar{x}_j)^2} \left(\frac{1}{1 - R_j^2} \right) \end{aligned}$$

We now make the following definition.

The Variance Inflation Factor

The *variance inflation factor* for b_j is

$$VIF_j = \frac{1}{1 - R_j^2}$$

From this definition it follows that

$$\sigma_{b_j}^2 = \frac{\sigma^2}{\sum_{i=1}^{n} (x_{ij} - \bar{x}_j)^2} VIF_j$$

If $R_j^2 = 0$, that is, if x_j is not related to the other independent variables through the above regression model, then the variance inflation factor VIF_j is equal to 1. This implies that

$$\sigma_{b_j}^2 = \frac{\sigma^2}{\sum_{i=1}^{n} (x_{ij} - \bar{x}_j)^2}$$

If $R_j^2 > 0$, x_j is related to the other independent variables through the above model. Then $1 - R_j^2$ is less than 1, and

$$VIF_j = \frac{1}{1 - R_j^2}$$

is greater than 1. This implies that

$$\sigma_{b_j}^2 = \frac{\sigma^2}{\sum_{i=1}^{n} (x_{ij} - \bar{x}_j)^2} VIF_j$$

is inflated beyond the value of $\sigma_{b_j}^2$ when $R_j^2 = 0$. Both the largest variance inflation factor among the independent variables and the mean of the variance inflation factors for the independent variables

$$\overline{VIF} = \frac{\sum_{j=1}^{p} VIF_j}{p}$$

are used as indicators of the severity of multicollinearity. If the largest variance inflation factor is greater than 10 (which means that the largest R_j^2 is greater than .9), or if the mean of the variance inflation factors is substantially greater than 1, then multicollinearity may be seriously influencing the least squares point estimates.

Example 10.3 Consider the two-variable fuel consumption model

$$y = \beta_0 + \beta_1 x_1 + \beta_2 x_2 + \varepsilon$$

Since

1. $R_1^2 = .5158$ is the multiple coefficient of determination for the model

$$x_1 = \beta_0 + \beta_1 x_2 + \varepsilon$$

it follows that

$$VIF_1 = \frac{1}{1 - R_1^2} = \frac{1}{1 - .5158} = 2.0653$$

2. $R_2^2 = .5158$ is the multiple coefficient of determination for the model

$$x_2 = \beta_0 + \beta_1 x_1 + \varepsilon$$

it follows that

$$VIF_2 = \frac{1}{1 - R_2^2} = \frac{1}{1 - .5158} = 2.0653$$

Next, consider the three-variable fuel consumption model

$$y = \beta_0 + \beta_1 x_1 + \beta_2 x_2 + \beta_3 x_3 + \varepsilon$$

Table 10.4 summarizes the calculation of the variance inflation factors for this model. Since the variance inflation factors for the three-variable model are larger than those for the two-variable model, the three-variable model exhibits more multicollinearity. Although none of the variance inflation factors in the three-variable model is greater than 10, the largest variance inflation factor in this model, $VIF_3 = 5.5340$, indicates that some of the information provided by x_3 is already provided by x_1 and x_2.

TABLE 10.4 **Variance inflation factors for the three-variable fuel consumption model**

Model	R_j^2	VIF_j
$x_1 = \beta_0 + \beta_1 x_2 + \beta_2 x_3 + \varepsilon$	$R_1^2 = .7510$	$VIF_1 = \frac{1}{1 - R_1^2} = \frac{1}{1 - .7510} = 4.0161$
$x_2 = \beta_0 + \beta_1 x_1 + \beta_2 x_3 + \varepsilon$	$R_2^2 = .6566$	$VIF_2 = \frac{1}{1 - R_2^2} = \frac{1}{1 - .6566} = 2.9121$
$x_3 = \beta_0 + \beta_1 x_1 + \beta_2 x_2 + \varepsilon$	$R_3^2 = .8193$	$VIF_3 = \frac{1}{1 - R_3^2} = \frac{1}{1 - .8193} = 5.5340$

We have seen that large simple correlation coefficients and large variance inflation factors indicate the existence of substantial multicollinearity. Another indication of substantial multicollinearity is when the overall F-test is significant while most (or all) of the t statistics are insignificant. That is, the prob-value related to the overall F statistic would be small, indicating that at least one independent variable is important and thus that the model does have significant overall utility. At the same time, most (or all) of the prob-values related to the individual t statistics would be large, indicating that many (or all) of the individual independent variables in the model do not have importance over and above the other independent variables. We illustrate this situation in the following example.

Example 10.4 In Table 10.5 we present data concerning the need for hospital labor in $n = 17$ U.S. Navy Hospitals. Here, x_1 denotes average daily patient load, x_2 denotes monthly X-ray exposures, x_3 denotes monthly occupied bed days, x_4 denotes eligible population in the area (divided by 1000), x_5 denotes average length of patients' stay in days, and y denotes monthly labor hours. Examining Figure 10.3, we see that data plots of y versus each of x_1, x_2, x_3, and x_4 indicate that y is related to each of these variables in a linear fashion. However, the data plot of y versus x_5 does not indicate a clear relationship between these variables. In general, although such individual data plots can be useful in tentatively specifying a regression model, they can also be misleading, particularly when strong multicollinearity exists. To investigate the multicollinearity between x_1, x_2, x_3, x_4, and x_5, we first use the data in Table 10.5 to calculate a *correlation matrix*.

TABLE 10.5 **Hospital labor needs data**

Hospital	x_1	x_2	x_3	x_4	x_5	y
1	15.57	2463	472.92	18.0	4.45	566.52
2	44.02	2048	1339.75	9.5	6.92	696.82
3	20.42	3940	620.25	12.8	4.28	1033.15
4	18.74	6505	568.33	36.7	3.90	1603.62
5	49.20	5723	1497.60	35.7	5.50	1611.37
6	44.92	11520	1365.83	24.0	4.60	1613.27
7	55.48	5779	1687.00	43.3	5.62	1854.17
8	59.28	5969	1639.92	46.7	5.15	2160.55
9	94.39	8461	2872.33	78.7	6.18	2305.58
10	128.02	20106	3655.08	180.5	6.15	3503.93
11	96.00	13313	2912.00	60.9	5.88	3571.89
12	131.42	10771	3921.00	103.7	4.88	3741.40
13	127.21	15543	3865.67	126.8	5.50	4026.52
14	252.90	36194	7684.10	157.7	7.00	10343.81
15	409.20	34703	12446.33	169.4	10.78	11732.17
16	463.70	39204	14098.40	331.4	7.05	15414.94
17	510.22	86533	15524.00	371.6	6.35	18854.45

Source: Procedures and Analysis for Staffing Standards Development: Regression Analysis Handbook (San Diego, CA: Navy Manpower and Material Analysis Center, 1979).

FIGURE 10.3 **SAS output of data plots of y versus each of x_1, x_2, x_3, x_4, and x_5 in the hospital labor needs problem**

(a) y versus x_1

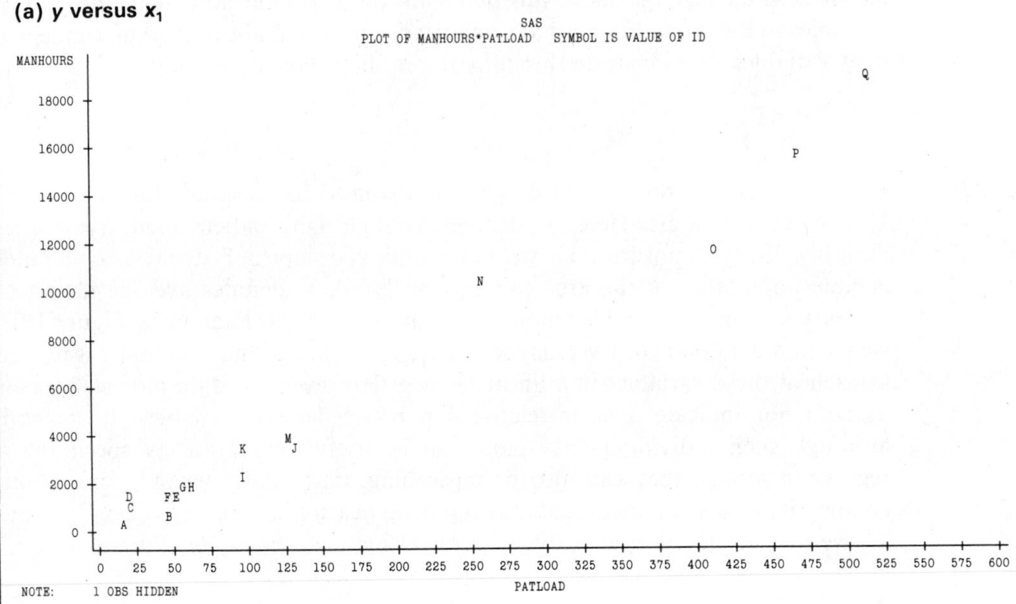

(b) y versus x_2

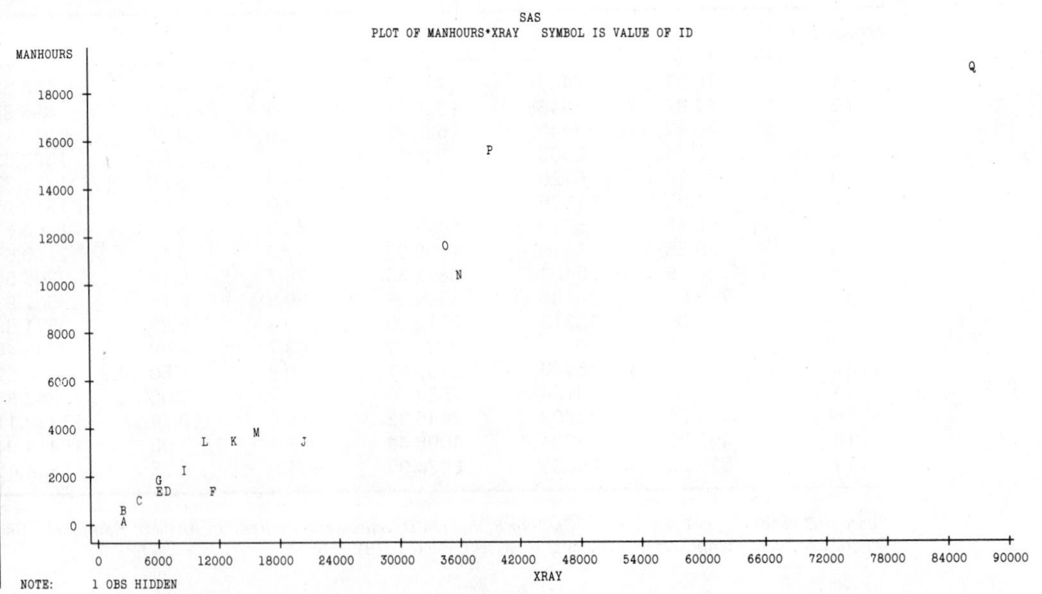

(c) y versus x_3

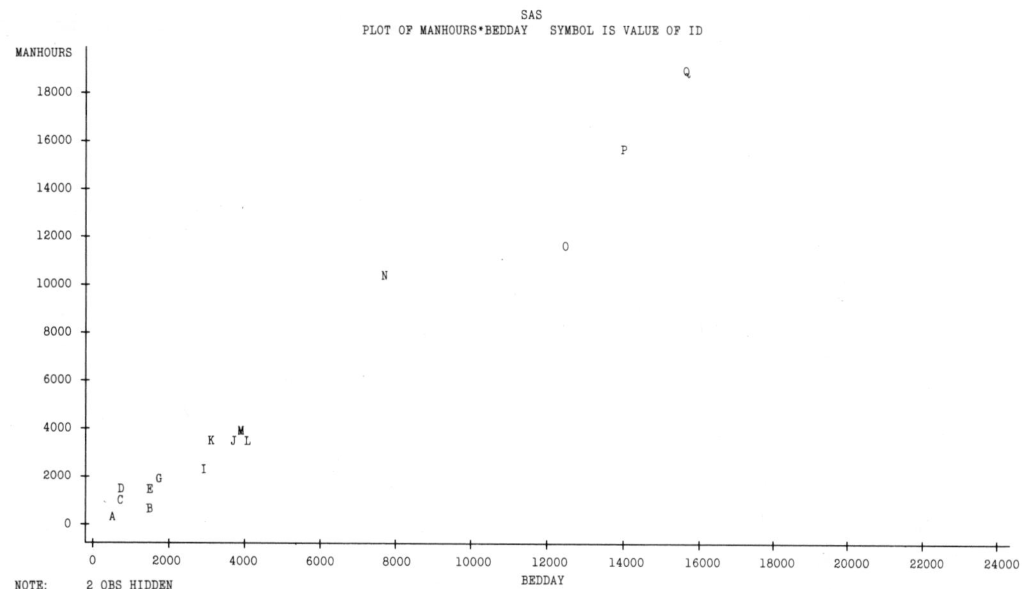

(d) y versus x_4

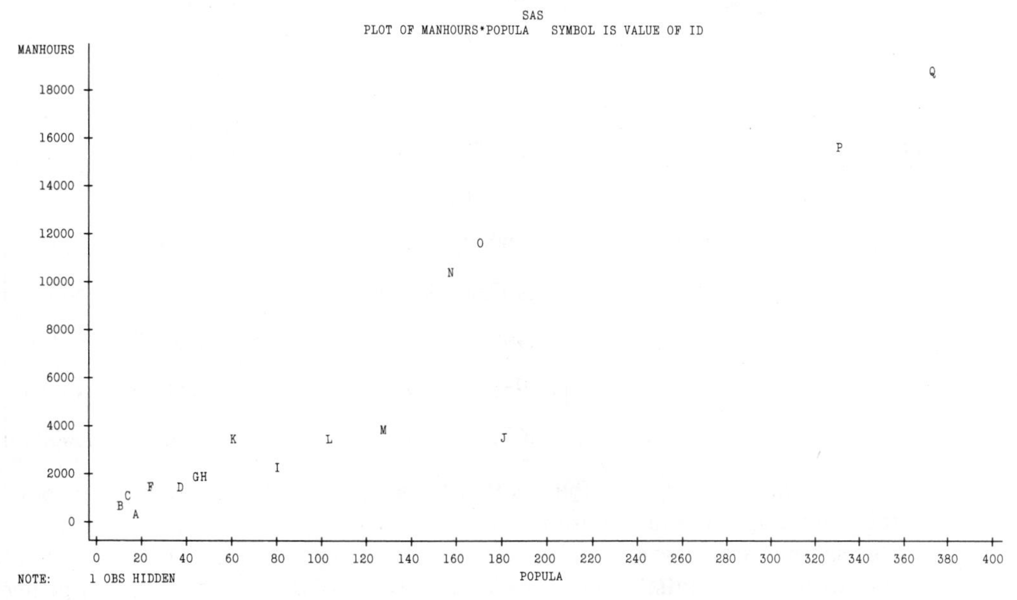

(continues)

FIGURE 10.3 Continued

(e) y versus x_5

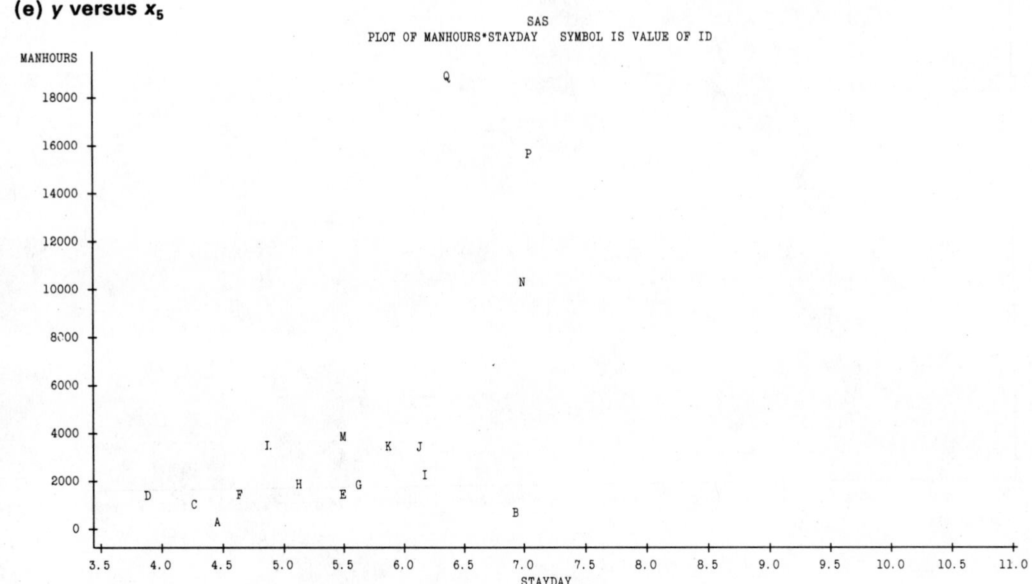

$$\text{Correlation matrix} = \begin{bmatrix} r_{x_1,x_1} & r_{x_1,x_2} & r_{x_1,x_3} & r_{x_1,x_4} & r_{x_1,x_5} \\ r_{x_2,x_1} & r_{x_2,x_2} & r_{x_2,x_3} & r_{x_2,x_4} & r_{x_2,x_5} \\ r_{x_3,x_1} & r_{x_3,x_2} & r_{x_3,x_3} & r_{x_3,x_4} & r_{x_3,x_5} \\ r_{x_4,x_1} & r_{x_4,x_2} & r_{x_4,x_3} & r_{x_4,x_4} & r_{x_4,x_5} \\ r_{x_5,x_1} & r_{x_5,x_2} & r_{x_5,x_3} & r_{x_5,x_4} & r_{x_5,x_5} \end{bmatrix}$$

$$= \begin{bmatrix} 1.00000 & .90738 & .99990 & .93569 & .67120 \\ .90738 & 1.00000 & .90715 & .91047 & .44665 \\ .99990 & .90715 & 1.00000 & .93317 & .67111 \\ .93569 & .91047 & .93317 & 1.00000 & .46286 \\ .67120 & .44665 & .67111 & .46286 & 1.00000 \end{bmatrix}$$

Noting that $r_{x_i,x_j} = r_{x_j,x_i}$, it follows that the simple correlation coefficients above the main diagonal equal those below the main diagonal. We see that six of the correlation coefficients above the main diagonal are greater than .9. This indicates that there is substantial multicollinearity between x_1, x_2, x_3, x_4, and x_5. Further evidence of multicollinearity is provided by Figure 10.4. This figure presents the

SAS output resulting from using the model

$$y = \beta_0 + \beta_1 x_1 + \beta_2 x_2 + \beta_3 x_3 + \beta_4 x_4 + \beta_5 x_5 + \varepsilon$$

to perform a regression analysis of the data in Table 10.5. Specifically, the prob-values related to testing the importance of x_1, x_3, x_4, and x_5 are greater than .05. This indicates that each of these variables is not important over and above the other independent variables in the model. However, the prob-value related to the overall F statistic is very small (.0001), much smaller than the prob-value related to x_2, which is .0234. This indicates that the model has significant overall utility, even though the model's individual independent variables do not appear to be very important. Another indication of substantial multicollinearity is the fact that the simple correlation coefficients between y and each of x_1, x_2, x_3, x_4, and x_5 (see Figure 10.4) are large, and the prob-values related to these correlation coefficients are small. We see that

$$r_{y,x_1} = .98565 \quad r_{y,x_2} = .94517 \quad r_{y,x_3} = .98599$$
$$(\text{prob-value} = .0001) \quad (\text{prob-value} = .0001) \quad (\text{prob-value} = .0001)$$

$$r_{y,x_4} = .94036 \quad r_{y,x_5} = .57858$$
$$(\text{prob-value} = .0001) \quad (\text{prob-value} = .0150)$$

Note that the prob-value related to r_{y,x_j} tests the null hypothesis $H_0: \rho_{y,x_j} = 0$, or, equivalently, the null hypothesis $H_0: \beta_1 = 0$ in the model $y = \beta_0 + \beta_1 x_j + \varepsilon$. This says that each of x_1, x_2, x_3, x_4, and x_5 is significantly related to y when not considered with any of the other independent variables. In contrast, when we employ all of x_1, x_2, x_3, x_4, and x_5 we have seen that each of x_1, x_3, x_4, and x_5 is not important over and above the other independent variables in the model.

Still further evidence suggesting substantial multicollinearity exists. Consider the signs of the least squares point estimates

$$b_1 = -15.8517 \quad \text{and} \quad b_4 = -4.2187$$

of the parameters β_1 and β_4 in the model

$$\begin{aligned} y &= \mu + \varepsilon \\ &= \beta_0 + \beta_1 x_1 + \beta_2 x_2 + \beta_3 x_3 + \beta_4 x_4 + \beta_5 x_5 + \varepsilon \end{aligned}$$

We see that these (negative) signs do not agree with our intuition. We would expect the true values of β_1 and β_4 to be positive because this would say that

1. The larger x_1 (average daily patient load) is, then (other things being equal) the larger is mean monthly labor hours

and

2. The larger x_4 (eligible population in the area) is, then (other things being equal) the larger is mean monthly labor hours.

Also, note that Myers (1986), who originally analyzed this data, states that the sign of $b_5 = -394.3141$ is different from the sign of the true value of β_5. We disagree

FIGURE 10.4 SAS output of a regression analysis of the hospital labor needs data by using the model $y = \beta_0 + \beta_1 x_1 + \beta_2 x_2 + \beta_3 x_3 + \beta_4 x_4 + \beta_5 x_5 + \varepsilon$ and of a correlation analysis of the hospital labor needs data

DEP VARIABLE: MANHOURS

ANALYSIS OF VARIANCE

SOURCE	DF	SUM OF SQUARES	MEAN SQUARE	F VALUE	PROB>F
MODEL	5	490177488	98035497.62	237.790[a]	0.0001[b]
ERROR	11	4535052.37	412277.49		
C TOTAL	16	494712540			

ROOT MSE	642.0884	R-SQUARE	0.9908
DEP MEAN	4978.48	ADJ R-SQ	0.9867
C.V.	12.89728		

PARAMETER ESTIMATES

VARIABLE	DF	PARAMETER[c] ESTIMATE	STANDARD ERROR	T FOR HO:[d] PARAMETER=0	PROB > |T|[e]	VARIANCE[f] INFLATION
INTERCEP	1	1962.94816	1071.36170	1.832	0.0941	0
PATLOAD	1	-15.85167473	97.65299018	-0.162	0.8740	9597.57076
XRAY	1	0.05593038	0.02125828	2.631	0.0234	7.94059253
BEDDAY	1	1.58962370	3.09208349	0.514	0.6174	8933.08650
POPULA	1	-4.21866799	7.17655737	-0.588	0.5685	23.29385611
STAYDAY	1	-394.31412	209.63954	-1.881	0.0867	4.27983530

CORRELATION COEFFICIENTS/PROB > |R| UNDER HO:RHO=0 / N = 17

	MANHOURS	PATLOAD	XRAY	BEDDAY	POPULA	STAYDAY
MANHOURS	1.00000 0.0000	0.98565 0.0001	0.94517 0.0001	0.98599 0.0001	0.94036 0.0001	0.57858 0.0150
PATLOAD	0.98565 0.0001	1.00000 0.0000	0.90738 0.0001	0.99990[h] 0.0001	0.93569 0.0001	0.67120 0.0032
XRAY	0.94517 0.0001	0.90738 0.0001	1.00000 0.0000	0.90715 0.0001	0.91047 0.0001	0.44665 0.0723
BEDDAY	0.98599[g] 0.0001	0.99990 0.0001	0.90715 0.0001	1.00000 0.0000	0.93317 0.0001	0.67111 0.0032
POPULA	0.94036 0.0001	0.93569 0.0001	0.91047 0.0001	0.93317 0.0001	1.00000 0.0000	0.46286 0.0614
STAYDAY	0.57858 0.0150	0.67120 0.0032	0.44665 0.0723	0.67111 0.0032	0.46286 0.0614	1.00000 0.0000

[a] *F (model)*
[b] *Prob-value for F (model)*
[c] b_j: b_0, b_1, b_2, b_3, b_4
[d] *t statistics*
[e] *Prob-values for the t statistics*
[f] VIF_j
[g] r_{y,x_3}
[h] r_{x_1,x_3}

because we think that the larger x_5 (average length of patients' stay in days) is, then (other things, such as average daily patient load, being equal) the less "startup" work that is required on patients, and thus the smaller is mean monthly labor hours. This means that the true value of β_5 is negative. Admittedly, however, this is debatable. It is difficult to theorize correctly about the true signs of the parameters in a regression model having so many redundant independent variables.

Finally, the variance inflation factors for x_1, x_2, x_3, x_4, and x_5 can be proven to be the diagonals of the inverse of the above correlation matrix. These variation inflation factors are

$$\begin{aligned} VIF_1 &= 9597.57 \\ VIF_2 &= 7.94 \\ VIF_3 &= 8933.09 \\ VIF_4 &= 23.29 \\ VIF_5 &= 4.28 \end{aligned}$$

Three of these variance inflation factors are greater than 10, and two of these variance inflation factors are very large.

To summarize the nature of the multicollinearity between x_1, x_2, x_3, x_4, and x_5, we might conclude that the values $r_{x_1,x_3} = .99990$, $VIF_1 = 9597.57$, and $VIF_3 = 8933.09$ indicate that x_1 and x_3 contribute very redundant information. Thus both x_1 and x_3 are probably not needed in the same model. Similarly, the values $r_{x_1,x_4} = .93569$, $r_{x_3,x_4} = .93317$, and $VIF_4 = 23.29$ indicate that x_4 might not be needed in a model utilizing x_1 or in a model utilizing x_3. In Chapter 11 we find that a good model describing the hospital labor needs data is

$$y = \beta_0 + \beta_1 x_2 + \beta_2 x_3 + \beta_3 x_5 + \varepsilon$$

Note that this model includes only one of the three highly collinear independent variables x_1, x_3, and x_4. Furthermore, the variance inflation factors for x_2, x_3, and x_5 in this model are 7.7373, 11.2693, and 2.4929. The prob-values related to the intercept, x_2, x_3, and x_5 in this model are .0749, .0205, .0001, and .0563. Therefore although this model still exhibits fairly strong multicollinearity, it has much less multicollinearity than the model containing all five independent variables.

EFFECTS OF MULTICOLLINEARITY ON THE EXPERIMENTAL REGION AND ON PREDICTION

Generally speaking, multicollinearity does not hinder the model's ability to predict the dependent variable on the basis of a combination of values of the independent variables in the experimental region. However, as illustrated in the following example, multicollinearity can cause combinations of values of the independent variables that one might expect to be in the experimental region to actually not be in the experimental region.

Example 10.5 Consider the fuel consumption model

$$y = \beta_0 + \beta_1 x_1 + \beta_2 x_2 + \varepsilon$$

In Figure 10.5 we illustrate the experimental region, which is the shaded region surrounded by the points representing the eight combinations of the observed values of x_1 and x_2. Consider the combination (40.0, 5). Since the average hourly temperature of 40.0 is in the range of observed average hourly temperatures (which is 28.0 to 62.5) and since the chill index of 5 is in the range of observed chill indices (which is 0 to 24), we might expect this combination to be in the experimental region. However, examination of Figure 10.5 indicates that the multicollinearity between x_1 and x_2 causes the experimental region to be a diagonal (or slanted) region that does not include the combination (40.0, 5).

FIGURE 10.5 **The effect of multicollinearity on the experimental region**

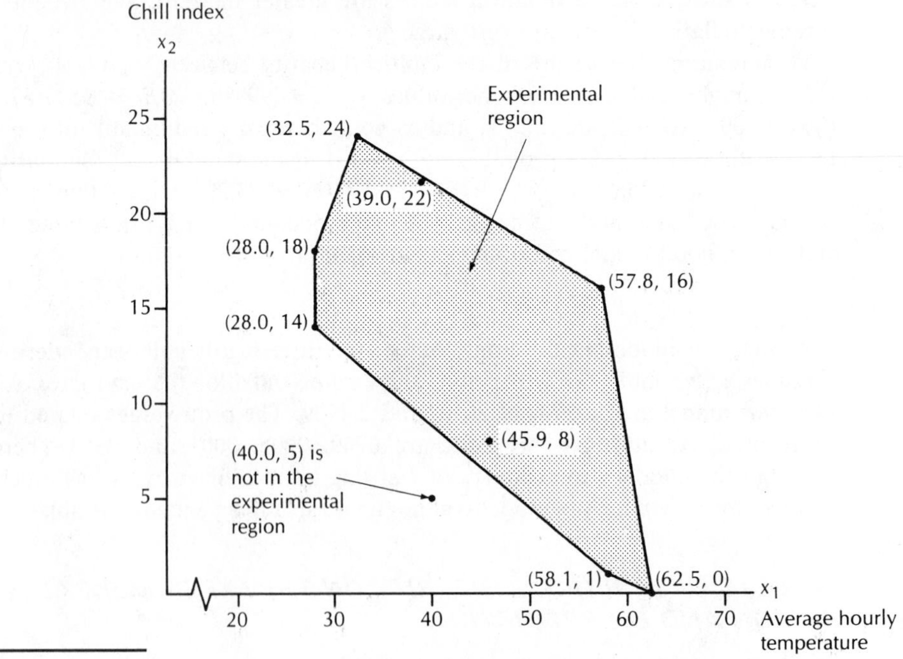

EFFECTS OF MULTICOLLINEARITY ON COMPUTATIONAL ACCURACY

The presence of strong multicollinearity can lead to serious rounding errors in the calculation of the least squares point estimates. This can occur even when a computer is being used to perform the calculations. The reason for this is that strong multicollinearity causes the columns of the matrix **X** to be nearly linearly dependent. This can cause substantial rounding errors in calculating $(\mathbf{X}'\mathbf{X})^{-1}$ and hence $\mathbf{b} = (\mathbf{X}'\mathbf{X})^{-1}\mathbf{X}'\mathbf{y}$. Many standard multiple regression computer programs are set up to minimize such errors, but serious rounding errors are more likely when strong multicollinearity exists. To lessen the chance of serious rounding errors, the analyst can "scale" the observed values of the independent and dependent variables. How to do this is discussed in optional Section 10.2.

DEALING WITH MULTICOLLINEARITY

The best solution to the multicollinearity problem is to try to avoid it by not including redundant independent variables in the regression model. If we can identify redundancy among independent variables already in the model, several remedies can be used to attempt to lessen the influence of the multicollinearity. One possible solution is to remove one or more of the highly correlated independent variables. However, one must be careful not to remove an independent variable that is important in describing and predicting the dependent variable. Such an independent variable should not be removed unless it can be replaced by another that is equally important and that does not cause the same strong multicollinearity. This approach, however, has two disadvantages. The first disadvantage is that when the independent variable is removed, the possibility of gathering any further information about the importance of that variable is eliminated. The second disadvantage is that the values of the least squares point estimates corresponding to the variables remaining in the model are influenced by the fact that there are correlated independent variables that are not included in the model. A second possible solution is to add more observations to the data used in building the regression model, so that the multicollinearity is (possibly) lessened. That is, sometimes the data we have collected makes it appear as though two or more independent variables are related when, in fact, no strong relationship exists. Collecting additional data may then lessen the (apparent) multicollinearity.

Finally, there are estimation procedures that are modifications of the least squares estimation procedure. When multicollinearity exists, these procedures are capable of producing point estimates that are "better" than the least squares point estimates in the sense that they are closer to the true values of the parameters. One such procedure is called *ridge regression*. This is discussed in optional Section 10.3.

*10.2 THE STANDARDIZED REGRESSION MODEL

As was stated in the preceding section, strong multicollinearity can cause serious rounding errors in calculating $(\mathbf{X}'\mathbf{X})^{-1}$. Such errors can also occur when the elements of $\mathbf{X}'\mathbf{X}$ have substantially different magnitudes. This occurs when the magnitudes of the independent variables differ substantially.

*This section is optional.

To control roundoff errors, we can transform the variables in the linear regression model

$$y_i = \beta_0 + \beta_1 x_{i1} + \cdots + \beta_p x_{ip} + \varepsilon_i$$

by using what is called a *correlation transformation.* The motivation for this terminology will be explained later. The correlation transformation of the dependent variable is

$$y_i' = \frac{1}{\sqrt{n-1}}\left(\frac{y_i - \bar{y}}{s_y}\right)$$

Here, $\bar{y}$ and s_y are the mean and standard deviation of the n observed values of the dependent variable. That is,

$$\bar{y} = \frac{\sum_{i=1}^{n} y_i}{n} \quad \text{and} \quad s_y = \sqrt{\frac{\sum_{i=1}^{n}(y_i - \bar{y})^2}{n-1}}$$

The correlation transformation of the jth independent variable ($j = 1, \ldots, p$) is

$$x_{ij}' = \frac{1}{\sqrt{n-1}}\left(\frac{x_{ij} - \bar{x}_j}{s_{x_j}}\right)$$

Here, $\bar{x}_j$ and s_{x_j} are the mean and standard deviation of the n observed values of the jth independent variable. That is,

$$\bar{x}_j = \frac{\sum_{i=1}^{n} x_{ij}}{n} \quad \text{and} \quad s_{x_j} = \sqrt{\frac{\sum_{i=1}^{n}(x_{ij} - \bar{x}_j)^2}{n-1}}$$

The *standardized regression model* utilizes the correlation transformations y_i' and x_{ij}' and is defined as

$$y_i' = \beta_1' x_{i1}' + \cdots + \beta_p' x_{ip}' + \varepsilon_i'$$

This model does not employ an intercept term. This is because the least squares calculations would always yield an estimated intercept of zero if an intercept parameter were used. The new parameters $\beta_1', \ldots, \beta_p'$ are related to the parameters $\beta_0, \beta_1, \ldots, \beta_p$ of the original model. It can be shown that

$$\beta_j = \left(\frac{s_y}{s_{x_j}}\right)\beta_j' \qquad j = 1, \ldots, p$$

$$\beta_0 = \bar{y} - \beta_1 \bar{x}_1 - \cdots - \beta_p \bar{x}_p$$

The least squares point estimates of the parameters $\beta'_1, \ldots, \beta'_p$ are calculated as

$$\begin{bmatrix} b'_1 \\ \cdot \\ \cdot \\ \cdot \\ b'_p \end{bmatrix} = (\mathring{\mathbf{X}}'\mathring{\mathbf{X}})^{-1}\mathring{\mathbf{X}}'\mathring{\mathbf{y}}$$

Here, the column vector $\mathring{\mathbf{y}}$ contains the n correlation transformed values of the dependent variable. Furthermore, the jth column of the matrix $\mathring{\mathbf{X}}$ consists of the n correlation transformed values of the jth independent variable ($j = 1, \ldots, p$). That is,

$$\mathring{\mathbf{y}} = \begin{bmatrix} y'_1 \\ y'_2 \\ \cdot \\ \cdot \\ \cdot \\ y'_n \end{bmatrix} \quad \text{and} \quad \mathring{\mathbf{X}} = \begin{bmatrix} x'_{11} & \ldots & x'_{1p} \\ x'_{21} & \ldots & x'_{2p} \\ \cdot & & \cdot \\ \cdot & & \cdot \\ \cdot & & \cdot \\ x'_{n1} & \ldots & x'_{np} \end{bmatrix}$$

It is easy to show that

$$\mathring{\mathbf{X}}'\mathring{\mathbf{X}} = \begin{bmatrix} 1 & r_{x_1,x_2} & \ldots & r_{x_1,x_p} \\ r_{x_2,x_1} & 1 & \ldots & r_{x_2,x_p} \\ \cdot & \cdot & & \cdot \\ \cdot & \cdot & & \cdot \\ \cdot & \cdot & & \cdot \\ r_{x_p,x_1} & r_{x_p,x_2} & \ldots & 1 \end{bmatrix} \quad \text{and} \quad \mathring{\mathbf{X}}'\mathring{\mathbf{y}} = \begin{bmatrix} r_{y,x_1} \\ r_{y,x_2} \\ \cdot \\ \cdot \\ \cdot \\ r_{y,x_p} \end{bmatrix}$$

Since $\mathring{\mathbf{X}}'\mathring{\mathbf{X}}$ consists of simple correlation coefficients, all of its elements are between -1 and 1. Therefore these elements have the same magnitudes. This can help to eliminate serious rounding errors in calculating $(\mathring{\mathbf{X}}'\mathring{\mathbf{X}})^{-1}$ and thus in calculating the least squares point estimates $b'_1, \ldots, b'_p$. By using these estimates, the least squares point estimates of the original model parameters are

$$b_j = \left(\frac{s_y}{s_{x_j}}\right) b'_j \qquad j = 1, \ldots, p$$

$$b_0 = \bar{y} - b_1\bar{x}_1 - \cdots - b_p\bar{x}_p$$

Some regression computer packages automatically use the correlation transformation to compute least squares point estimates. In addition, we will see in the next section that this transformation is used in *ridge regression.*

If we multiply both sides of the standardized regression model

$$\frac{1}{\sqrt{n-1}}\left(\frac{y_i - \bar{y}}{s_y}\right) = \beta'_1 \frac{1}{\sqrt{n-1}}\left(\frac{x_{i1} - \bar{x}_1}{s_{x_1}}\right) + \cdots + \beta'_p \frac{1}{\sqrt{n-1}}\left(\frac{x_{ip} - \bar{x}_p}{s_{x_p}}\right) + \varepsilon'_i$$

by $\sqrt{n-1}$, we obtain the model

$$\left(\frac{y_i - \bar{y}}{s_y}\right) = \beta_1''\left(\frac{x_{i1} - \bar{x}_1}{s_{x_1}}\right) + \cdots + \beta_p''\left(\frac{x_{ip} - \bar{x}_p}{s_{x_p}}\right) + \varepsilon_i''$$

It can be easily shown that the meaning and least squares point estimates of the parameters in the two models are the same. Therefore the estimate b_j' of β_j' reflects the change in the mean value of the dependent variable (in units of the standard deviation s_y) associated with an increase of one standard deviation s_{x_j} of x_j when the other independent variables are held constant. It might be tempting to compare the impact of two independent variables on the dependent variable by using the standardized estimates. However, one should be very cautious about doing this. This is because even the standardized estimates can be affected by multicollinearity.

Recall that in Chapter 9 we considered the Fresh Detergent model

$$y_i = \beta_0 + \beta_1 x_{i4} + \beta_2 x_{i3} + \beta_3 x_{i3}^2 + \beta_4 x_{i4} x_{i3} + \varepsilon_i$$

There is obviously strong multicollinearity in models using quadratic and/or interaction terms. For example, in the above model x_3 is related to x_3^2 and $x_4 x_3$, and x_4 is related to $x_4 x_3$. To eliminate serious rounding errors in calculating the least squares point estimates for such models, it is common practice to use either of two transformations of the independent variables. For the Fresh Detergent model the transformed model would be either

$$\begin{aligned} y_i = {} & \beta_0 + \beta_1(x_{i4} - \bar{x}_4) + \beta_2(x_{i3} - \bar{x}_3) + \beta_3(x_{i3} - \bar{x}_3)^2 \\ & + \beta_4(x_{i4} - \bar{x}_4)(x_{i3} - \bar{x}_3) + \varepsilon_i \end{aligned}$$

or

$$\begin{aligned} y_i = {} & \beta_0 + \beta_1\left(\frac{x_{i4} - \bar{x}_4}{s_{x_4}}\right) + \beta_2\left(\frac{x_{i3} - \bar{x}_3}{s_{x_3}}\right) + \beta_3\left(\frac{x_{i3} - \bar{x}_3}{s_{x_3}}\right)^2 \\ & + \beta_4\left(\frac{x_{i4} - \bar{x}_4}{s_{x_4}}\right)\left(\frac{x_{i3} - \bar{x}_3}{s_{x_3}}\right) + \varepsilon_i \end{aligned}$$

*10.3 RIDGE REGRESSION

When strong multicollinearity is present, we can sometimes use *ridge regression* to calculate point estimates that are "closer" to the true values of the model parameters than are the usual least squares point estimates. We first show how to calculate *ridge point estimates*. Then we discuss the advantage and disadvantages of these estimates.

To calculate the ridge estimates of the parameters in the model

$$y_i = \beta_0 + \beta_1 x_{i1} + \cdots + \beta_p x_{ip} + \varepsilon_i$$

*This section is optional.

we first consider the standardized regression model

$$y'_i = \beta'_1 x'_{i1} + \cdots + \beta'_p x'_{ip} + \varepsilon'_i$$

Here, we recall from Section 10.2 that

$$y'_i = \frac{1}{\sqrt{n-1}}\left(\frac{y_i - \bar{y}}{s_y}\right)$$

and

$$x'_{ij} = \frac{1}{\sqrt{n-1}}\left(\frac{x_{ij} - \bar{x}_j}{s_{x_j}}\right)$$

Then we form the matrices

$$\overset{*}{\mathbf{y}} = \begin{bmatrix} y'_1 \\ y'_2 \\ \vdots \\ y'_n \end{bmatrix} \qquad \overset{*}{\mathbf{X}} = \begin{bmatrix} x'_{11} & \cdots & x'_{1p} \\ x'_{21} & \cdots & x'_{2p} \\ \vdots & & \\ x'_{n1} & \cdots & x'_{np} \end{bmatrix}$$

$$\overset{*}{\mathbf{X}}{}'\overset{*}{\mathbf{X}} = \begin{bmatrix} 1 & r_{x_1,x_2} & \cdots & r_{x_1,x_p} \\ r_{x_2,x_1} & 1 & \cdots & r_{x_2,x_p} \\ \vdots & \vdots & & \vdots \\ r_{x_p,x_1} & r_{x_p,x_2} & \cdots & 1 \end{bmatrix} \qquad \overset{*}{\mathbf{X}}{}'\overset{*}{\mathbf{y}} = \begin{bmatrix} r_{y,x_1} \\ r_{y,x_2} \\ \vdots \\ r_{y,x_p} \end{bmatrix}$$

Ridge Estimation

The *ridge point estimates* of the parameters $\beta'_1, \ldots, \beta'_p$ of the standardized regression model are

$$\begin{bmatrix} b'_{1,R} \\ \vdots \\ b'_{p,R} \end{bmatrix} = (\overset{*}{\mathbf{X}}{}'\overset{*}{\mathbf{X}} + c\mathbf{I})^{-1}\overset{*}{\mathbf{X}}{}'\overset{*}{\mathbf{y}}$$

Here, we use a *biasing constant* $c \geqslant 0$. Then the ridge point estimates of the parameters $\beta_0, \beta_1, \ldots, \beta_p$ in the original regression model are

$$b_{j,R} = \left(\frac{s_y}{s_{x_j}}\right) b'_{j,R} \qquad j = 1, \ldots, p$$

$$b_{0,R} = \bar{y} - b_{1,R}\bar{x}_1 - b_{2,R}\bar{x}_2 - \cdots - b_{p,R}\bar{x}_p$$

To understand the biasing constant c, first note that if $c = 0$, then the ridge point estimates are the least squares point estimates. Recall that the least squares estimation procedure is unbiased. That is, $\mu_{b_j} = \beta_j$. If $c > 0$, the ridge estimation procedure is not unbiased. That is, $\mu_{b_{j,R}} \neq \beta_j$ if $c > 0$. We define the *bias* of the ridge estimation procedure to be $\{\mu_{b_{j,R}} - \beta_j\}$. To compare a biased estimation procedure with an unbiased estimation procedure, we employ *mean squared errors.* The mean squared error of an estimation procedure is defined to be the average of the squared deviations of the different possible point estimates from the unknown parameter. This can be proven to be equal to the sum of the *squared bias* of the procedure and the *variance* of the procedure. Here, the variance is the average of the squared deviations of the different possible point estimates from the mean of all possible point estimates. If the procedure is unbiased, the mean of all possible point estimates is the parameter we are estimating. In other words, when the bias is zero, the mean squared error and the variance of the procedure are the same.

Therefore the mean squared error of the (unbiased) least squares estimation procedure for estimating β_j is the variance $\sigma^2_{b_j}$. The mean squared error of the ridge estimation procedure is

$$[\mu_{b_{j,R}} - \beta_j]^2 + \sigma^2_{b_{j,R}}$$

It can be proved that as the biasing constant c increases from zero, the bias of the ridge estimation procedure increases, and the variance of this procedure decreases. It can further be proved that there is some $c > 0$ that makes $\sigma^2_{b_{j,R}}$ so much smaller than $\sigma^2_{b_j}$ that the mean squared error of the ridge estimation procedure is smaller than the mean squared error of the least squares estimation procedure. This is one advantage of ridge estimation. It implies that the ridge point estimates are less affected by multicollinearity than the least squares point estimates. Therefore, for example, they are less affected by small changes in the data. One problem is that the optimum value of c differs for different applications and is unknown.

One way to choose c is to calculate ridge point estimates for different values of c. We usually choose values between 0 and 1. Experience indicates that the ridge point estimates may fluctuate wildly as c is increased slightly from zero. The estimates may even change sign. Eventually, the values of the ridge point estimates begin to change slowly. It is reasonable to choose c to be the smallest value where all of the ridge point estimates begin to change slowly. Here, making a *ridge trace* can be useful. This is a simultaneous plot of the values of all of the ridge point estimates against values of c. Another way to choose c is to note that variance inflation factors related to the ridge point estimates of the parameters in the standardized regression model are the diagonal elements of the matrix

$$(\mathbf{\mathring{X}}'\mathbf{\mathring{X}} + c\mathbf{I})^{-1}\,\mathbf{\mathring{X}}'\mathbf{\mathring{X}}(\mathbf{\mathring{X}}'\mathbf{\mathring{X}} + c\mathbf{I})^{-1}$$

As c increases from zero, the variance inflation factors initially decrease quickly and then begin to change slowly. Therefore we might choose c to be a value where the variance inflation factors are sufficiently small. A related way to choose c is to consider the *trace* (the sum of the diagonal elements) of the matrix

$$\mathbf{H}_c = \mathbf{\mathring{X}}(\mathbf{\mathring{X}}'\mathbf{\mathring{X}} + c\mathbf{I})^{-1}\mathbf{\mathring{X}}'$$

It can be shown that as c increases from zero, this trace, denoted tr(H_c), initially decreases quickly and then begins to decrease slowly. We might choose c to be the smallest value where tr(H_c) begins to decrease slowly. This is because at this value the multicollinearity in the data begins to have a sufficiently small impact on the ridge point estimates.

One disadvantage of ridge regression is that the choice of c is somewhat subjective. Furthermore, the different ways to choose c often contradict each other. We have discussed only three such methods. Myers (1986) gives an excellent discussion of other methods for choosing c. Another major problem with ridge regression is that the exact probability distribution of all possible values of a ridge point estimate is unknown. This means that we cannot (easily) perform statistical inference. Ridge regression is very controversial. Our view is that before using ridge regression one should use the various model-building techniques of this book to eliminate severe multicollinearity by identifying redundant independent variables.

Example 10.6 If we use the hospital labor needs data in Table 10.5, the ridge point estimates of the parameters in the model

$$y = \beta_0 + \beta_1 x_1 + \beta_2 x_2 + \beta_3 x_3 + \beta_4 x_4 + \beta_5 x_5 + \varepsilon$$

are as given in Table 10.6. Here, we have ranged c from 0.00 to 0.20 and also include the values of tr(H_c). Noting the changes in sign in the ridge point estimates, it is certainly not easy to determine the value of c at which they begin to change slowly. We might arbitrarily choose $c = .16$. In contrast, the values of tr(H_c) seem to begin to change slowly at $c = .01$. If we do a finer search by ranging c in increments of .0001 from .0000 to .0010, the values of tr(H_c) begin to change slowly at $c = .0004$. The corresponding ridge point estimates can be calculated to be

$$b_{0,R} = 2053.33 \qquad b_{1,R} = 12.5411 \qquad b_{2,R} = .0565$$

$$b_{3,R} = .6849 \qquad b_{4,R} = -5.4249 \qquad b_{5,R} = -416.09$$

Experience indicates that various criteria for choosing c tend to differ when the data set has one or more observations that are considerably different from the others. Note from Table 10.5 that hospitals 15, 16, and 17 (and possibly 14) are considerably larger than hospitals 1 through 13. In Section 10.4 we carefully consider evaluating such outlying observations. At any rate, before using the results of ridge regression we should attempt to identify redundant independent variables. We reasoned in Example 10.4 that x_1, x_3, and x_4 are quite related. We will further find in Chapter 11 that one of the best models describing the hospital labor needs data is

$$y = \beta_0 + \beta_1 x_2 + \beta_3 x_3 + \beta_4 x_5 + \varepsilon$$

TABLE 10.6 **The ridge point estimates and values of tr(H_c) for the hospital labor needs model**

c	$b_{0,R}$	$b_{1,R}$	$b_{2,R}$	$b_{3,R}$	$b_{4,R}$	$b_{5,R}$	tr(H_c)
0.00	1962.95	−15.8517	0.0559	1.5896	−4.2187	−394.31	5.0000
0.01	1515.07	14.5765	0.0600	0.5104	−2.1732	−312.71	3.6955
0.02	1122.83	13.5101	0.0621	0.4664	0.2488	−236.20	3.4650
0.03	839.55	12.7104	0.0634	0.4358	1.9882	−180.25	3.2867
0.04	624.89	12.0993	0.0643	0.4130	3.2949	−137.25	3.1427
0.05	456.27	11.6180	0.0648	0.3951	4.3098	−102.94	3.0227
0.06	320.08	11.2286	0.0652	0.3808	5.1188	−74.75	2.9206
0.07	207.65	10.9066	0.0653	0.3690	5.7768	−51.05	2.8320
0.08	113.17	10.6353	0.0654	0.3591	6.3209	−30.75	2.7541
0.09	32.61	10.4031	0.0654	0.3507	6.7768	−13.07	2.6848
0.10	−36.91	10.2016	0.0654	0.3434	7.1632	2.50	2.6225
0.11	−97.52	10.0247	0.0653	0.3370	7.4937	16.39	2.5661
0.12	−150.81	9.8679	0.0652	0.3313	7.7787	28.88	2.5145
0.13	−198.00	9.7276	0.0651	0.3262	8.0261	40.21	2.4671
0.14	−240.04	9.6010	0.0649	0.3216	8.2422	50.56	2.4233
0.15	−277.70	9.4860	0.0648	0.3175	8.4319	60.07	2.3827
0.16	−311.58	9.3808	0.0646	0.3137	8.5990	68.85	2.3447
0.17	−342.18	9.2841	0.0644	0.3103	8.7469	77.00	2.3092
0.18	−369.91	9.1948	0.0642	0.3071	8.8782	84.59	2.2758
0.19	−395.10	9.1118	0.0640	0.3041	8.9950	91.69	2.2443
0.20	−418.03	9.0343	0.0638	0.3013	9.0992	98.35	2.2146

Source: Adapted from Raymond Myers, *Classical and Modern Regression with Applications* (Boston: PWS–KENT Publishing Company, 1986). See Tables 7.9 and 7.10.

This model uses only one of x_1, x_3, and x_4 and thus eliminates much multicollinearity. However, since we found in Example 10.4 that fairly strong multicollinearity still exists in this model, we could again use ridge regression.

10.4

DIAGNOSTIC STATISTICS FOR IDENTIFYING OUTLYING AND INFLUENTIAL OBSERVATIONS

In this section we present more sophisticated methods for identifying outliers with respect to their x or y values and for determining whether such outliers are influential. To identify outliers with respect to their x values, we can use what we call *leverage values.*

The Leverage Value h_{ii}

Consider the regression model

$$y_i = \beta_0 + \beta_1 x_{i1} + \beta_2 x_{i2} + \cdots + \beta_p x_{ip} + \varepsilon_i$$

which we assume has k parameters, and define the *hat matrix*

$$\mathbf{H} = \mathbf{X}(\mathbf{X}'\mathbf{X})^{-1}\mathbf{X}'$$

which has n rows and n columns. For $i = 1, 2, \ldots, n$ we define the *leverage value h_{ii} of the x values $x_{i1}, x_{i2}, \ldots, x_{ip}$* to be the ith diagonal element of $\mathbf{H}$. It can be shown that

$$h_{ii} = \mathbf{x}_i'(\mathbf{X}'\mathbf{X})^{-1}\mathbf{x}_i$$

where

$$\mathbf{x}_i' = [1 \ x_{i1} \ x_{i2} \ \ldots \ x_{ip}]$$

is a row vector containing the values of the independent variables in the ith observation.

It can be shown that

$$0 \leqslant h_{ii} \leqslant 1 \qquad \text{and} \qquad \sum_{i=1}^{n} h_{ii} = k$$

The leverage value h_{ii} indicates whether the x values $x_{i1}, x_{i2}, \ldots, x_{ip}$ are outlying because it can be shown that h_{ii} is a measure of the distance between these x values and the means

$$\bar{x}_1 = \frac{\sum_{i=1}^{n} x_{i1}}{n}, \ \bar{x}_2 = \frac{\sum_{i=1}^{n} x_{i2}}{n}, \ \ldots, \ \bar{x}_p = \frac{\sum_{i=1}^{n} x_{ip}}{n}$$

of the x values that have occurred in all n observations. If the leverage value h_{ii} is large, the ith observation is outlying with respect to its x values. Thus the ith observation will exert substantial leverage in determining the values of the least squares point estimates. To see why this is true, note that since observations 2 and 3 in Figure 6.22 are outlying with respect to their x values, they will exert substantial leverage in determining the values of the least squares point estimates. *A leverage value h_{ii} is generally considered to be large if it is substantially greater than most of the other leverage values or if it is greater than twice the average leverage value.* That is, h_{ii} is considered large if it is greater than

$$2\bar{h} = 2\frac{\sum_{i=1}^{n} h_{ii}}{n} = 2\frac{k}{n}$$

It should be noted, however, that an observation with a large leverage value is not necessarily influential. For example, both observation 2 and observation 3 in Figure 6.22 would have large leverage values because both observations are outliers with respect to their x values. However, as was previously stated, observation 3 is probably very influential because its y value is not consistent with the regression relationship displayed by the other observations. On the other hand, observation 2 is probably not influential because its y value is consistent with the regression relationship displayed by the nonoutlying observations.

We have discussed how to determine whether the ith observation is an outlier with respect to its y value by using the ith residual, $e_i = y_i - \hat{y}_i$, and standardized ith residual, e_i/s. We now present two modifications that will enable us to better identify outliers with respect to their y values. The first modification involves using:

The Studentized Residual

The *studentized residual* is

$$\frac{e_i}{s\sqrt{1 - h_{ii}}}$$

where $s\sqrt{1 - h_{ii}}$ is a point estimate of $\sigma_{e_i} = \sigma\sqrt{1 - h_{ii}}$.

It can be proved that σ_{e_i} is the standard deviation of the population of all possible values of the ith residual. The advantage of dividing e_i by $s\sqrt{1 - h_{ii}}$ is that doing so guarantees that the n populations of all possible studentized residuals have equal variances. For example, the population of all possible values of $e_1/s\sqrt{1 - h_{11}}$ has the same variance as the population of all possible values of $e_2/s\sqrt{1 - h_{22}}$. Therefore dividing e_i by $s\sqrt{1 - h_{ii}}$ implies that the n studentized residuals are of the same relative magnitude.

The second modification is to calculate:

The Deleted (or PRESS) Residual

The *deleted* (or *PRESS*) *residual* is defined to be

$$d_i = y_i - \hat{y}_{(i)}$$

where

$$\hat{y}_{(i)} = b_0^{(i)} + b_1^{(i)}x_{i1} + b_2^{(i)}x_{i2} + \cdots + b_p^{(i)}x_{ip}$$

is the point prediction of y_i calculated by using least squares point estimates $b_0^{(i)}$, $b_1^{(i)}, b_2^{(i)}, \ldots, b_p^{(i)}$ which are calculated by using all n observations except for the ith observation.

Note that we do not use the ith observation in computing $b_0^{(i)}, b_1^{(i)}, b_2^{(i)}, \ldots, b_p^{(i)}$ for the following reason. Even if the ith observation is an outlier with respect to its y value, using the ith observation to calculate $\hat{y}_i$ might draw $\hat{y}_i$ toward y_i and thus cause $e_i = y_i - \hat{y}_i$ to be small. This would falsely imply that the ith observation is not an outlier with respect to its y value. Next, let s_{d_i} denote the point estimate of σ_{d_i}, the standard deviation of the population of all possible values of d_i, and define:

The Studentized Deleted Residual

1. The *studentized deleted residual* is defined to be d_i/s_{d_i}, where it can be shown that

$$\frac{d_i}{s_{d_i}} = e_i\left[\frac{n-k-1}{SSE(1-h_{ii})-e_i^2}\right]^{1/2}$$

Here, $e_i = y_i - \hat{y}_i$, and the population of all possible values of d_i/s_{d_i} has a t-distribution with $n-k-1$ degrees of freedom.

2. If the absolute value of the studentized deleted residual d_i/s_{d_i} is greater than $t_{[.025]}^{(n-k-1)}$, this residual is considered to be large, and the ith observation is considered to be an outlier with respect to its y value.

We next define a statistic evaluating the difference between the point predictions of y_i made with and without using the ith observation.

The Difference in Fits Statistic

1. Let

$$f_i = \hat{y}_i - \hat{y}_{(i)}$$

If s_{f_i} denotes the standard error of this difference, then the *difference in fits statistic* is defined to be f_i/s_{f_i}. It can be shown that

$$\frac{f_i}{s_{f_i}} = \left[\frac{d_i}{s_{d_i}}\right]\left[\frac{h_{ii}}{1-h_{ii}}\right]^{1/2}$$

2. If the absolute value of f_i/s_{f_i} is greater than 2 (a commonly used critical value for this statistic), then removing the ith observation from the data set would substantially change the point prediction of y_i. For large data sets it might be appropriate to use critical values smaller than 2.*

If we have concluded that the ith observation is an outlier with respect to its x or y value, we can determine whether the ith observation is influential by calculating:

Cook's Distance Measure

Cook's distance measure is defined to be

$$D_i = \frac{(\mathbf{b} - \mathbf{b}^{(i)})' \mathbf{X}'\mathbf{X}(\mathbf{b} - \mathbf{b}^{(i)})}{ks^2}$$

*See Belsley, Kuh, and Welsch (1980) for an excellent discussion of choosing critical values related to this and the other statistics discussed in this section.

where $s^2 = SSE/(n - k)$ and

$$\mathbf{b} - \mathbf{b}^{(i)} = \begin{bmatrix} b_0 \\ b_1 \\ b_2 \\ \vdots \\ b_p \end{bmatrix} - \begin{bmatrix} b_0^{(i)} \\ b_1^{(i)} \\ b_2^{(i)} \\ \vdots \\ b_p^{(i)} \end{bmatrix} = \begin{bmatrix} b_0 - b_0^{(i)} \\ b_1 - b_1^{(i)} \\ b_2 - b_2^{(i)} \\ \vdots \\ b_p - b_p^{(i)} \end{bmatrix}$$

If D_i is large, the least squares point estimates $b_0, b_1, b_2, \ldots, b_p$ calculated by using all n observations differ substantially from the least squares point estimates $b_0^{(i)}, b_1^{(i)}, b_2^{(i)}, \ldots, b_p^{(i)}$ calculated by using all n observations except for the ith observation. Thus the ith observation is influential.

Although the population of all possible values of D_i does not have an F-distribution, practice has shown that

1. If D_i is less than $F_{[.80]}^{(k,n-k)}$ (the 20th percentile of the F-distribution having k and $n - k$ degrees of freedom), then the ith observation should not be considered influential.
2. If D_i is greater than $F_{[.50]}^{(k,n-k)}$ (the 50th percentile of the F-distribution having k and $n - k$ degrees of freedom), then the ith observation should be considered influential.
3. If $F_{[.80]}^{(k,n-k)} \leq D_i \leq F_{[.50]}^{(k,n-k)}$, then the nearer D_i is to $F_{[.50]}^{(k,n-k)}$, the greater the extent of the influence of the ith observation.

Finally, note that it can be shown that D_i can be calculated by the equation

$$D_i = \frac{e_i^2}{ks^2}\left[\frac{h_{ii}}{(1 - h_{ii})^2}\right]$$

If we determine that removing the ith observation from the data set would substantially change (as a group) the least squares point estimates, we might wish to determine whether the point estimate of a particular parameter β_j would change substantially.

The Difference in Estimate of β_j Statistic

1. Let

$$g_j^{(i)} = b_j - b_j^{(i)}$$

If $s_{g_j^{(i)}}$ denotes the standard error of this difference, then the *difference in estimate of the β_j statistic* is defined to be $g_j^{(i)}/s_{g_j^{(i)}}$. It can be shown that

$$\frac{g_j^{(i)}}{s_{g_j^{(i)}}} = \left[\frac{d_i}{s_{d_i}}\right]\left[\frac{r_{j,i}}{\sqrt{(\mathbf{r}_j'\mathbf{r}_j)(1 - h_{ii})}}\right]$$

Here, $r_{j,i}$ is the element in row j and column i of $\mathbf{R} = (\mathbf{X}'\mathbf{X})^{-1}\mathbf{X}'$, and $\mathbf{r}_j'$ is row j of $\mathbf{R}$.

2. If the absolute value of $g_j^{(i)}/s_{g_j^{(i)}}$ is greater than 2 (a commonly used critical value for this statistic), then removing the ith observation from the data set would substantially change the point estimate of β_j. For large data sets it might be appropriate to use critical values smaller than 2.

Once we have identified influential outlying observations, we must decide what to do about them. If an influential outlier has been caused by incorrect measurement (perhaps resulting from a faulty instrument) or erroneous recording (for example, an incorrect decimal point), the observation should be corrected (if it can be corrected), and the regression analysis should be rerun. If the observation cannot be corrected. then it should probably be dropped from the data set. If the influential outlying observation is accurate, it is possible that the regression model is inadequate. For example, it might not contain an important independent variable that would explain this observation, or it might not have the correct functional form. These possibilities should be investigated, and if need be, the model should be improved. Finally, if no explanation can be found, it might be appropriate to drop this observation from the data set. As an alternative, instead of calculating the least squares point estimates, we could dampen the effect of the influential outlier by calculating point estimates that minimize the sum of the *absolute values* of the residuals

$$\sum_{i=1}^{n} |e_i| = \sum_{i=1}^{n} |y_i - \hat{y}_i| = \sum_{i=1}^{n} |y_i - (b_0 + b_1 x_{i1} + \cdots + b_p x_{ip})|$$

The reader is referred to Kennedy and Gentle (1980) for a discussion of the computational aspects of such a minimization. Also, note that minimizing the sum of absolute residuals is only one of a variety of *robust regression* procedures. These procedures are intended to yield point estimates that are less sensitive than the least squares point estimates to both outlying observations and failures of the model assumptions. For example, if the populations sampled are not normal but are *heavy tailed*, then we are more likely to obtain a y_i value that is far from μ_i. This value will act much like an outlier, and its effect can be dampened by minimizing the sum of absolute residuals. An excellent discussion of robust regression procedures is given by Myers (1986).

Example 10.7 Home Real Estate Company has used the data in Figure 10.6 to develop the model

$$y_i = \beta_0 + \beta_1 x_{i1} + \beta_2 x_{i2} + \beta_3 x_{i2}^2 + \varepsilon_i$$

relating y_i (the selling price of a house, in thousands of dollars) to x_{i1} (the size of the house, in thousands of square feet) and x_{i2} (the age of the house, in years).

FIGURE 10.6 **SAS output of the Home Real Estate Company data and of diagnostics for detecting outlying and influential observations for the model $y_i = \beta_0 + \beta_1 x_{i1} + \beta_2 x_{i2} + \beta_3 x_{i2}^2 + \varepsilon_i$**

OBS	Y	X1	X2	HAT DIAG H	RESIDUAL	RSTUDENT	DFFITS
1	68.7	2.05	3.43	0.2728	-1.63733	-0.6669	-0.4085
2	54.9	1.70	11.61	0.1603	0.716355	0.2671	0.1167
3	51.5	1.47	8.31	0.1980	0.226113	0.0860	0.0427
4	71.6	1.75	0.00	0.1781	3.85264	1.6082	0.7487
5	58.4	1.94	7.41	0.2040	-4.96169	-2.3066	-1.1676
6	40.7	1.19	31.70	0.9587[a]	-1.19347[b]	-2.5079[c]	-12.0874[d]
7	51.7	1.56	16.10	0.2042	3.23091	1.3285	0.6730
8	71.9	1.95	2.05	0.1879	2.2038	0.8603	0.4136
9	57.1	1.60	1.74	0.1449	-4.62767	-1.9862	-0.8175
10	58.3	1.49	2.76	0.1982	0.535255	0.2039	0.1014
11	73.5	1.91	0.00	0.2307	1.92183	0.7657	0.4193
12	58.5	1.38	0.00	0.4231	-.388631	-0.1745	-0.1494
13	49.1	1.55	12.61	0.1829	-.849241	-0.3214	-0.1520
14	67.5	1.88	2.80	0.1286	0.448381	0.1638	0.0629
15	53.7	1.60	7.08	0.1111	-1.84092	-0.6786	-0.2399
16	50.0	1.55	18.00	0.2166	2.36366	0.9456	0.4972

OBS	COOK'S D	DFBETAS INTERCEP	DFBETAS X1	DFBETAS X2	DFBETAS X2SQ	STUDENT RESIDUAL
1	0.044	0.3069	-0.3432	0.0314	-0.0858	-0.683
2	0.004	-0.0235	0.0197	0.0896	-0.0758	0.278
3	0.000	0.0288	-0.0289	0.0184	-0.0264	0.090
4	0.124	0.1118	0.0190	-0.5538	0.4262	1.511
5	0.251	0.8156	-0.8515	-0.4547	0.2591	-1.978
6	25.351[e]	-0.0030[f]	-0.2241[g]	3.9352[h]	-7.4644[i]	-2.089[j]
7	0.108	-0.0014	-0.0392	0.5225	-0.4174	1.288
8	0.044	-0.2252	0.2797	-0.1347	0.1532	0.870
9	0.134	-0.5151	0.4020	0.3884	-0.1503	-1.780
10	0.003	0.0847	-0.0755	-0.0256	-0.0065	0.213
11	0.046	-0.1330	0.2003	-0.2658	0.2539	0.779
12	0.006	-0.1289	0.1137	0.0758	-0.0283	-0.182
13	0.006	-0.0350	0.0420	-0.1140	0.1111	-0.334
14	0.001	-0.0269	0.0354	-0.0160	0.0162	0.171
15	0.015	-0.1069	0.1005	-0.1040	0.1415	-0.694
16	0.062	-0.0407	0.0100	0.3627	-0.2539	0.950

[a] h_{66} [b] e_6 [c] d_6/s_{d_6} [d] f_6/s_{f_6} [e] D_6 [f] $g_0^{(6)}/s_{g_0^{(6)}}$ [g] $g_1^{(6)}/s_{g_1^{(6)}}$ [h] $g_2^{(6)}/s_{g_2^{(6)}}$ [i] $g_3^{(6)}/s_{g_3^{(6)}}$
[j] $e_6/s\sqrt{1-h_{66}}$

Examining the column labeled HAT DIAG H in Figure 10.6, we see that the leverage value $h_{66} = .9587$ is greater than $2(k/n) = 2(4/16) = .5$. This says that observation 6 is an outlier with respect to its x values. Intuitively, this is because house 6 is somewhat smaller and much older than the other houses. Examining the column labeled RSTUDENT, we see that the studentized deleted residual $d_6/s_{d_6} = -2.5079$ is (in absolute value) greater than $t_{[.025]}^{(n-k-1)} = 2.201$. This says that observation 6 is an outlier with respect to its y value. Next, we let $f_6 = \hat{y}_6 - \hat{y}_{(6)}$ denote the difference between the point predictions of y_6 made with and without using observation 6. Then, examining the column labeled DFFITS, we see that the absolute value of $f_6/s_{f_6} = -12.0874$ is greater than 2. This says that removing observation 6 from the data set would substantially change the point prediction of y_6. Examining the column labeled COOK'S D, we see that $D_6 = 25.351$ is greater than $F_{[.05]}^{(k,n-k)} = 5.41$ (which is itself greater than $F_{[.50]}^{(k,n-k)}$). This implies that removing observation 6 from the data set would substantially change (as a group) the least squares point estimates of the parameters β_0, β_1, β_2, and β_3. To determine whether the least squares point estimate of a particular parameter β_j would change substantially, we let $g_j^{(6)} = b_j - b_j^{(6)}$ denote the difference between the point estimates of β_j calculated with and without using observation 6, and we calculate $g_j^{(6)}/s_{g_j^{(6)}}$. Specifically, examining the columns labeled DFBETAS, we see that the absolute values of $g_0^{(6)}/s_{g_0^{(6)}} = -.0030$ and $g_1^{(6)}/s_{g_1^{(6)}} = -.2241$ are less than 2. This says that the least squares point estimates of β_0 and β_1 probably would not change substantially. The fact that the absolute values of $g_2^{(6)}/s_{g_2^{(6)}} = 3.9352$ and $g_3^{(6)}/s_{g_3^{(6)}} = -7.4644$ are greater than 2 indicates that the least squares point estimates of β_2 and β_3 probably would change substantially. The functional form of the above model can be verified to be correct. Therefore although we will lose information concerning smaller and older homes, we will remove outlying, influential observation 6 from the data set. It can be verified that doing this (1) changes the least squares point estimates of β_0, β_1, β_2, and β_3 from 25.8480, 23.9424, -1.4538, and .0335 to 25.8650, 24.6601, -2.2297, and .0847; (2) reduces the standard error for the above model from 2.8117 to 2.3425; and (3) changes the point prediction of and 95% prediction interval for the selling price of a 1700-square-foot, five-year-old house from 60.118 and [53.762, 66.474] to 58.756 and [53.274, 64.237]. Thus removing the influential outlier yields a shorter prediction interval.

A large value of $g_j^{(i)}/s_{g_j^{(i)}}$ implies that the ith observation significantly influences the value of the least squares point estimate b_j. However, a large value of $g_j^{(i)}/s_{g_j^{(i)}}$ does not shed any light on the effect that the ith observation has on $s\sqrt{c_{jj}}$. Note here that $s\sqrt{c_{jj}}$ is involved in the error bound of the $100(1-\alpha)\%$ confidence interval for β_j:

$$[b_j \pm t_{[\alpha/2]}^{(n-k)} s\sqrt{c_{jj}}]$$

Thus $s\sqrt{c_{jj}}$ is a measure of how precisely β_j is estimated. A statistic that is an overall measure of how the ith observation affects the precision of the least squares point estimates can be defined as follows. Notice that we use the notation "| |" to denote the *determinant** of the indicated matrix.

*See any matrix algebra text for a definition of the determinant. For our purposes it suffices to know that the determinant of a matrix is a single number that is a function of the elements of the matrix.

The Covariance Ratio

The *covariance ratio corresponding to the ith observation* is

$$CVR_i = \frac{|(\mathbf{X}'_{-i}\mathbf{X}_{-i})^{-1}s^2_{-i}|}{|(\mathbf{X}'\mathbf{X})^{-1}s^2|}$$

Here, $\mathbf{X}_{-i}$ and s^2_{-i} are the matrix of independent variable values (including the usual column of 1's) and the mean square error resulting from eliminating the ith observation.

Noting that it can be proven that

$$CVR_i = \frac{(s_{-i})^{2k}}{s^{2k}}\left(\frac{1}{1 - h_{ii}}\right)$$

Belsley, Kuh, and Welsch (1980) suggest that:

1. If

   $$CVR_i > 1 + 3k/n$$

   then eliminating the ith observation significantly damages the precision of at least some of the least squares point estimates.
2. If

   $$CVR_i < 1 - 3k/n$$

   then eliminating the ith observation significantly enhances the precision of at least some of the least squares point estimates.

Example 10.8 Reconsider the hospital labor needs problem. Figures 10.7 and 10.8 present SAS outputs of diagnostics for detecting outlying and influential observations that result from using

$$\text{Model 1: } y_i = \beta_0 + \beta_1 x_{i3} + \beta_2 x_{i5} + \varepsilon_i$$

and

$$\text{Model 2: } y_i = \beta_0 + \beta_1 x_{i2} + \beta_2 x_{i3} + \beta_3 x_{i5} + \varepsilon_i$$

to perform a regression analysis of the $n = 17$ observations in Table 10.5. Note that each of these models uses only one of the independent variables x_1, x_3, and x_4. This is reasonable, since we saw in Example 10.4 that these three variables contribute very redundant information. Furthermore, we will find in Chapter 11 that Models 1 and 2 are two of the best models describing the data in Table 10.5. The most significant diagnostics are labeled in Figures 10.7 and 10.8. We see that

$$\frac{d_{14}}{s_{d_{14}}} = \frac{y_{14} - \hat{y}_{(14)}}{s_{d_{14}}}$$

FIGURE 10.7 **SAS output of diagnostics for detecting outlying and influential observations for Model 1: $y_i = \beta_0 + \beta_1 x_{i3} + \beta_2 x_{i5} + \varepsilon_i$ in the hospital labor needs problem**

OBS	HAT DIAG H	RESIDUAL	RSTUDENT	COV RATIO	DFFITS
1	0.1155	-239.2	-0.3353	1.3761	-0.1212
2	0.2165	134.2	0.1994	1.5795	0.1048
3	0.1249	-44.3888	-0.0623	1.4259	-0.0235
4	0.1581	388.3	0.5623	1.3803	0.2436
5	0.0848	100.3	0.1378	1.3588	0.0419
6	0.1014	-213.2	-0.2963	1.3621	-0.0995
7	0.0841	173.4	0.2384	1.3459	0.0722
8	0.0805	288.3	0.3971	1.3101	0.1175
9	0.0838	-538.7	-0.7549	1.1986	-0.2283
10	0.0692	-321.0	-0.4400	1.2836	-0.1200
11	0.0700	519.4	0.7212	1.1939	0.1979
12	0.0957	-1085.5	-1.6478	0.7819	-0.5361
13	0.0629	-403.0	-0.5530	1.2430	-0.1432
14	0.0932	2004.7[d]	4.3040[e]	0.0966[f]	1.3796
15	0.6763[a]	-469.1	-1.1347	2.9061	-1.6399
16	0.3379[b]	-802.7	-1.3879	1.2463	-0.9916
17	0.5453[c]	508.2	1.0289	2.1719	1.1268

OBS	COOK'S D	INTERCEP DFBETAS	BEDDAY DFBETAS	STAYDAY DFBETAS
1	0.005	-0.0758	0.0247	0.0437
2	0.004	-0.0585	-0.0817	0.0818
3	0.000	-0.0168	0.0023	0.0110
4	0.021	0.2006	0.0098	-0.1495
5	0.001	0.0049	-0.0214	0.0077
6	0.004	-0.0663	0.0091	0.0412
7	0.002	0.0039	-0.0381	0.0176
8	0.005	0.0454	-0.0369	-0.0113
9	0.018	0.0460	0.1195	-0.1066
10	0.005	0.0119	0.0426	-0.0423
11	0.014	0.0029	-0.0792	0.0520
12	0.085	-0.4166	-0.1843	0.3291
13	0.007	-0.0644	-0.0080	0.0317
14	0.282	-0.2895	0.2806	0.3968
15	0.878	1.3401	0.3004	-1.3417
16	0.307	-0.2386	-0.8456	0.3367
17	0.421	0.5028	1.0586	-0.6290

[a] $h_{15,15}$ [b] $h_{16,16}$ [c] $h_{17,17}$ [d] e_{14} [e] $d_{14}/s_{d_{14}}$ [f] CVR_{14}

is very large (greater than $t_{[.025]}^{(n-k-1)}$) for both Model 1 ($d_{14}/s_{d_{14}} = 4.3040$) and Model 2 ($d_{14}/s_{d_{14}} = 4.5584$). This indicates that observation 14 is an outlier with respect to its y value. Specifically, the residuals corresponding to observation 14 in both figures indicate that both models greatly underpredict y_{14}. This indicates that the number of labor hours utilized in hospital 14 was much greater than would be expected for its values of x_2, x_3, and x_5. Moreover, CVR_{14} is very small (less than

FIGURE 10.8 **SAS output of diagnostics for detecting outlying and influential observations for Model 2: $y_i = \beta_0 + \beta_1 x_{i2} + \beta_2 x_{i3} + \beta_3 x_{i5} + \varepsilon_i$ in the hospital labor needs problem**

OBS	HAT DIAG H	RESIDUAL	RSTUDENT	COV RATIO	DFFITS
1	0.1207	-121.9	-0.2035	1.5451	-0.0754
2	0.2261	-25.0283	-0.0445	1.7787	-0.0240
3	0.1297	67.7570	0.1136	1.5758	0.0438
4	0.1588	431.2	0.7517	1.3620	0.3266
5	0.0849	84.5898	0.1383	1.4956	0.0421
6	0.1120	-380.6	-0.6419	1.3551	-0.2280
7	0.0841	177.6	0.2911	1.4621	0.0882
8	0.0830	369.1	0.6118	1.3284	0.1841
9	0.0846	-493.2	-0.8283	1.2046	-0.2518
10	0.1203	-687.4	-1.2136	0.9853	-0.4487
11	0.0773	380.9	0.6299	1.3107	0.1824
12	0.1771	-623.1	-1.1290	1.1177	-0.5237
13	0.0645	-337.7	-0.5526	1.3316	-0.1451
14	0.1465	1630.5[d]	4.5584[e]	0.0290[f]	1.8882
15	0.6818[a]	-348.7	-1.0059	3.1309	-1.4723
16	0.7855[b]	281.9	0.9892	4.6924	1.8930
17	0.8632[c]	-406.0	-1.9751	3.2670	-4.9523

OBS	COOK'S D	INTERCEP DFBETAS	XRAY DFBETAS	BEDDAY DFBETAS	STAYDAY DFBETAS
1	0.002	-0.0477	0.0157	-0.0083	0.0309
2	0.000	0.0138	-0.0050	0.0119	-0.0183
3	0.001	0.0307	-0.0084	0.0060	-0.0216
4	0.028	0.2416	-0.0217	0.0251	-0.1821
5	0.000	0.0035	0.0014	-0.0099	0.0074
6	0.014	-0.0881	-0.0703	0.0724	0.0401
7	0.002	0.0045	-0.0008	-0.0180	0.0179
8	0.009	0.0764	-0.0319	0.0063	-0.0314
9	0.016	0.0309	0.0243	0.0304	-0.0873
10	0.049	0.1787	-0.2924	0.3163	-0.2544
11	0.009	-0.0265	0.0560	-0.0792	0.0680
12	0.067	-0.4387	0.3549	-0.3782	0.3864
13	0.006	-0.0671	0.0230	-0.0243	0.0390
14	0.353	-0.8544	1.1389	-0.9198	0.9620
15	0.541	0.9616	0.1324	-0.0133	-0.9561
16	0.897	0.9880	-1.4289	1.7339	-1.1029
17	5.033[g]	0.0294	-3.0114[h]	1.2688	0.3155

[a] $h_{15,15}$ [b] $h_{16,16}$ [c] $h_{17,17}$ [d] e_{14} [e] $d_{14}/s_{d_{14}}$ [f] CVR_{14} [g] D_{16} [h] $g_{(1)}^{(14)}/s_{g_{(1)}^{(14)}}$

$1 - 3k/n$) for Model 1 ($CVR_{14} = .0966$) and Model 2 ($CVR_{14} = .0290$). This indicates that removing observation 14 from the data set significantly enhances the precision of at least some of the least squares point estimates of the parameters in Models 1 and 2. If the value $y_{14} = 10334$ was recorded erroneously or resulted from a situation that would be very unlikely to occur again, it would be reasonable to remove observation 14. If we do this and perform a regression analysis by using

FIGURE 10.9 **SAS output of diagnostics for detecting outlying and influential observations for Model 2: $y_i = \beta_0 + \beta_1 x_{i2} + \beta_2 x_{i3} + \beta_3 x_{i5} + \varepsilon_i$ when observation 14 is removed from the hospital labor needs data set**

OBS	HAT DIAG H	RESIDUAL	RSTUDENT	COV RATIO	DFFITS
1	0.1208	-125.6	-0.3330	1.5475	-0.1234
2	0.2351	141.7	0.4036	1.7458	0.2237
3	0.1297	60.5547	0.1607	1.6121	0.0620
4	0.1588	428.8	1.2336	1.0027	0.5359
5	0.0869	162.9	0.4249	1.4533	0.1311
6	0.1144	-294.3	-0.7953	1.2788	-0.2858
7	0.0861	256.3	0.6766	1.3164	0.2076
8	0.0835	409.8	1.1171	1.0055	0.3373
9	0.0876	-396.1	-1.0783	1.0386	-0.3342
10	0.1350	-473.0	-1.3591	0.8801	-0.5370
11	0.0833	517.7	1.4612	0.7600	0.4406
12	0.1780	-677.2	-2.2241	0.3901	-1.0350
13	0.0663	-262.2	-0.6851	1.2834	-0.1826
14	.	.	.	.	.
15	0.7144	-29.6792	-0.1375	4.9257	-0.2174
16	0.7868	219.0	1.2537	3.8927	2.4081
17	0.9334	61.2977	0.5966	18.7100[a]	2.2328

OBS	COOK'S D	INTERCEP DFBETAS	XRAY DFBETAS	BEDDAY DFBETAS	STAYDAY DFBETAS
1	0.004	-0.0768	0.0251	-0.0135	0.0495
2	0.013	-0.1320	0.0545	-0.1152	0.1722
3	0.001	0.0428	-0.0117	0.0085	-0.0300
4	0.069	0.3900	-0.0349	0.0407	-0.2927
5	0.005	0.0069	0.0090	-0.0338	0.0263
6	0.021	-0.0998	-0.0945	0.0962	0.0401
7	0.011	0.0045	0.0059	-0.0472	0.0474
8	0.028	0.1322	-0.0500	0.0060	-0.0506
9	0.028	0.0511	0.0157	0.0512	-0.1242
10	0.067	0.2311	-0.3635	0.3853	-0.3178
11	0.044	-0.0824	0.1552	-0.2041	0.1793
12	0.202	-0.8639	0.6970	-0.7456	0.7608
13	0.009	-0.0763	0.0204	-0.0235	0.0411
14	.	.	.	.	.
15	0.013	0.1449	0.0073	0.0073	-0.1446
16	1.384	1.2522	-1.7856	2.1796	-1.3917
17	1.317	-0.1253	1.4126	-0.6592	-0.0072

[a] CVR_{17}

Model 2 to analyze the remaining 16 observations, it can be verified that the standard error, s, is reduced from 614.779 to 387.160 and the prob-values corresponding to the intercept, x_2, x_3, and x_5 are reduced from .0749, .0205, .0001, and .0563 to .0023, .0120, .0001, and .0012. These results are consistent with the very small value of CVR_{14} for Model 2 in Figure 10.8. Figure 10.9 presents the diagnostics for detecting outlying and influential observations that result from

FIGURE 10.10 **SAS output of diagnostics for detecting outlying and influential observations for Model 1: $y_i = \beta_0 + \beta_1 x_{i3} + \beta_2 x_{i5} + \varepsilon_i$ when observations 14 and 17 are removed from the hospital labor needs data set**

OBS	HAT DIAG H	RESIDUAL	RSTUDENT	COV RATIO	DFFITS
1	0.1158	-213.1	-0.5693	1.3458	-0.2060
2	0.2889	-21.8576	-0.0642	1.8238	-0.0409
3	0.1258	13.4911	0.0357	1.4845	0.0136
4	0.1633	481.8	1.4186	0.9373	0.6267
5	0.0909	109.9	0.2864	1.3966	0.0906
6	0.1035	-120.2	-0.3158	1.4094	-0.1073
7	0.0906	187.8	0.4930	1.3370	0.1556
8	0.0818	348.3	0.9351	1.1240	0.2791
9	0.0903	-473.8	-1.3248	0.9154	-0.4175
10	0.0732	-180.3	-0.4683	1.3203	-0.1316
11	0.0732	619.9	1.8184	0.6367	0.5109
12	0.1332	-785.1	-2.6878	0.3294	-1.0537
13	0.0767	-173.6	-0.4516	1.3307	-0.1302
14	.	.	.	.	.
15	0.7161	-4.7409	-0.0220	4.5729	-0.0350
16	0.7767	211.6	1.1764	4.0744	2.1939
17	.	.	.	.	.

OBS	COOK'S D	INTERCEP DFBETAS	BEDDAY DFBETAS	STAYDAY DFBETAS
1	0.015	-0.1136	0.0308	0.0607
2	0.001	0.0267	0.0331	-0.0342
3	0.000	0.0091	-0.0001	-0.0059
4	0.121	0.5042	0.0984	-0.3798
5	0.003	-0.0003	-0.0446	0.0250
6	0.004	-0.0688	-0.0033	0.0438
7	0.009	-0.0094	-0.0788	0.0511
8	0.026	0.0894	-0.0658	-0.0146
9	0.055	0.1104	0.1950	-0.2054
10	0.006	0.0118	0.0274	-0.0392
11	0.073	-0.0098	-0.1497	0.1321
12	0.244	-0.8549	-0.6131	0.7412
13	0.006	-0.0678	-0.0417	0.0460
14	.	.	.	.
15	0.000	0.0246	0.0024	-0.0233
16	1.555	0.9978	2.0358	-1.2288
17	.	.	.	.

using Model 2 to analyze observations 1–13, 15, 16, and 17. The fact that $CVR_{17} = 18.7100$ is very large (greater than $1 + 3k/n$) indicates that removing observation 17 from the data set significantly damages the precision of at least some of the least squares point estimates of the parameters in Model 2. If, in addition to removing hospital 14, we remove hospital 17 (which has a particularly large value of x_2) from the data set, the prob-value corresponding to x_2 in Model 2 increases to .5802. Of course, we would remove hospital 17 only if we were willing to lose information about a hospital having a particularly large value of x_2 (monthly X-ray exposures). In this case—if we did not wish to predict labor requirements for hospitals having such large values of x_2—the large prob-value corresponding to x_2 in Model 2 would motivate us to eliminate x_2.

Figure 10.10 presents the diagnostics for detecting outlying and influential observations that result from using Model 1 (which results when x_2 is eliminated from Model 2) to analyze observations 1–13, 15, and 16. The hospital data set will be further analyzed in Chapters 11 and 12, and some final conclusions will be made.

*10.5

PARTIAL LEVERAGE RESIDUAL PLOTS

Suppose that we are attempting to relate the dependent variable y to the independent variables $x_1, \ldots, x_{j-1}, x_j, x_{j+1}, \ldots, x_p$. Let $b_0, b_1, \ldots, b_{j-1}, b_{j+1}, \ldots, b_p$ be the least squares point estimates of the parameters in the model

$$y = \beta_0 + \beta_1 x_1 + \cdots + \beta_{j-1} x_{j-1} + \beta_{j+1} x_{j+1} + \cdots + \beta_p x_p + \varepsilon$$

and let $b'_0, b'_1, \ldots, b'_{j-1}, b'_{j+1}, \ldots, b'_p$ be the least squares point estimates of the parameters in the model

$$x_j = \beta'_0 + \beta'_1 x_1 + \cdots + \beta'_{j-1} x_{j-1} + \beta'_{j+1} x_{j+1} + \cdots + \beta'_p x_p + \varepsilon$$

Then a *partial leverage residual plot* of

$$e_{(j)} = y - (b_0 + b_1 x_1 + \cdots + b_{j-1} x_{j-1} + b_{j+1} x_{j+1} + \cdots + b_p x_p)$$

versus

$$e'_{(j)} = x_j - (b'_0 + b'_1 x_1 + \cdots + b'_{j-1} x_{j-1} + b'_{j+1} x_{j+1} + \cdots + b'_p x_p)$$

represents a plot of y versus x_j, with the effects of the other independent variables $x_1, \ldots, x_{j-1}, x_{j+1}, \ldots, x_p$ removed. When strong multicollinearity exists between x_j and the other independent variables, a plot of y versus x_j can reveal an (apparent) significant relationship between y and x_j, while the partial leverage residual plot of $e_{(j)}$ versus $e'_{(j)}$ reveals very little or no relationship between $e_{(j)}$ and $e'_{(j)}$. This is a graphical illustration of the multicollinearity and says that there is very little or no relationship between y and x_j when the effects of the other independent variables are removed. In other words, x_j has little or no importance in describing y over and above the combined importance of the other independent variables.

Finally, note that the least squares point estimate of the slope parameter β_j in the simple linear model

$$e_{(j)} = \beta_0 + \beta_j e'_{(j)} + \varepsilon_{(j)}$$

equals the least squares point estimate of the parameter β_j in the model

$$y = \beta_0 + \beta_1 x_1 + \cdots + \beta_{j-1} x_{j-1} + \beta_j x_j + \beta_{j+1} x_{j+1} + \cdots + \beta_p x_p + \varepsilon$$

*This section is optional.

Example 10.9 Consider the hospital labor needs data in Table 10.5, and recall that we have previously shown that x_1, x_3, and x_4 are highly collinear. A graphical illustration of this multicollinearity results from comparing Figures 10.3(a), 10.3(c), and 10.3(d) with Figures 10.11(a), 10.11(c), and 10.11(d). Specifically, the plots in Figure 10.3 of y versus each of x_1, x_3, and x_4 suggest linear relationships between y and each of these variables. However, the partial leverage residual plots in Figures 10.11(a), 10.11(c), and 10.11(d) show very little or no relationship between

$$e_{(1)} = y - (b_0 + b_2x_2 + b_3x_3 + b_4x_4 + b_5x_5)$$

and

$$e'_{(1)} = x_1 - (b'_0 + b'_2x_2 + b'_3x_3 + b'_4x_4 + b'_5x_5)$$

or between

$$e_{(3)} = y - (b_0 + b_1x_1 + b_2x_2 + b_4x_4 + b_5x_5)$$

and

$$e'_{(3)} = x_3 - (b'_0 + b'_1x_1 + b'_2x_2 + b'_4x_4 + b'_5x_5)$$

or between

$$e_{(4)} = y - (b_0 + b_1x_1 + b_2x_2 + b_3x_3 + b_5x_5)$$

FIGURE 10.11 **Partial leverage residual plots in the hospital labor needs problem**

(a) $e_{(1)}$ versus $e'_{(1)}$

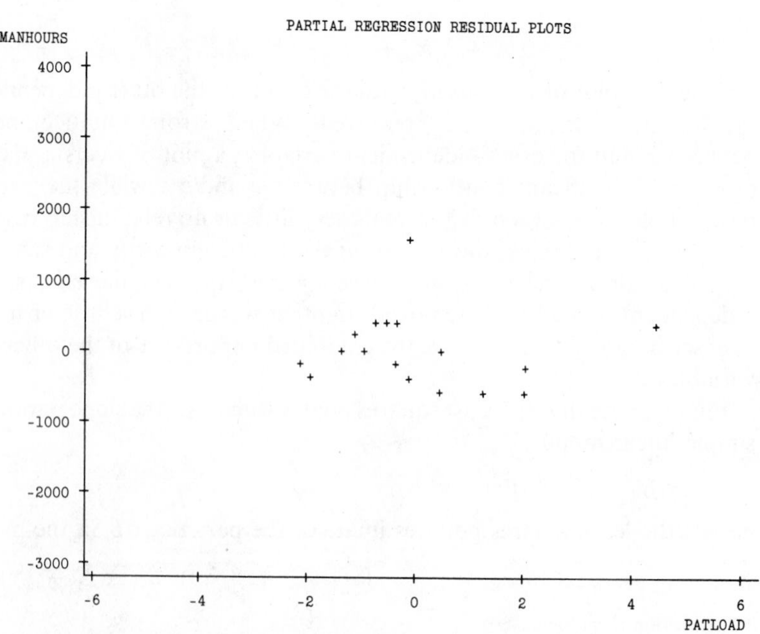

(b) $e_{(2)}$ versus $e'_{(2)}$

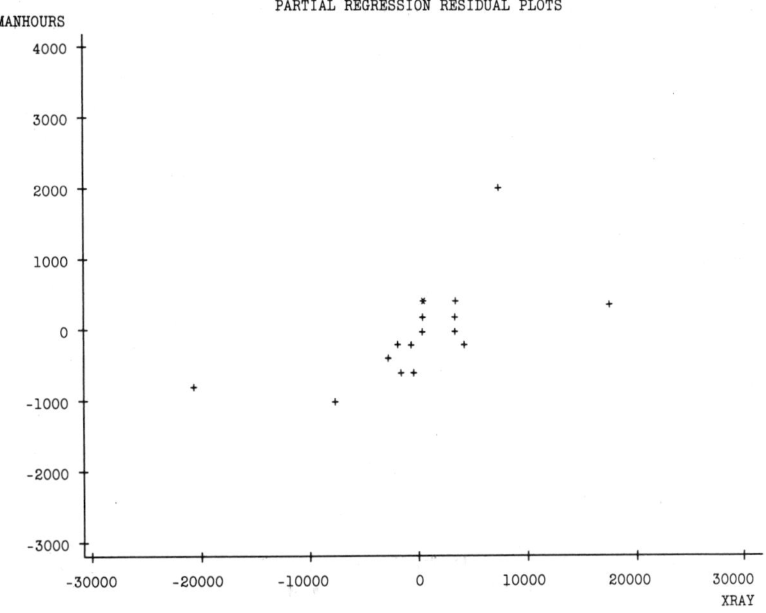

(c) $e_{(3)}$ versus $e'_{(3)}$

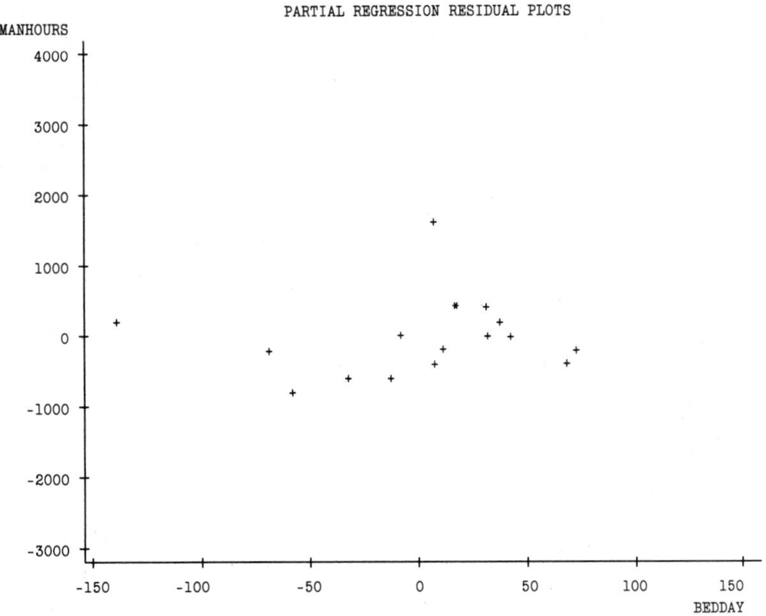

(*continues*)

FIGURE 10.11 **Continued**

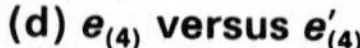

(d) $e_{(4)}$ **versus** $e'_{(4)}$

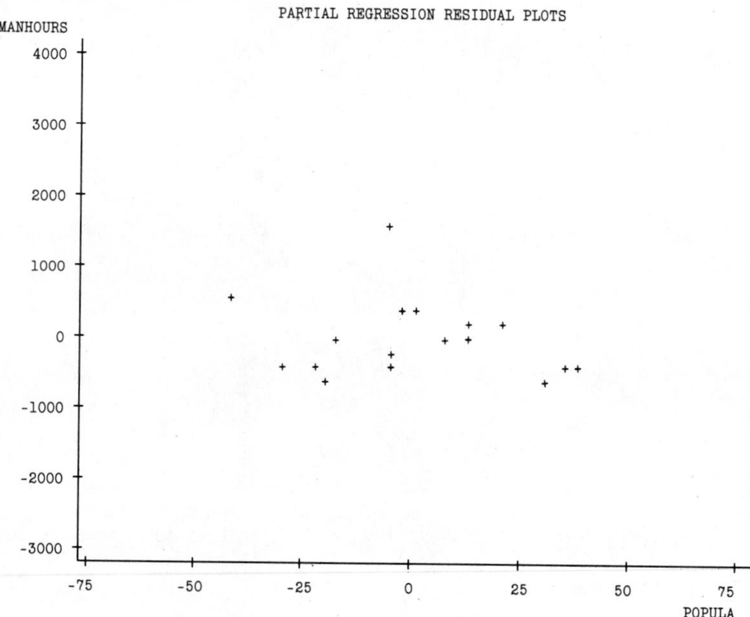

(e) $e_{(5)}$ **versus** $e'_{(5)}$

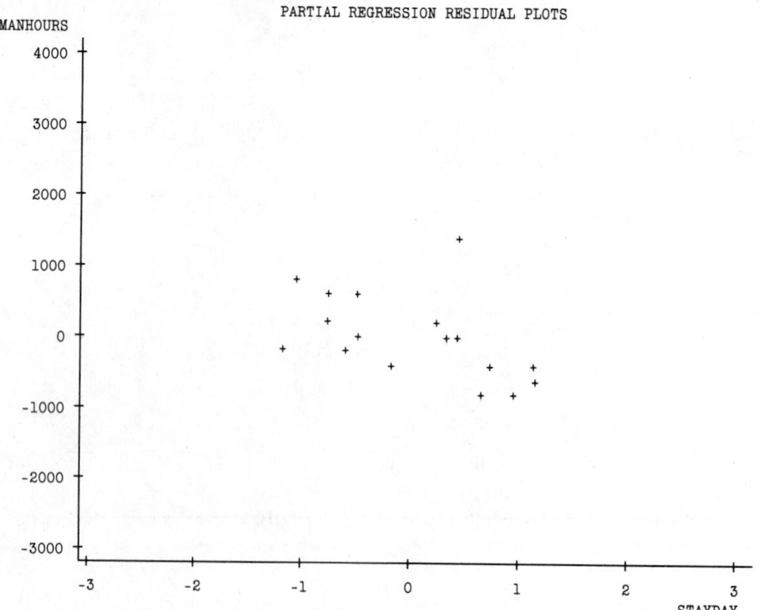

and

$$e'_{(4)} = x_4 - (b'_0 + b'_1x_1 + b'_2x_2 + b'_3x_3 + b'_5x_5)$$

On the other hand, the partial leverage residual plot in Figure 10.11(b) shows a relationship between

$$e_{(2)} = y - (b_0 + b_1x_1 + b_3x_3 + b_4x_4 + b_5x_5)$$

and

$$e'_{(2)} = x_2 - (b'_0 + b'_1x_1 + b'_3x_3 + b'_4x_4 + b'_5x_5)$$

that is as clear as the relationship between y and x_2 in Figure 10.3(b). Furthermore, the partial leverage residual plot in Figure 10.11(e) shows a clearer relationship between

$$e_{(5)} = y - (b_0 + b_1x_1 + b_2x_2 + b_3x_3 + b_4x_4)$$

and

$$e'_{(5)} = x_5 - (b_0 + b_1x_1 + b_2x_2 + b_3x_3 + b_4x_4)$$

than the plot in Figure 10.3(e) shows between y_5 and x_5.

Finally, Figure 10.4 tells us that $b_5 = -394.31412$ is the least squares point estimate of the parameter β_5 in the hospital labor needs model

$$y = \beta_0 + \beta_1x_1 + \beta_2x_2 + \beta_3x_3 + \beta_4x_4 + \beta_5x_5 + \varepsilon$$

It follows that $b_5 = -394.31412$ is the least squares point estimate of the slope parameter β_5 in the model

$$e_{(5)} = \beta_0 + \beta_5e'_{(5)} + \varepsilon_{(5)}$$

Here this model describes the values of $e_{(5)}$ and $e'_{(5)}$ plotted in Figure 10.11(e).

Example 10.9 illustrates that when an independent variable x_j is not highly correlated with the other independent variables in a regression problem (for example, x_2 and x_5 in the hospital labor needs problem are not highly correlated with the other independent variables), then the partial leverage residual plot of $e_{(j)}$ versus $e'_{(j)}$ can show a clearer pattern than does the plot of y versus x_j.

*10.6 USING SAS

In Figure 10.12 we present the SAS statements that implement many of the techniques of this chapter. Here these techniques are applied to the hospital labor needs data of Table 10.5.

In Figure 10.13 we present the SAS statements needed to obtain the SAS output of the Home Real Estate Company data analysis in Figure 10.6.

*This section is optional.

FIGURE 10.12 **SAS program to obtain the correlation matrix, variance inflation factors, and partial leverage residual plots in the hospital labor needs problem**

```
DATA HOSP;
INPUT Y X1 X2 X3 X4 X5; }  ----> Defines variables
CARDS;
  566.52    15.57    2463     479.92    18.0   4.45 )
  696.82    44.02    2048    1339.75     9.5   6.92 }  ----> Hospital labor needs data—see Table 10.5
                       :
18854.45   510.22   86533   15524.00   371.6   6.35 )

PROC PRINT;

PROC PLOT;                        }  ----> Plots y against each of
PLOT Y * (X1 X2 X3 X4 X5);        }        x1, x2, x3, x4, and x5

PROC CORR; } Prints correlation matrix
PROC REG;                          }  Gives partial leverage residual plots
MODEL Y = X1 X2 X3 X4 X5/PARTIAL;  }

PROC REG;                  }  ----> VIF calculates variance
MODEL Y = X2 X3 X5/P VIF;  }        inflation factors for model
```

Plots y against each of x_1, x_2, x_3, x_4, and x_5

VIF calculates variance inflation factors for model $y = \beta_0 + \beta_1 x_2 + \beta_2 x_3 + \beta_3 x_5 + \varepsilon$

FIGURE 10.13 **SAS program to compute diagnostics for outlying and influential observations for the model $y = \beta_0 + \beta_1 x_1 + \beta_2 x_2 + \beta_3 x_2^2 + \varepsilon$ in the Home Real Estate Company problem**

```
DATA HOUSE;  ----> Assigns name HOUSE to the data file

INPUT Y X1 X2;  ----> Defines variable names. Y = selling price,
                      X1 = size, X2 = age

X2SQ = X2 * X2;  ----> Transformation defines the squared term x2^2
                       and assigns the name X2SQ to this variable

CARDS;
68.7  2.05   3.43 )
54.9  1.70  11.61 }  ----> Home Real Estate data—see Figure 10.6
        :
50.0  1.55  18.00 )

  .   1.70   5.00  ----> Decimal point represents a missing value. Used
                         to obtain a point prediction when x01 = 1.70
                         and x02 = 5.00

PROC PRINT;  ----> Prints the data
PROC REG;    ----> Specifies regression procedure
MODEL Y = X1 X2 X2SQ/P R INFLUENCE CLM CLI;
```

Transformation defines the squared term x_2^2 and assigns the name X2SQ to this variable

Decimal point represents a missing value. Used to obtain a point prediction when $x_{01} = 1.70$ and $x_{02} = 5.00$

Specifies model $y = \beta_0 + \beta_1 x_1 + \beta_2 x_2 + \beta_3 x_2^2 + \varepsilon$

- P = predictions desired
- R, INFLUENCE: Diagnostics for outlying and influential observations desired
- CLM = 95% confidence interval desired
- CLI = 95% prediction interval desired

EXERCISES

10.1 Consider the residential sales data in Table 8.2. In Exercise 6.27 we found that residences 10, 50, and 61 in Table 4.3 (which are the same residences as those in Table 8.2) are possibly outlying and influential.

In Figure 10.14 we present the SAS output of the simple correlation coefficients between the values of the dependent and independent variables in Table 8.2 with residences 10, 50, and 61 removed. Do the correlation coefficients indicate that there is strong multicollinearity in the data? Justify your answer.

FIGURE 10.14 **SAS output of the simple correlation coefficients for the variables in Table 8.2 with residences 10, 50, and 61 removed from the data set**

SAS

PEARSON CORRELATION COEFFICIENTS / PROB > | R | UNDER HO:RHO=0 / N = 60

	SALEP	SQFT	ROOMS	BED	AGE
SALEP	1.00000 0.0000	0.83154 0.0001	0.58834 0.0001	0.45443 0.0003	-0.29483 0.0222
SQFT	0.83154 0.0001	1.00000 0.0000	0.66874 0.0001	0.68497 0.0001	-0.05572 0.6724
ROOMS	0.58834 0.0001	0.66874 0.0001	1.00000 0.0000	0.67771 0.0001	0.22258 0.0874
BED	0.45443 0.0003	0.68497 0.0001	0.67771 0.0001	1.00000 0.0000	0.07666 0.5605
AGE	-0.29483 0.0222	-0.05572 0.6724	0.22258 0.0874	0.07666 0.5605	1.00000 0.0000

10.2 Consider the hospital labor needs data in Table 10.5 and the model

$$y = \beta_0 + \beta_1 x_1 + \beta_2 x_2 + \beta_3 x_3 + \beta_4 x_4 + \beta_5 x_5 + \varepsilon$$

a. In Example 10.4 we used simple correlation coefficients and variance inflation factors to reason that there is very strong multicollinearity between x_1, x_3, and x_4. Discuss *intuitively* why the definitions of these independent variables indicate that there would be very strong multicollinearity among them.

b. In Example 10.4 we saw that $VIF_4 = 23.29$. Find R_4^2, the multiple coefficient of determination for the model

$$x_4 = \beta_0 + \beta_1 x_1 + \beta_2 x_2 + \beta_3 x_3 + \beta_4 x_5 + \varepsilon$$

10.3 In Chapter 11 we will find that a good model describing the hospital labor needs data in Table 10.5 is

$$y = \beta_0 + \beta_1 x_2 + \beta_2 x_3 + \beta_3 x_5 + \varepsilon$$

Note that this model includes only one of the three highly collinear independent variables x_1, x_3, and x_4. Figure 10.15 shows the SAS output of the t statistics, prob-values, and variance inflation factors associated with the independent variables in this model.

FIGURE 10.15 **SAS output of the *t* statistics, related prob-values, and variance inflation factors for the model $y = \beta_0 + \beta_1 x_2 + \beta_2 x_3 + \beta_3 x_5 + \varepsilon$ in the hospital labor needs problem**

VARIABLE	DF	PARAMETER ESTIMATE	STANDARD ERROR	T FOR H0: PARAMETER=0	PROB > |T|	VARIANCE INFLATION
INTERCEP	1	1523.389	786.898	1.936	0.0749	0
XRAY	1	0.052987	0.020092	2.637	0.0205	7.73733111
BEDDAY	1	0.978482	0.105154	9.305	0.0001	11.26934216
STAYDAY	1	-320.951	153.192	-2.095	0.0563	2.49290119

a. Compare the above variance inflation factors with the variance inflation factors given in Example 10.4 for the model containing all five independent variables x_1, x_2, x_3, x_4, and x_5. Which model has less multicollinearity?

b. The variance inflation factor VIF_3 for x_3 (BEDDAY) is

$$VIF_3 = \frac{1}{1 - R_3^2}$$

Discuss what R_3^2 measures in the model

$$y = \beta_0 + \beta_1 x_1 + \beta_2 x_2 + \beta_3 x_3 + \beta_4 x_4 + \beta_5 x_5 + \varepsilon$$

Discuss what R_3^2 measures in the model

$$y = \beta_0 + \beta_1 x_2 + \beta_2 x_3 + \beta_3 x_5 + \varepsilon$$

Explain why, logically, R_3^2 (and thus VIF_3) is larger for the five-independent-variable model than for the three-independent-variable model.

10.4 Below we summarize what we will see in Chapter 11 are six of the best models describing the hospital labor needs data in Table 10.5. Also presented are the variance inflation factors associated with the independent variables in the models.

Model 1: x_1 (1.8199), x_5 (1.8199)
Model 2: x_3 (1.8195), x_5 (1.8195)
Model 3: x_1 (5.6605), x_2 (5.6605)
Model 4: x_2 (5.6471), x_3 (5.6471)
Model 5: x_1 (11.3214), x_2 (7.7714), x_5 (2.4985)
Model 6: x_2 (7.7373), x_3 (11.2693), x_5 (2.4929)

Note that the variance inflation factors for the variables in Model 5 are very similar to the variance inflation factors for the variables in Model 6. The same can be said for Models 1 and 2 and for Models 3 and 4. Intuitively, why is this?

10.5 Consider the construction project data in Table 8.8 of Exercise 8.7. Two models describing this data that were considered in the exercises of Chapter 9 are

$$\underset{}{y_i} = \underset{(.0001)}{\beta_0} + \underset{(.2290)}{\beta_1 x_{i1}} + \underset{(.0001)}{\beta_2 x_{i2}} + \underset{(.0053)}{\beta_3 x_{i1}^2} + \underset{(.0001)}{\beta_4 x_{i1} x_{i2}} + \varepsilon_i$$

and

$$y_i = \underset{(.2696)}{\beta_0} + \underset{(.1306)}{\beta_1 x_{i1}} + \underset{(.7845)}{\beta_2 x_{i2}} + \underset{(.0032)}{\beta_3 x_{i1}^2} + \underset{(.4056)}{\beta_4 x_{i1} x_{i2}} + \underset{(.0471)}{\beta_5 x_{i1}^2 x_{i2}} + \varepsilon_i$$

The prob-values for the variables in these models are given under the variables. The second model might seem inferior because its variables have larger prob-values. However, in using quadratic and interaction terms, strong multicollinearity exists. For example, in the second model, x_{i1} is highly related to x_{i1}^2, $x_{i1}x_{i2}$, and $x_{i1}^2 x_{i2}$. The variance inflation factors associated with x_{i1}, x_{i2}, x_{i1}^2, $x_{i1}x_{i2}$, and $x_{i1}^2 x_{i2}$ in the second model are 411.6, 482.2, 387.8, 4082.3, and 1801.7, respectively. Intuitively, which model has more multicollinearity—the first or second? Why? In the exercises of Chapter 11 we will find that the second model is probably the better of the two models.

10.6 Measuring the percentage of fat (FAT) in pork bellies is an expensive procedure. Therefore it is important to see whether this percentage can be predicted from other, more easily measured properties of the pork carcass. In *SAS System for Regression*, 1986 Edition, Freund and Littell present data on FAT and ten predictor variables for a sample of 45 pork carcasses. The predictor variables are described by Freund and Littell as follows:

AVBF	is an average of three measures of back fat thickness.
MUS	is a muscling score for the carcass. The higher the number is, the more muscle and less fat.
LEA	is loin eye area.
DEP	is an average of three measures of fat depth opposite the tenth rib.
LWT	is live weight of the carcass.
CWT	is weight of the slaughtered carcass.
WTWAT	is a measure used to determine specific gravity.
DPSL	is the average of three determinations of depth of the belly.
LESL	is the average measure of leanness of three cross sections of the belly.
BELWT	is total weight of the belly.*

Their data are given in Table 10.7. The SAS output of the prob-values and variance inflation factors for the regression model relating FAT to all ten predictor variables is given in Figure 10.16.

a. Which four predictor variables have the largest variance inflation factors? Which of these variables are, logically, highly related to each other?
b. In the exercises of Chapter 11 we will find that a reasonable model describing the FAT data involves six of the ten predictor variables. A SAS output for this model is given in Figure 10.17. Which of the four variables from part (a) are involved in the six-variable model?

10.7 Use ridge regression and the hospital labor needs data in Table 10.5 to find point estimates of the parameters in the model

$$y = \beta_0 + \beta_1 x_2 + \beta_2 x_3 + \beta_3 x_5 + \varepsilon$$

Note: This a more difficult problem. SAS PROC MATRIX, which is described in SAS manuals, can be used to perform the matrix calculations.

*Source: Rudolph J. Freund and Ramon C. Littell, *SAS System for Regression*, 1986 Edition (Cary, N.C.: SAS Institute, 1986). Reprinted by permission.

TABLE 10.7 **Predictor variables and percentage of FAT in 45 pork carcasses**

AVBF	MUS	LEA	DEP	LWT	CWT	WTWAT	DPSL	LESL	BELWT	FAT
1.30	14	5.00	1.27	239	187	4.10	1.50000	10.0000	14.35	51.4
1.57	9	4.10	1.47	229	175	3.83	1.56667	6.0000	13.76	58.0
1.68	11	4.20	1.60	223	172	3.38	1.53333	5.0000	12.99	51.0
1.58	9	4.30	1.73	210	160	3.38	1.40000	8.6667	12.42	54.5
1.18	13	5.00	1.13	234	178	3.86	1.53333	8.0000	13.77	53.1
1.98	13	4.20	1.97	239	185	3.24	1.60000	4.3333	14.06	57.1
1.28	9	4.80	1.23	226	172	3.83	1.30000	5.6667	12.32	55.3
1.62	11	5.50	1.57	225	172	4.14	1.50000	9.3333	12.82	49.9
1.53	10	4.10	1.80	227	173	3.10	1.76667	4.6667	10.93	57.3
1.20	10	5.50	1.00	215	164	4.21	1.46667	12.0000	11.10	46.0
1.67	10	5.10	1.60	210	180	3.95	1.66667	7.6667	12.49	54.1
1.88	10	4.70	1.67	233	181	3.82	1.43333	5.3333	12.72	61.6
1.50	9	4.90	1.60	212	162	3.71	1.60000	6.6667	12.87	56.1
1.47	14	5.40	1.17	244	192	4.32	1.36667	8.3333	13.93	51.7
1.38	11	5.05	1.20	236	180	4.43	1.40000	12.0000	13.19	50.9
1.88	8	3.30	2.17	217	166	2.61	1.60000	4.0000	12.02	64.8
1.72	12	4.90	1.60	223	171	3.64	1.33333	8.3333	12.61	56.1
1.88	6	4.40	1.80	226	175	3.69	1.53333	8.0000	11.65	57.5
1.73	12	4.00	1.57	232	177	3.82	1.66667	5.6667	12.99	54.4
1.33	9	4.90	1.33	221	170	3.96	1.30000	7.0000	12.44	50.9
1.42	6	5.00	1.37	219	166	3.87	1.30000	10.0000	12.05	49.5
1.35	9	4.80	1.43	228	175	3.70	1.43333	10.0000	12.34	56.6
1.78	11	5.10	1.43	226	176	3.95	1.56667	9.6667	13.37	49.8
1.35	15	4.60	1.37	230	178	3.52	1.53333	7.0000	15.25	58.5
1.18	9	3.90	1.20	224	168	3.73	1.60000	7.3333	13.03	55.4
1.58	10	4.00	1.60	223	167	3.60	1.76667	7.0000	11.24	50.3
1.70	12	3.50	2.07	240	188	3.10	1.70000	5.0000	14.78	65.4
1.52	10	4.45	1.47	231	178	3.72	1.66667	6.3333	13.39	55.3
1.30	15	4.80	1.33	235	183	3.61	1.43333	6.6667	13.10	50.3
1.68	10	4.05	1.80	241	195	3.63	1.86667	6.6667	14.56	60.1
1.80	11	3.05	2.07	222	166	2.33	1.73333	3.6667	11.77	58.7
1.78	11	5.10	1.43	226	176	3.95	1.60000	7.0000	13.21	49.8
1.18	9	3.90	1.20	224	168	3.73	1.56667	5.6667	13.84	58.3
1.68	11	4.20	1.60	223	172	3.38	1.50000	9.6667	15.63	55.1
1.18	13	5.00	1.13	234	178	3.86	1.46667	8.0000	14.62	53.1
1.98	13	4.20	1.97	239	185	3.24	1.46667	6.3333	15.17	59.8
1.35	9	4.85	1.10	214	164	4.27	1.50000	9.6667	12.17	46.7
1.20	10	5.50	1.00	215	164	4.21	1.56667	12.6667	12.43	46.0
1.47	14	5.40	1.17	244	192	4.32	1.80000	7.6667	11.66	53.0
1.88	8	3.30	2.17	217	166	2.61	1.66667	4.6667	10.97	64.8
1.88	8	4.40	1.80	226	175	3.69	1.56667	6.6667	11.63	57.4
1.33	9	4.90	1.33	221	170	3.96	1.36667	7.0000	12.14	50.9
1.35	9	4.80	1.43	228	175	3.70	1.26667	6.3333	13.57	56.6
1.70	12	3.50	2.07	240	188	3.10	1.56667	5.3333	14.87	65.2
1.68	10	4.05	1.80	241	195	3.63	1.83333	6.3333	14.80	59.4

Source: Rudolph J. Freund and Ramon C. Littell, *SAS System for Regression*, 1986 Edition (Cary, N.C.: SAS Institute, 1986). Reprinted by permission.

FIGURE 10.16 SAS output of the prob-values and variance inflation factors for the regression model relating FAT to all ten predictor variables in Table 10.7

DEP VARIABLE: FAT

ANALYSIS OF VARIANCE

SOURCE	DF	SUM OF SQUARES	MEAN SQUARE	F VALUE	PROB>F
MODEL	10	861.78956	86.17895635	13.437	0.0001
ERROR	34	218.05844	6.41348343		
C TOTAL	44	1079.84800			

ROOT MSE	2.532486	R-SQUARE	0.7981
DEP MEAN	55.06	ADJ R-SQ	0.7387
C.V.	4.599502		

PARAMETER ESTIMATES

VARIABLE	DF	PARAMETER ESTIMATE	STANDARD ERROR	T FOR H0: PARAMETER=0	PROB > |T|	VARIANCE INFLATION
INTERCEP	1	24.8552840	21.01386546	1.183	0.2451	0
AVBF	1	-5.47255074	3.65597016	-1.497	0.1437	5.39256048
MUS	1	-0.58664327	0.30723799	-1.909	0.0647	3.13858275
LEA	1	-0.98927490	1.70794636	-0.579	0.5663	7.95643967
DEP	1	9.06495250	4.88651731	1.855	0.0723	17.00146970
LWT	1	0.15048631	0.20471925	0.735	0.4673	22.71827368
CWT	1	0.03036098	0.20963388	0.145	0.8857	24.57954351
WTWAT	1	-1.81247934	2.96541371	-0.611	0.5451	12.43644447
DPSL	1	-2.14189312	3.54748177	-0.604	0.5500	1.88110915
LESL	1	-0.43270697	0.28992712	-1.492	0.1448	2.64865964
BELWT	1	0.68668721	0.47218237	1.454	0.1550	2.28670951

FIGURE 10.17 SAS output of a regression relating FAT to six predictor variables in Table 10.7

	DF	SUM OF SQUARES	MEAN SQUARE	F	PROB>F
REGRESSION	6	850.72450159	141.78741693	23.52	0.0001
ERROR	38	229.12349841	6.02956575		
TOTAL	44	1079.84800000			

	B VALUE	STD ERROR	TYPE II SS	F	PROB>F
INTERCEP	16.10461148				
AVBF	-7.92198257	2.92430917	44.24938396	7.34	0.0101
MUS	-0.55034422	0.22980152	34.58191748	5.74	0.0217
DEP	13.19387520	2.60710596	154.42350816	25.61	0.0001
LWT	0.13245217	0.05875331	30.64358109	5.08	0.0300
LESL	-0.61409832	0.24034104	39.36461196	6.53	0.0147
BELWT	0.85301426	0.39015919	28.82139114	4.78	0.0350

10.8 Consider the residential sales data in Table 8.2. In Exercise 6.27 we found that residences 10, 50, and 61 in Table 4.3 (which are the same residences as those in Table 8.2) are possibly outlying and influential. In Figure 10.18 we present the SAS output of outlying and influential observation diagnostics for the $n = 63$ observations in Table 8.2. Here, we use the model relating SALEP to SQFT, ROOMS, BED, and AGE. Fully interpret the output. Be sure to interpret the diagnostics for residences 10, 50, and 61. Would you conclude that these residences are outlying and influential?

FIGURE 10.18 **SAS output of outlying and influential observation diagnostics for the $n = 63$ observations in Table 8.2 when using a model relating SALEP to SQFT, ROOMS, BED, and AGE**

SAS

DEP VARIABLE: SALEP

ANALYSIS OF VARIANCE

SOURCE	DF	SUM OF SQUARES	MEAN SQUARE	F VALUE	PROB>F
MODEL	4	55183.68778	13795.92195	38.371	0.0001
ERROR	58	20853.30206	359.53969		
C TOTAL	62	76036.98984			

ROOT MSE	18.96153	R-SQUARE	0.7257
DEP MEAN	78.80159	ADJ R-SQ	0.7068
C.V.	24.06237		

PARAMETER ESTIMATES

VARIABLE	DF	PARAMETER ESTIMATE	STANDARD ERROR	T FOR H0: PARAMETER=0	PROB > |T|
INTERCEP	1	10.36761556	11.49845952	0.902	0.3710
SQFT	1	0.05001119	0.008104134	6.171	0.0001
ROOMS	1	6.32177893	2.52798952	2.501	0.0152
BED	1	-11.10316277	5.86838064	-1.892	0.0635
AGE	1	-0.43186496	0.10970614	-3.937	0.0002

OBS	ACTUAL	PREDICT VALUE	STD ERR PREDICT	LOWER95% MEAN	UPPER95% MEAN	LOWER95% PREDICT	UPPER95% PREDICT	RESIDUAL	STD ERR RESIDUAL	STUDENT RESIDUAL	-2-1-0	1 2
1	53.5000	55.0662	4.2002	46.6586	63.4737	16.1905	93.9418	-1.5662	18.4905	-0.0847		
2	49.0000	63.9561	2.9738	58.0033	69.9089	25.5365	102.4	-14.9561	18.7269	-0.7986	*	
3	50.5000	66.1980	9.6129	46.9556	85.4404	23.6433	108.8	-15.6980	16.3441	-0.9605	*	
4	49.9000	36.5708	5.3003	25.9611	47.1804	-2.8398	75.9813	13.3292	18.2057	0.7321		*
5	52.0000	57.9277	3.3998	51.1222	64.7332	19.3668	96.4886	-5.9277	18.6542	-0.3178		
6	55.0000	64.5618	4.5354	55.4833	73.6404	25.5356	103.6	-9.5618	18.4111	-0.5194	*	
7	80.5000	77.1096	5.7708	65.5581	88.6610	37.4351	116.8	3.3904	18.0621	0.1877		
8	86.0000	93.1230	3.3420	86.4333	99.8128	54.5824	131.7	-7.1230	18.6647	-0.3816		
9	69.0000	64.5212	4.1229	56.2683	72.7741	25.6787	103.4	4.4788	18.5079	0.2420		
10	149.0	190.8	10.3532	170.0	211.5	147.5	234.0	-41.7682	15.8856	-2.6293	*****	
11	46.0000	35.8977	5.5280	24.8322	46.9632	-3.6380	75.4334	10.1023	18.1378	0.5570		*
12	38.0000	31.7500	4.7066	22.3286	41.1714	-7.3574	70.8574	6.2500	18.3681	0.3403		
13	49.5000	50.2848	4.5793	41.1183	59.4513	11.2380	89.3316	-0.7848	18.4003	-0.0427		
14	105.0	102.7	5.1358	92.4168	113.0	63.3740	142.0	2.3028	18.2528	0.1262		
15	152.5	120.5	4.8409	110.8	130.1	81.2785	159.6	32.0485	18.3332	1.7481		***
16	85.0000	90.6017	6.4282	77.7343	103.5	50.5243	130.7	-5.6017	17.8387	-0.3140		
17	60.0000	44.2050	8.2252	27.7405	60.6694	2.8322	85.5778	15.7950	17.0847	0.9245		*
18	58.5000	51.5853	6.4121	38.7501	64.4205	11.5182	91.6524	6.9147	17.8445	0.3875		
19	101.0	79.1951	4.8626	69.4615	88.9286	40.0112	118.4	21.8049	18.3274	1.1897		**
20	79.4000	75.0528	3.8992	67.2477	82.8579	36.3030	113.8	4.3472	18.5563	0.2343		
21	125.0	117.2	4.5367	108.2	126.3	78.2193	156.3	7.7539	18.4108	0.4212		

OBS	ACTUAL	PREDICT VALUE	STD ERR PREDICT	LOWER95% MEAN	UPPER95% MEAN	LOWER95% PREDICT	UPPER95% PREDICT	RESIDUAL	STD ERR RESIDUAL	STUDENT RESIDUAL	-2-1-0	1 2
22	87.9000	107.4	7.8955	91.6405	123.2	66.3304	148.6	-19.5450	17.2395	-1.1337	**	
23	80.0000	82.9342	4.2117	74.5036	91.3648	44.0535	121.8	-2.9342	18.4879	-0.1587		
24	94.0000	98.6424	5.6140	87.4047	109.9	59.0581	138.2	-4.6424	18.1114	-0.2563		
25	74.0000	98.5603	8.2736	81.9989	115.1	57.1488	140.0	-24.5603	17.0613	-1.4395	**	
26	69.0000	77.6451	2.5218	72.5971	82.6930	39.3552	115.9	-8.6451	18.7931	-0.4600		
27	63.0000	55.0662	4.2002	46.6586	63.4737	16.1905	93.9418	7.9338	18.4905	0.4291		
28	67.5000	64.1441	3.9468	56.2436	72.0445	25.3749	102.9	3.3559	18.5462	0.1809		
29	35.0000	44.5249	6.4818	31.5502	57.4996	4.4129	84.6369	-9.5249	17.8193	-0.5345	*	
30	142.5	136.4	5.5382	125.3	147.5	96.8584	175.9	6.1001	18.1347	0.3364		
31	92.2000	87.2581	4.5210	78.2084	96.3078	48.2385	126.3	4.9419	18.4147	0.2684		
32	56.0000	59.0904	5.1994	48.6826	69.4982	19.7336	98.4471	-3.0904	18.2347	-0.1695		
33	63.0000	62.0672	4.2865	53.4869	70.6475	23.1538	101.0	0.9328	18.4707	0.0505		
34	60.0000	90.1796	3.0705	84.0333	96.3260	51.7296	128.6	-30.1796	18.7113	-1.6129	***	
35	34.0000	20.8961	7.4112	6.0610	35.7312	-19.8557	61.6479	13.1039	17.4532	0.7508		*
36	52.0000	67.8950	4.8902	58.1062	77.6839	28.6974	107.1	-15.8950	18.3201	-0.8676	*	
37	75.0000	76.8496	2.4966	71.8521	81.8471	38.5664	115.1	-1.8496	18.7965	-0.0984		
38	93.0000	96.5081	4.5936	87.3131	105.7	57.4546	135.6	-3.5081	18.3967	-0.1907		
39	60.0000	91.6090	4.9552	81.6901	101.5	52.3788	130.8	-31.6090	18.3026	-1.7270	***	
40	73.0000	63.4175	4.2959	54.8184	72.0166	24.5000	102.3	9.5825	18.4685	0.5189		*
41	71.0000	72.9910	6.3486	60.2829	85.6991	32.9644	113.0	-1.9910	17.8671	-0.1114		
42	83.0000	84.1095	3.0375	78.0293	90.1897	45.6700	122.5	-1.1095	18.7167	-0.0593		
43	90.0000	101.2	3.4797	94.2547	108.2	62.6307	139.8	-11.2202	18.6395	-0.6020	*	
44	83.0000	94.5754	3.2831	88.0036	101.1	56.0550	133.1	-11.5754	18.6751	-0.6198	*	
45	115.0	120.7	4.8855	110.9	130.5	81.4761	159.9	-5.6713	18.3214	-0.3095		
46	50.0000	49.4967	4.2977	40.8940	58.0995	10.5784	88.4150	0.5033	18.4681	0.0273		
47	55.2000	58.4772	3.9728	50.5248	66.4297	19.6975	97.2570	-3.2772	18.5407	-0.1768		
48	61.0000	64.4311	3.8985	56.6275	72.2348	25.6816	103.2	-3.4311	18.5564	-0.1849		
49	147.0	128.7	5.6215	117.5	140.0	89.1325	168.3	18.2789	18.1091	1.0094		**
50	210.0	127.7	4.9880	117.7	137.7	88.4806	167.0	82.2725	18.2937	4.4973		******
51	60.0000	76.2588	2.6275	70.9994	81.5183	37.9406	114.6	-16.2588	18.7786	-0.8658	*	
52	100.0	110.3	4.8525	100.6	120.0	71.0966	149.5	-10.2754	18.3301	-0.5606	*	
53	44.5000	53.0192	4.3227	44.3665	61.6719	14.0898	91.9486	-8.5192	18.4622	-0.4614		
54	55.0000	70.4119	5.7391	58.9239	81.8999	30.7558	110.1	-15.4119	18.0722	-0.8528	*	
55	53.4000	46.2901	5.4478	35.3850	57.1951	6.7989	85.7812	7.1099	18.1621	0.3915		
56	65.0000	72.2177	4.7362	62.7372	81.6982	33.0960	111.3	-7.2177	18.3605	-0.3931		
57	73.0000	77.8190	4.5257	68.7598	86.8781	38.7972	116.8	-4.8190	18.4135	-0.2617		
58	40.0000	39.3221	8.1763	22.9555	55.6887	-2.0119	80.6561	0.6779	17.1081	0.0396		
59	141.0	117.0	10.7560	95.4329	138.5	73.3264	160.6	24.0365	15.6156	1.5393		***
60	68.0000	81.0823	2.5634	75.9511	86.2135	42.7814	119.4	-13.0823	18.7875	-0.6963	*	
61	139.0	88.3693	3.3328	81.6980	95.0406	49.8318	126.9	50.6307	18.6663	2.7124		*****
62	140.0	112.9	4.8629	103.2	122.7	73.7497	152.1	27.0664	18.3273	1.4768		**
63	55.0000	67.2137	4.4035	58.3992	76.0281	28.2480	106.2	-12.2137	18.4431	-0.6622	*	

OBS	RESIDUAL	RSTUDENT	HAT DIAG H	COV RATIO	DFFITS	COOK'S D	INTERCEP DFBETAS	SQFT DFBETAS	ROOMS DFBETAS	BED DFBETAS	AGE DFBETAS
1	-1.5662	-0.0840	0.0491	1.1464	-0.0191	0.000	-0.0150	-0.0003	-0.0026	0.0113	-0.0012
2	-14.9561	-0.7961	0.0246	1.0582	-0.1264	0.003	-0.0073	0.0631	0.0051	-0.0568	-0.0073
3	-15.6980	-0.9598	0.2570	1.3551	-0.5645	0.064	-0.0697	0.2545	-0.5073	0.2809	0.1426
4	13.3292	0.7292	0.0781	1.1296	0.2123	0.009	0.0105	-0.1324	-0.0608	0.1498	0.0251
5	-5.9277	-0.3153	0.0321	1.1173	-0.0575	0.001	0.0000	0.0347	-0.0006	-0.0281	-0.0068
6	-9.5618	-0.5161	0.0572	1.1304	-0.1271	0.003	-0.0321	0.0546	0.0401	-0.0744	0.0580
7	3.3904	0.1861	0.0926	1.1986	0.0595	0.001	-0.0394	-0.0173	0.0009	0.0354	0.0276
8	-7.1230	-0.3788	0.0311	1.1118	-0.0678	0.001	-0.0041	0.0122	-0.0393	0.0161	0.0358
9	4.4788	0.2400	0.0473	1.1392	0.0535	0.001	0.0144	-0.0174	-0.0236	0.0318	-0.0167
10	-41.7682	-2.7773	0.2981	0.8240	-1.8101	0.587	0.8524	-1.1751	0.3064	-0.0657	-0.0422
11	10.1023	0.5536	0.0850	1.1606	0.1687	0.006	0.0083	-0.1167	-0.0369	0.1200	0.0022
12	6.2500	0.3377	0.0616	1.1509	0.0865	0.002	0.0641	-0.0182	-0.0215	-0.0080	0.0193
13	-0.7848	-0.0423	0.0583	1.1582	-0.0105	0.000	0.0004	0.0087	-0.0020	-0.0056	0.0014
14	2.3028	0.1251	0.0734	1.1756	0.0352	0.000	-0.0039	0.0153	0.0143	-0.0215	0.0129
15	32.0485	1.7805	0.0652	0.8902	0.4701	0.043	0.0888	0.3018	0.0491	-0.2580	-0.1324
16	-5.6017	-0.3116	0.1149	1.2221	-0.1123	0.003	0.0762	0.0064	-0.0234	-0.0264	-0.0660
17	15.7950	0.9233	0.1882	1.2476	0.4445	0.040	-0.0397	0.1126	-0.1656	0.0313	0.4233
18	6.9147	0.3846	0.1144	1.2158	0.1382	0.004	0.0647	0.0645	-0.0302	-0.0732	0.0978
19	21.8049	1.1941	0.0658	1.0319	0.3168	0.020	-0.0292	0.1343	-0.0152	-0.0897	0.2494
20	4.3472	0.2324	0.0423	1.1336	0.0488	0.000	0.0075	-0.0282	0.0093	0.0168	-0.0320
21	7.7539	0.4182	0.0572	1.1395	0.1030	0.002	0.0194	0.0518	0.0210	-0.0524	-0.0434
22	-19.5450	-1.1366	0.1734	1.1798	-0.5205	0.054	-0.3111	-0.4222	0.1103	0.3825	-0.0031
23	-2.9342	-0.1574	0.0493	1.1450	-0.0358	0.000	-0.0135	-0.0096	0.0235	-0.0102	0.0090
24	-4.6424	-0.2542	0.0877	1.1889	-0.0788	0.001	-0.0274	-0.0559	0.0568	0.0016	-0.0013

(*continues*)

FIGURE 10.18 Continued

OBS	RESIDUAL	RSTUDENT	HAT DIAG H	COV RATIO	DFFITS	COOK'S D	INTERCEP DFBETAS	SQFT DFBETAS	ROOMS DFBETAS	BED DFBETAS	AGE DFBETAS
25	-24.5603	-1.4533	0.1904	1.1233	-0.7047	0.097	0.1052	0.3608	-0.6549	0.1738	0.3523
26	-8.6451	-0.4569	0.0177	1.0904	-0.0613	0.001	-0.0137	0.0046	0.0116	-0.0144	0.0084
27	7.9338	0.4260	0.0491	1.1290	0.0968	0.002	0.0761	0.0016	0.0133	-0.0574	0.0062
28	3.3559	0.1794	0.0433	1.1370	0.0382	0.000	0.0104	-0.0096	-0.0199	0.0227	-0.0076
29	-9.5249	-0.5312	0.1169	1.2051	-0.1932	0.008	-0.0833	-0.0703	0.0370	0.0902	-0.1446
30	6.1001	0.3338	0.0853	1.1810	0.1019	0.002	-0.0525	0.0147	0.0397	0.0017	-0.0364
31	4.9419	0.2662	0.0568	1.1494	0.0654	0.001	0.0241	0.0324	-0.0481	0.0106	-0.0047
32	-3.0904	-0.1681	0.0752	1.1766	-0.0479	0.000	-0.0017	0.0398	-0.0140	-0.0221	0.0246
33	0.9328	0.0501	0.0511	1.1493	0.0116	0.000	0.0096	0.0002	0.0022	-0.0072	-0.0023
34	-30.1796	-1.6361	0.0262	0.8905	-0.2685	0.014	-0.0767	-0.1522	0.1119	0.0297	-0.0112
35	13.1039	0.7479	0.1528	1.2262	0.3176	0.020	0.2810	0.0192	-0.0199	-0.1790	0.0562
36	-15.8950	-0.8658	0.0665	1.0947	-0.2311	0.011	-0.1788	0.0245	-0.0656	0.1219	0.1185
37	-1.8496	-0.0976	0.0173	1.1092	-0.0130	0.000	-0.0027	0.0000	0.0032	-0.0029	-0.0000
38	-3.5081	-0.1891	0.0587	1.1552	-0.0472	0.000	0.0328	0.0113	-0.0039	-0.0288	-0.0036
39	-31.6090	-1.7579	0.0683	0.8992	-0.4759	0.044	0.2371	0.0488	0.1718	-0.3735	-0.0313
40	9.5825	0.5156	0.0513	1.1234	0.1199	0.003	0.0999	0.0082	0.0201	-0.0769	-0.0214
41	-1.9910	-0.1105	0.1121	1.2273	-0.0393	0.000	0.0247	0.0082	0.0017	-0.0211	-0.0234
42	-1.1095	-0.0588	0.0257	1.1192	-0.0095	0.000	-0.0025	0.0015	0.0002	-0.0017	0.0054
43	-11.2202	-0.5986	0.0337	1.0940	-0.1118	0.003	-0.0134	-0.0129	-0.0506	0.0421	0.0567
44	-11.5754	-0.6165	0.0300	1.0878	-0.1084	0.002	-0.0355	-0.0336	0.0221	0.0064	0.0491
45	-5.6713	-0.3071	0.0664	1.1588	-0.0819	0.001	0.0406	-0.0008	-0.0079	-0.0314	0.0336
46	0.5033	0.0270	0.0514	1.1499	0.0063	0.000	0.0046	-0.0014	0.0015	-0.0030	-0.0002
47	-3.2772	-0.1753	0.0439	1.1379	-0.0376	0.000	-0.0008	0.0292	-0.0068	-0.0189	0.0095
48	-3.4311	-0.1834	0.0423	1.1356	-0.0385	0.000	-0.0098	-0.0038	0.0289	-0.0185	-0.0091
49	18.2789	1.0095	0.0879	1.0946	0.3134	0.020	0.0658	0.1908	0.0327	-0.1644	-0.1266
50	82.2725	5.5245	0.0692	0.1373	1.5063	0.301	-0.6854	0.4526	-0.0636	0.3700	-0.3854
51	-16.2588	-0.8639	0.0192	1.0422	-0.1209	0.003	-0.0217	-0.0235	0.0447	-0.0178	-0.0351
52	-10.2754	-0.5572	0.0655	1.1360	-0.1475	0.004	0.0088	-0.0583	-0.0773	0.0974	-0.0088
53	-8.5192	-0.4583	0.0520	1.1297	-0.1073	0.002	-0.0187	0.0515	0.0432	-0.0734	0.0101
54	-15.4119	-0.8508	0.0916	1.1274	-0.2702	0.015	0.0348	-0.0880	0.0193	0.0524	-0.2341
55	7.1099	0.3886	0.0825	1.1734	0.1166	0.003	0.0136	-0.0831	-0.0175	0.0795	-0.0352
56	-7.2177	-0.3902	0.0624	1.1480	-0.1007	0.002	-0.0294	0.0325	0.0305	-0.0498	0.0584
57	-4.8190	-0.2596	0.0570	1.1499	-0.0638	0.001	-0.0215	0.0087	0.0257	-0.0269	0.0349
58	0.6779	0.0393	0.1859	1.3398	0.0188	0.000	-0.0020	0.0032	-0.0064	0.0021	0.0177
59	24.0365	1.5581	0.3218	1.3056	1.0732	0.225	-0.5995	-0.3830	0.9304	-0.1433	-0.0144
60	-13.0823	-0.6932	0.0183	1.0655	-0.0946	0.002	-0.0234	-0.0212	0.0313	-0.0095	-0.0030
61	50.6307	2.8776	0.0309	0.5713	0.5138	0.047	0.1452	-0.0456	-0.0069	0.0597	-0.3356
62	27.0664	1.4924	0.0658	0.9640	0.3960	0.031	0.1282	0.2583	-0.1154	-0.1086	-0.1342
63	-12.2137	-0.6590	0.0539	1.1101	-0.1573	0.005	-0.1318	-0.0215	-0.0233	0.1051	0.0348

10.9 Figure 10.19 presents the SAS output of outlying and influential observation diagnostics for the residential sales observations in Table 8.2 with residences 10, 50, and 61 removed. Here, we use the model relating SALEP to SQFT, SQFTSQ, ROOMS, ROOMSSQ, BED, BEDSQ, and AGE. Recall that we found in the exercises of Chapter 8 that this is a reasonable model describing the $n = 60$ residences obtained by removing residences 10, 50, and 61. Fully interpret the ouput in Figure 10.19. Be sure to examine the influence of residence 57.

10.10 Interpret the SAS output in Figure 10.10 of outlying and influential observation diagnostics. Be sure to examine the influence of hospital 12.

10.11 If we wish to obtain a regression model describing labor needs for small to medium-sized hospitals, it might be reasonable to remove the larger hospitals 14, 15, 16, and 17 from Table 10.5 and use hospitals 1–13. We will find in Chapter 11 that one of the best

FIGURE 10.19 **SAS output of outlying and influential observation diagnostics for the residential sales observations in Table 8.2 with residences 10, 50, and 61 removed when using a model relating SALEP to SQFT, SQFTSQ, ROOMS, ROOMSSQ, BED, BEDSQ, and AGE**

SAS

ANALYSIS OF VARIANCE

SOURCE	DF	SUM OF SQUARES	MEAN SQUARE	F VALUE	PROB>F
MODEL	7	42692.82563	6098.97509	49.254	0.0001
ERROR	52	6438.94020	123.82577		
C TOTAL	59	49131.76583			

ROOT MSE	11.1277	R-SQUARE	0.8689
DEP MEAN	74.44167	ADJ R-SQ	0.8513
C.V.	14.94822		

PARAMETER ESTIMATES

VARIABLE	DF	PARAMETER ESTIMATE	STANDARD ERROR	T FOR HO: PARAMETER=0	PROB > |T|
INTERCEP	1	44.76848465	22.85500515	1.959	0.0555
SQFT	1	-0.05785964	0.02973061	-1.946	0.0571
SQFTSQ	1	0.000039090	.00000986478	3.963	0.0002
ROOMS	1	-11.88247653	6.99496011	-1.699	0.0953
ROOMSSQ	1	1.15033560	0.46272143	2.486	0.0162
BED	1	54.85589691	20.24428721	2.710	0.0091
BEDSQ	1	-11.31976861	3.51458392	-3.221	0.0022
AGE	1	-0.26334189	0.06943653	-3.793	0.0004

OBS	ACTUAL	PREDICT VALUE	STD ERR PREDICT	LOWER95% MEAN	UPPER95% MEAN	LOWER95% PREDICT	UPPER95% PREDICT	RESIDUAL	STD ERR RESIDUAL	STUDENT RESIDUAL	-2-1-0	1 2
1	53.5000	50.7254	2.6975	45.3125	56.1382	27.7493	73.7014	2.7746	10.7958	0.2570		
2	49.0000	58.5054	2.0781	54.3353	62.6755	35.7900	81.2208	-9.5054	10.9319	-0.8695	*	
3	50.5000	57.4340	6.6969	43.9957	70.8723	31.3728	83.4952	-6.9340	8.8869	-0.7802	*	
4	49.9000	45.7519	4.2163	37.2913	54.2125	21.8734	69.6303	4.1481	10.2980	0.4028		
5	52.0000	54.0438	2.3243	49.3798	58.7079	31.2326	76.8551	-2.0438	10.8822	-0.1878		
6	55.0000	61.1729	2.9157	55.3221	67.0237	38.0897	84.2560	-6.1729	10.7389	-0.5748	*	
7	80.5000	64.3544	4.8767	54.5687	74.1401	39.9749	88.7339	16.1456	10.0022	1.6142		***
8	86.0000	83.1380	2.6886	77.7429	88.5332	60.1662	106.1	2.8620	10.7980	0.2650		
9	69.0000	61.5442	2.7591	56.0077	67.0806	38.5387	84.5497	7.4558	10.7802	0.6916		*
10	46.0000	46.2502	4.6158	36.9880	55.5124	22.0761	70.4243	-0.2502	10.1252	-0.0247		
11	38.0000	47.8848	4.6813	38.4910	57.2786	23.6600	72.1096	-9.8848	10.0951	-0.9792	*	
12	49.5000	49.7536	3.4056	42.9197	56.5875	26.4019	73.1053	-0.2536	10.5937	-0.0239		
13	105.0	108.1	3.8227	100.5	115.8	84.5282	131.7	-3.1384	10.4505	-0.3003		
14	152.5	126.9	4.0022	118.9	134.9	103.2	150.6	25.6132	10.3831	2.4668		****
15	85.0000	91.0244	4.3214	82.3529	99.6958	67.0704	115.0	-6.0244	10.2543	-0.5875	*	
16	60.0000	49.8455	5.1365	39.5383	60.1527	25.2521	74.4389	10.1545	9.8713	1.0287		**
17	58.5000	48.4248	4.0554	40.2871	56.5625	24.6588	72.1908	10.0752	10.3624	0.9723		*
18	101.0	80.3636	3.3389	73.6636	87.0636	57.0507	103.7	20.6364	10.6150	1.9441		***

(*continues*)

FIGURE 10.19 Continued

SAS

OBS	ACTUAL	PREDICT VALUE	STD ERR PREDICT	LOWER95% MEAN	UPPER95% MEAN	LOWER95% PREDICT	UPPER95% PREDICT	RESIDUAL	STD ERR RESIDUAL	STUDENT RESIDUAL	-2-1-0	1 2
19	79.4000	65.3483	2.6697	59.9911	70.7055	42.3853	88.3113	14.0517	10.8027	1.3008		**
20	125.0	118.5	3.3239	111.9	125.2	95.2192	141.8	6.4766	10.6197	0.6099		*
21	87.9000	103.7	5.1717	93.3319	114.1	79.0865	128.3	-15.8096	9.8529	-1.6046	***	
22	80.0000	80.0728	3.1987	73.6540	86.4915	56.8392	103.3	-0.0728	10.6580	-.006828		
23	94.0000	106.1	4.1337	97.8323	114.4	82.3069	129.9	-12.1272	10.3314	-1.1738	**	
24	74.0000	86.6749	5.4340	75.7708	97.5790	61.8254	111.5	-12.6749	9.7107	-1.3053	**	
25	69.0000	70.8035	1.9689	66.8526	74.7544	48.1273	93.4797	-1.8035	10.9521	-0.1647		
26	63.0000	50.7254	2.6975	45.3125	56.1382	27.7493	73.7014	12.2746	10.7958	1.1370		**
27	67.5000	61.6062	2.7019	56.1845	67.0279	38.6281	84.5843	5.8938	10.7947	0.5460		*
28	35.0000	43.7147	3.9262	35.8362	51.5932	20.0362	67.3932	-8.7147	10.4120	-0.8370	*	
29	142.5	151.7	6.3235	139.0	164.3	126.0	177.3	-9.1540	9.1563	-0.9997	*	
30	92.2000	87.5370	3.3538	80.8071	94.2669	64.2156	110.9	4.6630	10.6103	0.4395		
31	56.0000	55.0141	3.5213	47.9481	62.0801	31.5934	78.4348	0.9859	10.5559	0.0934		
32	63.0000	54.6438	2.9371	48.7500	60.5376	31.5497	77.7379	8.3562	10.7331	0.7785		*
33	60.0000	87.4683	2.1282	83.1978	91.7389	64.7343	110.2	-27.4683	10.9223	-2.5149	*****	
34	34.0000	34.6449	8.6556	17.2762	52.0136	6.3559	62.9340	-0.6449	6.9933	-0.0922		
35	52.0000	58.2825	3.3822	51.4956	65.0695	34.9445	81.6205	-6.2825	10.6012	-0.5926	*	
36	75.0000	70.6005	1.9340	66.7197	74.4814	47.9365	93.2646	4.3995	10.9583	0.4015		
37	93.0000	85.8627	4.4052	77.0231	94.7023	61.8474	109.9	7.1373	10.2186	0.6985		*
38	60.0000	79.3817	4.9475	69.4539	89.3096	54.9449	103.8	-19.3817	9.9674	-1.9445	***	
39	73.0000	55.3228	3.0219	49.2690	61.3966	32.1948	78.4709	17.6672	10.7095	1.6497		***
40	71.0000	62.0418	4.9616	52.0855	71.9980	37.5933	86.4902	8.9582	9.9603	0.8994		*
41	83.0000	75.2296	2.3200	70.5741	79.8851	52.4201	98.0391	7.7704	10.8832	0.7140		*
42	90.0000	93.7940	2.5848	88.6071	98.9809	70.8701	116.7	-3.7940	10.8233	-0.3505		
43	83.0000	88.6489	2.3122	84.0092	93.2886	65.8427	111.5	-5.6489	10.8848	-0.5190	*	
44	115.0	118.8	4.8398	109.1	128.5	94.4401	143.1	-3.7900	10.0201	-0.3782		
45	50.0000	49.0382	2.8877	43.2435	54.8328	25.9692	72.1071	0.9618	10.7465	0.0895		
46	55.2000	54.1699	2.6956	48.7608	59.5790	31.1947	77.1450	1.0301	10.7963	0.0954		
47	61.0000	63.7263	2.8214	58.0647	69.3879	40.6904	86.7622	-2.7263	10.7641	-0.2533		
48	147.0	138.1	4.8418	128.4	147.8	113.7	162.4	8.9233	10.0191	0.8906		*
49	60.0000	71.4469	1.9509	67.5322	75.3616	48.7770	94.1168	-11.4469	10.9554	-1.0449	**	
50	100.0	114.2	3.6062	107.0	121.4	90.7158	137.7	-14.1884	10.5272	-1.3478	**	
51	44.5000	53.9253	2.7958	48.3151	59.5356	30.9020	76.9487	-9.4253	10.7708	-0.8751	*	
52	55.0000	71.8002	3.8278	64.1191	79.4813	48.1867	95.4137	-16.8002	10.4486	-1.6079	***	
53	53.4000	51.4634	4.9144	43.6086	59.3182	27.7928	75.1340	1.9366	10.4165	0.1859		
54	65.0000	66.6020	3.1886	60.2035	73.0004	43.3740	89.8300	-1.6020	10.6611	-0.1503		
55	73.0000	71.8899	3.2630	65.3423	78.4375	48.6204	95.1594	1.1101	10.6386	0.1043		
56	40.0000	44.8479	5.1318	34.5501	55.1456	20.2584	69.4374	-4.8479	9.8737	-0.4910		
57	141.0	134.2	10.4728	113.2	155.3	103.6	164.9	6.7539	3.7612	1.7957		***
58	68.0000	75.5810	1.9479	71.6722	79.4898	52.9121	98.2499	-7.5810	10.9559	-0.6920	*	
59	140.0	116.5	3.4297	109.6	123.4	93.1081	139.8	23.5260	10.5860	2.2224		****
60	55.0000	57.5493	3.2812	50.9650	64.1335	34.2694	80.8291	-2.5493	10.6329	-0.2398		

OBS	RESIDUAL	RSTUDENT	HAT DIAG H	COV RATIO	DFFITS	COOK'S D	INTERCEP DFBETAS	SQFT DFBETAS	SQFTSQ DFBETAS	ROOMS DFBETAS	ROOMSSQ DFBETAS	BED DFBETAS	BEDSQ DFBETAS
1	2.7746	0.2547	0.0588	1.2284	0.0636	0.001	0.0188	0.0113	-0.0130	0.0103	-0.0085	-0.0241	0.0181
2	-9.5054	-0.8674	0.0349	1.0765	-0.1649	0.003	0.0738	-0.0168	0.0317	-0.0409	0.0384	-0.0212	0.0147
3	-6.9340	-0.7773	0.3622	1.6667	-0.5857	0.043	0.1376	0.1266	-0.0696	-0.3673	0.2748	0.0635	-0.0159
4	4.1481	0.3995	0.1436	1.3302	0.1636	0.003	-0.0042	-0.1101	0.0985	-0.0255	0.0182	0.1038	-0.0885
5	-2.0438	-0.1861	0.0436	1.2147	-0.0397	0.000	0.0177	0.0057	-0.0014	-0.0114	0.0106	-0.0103	0.0083
6	-6.1729	-0.5711	0.0687	1.1918	-0.1551	0.003	-0.0215	-0.0201	0.0287	0.0585	-0.0518	-0.0245	0.0108
7	16.1456	1.6402	0.1921	0.9582	0.7997	0.077	0.0517	0.2804	-0.3139	0.2446	-0.2612	-0.4913	0.5555
8	2.8620	0.2627	0.0584	1.2271	0.0654	0.001	-0.0309	0.0234	-0.0259	0.0295	-0.0225	-0.0122	0.0086
9	7.4558	0.6881	0.0615	1.1559	0.1761	0.004	0.0243	0.0367	-0.0428	-0.0807	0.0700	0.0280	-0.0131
10	-0.2502	-0.0245	0.1721	1.4107	-0.0112	0.000	0.0003	0.0081	-0.0073	0.0012	-0.0008	-0.0071	0.0061
11	-9.8848	-0.9788	0.1770	1.2229	-0.4539	0.026	-0.2030	0.2930	-0.2894	0.1708	-0.1527	-0.2054	0.2004
12	-0.2536	-0.0237	0.0937	1.2887	-0.0076	0.000	0.0025	0.0042	-0.0032	-0.0026	0.0024	-0.0031	0.0026
13	-3.1384	-0.2977	0.1180	1.3061	-0.1089	0.002	0.0464	0.0335	-0.0413	-0.0189	0.0112	-0.0441	0.0541
14	25.6132	2.5998	0.1294	0.4957	1.0021	0.113	-0.1978	-0.4628	0.5639	0.1633	-0.1645	0.3687	-0.4305
15	-6.0244	-0.5838	0.1508	1.3041	-0.2460	0.008	-0.0273	-0.0007	-0.0015	-0.0233	0.0207	0.0656	-0.0791
16	10.1545	1.0293	0.2131	1.2592	0.5356	0.036	-0.1471	-0.0400	0.0621	-0.0428	0.0094	0.1603	-0.1604
17	10.0752	0.9718	0.1328	1.1631	0.3803	0.018	0.0608	0.1588	-0.1361	-0.0366	0.0245	-0.1165	0.0821
18	20.6364	1.9993	0.0900	0.7019	0.6289	0.047	-0.3029	-0.0398	0.0789	0.1094	-0.1084	0.1682	-0.2053
19	14.0517	1.3097	0.0576	0.9512	0.3237	0.013	-0.0837	0.0806	-0.1135	0.0876	-0.0712	-0.0380	0.0483
20	6.4766	0.6061	0.0892	1.2109	0.1897	0.005	-0.0480	-0.0519	0.0684	0.0438	-0.0394	0.0452	-0.0586
21	-15.8096	-1.6299	0.2160	0.9925	-0.8555	0.089	-0.2797	-0.3921	0.2780	0.1662	-0.1300	0.3211	-0.2293
22	-0.0728	-0.0068	0.0826	1.2733	-0.0020	0.000	-0.0004	-0.0009	0.0008	0.0012	-0.0010	-0.0000	-0.0000
23	-12.1272	-1.1782	0.1380	1.0931	-0.4714	0.028	-0.1088	0.0117	-0.0769	0.2681	-0.2102	-0.1414	0.1375
24	-12.6749	-1.3144	0.2385	1.1750	-0.7355	0.067	0.2245	-0.1672	0.2311	-0.1471	-0.0005	0.0305	0.0116
25	-1.8035	-0.1631	0.0313	1.2008	-0.0293	0.000	0.0103	-0.0136	0.0141	-0.0031	0.0033	0.0024	-0.0026
26	12.2746	1.1403	0.0588	1.0146	0.2849	0.010	0.0841	0.0506	-0.0581	0.0461	-0.0381	-0.1079	0.0812
27	5.8938	0.5423	0.0590	1.1854	0.1357	0.002	0.0187	0.0324	-0.0352	-0.0682	0.0584	0.0230	-0.0119
28	-8.7147	-0.8345	0.1245	1.1970	-0.3147	0.012	-0.0422	-0.0757	0.0617	0.0170	-0.0074	0.0618	-0.0358
29	-9.1540	-0.9997	0.3229	1.4771	-0.6904	0.060	-0.1531	0.4403	-0.4725	-0.0717	0.0750	-0.972	0.0729
30	4.6630	0.4360	0.0908	1.2471	0.1378	0.002	0.0312	0.0436	-0.0298	-0.0850	0.0698	0.0160	-0.0126
31	0.9859	0.0925	0.1001	1.2963	0.0309	0.000	-0.0083	-0.0125	0.0082	0.0115	-0.0098	0.0078	-0.0060
32	8.3562	0.7756	0.0697	1.1431	0.2122	0.006	0.0675	0.0798	-0.0851	0.0304	-0.0223	-0.1098	0.0901
33	-27.4683	-2.6574	0.0366	0.4296	-0.5178	0.030	0.1168	-0.1412	0.0903	0.0425	-0.0094	-0.0435	0.0583
34	-0.6449	-0.0913	0.6050	2.9535	-0.1130	0.002	-0.0994	0.0267	-0.0275	0.0178	-0.0128	0.0419	-0.0409
35	-6.2825	-0.5889	0.0924	1.2190	-0.1879	0.004	-0.0592	-0.0687	0.0763	-0.0325	0.0225	0.0994	-0.0849
36	4.3995	0.3982	0.0302	1.1749	0.0703	0.001	-0.0258	0.0316	-0.0321	0.0061	-0.0071	-0.0035	0.0037
37	7.1373	0.6950	0.1567	1.2845	0.2996	0.011	0.0474	0.0731	-0.0802	0.1015	-0.1090	-0.1900	0.2146
38	-19.3817	-1.9998	0.1977	0.7958	-0.9926	0.116	-0.2163	-0.2734	0.2834	-0.2742	0.3546	0.6379	-0.7350
39	17.6672	1.6782	0.0737	0.8203	0.4735	0.027	0.1496	0.2112	-0.2191	0.0557	-0.0391	-0.2595	0.2147
40	8.9582	0.8977	0.1988	1.2861	0.4472	0.025	0.0203	0.1372	-0.1520	0.1260	-0.1392	-0.2483	0.2833
41	7.7704	0.7106	0.0435	1.1286	0.1515	0.003	-0.0361	0.0749	-0.0791	0.0193	-0.0169	-0.0298	0.0303
42	-3.7940	-0.3476	0.0540	1.2115	-0.0830	0.001	0.0363	-0.0210	0.0202	-0.0331	0.0259	0.0084	-0.0025
43	-5.6489	-0.5153	0.0432	1.1711	-0.1095	0.002	0.0178	-0.0442	0.0383	0.0001	0.0026	0.0109	-0.0086
44	-3.7900	-0.3751	0.1892	1.4092	-0.1812	0.004	-0.0438	0.0386	-0.0420	-0.0558	0.0629	0.0615	-0.0742
45	0.9618	0.0886	0.0673	1.2509	0.0238	0.000	0.0059	-0.0071	0.0057	0.0063	-0.0054	-0.0022	0.0006
46	1.0301	0.0945	0.0587	1.2391	0.0236	0.000	-0.0087	-0.0066	0.0034	0.0084	-0.0074	0.0060	-0.0046
47	-2.7263	-0.2510	0.0643	1.2360	-0.0658	0.001	-0.0080	-0.0201	0.0178	0.0399	-0.0330	-0.0128	0.0088
48	8.9233	0.8888	0.1893	1.2741	0.4295	0.023	-0.0555	-0.2296	0.2705	0.0596	-0.0620	0.1654	-0.1878

(continues)

FIGURE 10.19 Continued

OBS	RESIDUAL	RSTUDENT	HAT DIAG H	COV RATIO	DFFITS	COOK'S D	INTERCEP DFBETAS	SQFT DFBETAS	SQFTSQ DFBETAS	ROOMS DFBETAS	ROOMSSQ DFBETAS	BED DFBETAS	BEDSQ DFBETAS
49	-11.4469	-1.0458	0.0307	1.0170	-0.1862	0.004	0.0700	-0.0746	0.0701	-0.0041	0.0112	-0.0054	0.0062
50	-14.1884	-1.3587	0.1050	0.9820	-0.4654	0.027	0.1894	0.1423	-0.1741	-0.0996	0.0584	-0.1726	0.2180
51	-9.4253	-0.8731	0.0631	1.1072	-0.2266	0.006	-0.0146	0.0418	-0.0276	0.0857	-0.0716	-0.0952	0.0715
52	-16.8002	-1.6335	0.1183	0.8810	-0.5984	0.043	0.2785	0.0282	-0.0556	-0.0911	0.0916	-0.1524	0.1807
53	1.9366	0.1842	0.1237	1.3259	0.0692	0.001	0.0017	-0.0411	0.0347	-0.0090	0.0072	0.0373	-0.0304
54	-1.6020	-0.1488	0.0821	1.2681	-0.0445	0.000	-0.0076	-0.0137	0.0153	0.0169	-0.0152	-0.0005	-0.0026
55	1.1101	0.1034	0.0860	1.2758	0.0317	0.000	0.0062	0.0135	-0.0136	-0.0140	0.0124	-0.0015	0.0032
56	-4.8479	-0.4874	0.2127	1.4295	-0.2533	0.008	0.0736	0.0293	-0.0357	0.0123	0.0018	-0.0803	0.0788
57	6.7539	1.8362	0.8858	6.1265	5.1126	3.125	1.5981	0.6969	-0.6776	-3.3899	4.0337	0.7545	-0.7651
58	-7.5810	-0.6885	0.0306	1.1190	-0.1224	0.002	0.0401	-0.0556	0.0520	-0.0027	0.0066	0.0045	-0.0033
59	23.5260	2.3135	0.0950	0.5809	0.7496	0.065	-0.0101	-0.1603	0.2526	-0.0685	0.0204	0.1965	-0.2203
60	-2.5493	-0.2376	0.0869	1.2680	-0.0733	0.001	-0.0228	-0.0405	0.0409	-0.0059	0.0035	0.0437	-0.0368

OBS	AGE DFBETAS	OBS	AGE DFBETAS	OBS	AGE DFBETAS
1	-0.0029	21	0.0980	41	-0.0846
2	-0.0035	22	0.0005	42	0.0366
3	0.1101	23	-0.0446	43	0.0492
4	0.0490	24	0.3444	44	0.0554
5	-0.0055	25	0.0062	45	0.0001
6	0.0643	26	-0.0129	46	-0.0035
7	0.0547	27	-0.0259	47	-0.0105
8	-0.0308	28	-0.1970	48	-0.0307
9	-0.0509	29	0.0364	49	-0.0272
10	-0.0026	30	-0.0108	50	-0.0936
11	-0.1502	31	-0.0105	51	-0.0010
12	-0.0003	32	-0.0681	52	-0.4758
13	-0.0477	33	-0.0057	53	-0.0039
14	-0.0145	34	-0.0048	54	0.0246
15	-0.0980	35	0.1094	55	-0.0165
16	0.4965	36	-0.0071	56	-0.2340
17	0.1945	37	-0.0541	57	0.0605
18	0.4523	38	0.1959	58	0.0086
19	-0.2049	39	-0.1521	59	-0.1208
20	-0.0390	40	0.0937	60	0.0273

FIGURE 10.20 SAS output of outlying and influential observation diagnostics for hospitals 1–13 in Table 10.5 when using the model $y = \beta_0 + \beta_1 x_1 + \beta_2 x_5 + \varepsilon$

SAS

ANALYSIS OF VARIANCE

SOURCE	DF	SUM OF SQUARES	MEAN SQUARE	F VALUE	PROB>F
MODEL	2	15373255.12	7686627.56	55.629	0.0001
ERROR	10	1381753.89	138175.39		
C TOTAL	12	16755009.01			

ROOT MSE	371.7195	R-SQUARE	0.9175
DEP MEAN	2176.061	ADJ R-SQ	0.9010
C.V.	17.08222		

PARAMETER ESTIMATES

VARIABLE	DF	PARAMETER ESTIMATE	STANDARD ERROR	T FOR HO: PARAMETER=0	PROB > |T|	VARIANCE INFLATION
INTERCEP	1	2070.44039	681.81832	3.037	0.0125	0
X1	1	29.21340629	2.85627863	10.228	0.0001	1.28383598
X5	1	-354.60309	139.99824	-2.533	0.0297	1.28383598

OBS	ACTUAL	PREDICT VALUE	STD ERR PREDICT	LOWER95% MEAN	UPPER95% MEAN	LOWER95% PREDICT	UPPER95% PREDICT	RESIDUAL	STD ERR RESIDUAL	STUDENT RESIDUAL	-2-1-0	1 2
1	566.5	947.3	174.9	557.5	1337.1	31.9289	1862.7	-380.8	328.0	-1.1610	**	
2	696.8	902.6	284.3	269.2	1535.9	-140.1	1945.2	-205.7	239.5	-0.8590	*	
3	1033.1	1149.3	177.3	754.1	1544.4	231.6	2067.0	-116.1	326.7	-0.3555		
4	1603.6	1234.9	207.9	771.7	1698.2	285.9	2184.0	368.7	308.1	1.1965		**
5	1611.4	1557.4	124.9	1279.1	1835.8	683.7	2431.2	53.9470	350.1	0.1541		
6	1613.3	1751.5	136.6	1447.1	2056.0	869.1	2634.0	-138.3	345.7	-0.4000		
7	1854.2	1698.3	123.7	1422.8	1973.9	825.5	2571.2	155.8	350.5	0.4446		
8	2160.5	1976.0	106.0	1739.9	2212.1	1114.8	2837.2	184.5	356.3	0.5179		*
9	2305.6	2636.4	150.1	2301.9	2971.0	1743.2	3529.7	-330.9	340.0	-0.9730	*	
10	3503.9	3629.5	186.7	3213.5	4045.6	2702.6	4556.4	-125.6	321.4	-0.3908		
11	3571.9	2789.9	131.9	2496.0	3083.8	1911.0	3668.7	782.0	347.5	2.2502		****
12	3741.4	4179.2	239.2	3646.3	4712.1	3194.3	5164.1	-437.8	284.6	-1.5385	***	
13	4026.5	3836.4	188.8	3415.7	4257.0	2907.4	4765.3	190.2	320.2	0.5939		*

OBS	RESIDUAL	RSTUDENT	HAT DIAG H	COV RATIO	DFFITS	COOK'S D	INTERCEP DFBETAS	X1 DFBETAS	X5 DFBETAS
1	-380.8	-1.1841	0.2215	1.1412	-0.6316	0.128	-0.3480	0.3372	0.1794
2	-205.7	-0.8467	0.5848	2.6251	-1.0049	0.346	0.7625	0.6177	-0.9116
3	-116.1	-0.3394	0.2276	1.7095	-0.1842	0.012	-0.1270	0.0710	0.0831
4	368.7	1.2262	0.3129	1.2558	0.8274	0.217	0.6847	-0.1916	-0.5212
5	53.9470	0.1464	0.1129	1.5354	0.0522	0.001	-0.0093	-0.0278	0.0218
6	-138.3	-0.3825	0.1351	1.5111	-0.1512	0.008	-0.1055	0.0215	0.0754
7	155.8	0.4260	0.1107	1.4528	0.1503	0.008	-0.0416	-0.0686	0.0735
8	184.5	0.4981	0.0812	1.3761	0.1481	0.008	0.0435	-0.0204	-0.0145
9	-330.9	-0.9701	0.1632	1.2163	-0.4284	0.062	0.2395	-0.0510	-0.2472
10	-125.6	-0.3736	0.2524	1.7520	-0.2171	0.017	0.0675	-0.1347	-0.0433
11	782.0	3.0384	0.1259	0.1888	1.1532	0.243	-0.3737	0.3690	0.3714
12	-437.8	-1.6706	0.4139	1.0409	-1.4040	0.557	-0.6698	-1.2281	0.8518
13	190.2	0.5736	0.2579	1.6598	0.3382	0.041	0.0508	0.2801	-0.0943

models describing hospitals 1–13 is the model

$$y = \beta_0 + \beta_1 x_1 + \beta_2 x_5 + \varepsilon$$

Figure 10.20 presents the SAS output of outlying and influential observation diagnostics for this model. Fully interpret this output.

10.12 (Optional) Use SAS to plot FAT against each of the predictor variables in Exercise 10.6. Also, use SAS to make the corresponding partial leverage residual plots. Interpret these plots.

CHAPTER 11

MODEL BUILDING

We saw in Chapter 10 that multicollinearity can hinder our ability to use the hypothesis test of $H_0: \beta_j = 0$ versus $H_1: \beta_j \neq 0$ to judge the importance of the independent variable x_j. Consequently, we need to use additional procedures to build an appropriate regression model. In this chapter we discuss several such procedures and how to combine many of the techniques of this book to determine an appropriate model. Sections 11.1 and 11.2 discuss various criteria that can be used to compare the utility of overall regression models: the multiple coefficient of determination, R^2; the mean square error, s^2 (and standard error, s); lengths of prediction intervals; corrected (or adjusted) R^2; the C statistic; and the PRESS statistic. In Section 11.3 we present several computerized screening procedures (stepwise regression, backward elimination, forward selection, and maximum R^2 improvement) that can be used to identify a set or sets of independent variables that might be important when there are a large number of potentially important independent variables. Once we have selected an appropriate model, it is important to validate this model by using it to analyze a data set that is different from the data set used to build the model. We discuss this procedure in Section 11.4. We complete this chapter with optional Section 11.5, which discusses using SAS to implement the techniques of this chapter.

11.1

COMPARING REGRESSION MODELS USING R^2, s, PREDICTION INTERVAL LENGTH, AND CORRECTED R^2

COMPARING REGRESSION MODELS USING R^2, s, AND PREDICTION INTERVAL LENGTH

When building a regression model, we wish to obtain a final model that is *reasonably easy and inexpensive to use* (in terms, for example, of being able to accurately and inexpensively determine future values of the independent variables). In addition, we want a model that *accurately describes, predicts, and controls* the dependent variable. In previous sections we saw that data plots, t statistics, and prob-values can help to determine an appropriate set of independent variables. However, since multicollinearity can cause the t statistics and prob-values to give misleading impressions of the importance of the independent variables, we need other procedures to help us find an appropriate final regression model.

One procedure is to consider all reasonable regression models and compare them on the basis of several criteria. One criterion is R^2, the multiple coefficient of determination. We have seen that the larger R^2 is, the larger is the proportion of the total variation that is explained by the regression model. However, we must balance the magnitude of R^2, or in general, the "goodness" of any criterion, against the difficulty and expense of using the regression model. Generally speaking, the size of R^2 does not necessarily indicate whether a regression model will accurately describe, predict, and control the dependent variable. For example, although a value of R^2 close to 1 indicates that the regression model explains a high proportion of the total variation, such a value of R^2 does not necessarily imply that the model will produce predictions accurate enough for the application at hand. Moreover, when using R^2 to compare overall regression models, we must be aware that (it can be proved that) *adding any independent variable to a regression model*, even an unimportant independent variable:

1. Will decrease the unexplained variation

$$SSE = \sum_{i=1}^{n} (y_i - \hat{y}_i)^2$$

2. Will leave the total variation unchanged, since

$$\text{Total variation} = \sum_{i=1}^{n} (y_i - \bar{y})^2$$

is a function of only the n observed values $y_1, y_2, \ldots, y_n$ and thus is independent of the regression model under consideration.

3. Will increase

$$R^2 = 1 - \frac{\text{Unexplained variation}}{\text{Total variation}}$$

Since adding any independent variable will increase R^2, it would be absurd to continue to add independent variables until R^2 decreases, because R^2 will never decrease!

One important consideration in predicting y_0, an individual value of the dependent variable, is the *length of the prediction interval*

$$[\hat{y}_0 \pm t^{(n-k)}_{[\alpha/2]} s\sqrt{1 + h_{00}}]$$

The length of this interval is

$$2t^{(n-k)}_{[\alpha/2]} s\sqrt{1 + h_{00}} = 2t^{(n-k)}_{[\alpha/2]} s\sqrt{1 + \mathbf{x}_0'(\mathbf{X}'\mathbf{X})^{-1}\mathbf{x}_0}$$

Since a shorter prediction interval provides a more precise guess of y_0 than does a longer prediction interval, a shorter prediction interval is more desirable. A criterion related to the length of the prediction interval is the standard error

$$s = \sqrt{\frac{SSE}{n - k}}$$

Since we want the length of the prediction interval to be small, we want s to be small. Consequently, we want R^2 to be large and s to be small. The standard error is a better criterion than R^2 for comparing regression models because adding *any* independent variable to a regression model will increase R^2 but adding an "unimportant" independent variable can increase s. As we will demonstrate in Example 11.1, this can happen when the decrease in the SSE caused by the addition of the independent variable is not enough to offset the decrease in the denominator $n - k$. If addition of an extra independent variable does increase s, this indicates that the length of the $100(1 - \alpha)\%$ prediction interval for y_0 will almost surely increase. This is because adding an independent variable also:

1. Increases k, the number of parameters in the regression model.
2. Decreases $n - k$, the number of degrees of freedom upon which the $t^{(n-k)}_{[\alpha/2]}$-point is based.
3. Increases the $t^{(n-k)}_{[\alpha/2]}$-point (as will be demonstrated in Example 11.1).
4. Usually does not decrease (but often increases) the matrix algebra expression $\sqrt{1 + h_{00}}$.

On the other hand, if adding an independent variable to a regression model decreases s, then the length of the prediction interval for y_0 will decrease if and only if the decrease in s is enough to offset the increase in $t^{(n-k)}_{[\alpha/2]}$ and the possible increase in $\sqrt{1 + h_{00}}$. To summarize:

1. The independent variable x_j should not be included in the final regression

model unless x_j reduces s enough to reduce the length of the $100(1 - \alpha)\%$ prediction interval for y_0.

2. Prediction of a value of the dependent variable requires knowledge of the corresponding value of the independent variable x_j. Thus we must decide whether the inclusion of x_j reduces s and the length of the $100(1 - \alpha)\%$ prediction interval for y_0 enough to offset the potential errors caused by possible inaccurate determination of values of x_j, or the possible expense of accurately (or inaccurately) determining values of x_j.

Since the key factor is the length of the $100(1 - \alpha)\%$ prediction interval for y_0, one might wonder why we do not directly compare prediction interval lengths. While it is useful to do this, the length of the $100(1 - \alpha)\%$ prediction interval for y_0 depends upon $\mathbf{x}_0'$, the row vector containing the values of the independent variables. We often wish to compute prediction intervals for several different combinations of values of the independent variables (and thus several different $\mathbf{x}_0'$ vectors). Thus we would compute prediction intervals having slightly different lengths. However, the standard error s is a constant factor with respect to the length of prediction intervals (as long as we are considering the same regression model). Thus it is common practice to compare regression models on the basis of s (and s^2).

Example 11.1 Recall that in the fuel consumption problem, x_1 is the average hourly temperature, x_2 is the chill index, x_3 is the average hourly wind velocity, and y is the weekly fuel consumption. In Table 11.1 we summarize the total variation, the unexplained variation, the explained variation, R^2, s^2, and s for eight regression models describing the fuel consumption data in Table 10.2. Note that the model

$$y = \beta_0 + \beta_1 x_1 + \varepsilon$$

has the largest R^2 and smallest s of the three one-variable models, and the model

$$y = \beta_0 + \beta_1 x_1 + \beta_2 x_2 + \varepsilon$$

has the largest R^2 and smallest s of the three two-variable models. Table 11.2 summarizes information concerning 95% prediction intervals given by four of the models in Table 11.1. This table contains, for each model, the 95% prediction interval for y_0, a future individual fuel consumption that will occur in a week having an average hourly temperature of $x_{01} = 40.0$, a chill index of $x_{02} = 10$, and an average hourly wind velocity of $x_{03} = 11$; and the length of this 95% prediction interval.

First, note from Tables 11.1 and 11.2 that adding the independent variable x_2 to the regression model

$$y = \beta_0 + \beta_1 x_1 + \varepsilon$$

TABLE 11.1 R^2, s^2, and s for eight regression models in the fuel consumption problem

Model	Total variation	Unexplained variation	Explained variation	$R^2 = \frac{\text{Explained variation}}{\text{Total variation}}$	k	$n - k = 8 - k$	$s^2 = \frac{\text{Unexplained variation}}{n - k}$	$s = \sqrt{s^2}$
$y = \beta_0 + \beta_1 x_1 + \varepsilon$	25.55	2.57	22.98	.90	2	6	.428	.654
$y = \beta_0 + \beta_1 x_2 + \varepsilon$	25.55	6.18	19.37	.758	2	6	1.03	1.015
$y = \beta_0 + \beta_1 x_3 + \varepsilon$	25.55	3.71	21.84	.855	2	6	.618	.786
$y = \beta_0 + \beta_1 x_1 + \beta_2 x_2 + \varepsilon$	25.55	.674	24.876	.974	3	5	.135	.367
$y = \beta_0 + \beta_1 x_1 + \beta_2 x_3 + \varepsilon$	25.55	1.48	24.07	.942	3	5	.295	.543
$y = \beta_0 + \beta_1 x_2 + \beta_2 x_3 + \varepsilon$	25.55	2.55	23.0	.90	3	5	.511	.715
$y = \beta_0 + \beta_1 x_1 + \beta_2 x_2 + \beta_3 x_1 x_2 + \varepsilon$	25.55	.660	24.89	.9742	4	4	.165	.406
$y = \beta_0 + \beta_1 x_1 + \beta_2 x_2 + \beta_3 x_3 + \varepsilon$	25.55	.533	25.017	.979	4	4	.133	.365

TABLE 11.2 **The lengths of 95% prediction intervals for four regression models in the fuel consumption problem**

Model	Point prediction when $x_{01} = 40.0$, $x_{02} = 10$, $x_{03} = 11$	k	$n - k = 8 - k$	$t^{(8-k)}_{[.025]}$	s
(1) $y = \beta_0 + \beta_1 x_1 + \varepsilon$	$\hat{y}_0 = b_0 + b_1(40) = 10.72$	2	6	2.447	.654
(2) $y = \beta_0 + \beta_1 x_1 + \beta_2 x_2 + \varepsilon$	$\hat{y}_0 = b_0 + b_1(40) + b_2(10) = 10.33$	3	5	2.571	.367
(3) $y = \beta_0 + \beta_1 x_1 + \beta_2 x_2 + \beta_3 x_1 x_2 + \varepsilon$	$\hat{y}_0 = b_0 + b_1(40) + b_2(10) + b_3(40)(10) = 10.27$	4	4	2.776	.406
(4) $y = \beta_0 + \beta_1 x_1 + \beta_2 x_2 + \beta_3 x_3 + \varepsilon$	$\hat{y}_0 = b_0 + b_1(40) + b_2(10) + b_3(11) = 10.60$	4	4	2.776	.365

Model	$\sqrt{1 + h_{00}}$	95% prediction interval	Length of 95% prediction interval $2t^{(n-k)}_{[\alpha/2]} s\sqrt{1 + h_{00}}$
(1)	$\sqrt{1.14}$	[9.01, 12.43]	3.42
(2)	$\sqrt{1.21}$	[9.29, 11.37]	2.08
(3)	$\sqrt{1.5}$	[8.89, 11.65]	2.76
(4)	$\sqrt{1.75}$	[9.26, 11.94]	2.68

to form the regression model

$$y = \beta_0 + \beta_1 x_1 + \beta_2 x_2 + \varepsilon$$

1. Increases R^2 from .90 to .974.
2. Decreases s^2 from .428 to .135 and s from .654 to .367.
3. Increases $t^{(8-k)}_{[.025]}$ from $t^{(6)}_{[.025]} = 2.447$ to $t^{(5)}_{[.025]} = 2.571$.
4. Increases $\sqrt{1 + h_{00}}$ from $\sqrt{1.14}$ to $\sqrt{1.21}$.
5. Reduces the length of the 95% prediction interval for y_0 from 3.42 to 2.08.

Here the decrease in s caused by the addition of x_2 has been enough to offset the increases in $t^{(8-k)}_{[.025]}$ and $\sqrt{1 + h_{00}}$. Recall that we have strong evidence that $H_0: \beta_2 = 0$ is false (because the prob-value related to this hypothesis is .01330 —see Table 8.6). Besides this, we now see that the addition of x_2 decreases s and the length of the 95% prediction interval for y_0. Thus we conclude that x_2 has substantial additional importance over the importance of x_1 in this two-variable model. Hence we conclude that the independent variable x_2 should be included in a final regression model.

We next consider including the cross-product term $x_1 x_2$, which should be utilized if interaction exists between x_1 and x_2. Table 11.1 tells us that adding $x_1 x_2$ to the model

$$y = \beta_0 + \beta_1 x_1 + \beta_2 x_2 + \varepsilon$$

to form the interaction model

$$y = \beta_0 + \beta_1 x_1 + \beta_2 x_2 + \beta_3 x_1 x_2 + \varepsilon$$

increases s^2 from .135 to .165. In this case, although the addition of $x_1 x_2$ has decreased the unexplained variation, this decrease (from .674 to .66) has not been enough to offset the change in the denominator of s^2, which decreases from 5 to 4. Examining Tables 11.1 and 11.2, we see that adding the cross-product term x_1x_2:

1. Increases R^2 from .974 to .9742.
2. Increases s^2 from .135 to .165 and increases s from .367 to .406.
3. Increases $t_{[.025]}^{(8-k)}$ from $t_{[.025]}^{(5)} = 2.571$ to $t_{[.025]}^{(4)} = 2.776$.
4. Increases $\sqrt{1 + h_{00}}$ from $\sqrt{1.21}$ to $\sqrt{1.5}$.
5. Increases the length of the 95% prediction interval from 2.08 to 2.76.

Here the addition of x_1x_2 increases $t_{[.025]}^{(8-k)}$, s, and $\sqrt{1 + h_{00}}$. Moreover, the t statistic and the prob-value for the hypothesis $H_0: \beta_3 = 0$ are

$$t = \frac{b_3}{s\sqrt{c_{33}}} = \frac{-.0006}{.406\sqrt{.000022}} = -.2894 \quad \text{and} \quad \text{prob-value} = .7866$$

Therefore we cannot reject $H_0: \beta_3 = 0$ with a small probability of a Type I error. Because of this, and because the addition of the cross-product term x_1x_2 increases s and the length of the 95% prediction interval for y_0, we conclude that x_1x_2 does not have substantial additional importance over the combined importance of x_1 and x_2 in the above interaction model. Therefore we should not include the independent variable x_1x_2 in a final regression model.

Last, note from Tables 11.1 and 11.2 that adding the independent variable x_3 to the regression model

$$y = \beta_0 + \beta_1x_1 + \beta_2x_2 + \varepsilon$$

to form the regression model

$$y = \beta_0 + \beta_1x_1 + \beta_2x_2 + \beta_3x_3 + \varepsilon$$

1. Increases R^2 from .974 to .979.
2. Decreases s from .367 to .365.
3. Increases $t_{[.025]}^{(8-k)}$ from $t_{[.025]}^{(5)} = 2.571$ to $t_{[.025]}^{(4)} = 2.776$.
4. Increases $\sqrt{1 + h_{00}}$ from $\sqrt{1.21}$ to $\sqrt{1.75}$.
5. Increases the length of the 95% prediction interval from 2.08 to 2.68.

Here the decrease in s (from .367 to .365) caused by the addition of x_3 has not been enough to offset the increases in $t_{[.025]}^{(8-k)}$ and $\sqrt{1 + h_{00}}$. It should be noted that the large increase in $\sqrt{1 + h_{00}}$ here is partly caused by the fact that $x_{03} = 11$ causes the point $(x_{01}, x_{02}, x_{03}) = (40.0, 10, 11)$ to be outside the experimental region. However, even if there had been no increase in $\sqrt{1 + h_{00}}$, the decrease in s caused by the addition of x_3 would not have been enough to offset the increase in $t_{[.025]}^{(8-k)}$. Recall that we cannot reject $H_0: \beta_3 = 0$ with a small probability of a Type I error (from Table 10.3 the prob-value for this hypothesis test is .3614). Besides this, we now see that the addition of x_3 does not decrease s enough to decrease the length of the 95% prediction interval for y_0. We conclude that x_3 does not have substantial additional importance over the combined importance of x_1 and x_2. In addition, inclusion of x_3 in a final regression model would make the model more difficult and possibly more expensive to use (since future values of x_3 would have to be predicted). Therefore the independent variable x_3 should not be included in a final regression model.

To summarize, we have decided that it is worthwhile to include the independent variables x_1 and x_2 in a final regression model but that it is not worthwhile to include x_1x_2 or x_3. Thus a reasonable final regression model for predicting weekly fuel consumption is

$$y = \beta_0 + \beta_1 x_1 + \beta_2 x_2 + \varepsilon$$

Example 11.2 Recall that in the Fresh Detergent problem x_1 is the price for Fresh, x_2 is the average industry price, x_4 is the price difference, x_3 is the advertising expenditure for Fresh, and y is the demand for Fresh. We now consider finding a final regression model to use to predict demand. Table 11.3 summarizes R^2 and s^2 for various possible models describing the Fresh Detergent data in Table 8.3. Consider using the term $x_4x_3^2$, in addition to the term x_4x_3, to model the interaction between x_4 and x_3. Examination of Table 11.3 reveals that adding $x_4x_3^2$ to the model

$$y = \beta_0 + \beta_1 x_4 + \beta_2 x_3 + \beta_3 x_3^2 + \beta_4 x_4 x_3 + \varepsilon$$

TABLE 11.3 **R^2 and s^2 for several regression models in the Fresh Detergent problem**

Model	R^2	s^2
$y = \beta_0 + \beta_1 x_1 + \varepsilon$	.2202	.3784
$y = \beta_0 + \beta_1 x_1 + \beta_2 x_1^2 + \varepsilon$	.2286	.3845
$y = \beta_0 + \beta_1 x_2 + \varepsilon$	.5490	.2168
$y = \beta_0 + \beta_1 x_2 + \beta_2 x_2^2 + \varepsilon$	.5590	.2198
$y = \beta_0 + \beta_1 x_3 + \varepsilon$	.7673	.1119
$y = \beta_0 + \beta_1 x_3 + \beta_2 x_3^2 + \varepsilon$	.8380	.0808
$y = \beta_0 + \beta_1 x_4 + \varepsilon$	.7915	.1002
$y = \beta_0 + \beta_1 x_4 + \beta_2 x_4^2 + \varepsilon$	.8043	.0975
$y = \beta_0 + \beta_1 x_1 + \beta_2 x_2 + \varepsilon$	.8288	.0854
$y = \beta_0 + \beta_1 x_1 + \beta_2 x_3 + \varepsilon$	.7717	.1138
$y = \beta_0 + \beta_1 x_2 + \beta_2 x_3 + \varepsilon$	.8377	.0809
$y = \beta_0 + \beta_1 x_1 + \beta_2 x_2 + \beta_3 x_3 + \varepsilon$	.8936	.0551
$y = \beta_0 + \beta_1 x_1 + \beta_2 x_2 + \beta_3 x_3 + \beta_4 x_3^2 + \varepsilon$	.9084	.0493
$y = \beta_0 + \beta_1 x_1 + \beta_2 x_1^2 + \beta_3 x_2 + \beta_4 x_2^2 + \beta_5 x_3 + \beta_6 x_3^2 + \varepsilon$	.9161	.0491
$y = \beta_0 + \beta_1 x_4 + \beta_2 x_3 + \varepsilon$	.8860	.0568
$y = \beta_0 + \beta_1 x_4 + \beta_2 x_3 + \beta_3 x_3^2 + \varepsilon$	.9054	.0490
$y = \beta_0 + \beta_1 x_4 + \beta_2 x_4^2 + \beta_3 x_3 + \beta_4 x_3^2 + \varepsilon$	.9106	.0481
$y = \beta_0 + \beta_1 x_4 + \beta_2 x_3 + \beta_3 x_3^2 + \beta_4 x_4 x_3 + \varepsilon$	.9209	.0426
$y = \beta_0 + \beta_1 x_4 + \beta_2 x_3 + \beta_3 x_3^2 + \beta_4 x_4 x_3 + \beta_5 x_4 x_3^2 + \varepsilon$	.9225	.0434

to form the model

$$y = \beta_0 + \beta_1 x_4 + \beta_2 x_3 + \beta_3 x_3^2 + \beta_4 x_4 x_3 + \beta_5 x_4 x_3^2 + \varepsilon$$

increases s^2 from .0426 to .0434. Thus the addition of $x_4 x_3^2$ is not warranted. It should also be noted that since the independent variables $x_4 x_3$ and $x_4 x_3^2$ are related to each other (and to the independent variables x_4, x_3, and x_3^2), substantial multicollinearity exists between the independent variables in this model. In fact, so much multicollinearity exists that the prob-values related to the hypotheses $H_0: \beta_4 = 0$ and $H_0: \beta_5 = 0$ in this model are both larger than .10. This makes the interaction terms $x_4 x_3$ and $x_4 x_3^2$ appear to be unimportant. If we were to conclude from this that little or no interaction exists between x_4 and x_3, we would be making a mistake. This is because we saw in Figure 9.5 that the prob-value related to the hypothesis $H_0: \beta_4 = 0$ in the model

$$y = \beta_0 + \beta_1 x_4 + \beta_2 x_3 + \beta_3 x_3^2 + \beta_4 x_4 x_3 + \varepsilon$$

is less than .05. This says that the interaction term $x_4 x_3$ really is important. Thus we should include the interaction term $x_4 x_3$ in the Fresh Detergent demand model. This illustrates that when we are considering using squared terms and interaction terms, multicollinearity can hinder the ability of t statistics and prob-values to indicate accurately which of these terms are important.

To decide upon a final model, note that the model

$$y = \beta_0 + \beta_1 x_4 + \beta_2 x_3 + \beta_3 x_3^2 + \beta_4 x_4 x_3 + \varepsilon$$

has the smallest s^2 of any model in Table 11.3. It also yields the shortest prediction interval of any model in Table 11.4. In addition, the prob-values in Figure 9.5 indicate that the intercept β_0 and the independent variables x_4, x_3, x_3^2, and $x_4 x_3$ are important in describing demand for Fresh. All these considerations imply that this model is a reasonable final model to use to predict demand for Fresh Detergent.

To conclude this example, we should note that, although we have reasoned that it is not appropriate to include the cross-product term $x_4 x_3^2$ in the Fresh Detergent model, in some situations it is appropriate to model interaction by using such terms. For example, consider the construction project model of the exercises of Chapter 9:

$$y = \beta_0 + \beta_1 x_1 + \beta_2 x_1^2 + \beta_3 x_2 + \beta_4 x_1 x_2 + \varepsilon$$

It can be shown that this model has a mean square error equal to .9422. The reader will see in the exercises at the end of this chapter that if we add the term $x_1^2 x_2$ to this model to form the model

$$y = \beta_0 + \beta_1 x_1 + \beta_2 x_1^2 + \beta_3 x_2 + \beta_4 x_1 x_2 + \beta_5 x_1^2 x_2 + \varepsilon$$

the mean square error is reduced to .7251. This occurs even though (because of added multicollinearity) the prob-values related to the t statistics corresponding to the individual independent variables in the expanded model do not make these variables seem important. This and other considerations to be analyzed in the

TABLE 11.4 The lengths of 95% prediction intervals for three regression models in the Fresh Detergent problem

Model	Point prediction when $x_{01} = 3.80$, $x_{02} = 3.90$, $x_{04} = .10$, $x_{03} = 6.80$	95% prediction interval	Length of 95% prediction interval
(1) $y = \beta_0 + \beta_1 x_1 + \beta_2 x_2 + \beta_3 x_3 + \varepsilon$	$\hat{y}_0 = b_0 + b_1(3.80) + b_2(3.90) + b_3(6.80) = 8.325$	[7.80, 8.85]	1.05
(2) $y = \beta_0 + \beta_1 x_4 + \beta_2 x_3 + \beta_3 x_3^2 + \varepsilon$	$\hat{y}_0 = b_0 + b_1(.10) + b_2(6.80) + b_3(6.80)^2 = 8.445$	[7.96, 8.93]	.97
(3) $y = \beta_0 + \beta_1 x_4 + \beta_2 x_3 + \beta_3 x_3^2 + \beta_4 x_4 x_3 + \varepsilon$	$\hat{y}_0 = b_0 + b_1(.10) + b_2(6.80) + b_3(6.80)^2 + b_4(.10)(6.80) = 8.526$	[8.07, 8.98]	.91

exercises at the end of this chapter imply that it is appropriate to include the term $x_1^2x_2$ in the model.

Example 11.3 In Table 11.5 we repeat the data in Table 10.5 concerning the need for hospital labor in $n = 17$ U.S. Navy Hospitals. Here, x_1 denotes average daily patient load, x_2 denotes monthly X-ray exposures, x_3 denotes monthly occupied bed days, x_4 denotes eligible population in the area (divided by 1000), x_5 denotes average length of patients' stay in days, and y denotes monthly labor hours. In Table 11.6 we present s^2 and R^2 for all reasonable regression models describing this data. We see that the model

$$y = \beta_0 + \beta_1x_2 + \beta_2x_3 + \beta_3x_5 + \varepsilon$$

has the smallest s^2 of any model in the table. If we calculate the least squares point estimates of the model parameters, we obtain the prediction equation

$$\hat{y} = 1523.389 + .0530x_2 + .9785x_3 - 320.951x_5$$

If we use this prediction equation to calculate the residuals for all 17 hospitals, we find that 16 of the 17 residuals are reasonably small. However, e_{14}, the residual corresponding to hospital 14, is quite large. Specifically, since Table 11.5 tells us that $x_{14,2} = 36194$, $x_{14,3} = 7684.10$, and $x_{14,5} = 7.00$, the above prediction equation tells us that

$$\begin{aligned}\hat{y}_{14} &= 1523.389 + .0530(36194) + .9785(7684.10) - 320.951(7) \\ &= 8713\end{aligned}$$

TABLE 11.5 **Hospital labor data**

Hospital	x_1	x_2	x_3	x_4	x_5	y
1	15.57	2,463	472.92	18.0	4.45	566.52
2	44.02	2,048	1,339.75	9.5	6.92	696.82
3	20.42	3,940	620.25	12.8	4.28	1,033.15
4	18.74	6,505	568.33	36.7	3.90	1,603.62
5	49.20	5,723	1,497.60	35.7	5.50	1,611.37
6	44.92	11,520	1,365.83	24.0	4.60	1,613.27
7	55.48	5,779	1,687.00	43.3	5.62	1,854.17
8	59.28	5,969	1,639.92	46.7	5.15	2,160.55
9	94.39	8,461	2,872.33	78.7	6.18	2,305.58
10	128.02	20,106	3,655.08	180.5	6.15	3,503.93
11	96.00	13,313	2,912.00	60.9	5.88	3,571.89
12	131.42	10,771	3,921.00	103.7	4.88	3,741.40
13	127.21	15,543	3,865.67	126.8	5.50	4,026.52
14	252.90	36,194	7,684.10	157.7	7.00	10,343.81
15	409.20	34,703	12,446.33	169.4	10.78	11,732.17
16	463.70	39,204	14,098.40	331.4	7.05	15,414.94
17	510.22	86,533	15,524.00	371.6	6.35	18,854.45

Source: Procedures and Analysis for Staffing Standards Development: Regression Analysis Handbook (San Diego, CA: Navy Manpower and Material Analysis Center, 1979).

TABLE 11.6 **s^2 and R^2 for all reasonable regression models describing the hospital labor data**

Variables in model	s^2	R^2
x_5	21,940,359	.3348
x_4	3,816,879	.8843
x_2	3,517,337	.8934
x_1	939,990	.9715
x_3	917,487	.9722
x_4, x_5	3,165,704	.9104
x_2, x_5	2,688,542	.9239
x_2, x_4	2,452,812	.9306
x_1, x_3	971,399	.9725
x_1, x_4	914,181	.9741
x_3, x_4	870,761	.9754
x_1, x_5	564,291	.9840
x_3, x_5	538,720	.9848
x_1, x_2	490,585	.9861
x_2, x_3	469,456	.9867
x_2, x_4, x_5	1,816,626	.9523
x_1, x_3, x_4	818,290	.9785
x_1, x_4, x_5	583,845	.9846
x_3, x_4, x_5	570,794	.9850
x_1, x_3, x_5	569,830	.9850
x_1, x_2, x_4	528,257	.9861
x_2, x_3, x_4	504,218	.9868
x_1, x_2, x_3	482,286	.9873
x_1, x_2, x_5	403,215	.9894
x_2, x_3, x_5	377,954	.9901
x_1, x_3, x_4, x_5	615,741	.9851
x_1, x_2, x_3, x_4	499,469	.9879
x_1, x_2, x_3, x_5	389,793	.9905
x_1, x_2, x_4, x_5	387,001	.9906
x_2, x_3, x_4, x_5	378,826	.9908
x_1, x_2, x_3, x_4, x_5	412,277	.9908

is the point prediction of $y_{14} = 10{,}344$. This implies that

$$e_{14} = y_{14} - \hat{y}_{14} = 10{,}344 - 8713 = 1631$$

The fact that the above prediction equation greatly underpredicts y_{14} indicates that the number of labor hours utilized in hospital 14 was much greater than would be expected for its values of x_2, x_3, and x_5. If the value $y_{14} = 10{,}344$ was recorded erroneously or resulted from a situation that would be very unlikely to occur again, it would be reasonable to remove observation 14 from the data set. It can be verified that removing observation 14 and performing a regression analysis by using the remaining 16 observations (1) reduces the standard error s (which is a

FIGURE 11.1 **SAS output obtained by using all observations in Table 11.5**

VARIABLE	DF	PARAMETER ESTIMATE	STANDARD ERROR	T FOR H0: PARAMETER=0	PROB > |T|	VARIANCE INFLATION
INTERCEP	1	1523.389	786.898	1.936	0.0749	0
XRAY	1	0.052987	0.020092	2.637	0.0205	7.73733111
BEDDAY	1	0.978482	0.105154	9.305	0.0001	11.26934216
STAYDAY	1	-320.951	153.192	-2.095	0.0563	2.49290119

FIGURE 11.2 **SAS output obtained by using all observations except observation 14 in Table 11.5**

VARIABLE	DF	PARAMETER ESTIMATE	STANDARD ERROR	T FOR H0: PARAMETER=0	PROB > |T|	VARIANCE INFLATION
INTERCEP	1	1946.802	504.182	3.861	0.0023	0
XRAY	1	0.038577	0.013042	2.958	0.0120	7.82831925
BEDDAY	1	1.039392	0.067556	15.386	0.0001	11.39619473
STAYDAY	1	-413.758	98.598280	-4.196	0.0012	2.51955937

function of the residuals) from 614.779 to 387.160 and (2) reduces the prob-values corresponding to the intercept, x_2, x_3, and x_5 from .0749, .0205, .0001, and .0563 to .0023, .0120, .0001, and .0012 (see Figures 11.1 and 11.2). Moreover, the model

$$y = \beta_0 + \beta_1 x_2 + \beta_2 x_3 + \beta_3 x_5 + \varepsilon$$

still has the smallest s^2 of any model in Table 11.6. To illustrate utilizing this model to predict future hospital labor requirements, consider a hospital for which $x_{02} = 54{,}000$, $x_{03} = 14{,}400$, and $x_{05} = 6.9$. Note that this combination of values is roughly in the range of "large hospitals" 15, 16, and 17. In the next section we discuss the implications of the fact that hospitals 15, 16, and 17 are substantially larger than the other hospitals in Table 11.5. Doing all calculations with observation 14 removed from the data set, we find that

$$\begin{aligned} \hat{y}_0 &= b_0 + b_1 x_{02} + b_2 x_{03} + b_3 x_{05} \\ &= 1946.802 + .0386(54000) + 1.0394(14400) - 413.758(6.9) \\ &= 16142 \end{aligned}$$

is a point prediction of y_0, the actual labor hour requirement at the hospital in the future month. Moreover, using $\mathbf{x}_0' = [1 \quad 54000 \quad 14400 \quad 6.9]$ implies that $h_{00} = \mathbf{x}_0'(\mathbf{X}'\mathbf{X})^{-1}\mathbf{x}_0 = .4415$. It follows that a 95% prediction interval for y_0 is (since $n - k = 16 - 4 = 12$)

$$\begin{aligned} [\hat{y}_0 \pm t_{[.025]}^{(12)} s\sqrt{1 + h_{00}}] &= [16142 \pm 2.179(387.160)\sqrt{1.4415}] \\ &= [15130, 17155] \end{aligned}$$

CORRECTED R^2

Since adding an unimportant independent variable will increase R^2 to some extent, it is sometimes useful to correct for this by reducing R^2 appropriately. We do this by considering a corrected R^2, denoted by $\bar{R}^2$ and defined as follows.

Corrected (Adjusted) R^2

Assume that the regression model under consideration

$$y = \beta_0 + \beta_1 x_1 + \cdots + \beta_{k-1} x_{k-1} + \varepsilon$$

includes $k - 1$ independent variables (and thus, because of the intercept β_0, utilizes k parameters). Then the *corrected multiple coefficient of determination* (*corrected* R^2) is

$$\bar{R}^2 = \left(R^2 - \frac{k-1}{n-1}\right)\left(\frac{n-1}{n-k}\right)$$

where R^2 is the multiple coefficient of determination, and n is the number of observations.

To understand $\bar{R}^2$, first suppose that the values of the $k - 1$ independent variables are completely random (that is, randomly chosen from a population of numbers). It can be shown that these independent variables will still explain enough of the total variation in the observed values of the dependent variable to make R^2 equal to, on the average, $(k - 1)/(n - 1)$. Therefore our first step in correcting R^2 is to subtract this random explanation and form the quantity

$$R^2 - \frac{k-1}{n-1}$$

If the values of the independent variables are completely random, then this corrected version of R^2 is equal to zero. However, if the values of the independent variables are not completely random, then this quantity reduces R^2 too much. To see why, note that if R^2 is equal to 1, then the preceding corrected version of R^2 is not equal to 1 but is equal to

$$1 - \frac{k-1}{n-1} = \frac{n-k}{n-1}$$

which is less than 1, since $n - k < n - 1$. To define a corrected R^2 that is equal to 1 if R^2 is equal to 1, we multiply the above corrected version of R^2 by the factor

$$\frac{n-1}{n-k}$$

to form the quantity

$$\left(R^2 - \frac{k-1}{n-1}\right)\left(\frac{n-1}{n-k}\right)$$

This is the previously defined $\bar{R}^2$. Before presenting an example, note that it can be shown that the mean square error, s^2, and $\bar{R}^2$ are related by the equation

$$s^2 = (1 - \bar{R}^2)s_y^2$$

where

$$s_y^2 = \frac{\sum_{i=1}^{n}(y_i - \bar{y})^2}{n - 1}$$

Here s_y^2 is independent of the model under consideration because this quantity depends only on the observed values of the dependent variable $y_1, y_2, \ldots, y_n$. It follows that s^2 *decreases if and only if* $\bar{R}^2$ *increases.*

Example 11.4 In Table 11.7 we summarize R^2, $\bar{R}^2$, and s^2 for three regression models in the fuel consumption problem and present the SAS output of these quantities for the model

$$y = \beta_0 + \beta_1 x_1 + \beta_2 x_2 + \varepsilon$$

TABLE 11.7 R^2, $\bar{R}^2$, and s^2 for three regression models in the fuel consumption problem and the SAS output of R^2, $\bar{R}^2$, and s^2 for the model $y = \beta_0 + \beta_1 x_1 + \beta_2 x_2 + \varepsilon$

Model	k	R^2	$\bar{R}^2 = \left(R^2 - \frac{k-1}{n-1}\right)\left(\frac{n-1}{n-k}\right) = \left(R^2 - \frac{k-1}{7}\right)\left(\frac{7}{8-k}\right)$	s^2
$y = \beta_0 + \beta_1 x_1 + \varepsilon$	2	.90	$\bar{R}^2 = \left(.90 - \frac{1}{7}\right)\left(\frac{7}{6}\right) = .883$	.428
$y = \beta_0 + \beta_1 x_1 + \beta_2 x_2 + \varepsilon$	3	.974	$\bar{R}^2 = \left(.974 - \frac{2}{7}\right)\left(\frac{7}{5}\right) = .963$	.135
$y = \beta_0 + \beta_1 x_1 + \beta_2 x_2 + \beta_3 x_1 x_2 + \varepsilon$	4	.9742	$\bar{R}^2 = \left(.9742 - \frac{3}{7}\right)\left(\frac{7}{4}\right) = .955$	.165

DEP VARIABLE: Y

SOURCE	DF	SUM OF SQUARES	MEAN SQUARE	F VALUE	PROB>F
MODEL	2	24.875018[a]	12.437509	92.303	0.0001
ERROR	5	0.673732[b]	0.134746[e]		
C TOTAL	7	25.548750[c]			
ROOT MSE		0.367078[d]	R-SQUARE	0.9736[f]	
DEP MEAN		10.212500	ADJ R-SQ	0.9631[g]	
C.V.		3.594401			

[a]Explained variation [b]SSE [c]Total variation [d]s
[e]s^2 [f]R^2 [g]$\bar{R}^2$

Examining this table, we see that adding the independent variable x_2 to the regression model

$$y = \beta_0 + \beta_1 x_1 + \varepsilon$$

to form the model

$$y = \beta_0 + \beta_1 x_1 + \beta_2 x_2 + \varepsilon$$

increases R^2, increases $\bar{R}^2$, and decreases s^2. However, adding the cross-product term $x_1 x_2$ to this model to form the model

$$y = \beta_0 + \beta_1 x_1 + \beta_2 x_2 + \beta_3 x_1 x_2 + \varepsilon$$

increases R^2, decreases $\bar{R}^2$, and increases s^2. This indicates that adding $x_1 x_2$ is not warranted.

11.2

ADVANCED MODEL COMPARISON METHODS: THE *C* STATISTIC AND THE PRESS STATISTIC

THE C STATISTIC

Another criterion for comparing overall regression models is called the C statistic. This statistic is often called the C_k statistic in the statistical literature, but we omit the k for notational simplicity.

The C Statistic

Suppose we are attempting to choose an appropriate set of independent variables (to form a final regression model) from p potential independent variables. Then if the regression model

$$y = \beta_0 + \beta_1 x_1 + \cdots + \beta_{k-1} x_{k-1} + \varepsilon$$

includes $k - 1$ independent variables (and, because of the intercept β_0, utilizes k parameters), we define the *C statistic* to be

$$C = \frac{SSE}{s_p^2} - (n - 2k)$$

Here, n is the number of observations, SSE is the unexplained variation calculated from the above model, and s_p^2 is the mean square error calculated from the model

$$y = \beta_0 + \beta_1 x_1 + \cdots + \beta_p x_p + \varepsilon$$

which includes all p independent variables (and thus utilizes $p + 1$ parameters).

Since the C statistic for a given model is a function of the model's SSE, and since we want SSE to be small, *we want C to be small.* Although adding an unimportant independent variable to a regression model will decrease SSE, adding such a variable can increase C. As we show in Example 11.5, this can happen when the decrease in SSE caused by the addition of the extra independent variable is not enough to offset the decrease in $n - 2k$ caused by the addition of the extra independent variable (which increases k by 1). It should be noted that although adding an unimportant independent variable to a regression model can increase both s^2 and C, there is no exact relationship between s^2 and C. For example, as we show in Example 11.5, adding an independent variable to a regression model can decrease s^2 and increase C. While we want C to be small, in addition it can be shown from the theory behind the C statistic that *we also wish to find a model for which the C statistic roughly equals k* (k equals the number of parameters in the model). If a model has a C statistic substantially greater than k, it can be shown that this model has substantial *bias* and is undesirable. Thus although we want to find a model for which C is as small as possible, if C for such a model is substantially greater than k, we may prefer to choose a different model for which C is slightly larger and more nearly equal to the number of parameters in that (different) model. If a particular model has a small value of C and C for this model is less than k, then the model should be considered desirable. Finally, it should be noted that for the model that includes all p potential independent variables (and thus utilizes $p + 1$ parameters)

$$\begin{aligned} C &= \frac{SSE}{s_p^2} - (n - 2k) \\ &= \frac{SSE}{SSE/(n - (p + 1))} - (n - 2(p + 1)) \\ &= n - (p + 1) - (n - 2(p + 1)) \\ &= p + 1 \end{aligned}$$

Example 11.5 For the fuel consumption problem, assume that we are attempting to choose an appropriate set of independent variables (to form a final regression model) from the $p = 3$ potential independent variables x_1, the average hourly temperature, x_2, the chill index, and x_3, the average hourly wind velocity. The mean square error calculated from the model

$$y = \beta_0 + \beta_1 x_1 + \beta_2 x_2 + \beta_3 x_3 + \varepsilon$$

which includes all $p = 3$ independent variables is

$$s_p^2 = \frac{SSE}{n - (p + 1)} = \frac{.532}{8 - (3 + 1)} = .133$$

Thus the C statistics for three possible models in the fuel consumption problem are

TABLE 11.8 **The C statistics for three regression models in the fuel consumption problem**

Model	SSE	k	s^2	$C = \frac{SSE}{s_p^2} - (n - 2k) = \frac{SSE}{.133} - (8 - 2k)$
$y = \beta_0 + \beta_1 x_1 + \varepsilon$	2.57	2	.428	$C = \frac{2.57}{.133} - (8 - 2(2)) = 19.323 - 4 = 15.323$
$y = \beta_0 + \beta_1 x_1 + \beta_2 x_2 + \varepsilon$	.674	3	.135	$C = \frac{.674}{.133} - (8 - 2(3)) = 5.068 - 2 = 3.068$
$y = \beta_0 + \beta_1 x_1 + \beta_2 x_2 + \beta_3 x_3 + \varepsilon$	.532	4	.133	$C = \frac{.532}{.133} - (8 - 2(4)) = 4.0 - 0 = 4.0$

as summarized in Table 11.8. We see that adding x_2 to the model

$$y = \beta_0 + \beta_1 x_1 + \varepsilon$$

to form the model

$$y = \beta_0 + \beta_1 x_1 + \beta_2 x_2 + \varepsilon$$

has decreased both s^2 and C. In addition, we see that adding x_3 to the above two-variable model to form the model

$$y = \beta_0 + \beta_1 x_1 + \beta_2 x_2 + \beta_3 x_3 + \varepsilon$$

has decreased s^2 (from .135 to .133) but has increased C (from 3.068 to 4.0). In this case, note from Table 11.8 that although the addition of x_3 has decreased the unexplained variation, this decrease (from .674 to .532) has not been enough to offset the change in $n - 2k$, which decreases from $n - 2k = 2$ to $n - 2k = 0$. Thus the model

$$y = \beta_0 + \beta_1 x_1 + \beta_2 x_2 + \varepsilon$$

has the smallest C statistic, $C = 3.068$, which roughly equals the number of model parameters ($k = 3$). It follows that this model is the best of the three models in Table 11.8 on the basis of the C statistic.

Example 11.6 Table 11.9 presents the sales price y, square footage x_1, number of rooms x_2, number of bedrooms x_3, age x_4, and number of bathrooms x_5 for each of 63 single-family residences sold during 1988 in Oxford, Ohio. Figure 11.3 presents the SAS output of the values of R^2 and C for all reasonable regression models describing this data. The values of C have been computed by using the fact that the mean square error s_p^2 for the model containing all $p = 5$ potential independent variables

$$y = \beta_0 + \beta_1 x_1 + \beta_2 x_2 + \beta_3 x_3 + \beta_4 x_4 + \beta_5 x_5 + \varepsilon$$

TABLE 11.9 Measurements taken on 63 single-family residences

Residence, i	Sales price, y(×$1000)	Square feet, x_1	Rooms, x_2	Bedrooms, x_3	Age, x_4	Bathrooms, x_5
1	53.5	1008	5	2	35	1.0
2	49.0	1290	6	3	36	1.0
3	50.5	860	8	2	36	1.0
4	49.9	912	5	3	41	1.0
5	52.0	1204	6	3	40	1.0
6	55.0	1204	5	3	10	1.5
7	80.5	1764	8	4	64	1.5
8	86.0	1600	7	3	19	2.0
9	69.0	1255	5	3	16	2.0
10	149.0	3600	10	5	17	2.5
11	46.0	864	5	3	37	1.0
12	38.0	720	4	2	41	1.0
13	49.5	1008	6	3	35	2.0
14	105.0	1950	8	3	52	1.5
15	152.5	2086	7	3	12	2.0
16	85.0	2011	9	4	76	1.5
17	60.0	1465	6	3	102	1.0
18	58.5	1232	5	2	69	1.5
19	101.0	1736	7	3	67	1.0
20	79.4	1296	6	3	11	1.5
21	125.0	1996	7	3	9	2.5
22	87.9	1874	5	2	14	2.0
23	80.0	1580	5	3	11	1.0
24	94.0	1920	5	3	14	2.5
25	74.0	1430	9	3	16	2.0
26	69.0	1486	6	3	27	2.0
27	63.0	1008	5	2	35	1.0
28	67.5	1282	5	3	20	2.0
29	35.0	1134	5	2	74	1.0
30	142.5	2400	9	4	15	2.5
31	92.2	1701	5	3	15	2.0
32	56.0	1020	6	3	16	1.0
33	63.0	1053	5	2	24	2.0
34	60.0	1728	6	3	26	1.5
35	34.0	416	3	1	42	1.0
36	52.0	1040	5	2	9	1.5
37	75.0	1496	6	3	30	2.0
38	93.0	1936	8	4	39	1.5
39	60.0	1904	7	4	32	1.0
40	73.0	1080	5	2	24	1.5
41	71.0	1768	8	4	74	1.5
42	83.0	1503	6	3	14	2.0
43	90.0	1736	7	3	16	2.5
44	83.0	1695	6	3	12	2.0
45	115.0	2186	8	4	12	2.5
46	50.0	888	5	2	34	1.0
47	55.2	1120	6	3	29	1.0
48	61.0	1400	5	3	33	1.0
49	147.0	2165	7	3	2	2.0

(continues)

TABLE 11.9 Continued

Residence, i	Sales price, $y(\times \$1000)$	Square feet, x_1	Rooms, x_2	Bedrooms, x_3	Age, x_4	Bathrooms, x_5
50	210.0	2353	8	4	15	2.5
51	60.0	1536	6	3	36	2.0
52	100.0	1972	8	3	37	2.0
53	44.5	1120	5	3	27	1.0
54	55.0	1664	7	3	79	2.0
55	53.4	925	5	3	20	1.0
56	65.0	1288	5	3	2	1.5
57	73.0	1400	5	3	2	1.5
58	40.0	1376	6	3	103	1.0
59	141.0	2038	12	4	62	2.0
60	68.0	1572	6	3	29	1.0
61	139.0	1545	6	3	9	2.0
62	140.0	1993	6	3	4	2.0
63	55.0	1130	5	2	21	2.0

Source: Reprinted by permission of Alpha, Inc., Oxford, Ohio.

FIGURE 11.3 SAS output of values of R^2 and C for all reasonable residential sales models

NUMBER IN MODEL	R-SQUARE	C(P)	VARIABLES IN MODEL
1	0.10635858	130.881	AGE
1	0.28084477	93.806110	BED
1	0.35143113	78.807918	ROOMS
1	0.44194830	59.574830	BATH
1	0.63320895	18.935765	SQFT
2	0.37317349	76.188102	ROOMS BED
2	0.39330465	71.910633	BED AGE
2	0.44547115	60.826295	BATH AGE
2	0.53902497	40.947978	ROOMS AGE
2	0.54199971	40.315906	BATH BED
2	0.56831004	34.725486	ROOMS BATH
2	0.63472579	20.613466	SQFT ROOMS
2	0.64449803	18.537062	SQFT BED
2	0.67830492	11.353773	SQFT BATH
2	0.69122057	8.609456	SQFT AGE
3	0.54567725	41.534501	ROOMS BED AGE
3	0.55669758	39.192902	BATH BED AGE
3	0.57886699	34.482345	ROOMS BATH BED
3	0.61672255	26.438797	ROOMS BATH AGE
3	0.65247290	18.842560	SQFT ROOMS BED
3	0.68145441	12.684570	SQFT ROOMS BATH
3	0.68161437	12.650582	SQFT BATH BED
3	0.69617802	9.556097	SQFT BED AGE
3	0.70524070	7.630458	SQFT BATH AGE
3	0.70882097	6.869722	SQFT ROOMS AGE

4	0.62257055	27.196213	ROOMS BATH BED AGE
4	0.68900317	13.080609	SQFT ROOMS BATH BED
4	0.70760346	9.128420	SQFT BATH BED AGE
4	0.72024721	6.441875	SQFT ROOMS BATH AGE
4	0.72574793	5.273083	SQFT ROOMS BED AGE
5	0.73173947	6.000000	SQFT ROOMS BATH BED AGE

is 357.9074. For example, the C statistic for the model

$$y = \beta_0 + \beta_1 x_1 + \beta_2 x_2 + \beta_3 x_3 + \beta_4 x_4 + \varepsilon$$

which has an unexplained variation of $SSE = 20{,}853.3021$ and $k = 5$, is

$$\begin{aligned} C &= \frac{SSE}{s_p^2} - (n - 2k) \\ &= \frac{20{,}853.3021}{357.9074} - (63 - 2(5)) \\ &= 5.273083 \end{aligned}$$

Examining Figure 11.3, we conclude that this latter model is the best model with respect to the C statistic. This is because this model has the smallest value of C and because, since $C = 5.27$ is roughly equal to $k = 5$ (the number of parameters in the model), the C statistic indicates that the model is not biased. We discussed this model previously in Examples 8.16 and 8.21.

THE PRESS STATISTIC

We now discuss another statistic that can be used for model comparisons—the *PRESS statistic*.

PRESS Residuals and the PRESS Statistic

Consider the regression model

$$y_i = \beta_0 + \beta_1 x_{i1} + \beta_2 x_{i2} + \cdots + \beta_p x_{ip} + \varepsilon_i$$

Then

1. The *ith PRESS* (or *deleted*) *residual* is

   $$d_i = y_i - \hat{y}_{(i)}$$

   where

   $$\hat{y}_{(i)} = b_0^{(i)} + b_1^{(i)} x_{i1} + b_2^{(i)} x_{i2} + \cdots + b_p^{(i)} x_{ip}$$

   is the point prediction of y_i calculated by using least squares point estimates $b_0^{(i)}, b_1^{(i)}, b_2^{(i)}, \ldots, b_p^{(i)}$ that are computed by utilizing all n observations except for observation i.

2. Recalling that the usual residual is $e_i = y_i - \hat{y}_i$, where $\hat{y}_i$ is the point prediction of y_i calculated by using least squares point estimates b_0, b_1, $b_2, \ldots, b_p$ that are computed by using all n observations, it can be proven that

$$d_i = \frac{e_i}{1 - h_{ii}}$$

Here, $h_{ii} = \mathbf{x}_i'(\mathbf{X}'\mathbf{X})^{-1}\mathbf{x}_i$ is called a *leverage value*, where the matrix $\mathbf{X}$ for the model above includes all n observations and

$$\mathbf{x}_i' = [1 \quad x_{i1} \quad x_{i2} \quad \ldots \quad x_{ip}]$$

is a row vector containing the independent variable values for the ith observation.

3. The *PRESS statistic* is the sum of squared press residuals. That is,

$$\text{PRESS} = \sum_{i=1}^{n} d_i^2 = \sum_{i=1}^{n} (y_i - \hat{y}_{(i)})^2 = \sum_{i=1}^{n} \left(\frac{e_i}{1 - h_{ii}}\right)^2$$

Note that it can be shown that $0 < h_{ii} \leqslant 1$ for any regression model employing an intercept β_0. Therefore for such a model the PRESS residual

$$d_i = \frac{e_i}{1 - h_{ii}}$$

is of larger magnitude than the usual residual e_i.

We now present an example illustrating both the use of the C statistic and the PRESS statistic.

Example 11.7 In Table 11.10 we present several statistics related to all reasonable regression models describing all $n = 17$ observations comprising the hospital labor data set in Table 11.5. In Table 11.11 we present several statistics related to what we consider to be the six best models. Note that the numbers in parentheses corresponding to the variables are the variance inflation factors for the specified model. The C statistics in these tables have been computed by using the fact that the mean square error s_p^2 for the model containing all $p = 5$ potential independent variables

$$y = \beta_0 + \beta_1 x_1 + \beta_2 x_2 + \beta_3 x_3 + \beta_4 x_4 + \beta_5 x_5 + \varepsilon$$

is 412,277. For example, the C statistic for

$$\text{Model 6:} \quad y = \beta_0 + \beta_1 x_2 + \beta_2 x_3 + \beta_3 x_5 + \varepsilon$$

TABLE 11.10 **Model comparison statistics for all reasonable regression models describing the hospital labor data based on using all 17 observations in Table 11.5**

Variables in model	s^2	R^2	C	*PRESS*
x_5	21,940,359	.3348	785.26	455,789,688
x_4	3,816,879	.8843	125.87	70,075,727
x_2	3,517,337	.8934	114.97	158,243,072
x_1	939,990	.9715	21.20	22,243,776
x_3	917,487	.9722	20.38	21,841,500
x_4, x_5	3,165,704	.9104	96.50	71,164,912
x_2, x_5	2,688,542	.9239	80.30	128,703,914
x_2, x_4	2,452,812	.9306	72.29	140,743,693
x_1, x_3	971,399	.9725	21.99	22,564,832
x_1, x_4	914,181	.9741	20.04	35,786,830
x_3, x_4	870,761	.9754	18.57	32,454,824
x_1, x_5	564,291	.9840	8.16	12,977,100
x_3, x_5	538,720	.9848	7.29	12,628,470
x_1, x_2	490,585	.9861	5.66	18,036,834
x_2, x_3	469,456	.9867	4.94	17,853,441
x_2, x_4, x_5	1,816,626	.9523	48.28	107,102,405
x_1, x_3, x_4	818,290	.9785	16.80	34,400,018
x_1, x_4, x_5	583,845	.9846	9.41	17,828,243
x_3, x_4, x_5	570,794	.9850	9.00	16,275,120
x_1, x_3, x_5	569,830	.9850	8.97	13,036,635
x_1, x_2, x_4	528,257	.9861	7.66	32,814,300
x_2, x_3, x_4	504,218	.9868	6.90	30,139,304
x_1, x_2, x_3	482,286	.9873	6.21	22,794,229
x_1, x_2, x_5	403,215	.9894	3.71	18,051,730
x_2, x_3, x_5	377,954	.9901	2.92	17,846,717
x_1, x_3, x_4, x_5	615,741	.9851	10.92	18,621,398
x_1, x_2, x_3, x_4	499,469	.9879	7.54	37,719,883
x_1, x_2, x_3, x_5	389,793	.9905	4.34	22,464,723
x_1, x_2, x_4, x_5	387,001	.9906	4.26	30,255,902
x_2, x_3, x_4, x_5	378,826	.9908	4.03	28,629,419
x_1, x_2, x_3, x_4, x_5	412,277	.9908	6.00	32,195,222

which has an unexplained variation of $SSE = 4{,}913{,}402$ and $k = 4$, is

$$\begin{aligned} C &= \frac{SSE}{s_p^2} - (n - 2k) \\ &= \frac{4{,}913{,}402}{412{,}277} - [17 - 2(4)] \\ &= 2.92 \end{aligned}$$

Examining Table 11.10, we conclude that Model 6 is the best model with respect to the C statistic. This is because Model 6 has the smallest C statistic and because,

TABLE 11.11 **Summary statistics for the six best hospital labor models based on using all 17 observations in Table 11.5**

Model, variables, and variance inflation factors	*C*	s^2	R^2	*PRESS*
Model 1: $x_1(1.8199)$, $x_5(1.8199)$	8.16	564,291	.9840	12,977,100
Model 2: $x_3(1.8195)$, $x_5(1.8195)$	7.29	538,720	.9848	12,628,470
Model 3: $x_1(5.6605)$, $x_2(5.6605)$	5.66	490,585	.9861	18,036,834
Model 4: $x_2(5.6471)$, $x_3(5.6471)$	4.94	469,456	.9867	17,853,441
Model 5: $x_1(11.3214)$, $x_2(7.7714)$, $x_5(2.4985)$	3.71	403,215	.9894	18,051,730
Model 6: $x_2(7.7373)$, $x_3(11.2693)$, $x_5(2.4929)$	2.92	377,954	.9901	17,846,717

since $C = 2.92$ is not greater than $k = 4$ (the number of parameters in Model 6), the C statistic indicates that Model 6 is not biased.

Since Model 6 has not only the smallest C statistic but also the smallest s^2 of any model in Table 11.10, we might be tempted to conclude that this model is the best model to use to predict hospital labor needs. However, the hospital labor data set in Table 11.5 contains three hospitals (15, 16, and 17) that are substantially larger than the other 14 small to medium-sized hospitals (note that classifying hospital 14 is difficult). Therefore it is important to compute the PRESS statistic for each model under consideration.

For example, consider

$$\text{Model 2:} \quad y_i = \beta_0 + \beta_1 x_{i3} + \beta_2 x_{i5} + \varepsilon_i$$

When we use all $n = 17$ observations in Table 11.5 to calculate the least squares point estimates of the model parameters, we obtain the prediction equation

$$\begin{aligned} \hat{y}_i &= b_0 + b_1 x_{i3} + b_2 x_{i5} \\ &= 2585.520 + 1.2324 x_{i3} - 530.933 x_{i5} \end{aligned}$$

To illustrate the calculation of PRESS residuals for Model 2, consider (for example) hospital 8 in Table 11.5. Since $x_{83} = 1639.92$ and $x_{85} = 5.15$, it follows that

$$\begin{aligned} \hat{y}_8 &= b_0 + b_1 x_{83} + b_2 x_{85} \\ &= 2585.520 + 1.2324(1639.92) - 530.933(5.15) \\ &= 1872.295 \end{aligned}$$

$$\begin{aligned} \mathbf{x}_8' &= [1 \quad x_{83} \quad x_{85}] \\ &= [1 \quad 1639.92 \quad 5.15] \end{aligned}$$

and

$$h_{88} = \mathbf{x}_8'(\mathbf{X}'\mathbf{X})^{-1}\mathbf{x}_8 = .0805$$

Therefore the usual residual is

$$\begin{aligned} e_8 &= y_8 - \hat{y}_8 \\ &= 2160.55 - 1872.295 \\ &= 288.255 \end{aligned}$$

TABLE 11.12 **The usual and PRESS residuals for Model 2 corresponding to all $n = 17$ hospitals in Table 11.5**

Hospital	*Usual residuals*	h_{ii}	*PRESS residuals*
1	−239.187	.1155	−270.425
2	134.214	.2165	171.296
3	−44.389	.1249	−50.723
4	388.314	.1581	461.216
5	100.301	.0848	109.597
6	−213.242	.1014	−237.293
7	173.391	.0841	189.307
8	288.255	.0805	313.497
9	−538.707	.0838	−587.985
10	−320.967	.0692	−344.822
11	519.432	.0700	558.552
12	−1085.508	.0957	−1200.408
13	−403.020	.0629	−430.060
14	2004.738	.0932	2210.715
15	−469.070	.6763	−1448.866
16	−802.734	.3379	−1212.467
17	508.177	.5453	1117.665

and the PRESS residual is

$$d_8 = \frac{e_8}{1 - h_{88}} = \frac{288.255}{1 - .0805} = 313.497$$

The usual and PRESS residuals for Model 2 corresponding to all $n = 17$ hospitals in Table 11.5 are summarized in Table 11.12. To see how the PRESS residuals can help us, note that each of the large hospitals 15, 16, and 17 is somewhat outside of the experimental region defined by the other 16 hospitals. This is because each of the large hospitals is different from the 14 small to medium-sized hospitals and is different (with respect to at least some of the values of x_1, x_2, x_3, x_4, and x_5) from the other two large hospitals. Suppose that we use a particular regression model to calculate for large hospital i the PRESS residual, $d_i = y_i - \hat{y}_{(i)}$. Then, since the point prediction $\hat{y}_{(i)}$ of y_i is made by using all 16 hospitals except for large hospital i, it follows that we are predicting y_i by extrapolating the regression model somewhat outside of the experimental region defined by the other 16 hospitals. The smaller the magnitude of d_i, the less dangerous it is to extrapolate the particular regression model to predict y_i. It can be verified that, while the magnitudes of the PRESS residuals corresponding to hospitals 1–14 are of roughly the same sizes for all six models, the magnitudes of the PRESS residuals corresponding to hospitals 15, 16, and 17 are smallest for Model 2. Therefore the PRESS statistic is the smallest for Model 2. This might be interpreted to mean that it is less dangerous to extrapolate Model 2 to predict labor needs for large hospitals. However, recall from Example 11.3 that $y_{14} = 10{,}344$ was much greater than would be expected for hospital 14's values of x_2, x_3, and x_5. Therefore if y_{14} was recorded erroneously or

TABLE 11.13 **Summary statistics for the six best hospital labor models based on excluding observation 14 and using the other 16 observations in Table 11.5**

Model, variables, and variance inflation factors	*C*	s^2	R^2	*PRESS*
Model 1: $x_1(1.7761)$, $x_5(1.7761)$	11.15	259,918	.9927	6,915,443
Model 2: $x_3(1.7755)$, $x_5(1.7755)$	9.47	239,245	.9933	6,488,227
Model 3: $x_1(5.5329)$, $x_2(5.5329)$	19.38	361,059	.9899	12,257,014
Model 4: $x_2(5.5164)$, $x_3(5.5164)$	17.78	341,407	.9904	11,291,921
Model 5: $x_1(11.4687)$, $x_2(7.8753)$, $x_5(2.5280)$	4.96	172,617	.9955	5,605,173
Model 6: $x_2(7.8283)$, $x_3(11.3962)$, $x_5(2.5196)$	3.26	149,893	.9961	4,215,645

resulted from a situation that would be very unlikely to occur again, it would be reasonable to remove observation 14 from the data set. It can be verified that when observation 14 is removed from the data set and a regression analysis is performed by using the remaining 16 observations, the magnitudes of the usual and PRESS residuals corresponding to hospitals 1–13 for Model 6 are roughly equal to the magnitudes of these residuals for Model 2. However, the magnitudes of the usual and PRESS residuals corresponding to hospitals 15, 16, and 17, and also the values of C, s^2, and PRESS, are smaller for Model 6 than for Model 2 or any other reasonable model (see Table 11.13). This might be interpreted to mean that it is best to use Model 6 (based on hospitals 1–13, 15, 16, and 17) to predict labor needs for large hospitals. Moreover, it can be verified that the magnitudes of the usual and PRESS residuals corresponding to hospital 17 (for which the value of x_2, monthly X-ray exposures, is particularly large) are much smaller for Model 6 than for Model 2. Thus it seems very important to use Model 6 to predict labor needs for large hospitals having large values of x_2. This analysis implies that it was appropriate in Example 11.3 to use Model 6 to predict labor needs for a large hospital for which the values of x_2, x_3, and x_5 in a future month will be $x_{02} = 54{,}000$, $x_{03} = 14{,}400$, and $x_{05} = 6.9$.

Suppose that in addition to removing hospital 14, we remove hospital 17 (which has a particularly large value of x_2) from the data set. Then the prob-value corresponding to x_2 in Model 6 increases to .5802, and Model 2 (which results when x_2 is eliminated from Model 6) has slightly smaller (in magnitude) usual and PRESS residuals corresponding to all hospitals remaining in the data set than does Model 6. Moreover, the SAS output in Figure 11.4 shows that Model 2 has the smallest value of C of all reasonable regression models based on observations 1–13, 15, and 16. It can be further verified that Model 2 has the smallest values of s^2 and PRESS and the smallest variance inflation factors of any model in Table 11.13, based on observations 1–13, 15, and 16. Therefore it seems reasonable to use Model 2 to predict labor needs for small, medium-sized, and possibly large hospitals (that are not much different from large hospitals 15 or 16), as long as these hospitals do not have particularly large values of x_2. Note that Model 2 also has the advantage

FIGURE 11.4 **SAS output of values of R^2 and C for all reasonable hospital labor models based on using observations 1–13, 15, and 16**

NUMBER IN MODEL	R-SQUARE	C(P)	VARIABLES IN MODEL
1	0.47189469	729.194	STAYDAY
1	0.82089803	240.030	POPULA
1	0.92232790	97.865438	XRAY
1	0.98061790	16.166001	PATLOAD
1	0.98114543	15.426617	BEDDAY
2	0.88462038	152.716	POPULA STAYDAY
2	0.92296998	98.965495	XRAY STAYDAY
2	0.92543890	95.505061	XRAY POPULA
2	0.98118884	17.365778	PATLOAD BEDDAY
2	0.98155004	16.859523	PATLOAD XRAY
2	0.98221834	15.922833	PATLOAD POPULA
2	0.98241996	15.640239	XRAY BEDDAY
2	0.98348289	14.150427	BEDDAY POPULA
2	0.99204568	2.148793	PATLOAD STAYDAY
2	0.99278739	1.109214	BEDDAY STAYDAY
3	0.92844177	93.296221	XRAY POPULA STAYDAY
3	0.98238554	17.688482	PATLOAD XRAY POPULA
3	0.98299192	16.838576	PATLOAD XRAY BEDDAY
3	0.98365279	15.912297	XRAY BEDDAY POPULA
3	0.98525049	13.672957	PATLOAD BEDDAY POPULA
3	0.99211198	4.055875	PATLOAD XRAY STAYDAY
3	0.99274866	3.163505	PATLOAD POPULA STAYDAY
3	0.99287289	2.989374	PATLOAD BEDDAY STAYDAY
3	0.99298224	2.836117	BEDDAY POPULA STAYDAY
3	0.99299428	2.819232	XRAY BEDDAY STAYDAY
4	0.98570214	15.039917	PATLOAD XRAY BEDDAY POPULA
4	0.99298230	4.836031	PATLOAD BEDDAY POPULA STAYDAY
4	0.99325018	4.460573	PATLOAD XRAY POPULA STAYDAY
4	0.99328342	4.413977	PATLOAD XRAY BEDDAY STAYDAY
4	0.99356644	4.017293	XRAY BEDDAY POPULA STAYDAY
5	0.99357878	6.000000	PATLOAD XRAY BEDDAY POPULA STAYDAY

of not requiring that we know exact future values of x_2. Of course, for the purposes of developing a prediction model for small to medium-sized hospitals, it might be reasonable to remove all of the large hospitals (15, 16, and 17), as well as hospital 14, from the data set. If this is done, both Model 2 and Model 6 give very similar results, and thus we might choose the simpler Model 2. Finally, note that in Chapter 12 we will use what is called a *dummy variable* to further analyze the hospital labor data.

Example 11.8 In Table 11.14 we present summary statistics for four of the best models describing the sales territory performance data presented in Table 8.5.

TABLE 11.14 Summary statistics for four of the best models describing the sales territory performance data

Model	Independent variables and related prob-values		R^2	s^2	C	PRESS
1	POTEN	0.0001	0.90	205,967	5.4	5,794,837.54
	ADV	0.0001				
	SHARE	0.0011				
	ACCTS	0.0043				
2	TIME	0.0065	0.92	185,099	4.4	5,440,757.72
	POTEN	0.0001				
	ADV	0.0025				
	SHARE	0.0001				
	CHANGE	0.0530				
3	POTEN	0.0001	0.91	190,747	5.0	5,338,063.76
	ADV	0.0006				
	SHARE	0.0006				
	CHANGE	0.1236				
	ACCTS	0.0089				
4	TIME	0.1983	0.92	183,187	5.3	6,089,047.96
	POTEN	0.0001				
	ADV	0.0018				
	SHARE	0.0004				
	CHANGE	0.0927				
	ACCTS	0.2881				

11.3

STEPWISE REGRESSION, FORWARD SELECTION, BACKWARD ELIMINATION, AND MAXIMUM R^2 IMPROVEMENT

When the number of potential independent variables is not large, we can fairly easily compare all reasonable regression models with respect to various criteria (such as R^2, s, and C). However, if we are attempting to choose an appropriate set of independent variables from a large number of potential variables, comparing all reasonable regression models can be quite unwieldy. In this case it is useful to employ a screening procedure that can be used to identify one set (or several sets) of the "most important" independent variables. We now present four such screening procedures.

STEPWISE REGRESSION

Stepwise regression is generally carried out on a computer and is available in most standard regression computer packages. There are slight variations in the way in which different computer packages carry out stepwise regression. Assuming that y is the dependent variable and $x_1, x_2, \ldots, x_p$ are the p potential independent variables (where p will generally be large), we explain how most of the computer packages carry out stepwise regression. To make our description as concise as possible, we need to introduce some new terminology. Stepwise regression uses t statistics (and related prob-values) to determine the importance (or significance) of the independent variables in various regression models. In this context the *t statistic indicates that the independent variable x_j is significant at the α level if and only if the related prob-value is less than α*. This implies that we can reject $H_0: \beta_j = 0$ in favor of $H_1: \beta_j \neq 0$ with the probability of a Type I error equal to α. Then stepwise regression is performed as follows.

Choice of α_{entry} and α_{stay}

Before beginning the stepwise procedure we choose a value of α_{entry}, which we call "the probability of a Type I error related to entering an independent variable into the regression model." We also choose a value of α_{stay}, which we call "the probability of a Type I error related to retaining an independent variable that was previously entered into the model." We discuss the considerations involved in choosing these values after our description of the stepwise procedure. For now, suffice it to say that it is common practice to set both α_{entry} and α_{stay} equal to .05.

Step 1 The stepwise procedure considers the p possible one-independent-variable regression models of the form

$$y = \beta_0 + \beta_1 x_j + \varepsilon$$

Each different model includes a different potential independent variable. For each model the t statistic (and prob-value) related to testing H_0: $\beta_1 = 0$ versus H_1: $\beta_1 \neq 0$ is calculated. Denoting the independent variable giving the largest absolute value of the t statistic (and the smallest prob-value) by the symbol $x_{[1]}$, we consider the model

$$y = \beta_0 + \beta_1 x_{[1]} + \varepsilon$$

If the t statistic does not indicate that $x_{[1]}$ is significant at the α_{entry} level, then the stepwise procedure terminates by choosing the model

$$y = \beta_0 + \varepsilon$$

If the t statistic indicates that the independent variable $x_{[1]}$ is significant at the α_{entry} level, then $x_{[1]}$ is retained for use in Step 2.

Step 2 The stepwise procedure considers the $p - 1$ possible two-independent-variable regression models of the form

$$y = \beta_0 + \beta_1 x_{[1]} + \beta_2 x_j + \varepsilon$$

Each different model includes $x_{[1]}$, the independent variable chosen in Step 1, and a different potential independent variable chosen from the remaining $p - 1$ independent variables that were not chosen in Step 1. For each model the t statistic (and prob-value) related to testing $H_0: \beta_2 = 0$ versus $H_1: \beta_2 \neq 0$ is calculated. Denoting the independent variable giving the largest absolute value of the t statistic (and the smallest prob-value) by the symbol $x_{[2]}$, we consider the model

$$y = \beta_0 + \beta_1 x_{[1]} + \beta_2 x_{[2]} + \varepsilon$$

If the t statistic indicates that $x_{[2]}$ is significant at the α_{entry} level, then $x_{[2]}$ is retained in this model, and the stepwise procedure checks to see whether $x_{[1]}$ should be allowed to stay in the model. This check should be made because multicollinearity will probably cause the t statistic related to the importance of $x_{[1]}$ to change when $x_{[2]}$ is added to the model. If the t statistic does not indicate that $x_{[1]}$ is significant at the α_{stay} level, then the stepwise procedure returns to the beginning of Step 2. Starting with a new one-independent-variable model that uses the new significant independent variable $x_{[2]}$, the stepwise procedure attempts to find a new two-independent-variable model

$$y = \beta_0 + \beta_1 x_{[2]} + \beta_2 x_j + \varepsilon$$

If the t statistic indicates that $x_{[1]}$ is significant at the α_{stay} level in the model

$$y = \beta_0 + \beta_1 x_{[1]} + \beta_2 x_{[2]} + \varepsilon$$

then both the independent variables $x_{[1]}$ and $x_{[2]}$ are retained for use in further steps.

Further steps The stepwise procedure continues by adding independent variables one at a time to the model. At each step an independent variable is added to the model if and only if it has the largest (in absolute value) t statistic of the independent variables not in the model and if its t statistic indicates that it is significant at the α_{entry} level. After adding an independent variable the stepwise procedure checks all the independent variables already included in the model and removes any independent variable that is not significant at the α_{stay} level. Only after the necessary removals are made does the stepwise procedure attempt to add another independent variable to the model. The stepwise procedure terminates when all the independent variables not in the model are insignificant at the α_{entry} level or when the variable to be added to the model is the one just removed from it.

Regarding the choice of α_{entry} and α_{stay}, Draper and Smith (1981) state that it is usually best to choose α_{entry} equal to α_{stay}. It is not recommended that α_{stay} be chosen less than α_{entry} because this makes it too likely that an independent variable that has just been added to the model will (in subsequent steps) be removed from the model. Sometimes, however, it is reasonable to choose α_{stay} to be greater than α_{entry}. This makes it more likely that an independent variable whose significance decreases as new independent variables are added to the model will be allowed to stay in the model. Draper and Smith go on to suggest that α_{entry} and α_{stay} be set equal to .05 or .10.

However, we should point out that setting α_{entry} and α_{stay} higher than .10 is also reasonable. This will cause more independent variables to be included in the model and will give the analyst an opportunity to consider additional independent variables. Indeed, though the model obtained by the stepwise procedure may be reasonable, it should not necessarily be regarded as the best final regression model. First, the choices of α_{entry} and α_{stay} are arbitrary, and the many hypothesis tests performed by the stepwise procedure imply that Type I and Type II errors might be committed. It follows that the stepwise procedure might include some unimportant independent variables in the model and exclude some important independent variables from the model. Second, while higher-order terms and interaction terms can be included in the set of potential independent variables to be considered by stepwise regression, we often omit them. This is done so that the (probably) already large list of potential independent variables is not unduly increased. Thus it is possible that some important terms will be excluded. In general, then, stepwise regression should be regarded as a screening procedure that can be used to find at least some of the most important independent variables. Once the stepwise procedure identifies these variables, we should then carefully use the other model-building techniques discussed in this book to arrive at an appropriate final regression model.

Example 11.9 In the fuel consumption problem we let x_1, x_2, and x_3 be the $p = 3$ potential independent variables to be considered in stepwise regression, and we set both α_{entry} and α_{stay} equal to .05. Note that in this example we place the prob-value for testing the importance of an independent variable in parentheses below the appropriate model term.

Step 1 The $p = 3$ possible one-independent-variable regression models of the form

$$y = \beta_0 + \beta_1 x_j + \varepsilon$$

are

$$y = \beta_0 + \underset{(.00033)}{\beta_1 x_1} + \varepsilon$$

$$y = \beta_0 + \underset{(.00490)}{\beta_1 x_2} + \varepsilon$$

$$y = \beta_0 + \underset{(.00101)}{\beta_1 x_3} + \varepsilon$$

Since the independent variable giving the smallest prob-value related to testing $H_0: \beta_1 = 0$ is x_1, we consider the model

$$y = \beta_0 + \underset{(.00033)}{\beta_1 x_1} + \varepsilon$$

Since .00033, the prob-value related to $H_0: \beta_1 = 0$, is less than $\alpha_{entry} = .05$, x_1 is significant at the α_{entry} level and is retained for use in Step 2.

Step 2 The $p - 1 = 2$ possible two-independent-variable regression models of the form

$$y = \beta_0 + \beta_1 x_{[1]} + \beta_2 x_j + \varepsilon$$

are

$$y = \beta_0 + \underset{(.00139)}{\beta_1 x_1} + \underset{(.01330)}{\beta_2 x_2} + \varepsilon$$

$$y = \beta_0 + \underset{(.04040)}{\beta_1 x_1} + \underset{(.11250)}{\beta_2 x_3} + \varepsilon$$

Since the independent variable giving the smallest prob-value related to $H_0: \beta_2 = 0$ is x_2, we consider the model

$$y = \beta_0 + \underset{(.00139)}{\beta_1 x_1} + \underset{(.01330)}{\beta_2 x_2} + \varepsilon$$

Since .01330, the prob-value related to $H_0: \beta_2 = 0$, is less than $\alpha_{entry} = .05$, x_2 is significant at the α_{entry} level. Since .00139, the prob-value related to $H_0: \beta_1 = 0$, is less than $\alpha_{stay} = .05$, x_1 is significant at the α_{stay} level. Thus both x_1 and x_2 are retained for use in Step 3.

Step 3 The three-independent-variable regression model of the form

$$y = \beta_0 + \beta_1 x_{[1]} + \beta_2 x_{[2]} + \beta_3 x_j + \varepsilon$$

is

$$y = \beta_0 + \underset{(.01761)}{\beta_1 x_1} + \underset{(.05627)}{\beta_2 x_2} + \underset{(.36145)}{\beta_3 x_3} + \varepsilon$$

Since .36145, the prob-value related to $H_0: \beta_3 = 0$, is not less than $\alpha_{entry} = .05$, x_3 is not significant at the α_{entry} level. Thus the stepwise procedure terminates by choosing the model

$$y = \beta_0 + \beta_1 x_1 + \beta_2 x_2 + \varepsilon$$

Note that this model was chosen by the other model-building techniques presented in this chapter.

Although we have used only $p = 3$ potential independent variables in the fuel consumption problem, stepwise regression is most profitably used as a screening procedure in a regression problem having a large number of potential independent variables. We now illustrate such a use of stepwise regression.

Example 11.10 In the hospital labor problem we let x_1, x_2, x_3, x_4, and x_5 (defined in Example 11.3) be the $p = 5$ potential independent variables to be considered in stepwise regression. When both α_{entry} and α_{stay} are set equal to .10, the stepwise procedure that we have described (when applied to all $n = 17$ observations in Table 11.5) (1) adds x_3 on the first step, (2) adds x_2 (and retains x_3) on the second step, (3) adds x_5 (and retains x_2 and x_3) on the third step, and (4) terminates after Step 3 when no more independent variables can be added. The SAS output of this stepwise procedure is given in Figure 11.5. Note that the stepwise procedure arrives at the final model

$$y = \beta_0 + \beta_1 x_2 + \beta_2 x_3 + \beta_3 x_5 + \varepsilon$$

This is the model that the statistics of Tables 11.10 and 11.11 indicate is (possibly) best.

FIGURE 11.5 **The SAS output of the stepwise procedure for the hospital labor problem**

```
STEP 1   VARIABLE X3 ENTERED          R SQUARE = 0.97218120          C(P) = 20.38117958

                        DF           SUM OF SQUARES          MEAN SQUARE          F      PROB > F

      REGRESSION         1       480950231.62604150   480950231.62604150     524.20        0.0001
      ERROR             15        13762308.86295839      917487.25753056
      TOTAL             16       494712540.48899990

                   B VALUE                STD ERROR       PARTIAL REG SS          F      PROB > F

      INTERCEPT -28.12861560
      X3          1.11739237             0.04880403   480950231.62604150     524.20        0.0001
-------------------------------------------------------------------------------------------------

STEP 2   VARIABLE X2 ENTERED          R SQUARE = 0.98671474          C(P) = 4.94164787

                        DF           SUM OF SQUARES          MEAN SQUARE          F      PROB > F

      REGRESSION         2       488140157.95096330   244070078.97548168     519.90        0.0001
      ERROR             14         6572382.53803656      469455.89557404
      TOTAL             16       494712540.48899990

                   B VALUE                STD ERROR       PARTIAL REG SS          F      PROB > F

      INTERCEPT -68.31395896
      X2          0.07486591             0.01913019     7189926.32492182      15.32        0.0016
      X3          0.82287456             0.08295986    46187674.54075647      98.39        0.0001
-------------------------------------------------------------------------------------------------

STEP 3   VARIABLE X5 ENTERED          R SQUARE = 0.99006817          C(P) = 2.91769778

                        DF           SUM OF SQUARES          MEAN SQUARE          F      PROB > F

      REGRESSION         3       489799141.98626880   163266380.66208962     431.97        0.0001
      ERROR             13         4913398.50273108      377953.73097931
      TOTAL             16       494712540.48899990

                   B VALUE                STD ERROR       PARTIAL REG SS          F      PROB > F

      INTERCEPT 1523.38923568
      X2          0.05298733             0.02009194     2628687.59792946       6.96        0.0205
      X3          0.97848162             0.10515362    32726194.93174630      86.59        0.0001
      X5       -320.95082518           153.19222065     1658984.03530548       4.39        0.0563
```

FORWARD SELECTION

This procedure works the same way as stepwise regression *except that once an independent variable is entered into the model, it is never removed.* Forward selection is generally considered to be less effective than stepwise regression but to be useful in some problems.

BACKWARD ELIMINATION

A regression analysis is performed by using a regression model containing all the p potential independent variables. Then the independent variable having the smallest (in absolute value) t statistic is chosen. If the t statistic indicates that this independent variable is significant at the α_{stay} level (α_{stay} is chosen prior to the beginning of the procedure), then the procedure terminates by choosing the regression model containing all p independent variables. If this independent variable is not significant at the α_{stay} level, then it is removed from the model, and a regression analysis is performed by using a regression model containing all the remaining independent variables. The procedure continues by removing independent variables one at a time from the model. At each step an independent variable is removed from the model if it has the smallest (in absolute value) t statistic of the independent variables remaining in the model and if it is not significant at the α_{stay} level. The procedure terminates when no independent variable remaining in the model can be removed. Backward elimination is generally considered to be a reasonable procedure, especially for analysts who like to start with all possible independent variables in the model so that they will not "miss any important variables." For example, Figure 11.6 presents the SAS output of the backward elimination procedure for the hospital labor problem (using all $n = 17$ observations in Table 11.5). Here α_{stay} has been set equal to .10. Note that the backward elimination procedure arrives at the final model

$$y = \beta_0 + \beta_1 x_2 + \beta_2 x_3 + \beta_3 x_5 + \varepsilon$$

MAXIMUM R^2 IMPROVEMENT

The following description of this procedure is quoted from the *SAS User's Guide*, which calls this procedure the MAXR method.

> This method does not settle on a single model. Instead, it looks for the "best" one-variable model, the "best" two-variable model, and so forth.
>
> The MAXR method begins by finding the one-variable model producing the highest R^2. Then another variable, the one that would yield the greatest increase in R^2, is added.
>
> Once the two-variable model is obtained, each of the variables in the model is compared to each variable not in the model. For each comparison, MAXR determines if removing one variable and replacing it with the other variable would increase R^2.

FIGURE 11.6 **The SAS output of the backward elimination procedure for the hospital labor problem**

STEP 0 ALL VARIABLES ENTERED R SQUARE = 0.99083295 C(P) = 6.00000000

	DF	SUM OF SQUARES	MEAN SQUARE	F	PROB>F
REGRESSION	5	490177488.12165090	98035497.62433018	237.79	0.0001
ERROR	11	4535052.36734900	412277.48794082		
TOTAL	16	494712540.48899990			

	B VALUE	STD ERROR	PARTIAL REG SS	F	PROB>F
INTERCEPT	1962.94815647				
X1	-15.85167473	97.65299018	10863.47659691	0.03	0.8740
X2	0.05593038	0.02125828	2853834.33814818	6.92	0.0234
X3	1.58962370	3.09208349	108962.19771764	0.26	0.6174
X4	-4.21866799	7.17655737	142464.93826739	0.35	0.5685
X5	-394.31411702	209.63954082	1458572.00281510	3.54	0.0867

STEP 1 VARIABLE X1 REMOVED R SQUARE = 0.99081100 C(P) = 4.02634991

	DF	SUM OF SQUARES	MEAN SQUARE	F	PROB>F
REGRESSION	4	490166624.64505400	122541656.16126351	323.48	0.0001
ERROR	12	4545915.84394591	378826.32032883		
TOTAL	16	494712540.48899990			

	B VALUE	STD ERROR	PARTIAL REG SS	F	PROB>F
INTERCEPT	2032.18806215				
X2	0.05607934	0.02035863	2874411.76901339	7.59	0.0175
X3	1.08836904	0.15339754	19070221.63728595	50.34	0.0001
X4	-5.00406579	5.08071295	367482.65878517	0.97	0.3441
X5	-410.08295954	178.07810366	2008919.55996109	5.30	0.0400

STEP 2 VARIABLE X4 REMOVED R SQUARE = 0.99006817 C(P) = 2.91769778

	DF	SUM OF SQUARES	MEAN SQUARE	F	PROB>F
REGRESSION	3	489799141.98626880	163266380.66208962	431.97	0.0001
ERROR	13	4913398.50273108	377953.73097931		
TOTAL	16	494712540.48899990			

	B VALUE	STD ERROR	PARTIAL REG SS	F	PROB>F
INTERCEPT	1523.38923568				
X2	0.05298733	0.02009194	2628687.59792946	6.96	0.0205
X3	0.97848162	0.10515362	32726194.93174914	86.59	0.0001
X5	-320.95082518	153.19222065	1658984.03530548	4.39	0.0563

After comparing all possible switches, the one that produces the largest increase in R^2 is made.

Comparisons begin again, and the process continues until MAXR finds that no switch could increase R^2. The two-variable model thus achieved is considered the "best" two-variable model the technique can find.

Another variable is then added to the model, and the comparing and switching process is repeated to find the "best" three-variable model, and so forth.

The difference between the stepwise technique and the maximum R^2 improvement method is that all switches are evaluated before any switch is made in the MAXR method. In the stepwise method, the "worst" variable may be removed without considering what adding the "best" remaining variable might accomplish.*

*Reprinted by permission from SAS Institute, Cary, N.C.

In Figure 11.7 we present the SAS output of the maximum R^2 improvement procedure for the sales territory performance data in Table 8.5.

FIGURE 11.7 **The SAS output of the maximum R^2 improvement procedure for the sales territory performance data**

```
MAXIMUM R-SQUARE IMPROVEMENT FOR DEPENDENT VARIABLE SALES

STEP 1   VARIABLE ACCTS ENTERED       R SQUARE = 0.56849518      C(P) = 67.55826191

                 DF            SUM OF SQUARES       MEAN SQUARE            F      PROB > F

REGRESSION        1            23524074.19189501    23524074.19189501    30.30    0.0001
ERROR            23            17855474.73496109      776324.98847657
TOTAL            24            41379548.92685610

                 B VALUE       STD ERROR            TYPE II SS             F      PROB > F

INTERCEPT      709.32383372
ACCTS           21.72176971    3.94603304           23524074.19189501    30.30    0.0001

THE ABOVE MODEL IS THE BEST 1 VARIABLE MODEL FOUND.
-----------------------------------------------------------------------------------------
STEP 2   VARIABLE ADV ENTERED       R SQUARE = 0.77510077      C(P) = 27.15634512

                 DF            SUM OF SQUARES       MEAN SQUARE            F      PROB > F

REGRESSION        2            32073320.05178234    16036660.02589117    37.91    0.0001
ERROR            22             9306228.87507376      423010.40341244
TOTAL            24            41379548.92685610

                 B VALUE       STD ERROR            TYPE II SS             F      PROB > F

INTERCEPT       50.29906160
ADV              0.22652657    0.05038842            8549245.85988733    20.21    0.0002
ACCTS           19.04824598    2.97291380           17365864.05826321    41.05    0.0001

THE ABOVE MODEL IS THE BEST 2 VARIABLE MODEL FOUND.
-----------------------------------------------------------------------------------------
STEP 3   VARIABLE POTEN ENTERED       R SQUARE = 0.82772280      C(P) = 18.35666096

                 DF            SUM OF SQUARES       MEAN SQUARE            F      PROB > F

REGRESSION        3            34250796.02722474    11416932.00907491    33.63    0.0001
ERROR            21             7128752.89963136      339464.42379197
TOTAL            24            41379548.92685610

                 B VALUE       STD ERROR            TYPE II SS             F      PROB > F

INTERCEPT     -327.23338939
POTEN            0.02192192    0.00865564            2177475.97544240     6.41    0.0194
ADV              0.21607079    0.04532744            7713721.79142546    22.72    0.0001
ACCTS           15.55392158    2.99936692            9128825.32465885    26.89    0.0001

STEP 3   ACCTS REPLACED BY SHARE       R SQUARE = 0.84897591      C(P) = 13.99485796

                 DF            SUM OF SQUARES       MEAN SQUARE            F      PROB > F

REGRESSION        3            35130240.37743155    11710080.12581052    39.35    0.0001
ERROR            21             6249308.54942455      297586.12140117
TOTAL            24            41379548.92685610
```

```
             B VALUE          STD ERROR            TYPE II SS              F        PROB > F

INTERCEPT    -1603.58091828
POTEN            0.05428605    0.00747411          15698916.49197692    52.75     0.0001
ADV              0.16748034    0.04427318           4258521.38246777    14.31     0.0011
SHARE          282.74666591   48.75558026          10008269.67486566    33.63     0.0001

THE ABOVE MODEL IS THE BEST 3 VARIABLE MODEL FOUND.
---------------------------------------------------------------------------------------------
STEP 4  VARIABLE ACCTS ENTERED     R SQUARE = 0.90044970       C(P) = 5.43082968

             DF               SUM OF SQUARES       MEAN SQUARE             F        PROB > F

REGRESSION    4               37260202.46282121    9315050.61570530     45.23     0.0001
ERROR        20                4119346.46403489     205967.32320174
TOTAL        24               41379548.92685610

             B VALUE          STD ERROR            TYPE II SS              F        PROB > F

INTERCEPT    -1441.93182868
POTEN            0.03821753    0.00797694          4727717.33107410     22.95     0.0001
ADV              0.17499004    0.03690666          4630368.54356449     22.48     0.0001
SHARE          190.14429731   49.74415347          3009406.43559647     14.61     0.0011
ACCTS            9.21389567    2.86521038          2129962.08538966     10.34     0.0043

THE ABOVE MODEL IS THE BEST 4 VARIABLE MODEL FOUND.
---------------------------------------------------------------------------------------------
STEP 5  VARIABLE CHANGE ENTERED    R SQUARE = 0.91241574       C(P) = 4.97502565

             DF               SUM OF SQUARES       MEAN SQUARE             F        PROB > F

REGRESSION    5               37755351.59771667    7551070.31954333     39.59     0.0001
ERROR        19                3624197.32913943     190747.22784944
TOTAL        24               41379548.92685610

             B VALUE          STD ERROR            TYPE II SS              F        PROB > F

INTERCEPT    -1285.94337067
POTEN            0.03763121     0.00768517         4573489.56079583     23.98     0.0001
ADV              0.15443602     0.03773852         3194373.73424488     16.75     0.0006
SHARE          196.94952750    48.05692351         3203732.09272627     16.80     0.0006
CHANGE         262.50049338   162.92631286          495149.13489546      2.60     0.1236
ACCTS            8.23411280     2.82357962         1622152.88530014      8.50     0.0089


STEP 5  ACCTS REPLACED BY TIME     R SQUARE = 0.91500898       C(P) = 4.44281083

             DF               SUM OF SQUARES       MEAN SQUARE             F        PROB > F

REGRESSION    5               37862658.90021934    7572531.78004387     40.91     0.0001
ERROR        19                3516890.02663676     185099.47508615
TOTAL        24               41379548.92685610

             B VALUE          STD ERROR            TYPE II SS              F        PROB > F

INTERCEPT    -1113.78787943
TIME             3.61210118     1.18169995         1729460.18780281      9.34     0.0065
POTEN            0.04208812     0.00673122         7236631.11940864     39.10     0.0001
ADV              0.12885675     0.03703609         2240625.11884974     12.10     0.0025
SHARE          256.95554016    39.13606967         7979335.79655744     43.11     0.0001
CHANGE         324.53344995   157.28308421          788061.45599777      4.26     0.0530

THE ABOVE MODEL IS THE BEST 5 VARIABLE MODEL FOUND.
---------------------------------------------------------------------------------------------
```

(continues)

FIGURE 11.7 Continued

STEP 6 VARIABLE ACCTS ENTERED R SQUARE = 0.92031405 C(P) = 5.35404547

	DF	SUM OF SQUARES	MEAN SQUARE	F	PROB > F
REGRESSION	6	38082180.17286742	6347030.02881124	34.65	0.0001
ERROR	18	3297368.75398868	183187.15299937		
TOTAL	24	41379548.92685610			

	B VALUE	STD ERROR	TYPE II SS	F	PROB > F
INTERCEPT	-1165.47855369				
TIME	2.26935112	1.69898362	326828.57515075	1.78	0.1983
POTEN	0.03827800	0.00754688	4712573.84321485	25.73	0.0001
ADV	0.14067029	0.03839221	2459312.09671656	13.43	0.0018
SHARE	221.60469221	50.58309112	3515945.91528346	19.19	0.0004
CHANGE	285.10928426	160.55965553	577623.53145916	3.15	0.0927
ACCTS	4.37770296	3.99903763	219521.27264808	1.20	0.2881

THE ABOVE MODEL IS THE BEST 6 VARIABLE MODEL FOUND.

STEP 7 VARIABLE WORK ENTERED R SQUARE = 0.92201937 C(P) = 7.00406114

	DF	SUM OF SQUARES	MEAN SQUARE	F	PROB > F
REGRESSION	7	38152745.42925927	5450392.20417990	28.71	0.0001
ERROR	17	3226803.49759683	189811.97044687		
TOTAL	24	41379548.92685610			

	B VALUE	STD ERROR	TYPE II SS	F	PROB > F
INTERCEPT	-1485.88075785				
TIME	1.97454330	1.79574963	229490.83764567	1.21	0.2868
POTEN	0.03729049	0.00785101	4282222.67815504	22.56	0.0002
ADV	0.15196094	0.04324545	2343725.58802753	12.35	0.0027
SHARE	198.30848767	64.11718234	1815755.40273993	9.57	0.0066
CHANGE	295.86609393	164.38654930	614867.35490257	3.24	0.0897
ACCTS	5.61018822	4.54495654	289213.78714048	1.52	0.2339
WORK	19.89903131	32.63610128	70565.25639185	0.37	0.5501

THE ABOVE MODEL IS THE BEST 7 VARIABLE MODEL FOUND.

STEP 8 VARIABLE RATE ENTERED R SQUARE = 0.92203915 C(P) = 9.00000000

	DF	SUM OF SQUARES	MEAN SQUARE	F	PROB > F
REGRESSION	8	38153564.25210440	4769195.53151305	23.65	0.0001
ERROR	16	3225984.67475170	201624.04217198		
TOTAL	24	41379548.92685610			

	B VALUE	STD ERROR	TYPE II SS	F	PROB > F
INTERCEPT	-1507.81372984				
TIME	2.00956615	1.93065421	218442.90153751	1.08	0.3134
POTEN	0.03720491	0.00820230	4148313.42001816	20.57	0.0003
ADV	0.15098890	0.04710851	2071255.37925263	10.27	0.0055
SHARE	199.02353635	67.02792230	1777623.07698218	8.82	0.0090
CHANGE	290.85513399	186.78199574	488906.44513135	2.42	0.1390
ACCTS	5.55096065	4.77554962	272415.54793787	1.35	0.2621
WORK	19.79389189	33.67669223	69653.96590204	0.35	0.5649
RATE	8.18928366	128.50561301	818.82284513	0.00	0.9500

THE ABOVE MODEL IS THE BEST 8 VARIABLE MODEL FOUND.

11.4

MODEL VALIDATION

When we have selected an appropriate model, it is important to validate this model by using it to analyze a data set that is different from the data set used to build the model. Generally, we check the model's predictive ability on the different data set, which might be newly collected data or a hold-out sample from the original data. Another way to validate the model is to compare it to theoretical expectations and earlier empirical findings.

For example, in the article "A Model of MBA Student Performance," which was published in *Research in Higher Education* (Vol. 25, No. 2, 1986), McClure, Wells, and Bowerman developed a model to predict the graduate grade point average (GGPA) of an MBA student. Using a sample of 89 students, they found that important independent variables included undergraduate grade point average, Graduate Management Admission Test score, the level of competitiveness of a student's undergraduate institution, and the type of undergraduate major (business, quantitative, education, social science, liberal arts, and fine arts). Note that level of competitiveness and type of undergraduate major are qualitative variables that are modeled by using *dummy variables*, which are discussed in Chapter 12. By employing interaction terms between undergraduate grade point average and the qualitative variables the authors obtained an R^2 of .61 and an s^2 of .04. These results were substantially better than had previously been reported in the literature. To validate their model, the authors predicted the GGPA of 100 students in a new data set. The mean square error for the difference between the observed and predicted values for the new data was .075. Of the 91 students who had graduated, eight students had predicted GGPA's less than 3.0 (the minimum required for graduation). Five of these eight students had graduated with a GGPA of exactly 3.0, while the average GGPA for all eight students was 3.071. The average GGPA for the remaining 83 students who had graduated was 3.362. Of the nine students who did not graduate, two had predicted GGPA's less than 3.0. The actual average GGPA for these two students was 2.576. For the remaining seven students, who had predicted GGPA's above 3.0, the average GGPA was 2.768. Two of these seven students had GGPA's above 2.9 and therefore possibly withdrew for reasons other than grades. In summary, the model did a fairly good job in identifying potential grade problems with the MBA students.

Finally, we note that if we cannot obtain new data or if the original data set is too small to use a hold-out sample (this is called *data-splitting*), the PRESS residuals and statistic can be used in validation. Recall that a PRESS residual is calculated by predicting y_i without using the ith observation. Therefore the size of the PRESS residuals and statistic give some indication of the model's ability to predict new data.

*11.5

USING SAS

In Figure 11.8 we present the SAS statements that implement many of the model-building techniques of this chapter. Here these techniques are applied to the hospital labor data of Table 11.5.

FIGURE 11.8 SAS program to implement model building for the hospital labor data

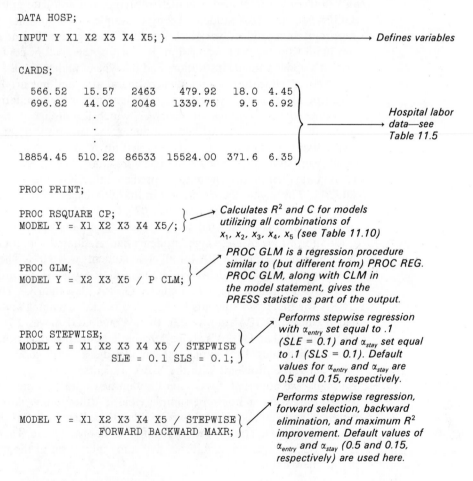

```
DATA HOSP;
INPUT Y X1 X2 X3 X4 X5; }  ——————————→ Defines variables

CARDS;
   566.52   15.57   2463     479.92   18.0  4.45 )
   696.82   44.02   2048    1339.75    9.5  6.92 |
                                                  |
              .                                   }——→ Hospital labor
              .                                   |    data—see
              .                                   |    Table 11.5
18854.45  510.22  86533  15524.00  371.6  6.35   )

PROC PRINT;

PROC RSQUARE CP;               }  ——→ Calculates R² and C for models
MODEL Y = X1 X2 X3 X4 X5/;     }      utilizing all combinations of
                                      x1, x2, x3, x4, x5 (see Table 11.10)

PROC GLM;                      }  ——→ PROC GLM is a regression procedure
MODEL Y = X2 X3 X5 / P CLM;    }      similar to (but different from) PROC REG.
                                      PROC GLM, along with CLM in
                                      the model statement, gives the
                                      PRESS statistic as part of the output.

PROC STEPWISE;                          }  ——→ Performs stepwise regression
MODEL Y = X1 X2 X3 X4 X5 / STEPWISE     }      with α_entry set equal to .1
                   SLE = 0.1 SLS = 0.1; }      (SLE = 0.1) and α_stay set equal
                                               to .1 (SLS = 0.1). Default
                                               values for α_entry and α_stay are
                                               0.5 and 0.15, respectively.

MODEL Y = X1 X2 X3 X4 X5 / STEPWISE     }  ——→ Performs stepwise regression,
               FORWARD BACKWARD MAXR;   }      forward selection, backward
                                               elimination, and maximum R²
                                               improvement. Default values of
                                               α_entry and α_stay (0.5 and 0.15,
                                               respectively) are used here.
```

*This section is optional.

EXERCISES

11.1 Consider the construction project data in Table 8.8. Table 11.15 summarizes the values of R^2, $\bar{R}^2$, and s^2, and the prediction intervals (for the future combination $x_{01} = 4.8$ and $x_{02} = 6$) that result when various models are used to perform regression analyses of the data.

TABLE 11.15 Summary measures for various construction project models

Model	R^2	$\bar{R}^2$	s^2	Prediction interval
$y = \beta_0 + \beta_1 x_1 + \beta_2 x_2 + \varepsilon$	.3284	.2389	3.9971	[−.2618, 8.834]
$y = \beta_0 + \beta_1 x_1 + \beta_2 x_2 + \beta_3 x_1^2 + \varepsilon$	.3797	.2468	3.9556	[−.1358, 8.986]
$y = \beta_0 + \beta_1 x_1 + \beta_2 x_2 + \beta_3 x_1^2 + \beta_4 x_1 x_2 + \varepsilon$	.8628	.8206	.9422	[3.758, 8.362]
$y = \beta_0 + \beta_1 x_1 + \beta_2 x_2 + \beta_3 x_1^2 + \beta_4 x_1 x_2 + \beta_5 x_1^2 x_2 + \varepsilon$	.9025	.8619	.7251	[3.968, 8.043]
$y = \beta_0 + \beta_1 x_1 + \beta_2 x_2 + \beta_3 x_1^2 + \beta_4 x_1 x_2 + \beta_5 x_2^2 + \varepsilon$	.8647	.8084	1.0063	[3.620, 8.432]

a. Consider the model

$$y = \beta_0 + \beta_1 x_1 + \beta_2 x_2 + \beta_3 x_1^2 + \beta_4 x_1 x_2 + \varepsilon$$

Using the fact that $R^2 = .8628$, demonstrate how $\bar{R}^2 = .8206$ has been calculated.

b. On the basis of Table 11.15, which model seems best? Justify your answer. Discuss why the model you have chosen is not the model you would (probably) choose by examining the prob-values (given under the appropriate variables) in the following models:

$$\underset{}{y_i} = \underset{(.0001)}{\beta_0} + \underset{(.2290)}{\beta_1 x_{i1}} + \underset{(.0001)}{\beta_2 x_{i2}} + \underset{(.0053)}{\beta_3 x_{i1}^2} + \underset{(.0001)}{\beta_4 x_{i1} x_{i2}} + \varepsilon_i$$

$$y_i = \underset{(.2696)}{\beta_0} + \underset{(.1306)}{\beta_1 x_{i1}} + \underset{(.7845)}{\beta_2 x_{i2}} + \underset{(.0032)}{\beta_3 x_{i1}^2} + \underset{(.4056)}{\beta_4 x_{i1} x_{i2}} + \underset{(.0471)}{\beta_5 x_{i1}^2 x_{i2}} + \varepsilon_i$$

11.2 Figure 11.9 shows the SAS output of the values of R^2 and C that result from regression models representing all possible combinations of the independent variables describing the residential sales data in Table 11.9. Here, outlying and influential residences 10, 50, and 61 have been removed from the data, and thus the values are based on the remaining 60 observations.

a. Which model has the smallest value of C? Is this model biased?

b. It can be verified that the model having the smallest value of C has an s^2 equal to 171.73. On the other hand, Figure 8.29 tells us that s^2 equals 123.83 for the model relating SALEP to SQFT, SQFTSQ, ROOMS, ROOMSSQ, BED, BEDSQ, and AGE. Which of the two models seems best? Explain.

11.3 Figure 11.10 shows the SAS output of the values of R^2 and C that result from regression models representing all possible combinations of the independent variables describing the hospital labor data in Table 11.5 when hospital 14 is removed.

FIGURE 11.9 SAS output of values of R^2 and C when residences 10, 50, and 61 are removed from Table 11.9

SAS

N=60 REGRESSION MODELS FOR DEPENDENT VARIABLE: SALEP

NUMBER IN MODEL	R-SQUARE	C(P)	VARIABLES IN MODEL
1	0.08692707	200.535	AGE
1	0.20650257	166.940	BED
1	0.34614282	127.706	ROOMS
1	0.39472538	114.057	BATH
1	0.69146379	30.685749	SQFT
2	0.31582728	138.224	BED AGE
2	0.35188062	128.094	ROOMS BED
2	0.39795143	115.150	BATH AGE
2	0.48575258	90.481980	BATH BED
2	0.53688401	76.116188	ROOMS AGE
2	0.55296605	71.597810	ROOMS BATH
2	0.69334520	32.157151	SQFT ROOMS
2	0.71644449	25.667218	SQFT BED
2	0.71906684	24.930447	SQFT BATH
2	0.75340716	15.282256	SQFT AGE
3	0.49992297	88.500693	BATH BED AGE
3	0.53782767	77.851060	ROOMS BED AGE
3	0.55624058	72.677805	ROOMS BATH BED
3	0.61070308	57.376124	ROOMS BATH AGE
3	0.72259172	25.940104	SQFT ROOMS BATH
3	0.73026524	23.784167	SQFT ROOMS BED
3	0.73345692	22.887437	SQFT BATH BED
3	0.75739895	16.160731	SQFT BATH AGE
3	0.76787299	13.217963	SQFT BED AGE
3	0.77281301	11.830026	SQFT ROOMS AGE
4	0.61187959	59.045574	ROOMS BATH BED AGE
4	0.74700939	21.079763	SQFT ROOMS BATH BED
4	0.76940685	14.787013	SQFT BATH BED AGE
4	0.77567695	13.025380	SQFT ROOMS BATH AGE
4	0.80775853	4.011799	SQFT ROOMS BED AGE
5	0.80780052	6.000000	SQFT ROOMS BATH BED AGE

FIGURE 11.10 SAS output of values of R^2 and C when hospital 14 is removed from Table 11.5

SAS

N=16 REGRESSION MODELS FOR DEPENDENT VARIABLE: Y

NUMBER IN MODEL	R-SQUARE	C(P)	VARIABLES IN MODEL
1	0.31391746	1981.054	X5
1	0.88832232	312.421	X2
1	0.89587519	290.480	X4
1	0.97729076	53.969806	X1
1	0.97786107	52.313060	X3

NUMBER IN MODEL	R-SQUARE	C(P)	VARIABLES IN MODEL
2	0.91806856	228.009	X4 X5
2	0.91967400	223.345	X2 X5
2	0.93303588	184.529	X2 X4
2	0.97804641	53.774669	X1 X3
2	0.98111090	44.872390	X1 X4
2	0.98233794	41.307859	X3 X4
2	0.98988689	19.378342	X1 X2
2	0.99043733	17.779321	X2 X3
2	0.99271981	11.148785	X1 X5
2	0.99329885	9.466671	X3 X5
3	0.95334848	127.522	X2 X4 X5
3	0.98557865	33.893679	X1 X3 X4
3	0.99008577	20.800604	X1 X2 X4
3	0.99079427	18.742433	X2 X3 X4
3	0.99087491	18.508171	X1 X2 X3
3	0.99315453	11.885927	X1 X4 X5
3	0.99342170	11.109802	X1 X3 X5
3	0.99342344	11.104740	X3 X4 X5
3	0.99553699	4.964935	X1 X2 X5
3	0.99612452	3.258171	X2 X3 X5
4	0.99220851	16.634089	X1 X2 X3 X4
4	0.99344662	13.037408	X1 X3 X4 X5
4	0.99637228	4.538455	X1 X2 X4 X5
4	0.99637896	4.519035	X1 X2 X3 X5
4	0.99654958	4.023375	X2 X3 X4 X5
5	0.99655763	6.000000	X1 X2 X3 X4 X5

a. What is the best two-variable model with respect to C?
b. What is the best three-variable model with respect to C?
c. In Figure 11.11 we present the SAS output for the model using x_3 and x_5, and in Figure 11.12 we present the SAS output for the model using x_2, x_3, and x_5. Compare the residuals representing observations 14, 15, and 16 (which are "large hospitals" 15, 16, and 17 in Table 11.5) for the two models. Which model should be used to predict labor requirements for large hospitals?

11.4 If we wish to obtain a regression model describing labor needs for small to medium-sized hospitals, it might be reasonable to remove the larger hospitals 14, 15, 16, and 17 from Table 11.5 and to use hospitals 1–13. Figure 11.13 shows the SAS output of the values of R^2 and C that result from regression models representing all possible combinations of the independent variables describing the hospital labor data in Table 11.5 when we use hospitals 1–13.

a. What model has the smallest value of C?
b. In Figure 11.14 we present the SAS output for the model using x_1 and x_5. What is the standard error, s, for this model? What is s for the model using x_2, x_3, and x_5 and based on hospitals 1–13, 15, 16, and 17 in Figure 11.12? Which of the two models would you use to predict labor needs for small to medium-sized hospitals?

FIGURE 11.11 SAS output for the hospital labor model using x_3 and x_5 when hospital 14 is removed from Table 11.5

SAS

ANALYSIS OF VARIANCE

SOURCE	DF	SUM OF SQUARES	MEAN SQUARE	F VALUE	PROB>F
MODEL	2	461016421	230508211	963.483	0.0001
ERROR	13	3110180.18	239244.63		
C TOTAL	15	464126602			

ROOT MSE	489.1264	R-SQUARE	0.9933
DEP MEAN	4643.147	ADJ R-SQ	0.9923
C.V.	10.53437		

PARAMETER ESTIMATES

VARIABLE	DF	PARAMETER ESTIMATE	STANDARD ERROR	T FOR HO: PARAMETER=0	PROB > |T|	VARIANCE INFLATION
INTERCEP	1	2741.24376	539.06763	5.085	0.0002	0
X3	1	1.22299232	0.03368758	36.304	0.0001	1.77547852
X5	1	-572.24921	104.56724	-5.473	0.0001	1.77547852

OBS	ACTUAL	PREDICT VALUE	STD ERR PREDICT	LOWER95% MEAN	UPPER95% MEAN	LOWER95% PREDICT	UPPER95% PREDICT	RESIDUAL
1	566.5	773.1	166.4	413.6	1132.6	-343.1	1889.3	-206.6
2	696.8	419.8	230.0	-77.0688	916.6	-747.9	1587.5	277.0
3	1033.1	1050.6	173.0	676.9	1424.2	-70.2357	2171.4	-17.4281
4	1603.6	1204.5	194.5	784.4	1624.7	67.3797	2341.7	399.1
5	1611.4	1425.4	143.8	1114.7	1736.2	324.0	2526.9	185.9
6	1613.3	1779.3	156.1	1442.1	2116.5	670.1	2888.5	-166.0
7	1854.2	1588.4	143.4	1278.5	1898.3	487.2	2689.6	265.8
8	2160.5	1799.8	139.8	1497.7	2101.8	700.8	2898.8	360.8
9	2305.6	2717.6	144.6	2405.1	3030.0	1615.7	3819.5	-412.0
10	3503.9	3692.0	132.3	3406.2	3977.9	2597.4	4786.7	-188.1
11	3571.9	2937.8	132.2	2652.3	3223.3	1843.2	4032.4	634.1
12	3741.4	4744.0	152.5	4414.5	5073.6	3637.1	5850.9	-1002.6
13	4026.5	4321.6	125.2	4051.1	4592.0	3230.8	5412.3	-295.0
14	11732.2	11794.2	413.2	10901.5	12686.8	10410.9	13177.4	-61.9933
15	15414.9	15949.1	291.1	15320.2	16578.0	14719.4	17178.8	-534.2
16	18854.4	18093.2	366.0	17302.6	18883.8	16773.5	19412.9	761.3

FIGURE 11.12 SAS output for the hospital labor model using x_2, x_3, and x_5 when hospital 14 is removed from Table 11.5

SAS

ANALYSIS OF VARIANCE

SOURCE	DF	SUM OF SQUARES	MEAN SQUARE	F VALUE	PROB>F
MODEL	3	462327889	154109296	1028.131	0.0001
ERROR	12	1798712.22	149892.68		
C TOTAL	15	464126602			

ROOT MSE	387.1598	R-SQUARE	0.9961
DEP MEAN	4643.147	ADJ R-SQ	0.9952
C.V.	8.338305		

PARAMETER ESTIMATES

VARIABLE	DF	PARAMETER ESTIMATE	STANDARD ERROR	T FOR HO: PARAMETER=0	PROB > |T|	VARIANCE INFLATION
INTERCEP	1	1946.80204	504.18193	3.861	0.0023	0
X2	1	0.03857709	0.01304190	2.958	0.0120	7.82831925
X3	1	1.03939197	0.06755556	15.386	0.0001	11.39619473
X5	1	-413.75780	98.59827976	-4.196	0.0012	2.51955937

OBS	ACTUAL	PREDICT VALUE	STD ERR PREDICT	LOWER95% MEAN	UPPER95% MEAN	LOWER95% PREDICT	UPPER95% PREDICT	RESIDUAL
1	566.5	692.1	134.5	399.0	985.3	-200.9	1585.2	-125.6
2	696.8	555.1	187.7	146.2	964.1	-382.3	1492.6	141.7
3	1033.1	972.6	139.4	668.8	1276.4	76.0159	1869.2	60.5547
4	1603.6	1174.8	154.3	838.7	1510.9	266.8	2082.9	428.8
5	1611.4	1448.5	114.1	1199.9	1697.1	569.1	2327.9	162.9
6	1613.3	1907.6	131.0	1622.2	2192.9	1017.1	2798.1	-294.3
7	1854.2	1597.9	113.6	1350.4	1845.3	718.8	2477.0	256.3
8	2160.5	1750.7	111.9	1506.9	1994.5	872.7	2628.8	409.8
9	2305.6	2701.7	114.6	2452.0	2951.4	1821.9	3581.4	-396.1
10	3503.9	3976.9	142.3	3666.9	4286.9	3078.2	4875.6	-473.0
11	3571.9	3054.2	111.8	2810.7	3297.7	2176.2	3932.2	517.7
12	3741.4	4418.6	163.3	4062.7	4774.5	3503.1	5334.2	-577.2
13	4026.5	4288.7	99.7120	4071.4	4505.9	3417.6	5159.8	-262.2
14	11732.2	11761.8	327.2	11048.8	12474.9	10657.3	12866.4	-29.6792
15	15414.9	15195.9	343.4	14447.7	15944.2	14068.4	16323.5	219.0
16	18854.4	18793.2	374.0	17978.2	19608.1	17620.2	19966.1	61.2977

FIGURE 11.13 SAS output of values of R^2 and C based on hospitals 1–13 in Table 11.5

SAS

N=13 REGRESSION MODELS FOR DEPENDENT VARIABLE: Y

NUMBER IN MODEL	R-SQUARE	C(P)	VARIABLES IN MODEL
1	0.05485277	97.133560	X5
1	0.71194981	23.346063	X4
1	0.73146110	21.155079	X2
1	0.86357034	6.320116	X3
1	0.86462334	6.201872	X1
2	0.71307008	25.220265	X4 X5
2	0.73412283	22.856185	X2 X5
2	0.77016642	18.808737	X2 X4
2	0.86484201	8.177317	X1 X3
2	0.86636542	8.006248	X1 X4
2	0.86854227	7.761802	X3 X4
2	0.89567538	4.714941	X1 X2
2	0.89898751	4.343011	X2 X3
2	0.91728347	2.288500	X3 X5
2	0.91753189	2.260604	X1 X5
3	0.77016840	20.808514	X2 X4 X5
3	0.86871683	9.742201	X1 X3 X4
3	0.89983334	6.248029	X1 X2 X3
3	0.90131328	6.081842	X1 X2 X4
3	0.90206258	5.997702	X2 X3 X4

(continues)

FIGURE 11.13 Continued

NUMBER IN MODEL	R-SQUARE	C(P)	VARIABLES IN MODEL
3	0.91767882	4.244105	X3 X4 X5
3	0.91779066	4.231545	X1 X4 X5
3	0.91822776	4.182462	X1 X3 X5
3	0.92937010	2.931254	X1 X2 X5
3	0.93251479	2.578127	X2 X3 X5
4	0.90208590	7.995083	X1 X2 X3 X4
4	0.91822805	6.182430	X1 X3 X4 X5
4	0.93267449	4.560194	X1 X2 X3 X5
4	0.93661280	4.117948	X2 X3 X4 X5
4	0.93758890	4.008339	X1 X2 X4 X5
5	0.93766316	6.000000	X1 X2 X3 X4 X5

FIGURE 11.14 SAS output for the hospital labor model using x_1 and x_5 based on hospitals 1–13 in Table 11.5

SAS

ANALYSIS OF VARIANCE

SOURCE	DF	SUM OF SQUARES	MEAN SQUARE	F VALUE	PROB>F
MODEL	2	15373255.12	7686627.56	55.629	0.0001
ERROR	10	1381753.89	138175.39		
C TOTAL	12	16755009.01			

ROOT MSE	371.7195	R-SQUARE	0.9175
DEP MEAN	2176.061	ADJ R-SQ	0.9010
C.V.	17.08222		

PARAMETER ESTIMATES

VARIABLE	DF	PARAMETER ESTIMATE	STANDARD ERROR	T FOR HO: PARAMETER=0	PROB > |T|	VARIANCE INFLATION
INTERCEP	1	2070.44039	681.81832	3.037	0.0125	0
X1	1	29.21340629	2.85627863	10.228	0.0001	1.28383598
X5	1	-354.60309	139.99824	-2.533	0.0297	1.28383598

OBS	ACTUAL	PREDICT VALUE	STD ERR PREDICT	LOWER95% MEAN	UPPER95% MEAN	LOWER95% PREDICT	UPPER95% PREDICT	RESIDUAL
1	566.5	947.3	174.9	557.5	1337.1	31.9289	1862.7	-380.8
2	696.8	902.6	284.3	269.2	1535.9	-140.1	1945.2	-205.7
3	1033.1	1149.3	177.3	754.1	1544.4	231.6	2067.0	-116.1
4	1603.6	1234.9	207.9	771.7	1698.2	285.9	2184.0	368.7
5	1611.4	1557.4	124.9	1279.1	1835.8	683.7	2431.2	53.9470
6	1613.3	1751.5	136.6	1447.1	2056.0	869.1	2634.0	-138.3
7	1854.2	1698.3	123.7	1422.8	1973.9	825.5	2571.2	155.8
8	2160.5	1976.0	106.0	1739.9	2212.1	1114.8	2837.2	184.5
9	2305.6	2636.4	150.1	2301.9	2971.0	1743.2	3529.7	-330.9
10	3503.9	3629.5	186.7	3213.5	4045.6	2702.6	4556.4	-125.6
11	3571.9	2789.9	131.9	2496.0	3083.8	1911.0	3668.7	782.0
12	3741.4	4179.2	239.2	3646.3	4712.1	3194.3	5164.1	-437.8
13	4026.5	3836.4	188.8	3415.7	4257.0	2907.4	4765.3	190.2

11.5 Market Planning, Inc., a marketing research firm, has obtained the prescription sales data in Table 11.16 for $n = 20$ independent pharmacies.*

TABLE 11.16 **Prescription sales data**

Pharmacy	Sales, y	Floor space, x_1	Prescription %, x_2	Parking, x_3	Income, x_4	Shopping center, x_5
1	22	4900	9	40	18	1
2	19	5800	10	50	20	1
3	24	5000	11	55	17	1
4	28	4400	12	30	19	0
5	18	3850	13	42	10	0
6	21	5300	15	20	22	1
7	29	4100	20	25	8	0
8	15	4700	22	60	15	1
9	12	5600	24	45	16	1
10	14	4900	27	82	14	1
11	18	3700	28	56	12	0
12	19	3800	31	38	8	0
13	15	2400	36	35	6	0
14	22	1800	37	28	4	0
15	13	3100	40	43	6	0
16	16	2300	41	20	5	0
17	8	4400	42	46	7	1
18	6	3300	42	15	4	0
19	7	2900	45	30	9	1
20	17	2400	46	16	3	0

These variables can be described precisely as follows:

y = average weekly prescription sales over the past year (in units of \$1000)

x_1 = floor space (in square feet)

x_2 = percent of floor space allocated to the prescription department

x_3 = number of parking spaces available for the store

x_4 = monthly per capita income for the surrounding community (in units of \$100)

x_5 is an independent variable that equals 1 if the pharmacy is located in a shopping center and equals 0 otherwise (x_5 is called a *dummy variable*; such variables will be discussed in detail in Chapter 12).

Figure 11.15 is the SAS output of the values of R^2 and C that result from regression models representing all possible combinations of the above independent variables. On the basis of the C statistic, which model seems best? Why?

*This problem is taken from an example in Lyman Ott, *An Introduction to Statistical Methods and Data Analysis*, 2nd ed., PWS–KENT Publishing Company, Boston. © 1987. Used with permission.

FIGURE 11.15 SAS output of values of R^2 and C based on the prescription sales data in Table 11.16

```
                            REGRESSION ANALYSES
                  PROC RSQUARE—ALL POSSIBLE SUBSETS ANALYSIS

  N=20                                  REGRESSION MODELS FOR DEPENDENT VARIABLE VOLUME

NUMBER IN
  MODEL        R-SQUARE          C(P)                      VARIABLES IN MODEL

    1        0.00480421     30.45388047     PARKING
    1        0.03353172     29.11293360     FLOOR_SP
    1        0.04105340     28.76183600     SHOPCNTR
    1        0.14798995     23.77023759     INCOME
    1        0.43933184     10.17094219     PRESC_RX

    2        0.04210776     30.71262010     PARKING SHOPCNTR
    2        0.06855667     29.47803470     FLOOR_SP PARKING
    2        0.20543099     23.08899693     PARKING INCOME
    2        0.23487329     21.71468547     FLOOR_SP INCOME
    2        0.25653635     20.70349407     FLOOR_SP SHOPCNTR
    2        0.49576794      9.53661080     SHOPCNTR INCOME
    2        0.53142435      7.87223587     PRESC_RX PARKING
    2        0.54748785      7.12242198     PRESC_RX INCOME
    2        0.64706473      2.47435928     PRESC_RX SHOPCNTR
    2        0.66566267      1.60624219     FLOOR_SP PRESC_RX

    3        0.25569607     22.74271718     FLOOR_SP PARKING INCOME
    3        0.26507110     22.30510820     FLOOR_SP PARKING SHOPCNTR
    3        0.49828073     11.41931841     PARKING SHOPCNTR INCOME
    3        0.50012580     11.33319388     FLOOR_SP SHOPCNTR INCOME
    3        0.60243233      6.55771633     PRESC_RX PARKING INCOME
    3        0.64711563      4.47198330     PRESC_RX SHOPCNTR INCOME
    3        0.66259120      3.74961255     PRESC_RX PARKING SHOPCNTR
    3        0.66641145      3.57129027     FLOOR_SP PRESC_RX INCOME
    3        0.67943313      2.96346249     FLOOR_SP PRESC_RX PARKING
    3        0.69072432      2.43641080     FLOOR_SP PRESC_RX SHOPCNTR

    4        0.50128901     13.27889728     FLOOR_SP PARKING SHOPCNTR INCOME
    4        0.66300855      5.73013127     PRESC_RX PARKING SHOPCNTR INCOME
    4        0.68058567      4.90966443     FLOOR_SP PRESC_RX PARKING INCOME
    4        0.69326657      4.31774327     FLOOR_SP PRESC_RX SHOPCNTR INCOME
    4        0.69873952      4.06227626     FLOOR_SP PRESC_RX PARKING SHOPCNTR

    5        0.70007369      6.00000000     FLOOR_SP PRESC_RX PARKING SHOPCNTR INCOME
```

11.6 (Optional) Write a SAS program to calculate the values of R^2 and C that result from regression models representing all possible combinations of the ten predictor variables describing FAT in the data of Exercise 10.6.

11.7 Show exactly how the usual and PRESS residuals corresponding to hospital 15 in Table 11.12 have been calculated. By working backward, determine the leverage value

$$h_{15,15} = \mathbf{x}_{15}'(\mathbf{X}'\mathbf{X})^{-1}\mathbf{x}_{15}$$

used to calculate the PRESS residual corresponding to hospital 15.

11.8 Figure 11.16 is the SAS output of a stepwise regression of the data in Exercise 11.5. Here, both α_{entry} and α_{stay} have been set equal to .15.

a. What is the first independent variable entered?

FIGURE 11.16 **SAS output of a stepwise regression of the data in Exercise 11.5**

```
STEPWISE REGRESSION PROCEDURE FOR DEPENDENT VARIABLE VOLUME

STEP 1  VARIABLE PRESC PCT
        ENTERED             R SQUARE=0.43933184          C(P)=10.17094219

                                  DF     SUM OF SQUARES   MEAN SQUARE       F  PROB>F

        REGRESSION                 1       329.74051403  329.74051403   14.10  0.0014
        ERROR                     18       420.80948597   23.37830478
        TOTAL                     19       750.55000000

                                  B VALUE       STD ERROR   TYPE II SS        F  PROB>F

        INTERCEPT             25.98133346
        PRESC PCT             -0.32055657      0.08535423  329.74051403   14.10  0.0014
--------------------------------------------------------------------------------------
STEP 2  VARIABLE FLOOR SP
        ENTERED             R SQUARE=0.66566267          C(P)=1.60624219

                                  DF     SUM OF SQUARES   MEAN SQUARE       F  PROB>F

        REGRESSION                 2       499.61311336  249.80655668   16.92  0.0001
        ERROR                     17       250.93688664   14.76099333
        TOTAL                     19       750.55000000

                                  B VALUE       STD ERROR   TYPE II SS        F  PROB>F

        INTERCEPT             48.29085530
        FLOOR SP               0.00384220      0.00113262  169.87259933   11.51  0.0035
        PRESC PCT             -0.58189034      0.10263739  474.44587802   32.14  0.0001
--------------------------------------------------------------------------------------
NO OTHER VARIABLES MET THE 0.1500 SIGNIFICANCE LEVEL
FOR ENTRY INTO THE MODEL
```

b. What is the second independent variable entered? Has the first independent entered been retained? Why?
c. What is the final model arrived at by stepwise regression? Is this model the same model that you chose in Exercise 11.5?

11.9 Figure 11.17 is the SAS output of a backward elimination procedure applied to the data in Exercise 11.5. Here, α_{stay} has been set equal to .10.

a. Describe the order in which the independent variables are removed by the backward elimination procedure.
b. Is the model arrived at by backward elimination the same as the model arrived at by stepwise regression in Exercise 11.8?

FIGURE 11.17 **SAS output of a backward elimination of the data in Exercise 11.5**

```
                       REGRESSION ANALYSIS, USING BACKWARD ELIMINATION

                 BACKWARD ELIMINATION PROCEDURE FOR DEPENDENT VARIABLE VOLUME

STEP 0     ALL VARIABLES ENTERED     R SQUARE = 0.70007369     C(P) = 6.00000000

                   DF          SUM OF SQUARES      MEAN SQUARE          F       PROB>F

REGRESSION          5            525.44030541     105.08806108       6.54       0.0025
ERROR              14            225.10969459      16.07926390
TOTAL              19            750.55000000
```

(continues)

FIGURE 11.17 Continued

```
                 B VALUE          STD ERROR        TYPE II SS          F     PROB>F

INTERCEPT      42.08710826
FLOOR_SP       -0.00241878       0.00183889       27.81923726       1.73     0.2095
PRESC_RX       -0.50046955       0.16429694      149.19783807       9.28     0.0087
PARKING        -0.03690284       0.06546687        5.10907792       0.32     0.5819
SHOPCNTR       -3.09957355       3.24983522       14.62673442       0.91     0.3564
INCOME          0.10666360       0.42742012        1.00135642       0.06     0.8066

BOUNDS ON CONDITION NUMBER:    7.823107,     117.1991
-----------------------------------------------------------------------------------
STEP 1    VARIABLE INCOME REMOVED     R SQUARE = 0.69873952    C(P) = 4.06227626

                    DF          SUM OF SQUARES      MEAN SQUARE          F     PROB>F

REGRESSION           4           524.43894899     131.10973725       8.70     0.0008
ERROR               15           226.11105101      15.07407007
TOTAL               19           750.55000000

                 B VALUE          STD ERROR        TYPE II SS          F     PROB>F

INTERCEPT      43.46782063
FLOOR_SP       -0.00228513       0.00170330       27.13112543       1.80     0.1997
PRESC_RX       -0.52910174       0.11386382      325.48983690      21.59     0.0003
PARKING        -0.03952477       0.06256589        6.01580808       0.40     0.5371
SHOPCNTR       -2.71387948       2.76799605       14.49041122       0.96     0.3424

BOUNDS ON CONDITION NUMBER:    5.071729,     46.98862
-----------------------------------------------------------------------------------
STEP 2    VARIABLE PARKING REMOVED    R SQUARE = 0.69072432    C(P) = 2.43641080

                    DF          SUM OF SQUARES      MEAN SQUARE          F     PROB>F

REGRESSION           3           518.42314091     172.80771364      11.91     0.0002
ERROR               16           232.12685909      14.50792869
TOTAL               19           750.55000000

                 B VALUE          STD ERROR        TYPE II SS          F     PROB>F

INTERCEPT      42.82702645
FLOOR_SP       -0.00247284       0.00164539       32.76871130       2.26     0.1523
PRESC_RX       -0.52941361       0.11170410      325.87978038      22.46     0.0002
SHOPCNTR       -3.03834296       2.66836223       18.81002755       1.30     0.2716

BOUNDS ON CONDITION NUMBER:    4.917388,     30.31995
-----------------------------------------------------------------------------------
STEP 3    VARIABLE SHOPCNTR REMOVED   R SQUARE = 0.66566267    C(P) = 1.60624219

                    DF          SUM OF SQUARES      MEAN SQUARE          F     PROB>F

REGRESSION           2           499.61311336     249.80655668      16.92     0.0001
ERROR               17           250.93688664      14.76099333
TOTAL               19           750.55000000

                 B VALUE          STD ERROR        TYPE II SS          F     PROB>F

INTERCEPT      48.29085530
FLOOR_SP       -0.00384228       0.00113262      169.87259933      11.51     0.0035
PRESC_RX       -0.58189034       0.10263739      474.44587802      32.14     0.0001

BOUNDS ON CONDITION NUMBER:    2.290122,     9.160487
-----------------------------------------------------------------------------------
ALL VARIABLES IN THE MODEL ARE SIGNIFICANT AT THE 0.1000 LEVEL.
```

11.10 Figure 11.18 is the SAS output of a stepwise regression of the residential sales data in Table 11.9 with outlying and influential residences 10, 50, and 61 removed. Here, both

α_{entry} and α_{stay} have been set equal to .10. Does the stepwise regression arrive at the model having the smallest value of C in Exercise 11.2?

FIGURE 11.18 **SAS output of a stepwise regression of the data in Table 11.9 with residences 10, 50, and 61 removed**

```
                                      SAS

              STEPWISE REGRESSION PROCEDURE FOR DEPENDENT VARIABLE SALEP

STEP 1     VARIABLE SQFT ENTERED      R SQUARE = 0.69146379      C(P) = 30.685749

                    DF          SUM OF SQUARES        MEAN SQUARE            F      PROB>F

REGRESSION           1       33972.83686517      33972.83686517       129.98      0.0001
ERROR               58       15158.92896816        261.36084428
TOTAL               59       49131.76583333

                     B VALUE          STD ERROR         TYPE II SS            F      PROB>F

INTERCEPT        -7.26896622
SQFT              0.05607886         0.00491874     33972.83686517       129.98      0.0001

BOUNDS ON CONDITION NUMBER:            1,          1

STEP 2     VARIABLE AGE ENTERED       R SQUARE = 0.75340716      C(P) = 15.282256

                    DF          SUM OF SQUARES        MEAN SQUARE            F      PROB>F

REGRESSION           2       37016.22425589      18508.11212794        87.08      0.0001
ERROR               57       12115.54157745        212.55336101
TOTAL               59       49131.76583333

                     B VALUE          STD ERROR         TYPE II SS            F      PROB>F

INTERCEPT         3.84771042
SQFT              0.05514210         0.00444265     32745.34364668       154.06      0.0001
AGE              -0.30175654         0.07974658      3043.38739071        14.32      0.0004

BOUNDS ON CONDITION NUMBER:     1.003115,      4.012459

STEP 3     VARIABLE ROOMS ENTERED     R SQUARE = 0.77281301      C(P) = 11.830026

                    DF          SUM OF SQUARES        MEAN SQUARE            F      PROB>F

REGRESSION           3       37969.66777032      12656.55592344        63.50      0.0001
ERROR               56       11162.09806301        199.32317970
TOTAL               59       49131.76583333

                     B VALUE          STD ERROR         TYPE II SS            F      PROB>F

INTERCEPT        -4.26870998
SQFT              0.04592536         0.00602226     11591.60813025        58.15      0.0001
ROOMS             3.83502512         1.75347522       953.44351443         4.78      0.0329
AGE              -0.36486863         0.08244012      3904.39379619        19.59      0.0001

BOUNDS ON CONDITION NUMBER:     2.061629,      15.51122

STEP 4     VARIABLE BED ENTERED       R SQUARE = 0.80775853      C(P) = 4.011799

                    DF          SUM OF SQUARES        MEAN SQUARE            F      PROB>F

REGRESSION           4       39686.60289604       9921.65072401        57.77      0.0001
ERROR               55        9445.16293729        171.73023522
TOTAL               59       49131.76583333

                     B VALUE          STD ERROR         TYPE II SS            F      PROB>F

INTERCEPT         8.24325500
SQFT              0.05395555         0.00613977     13262.17971605        77.23      0.0001
ROOMS             5.94033538         1.75851248      1959.64672577        11.41      0.0013
BED             -12.86398180         4.06838324      1716.93512572        10.00      0.0026
AGE              -0.36036951         0.07653466      3807.38231081        22.17      0.0001

BOUNDS ON CONDITION NUMBER:     2.406651,      32.70757
```

11.11 Figure 11.19 is the SAS output of a stepwise regression of the hospital labor data in Table 11.5 with hospital 14 removed. Here, both α_{entry} and α_{stay} have been set equal to .10. Does the stepwise regression arrive at the model having the smallest value of C in Exercise 11.3?

FIGURE 11.19 **SAS output of a stepwise regression of the data in Table 11.5 with hospital 14 removed**

```
                                        SAS

                  STEPWISE REGRESSION PROCEDURE FOR DEPENDENT VARIABLE Y

STEP 1      VARIABLE X3 ENTERED      R SQUARE = 0.97786107        C(P) = 52.31305986

                     DF                SUM OF SQUARES            MEAN SQUARE          F   PROB>F

REGRESSION            1            453851336.67208900     453851336.67208900     618.37   0.0001
ERROR                14             10275264.93245420        733947.49517530
TOTAL                15            464126601.60454400

                         B VALUE              STD ERROR            TYPE II SS          F   PROB>F

INTERCEPT       -70.23024595
X3                1.10115326             0.04428162     453851336.67208900     618.37   0.0001

BOUNDS ON CONDITION NUMBER:              1,          1
-----------------------------------------------------------------------------------------------
STEP 2      VARIABLE X5 ENTERED      R SQUARE = 0.99329885        C(P) = 9.46667119

                     DF                SUM OF SQUARES            MEAN SQUARE          F   PROB>F

REGRESSION            2            461016421.42888300     230508210.71444200     963.48   0.0001
ERROR                13              3110180.17566051        239244.62889696
TOTAL                15            464126601.60454400

                         B VALUE              STD ERROR            TYPE II SS          F   PROB>F

INTERCEPT      2741.24375965
X3                1.22299232             0.03368758     315318979.25576300    1317.98   0.0001
X5             -572.24920511           104.56723642       7165084.75679367      29.95   0.0001

BOUNDS ON CONDITION NUMBER:       1.775479,       7.101914
-----------------------------------------------------------------------------------------------
STEP 3      VARIABLE X2 ENTERED      R SQUARE = 0.99612452        C(P) = 3.25817070

                     DF                SUM OF SQUARES            MEAN SQUARE          F   PROB>F

REGRESSION            3            462327889.38661900     154109296.46220600    1028.13   0.0001
ERROR                12              1798712.21792442        149892.68482704
TOTAL                15            464126601.60454400

                         B VALUE              STD ERROR            TYPE II SS          F   PROB>F

INTERCEPT      1946.80203866
X2                0.03857709             0.01304190       1311467.95773609       8.75   0.0120
X3                1.03939197             0.06755556      35482721.96191110     236.72   0.0001
X5             -413.75779647            98.59827976       2639575.83272846      17.61   0.0012

BOUNDS ON CONDITION NUMBER:       11.39619,       65.23222
```

11.12 Figure 11.20 is the SAS output of a stepwise regression of the hospital labor data in Table 11.5 with large hospitals 14, 15, 16, and 17 removed. Here, both α_{entry} and α_{stay} have been set equal to .10. Does the stepwise regression arrive at the model having the smallest value of C in Exercise 11.4?

FIGURE 11.20 **SAS output of a stepwise regression of the data in Table 11.5 with hospitals 14, 15, 16, and 17 removed**

```
                                     SAS

              STEPWISE REGRESSION PROCEDURE FOR DEPENDENT VARIABLE Y

STEP 1     VARIABLE X1 ENTERED     R SQUARE = 0.86462334     C(P) = 6.20187207

                  DF          SUM OF SQUARES           MEAN SQUARE         F    PROB>F

REGRESSION         1       14486771.77878970     14486771.77878970     70.25    0.0001
ERROR             11        2268237.22790266       206203.38435479
TOTAL             12       16755009.00669230

                   B VALUE             STD ERROR            TYPE II SS         F    PROB>F

INTERCEPT     419.53628641
X1             25.81167924            3.07948679     14486771.77878970     70.25    0.0001

BOUNDS ON CONDITION NUMBER:           1,          1
--------------------------------------------------------------------------------------
STEP 2     VARIABLE X5 ENTERED     R SQUARE = 0.91753189     C(P) = 2.26060361

                  DF          SUM OF SQUARES           MEAN SQUARE         F    PROB>F

REGRESSION         2       15373255.11680240      7686627.55840120     55.63    0.0001
ERROR             10        1381753.88988993       138175.38898899
TOTAL             12       16755009.00669230

                   B VALUE             STD ERROR            TYPE II SS         F    PROB>F

INTERCEPT    2070.44039282
X1             29.21340629            2.85627863     14454196.47419310    104.61    0.0001
X5           -354.60309015          139.99823977       886483.33801274      6.42    0.0297

BOUNDS ON CONDITION NUMBER:      1.283836,      5.135344
```

11.13 The SAS output in part (b) of Exercise 10.6 is the result of a backward elimination procedure applied to the FAT data in Exercise 10.6. Here, α_{stay} has been set equal to .05. Try to improve the six-variable model in the SAS output by adding quadratic and interaction terms.

11.14 Consider the hotel rate data of Exercise 5.41 (see Table 5.11). Using the model-building techniques you have studied, develop a model for predicting hotel room rates.

11.15 Consider the economic data for selected cities in Table 5.12. Using the model-building techniques you have studied, develop models for predicting (1) HCOST (indexed housing cost), (2) JG (jobs growth), and (3) IG (personal income growth).

11.16 An analyst at National Motors is interested in developing a regression model for predicting automobile sales in Ohio for the standard and luxury models of the Lance.* To do this, monthly data have been observed over the past 18 months on the following variables:

y_i = number of Lances sold in month i (measured in thousands of cars)

x_{i1} = (average) price per gallon of gasoline in month i

x_{i2} = (average) interest rate in month i

*This problem is taken from an example in Lyman Ott, *An Introduction to Statistical Methods and Data Analysis*, 2nd ed., PWS–KENT Publishing Company, Boston. © 1987. Used with permission.

$$x_{i3} = \begin{cases} 1 & \text{if we are considering standard model sales} \\ 0 & \text{if we are considering luxury model sales} \end{cases}$$

The data are given in Table 11.17.

TABLE 11.17 **Lance sales data**

i	y_i	x_{i1}	x_{i2}	x_{i3}
1	22.1	1.39	12.1	1
1	7.2	1.39	12.1	0
2	15.4	1.44	12.2	1
2	5.4	1.44	12.2	0
3	11.7	1.45	12.3	1
3	7.6	1.45	12.3	0
4	10.3	1.32	14.2	1
4	2.5	1.32	14.2	0
5	11.4	1.35	15.8	1
5	2.4	1.35	15.8	0
6	7.5	1.28	16.3	1
6	1.7	1.28	16.3	0
7	13.0	1.26	16.5	1
7	4.3	1.26	16.5	0
8	12.8	1.26	14.7	1
8	3.7	1.26	14.7	0
9	14.6	1.25	13.4	1
9	3.9	1.25	13.4	0
10	18.9	1.24	12.9	1
10	7.0	1.24	12.9	0
11	19.3	1.20	11.2	1
11	6.8	1.20	11.2	0
12	30.1	1.20	10.9	1
12	10.1	1.20	10.9	0
13	28.2	1.18	10.3	1
13	9.4	1.18	10.3	0
14	25.6	1.10	9.7	1
14	7.9	1.10	9.7	0
15	37.5	1.11	9.6	1
15	14.1	1.11	9.6	0
16	36.1	1.14	9.1	1
16	14.5	1.14	9.1	0
17	39.8	1.17	7.8	1
17	14.9	1.17	7.8	0
18	44.3	1.18	8.3	1
18	15.6	1.18	8.3	0

a. Using all the model-building techniques discussed in this book, develop a regression model relating y_i to x_{i1} and x_{i2} for the standard model (that is, when $x_{i3} = 1$).
b. Using all the model-building techniques discussed in this book, develop a regression model relating y_i to x_{i1} and x_{i2} for the luxury model (that is, when $x_{i3} = 0$).
c. Compare the regression models that you developed in parts (a) and (b). Note that in the exercises at the end of Chapter 12 you will combine the two models obtained in parts (a) and (b).

11.17 In *SAS System for Regression*, 1986 Edition, Rudolph J. Freund and Ramon C. Littell consider data relating to the cost of providing air service. The authors take the data from a Civil Aeronautics Board report, "Aircraft Operation Costs and Performance Report," August 1972. They describe the variables considered as follows:

CPM is cost per passenger mile (cents).
UTL is average hours per day use of aircraft.
ASL is average length of nonstop legs of flights (1000 miles).
SPA is average number of seats per aircraft (100 seats).
ALF is average load factor (percentage of seats occupied by passengers).*

The data represent 33 U.S. airlines with average nonstop lengths of flights greater than 800 miles. An additional dummy variable, TYPE, has been used. This variable equals 0 for airlines with ASL < 1200 miles and 1 for airlines with ASL ⩾ 1200 miles. Develop a model to predict CPM by using the data in Table 11.18.

TABLE 11.18 Air service cost data

OBS	ALF	UTL	ASL	SPA	TYPE	CPM
1	0.287	8.09	1.528	0.3522	1	3.306
2	0.349	9.56	2.189	0.3279	1	3.527
3	0.362	10.80	1.518	0.1356	1	3.959
4	0.378	5.65	0.821	0.1290	0	4.737
5	0.381	10.20	1.692	0.3007	1	3.096
6	0.394	7.94	0.949	0.1488	0	3.689
7	0.397	13.30	3.607	0.3390	1	2.357
8	0.400	8.42	1.495	0.3597	1	2.833
9	0.405	9.57	0.863	0.1390	0	3.313
10	0.409	9.00	0.845	0.1390	0	3.044
11	0.410	9.62	0.840	0.1390	0	2.846
12	0.412	7.91	1.350	0.1920	1	2.341
13	0.417	8.83	2.377	0.3287	1	2.780
14	0.422	8.35	1.031	0.1365	0	3.392
15	0.425	10.60	2.780	0.1282	1	3.856
16	0.426	7.52	0.975	0.2025	0	3.462
17	0.434	8.36	1.912	0.3148	1	2.711
18	0.439	8.43	1.584	0.1607	1	2.743
19	0.452	7.55	1.164	0.1270	0	3.760
20	0.455	7.70	1.236	0.1221	1	3.311
21	0.466	9.38	1.123	0.1481	0	2.404
22	0.476	8.91	0.961	0.1236	0	2.962
23	0.476	7.27	1.416	0.1145	1	3.437
24	0.478	8.71	1.392	0.1148	1	2.906
25	0.486	8.29	0.877	0.1060	0	3.140
26	0.488	9.50	2.515	0.3546	1	2.275
27	0.495	8.44	0.871	0.1186	0	2.954
28	0.504	9.47	1.408	0.1345	1	3.306
29	0.535	10.80	1.576	0.1361	1	2.425
30	0.539	6.84	1.008	0.1150	0	2.971
31	0.541	6.31	0.823	0.0943	0	4.024
32	0.582	8.48	1.963	0.1381	1	2.363
33	0.591	7.87	1.790	0.1375	1	2.258

**Source:* Rudolph J. Freund and Ramon C. Littell, *SAS System for Regression*, 1986 Edition (Cary, N.C.: SAS Institute, 1986).

11.18 Consider the dependent variable

y = sales price of a home (measured in thousands of dollars)

and the independent variables

x_1 = taxes (local, school, county) (in hundreds of dollars)

x_2 = number of baths

x_3 = lot size (in thousands of square feet)

x_4 = living space (in thousands of square feet)

x_5 = number of garages

x_6 = number of rooms

x_7 = number of bedrooms

x_8 = age of the home (in years)

x_9 = number of fireplaces

Twenty-eight observations on these variables are listed in Table 11.19. These variables

TABLE 11.19 **Residential sales data**

y	x_1	x_2	x_3	x_4	x_5	x_6	x_7	x_8	x_9
25.9	4.9176	1.0	3.4720	0.9980	1.0	7	4	42	0
29.5	5.0208	1.0	3.5310	1.5000	2.0	7	4	62	0
27.9	4.5429	1.0	2.2750	1.1750	1.0	6	3	40	0
25.9	4.5573	1.0	4.0500	1.2320	1.0	6	3	54	0
29.9	5.0597	1.0	4.4550	1.1210	1.0	6	3	42	0
29.9	3.8910	1.0	4.4550	0.9880	1.0	6	3	56	0
30.9	5.8980	1.0	5.8500	1.2400	1.0	7	3	51	1
28.9	5.6039	1.0	9.5200	1.5010	0.0	6	3	32	0
84.9	15.4202	2.5	9.8000	3.4200	2.0	10	5	42	1
82.9	14.4598	2.5	12.8000	3.0000	2.0	9	5	14	1
35.9	5.8282	1.0	6.4350	1.2250	2.0	6	3	32	0
31.5	5.3003	1.0	4.9883	1.5520	1.0	6	3	30	0
31.0	6.2712	1.0	5.5200	0.9750	1.0	5	2	30	0
30.9	5.9592	1.0	6.6660	1.1210	2.0	6	3	32	0
30.0	5.0500	1.0	5.0000	1.0200	0.0	5	2	46	1
28.9	5.6039	1.0	9.5200	1.5010	0.0	6	3	32	0
36.9	8.2464	1.5	5.1500	1.6640	2.0	8	4	50	0
41.9	6.6969	1.5	6.9020	1.4880	1.5	7	3	22	1
40.5	7.7841	1.5	7.1020	1.3760	1.0	6	3	17	0
43.9	9.0384	1.0	7.8000	1.5000	1.5	7	3	23	0
37.5	5.9894	1.0	5.5200	1.2560	2.0	6	3	40	1
37.9	7.5422	1.5	4.0000	1.6900	1.0	6	3	22	0
44.5	8.7951	1.5	9.8900	1.8200	2.0	8	4	50	1
37.9	6.0931	1.5	6.7265	1.6520	1.0	6	3	44	0
38.9	8.3607	1.5	9.1500	1.7770	2.0	8	4	48	1
36.9	8.1400	1.0	8.0000	1.5040	2.0	7	3	3	0
45.8	9.1416	1.5	7.3262	1.8310	1.5	8	4	31	0
41.0	12.0000	1.5	5.0000	1.2000	2.0	6	3	30	1

Source: Reprinted by permission from *Technometrics*, Vol. 19, No. 2. © 1977.

were obtained from *Multiple Listing*, Vol. 87 for area 12 (Erie, PA). The data was originally presented by Narula and Wellington in "Prediction, Linear Regression and the Minimum Sum of Relative Errors." Using this data, develop a model to predict y.

11.19 Using the model-building techniques you have studied and the data in Table 8.10, find a model for predicting sales of Crest toothpaste.

11.20 In an article in the *Journal of Petroleum Technology*, G. C. Wang relates

y = percentage of oil recovery

to

x_1 = pressure (in pounds per square inch) of carbon dioxide flooded into oil pockets

and

x_2 = dipping angle of tubes doing the flooding.

The data in Table 11.20 was analyzed. Develop a model for predicting y on the basis of x_1 and x_2.

TABLE 11.20 Oil recovery data

x_1	x_2	y
1000	0	60.58
1000	15	72.72
1000	30	79.99
1500	0	66.83
1500	15	80.78
1500	30	89.78
2000	0	69.18
2000	15	80.31
2000	30	91.99

Source: G. C. Wang, "Microscopic Investigation of CO_2 Flooding Process," *Journal of Petroleum Technology*, Vol. 34, No. 8, August 1982.

11.21 In Appendix F, consider Data Base 1: Manufacturing Data. Using the model-building techniques you have learned, find a model to predict NCAPEX.

11.22 In Appendix F, consider Data Base 2: Ohio Local Government and Payroll Data. Using the model-building techniques you have learned, find a model to predict OCTPAY.

11.23 In Appendix F, consider Data Base 3: Population Data. Using the model-building techniques you have learned, find a model to predict PCTUNEM.

CHAPTER 12

DUMMY VARIABLES AND ADVANCED STATISTICAL INFERENCES

This chapter discusses using dummy variables in regression models and also presents some advanced inference procedures that are often useful in analyzing dummy variable (and other) models. We begin in Section 12.1 by illustrating how dummy variables can be used to model the effects of the different levels of a qualitative (that is, nonnumerical) independent variable. Then Section 12.2 explains how to carry out hypothesis tests concerning linear combinations of regression parameters. We also show how to compute confidence intervals for such linear combinations. In addition, we demonstrate how these inference techniques can be used to compare the effects of the different levels of a qualitative independent variable. Section 12.3 describes how to compute simultaneous confidence intervals for linear combinations of parameters. Here we discuss both Scheffé and Bonferroni simultaneous intervals. We complete this chapter with optional Section 12.4, which illustrates the use of SAS to implement dummy variable techniques.

12.1

USING DUMMY VARIABLES TO MODEL QUALITATIVE INDEPENDENT VARIABLES

Whereas the levels of a quantitative independent variable are numerical, the levels of a *qualitative* independent variable are defined by describing them. For example, we might describe three different levels of the qualitative independent variable door-to-door sales technique: high pressure, medium pressure, and low pressure. We can model the effects of the different levels of a qualitative independent variable by using *dummy variables*. We begin with an example.

Example 12.1 ***Part 1: The Data and Data Plots***

Suppose that Electronics World, a chain of stores that sells radio and video equipment, has gathered the data in Table 12.1. The company wishes to study the relationship between store sales volume in July 1988 (y, measured in thousands of dollars) and the number of households in the store's area (x, measured in thousands) and the location of the store (on a suburban street, in a suburban shopping mall, or downtown—a qualitative independent variable). The data plot in Figure 12.1 indicates that y tends to increase in a straight-line fashion as x increases. Furthermore, the line relating the (x, y) combinations for mall locations (the asterisks in the figure) has a greater y-intercept than the line relating the (x, y) combinations for downtown locations (the open circles in the figure). In

TABLE 12.1 **Electronics World sales volume data**

Store, i	Number of households, x_i	Location	Sales volume, y_i
1	161	Street	157.27
2	99	Street	93.28
3	135	Street	136.81
4	120	Street	123.79
5	164	Street	153.51
6	221	Mall	241.74
7	179	Mall	201.54
8	204	Mall	206.71
9	214	Mall	229.78
10	101	Mall	135.22
11	231	Downtown	224.71
12	206	Downtown	195.29
13	248	Downtown	242.16
14	107	Downtown	115.21
15	205	Downtown	197.82

addition, both of these lines have a greater y-intercept than the line relating the (x, y) combinations for street locations (the solid dots in the figure).

Part 2: A Dummy Variable Model

A reasonable model describing the store sales volumes is

$$\begin{aligned} y_i &= \mu_i + \varepsilon_i \\ &= \beta_0 + \beta_1 x_i + \beta_2 D_{i,M} + \beta_3 D_{i,D} + \varepsilon_i \end{aligned}$$

Here, we are modeling the effects of the street, shopping mall, and downtown locations by using the dummy variables $D_{i,M}$ and $D_{i,D}$. These variables are defined as follows.

$$D_{i,M} = \begin{cases} 1 & \text{if store } i \text{ is in a mall location} \\ 0 & \text{otherwise} \end{cases}$$

$$D_{i,D} = \begin{cases} 1 & \text{if store } i \text{ is in a downtown location} \\ 0 & \text{otherwise} \end{cases}$$

With these definitions the equation

$$\mu_i = \beta_0 + \beta_1 x_i + \beta_2 D_{i,M} + \beta_3 D_{i,D}$$

has the following implications.

1. For a street location,

$$\begin{aligned} \mu_i &= \beta_0 + \beta_1 x_i + \beta_2 D_{i,M} + \beta_3 D_{i,D} \\ &= \beta_0 + \beta_1 x_i + \beta_2(0) + \beta_3(0) \\ &= \beta_0 + \beta_1 x_i \end{aligned}$$

2. For a mall location,

$$\begin{aligned} \mu_i &= \beta_0 + \beta_1 x_i + \beta_2 D_{i,M} + \beta_3 D_{i,D} \\ &= \beta_0 + \beta_1 x_i + \beta_2(1) + \beta_3(0) \\ &= (\beta_0 + \beta_2) + \beta_1 x_i \end{aligned}$$

3. For a downtown location,

$$\begin{aligned} \mu_i &= \beta_0 + \beta_1 x_i + \beta_2 D_{i,M} + \beta_3 D_{i,D} \\ &= \beta_0 + \beta_1 x_i + \beta_2(0) + \beta_3(1) \\ &= (\beta_0 + \beta_3) + \beta_1 x_i \end{aligned}$$

Thus the dummy variables allow us to model the situation illustrated in Figure 12.1. Here, the lines relating μ_i to x_i for the street, mall, and downtown locations have different y-intercepts and the same slope β_1. It follows that the equation

$$\mu_i = \beta_0 + \beta_1 x_i + \beta_2 D_{i,M} + \beta_3 D_{i,D}$$

assumes that there is no interaction between x_i (number of households) and store location.

Using

$$\mathbf{y} = \begin{bmatrix} 157.27 \\ 93.28 \\ 136.81 \\ 123.79 \\ 153.51 \\ 241.74 \\ 201.54 \\ 206.71 \\ 229.78 \\ 135.22 \\ 224.71 \\ 195.29 \\ 242.16 \\ 115.21 \\ 197.82 \end{bmatrix} \quad \text{and} \quad \mathbf{X} = \begin{array}{c} \begin{array}{cccc} 1 & x_i & D_{i,\text{M}} & D_{i,\text{D}} \end{array} \\ \begin{bmatrix} 1 & 161 & 0 & 0 \\ 1 & 99 & 0 & 0 \\ 1 & 135 & 0 & 0 \\ 1 & 120 & 0 & 0 \\ 1 & 164 & 0 & 0 \\ 1 & 221 & 1 & 0 \\ 1 & 179 & 1 & 0 \\ 1 & 204 & 1 & 0 \\ 1 & 214 & 1 & 0 \\ 1 & 101 & 1 & 0 \\ 1 & 231 & 0 & 1 \\ 1 & 206 & 0 & 1 \\ 1 & 248 & 0 & 1 \\ 1 & 107 & 0 & 1 \\ 1 & 205 & 0 & 1 \end{bmatrix} \end{array}$$

We can calculate the least squares point estimates of β_0, β_1, β_2, and β_3 to be

$$\begin{bmatrix} b_0 \\ b_1 \\ b_2 \\ b_3 \end{bmatrix} = (\mathbf{X}'\mathbf{X})^{-1}\mathbf{X}'\mathbf{y} = \begin{bmatrix} 14.9777 \\ .8686 \\ 28.3738 \\ 6.8638 \end{bmatrix}$$

Furthermore, the standard error is $s = 6.3494$, and $R^2 = .9868$. The SAS output in Figure 12.2 presents these and other related quantities.

Part 3: Comparing the Locations

To compare the effects of the street, shopping mall, and downtown locations, consider comparing three means, which we denote as $\mu_{[h,\text{S}]}$, $\mu_{[h,\text{M}]}$, and $\mu_{[h,\text{D}]}$. These means represent the mean sales volumes at stores having h households in the area and located on streets, in shopping malls, and downtown, respectively. If we set

FIGURE 12.1 **Plot of the sales volume data and geometrical interpretation of the model $y_i = \mu_i + \varepsilon_i = \beta_0 + \beta_1 x_i + \beta_2 D_{i,M} + \beta_3 D_{i,D} + \varepsilon_i$**

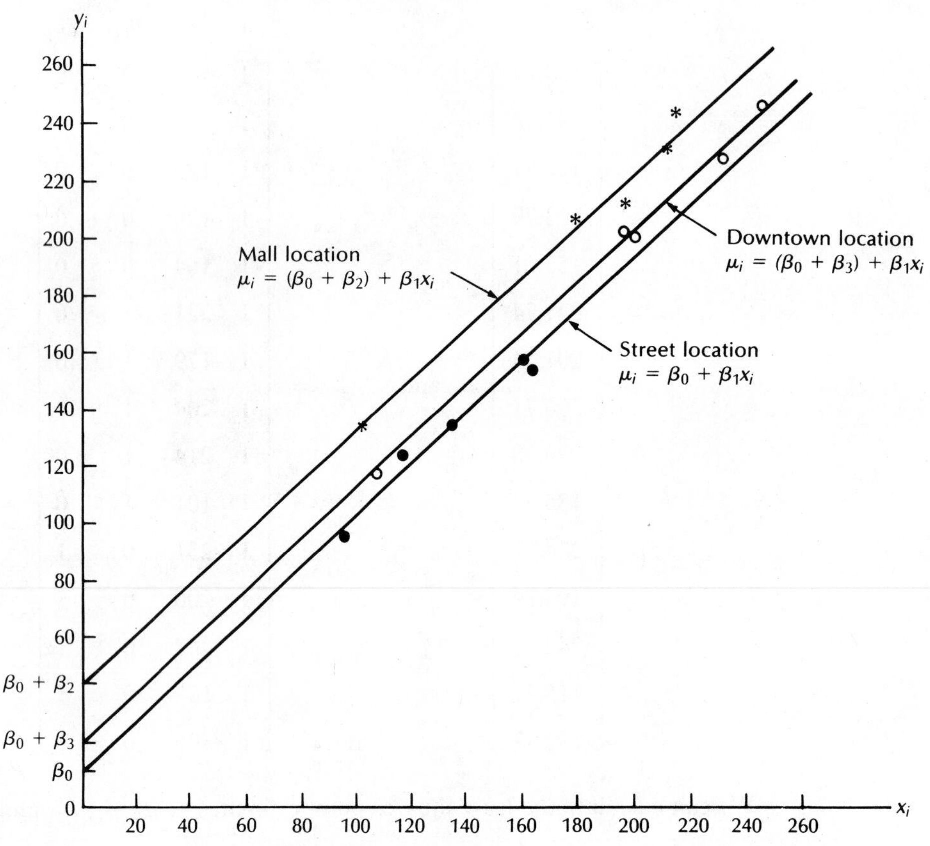

$x_i = h$ in the above equation for μ_i, it follows that

$$\begin{aligned}\mu_{[h,S]} &= \beta_0 + \beta_1 h + \beta_2(0) + \beta_3(0)\\ &= \beta_0 + \beta_1 h\end{aligned}$$

$$\begin{aligned}\mu_{[h,M]} &= \beta_0 + \beta_1 h + \beta_2(1) + \beta_3(0)\\ &= \beta_0 + \beta_1 h + \beta_2\end{aligned}$$

and

$$\begin{aligned}\mu_{[h,D]} &= \beta_0 + \beta_1 h + \beta_2(0) + \beta_3(1)\\ &= \beta_0 + \beta_1 h + \beta_3\end{aligned}$$

FIGURE 12.2 SAS output of a regression analysis of the sales volume data using the model $y_i = \mu_i + \varepsilon_i = \beta_0 + \beta_1 x_i + \beta_2 D_{i,M} + \beta_3 D_{i,D} + \varepsilon_i$

SAS

ANALYSIS OF VARIANCE

SOURCE	DF	SUM OF SQUARES	MEAN SQUARE	F VALUE	PROB>F
MODEL	3	33268.69529	11089.56510	275.073	0.0001
ERROR	11	443.46500	40.31500010		
C TOTAL	14	33712.16029			

ROOT MSE	6.349409	R-SQUARE	0.9868
DEP MEAN	176.9893	ADJ R-SQ	0.9833
C.V.	3.587453		

PARAMETER ESTIMATES

VARIABLE	DF	PARAMETER ESTIMATE	STANDARD ERROR	T FOR HO: PARAMETER=0	PROB > |T|
INTERCEP	1	14.97769322	6.18844540	2.420	0.0340
X	1	0.86858842	0.04048993	21.452	0.0001
DM	1	28.37375607	4.46130660	6.360	0.0001
DD	1	6.86377679	4.77047650	1.439	0.1780

OBS	ACTUAL	PREDICT VALUE	STD ERR PREDICT	LOWER95% MEAN	UPPER95% MEAN	LOWER95% PREDICT	UPPER95% PREDICT	RESIDUAL
1	157.3	154.8	3.0173	148.2	161.5	139.3	170.3	2.4496
2	93.2800	101.0	3.2067	93.9099	108.0	85.3118	116.6	-7.6879
3	136.8	132.2	2.8397	126.0	138.5	116.9	147.5	4.5729
4	123.8	119.2	2.9107	112.8	125.6	103.8	134.6	4.5817
5	153.5	157.4	3.0605	150.7	164.2	141.9	172.9	-3.9162
6	241.7	235.3	3.2143	228.2	242.4	219.6	251.0	6.4305
7	201.5	198.8	2.8462	192.6	205.1	183.5	214.1	2.7112
8	206.7	220.5	2.9550	214.0	227.0	205.1	236.0	-13.8335
9	229.8	229.2	3.0916	222.4	236.0	213.7	244.8	0.5506
10	135.2	131.1	4.3935	121.4	140.7	114.1	148.1	4.1411
11	224.7	222.5	3.1145	215.6	229.3	206.9	238.1	2.2246
12	195.3	200.8	2.8521	194.5	207.0	185.5	216.1	-5.4807
13	242.2	237.3	3.4547	229.6	244.9	221.3	253.2	4.9086
14	115.2	114.8	4.6968	104.4	125.1	97.3974	132.2	0.4296
15	197.8	199.9	2.8486	193.6	206.2	184.6	215.2	-2.0821
16	.	217.1[a]	2.9143[b]	210.7	223.5	201.7	232.4	

95% prediction interval

SUM OF RESIDUALS	3.55271E-13
SUM OF SQUARED RESIDUALS	443.465
PREDICTED RESID SS (PRESS)	782.6423

[a] $\hat{y}_0$ [b] $s\sqrt{\mathbf{x}_0'(\mathbf{X}'\mathbf{X})^{-1}\mathbf{x}_0}$

These equations imply that

$$\begin{aligned}\mu_{[h,\text{M}]} - \mu_{[h,\text{S}]} &= [\beta_0 + \beta_1 h + \beta_2] - [\beta_0 + \beta_1 h] \\ &= \beta_2\end{aligned}$$

We have seen that the least squares point estimate of β_2 is $b_2 = 28.3738$. This says that for any given number of households in a store's area we estimate that the mean monthly sales volume in a mall location is \$28,374 greater than the mean monthly sales volume in a street location. Furthermore, the 95% confidence interval for β_2 is

$$\begin{aligned}[b_2 \pm t_{[.025]}^{(15-4)} s\sqrt{c_{22}}] &= [28.3738 \pm 2.201(4.4613)] \\ &= [28.3738 \pm 9.8193] \\ &= [18.5545, 38.1931]\end{aligned}$$

This says that we are 95% confident that for any given number of households in a store's area the mean monthly sales volume in a mall location is between \$18,554 and \$38,193 greater than the mean monthly sales volume in a street location. Also, note that

$$\begin{aligned}\mu_{[h,D]} - \mu_{[h,S]} &= [\beta_0 + \beta_1 h + \beta_3] - [\beta_0 + \beta_1 h] \\ &= \beta_3\end{aligned}$$

We have seen that the least squares point estimate of β_3 is $b_3 = 6.8638$. This says that for any given number of households in a store's area we estimate that the mean monthly sales volume in a downtown location is \$6,864 greater than the mean monthly sales volume in a street location. Furthermore, note that

$$\begin{aligned}\mu_{[h,M]} - \mu_{[h,D]} &= [\beta_0 + \beta_1 h + \beta_2] - [\beta_0 + \beta_1 h + \beta_3] \\ &= \beta_2 - \beta_3\end{aligned}$$

Therefore the least squares point estimates imply that an estimate of $\beta_2 - \beta_3$ is $b_2 - b_3 = 28.3738 - 6.8638 = 21.51$. This says that for any given number of households in a store's area we estimate that the mean monthly sales volume in a mall location is \$21,510 greater than the mean monthly sales volume in a downtown location. Of course, the above interpretations (and other interpretations in this example) assume that the regression relationships between y and x and the store locations apply to future months and other stores. Thus we assume that there are no trend, seasonal, or other time-related influences affecting store sales volume. In the exercises at the end of this chapter the reader will continue the above analysis by calculating 95% confidence intervals for β_3 and $(\beta_2 - \beta_3)$. Calculation of the latter interval requires techniques to be discussed in Section 12.2.

Part 4: Predicting a Future Sales Volume

Next, we will predict

$$\begin{aligned}y_0 &= \mu_0 + \varepsilon_0 \\ &= \beta_0 + \beta_1(200) + \beta_2(1) + \beta_3(0) + \varepsilon_0\end{aligned}$$

the actual sales volume in a future month for a store that has 200,000 households in its area and is located in a shopping mall. A point prediction of y_0 is

$$\begin{aligned}\hat{y}_0 &= b_0 + b_1(200) + b_2(1) + b_3(0) \\ &= 14.9777 + .8686(200) + 28.3738(1) \\ &= 217.1\end{aligned}$$

Furthermore, note that $\mathbf{x}_0' = [1 \;\; 200 \;\; 1 \;\; 0]$ and that the SAS output in Figure 12.2 tells us that

$$s\sqrt{\mathbf{x}_0'(\mathbf{X}'\mathbf{X})^{-1}\mathbf{x}_0} = 2.9143 \qquad \text{and} \qquad s = 6.3494$$

This implies that

$$\mathbf{x}_0'(\mathbf{X}'\mathbf{X})^{-1}\mathbf{x}_0 = \left(\frac{2.9143}{6.3494}\right)^2 = .2107$$

It follows that a 95% prediction interval for y_0 is

$$\begin{aligned}[\hat{y}_0 \pm t_{[.025]}^{(15-4)} s\sqrt{1 + \mathbf{x}_0'(\mathbf{X}'\mathbf{X})^{-1}\mathbf{x}_0}] &= [217.1 \pm 2.201(6.3494)\sqrt{1.2107}] \\ &= [201.7, 232.4]\end{aligned}$$

FIGURE 12.3 **Geometrical interpretation of the model** $y_i = \mu_i + \varepsilon_i = \beta_0 + \beta_1 x_i + \beta_2 D_{i,M} + \beta_3 D_{i,D} + \beta_4 x_i D_{i,M} + \beta_5 x_i D_{i,D} + \varepsilon_i$

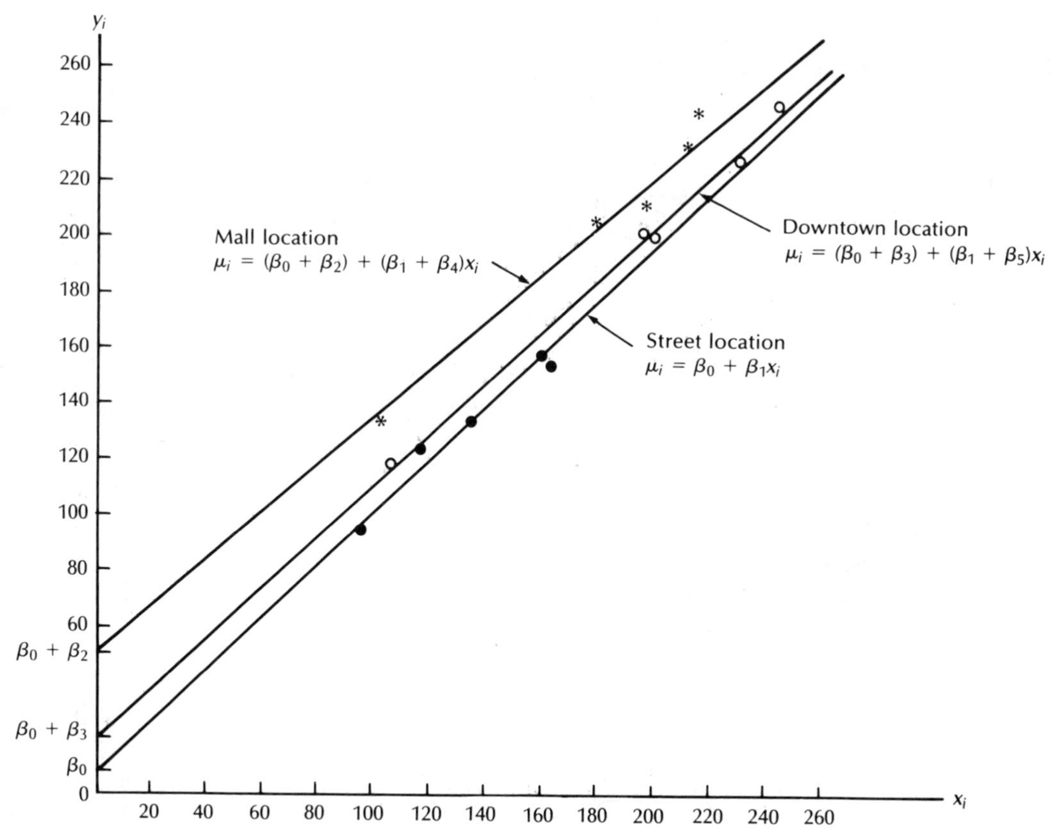

FIGURE 12.4 SAS output of a regression analysis of the sales volume data using the model $y_i = \mu_i + \varepsilon_i = \beta_0 + \beta_1 x_i + \beta_2 D_{i,M} + \beta_3 D_{i,D} + \beta_4 x_i D_{i,M} + \beta_5 x_i D_{i,D} + \varepsilon_i$

SAS

ANALYSIS OF VARIANCE

SOURCE	DF	SUM OF SQUARES	MEAN SQUARE	F VALUE	PROB>F
MODEL	5	33296.05762	6659.21152	144.034	0.0001
ERROR	9	416.10268	46.23363057		
C TOTAL	14	33712.16029			

ROOT MSE	6.799532	R-SQUARE	0.9877
DEP MEAN	176.9893	ADJ R-SQ	0.9808
C.V.	3.841775		

PARAMETER ESTIMATES

VARIABLE	DF	PARAMETER ESTIMATE	STANDARD ERROR	T FOR HO: PARAMETER=0	PROB > |T|
INTERCEP	1	7.90041914	17.03513159	0.464	0.6538
X	1	0.92070384	0.12342808	7.459	0.0001
DM	1	42.72974386	21.50420409	1.987	0.0782
DD	1	10.25503220	21.28318910	0.482	0.6414
XDM	1	-0.09171669	0.14163028	-0.648	0.5334
XDD	1	-0.03362987	0.13818798	-0.243	0.8132

OBS	ACTUAL	PREDICT VALUE	STD ERR PREDICT	LOWER95% MEAN	UPPER95% MEAN	LOWER95% PREDICT	UPPER95% PREDICT	RESIDUAL
1	157.3	156.1	4.3499	146.3	166.0	137.9	174.4	1.1363
2	93.2800	99.0501	5.4661	86.6849	111.4	79.3144	118.8	-5.7701
3	136.8	132.2	3.0424	125.3	139.1	115.3	149.0	4.6146
4	123.8	118.4	3.6125	110.2	126.6	101.0	135.8	5.4051
5	153.5	158.9	4.6219	148.4	169.4	140.3	177.5	-5.3858
6	241.7	233.8	3.9904	224.8	242.9	216.0	251.7	7.9037
7	201.5	199.0	3.0591	192.1	205.9	182.2	215.9	2.5211
8	206.7	219.7	3.3489	212.2	227.3	202.6	236.9	-13.0335
9	229.8	228.0	3.6942	219.7	236.4	210.5	245.5	1.7466
10	135.2	134.4	6.5057	119.6	149.1	113.1	155.6	0.8621
11	224.7	223.1	3.6198	214.9	231.3	205.6	240.5	1.6405
12	195.3	200.9	3.0684	194.0	207.8	184.0	217.8	-5.6027
13	242.2	238.1	4.2857	228.5	247.8	220.0	256.3	4.0102
14	115.2	113.1	6.4973	98.3744	127.8	91.7973	134.3	2.1376
15	197.8	200.0	3.0607	193.1	206.9	183.1	216.9	-2.1856
16	.	216.4	3.2424	209.1	223.8	199.4	233.5	.

SUM OF RESIDUALS	4.29878E-13
SUM OF SQUARED RESIDUALS	416.1027
PREDICTED RESID SS (PRESS)	1732.719

Therefore we are 95% confident that the actual sales volume in a future month for a store that has 200,000 households in its area and is located in a shopping mall will be between \$201,700 and \$232,400.

Part 5: An Interaction Model

In modeling the sales volume data we might consider using the equation

$$\mu_i = \beta_0 + \beta_1 x_i + \beta_2 D_{i,M} + \beta_3 D_{i,D} + \beta_4 x_i D_{i,M} + \beta_5 x_i D_{i,D}$$

This equation implies that

1. For a street location,

$$\begin{aligned}\mu_i &= \beta_0 + \beta_1 x_i + \beta_2(0) + \beta_3(0) + \beta_4 x_i(0) + \beta_5 x_i(0)\\ &= \beta_0 + \beta_1 x_i\end{aligned}$$

2. For a mall location,

$$\begin{aligned}\mu_i &= \beta_0 + \beta_1 x_i + \beta_2(1) + \beta_3(0) + \beta_4 x_i(1) + \beta_5 x_i(0)\\ &= (\beta_0 + \beta_2) + (\beta_1 + \beta_4)x_i\end{aligned}$$

3. For a downtown location,

$$\begin{aligned}\mu_i &= \beta_0 + \beta_1 x_i + \beta_2(0) + \beta_3(1) + \beta_4 x_i(0) + \beta_5 x_i(1)\\ &= (\beta_0 + \beta_3) + (\beta_1 + \beta_5)x_i\end{aligned}$$

Figure 12.3 illustrates that, if we use the above equation, then the straight lines relating μ_i to x_i for the street, mall, and downtown locations have different y-intercepts and different slopes. It follows that the model

$$\begin{aligned}y_i &= \mu_i + \varepsilon_i\\ &= \beta_0 + \beta_1 x_i + \beta_2 D_{i,\mathrm{M}} + \beta_3 D_{i,\mathrm{D}} + \beta_4 x_i D_{i,\mathrm{M}} + \beta_5 x_i D_{i,\mathrm{D}} + \varepsilon_i\end{aligned}$$

assumes that there is interaction between x_i and store location. However, the SAS output in Figure 12.4 indicates that the prob-values for testing H_0: $\beta_4 = 0$ and H_0: $\beta_5 = 0$ are quite large. This implies that the interaction terms $x_i D_{i,\mathrm{M}}$ and $x_i D_{i,\mathrm{D}}$ are not important. Furthermore, Figure 12.4 tells us that the standard error for the interaction model is $s = 6.7995$. This is larger than the standard error for the no-interaction model. Hence the no-interaction model seems best.

We next consider a more detailed example using dummy variables.

Example 12.2 ***Part 1: The Problem and Data***

Again consider the Fresh Detergent demand example. Recall that Enterprise Industries has employed the historical data of Table 8.3 to develop the regression model

$$\begin{aligned}y_i &= \mu_i + \varepsilon_i\\ &= \beta_0 + \beta_1 x_{i4} + \beta_2 x_{i3} + \beta_3 x_{i3}^2 + \beta_4 x_{i4} x_{i3} + \varepsilon_i\end{aligned}$$

This model relates the dependent variable y (demand for Fresh) to the independent variables x_3 (Enterprise Industries' advertising expenditure) and x_4 (the price difference).

To ultimately increase the demand for Fresh, Enterprise Industries' marketing department is conducting a study comparing the effectiveness of three different advertising campaigns. These campaigns are denoted as campaigns A, B, and C. Here, campaign A consists entirely of television commercials, campaign B consists

TABLE 12.2 Advertising campaigns used by Enterprise Industries

Sales region, i	*Advertising campaign*	*Sales region, i*	*Advertising campaign*
1	B	16	B
2	B	17	B
3	B	18	A
4	A	19	B
5	C	20	B
6	A	21	C
7	C	22	A
8	C	23	A
9	B	24	A
10	C	25	A
11	A	26	B
12	C	27	C
13	C	28	B
14	A	29	C
15	B	30	C

of a balanced mixture of television and radio commercials, and campaign C consists of a balanced mixture of television, radio, newspaper, and magazine ads. To conduct this study, Enterprise Industries has randomly selected an advertising campaign to be used in each of the sales regions in Table 8.3. Table 12.2 lists the campaigns used in these sales regions.

Part 2: The Model and Interpretation of the Model Parameters

In this example we explain how Enterprise Industries can study and compare the effectiveness of advertising campaigns A, B, and C. To do this, we define two dummy variables:

$$D_{i,\mathrm{B}} = \begin{cases} 1 & \text{if campaign B is used in sales region } i \\ 0 & \text{otherwise} \end{cases}$$

$$D_{i,\mathrm{C}} = \begin{cases} 1 & \text{if campaign C is used in sales region } i \\ 0 & \text{otherwise} \end{cases}$$

Using these dummy variables, we now consider a regression model relating y_i to x_{i3}, x_{i4}, $D_{i,\mathrm{B}}$, and $D_{i,\mathrm{C}}$. This model is

$$\begin{aligned} y_i &= \mu_i + \varepsilon_i \\ &= \beta_0 + \beta_1 x_{i4} + \beta_2 x_{i3} + \beta_3 x_{i3}^2 + \beta_4 x_{i4} x_{i3} + \beta_5 D_{i,\mathrm{B}} + \beta_6 D_{i,\mathrm{C}} + \varepsilon_i \end{aligned}$$

To interpret the meaning of the parameters β_5 and β_6, consider comparing three means, denoted $\mu_{[d,a,\mathrm{A}]}$, $\mu_{[d,a,\mathrm{B}]}$, and $\mu_{[d,a,\mathrm{C}]}$. These means represent the mean (quarterly) demands for Fresh when the price difference is d, the advertising expenditure is a, and we use advertising campaigns A, B, and C, respectively. If we set $x_{i4} = d$ and $x_{i3} = a$ in the equation

$$\mu_i = \beta_0 + \beta_1 x_{i4} + \beta_2 x_{i3} + \beta_3 x_{i3}^2 + \beta_4 x_{i4} x_{i3} + \beta_5 D_{i,\mathrm{B}} + \beta_6 D_{i,\mathrm{C}}$$

it follows that

$$\begin{aligned}\mu_{[d,a,\mathrm{A}]} &= \beta_0 + \beta_1 d + \beta_2 a + \beta_3 a^2 + \beta_4 da + \beta_5(0) + \beta_6(0)\\ &= \beta_0 + \beta_1 d + \beta_2 a + \beta_3 a^2 + \beta_4 da\end{aligned}$$

$$\begin{aligned}\mu_{[d,a,\mathrm{B}]} &= \beta_0 + \beta_1 d + \beta_2 a + \beta_3 a^2 + \beta_4 da + \beta_5(1) + \beta_6(0)\\ &= \beta_0 + \beta_1 d + \beta_2 a + \beta_3 a^2 + \beta_4 da + \beta_5\end{aligned}$$

and

$$\begin{aligned}\mu_{[d,a,\mathrm{C}]} &= \beta_0 + \beta_1 d + \beta_2 a + \beta_3 a^2 + \beta_4 da + \beta_5(0) + \beta_6(1)\\ &= \beta_0 + \beta_1 d + \beta_2 a + \beta_3 a^2 + \beta_4 da + \beta_6\end{aligned}$$

These equations imply that

$$\begin{aligned}\mu_{[d,a,\mathrm{B}]} - \mu_{[d,a,\mathrm{A}]} &= (\beta_0 + \beta_1 d + \beta_2 a + \beta_3 a^2 + \beta_4 da + \beta_5)\\ &\quad - (\beta_0 + \beta_1 d + \beta_2 a + \beta_3 a^2 + \beta_4 da)\\ &= \beta_5\end{aligned}$$

$$\begin{aligned}\mu_{[d,a,\mathrm{C}]} - \mu_{[d,a,\mathrm{A}]} &= (\beta_0 + \beta_1 d + \beta_2 a + \beta_3 a^2 + \beta_4 da + \beta_6)\\ &\quad - (\beta_0 + \beta_1 d + \beta_2 a + \beta_3 a^2 + \beta_4 da)\\ &= \beta_6\end{aligned}$$

and

$$\begin{aligned}\mu_{[d,a,\mathrm{C}]} - \mu_{[d,a,\mathrm{B}]} &= (\beta_0 + \beta_1 d + \beta_2 a + \beta_3 a^2 + \beta_4 da + \beta_6)\\ &\quad - (\beta_0 + \beta_1 d + \beta_2 a + \beta_3 a^2 + \beta_4 da + \beta_5)\\ &= \beta_6 - \beta_5\end{aligned}$$

In summary, the parameters

$$\begin{aligned}\beta_5 &= \mu_{[d,a,\mathrm{B}]} - \mu_{[d,a,\mathrm{A}]}\\ \beta_6 &= \mu_{[d,a,\mathrm{C}]} - \mu_{[d,a,\mathrm{A}]}\\ \beta_6 - \beta_5 &= \mu_{[d,a,\mathrm{C}]} - \mu_{[d,a,\mathrm{B}]}\end{aligned}$$

measure the effects on mean demand for Fresh of (1) changing from advertising campaign A to advertising campaign B, (2) changing from advertising campaign A to advertising campaign C, and (3) changing from advertising campaign B to advertising campaign C, respectively.

We are able to compare the effects of the three advertising campaigns by using two dummy variables because β_5 and β_6 (the parameters multiplied by $D_{i,\mathrm{B}}$ and $D_{i,\mathrm{C}}$) express the effects of campaigns B and C with respect to the effect of campaign A. We do not employ three dummy variables ($D_{i,\mathrm{A}}$, $D_{i,\mathrm{B}}$, and $D_{i,\mathrm{C}}$, where $D_{i,\mathrm{A}} = 1$ if campaign A is used in sales region i, and $D_{i,\mathrm{A}} = 0$ otherwise). This is because, if we use three dummy variables, the columns of the **X** matrix can be shown to be linearly dependent and the least squares point estimates of the model parameters cannot be computed by using the methods presented in this book. Instead of using $D_{i,B}$ and $D_{i,C}$, we could use any two of the dummy variables $D_{i,A}$, $D_{i,B}$, and $D_{i,C}$. If we used $D_{i,A}$ and $D_{i,B}$, the parameters β_5 and β_6 would express the effects of campaigns A and B with respect to the effect of campaign C. If we used $D_{i,A}$ and $D_{i,C}$, the parameters β_5 and β_6 would express the effects of campaigns A and C with respect to the effect of campaign B.

Part 3: Least Squares Point Estimates and Statistical Inference for the Model Parameters

The least squares point estimates of the parameters in the previously given dummy variable model can be obtained by employing the column vector

$$\mathbf{y} = \begin{bmatrix} y_1 \\ y_2 \\ y_3 \\ \cdot \\ \cdot \\ \cdot \\ y_{30} \end{bmatrix} = \begin{bmatrix} 7.38 \\ 8.51 \\ 9.52 \\ \cdot \\ \cdot \\ \cdot \\ 9.26 \end{bmatrix}$$

and the matrix (see Tables 8.3 and 12.2)

$$\mathbf{X} = \begin{array}{c} \begin{array}{ccccccc} 1 & x_4 & x_3 & x_3^2 & x_4x_3 & D_{i,B} & D_{i,C} \end{array} \\ \begin{bmatrix} 1 & -.05 & 5.50 & (5.50)^2 & (-.05)(5.50) & 1 & 0 \\ 1 & .25 & 6.75 & (6.75)^2 & (.25)(6.75) & 1 & 0 \\ 1 & .60 & 7.25 & (7.25)^2 & (.60)(7.25) & 1 & 0 \\ 1 & 0 & 5.50 & (5.50)^2 & (0)(5.50) & 0 & 0 \\ 1 & .25 & 7.00 & (7.00)^2 & (.25)(7.00) & 0 & 1 \\ 1 & .20 & 6.50 & (6.50)^2 & (.20)(6.50) & 0 & 0 \\ \cdot & \cdot & \cdot & \cdot & \cdot & \cdot & \cdot \\ \cdot & \cdot & \cdot & \cdot & \cdot & \cdot & \cdot \\ \cdot & \cdot & \cdot & \cdot & \cdot & \cdot & \cdot \\ 1 & .55 & 6.80 & (6.80)^2 & (.55)(6.80) & 0 & 1 \end{bmatrix} \end{array}$$

When the appropriate calculations are done, we obtain the prediction equation

$$\begin{aligned} \hat{y}_i &= b_0 + b_1x_{i4} + b_2x_{i3} + b_3x_{i3}^2 + b_4x_{i4}x_{i3} + b_5D_{i,B} + b_6D_{i,C} \\ &= 25.6127 + 9.0587x_{i4} - 6.5377x_{i3} + .5844x_{i3}^2 - 1.1565x_{i4}x_{i3} \\ &\quad + .2137D_{i,B} + .3818D_{i,C} \end{aligned}$$

In addition, the unexplained variation is $SSE = .3936$, the standard error is $s = .1308$, and $R^2 = .9708$. Figure 12.5 presents the SAS output of the t statistics and the prob-values for testing the importance of the independent variables. Since each prob-value is less than .05, we are confident that the intercept β_0 and each of x_{i4}, x_{i3}, x_{i3}^2, $x_{i4}x_{i3}$, $D_{i,B}$, and $D_{i,C}$ has substantial importance over the combined importance of the other independent variables in the model. Therefore each of these independent variables should be included in a final regression model.

Part 4: Comparing the Effects of Campaigns A, B, and C

We can now employ the dummy variable regression model to compare the effects of advertising campaigns A, B, and C on mean demand for Fresh. Figure 12.5 tells us that a 95% confidence interval for $\mu_{[d,a,B]} - \mu_{[d,a,A]} = \beta_5$ is

$$\begin{aligned} &[b_5 \pm 2.069s\sqrt{c_{55}}] \\ &= [.0851, .3423] \end{aligned}$$

This interval makes Enterprise Industries 95% confident that the effect of changing from advertising campaign A to advertising campaign B is to increase mean demand for Fresh by between .0851 (8,510 bottles) and .3423 (34,230 bottles). Figure 12.5 also tells us that a 95% confidence interval for $\mu_{[d,a,C]} - \mu_{[d,a,A]} = \beta_6$ is

$$\begin{aligned} &[b_6 \pm 2.069s\sqrt{c_{66}}] \\ &= [.2550, .5085] \end{aligned}$$

This interval makes Enterprise Industries 95% confident that the effect of changing from advertising campaign A to advertising campaign C is to increase mean demand for Fresh by between 25,500 bottles and 50,850 bottles. In Section 12.2 we will discuss how to compute a 95% confidence interval for $\mu_{[d,a,C]} - \mu_{[d,a,B]} = \beta_6 - \beta_5$. However, if we consider the dummy variable model

$$\begin{aligned} y_i &= \mu_i + \varepsilon_i \\ &= \beta_0 + \beta_1 x_{i4} + \beta_2 x_{i3} + \beta_3 x_{i3}^2 + \beta_4 x_{i4} x_{i3} + \beta_5 D_{i,A} + \beta_6 D_{i,C} + \varepsilon_i \end{aligned}$$

FIGURE 12.5 **SAS output of the *t* statistics and prob-values for testing the importance of the independent variables and the 95% confidence intervals for β_5 and β_6**

VARIABLE	PARAMETER ESTIMATE[a]	T FOR HO: PARAMETER=0[b]	PROB > | T |[c]	STD ERROR OF ESTIMATE[d]	$[b_j \pm 2.069s\sqrt{c_{jj}}]$ = 95% Confidence Interval for β_j
INTERCEP	25.61269602	5.34	0.0001	4.79378249	
X4	9.05868432	2.99	0.0066	3.03170457	
X3	-6.53767133	-4.13	0.0004	1.58136655	
X3SQ	0.58444394	4.50	0.0002	0.12987222	
X43	-1.15648054	-2.54	0.0184	0.45573648	
DB	0.21368626	3.44	0.0023	0.06215362	[.0851, .3423]
DC	0.38177617	6.23	0.0001	0.06125253	[.2550, .5085]

[a] b_j: $b_0, b_1, b_2, b_3, b_4, b_5, b_6$ [b] t [c] *Prob-value* [d] $s\sqrt{c_{jj}}$

it can be verified that

$$\mu_{[d,a,\text{C}]} - \mu_{[d,a,\text{B}]} = \beta_6 .$$

Thus by using this model we can calculate a 95% confidence interval for

$$\beta_6 = \mu_{[d,a,\text{C}]} - \mu_{[d,a,\text{B}]}$$

The reader will have an opportunity to do this in the exercises at the end of this chapter. We find that a 95% confidence interval for β_6 is [.0363, .2999]. This interval makes Enterprise Industries 95% confident that the effect of changing from advertising campaign B to advertising campaign C is to increase mean demand for Fresh by between 3,630 bottles and 29,990 bottles.

Part 5: Using the Model to Predict Demand for Fresh

Suppose that on the basis of the preceding analyses, Enterprise Industries concludes that campaign C is the most effective advertising campaign. Therefore the company decides to employ campaign C in its sales regions in future quarters. Further, suppose that Enterprise Industries wishes to predict an individual demand, y_0, when the price of Fresh will be $x_{01} = \$3.80$, the average industry price will be $x_{02} = \$3.90$, the price difference will be $x_{04} = x_{02} - x_{01} = \$.10$, and the advertising expenditure for Fresh will be $x_{03} = 6.80$ (or \$680,000). We can express the individual demand as

$$\begin{aligned} y_0 &= \mu_0 + \varepsilon_0 \\ &= \beta_0 + \beta_1 x_{04} + \beta_2 x_{03} + \beta_3 x_{03}^2 + \beta_4 x_{04} x_{03} + \beta_5 D_{0,\text{B}} + \beta_6 D_{0,\text{C}} + \varepsilon_0 \end{aligned}$$

Therefore a point prediction of y_0 is

$$\begin{aligned} \hat{y}_0 &= b_0 + b_1 x_{04} + b_2 x_{03} + b_3 x_{03}^2 + b_4 x_{04} x_{03} + b_5 D_{0,\text{B}} + b_6 D_{0,\text{C}} \\ &= 25.6127 + 9.0587 x_{04} - 6.5377 x_{03} + .5844 x_{03}^2 \\ &\quad - 1.1565 x_{04} x_{03} + .2137 D_{0,\text{B}} + .3818 D_{0,\text{C}} \\ &= 25.6127 + 9.0587(.10) - 6.5377(6.80) + .5844(6.80)^2 \\ &\quad - 1.1565(.10)(6.80) + .2137(0) + .3818(1) \\ &= 8.6825 \text{ (or 868,250 bottles)} \end{aligned}$$

Here, $D_{0,\text{B}} = 0$ and $D_{0,\text{C}} = 1$, since Enterprise Industries will use advertising campaign C. A 95% prediction interval for y_0 is [8.385, 8.980], or [838,500 bottles, 898,000 bottles]. This prediction interval is computed by using the formula

$$\left[\hat{y}_0 \pm t_{[.025]}^{(n-k)} s \sqrt{1 + \mathbf{x}_0'(\mathbf{X}'\mathbf{X})^{-1}\mathbf{x}_0}\right]$$

where

$$\hat{y}_0 = 8.6825$$

$$t_{[.025]}^{(n-k)} = t_{[.025]}^{(30-7)} = t_{[.025]}^{(23)} = 2.069$$

and

$$\begin{aligned}\mathbf{x}_0' &= [1 \quad .10 \quad 6.80 \quad (6.80)^2 \quad (.10)(6.80) \quad 0 \quad 1] \\ &= [1 \quad .10 \quad 6.80 \quad 46.24 \quad .68 \quad 0 \quad 1]\end{aligned}$$

This prediction interval says that if $x_{04} = \$.10$, if $x_{03} = \$680{,}000$, and if Enterprise Industries uses advertising campaign C, then the company can be 95% confident that quarterly demand for Fresh in a sales region will be between 838,500 bottles and 898,000 bottles.

Part 6: Investigating Interactions Involving Advertising Campaign

The regression model we have used,

$$y_i = \beta_0 + \beta_1 x_{i4} + \beta_2 x_{i3} + \beta_3 x_{i3}^2 + \beta_4 x_{i4} x_{i3} + \beta_5 D_{i,\mathrm{B}} + \beta_6 D_{i,\mathrm{C}} + \varepsilon_i$$

does not contain any interaction terms involving the dummy variables $D_{i,\mathrm{B}}$ and $D_{i,\mathrm{C}}$. However, it is possible that interaction exists between the qualitative independent variable advertising campaign and either or both of the quantitative independent variables price difference and advertising expenditure. If interaction exists between advertising campaign and price difference, we would model this interaction by using cross-product terms such as $x_{i4}D_{i,\mathrm{B}}$ and $x_{i4}D_{i,\mathrm{C}}$. If interaction exists between advertising campaign and advertising expenditure, we would model this interaction by using cross-product terms such as $x_{i3}D_{i,\mathrm{B}}$ and $x_{i3}D_{i,\mathrm{C}}$ and possibly $x_{i3}^2D_{i,\mathrm{B}}$ and $x_{i3}^2D_{i,\mathrm{C}}$.

To discuss the meaning of interaction between advertising campaign and advertising expenditure, consider Figure 12.6. This figure illustrates one hypothetical type of interaction between advertising campaign and advertising expenditure.

FIGURE 12.6 **One hypothetical type of interaction between advertising campaign and advertising expenditure**

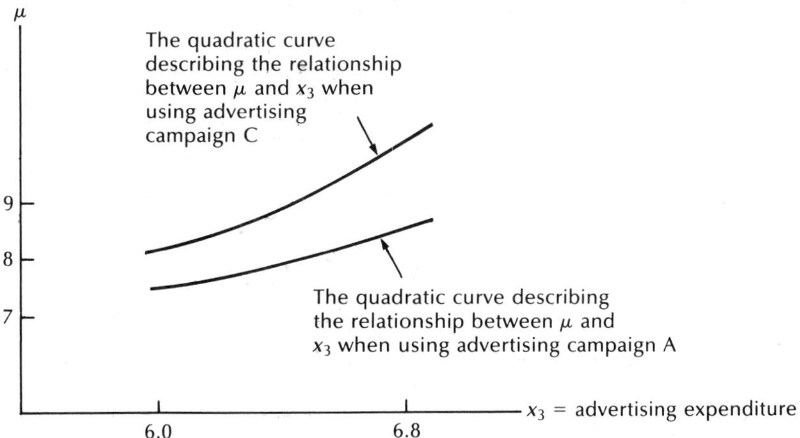

Specifically, Figure 12.6 first says that advertising campaign C is more effective at all advertising expenditure levels than is advertising campaign A. In addition, this figure says that as advertising expenditure increases, mean demand increases at a faster rate when advertising campaign C is used than when advertising campaign A is used.

To determine whether interaction involving advertising campaign does indeed exist, first consider the model

$$\begin{aligned} y_i = {} & \beta_0 + \beta_1 x_{i4} + \beta_2 x_{i3} + \beta_3 x_{i3}^2 + \beta_4 x_{i4}x_{i3} + \beta_5 D_{i,B} + \beta_6 D_{i,C} + \beta_7 x_{i4} D_{i,B} \\ & + \beta_8 x_{i4} D_{i,C} + \beta_9 x_{i3} D_{i,B} + \beta_{10} x_{i3} D_{i,C} + \beta_{11} x_{i4} x_{i3} D_{i,B} \\ & + \beta_{12} x_{i4} x_{i3} D_{i,C} + \varepsilon_i \end{aligned}$$

This model contains the two-factor interaction terms $x_{i4}D_{i,B}$, $x_{i4}D_{i,C}$, $x_{i3}D_{i,B}$, and $x_{i3}D_{i,C}$, as well as the three-factor interaction terms $x_{i4}x_{i3}D_{i,B}$ and $x_{i4}x_{i3}D_{i,C}$. Three-factor interaction terms can be very difficult to interpret. However, as an intuitive example, the term $x_{i4}x_{i3}D_{i,B}$ measures the amount of interaction between the interaction variable $x_{i4}x_{i3}$ and the dummy variable $D_{i,B}$, which represents the effect of advertising campaign B. Figure 12.7 gives the SAS output of the least squares point estimates of the parameters in the expanded model along with the t statistics and associated prob-values. Looking at this figure, we see that the prob-values for all the interaction terms in the expanded model are large. Most are substantially greater than .05. In addition, the standard error for this expanded model is $s = .1311$, which is larger than the standard error (.1308) for our previous model. This indicates that the inclusion of the additional interaction terms is not warranted. To reduce the degree of multicollinearity, we might wish to try some other models with fewer interaction terms. For example, we might consider the

FIGURE 12.7 **SAS output of the least squares point estimates, t statistics, and prob-values for the model $y_i = \beta_0 + \beta_1 x_{i4} + \beta_2 x_{i3} + \beta_3 x_{i3}^2 + \beta_4 x_{i4}x_{i3} + \beta_5 D_{i,B} + \beta_6 D_{i,C} + \beta_7 x_{i4} D_{i,B} + \beta_8 x_{i4} D_{i,C} + \beta_9 x_{i3} D_{i,B} + \beta_{10} x_{i3} D_{i,C} + \beta_{11} x_{i4} x_{i3} D_{i,B} + \beta_{12} x_{i4} x_{i3} D_{i,C} + \varepsilon_i$**

VARIABLE	DF	PARAMETER ESTIMATE	STANDARD ERROR	T FOR HO: PARAMETER=0	PROB > |T|
INTERCEP	1	32.590707	5.740052	5.678	0.0001
X4	1	5.887626	5.658572	1.040	0.3127
X3	1	-8.603161	1.838238	-4.680	0.0002
X3SQ	1	0.735327	0.147080	5.000	0.0001
X43	1	-0.665556	0.831766	-0.800	0.4347
DB	1	-0.761000	1.169902	-0.650	0.5241
DC	1	-2.415693	1.398314	-1.728	0.1022
X4DB	1	8.277369	6.093695	1.358	0.1921
X4DC	1	11.961760	9.746570	1.227	0.2364
X3DB	1	0.156574	0.192191	0.815	0.4265
X3DC	1	0.447779	0.224349	1.996	0.0622
X43DB	1	-1.231316	0.887758	-1.387	0.1834
X43DC	1	-1.836136	1.445220	-1.270	0.2210

FIGURE 12.8 **SAS output of the least squares point estimates, *t* statistics, and prob-values for the model $y_i = \beta_0 + \beta_1 x_{i4} + \beta_2 x_{i3} + \beta_3 x_{i3}^2 + \beta_4 x_{i4} x_{i3} + \beta_5 D_{i,B} + \beta_6 D_{i,C} + \beta_7 x_{i4} D_{i,B} + \beta_8 x_{i4} D_{i,C} + \beta_9 x_{i3} D_{i,B} + \beta_{10} x_{i3} D_{i,C} + \varepsilon_i$**

VARIABLE	DF	PARAMETER ESTIMATE	STANDARD ERROR	T FOR HO: PARAMETER=0	PROB > |T|
INTERCEP	1	30.727132	5.611330	5.476	0.0001
X4	1	12.100134	3.600116	3.361	0.0033
X3	1	-8.016542	1.797726	-4.459	0.0003
X3SQ	1	0.689466	0.143943	4.790	0.0001
X43	1	-1.575823	0.531068	-2.967	0.0079
DB	1	-0.722749	1.176255	-0.614	0.5462
DC	1	-1.680605	1.139724	-1.475	0.1567
X4DB	1	-0.132349	0.454120	-0.291	0.7739
X4DC	1	-0.443757	0.442450	-1.003	0.3285
X3DB	1	0.149000	0.193337	0.771	0.4504
X3DC	1	0.331765	0.184911	1.794	0.0887

model

$$\begin{aligned} y_i &= \beta_0 + \beta_1 x_{i4} + \beta_2 x_{i3} + \beta_3 x_{i3}^2 + \beta_4 x_{i4} x_{i3} + \beta_5 D_{i,B} + \beta_6 D_{i,C} + \beta_7 x_{i4} D_{i,B} \\ &\quad + \beta_8 x_{i4} D_{i,C} + \beta_9 x_{i3} D_{i,B} + \beta_{10} x_{i3} D_{i,C} + \varepsilon_i \end{aligned}$$

Figure 12.8 presents the SAS output of key quantities related to this model. The output indicates that inclusion of the additional interaction terms in this model is not warranted. Moreover, when other possible models are analyzed, we find that the inclusion of interaction terms beyond $x_{i4}x_{i3}$ is not warranted.

Finally, however, suppose that interaction between advertising campaign and advertising expenditure did exist. Also suppose that this is the only type of interaction involving advertising campaign that exists. Then the model

$$\begin{aligned} y_i &= \mu_i + \varepsilon_i \\ &= \beta_0 + \beta_1 x_{i4} + \beta_2 x_{i3} + \beta_3 x_{i3}^2 + \beta_4 x_{i4} x_{i3} \\ &\quad + \beta_5 D_{i,B} + \beta_6 D_{i,C} + \beta_7 x_{i3} D_{i,B} + \beta_7 x_{i3} D_{i,C} + \varepsilon_i \end{aligned}$$

might be a reasonable final model. This model implies that

$$\begin{aligned} \mu_{[d,a,C]} &= \beta_0 + \beta_1 d + \beta_2 a + \beta_3 a^2 + \beta_4 da \\ &\quad + \beta_5(0) + \beta_6(1) + \beta_7 a(0) + \beta_8 a(1) \end{aligned}$$

and

$$\begin{aligned} \mu_{[d,a,A]} &= \beta_0 + \beta_1 d + \beta_2 a + \beta_3 a^2 + \beta_4 da \\ &\quad + \beta_5(0) + \beta_6(0) + \beta_7 a(0) + \beta_8 a(0) \end{aligned}$$

Therefore

$$\mu_{[d,a,C]} - \mu_{[d,a,A]} = \beta_6 + \beta_8 a$$

In other words, this interaction model says that the change in mean demand associated with changing from advertising campaign A to advertising campaign C, while the price difference is held constant at d and the advertising expenditure is held constant at a, depends on the advertising expenditure a. Furthermore, statistical inferences concerning this change can be obtained by making statistical inferences for $(\beta_6 + \beta_8 a)$. We will see how to do this in Section 12.2.

Part 7 (Optional): Comparing the Effects of Campaigns A, B, and C by Using a Partial F-Test

There is a theoretical problem with using the previously discussed 95% confidence intervals to make pairwise comparisons of campaigns A, B and C. Doing this is equivalent to using t statistics (and related prob-values) to perform individual tests of

$$H_0: \beta_5 = \mu_{[d,a,B]} - \mu_{[d,a,A]} = 0 \quad \text{versus} \quad H_1: \beta_5 = \mu_{[d,a,B]} - \mu_{[d,a,A]} \neq 0$$

$$H_0: \beta_6 = \mu_{[d,a,C]} - \mu_{[d,a,A]} = 0 \quad \text{versus} \quad H_1: \beta_6 = \mu_{[d,a,C]} - \mu_{[d,a,A]} \neq 0$$

$$H_0: \beta_6 - \beta_5 = \mu_{[d,a,C]} - \mu_{[d,a,B]} = 0 \quad \text{versus}$$

$$H_1: \beta_6 - \beta_5 = \mu_{[d,a,C]} - \mu_{[d,a,B]} \neq 0$$

We can set the probability of a Type I error equal to .05 for each individual t-test. However, it can be shown that the probability of falsely rejecting H_0 in at least one of these tests is greater than .05. Because of this problem, some practitioners feel that we should test for differences in the effect of advertising campaigns A, B, and C by testing the single hypothesis

$$H_0: \mu_{[d,a,A]} = \mu_{[d,a,B]} = \mu_{[d,a,C]}$$

before we make pairwise comparisons of advertising campaigns. We do this by using a partial F statistic and by considering the following complete model:

$$\text{Complete model:} \quad y_i = \beta_0 + \beta_1 x_{i4} + \beta_2 x_{i3} + \beta_3 x_{i3}^2 + \beta_4 x_{i4} x_{i3} + \beta_5 D_{i,B} + \beta_6 D_{i,C} + \varepsilon_i$$

This model has $k = 7$ parameters and an unexplained variation equal to $SSE_C = .3936$. For this model,

$$\beta_5 = \mu_{[d,a,B]} - \mu_{[d,a,A]}$$

$$\beta_6 = \mu_{[d,a,C]} - \mu_{[d,a,A]}$$

Thus the null hypothesis

$$H_0: \mu_{[d,a,A]} = \mu_{[d,a,B]} = \mu_{[d,a,C]}$$

is equivalent to $H_0: \beta_5 = \beta_6 = 0$. We test H_0 versus the alternative hypothesis

H_1: At least two of $\mu_{[d,a,A]}$, $\mu_{[d,a,B]}$, and $\mu_{[d,a,C]}$ differ from each other or

H_1: At least one of β_5 and β_6 does not equal zero

Here, H_1 says that at least two of advertising campaigns A, B, and C have different effects on mean demand for Fresh.

To carry out a partial F-test of H_0 versus H_1, first note that $p - g = 2$. This is because two parameters (β_5 and β_6) are set equal to zero in the statement of H_0. Also note that under the assumption that H_0 is true, the complete model becomes the following reduced model:

$$\text{Reduced model:} \quad y_i = \beta_0 + \beta_1 x_{i4} + \beta_2 x_{i3} + \beta_3 x_{i3}^2 + \beta_4 x_{i4} x_{i3} + \varepsilon_i$$

For this model the unexplained variation is $SSE_R = 1.0644$. Thus we use the following partial F statistic and prob-value:

$$\begin{aligned} F(D_{i,\text{B}}, D_{i,\text{C}} \mid x_{i4}, x_{i3}, x_{i3}^2, x_{i4}x_{i3}) &= \frac{MS_{\text{drop}}}{MSE_C} \\ &= \frac{SS_{\text{drop}}/(p - g)}{SSE_C/(n - k)} \\ &= \frac{\{SSE_R - SSE_C\}/(p - g)}{SSE_C/(n - k)} \\ &= \frac{\{1.0644 - .3936\}/2}{.3936/(30 - 7)} \\ &= \frac{.6708/2}{.3936/23} \\ &= \frac{.3354}{.0171} \\ &= 19.614 \end{aligned}$$

and

$$\text{prob-value} = .0001$$

Since the prob-value is so small, we have very strong evidence that H_0 should be rejected. Thus we have substantial statistical evidence that at least two of advertising campaigns A, B, and C have different effects on mean demand for Fresh. Having reached this conclusion, it makes sense to compare the effects of specific pairs of campaigns. We have already done this in Part 4 of this example. It should also be noted that even if H_0 were not rejected, some practitioners feel that pairwise comparisons should still be made. This is because there is always a possibility that we have erroneously decided to not reject H_0.

In the Fresh Detergent model of Example 12.2,

$$y_i = \beta_0 + \beta_1 x_{i4} + \beta_2 x_{i3} + \beta_3 x_{i3}^2 + \beta_4 x_{i4} x_{i3} + \beta_5 D_{i,\text{B}} + \beta_6 D_{i,\text{C}} + \varepsilon_i$$

the dummy variables $D_{i,\text{B}}$ and $D_{i,\text{C}}$ model the effects of advertising campaigns A, B, and C. Suppose that, in addition to being affected by the qualitative independent variable advertising campaign, we feel that demand for Fresh is also affected by another qualitative independent variable: competitors' advertising activity. This

TABLE 12.3 **Advertising campaigns and competitors' advertising activity levels in the Fresh Detergent problem**

Sales region, i	*Advertising campaign*	*Competitors' advertising activity level*	*Sales region, i*	*Advertising campaign*	*Competitors' advertising activity level*
1	B	V	16	B	H
2	B	M	17	B	L
3	B	M	18	A	H
4	A	L	19	B	M
5	C	V	20	B	H
6	A	V	21	C	M
7	C	H	22	A	L
8	C	L	23	A	H
9	B	H	24	A	H
10	C	H	25	A	V
11	A	M	26	B	M
12	C	L	27	C	V
13	C	L	28	B	L
14	A	V	29	C	V
15	B	M	30	C	H

variable is defined to have four levels: very high (V), high (H), medium (M), and low (L). Moreover, suppose that Table 12.3 lists the campaigns (A, B, or C) and the competitors' advertising activity levels (V, H, M, or L) in the 30 sales regions. Also suppose first that *no interaction* exists between advertising campaign and competitors' advertising activity. Then we would use the previously discussed dummy variables $D_{i,B}$ and $D_{i,C}$ to measure the effects of advertising campaigns A, B, and C. Similarly, we would use a separate set of dummy variables, which we denote $D_{i,H}$, $D_{i,M}$, and $D_{i,L}$, to model the effects of competitors' advertising activity levels V, H, M, and L. Here,

$$D_{i,H} = \begin{cases} 1 & \text{if there was a high level of competitors' advertising activity in sales region } i \\ 0 & \text{otherwise} \end{cases}$$

$$D_{i,M} = \begin{cases} 1 & \text{if there was a medium level of competitors' advertising activity in sales region } i \\ 0 & \text{otherwise} \end{cases}$$

$$D_{i,L} = \begin{cases} 1 & \text{if there was a low level of competitors' advertising activity in sales region } i \\ 0 & \text{otherwise} \end{cases}$$

Then a reasonable model describing demand for Fresh detergent would be

$$\begin{aligned} y_i &= \mu_i + \varepsilon_i \\ &= \beta_0 + \beta_1 x_{i4} + \beta_2 x_{i3} + \beta_3 x_{i3}^2 + \beta_4 x_{i4} x_{i3} + \beta_5 D_{i,\mathrm{B}} + \beta_6 D_{i,\mathrm{C}} \\ &\quad + \beta_7 D_{i,\mathrm{H}} + \beta_8 D_{i,\mathrm{M}} + \beta_9 D_{i,\mathrm{L}} + \varepsilon_i \end{aligned}$$

This model would say, for example, that

$$\begin{aligned} \mu_{[d,a,\mathrm{C},\mathrm{L}]} &= \beta_0 + \beta_1 d + \beta_2 a + \beta_3 a^2 + \beta_4 da + \beta_5(0) + \beta_6(1) \\ &\quad + \beta_7(0) + \beta_8(0) + \beta_9(1) \end{aligned}$$

and

$$\begin{aligned} \mu_{[d,a,\mathrm{C},\mathrm{H}]} &= \beta_0 + \beta_1 d + \beta_2 a + \beta_3 a^2 + \beta_4 da + \beta_5(0) + \beta_6(1) \\ &\quad + \beta_7(1) + \beta_8(0) + \beta_9(0) \end{aligned}$$

Therefore

$$\mu_{[d,a,\mathrm{C},\mathrm{H}]} - \mu_{[d,a,\mathrm{C},\mathrm{L}]} = \beta_7 - \beta_9$$

This difference is the change in mean demand for Fresh associated with changing from a low level to a high level of competitors' advertising activity while the price difference, advertising expenditure, and advertising campaign remain constant.

If interaction exists between advertising campaign and competitors' advertising activity, one might attempt to model this interaction by multiplying the appropriate dummy variables together. We might consider the model

$$\begin{aligned} y_i &= \beta_0 + \beta_1 x_{i4} + \beta_2 x_{i3} + \beta_3 x_{i3}^2 + \beta_4 x_{i4} x_{i3} + \beta_5 D_{i,\mathrm{B}} + \beta_6 D_{i,\mathrm{C}} \\ &\quad + \beta_7 D_{i,\mathrm{H}} + \beta_8 D_{i,\mathrm{M}} + \beta_9 D_{i,\mathrm{L}} + \beta_{10} D_{i,\mathrm{B}} D_{i,\mathrm{H}} + \beta_{11} D_{i,\mathrm{B}} D_{i,\mathrm{M}} \\ &\quad + \beta_{12} D_{i,\mathrm{B}} D_{i,\mathrm{L}} + \beta_{13} D_{i,\mathrm{C}} D_{i,\mathrm{H}} + \beta_{14} D_{i,\mathrm{C}} D_{i,\mathrm{M}} + \beta_{15} D_{i,\mathrm{C}} D_{i,\mathrm{L}} + \varepsilon_i \end{aligned}$$

We have demonstrated that multiplying independent variables together is appropriate in modeling interaction between two quantitative independent variables or between one quantitative independent variable and one qualitative independent variable. However, this multiplication procedure is not always appropriate in modeling interaction between two qualitative independent variables. Specifically, the multiplication procedure is appropriate if at least one value of the dependent variable has been observed for each and every combination of levels of the two qualitative independent variables. For example, the different combinations of advertising campaigns and competitors' advertising activity levels are

AV	AH	AM	AL
BV	BH	BM	BL
CV	CH	CM	CL

Table 12.3 indicates that at least one demand has been observed for each and every one of these combinations. Therefore the above model is appropriate.

However, if no data has been observed for at least one combination of levels then the multiplication procedure might not be appropriate. This is because it might produce a model whose **X**-matrix does not have linearly independent columns, which means that $(\mathbf{X}'\mathbf{X})^{-1}$ does not exist. In such a case, interaction can be appropriately modeled by using a different dummy variable to represent the effect of each *combination* of levels for which data has been observed. Note here that a dummy variable will not be defined for one arbitrarily selected combination of levels for which data has been observed. For example, suppose that in the Fresh Detergent problem at least one Fresh demand had been observed for all combinations except for the combination BM. Then we could model interaction between advertising campaign and competitors' advertising activity by using the model

$$\begin{aligned} y_i = {} & \beta_0 + \beta_1 x_{i4} + \beta_2 x_{i3} + \beta_3 x_{i3}^2 + \beta_4 x_{i4} x_{i3} + \beta_5 D_{i,\mathrm{AH}} + \beta_6 D_{i,\mathrm{AM}} \\ & + \beta_7 D_{i,\mathrm{AL}} + \beta_8 D_{i,\mathrm{BV}} + \beta_9 D_{i,\mathrm{BH}} + \beta_{10} D_{i,\mathrm{BL}} \\ & + \beta_{11} D_{i,\mathrm{CV}} + \beta_{12} D_{i,\mathrm{CH}} + \beta_{13} D_{i,\mathrm{CM}} + \beta_{14} D_{i,\mathrm{CL}} + \varepsilon_i \end{aligned}$$

Here, for example,

$$D_{i,\mathrm{BH}} = \begin{cases} 1 & \text{if advertising campaign B was used in sales region } i \text{ and there was a high level of competitors' advertising activity in sales region } i \\ 0 & \text{otherwise} \end{cases}$$

Note that we have not defined a dummy variable for the combination BM. Also, note that we have not defined a dummy variable for the arbitrarily selected combination AV. This means that the parameters multiplied by the dummy variables in the model compare the corresponding combinations to the combination AV. In Chapters 15 and 16 we discuss in detail how to assess the effects of two qualitative independent variables on a dependent variable.

Example 12.3 We found in Chapter 11 that one of the best models describing all 17 observations in the hospital labor data set of Table 11.5 is

$$y_i = \beta_0 + \beta_1 x_{i2} + \beta_2 x_{i3} + \beta_3 x_{i5} + \varepsilon_i$$

This model has a standard error, s, equal to 614.78. Furthermore, it yields a residual corresponding to hospital 14 equal to $e_{14} = 1631$. This implies that the model greatly underpredicts labor requirements for hospital 14 and suggests that we might remove observation 14 from the data set.

We now consider using a dummy variable to divide the large hospitals (14, 15, 16, and 17) into a class by themselves. A possible model would be

$$y_i = \beta_0 + \beta_1 x_{i2} + \beta_2 x_{i3} + \beta_3 x_{i5} + \beta_4 D_{i,\mathrm{L}} + \varepsilon_i$$

Here,

$$D_{i,L} = \begin{cases} 1 & \text{if hospital } i \text{ is a large hospital} \\ 0 & \text{otherwise} \end{cases}$$

If we calculate the least squares point estimates of the parameters in this model by using all 17 observations, we obtain the prediction equation

$$\hat{y}_i = 2462.22 + .0482x_{i2} + .7843x_{i3} - 432.41x_{i5} + 2871.78D_{i,L}$$

This model has a substantially smaller standard error, s, equal to 363.85. Furthermore, it yields a largest residual of $e_{11} = 727.2$ and a next largest residual of 485.9. As an exercise, the reader can verify that removing observation 11 from the data set does not yield a much improved model. The reader can also verify that there seems to be little or no need for the interaction terms $x_{i2}D_{i,L}$, $x_{i3}D_{i,L}$, and $x_{i5}D_{i,L}$. Of course, we might find such a need if we could add more large hospitals to the data set. The lack of interaction for the available data suggests that the large hospitals respond to increases in x_{i2}, x_{i3}, and x_{i5} at the same rate as small hospitals, while requiring a consistent incremental increase in labor hours. This increment is estimated to be $b_4 = 2871.78$ and may reflect a certain inefficiency related to large organizations.

Example 12.4 In Table 11.9 we presented the sales price y, square footage x_1, number of rooms x_2, number of bedrooms x_3, age x_4, and number of bathrooms x_5 for each of 63 single-family residences sold during 1988 in Oxford, Ohio. Consider the regression model

$$y_i = \beta_0 + \beta_1 x_{i1} + \beta_2 x_{i2} + \beta_3 x_{i3} + \beta_4 x_{i4} + \beta_5 x_{i5} + \beta_6 x_{i6} + \beta_7 x_{i7} + \varepsilon_i$$

Here, x_{i6} and x_{i7} are dummy variables defined as follows:

$$x_{i6} = \begin{cases} 1 & \text{if residence } i \text{ has a garage} \\ 0 & \text{otherwise} \end{cases}$$

$$x_{i7} = \begin{cases} 1 & \text{if residence } i \text{ has a porch} \\ 0 & \text{otherwise} \end{cases}$$

Therefore β_6 represents the change in mean sales price associated with a house having a garage when all other independent variables are held constant. Similarly, β_7 represents the change in mean sales price associated with a house having a porch when all other independent variables are held constant. Table 12.4 provides information concerning the independent variables x_{i6} and x_{i7} for the 63 residences in Table 11.9. Using the above model as the model containing all $p = 7$ potential independent variables, we present in Figure 12.9 the SAS output of the values of R^2 and C for all reasonable models describing the data in Tables 11.9 and 12.4. The

TABLE 12.4 Measurements taken on 63 single-family residences

Residence, i	Garage, x_6	Porch, x_7	Residence, i	Garage, x_6	Porch, x_7
1	0	0	33	1	0
2	0	1	34	0	0
3	0	1	35	0	0
4	1	0	36	1	0
5	1	1	37	0	0
6	1	0	38	1	1
7	1	0	39	1	0
8	1	1	40	1	0
9	1	0	41	1	0
10	1	0	42	1	1
11	1	0	43	1	0
12	0	0	44	1	0
13	0	0	45	1	0
14	1	0	46	1	1
15	1	0	47	0	0
16	0	0	48	1	0
17	1	0	49	1	0
18	1	0	50	1	0
19	0	1	51	1	1
20	1	0	52	1	1
21	1	0	53	1	0
22	1	1	54	1	0
23	0	0	55	1	0
24	1	0	56	1	0
25	1	0	57	1	1
26	1	0	58	1	0
27	1	1	59	1	1
28	1	1	60	1	0
29	1	1	61	1	0
30	1	1	62	1	0
31	1	0	63	1	0
32	1	0			

Source: Reprinted by permission of Alpha, Inc., Oxford, Ohio.

FIGURE 12.9 SAS output of values of R^2 and C for all reasonable residential sales models

```
NUMBER IN
  MODEL      R-SQUARE        C(P)    VARIABLES IN MODEL

    1       0.00002895     147.937   PORCH
    1       0.06253369     135.002   GARAGE
    1       0.10635858     125.932   AGE
    1       0.28084477   89.823829   BED
    1       0.35143113   75.216505   ROOMS
    1       0.44194830   56.484651   BATH
    1       0.63320895   16.904681   SQFT

    2       0.06265043     136.978   PORCH GARAGE
    2       0.10666225     127.870   AGE PORCH
    2       0.14776102     119.365   AGE GARAGE
```

NUMBER IN MODEL	R-SQUARE	C(P)	VARIABLES IN MODEL
2	0.28162931	91.661474	BED PORCH
2	0.30175703	87.496192	BED GARAGE
2	0.36178651	75.073538	ROOMS PORCH
2	0.37317349	72.717087	ROOMS BED
2	0.38807272	69.633801	ROOMS GARAGE
2	0.39330465	68.551093	BED AGE
2	0.44279414	58.309610	BATH PORCH
2	0.44304931	58.256806	BATH GARAGE
2	0.44547115	57.755623	BATH AGE
2	0.53902497	38.395356	ROOMS AGE
2	0.54199971	37.779756	BATH BED
2	0.56831004	32.335028	ROOMS BATH
2	0.63327355	18.891313	SQFT PORCH
2	0.63472579	18.590782	SQFT ROOMS
2	0.63609600	18.307228	SQFT GARAGE
2	0.64449803	16.568490	SQFT BED
2	0.67830492	9.572406	SQFT BATH
2	0.69122057	6.899608	SQFT AGE
3	0.14787766	121.340	AGE PORCH GARAGE
3	0.30228178	89.387598	BED PORCH GARAGE
3	0.37746104	73.829810	ROOMS BED PORCH
3	0.39600497	69.992282	BED AGE PORCH
3	0.39865000	69.444912	ROOMS PORCH GARAGE
3	0.40079757	69.000489	ROOMS BED GARAGE
3	0.40198696	68.754353	BED AGE GARAGE
3	0.44381906	60.097511	BATH PORCH GARAGE
3	0.44650361	59.541962	BATH AGE GARAGE
3	0.44651418	59.539776	BATH AGE PORCH
3	0.54200147	39.779391	BATH BED GARAGE
3	0.54395811	39.374479	BATH BED PORCH
3	0.54567725	39.018715	ROOMS BED AGE
3	0.54589565	38.973519	ROOMS AGE PORCH
3	0.55327988	37.445408	ROOMS AGE GARAGE
3	0.55669758	36.738141	BATH BED AGE
3	0.57009038	33.966601	ROOMS BATH PORCH
3	0.57045084	33.892006	ROOMS BATH GARAGE
3	0.57886699	32.150345	ROOMS BATH BED
3	0.61672255	24.316420	ROOMS BATH AGE
3	0.63502237	20.529407	SQFT ROOMS PORCH
3	0.63618057	20.289727	SQFT PORCH GARAGE
3	0.63814292	19.883633	SQFT ROOMS GARAGE
3	0.64484797	18.496073	SQFT BED PORCH
3	0.64770433	17.904970	SQFT BED GARAGE
3	0.65247290	16.918149	SQFT ROOMS BED
3	0.67833190	11.566822	SQFT BATH GARAGE
3	0.67834949	11.563181	SQFT BATH PORCH
3	0.68145441	10.920643	SQFT ROOMS BATH
3	0.68161437	10.887540	SQFT BATH BED
3	0.69130132	8.882893	SQFT AGE PORCH
3	0.69175883	8.788219	SQFT AGE GARAGE
3	0.69617802	7.873701	SQFT BED AGE
3	0.70524070	5.998245	SQFT BATH AGE
3	0.70882097	5.257335	SQFT ROOMS AGE

(continues)

FIGURE 12.9 **Continued**

NUMBER IN MODEL	R-SQUARE	C(P)	VARIABLES IN MODEL
4	0.40425439	70.285126	BED AGE PORCH GARAGE
4	0.40648449	69.823623	ROOMS BED PORCH GARAGE
4	0.44746285	61.343456	BATH AGE PORCH GARAGE
4	0.54395862	41.374373	BATH BED PORCH GARAGE
4	0.54977422	40.170878	ROOMS BED AGE PORCH
4	0.55670805	38.735973	BATH BED AGE GARAGE
4	0.55714261	38.646044	ROOMS BED AGE GARAGE
4	0.55940868	38.177099	BATH BED AGE PORCH
4	0.56041794	37.968240	ROOMS AGE PORCH GARAGE
4	0.57243390	35.431627	ROOMS BATH PORCH GARAGE
4	0.57914972	34.091837	ROOMS BATH BED PORCH
4	0.57971666	33.974513	ROOMS BATH BED GARAGE
4	0.61879575	25.887386	ROOMS BATH AGE GARAGE
4	0.61908203	25.828143	ROOMS BATH AGE PORCH
4	0.62257055	25.106219	ROOMS BATH BED AGE
4	0.63854536	21.800350	SQFT ROOMS PORCH GARAGE
4	0.64810799	19.821436	SQFT BED PORCH GARAGE
4	0.65465062	18.467455	SQFT ROOMS BED PORCH
4	0.65726622	17.926208	SQFT ROOMS BED GARAGE
4	0.67837390	13.558130	SQFT BATH PORCH GARAGE
4	0.68148834	12.913620	SQFT ROOMS BATH PORCH
4	0.68156793	12.897149	SQFT ROOMS BATH GARAGE
4	0.68161441	12.887531	SQFT BATH BED PORCH
4	0.68171808	12.866078	SQFT BATH BED GARAGE
4	0.68900317	11.358482	SQFT ROOMS BATH BED
4	0.69182729	10.774052	SQFT AGE PORCH GARAGE
4	0.69617926	9.873443	SQFT BED AGE PORCH
4	0.69688379	9.727647	SQFT BED AGE GARAGE
4	0.70524074	7.998237	SQFT BATH AGE GARAGE
4	0.70543784	7.957450	SQFT BATH AGE PORCH
4	0.70760346	7.509291	SQFT BATH BED AGE
4	0.70921493	7.175808	SQFT ROOMS AGE PORCH
4	0.70989928	7.034187	SQFT ROOMS AGE GARAGE
4	0.72024721	4.892760	SQFT ROOMS BATH AGE
4	0.72574793	3.754428	SQFT ROOMS BED AGE
5	0.55944207	40.170188	BATH BED AGE PORCH GARAGE
5	0.56212312	39.615365	ROOMS BED AGE PORCH GARAGE
5	0.58010608	35.893926	ROOMS BATH BED PORCH GARAGE
5	0.62138450	27.351664	ROOMS BATH AGE PORCH GARAGE
5	0.62346247	26.921644	ROOMS BATH BED AGE PORCH
5	0.62365251	26.882314	ROOMS BATH BED AGE GARAGE
5	0.65990874	19.379359	SQFT ROOMS BED PORCH GARAGE
5	0.68161010	14.888424	SQFT ROOMS BATH PORCH GARAGE
5	0.68171858	14.865973	SQFT BATH BED PORCH GARAGE
5	0.68951766	13.252012	SQFT ROOMS BATH BED GARAGE
5	0.68971183	13.211831	SQFT ROOMS BATH BED PORCH
5	0.69688381	11.727642	SQFT BED AGE PORCH GARAGE
5	0.70543797	9.957422	SQFT BATH AGE PORCH GARAGE
5	0.70762357	9.505129	SQFT BATH BED AGE GARAGE
5	0.70766990	9.495540	SQFT BATH BED AGE PORCH
5	0.71035860	8.939135	SQFT ROOMS AGE PORCH GARAGE
5	0.72037865	6.865559	SQFT ROOMS BATH AGE GARAGE
5	0.72041831	6.857353	SQFT ROOMS BATH AGE PORCH

NUMBER IN MODEL	R-SQUARE	C(P)	VARIABLES IN MODEL
5	0.72763857	5.363174	SQFT ROOMS BED AGE GARAGE
5	0.72811603	5.264367	SQFT ROOMS BED AGE PORCH
5	0.73173947	4.514523	SQFT ROOMS BATH BED AGE
6	0.62475864	28.653411	ROOMS BATH BED AGE PORCH GARAGE
6	0.69035536	15.078657	SQFT ROOMS BATH BED PORCH GARAGE
6	0.70768649	11.492109	SQFT BATH BED AGE PORCH GARAGE
6	0.72056949	8.826065	SQFT ROOMS BATH AGE PORCH GARAGE
6	0.73030776	6.810804	SQFT ROOMS BED AGE PORCH GARAGE
6	0.73243712	6.370149	SQFT ROOMS BATH BED AGE GARAGE
6	0.73330033	6.191513	SQFT ROOMS BATH BED AGE PORCH
7	0.73422578	8.000000	SQFT ROOMS BATH BED AGE PORCH GARAGE

FIGURE 12.10 **SAS output of a stepwise regression of the residential data**

STEPWISE REGRESSION PROCEDURE FOR DEPENDENT VARIABLE SALEP

STEP 1 VARIABLE SQFT ENTERED R SQUARE = 0.63320895 C(P) = 16.90468097

	DF	SUM OF SQUARES	MEAN SQUARE	F	PROB>F
REGRESSION	1	48147.30258391	48147.30258391	105.31	0.0001
ERROR	61	27889.68725736	457.20798783		
TOTAL	62	76036.98984127			

	B VALUE	STD ERROR	TYPE II SS	F	PROB>F
INTERCEPT	-3.72102928				
SQFT	0.05477049	0.00533725	48147.30258391	105.31	0.0001

BOUNDS ON CONDITION NUMBER: 1, 1

STEP 2 VARIABLE AGE ENTERED R SQUARE = 0.69122057 C(P) = 6.89960812

	DF	SUM OF SQUARES	MEAN SQUARE	F	PROB>F
REGRESSION	2	52558.33129612	26279.16564806	67.16	0.0001
ERROR	60	23478.65854515	391.31097575		
TOTAL	62	76036.98984127			

	B VALUE	STD ERROR	TYPE II SS	F	PROB>F
INTERCEPT	10.31301793				
SQFT	0.05295333	0.00496724	44471.14528680	113.65	0.0001
AGE	-0.35942252	0.10705239	4411.02871221	11.27	0.0014

BOUNDS ON CONDITION NUMBER: 1.012015, 4.04806

STEP 3 VARIABLE ROOMS ENTERED R SQUARE = 0.70882097 C(P) = 5.25733481

	DF	SUM OF SQUARES	MEAN SQUARE	F	PROB>F
REGRESSION	3	53896.61323550	17965.53774517	47.87	0.0001
ERROR	59	22140.37660577	375.26062044		
TOTAL	62	76036.98984127			

(continues)

FIGURE 12.10 Continued

	B VALUE	STD ERROR	TYPE II SS	F	PROB>F
INTERCEPT	-0.08788207				
SQFT	0.04271768	0.00728279	12910.77698346	34.40	0.0001
ROOMS	4.51653219	2.39165096	1338.28193937	3.57	0.0639
AGE	-0.43424655	0.11207156	5633.97454061	15.01	0.0003

BOUNDS ON CONDITION NUMBER: 2.308184, 17.19982

STEP 4 VARIABLE BED ENTERED R SQUARE = 0.72574793 C(P) = 3.75442773

	DF	SUM OF SQUARES	MEAN SQUARE	F	PROB>F
REGRESSION	4	55183.68778283	13795.92194571	38.37	0.0001
ERROR	58	20853.30205844	359.53969066		
TOTAL	62	76036.98984127			

	B VALUE	STD ERROR	TYPE II SS	F	PROB>F
INTERCEPT	10.36761556				
SQFT	0.05001119	0.00810413	13692.03206797	38.08	0.0001
ROOMS	6.32177893	2.52798952	2248.40691315	6.25	0.0152
BED	-11.10316277	5.86838064	1287.07454734	3.58	0.0635
AGE	-0.43186496	0.10970614	5571.61231461	15.50	0.0002

BOUNDS ON CONDITION NUMBER: 2.931872, 38.17696

NO OTHER VARIABLES MET THE 0.10000 SIGNIFICANCE LEVEL FOR ENTRY INTO THE MODEL.

model having the smallest value of C is

$$y_i = \beta_0 + \beta_1 x_{i1} + \beta_2 x_{i2} + \beta_3 x_{i3} + \beta_4 x_{i4} + \varepsilon_i$$

As further evidence that this model is reasonable, note from Figure 12.10 that a stepwise regression of the residential data (with $\alpha_{\text{entry}} = .1$ and $\alpha_{\text{stay}} = .1$) arrives at this model. We discussed this model previously in Examples 8.16 and 8.21.

12.2

STATISTICAL INFERENCES FOR A LINEAR COMBINATION OF REGRESSION PARAMETERS

A CONFIDENCE INTERVAL FOR $\lambda'\beta$

In Example 12.2 we considered the Fresh Detergent model

$$y_i = \beta_0 + \beta_1 x_{i4} + \beta_2 x_{i3} + \beta_3 x_{i3}^2 + \beta_4 x_{i4} x_{i3} + \beta_5 D_{i,\text{B}} + \beta_6 D_{i,\text{C}} + \varepsilon_i$$

We said that a 95% confidence interval for

$$\mu_{[d,a,\text{C}]} - \mu_{[d,a,\text{B}]} = \beta_6 - \beta_5$$

is [.0363, .2999]. To see how we calculate this interval, note that $\beta_6 - \beta_5$ can be expressed as a linear combination of the parameters in this model. That is, we can express $\beta_6 - \beta_5$ as

$$\begin{aligned}
\beta_6 - \beta_5 &= 0 \cdot \beta_0 + 0 \cdot \beta_1 + 0 \cdot \beta_2 + 0 \cdot \beta_3 + 0 \cdot \beta_4 + (-1) \cdot \beta_5 + 1 \cdot \beta_6 \\
&= [0 \quad 0 \quad 0 \quad 0 \quad 0 \quad -1 \quad 1] \begin{bmatrix} \beta_0 \\ \beta_1 \\ \beta_2 \\ \beta_3 \\ \beta_4 \\ \beta_5 \\ \beta_6 \end{bmatrix} \\
&= \boldsymbol{\lambda}'\boldsymbol{\beta}
\end{aligned}$$

where

$$\boldsymbol{\lambda}' = [0 \quad 0 \quad 0 \quad 0 \quad 0 \quad -1 \quad 1] \quad \text{and} \quad \boldsymbol{\beta} = \begin{bmatrix} \beta_0 \\ \beta_1 \\ \beta_2 \\ \beta_3 \\ \beta_4 \\ \beta_5 \\ \beta_6 \end{bmatrix}$$

To compute a confidence interval for a linear combination of the parameters $\beta_0, \beta_1, \beta_2, \ldots, \beta_p$ in the model

$$y = \beta_0 + \beta_1 x_1 + \beta_2 x_2 + \cdots + \beta_p x_p + \varepsilon$$

consider the linear combination of regression parameters

$$\boldsymbol{\lambda}'\boldsymbol{\beta} = \sum_{j=0}^{p} \lambda_j \beta_j = [\lambda_0 \quad \lambda_1 \quad \lambda_2 \quad \ldots \quad \lambda_p] \begin{bmatrix} \beta_0 \\ \beta_1 \\ \beta_2 \\ \vdots \\ \beta_p \end{bmatrix}$$

where

$$\boldsymbol{\lambda}' = [\lambda_0 \quad \lambda_1 \quad \lambda_2 \quad \ldots \quad \lambda_p] \quad \text{and} \quad \boldsymbol{\beta} = \begin{bmatrix} \beta_0 \\ \beta_1 \\ \beta_2 \\ \vdots \\ \beta_p \end{bmatrix}$$

A *point estimate* of the linear combination

$$\boldsymbol{\lambda}'\boldsymbol{\beta} = \sum_{j=0}^{p} \lambda_j \beta_j = \lambda_0\beta_0 + \lambda_1\beta_1 + \lambda_2\beta_2 + \cdots + \lambda_p\beta_p$$

is

$$\boldsymbol{\lambda}'\mathbf{b} = \sum_{j=0}^{p} \lambda_j b_j = \lambda_0 b_0 + \lambda_1 b_1 + \lambda_2 b_2 + \cdots + \lambda_p b_p$$

A confidence interval for $\boldsymbol{\lambda}'\boldsymbol{\beta}$ is obtained as follows.

A Confidence Interval for $\boldsymbol{\lambda}'\boldsymbol{\beta}$

1. *The population of all possible point estimates of* $\boldsymbol{\lambda}'\boldsymbol{\beta}$ (that is, values of $\boldsymbol{\lambda}'\mathbf{b}$)
   a. has *mean* $\mu_{\boldsymbol{\lambda}'\mathbf{b}} = \boldsymbol{\lambda}'\boldsymbol{\beta}$
   b. has *variance* $\sigma^2_{\boldsymbol{\lambda}'\mathbf{b}} = \sigma^2\boldsymbol{\lambda}'(\mathbf{X}'\mathbf{X})^{-1}\boldsymbol{\lambda}$ (if inference assumptions 1 and 2 hold)
   c. has *standard deviation* $\sigma_{\boldsymbol{\lambda}'\mathbf{b}} = \sigma\sqrt{\boldsymbol{\lambda}'(\mathbf{X}'\mathbf{X})^{-1}\boldsymbol{\lambda}}$ (if inference assumptions 1 and 2 hold)
   d. has a *normal distribution* (if inference assumptions 1, 2, and 3 hold)
2. A point estimate of $\sigma_{\boldsymbol{\lambda}'\mathbf{b}} = \sigma\sqrt{\boldsymbol{\lambda}'(\mathbf{X}'\mathbf{X})^{-1}\boldsymbol{\lambda}}$, is $s_{\boldsymbol{\lambda}'\mathbf{b}} = s\sqrt{\boldsymbol{\lambda}'(\mathbf{X}'\mathbf{X})^{-1}\boldsymbol{\lambda}}$, which is called the *standard error of the estimate* $\boldsymbol{\lambda}'\mathbf{b}$.
3. If the inference assumptions hold, the population of all possible values of

   $$\frac{\boldsymbol{\lambda}'\mathbf{b} - \boldsymbol{\lambda}'\boldsymbol{\beta}}{s\sqrt{\boldsymbol{\lambda}'(\mathbf{X}'\mathbf{X})^{-1}\boldsymbol{\lambda}}}$$

   has a *t-distribution with n − k degrees of freedom.*
4. If the inference assumptions hold, a *100(1 − α)% confidence interval for* $\boldsymbol{\lambda}'\boldsymbol{\beta}$ is

   $$[\boldsymbol{\lambda}'\mathbf{b} \pm t^{(n-k)}_{[\alpha/2]}\, s\sqrt{\boldsymbol{\lambda}'(\mathbf{X}'\mathbf{X})^{-1}\boldsymbol{\lambda}}]$$

Example 12.5 Consider the Fresh Detergent model

$$y_i = \beta_0 + \beta_1 x_{i4} + \beta_2 x_{i3} + \beta_3 x_{i3}^2 + \beta_4 x_{i4} x_{i3} + \beta_5 D_{i,\text{B}} + \beta_6 D_{i,\text{C}} + \varepsilon_i$$

Here we wish to calculate a 95% confidence interval for

$$\begin{aligned} \mu_{[d,a,\text{C}]} - \mu_{[d,a,\text{B}]} &= \beta_6 - \beta_5 \\ &= 0 \cdot \beta_0 + 0 \cdot \beta_1 + 0 \cdot \beta_2 + 0 \cdot \beta_3 \\ &\quad + 0 \cdot \beta_4 + (-1) \cdot \beta_5 + 1 \cdot \beta_6 \\ &= [0 \;\; 0 \;\; 0 \;\; 0 \;\; 0 \;\; -1 \;\; 1] \begin{bmatrix} \beta_0 \\ \beta_1 \\ \beta_2 \\ \beta_3 \\ \beta_4 \\ \beta_5 \\ \beta_6 \end{bmatrix} \\ &= \boldsymbol{\lambda}'\boldsymbol{\beta} \end{aligned}$$

where

$$\boldsymbol{\lambda}' = [0 \;\; 0 \;\; 0 \;\; 0 \;\; 0 \;\; -1 \;\; 1]$$

We saw in Example 12.2 that the least squares point estimates of β_5 and β_6 are $b_5 = .2137$ and $b_6 = .3818$ and the standard error is $s = .1308$. It follows that a point estimate of $\mu_{[d,a,\text{C}]} - \mu_{[d,a,\text{B}]} = \beta_6 - \beta_5 = \boldsymbol{\lambda}'\boldsymbol{\beta}$ is

$$\begin{aligned} \boldsymbol{\lambda}'\mathbf{b} &= [0 \;\; 0 \;\; 0 \;\; 0 \;\; 0 \;\; -1 \;\; 1] \begin{bmatrix} b_0 \\ b_1 \\ b_2 \\ b_3 \\ b_4 \\ b_5 \\ b_6 \end{bmatrix} \\ &= b_6 - b_5 = .3818 - .2137 = .1681 \end{aligned}$$

Furthermore, a 95% confidence interval for $\mu_{[d,a,\text{C}]} - \mu_{[d,a,\text{B}]} = \beta_6 - \beta_5 = \boldsymbol{\lambda}'\boldsymbol{\beta}$ is

$$\begin{aligned} [\boldsymbol{\lambda}'\mathbf{b} \pm t_{[.025]}^{(n-k)} s\sqrt{\boldsymbol{\lambda}'(\mathbf{X}'\mathbf{X})^{-1}\boldsymbol{\lambda}}] &= [\boldsymbol{\lambda}'\mathbf{b} \pm t_{[.025]}^{(30-7)} s\sqrt{\boldsymbol{\lambda}'(\mathbf{X}'\mathbf{X})^{-1}\boldsymbol{\lambda}}] \\ &= [.1681 \pm 2.069(.1308)\sqrt{.2372}] \\ &= [.1681 \pm .1318] \\ &= [.0363, .2999] \end{aligned}$$

This interval makes Enterprise Industries 95% confident that the effect of changing from advertising campaign B to advertising campaign C is to increase mean demand for Fresh by between 3,630 bottles and 29,990 bottles.

A HYPOTHESIS TEST CONCERNING $\lambda'\beta$

Next we discuss hypothesis tests concerning linear combinations of regression parameters. Suppose that $\boldsymbol{\lambda}'\boldsymbol{\beta}$ is a linear combination of the parameters in the regression model

$$y = \beta_0 + \beta_1 x_1 + \beta_2 x_2 + \cdots + \beta_p x_p + \varepsilon$$

It is often useful to test the null hypothesis $H_0: \boldsymbol{\lambda}'\boldsymbol{\beta} = 0$ versus the alternative hypothesis $H_1: \boldsymbol{\lambda}'\boldsymbol{\beta} \neq 0$. For example, in the Fresh Detergent problem it would be useful to test

$$H_0: \beta_6 - \beta_5 = \mu_{[d,a,\mathrm{C}]} - \mu_{[d,a,\mathrm{B}]} = 0$$

which says that advertising campaigns B and C have the same effect on mean demand for Fresh, versus

$$H_1: \beta_6 - \beta_5 = \mu_{[d,a,\mathrm{C}]} - \mu_{[d,a,\mathrm{B}]} \neq 0$$

which says that advertising campaigns B and C have different effects on mean demand for Fresh.

Testing $H_0: \boldsymbol{\lambda}'\boldsymbol{\beta} = 0$ Versus $H_1: \boldsymbol{\lambda}'\boldsymbol{\beta} \neq 0$

Define the test statistic

$$t = \frac{\boldsymbol{\lambda}'\mathbf{b}}{s\sqrt{\boldsymbol{\lambda}'(\mathbf{X}'\mathbf{X})^{-1}\boldsymbol{\lambda}}}$$

Also, define the prob-value to be twice the area under the curve of the t-distribution having $n - k$ degrees of freedom to the right of $|t|$. If the inference assumptions are satisfied, we can reject $H_0: \boldsymbol{\lambda}'\boldsymbol{\beta} = 0$ in favor of $H_1: \boldsymbol{\lambda}'\boldsymbol{\beta} \neq 0$ by setting the probability of a Type I error equal to α if and only if either of the following equivalent conditions holds:

1. $|t| > t_{[\alpha/2]}^{(n-k)}$—that is, if $t > t_{[\alpha/2]}^{(n-k)}$ or $t < -t_{[\alpha/2]}^{(n-k)}$.
2. Prob-value $< \alpha$.

Example 12.6 We consider the Fresh Detergent model

$$y_i = \beta_0 + \beta_1 x_{i4} + \beta_2 x_{i3} + \beta_3 x_{i3}^2 + \beta_4 x_{i4} x_{i3} + \beta_5 D_{i,\mathrm{B}} + \beta_6 D_{i,\mathrm{C}} + \varepsilon_i$$

and test

$$H_0: \beta_6 - \beta_5 = \mu_{[d,a,\mathrm{C}]} - \mu_{[d,a,\mathrm{B}]} = 0$$

FIGURE 12.11 **Statistical inference in the Fresh Detergent dummy variable model**

(a) SAS output of the t statistics and prob-values for testing $H_0: \lambda'\beta = 0$ in the Fresh Detergent dummy variable model

PARAMETER[a]	ESTIMATE[b]	T FOR HO: PARAMETER=0[c]	PR > | T |[d]	STD ERROR OF ESTIMATE[e]
INTERCEPT	25.61269602	5.34	0.0001	4.79378249
X4	9.05868432	2.99	0.0066	3.03170457
X3	-6.53767133	-4.13	0.0004	1.58136655
X3SQ	0.58444394	4.50	0.0002	0.12987222
X43	-1.15648054	-2.54	0.0184	0.45573648
DB	0.21368626	3.44	0.0022	0.06215362
DC	0.38177617	6.23	0.0001	0.06125253
MUDAB-MUDAA	0.21368626	3.44	0.0022	0.06215362
MUDAC-MUDAA	0.38177617	6.23	0.0001	0.06125253
MUDAC-MUDAB	0.16808991	2.64	0.0147	0.06370664

[a] $\lambda'\beta$ [b] $\lambda'\mathbf{b}$ [c] t [d] *Prob-value* [e] $s_{\lambda'\mathbf{b}} = s\sqrt{\lambda'(\mathbf{X}'\mathbf{X})^{-1}\lambda}$

(b) 95% confidence intervals for $\mu_{[d,a,B]} - \mu_{[d,a,A]} = \beta_5$, $\mu_{[d,a,C]} - \mu_{[d,a,A]} = \beta_6$ and $\mu_{[d,a,C]} - \mu_{[d,a,B]} = \beta_6 - \beta_5$

Parameter $\lambda'\beta$	***95% confidence interval for $\lambda'\beta$: $[\lambda'b \pm 2.069s\sqrt{\lambda'(X'X)^{-1}\lambda}]$***
$\mu_{[d,a,B]} - \mu_{[d,a,A]} = \beta_5$	$[.21368626 \pm 2.069(.06215362)] = [.0851, .3423]$
$\mu_{[d,a,C]} - \mu_{[d,a,A]} = \beta_6$	$[.38177617 \pm 2.069(.06125253)] = [.2550, .5085]$
$\mu_{[d,a,C]} - \mu_{[d,a,B]} = \beta_6 - \beta_5$	$[.16808991 \pm 2.069(.06370664)] = [.0363, .2999]$

versus

$$H_1: \beta_6 - \beta_5 = \mu_{[d,a,C]} - \mu_{[d,a,B]} \neq 0$$

From condition (1) we can test H_0 versus H_1 by setting α, the probability of a Type I error, equal to .05 by calculating

$$\begin{aligned} t &= \frac{\lambda'\mathbf{b}}{s\sqrt{\lambda'(\mathbf{X}'\mathbf{X})^{-1}\lambda}} \\ &= \frac{b_6 - b_5}{s\sqrt{\lambda'(\mathbf{X}'\mathbf{X})^{-1}\lambda}} \\ &= \frac{.1681}{.1308\sqrt{.2372}} \\ &= 2.6389 \end{aligned}$$

We use the rejection points

$$t_{[\alpha/2]}^{(n-k)} = t_{[.05/2]}^{(30-7)} = t_{[.025]}^{(23)} = 2.069 \quad \text{and} \quad -t_{[\alpha/2]}^{(n-k)} = -2.069$$

Since $t = 2.6389 > 2.069 = t_{[.025]}^{(23)}$, we can reject H_0 in favor of H_1 by setting α equal to .05.

Next, we note that the prob-value is twice the area under the curve of the t-distribution having $n - k = 23$ degrees of freedom to the right of $|t| = 2.6389$. The prob-value can be computer calculated to be .0147. Since this prob-value is less than .05, we can reject H_0 in favor of H_1 by setting α equal to .05.

In Figure 12.11(a) we present the SAS output of the least squares point estimates of the parameters β_0, β_1, β_2, β_3, β_4, β_5, β_6, and $\beta_6 - \beta_5$ in the above model. Note that on this output the differences $\mu_{[d,a,\mathrm{B}]} - \mu_{[d,a,\mathrm{A}]} = \beta_5$, $\mu_{[d,a,\mathrm{C}]} - \mu_{[d,a,\mathrm{A}]} = \beta_6$, and $\mu_{[d,a,\mathrm{C}]} - \mu_{[d,a,\mathrm{B}]} = \beta_6 - \beta_5$ are denoted MUDAB − MUDAA, MUDAC − MUDAA, and MUDAC − MUDAB, respectively. Moreover, if we let $\boldsymbol{\lambda}'\boldsymbol{\beta}$ denote any one of the parameters, Figure 12.11(a) also presents the t statistic and related prob-value for testing $H_0: \boldsymbol{\lambda}'\boldsymbol{\beta} = 0$. Although SAS does not calculate confidence intervals for $\boldsymbol{\lambda}'\boldsymbol{\beta}$, SAS does present $\boldsymbol{\lambda}'\mathbf{b}$ and $s_{\boldsymbol{\lambda}'\mathbf{b}} = s\sqrt{\boldsymbol{\lambda}'(\mathbf{X}'\mathbf{X})^{-1}\boldsymbol{\lambda}}$. Thus we can calculate a $100(1 - \alpha)\%$ confidence interval for $\boldsymbol{\lambda}'\boldsymbol{\beta}$ by using the formula

$$\left[\boldsymbol{\lambda}'\mathbf{b} \pm t_{[\alpha/2]}^{(n-k)}\, s\sqrt{\boldsymbol{\lambda}'(\mathbf{X}'\mathbf{X})^{-1}\boldsymbol{\lambda}}\right]$$

In Figure 12.11(b) we use this formula to calculate 95% confidence intervals for β_5, β_6, and $\beta_6 - \beta_5$.

RELATIONSHIPS BETWEEN THE INTERVALS FOR $\boldsymbol{\lambda}'\boldsymbol{\beta}$, β_j, AND μ_0

We wish to show that the previously discussed confidence intervals for β_j and μ_0 are special cases of the confidence interval for $\boldsymbol{\lambda}'\boldsymbol{\beta}$. Note that the single parameter β_j in the regression model

$$\begin{aligned} y = {} & \beta_0 + \beta_1 x_1 + \cdots + \beta_{j-1}x_{j-1} + \beta_j x_j + \beta_{j+1}x_{j+1} \\ & + \cdots + \beta_p x_p + \varepsilon \end{aligned}$$

can be written as

$$\begin{aligned} \beta_j &= 0 \cdot \beta_0 + 0 \cdot \beta_1 + \cdots + 0 \cdot \beta_{j-1} + 1 \cdot \beta_j + 0 \cdot \beta_{j+1} + \cdots + 0 \cdot \beta_p \\ &= [0 \;\; 0 \;\; \cdots \;\; 0 \;\; 1 \;\; 0 \;\; \cdots \;\; 0] \begin{bmatrix} \beta_0 \\ \beta_1 \\ \vdots \\ \beta_{j-1} \\ \beta_j \\ \beta_{j+1} \\ \vdots \\ \beta_p \end{bmatrix} \\ &= \boldsymbol{\lambda}'\boldsymbol{\beta}, \qquad \text{where } \boldsymbol{\lambda}' = [0 \;\; 0 \;\; \cdots \;\; 0 \;\; 1 \;\; 0 \;\; \cdots \;\; 0] \end{aligned}$$

Thus the single parameter β_j can be considered a special case of the linear combination $\boldsymbol{\lambda}'\boldsymbol{\beta}$. This implies that the formula for a $100(1 - \alpha)\%$ confidence interval for β_j,

$$[b_j \pm t_{[\alpha/2]}^{(n-k)} s\sqrt{c_{jj}}]$$

is a special case of the formula for a $100(1 - \alpha)\%$ confidence interval for $\boldsymbol{\lambda}'\boldsymbol{\beta}$,

$$[\boldsymbol{\lambda}'\mathbf{b} \pm t_{[\alpha/2]}^{(n-k)} s\sqrt{\boldsymbol{\lambda}'(\mathbf{X}'\mathbf{X})^{-1}\boldsymbol{\lambda}}]$$

In this case, $\boldsymbol{\lambda}'$ is a row vector containing the number 1 in position $(j + 1)$ and containing zeroes elsewhere. Thus

$$\boldsymbol{\lambda}'\mathbf{b} = b_j \quad \text{and} \quad \boldsymbol{\lambda}'(\mathbf{X}'\mathbf{X})^{-1}\boldsymbol{\lambda} = c_{jj}$$

Next

$$\begin{aligned} \mu_0 &= \beta_0 + \beta_1 x_{01} + \beta_2 x_{02} + \cdots + \beta_p x_{0p} \\ &= [1 \quad x_{01} \quad x_{02} \quad \ldots \quad x_{0p}] \begin{bmatrix} \beta_0 \\ \beta_1 \\ \cdot \\ \cdot \\ \cdot \\ \beta_p \end{bmatrix} \\ &= \boldsymbol{\lambda}'\boldsymbol{\beta}, \quad \text{where } \boldsymbol{\lambda}' = [1 \quad x_{01} \quad x_{02} \quad \ldots \quad x_{0p}] \end{aligned}$$

It follows that μ_0 is also a special case of the linear combination $\boldsymbol{\lambda}'\boldsymbol{\beta}$. This implies that the formula for a $100(1 - \alpha)\%$ confidence interval for μ_0

$$[\hat{y}_0 \pm t_{[\alpha/2]}^{(n-k)} s\sqrt{\mathbf{x}_0'(\mathbf{X}'\mathbf{X})^{-1}\mathbf{x}_0}]$$

where

$$\mathbf{x}_0' = [1 \quad x_{01} \quad x_{02} \quad \ldots \quad x_{0p}]$$

is a special case of the formula for a $100(1 - \alpha)\%$ confidence interval for $\boldsymbol{\lambda}'\boldsymbol{\beta}$. This is because

$$\begin{aligned} \boldsymbol{\lambda}'\mathbf{b} &= [1 \quad x_{01} \quad x_{02} \quad \ldots \quad x_{0p}] \begin{bmatrix} b_0 \\ b_1 \\ b_2 \\ \cdot \\ \cdot \\ \cdot \\ b_p \end{bmatrix} \\ &= b_0 + b_1 x_{01} + b_2 x_{02} + \cdots + b_p x_{0p} \\ &= \hat{y}_0 \end{aligned}$$

and

$$\boldsymbol{\lambda}'(\mathbf{X}'\mathbf{X})^{-1}\boldsymbol{\lambda} = \mathbf{x}_0'(\mathbf{X}'\mathbf{X})^{-1}\mathbf{x}_0$$

12.3

SIMULTANEOUS CONFIDENCE INTERVALS

From now on we sometimes refer to a $100(1 - \alpha)\%$ confidence interval for a linear combination $\boldsymbol{\lambda}'\boldsymbol{\beta}$ computed by using the formula

$$[\boldsymbol{\lambda}'\mathbf{b} \pm t_{[\alpha/2]}^{(n-k)} s\sqrt{\boldsymbol{\lambda}'(\mathbf{X}'\mathbf{X})^{-1}\boldsymbol{\lambda}}]$$

as an *individual* $100(1 - \alpha)\%$ *confidence interval for* $\boldsymbol{\lambda}'\boldsymbol{\beta}$. To clarify what we mean by this, we say that a sample in the population of all possible samples is *individually successful* for $\boldsymbol{\lambda}'\boldsymbol{\beta}$ if a $100(1 - \alpha)\%$ confidence interval for $\boldsymbol{\lambda}'\boldsymbol{\beta}$ computed by using the sample contains the true value of $\boldsymbol{\lambda}'\boldsymbol{\beta}$. When we use the term *individual* $100(1 - \alpha)\%$ *confidence*, we are saying that, if we were to compute individual $100(1 - \alpha)\%$ confidence intervals for $\boldsymbol{\lambda}'\boldsymbol{\beta}$ by using each of the samples in the population of all possible samples, then $100(1 - \alpha)\%$ of these samples would be individually successful for $\boldsymbol{\lambda}'\boldsymbol{\beta}$.

In contrast to *an individual* $100(1 - \alpha)\%$ *confidence interval*, there is *a simultaneous* $100(1 - \alpha)\%$ *confidence interval*. There are several formulas for simultaneous $100(1 - \alpha)\%$ confidence intervals. We say that a sample in the population of all possible samples is *simultaneously successful* for all the linear combinations of regression parameters in a set of linear combinations if all the $100(1 - \alpha)\%$ confidence intervals for the linear combinations in the set computed by using the sample contain their respective linear combinations.

To explain more fully the difference between individual and simultaneous success, we look at a very simple example. Suppose we wish to compute $\frac{2}{3} \times 100\% = 66.67\%$ confidence intervals for two parameters β_1 and β_2. Assume here that (a supernatural power knows that) $\beta_1 = 2$ and $\beta_2 = 5$, and consider three possible samples. The upper part of Table 12.5 lists hypothetical *individual* 66.67% confidence intervals for β_1 and β_2 computed by using these three samples. Here we see that:

1. 66.67 percent (two out of three) of these samples are individually successful for β_1.
2. 66.67 percent (two out of three) of these samples are individually successful for β_2.
3. Only 33.33 percent (one out of three) of these samples are simultaneously successful for β_1 and β_2.

Generalizing, suppose that we were to compute *individual* $100(1 - \alpha)\%$ confidence intervals for each parameter in a set of parameters by applying the individual confidence interval formula to each of the samples in the population of all possible samples. Then $100(1 - \alpha)\%$ of these samples would be individually successful for each parameter in the set, but less than $100(1 - \alpha)\%$ of these samples would be simultaneously successful for all of the parameters in the set. The lower part of

TABLE 12.5 **Individual 66.67% confidence intervals and simultaneous 66.67% confidence intervals for β_1 and β_2**

Sample used to calculate intervals	*Individual 66.67% confidence intervals for*		*Result*
	$\beta_1 = 2$	$\beta_2 = 5$	
First sample	*[1.1, 3.1]	*[3.9, 5.9]	Simultaneously successful for β_1 and β_2
Second sample	*[1.5, 3.5]	[5.4, 7.4]	Not simultaneously successful for β_1 and β_2
Third sample	[2.1, 4.1]	*[4.2, 6.2]	Not simultaneously successful for β_1 and β_2
	$\frac{2}{3} \times 100\% = 66.67\%$ of the samples are individually successful for β_1	$\frac{2}{3} \times 100\% = 66.67\%$ of the samples are individually successful for β_2	$\frac{1}{3} \times 100\% = 33.33\%$ of the samples are simultaneously successful for β_1 and β_2

Sample used to calculate intervals	*Simultaneous 66.67% confidence intervals for*		*Result*
	$\beta_1 = 2$	$\beta_2 = 5$	
First sample	*[.8, 3.4]	*[3.6, 6.2]	Simultaneously successful for β_1 and β_2
Second sample	*[1.2, 3.8]	[5.1, 7.7]	Not simultaneously successful for β_1 and β_2
Third sample	*[1.8, 4.4]	*[3.9, 6.5]	Simultaneouly successful for β_1 and β_2
	$\frac{3}{3} \times 100\% = 100\%$ of the samples are individually successful for β_1	$\frac{2}{3} \times 100\% = 66.67\%$ of the samples are individually successful for β_2	$\frac{2}{3} \times 100\% = 66.67\%$ of the samples are simultaneously successful for β_1 and β_2

Note: An asterisk beside an interval indicates that the interval contains the parameter of interest.

Table 12.5 lists hypothetical *simultaneous* 66.67% confidence intervals for β_1 and β_2 computed by using the three samples. Here we see that:

1. 66.67 percent (two out of three) of these samples are simultaneously successful for β_1 and β_2.
2. 100 percent (all three) of these samples are individually successful for β_1.

3. 66.67 percent (two out of three) of these samples are individually successful for β_2.

Generalizing, suppose that we were to compute simultaneous $100(1 - \alpha)\%$ confidence intervals for all the parameters in a set of parameters by applying a simultaneous $100(1 - \alpha)\%$ confidence interval formula for each of the samples in the population of all possible samples. Then $100(1 - \alpha)\%$ of these samples would be simultaneously successful for all the parameters in the set, and at least $100(1 - \alpha)\%$ of these samples would be individually successful for each parameter in the set.

Comparing the individual and simultaneous intervals in Table 12.5, we see that the simultaneous 66.67% confidence intervals are longer (less precise) than the individual 66.67% confidence intervals. In general, simultaneous $100(1 - \alpha)\%$ confidence intervals are less precise than individual $100(1 - \alpha)\%$ confidence intervals computed for the same linear combinations of parameters. Thus intuitively, we are "paying for" simultaneous confidence by obtaining less precise intervals. Some analysts argue that less precision is too high a price to pay for simultaneous confidence, particularly when the simultaneous confidence intervals are so imprecise that they provide little or no meaningful information. Others are willing to pay this price for simultaneous confidence. We suggest calculating both individual and simultaneous confidence intervals. Conclusions (about, for example, differences in population means) can be drawn by examining both intervals.

We now explain how to calculate *Scheffé simultaneous $100(1 - \alpha)\%$ confidence intervals*. Consider the linear regression model

$$y = \beta_0 + \beta_1 x_1 + \beta_2 x_2 + \cdots + \beta_p x_p + \varepsilon$$

Suppose that we wish to calculate Scheffé simultaneous $100(1 - \alpha)\%$ confidence intervals for all the linear combinations of parameters in a set of linear combinations. We employ q to denote the number of parameters that are *involved nontrivially* in at least one of the linear combinations in the set. Here, any parameter β_j is said to be involved nontrivially in a linear combination.

$$\boldsymbol{\lambda}'\boldsymbol{\beta} = \sum_{j=0}^{p} \lambda_j \beta_j$$

if $\lambda_j \neq 0$. Then, the Scheffé simultaneous $100(1 - \alpha)\%$ confidence interval for any linear combination in the set is as follows.

Scheffé Simultaneous Confidence Intervals

If the inference assumptions are satisfied, the Scheffé simultaneous $100(1 - \alpha)\%$ confidence interval for $\boldsymbol{\lambda}'\boldsymbol{\beta}$ is

$$\left[\boldsymbol{\lambda}'\mathbf{b} \pm \sqrt{qF_{[\alpha]}^{(q,n-k)}}\, s\sqrt{\boldsymbol{\lambda}'(\mathbf{X}'\mathbf{X})^{-1}\boldsymbol{\lambda}}\right]$$

Here:

1. $F_{[\alpha]}^{(q,n-k)}$ is the point on the scale of the F-distribution having q and $(n - k)$ degrees of freedom so that the area under this curve to the right of $F_{[\alpha]}^{(q,n-k)}$ is α.
2. $\mathbf{b}$ is a column vector containing the least squares estimates $b_0, b_1, b_2, \ldots, b_p$.
3. s is the standard error.
4. n is the number of observations.
5. k is the number of parameters in the regression model.
6. q is the number of parameters that are *involved nontrivially* in at least one of the linear combinations for which simultaneous intervals are being computed.

Example 12.7 We now demonstrate how Enterprise Industries can calculate Scheffé simultaneous 95% confidence intervals for all the linear combinations of regression parameters in the following set:

Set I

$$\mu_{[d,a,\mathrm{B}]} - \mu_{[d,a,\mathrm{A}]} = \beta_5$$

$$\mu_{[d,a,\mathrm{C}]} - \mu_{[d,a,\mathrm{A}]} = \beta_6$$

$$\mu_{[d,a,\mathrm{C}]} - \mu_{[d,a,\mathrm{B}]} = \beta_6 - \beta_5$$

There are two parameters, β_5 and β_6, from the model

$$y_i = \beta_0 + \beta_1 x_{i4} + \beta_2 x_{i3} + \beta_3 x_{i3}^2 + \beta_4 x_{i4} x_{i3} + \beta_5 D_{i,\mathrm{B}} + \beta_6 D_{i,\mathrm{C}} + \varepsilon_i$$

involved nontrivially in at least one of the linear combinations in Set I. Therefore $q = 2$, and

$$\begin{aligned} \sqrt{qF_{[\alpha]}^{(q,n-k)}} &= \sqrt{2F_{[.05]}^{(2,30-7)}} \\ &= \sqrt{2F_{[.05]}^{(2,23)}} \\ &= \sqrt{2(3.42)} \\ &= 2.6153 \end{aligned}$$

Thus

$$\left[\boldsymbol{\lambda}'\mathbf{b} \pm \sqrt{qF_{[\alpha]}^{(q,n-k)}}\, s\sqrt{\boldsymbol{\lambda}'(\mathbf{X}'\mathbf{X})^{-1}\boldsymbol{\lambda}}\right] = \left[\boldsymbol{\lambda}'\mathbf{b} \pm (2.6153)\, s\sqrt{\boldsymbol{\lambda}'(\mathbf{X}'\mathbf{X})^{-1}\boldsymbol{\lambda}}\right]$$

is a Scheffé simultaneous 95% confidence interval for any linear combination $\boldsymbol{\lambda}'\boldsymbol{\beta}$ in Set I. Note that this simultaneous interval is longer than

$$\left[\boldsymbol{\lambda}'\mathbf{b} \pm t_{[\alpha/2]}^{(n-k)}\, s\sqrt{\boldsymbol{\lambda}'(\mathbf{X}'\mathbf{X})^{-1}\boldsymbol{\lambda}}\right] = \left[\boldsymbol{\lambda}'\mathbf{b} \pm (2.069)\, s\sqrt{\boldsymbol{\lambda}'(\mathbf{X}'\mathbf{X})^{-1}\boldsymbol{\lambda}}\right]$$

which is the individual 95% confidence interval for $\boldsymbol{\lambda}'\boldsymbol{\beta}$. Therefore we are "paying for" simultaneous confidence by obtaining less precise intervals.

TABLE 12.6 Scheffé simultaneous 95% confidence intervals

Linear combination, $\lambda'\beta$	*Point estimate,* $\lambda' b$	$s\sqrt{\lambda'(X'X)^{-1}\lambda}$	*Individual 95% confidence interval for* $\lambda'\beta$, $[\lambda' b \pm (2.069) \times s\sqrt{\lambda'(X'X)^{-1}\lambda}]$	*Scheffé simultaneous 95% confidence interval for* $\lambda'\beta$ *(in Set I),* $[\lambda' b \pm (2.6153) \times s\sqrt{\lambda'(X'X)^{-1}\lambda}]$	*Scheffé simultaneous 95% confidence interval for* $\lambda'\beta$ *(in Set II),* $[\lambda' b \pm (4.1328) \times s\sqrt{\lambda'(X'X)^{-1}\lambda}]$
$\mu_{[d,a,B]} - \mu_{[d,a,A]} = \beta_5$	$b_5 = .2137$	.0622	$[.2137 \pm 2.069(.0622)] = [.0851, .3423]$	$[.2137 \pm 2.6153(.0622)] = [.051, .3764]$	$[.2137 \pm 4.1328(.0622)] = [-.0434, .4708]$
$\mu_{[d,a,C]} - \mu_{[d,a,A]} = \beta_6$	$b_6 = .3818$	.0613	$[.3818 \pm 2.069(.0613)] = [.2550, .5085]$	$[.3818 \pm 2.6153(.0613)] = [.2215, .5421]$	$[.3818 \pm 4.1328(.0613)] = [.1285, .6351]$
$\mu_{[d,a,C]} - \mu_{[d,a,B]} = \beta_6 - \beta_5$	$b_6 - b_5 = .1681$	.0637	$[.1681 \pm 2.069(.0637)] = [.0363, .2999]$	$[.1681 \pm 2.6153(.0637)] = [-.0048, .3284]$	$[.1681 \pm 4.1328(.0637)] = [-.1015, .4251]$
$\mu_0 = \beta_0 + \beta_1(.10) + \beta_2(6.80) + \beta_3(6.80)^2 + \beta_4(.10)(6.80) + \beta_5(0) + \beta_6(1)$	$b_0 + b_1(.10) + b_2(6.80) + b_3(6.80)^2 + b_4(.10)(6.80) + b_5(0) + b_6(1) = 8.6825$	.0597	$[8.6825 \pm 2.069(.0597)] = [8.559, 8.806]$		$[8.6825 \pm 4.1328(.0597)] = [8.4358, 8.9292]$

In Table 12.6 we calculate both individual 95% confidence intervals and Scheffé simultaneous 95% confidence intervals for the three linear combinations in Set I. We are 95 percent confident that each of the individual 95% confidence intervals contains the linear combination it is meant to contain. However, we are less than 95 percent confident that all three of the individual 95% confidence intervals contain the three linear combinations that they are meant to contain. In contrast, we are 95 percent confident that all three of the longer Scheffé simultaneous 95% confidence intervals contain the three linear combinations that they are meant to contain. To illustrate that the simultaneous intervals are less precise than the individual intervals, note that the individual 95% confidence interval for $\beta_6 - \beta_5$ makes us 95% (individually) confident that $\mu_{[d,a,C]} - \mu_{[d,a,B]}$ does not equal zero. However, the simultaneous 95% confidence interval for $\beta_6 - \beta_5$ does not make us 95 percent (simultaneously) confident that $\mu_{[d,a,C]} - \mu_{[d,a,B]}$ does not equal zero.

We next demonstrate how Enterprise Industries can calculate Scheffé simultaneous 95% confidence intervals for all the linear combinations of regression parameters in the following set:

Set II

$$
\begin{aligned}
\mu_{[d,a,B]} - \mu_{[d,a,A]} &= \beta_5 \\
\mu_{[d,a,C]} - \mu_{[d,a,A]} &= \beta_6 \\
\mu_{[d,a,C]} - \mu_{[d,a,B]} &= \beta_6 - \beta_5 \\
\mu_0 &= \beta_0 + \beta_1 x_{04} + \beta_2 x_{03} + \beta_3 x_{03}^2 + \beta_4 x_{04} x_{03} + \beta_5 D_{0,B} \\
&\quad + \beta_6 D_{0,C} \\
&= \beta_0 + \beta_1(.10) + \beta_2(6.80) + \beta_3(6.80)^2 + \beta_4(.10)(6.80) \\
&\quad + \beta_5(0) + \beta_6(1)
\end{aligned}
$$

All seven parameters from the above model are involved nontrivially in at least one of the linear combinations in Set II. It follows that $q = 7$ and that

$$
\begin{aligned}
\sqrt{qF_{[\alpha]}^{(q,n-k)}} &= \sqrt{7F_{[.05]}^{(7,30-7)}} \\
&= \sqrt{7F_{[.05]}^{(7,23)}} \\
&= \sqrt{7(2.44)} \\
&= 4.1328
\end{aligned}
$$

Thus

$$
\left[\boldsymbol{\lambda}'\mathbf{b} \pm \sqrt{qF_{[\alpha]}^{(q,n-k)}}\, s\sqrt{\boldsymbol{\lambda}'(\mathbf{X}'\mathbf{X})^{-1}\boldsymbol{\lambda}}\right] = \left[\boldsymbol{\lambda}'\mathbf{b} \pm (4.1328)\, s\sqrt{\boldsymbol{\lambda}'(\mathbf{X}'\mathbf{X})^{-1}\boldsymbol{\lambda}}\right]
$$

is a Scheffé simultaneous 95% confidence interval for any linear combination $\boldsymbol{\lambda}'\boldsymbol{\beta}$ in Set II. Since $q = 7$ for Set II, while $q = 2$ for Set I, this simultaneous 95% interval is longer than the corresponding Scheffé simultaneous 95% interval for $\boldsymbol{\lambda}'\boldsymbol{\beta}$ in Set I. In Table 12.6 we calculate the Scheffé simultaneous 95% confidence intervals for the linear combinations in Set II.

In using the Scheffé formula, suppose that we prespecify (before the data is observed) a particular set of linear combinations for which we wish to calculate simultaneous confidence intervals. Then it is permissible to set q equal to a value that is less than the total number of parameters in the model. For example, suppose that we know (before we observe the Fresh Detergent data) that we wish to find simultaneous confidence intervals for the linear combinations β_5, β_6, and $\beta_6 - \beta_5$ and only for these linear combinations. Then it is permissible to let $q = 2$, which is less than 7, the total number of parameters in the Fresh Detergent model. However, sometimes we do not prespecify (before we observe the data) which linear combinations will be investigated. Instead, we sometimes let the data suggest the linear combinations we will investigate. This is called *data snooping*. When we data snoop, we should set q equal to the largest possible value that might be suggested by the data. Furthermore, we should use this "largest possible value of q" even if, after we observe the data, the data suggests that we investigate a set of linear combinations for which q would be smaller. For example, in some problems the data could potentially suggest a set of linear combinations involving each and every regression parameter. In such a case we should set q equal to the total number of parameters in the model.

In addition to Scheffé intervals there are other types of simultaneous confidence intervals. Another useful simultaneous confidence interval is called a *Bonferroni interval*, which is computed as follows.

Bonferroni Simultaneous Confidence Intervals

If the inference assumptions are satisfied, the *Bonferroni simultaneous 100(1 − α)% confidence interval* for $\boldsymbol{\lambda}'\boldsymbol{\beta}$ in a *prespecified set* of g linear combinations of the regression parameters $(\beta_0, \beta_1, \ldots, \beta_p)$ is

$$\left[\boldsymbol{\lambda}'\mathbf{b} \pm t_{[\alpha/2g]}^{(n-k)}\, s\sqrt{\boldsymbol{\lambda}'(\mathbf{X}'\mathbf{X})^{-1}\boldsymbol{\lambda}}\right]$$

The Bonferroni and Scheffé formulas are alike except that the Bonferroni formula employs $t_{[\alpha/2g]}^{(n-k)}$ while the Scheffé formula employs $\sqrt{qF_{[\alpha]}^{(q,n-k)}}$. It follows that the Bonferroni formula gives more precise simultaneous intervals than the Scheffé formula for a prespecified set of linear combinations if and only if $t_{[\alpha/2g]}^{(n-k)}$ is less than $\sqrt{qF_{[\alpha]}^{(q,n-k)}}$.

Example 12.8 In the Fresh Detergent problem, consider finding simultaneous 95% confidence intervals for the linear combinations in the following prespecified set:

Set I: $\beta_5, \beta_6, \beta_6 - \beta_5$

Suppose that we wish to compute Bonferroni intervals. Then, since there are $g = 3$ linear combinations in Set I, we use

$$t_{[\alpha/2g]}^{(n-k)} = t_{[.05/2(3)]}^{(30-7)} = t_{[.0083]}^{(23)}$$

Although this t-point is not given in Table E.2 in Appendix E, this table indicates that $t_{[.0083]}^{(23)}$ is between $t_{[.01]}^{(23)} = 2.5$ and $t_{[.0075]}^{(23)} = 2.629$. To determine an approximate value for $t_{[.0083]}^{(23)}$, we could linearly interpolate between these known t-points. Alternatively, to be conservative (calculate intervals that are at least as long as would be given by the true value of $t_{[.0083]}^{(23)}$), we could use $t_{[.0075]}^{(23)} = 2.629$. If we wish to compute Scheffé intervals, there are $q = 2$ parameters (β_5 and β_6) involved in the three linear combinations in Set I. Thus we use

$$\begin{aligned}\sqrt{qF_{[\alpha]}^{(q,n-k)}} &= \sqrt{2F_{[.05]}^{(2,30-7)}} \\ &= 2.6153\end{aligned}$$

Therefore for this set of linear combinations the Bonferroni and Scheffé intervals are of roughly the same precision.

Next, suppose that we wish to find simultaneous 95% confidence intervals for the linear combinations in the following prespecified set:

$$\text{Set II:} \quad \beta_5,\ \beta_6,\ \beta_6 - \beta_5$$

$$\mu_0 = \beta_0 + \beta_1(.10) + \beta_2(6.8) + \beta_3(6.8)^2 + \beta_4(.10)(6.8) + \beta_6$$

If we wish to compute Bonferroni intervals, there are $g = 4$ linear combinations in Set II. Thus we use

$$t_{[\alpha/2g]}^{(n-k)} = t_{[.05/2(4)]}^{(30-7)} = t_{[.00625]}^{(23)}$$

Table E.2 in Appendix E indicates that this t-point is between $t_{[.0075]}^{(23)} = 2.629$ and $t_{[.005]}^{(23)} = 2.807$. If we wish to compute Scheffé intervals, there are $q = 7$ parameters (β_0, β_1, β_2, β_3, β_4, β_5, and β_6) involved in the linear combinations in Set II. Thus we use

$$\sqrt{qF_{[\alpha]}^{(q,n-k)}} = \sqrt{7F_{[.05]}^{(7,30-7)}} = 4.1328$$

Therefore the Bonferroni intervals are substantially more precise than the Scheffé intervals. For example, if we use the information in Table 12.6 and use (to be conservative) $t_{[.005]}^{(23)} = 2.807$ in place of $t_{[.00625]}^{(23)}$, it follows that a Bonferroni simultaneous 95% confidence interval for

$$\mu_{[d,a,\mathrm{C}]} - \mu_{[d,a,\mathrm{B}]} = \beta_6 - \beta_5$$

is

$$\begin{aligned}\left[b_6 - b_5 \pm t_{[.00625]}^{(23)} s\sqrt{\boldsymbol{\lambda}'(\mathbf{X}'\mathbf{X})^{-1}\boldsymbol{\lambda}}\right] &= [.1618 \pm 2.807(.0637)] \\ &= [.1618 \pm .1788] \\ &= [-.017, .3406]\end{aligned}$$

This interval is more precise than the Scheffé simultaneous 95% confidence interval for $\beta_6 - \beta_5$, which is $[-.1015, .4251]$ (see Table 12.6).

As a rule of thumb, the Bonferroni formula generally yields more precise simultaneous intervals than the Scheffé formula (for a given set of linear combinations) if g, the number of linear combinations in the set, is substantially less than q, the number of parameters involved in the g linear combinations. In any given problem we can calculate both the Bonferroni and Scheffé intervals and choose the intervals that are the most precise.

Finally, we emphasize that Bonferroni intervals require that we prespecify a set of linear combinations. Therefore we cannot use Bonferroni intervals to estimate linear combinations suggested by the observed data (unless these linear combinations were prespecified). On the other hand, we can use Scheffé intervals when data snooping. Here we must set q equal to the largest possible value that might be suggested by the data. Specifically, if we set q equal to the total number of parameters in the model, we can investigate any or all linear combinations of parameters that might be suggested by the data.

*12.4 USING SAS

In Figure 12.12 we present the SAS program that produces the Fresh Detergent dummy variable model output in Figure 12.11. The ESTIMATE statement gives $\boldsymbol{\lambda}'\mathbf{b}$, $s_{\boldsymbol{\lambda}'\mathbf{b}} = s\sqrt{\boldsymbol{\lambda}'(\mathbf{X}'\mathbf{X})^{-1}\boldsymbol{\lambda}}$, t, and the related prob-value for each specified linear combination $\boldsymbol{\lambda}'\boldsymbol{\beta}$. Expressions in single quotes are labels (which may be up to 16 characters in length). The linear combinations β_5, β_6, and $\beta_6 - \beta_5$ are specified as follows:

1. Since $\mu_{[d,a,\mathrm{B}]} - \mu_{[d,a,\mathrm{A}]} = \beta_5$, we estimate this difference between means by specifying the linear combination $\boldsymbol{\lambda}'\boldsymbol{\beta} = \beta_5$. Here, DB 1 specifies the coefficient (1) multiplied by β_5 (the parameter that corresponds to DB) in $\boldsymbol{\lambda}'\boldsymbol{\beta} = \beta_5$. Coefficients for parameters other than β_5 are assumed to be zero.
2. Since $\mu_{[d,a,\mathrm{C}]} - \mu_{[d,a,\mathrm{A}]} = \beta_6$, we estimate this difference between means by specifying the linear combination $\boldsymbol{\lambda}'\boldsymbol{\beta} = \beta_6$. Here, DC 1 specifies the coefficient (1) multiplied by β_6 (the parameter that corresponds to DC) in $\boldsymbol{\lambda}'\boldsymbol{\beta} = \beta_6$. Coefficients for parameters other than β_6 are assumed to be zero.
3. Since $\mu_{[d,a,\mathrm{C}]} - \mu_{[d,a,\mathrm{B}]} = \beta_6 - \beta_5$, we estimate this difference between means by specifying the linear combination $\boldsymbol{\lambda}'\boldsymbol{\beta} = \beta_6 - \beta_5$. Here, DB $-$ 1 specifies the coefficient (-1) multiplied by β_5 (the parameter that corresponds to DB) in the expression $\beta_6 - \beta_5$, and DC 1 specifies the coefficient (1) multiplied by β_6 (the parameter that corresponds to DC) in the expression $\beta_6 - \beta_5$. Coefficients for parameters other than β_5 (and β_6 are assumed to be zero.

*This section is optional.

FIGURE 12.12 **SAS program to produce the Fresh Detergent dummy variable model output in Figure 12.11**

```
DATA DETR;
INPUT Y X4 X3 DB DC; }
X3SQ = X3*X3; }
X43 = X4*X3;  }
CARDS;
7.38  -0.05  5.50  1  0
8.51   0.25  6.75  1  0
9.52   0.60  7.25  1  0
7.50   0.00  5.50  0  0
  .
  .
  .
9.26   0.55  6.80  0  1

 .     0.10  6.80  0  1 }

PROC GLM DATA = DETR; }
MODEL Y = X4 X3 X3SQ X43 DB DC/CLI;

ESTIMATE 'MUDAB-MUDAA' DB 1; }
ESTIMATE 'MUDAC-MUDAA' DC 1; }
ESTIMATE 'MUDAC-MUDAB' DB -1 DC 1; }
```

Defines Y = demand, X4 = price difference, X3 = advertising expenditure. DB and DC are names assigned to the dummy variables for advertising campaigns B and C

→ *Transformations defining x_3^2 and x_4x_3*

→ *Dummy variable input*

→ *Data (see Tables 8.3 and 12.2) including dummy variables*

Generates prediction when $x_{04} = .10$, $x_{03} = 6.80$, and advertising campaign C is used

→ *General linear models procedure (PROC GLM is needed for "estimation")*

Specifies model $y_i = \beta_0 + \beta_1 x_{i4} + \beta_2 x_{i3} + \beta_3 x_{i3}^2 + \beta_4 x_{i4} x_{i3} + \beta_5 D_{i,B} + \beta_6 D_{i,C} + \varepsilon_i$

→ *Estimates $\mu_{[d,a,B]} - \mu_{[d,a,A]} = \beta_5$*

→ *Estimates $\mu_{[d,a,C]} - \mu_{[d,a,A]} = \beta_6$*

→ *Estimates $\mu_{[d,a,C]} - \mu_{[d,a,B]} = \beta_6 - \beta_5$*

MUDAB − MUDAA, MUDAC − MUDAA, and MUDAC − MUDAB are labels denoting the differences in 1, 2, and 3 above, respectively. Note that the ESTIMATE statement can be used only with PROC GLM.

EXERCISES

Exercises 12.1–12.6 refer to the following situation. Panasound, Inc. wishes to use the regression model

$$\begin{aligned} y_i &= \mu_i + \varepsilon_i \\ &= \beta_0 + \beta_1 x_i + \beta_2 x_i^2 + \beta_3 D_{i,B} + \beta_4 D_{i,C} + \varepsilon_i \end{aligned}$$

to relate the dependent variable

y_i = sales of the Panasound Video Recorder (in thousands of units) in sales region i

to the independent variables

x_i = advertising expenditure (in hundreds of thousands of dollars) in sales region i

and the advertising agency (A, B, or C) that handles the Panasound account in sales region i. To model the effect of the advertising agencies, we use dummy variables defined as follows:

$$D_{i,B} = \begin{cases} 1 & \text{if advertising agency B handles the Panasound account in sales region } i \\ 0 & \text{otherwise} \end{cases}$$

$$D_{i,C} = \begin{cases} 1 & \text{if advertising agency C handles the Panasound account in sales region } i \\ 0 & \text{otherwise} \end{cases}$$

By using data observed on these variables for $n = 25$ sales regions of equal sales potential, the following calculations are made.

$$(\mathbf{X'X})^{-1} = \begin{bmatrix} .02 & 0 & 0 & 0 & 0 \\ 0 & .01 & 0 & 0 & 0 \\ 0 & 0 & .05 & 0 & 0 \\ 0 & 0 & 0 & .05 & 0 \\ 0 & 0 & 0 & 0 & .10 \end{bmatrix} \quad \text{and} \quad \mathbf{X'y} = \begin{bmatrix} 1000 \\ 300 \\ 50 \\ 60 \\ 50 \end{bmatrix}$$

$$\mathbf{b} = (\mathbf{X'X})^{-1}\mathbf{X'y} = \begin{bmatrix} b_0 \\ b_1 \\ b_2 \\ b_3 \\ b_4 \end{bmatrix} = \begin{bmatrix} 20 \\ 3 \\ 2.5 \\ 3 \\ 5 \end{bmatrix}$$

$$SSE = \sum_{i=1}^{25} y_i^2 - \mathbf{b'X'y} = 21{,}500 - 21{,}455 = 45$$

$$s = \sqrt{\frac{SSE}{n-k}} = \sqrt{\frac{45}{25-5}} = 1.5$$

$$\text{Total variation} = \sum_{i=1}^{25} (y_i - \bar{y})^2 = 500$$

12.1 Consider the Panasound, Inc. situation. For the Panasound dummy variable model,

a. Calculate R^2 and interpret its value.
b. Set up the appropriate null and alternative hypotheses needed to carry out an overall F-test for the model.
c. Compute F(model), the overall F statistic.
d. By comparing F(model) with the appropriate rejection point, decide whether or not the null hypothesis you set up in part (b) can be rejected. Interpret the results of this hypothesis test. Use $\alpha = .05$.

12.2 Consider the Panasound, Inc. situation.

a. Show that the Panasound dummy variable model

$$\begin{aligned} y_i &= \mu_i + \varepsilon_i \\ &= \beta_0 + \beta_1 x_i + \beta_2 x_i^2 + \beta_3 D_{i,\mathrm{B}} + \beta_4 D_{i,\mathrm{C}} + \varepsilon_i \end{aligned}$$

implies that

$$\begin{aligned} \mu_{[a,\mathrm{C}]} - \mu_{[a,\mathrm{A}]} &= \beta_4 \\ \mu_{[a,\mathrm{B}]} - \mu_{[a,\mathrm{A}]} &= \beta_3 \\ \mu_{[a,\mathrm{C}]} - \mu_{[a,\mathrm{B}]} &= \beta_4 - \beta_3 \end{aligned}$$

where

$\mu_{[a,\mathrm{A}]}$ = mean sales when advertising expenditure is a and agency A handles the Panasound account

$\mu_{[a,\mathrm{C}]}$ = mean sales when advertising expenditure is a and agency C handles the Panasound account

$\mu_{[a,\mathrm{B}]}$ = mean sales when advertising expenditure is a and agency B handles the Panasound account

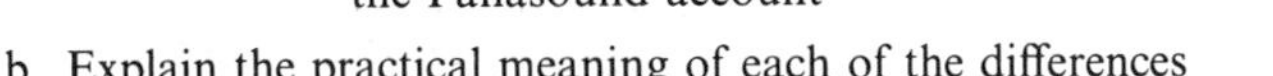

b. Explain the practical meaning of each of the differences

$$\begin{aligned} \mu_{[a,\mathrm{C}]} - \mu_{[a,\mathrm{A}]} &= \beta_4 \\ \mu_{[a,\mathrm{B}]} - \mu_{[a,\mathrm{A}]} &= \beta_3 \\ \mu_{[a,\mathrm{C}]} - \mu_{[a,\mathrm{B}]} &= \beta_4 - \beta_3 \end{aligned}$$

12.3 Consider the Panasound, Inc. situation.

a. Set up the appropriate null and alternative hypotheses required to employ a partial F-test that will test to see whether or not there are any differences between the effects of advertising agencies A, B, and C.
b. Specify the appropriate complete and reduced models for the partial F-test needed to test the hypotheses you set up in part (a).
c. For the appropriate reduced model it can be shown that

$$(\mathbf{X}'\mathbf{X})^{-1} = \begin{bmatrix} .02 & 0 & 0 \\ 0 & .01 & 0 \\ 0 & 0 & .05 \end{bmatrix} \qquad \mathbf{X}'\mathbf{y} = \begin{bmatrix} 1000 \\ 300 \\ 50 \end{bmatrix}$$

and

$$\sum_{i=1}^{25} y_i^2 = 21{,}500$$

Use this information to calculate SSE_R, the unexplained variation for the reduced model.
d. Calculate the partial F statistic needed to test the hypotheses you set up in part (a).
e. By comparing the partial F statistic computed in part (d) with the appropriate rejection point, decide whether or not the null hypothesis you set up in part (a) can be rejected with the probability of a Type I error α set equal to .05. What is the practical meaning of the result of this test?

12.4 Consider the Panasound, Inc. situation.

a. Using the appropriate t statistic and rejection point, test $H_0: \beta_3 = 0$ versus $H_1: \beta_3 \neq 0$ by setting $\alpha = .05$. What does the result of this test say about the difference between $\mu_{[a,B]}$ and $\mu_{[a,A]}$?
b. Using the appropriate t statistic and rejection point, test $H_0: \beta_4 = 0$ versus $H_1: \beta_4 \neq 0$ by setting $\alpha = .05$. What does the result of this test say about the difference between $\mu_{[a,C]}$ and $\mu_{[a,A]}$?
c. Calculate a 95% confidence interval for β_3. Interpret what this interval says about the effect of changing from agency A to agency B.
d. Calculate a 95% confidence interval for β_4. Interpret what this interval says about the effect of changing from agency A to agency C.

12.5 Consider the Panasound, Inc. situation.

a. Suppose that we wish to express $\beta_4 - \beta_3$ as a linear combination $\boldsymbol{\lambda}'\boldsymbol{\beta}$, where $\boldsymbol{\beta}' = [\beta_0\ \beta_1\ \beta_2\ \beta_3\ \beta_4]$. Specify the row vector $\boldsymbol{\lambda}'$.
b. Using $(\mathbf{X}'\mathbf{X})^{-1}$ for the Panasound dummy variable model and the vector $\boldsymbol{\lambda}'$ specified in part (a), compute $\boldsymbol{\lambda}'(\mathbf{X}'\mathbf{X})^{-1}\boldsymbol{\lambda}$.
c. Set up the null and alternative hypotheses needed to test to see whether or not advertising agencies B and C have different effects on sales of the Panasound Video Recorder.
d. Calculate the t statistic needed to test the hypotheses you set up in part (c).
e. By comparing the t statistic you computed in part (d) with the appropriate rejection point, decide whether or not the null hypotheses you set up in part (c) can be rejected. Can we conclude that the effects of advertising agencies B and C differ? Use $\alpha = .05$.
f. Calculate a 95% confidence interval for $\beta_4 - \beta_3$. Interpret what this interval says about the effect of changing from agency B to agency C.

12.6 Consider the Panasound, Inc. situation.

a. Calculate a point prediction of sales of the Panasound Video Recorder when advertising expenditure will be \$500,000 (that is, $x_0 = 5$) and when advertising agency C will handle the Panasound account.
b. Calculate a 95% confidence interval for mean sales of the Panasound Video Recorder when advertising expenditure will be \$500,000 and advertising agency C will handle the Panasound account. Interpret this interval.
c. Calculate a 95% prediction interval for individual sales of the Panasound Video Recorder when advertising expenditure will be \$500,000 and advertising agency C will handle the Panasound account. Interpret this interval.

12.7 Consider the model

$$\begin{aligned} y_i &= \mu_i + \varepsilon_i \\ &= \beta_0 + \beta_1 x_i + \beta_2 x_i^2 + \beta_3 D_{i,B} + \beta_4 D_{i,C} + \varepsilon_i \end{aligned}$$

where

y_i = sales in sales region i
x_i = advertising expenditure in sales region i

$$D_{i,B} = \begin{cases} 1 & \text{if we use advertising campaign B in sales region } i \\ 0 & \text{otherwise} \end{cases}$$

$$D_{i,C} = \begin{cases} 1 & \text{if we use advertising campaign C in sales region } i \\ 0 & \text{otherwise} \end{cases}$$

Using data observed on these variables in $n = 25$ sales regions of equal sales potential, we obtain the following results:

$$(\mathbf{X}'\mathbf{X})^{-1} = \begin{bmatrix} .02 & 0 & 0 & 0 & 0 \\ 0 & .01 & 0 & 0 & 0 \\ 0 & 0 & .05 & 0 & 0 \\ 0 & 0 & 0 & .04 & 0 \\ 0 & 0 & 0 & 0 & .08 \end{bmatrix} \qquad \mathbf{X}'\mathbf{y} = \begin{bmatrix} 1000 \\ 300 \\ 50 \\ 80 \\ 25 \end{bmatrix}$$

and

$$\sum_{i=1}^{25} y_i^2 = 21{,}371$$

a. Calculate the least squares estimates b_0, b_1, b_2, b_3, and b_4.
b. Calculate SSE and s.
c. Write $\beta_3 - \beta_4$ as a linear combination $\boldsymbol{\lambda}'\boldsymbol{\beta}$.
d. Calculate a point estimate of $\beta_3 - \beta_4$.
e. Calculate a 95% confidence interval for $\beta_3 - \beta_4$.
f. Consider testing the null hypothesis

$$H_0: \beta_3 - \beta_4 = 0$$

versus the alternative hypothesis

$$H_1: \beta_3 - \beta_4 \neq 0$$

Calculate the t statistic needed to test these hypotheses.
g. If we set $\alpha = .05$, can the null hypothesis of part (f) be rejected? Explain your answer by using an appropriate rejection point.

12.8 Consider the sales model of Exercise 12.7. Using this model:

a. Calculate Scheffé simultaneous 95% confidence intervals for each of the parameters in the following set: β_3, $\beta_3 - \beta_4$, and $\beta_1 + \beta_2 + \beta_3$. *Hint*: Write each of the parameters above as a linear combination $\boldsymbol{\lambda}'\boldsymbol{\beta}$.
b. Compare the individual 95% confidence interval for $\beta_3 - \beta_4$ you calculated in Exercise 12.7 with the Scheffé simultaneous 95% confidence interval for $\beta_3 - \beta_4$ you calculated in part (a). Which interval is more precise (that is, which is shorter)?

12.9 Consider the sales model of Exercise 12.7. Using this model:

a. Calculate Bonferroni simultaneous 95% confidence intervals for each of the following prespecified linear combinations: β_3, $\beta_3 - \beta_4$, and $\beta_1 + \beta_2 + \beta_3$.
b. Compare the Scheffé simultaneous 95% confidence intervals you calculated in Exercise 12.8 with the Bonferroni 95% confidence intervals you calculated in part (a). Which intervals are more precise (that is, which are shorter)?

12.10 Consider the Panasound, Inc. problem of Exercise 12.1 and the set of regression parameters

$$\beta_3 = \mu_{[a,B]} - \mu_{[a,A]}$$
$$\beta_4 = \mu_{[a,C]} - \mu_{[a,A]}$$
$$\beta_4 - \beta_3 = \mu_{[a,C]} - \mu_{[a,B]}$$

a. Compute Scheffé simultaneous 95% confidence intervals for each of β_3, β_4, and $\beta_4 - \beta_3$.

b. What do the Scheffé simultaneous 95% confidence intervals computed in part (a) say about the effects of
(1) Changing from advertising agency A to advertising agency B?
(2) Changing from advertising agency A to advertising agency C?
(3) Changing from advertising agency B to advertising agency C?

c. Consider

$$\mu_0 = \beta_0 + \beta_1(2) + \beta_2(2)^2 + \beta_3(1) + \beta_4(0)$$

which is the mean sales of the Panasound Video Recorder when Panasound, Inc. spends \$200,000 on advertising and when advertising agency B handles the Panasound account. Calculate Scheffé simultaneous 95% confidence intervals for each parameter in the set β_3, β_4, $\beta_4 - \beta_3$, and

$$\mu_0 = \beta_0 + \beta_1(2) + \beta_2(2)^2 + \beta_3(1) + \beta_4(0)$$

12.11 Consider the Pansound, Inc. problem of Exercise 12.1.

a. Compute Bonferroni simultaneous 95% confidence intervals for each of the (prespecified) linear combinations β_3, β_4, and $\beta_4 - \beta_3$. Compare these intervals to the Scheffé simultaneous confidence intervals you computed in part (a) of Exercise 12.10. Which intervals are more precise?

b. Compute Bonferroni simultaneous 95% confidence intervals for each of the (prespecified) linear combinations β_3, β_4, $\beta_4 - \beta_3$, and

$$\mu_0 = \beta_0 + \beta_1(2) + \beta_2(2)^2 + \beta_3(1) + \beta_4(0)$$

Compare these intervals to the Scheffé simultaneous confidence intervals you computed in part (c) of Exercise 12.10. Which intervals are more precise?

Exercises 12.12–12.19 refer to the following situation: International Oil, Inc. is attempting to develop a reasonably priced regular unleaded gasoline that will deliver higher gasoline mileages than can be achieved by its current regular unleaded gasoline. As part of the development process, the company wishes to study the effect of one qualitative independent variable—x_1, regular unleaded gasoline type (A, B, or C)—and of one quantitative independent variable—x_2, amount of gasoline additive VST (1, 2, 3, or 4 units)—on the gasoline mileage (y) obtained by an automobile called the GT-1100. For testing purposes a sample of $n = 22$ GT-1100s is randomly selected and driven under normal driving conditions.

The combinations of x_1 and x_2 used in the experiment, along with the corresponding observed values of y, are given in Table 12.7.

12.12 Consider the International Oil regular unleaded gasoline mileage data in Table 12.7.

a. Plot y against x_1 (= A, B, C).

b. Plot y against x_1 when $x_2 = 1$.

c. Plot y against x_1 when $x_2 = 2$.

TABLE 12.7 International oil regular unleaded gasoline mileage data

Gasoline mileage, y (mpg)	Regular unleaded gasoline type, x_1	Amount of gasoline additive VST, x_2
29.4	A	1
30.6	A	1
30.0	A	1
35.0	A	2
34.0	A	2
34.3	A	3
35.5	A	3
32.6	B	1
31.6	B	1
35.3	B	2
36.5	B	2
37.6	B	3
36.4	B	3
37.0	B	3
35.4	B	4
28.6	C	1
29.8	C	1
34.0	C	3
34.7	C	3
33.3	C	3
32.0	C	4
33.0	C	4

d. Plot y against x_1 when $x_2 = 3$.
e. Plot y against x_1 when $x_2 = 4$.
f. Combine the plots you made in parts (b), (c), (d), and (e) by drawing these plots on the same set of axes. Use a different color for each level of x_2 ($=1$, 2, 3, and 4). What do these plots say about whether interaction exists between x_1 and x_2?
g. Plot y against x_2.
h. Plot y against x_2 when $x_1 = $ A.
i. Plot y against x_2 when $x_1 = $ B.
j. Plot y against x_2 when $x_1 = $ C.
k. Combine the plots you made in parts (h), (i), and (j) by drawing these plots on the same set of axes. Use a different color for each level of x_1 ($=$A, B, and C). What do these plots say about whether interaction exists between x_1 and x_2?

12.13 Consider the International Oil regular unleaded gasoline mileage situation and the data in Table 12.7.

a. Define appropriate dummy variables that will enable us to compare the effects of gasoline types B and C on mean gasoline mileage to the effect of gasoline type A. Denote these dummy variables as $D_{i,B}$ and $D_{i,C}$.
b. Explain why the graphical analysis of parts (a)–(k) in Exercise 12.12 suggests that an appropriate model relating y to x_1 and x_2 is

$$\begin{aligned} y_i &= \mu_i + \varepsilon_i \\ &= \beta_0 + \beta_1 x_{i2} + \beta_2 x_{i2}^2 + \beta_3 D_{i,B} + \beta_4 D_{i,C} + \varepsilon_i \end{aligned}$$

c. Specify the vector **y** and matrix **X** that should be used to calculate the least squares point estimates of the parameters in the above model.

12.14 Figure 12.13 gives the SAS output that is obtained when each of the models

$$y_i = \beta_0 + \beta_1 x_{i2} + \beta_2 x_{i2}^2 + \beta_3 D_{i,B} + \beta_4 D_{i,C} + \varepsilon_i$$
$$y_i = \beta_0 + \beta_1 x_{i2} + \beta_2 x_{i2}^2 + \beta_3 D_{i,B} + \beta_4 D_{i,C} + \beta_5 x_{i2} D_{i,B} + \beta_6 x_{i2} D_{i,C} + \varepsilon_i$$

is employed to analyze the gasoline mileage data in Table 12.7.

FIGURE 12.13 **SAS output of regression analysis of the regular unleaded gasoline mileage data using two models**

(a) $y_i = \beta_0 + \beta_1 x_{i2} + \beta_2 x_{i2}^2 + \beta_3 D_{i,B} + \beta_4 D_{i,C} + \varepsilon_i$

ANALYSIS OF VARIANCE

SOURCE	DF	SUM OF SQUARES	MEAN SQUARE	F VALUE	PROB>F
MODEL	4	137.51374	34.37843561	86.468	0.0001
ERROR	17	6.75898482	0.39758734		
C TOTAL	21	144.27273			

ROOT MSE	0.6305453	R-SQUARE	0.9532
DEP MEAN	33.48182	ADJ R-SQ	0.9421
C.V.	1.883247		

PARAMETER ESTIMATES

VARIABLE	DF	PARAMETER ESTIMATE	STANDARD ERROR	T FOR H0: PARAMETER=0	PROB > |T|
INTERCEP	1	23.48733397	0.74509730	31.523	0.0001
X2	1	8.06795541	0.70510295	11.442	0.0001
X2SQ	1	-1.39637097	0.15085150	-9.257	0.0001
DB	1	1.90222960	0.33400914	5.695	0.0001
DC	1	-1.01769450	0.36883510	-2.759	0.0134

(b) $y_i = \beta_0 + \beta_1 x_{i2} + \beta_2 x_{i2}^2 + \beta_3 D_{i,B} + \beta_4 D_{i,C} + \beta_5 x_{i2} D_{i,B} + \beta_6 x_{i2} D_{i,C} + \varepsilon_i$

ANALYSIS OF VARIANCE

SOURCE	DF	SUM OF SQUARES	MEAN SQUARE	F VALUE	PROB>F
MODEL	6	137.51409	22.91901478	50.866	0.0001
ERROR	15	6.75863861	0.45057591		
C TOTAL	21	144.27273			

ROOT MSE	0.6712495	R-SQUARE	0.9532
DEP MEAN	33.48182	ADJ R-SQ	0.9344
C.V.	2.004818		

PARAMETER ESTIMATES

VARIABLE	DF	PARAMETER ESTIMATE	STANDARD ERROR	T FOR H0: PARAMETER=0	PROB > |T|
INTERCEP	1	23.48180693	0.83138382	28.244	0.0001
X2	1	8.06701733	0.75179443	10.730	0.0001
X2SQ	1	-1.39461634	0.17439840	-7.997	0.0001
DB	1	1.91423267	0.90694822	2.111	0.0520
DC	1	-0.99467822	0.91848876	-1.083	0.2959
DBX2	1	-0.006683168	0.41007375	-0.016	0.9872
DCX2	1	-0.01113861	0.40490391	-0.028	0.9784

a. Which of the above models seems to best relate y to x_1 and x_2? Fully explain your answer.

b. Is your answer to part (a) consistent with the graphical analysis you carried out in Exercise 12.12? Fully explain.

12.15 Consider the International Oil regular unleaded gasoline mileage situation.

a. Define $\mu_{[A,x_2]}$, $\mu_{[B,x_2]}$, and $\mu_{[C,x_2]}$ to be the mean gasoline mileages obtained by the GT-1100 when using additive VST amount x_2 and regular unleaded gasoline types A, B, and C, respectively. Express each of the differences $\mu_{[B,x_2]} - \mu_{[A,x_2]}$, $\mu_{[C,x_2]} - \mu_{[A,x_2]}$, $\mu_{[C,x_2]} - \mu_{[B,x_2]}$, and $(\mu_{[A,x_2]} + \mu_{[C,x_2]})/2 - \mu_{[B,x_2]}$ as linear combinations of the regression parameters in the model

$$y_i = \beta_0 + \beta_1 x_{i2} + \beta_2 x_{i2}^2 + \beta_3 D_{i,B} + \beta_4 D_{i,C} + \varepsilon_i$$

b. Explain the practical interpretation of each of the differences given in part (a).

12.16 Figure 12.14 gives the SAS output that is obtained when PROC GLM is employed to run a regression analysis of the regular unleaded gasoline mileage data in Table 12.7 using the model

$$y_i = \beta_0 + \beta_1 x_{i2} + \beta_2 x_{i2}^2 + \beta_3 D_{i,B} + \beta_4 D_{i,C} + \varepsilon_i$$

On this output the difference $\mu_{[B,x_2]} - \mu_{[A,x_2]}$ is denoted MUBX2 – MUAX2, and other differences are denoted similarly.

a. Use the SAS output in Figure 12.14 to test the significance of each of the differences $\mu_{[B,x_2]} - \mu_{[A,x_2]}$, $\mu_{[C,x_2]} - \mu_{[A,x_2]}$, $\mu_{[C,x_2]} - \mu_{[B,x_2]}$, and $(\mu_{[A,x_2]} + \mu_{[C,x_2]})/2 - \mu_{[B,x_2]}$ by setting $\alpha = .05$ for each test.

b. Explain the practical interpretation of the results of each of these tests.

c. Use the SAS output in Figure 12.14 to compute an individual 95% confidence interval for each of the above differences. Interpret each interval.

FIGURE 12.14 **SAS output of a regression analysis of the regular unleaded gasoline mileage data using the model $y_i = \beta_0 + \beta_1 x_{i2} + \beta_2 x_{i2}^2 + \beta_3 D_{i,B} + \beta_4 D_{i,C} + \varepsilon_i$**

SAS

GENERAL LINEAR MODELS PROCEDURE

DEPENDENT VARIABLE: MILEAGE

SOURCE	DF	SUM OF SQUARES	MEAN SQUARE	F VALUE	PR > F	R-SQUARE	C.V.
MODEL	4	137.51374245	34.37843561	86.47	0.0001	0.953151	1.8832
ERROR	17	6.75898482	0.39758734		ROOT MSE		MILEAGE MEAN
CORRECTED TOTAL	21	144.27272727			0.63054527		33.48181818

SOURCE	DF	TYPE I SS	F VALUE	PR > F	DF	TYPE III SS	F VALUE	PR > F
X2	1	54.15459883	136.21	0.0001	1	52.05404309	130.92	0.0001
X2SQ	1	51.48026136	129.48	0.0001	1	34.06707941	85.68	0.0001
DB	1	28.85194857	72.57	0.0001	1	12.89560521	32.43	0.0001
DC	1	3.02693369	7.61	0.0134	1	3.02693369	7.61	0.0134

(*continues*)

FIGURE 12.14 Continued

PARAMETER	ESTIMATE	T FOR HO: PARAMETER=0	PR > |T|	STD ERROR OF ESTIMATE
INTERCEPT	23.48733397	31.52	0.0001	0.74509730
X2	8.06795541	11.44	0.0001	0.70510295
X2SQ	-1.39637097	-9.26	0.0001	0.15085150
DB	1.90222960	5.70	0.0001	0.33400914
DC	-1.01769450	-2.76	0.0134	0.36883510
MUBX2-MUAX2	1.90222960	5.70	0.0001	0.33400914
MUCX2-MUAX2	-1.01769450	-2.76	0.0134	0.36883510
MUCX2-MUBX2	-2.91992410	-8.61	0.0001	0.33913044
(AX2+CX2)/2-BX2	-2.41107685	-8.56	0.0001	0.28155984

OBSERVATION	OBSERVED VALUE	PREDICTED VALUE	RESIDUAL	LOWER95% CL FOR MEAN	UPPER95% CL FOR MEAN
1	29.40000000	30.15891841	-0.75891841	29.54669967	30.77113714
2	30.60000000	30.15891841	0.44108159	29.54669967	30.77113714
3	30.00000000	30.15891841	-0.15891841	29.54669967	30.77113714
4	35.00000000	34.03776091	0.96223909	33.47293488	34.60258694
5	34.00000000	34.03776091	-0.03776091	33.47293488	34.60258694
6	34.30000000	35.12386148	-0.82386148	34.51366497	35.73405799
7	35.50000000	35.12386148	0.37613852	34.51366497	35.73405799
8	32.60000000	32.06114801	0.53885199	31.41328411	32.70901191
9	31.60000000	32.06114801	-0.46114801	31.41328411	32.70901191
10	35.30000000	35.93999051	-0.63999051	35.37851880	36.50146222
11	36.50000000	35.93999051	0.56000949	35.37851880	36.50146222
12	37.60000000	37.02609108	0.57390892	36.49973472	37.55244744
13	36.40000000	37.02609108	-0.62609108	36.49973472	37.55244744
14	37.00000000	37.02609108	-0.02609108	36.49973472	37.55244744
15	35.40000000	35.31944972	0.08055028	34.42242093	36.21647850
16	28.60000000	29.14122391	-0.54122391	28.44131207	29.84113574
17	29.80000000	29.14122391	0.65877609	28.44131207	29.84113574
18	34.00000000	34.10616698	-0.10616698	33.50130617	34.71102779
19	34.70000000	34.10616698	0.59383302	33.50130617	34.71102779
20	33.30000000	34.10616698	-0.80616698	33.50130617	34.71102779
21	32.00000000	32.39952562	-0.39952562	31.63841557	33.16063566
22	33.00000000	32.39952562	0.60047438	31.63841557	33.16063566
23*	.	37.02609108	.	36.49973472	37.55244744

*OBSERVATION WAS NOT USED IN THIS ANALYSIS

SUM OF RESIDUALS	0.00000000
SUM OF SQUARED RESIDUALS	6.75898482
SUM OF SQUARED RESIDUALS - ERROR SS	-0.00000000
PRESS STATISTIC	11.09877850
FIRST ORDER AUTOCORRELATION	-0.28633215
DURBIN-WATSON D	2.43410406

12.17 Consider the International Oil regular unleaded gasoline mileage situation and the SAS output in Figure 12.14. Use this SAS output to

a. Compute 95% Scheffé simultaneous confidence intervals for each of the differences given in Exercise 12.16.
b. Compute 95% Bonferroni simultaneous confidence intervals for each of the differences given in Exercise 12.16.

12.18 Consider the International Oil regular unleaded gasoline mileage situation. Note from Table 12.7 that gasoline type B seems to maximize gasoline mileage and assume that International Oil will produce this gasoline type. Also note that "observation 23" on the SAS output in Figure 12.14 corresponds to using gasoline type B and three units of additive VST.

a. Use the SAS output in Figure 12.14 to write the least squares prediction equation for predicting regular unleaded gasoline mileage.
b. Use the SAS output in Figure 12.14 to find a point estimate of and a 95% confidence interval for mean gasoline mileage when gasoline type B is produced by using three units of additive VST.
c. Use the SAS output in Figure 12.14 to find a point prediction of and a 95% prediction interval for an individual gasoline mileage when gasoline type B is produced by using three units of additive VST.
d. Let $\hat{y}_{[B,x_2]}$ denote predicted gasoline mileage when gasoline type B is produced using x_2 units of additive VST. Use differential calculus to show that the value of x_2 maximizing $\hat{y}_{[B,x_2]}$ is 2.8889.
e. Calculate a point estimate of the mean regular unleaded gasoline mileage obtained when gasoline type B is produced by using 2.889 units of additive VST.

12.19 Consider the International Oil regular unleaded gasoline mileage situation and the model

$$y_i = \beta_0 + \beta_1 x_{i2} + \beta_2 x_{i2}^2 + \beta_3 D_{i,B} + \beta_4 D_{i,C} + \varepsilon_i$$

a. Figure 12.15 gives residual plots for this model versus x_{i2}, $D_{i,B}$, $D_{i,C}$, and $\hat{y}_i$. Fully interpret these plots.
b. Use the residuals given in Figure 12.14 to check the normality assumption for this model.

Exercises 12.20–12.24 refer to the following situation. Consider analyzing the Fresh Detergent data of Tables 8.3 and 12.2 using the model

$$\begin{aligned} y_i &= \mu_i + \varepsilon_i \\ &= \beta_0 + \beta_1 x_{i4} + \beta_2 x_{i3} + \beta_3 x_{i3}^2 + \beta_4 x_{i4} x_{i3} + \beta_5 D_{i,A} + \beta_6 D_{i,C} + \varepsilon_i \end{aligned}$$

where $D_{i,A}$ and $D_{i,C}$ are dummy variables defined as follows:

$$D_{i,A} = \begin{cases} 1 & \text{if campaign A is used in sales region } i \\ 0 & \text{otherwise} \end{cases}$$

$$D_{i,C} = \begin{cases} 1 & \text{if campaign C is used in sales region } i \\ 0 & \text{otherwise} \end{cases}$$

Figure 12.16 presents the SAS output that is obtained when PROC GLM is employed to analyze the data of Tables 8.3 and 12.2 using the above model.

12.20 Consider the Fresh Detergent situation and the SAS output of Figure 12.16.

a. We can employ the overall F-test to test the adequacy of the dummy variable model. Set up the appropriate null and alternative hypotheses for the overall F-test.
b. Using the overall F statistic and the associated prob-value on the SAS output of Figure 12.16, perform the overall F test. What do the results say about the adequacy of the model? Explain in detail. Use $\alpha = .05$.

FIGURE 12.15 **SAS output of residual plots for the regular unleaded gasoline mileage model $y_i = \beta_0 + \beta_1 x_{i2} + \beta_2 x_{i2}^2 + \beta_3 D_{i,B} + \beta_4 D_{i,C} + \varepsilon_i$**

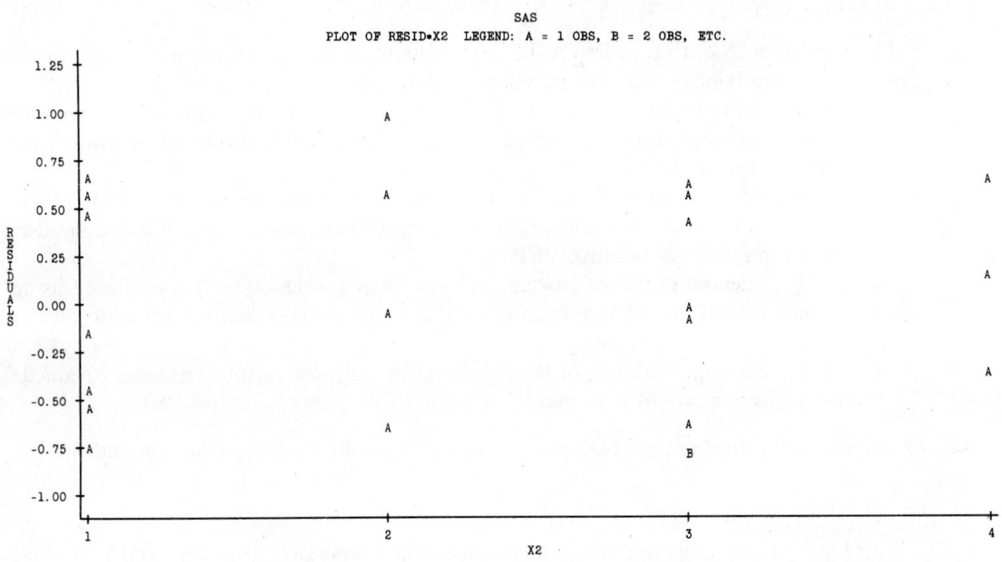

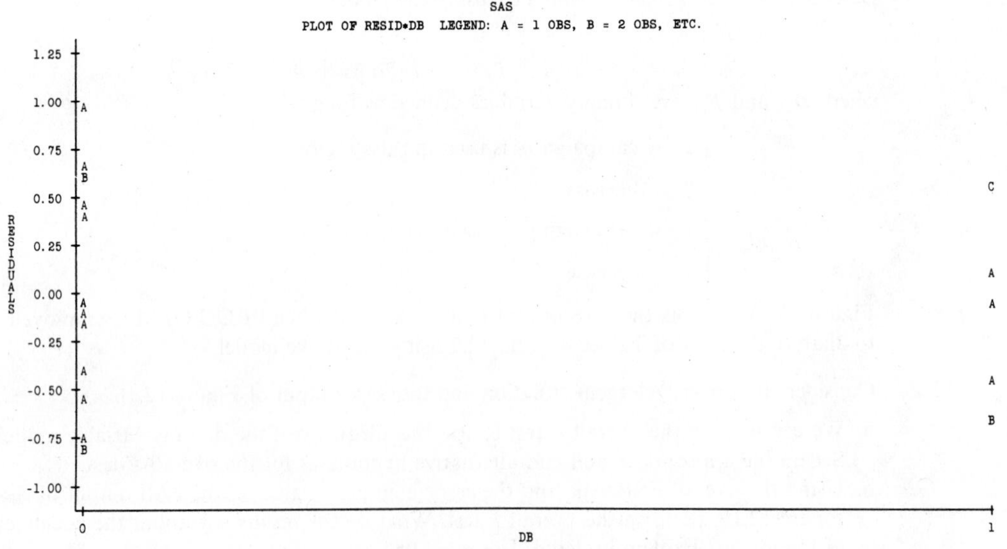

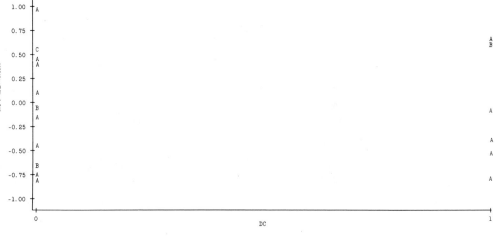

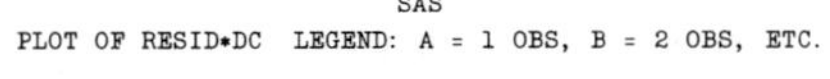
SAS
PLOT OF RESID*DC LEGEND: A = 1 OBS, B = 2 OBS, ETC.

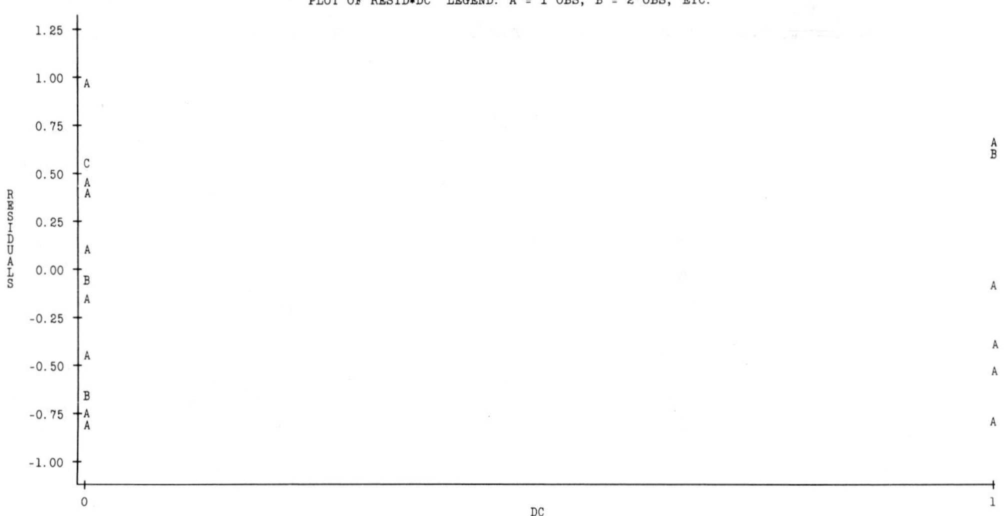
RESIDUALS
1.25
1.00
0.75
0.50
0.25
0.00
-0.25
-0.50
-0.75
-1.00
0
1
DC

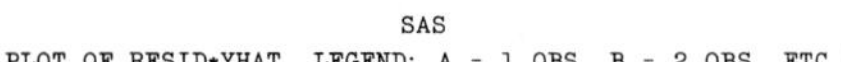
SAS
PLOT OF RESID*YHAT LEGEND: A = 1 OBS, B = 2 OBS, ETC.

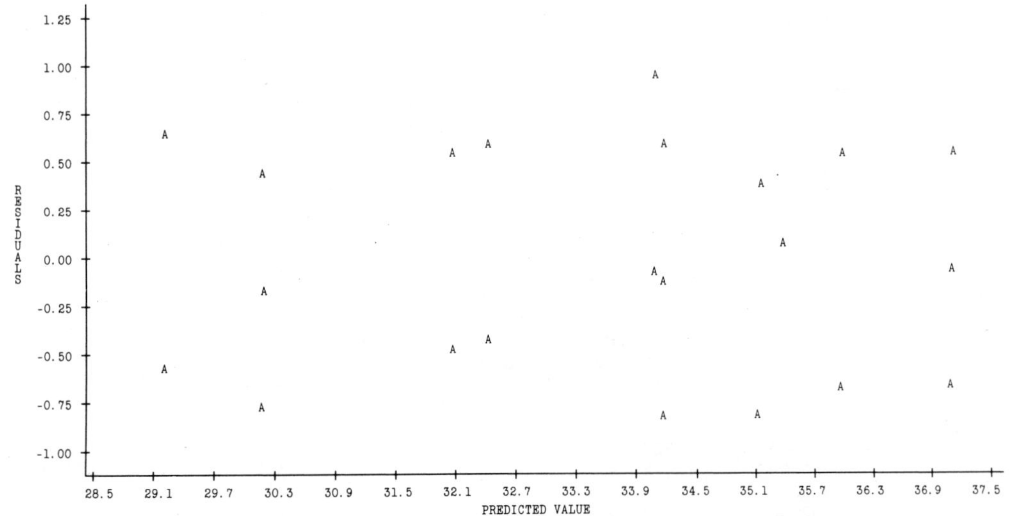
RESIDUALS
1.25
1.00
0.75
0.50
0.25
0.00
-0.25
-0.50
-0.75
-1.00
28.5
29.1
29.7
30.3
30.9
31.5
32.1
32.7
33.3
33.9
34.5
35.1
35.7
36.3
36.9
37.5
PREDICTED VALUE

FIGURE 12.16 SAS output of a regression analysis of the Fresh demand data using the model $y_i = \beta_0 + \beta_1 x_{i4} + \beta_2 x_{i3} + \beta_3 x_{i3}^2 + \beta_4 x_{i4} x_{i3} + \beta_5 D_{i,A} + \beta_6 D_{i,C} + \varepsilon_i$

SAS

GENERAL LINEAR MODELS PROCEDURE

DEPENDENT VARIABLE: Y

SOURCE	DF	SUM OF SQUARES	MEAN SQUARE	F VALUE	PR > F	R-SQUARE	C.V.
MODEL	6	13.06501880	2.17750313	127.25	0.0001	0.970757	1.5605
ERROR	23	0.39356786	0.01711165		ROOT MSE		Y MEAN
CORRECTED TOTAL	29	13.45858667			0.13081149		8.38266667

SOURCE	DF	TYPE I SS	F VALUE	PR > F	DF	TYPE III SS	F VALUE	PR > F
X4	1	10.65268464	622.54	0.0001	1	0.15277359	8.93	0.0066
X3	1	1.27222152	74.35	0.0001	1	0.29246430	17.09	0.0004
X3SQ	1	0.26041192	15.22	0.0007	1	0.34653331	20.25	0.0002
X43	1	0.20887206	12.21	0.0020	1	0.11018966	6.44	0.0184
DA	1	0.55170291	32.24	0.0001	1	0.20226102	11.82	0.0022
DC	1	0.11912574	6.96	0.0147	1	0.11912574	6.96	0.0147

PARAMETER	ESTIMATE	T FOR HO: PARAMETER=0	PR > |T|	STD ERROR OF ESTIMATE
INTERCEPT	25.82638229	5.39	0.0001	4.79455950
X4	9.05868432	2.99	0.0066	3.03170457
X3	-6.53767133	-4.13	0.0004	1.58136655
X3SQ	0.58444394	4.50	0.0002	0.12987222
X43	-1.15648054	-2.54	0.0184	0.45573648
DA	-0.21368626	-3.44	0.0022	0.06215362
DC	0.16808991	2.64	0.0147	0.06370664
MUDAA-MUDAB	-0.21368626	-3.44	0.0022	0.06215362
MUDAC-MUDAA	0.38177617	6.23	0.0001	0.06125253
MUDAC-MUDAB	0.16808991	2.64	0.0147	0.06370664

OBSERVATION	OBSERVED VALUE	PREDICTED VALUE	RESIDUAL	LOWER 95% CL INDIVIDUAL	UPPER 95% CL INDIVIDUAL
1	7.38000000	7.41371707	-0.03371707	7.11760477	7.70982936
2	8.51000000	8.63893796	-0.12893796	8.34769100	8.93018491
3	9.52000000	9.55261994	-0.03261994	9.23351784	9.87172203
4	7.50000000	7.33493287	0.16506713	7.03257634	7.63728940
5	9.33000000	9.10935604	0.22064396	8.80862379	9.41008829
6	8.28000000	8.11890097	0.16109903	7.82974740	8.40805453
7	8.75000000	8.68178380	0.06821620	8.39191462	8.97165297
8	7.87000000	7.92979186	-0.05979186	7.58321921	8.27636451
9	7.10000000	7.16426966	-0.06426966	6.83394925	7.49459006
10	8.00000000	8.12639619	-0.12639619	7.83076415	8.42202822
11	7.89000000	8.11890097	-0.22890097	7.82974740	8.40805453
12	8.15000000	8.14693585	0.00306415	7.85714860	8.43672310
13	9.10000000	9.25385412	-0.15385412	8.95881712	9.54889112
14	8.86000000	8.81367565	0.04632435	8.52334195	9.10400936
15	8.90000000	8.81302081	0.08697919	8.52467584	9.10136578
16	8.87000000	8.75328998	0.11671002	8.46396165	9.04261831
17	9.26000000	9.29457106	-0.03457106	9.00148035	9.58766177
18	9.00000000	8.96841000	0.03159000	8.67148190	9.26533809
19	8.75000000	8.87275165	-0.12275165	8.58399515	9.16150814
20	7.95000000	7.94719703	0.00280297	7.64336403	8.25103003
21	7.65000000	7.87233371	-0.22233371	7.57110389	8.17356353
22	7.27000000	7.21466975	0.05533025	6.90116423	7.52817527

OBSERVATION	OBSERVED VALUE	PREDICTED VALUE	RESIDUAL	LOWER 95% CL INDIVIDUAL	UPPER 95% CL INDIVIDUAL
23	8.00000000	8.11890097	-0.11890097	7.82974740	8.40805453
24	8.50000000	8.58308179	-0.08308179	8.25237317	8.91379040
25	8.75000000	8.77852705	-0.02852705	8.48377953	9.07327456
26	9.21000000	9.11167497	0.09832503	8.80772468	9.41562527
27	8.27000000	8.11528694	0.15471306	7.81663778	8.41393610
28	7.67000000	7.55794988	0.11205012	7.26478943	7.85111033
29	7.93000000	7.85422745	0.07577255	7.56353424	8.14492066
30	9.26000000	9.22003405	0.03996595	8.91598369	9.52408441
31*		8.50067714		8.21322415	8.78813013

*OBSERVATION WAS NOT USED IN THIS ANALYSIS

SUM OF RESIDUALS	0.00000000
SUM OF SQUARED RESIDUALS	0.39356786
SUM OF SQUARED RESIDUALS - ERROR SS	0.00000000
PRESS STATISTIC	0.65774226
FIRST ORDER AUTOCORRELATION	0.38997826
DURBIN-WATSON D	1.21309648

c. Using the explained variation (SS_{model}) and unexplained variation (SSE) on the SAS output, demonstrate the calculation of (that is, hand calculate) the overall F statistic for the model.

12.21 Consider the Fresh Detergent situation and the SAS output of Figure 12.16.

a. Using the fact that

$$\mu_i = \beta_0 + \beta_1 x_{i4} + \beta_2 x_{i3} + \beta_3 x_{i3}^2 + \beta_4 x_{i4} x_{i3} + \beta_5 D_{i,A} + \beta_6 D_{i,C}$$

show that

$$\mu_{[d,a,A]} - \mu_{[d,a,B]} = \beta_5$$
$$\mu_{[d,a,C]} - \mu_{[d,a,B]} = \beta_6$$
$$\mu_{[d,a,C]} - \mu_{[d,a,A]} = \beta_6 - \beta_5$$

Explain the practical interpretations of the three differences.

b. We can test to see whether or not there are any differences between the effects of advertising campaigns A, B, and C by using a partial F-test. Set up the appropriate null and alternative hypotheses for this partial F-test. *Hint*: The null hypothesis should state that there are no differences between the effects of campaigns A, B, and C.

c. Define the appropriate complete model and reduced model for the partial F-test.

d. For the appropriate reduced model it can be shown that $SSE_R = 1.06439652$. Using SSE_C and SSE_R, calculate the partial F statistic needed for the partial F-test you set up in part (b).

e. Using the partial F statistic, carry out the partial F-test. Can the null hypothesis you set up be rejected? What do the results of this test say about differences between the effects of campaigns A, B, and C? Use $\alpha = .05$.

12.22 Consider the Fresh Detergent situation and the SAS output of Figure 12.16.

a. Set up the appropriate null and alternative hypotheses needed to test to see whether or not there is any difference between the effects of campaigns A and B.

b. Use the appropriate t statistic and associated prob-value to test the hypotheses in part (a). Can the null hypothesis be rejected? What do the results of this test say about whether or not there is a difference between campaigns A and B? Use $\alpha = .05$.
c. Calculate a 95% confidence interval for $\mu_{[d,a,\text{A}]} - \mu_{[d,a,\text{B}]}$. Interpret the practical meaning of this interval.
d. Set up the appropriate null and alternative hypotheses needed to test to see whether or not there is any difference between the effects of campaigns C and B.
e. Use the appropriate t statistic and associated prob-value to test the hypotheses in part (d). Can the null hypothesis be rejected? What do the results of this test say about whether or not there is a difference between campaigns C and B? Use $\alpha = .05$.
f. Calculate a 95% confidence interval for $\mu_{[d,a,\text{C}]} - \mu_{[d,a,\text{B}]}$. Interpret the practical meaning of this interval.
g. Set up the appropriate null and alternative hypotheses needed to test to see whether or not there is any difference between the effects of campaigns C and A.
h. Use the appropriate t statistic and associated prob-value to test the hypotheses in part (g). Can the null hypothesis be rejected? What do the results of this test say about whether or not there is a difference between campaigns C and A? Use $\alpha = .05$.
i. Calculate a 95% confidence interval for $\mu_{[d,a,\text{C}]} - \mu_{[d,a,\text{A}]}$. Interpret the practical meaning of this interval.

12.23 Consider the Fresh Detergent situation and the SAS output of Figure 12.16. Noting that "observation 31" corresponds to an advertising expenditure of $650,000, a price difference of $.20, and using advertising campaign C:

a. Find a point prediction for Fresh demand when advertising expenditure is $650,000, the price difference is $.20, and Enterprise Industries employs advertising campaign C.
b. Find a 95% prediction interval for Fresh demand when advertising expenditure is $650,000, the price difference is $.20, and Enterprise Industries employs advertising campaign C.
c. Find a 95% confidence interval for mean Fresh demand when advertising expenditure is $650,000, the price difference is $.20, and Enterprise Industries employs advertising campaign C. *Hint*: Solve for $\mathbf{x}_0'(\mathbf{X}'\mathbf{X})^{-1}\mathbf{x}_0$.
d. With regard to part (c), what is $\mathbf{x}_0'$?

12.24 Consider the Fresh Detergent situation and the SAS output of Figure 12.16.

a. Calculate Bonferroni simultaneous 95% confidence intervals for each of the linear combinations in the (prespecified) set β_5, β_6, and $\beta_6 - \beta_5$.
b. Calculate Scheffé simultaneous 95% confidence intervals for each of the linear combinations in the above set.

12.25 Consider Example 12.7 and the linear combinations in

$$\text{Set II:}\quad \beta_5,\ \beta_6,\ \beta_6 - \beta_5$$
$$\mu_0 = \beta_0 + \beta_1(.10) + \beta_2(6.8) + \beta_3(6.8)^2 + \beta_4(.10)(6.8) + \beta_6$$

a. Calculate Bonferroni simultaneous 95% confidence intervals for each of the parameters β_5, β_6, and μ_0 (recall that the Bonferroni interval for $\beta_6 - \beta_5$ was computed in Example 12.8) in the prespecified Set II.

TABLE 12.8 International Oil premium unleaded gasoline mileage data

Gasoline mileage, y (mpg)	Premium unleaded gasoline type, x_1	Amount of gasoline additive VST, x_2
28.0	A	0
28.6	A	0
27.4	A	0
33.3	B	0
34.5	B	0
33.0	A	1
32.0	A	1
35.6	B	1
34.4	B	1
35.0	B	1
34.0	C	1
33.3	C	1
34.7	C	1
33.5	A	2
32.3	A	2
33.4	B	2
33.0	C	2
32.0	C	2
29.6	B	3
30.6	B	3
28.6	C	3
29.8	C	3

b. Compare the intervals you computed in part (a) with the corresponding Scheffé simultaneous 95% confidence intervals given in Table 12.6.
c. Interpret the practical meaning of each of the intervals you computed in part (a).

Exercises 12.26–12.35 refer to the following situation. International Oil, Inc. is attempting to develop a reasonably priced premium unleaded gasoline that will deliver higher gasoline mileages than can be achieved by its current premium unleaded gasolines. As part of its development process, International Oil wishes to study the effect of one qualitative independent variable—x_1, premium unleaded gasoline type (A, B, or C)—and of one quantitative independent variable—x_2, amount of gasoline additive VST (0, 1, 2, or 3 units)—on the gasoline mileage y obtained by an automobile called the Encore. For testing purposes a sample of $n = 22$ Encores is randomly selected and driven under normal driving conditions. The combinations of x_1 and x_2 used in the experiment, along with the corresponding values of y, are given in Table 12.8.

12.26 Consider the International Oil premium unleaded gasoline mileage data in Table 12.8.

a. Plot y against x_1 (= A, B, and C).
b. Plot y against x_1 when $x_2 = 0$.
c. Plot y against x_1 when $x_2 = 1$.
d. Plot y against x_1 when $x_2 = 2$.
e. Plot y against x_1 when $x_2 = 3$.
f. Combine the plots you made in parts (b), (c), (d), and (e) by drawing these plots on

the same set of axes. Use a different color for each level of x_2 (=0, 1, 2, and 3). What do these plots say about whether interaction exists between x_1 and x_2?

g. Plot y against x_2.
h. Plot y against x_2 when x_1 = A.
i. Plot y against x_2 when x_1 = B.
j. Plot y against x_2 when x_1 = C.
k. Combine the plots you made in parts (h), (i), and (j) by drawing these plots on the same set of axes. Use a different color for each level of x_1 (= A, B, and C). What do these plots say about whether interaction exists between x_1 and x_2?

12.27 Consider the International Oil premium unleaded gasoline mileage situation and the data in Table 12.8.

a. Define appropriate dummy variables that will enable us to compare the effects of premium unleaded gasoline types B and C on mean gasoline mileage to the effect of premium unleaded gasoline type A. Denote these dummy variables as $D_{i,\text{B}}$ and $D_{i,\text{C}}$.
b. Explain why the graphical analysis of parts (a)–(k) in Exercise 12.26 suggests that an appropriate model relating y to x_1 and x_2 is

$$\begin{aligned} y_i &= \mu_i + \varepsilon_i \\ &= \beta_0 + \beta_1 D_{i,\text{B}} + \beta_2 D_{i,\text{C}} + \beta_3 x_{i2} + \beta_4 x_{i2}^2 + \beta_5 D_{i,\text{B}} x_{i2} + \beta_6 D_{i,\text{C}} x_{i2} \\ &\quad + \beta_7 D_{i,\text{B}} x_{i2}^2 + \beta_8 D_{i,\text{C}} x_{i2}^2 + \varepsilon_i \end{aligned}$$

c. Specify the vector **y** and matrix **X** that should be used to calculate the least squares point estimates of the parameters in the above model.

12.28 Consider the International Oil premium unleaded gasoline mileage situation.

a. Define $\mu_{[\text{A},x_2]}$, $\mu_{[\text{B},x_2]}$, and $\mu_{[\text{C},x_2]}$ to be the mean premium unleaded gasoline mileages obtained by the Encore when using additive VST amount x_2 and premium unleaded gasline types A, B, and C, respectively, and consider the model

$$\begin{aligned} y_i &= \mu_i + \varepsilon_i \\ &= \beta_0 + \beta_1 D_{i,\text{B}} + \beta_2 D_{i,\text{C}} + \beta_3 x_{i2} + \beta_4 x_{i2}^2 + \beta_5 D_{i,\text{B}} x_{i2} \\ &\quad + \beta_6 D_{i,\text{C}} x_{i2} + \beta_7 D_{i,\text{B}} x_{i2}^2 + \beta_8 D_{i,\text{C}} x_{i2}^2 + \varepsilon_i \end{aligned}$$

Express the following differences in terms of $\beta_1, \beta_2, \ldots, \beta_8$ and x_2:

$$\mu_{[\text{B},x_2]} - \mu_{[\text{A},x_2]}$$

$$\mu_{[\text{C},x_2]} - \mu_{[\text{A},x_2]}$$

$$\mu_{[\text{C},x_2]} - \mu_{[\text{B},x_2]}$$

$$\frac{\mu_{[\text{C},x_2]} + \mu_{[\text{B},x_2]}}{2} - \mu_{[\text{A},x_2]}$$

b. Explain the practical interpretation of each of the differences given in part (a).

12.29 Figure 12.17 gives the SAS output that is obtained when PROC GLM is employed to run a regression analysis of the premium unleaded gasoline mileage data in Table 12.8 using the model

$$\begin{aligned} y_i = \beta_0 &+ \beta_1 D_{i,\text{B}} + \beta_2 D_{i,\text{C}} + \beta_3 x_{i2} + \beta_4 x_{i2}^2 + \beta_5 D_{i,\text{B}} x_{i2} + \beta_6 D_{i,\text{C}} x_{i2} \\ &+ \beta_7 D_{i,\text{B}} x_{i2}^2 + \beta_8 D_{i,\text{C}} x_{i2}^2 + \varepsilon_i \end{aligned}$$

Note here that "observation 23" on the output corresponds to using gasoline type B and .93 unit of additive VST.

FIGURE 12.17 SAS output of a regression analysis of the premium unleaded mileage data using the model $y_i = \beta_0 + \beta_1 D_{i,B} + \beta_2 D_{i,C} + \beta_3 x_{i2} + \beta_4 x_{i2}^2 + \beta_5 D_{i,B} x_{i2} + \beta_6 D_{i,C} x_{i2} + \beta_7 D_{i,B} x_{i2}^2 + \beta_8 D_{i,C} x_{i2}^2 + \varepsilon_i$

SAS

GENERAL LINEAR MODELS PROCEDURE

DEPENDENT VARIABLE: MILEAGE

SOURCE	DF	SUM OF SQUARES	MEAN SQUARE	F VALUE	PR > F	R-SQUARE	C.V.
MODEL	8	121.31580420	15.16447552	32.02	0.0001	0.951700	2.1427
ERROR	13	6.15692308	0.47360947		ROOT MSE		MILEAGE MEAN
CORRECTED TOTAL	21	127.47272727			0.68819290		32.11818182

SOURCE	DF	TYPE I SS	F VALUE	PR > F	DF	TYPE III SS	F VALUE	PR > F
DB	1	17.55844156	37.07	0.0001	1	43.31803755	91.46	0.0001
DC	1	8.02571429	16.95	0.0012	1	3.89880000	8.23	0.0132
X2	1	8.38243521	17.70	0.0010	1	14.92260870	31.51	0.0001
X2SQ	1	71.09377466	150.11	0.0001	1	5.93294118	12.53	0.0036
DBX2	1	7.35368820	15.53	0.0017	1	4.67288204	9.87	0.0078
DCX2	1	7.73264395	16.33	0.0014	1	2.03236686	4.29	0.0588
DBX2SQ	1	0.23557692	0.50	0.4931	1	0.99516905	2.10	0.1709
DCX2SQ	1	0.93352941	1.97	0.1838	1	0.93352941	1.97	0.1838

PARAMETER	ESTIMATE	T FOR HO: PARAMETER=0	PR > |T|	STD ERROR OF ESTIMATE
INTERCEPT	28.00000000	70.47	0.0001	0.39732836
DB	5.93846154	9.56	0.0001	0.62094016
DC	5.70000000	2.87	0.0132	1.98664178
X2	6.55000000	5.61	0.0001	1.16688783
X2SQ	-2.05000000	-3.54	0.0036	0.57920063
DBX2	-4.42692308	-3.14	0.0078	1.40935290
DCX2	-5.35000000	-2.07	0.0588	2.58263431
DBX2SQ	0.91153846	1.45	0.1709	0.62883507
DCX2SQ	1.15000000	1.40	0.1838	0.81911339
MUBX2-MUAX2	2.60981269	4.41	0.0007	0.59169048
MUCX2-MUAX2	1.71913500	2.61	0.0216	0.65869936
MUCX2-MUBX2	-0.89067769	-1.59	0.1360	0.56040913
(CX2+BX2)/2-AX2	2.16400000	3.86	0.0019	0.55980000

OBSERVATION	OBSERVED VALUE	PREDICTED VALUE	RESIDUAL	LOWER 95% CL INDIVIDUAL	UPPER 95% CL INDIVIDUAL
1	28.00000000	28.00000000	0.00000000	27.14162479	28.85837521
2	28.60000000	28.00000000	0.60000000	27.14162479	28.85837521
3	27.40000000	28.00000000	-0.60000000	27.14162479	28.85837521
4	33.30000000	33.93846154	-0.63846154	32.90758625	34.96933682
5	34.50000000	33.93846154	0.56153846	32.90758625	34.96933682
6	33.00000000	32.50000000	0.50000000	31.44870936	33.55129064
7	32.00000000	32.50000000	-0.50000000	31.44870936	33.55129064
8	35.60000000	34.92307692	0.67692308	34.17023206	35.67592179
9	34.40000000	34.92307692	-0.52307692	34.17023206	35.67592179
10	35.00000000	34.92307692	0.07692308	34.17023206	35.67592179
11	34.00000000	34.00000000	-0.00000000	33.14162479	34.85837521

(continues)

FIGURE 12.17 Continued

SAS

GENERAL LINEAR MODELS PROCEDURE

DEPENDENT VARIABLE: MILEAGE

OBSERVATION	OBSERVED VALUE	PREDICTED VALUE	RESIDUAL	LOWER 95% CL FOR MEAN	UPPER 95% CL FOR MEAN
12	33.30000000	34.00000000	-0.70000000	33.14162479	34.85837521
13	34.70000000	34.00000000	0.70000000	33.14162479	34.85837521
14	33.50000000	32.90000000	0.60000000	31.84870936	33.95129064
15	32.30000000	32.90000000	-0.60000000	31.84870936	33.95129064
16	33.40000000	33.63076923	-0.23076923	32.80606900	34.45546946
17	33.00000000	32.50000000	0.50000000	31.44870936	33.55129064
18	32.00000000	32.50000000	-0.50000000	31.44870936	33.55129064
19	29.60000000	30.06153846	-0.46153846	29.03066318	31.09241375
20	30.60000000	30.06153846	0.53846154	29.03066318	31.09241375
21	28.60000000	29.20000000	-0.60000000	28.14870936	30.25129064
22	29.80000000	29.20000000	0.60000000	28.14870936	30.25129064
23*	.	34.92826769	.	34.19521973	35.66131566

* OBSERVATION WAS NOT USED IN THIS ANALYSIS

SUM OF RESIDUALS	0.00000000
SUM OF SQUARED RESIDUALS	6.15692308
SUM OF SQUARED RESIDUALS - ERROR SS	-0.00000000
PRESS STATISTIC	19.57754033
FIRST ORDER AUTOCORRELATION	-0.38902664
DURBIN-WATSON D	2.71958252

a. Use the SAS output of Figure 12.17 to write the least squares prediction equation for predicting premium unleaded gasoline mileage.

b. Note from Table 12.8 that premium unleaded gasoline type B seems to maximize gasoline mileage and assume that International Oil will produce this gasoline type. Let $\hat{y}_{[B,x_2]}$ denote predicted gasoline mileage when gasoline type B is produced using x_2 units of additive VST. Use differential calculus to show that the value of x_2 maximizing $\hat{y}_{[B,x_2]}$ is .93.

c. Calculate a point estimate of the mean premium unleaded gasoline mileage obtained when gasoline type B is produced by using .93 unit of additive VST.

d. Use the SAS output in Figure 12.17 to find a 95% confidence interval for the mean premium unleaded gasoline mileage obtained when gasoline type B is produced by using .93 unit of additive VST.

e. Use the SAS output in Figure 12.17 to find a 95% prediction interval for an individual premium unleaded gasoline mileage obtained when gasoline type B is produced by using .93 unit of additive VST. *Hint*: Solve for $\mathbf{x}_0'(\mathbf{X}'\mathbf{X})^{-1}\mathbf{x}_0$.

f. With regard to part (e), what is $\mathbf{x}_0'$?

12.30 Consider the International Oil premium gasoline mileage situation. Suppose that the company will produce premium unleaded gasoline type B by using .93 unit of additive VST.

a. Express each of the following:

$$\mu_{[B,.93]} - \mu_{[A,.93]}$$
$$\mu_{[C,.93]} - \mu_{[A,.93]}$$
$$\mu_{[C,.93]} - \mu_{[B,.93]}$$
$$\frac{\mu_{[C,.93]} + \mu_{[B,.93]}}{2} - \mu_{[A,.93]}$$
$$\mu_{[B,.93]}$$

as a linear combination of parameters in the model

$$y_i = \beta_0 + \beta_1 D_{i,B} + \beta_2 D_{i,C} + \beta_3 x_{i2} + \beta_4 x_{i2}^2 + \beta_5 D_{i,B} x_{i2} + \beta_6 D_{i,C} x_{i2} + \beta_7 D_{i,B} x_{i2}^2 + \beta_8 D_{i,C} x_{i2}^2 + \varepsilon_i$$

b. For each of the linear combinations in part (a), specify the appropriate $\boldsymbol{\lambda}'$ vector.

c. The SAS output of Figure 12.17 gives the point estimate, t statistic and associated prob-value, and standard error of the estimate for each of the differences

$$\mu_{[B,.93]} - \mu_{[A,.93]}, \qquad \mu_{[C,.93]} - \mu_{[A,.93]}, \qquad \mu_{[C,.93]} - \mu_{[B,.93]}$$

and

$$\frac{\mu_{[C,.93]} + \mu_{[B,.93]}}{2} - \mu_{[A,.93]}$$

Here, for example, the difference $\mu_{[B,.93]} - \mu_{[A,.93]}$ is denoted MUBX2 – MUAX2 on the SAS output and other differences are denoted similarly. Use the SAS output to test the significance of each of the above differences by setting $\alpha = .05$ for each test.

d. Explain the practical interpretation of the results of each of these tests.

e. Use the SAS output in Figure 12.17 to compute an individual 95% confidence interval for each of the above differences. Interpret each interval.

12.31 Consider the International Oil premium unleaded gasoline mileage situation and the SAS output in Figure 12.17. Use this SAS output to:

a. Compute 95% Scheffé simultaneous confidence intervals for each of the differences

$$\mu_{[B,.93]} - \mu_{[A,.93]}, \qquad \mu_{[C,.93]} - \mu_{[A,.93]}, \qquad \mu_{[C,.93]} - \mu_{[B,.93]}$$

and

$$\frac{\mu_{[C,.93]} + \mu_{[B,.93]}}{2} - \mu_{[A,.93]}$$

Recall from Exercise 12.30 that the difference $\mu_{[B,.93]} - \mu_{[A,.93]}$ is denoted as MUBX2 – MUAX2 on the SAS output of Figure 12.17 and that other differences are denoted similarly.

b. Compute 95% Bonferroni simultaneous confidence intervals for each of the differences given in part (a).

12.32 Consider the International Oil premium unleaded gasoline mileage situation and the SAS output in Figure 12.17. Use the SAS output to:

a. Compute 95% Scheffé simultaneous confidence intervals for each of the linear combinations

$$\mu_{[B,.93]} - \mu_{[A,.93]} \qquad \mu_{[C,.93]} - \mu_{[A,.93]} \qquad \mu_{[C,.93]} - \mu_{[B,.93]}$$

$$\frac{\mu_{[C,.93]} + \mu_{[B,.93]}}{2} - \mu_{[A,.93]} \quad \text{and} \quad \mu_{[B,.93]}$$

Recall from Exercise 12.30 that the difference $\mu_{[B,.93]} - \mu_{[A,.93]}$ is denoted as MUBX2 − MUAX2 on the SAS output of Figure 12.17 and that other linear combinations are denoted similarly.

b. Compute 95% Bonferroni simultaneous confidence intervals for each of the linear combinations given in part (a).

12.33 Consider the International Oil premium unleaded gasoline mileage situation. Figure 12.18 gives the SAS output obtained when a regression analysis of the gasoline mileage data in Table 12.8 is performed using each of the models

$$y_i = \beta_0 + \beta_1 D_{i,B} + \beta_2 D_{i,C} + \beta_3 x_{i2} + \beta_4 x_{i2}^2 + \varepsilon_i \quad \text{(Model 1)}$$

$$\begin{aligned} y_i = {} & \beta_0 + \beta_1 D_{i,B} + \beta_2 D_{i,C} + \beta_3 x_{i2} + \beta_4 x_{i2}^2 \\ & + \beta_5 D_{i,B} x_{i2} + \beta_6 D_{i,C} x_{i2} + \varepsilon_i \end{aligned} \quad \text{(Model 2)}$$

$$\begin{aligned} y_i = {} & \beta_0 + \beta_1 D_{i,B} + \beta_2 D_{i,C} + \beta_3 x_{i2} + \beta_4 x_{i2}^2 + \beta_5 D_{i,B} x_{i2} \\ & + \beta_6 D_{i,C} x_{i2} + \beta_7 D_{i,B} x_{i2}^2 + \beta_8 D_{i,C} x_{i2}^2 + \varepsilon_i \end{aligned} \quad \text{(Model 3)}$$

Using all of the information at your disposal, decide which of these models seems best. Carefully justify your model choice.

12.34 Consider the International Oil premium unleaded gasoline mileage situation and the model

$$\begin{aligned} y_i = {} & \beta_0 + \beta_1 D_{i,B} + \beta_2 D_{i,C} + \beta_3 x_{i2} + \beta_4 x_{i2}^2 + \beta_5 D_{i,B} x_{i2} \\ & + \beta_6 D_{i,C} x_{i2} + \beta_7 D_{i,B} x_{i2}^2 + \beta_8 D_{i,C} x_{i2}^2 + \varepsilon_i \end{aligned}$$

FIGURE 12.18 **SAS output of a regression analysis of the premium unleaded mileage data using three possible models**

(a) $y_i = \beta_0 + \beta_1 D_{i,B} + \beta_2 D_{i,C} + \beta_3 x_{i2} + \beta_4 x_{i2}^2 + \varepsilon_i$

SAS

DEP VARIABLE: MILEAGE

ANALYSIS OF VARIANCE

SOURCE	DF	SUM OF SQUARES	MEAN SQUARE	F VALUE	PROB>F
MODEL	4	105.06037	26.26509148	19.922	0.0001
ERROR	17	22.41236155	1.31837421		
C TOTAL	21	127.47273			

ROOT MSE	1.148205	R-SQUARE	0.8242
DEP MEAN	32.11818	ADJ R-SQ	0.7828
C.V.	3.574937		

PARAMETER ESTIMATES

VARIABLE	DF	PARAMETER ESTIMATE	STANDARD ERROR	T FOR H0: PARAMETER=0	PROB > |T|
INTERCEP	1	29.08646704	0.56947334	51.076	0.0001
DB	1	3.19760855	0.61015866	5.241	0.0001
DC	1	1.59648376	0.67085424	2.380	0.0293
X2	1	4.96485788	0.80796017	6.145	0.0001
X2SQ	1	-1.85944166	0.25321307	-7.343	0.0001

(b) $y_i = \beta_0 + \beta_1 D_{i,B} + \beta_2 D_{i,C} + \beta_3 x_{i2} + \beta_4 x_{i2}^2 + \beta_5 D_{i,B} x_{i2} + \beta_6 D_{i,C} x_{i2} + \varepsilon_i$

SAS

DEP VARIABLE: MILEAGE

ANALYSIS OF VARIANCE

SOURCE	DF	SUM OF SQUARES	MEAN SQUARE	F VALUE	PROB>F
MODEL	6	120.14670	20.02444964	41.000	0.0001
ERROR	15	7.32602941	0.48840196		
C TOTAL	21	127.47273			

ROOT MSE	0.6988576	R-SQUARE	0.9425
DEP MEAN	32.11818	ADJ R-SQ	0.9195
C.V.	2.175894		

PARAMETER ESTIMATES

VARIABLE	DF	PARAMETER ESTIMATE	STANDARD ERROR	T FOR H0: PARAMETER=0	PROB > |T|
INTERCEP	1	28.19362745	0.38232113	73.743	0.0001
DB	1	5.64387255	0.57915905	9.745	0.0001
DC	1	4.46740196	0.97686193	4.573	0.0004
X2	1	4.95257353	0.52172245	9.493	0.0001
X2SQ	1	-1.22708333	0.21342508	-5.749	0.0001
DBX2	1	-2.55465686	0.45968865	-5.557	0.0001
DCX2	1	-2.46348039	0.61911853	-3.979	0.0012

(c) $y_i = \beta_0 + \beta_1 D_{i,B} + \beta_2 D_{i,C} + \beta_3 x_{i2} + \beta_4 x_{i2}^2 + \beta_5 D_{i,B} x_{i2} + \beta_6 D_{i,C} x_{i2} + \beta_7 D_{i,B} x_{i2}^2 + \beta_8 D_{i,C} x_{i2}^2 + \varepsilon_i$

SAS

DEP VARIABLE: MILEAGE

ANALYSIS OF VARIANCE

SOURCE	DF	SUM OF SQUARES	MEAN SQUARE	F VALUE	PROB>F
MODEL	8	121.31580	15.16447552	32.019	0.0001
ERROR	13	6.15692308	0.47360947		
C TOTAL	21	127.47273			

ROOT MSE	0.6881929	R-SQUARE	0.9517
DEP MEAN	32.11818	ADJ R-SQ	0.9220
C.V.	2.142689		

PARAMETER ESTIMATES

VARIABLE	DF	PARAMETER ESTIMATE	STANDARD ERROR	T FOR H0: PARAMETER=0	PROB > |T|
INTERCEP	1	28.00000000	0.39732836	70.471	0.0001
DB	1	5.93846154	0.62094016	9.564	0.0001
DC	1	5.70000000	1.98664178	2.869	0.0132
X2	1	6.55000000	1.16688783	5.613	0.0001
X2SQ	1	-2.05000000	0.57920063	-3.539	0.0036
DBX2	1	-4.42692308	1.40935290	-3.141	0.0078
DCX2	1	-5.35000000	2.58263431	-2.072	0.0588
DBX2SQ	1	0.91153846	0.62883507	1.450	0.1709
DCX2SQ	1	1.15000000	0.81911339	1.404	0.1838

a. Using the SAS output in Figure 12.18, hand calculate the overall F statistic. Using this statistic and the appropriate rejection point, determine whether we can, by setting α equal to .05, reject

$$H_0: \beta_1 = \beta_2 = \beta_3 = \beta_4 = \beta_5 = \beta_6 = \beta_7 = \beta_8 = 0$$

in favor of

H_1: At least one of $\beta_1, \beta_2, \beta_3, \beta_4, \beta_5, \beta_6, \beta_7$, and β_8 does not equal zero

b. The prob-value for testing H_0 versus H_1 is calculated to be .0001. Demonstrate how this prob-value has been calculated, and use it to determine whether we can reject H_0 in favor of H_1 by setting α equal to .05 or .01.

c. Using the SAS output of Figure 12.18, calculate the partial F statistics used to test

$$H_0: \beta_5 = \beta_6 = \beta_7 = \beta_8 = 0$$

versus

H_1: At least one of $\beta_5, \beta_6, \beta_7$, and β_8 does not equal zero

and to test

$$H_0: \beta_7 = \beta_8 = 0$$

versus

H_1: At least one of β_7 and β_8 does not equal zero

Then use these partial F statistics and the appropriate rejection points to determine whether we can reject the preceding null hypotheses by setting α equal to .05.

d. The prob-values related to the partial F statistics computed in part (c) can be calculated to be .0013 and .3230, respectively. Demonstrate how these prob-values have been calculated, and use them to determine whether we can reject the null hypotheses by setting α equal to .05 or .01.

12.35 Consider the International Oil premium unleaded gasoline mileage situation and the model

$$y_i = \beta_0 + \beta_1 D_{i,B} + \beta_2 D_{i,C} + \beta_3 x_{i2} + \beta_4 x_{i2}^2 + \beta_5 D_{i,B} x_{i2} + \beta_6 D_{i,C} x_{i2} + \beta_7 D_{i,B} x_{i2}^2 + \beta_8 D_{i,C} x_{i2}^2 + \varepsilon_i$$

a. Figure 12.19 gives residual plots for this model versus x_{i2}, $D_{i,B}$, $D_{i,C}$, and $\hat{y}_i$. Fully interpret these plots.

b. Use the residuals given in Figure 12.17 to check the normality assumption for this model.

12.36 Figure 12.20 gives the SAS output that results from using the model

$$y_i = \mu_i + \varepsilon_i = \beta_0 + \beta_1 x_i + \beta_2 D_{i,M} + \beta_3 D_{i,D} + \varepsilon_i$$

to analyze the sales volume data in Table 12.1. Using this output:

a. Test the significance of the differences

$$\mu_{[h,M]} - \mu_{[h,D]} = \beta_2 - \beta_3 \quad \text{and} \quad \mu_{[h,D]} - \mu_{[h,S]} = \beta_3$$

Note that these differences are denoted as MUMALL – MUDOWNTN and MUDOWNTN – MUSRT on the SAS output. Set α = .05 for each test of significance.

b. Interpret the practical meaning of the results of these tests.

FIGURE 12.19 **SAS output of residual plots for the premium unleaded gasoline mileage model $y_i = \beta_0 + \beta_1 D_{i,B} + \beta_2 D_{i,C} + \beta_3 x_{i2} + \beta_4 x_{i2}^2 + \beta_5 D_{i,B} x_{i2} + \beta_6 D_{i,C} x_{i2} + \beta_7 D_{i,B} x_{i2}^2 + \beta_8 D_{i,C} x_{i2}^2 + \varepsilon_i$**

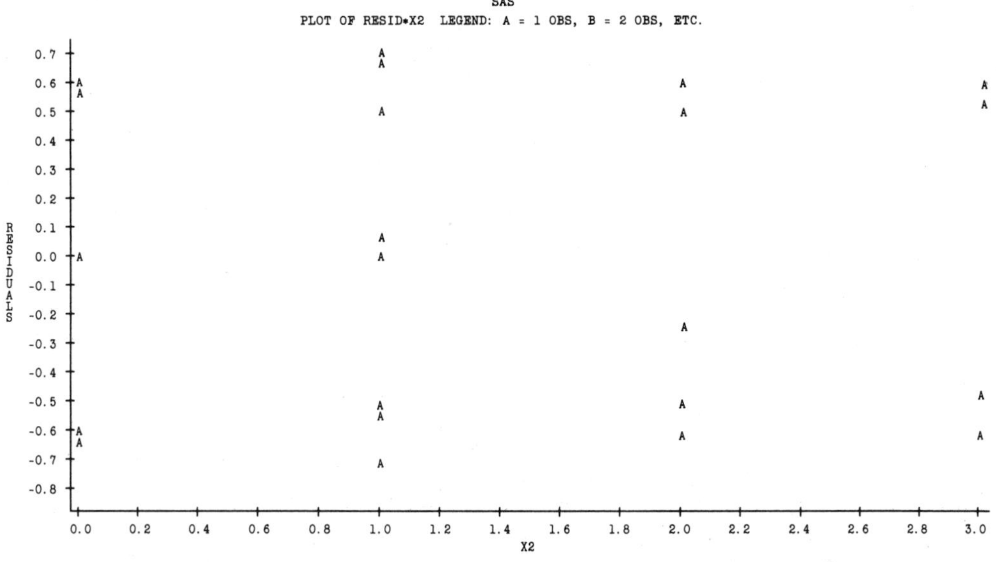

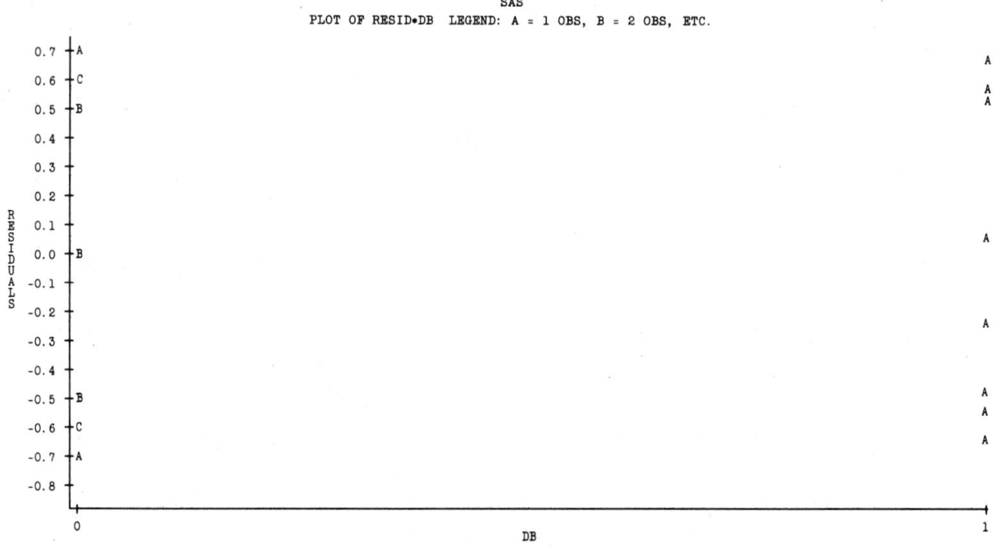

(continues)

FIGURE 12.19 Continued

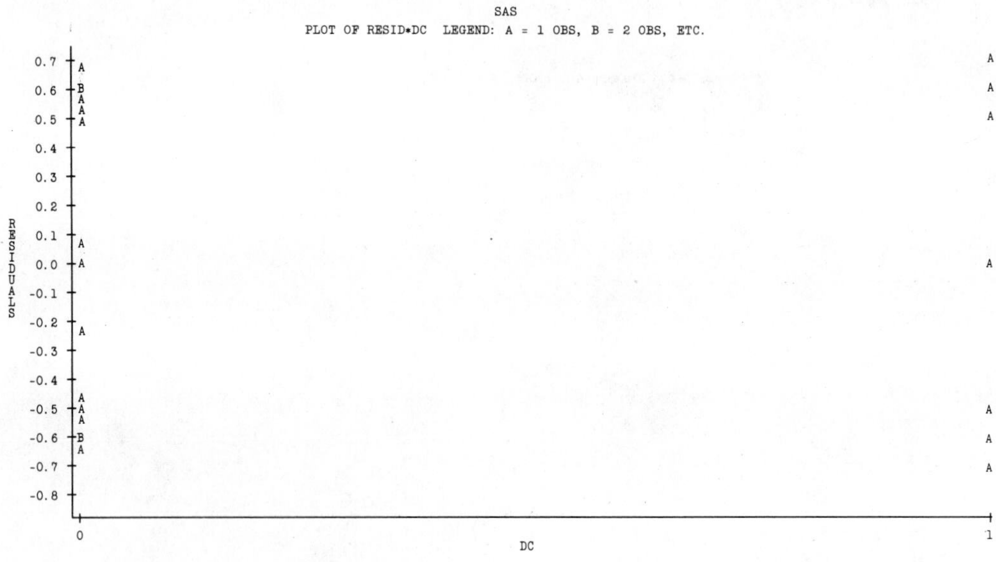

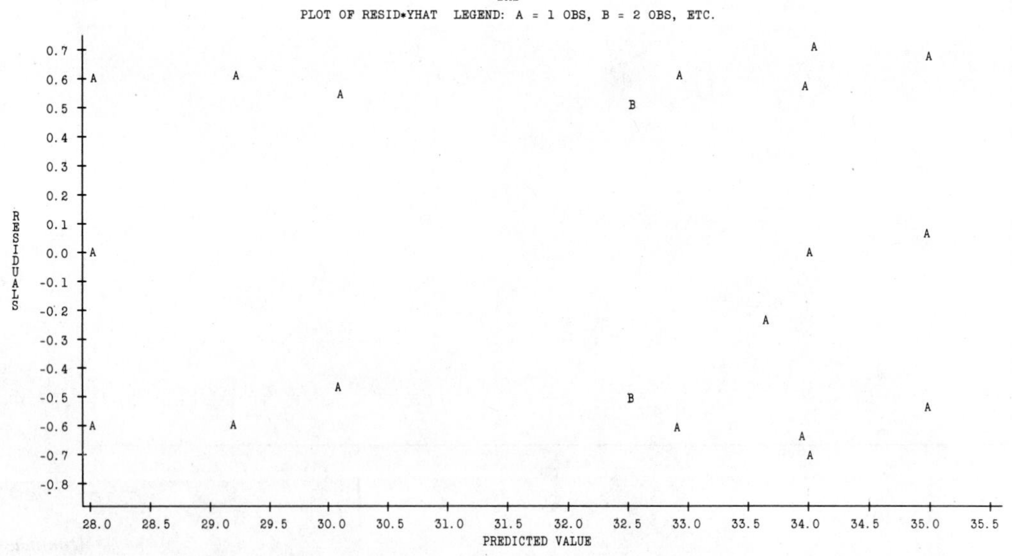

FIGURE 12.20 **SAS output of a regression analysis of the sales volume data in Table 12.1 using the model $y_i = \beta_0 + \beta_1 x_i + \beta_2 D_{i,M} + \beta_3 D_{i,D} + \varepsilon_i$**

```
                                         SAS
                          GENERAL LINEAR MODELS PROCEDURE
DEPENDENT VARIABLE: Y
SOURCE              DF     SUM OF SQUARES        MEAN SQUARE    F VALUE          PR > F
MODEL                3     33268.69529227     11089.56509742     275.07          0.0001
ERROR               11       443.46500107        40.31500010                   ROOT MSE
CORRECTED TOTAL     14     33712.16029333                                     6.34940943

SOURCE          DF              TYPE I SS          F VALUE          PR > F            DF
X                1         31244.35358664           775.01          0.0001             1
DM               1          1940.88342229            48.14          0.0001             1
DD               1            83.45828333             2.07          0.1780             1

                                            T FOR HO:                     STD ERROR OF
PARAMETER                ESTIMATE         PARAMETER=0       PR > |T|        ESTIMATE
INTERCEPT             14.97769322                2.42         0.0340      6.18844540
X                      0.86858842               21.45         0.0001      0.04048993
DM                    28.37375607                6.36         0.0001      4.46130660
DD                     6.86377679                1.44         0.1780      4.77047650

MUMALL-MUSRT          28.37375607                6.36         0.0001      4.46130660
MUMALL-MUDOWNTN       21.50997928                5.29         0.0003      4.06509197
MUDOWNTN-MUSRT         6.86377679                1.44         0.1780      4.77047650
```

c. Compute 95% (individual) confidence intervals for each of

$$\mu_{[h,\mathrm{M}]} - \mu_{[h,\mathrm{D}]} \quad \text{and} \quad \mu_{[h,\mathrm{D}]} - \mu_{[h,\mathrm{S}]}$$

Interpret the meaning of these intervals.

d. Compute 95% Scheffé simultaneous confidence intervals for each of

$$\mu_{[h,\mathrm{M}]} - \mu_{[h,\mathrm{D}]}, \quad \mu_{[h,\mathrm{D}]} - \mu_{[h,\mathrm{S}]}, \quad \text{and} \quad \mu_{[h,\mathrm{M}]} - \mu_{[h,\mathrm{S}]}$$

e. Compute 95% Bonferroni simultaneous confidence intervals for each of the differences given in part (d).

12.37 Reconsider the standard and luxury model automobile sales data in Exercise 11.16. Using all the model-building techniques discussed in this book, develop one regression model relating y to x_1 and x_2 for both the standard and luxury models. *Hint*: Use an appropriate dummy variable.

12.38 Consider the Andrews and Ferguson data of Exercise 8.19. Suppose that the home condition rating was either "poor," "fair," or "excellent." Also, suppose that the data in Table 12.9 is observed.

a. Using an appropriate set of dummy variables, determine an appropriate model relating y to x_1 and home condition rating.
b. Use the model that you developed in part (a) to estimate the pairwise differences between mean sales prices of homes having poor, fair, and excellent home condition ratings. Carry out appropriate tests of significance and compute appropriate individual and simultaneous confidence intervals.

TABLE 12.9 **Real estate sales data with qualitative home condition ratings**

Home size, x_1 (hundreds of square feet)	*Home condition rating, x_2*	*Sales price, y ($1000s)*
23	Fair	60.0
11	Poor	32.7
20	Excellent	57.7
17	Poor	45.5
15	Excellent	47.0
21	Fair	55.3
24	Fair	64.5
13	Fair	42.6
19	Fair	54.5
25	Poor	57.5

Source: R.L. Andrews and J.T. Ferguson, "Integrating Judgment with a Regression Appraisal," *The Real Estate Appraiser and Analyst,* Vol. 52, No. 2, Spring 1986 (Table 1).

c. Use the model to find a point prediction and 95% prediction interval for the selling price of a home having 2000 square feet and an "excellent" home condition rating.

12.39 Consider Exercise 11.18. Table 12.10 lists 28 observations (corresponding to the observations in Exercise 11.18) on two additional independent variables:

x_{10} = construction type (brick = 1, brick and frame = 2, aluminum and frame = 3, frame = 4)

x_{11} = style (two-story = 1, one-and-a-half-story = 2, ranch = 3)

TABLE 12.10 **Additional home data: Construction type and style**

Construction type, x_{10}	*Style, x_{11}*	*Construction type, x_{10}*	*Style, x_{11}*
3	1	1	1
1	1	4	1
2	1	1	1
4	1	2	1
3	1	3	3
2	1	4	1
2	1	1	1
1	1	1	1
2	1	4	1
4	1	1	1
1	1	1	3
1	2	4	1
1	2	3	1
2	1		
4	1		

a. Define a set of dummy variables to model the effect of x_{10} (construction type) and a set of dummy variables to model the effect of x_{11} (style).
b. Using the independent variables $x_1, x_2, \ldots, x_9$ from Exercise 11.18 and the sets of dummy variables from part (a), develop a model to predict sales price (y).
c. How can we determine whether there is interaction between construction type and style? Is there interaction between these variables?
d. Fully analyze the effects of different construction types and styles by using your model.

12.40 (Optional) Consider the regular unleaded gasoline mileage data in Table 12.7. Write a SAS program that will generate the output given in Figure 12.14, which analyzes the data using the model

$$y_i = \beta_0 + \beta_1 x_{i2} + \beta_2 x_{i2}^2 + \beta_3 D_{i,\mathrm{B}} + \beta_4 D_{i,\mathrm{C}} + \varepsilon_i$$

Be sure to use PROC GLM to estimate the differences between means that are given in Figure 12.14.

12.41 (Optional) Consider the premium unleaded gasoline mileage data in Table 12.8. Write a SAS program that will generate the output given in Figure 12.17, which analyzes the data using the model

$$y_i = \beta_0 + \beta_1 D_{i,\mathrm{B}} + \beta_2 D_{i,\mathrm{C}} + \beta_3 x_{i2} + \beta_4 x_{i2}^2 + \beta_5 D_{i,\mathrm{B}} x_{i2} + \beta_6 D_{i,\mathrm{C}} x_{i2} + \beta_7 D_{i,\mathrm{B}} x_{i2}^2 + \beta_8 D_{i,\mathrm{C}} x_{i2}^2 + \varepsilon_i$$

Be sure to use PROC GLM to estimate the differences between means that are given in Figure 12.17.

12.42 (Optional) Write a SAS program that will completely analyze the premium unleaded gasoline mileage data in Table 12.8 by using the model

$$y_i = \beta_0 + \beta_1 D_{i,\mathrm{B}} + \beta_2 D_{i,\mathrm{C}} + \beta_3 x_{i2} + \beta_4 x_{i2}^2 + \beta_5 D_{i,\mathrm{B}} x_{i2} + \beta_6 D_{i,\mathrm{C}} x_{i2} + \varepsilon_i$$

Compare the results for this model with the results of Figure 12.17.

12.43 (Optional) Write a SAS program that will completely analyze the real estate sales data given in Table 12.9. Be sure to estimate and analyze all relevant differences between means.

12.44 (Optional) Write a SAS program that will completely analyze the home data (including construction type and home style) by using a model developed using the independent variables $x_1, x_2, \ldots, x_{11}$. Fully analyze the effects of different construction types and home styles. Recall that the data for this exercise is given in Table 12.10 and in Table 11.19 of Exercise 11.18.

12.45 (Optional) Consider analyzing the International Oil premium unleaded gasoline mileage data in Table 12.8 by using the model

$$y_i = \beta_0 + \beta_1 D_{i,\mathrm{B}} + \beta_2 D_{i,\mathrm{C}} + \beta_3 x_{i2} + \beta_4 x_{i2}^2 + \beta_5 D_{i,\mathrm{B}} x_{i2} + \beta_6 D_{i,\mathrm{C}} x_{i2} + \beta_7 D_{i,\mathrm{B}} x_{i2}^2 + \beta_8 D_{i,\mathrm{C}} x_{i2}^2 + \varepsilon_i$$

a. Use the SAS output of Figure 12.17 to write the least squares prediction equation for this model. Using the prediction equation and differential calculus, compute the values of x_2 that maximize mean gasoline mileage when using each of premium unleaded gasoline types A, B, and C. Define these values of x_2 to be ${}_{\mathrm{A}}x_2$, ${}_{\mathrm{B}}x_2$, and ${}_{\mathrm{C}}x_2$.

b. Let

$\mu_{[A,_A x_2]}$ = the mean gasoline mileage obtained when using premium unleaded gasoline type A and the value of x_2 that maximizes mean gasoline mileage when using gasoline type A

and let $\mu_{[B,_B x_2]}$ and $\mu_{[C,_C x_2]}$ be defined similarly for premium unleaded gasoline types B and C.

Write a SAS program that will estimate each of the differences

$$\mu_{[B,_B x_2]} - \mu_{[A,_A x_2]}$$

$$\mu_{[C,_C x_2]} - \mu_{[A,_A x_2]}$$

$$\mu_{[C,_C x_2]} - \mu_{[B,_B x_2]}$$

$$\frac{\mu_{[C,_C x_2]} + \mu_{[A,_A x_2]}}{2} - \mu_{[B,_B x_2]}$$

Using the output from your SAS program, test the significance of each of the differences and compute 95% individual, Scheffé, and Bonferroni confidence intervals for each difference. Carefully interpret your results.

12.46 In the article "A Model of MBA Student Performance," which was published in *Reseach in Higher Education* (Vol. 25, No. 2, 1986), McClure, Wells, and Bowerman developed a model to predict the graduate grade point average (GGPA) of an MBA student. Using a sample of 89 students, they found that important independent variables included undergraduate grade point average (UGGPA), Graduate Management Admission Test (GMAT) score, the level of competitiveness of a student's undergraduate institution, and the type of undergraduate major (business, quantitative, education, social science, liberal arts, and fine arts). Note that level of competitiveness and type of undergraduate major are qualitative variables that are modeled by using the following dummy variables:

$$D123 = \begin{cases} 1 & \text{if degree was received from an academic institution in one of the three most competitive categories,} \\ 0 & \text{otherwise;} \end{cases}$$

$$D89 = \begin{cases} 1 & \text{if degree was received from an academic institution in one of the two least competitive categories,} \\ 0 & \text{otherwise;} \end{cases}$$

$$DQUANT = \begin{cases} 1 & \text{if undergraduate degree was from a quantitative area,} \\ 0 & \text{otherwise;} \end{cases}$$

$$DEDC = \begin{cases} 1 & \text{if undergraduate degree was in education,} \\ 0 & \text{otherwise;} \end{cases}$$

$$DSOC = \begin{cases} 1 & \text{if undergraduate degree was in a social science area,} \\ 0 & \text{otherwise;} \end{cases}$$

$$DLIB = \begin{cases} 1 & \text{if undergraduate degree was in liberal arts,} \\ 0 & \text{otherwise;} \end{cases}$$

TABLE 12.11 **Results from the MBA student performance model**

Source of variation	*Sum of squares*	*df*	*Mean square*	*F*	*Prob value*
Model	4.29	17	0.252	6.32	0.0001
Error	2.75	69	0.040		
Total	7.04	86			

R^2 = .61
Adjusted R^2 = 0.51

Variable	*Parameter estimate*	*Standard error*	*Prob value*[a]
Intercept	2.07	0.24	0.0001
Total GMAT score, GMAT	0.0011	0.0003	0.001
Undergraduate grade point, UGGPA	0.219	0.07	0.006
Years since undergraduate degree, YRS	0.013	0.01	0.237
Student from 3 most competitive schools, D123	−2.11	1.12	0.063
Student from 2 least competitive schools, D89	−1.25	0.61	0.043
Undergraduate quantitative major, DQUANT	0.384	0.55	0.487
Undergraduate education major, DEDC	−1.16	1.05	0.270
Undergraduate social science major, DSOC	−2.00	0.75	0.010
Undergraduate liberal arts major, DLIB	−0.660	0.56	0.240
Undergraduate fine arts major, DFIN	−2.65	1.22	0.034
UGGPA*D123	0.744	0.40	0.064
UGGPA*D89	0.350	0.19	0.064
UGGPA*DQUANT	−0.091	0.18	0.614
UGGPA*DEDC	0.382	0.32	0.241
UGGPA*DSOC	0.654	0.24	0.008
UGGPA*DLIB	0.232	0.19	0.224
UGGPA*DFIN	0.893	0.42	0.038

[a] For t-value in testing H_0: Parameter = 0.

$$\text{DFIN} = \begin{cases} 1 & \text{if undergraduate degree was in fine arts,} \\ 0 & \text{otherwise.} \end{cases}$$

Note that if all of the major related dummy variables equal zero, then the student was an undergraduate business major. Table 12.11 presents a summary of some of the statistical results for one of the authors' best models. Interpret the above results. Specifically, discuss the meaning of the interaction terms between undergraduate grade point average and the qualitative independent variables.

12.47 In *The Miami University Report*, Provost Carlisle wrote the following in a report concerning the pursuit of salary equity:

> In the most recent study of faculty salaries, conducted this past month in preparation for 1989–90 salary improvements, no statistically significant group differences have been identified that call for redress in 1989–90. More specifically, at no rank are there categorical differences—by gender, ethnic classification or campus of assignment

—which cannot be attributed to relative market values between disciplines or to various indicators of professional experience.

In each of these recent salary studies, the Office of Budgeting, Planning and Analysis has relied on regression analyses. The variables that enter the regression equation include a measure of the relative market value of individual's discipline, and several measures of professional experience—age, years in rank, years of Miami service and years since completing last degree. The regression equation is used to calculate an expected or predicted salary, and predicted salaries are compared to actual salaries.*

Formulate one or more possible regression models employing the types of independent variables discussed by Carlisle. Discuss how one might use the model(s) to investigate the existence of the group differences discussed by Carlisle.

*Source: *The Miami University Report* (Miami University, Oxford, Ohio), Vol. 8, No. 26, 1989.

CHAPTER 13

REMEDIES FOR VIOLATIONS OF THE REGRESSION ASSUMPTIONS

What can be done when the assumptions behind regression analysis are violated? This chapter provides some answers to this question.

We begin Chapter 13 by considering the linearity assumption. We find that some useful regression models are not linear in the parameters. However, we can often use data transformations that will produce a linear regression model. The use of these *transformations to achieve linearity* is the subject of Section 13.1. In Section 13.2 we turn to a discussion of remedies that can be employed when the constant variance assumption is violated. Remedies here also involve the use of data transformation techniques. These *variance-equalizing transformations* and *weighted least squares* are explained in Section 13.2. Section 13.3 presents a related regression problem involving what we call *binary dependent variables*. Also discussed is *logistic regression*. When the independence assumption is violated, remedies often involve the use of *deterministic time series components* and *autoregressive terms*. These topics are discussed in Sections 13.4 and 13.5. In particular, a technique called the *Cochran-Orcutt procedure*, which can be employed when an autoregressive error structure exists, is explained in detail in Section 13.5. Then, in Section 13.6 we discuss remedies that can be used when the normality assumption is violated. We complete this chapter with optional Section 13.7, which shows how to use SAS to implement the techniques of this chapter.

13.1

TRANSFORMATIONS TO ACHIEVE LINEARITY

Up to this point, we have studied models of the form

$$y = \beta_0 + \beta_1 x_1 + \beta_2 x_2 + \cdots + \beta_p x_p + \varepsilon$$

Such models are said to be *linear in the parameters* $\beta_0, \beta_1, \beta_2, \ldots, \beta_p$. This is because the model contains $p + 1$ terms, each of which is a parameter β_j multiplied by a value determined from the data. Models including higher-order terms and interaction terms, such as

$$y = \beta_0 + \beta_1 x_1 + \beta_2 x_2 + \beta_3 x_2^2 + \beta_4 x_1 x_2 + \varepsilon$$

are also linear in the parameters. Again, this is because each term in the model is a parameter β_j multiplied by a value determined from the data.

Sometimes, however, useful models are not linear in the parameters. For instance, the model

$$y = \beta_0(\beta_1^x)\varepsilon$$

is not linear in the parameters. Here, the independent variable x enters as an exponent, and β_0 is multiplied by β_1^x. This model further departs from the usual linear model because the error term ε is multiplicative rather than additive. That is, ε is multiplied by, rather than added to, $\beta_0(\beta_1^x)$. To apply the techniques of estimation and prediction we presented in previous chapters, we must *transform* such a nonlinear model into a model that is linear in the parameters.

Some nonlinear models can be transformed into a linear model by taking logarithms on both sides. Either base 10 logarithms (denoted log) or natural (base *e*) logarithms (denoted ln) can be used. Both base 10 and natural logarithms can easily be obtained on many modern pocket calculators or by using computer routines. Two important properties of logarithms are

$$\log AB = \log A + \log B$$

$$\log A^r = r \log A$$

where A and B are positive numbers. These properties allow us to transform some nonlinear models into linear models.

If $\beta_0 > 0$ and $\beta_1 > 0$, applying a logarithmic transformation to the model

$$y = \beta_0(\beta_1^x)\varepsilon$$

yields

$$\log y = \log \beta_0 + x \log \beta_1 + \log \varepsilon$$

If we let $\alpha_0 = \log \beta_0$, $\alpha_1 = \log \beta_1$, and $U = \log \varepsilon$, the transformed version of the

FIGURE 13.1 Exponential curves of the form $y = \beta_0(\beta_1^x)$

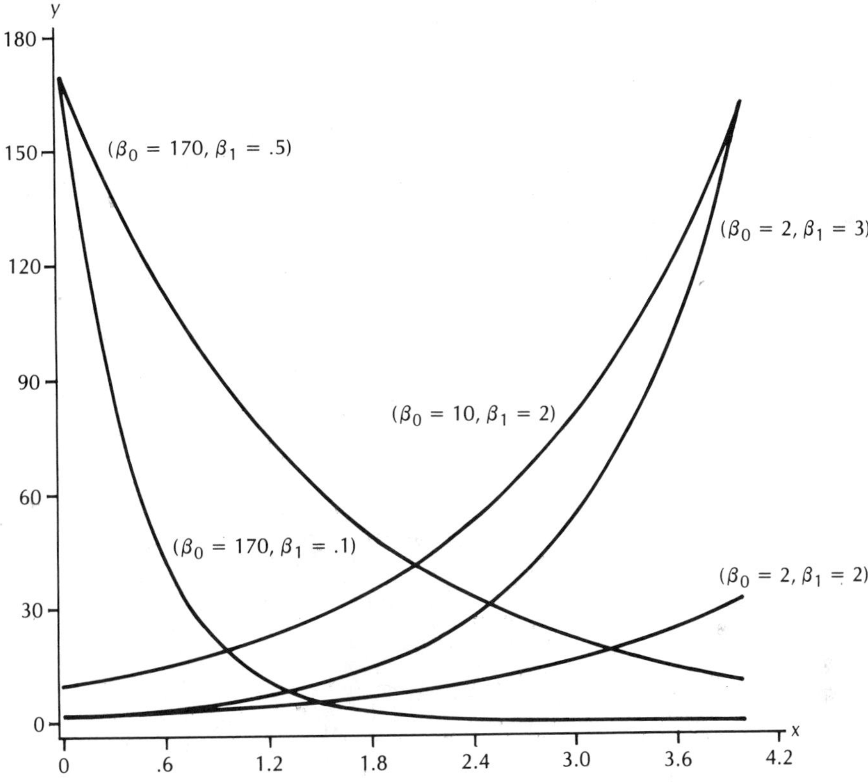

model becomes

$$\log y = \alpha_0 + \alpha_1 x + U$$

Thus we see that the model with dependent variable $\log y$ is linear in the parameters α_0 and α_1. If, in addition, we can assume that the inference assumptions are satisfied, we are justified in using the regression procedures we have presented to estimate α_0 and α_1.

Cases in which a model

$$y = \beta_0(\beta_1^x)\varepsilon$$

may be appropriate can be identified by data plots of y versus x. Plots of the expression $\beta_0(\beta_1^x)$ for several combinations of β_0 and β_1 are shown in Figure 13.1. Plots of observed data would have points scattered about such a function. The multiplicative error term would cause more variation around the high parts of the curve and less variation around the low parts. This is because the variation in y

depends on the level of $\beta_0(\beta_1^x)$. That is, given the same error term ε, the larger $\beta_0(\beta_1^x)$ is, the larger will be the variation in y.

Figure 13.1 shows that the curves described by this functional form may be increasing ($\beta_1 > 1$) or decreasing ($0 < \beta_1 < 1$) functions of x. We can see that β_0 is the intercept and that β_1 determines the amount of curvature in the plot. The curvature gets more pronounced as β_1 moves away from 1 in either direction. To illustrate our discussion thus far, we present an example.

Example 13.1 Wild Bill's Steakhouses, a fast food chain, opened in 1974. Each year from 1974 to 1988 the number of steakhouses in operation, y_t, is recorded. For convenience, we let $t = 0$ for the year 1974, $t = 1$ for the year 1975, and so on, where t is a time index. An analyst for the firm wishes to use this data (presented in Table 13.1) to estimate the growth rate for the company. Here, the analyst believes that the model

$$y_t = \beta_0(\beta_1^t)\varepsilon_t \qquad \text{for } t = 0, 1, \ldots, 14$$

may be appropriate. This model implies that

$$y_t = [\beta_0(\beta_1^{t-1})]\beta_1\varepsilon_t \approx (y_{t-1})\beta_1\varepsilon_t$$

This says that y_t is expected to be approximately β_1 times y_{t-1}. For example, if the true value of β_1 is 1.3, then y_t is expected to be approximately 1.3 times y_{t-1}. This says that y_t is expected to be approximately

$$100(\beta_1 - 1)\% = 100(1.3 - 1)\% = 30\%$$

greater than y_{t-1}. We call $100(\beta_1 - 1)\%$ the *growth rate* of the company. Since we do not know the true value of β_1, we wish to obtain both a point estimate and a 95% confidence interval for this growth rate.

The growth data are given in Table 13.1, along with the natural logarithm of y_t for each year. The original steakhouse data (y_t) are plotted against time (t) in Figure 13.2. This plot shows an exponential increase reminiscent of the plots in

TABLE 13.1 **Number of Wild Bill's Steakhouses (y_t) in operation for the years 1974–1988**

Year	t	y_t	$\ln y_t$	Year	t	y_t	$\ln y_t$
1974	0	11	2.398	1982	8	82	4.407
1975	1	14	2.639	1983	9	99	4.595
1976	2	16	2.773	1984	10	119	4.779
1977	3	22	3.091	1985	11	156	5.050
1978	4	28	3.332	1986	12	257	5.549
1979	5	36	3.584	1987	13	284	5.649
1980	6	46	3.829	1988	14	403	5.999
1981	7	67	4.205				

FIGURE 13.2 **Number of Wild Bill's Steakhouses in operation for the years 1974–1988 plotted versus time**

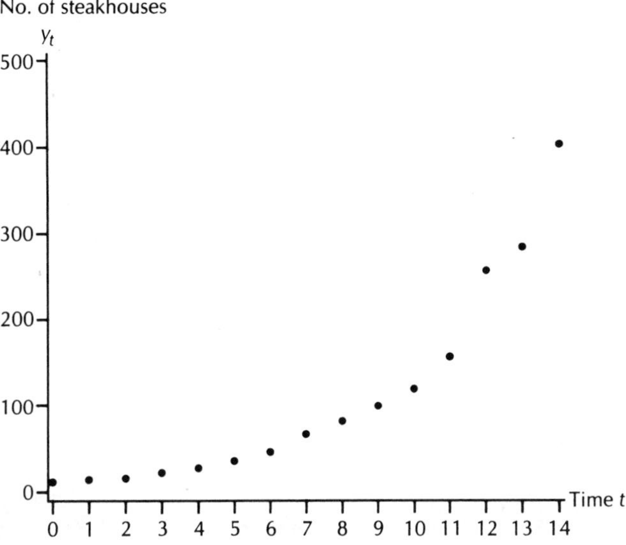

Figure 13.1 where β_1 is greater than 1. This suggests that the model

$$y_t = \beta_0(\beta_1^t)\varepsilon_t$$

where the independent variable is the time period t, may be appropriate. The natural logarithms of the steakhouse data ($\ln y_t$) are plotted in Figure 13.3. This plot suggests that the relationship between $\ln y_t$ and t is linear. Applying the logarithmic transformation to the above model, we obtain

$$\ln y_t = \ln \beta_0 + t \ln \beta_1 + \ln \varepsilon_t$$

Defining $\alpha_0 = \ln \beta_0$, $\alpha_1 = \ln \beta_1$, and $U_t = \ln \varepsilon_t$, we have

$$\ln y_t = \alpha_0 + \alpha_1 t + U_t$$

To estimate α_0 and α_1, we let

$$\mathbf{y} = \begin{bmatrix} 2.398 \\ 2.639 \\ 2.773 \\ \vdots \\ 5.999 \end{bmatrix} \quad \text{and} \quad \mathbf{X} = \begin{bmatrix} 1 & 0 \\ 1 & 1 \\ 1 & 2 \\ \vdots & \vdots \\ 1 & 14 \end{bmatrix}$$

FIGURE 13.3 Natural logarithms of the steakhouse data plotted versus time

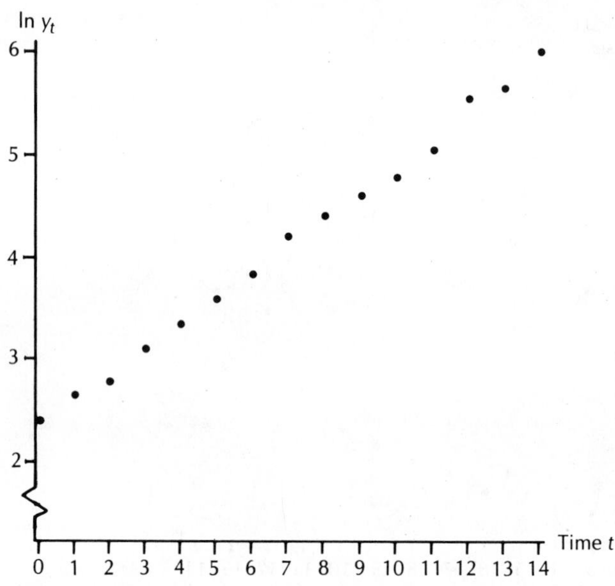

We compute

$$\mathbf{X}'\mathbf{X} = \begin{bmatrix} 15 & 105 \\ 105 & 1015 \end{bmatrix} \qquad (\mathbf{X}'\mathbf{X})^{-1} = \begin{bmatrix} .2417 & -.0250 \\ -.0250 & .0036 \end{bmatrix}$$

$$\mathbf{X}'\mathbf{y} = \begin{bmatrix} 61.8774 \\ 505.0686 \end{bmatrix} \qquad \mathbf{b} = (\mathbf{X}'\mathbf{X})^{-1}\mathbf{X}'\mathbf{y} = \begin{bmatrix} 2.3270 \\ .2569 \end{bmatrix}$$

Thus point estimates of α_0 and α_1 are $\hat{\alpha}_0 = 2.3270$ and $\hat{\alpha}_1 = .2569$, respectively. Moreover, for the logarithmic model, $SSE = .0741$ and $s = .0754983$. To estimate the rate of growth $100(\beta_1 - 1)\%$, we can use $\hat{\alpha}_1 = .2569$, the estimate of $\alpha_1 = \ln \beta_1$ in the logarithmic model.

Since $e^{\alpha_1} = e^{\ln \beta_1} = \beta_1$, a point estimate of β_1 is $e^{.2569} = 1.293$. This says that a point estimate of the growth rate $100(\beta_1 - 1)\%$ is

$$100(1.293 - 1)\% = 29.3\%$$

Since $t_{[\alpha/2]}^{(n-k)} = t_{[.025]}^{(13)} = 2.160$, a 95% confidence interval for $\alpha_1 = \ln \beta_1$ is

$$\begin{aligned}[.2569 \pm t_{[\alpha/2]}^{(n-k)} s\sqrt{c_{11}}] &= [.2569 \pm 2.160(.0754983)\sqrt{.0036}] \\ &= [.24715, .26665]\end{aligned}$$

Since $e^{.24715} = 1.2804$ and $e^{.26665} = 1.3056$, we are 95 percent confident that $\beta_1 = e^{\alpha_1}$ is contained in the interval [1.2804, 1.3056]. Thus we are 95 percent confident that the growth rate is between

$$100[1.2804 - 1]\% = 28.04\% \quad \text{and} \quad 100[1.3056 - 1]\% = 30.56\%$$

per year. Our confidence interval here is very narrow because there is little variation in the data points around the estimate of the function $\beta_0(\beta_1^t)$ (see Figure 13.2). It is probably reasonable to use this model to predict growth of Wild Bill's Steakhouses for the next year. However, one should be hesitant to use it as a predictor for growth over, say, the next five years. This is because, though the exponential growth model describes the data for 1974–1988 very well, there is no guarantee that growth will continue to be exponential so far outside the experimental region (so far into the future).

To conclude this example, we calculate a point prediction of and a 95% prediction interval for y_{15}, the number of steakhouses that will be in operation in 1989. A point prediction of $\ln y_{15}$ is

$$\begin{aligned}\widehat{\ln y_{15}} &= \hat{\alpha}_0 + \hat{\alpha}_1(15) \\ &= 2.3270 + .2569(15) \\ &= 6.1850\end{aligned}$$

Thus a point prediction of y_{15} is

$$\hat{y}_{15} = e^{6.1850} = 485.41 \quad \text{(steakhouses)}$$

Furthermore, since $\mathbf{x}_0' = [1 \quad 15]$, then

$$\begin{aligned}\mathbf{x}_0'(\mathbf{X}'\mathbf{X})^{-1}\mathbf{x}_0 &= [1 \quad 15]\begin{bmatrix} .2417 & -.0250 \\ -.0250 & .0036 \end{bmatrix}\begin{bmatrix} 1 \\ 15 \end{bmatrix} \\ &= .3017\end{aligned}$$

It follows that a 95% prediction interval for $\ln y_{15}$ is

$$\begin{aligned}\left[\widehat{\ln y_{15}} \pm t_{[.025]}^{(13)} s\sqrt{1 + \mathbf{x}_0'(\mathbf{X}'\mathbf{X})^{-1}\mathbf{x}_0}\right] &= [6.1850 \pm 2.16(.0755)\sqrt{1.3017}] \\ &= [5.9989, 6.3711]\end{aligned}$$

Therefore a 95% prediction interval for y_{15} is

$$[e^{5.9989}, e^{6.3711}] = [402.99 \text{ steakhouses}, 584.70 \text{ steakhouses}]$$

This says that we can be 95% confident that between approximately 403 and 585 steakhouses will be in operation in 1989.

Another model that can be linearized by a transformation is the model

$$y = \beta_0 x^{\beta_1} \varepsilon$$

This model also employs a multiplicative error term. Taking logarithms (natural or base 10) on both sides, we obtain (with base 10 logs)

$$\log y = \log \beta_0 + \beta_1 \log x + \log \varepsilon$$

FIGURE 13.4 Plots of $y = \beta_0 x^{\beta_1}$

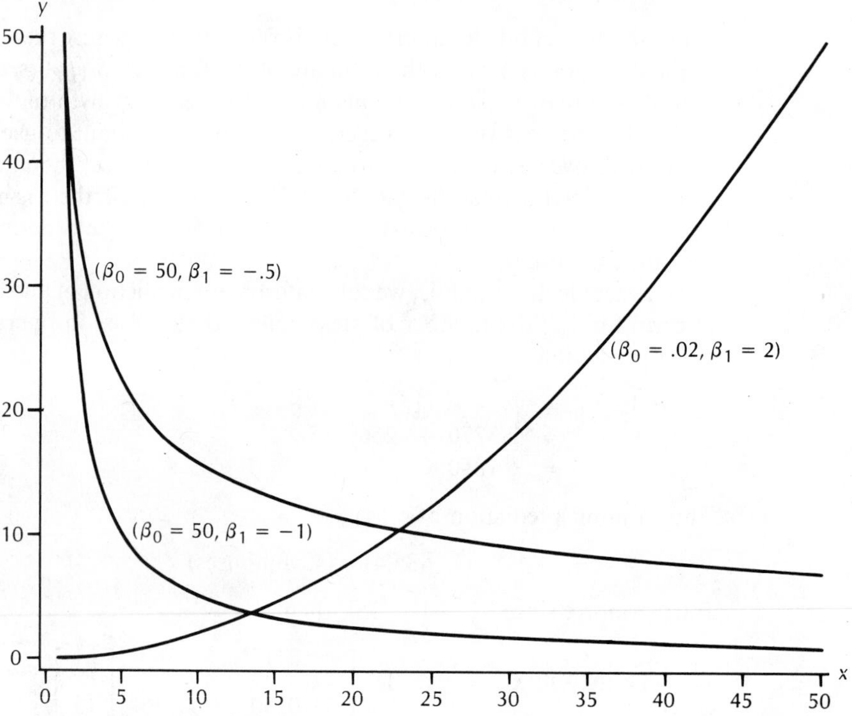

Letting $\alpha_0 = \log \beta_0$, $\alpha_1 = \beta_1$, and $U = \log \varepsilon$, we obtain

$$\log y = \alpha_0 + \alpha_1 \log x + U$$

which is linear in the parameters α_0 and α_1. If we can assume that the inference assumptions are satisfied, we can use our standard regression formulas to estimate $\alpha_0 = \log(\beta_0)$ and $\alpha_1 = \beta_1$. This model is sometimes used by economists to estimate the relationship between y, the quantity of a good demanded, and x, the per unit price for the good. In this application, $|\beta_1|$ is often referred to as the elasticity of demand for the good.

Plots of the expression $\beta_0 x^{\beta_1}$ are given in Figure 13.4 for several combinations of β_0 and β_1. We can see that if $\beta_0 > 0$ and $\beta_1 < 0$, the graph has the y and x axes as asymptotes. As can be seen by comparing Figures 13.4 and 13.1, the shapes obtainable with this model are similar to those obtainable with our previous model.

The choice of a model by visual inspection of a data plot can therefore be difficult. This is especially true when a data plot displays a high degree of variability. Clearly, a statistician would be more comfortable when theoretical considerations (economic theory, for example) suggest a model. Such a situation more often

FIGURE 13.5 Plots of functions involving reciprocals of y and x

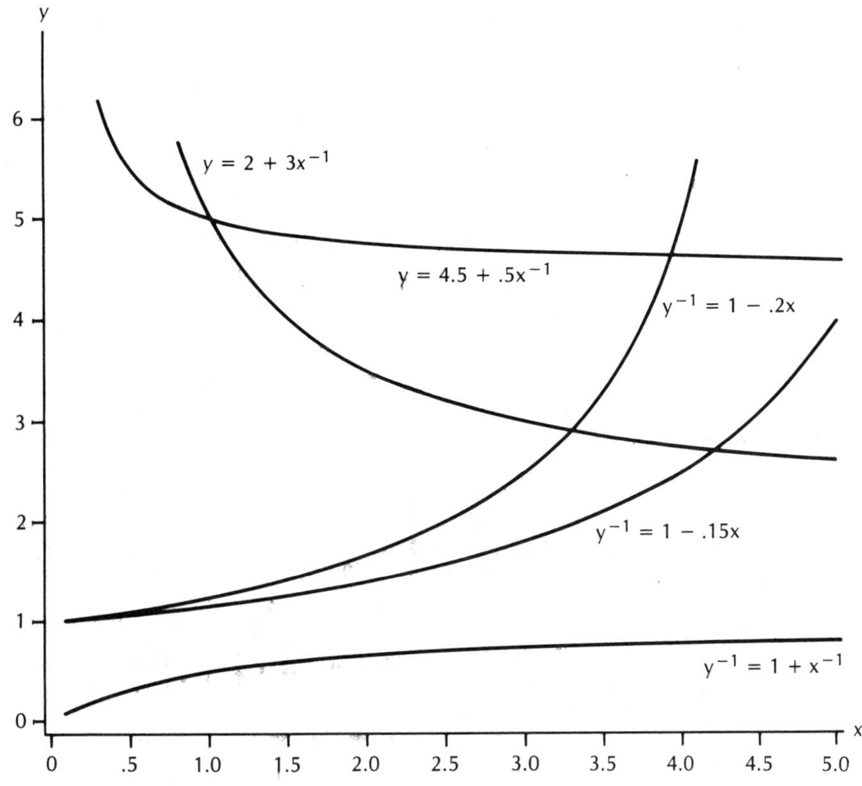

involves regression coefficients with meaningful interpretations (like elasticity). We will illustrate the use of the model

$$y = \beta_0 x^{\beta_1} \varepsilon$$

in a demand and price problem at the end of this chapter.

So far we have seen that the regression of log y on x and the regression of log y on log x can both be useful. In addition, although we do not present an example here, we can also regress y on log x if this seems appropriate for a given data set.

When data plots suggest that the usual linear model may not be appropriate, the reciprocal transformation may be useful. Examples of models involving the reciprocals of y, x, or both, are

$$y = \beta_0 + \beta_1 \left(\frac{1}{x}\right) + \varepsilon$$

$$\frac{1}{y} = \beta_0 + \beta_1 x + \varepsilon$$

$$\frac{1}{y} = \beta_0 + \beta_1 \left(\frac{1}{x}\right) + \varepsilon$$

Figure 13.5 shows several graphs that are helpful in identifying the general shapes of data plots that suggest these models. As with the logarithmic transformations, plots of the transformed data should appear linear. This class of models is easily enlarged by adding higher-order polynomial terms in $1/x$. For example, the model

$$y = \beta_0 + \beta_1 \left(\frac{1}{x}\right) + \beta_2 \left(\frac{1}{x^2}\right) + \varepsilon$$

might be useful in some situations.

Looking at Figure 13.5, we see that in the equation

$$y = \beta_0 + \beta_1 \left(\frac{1}{x}\right)$$

y approaches β_0 as x gets larger. This model might be considered in a situation in which learning over time (x) causes the time required to perform a task (y) to decrease toward an asymptotic value (β_0). We now present such an example.

Example 13.2 The State Department of Taxation wishes to investigate the effect of experience, x, on the amount of time, y, required to fill out Form ST 1040AVG, the state income-averaging form. In order to do this, nine people whose financial status makes income averaging advantageous are chosen at random. Each is asked to fill out Form ST 1040AVG and to report (1) the time y (in hours) required to complete the form and (2) the number of times x (including this one) that he or she has filled out this form. We might anticipate a model like

$$y = \beta_0 + \beta_1 \left(\frac{1}{x}\right) + \varepsilon$$

because learning over time might logically cause the time required to fill out Form ST 1040AVG to decrease toward an asymptotic value. Thus we have defined x in such a way that a zero value for x is impossible.

TABLE 13.2 **Completion times for Tax Form ST 1040AVG**

Person	1	2	3	4	5	6	7	8	9
Completion time, y (in hours)	8.0	4.7	3.7	2.8	8.9	5.8	2.0	1.9	3.3
Experience, x	1	8	4	16	1	2	12	5	3

FIGURE 13.6 **Completion time versus experience for Tax Form ST 1040AVG**

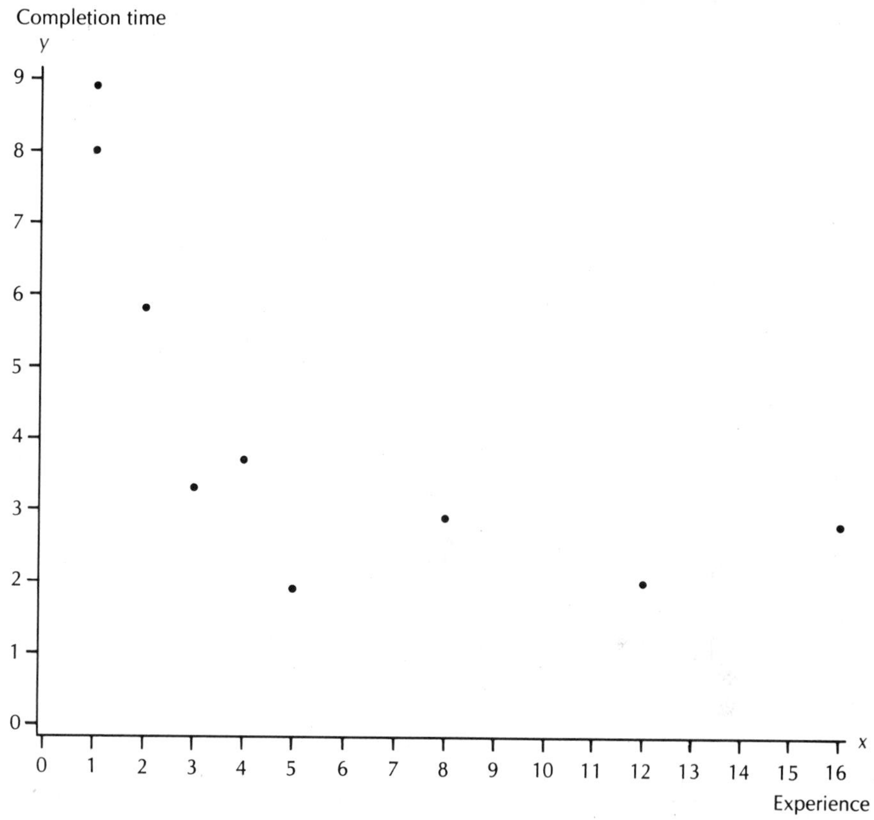

The data obtained in the tax form experiment are given in Table 13.2. We plot y versus x in Figure 13.6. We see a tendency for a decrease in y with increasing values of x in a pattern consistent with the proposed model.

A plot of y versus $1/x$ is shown in Figure 13.7. Notice that this plot has a linear appearance. We estimate β_0 and β_1 by using

$$\mathbf{y} = \begin{bmatrix} 8.0 \\ 4.7 \\ 3.7 \\ \vdots \\ 3.3 \end{bmatrix} \quad \text{and} \quad \mathbf{X} = \begin{bmatrix} 1 & 1.0000 \\ 1 & .1250 \\ 1 & .2500 \\ \vdots & \vdots \\ 1 & .3333 \end{bmatrix}$$

Here, the second column of $\mathbf{X}$ contains the reciprocals of the observations on

FIGURE 13.7 **Completion time versus reciprocal of experience for Tax Form ST 1040AVG**

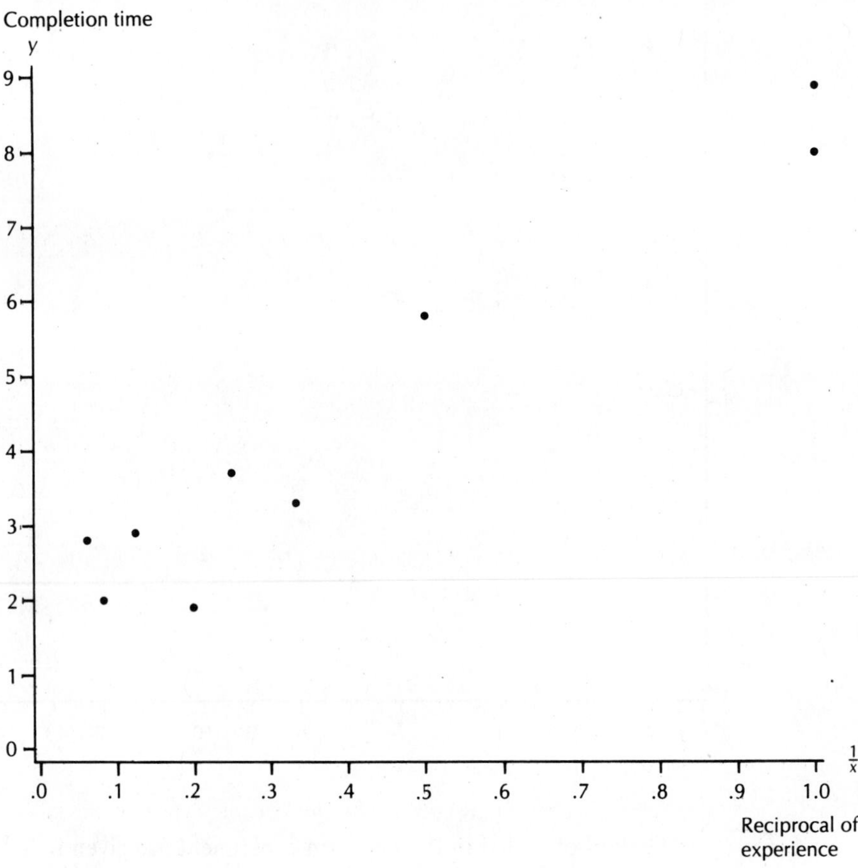

experience. We find that

$$\mathbf{X'X} = \begin{bmatrix} 9.0 & 3.5541 \\ 3.5541 & 2.4901 \end{bmatrix} \qquad (\mathbf{X'X})^{-1} = \begin{bmatrix} .2546 & -.3634 \\ -.3634 & .9203 \end{bmatrix}$$

$$\mathbf{X'y} = \begin{bmatrix} 41.1000 \\ 23.1340 \end{bmatrix} \qquad \mathbf{b} = (\mathbf{X'X})^{-1}\mathbf{X'y} = \begin{bmatrix} 2.0572 \\ 6.3545 \end{bmatrix}$$

Moreover, for this model, $SSE = 7.4141$ and $s = 1.029$.

The first goal of the experiment is to see whether there is significant evidence suggesting that the mean completion time decreases with increasing experience. To investigate this, we can test the null hypothesis $H_0: \beta_1 = 0$ against the alternative hypothesis $H_1: \beta_1 \neq 0$. The least squares point estimate of β_1 is $b_1 = 6.3545$, and the t statistic for testing $H_0: \beta_1 = 0$ is

$$t = \frac{b_1}{s\sqrt{c_{11}}} = \frac{6.3545}{1.029\sqrt{.9203}} = 6.44$$

Comparing $t = 6.44$ with the rejection point $t_{[\alpha/2]}^{(n-k)} = t_{[.025]}^{(7)} = 2.365$, we reject $H_0: \beta_1 = 0$ with $\alpha = .05$. Furthermore, since b_1 is positive, we have significant evidence that $\beta_1 > 0$. Note that if $\beta_1 > 0$, the equation

$$\mu = \beta_0 + \beta_1\left(\frac{1}{x}\right)$$

says that μ, mean completion time, will decrease towards β_0 as x, experience, increases. Thus we have significant evidence to suggest that the mean time to complete Form ST 1040AVG decreases with increasing experience.

The second goal of the experiment is to predict the amount of time it will take an individual with a given amount of experience to complete Form ST 1040AVG. We illustrate the procedure for an individual who has completed four such forms and who is thus filling out the fifth form. A point prediction for an individual completion time when $x = 5$ is

$$\begin{aligned}\hat{y} &= b_0 + b_1\left(\frac{1}{x}\right) \\ &= 2.0572 + 6.3545\left(\frac{1}{5}\right) = 3.3281 \text{ hours}\end{aligned}$$

Since we are predicting an *individual* completion time, we wish to compute a 95% prediction interval when $x = 5$. Letting

$$\mathbf{x}_0' = \begin{bmatrix} 1 & \frac{1}{x} \end{bmatrix} = \begin{bmatrix} 1 & \frac{1}{5} \end{bmatrix}$$

we obtain

$$\begin{aligned}&[3.3281 \pm t_{[\alpha/2]}^{(n-k)} s\sqrt{1 + \mathbf{x}_0'(\mathbf{X}'\mathbf{X})^{-1}\mathbf{x}_0}] \\ &= [3.3281 \pm 2.365(1.029)\sqrt{1 + .2138}] \\ &= [.7225, 5.9337]\end{aligned}$$

This says that we can be 95 percent confident that an individual who has completed Form ST 1040AVG four times will require between .7225 and 5.9337 hours to complete the fifth such form.

13.2

HANDLING NONCONSTANT ERROR VARIANCES AND WEIGHTED LEAST SQUARES

VARIANCE-EQUALIZING TRANSFORMATIONS

When an increasing or decreasing error variance exists, inference assumption 1 is violated. When this constant variance assumption is seriously violated, we must equalize the variances of the populations of potential error terms. If we do not, the statistical inference formulas (confidence and prediction intervals, hypothesis tests, and so on) that we have presented will not be valid.

To equalize variances, we use transformations. For instance, consider the general linear regression model

$$y_i = \beta_0 + \beta_1 x_{i1} + \beta_2 x_{i2} + \cdots + \beta_p x_{ip} + \varepsilon_i$$

If the variances $\sigma_1^2, \sigma_2^2, \ldots, \sigma_n^2$ are unequal and *known*, then the variances can be equalized by using the transformed model

$$\frac{y_i}{\sigma_i} = \beta_0\left(\frac{1}{\sigma_i}\right) + \beta_1\left(\frac{x_{i1}}{\sigma_i}\right) + \beta_2\left(\frac{x_{i2}}{\sigma_i}\right) + \cdots + \beta_p\left(\frac{x_{ip}}{\sigma_i}\right) + \eta_i$$

where $\eta_i = \varepsilon_i/\sigma_i$. This transformed model has the same parameters as the original model and also satisfies the constant variance assumption. This is because the properties of the variance tell us that the variance of the potential η_i values for the transformed model is $\sigma_i^2(1/\sigma_i^2) = 1$. This follows because, if we divide every value in a population by a constant k, the variance of the new population is $1/k^2$ multiplied by the variance of the original population. The parameters $\beta_0, \beta_1, \beta_2, \ldots, \beta_p$ in the transformed model are estimated by using

$$\mathbf{y} = \begin{bmatrix} \frac{y_1}{\sigma_1} \\ \frac{y_2}{\sigma_2} \\ \vdots \\ \frac{y_n}{\sigma_n} \end{bmatrix} \quad \text{and} \quad \mathbf{X} = \begin{bmatrix} \frac{1}{\sigma_1} & \frac{x_{11}}{\sigma_1} & \frac{x_{12}}{\sigma_1} & \cdots & \frac{x_{1p}}{\sigma_1} \\ \frac{1}{\sigma_2} & \frac{x_{21}}{\sigma_2} & \frac{x_{22}}{\sigma_2} & \cdots & \frac{x_{2p}}{\sigma_2} \\ \vdots & \vdots & \vdots & & \vdots \\ \frac{1}{\sigma_n} & \frac{x_{n1}}{\sigma_n} & \frac{x_{n2}}{\sigma_n} & \cdots & \frac{x_{np}}{\sigma_n} \end{bmatrix}$$

Notice here that some computer packages by default supply a column of 1's for an intercept term. We would have to override such a default here, since we wish to include a transformed intercept column containing $1/\sigma_i$.

The trouble with this approach is, of course, that the variances $\sigma_1^2, \sigma_2^2, \ldots, \sigma_n^2$ are not precisely known in a practical regression problem. In such a case these variances can be estimated when several observations have been randomly selected from each population of values of the dependent variable. Defining s_i^2 to be the sample variance of the observations drawn from the ith population, the estimates of $\sigma_1^2, \sigma_2^2, \ldots, \sigma_n^2$ are $s_1^2, s_2^2, \ldots, s_n^2$. Then if the usual linear model has a nonconstant error variance, the variances can be approximately equalized by using the transformed model

$$\left(\frac{y_i}{s_i}\right) = \beta_0\left(\frac{1}{s_i}\right) + \beta_1\left(\frac{x_{i1}}{s_i}\right) + \beta_2\left(\frac{x_{i2}}{s_i}\right) + \cdots + \beta_p\left(\frac{x_{ip}}{s_i}\right) + \eta_i$$

where $\eta_i = \varepsilon_i/s_i$. The parameters in this transformed model can be estimated by using the **y** vector and **X** matrix specified previously with $\sigma_1, \sigma_2, \ldots, \sigma_n$ replaced by $s_1, s_2, \ldots, s_n$.

Sometimes the variances $\sigma_1^2, \sigma_2^2, \ldots, \sigma_n^2$ cannot be estimated. For example, there might be only one observation from each population of values of the dependent variable. Then other kinds of transformations can be used to equalize variances. Often, the error variance is a function of one of the independent variables. As we have seen, such behavior would be characterized by fanning out or funneling in patterns in a residual plot against the independent variable.

Specifically, consider the regression model

$$y_i = \beta_0 + \beta_1 x_{i1} + \cdots + \beta_j x_{ij} + \cdots + \beta_p x_{ip} + \varepsilon_i$$

If a residual plot against x_{ij} fans out, it implies that the error variance σ_i^2 and standard deviation σ_i increase as x_{ij} increases. In such a situation it might be reasonable to conclude that $\sigma_i = x_{ij}^c\sigma$. This says that σ_i is proportional to some power of x_{ij}, where σ is a proportionality constant. Frequently, either $c = 1/2$, in which case $\sigma_i = x_{ij}^{1/2}\sigma$, or $c = 1$, in which case $\sigma_i = x_{ij}\sigma$. If $\sigma_i = x_{ij}^c\sigma$, then we can obtain an appropriate transformed model by dividing all terms in the model by x_{ij}^c. The transformed model is

$$\left(\frac{y_i}{x_{ij}^c}\right) = \beta_0\left(\frac{1}{x_{ij}^c}\right) + \beta_1\left(\frac{x_{i1}}{x_{ij}^c}\right) + \cdots + \beta_j\left(\frac{x_{ij}}{x_{ij}^c}\right) + \cdots + \beta_p\left(\frac{x_{ip}}{x_{ij}^c}\right) + \eta_i$$

where $\eta_i = \varepsilon_i/x_{ij}^c$. Here, the properties of the variance tell us that for any i the variance of the potential η_i values in the population of error terms for the transformed model is

$$\sigma_i^2\left(\frac{1}{x_{ij}^c}\right)^2 = (x_{ij}^c\sigma)^2\left(\frac{1}{x_{ij}^c}\right)^2 = \sigma^2$$

Therefore the transformed model satisfies the constant variance assumption. The parameters $\beta_0, \beta_1, \ldots, \beta_p$ in the transformed model (which are the same parameters as in the original model) are estimated by using

$$\mathbf{y} = \begin{bmatrix} \dfrac{y_1}{x_{1j}^c} \\ \dfrac{y_2}{x_{2j}^c} \\ \vdots \\ \dfrac{y_n}{x_{nj}^c} \end{bmatrix} \quad \text{and} \quad \mathbf{X} = \begin{bmatrix} \dfrac{1}{x_{1j}^c} & \dfrac{x_{11}}{x_{1j}^c} & \cdots & \dfrac{x_{1j}}{x_{1j}^c} & \cdots & \dfrac{x_{1p}}{x_{1j}^c} \\ \dfrac{1}{x_{2j}^c} & \dfrac{x_{21}}{x_{2j}^c} & \cdots & \dfrac{x_{2j}}{x_{2j}^c} & \cdots & \dfrac{x_{2p}}{x_{2j}^c} \\ \vdots & \vdots & & \vdots & & \vdots \\ \dfrac{1}{x_{nj}^c} & \dfrac{x_{n1}}{x_{nj}^c} & \cdots & \dfrac{x_{nj}}{x_{nj}^c} & \cdots & \dfrac{x_{np}}{x_{nj}^c} \end{bmatrix}$$

Again, some computer packages by default supply a column of 1's for an intercept term. We would have to override this default here, since we wish to include an intercept column containing $1/x_{ij}^c$ values.

We now (without going into a detailed discussion of each case) list some other transformations that can be used to equalize variances.

1. If the *variance* σ_i^2 appears to be an increasing linear function of the $\hat{y}_i$ values, an appropriate transformed model can often be obtained by replacing y by its square root, $\sqrt{y}$.
2. If the *standard deviation* σ_i appears to be an increasing linear function of the $\hat{y}_i$ values, an appropriate transformed model can often be obtained by replacing y by the logarithm (natural or base 10) of y.
3. If the data display a multiplicative error structure, then the variance σ_i^2 will often appear to increase or decrease with increasing values of y. An appropriate transformation that will often equalize the variances is obtained by taking logarithms (natural or base 10) of both sides of the regression equation (for instance, see Example 13.1).
4. When the values of y are "count data," an appropriate transformation that will often equalize the variances is obtained by replacing y by its square root, $\sqrt{y}$. Here, by count data we mean, for example, the number of occurrences of an event in a given period of time. Such data often can be described by a *Poisson distribution.* For a discussion of this distribution, see any basic statistics book. We present an example of this kind of problem in Example 13.5.

Example 13.3 The National Association of Retail Hardware Stores (NARHS) is a nationally known trade association. It wishes to investigate the relationship between x, home value (in thousands of dollars), and y, yearly expenditure on upkeep like lawn care, painting, and repairs (in dollars). A random sample of 40 homeowners is taken, and the results are given in Table 13.3. Figure 13.8 gives a SAS plot of y versus x. From this plot it appears that y is increasing (probably in a quadratic fashion) as x increases. Also, the variance of the y values is increasing as x increases. Increasing

TABLE 13.3 NARHS upkeep expenditure data

House	Value of house, x (thousands of dollars)	Upkeep expenditure, y (dollars)
1	118.50	706.04
2	76.54	398.60
3	92.43	436.24
4	111.03	501.71
5	80.34	426.45
6	49.84	144.24
7	114.52	644.23
8	50.89	211.54
9	128.93	675.87
10	48.14	189.02
11	85.50	459.04
12	115.51	813.62
13	114.16	602.39
14	102.95	428.52
15	92.86	387.50
16	84.39	434.63
17	123.53	698.00
18	77.77	355.75
19	112.10	737.59
20	101.02	706.66
21	76.52	424.57
22	116.09	656.92
23	62.72	301.03
24	84.91	321.07
25	88.64	519.40
26	81.41	348.50
27	60.22	162.17
28	95.55	482.55
29	79.39	460.07
30	89.25	475.45
31	136.10	835.16
32	24.45	62.70
33	52.28	239.89
34	143.09	1005.32
35	41.86	184.18
36	43.10	212.80
37	66.79	313.45
38	106.43	658.47
39	61.01	195.08
40	99.01	545.42

variance makes some intuitive sense here, since people with more expensive homes generally have higher incomes and can afford to pay to have upkeep done or can perform upkeep chores themselves if they wish. This would cause a relatively large variation in upkeep expenses for people owning expensive homes.

Since y appears to increase in a quadratic fashion as x increases, we begin by trying the model

$$y_i = \beta_0 + \beta_1 x_i + \beta_2 x_i^2 + \varepsilon_i$$

FIGURE 13.8 SAS plot of upkeep expenditure versus value of house for the NARHS data

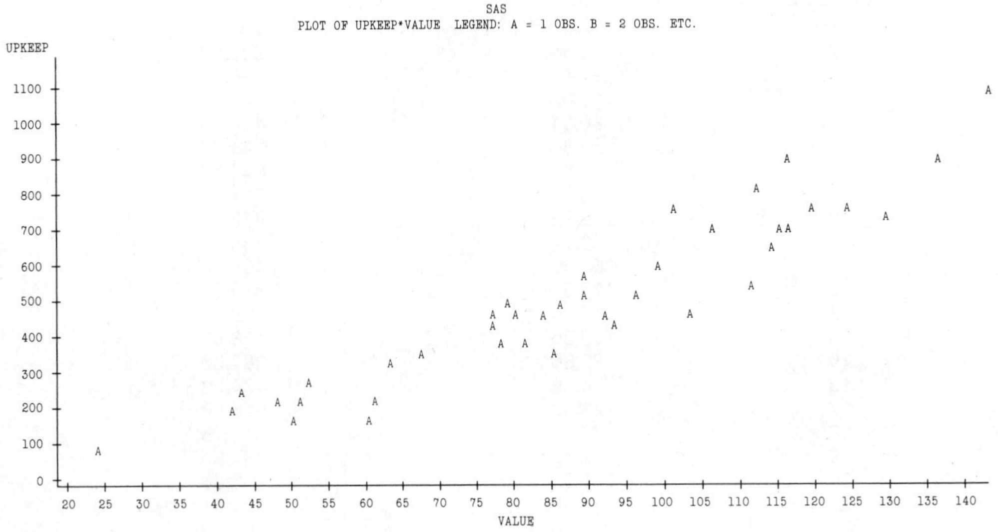

When we estimate β_0, β_1, and β_2 using the data in Table 13.3, we obtain the prediction equation

$$\hat{y}_i = -14.4436 + 3.0585x_i + .0246x_i^2$$

The t statistic for the x_i^2 term is $t = 1.98$, which has an associated prob-value of .0554. Figure 13.9 shows a SAS plot of the residuals for this model against x_i. We see that the residuals appear to fan out as x_i increases, indicating the existence of an increasing error variance. Therefore the t-test and prob-value for the x_i^2 term are not valid. This is because the constant variance assumption is not met.

Since both Figures 13.8 and 13.9 suggest the existence of an increasing error variance, we consider using a variance-equalizing transformation. Here we will divide all terms in the model by x_i^c, although the value of c is not at all obvious. Our strategy will be to choose a transformation and then check it, using a plot of the residuals from the transformed model. If this plot fails to suggest the existence of a nonconstant error variance, then we will be satisfied even though some other transformation may also solve the unequal variances problem.

We will start by assuming that $c = 1$. This implies that $\sigma_i = x_i\sigma$. Under this assumption the appropriate transformation is division of each term in the model by x_i. The transformed model is

$$\frac{y_i}{x_i} = \beta_0\left(\frac{1}{x_i}\right) + \beta_1\left(\frac{x_i}{x_i}\right) + \beta_2\left(\frac{x_i^2}{x_i}\right) + \eta_i$$

FIGURE 13.9 SAS plot of the residuals for the quadratic NARHS model versus *x* (value of house)

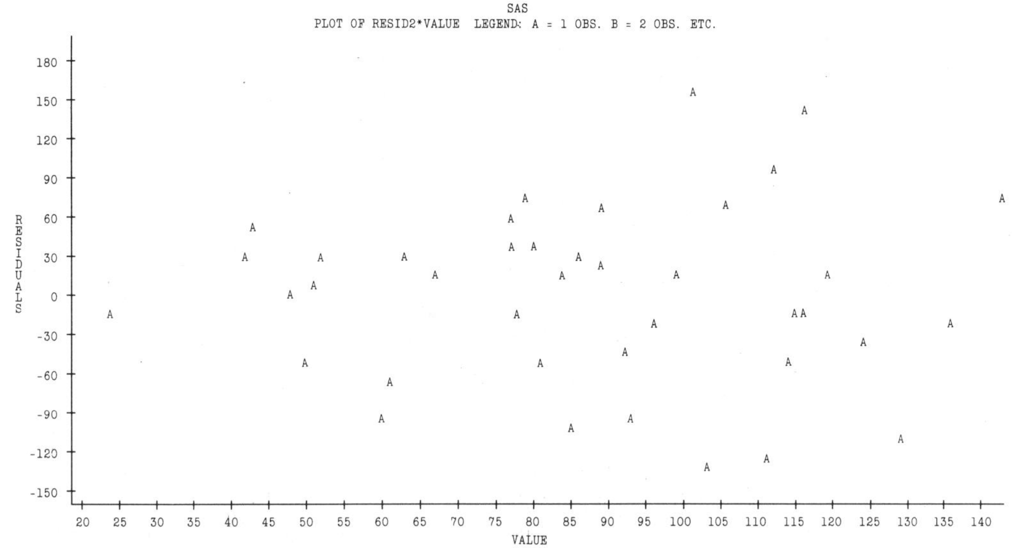

$$\frac{y_i}{x_i} = \beta_0\left(\frac{1}{x_i}\right) + \beta_1 + \beta_2 x_i + \eta_i$$

where $\eta_i = \varepsilon_i/x_i$. If this transformation remedies the unequal variances problem, a residual plot against x_i should appear as a horizontal band. However, if this transformation is too extreme, a residual plot for the transformed model should show the residuals funneling in as x_i increases.

We estimate β_0, β_1, and β_2 in the transformed model by using (see Table 13.3)

$$\mathbf{y} = \begin{bmatrix} \dfrac{706.04}{118.50} \\ \dfrac{398.60}{76.54} \\ \vdots \\ \dfrac{545.42}{99.01} \end{bmatrix} = \begin{bmatrix} 5.9581 \\ 5.2007 \\ \vdots \\ 5.5087 \end{bmatrix}$$

and

$$\mathbf{X} = \begin{bmatrix} \frac{1}{118.50} & 1 & 118.50 \\ \frac{1}{76.54} & 1 & 76.54 \\ \vdots & \vdots & \vdots \\ \frac{1}{99.01} & 1 & 99.01 \end{bmatrix} = \begin{bmatrix} .008439 & 1 & 118.50 \\ .013065 & 1 & 76.54 \\ \vdots & \vdots & \vdots \\ .0101 & 1 & 99.01 \end{bmatrix}$$

We obtain the prediction equation

$$\frac{\hat{y}_i}{x_i} = \frac{-26.7501}{x_i} + 3.4089 + 0.0224x_i$$

Figure 13.10 presents the SAS output of the least squares point estimates, the standard error $s = .7935$, and other quantities related to the transformed model.

We calculate the residuals for the transformed model by the equation

$$e_i = \frac{y_i}{x_i} - \left(\frac{-26.7501}{x_i} + 3.4089 + 0.0224x_i\right)$$

These residuals are given in Figure 13.10 and plotted in Figure 13.11. Since the plot has a horizontal band appearance, we can assume that our transformation has

FIGURE 13.10 **SAS output of a regression analysis of the NARHS data by using the transformed model** $\frac{y_i}{x_i} = \beta_0\left(\frac{1}{x_i}\right) + \beta_1 + \beta_2 x_i + \eta_i$

ANALYSIS OF VARIANCE

SOURCE	DF	SUM OF SQUARES	MEAN SQUARE	F VALUE	PROB>F
MODEL	3	1029.79657	343.26552	545.233	0.0001
ERROR	37	23.29428654	0.62957531		
U TOTAL	40	1053.09086			

ROOT MSE	0.7934578	R-SQUARE	0.9779
DEP MEAN	5.014115	ADJ R-SQ	0.9761
C.V.	15.82448		

PARAMETER ESTIMATES

VARIABLE	DF	PARAMETER ESTIMATE	STANDARD ERROR	T FOR H0: PARAMETER=0	PROB > |T|
RVALUE	1	-26.75007521	41.59970494	-0.643	0.5242
ONE	1	3.40891448	1.32081782	2.581	0.0140
VALUE	1	0.02244715	0.009253194	2.426	0.0203

OBS	ACTUAL	PREDICT VALUE	STD ERR PREDICT	RESIDUAL
1	5.9581	5.8432	0.1960	0.1150
2	5.2077	4.7775	0.1630	0.4302
3	4.7197	5.1943	0.1398	-0.4746
4	4.5187	5.6603	0.1657	-1.1416
5	5.3081	4.8794	0.1560	0.4287
6	2.8941	3.9910	0.2103	-1.0969
7	5.6255	5.7460	0.1787	-0.1205
8	4.1568	4.0256	0.2075	0.1312
9	5.2421	6.0955	0.2510	-0.8534
10	3.9265	3.9338	0.2158	-.007383
11	5.3689	5.0153	0.1473	0.3536
12	7.0437	5.7702	0.1828	1.2735
13	5.2767	5.7372	0.1773	-0.4604
14	4.1624	5.4600	0.1450	-1.2976
15	4.1729	5.2053	0.1396	-1.0323
16	5.1503	4.9862	0.1490	0.1640
17	5.6504	5.9653	0.2210	-0.3148
18	4.5744	4.8107	0.1607	-0.2363
19	6.5798	5.6866	0.1695	0.8931
20	6.9952	5.4117	0.1423	1.5835
21	5.5485	4.7770	0.1631	0.7715
22	5.6587	5.7844	0.1852	-0.1257
23	4.7996	4.3903	0.1870	0.4093
24	3.7813	4.9999	0.1482	-1.2186
25	5.8597	5.0968	0.1431	0.7628
26	4.2808	4.9078	0.1540	-0.6270
27	2.6930	4.3165	0.1908	-1.6235
28	5.0502	5.2738	0.1390	-0.2235
29	5.7951	4.8540	0.1577	0.9410
30	5.3272	5.1126	0.1425	0.2146
31	6.1364	6.2674	0.2947	-0.1311
32	2.5644	2.8637	0.7142	-0.2993
33	4.5886	4.0708	0.2042	0.5178
34	7.0258	6.4339	0.3406	0.5919
35	4.3999	3.7095	0.2515	0.6904
36	4.9374	3.7557	0.2418	1.1816
37	4.6931	4.5076	0.1806	0.1854
38	6.1869	5.5466	0.1522	0.6403
39	3.1975	4.3400	0.1896	-1.1425
40	5.5087	5.3612	0.1403	0.1475

suitably equalized the variances. If the residuals had fanned out after division by x_i, we could have tried a stronger transformation, such as division by x_i^2. A funneling in of this residual plot would lead us to try a milder transformation, such as division by $\sqrt{x_i}$.

Since we have concluded that $\sigma_i = x_i\sigma$, which implies that $\sigma_i^2 = x_i^2\sigma^2$, we would report that the mean square error here is $s^2x_i^2 = (.7935)^2x_i^2 = .6296x_i^2$. This emphasizes that the error variance depends on x_i.

Multiplying the transformed model through by x_i, we get the following final prediction equation:

$$\hat{y}_i = -26.7501 + 3.4089x_i + .0224x_i^2$$

FIGURE 13.11 SAS plot of residuals versus x for the model $\frac{y}{x} = \beta_0\left(\frac{1}{x}\right) + \beta_1 + \beta_2 x + \eta$

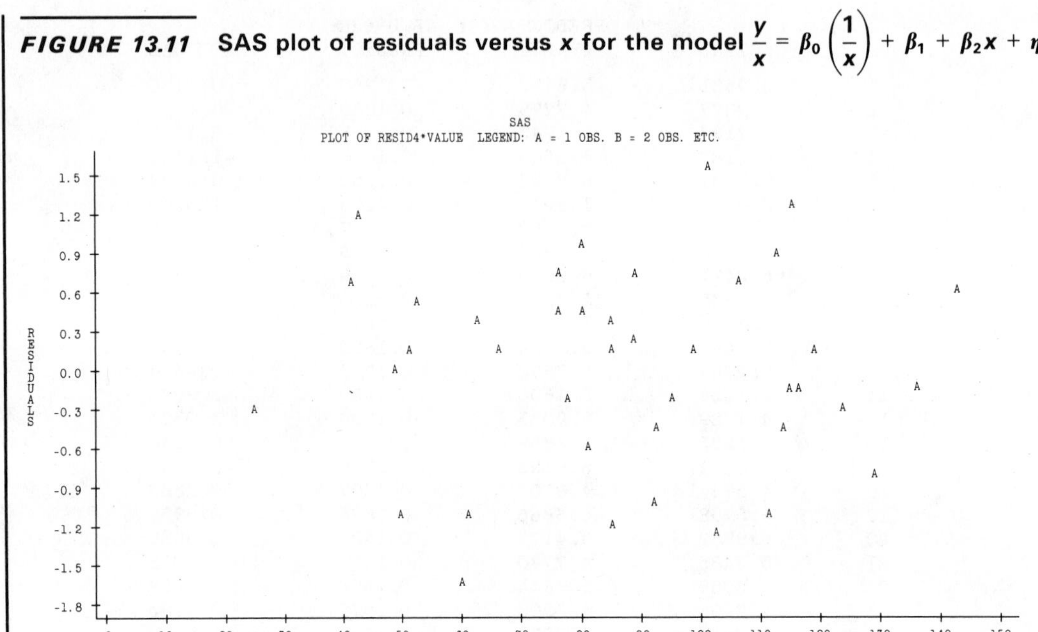

For example, since the value of house 1 in Table 13.3 is $x_1 = 118.5$, the predicted upkeep expenditure is

$$\begin{aligned}\hat{y}_1 &= -26.7501 + 3.4089(118.5) + .0224(118.5)^2 \\ &= 692.4\end{aligned}$$

If we conclude that $\sigma_i = x_{ij}^c \sigma$, procedures for computing confidence and prediction intervals are modifications of the standard procedures. The calculation of confidence intervals for the regression parameters and the calculation of t statistics present no new problems. We simply note that the transformed data satisfy the inference assumptions and that the transformed model involves the same parameters as does the original model. It follows that t statistics and their associated prob-values as well as confidence intervals for regression parameters are calculated by using the usual formulas. However, all quantities involved in the calculations (least squares point estimates, the standard error, and elements on the diagonal of $(\mathbf{X}'\mathbf{X})^{-1}$) *must be calculated by using the transformed data.*

Next, we consider calculating a confidence interval for the mean μ_0 and a prediction interval for the individual value y_0.

Confidence and Prediction Intervals When $\sigma_i = x_{ij}^c \sigma$

If $\sigma_i = x_{ij}^c \sigma$, then

1. An appropriate transformed model is obtained by dividing each term in the original model by x_{ij}^c.
2. If $b_0, b_1, \ldots, b_p$ are the least squares point estimates calculated by using the transformed model, then

   $$\hat{y}_0 = b_0 + b_1 x_{01} + \cdots + b_p x_{0p}$$

   is a *point estimate of* μ_0 and is a *point prediction of* y_0.
3. A *100(1 − α)% confidence interval for* μ_0 is

   $$[\hat{y}_0 \pm t_{[\alpha/2]}^{(n-k)} s \sqrt{\mathbf{x}_0'(\mathbf{X}'\mathbf{X})^{-1}\mathbf{x}_0}]$$

   where $(\mathbf{X}'\mathbf{X})^{-1}$ and s *are computed from the transformed data* and $\mathbf{x}_0' = [1 \quad x_{01} \quad \ldots \quad x_{0p}]$ is a row vector containing the values of the independent variables.
4. A *100(1 − α)% prediction interval for* y_0 is

   $$[\hat{y}_0 \pm t_{[\alpha/2]}^{(n-k)} s \sqrt{x_{0j}^{2c} + \mathbf{x}_0'(\mathbf{X}'\mathbf{X})^{-1}\mathbf{x}_0}]$$

   where x_{0j} is the value of the independent variable x_{ij}.

Looking at these formulas, we see that correct prediction intervals will tend to be wider at larger values of x_{0j} and will tend to be narrower at small values of x_{0j}. Ignoring the unequal variances problem would produce prediction intervals that are too narrow for large values of x_{0j} and too wide for small values of x_{0j}.

We now illustrate the calculation of confidence and prediction intervals.

Example 13.4 Consider Example 13.3 and the NARHS problem. In this problem the transformed model

$$\frac{y_i}{x_i} = \beta_0\left(\frac{1}{x_i}\right) + \beta_1 + \beta_2(x_i) + \eta_i$$

involves the same parameters (β_0, β_1, and β_2) as does the original model

$$y_i = \beta_0 + \beta_1 x_i + \beta_2 x_i^2 + \varepsilon_i$$

Suppose that we wish to find a 95% confidence interval for the parameter β_1. Then we use the least squares point estimate $b_1 = 3.4089$ and the standard error of the estimate $s\sqrt{c_{11}} = 1.3208$ that we calculated using the transformed model (see Figure 13.10). Here, c_{11} is obtained from the $(\mathbf{X}'\mathbf{X})^{-1}$ matrix calculated by using the transformed data, and s is also computed by using the transformed data. Hence the

95% confidence interval for β_1 is

$$\begin{aligned}[b_1 \pm t_{[\alpha/2]}^{(n-k)} s\sqrt{c_{11}}] &= [3.4089 \pm t_{[.05/2]}^{(40-3)}(1.3208)] \\ &= [3.4089 \pm 2.026(1.3208)] \\ &= [.7330, 6.0848]\end{aligned}$$

Figure 13.10 presents the t statistics and related prob-values for the transformed model. These results also apply to the original model. Notice that the t statistic for the quadratic term is now $t = 2.426$. Unlike the t statistic calculated from the untransformed data, this t statistic has a prob-value (.0203) less than .05. This t statistic and associated prob-value are valid, since the variance has been equalized by our transformation (assuming that inference assumptions 2 and 3 also hold). We see here that failure to recognize the nonconstant error variance could cause us to incorrectly omit the quadratic term.

Next, we compute a 95% confidence interval for μ_0, the mean upkeep expenditure for all houses that are worth \$110,000. A point estimate of $\mu_0 = \beta_0 + \beta_1(110) + \beta_2(110)^2$ is

$$\begin{aligned}\hat{y}_0 &= b_0 + b_1(110) + b_2(110)^2 \\ &= -26.7501 + 3.4089(110) + .0224(110)^2 = \$619.80\end{aligned}$$

Using $(\mathbf{X}'\mathbf{X})^{-1}$ from the transformed data and $\mathbf{x}_0' = [1 \quad 110 \quad (110)^2]$, we find that for this problem,

$$\mathbf{x}_0'(\mathbf{X}'\mathbf{X})^{-1}\mathbf{x}_0 = 506.354$$

Thus a 95% confidence interval for μ_0 (using s calculated from the transformed

FIGURE 13.12 **SAS output of predictions using the equation $\hat{y} = -26.7501 + 3.4089x + .0224x^2$**

ANALYSIS OF VARIANCE

SOURCE	DF	SUM OF SQUARES	MEAN SQUARE	F VALUE	PROB>F
MODEL	2	313.39579	156.69790	248.895	0.0001
ERROR	37	23.29428654	0.62957531		
C TOTAL	39	336.69008			

ROOT MSE	0.7934578	R-SQUARE	0.9308
DEP MEAN	289.9999	ADJ R-SQ	0.9271
C.V.	0.2736062		

PARAMETER ESTIMATES

VARIABLE	DF	PARAMETER ESTIMATE	STANDARD ERROR	T FOR H0: PARAMETER=0	PROB > |T|
INTERCEP	1	-26.75007521	41.59970494	-0.643	0.5242
VALUE	1	3.40891448	1.32081782	2.581	0.0140
VALUESQ	1	0.02244715	0.009253194	2.426	0.0203

OBS	ACTUAL	PREDICT VALUE	STD ERR PREDICT	LOWER95% MEAN	UPPER95% MEAN	LOWER95% PREDICT	UPPER95% PREDICT
1	706.0	692.4	23.2275	645.4	739.5	496.2	888.7
2	398.6	365.7	12.4781	340.4	391.0	240.0	491.3
3	436.2	480.1	12.9255	453.9	506.3	329.2	631.0
4	501.7	628.5	18.3994	591.2	665.7	446.1	810.8
5	426.4	392.0	12.5299	366.6	417.4	260.4	523.6
6	144.2	198.9	10.4822	177.7	220.1	116.0	281.8
7	644.2	658.0	20.4660	616.6	699.5	469.3	846.8
8	211.5	204.9	10.5588	183.5	226.3	120.3	289.4
9	675.9	785.9	32.3650	720.3	851.5	568.5	1003.3
10	189.0	189.4	10.3876	168.3	210.4	109.2	269.6
11	459.0	428.8	12.5952	403.3	454.3	289.0	568.6
12	813.6	666.5	21.1133	623.7	709.3	475.9	857.1
13	602.4	655.0	20.2372	613.9	696.0	466.9	843.0
14	428.5	562.1	14.9324	531.9	592.4	393.9	730.4
15	387.5	483.4	12.9643	457.1	509.6	331.8	634.9
16	434.6	420.8	12.5758	395.3	446.3	282.7	558.8
17	698.0	736.9	27.3037	681.6	792.2	530.7	943.0
18	355.8	374.1	12.4987	348.8	399.5	246.6	501.7
19	737.6	637.5	18.9969	599.0	676.0	453.2	821.8
20	706.7	546.7	14.3716	517.6	575.8	381.7	711.7
21	424.6	365.5	12.4777	340.3	390.8	239.9	491.1
22	656.9	671.5	21.5048	627.9	715.1	479.9	863.2
23	301.0	275.4	11.7316	251.6	299.1	171.8	379.0
24	321.1	424.5	12.5842	399.0	450.0	285.7	563.4
25	519.4	451.8	12.6883	426.1	477.5	307.0	596.6
26	348.5	399.5	12.5406	374.1	424.9	266.2	532.9
27	162.2	259.9	11.4899	236.7	283.2	160.4	359.5
28	482.5	503.9	13.2770	477.0	530.8	348.0	659.9
29	460.1	385.4	12.5197	360.0	410.7	255.2	515.5
30	475.4	456.3	12.7155	430.5	482.1	310.5	602.1
31	835.2	853.0	40.1117	771.7	934.3	619.6	1086.4
32	62.7000	70.0168	17.4619	34.6358	105.4	17.1308	122.9
33	239.9	212.8	10.6766	191.2	234.5	126.0	299.6
34	1005.3	920.6	48.7293	821.9	1019.4	670.3	1171.0
35	184.2	155.3	10.5260	134.0	176.6	84.6838	225.9
36	212.8	161.9	10.4216	140.8	183.0	89.4347	234.3
37	313.4	301.1	12.0624	276.6	325.5	190.9	411.2
38	658.5	590.3	16.1997	557.5	623.2	416.1	764.6
39	195.1	264.8	11.5688	241.3	288.2	163.9	365.6
40	545.4	530.8	13.8889	502.7	559.0	369.2	692.5
41	.	619.8	17.8546	583.7	656.0	439.3	800.3
		$\hat{y}_0$		95% confidence interval for μ_0		95% prediction interval for y_0	

data) is

$$[\hat{y}_0 \pm t_{[\alpha/2]}^{(n-k)} s\sqrt{\mathbf{x}_0'(\mathbf{X}'\mathbf{X})^{-1}\mathbf{x}_0}] = [619.8 \pm 2.026(.7935)\sqrt{506.354}]$$
$$= [583.7, 656.0]$$

Therefore the NARHS can be 95% confident that the mean yearly upkeep expenditure for all houses worth \$110,000 is between \$583.70 and \$656.00.

Finally, we compute a 95% prediction interval for y_0, the individual yearly expenditure on a house worth \$110,000. This interval is

$$[\hat{y}_0 \pm t_{[\alpha/2]}^{(n-k)} s\sqrt{x_0^2 + \mathbf{x}_0'(\mathbf{X}'\mathbf{X})^{-1}\mathbf{x}_0}] = [619.8 \pm 2.026(.7935)\sqrt{(110)^2 + 506.354}]$$
$$= [439.3, 800.3]$$

Here we again emphasize that $(\mathbf{X}'\mathbf{X})^{-1}$ and s are both computed by using the transformed data. Also, the "extra" x_0^2 under the radical is employed because we are assuming that (and residual plots verify that) $\sigma_i = x_i\sigma$ (see Example 13.3). This interval says that the NARHS can be 95 percent confident that the yearly upkeep expenditure for an individual house valued at \$110,000 will be between \$439.30 and \$800.30. Note that $\hat{y}_0$, the 95% confidence interval for μ_0, and the 95% prediction interval for y_0 are given in the SAS output of Figure 13.12.

Example 13.5 Republic Wholesalers, Inc., wants to investigate its telephone orders. The company believes that there is a repeating daily pattern for the volume of phone business. The goal of the analysis is to describe this daily pattern as a mathematical function of the time of day. More specifically, the company wishes to find 95% prediction intervals for the number of orders coming into the company switchboard at various times of day. The results obtained will be used to help in the scheduling of operators who have been trained to process orders.

Republic Wholesalers takes phone orders from 9 a.m. (hour 1 is defined to be 9 a.m. to 10 a.m.) through 5 p.m. (hour 8 is defined to be 4 p.m. to 5 p.m.). For each hour, 1 through 8, the number of telephone orders is recorded over a span of ten business days. This telephone order data is given in Table 13.4. Considering this table, we let y_t denote the number of telephone orders occurring in time period t

TABLE 13.4 Republic Wholesalers data: Hourly number of telephone orders

	Hour, H_t							
Day	1	2	3	4	5	6	7	8
1	10(=y_1)	34(=y_2)	29	44	36	17	8	11(=y_8)
2	31(=y_9)	19(=y_{10})	18	16	28	12	14	8(=y_{16})
3	8	9	12	14	23	28	15	1
4	0	23	34	53	56	15	7	2
5	8	13	30	69	26	69	22	0
6	2	35	15	6	19	23	7	3
7	12	16	13	19	7	4	18	8
8	16	17	28	73	35	15	6	14
9	1	16	7	28	19	10	10	0
10	12(=y_{73})	7(=y_{74})	46	4	13	13	5	10(=y_{80})
Hour Avg., $\bar{y}_{H_t}$	10.0	18.9	23.2	32.6	26.2	20.6	11.2	5.7
Hour Var., $s_{H_t}^2$	$(9.1)^2$	$(9.4)^2$	$(12.1)^2$	$(25.5)^2$	$(13.8)^2$	$(18.3)^2$	$(5.8)^2$	$(5.1)^2$

FIGURE 13.13 Number of telephone orders versus hour of day

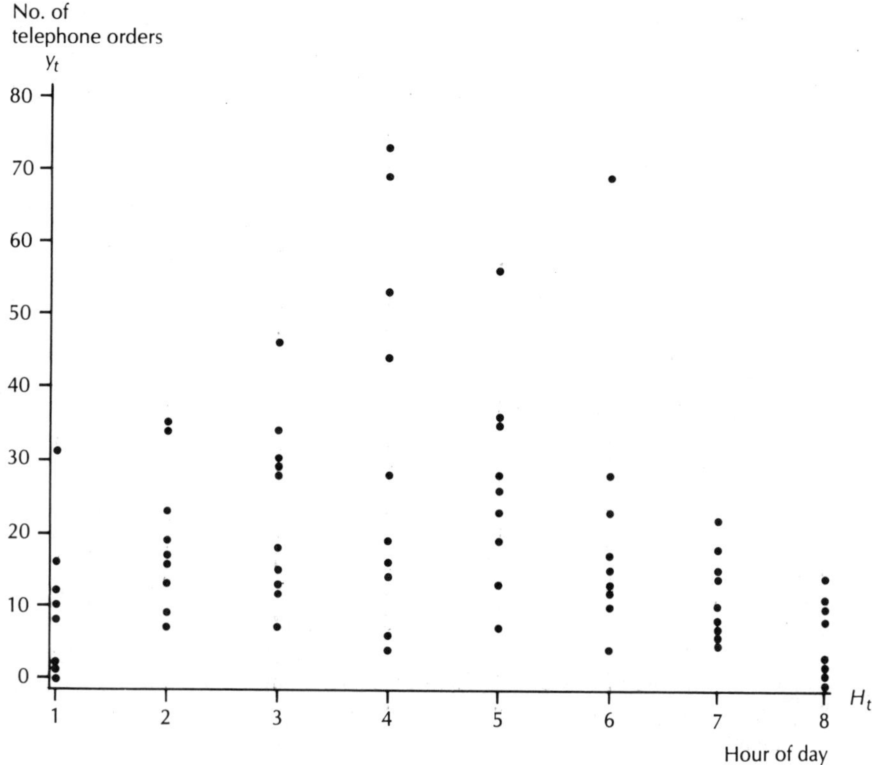

(note that there are $10 \times 8 = 80$ different time periods and thus 80 different y_t values). We let H_t (for $t = 1, 2, \ldots, 80$) denote the hour number corresponding to time period t. Here, H_t can take on any of the values 1, 2, 3, 4, 5, 6, 7, or 8. Thus, for example, $H_{10} = 2$ (see Table 13.4). The different values of y_t are plotted against H_t in Figure 13.13. Two things are apparent from this plot. First, the average number of telephone orders is larger in the middle of the day. Second, the variance of the number of telephone orders is also larger in the middle of the day. This second fact indicates that a nonconstant error variance exists. As further evidence of this, let $\bar{y}_{H_t}$ and $s^2_{H_t}$ denote the mean and variance, respectively, of the different numbers of telephone orders occurring in hour H_t (for $H_t = 1, 2, \ldots, 8$). These quantities are calculated in Table 13.4. The different values of $s^2_{H_t}$ are plotted versus $\bar{y}_{H_t}$ in Figure 13.14. We see that $s^2_{H_t}$ is probably a linear function of the average level of telephone orders as measured by $\bar{y}_{H_t}$. This indicates that the error variance corresponding to hour H_t is an increasing linear function of the true mean number of telephone orders that could occur in hour H_t. This fact, along with the fact that y represents count data (the number of telephone orders) that might be described

FIGURE 13.14 Plot of $s^2_{H_t}$ versus $\bar{y}_{H_t}$ for the Republic Wholesalers data

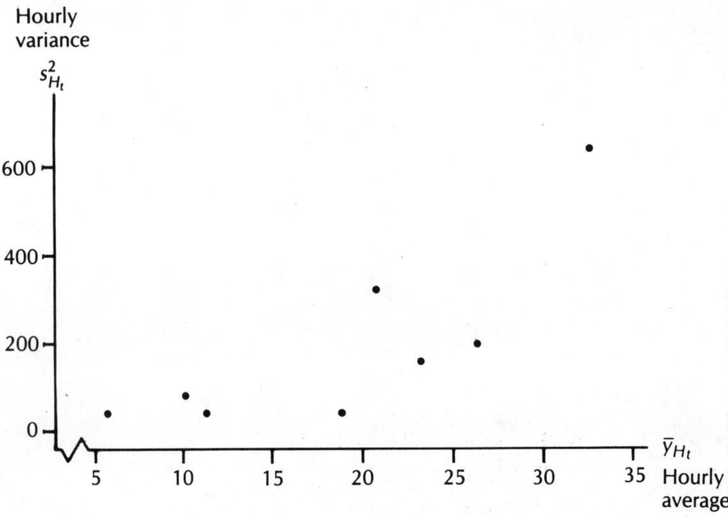

by a Poisson distribution, motivates us to use a variance-equalizing transformation in which we replace y by its square root.

We now let g_t denote the square root of the number of telephone orders at time t. A plot of the square root g_t against the hour number H_t is given in Figure 13.15. Notice that the square root transformation seems to have equalized the error variance. The variance of the number of telephone orders is larger in the middle of the day (see Figure 13.13), while the variance of the square roots is not dramatically larger in the middle of the day (see Figure 13.15). To account for the apparent curvature in Figure 13.15 (which is still present after taking square roots), we try the model

$$g_t = \beta_0 + \beta_1 H_t + \beta_2 H_t^2 + \varepsilon_t$$

To find the least squares point estimates of the parameters β_0, β_1, and β_2, we use (see Table 13.4)

$$\mathbf{y} = \begin{bmatrix} \sqrt{10} \\ \sqrt{34} \\ \sqrt{29} \\ \cdot \\ \cdot \\ \cdot \\ \sqrt{5} \\ \sqrt{10} \end{bmatrix} \quad \text{and} \quad \mathbf{X} = \begin{bmatrix} 1 & 1 & 1 \\ 1 & 2 & 4 \\ 1 & 3 & 9 \\ \cdot & \cdot & \cdot \\ \cdot & \cdot & \cdot \\ \cdot & \cdot & \cdot \\ 1 & 7 & 49 \\ 1 & 8 & 64 \end{bmatrix}$$

FIGURE 13.15 Square roots of telephone orders versus hour of day

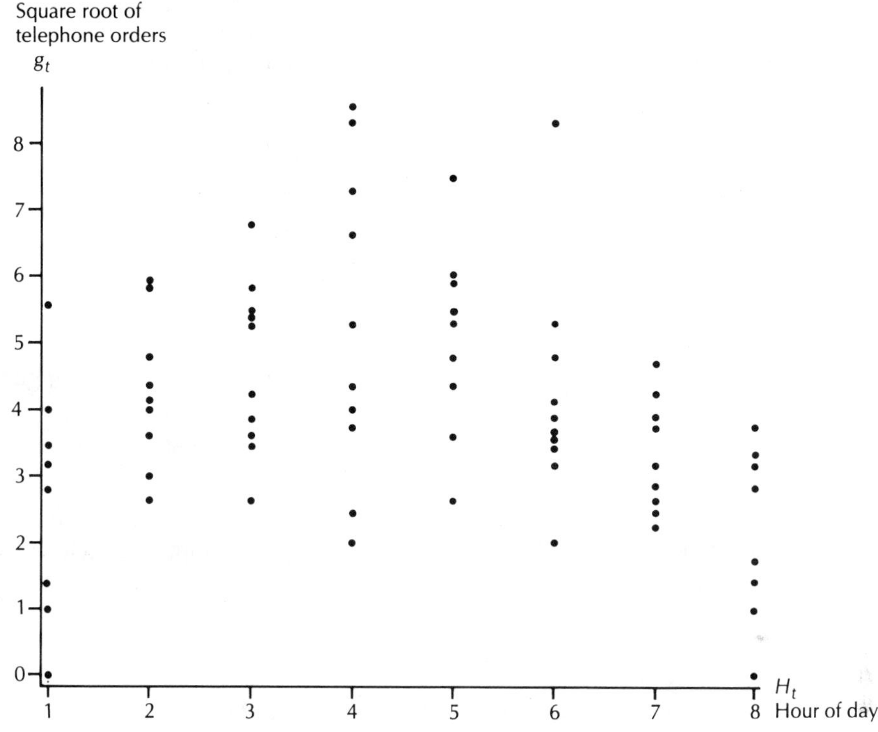

We obtain the prediction equation

$$\hat{g}_t = 1.2604 + 1.8363H_t - .2197H_t^2$$

For this transformed model we also obtain $s = 1.4584$.

Table 13.5 gives the t statistics for this model. Since Figure 13.15 indicates that our transformed model displays a constant error variance, these t statistics

TABLE 13.5 The t statistics for the model $g_t = \beta_0 + \beta_1 H_t + \beta_2 H_t^2 + \varepsilon_t$

Parameter, β_j	*Least squares point estimate,* b_j	*Standard error of the estimate,* $s\sqrt{c_{jj}}$	*t statistic,* $t = \frac{b_j}{s\sqrt{c_{jj}}}$
β_0	$b_0 = 1.2604$	.6434	$t = 1.96$
β_1	$b_1 = 1.8363$	.3280	$t = 5.60$
β_2	$b_2 = -.2197$	.0356	$t = -6.17$

are valid (assuming that inference assumptions 2 and 3 also hold). We see that the quadratic term is important, as the plot in Figure 13.15 indicated it might be.

We now calculate a 95% prediction interval for the number of telephone orders received on a future day between noon and 1 p.m. (hour 4). First, a point prediction of the square root of the number of telephone orders is

$$\begin{aligned}\hat{g}_0 &= 1.2604 + 1.8363H_0 - .2197H_0^2 \\ &= 1.2604 + 1.8363(4) - .2197(16) = 5.0904\end{aligned}$$

Therefore a point prediction of the number of telephone orders received is $(5.0904)^2 = 25.91$. Next, a 95% prediction interval for the square root of the number of telephone orders is

$$\begin{aligned}[\hat{g}_0 \pm t_{[\alpha/2]}^{(n-k)} s\sqrt{1 + \mathbf{x}_0'(\mathbf{X}'\mathbf{X})^{-1}\mathbf{x}_0}] &= [5.0904 \pm t_{[.025]}^{(80-3)}(1.4584)\sqrt{1 + .0287}] \\ &= [5.0904 \pm 1.990(1.4584)\sqrt{1.0287}] \\ &= [2.1468, 8.0340]\end{aligned}$$

Here $\mathbf{x}_0' = [1 \quad 4 \quad 16]$ (we have omitted a detailed calculation of $\mathbf{x}_0'(\mathbf{X}'\mathbf{X})^{-1}\mathbf{x}_0$). It follows that a 95% prediction interval for the number of telephone orders between noon and 1 p.m. is

$$[(2.1468)^2, (8.0340)^2] \quad \text{or} \quad [4.61, 64.55]$$

Thus Republic Wholesalers, Inc., can be 95 percent confident that between 4.61 and 64.55 (or about 5 to 65) telephone orders will be received between noon and 1 p.m. This interval is quite wide. For this problem we have $\mathbf{x}_0'(\mathbf{X}'\mathbf{X})^{-1}\mathbf{x}_0 = .0287$, which could be reduced by collecting more data. However, even if we were to set $\mathbf{x}_0'(\mathbf{X}'\mathbf{X})^{-1}\mathbf{x}_0$ equal to zero, we would only reduce

$$s\sqrt{1 + \mathbf{x}_0'(\mathbf{X}'\mathbf{X})^{-1}\mathbf{x}_0} = 1.4584\sqrt{1 + .0287} = 1.4792$$

to

$$s\sqrt{1 + 0} = s = 1.4584$$

This implies that the large width of this prediction interval is due mostly to the variability that is inherent in the ordering process. In this case, Republic Wholesalers would have to decide whether or not it is worthwhile to staff their switchboard with enough trained operators to handle the maximum number of calls (65) that might be received. Prediction intervals for the other hours in the day can be found in a similar manner.

WEIGHTED LEAST SQUARES

Consider the regression model

$$\begin{aligned} y_i &= \mu_i + \varepsilon_i \\ &= \beta_0 + \beta_1 x_{i1} + \cdots + \beta_p x_{ip} + \varepsilon_i \end{aligned}$$

If $\sigma_i^2 = \sigma^2/w_i$, then it is intuitively reasonable to choose point estimates $b_0, b_1, \ldots, b_p$ that minimize the sum of weighted squared residuals

$$SSE_w = \sum_{i=1}^{n} w_i(y_i - (b_0 + b_1 x_{i1} + \cdots + b_p x_{ip}))^2$$

Intuitively, this is because the smaller w_i is, the larger $\sigma_i^2 = \sigma^2/w_i$ is. This says that y_i (being more variable) provides less information about μ_i and should count less in determining the point estimates. It can be proven that the point estimates that minimize SSE_w are calculated by the equation

$$\begin{bmatrix} b_0 \\ b_1 \\ \cdot \\ \cdot \\ \cdot \\ b_p \end{bmatrix} = (\mathbf{X}'\mathbf{W}\mathbf{X})^{-1}\mathbf{X}'\mathbf{W}\mathbf{y}$$

Here,

$$\mathbf{W} = \begin{bmatrix} w_1 & 0 & \cdots & 0 \\ 0 & w_2 & \cdots & 0 \\ \vdots & \vdots & \ddots & \vdots \\ 0 & 0 & \cdots & w_n \end{bmatrix}$$

In this case we say that we are using *weighted least squares.*

It can be proven that the point estimates that minimize SSE_w are the same as the least squares point estimates when using the transformed model

$$\left(\frac{y_i}{(1/\sqrt{w_i})}\right) = \beta_0\left(\frac{1}{(1/\sqrt{w_i})}\right) + \beta_1\left(\frac{x_{i1}}{(1/\sqrt{w_i})}\right) + \cdots + \beta_p\left(\frac{x_{ip}}{(1/\sqrt{w_i})}\right) + \eta_i$$

Therefore when $\sigma_i = x_{ij}^c\sigma$, and we therefore use the transformed model

$$\left(\frac{y_i}{x_{ij}^c}\right) = \beta_0\left(\frac{1}{x_{ij}^c}\right) + \beta_1\left(\frac{x_{i1}}{x_{ij}^c}\right) + \cdots + \beta_p\left(\frac{x_{ip}}{x_{ij}^c}\right) + \eta_i$$

we are using weighted least squares, where $w_i = 1/x_{ij}^{2c}$.

13.3

BINARY DEPENDENT VARIABLES AND LOGISTIC REGRESSION

A *binary* dependent variable is a dependent variable that takes the value 0 or 1.

Example 13.6 Suppose that we wish to investigate the attitudes of people in various income brackets toward a new tax cut. To do this, we take a sample survey of n people and ask each person to report his or her income (x) and whether or not a tax cut is favored. We now let $y = 1$ if a tax cut is favored and $y = 0$ otherwise. By defining y in this way we have set up the problem so that the mean value of y in the population of all people at a particular income level x is simply the proportion of people with that income who favor a tax cut. Therefore if we consider a regression in which y is the dependent variable and x is the independent variable, we can try to establish a relationship between income x and the probability that a person selected at random with income x will favor a tax cut. This can be done by estimating the mean value of y as a function of income x.

When a binary variable y is expressed as a function of x, the mean value of y, denoted $p(x)$, is interpreted as a probability. This being the case, a prediction equation obtained from such a regression makes sense only when it produces probabilities between zero and 1. For example, the prediction equation $\hat{y} = .01 + .00002x$ would not be sensible for an income of \$50,000. This is because this equation produces a probability larger than 1, since $.01 + .00002(50{,}000) = 1.01$. Another feature of such a regression is that for a given x the dependent variable y is 1 with some probability $p(x)$ and is zero with probability $1 - p(x)$. The variance of y at this value of x can be shown to be $p(x)[1 - p(x)]$. This variance will of course not be the same for all values of x (unless $p(x)$ is a constant function of x, in which case our regression makes little sense in the first place). Thus a regression of this type displays a nonconstant error variance. Our remedy for this problem will involve a transformation of the data. The effect of unequal variances in this kind of problem is most pronounced when $p(x)$ takes on values near zero or 1. For example, if $.3 \leqslant p(x) \leqslant .7$ for all the values of x we are considering, then $.21 \leqslant p(x)[1 - p(x)] \leqslant .25$. Note that the maximum value that $p(x)[1 - p(x)]$ attains for values of $p(x)$ between zero and 1 is .25, which occurs when $p(x) = .5$. Since the variance of y is practically constant here, little would be gained by transformation.

A STRAIGHT-LINE MODEL

Example 13.7 Happy Valley Real Estate is a company that sells recreational property. The company wishes to investigate the chances that a randomly selected person with a given income will purchase property. As part of its marketing strategy, Happy Valley Real Estate offers incentives (free gifts and vacation trips) to encourage prospective buyers to visit sites at Happy Valley. After a tour the prospective buyer is asked to fill out a questionnaire, which, among other things, includes yearly income (denoted x). If the customer purchases property from Happy Valley within a month of the visit, $y = 1$ is recorded for the customer. If the customer does not purchase property within a month, $y = 0$ is recorded. The goal of the analysis is to compute a 95% confidence interval for the probability $p(x)$ that a randomly selected person with income x will purchase property.

Happy Valley Real Estate collects data as described from 30 prospective buyers. The data is given in Table 13.6 and plotted in Figure 13.16. Looking at the data, we see that eight people in this sample have incomes over \$35,000, and of those, six purchased property. On the other hand, of the eight people with incomes less than \$21,000, only one purchased property. This leads us to suspect that income does indeed affect the buying decision.

To establish a relationship between income x and $p(x)$, first notice that for most incomes x we have only one observation of y. This means that we cannot satisfactorily estimate the probability $p(x)$ by using only the data at a given x value. Instead, we use the regression model

$$\begin{aligned} y &= \mu + \varepsilon \\ &= p(x) + \varepsilon \\ &= \beta_0 + \beta_1 x + \varepsilon \end{aligned}$$

This model relates the purchase decision y to income x. Here, the mean value of y in the population of all people at a particular income level x is the proportion $p(x)$ of people with that income who will purchase Happy Valley property. Therefore we use the data in Table 13.6 to estimate the parameters β_0 and β_1 in the preceding

TABLE 13.6 Happy Valley Real Estate purchase data

Income, x (thousands of dollars)			50	25	30	28	42	17	20	25	38
Purchase decision, y (1 = buy, 0 = not buy)			1	0	0	1	1	0	0	0	1
Income, x	15	30	28	48	16	21	13	36	34	22	20
Purchase decision, y	0	1	0	1	0	1	0	1	0	0	0
Income, x	37	45	17	24	28	25	22	18	26	44	
Purchase decision, y	0	1	1	0	1	1	0	0	0	0	

FIGURE 13.16 **Purchase decision versus income for the Happy Valley Real Estate data**

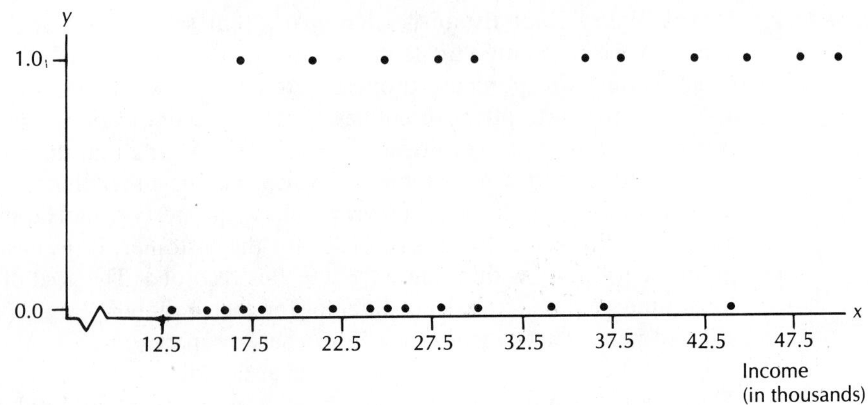

model. Then, since the mean value of y at a particular income level x is

$$\mu = \beta_0 + \beta_1 x$$

our estimate of $p(x)$ is $b_0 + b_1 x$.

However, we are now in a dilemma. We know that the variance of y at a particular value of x is $p(x)[1 - p(x)]$. We also know that this variance varies as x varies, since $p(x)$ changes as x changes. This says that a nonconstant error variance exists. If we could divide each term in the preceding model by the quantity $\sqrt{p(x)[1 - p(x)]}$ (call this v), we would have a model

$$\frac{y}{v} = \beta_0\left(\frac{1}{v}\right) + \beta_1\left(\frac{x}{v}\right) + \frac{\varepsilon}{v}$$

Here, ε/v has variance 1 at all x values (this is analogous to what we did in Section 13.2). The dilemma arises because $p(x)$, and therefore v, are unknown. Of course, if we knew $p(x)$, we would not be trying to estimate it!

Fortunately, the least squares point estimates b_0 and b_1 are not totally unreasonable, even though inference assumption 1 is violated. This means that an iterative scheme can be employed in which b_0 and b_1 are used to estimate $p(x)$, and then this estimate is used to transform the model. We then use the transformed model to reestimate β_0 and β_1. Following this, a new $p(x)$ estimate can be computed and can be used to obtain a new transformed model. This model is in turn used to find yet a third set of estimates of β_0 and β_1. Often, iteration produces only small changes in the estimates, and usually only a few rounds of iteration are necessary.

To begin the iteration procedure, we use the data in Table 13.6 to calculate the least squares point estimates b_0 and b_1 of the parameters β_0 and β_1 in the model

$$\begin{aligned} y &= \mu + \varepsilon \\ &= p(x) + \varepsilon \\ &= \beta_0 + \beta_1 x + \varepsilon \end{aligned}$$

We obtain $b_0 = -.2411$ and $b_1 = .0228$. It follows that our estimate of the probability $p(x)$ is

$$\hat{p}(x) = -.2411 + .0228x$$

Looking at Table 13.6, we see that the observed range of incomes is from 13 (thousand dollars) to 50 (thousand dollars). This implies that our estimated probabilities at various income levels range from

$$\hat{p}(13) = -.2411 + .0228(13) = .0553$$

to

$$\hat{p}(50) = -.2411 + .0228(50) = .8983$$

Since

$$\hat{p}(13)[1 - \hat{p}(13)] = .0553[1 - .0553] = .0522$$

and

$$\hat{p}(50)[1 - \hat{p}(50)] = .8983[1 - .8983] = .0914$$

the estimated variance $\hat{p}(x)[1 - \hat{p}(x)]$ ranges from .0522 to .25. Recall that $p(x)[1 - p(x)]$ attains a maximum value of $.5[1 - .5] = .25$ when $p(x) = .5$. Table 13.7 shows the original Happy Valley data, the estimated probability $\hat{p}(x)$, and the transformed data $y/\hat{v}$, $1/\hat{v}$, and $x/\hat{v}$ for the transformed model

$$\frac{y}{\hat{v}} = \beta_0\left(\frac{1}{\hat{v}}\right) + \beta_1\left(\frac{x}{\hat{v}}\right) + \frac{\varepsilon}{\hat{v}}$$

Here, $\hat{v} = \sqrt{\hat{p}(x)[1 - \hat{p}(x)]}$ is the estimated standard deviation of y computed by using the estimated probability $\hat{p}(x)$. The least squares point estimates of the parameters in the transformed model are $\tilde{b}_0 = -.2676$ and $\tilde{b}_1 = .0239$. These estimates are not very different from the original estimates $b_0 = -.2411$ and $b_1 = .0228$. Therefore further iteration (yielding a second set of revised estimates) will not be done. The t statistics associated with the revised estimates are

$$t = \frac{\tilde{b}_0}{s\sqrt{c_{00}}} = \frac{-.2676}{.1861} = -1.438$$

$$t = \frac{\tilde{b}_1}{s\sqrt{c_{11}}} = \frac{.0239}{.0064} = 3.734$$

These t statistics are valid because the transformed model approximately satisfies the constant variance assumption.

TABLE 13.7 The transformed Happy Valley Real Estate purchase data

Purchase decision, y	Income, x (thousands of dollars)	$\hat{p}(x) = -.2411 + .0228x$	$y/\hat{v} = y/\sqrt{\hat{p}(x)[1-\hat{p}(x)]}$	$1/\hat{v} = 1/\sqrt{\hat{p}(x)[1-\hat{p}(x)]}$	$x/\hat{v} = x/\sqrt{\hat{p}(x)[1-\hat{p}(x)]}$
1	50	0.898276	3.30814	3.30814	165.407
0	25	0.328601	0.00000	2.12900	53.225
0	30	0.442536	0.00000	2.01334	60.400
1	28	0.396962	2.04387	2.04387	57.228
1	42	0.715980	2.21756	2.21756	93.137
0	17	0.146305	0.00000	2.82956	48.103
0	20	0.214666	0.00000	2.43552	48.710
0	25	0.328601	0.00000	2.12900	53.225
1	38	0.624832	2.06541	2.06541	78.485
0	15	0.100731	0.00000	3.32257	49.838
1	30	0.442536	2.01334	2.01334	60.400
0	28	0.396962	0.00000	2.04387	57.228
1	48	0.852702	2.82165	2.82165	135.439
0	16	0.123518	0.00000	3.03923	48.628
1	21	0.237453	2.35005	2.35005	49.351
0	13	0.055157	0.00000	4.38046	56.946
1	36	0.579258	2.02561	2.02561	72.922
0	34	0.533684	0.00000	2.00455	68.155
0	22	0.260240	0.00000	2.27912	50.141
0	20	0.214666	0.00000	2.43552	48.710
0	37	0.602045	0.00000	2.04300	75.591
1	45	0.784341	2.43144	2.43144	109.415
1	17	0.146305	2.82956	2.82956	48.103
0	24	0.305814	0.00000	2.17037	52.089
1	28	0.396962	2.04387	2.04387	57.228
1	25	0.328601	2.12900	2.12900	53.225
0	22	0.260240	0.00000	2.27912	50.141
0	18	0.169092	0.00000	2.66785	48.021
0	26	0.351388	0.00000	2.09466	54.461
0	44	0.761554	0.00000	2.34668	103.254

Because the transformed model has the same parameters β_0 and β_1 as the original model, we use $\tilde{b}_0$ and $\tilde{b}_1$ to estimate $p(x)$. Calling this estimate $\tilde{p}(x)$, we obtain

$$\tilde{p}(x) = -.2676 + .0239x$$

We now use our results to find point and interval estimates of $p(x)$ when, for example, $x = 23$. The point estimate of $p(23)$ is

$$\tilde{p}(23) = -.2676 + .0239(23) = .2821$$

Thus we estimate that 28.21 percent of all people with a yearly income of \$23,000 would purchase property within one month of a Happy Valley tour. To compute a 95% confidence interval for $p(x)$, we use the standard error, s, and the $(\mathbf{X}'\mathbf{X})^{-1}$ matrix *computed from the transformed data.* Letting $\mathbf{x}_0' = [1 \quad 23]$, we obtain

$$\mathbf{x}_0'(\mathbf{X}'\mathbf{X})^{-1}\mathbf{x}_0 = .0059217$$

Also, for the transformed model s equals 1.01494. It follows that a 95% confidence interval for $p(23)$ is

$$\begin{aligned}
[\tilde{p}(23) \pm t_{[\alpha/2]}^{(n-k)} s\sqrt{\mathbf{x}_0'(\mathbf{X}'\mathbf{X})^{-1}\mathbf{x}_0}] &= [\tilde{p}(23) \pm t_{[.025]}^{(30-2)} s\sqrt{\mathbf{x}_0'(\mathbf{X}'\mathbf{X})^{-1}\mathbf{x}_0}] \\
&= [.2821 \pm 2.048(1.01494)\sqrt{.0059217}] \\
&= [.2821 \pm .1600] \\
&= [.1221, .4421]
\end{aligned}$$

Notice that we compute a 95% *confidence* interval here rather than a prediction interval. This is because $p(23)$ is the *mean* of the population of y values (0's and 1's) when yearly income is \$23,000 ($x = 23$). The interval says that we can be 95 percent confident that between 12.21 and 44.21 percent of all people with a yearly income of \$23,000 would purchase property within one month of a Happy Valley tour. Point and interval estimates of $p(x)$ at other income levels can be established in a similar fashion.

LOGISTIC REGRESSION

In Example 13.7 we employed a simple linear regression model in which $p(x) = \beta_0 + \beta_1 x$. Fortunately, for the range of x values (yearly incomes) in which we wished to estimate $p(x)$ we did not obtain any $\tilde{p}(x)$ values (estimated probabilities) less than zero or greater than 1. For a verification of this, see Table 13.7. Another function that can be used is the *logistic function.* This function, unlike the linear function, can be forced to give estimates of $p(x)$ that are between zero and 1 for all values of x.

Example 13.8 An advertising firm has decided to use six groups of 50 people each in an experiment. The experiment is designed to estimate $p(x)$, the probability that a person will remember a product name one week after viewing a 60-second commercial that mentions the product name x times. The firm shows the $n_i = 50$ people in group

TABLE 13.8 **Product name data**

i	x_i	y_i	$\hat{p}_i$	$l_i = \ln[\hat{p}_i/(1-\hat{p}_i)]$	$\hat{v}_i = 1/[n_i\hat{p}_i(1-\hat{p}_i)]^{1/2}$
1	1	4	.08	−2.4423	.5212
2	2	7	.14	−1.8153	.4076
3	3	20	.40	−0.4055	.2886
4	4	35	.70	.8473	.3085
5	5	44	.88	1.9924	.4352
6	6	46	.92	2.4423	.5212

i (i = 1, 2, 3, 4, 5, 6) a 60-second commercial that mentions the product name $x_i = i$ times. One week later, the number, y_i, and the proportion, $\hat{p}_i = y_i/50$, of the 50 people who recall the product name are recorded. Table 13.8 gives the data that is obtained, and Figure 13.17 presents a plot of the observed $\hat{p}_i$ values (represented by the dots in the figure) against x_i.

A curve having the shape of the curve in Figure 13.17 is called a *logistic curve.* It is defined by the equation

$$p(x_i) = \frac{e^{(\beta_0 + \beta_1 x_i)}}{1 + e^{(\beta_0 + \beta_1 x_i)}}$$

Note that the observed $\hat{p}_i$ values seem to follow the pattern defined by the logistic

FIGURE 13.17 **Observed proportions and the logistic curve**

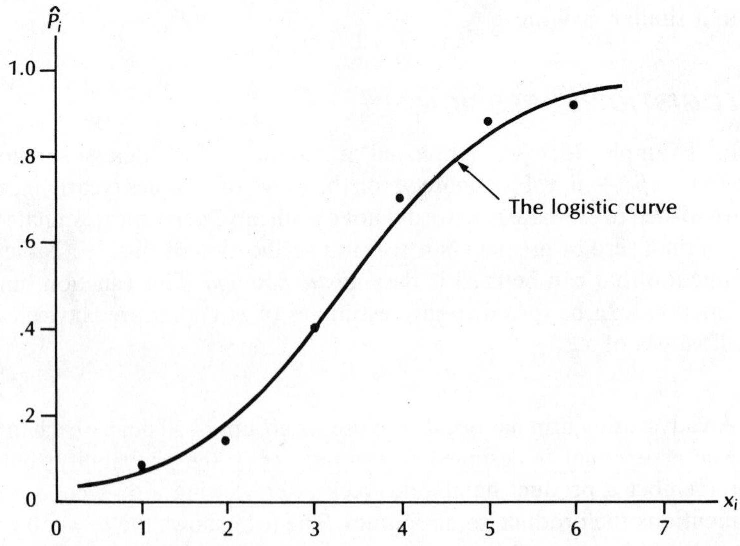

curve. We can fit this curve to the $\hat{p}_i$ values by noting that the equation defining $p(x_i)$ implies that

$$\frac{p(x_i)}{1 - p(x_i)} = e^{(\beta_0 + \beta_1 x_i)}$$

Therefore

$$\ln\left(\frac{p(x_i)}{1 - p(x_i)}\right) = \beta_0 + \beta_1 x_i$$

We can estimate $p(x_i)$ by $\hat{p}_i$ (see Table 13.8). Thus we consider the regression model

$$l_i = \beta_0 + \beta_1 x_i + \varepsilon_i$$

where

$$l_i = \ln\left(\frac{\hat{p}_i}{1 - \hat{p}_i}\right)$$

It can be shown that since n_i $(= 50)$ is large, the approximate estimated standard deviation of l_i is

$$\hat{v}_i = \frac{1}{(n_i \hat{p}_i(1 - \hat{p}_i))^{1/2}}$$

Therefore we form the transformed logistic regression model

$$\frac{l_i}{\hat{v}_i} = \beta_0\left(\frac{1}{\hat{v}_i}\right) + \beta_1\left(\frac{x_i}{\hat{v}_i}\right) + \frac{\varepsilon_i}{\hat{v}_i}$$

Using the values of l_i and $\hat{v}_i$ calculated in Table 13.8, we compute point estimates $\tilde{b}_0$ and $\tilde{b}_1$ of the parameters β_0 and β_1 as follows:

$$\mathbf{y} = \begin{bmatrix} \dfrac{l_1}{\hat{v}_1} \\ \vdots \\ \dfrac{l_6}{\hat{v}_6} \end{bmatrix} = \begin{bmatrix} \dfrac{-2.4423}{.5212} \\ \vdots \\ \dfrac{2.4423}{.5212} \end{bmatrix} = \begin{bmatrix} -4.6859 \\ \vdots \\ 4.6859 \end{bmatrix}$$

$$\mathbf{X} = \begin{bmatrix} \dfrac{1}{\hat{v}_1} & \dfrac{x_1}{\hat{v}_1} \\ \vdots & \vdots \\ \dfrac{1}{\hat{v}_6} & \dfrac{x_6}{\hat{v}_6} \end{bmatrix} = \begin{bmatrix} \dfrac{1}{.5212} & \dfrac{1}{.5212} \\ \vdots & \vdots \\ \dfrac{1}{.5212} & \dfrac{6}{.5212} \end{bmatrix} = \begin{bmatrix} 1.9186 & 1.9186 \\ \vdots & \vdots \\ 1.9186 & 11.5119 \end{bmatrix}$$

FIGURE 13.18 **SAS output for the transformed logistic model** $\frac{l_i}{\hat{v}_i} = \beta_0\left(\frac{1}{\hat{v}_i}\right) + \beta_1\left(\frac{x_i}{\hat{v}_i}\right) + \frac{\varepsilon_i}{\hat{v}_i}$

ANALYSIS OF VARIANCE

SOURCE	DF	SUM OF SQUARES	MEAN SQUARE	F VALUE	PROB>F
MODEL	1	92.01801757	92.01801757	196.324	0.0002
ERROR	4	1.87482246	0.46870562		
C TOTAL	5	93.89284003			

ROOT MSE	0.6846208[c]	R-SQUARE	0.9800
DEP MEAN	0.08802313	ADJ R-SQ	0.9750
C.V.	777.7738		

PARAMETER ESTIMATES

VARIABLE	DF	PARAMETER ESTIMATE	STANDARD ERROR	T FOR H0: PARAMETER=0	PROB > |T|
INTERCEP	1	-3.68953675[a]	0.28995392	-12.725	0.0002
MENTION	1	1.09342028[b]	0.07803704	14.012	0.0002

OBS	ACTUAL	PREDICT VALUE	STD ERR PREDICT	LOWER95% MEAN	UPPER95% MEAN	RESIDUAL
1	-2.4423	-2.5961	0.2193	-3.2049	-1.9873	0.1538
2	-1.8153	-1.5027	0.1558	-1.9353	-1.0701	-0.3126
3	-0.4055	-0.4093	0.1125	-0.7215	-0.0970	.0038108
4	0.8473	0.6841[d]	0.1149[e]	0.3652[f]	1.0031[g]	0.1632
5	1.9924	1.7776	0.1610	1.3305	2.2246	0.2149
6	2.4423	2.8710	0.2255	2.2450	3.4970	-0.4286

[a] $\tilde{b}_0$ [b] $\tilde{b}_1$ [c] s [d] $\tilde{b}_0 + \tilde{b}_1(4)$ [e] $s\sqrt{\mathbf{x}'(\mathbf{X}'\mathbf{X})^{-1}\mathbf{x}}$ where $\mathbf{x}' = [1 \quad 4]$ [f] $\tilde{b}_0 + \tilde{b}_1(4) - \beta(4)$ [g] $\tilde{b}_0 + \tilde{b}_1(4) + \beta(4)$

$$(\mathbf{X}'\mathbf{X})^{-1} = \begin{bmatrix} .1794 & -.0448 \\ -.0448 & .0130 \end{bmatrix}$$

$$\begin{bmatrix} \tilde{b}_0 \\ \tilde{b}_1 \end{bmatrix} = \tilde{\mathbf{b}} = (\mathbf{X}'\mathbf{X})^{-1}\mathbf{X}'\mathbf{y} = \begin{bmatrix} -3.6895 \\ 1.0934 \end{bmatrix}$$

Moreover, the standard error, s, for this model is .6846. The SAS output in Figure 13.18 gives $\tilde{b}_0$, $\tilde{b}_1$, s, and other quantities related to the following analysis.

It can be shown that

1. A point estimate $\tilde{p}(x)$ of $p(x)$ is given by the *estimated logistic function*

$$\tilde{p}(x) = \frac{e^{(\tilde{b}_0 + \tilde{b}_1 x)}}{1 + e^{(\tilde{b}_0 + \tilde{b}_1 x)}} = \frac{e^{(-3.6895 + 1.0934x)}}{1 + e^{(-3.6895 + 1.0934x)}}$$

For example, a point estimate of $p(4)$ is calculated as follows:

$$\tilde{b}_0 + \tilde{b}_1(4) = -3.6895 + 1.0934(4) = .6841$$

$$\tilde{p}(4) = \frac{e^{.6841}}{1 + e^{.6841}} = \frac{1.982}{2.982} = .6647$$

Furthermore, we can use the estimated logistic function to compute point estimates $\tilde{p}(1)$ through $\tilde{p}(6)$ of $p(1)$ through $p(6)$. If we plot these point estimates against x (1 through 6), then we obtain a plot of the estimated logistic function. This plot is the logistic curve in Figure 13.17.

2. If we let

$$\beta(x) = t_{[\alpha/2]}^{(6-2)} s\sqrt{\mathbf{x}'(\mathbf{X}'\mathbf{X})^{-1}\mathbf{x}}$$

where

$$\mathbf{x}' = [1 \quad x]$$

then a $100(1 - \alpha)\%$ confidence interval for $p(x)$ is

$$\left[\frac{e^{(\tilde{b}_0 + \tilde{b}_1 x - \beta(x))}}{1 + e^{(\tilde{b}_0 + \tilde{b}_1 x - \beta(x))}}, \frac{e^{(\tilde{b}_0 + \tilde{b}_1 x + \beta(x))}}{1 + e^{(\tilde{b}_0 + \tilde{b}_1 x + \beta(x))}}\right]$$

For example, a 95% confidence interval for $p(4)$ is calculated as follows:

$$\tilde{b}_0 + \tilde{b}_1(4) = -3.6895 + 1.0934(4) = .6841$$

$$\mathbf{x}' = [1 \quad 4]$$

$$\mathbf{x}'(\mathbf{X}'\mathbf{X})^{-1}\mathbf{x} = [1 \quad 4]\begin{bmatrix} .1794 & -.0448 \\ -.0448 & .0130 \end{bmatrix}\begin{bmatrix} 1 \\ 4 \end{bmatrix} = .0282$$

$$s = .6846$$

$$\beta(4) = t_{[.025]}^{(6-2)} s\sqrt{\mathbf{x}'(\mathbf{X}'\mathbf{X})^{-1}\mathbf{x}} = 2.776(.1149) = .3190$$

$$\tilde{b}_0 + \tilde{b}_1(4) - \beta(4) = .6841 - .3190 = .3652$$

$$\tilde{b}_0 + \tilde{b}_1(4) + \beta(4) = .6841 + .3190 = 1.0031$$

$$\left[\frac{e^{.3652}}{1 + e^{.3652}}, \frac{e^{1.0031}}{1 + e^{1.0031}}\right] = \left[\frac{1.440802}{2.440802}, \frac{2.726722}{3.726722}\right] = [.5903, \quad .7317]$$

This interval says that the advertising firm can be 95 percent confident that $p(4)$, the probability that a person will remember the product name one week after viewing a 60-second commercial that mentions the product name four times, is between .5903 and .7317.

The logistic function can be extended to relate a binary dependent variable to any number of independent variables $x_1, \ldots, x_p$. Let $p(x_{i1}, \ldots, x_{ip})$ denote the probability that y_i equals 1 when $x_{i1}, \ldots, x_{ip}$ are the values of the independent variables. Then the *multiple logistic function* is

$$p(x_{i1}, \ldots, x_{ip}) = \frac{e^{(\beta_0 + \beta_1 x_{i1} + \cdots + \beta_p x_{ip})}}{1 + e^{(\beta_0 + \beta_1 x_{i1} + \cdots + \beta_p x_{ip})}}$$

Suppose that we have observed a large number (n_i) of values of the dependent variable at each combination of values $x_{i1}, \ldots, x_{ip}$ of the independent variables. Then we can estimate the parameters of the multiple logistic function by using the transformed logistic multiple regression model

$$\frac{l_i}{\hat{v}_i} = \beta_0\left(\frac{1}{\hat{v}_i}\right) + \beta_1\left(\frac{x_{i1}}{\hat{v}_i}\right) + \cdots + \beta_p\left(\frac{x_{ip}}{\hat{v}_i}\right) + \frac{\varepsilon_i}{\hat{v}_i}$$

Here,

$$l_i = \ln\left(\frac{\hat{p}_i}{1 - \hat{p}_i}\right)$$

where $\hat{p}_i$ is the proportion of the n_i values of the dependent variable that equaled 1 when $x_{i1}, \ldots, x_{ip}$ were the values of the independent variables. Furthermore

$$\hat{v}_i = \frac{1}{(n_i\hat{p}_i(1 - \hat{p}_i))^{1/2}}$$

is the approximate estimated standard deviation of l_i (assuming that n_i is large). Then if $\tilde{b}_0, \tilde{b}_1, \ldots, \tilde{b}_p$ denote the least squares point estimates of $\beta_0, \beta_1, \ldots, \beta_p$ in the transformed model, the point estimate of $p(x_{i1}, \ldots, x_{ip})$ is

$$\hat{p}(x_{i1}, \ldots, x_{ip}) = \frac{e^{(\tilde{b}_0 + \tilde{b}_1 x_{i1} + \cdots + \tilde{b}_p x_{ip})}}{1 + e^{(\tilde{b}_0 + \tilde{b}_1 x_{i1} + \cdots + \tilde{b}_p x_{ip})}}$$

Furthermore, a confidence interval for $p(x_{i1}, \ldots, x_{ip})$ can be calculated by using the methods illustrated in Example 13.8.

Sometimes some or all of the n_i values are not large. This was the case for the purchase data in Table 13.6. In such a situation we can estimate the parameters of the logistic function by using the method of *maximum likelihood estimation.* Such estimation can be done by using standard computer packages. For example, the maximum likelihood estimates of β_0 and β_1 in the logistic function

$$p(x) = \frac{e^{(\beta_0 + \beta_1 x)}}{1 + e^{(\beta_0 + \beta_1 x)}}$$

describing the purchase data of Table 13.6 are $b_0 = -3.35117$ and $b_1 = .103971$. It follows that a point estimate of $p(23)$, the probability that a randomly selected person with an income of \$23,000 will purchase property, is

$$\begin{aligned}\hat{p}(23) &= \frac{e^{(-3.35117+.103971(23))}}{1 + e^{(-3.35117+.103971(23))}} \\ &= .2769\end{aligned}$$

It is interesting to compare this point estimate with $\hat{p}(23) = .2821$, which is the point estimate of $p(23)$ obtained by using the straight line model in Example 13.7. Finally, note that there are confidence interval and hypothesis-testing procedures based on maximum likelihood estimation of the logistic function parameters. These procedures are based on large samples. The interested reader is referred to Neter, Wasserman, and Kutner (1989).

13.4

MODELING DETERMINISTIC TIME SERIES COMPONENTS

If the appropriate residual plots or statistical tests indicate that the independence assumption is violated, remedies are available. A common remedy is to use *time series models*.

In general, time series data can be described by *trends*, *seasonal effects*, and *cyclical effects* as well as *causal variables* (variables like advertising expenditure and the price difference).

Trend refers to the upward or downward movement in a time series over a period of time. Thus trend reflects the long-run growth or decline in the time series. Trends can represent a variety of factors. For example, trends in sales might be determined by (1) technological change, (2) changes in consumer tastes, (3) increases in per capita income, and (4) market growth. If we denote the trend in time period t as TR_t, some common trends are

1. Linear trend, which is modeled as $TR_t = \beta_0 + \beta_1 t$.
2. Quadratic trend, which is modeled as $TR_t = \beta_0 + \beta_1 t + \beta_2 t^2$.
3. Exponential trends (such as the trend in Example 13.1, which we modeled as $TR_t = \beta_0(\beta_1)^t$).

Although these trends are probably most common, other, more complicated trends also exist.

Time series data also sometimes show *seasonal effects*. Seasonal variations are periodic patterns in a time series that complete themselves within a calendar year and are then repeated on a yearly basis. Such variations are usually caused by factors like the weather and customs (such as holidays). For instance, air conditioner sales and soft drink sales are seasonal in nature, with highest sales occurring during the summer months. As another example, monthly sales volume for a department store might be seasonal. Here, the seasonal variation would be caused by the observance of various holidays.

Seasonal patterns can be modeled by using dummy variables. Let y_t denote the value of the dependent variable in time period t, TR_t denote the trend in time period t, SN_t denote the seasonal factor in time period t, and ε_t denote the error term in time period t. Then we use the model

$$y_t = TR_t + SN_t + \varepsilon_t$$

Moreover, suppose that there are L seasons (months, quarters, and so on) per year. Then we assume that SN_t is given by the equation

$$SN_t = \beta_{S1}x_{S1,t} + \beta_{S2}x_{S2,t} + \cdots + \beta_{S(L-1)}x_{S(L-1),t}$$

Here, $x_{S1,t}, x_{S2,t}, \ldots, x_{S(L-1),t}$ are dummy variables, which are defined as follows:

$$x_{S1,t} = \begin{cases} 1 & \text{if time period } t \text{ is season 1} \\ 0 & \text{otherwise} \end{cases}$$

$$x_{S2,t} = \begin{cases} 1 & \text{if time period } t \text{ is season 2} \\ 0 & \text{otherwise} \end{cases}$$

$$\vdots \qquad\qquad \vdots$$

$$x_{S(L-1),t} = \begin{cases} 1 & \text{if time period } t \text{ is season } L-1 \\ 0 & \text{otherwise} \end{cases}$$

For example, if $L = 12$ (monthly data) and period t is season 2 (February), we have

$$\begin{aligned} y_t &= TR_t + SN_t + \varepsilon_t \\ &= TR_t + \beta_{S1}x_{S1,t} + \beta_{S2}x_{S2,t} + \beta_{S3}x_{S3,t} + \cdots + \beta_{S11}x_{S11,t} + \varepsilon_t \\ &= TR_t + \beta_{S1}(0) + \beta_{S2}(1) + \beta_{S3}(0) + \cdots + \beta_{S11}(0) + \varepsilon_t \\ &= TR_t + \beta_{S2} + \varepsilon_t \end{aligned}$$

The use of the dummy variables ensures that a seasonal parameter for season 2 is added to the trend in each time period that is season 2. This seasonal parameter, β_{S2}, accounts for the seasonality in season 2.

In general, the purpose of the dummy variables is to ensure that the appropriate seasonal parameter is included in the model in each time period. Also note that SN_t is a linear function of the parameters $\beta_{S1}, \beta_{S2}, \beta_{S3}, \ldots, \beta_{S(L-1)}$. Thus if the trend is also a linear function of its parameters, least squares point estimates of the parameters in the model

$$y_t = TR_t + SN_t + \varepsilon_t$$

can be computed. Letting tr_t be the estimate of the trend, we have, for example,

$$tr_t = b_0 + b_1 t$$

if there is a linear trend or

$$tr_t = b_0 + b_1 t + b_2 t^2$$

if there is a quadratic trend. Furthermore, sn_t, the estimate of SN_t, is

$$sn_t = b_{S1}x_{S1,t} + b_{S2}x_{S2,t} + \cdots + b_{S(L-1)}x_{S(L-1),t}$$

Here, $b_{S1}, b_{S2}, \ldots, b_{S(L-1)}$ are the least squares point estimates of the parameters $\beta_{S1}, \beta_{S2}, \ldots, \beta_{S(L-1)}$. Therefore the point prediction of y_t is

$$\hat{y}_t = tr_t + sn_t$$

For example, if period t is season 1, then

$$y_t = TR_t + \beta_{S1} + \varepsilon_t$$

and a point prediction of y_t is

$$\hat{y}_t = tr_t + b_{S1}$$

If period t is season 2, then

$$y_t = TR_t + \beta_{S2} + \varepsilon_t$$

and a point prediction of y_t is

$$\hat{y}_t = tr_t + b_{S2}$$

If period t is season $L - 1$, then

$$y_t = TR_t + \beta_{S(L-1)} + \varepsilon_t$$

and a point prediction of y_t is

$$\hat{y}_t = tr_t + b_{S(L-1)}$$

If period t is season L, then

$$y_t = TR_t + \varepsilon_t$$

and a point prediction of y_t is

$$\hat{y}_t = tr_t$$

We have, quite arbitrarily, set the seasonal parameter for season L equal to zero. Thus the other seasonal parameters—$\beta_{S1}, \beta_{S2}, \ldots, \beta_{S(L-1)}$—are defined with respect to season L. Intuitively, β_{Sj} is the difference, excluding trend, between the level of the time series in season j and the level of the time series in season L. If β_{Sj} is positive, this implies that, excluding trend, the value of the time series in season j can be expected to be greater than the value of the time series in season L. If β_{Sj} is negative, this implies that, excluding trend, the value of the time series in season j can be expected to be smaller than the value of the time series in season L. We do not have to set the seasonal parameter for season L equal to zero. We can set the seasonal parameter for any particular season equal to zero and thus define the other seasonal parameters with respect to that particular season. However, we must arbitrarily set one of the seasonal parameters equal to zero. If we do not, the columns of the **X** matrix can be shown to be linearly dependent. This would imply that the least squares point estimates of the model parameters cannot be computed using the methods we have presented in this book.

Example 13.9 Farmers' Bureau Coop, a small agricultural cooperative, would like to predict its propane gas bills for the four quarters of next year. In addition, the coop would like to estimate the trend in its propane gas bills after adjusting for seasonal effects. To do this, Farmers' Bureau has compiled its propane gas bills for the last 10 years. The quarterly propane gas bills for this 10-year period (40 quarters) are given in Table 13.9.

A plot of the gas bill data is displayed in Figure 13.19. If we look at this data plot, there appears to be an upward trend in the gas bills. Large seasonal (quarterly) fluctuations, however, make it difficult to ascertain the exact nature of this trend. In Figure 13.20 we plot the yearly mean propane gas bills against the year number (1 to 10). We do this to average out the quarterly effects in the gas bill data. Looking at this figure, we see that the trend is possibly increasing at an increasing rate. On the basis of this plot we assume that the trend here is quadratic. That is, the trend is modeled as

$$TR_t = \beta_0 + \beta_1 t + \beta_2 t^2$$

To model the seasonality of the gas bill data, we use dummy variables $x_{S1,t}$, $x_{S2,t}$, and $x_{S3,t}$. These are defined as follows:

$$x_{S1,t} = \begin{cases} 1 & \text{if time period } t \text{ is quarter 1} \\ 0 & \text{otherwise} \end{cases}$$

$$x_{S2,t} = \begin{cases} 1 & \text{if time period } t \text{ is quarter 2} \\ 0 & \text{otherwise} \end{cases}$$

TABLE 13.9 **Quarterly propane gas bills for Farmers' Bureau Coop**

Year	*Quarter 1*	*Quarter 2*	*Quarter 3*	*Quarter 4*
1	344.39 ($=y_1$)	246.63 ($=y_2$)	131.53 ($=y_3$)	288.87 ($=y_4$)
2	313.45 ($=y_5$)	189.76 ($=y_6$)	179.10 ($=y_7$)	221.10 ($=y_8$)
3	246.84	209.00	51.21	133.89
4	277.01	197.98	50.68	218.08
5	365.10	207.51	54.63	214.09
6	267.00	230.28	230.32	426.41
7	467.06	306.03	253.23	279.46
8	336.56	196.67	152.15	319.67
9	440.00	315.04	216.42	339.78
10	434.66 ($=y_{37}$)	399.66 ($=y_{38}$)	330.80 ($=y_{39}$)	539.78 ($=y_{40}$)

FIGURE 13.19 Quarterly propane gas bill versus time for Farmers' Bureau Coop

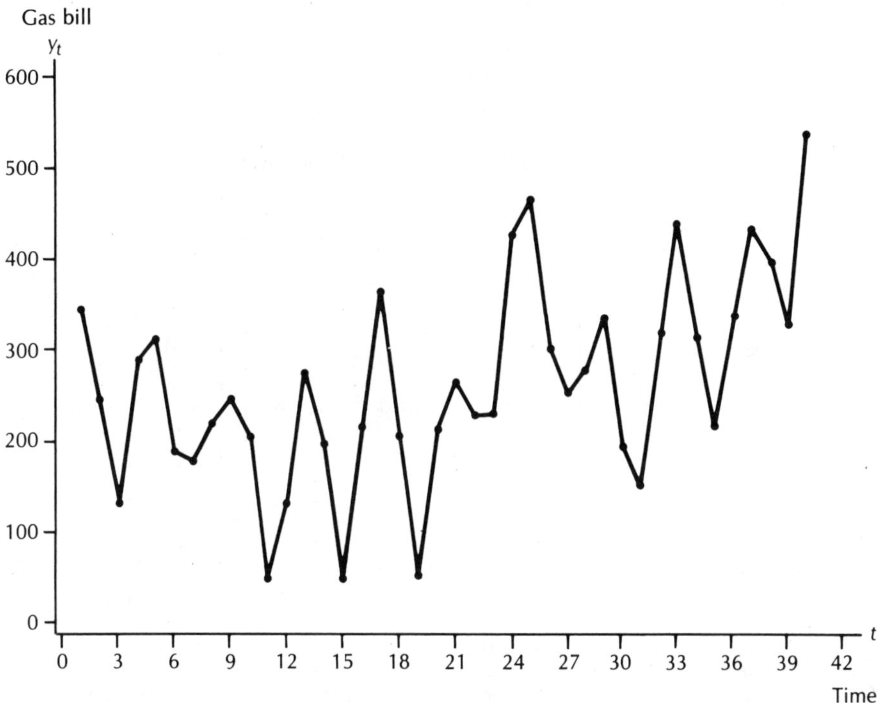

FIGURE 13.20 Mean yearly propane gas bill versus year for Farmers' Bureau Coop

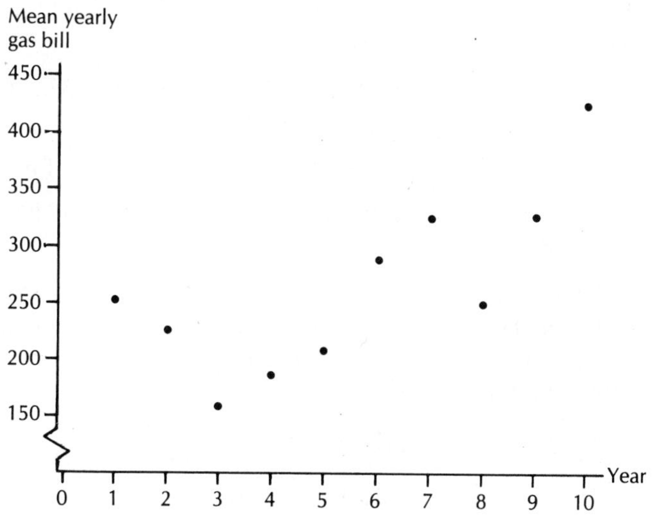

$$x_{S3,t} = \begin{cases} 1 & \text{if time period } t \text{ is quarter 3} \\ 0 & \text{otherwise} \end{cases}$$

Our regression model is

$$\begin{aligned} y_t &= TR_t + SN_t + \varepsilon_t \\ &= \beta_0 + \beta_1 t + \beta_2 t^2 + \beta_3 x_{S1,t} + \beta_4 x_{S2,t} + \beta_5 x_{S3,t} + \varepsilon_t \end{aligned}$$

The least squares point estimates of the model parameters are calculated by using

$$\mathbf{y} = \begin{bmatrix} 344.39 \\ 246.63 \\ 131.53 \\ 288.87 \\ \vdots \\ 434.66 \\ 399.66 \\ 330.80 \\ 539.78 \end{bmatrix} \quad \text{and} \quad \mathbf{X} = \begin{bmatrix} 1 & 1 & 1 & 1 & 0 & 0 \\ 1 & 2 & 4 & 0 & 1 & 0 \\ 1 & 3 & 9 & 0 & 0 & 1 \\ 1 & 4 & 16 & 0 & 0 & 0 \\ \vdots & \vdots & \vdots & \vdots & \vdots & \vdots \\ 1 & 37 & 1369 & 1 & 0 & 0 \\ 1 & 38 & 1444 & 0 & 1 & 0 \\ 1 & 39 & 1521 & 0 & 0 & 1 \\ 1 & 40 & 1600 & 0 & 0 & 0 \end{bmatrix}$$

We obtain the prediction equation

$$\hat{y}_t = 276.64 - 7.46t + .30t^2 + 65.77x_{S1,t} - 37.87x_{S2,t} - 127.61x_{S3,t}$$

Figure 13.21 presents the SAS output for this model. Using the prediction equation, we obtain point predictions of the quarterly propane gas bills for the four quarters of next year (time periods 41, 42, 43, and 44).

For quarter 1 ($x_{S1,41} = 1$, $x_{S2,41} = 0$, and $x_{S3,41} = 0$) a point forecast of y_{41} (the propane gas bill in time period 41) is

$$\begin{aligned} \hat{y}_{41} &= 276.64 - 7.46(41) + .30(41)^2 + 65.77(1) - 37.87(0) - 127.61(0) \\ &= 543.0 \end{aligned}$$

For quarter 2 ($x_{S1,42} = 0$, $x_{S2,42} = 1$, and $x_{S3,42} = 0$) a point forecast of y_{42} is

$$\begin{aligned} \hat{y}_{42} &= 276.64 - 7.46(42) + .30(42)^2 + 65.77(0) - 37.87(1) - 127.61(0) \\ &= 456.9 \end{aligned}$$

Similarly, point forecasts of y_{43} and y_{44} are $\hat{y}_{43} = 385.3$ and $\hat{y}_{44} = 531.7$, respectively. Figure 13.21 gives these point forecasts, along with the 95% prediction intervals for y_{41}, y_{42}, y_{43}, and y_{44}. For example, the 95% prediction interval for y_{42} is calculated by using the formula

$$[\hat{y}_{42} \pm t_{[.025]}^{(40-6)} s\sqrt{1 + \mathbf{x}'_{42}(\mathbf{X}'\mathbf{X})^{-1}\mathbf{x}_{42}}]$$

Here, $\mathbf{x}'_{42} = [1 \quad 42 \quad (42)^2 \quad 0 \quad 1 \quad 0]$.

FIGURE 13.21 **SAS output for the Farmers' Bureau Coop dummy variable model**

ANALYSIS OF VARIANCE

SOURCE	DF	SUM OF SQUARES	MEAN SQUARE	F VALUE	PROB>F
MODEL	5	361966.91	72393.38111	19.796	0.0001
ERROR	34	124335.67	3656.93152		
C TOTAL	39	486302.58			

ROOT MSE	60.47257	R-SQUARE	0.7443
DEP MEAN	265.5457	ADJ R-SQ	0.7067
C.V.	22.77294		

PARAMETER ESTIMATES

VARIABLE	DF	PARAMETER ESTIMATE	STANDARD ERROR	T FOR H0: PARAMETER=0	PROB > |T|
INTERCEP	1	276.63631	35.04850316	7.893	0.0001
TIME	1	-7.45825468	3.39603097	-2.196	0.0350
TIMESQ	1	0.30123099	0.08030439	3.751	0.0007
Q1	1	65.77064773	27.15915517	2.422	0.0209
Q2	1	-37.87010620	27.09580229	-1.398	0.1713
Q3	1	-127.61132	27.05743335	-4.716	0.0001

OBS	ACTUAL	PREDICT VALUE	STD ERR PREDICT	LOWER95% MEAN	UPPER95% MEAN	LOWER95% PREDICT	UPPER95% PREDICT	RESIDUAL
1	344.4	335.2	31.3106	271.6	398.9	196.9	473.6	9.1401
2	246.6	225.1	29.5717	165.0	285.2	88.2533	361.9	21.5754
3	131.5	129.4	28.0239	72.4102	186.3	-6.0877	264.8	2.1687
4	288.9	251.6	26.7003	197.4	305.9	117.3	386.0	37.2470
5	313.4	312.6	23.9562	264.0	361.3	180.5	444.8	0.8035
6	189.8	204.9	23.2525	157.6	252.1	73.1948	336.5	-15.1010
7	179.1	111.6	22.7501	65.3441	157.8	-19.7257	242.9	67.5225
8	221.1	236.2	22.4625	190.6	281.9	105.2	367.3	-15.1491
9	246.8	299.7	20.8562	257.3	342.1	169.7	429.7	-52.8424
10	209.0	194.3	20.9302	151.8	236.8	64.2597	324.4	14.6932
11	51.2100	103.4	21.1265	60.4992	146.4	-26.7449	233.6	-52.2231
12	133.9	230.5	21.4416	186.9	274.1	100.1	360.9	-96.6245
13	277.0	296.4	20.7331	254.2	338.5	166.4	426.3	-19.3477
14	198.0	193.4	21.0443	150.6	236.2	63.2689	323.5	4.5881
15	50.6800	104.9	21.3944	61.4498	148.4	-25.4305	235.3	-54.2481
16	218.1	234.4	21.7815	190.2	278.7	103.8	365.0	-16.3394
17	365.1	302.7	21.5510	258.9	346.5	172.2	433.1	62.4276
18	207.5	202.1	21.6990	158.0	246.2	71.5501	332.7	5.3935
19	54.6300	116.1	21.8506	71.6571	160.5	-14.6082	246.7	-61.4325
20	214.1	248.0	22.0059	203.2	292.7	117.2	378.7	-33.8736
21	267.0	318.6	22.0059	273.9	363.3	187.8	449.4	-51.6265
22	230.3	220.5	21.8506	176.1	264.9	89.8096	351.2	9.7996
23	230.3	136.8	21.6990	92.7390	180.9	6.2700	267.4	93.4837
24	426.4	271.1	21.5510	227.4	314.9	140.7	401.6	155.3
25	467.1	344.2	21.7815	300.0	388.5	213.6	474.8	122.8
26	306.0	248.5	21.3944	205.0	292.0	118.1	378.8	57.5463
27	253.2	167.2	21.0443	124.5	210.0	37.1264	297.4	85.9805
28	279.5	304.0	20.7331	261.8	346.1	174.1	433.9	-24.5103
29	336.6	379.5	21.4416	335.9	423.0	249.1	509.8	-42.8928
30	196.7	286.1	21.1265	243.2	329.1	155.9	416.3	-89.4565
31	152.1	207.3	20.9302	164.8	249.8	77.2550	337.3	-55.1521
32	319.7	346.4	20.8562	304.0	388.8	216.4	476.4	-26.7627

(continues)

FIGURE 13.21 Continued

OBS	ACTUAL	PREDICT VALUE	STD ERR PREDICT	LOWER95% MEAN	UPPER95% MEAN	LOWER95% PREDICT	UPPER95% PREDICT	RESIDUAL
33	440.0	424.3	22.4625	378.7	470.0	293.2	555.4	15.6749
34	315.0	333.4	22.7501	287.2	379.6	202.1	464.7	-18.3686
35	216.4	257.0	23.2525	209.7	304.2	125.3	388.7	-40.5740
36	339.8	398.5	23.9562	349.8	447.2	266.3	530.7	-58.7545
37	434.7	478.8	26.7003	424.6	533.1	344.5	613.2	-44.1768
38	399.7	390.3	28.0239	333.4	447.3	254.9	525.8	9.3299
39	330.8	316.3	29.5717	256.2	376.4	179.5	453.1	14.4746
40	539.8	460.3	31.3106	396.6	523.9	321.9	598.7	79.5043
41	.	543.0	35.0485	471.8	614.2	400.9	685.0	.
42	.	456.9[a]	37.4591	380.8	533.0	312.3[b]	601.5[b]	.
43	.	385.3	40.0368	303.9	466.7	237.9	532.7	.
44	.	531.7	42.7513	444.8	618.5	381.2	682.2	.

```
SUM OF RESIDUALS                 1.35323E-11
SUM OF SQUARED RESIDUALS            124335.7
PREDICTED RESID SS (PRESS)          168189.9
DURBIN-WATSON D                  0.840[c]
(FOR NUMBER OF OBS.)                40
1ST ORDER AUTOCORRELATION        0.554
```

[a] $\hat{y}_{42}$ [b] *95% prediction interval for* y_{42} [c] *Durbin-Watson statistic*

FIGURE 13.22 Residuals versus time for Farmers' Bureau Coop dummy variable model

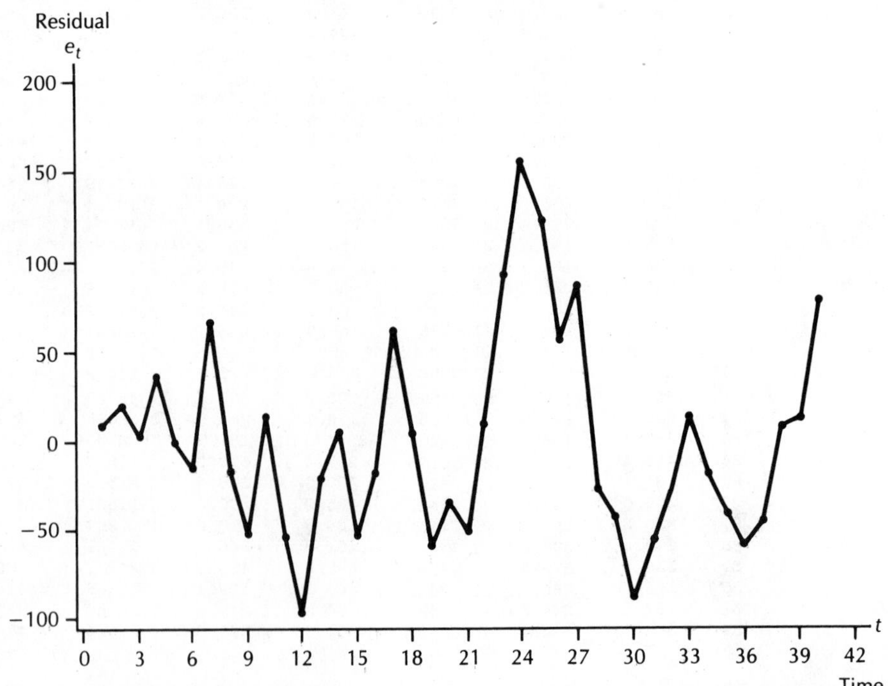

The residuals for the model are plotted in time order in Figure 13.22. The plotted residuals have these signs: + + + + + − + − − + − − − + − − + + − − − + + + + + + − − − − − − + − − − − − + + +. Here, we see a tendency for residuals to be followed by residuals of the same sign. This suggests that the error terms for this model display positive autocorrelation and that this model therefore violates the independence assumptions. The Durbin–Watson statistic is

$$d = \frac{\sum_{t=2}^{n} (e_t - e_{t-1})^2}{\sum_{t=1}^{n} e_t^2} = .840$$

Our model has $k - 1 = 5$ independent variables (we exclude the intercept), and we have $n = 40$ observations in the Farmers' Bureau Coop data. Therefore letting $\alpha = .05$, we obtain $d_{L,.05} = 1.23$ and $d_{U,.05} = 1.79$ (see Table E.5 in Appendix E). We test the null hypothesis

H_0: The error terms are not autocorrelated

versus the alternative hypothesis

H_1: The error terms are positively autocorrelated

We see that since $d = .84 < d_{L,.05} = 1.23$, we reject H_0. Thus we conclude that the error terms for the dummy variable model are positively autocorrelated. In the next section we will complete the analysis of the Farmers' Bureau Coop data by building a model that will remedy this problem.

The dummy variable regression model

$$y_t = TR_t + SN_t + \varepsilon_t$$

where

$$SN_t = \beta_{S1}x_{S1,t} + \beta_{S2}x_{S2,t} + \cdots + \beta_{S(L-1)}x_{S(L-1),t}$$

assumes that we have *additive seasonal variation*. If a time series displays additive seasonal variation, the magnitude of the seasonal swing is independent of the level of the trend. Additive seasonal variation is illustrated in Figure 13.23. This is not the only kind of seasonal variation, however. Sometimes, seasonal variation is multiplicative. If a time series displays *multiplicative seasonal variation*, the magnitude of the seasonal swing is proportional to the level of the trend. Thus if the trend is increasing, then so is the magnitude of the seasonal swing. On the other hand, if the trend is decreasing, the magnitude of the seasonal swing is also decreasing. Multiplicative seasonal variation is illustrated in Figure 13.24. Very few actual time series possess seasonal variation that is precisely additive or precisely multiplicative. However, it is useful to try to classify seasonal variation as either additive or multiplicative in order to model a time series.

FIGURE 13.23 **Additive seasonal variation: The seasonal swing is the same as the trend increases or decreases**

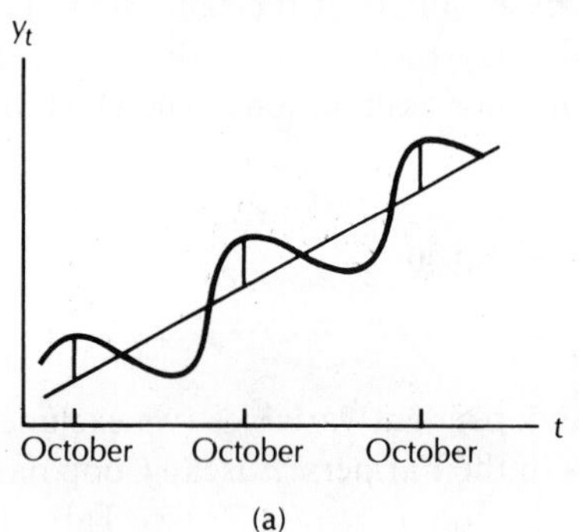

(a)

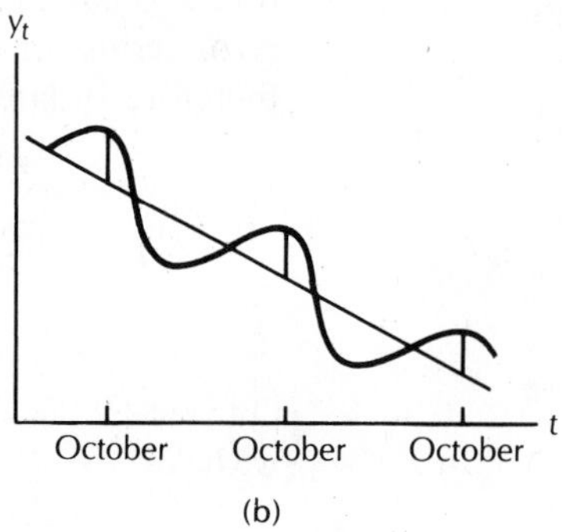

(b)

As we have stated, the dummy variable regression model is useful for modeling time series with additive seasonal variation. Other regression models can also be used, however. Two models involving *trigonometric terms* that are useful for modeling additive seasonal variation are the following (here, L is the number of seasons in a year):

$$y_t = \beta_0 + \beta_1 t + \beta_2 \sin \frac{2\pi t}{L} + \beta_3 \cos \frac{2\pi t}{L} + \varepsilon_t$$

and

$$y_t = \beta_0 + \beta_1 t + \beta_2 \sin \frac{2\pi t}{L} + \beta_3 \cos \frac{2\pi t}{L} + \beta_4 \sin \frac{4\pi t}{L} + \beta_5 \cos \frac{4\pi t}{L} + \varepsilon_t$$

FIGURE 13.24 **Multiplicative seasonal variation: The magnitude of the seasonal swing is proportional to the trend**

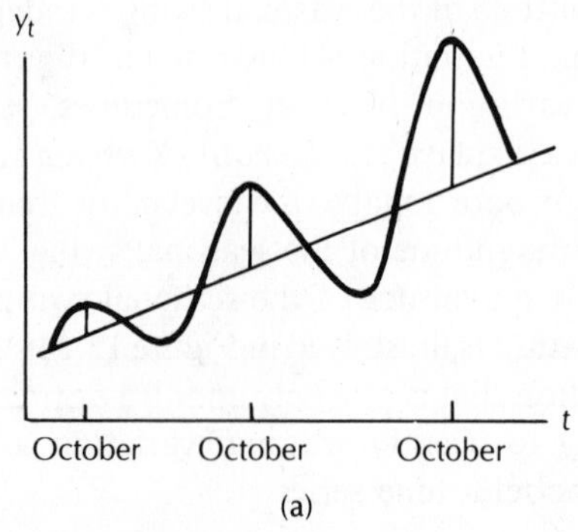

(a)

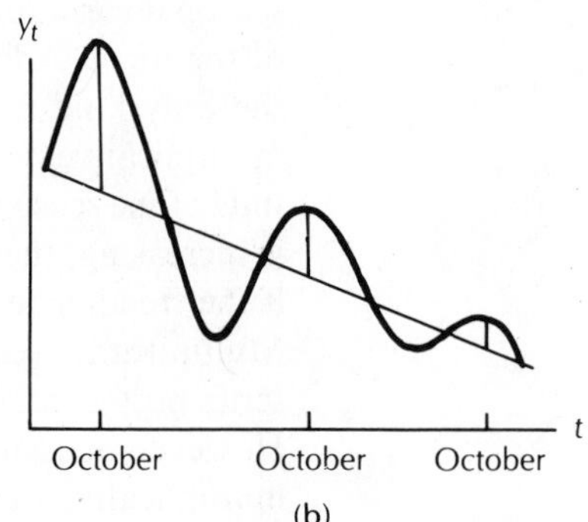

(b)

These models assume a linear trend, but they can be altered to handle other trends. The first is useful in modeling a very regular additive seasonal pattern. The second possesses terms that allow modeling of a more complicated additive seasonal pattern.

Regression models can also be employed to model multiplicative seasonal variation. Two useful models that involve trigonometric terms are

$$y_t = \beta_0 + \beta_1 t + \beta_2 \sin \frac{2\pi t}{L} + \beta_3 t \sin \frac{2\pi t}{L} + \beta_4 \cos \frac{2\pi t}{L} + \beta_5 t \cos \frac{2\pi t}{L} + \varepsilon_t$$

and

$$\begin{aligned} y_t = {} & \beta_0 + \beta_1 t + \beta_2 \sin \frac{2\pi t}{L} + \beta_3 t \sin \frac{2\pi t}{L} + \beta_4 \cos \frac{2\pi t}{L} + \beta_5 t \cos \frac{2\pi t}{L} \\ & + \beta_6 \sin \frac{4\pi t}{L} + \beta_7 t \sin \frac{4\pi t}{L} + \beta_8 \cos \frac{4\pi t}{L} + \beta_9 t \cos \frac{4\pi t}{L} + \varepsilon_t \end{aligned}$$

Again, these models assume a linear trend but can be altered to handle other trends. The first is useful in modeling a very regular multiplicative seasonal pattern. The second possesses terms that allow modeling of a more complicated multiplicative seasonal pattern. For several examples demonstrating the use of these trigonometric models, see Bowerman and O'Connell (1987). Other models (such as *Winters' models* and *decomposition models*) can be useful in modeling additive or multiplicative seasonal time series. For a discussion of some of these models, see Bowerman and O'Connell (1987). Included is a discussion of how to handle *cyclical effects*, which are recurring up and down movements around the trend that have a duration of more than a year. For instance, such cycles might be caused by the business cycle or by cyclical weather patterns.

Finally, we can combine *causal variables* and *time series variables* in the same regression model. As an example of this, consider forecasting monthly sales, y_t, of a fishing lure. We suppose that the lure has been sold at a constant price for the past several years. Also, assume that monthly sales of this lure have shown an increasing linear trend. In addition, suppose that monthly sales of this lure are affected by the monthly advertising expenditure, x_t, devoted to promoting the lure and that sales are seasonal. This would suggest using the model

$$y_t = \beta_0 + \beta_1 x_t + \beta_2 t + \beta_3 x_{S1,t} + \beta_4 x_{S2,t} + \cdots + \beta_{13} x_{S11,t} + \varepsilon_t$$

Here, $x_{S1,t}, x_{S2,t}, \ldots, x_{S11,t}$ are monthly dummy variables defined so that

$$x_{Sj,t} = \begin{cases} 1 & \text{if time period } t \text{ is month } j \\ 0 & \text{otherwise} \end{cases}$$

In this model the $\beta_1 x_t$ term (assuming that the functional form is correct) accounts for the effect on monthly sales of monthly advertising expenditure. The $\beta_2 t$ term accounts for the linear trend in sales. The dummy variables account for the seasonal pattern in sales. Thus in this model we are including a causal variable, x_t, along with the time series variables t, $x_{S1,t}, x_{S2,t}, \ldots,$ and $x_{S11,t}$.

13.5

AUTOREGRESSIVE ERROR STRUCTURES

When the error terms for a regression model are autocorrelated, we can remedy the problem by modeling the autocorrelation. In general, if the error terms are autocorrelated and we ignore this fact, the least squares procedure tends to produce values of $s\sqrt{c_{jj}}$ (the standard error of the estimate b_j) that are too small. Consequently, values of the t statistic are too large. It should be noted that the opposite behavior is possible but less common. Therefore if we ignore autocorrelated error terms, we tend to get spurious declarations of significance when variables are really not important. For example, further analysis in this section will reveal that the linear term $(-7.46t)$ in the Farmers' Bureau Coop dummy variable model is less significant than the t statistic in Figure 13.21 ($t = -2.2$) indicates. The reason for this misleading t statistic is the autocorrelation in the error terms for this model. Another consequence of ignoring autocorrelated error terms is that we pay a penalty in terms of wider prediction intervals. By taking autocorrelation into account we can achieve more accurate prediction intervals.

One autocorrelated error structure that is frequently encountered is the *first-order autoregressive process*. To define this process, we consider the model

$$\varepsilon_t = \rho\varepsilon_{t-1} + U_t$$

This model relates ε_t, the error term in time period t, to ε_{t-1}, the error term in time period $t - 1$. Here, we assume that $U_1, U_2, \ldots, U_n$ each have mean zero and satisfy the inference assumptions, and we define ρ to be the correlation coefficient between ε_t and ε_{t-1}. That is, ρ is defined to be the correlation coefficient between error terms separated by one time period. If $\rho > 0$, this indicates that the error terms are positively autocorrelated. It is easy to see that the above equation implies that if $\rho > 0$, then a positive error term ε_{t-1} will tend to produce another positive error term ε_t. Likewise, a negative error term ε_{t-1} will tend to produce another negative error term ε_t. On the other hand, if $\rho < 0$, the error terms are negatively autocorrelated. Again, if $\rho < 0$, then a positive error term ε_{t-1} will tend to produce a negative error term ε_t. However, a negative error term ε_{t-1} will tend to produce a positive error term ε_t.

Now consider the general linear regression model

$$y_t = \beta_0 + \beta_1 x_{t1} + \beta_2 x_{t2} + \cdots + \beta_p x_{tp} + \varepsilon_t$$

in which we assume that

$$\varepsilon_t = \rho\varepsilon_{t-1} + U_t$$

and that $\rho \neq 0$. This implies that the error terms are autocorrelated according to a first-order autoregressive process. It follows that the independence assumption is violated. To obtain a transformed model that satisfies the inference assumptions, consider the model

$$\rho y_{t-1} = \rho\beta_0 + \rho\beta_1 x_{t-1,1} + \rho\beta_2 x_{t-1,2} + \cdots + \rho\beta_p x_{t-1,p} + \rho\varepsilon_{t-1}$$

For $t = 2, 3, \ldots, n$, subtract ρy_{t-1} from y_t, which yields the model

$$\begin{aligned} y_t - \rho y_{t-1} &= \beta_0(1 - \rho) + \beta_1(x_{t1} - \rho x_{t-1,1}) + \beta_2(x_{t2} - \rho x_{t-1,2}) \\ &\quad + \cdots + \beta_p(x_{tp} - \rho x_{t-1,p}) + [\varepsilon_t - \rho\varepsilon_{t-1}] \end{aligned}$$

Since $[\varepsilon_t - \rho\varepsilon_{t-1}] = U_t$, and since $U_1, U_2, \ldots, U_n$ satisfy the inference assumptions, this new transformed model satisfies the inference assumptions.

Notice that when we subtract ρy_{t-1} from y_t we lose information concerning y_1 (that is, $y_t - \rho y_{t-1}$ cannot be computed for $t = 1$). This is not serious if n is large, but this might be serious if n is small. However, we can regain this loss by multiplying both sides of the regression model describing y_1 by $\sqrt{1 - \rho^2}$. This yields

$$\begin{aligned} \sqrt{1 - \rho^2}\, y_1 &= \sqrt{1 - \rho^2}\,\beta_0 + \beta_1(\sqrt{1 - \rho^2}\, x_{11}) + \beta_2(\sqrt{1 - \rho^2}\, x_{12}) \\ &\quad + \cdots + \beta_p(\sqrt{1 - \rho^2}\, x_{1p}) + \sqrt{1 - \rho^2}\,\varepsilon_1 \end{aligned}$$

Note that it can be shown that this equation, along with the equations

$$\begin{aligned} y_t - \rho y_{t-1} &= \beta_0(1 - \rho) + \beta_1(x_{t1} - \rho x_{t-1,1}) + \beta_2(x_{t2} - \rho x_{t-1,2}) \\ &\quad + \cdots + \beta_p(x_{tp} - \rho x_{t-1,p}) + U_t \end{aligned}$$

for $t = 2, 3, \ldots, n$, satisfy the inference assumptions.

We are faced with a dilemma here. In practice, the correlation coefficient ρ must be estimated in order to transform the data. However, to estimate ρ, we must use the untransformed data. Therefore an iterative procedure is suggested. This procedure is called the *Cochran-Orcutt procedure*, and it works as follows.

First, for the untransformed model

$$y_t = \beta_0 + \beta_1 x_{t1} + \beta_2 x_{t2} + \cdots + \beta_p x_{tp} + \varepsilon_t$$

we calculate the least squares point estimates using the *untransformed data*. Using these estimates, we compute the residuals $e_1, e_2, \ldots, e_n$.

Second, we perform regression analysis on the residuals using the model

$$e_t = \rho e_{t-1} + U_t$$

Here we let

$$\mathbf{y} = \begin{bmatrix} e_2 \\ e_3 \\ \cdot \\ \cdot \\ \cdot \\ e_n \end{bmatrix} \quad \text{and} \quad \mathbf{X} = \begin{bmatrix} e_1 \\ e_2 \\ \cdot \\ \cdot \\ \cdot \\ e_{n-1} \end{bmatrix}$$

We calculate the least squares estimate of ρ (denoted r) to be

$$r = (\mathbf{X}'\mathbf{X})^{-1}\mathbf{X}'\mathbf{y} = \frac{\sum_{t=2}^{n} e_t e_{t-1}}{\sum_{t=2}^{n} e_{t-1}^2}$$

Third, we make the transformation discussed previously, using r as an estimate of the correlation coefficient ρ. Thus we let

$$\mathbf{y} = \begin{bmatrix} \sqrt{1-r^2}\,y_1 \\ y_2 - ry_1 \\ y_3 - ry_2 \\ \vdots \\ y_n - ry_{n-1} \end{bmatrix}$$

and

$$\mathbf{X} = \begin{bmatrix} \sqrt{1-r^2} & \sqrt{1-r^2}\,x_{11} & \cdots & \sqrt{1-r^2}\,x_{1j} & \cdots & \sqrt{1-r^2}\,x_{1p} \\ 1-r & x_{21} - rx_{11} & \cdots & x_{2j} - rx_{1j} & \cdots & x_{2p} - rx_{1p} \\ 1-r & x_{31} - rx_{21} & \cdots & x_{3j} - rx_{2j} & \cdots & x_{3p} - rx_{2p} \\ \vdots & \vdots & & \vdots & & \vdots \\ 1-r & x_{n1} - rx_{n-1,1} & \cdots & x_{nj} - rx_{n-1,j} & \cdots & x_{np} - rx_{n-1,p} \end{bmatrix}$$

Then, utilizing the new **X** and **y**, we compute new least squares point estimates.

Fourth, using the new least squares estimates, we recompute the residuals and return to the second step, in which we calculate a revised estimate of ρ using the newest residuals. We then use the revised estimate of ρ to compute newly transformed data (as in the third step). Then we use these newly transformed data to calculate revised least squares point estimates.

The iterative process ends when the new least squares point estimates change little between iterations. Often, one or two iterations are sufficient. When we obtain the final regression, the least squares point estimates apply to the untransformed model

$$y_t = \beta_0 + \beta_1 x_{t1} + \beta_2 x_{t2} + \cdots + \beta_p x_{tp} + \varepsilon_t$$

where $\varepsilon_t = \rho\varepsilon_{t-1} + U_t$. This is because the parameters in this model are the same as the parameters in the transformed model. In addition, the standard error and the t statistics obtained for the transformed model also apply to the untransformed model. Moreover, these values are correct because the transformed model satisfies the inference assumptions.

Point predictions and prediction intervals are found as follows. First, denote the predictions of y_t, μ_t, and ε_t as $\hat{y}_t$, $\hat{\mu}_t$, and $\hat{\varepsilon}_t$. A point prediction of the future value

$$\begin{aligned} y_{n+\tau} &= \mu_{n+\tau} + \varepsilon_{n+\tau} \\ &= \beta_0 + \beta_1 x_{n+\tau,1} + \beta_2 x_{n+\tau,2} + \cdots + \beta_p x_{n+\tau,p} + \varepsilon_{n+\tau} \end{aligned}$$

where $\varepsilon_{n+\tau} = \rho\varepsilon_{n+\tau-1} + U_{n+\tau}$ is given by

$$\begin{aligned} \hat{y}_{n+\tau} &= \hat{\mu}_{n+\tau} + \hat{\varepsilon}_{n+\tau} = \hat{\mu}_{n+\tau} + r\hat{\varepsilon}_{n+\tau-1} + \hat{U}_{n+\tau} \\ &= b_0 + b_1 x_{n+\tau,1} + b_2 x_{n+\tau,2} + \cdots + b_p x_{n+\tau,p} + r\hat{\varepsilon}_{n+\tau-1} \end{aligned}$$

Here, $\hat{\varepsilon}_{n+\tau}$ is obtained by substituting the estimate r for ρ and by predicting $U_{n+\tau}$ to be zero. If $\tau = 1$ (the prediction is being made for one time period ahead), then

$$\begin{aligned} \hat{\varepsilon}_{n+\tau-1} &= \hat{\varepsilon}_n = y_n - \hat{\mu}_n \\ &= y_n - [b_0 + b_1 x_{n1} + b_2 x_{n2} + \cdots + b_p x_{np}] \end{aligned}$$

since y_n has been observed. If $\tau > 1$ (the prediction is for more than one time period ahead), then, since $y_{n+\tau-1}$ has not been observed,

$$\begin{aligned} \hat{\varepsilon}_{n+\tau-1} &= \hat{y}_{n+\tau-1} - \hat{\mu}_{n+\tau-1} \\ &= \hat{y}_{n+\tau-1} - [b_0 + b_1 x_{n+\tau-1,1} + b_2 x_{n+\tau-1,2} + \cdots + b_p x_{n+\tau-1,p}] \end{aligned}$$

Furthermore, approximate $100(1 - \alpha)\%$ prediction intervals are obtained as follows.

1. If $\tau = 1$, then an approximate $100(1 - \alpha)\%$ prediction interval for y_{n+1} is

$$[\hat{y}_{n+1} \pm t_{[\alpha/2]}^{(n-k)} s]$$

2. If $\tau = 2$, then an approximate $100(1 - \alpha)\%$ prediction interval for y_{n+2} is

$$[\hat{y}_{n+2} \pm t_{[\alpha/2]}^{(n-k)} s\sqrt{1 + r^2}]$$

3. If $\tau \geqslant 3$, then an approximate $100(1 - \alpha)\%$ prediction interval for $y_{n+\tau}$ is

$$[\hat{y}_{n+\tau} \pm t_{[\alpha/2]}^{(n-k)} s\sqrt{1 + r^2 + \cdots + r^{2(\tau-1)}}]$$

In these formulas, $t_{[\alpha/2]}^{(n-k)}$ is defined as usual, s is the standard error computed from the transformed data, and r is the final estimate of ρ. We demonstrate the use of these formulas in the following example.

Example 13.10 Reconsider the Farmers' Bureau Coop problem, in which the coop wishes to predict its quarterly propane gas bill, y_t. We saw in Example 13.9 that the dummy variable model

$$\begin{aligned} y_t &= TR_t + SN_t + \varepsilon_t \\ &= \beta_0 + \beta_1 t + \beta_2 t^2 + \beta_3 x_{S1,t} + \beta_4 x_{S2,t} + \beta_5 x_{S3,t} + \varepsilon_t \end{aligned}$$

has a positively autocorrelated error structure. We now continue to analyze the data by assuming a first-order autoregressive error structure. That is, we suppose that we can express ε_t in the above model as

$$\varepsilon_t = \rho\varepsilon_{t-1} + U_t$$

To use the Cochran-Orcutt procedure, we note that we computed (in Example 13.9) least squares point estimates using the untransformed Farmers' Bureau Coop data. We obtained the residuals $e_1, e_2, \ldots, e_{40}$ that are given in Figure 13.21. Using these residuals and the model

$$e_t = \rho e_{t-1} + U_t$$

we perform a regression analysis on the residuals. Here we let

$$\mathbf{y} = \begin{bmatrix} e_2 \\ e_3 \\ \cdot \\ \cdot \\ \cdot \\ e_{40} \end{bmatrix} \quad \text{and} \quad \mathbf{X} = \begin{bmatrix} e_1 \\ e_2 \\ \cdot \\ \cdot \\ \cdot \\ e_{39} \end{bmatrix}$$

We calculate the least squares estimate of ρ to be

$$r = (\mathbf{X}'\mathbf{X})^{-1}\mathbf{X}'\mathbf{y} = .5841$$

Next, we compute transformed data. We find that

$$\mathbf{y} = \begin{bmatrix} \sqrt{1 - r^2}y_1 \\ y_2 - ry_1 \\ y_3 - ry_2 \\ \cdot \\ \cdot \\ \cdot \\ y_{40} - ry_{39} \end{bmatrix} = \begin{bmatrix} \sqrt{1 - (.5841)^2}(344.39) \\ 246.63 - .5841(344.39) \\ 131.53 - .5841(246.63) \\ \cdot \\ \cdot \\ \cdot \\ 539.78 - .5841(330.8) \end{bmatrix}$$

and that

$$\mathbf{X} = \begin{bmatrix} \sqrt{1 - r^2} & \sqrt{1 - r^2}x_{11} & \cdots & \sqrt{1 - r^2}x_{1j} & \cdots & \sqrt{1 - r^2}x_{1p} \\ 1 - r & x_{21} - rx_{11} & \cdots & x_{2j} - rx_{1j} & \cdots & x_{2p} - rx_{1p} \\ 1 - r & x_{31} - rx_{21} & \cdots & x_{3j} - rx_{2j} & \cdots & x_{3p} - rx_{2p} \\ \vdots & \vdots & & \vdots & & \vdots \\ 1 - r & x_{n1} - rx_{n-1,1} & \cdots & x_{nj} - rx_{n-1,j} & \cdots & x_{np} - rx_{n-1,p} \end{bmatrix}$$

$$= \begin{bmatrix} & t & t^2 & x_{S1,t} & x_{S2,t} & x_{S3,t} \\ \sqrt{1 - (.5841)^2} & \sqrt{1 - (.5841)^2}(1) & \sqrt{1 - (.5841)^2}(1)^2 & \sqrt{1 - (.5841)^2}(1) & \sqrt{1 - (.5841)^2}(0) & \sqrt{1 - (.5841)^2}(0) \\ 1 - .5841 & 2 - (.5841)(1) & (2)^2 - .5841(1)^2 & 0 - (.5841)(1) & 1 - (.5841)(0) & 0 - (.5841)(0) \\ 1 - .5841 & 3 - (.5841)(2) & (3)^2 - .5841(2)^2 & 0 - (.5841)(0) & 0 - (.5841)(0) & 1 - (.5841)(0) \\ \vdots & \vdots & \vdots & \vdots & \vdots & \vdots \\ 1 - .5841 & 40 - (.5841)(39) & (40)^2 - .5841(39)^2 & 0 - (.5841)(0) & 0 - (.5841)(0) & 0 - (.5841)(1) \end{bmatrix}$$

TABLE 13.10 The *t* statistics for the Farmers' Bureau Coop first-order autoregressive model

Parameter	Least squares point estimate	Standard error of the estimate, $s\sqrt{c_{jj}}$	t statistic
β_0	$b_0 = 283.9803$	51.9057	$t = 5.47$
β_1	$b_1 = -9.1870$	5.6835	$t = -1.62$
β_2	$b_2 = .3522$	.1338	$t = 2.63$
β_3	$b_3 = 70.0615$	17.2703	$t = 4.06$
β_4	$b_4 = -35.4887$	19.2970	$t = -1.84$
β_5	$b_5 = -126.5643$	16.7836	$t = -7.54$

Using this **y** and **X**, we compute updated least squares point estimates. These estimates, which are given in Table 13.10 along with the appropriate t statistics, yield the equation

$$\begin{aligned}\hat{\mu}_t &= 283.9803 - 9.1870t + .3522t^2 + 70.0615x_{S1,t} \\ &\quad - 35.4887x_{S2,t} - 126.5643x_{S3,t}\end{aligned}$$

For the transformed model we obtain $s = 49.5102$.

Note that since the transformed model utilizes r, the point estimate of ρ, some computer packages (including SAS) recommend calculating s by the more conservative equation $s = \sqrt{SSE/(n - (k + 1))} = \sqrt{83342.925/(40 - 7)} = 50.2548$. Comparing Figure 13.21 and Table 13.10, we see that the least squares point estimates obtained from the transformed data do not differ a great deal from those obtained by using the untransformed data. Therefore further iterations will not be done.

The t statistics in Table 13.10 are valid because they come from the transformed regression that satisfies the inference assumptions. Recall that the t statistics in Figure 13.21 are invalid and should be ignored. For example, Table 13.10 indicates that the variable t (with t statistic -1.62) is less significant in our new correct analysis than the incorrect t statistic ($= -2.2$) in Figure 13.21 had indicated. The t term would usually be retained anyway, since Table 13.10 indicates that the t^2 term is important.

Recall that Farmers' Bureau Coop wishes to find out whether the seasonally adjusted trend in the propane gas bills is increasing. One way to remove the seasonality is to ignore the seasonal dummy variables $x_{S1,t}$, $x_{S2,t}$, and $x_{S3,t}$. Thus we look only at the estimated trend

$$tr_t = 283.9803 - 9.1870t + .3522t^2$$

This essentially puts all trend estimates on a fourth-quarter basis. We see that there is an initial decline in tr_t because of the negative coefficient on t. However, as t gets large (that is, as time advances) the positive term $.3522t^2$ dominates, and the estimated trend rises. For any quadratic function

$$tr_t = a + bt + dt^2$$

where $d > 0$, it can easily be shown by using calculus that the minimum value of tr_t occurs at $t = -b/2d$. Therefore in our case the estimated propane gas bill trend will increase after

$$t = \frac{-(-9.1870)}{2(.3522)} = 13$$

Since our last observation is at time $t = 40$, the seasonally adjusted propane gas bill is increasing. Also, by differentiating the above quadratic function it can easily be shown that the instantaneous rate of growth is $b + 2dt$. So at $t = 40$ we estimate that the propane gas bill trend is rising at a rate of

$$-9.1870 + 2(.3522)(40) = \$18.99 \text{ per quarter}$$

The coop now wishes to forecast its quarterly propane gas bills for the four quarters of next year (time periods 41, 42, 43, and 44). For time period 41,

$$y_{41} = \mu_{41} + \varepsilon_{41} = \mu_{41} + [\rho\varepsilon_{40} + U_{41}]$$

Therefore the predicted propane gas bill for time period 41 is

$$\hat{y}_{41} = \hat{\mu}_{41} + r\hat{\varepsilon}_{40}$$

where $\hat{\varepsilon}_{40} = y_{40} - \hat{\mu}_{40}$. Here,

$$\begin{aligned}\hat{\varepsilon}_{40} &= 539.78 - [283.9803 - 9.1870(40) + .3522(40^2) + 70.0615(0) \\ &\quad - 35.4887(0) - 126.5643(0)] \\ &= 539.78 - 480.02 \\ &= 59.76\end{aligned}$$

and

$$\begin{aligned}\hat{y}_{41} &= \hat{\mu}_{41} + r\hat{\varepsilon}_{40} \\ &= [283.9803 - 9.1870(41) + .3522(41)^2 + 70.0615(1) \\ &\quad - 35.4887(0) - 126.5643(0)] + .5841(59.76) \\ &= 569.42 + .5841(59.76) = 604.33\end{aligned}$$

Thus the quarterly propane gas bill for time period 41 is predicted to be \$604.33. Furthermore, an approximate 95% prediction interval for y_{41} is

$$\begin{aligned}[\hat{y}_{41} \pm t_{[\alpha/2]}^{(n-k)} s] &= [604.33 \pm t_{[.025]}^{(40-6)} s] \\ &= [604.33 \pm 2.034(49.5102)] \\ &= [503.63, 705.03]\end{aligned}$$

This interval says that Farmers' Bureau Coop is 95 percent confident that the propane gas bill for period 41 will be between \$503.63 and \$705.03.

For time period 42,

$$y_{42} = \mu_{42} + \varepsilon_{42} = \mu_{42} + [\rho\varepsilon_{41} + U_{42}]$$

FIGURE 13.25 **PROC ARIMA output for the Farmers' Bureau Coop first-order autoregressive model**

PARAMETER	ESTIMATE	APPROX. STD ERROR	T RATIO	LAG	VARIABLE
MU	283.949	47.5501	5.97	0	BILL
AR1,1	0.594084	0.147014	4.04	1	BILL
NUM1	-9.21968	5.46548	-1.69	0	TIME
NUM2	0.353478	0.133145	2.65	0	TIMESQ
NUM3	70.1069	17.427	4.02	0	Q1
NUM4	-35.4286	19.5222	-1.81	0	Q2
NUM5	-126.525	16.9569	-7.46	0	Q3

```
CONSTANT ESTIMATE    =  115.26

VARIANCE ESTIMATE    =  2525.2
STD ERROR ESTIMATE   = 50.2513
AIC                  = 433.183*
SBC                  = 445.005*
NUMBER OF RESIDUALS  =      40

FORECASTS FOR VARIABLE BILL
```

OBS	FORECAST	STD ERROR	LOWER 95%	UPPER 95%
	FORECAST BEGINS			
41	605.3285	50.2513	506.8380	703.8191
42	505.6717	58.4502	391.1116	620.2318
43	426.9411	61.0817	307.2235	546.6588
44	569.9732	61.9838	448.4875	691.4589

Note: The values given under ESTIMATE are point estimates of, in order, the parameters β_0, ρ, β_1, β_2, β_3, β_4, and β_5.

Therefore the predicted propane gas bill for time period 42 is

$$\hat{y}_{42} = \hat{\mu}_{42} + r\hat{\varepsilon}_{41}$$

where $\hat{\varepsilon}_{41} = \hat{y}_{41} - \hat{\mu}_{41}$. Here, $\hat{\varepsilon}_{41} = \hat{y}_{41} - \hat{\mu}_{41} = 604.33 - 569.42 = 34.91$, and

$$\begin{aligned}\hat{y}_{42} &= \hat{\mu}_{42} + r\hat{\varepsilon}_{41} \\ &= [283.9803 - 9.1870(42) + .3522(42)^2 + 70.0615(0) - 35.4887(1) \\ &\quad - 126.5643(0)] + .5841(34.91) \\ &= 483.92 + .5841(34.91) = 504.31\end{aligned}$$

Thus the quarterly propane gas bill for time period 42 is predicted to be \$504.31. Moreover, an approximate 95% prediction interval for y_{42} is

$$\begin{aligned}[\hat{y}_{42} \pm t_{[.025]}^{(40-6)} s\sqrt{1 + r^2}] &= [504.31 \pm 2.034(49.5102)\sqrt{1 + (.5841)^2}] \\ &= [504.31 \pm 116.62] \\ &= [387.69, 620.93]\end{aligned}$$

In a similar fashion we find that a point prediction of y_{43} is $\hat{y}_{43} = \$425.50$. We also find that an approximate 95% prediction interval for y_{43} is

$$\begin{aligned}[\hat{y}_{43} \pm t_{[.025]}^{(40-6)} s\sqrt{1 + r^2 + r^4}] &= [425.50 \pm 2.034(49.5102)\sqrt{1 + (.5841)^2 + (.5841)^4}] \\ &= [425.50 \pm 121.58] \\ &= [303.92, 547.08]\end{aligned}$$

Finally, we find that a point prediction of y_{44} is $\hat{y}_{44} = \$568.57$. Furthermore, an approximate 95% prediction interval for y_{44} is

$$\begin{aligned}[\hat{y}_{44} \pm t_{[.025]}^{(40-6)} s\sqrt{1 + r^2 + r^4 + r^6}] &= [568.57 \pm 2.034(49.5102)\sqrt{1 + (.5841)^2 + (.5841)^4 + (.5841)^6}] \\ &= [568.57 \pm 123.22] \\ &= [445.35, 691.79]\end{aligned}$$

SAS PROC AUTOREG can be used to calculate the point estimates of the parameters in a model having an autoregressive error structure. However, this procedure does not (at the time of this writing) calculate prediction intervals as shown above. Therefore we recommend using SAS PROC ARIMA, which does calculate prediction intervals very similar to those calculated above. In optional Section 13.7 we show how to use PROC ARIMA. The resulting output for the model currently being discussed is shown in Figure 13.25. PROC ARIMA uses a more complicated estimation procedure than that described above. However, note from Figure 13.25 that the point predictions of and 95% prediction intervals for y_{41}, y_{42}, y_{43}, and y_{44} that are provided by PROC ARIMA are quite similar to those calculated above.

In addition to the first-order autoregressive process, other autocorrelated error structures exist. For instance, error terms can follow the *autoregressive process of order q*. This process is written as

$$\varepsilon_t = \phi_1\varepsilon_{t-1} + \phi_2\varepsilon_{t-2} + \cdots + \phi_q\varepsilon_{t-q} + U_t$$

It relates ε_t, the error term in time period t, to the previous error terms ε_{t-1}, ε_{t-2}, . . . , ε_{t-q}. Here $\phi_1, \phi_2, \ldots, \phi_q$ are parameters, and we assume that $U_1, U_2, \ldots, U_n$ each have mean zero and satisfy the inference assumptions. When this model is appropriate, an iterative procedure known as the *Cochran-Orcutt procedure for an autoregressive process of order q* can be employed. This procedure works as follows.

First, for the untransformed model

$$y_t = \beta_0 + \beta_1 x_{t1} + \beta_2 x_{t2} + \cdots + \beta_p x_{tp} + \varepsilon_t$$

we calculate least squares point estimates $b_0, b_1, \ldots, b_p$ using the *untransformed data.* Using these estimates, we compute residuals $e_1, e_2, \ldots, e_n$.

Second, we model e_t as an autoregressive process

$$e_t = \phi_1 e_{t-1} + \phi_2 e_{t-2} + \cdots + \phi_q e_{t-q} + U_t$$

and we compute least squares point estimates of the parameters $\phi_1, \phi_2, \ldots, \phi_q$. We denote these estimates as $\hat{\phi}_1, \hat{\phi}_2, \ldots, \hat{\phi}_q$.

Third, we transform y_t to

$$y_t^* = y_t - \hat{\phi}_1 y_{t-1} - \cdots - \hat{\phi}_q y_{t-q}$$

for $t = q + 1, q + 2, \ldots, n$. Similarly, we transform each column of the **X** matrix. For example, the intercept column of n 1's becomes a column of $n - q$ entries, each entry being $1 - \hat{\phi}_1 - \hat{\phi}_2 - \cdots - \hat{\phi}_q$. The other columns of the **X** matrix are transformed by replacing, for $t = q + 1, q + 2, \ldots, n$, x_{tj} by

$$x_{tj}^* = x_{tj} - \hat{\phi}_1 x_{t-1,j} - \hat{\phi}_2 x_{t-2,j} - \cdots - \hat{\phi}_q x_{t-q,j}$$

We then compute new least squares point estimates $b_0, b_1, \ldots, b_p$ using the *transformed data.*

Fourth, using the new least squares point estimates, we recompute residuals and return to the second step, in which we calculate revised estimates of $\phi_1, \phi_2, \ldots, \phi_q$ using the newest residuals. We then use the revised estimates of $\phi_1, \phi_2, \ldots, \phi_q$ to compute newly transformed data, which are in turn used to compute revised least squares point estimates $b_0, b_1, \ldots, b_p$.

Again, the iterative procedure ends when the least squares point estimates $b_0, b_1, \ldots, b_p$ change little between iterations. Usually, one or two iterations are sufficient. When we obtain the final regression, the final least squares point estimates apply to the untransformed model. In addition, the standard error and the t statistics obtained for the transformed model also apply to the untransformed model. Furthermore, these values are correct because the transformed model satisfies the inference assumptions.

This procedure loses information from the first q observations. A method similar to that given in the context of the first-order autoregressive process is available to recoup this information. Fortunately, the loss of information is not very severe for large n. Therefore since the method used to recoup this information is somewhat complicated, we do not present it here. PROC ARIMA allows one to run regressions with error terms following an autoregressive process of order q.

Another common autoregressive process is the *p-th order autoregressive process in the observations* $y_1, y_2, \ldots, y_n$. This process is written as

$$y_t = \beta_0 + \beta_1 y_{t-1} + \beta_2 y_{t-2} + \cdots + \beta_p y_{t-p} + \varepsilon_t$$

It expresses y_t in terms of the previous observations $y_{t-1}, y_{t-2}, \ldots, y_{t-p}$ and an error term ε_t. Although the independence assumption does not hold for such a model, it can be shown that for large samples the least squares procedure is appropriate. As an example, the second-order autoregressive process in y_t,

$$y_t = \beta_0 + \beta_1 y_{t-1} + \beta_2 y_{t-2} + \varepsilon_t$$

expresses the observation y_t in terms of the previous observations y_{t-1} and y_{t-2}.

Here, estimation of the parameters β_0, β_1, and β_2 would be accomplished by using

$$\mathbf{y} = \begin{bmatrix} y_3 \\ y_4 \\ y_5 \\ \cdot \\ \cdot \\ \cdot \\ y_n \end{bmatrix} \quad \text{and} \quad \mathbf{X} = \begin{bmatrix} 1 & y_2 & y_1 \\ 1 & y_3 & y_2 \\ 1 & y_4 & y_3 \\ \cdot & \cdot & \cdot \\ \cdot & \cdot & \cdot \\ \cdot & \cdot & \cdot \\ 1 & y_{n-1} & y_{n-2} \end{bmatrix}$$

Finally, *Box-Jenkins models* express y_t in terms of previous observations y_{t-1}, $y_{t-2}, \ldots, y_{t-p}$ and previous error terms $\varepsilon_{t-1}, \varepsilon_{t-2}, \ldots, \varepsilon_{t-q}$. Such a model can be written as

$$\begin{aligned} y_t &= \beta_0 + \beta_1 y_{t-1} + \beta_2 y_{t-2} + \cdots + \beta_p y_{t-p} + \phi_1 \varepsilon_{t-1} \\ &\quad + \phi_2 \varepsilon_{t-2} + \cdots + \phi_q \varepsilon_{t-q} + U_t \end{aligned}$$

The use of Box-Jenkins models involves four steps:

1. The *identification* of an appropriate model.
2. The *estimation* of model parameters.
3. *Diagnostic checking* (checking the adequacy of the model).
4. *Forecasting.*

A detailed presentation of the Box-Jenkins methodology can be found in Box and Jenkins (1977) and Bowerman and O'Connell (1987).

13.6

REMEDIES FOR NON-NORMALITY

Mild departures from the normality assumption are not serious. However, if examination of the residuals indicates a pronounced departure from the normality assumption, there are remedies that can be employed. Generally, these remedies involve transformation of the data. In particular, incorrect functional forms, omitted variables, and violations of the constant variance assumption can cause the error terms to look non-normal. Fortunately, remedies for these problems (for example, transformations to achieve equal variances) often correct the non-normality problem. We demonstrate this phenomenon in the following example.

Example 13.11 Reconsider the Republic Wholesalers telephone order data. A frequency distribution and histogram of the residuals for the telephone order data are given in Table 13.11 and Figure 13.26. The residuals are calculated as deviations from the column (hour) means in Table 13.4. This is reasonable, since we have several

TABLE 13.11 **Frequency distribution of the residuals (from hour means) for the Republic Wholesalers telephone order data**

Subinterval	*Frequency (no. in subinterval)*	*Subinterval*	*Frequency (no. in subinterval)*
−29.99 to −25	2	10 to 14.99	3
−24.99 to −20	0	15 to 19.99	2
−19.99 to −15	5	20 to 24.99	3
−14.99 to −10	6	25 to 29.99	1
−9.99 to −5	16	30 to 34.99	0
−4.99 to 0	16	35 to 39.99	1
0 to 4.99	14	40 to 44.99	1
5 to 9.99	9	45 to 49.99	1

observations for each hour. The histogram in Figure 13.26 appears to be skewed to the right. Notice that three positive residuals exceed 30, while no negative residuals are more than 30 units away from zero. This suggests a violation of the normality assumption, although the decision as to whether or not the histogram looks normal is a subjective judgment. The normal plot of these residuals is given

FIGURE 13.26 **Histogram of the residuals (from hour means) for the Republic Wholesalers telephone order data**

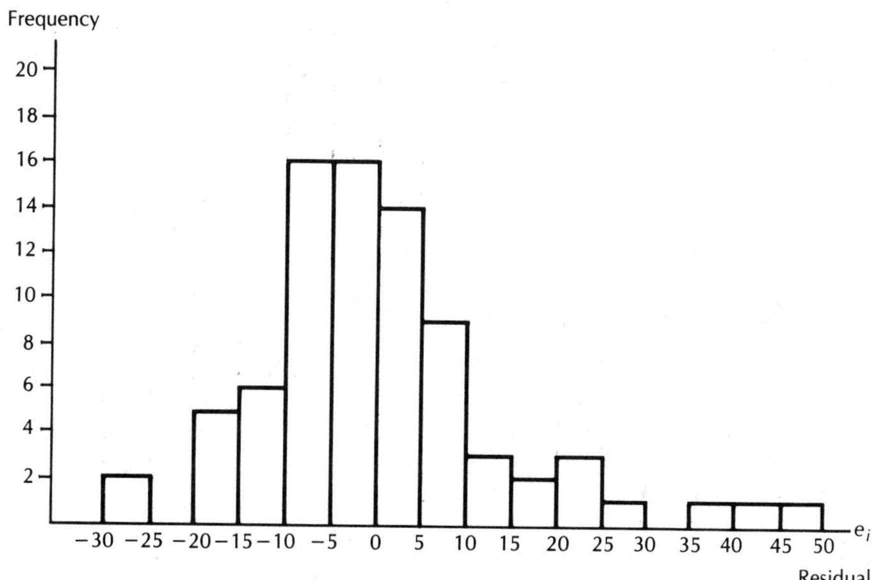

FIGURE 13.27 Normal plot of the residuals (from hour means) for the Republic Wholesalers telephone order data

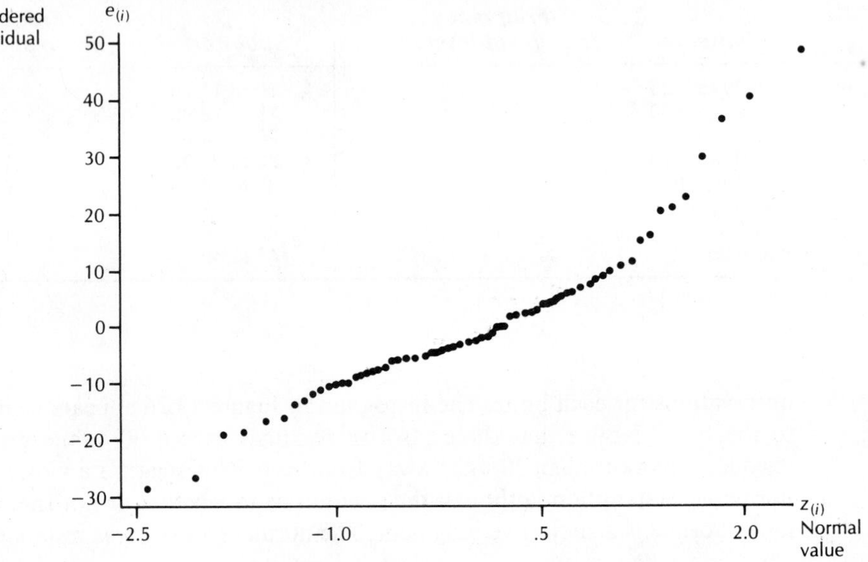

FIGURE 13.28 Normal plot of the residuals (from hour means) for the transformed Republic Wholesalers telephone order data

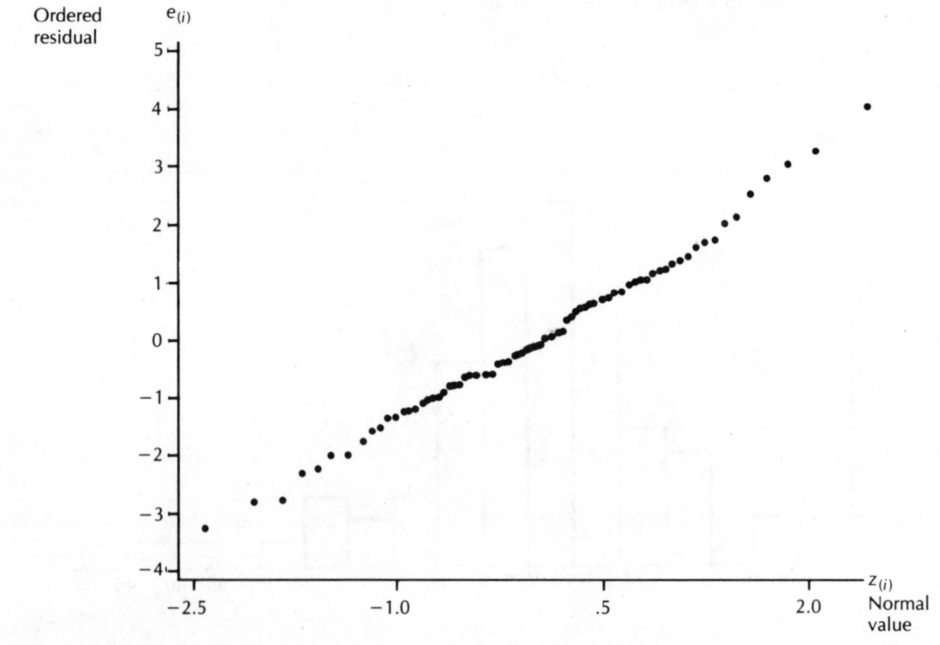

in Figure 13.27. The plot appears to be somewhat nonlinear (although this is a matter of judgment), again suggesting a violation of the normality assumption.

As we have said, when data appears to be non-normal, the technique of data transformation is often applied. Recall that in Example 13.5 we saw that the telephone order data appears to violate the constant variance assumption. To remedy this problem, we used a variance-equalizing transformation. We replaced y (the number of hourly telephone orders) by its square root. Figure 13.28 gives a normal plot of the residuals for the transformed telephone order data. Here the entries in Table 13.4 have been replaced by their square roots before computation of column (hour) means and residuals. Again, the residuals have been computed as deviations from column means, since we have several observations for each hour. If we look at Figure 13.28, the normal plot for this transformed data set seems straighter than the normal plot for the original data.

Notice that in Example 13.11 the transformation that appeared to make the variances roughly equal also made the normal plot appear straighter. It is common to observe this phenomenon. In the Republic Wholesalers problem the normal plot of Figure 13.27 and the histogram in Figure 13.26 do not indicate a terribly serious violation of the normality assumption. Here, the most convincing evidence indicating that a transformation is needed is the plot of $s^2_{H_t}$ versus $\bar{y}_{H_t}$ in Figure 13.14. This plot shows that the error variance tends to increase as $\bar{y}_{H_t}$ increases. However, we see that using an appropriate variance-equalizing transformation simultaneously seems to correct any non-normality problem that might exist. The main point we are making is that when other problems are remedied, any non-normality problem that might exist is often also corrected simultaneously. If it is not, then additional data transformation might be necessary.

We emphasize that histograms and normal plots, like checks for autocorrelation, are sensitive to incorrect functional forms. It is crucial that checks for omitted variables and incorrect functional form be made and appropriate changes be implemented before validation of the normality assumption.

*13.7 USING SAS

The SAS program in Figure 13.29 performs a regression analysis of the steakhouse data in Table 13.1 by using the model

$$\ln y_t = \alpha_0 + \alpha_1 t + \varepsilon_t$$

In Figure 13.30 we present the SAS program to analyze the NARHS data in Table 13.3.

*This section is optional.

FIGURE 13.29 SAS program to perform a regression analysis of the steakhouse data using the model $\ln y_t = \alpha_0 + \alpha_1 t + \varepsilon_t$

```
DATA STEAK;
INPUT Y T ;
LY = LOG(Y);        } → Transformation to obtain logged data
CARDS ;
11   0  )
14   1  |
 .   .  |—→ Data—see Table 13.1
 .   .  |
 .   .  |
403  14 )
 .    15     } → Generates prediction for period 15
PROC PRINT ;
PROC REG ;
MODEL LY = T / P CLI ; } → Specifies model ln y_t = α_0 + α_1 t + ε_t
```

FIGURE 13.30 SAS program to analyze the NARHS data

```
DATA HARDWARE;
INPUT VALUE UPKEEP;
VALUESQ = VALUE*VALUE;            )
VALUEINV = 1/(VALUE*VALUE);       |
TRANSY = UPKEEP/VALUE;            |—→ Transformations
RVALUE = 1/VALUE;                 |
ONE = 1;                          )

CARDS;
118.50 706.04 )
 76.54 398.60 |
   .     .    |—→ Data—see Table 13.3
   .     .    |
   .     .    |
 99.01 545.42 )
110      .    } —→ Predicts upkeep for house value of 110

PROC PRINT;

PROC REG;
MODEL TRANSY = RVALUE ONE VALUE/NOINT P; } —→ Produces output in Figure 13.10
                                              by fitting transformed model
                                              y_i/x_i = β_0(1/x_i) + β_1 + β_2 x_i + η_i
                                              NOINT omits the usual intercept
```

```
OUTPUT OUT = ONE PREDICTED = YHAT RESIDUAL = RESID;  }
PROC PLOT DATA = ONE;                                 }—→ Produces residual
PLOT RESID*(VALUE YHAT);                              }   plot in Figure 13.11
```

Produces predictions in Table 13.12 by using original prediction equation $\hat{y}_i = b_0 + b_1x_i + b_2x_i^2$ Here, b_0, b_1, and b_2 are calculated from the transformed model. Weighted least squares is being used. Since $\sigma_i = x_i\sigma$, the weight is $w_i = 1/x_i^2$

```
PROC REG;
MODEL UPKEEP = VALUE VALUESQ/P CLM CLI;
WEIGHT VALUEINV;
```

In Figure 13.31 we present the SAS program to fit the transformed logistic regression model

$$\frac{l_i}{\hat{v}_i} = \beta_0\left(\frac{1}{\hat{v}_i}\right) + \beta_1\left(\frac{x_i}{\hat{v}_i}\right) + \frac{\varepsilon_i}{\hat{v}_i}$$

to the product name recognition data of Table 13.8.

In Figure 13.32 we present the SAS program to fit the Farmers' Bureau Coop dummy variable model

$$y_t = \beta_0 + \beta_1 t + \beta_2 t^2 + \beta_3 x_{S1,t} + \beta_4 x_{S2,t} + \beta_5 x_{S3,t} + \varepsilon_t$$

FIGURE 13.31 **SAS program to analyze the product name data of Table 13.8 by using logistic regression**

```
DATA COMM;
INPUT NRECALL MENTION;
PROP = NRECALL/50;          }
R = PROP/(1 - PROP);        }
L = LOG(R);                 } Transformations
M = PROP*(1 - PROP);        }
NM = 50*M;                  }
CARDS;
 4 1 }
 7 2 }
20 3 }
35 4 }—→ Data—see Table 13.8
44 5 }
46 6 }
PROC PRINT;
```

Produces output in Figure 13.18. Weighted least squares is being used. Since the approximate estimated standard deviation of l_i is

$$\hat{v}_i = \frac{1}{(n_i\hat{p}_i(1-\hat{p}_i))^{1/2}}$$

the weight is

$$w_i = \frac{1}{\hat{v}_i^2} = n_i\hat{p}_i(1-\hat{p}_i) = 50\hat{p}_i(1-\hat{p}_i)$$

```
PROC REG;
MODEL L = MENTION/P CLM;
WEIGHT NM;
```

FIGURE 13.32 **SAS program to analyze the Farmers' Bureau Coop data by using dummy variable regression**

```
DATA GAS;
INPUT BILL TIME Q1 Q2 Q3;
TIMESQ = TIME*TIME;

CARDS;
                    → Dummy variable input data
344.39  1 1 0 0
246.63  2 0 1 0
131.53  3 0 0 1
288.87  4 0 0 0     → Data—see Table 13.9
   .
   .
   .
539.78 40 0 0 0
  .    41 1 0 0
  .    42 0 1 0     Calculates predictions for periods 41, 42, 43, 44
  .    43 0 0 1
  .    44 0 0 0

PROC PRINT;

PROC REG;
MODEL BILL = TIME TIMESQ Q1 Q2 Q3/P DW CLM CLI;
```

Fits the dummy variable model assuming independent errors. Produces the output in Figure 13.21

```
PROC ARIMA;
IDENTIFY VAR = BILL NOPRINT CROSSCOR = (TIME TIMESQ Q1 Q2 Q3);
ESTIMATE INPUT = (TIME TIMESQ Q1 Q2 Q3) P = 1 PRINTALL PLOT;
FORECAST LEAD = 4;
```

Fits the dummy variable model assuming that $\varepsilon_t = \rho\varepsilon_{t-1} + U_t$. Produces the output in Figure 13.25. Note that the autoregressive process $\varepsilon_t = \rho\varepsilon_{t-1} + U_t$ is specified by P = 1. As another example, the autoregressive process

$$\varepsilon_t = \rho_1\varepsilon_{t-1} + \rho_2\varepsilon_{t-2} + \rho_4\varepsilon_{t-4} + U_t$$

would be specified by P = (1, 2, 4). One way to choose an appropriate autoregressive process would be to use the commands

```
PROC AUTOREG;
MODEL BILL = TIME TIMESQ Q1 Q2 Q3/NLAG = 6 BACKSTEP SLSTAY = 0.05;
```

Starting with the model $\varepsilon_t = \rho_1\varepsilon_{t-1} + \rho_2\varepsilon_{t-2} + \rho_3\varepsilon_{t-3} + \rho_4\varepsilon_{t-4} + \rho_5\varepsilon_{t-5} + \rho_6\varepsilon_{t-6} + U_t$, an appropriate autoregressive process is chosen by performing backward elimination with $\alpha_{\text{stay}} = .05$.

EXERCISES

Exercises 13.1 through 13.5 refer to the following situation. The Natugrain Cereal Company wants to set a price for its new cereal, Bran-Nu. To do this, Bran-Nu is marketed in a test area at a regular store price of \$1.50 per box. To estimate the demand for Bran-Nu at various price levels, the company mails coupons for discounts from 5¢ to \$1.10 in 5¢ increments. For each of the 22 discount prices (from \$1.45 to 40¢), 500 coupons are mailed. The company will measure demand by the number of coupons, y, (out of 500) redeemed at each discounted price.

The results of the Bran-Nu marketing experiment are given in Table 13.12. This table presents y, the number of coupons redeemed (out of 500), for each discount price x (expressed in cents). For future reference the table also gives the natural logarithms of y and x. Using this data, the Natugrain Cereal Company wishes to estimate the elasticity of demand for Bran-Nu and the demand for Bran-Nu at various prices, where demand will be measured as the proportion of the cereal-buying population who would buy Bran-Nu at a given price.

13.1 Consider the Bran-Nu Cereal demand data in Table 13.12.

a. Plot y (number of coupons redeemed) versus x (discount price).

TABLE 13.12 **Bran-Nu Cereal demand data**

Number of coupons redeemed, y	*Discount price, x (cents)*	ln x	ln y
43	145	4.97673	3.76120
64	140	4.94164	4.15888
62	135	4.90527	4.12713
54	130	4.86753	3.98898
81	125	4.82831	4.39445
93	120	4.78749	4.53260
53	115	4.74493	3.97029
80	110	4.70048	4.38203
87	105	4.65396	4.46591
134	100	4.60517	4.89784
120	95	4.55388	4.78749
166	90	4.49981	5.11199
118	85	4.44265	4.77088
189	80	4.38203	5.24175
174	75	4.31749	5.15906
249	70	4.24850	5.51745
317	65	4.17439	5.75890
217	60	4.09434	5.37990
297	55	4.00733	5.69373
421	50	3.91202	6.04263
488	45	3.80666	6.19032
438	40	3.68888	6.08222

b. Is the plot of the Bran-Nu data consistent with a model of the form

$$y = \beta_0 + \beta_1 x + \varepsilon$$

Explain why or why not?

c. Is the plot of the Bran-Nu data consistent with a model of the form

$$y = \beta_0 x^{\beta_1} \varepsilon$$

Explain why or why not.

d. Is the plot of the Bran-Nu data consistent with a model of the form

$$y = \beta_0 \beta_1^x \varepsilon$$

Explain why or why not.

13.2 Consider the Bran-Nu Cereal demand situation. Suppose that economists working for the Natugrain Cereal Company have theoretical reasons for believing that an appropriate model relating y to x is

$$y = \beta_0 x^{\beta_1} \varepsilon$$

a. Show that by taking natural logarithms of both sides of the model

$$y = \beta_0 x^{\beta_1} \varepsilon$$

we obtain the model

$$\ln y = \alpha_0 + \alpha_1 \ln x + U$$

where $\alpha_0 = \ln \beta_0$, $\alpha_1 = \beta_1$, and $U = \ln \varepsilon$.

b. Plot the natural logarithms of y versus the natural logarithms of x. Is the plot fairly linear? Explain why we would expect this plot to be linear if the model

$$y = \beta_0 x^{\beta_1} \varepsilon$$

appropriately describes the original Bran-Nu data.

13.3 Consider the Bran-Nu Cereal demand situation and the transformed model

$$\ln y = \alpha_0 + \alpha_1 \ln x + U$$

a. Specify the vector **y** and the matrix **X** used to calculate the least squares point estimates of α_0 and α_1 in the model

$$\ln y = \alpha_0 + \alpha_1 \ln x + U$$

b. When we use the model in part (a) to perform a regression analysis of the Bran-Nu data, we obtain

$$(\mathbf{X}'\mathbf{X})^{-1} = \begin{bmatrix} 6.4078 & -1.4262 \\ -1.4262 & 0.3197 \end{bmatrix} \qquad \mathbf{X}'\mathbf{y} = \begin{bmatrix} 108.4154 \\ 477.8322 \end{bmatrix}$$

$$\begin{bmatrix} \hat{\alpha}_0 \\ \hat{\alpha}_1 \end{bmatrix} = (\mathbf{X}'\mathbf{X})^{-1}\mathbf{X}'\mathbf{y} = \begin{bmatrix} 13.1956 \\ -1.8543 \end{bmatrix} \qquad SSE = .8300$$

Calculate the standard error for this model. Then, using the fact that $\alpha_0 = \ln \beta_0$ and $\alpha_1 = \beta_1$ (see Exercise 13.2), find point estimates of β_0 and β_1.

c. Noting that $|\beta_1| = |\alpha_1|$ is called the elasticity of demand, calculate a point estimate of $|\beta_1|$. Then calculate a 95% confidence interval for $|\beta_1|$. Interpret this interval.

13.4 Consider the Bran-Nu Cereal demand situation and the transformed model

$$\ln y = \alpha_0 + \alpha_1 \ln x + U$$

We now turn to the problem of estimating demand (the proportion of the cereal-buying population who would buy Bran-Nu) at a given price.

a. Using the above model, calculate a point estimate of the natural logarithm of the number of people per 500 who would buy Bran-Nu at a price of \$1.00.
b. Calculate a 95% confidence interval for the natural logarithm of the number of people per 500 who would buy Bran-Nu at the \$1.00 price.
c. Calculate a point estimate of and a 95% confidence interval for the number of people per 500 who would buy Bran-Nu at the \$1.00 price.
d. Calculate a point estimate of and a 95% confidence interval for the percentage of the cereal-buying population who would buy Bran-Nu at the \$1.00 price.

13.5 Consider the Bran-Nu Cereal demand data in Table 13.12. Suppose that we wish to analyze this data by using the model

$$y = \beta_0 \beta_1^x \varepsilon$$

a. Define an appropriate transformed model that is linear in the parameters.
b. Use the Bran-Nu data in Table 13.12 and a computer to calculate the least squares point estimates for the transformed model.
c. Use the transformed model to calculate point estimates of and 95% confidence intervals for both the number and percentage of the cereal-buying population who would buy Bran-Nu at the \$1.00 price.
d. Compare the confidence intervals you calculated in part (c) to the confidence intervals you calculated in Exercise 13.4. Which intervals are more precise?

Exercises 13.6 through 13.13 refer to the following situation. Reconsider Exercise 6.21. Recall that an economist studied the relationship between x, 1970 yearly income (in thousands of dollars) for a family of four, and y, 1970 yearly clothing expenditure (in hundreds of dollars) for the family. The data in Table 13.13 were observed. Next, recall that when the regression model

$$y_i = \beta_0 + \beta_1 x_i + \varepsilon_i$$

is used to perform a regression analysis of these data, we find that the least squares point estimates of β_0 and β_1 are $b_0 = .2968$ and $b_1 = .6677$ and that the standard error, s, is equal to 1.3434. Moreover, the plot of the standardized residuals versus x fans out. This indicates that the constant variance assumption is violated.

13.6 Consider the clothing expenditure data in Table 13.13.

a. Assuming that the standard deviation σ_i is proportional to x_i, specify a transformed model that will remedy the unequal variances problem.
b. Specify the **y** vector and **X** matrix used to calculate the least squares point estimates b_0 and b_1 of the parameters in the transformed model.
c. Using **y** and **X**, calculate **X′X**, $(\mathbf{X'X})^{-1}$, **X′y**, b_0, b_1, and the standard error for the

TABLE 13.13 **1970 clothing expenditure data**

x_i:	8	10	12	14	16	18	20
y_i:	6.47	6.17	7.4	10.57	11.93	10.3	14.67

transformed model. *Hint*: To calculate $(\mathbf{X}'\mathbf{X})^{-1}$, it is useful to know that if

$$\mathbf{A} = \begin{bmatrix} a & b \\ c & d \end{bmatrix}$$

is a 2×2 matrix, then the inverse of $\mathbf{A}$ is

$$\mathbf{A}^{-1} = \begin{bmatrix} \frac{d}{D} & -\frac{b}{D} \\ -\frac{c}{D} & \frac{a}{D} \end{bmatrix}$$

where $D = ad - bc$.

d. Calculate the residuals and standardized residuals for the transformed model.
e. Plot the standardized residuals for the transformed model versus x. Has the transformation suitably equalized the variances? Explain why or why not.

13.7 Consider the clothing expenditure data in Table 13.13. Use the transformed model you defined in Exercise 13.6 to:

a. Calculate a 95% confidence interval for β_1.
b. Calculate a 95% confidence interval for the mean 1970 yearly clothing expenditure for all families of four that had a 1970 yearly income of \$16,000.
c. Calculate a 95% prediction interval for the 1970 yearly clothing expenditure of an individual family of four that had a 1970 yearly income of \$16,000.

13.8 Repeat parts (a) through (e) of Exercise 13.6, assuming that the standard deviation σ_i is proportional to x_i^2. Does the appropriate transformation here or the transformation of Exercise 13.6 seem to more suitably equalize the variances? Explain.

13.9 Repeat parts (a), (b), and (c) of Exercise 13.7, assuming that the standard deviation σ_i is proportional to x_i^2. Compare the confidence interval and prediction interval obtained here with those calculated in Exercise 13.7. Which intervals are more precise?

13.10 Consider the clothing expenditure data in Table 13.13.

a. Calculate the square roots of the y values and consider the model

$$\sqrt{y_i} = \beta_0 + \beta_1 x_i + \varepsilon_i$$

Calculate the least squares point estimates b_0 and b_1 of the parameters in this model.

b. Using the prediction equation obtained by using the least squares point estimates b_0 and b_1, calculate the predicted square roots

$$\widehat{\sqrt{y_i}} = b_0 + b_1 x_i$$

and the transformed residuals $\sqrt{y_i} - \widehat{\sqrt{y_i}}$.

c. Plot the transformed residuals versus x and $\widehat{\sqrt{y}}$. Do the plots indicate that the variances have been equalized? Explain.

13.11 Consider the clothing expenditure data in Table 13.13. Using the transformed model of Exercise 13.10,

$$\sqrt{y_i} = \beta_0 + \beta_1 x_i + \varepsilon_i$$

a. Compute a point prediction of the square root of the yearly clothing expenditure for a family of four whose 1970 yearly income was \$16,000.

b. Compute a 95% prediction interval for the square root of the yearly clothing expenditure for a family of four whose 1970 yearly income was \$16,000.
c. Compute a point prediction of the yearly clothing expenditure for a family of four whose 1970 yearly income was \$16,000.
d. Compute a 95% prediction interval for the yearly clothing expenditure for a family of four whose 1970 yearly income was \$16,000.

13.12 Consider the clothing expenditure data in Table 13.13.

a. Calculate the natural logarithms of the y values and consider the model

$$\ln y_i = \beta_0 + \beta_1 x_i + \varepsilon_i$$

Calculate the least squares point estimates b_0 and b_1 of the parameters in this model.

b. Using the prediction equation obtained by utilizing the least squares point estimates b_0 and b_1, calculate the predicted natural logarithms

$$\widehat{\ln y_i} = b_0 + b_1 x_i$$

and the transformed residuals $\ln y_i - \widehat{\ln y_i}$.

c. Plot the transformed residuals versus x and $\widehat{\ln y_i}$. Do the plots indicate that the variances have been equalized? Explain.
d. Compare your results of this exercise to the results of Exercise 13.10. Which transformation—the square root or logarithmic—seems to better equalize the variances?

13.13 Consider the clothing expenditure data in Table 13.13. Using the transformed model of Exercise 13.12,

$$\ln y_i = \beta_0 + \beta_1 x_i + \varepsilon_i$$

a. Compute a point prediction of the natural logarithm of the yearly clothing expenditure for a family of four whose 1970 yearly income was \$16,000.
b. Compute a 95% prediction interval for the natural logarithm of the yearly clothing expenditure for a family of four whose 1970 yearly income was \$16,000.
c. Compute a point prediction of the yearly clothing expenditure for a family of four whose 1970 yearly income was \$16,000.
d. Compute a 95% prediction interval for the yearly clothing expenditure for a family of four whose 1970 yearly income was \$16,000.
e. Compare the intervals you computed in parts (b) and (d) with the intervals you computed in Exercise 13.11. Which transformed model yields more precise prediction intervals?

13.14 Consider the NARHS data given in Table 13.3. Figure 13.33 gives SAS output of residual plots versus x (value of house) and $\hat{y}$ (predicted upkeep expenditure) for each of the models

$$\begin{aligned}\frac{y}{x} &= \beta_0\left(\frac{1}{x}\right) + \beta_1\left(\frac{x}{x}\right) + \beta_2\left(\frac{x^2}{x}\right) + \eta \\ &= \beta_0\left(\frac{1}{x}\right) + \beta_1 + \beta_2 x + \eta\end{aligned}$$

FIGURE 13.33 (a) SAS plot of residuals versus x for the model $\frac{y}{x} = \beta_0\left(\frac{1}{x}\right) + \beta_1 + \beta_2 x + \eta$

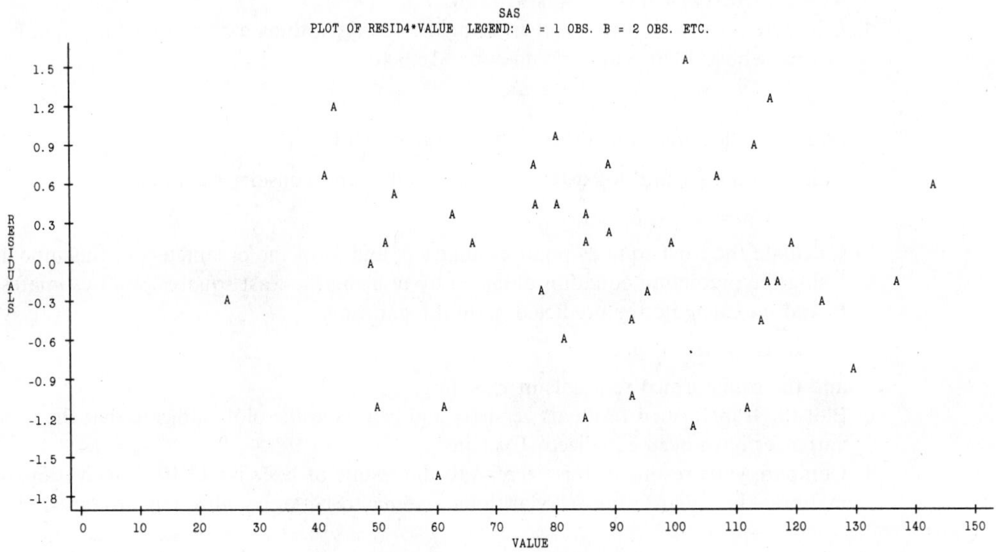

FIGURE 13.33 (b) SAS plot of residuals versus $\hat{y}$ for the model $\frac{y}{x} = \beta_0\left(\frac{1}{x}\right) + \beta_1 + \beta_2 x + \eta$

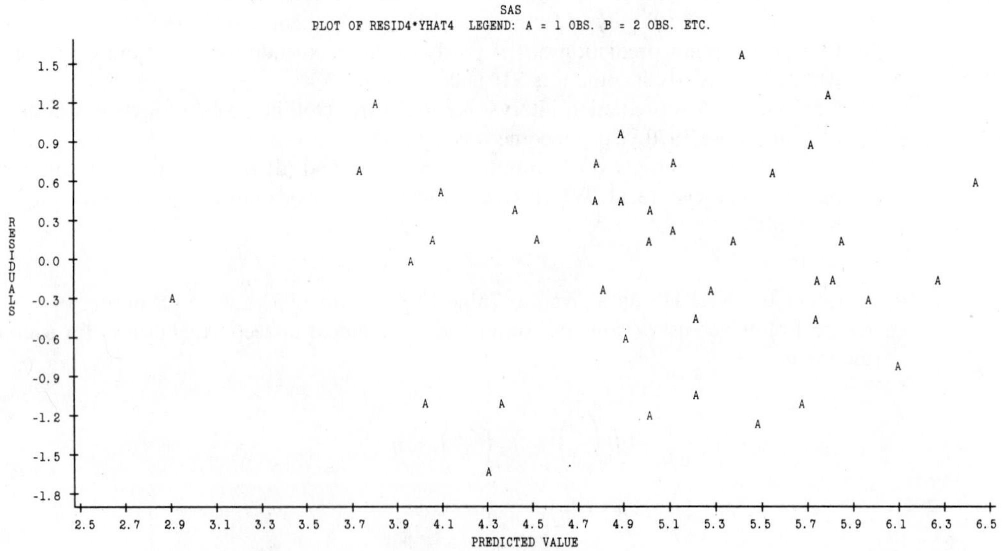

FIGURE 13.33 (c) SAS plot of residuals versus x for the model

$$\frac{y}{x^2} = \beta_0\left(\frac{1}{x^2}\right) + \beta_1\left(\frac{1}{x}\right) + \beta_2 + \eta$$

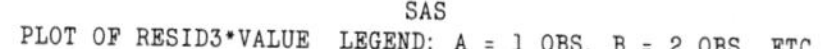

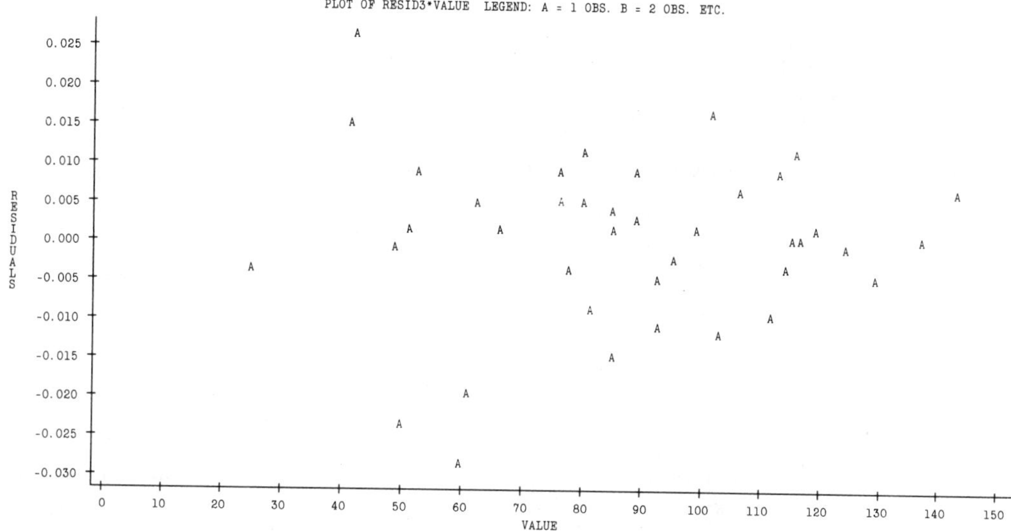

FIGURE 13.33 (d) SAS plot of residuals versus $\hat{y}$ for the model

$$\frac{y}{x^2} = \beta_0\left(\frac{1}{x^2}\right) + \beta_1\left(\frac{1}{x}\right) + \beta_2 + \eta$$

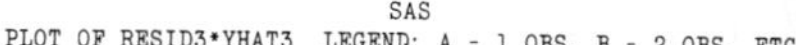

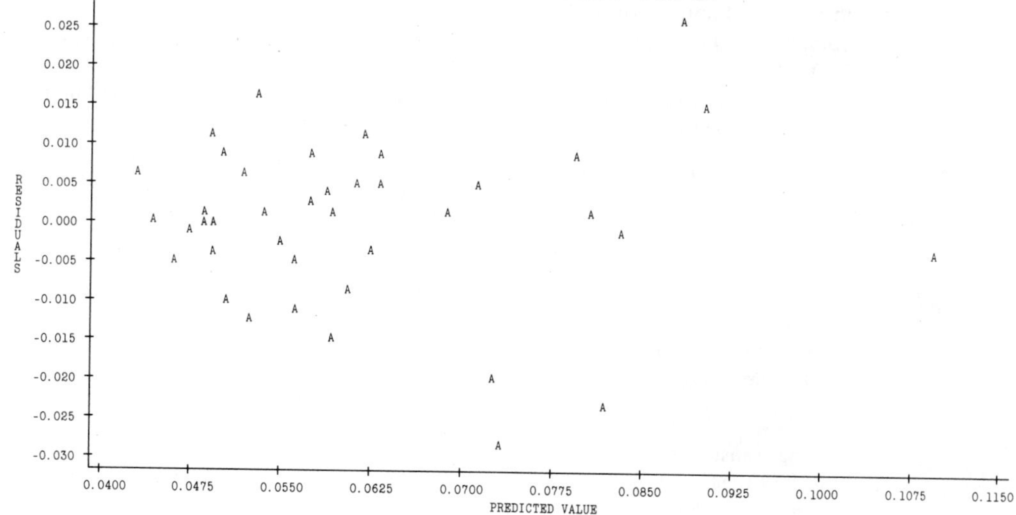

and

$$\frac{y}{x^2} = \beta_0\left(\frac{1}{x^2}\right) + \beta_1\left(\frac{x}{x^2}\right) + \beta_2\left(\frac{x^2}{x^2}\right) + \eta$$

$$= \beta_0\left(\frac{1}{x^2}\right) + \beta_1\left(\frac{1}{x}\right) + \beta_2 + \eta$$

By examining the residual plots, decide which transformation most suitably equalizes the variances. That is, which transformed model best satisfies the equal variances assumption? Carefully explain your answer.

13.15 Consider the NARHS data given in Table 13.3.

a. Calculate the square roots of the y values and consider the model

$$\sqrt{y_i} = \beta_0 + \beta_1 x_i + \beta_2 x_i^2 + \varepsilon_i$$

Using a computer, calculate the least squares point estimates b_0, b_1, and b_2 of the parameters in this model.

b. Using a computer and the prediction equation obtained by utilizing the least squares point estimates b_0, b_1, and b_2, calculate the predicted square roots

$$\widehat{\sqrt{y_i}} = b_0 + b_1 x_i + b_2 x_i^2$$

and the transformed residuals $\sqrt{y_i} - \widehat{\sqrt{y_i}}$.

c. Plot the transformed residuals versus x and $\widehat{\sqrt{y_i}}$. Do the plots indicate that the variances have been equalized? Explain.

13.16 Consider the NARHS data of Table 13.3. Using the transformed model of Exercise 13.15,

$$\sqrt{y_i} = \beta_0 + \beta_1 x_i + \beta_2 x_i^2 + \varepsilon_i$$

a. Compute a point prediction of the square root of the upkeep expenditure for an individual house worth $110,000.

b. Compute a 95% prediction interval for the square root of the upkeep expenditure for an individual house worth $110,000.

c. Compute a point prediction of the upkeep expenditure for an individual house worth $110,000.

d. Compute a 95% prediction interval for the upkeep expenditure for an individual house worth $110,000.

13.17 Consider the NARHS data of Table 13.3.

a. Calculate the natural logarithms of the y values and consider the model

$$\ln y_i = \beta_0 + \beta_1 x_i + \beta_2 x_i^2 + \varepsilon_i$$

Using a computer, calculate the least squares point estimates b_0, b_1, and b_2 of the parameters in this model.

b. Using a computer and the prediction equation obtained by utilizing the least squares point estimates b_0, b_1, and b_2, calculate the predicted natural logarithms

$$\widehat{\ln y_i} = b_0 + b_1 x_i + b_2 x_i^2$$

and the transformed residuals $\ln y_i - \widehat{\ln y_i}$.

c. Plot the transformed residuals versus x and $\widehat{\ln y_i}$. Do the plots indicate that the variances have been equalized? Explain.

d. Compare your results of this exercise to the results of Exercise 13.15. Which transformation—the square root or logarithmic—seems to better equalize the variances?

13.18 Consider the NARHS data of Table 13.3. Using the transformed model of Exercise 13.17,

$$\ln y_i = \beta_0 + \beta_1 x_i + \beta_2 x_i^2 + \varepsilon_i$$

a. Compute a point prediction of the natural logarithm of the upkeep expenditure for an individual house worth \$110,000.
b. Compute a 95% prediction interval for the natural logarithm of the upkeep expenditure for an individual house worth \$110,000.
c. Compute a point prediction of the upkeep expenditure for an individual house worth \$110,000.
d. Compute a 95% prediction interval for the upkeep expenditure for an individual house worth \$110,000.
e. Compare the intervals you computed in parts (b) and (d) with the intervals you computed in Exercise 13.16. Which transformed model yields more precise prediction intervals?

13.19 Consider the simple linear regression model in Exercise 6.25, which relates time (y) required to perform service on a service call to the number of microcomputers serviced (x). Recall that Figure 6.31 indicates that this model violates the equal variances assumption. Figure 13.34 gives the SAS output of residual plots versus x and $\hat{y}$ for each

FIGURE 13.34 **(a) SAS plot of residuals versus x for the model** $\frac{y}{x} = \beta_0\left(\frac{1}{x}\right) + \beta_1 + \eta$

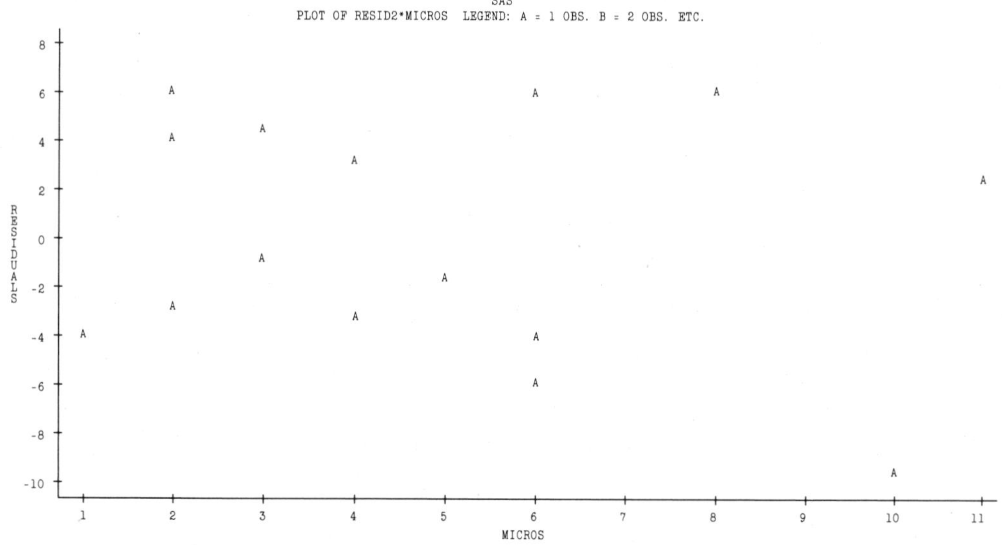

(continues)

FIGURE 13.34 (b) SAS plot of residuals versus $\hat{y}$ for the model $\frac{y}{x} = \beta_0\left(\frac{1}{x}\right) + \beta_1 + \eta$

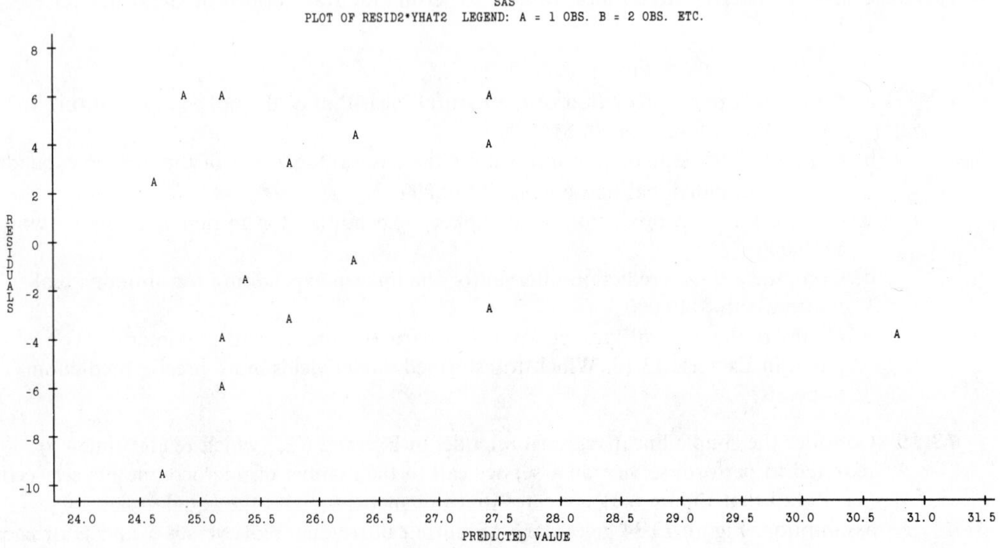

FIGURE 13.34 (c) SAS plot of residuals versus x for the model $\frac{y}{x^2} = \beta_0\left(\frac{1}{x^2}\right) + \beta_1\left(\frac{1}{x}\right) + \eta$

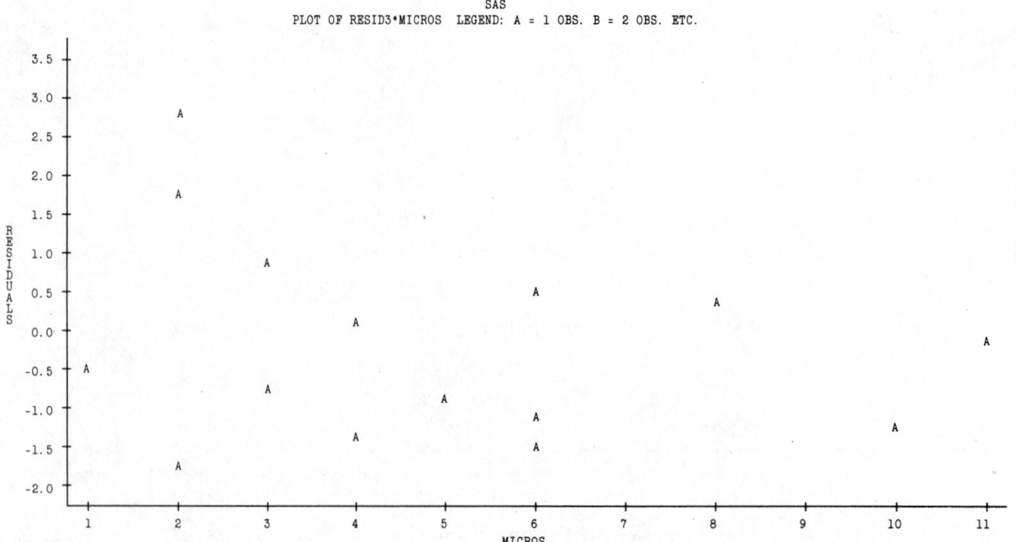

FIGURE 13.34 **(d) SAS plot of residuals versus $\hat{y}$ for the model $\frac{y}{x^2} = \beta_0\left(\frac{1}{x^2}\right) + \beta_1\left(\frac{1}{x}\right) + \eta$**

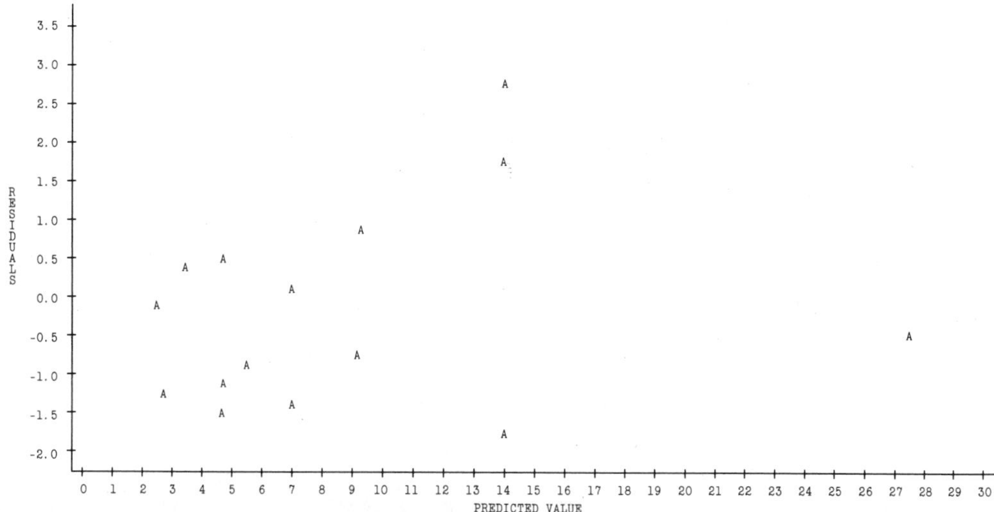

of the models

$$\frac{y}{x} = \beta_0\left(\frac{1}{x}\right) + \beta_1\left(\frac{x}{x}\right) + \eta$$
$$= \beta_0\left(\frac{1}{x}\right) + \beta_1 + \eta$$

and

$$\frac{y}{x^2} = \beta_0\left(\frac{1}{x^2}\right) + \beta_1\left(\frac{x}{x^2}\right) + \eta$$
$$= \beta_0\left(\frac{1}{x^2}\right) + \beta_1\left(\frac{1}{x}\right) + \eta$$

By examining the residual plots, decide which transformation most suitably equalizes the variances. That is, which transformed model best satisfies the equal variances assumption? Carefully explain your answer.

13.20 Consider Exercise 13.19 and an analysis of the service call data of Exercise 6.25 by using the transformed model

$$\frac{y}{x} = \beta_0\left(\frac{1}{x}\right) + \beta_1 + \eta$$

Figure 13.35 gives the SAS output of this regression analysis. Note that on this output the variable $(1/x)$ is denoted as INVMICRO. Using this SAS output,

a. Write the least squares prediction equation for predicting y/x.

FIGURE 13.35 **SAS output of a regression analysis of the service call data of Exercise 6.25 using the model** $\frac{y}{x} = \beta_0\left(\frac{1}{x}\right) + \beta_1 + \eta$

```
                                        SAS
DEP VARIABLE: TSERTIME
                                 ANALYSIS OF VARIANCE

                               SUM OF              MEAN
      SOURCE      DF          SQUARES            SQUARE      F VALUE      PROB>F

      MODEL        2      10295.30974        5147.65487      193.473      0.0001
      ERROR       13        345.88574       26.60659557
      U TOTAL     15      10641.19548

             ROOT MSE        5.158158         R-SQUARE        0.9675
             DEP MEAN        26.15217         ADJ R-SQ        0.9625
             C.V.            19.72363

NOTE: NO INTERCEPT TERM IS USED, R-SQUARE IS REDEFINED.

                                 PARAMETER ESTIMATES

                             PARAMETER           STANDARD        T FOR HO:
   VARIABLE      DF          ESTIMATE              ERROR      PARAMETER=0      PROB > |T|

   INVMICRO       1        6.76419808         5.79357245            1.168          0.2640
   ONE            1       24.04058039         2.24605629           10.703          0.0001

                  PREDICT   STD ERR  LOWER95%  UPPER95%  LOWER95%  UPPER95%
   OBS   ACTUAL     VALUE   PREDICT      MEAN      MEAN   PREDICT   PREDICT   RESIDUAL

     1  30.6667   26.2953    1.3375   23.4059   29.1847   14.7833   37.8073     4.3714
     2  31.5000   27.4227    1.7199   23.7071   31.1382   15.6761   39.1693     4.0773
     3  21.0000   25.1679    1.5762   21.7628   28.5731   13.5158   36.8201    -4.1679
     4  30.8750   24.8861    1.7175   21.1758   28.5965   13.1411   36.6311     5.9889
     5  24.5000   27.4227    1.7199   23.7071   31.1382   15.6761   39.1693    -2.9227
     6  22.5000   25.7316    1.3797   22.7510   28.7122   14.1964   37.2669    -3.2316
     7  23.8000   25.3934    1.4819   22.1919   28.5949   13.7991   36.9877    -1.5934
     8  19.0000   25.1679    1.5762   21.7628   28.5731   13.5158   36.8201    -6.1679
     9  33.5000   27.4227    1.7199   23.7071   31.1382   15.6761   39.1693     6.0773
    10  28.7500   25.7316    1.3797   22.7510   28.7122   14.1964   37.2669     3.0184
    11  31.3333   25.1679    1.5762   21.7628   28.5731   13.5158   36.8201     6.1654
    12  27.0909   24.6555    1.8485   20.6620   28.6490   12.8180   36.4930     2.4354
    13  25.6667   26.2953    1.3375   23.4059   29.1847   14.7833   37.8073    -0.6286
    14  15.1000   24.7170    1.8124   20.8016   28.6324   12.9056   36.5284    -9.6170
    15  27.0000   30.8048    4.2016   21.7277   39.8819   16.4322   45.1774    -3.8048

SUM OF RESIDUALS              1.54765E-13
SUM OF SQUARED RESIDUALS         345.8857
PREDICTED RESID SS (PRESS)       541.7949
DURBIN-WATSON D               1.681
(FOR NUMBER OF OBS.)             15
1ST ORDER AUTOCORRELATION     0.111
```

b. Find the point prediction of y/x for a service call on which ten microcomputers are serviced. *Hint*: See Table 6.10.

c. Find a 95% confidence interval for the mean value of y/x when five microcomputers are serviced.

d. Find a 95% prediction interval for y/x for a service call on which five microcomputers are serviced.

e. Write the least squares prediction equation for predicting y.

f. Find the point prediction of y for a service call on which ten microcomputers are serviced.
g. Find a 95% confidence interval for the mean value of y when five microcomputers are serviced.
h. Find a 95% prediction interval for y for a service call on which ten microcomputers are serviced.
i. Compute a 99% confidence interval for the mean value of y when ten microcomputers are serviced.

13.21 Consider the steakhouse data of Table 13.1. Using any needed information in Example 13.1,

a. Calculate a point prediction of and a 95% prediction interval for y_{16}, the number of steakhouses that will be in operation in 1990.
b. Calculate a point prediction of and a 95% prediction interval for y_{17}, the number of steakhouses that will be in operation in 1991.
c. Compare the prediction intervals for y_{16} and y_{17} calculated in parts (a) and (b). Which interval is more precise? Does this make intuitive sense? Explain.

13.22 Consider the tax form completion data of Table 13.2. Using any needed information in Example 13.2:

a. Calculate a point prediction of and a 95% prediction interval for the amount of time it will take an individual who has completed the tax form seven times to fill out the form an eighth time.
b. Calculate a point estimate of and a 99% confidence interval for the mean completion time for all individuals who have completed the tax form three times and are filling out their fourth form.

Exercises 13.23 and 13.24 refer to the following situation. A professional organization wishes to study the relationship between the ability of its members to pass a certification examination and the number of organization-sponsored preparation sessions members have attended before taking the examination. Twenty-five members who are taking the certification examination for the first time are randomly selected to participate in the study. For each, the number of preparation sessions attended before taking the examination is recorded along with the examination result. Here the data is coded so that $y = 1$ if a member passes and $y = 0$ if a member does not pass. (The state certification board provides only pass/fail results; raw test scores are not released.) The data in Table 13.14 is observed.

13.23 Consider the certification examination data in Table 13.14.

a. Plot y versus x.
b. Consider a randomly selected member taking the certification examination for the first time who has attended x preparation sessions, and let $p(x)$ be the probability that this member passes the examination. On the basis of your data plot, does $p(x)$ seem to be related to x? Explain.
c. Consider the model

$$\begin{aligned} y &= \mu + \varepsilon \\ &= p(x) + \varepsilon \\ &= \beta_0 + \beta_1 x + \varepsilon \end{aligned}$$

TABLE 13.14 **Certification examination data**

Member	*Number of preparation sessions attended, x*	*Test result, y (pass = 1, fail = 0)*
1	4	1
2	1	0
3	7	1
4	5	0
5	3	1
6	2	0
7	8	1
8	5	1
9	3	0
10	2	1
11	10	1
12	6	1
13	4	0
14	2	0
15	7	1
16	5	0
17	7	0
18	1	0
19	3	1
20	9	1
21	3	0
22	1	0
23	5	1
24	6	1
25	8	1

Use a computer and the data in Table 13.14 to calculate the least squares point estimates b_0 and b_1 of the parameters in the model.

d. Use the least squares point estimates b_0 and b_1 to calculate $\hat{p}(x)$, the point estimate of $p(x)$, $\hat{v} = \sqrt{\hat{p}(x)[1 - \hat{p}(x)]}$, $y/\hat{v}$, $1/\hat{v}$, and $x/\hat{v}$ for each observation in Table 13.14. Use a computer to aid in the calculations.

e. Consider the transformed model

$$\frac{y}{\hat{v}} = \beta_0\left(\frac{1}{\hat{v}}\right) + \beta_1\left(\frac{x}{\hat{v}}\right) + \frac{\varepsilon}{\hat{v}}$$

Using a computer, calculate the least squares point estimates of the parameters in the transformed model. If we denote these estimates as $\tilde{b}_0$ and $\tilde{b}_1$, are these estimates much different from the estimates b_0 and b_1 computed in part (c)?

f. Assess the importance of the intercept and x in the original model of part (c) by using appropriate t statistics and associated prob-values.

13.24 Consider the certification examination data in Table 13.14 and the transformed model

$$\frac{y}{\hat{v}} = \beta_0\left(\frac{1}{\hat{v}}\right) + \beta_1\left(\frac{x}{\hat{v}}\right) + \frac{\varepsilon}{\hat{v}}$$

of Exercise 13.23(e).

a. Use the least squares point estimates $\tilde{b}_0$ and $\tilde{b}_1$ of the parameters in the transformed model to write a prediction equation for $p(x)$. (Recall that $p(x)$ is defined in Exercise 13.23(b).)
b. Calculate point estimates of $p(3)$ and $p(7)$ by using your prediction equation.
c. Using a computer, find 95% confidence intervals for each of $p(3)$ and $p(7)$. Interpret these intervals.
d. Calculate 99% confidence intervals for each of $p(3)$ and $p(7)$. Interpret these intervals.

Exercises 13.25 and 13.26 refer to the following situation. Happy Valley Real Estate has decided to use eight groups of people in an experiment. The experiment is designed to estimate $p(x)$, the probability that a person will remember the Happy Valley name one day after viewing a 1-minute media presentation that mentions the Happy Valley name x times. Happy Valley shows the $n_i = 50$ people in group i ($i = 1, 2, \ldots, 8$) a 1-minute media presentation that mentions Happy Valley $x_i = i$ times. One day later, the number, y_i, and the proportion, $\hat{p}_i = y_i/50$, of the 50 people who remember the Happy Valley name are recorded. Table 13.15 gives the data that is obtained.

13.25 Consider the Happy Valley name recognition data in Table 13.15.

a. Plot $\hat{p}_i$ versus x_i. Does the logistic curve model seem reasonable for this data? Explain.
b. Consider the regression model

$$l_i = \beta_0 + \beta_1 x_i + \varepsilon_i \qquad \text{where} \qquad l_i = \ln\left(\frac{\hat{p}_i}{1 - \hat{p}_i}\right)$$

and the transformed model

$$\frac{l_i}{\hat{v}_i} = \beta_0\left(\frac{1}{\hat{v}_i}\right) + \beta_1\left(\frac{x_i}{\hat{v}_i}\right) + \frac{\varepsilon_i}{\hat{v}_i} \qquad \text{where} \qquad \hat{v}_i = \frac{1}{[n_i\hat{p}_i(1 - \hat{p}_i)]^{1/2}}$$

Hand calculate (or use a computer to calculate) point estimates $\tilde{b}_0$ and $\tilde{b}_1$ of the parameters in this model. *Hint*: Compute l_i, $l_i/\hat{v}_i$, $1/\hat{v}_i$, and $x_i/\hat{v}_i$ for each observation.
c. Hand calculate (or use a computer to calculate) the standard error for this transformed model.

13.26 Consider the Happy Valley name recognition data in Table 13.15 and the transformed logistic regression model as given in Exercise 13.25(b).

a. Using the point estimates $\tilde{b}_0$ and $\tilde{b}_1$, compute point estimates of $p(1)$, $p(2)$, $p(3)$, $p(4)$, $p(5)$, $p(6)$, $p(7)$, and $p(8)$. Denote the point estimate of $p(x_i)$ as $\tilde{p}(x_i)$.
b. Plot the estimated logistic function. That is, plot the estimates $\tilde{p}(1), \tilde{p}(2), \ldots, \tilde{p}(8)$ versus x.

TABLE 13.15 **Happy Valley name recognition data**

i	x_i	y_i	$\hat{p}_i$	i	x_i	y_i	$\hat{p}_i$
1	1	2	.04	5	5	36	.72
2	2	8	.16	6	6	41	.82
3	3	15	.30	7	7	47	.94
4	4	29	.58	8	8	48	.96

c. Hand calculate (or use a computer to calculate) 95% confidence intervals for $p(3)$ and $p(6)$. Interpret these intervals.

13.27 Consider the Happy Valley Real Estate situation of Example 13.7. Use a computer to calculate 95% confidence intervals for $p(28)$, $p(35)$, and $p(48)$. Interpret these intervals.

13.28 Consider the product name data in Table 13.8 and the transformed logistic regression model in Example 13.8.

a. Calculate the point estimates $\tilde{p}(1)$, $\tilde{p}(2)$, $\tilde{p}(3)$, $\tilde{p}(5)$, and $\tilde{p}(6)$. Interpret these point estimates.
b. Calculate 95% confidence intervals for $p(2)$ and $p(5)$. Interpret these intervals.

13.29 An advertising agency employs ten groups of 100 consumers each in a marketing experiment. The experiment is designed to estimate $p(x)$, the probability that a consumer will recall an advertising claim after receiving x phone calls in which the claim is made. Each consumer in group i receives $x_i = i$ phone calls making the claim over a four-week period of time. One week later, the number, y_i, and the proportion, $\hat{p}_i = y_i/100$, of the 100 consumers in each group who recall the claim is recorded. The data obtained is given in Table 13.16. Using logistic regression, completely analyze the data.

TABLE 13.16 **Advertising claim experiment data**

Group, i	x_i	y_i	$\hat{p}_i = y_i/100$
1	1	4	.04
2	2	10	.10
3	3	16	.16
4	4	28	.28
5	5	42	.42
6	6	77	.77
7	7	85	.85
8	8	92	.92
9	9	95	.95
10	10	99	.99

13.30 Consider the advertising experiment in Example 13.8. Suppose that the firm wishes to estimate $p(x_1, x_2)$, the probability that a consumer will remember a product name one week after viewing a 60-second commercial that mentions the product name x_1 times and in which the product name appears on the screen x_2 times. The firm shows the $n_i = 50$ people in group i ($i = 1, 2, \ldots, 12$) a 60-second commercial that mentions the product name x_{i1} times and in which the product name appears on the screen x_{i2} times. One week later, the number, y_i, and the proportion, $\hat{p}_i = y_i/50$, of the 50 people who recall the product name are recorded. Table 13.17 gives the data that is obtained. Completely analyze this data using the transformed multiple logistic regression model

$$\frac{l_i}{\hat{v}_i} = \beta_0\left(\frac{1}{\hat{v}_i}\right) + \beta_1\left(\frac{x_{i1}}{\hat{v}_i}\right) + \beta_2\left(\frac{x_{i2}}{\hat{v}_i}\right) + \frac{\varepsilon_i}{\hat{v}_i}$$

TABLE 13.17 **Two-variable advertising experiment data**

Group, i	x_{i1}	x_{i2}	y_i	$\hat{p}_i = y_i/50$
1	1	1	5	.10
2	1	2	12	.24
3	1	3	21	.42
4	2	1	10	.20
5	2	2	24	.48
6	2	3	37	.74
7	3	1	23	.46
8	3	2	39	.78
9	3	3	45	.90
10	4	1	35	.70
11	4	2	43	.86
12	4	3	49	.98

where

$$l_i = \ln\left(\frac{\hat{p}_i}{1 - \hat{p}_i}\right) \quad \text{and} \quad \hat{v}_i = \frac{1}{[n_i\hat{p}_i(1 - \hat{p}_i)]^{1/2}}$$

13.31 Value City, a small department store, sells a particular brand of microwave oven. Table 13.18 contains four years of quarterly sales data for this microwave oven.

a. Plot the microwave sales data versus time.

b. Consider the dummy variable regression model

$$\begin{aligned} y_t &= TR_t + SN_t + \varepsilon_t \\ &= \beta_0 + \beta_1 t + \beta_{S2}x_{S2,t} + \beta_{S3}x_{S3,t} + \beta_{S4}x_{S4,t} + \varepsilon_t \end{aligned}$$

where $x_{S2,t}$, $x_{S3,t}$, and $x_{S4,t}$ are dummy variables defined as follows:

$$x_{S2,t} = \begin{cases} 1 & \text{if time period } t \text{ is quarter 2} \\ 0 & \text{otherwise} \end{cases}$$

$$x_{S3,t} = \begin{cases} 1 & \text{if time period } t \text{ is quarter 3} \\ 0 & \text{otherwise} \end{cases}$$

$$x_{S4,t} = \begin{cases} 1 & \text{if time period } t \text{ is quarter 4} \\ 0 & \text{otherwise} \end{cases}$$

Using your data plot from part (a), explain why this model may be an appropriate model relating sales (y_t) to time (t).

c. Explain the meaning of the seasonal regression parameters β_{S2}, β_{S3}, and β_{S4}.

d. Write out the vector **y** and matrix **X** that are used to calculate the least squares point estimates of the model parameters.

13.32 Consider the microwave oven sales data in Table 13.18 and the dummy variable regression model of Exercise 13.31. When this model is fit to the data, the least squares estimates of the model parameters are found to be

$$b_0 = 8.75, \quad b_1 = .5, \quad b_{S2} = 21, \quad b_{S3} = 33.5, \quad \text{and} \quad b_{S4} = 4.5$$

TABLE 13.18 Microwave oven sales data

Year	Quarter	t	Demand, y_t	Year	Quarter	t	Demand, y_t
1	1 (winter)	1	10	3	1	9	13
	2 (spring)	2	31		2	10	34
	3 (summer)	3	43		3	11	48
	4 (fall)	4	16		4	12	19
2	1	5	11	4	1	13	15
	2	6	33		2	14	37
	3	7	45		3	15	51
	4	8	17		4	16	21

Furthermore, the squared residuals obtained from the prediction equation

$$\begin{aligned}\hat{y}_t &= tr_t + sn_t \\ &= b_0 + b_1 t + b_{S2}x_{S2,t} + b_{S3}x_{S3,t} + b_{S4}x_{S4,t}\end{aligned}$$

are given in Table 13.19.

a. Calculate the standard error s.

b. Plot the residuals versus time. Is there any evidence of positive or negative autocorrelation? Explain.

c. Calculate the Durbin-Watson statistic. Use this statistic to test for positive autocorrelation with $\alpha = .05$. Also test for negative autocorrelation with $\alpha = .05$.

d. Assuming the regression assumptions are satisfied, calculate point predictions of and 95% prediction intervals for y_{17}, y_{18}, y_{19}, and y_{20}, the sales of the microwave oven in

TABLE 13.19 Residuals for the dummy variable microwave oven sales model

t	y_t	$tr_t = 8.75 + .5t$	sn_t	$\hat{y}_t = tr_t + sn_t$	$(y_t - \hat{y}_t)^2$
1	10	9.25	0	9.25	$(.75)^2$
2	31	9.75	21.0	30.75	$(.25)^2$
3	43	10.25	33.5	43.75	$(-.75)^2$
4	16	10.75	4.5	15.25	$(.75)^2$
5	11	11.25	0	11.25	$(-.25)^2$
6	33	11.75	21.0	32.75	$(.25)^2$
7	45	12.25	33.5	45.75	$(-.75)^2$
8	17	12.75	4.5	17.25	$(-.25)^2$
9	13	13.25	0	13.25	$(-.25)^2$
10	34	13.75	21.0	34.75	$(-.75)^2$
11	48	14.25	33.5	47.75	$(.25)^2$
12	19	14.75	4.5	19.25	$(-.25)^2$
13	15	15.25	0	15.25	$(-.25)^2$
14	37	15.75	21.0	36.75	$(.25)^2$
15	51	16.25	33.5	49.75	$(1.25)^2$
16	21	16.75	4.5	21.25	$(-.25)^2$

$$\sum_{t=1}^{16} (y_t - \hat{y}_t)^2 = 5.00$$

periods 17, 18, 19, and 20, respectively. *Hint*: If $\mathbf{x}'_{17}$, $\mathbf{x}'_{18}$, $\mathbf{x}'_{19}$, and $\mathbf{x}'_{20}$ denote the row vectors containing the numbers multiplied by b_0, b_1, b_{S2}, b_{S3}, and b_{S4} in $\hat{y}_{17}$, $\hat{y}_{18}$, $\hat{y}_{19}$, and $\hat{y}_{20}$ respectively, then each of $\mathbf{x}'_{17}(\mathbf{X}'\mathbf{X})^{-1}\mathbf{x}_{17}$, $\mathbf{x}'_{18}(\mathbf{X}'\mathbf{X})^{-1}\mathbf{x}_{18}$, $\mathbf{x}'_{19}(\mathbf{X}'\mathbf{X})^{-1}\mathbf{x}_{19}$, and $\mathbf{x}'_{20}(\mathbf{X}'\mathbf{X})^{-1}\mathbf{x}_{20}$ can be calculated to be .5543.

e. With regard to part (d), what are $\mathbf{x}'_{17}$, $\mathbf{x}'_{18}$, $\mathbf{x}'_{19}$, and $\mathbf{x}'_{20}$?

13.33 Bargain Department Stores is a chain of department stores in the Midwest. Quarterly sales of the Bargain 8000-Btu air conditioner over the past three years are given in Table 13.20.

a. Plot sales (y_t) versus time (t).
b. From your data plot, what kind of trend appears to exist?
c. Figure 13.36 gives the SAS output that is obtained when the regression model

$$y_t = \beta_0 + \beta_1 t + \beta_2 t^2 + \beta_3 x_{S2,t} + \beta_4 x_{S3,t} + \beta_5 x_{S4,t} + \varepsilon_t$$

FIGURE 13.36 **SAS output of a regression analysis of the air conditioner sales data in Table 13.20 using the model $y_t = \beta_0 + \beta_1 t + \beta_2 t^2 + \beta_3 x_{S2,t} + \beta_4 x_{S3,t} + \beta_5 x_{S4,t} + \varepsilon_t$**

DEP VARIABLE: Y

SOURCE	DF	SUM OF SQUARES	MEAN SQUARE	F VALUE	PROB > F
MODEL	5	142952179	28590436	3346.937	0.0001
ERROR	6	51253.616	8542.269		
C TOTAL	11	143003433			

ROOT MSE	92.424398	R-SQUARE	0.9996
DEP MEAN	6948.667	ADJ R-SQ	0.9993
C.V.	1.330103		

VARIABLE	DF	PARAMETER ESTIMATE	STANDARD ERROR	T FOR HO: PARAMETER=0	PROB > |T|
INTERCEP	1	2624.527	100.364	26.150	0.0001
T	1	382.819	34.032350	11.249	0.0001
TSQ	1	-11.353831	2.541332	-4.468	0.0042
D2	1	4629.740	76.075069	60.858	0.0001
D3	1	6738.855	77.379749	87.088	0.0001
D4	1	-1565.323	79.344026	-19.728	0.0001

OBS	ACTUAL	PREDICT VALUE	STD ERR PREDICT	LOWER95% MEAN	UPPER95% MEAN	LOWER95% PREDICT	UPPER95% PREDICT	RESIDUAL
1	2915	2996	76.522	2809	3183	2702	3290	-80.992
2	8032	7974	66.916	7811	8138	7695	8254	57.511
3	10411	10410	62.663	10256	10563	10136	10683	1.347
4	2427	2409	64.825	2250	2567	2133	2685	18.183
5	4381	4255	59.852	4108	4401	3985	4524	126.226
6	9138	9142	59.852	8996	9289	8873	9412	-4.441
7	11386	11487	59.852	11340	11633	11217	11756	-100.774
8	3382	3395	59.852	3249	3542	3126	3665	-13.108
9	5105	5150	64.825	4992	5309	4874	5426	-45.234
10	9894	9947	62.663	9794	10100	9674	10220	-53.070
11	12300	12201	66.916	12037	12364	11921	12480	99.427
12	4013	4018	76.522	3831	4205	3724	4312	-5.075
13	.	5682	112.586	5407	5958	5326	6039	.
14	.	10388	142.798	10039	10738	9972	10805	.
15	.	12551	177.239	12117	12985	12062	13040	.
16	.	4278	213.875	3754	4801	3708	4848	.

TABLE 13.20 Quarterly air conditioner sales

Year	Quarter	Period, t	Sales, y_t
1	1	1	2915
	2	2	8032
	3	3	10411
	4	4	2427
2	1	5	4381
	2	6	9138
	3	7	11386
	4	8	3382
3	1	9	5105
	2	10	9894
	3	11	12300
	4	12	4013

is used to analyze the data in Table 13.20. Here, $x_{S2,t}$, $x_{S3,t}$, and $x_{S4,t}$ are appropriately defined dummy variables for quarters 2, 3, and 4. Define these dummy variables.

d. Noting that observations 13, 14, 15, and 16 correspond to time periods 13, 14, 15, and 16, answer the following:
   1. Do all of the independent variables in the model seem important? Justify your answer.
   2. Find and identify $\hat{y}_{13}$, $\hat{y}_{14}$, $\hat{y}_{15}$, and $\hat{y}_{16}$, the point predictions of y_{13}, y_{14}, y_{15}, and y_{16}.
   3. Calculate $\hat{y}_{15}$ by using the appropriate least squares point estimates.
   4. Calculate $\hat{y}_{16}$ by using the appropriate least squares point estimates.
   5. Find and identify the 95% prediction intervals for y_{13}, y_{14}, y_{15}, and y_{16}. Interpret these intervals.

13.34 The data in Table 13.21 gives the number of reported cases y_t of a newly discovered disease over the last 11 months.

a. Plot y_t versus time (t). Explain why the model

$$y_t = \beta_0(\beta_1^t)\varepsilon_t$$

might be an appropriate model relating y_t to t. In this context, such a model is called a *growth curve model.*

b. Using natural logarithms, define a transformed growth curve model that will be linear in its parameters.

c. Plot the natural logarithms of the y_t values versus time. Has the logarithmic transformation linearized the data?

d. Using a computer, calculate least squares point estimates of the parameters in the transformed model.

e. Using a computer, calculate a point prediction and 95% prediction interval for $\ln y_{12}$ (the natural logarithm of next month's number of reported cases of the disease).

f. Since the above model implies that

$$y_t = \beta_0(\beta_1^t)\varepsilon_t = [\beta_0(\beta_1^{t-1})]\beta_1\varepsilon_t \doteq (y_{t-1})\beta_1\varepsilon_t$$

TABLE 13.21 Reported cases of a new disease

Month, t	*Number of reported cases, y_t*
1	1
2	1
3	2
4	3
5	4
6	6
7	8
8	13
9	21
10	27
11	45

the parameter β_1 is called the *growth rate* of the disease. Compute a point estimate of and a 95% confidence interval for the growth rate. Interpret the point estimate and the confidence interval.

g. Calculate a point forecast of y_{12}, the number of reported cases of the disease in month 12.

h. Using a computer, calculate a 95% prediction interval for y_{12}. Interpret this interval.

13.35 Consider the Farmers' Bureau Coop situation in Example 13.10. Using any needed information in this example, compute the point predictions $\hat{y}_{43}$ and $\hat{y}_{44}$ of y_{43} and y_{44}, the propane gas bills in time periods 43 and 44.

Exercises 13.36–13.38 refer to the following situation. Nite's Rest, Inc., operates four hotels in Central City. The analysts in the operating division of the corporation were asked to develop a model that could be used to obtain short-term forecasts (up to one year) of the number of occupied rooms in the hotels. These forecasts were needed by various personnel to assist in decision making with regard to hiring additional help during the summer months, ordering supplies that have long delivery lead times, budgeting of local advertising expenditures, and so on.

The available historical data consisted of the number of occupied rooms during each day for the 15 years from 1976 to 1989. Because it was desired to obtain k-step-ahead monthly forecasts for $k = 1, 2, \ldots, 12$, the data was reduced to monthly averages by dividing each monthly total by the number of days in the month. The monthly room averages denoted by $y_1, y_2, \ldots, y_{168}$, are given in Table 13.22 and plotted in Figure 13.37. This figure shows that the monthly room averages follow a strong trend and that they have a seasonal pattern with one major and several minor peaks during the year. It also appears that the amount of seasonal variation is increasing with the level of the time series, indicating that the use of a transformation (such as the natural logarithms of the observations) might be warranted. The natural logarithms of the room averages are denoted by $y_1^*, y_2^*, \ldots, y_{168}^*$ and are plotted in Figure 13.38. This figure indicates that the log transformation has equalized the amount of seasonal variation over the range of the data.

TABLE 13.22 Nites' Rest hotel room occupancy data

t	y_t	t	y_t	t	y_t	t	y_t
1	501	43	785	85	645	127	1067
2	488	44	830	86	602	128	1038
3	504	45	645	87	601	129	812
4	578	46	643	88	709	130	790
5	545	47	551	89	706	131	692
6	632	48	606	90	817	132	782
7	728	49	585	91	930	133	758
8	725	50	553	92	983	134	709
9	585	51	576	93	745	135	715
10	542	52	665	94	735	136	788
11	480	53	656	95	620	137	794
12	530	54	720	96	698	138	893
13	518	55	826	97	665	139	1046
14	489	56	838	98	626	140	1075
15	528	57	652	99	649	141	812
16	599	58	661	100	740	142	822
17	572	59	584	101	729	143	714
18	659	60	644	102	824	144	802
19	739	61	623	103	937	145	748
20	758	62	553	104	994	146	731
21	602	63	599	105	781	147	748
22	587	64	657	106	759	148	827
23	497	65	680	107	643	149	788
24	558	66	759	108	728	150	937
25	555	67	878	109	691	151	1076
26	523	68	881	110	649	152	1125
27	532	69	705	111	656	153	840
28	623	70	684	112	735	154	864
29	598	71	577	113	743	155	717
30	683	72	656	114	837	156	813
31	774	73	645	115	995	157	811
32	780	74	593	116	1040	158	732
33	609	75	617	117	809	159	745
34	604	76	686	118	793	160	844
35	531	77	679	119	692	161	833
36	592	78	773	120	763	162	935
37	578	79	906	121	723	163	1110
38	543	80	934	122	655	164	1124
39	565	81	713	123	658	165	868
40	648	82	710	124	761	166	860
41	615	83	600	125	768	167	762
42	697	84	676	126	885	168	877

13.36 Consider the hotel room data in Table 13.22 and the model

$$\begin{aligned} y_t^* &= \mu_t^* + \varepsilon_t^* \\ &= \beta_0 + \beta_1 t + \beta_{S1}x_{S1,t} + \beta_{S2}x_{S2,t} + \beta_{S3}x_{S3,t} + \beta_{S4}x_{S4,t} + \beta_{S5}x_{S5,t} + \beta_{S6}x_{S6,t} \\ &\quad + \beta_{S7}x_{S7,t} + \beta_{S8}x_{S8,t} + \beta_{S9}x_{S9,t} + \beta_{S10}x_{S10,t} + \beta_{S11}x_{S11,t} + \varepsilon_t^* \end{aligned}$$

FIGURE 13.37 Plot of the hotel room averages versus time

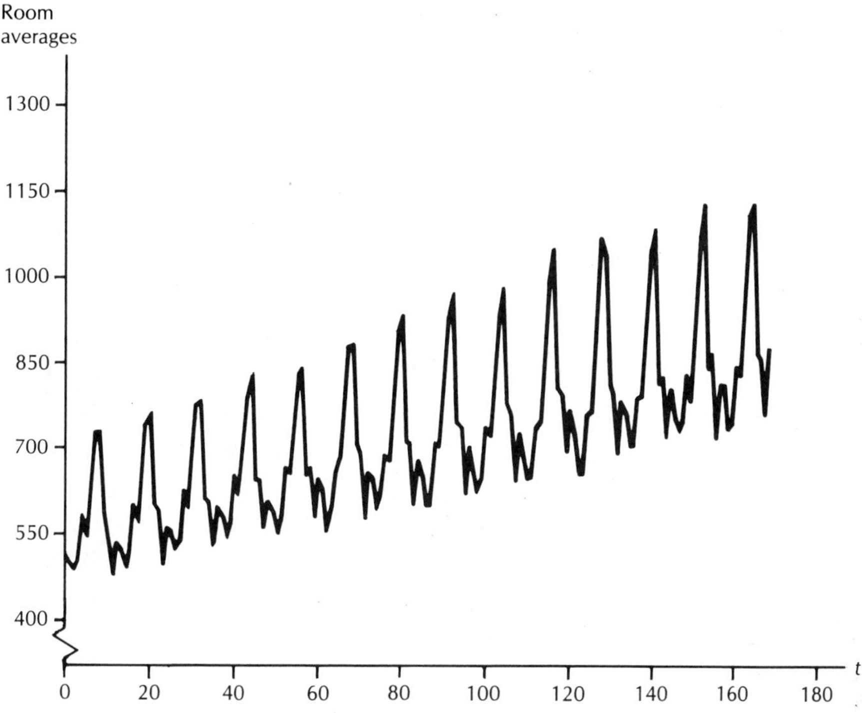

Here, $x_{S1,t}, x_{S2,t}, \ldots, x_{S11,t}$ are dummy variables, where

$$x_{S1,t} = \begin{cases} 1 & \text{if time period } t \text{ is season 1 (January)} \\ 0 & \text{otherwise} \end{cases}$$

$$x_{S2,t} = \begin{cases} 1 & \text{if time period } t \text{ is season 2 (February)} \\ 0 & \text{otherwise} \end{cases}$$

and the other dummy variables are defined similarly.

a. Explain why this model is a reasonable model to use to describe the logarithms of the room averages.
b. Explain the meanings of the seasonal parameters $\beta_{S1}, \beta_{S2}, \ldots, \beta_{S11}$.
c. Specify the first 12 elements of the vector **y** and the first 12 elements of each column of the matrix **X** used to calculate the least squares point estimates of the parameters in the above model.

13.37 Consider the hotel room data in Table 13.22 and the dummy variable regression model of Exercise 13.36. When this model is used to perform a regression analysis of the logarithms of the room averages, the least squares point estimates of the parameters in

FIGURE 13.38 **Plot of the natural logarithms of the hotel room averages versus time**

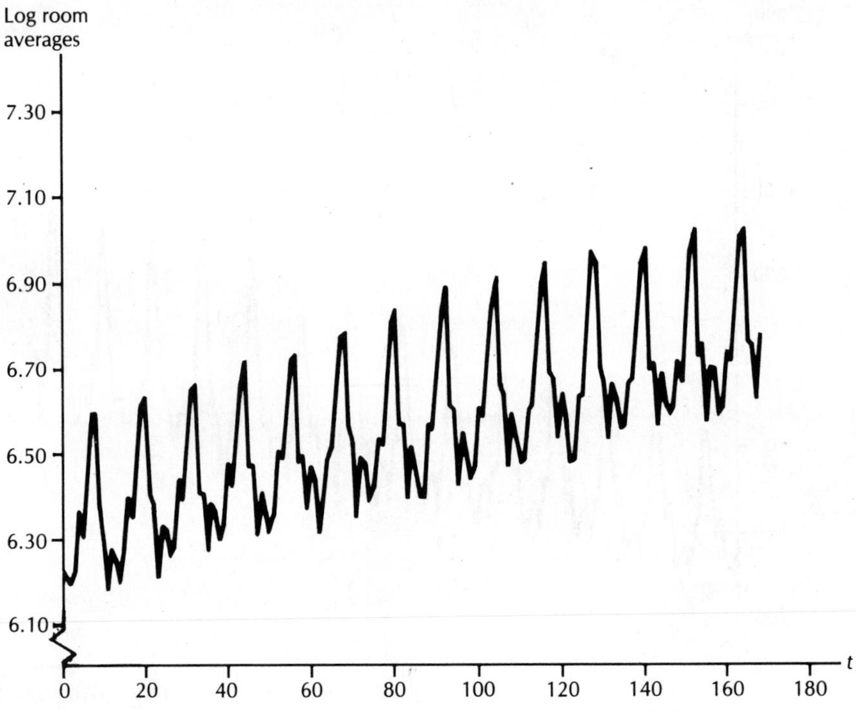

this model are found to be

$$
\begin{aligned}
b_0 &= 6.287557 & b_{S6} &= .1469094 \\
b_1 &= .00272528 & b_{S7} &= .2890226 \\
b_{S1} &= -.0416063 & b_{S8} &= .3111946 \\
b_{S2} &= -.112079 & b_{S9} &= .05598723 \\
b_{S3} &= -.084459 & b_{S10} &= .03954382 \\
b_{S4} &= .03983306 & b_{S11} &= -.112215 \\
b_{S5} &= .02039515 & &
\end{aligned}
$$

Moreover, the observed hotel room averages and the logarithms of these averages for periods 144 through 168 are given in Table 13.23.

a. Calculate the residuals $e^*_{167} = y^*_{167} - \hat{y}^*_{167}$ and $e^*_{168} = y^*_{168} - \hat{y}^*_{168}$.
b. The Durbin-Watson statistic calculated by using the residuals $e^*_1, e^*_2, \ldots, e^*_{168}$ indicates that positive autocorrelation exists. We continue to analyze $y^*_1, y^*_2, \ldots, y^*_{168}$ by assuming that these data have a first-order autoregressive error structure. That is, we

TABLE 13.23 **Hotel room averages and natural logs of hotel room averages for periods 144 through 168**

t	y_t	y_t^*	t	y_t	y_t^*
144	802	6.68711	157	811	6.69827
145	748	6.61740	158	732	6.59578
146	731	6.59441	159	745	6.61338
147	748	6.61740	160	844	6.73815
148	827	6.71780	161	833	6.72503
149	788	6.66950	162	935	6.84055
150	937	6.84268	163	1110	7.01212
151	1076	6.98101	164	1124	7.02465
152	1125	7.02554	165	868	6.76619
153	840	6.73340	166	860	6.75693
154	864	6.76157	167	762	6.63595
155	717	6.57508	168	877	6.77651
156	813	6.70073			

consider the model

$$\begin{aligned} y_t^* &= \mu_t^* + \varepsilon_t^* \\ &= \beta_0 + \beta_1 t + \beta_{S1}x_{S1,t} + \beta_{S2}x_{S2,t} + \cdots + \beta_{S11}x_{S11,t} + \varepsilon_t^* \end{aligned}$$

where $\varepsilon_t^* = \rho\varepsilon_{t-1}^* + U_t$. Describe the vector $\mathbf{y}$ and matrix $\mathbf{X}$ used to calculate r, the least squares point estimate of ρ, by using the model $e_t^* = \rho e_{t-1}^* + U_t$ to perform a regression analysis of the residuals $e_1^*, e_2^*, \ldots, e_{168}^*$ obtained from the model in part (a).

c. When we perform the regression analysis using the model $e_t^* = \rho e_{t-1}^* + U_t$ we find that $r = .390677$. Specify the first element and the last 12 elements of the vector $\mathbf{y}$ and the first element and the last 12 elements of each column of the matrix $\mathbf{X}$ used to calculate the revised least squares point estimates of $\beta_0, \beta_1, \beta_{S1}, \beta_{S2}, \ldots, \beta_{S11}$ by using the appropriate transformed model. *Hint*: first specify the appropriate transformed model.

d. When we calculate the revised least squares point estimates, we obtain

$b_0 = 6.28581951250$		$b_{S6} = .14786905558$
$b_1 = .00273459989$		$b_{S7} = .28996311620$
$b_{S1} = -.039732125$		$b_{S8} = .31211113045$
$b_{S2} = -.1107463$		$b_{S9} = .05686103800$
$b_{S3} = -.083343147$		$b_{S10} = .04032441500$
$b_{S4} = .04085815100$		$b_{S11} = -.1116581$
$b_{S5} = .02137845045$		

We also find that the unexplained variation, SSE, equals .05860552. Since we are estimating $k = 13$ parameters—$\beta_0, \beta_1, \beta_{S1}, \beta_{S2}, \ldots, \beta_{S11}$—and since the transformed model utilizes r, the point estimate of ρ, we compute the standard error to be

$$s = \sqrt{\frac{SSE}{n - (k + 1)}} = \sqrt{\frac{.05860552}{168 - (13 + 1)}} = .0195078$$

Using the revised least squares point estimates, the standard error, any other pertinent information, and the methods discussed in Section 13.5, calculate point predictions of, and 95% prediction intervals for, y^*_{169}, y_{169}, y^*_{170}, y_{170}, y^*_{171}, and y_{171}. *Hint*: Since $y^*_t = \ln y_t$, which implies that $y_t = e^{y^*_t}$, it follows that if $\hat{y}^*_t$ is a point prediction of y^*_t, then $e^{\hat{y}^*_t}$ is a point prediction of y_t, and if $[a_t, b_t]$ is a 95% prediction interval for y^*_t, then $[e^{a_t}, e^{b_t}]$ is a 95% prediction interval for y_t.

13.38 Consider the hotel room data in Table 13.22 and the model in Exercise 13.37.

a. Although we have assumed that $y^*_1, y^*_2, \ldots, y^*_{168}$ has a first-order autoregressive error structure, further analysis shows that the model

$$\begin{aligned} y^*_t &= \mu^*_t + \varepsilon^*_t \\ &= \beta_0 + \beta_1 t + \beta_{S1} x_{S1,t} + \beta_{S2} x_{S2,t} + \cdots + \beta_{S11} x_{S11,t} + \varepsilon^*_t \end{aligned}$$

where

$$\varepsilon^*_t = \phi_1 \varepsilon^*_{t-1} + \phi_2 \varepsilon^*_{t-2} + \phi_3 \varepsilon^*_{t-3} + \phi_{10} \varepsilon^*_{t-10} + \phi_{13} \varepsilon^*_{t-13} + U_t$$

is perhaps the best model describing $y^*_1, y^*_2, \ldots, y^*_{168}$. Describe the vector **y** and the matrix **X** used to calculate $\hat{\phi}_1$, $\hat{\phi}_2$, $\hat{\phi}_3$, $\hat{\phi}_{10}$, and $\hat{\phi}_{13}$, the least squares point estimates of ϕ_1, ϕ_2, ϕ_3, ϕ_{10}, and ϕ_{13}, by using the model

$$e^*_t = \phi_1 e^*_{t-1} + \phi_2 e^*_{t-2} + \phi_3 e^*_{t-3} + \phi_{10} e^*_{t-10} + \phi_{13} e^*_{t-13} + U_t$$

to perform a regression analysis of the residuals $e^*_1, e^*_2, \ldots, e^*_{168}$. Note here that the above model describing ε^*_t has been found by using SAS (which employs a backward elimination technique to determine an appropriate model).

b. When we perform the regression analysis using the model

$$e^*_t = \phi_1 e^*_{t-1} + \phi_2 e^*_{t-2} + \phi_3 e^*_{t-3} + \phi_{10} e^*_{t-10} + \phi_{13} e^*_{t-13} + U_t$$

we obtain

$\hat{\phi}_1 = -.32473663$ $\quad\hat{\phi}_{10} = -.14454908$

$\hat{\phi}_2 = -.17161490$ $\quad\hat{\phi}_{13} = -.18268925$

$\hat{\phi}_3 = .23998988$

Specify the last 12 elements of the vector **y** and the last 12 elements of each column of the matrix **X** used to calculate the revised least squares point estimates of β_0, β_1, β_{S1}, $\beta_{S2}, \ldots, \beta_{S11}$ by using the appropriate transformed model. *Hint*: First specify the appropriate transformed model.

c. When we calculate the revised least squares point estimates, we obtain

$b_0 = 6.28375470060$ $\quad b_{S6} = .14814188966$

$b_1 = .00274053113$ $\quad b_{S7} = .28940666366$

$b_{S1} = -.038092682$ $\quad b_{S8} = .31231859565$

$b_{S2} = -.1089287$ $\quad b_{S9} = .05624999334$

$b_{S3} = -.082511866$ $\quad b_{S10} = .04000362601$

$b_{S4} = .04115035831$ $\quad b_{S11} = -.11246398$

$b_{S5} = .02156220460$

We find that the unexplained variation, *SSE*, equals .04981187. Since we are estimating $k = 13$ parameters—β_0, β_1, β_{S1}, $\beta_{S2}, \ldots, \beta_{S11}$—and since the appropriate transformed model utilizes $\hat{\phi}_1$, $\hat{\phi}_2$, $\hat{\phi}_3$, $\hat{\phi}_{10}$, and $\hat{\phi}_{13}$, we calculate s by the (conservative) equation

$$s = \sqrt{\frac{SSE}{n - (k + 5)}} = \sqrt{\frac{.04981187}{168 - (13 + 5)}} = .018223$$

Since this model has a smaller standard error, we have some evidence to suggest that the model of this exercise is more appropriate than the model of Exercise 13.37. Using the revised least squares point estimates and any other pertinent information, calculate point predictions of y^*_{169}, y_{169}, y^*_{170}, y_{170}, y^*_{171}, y_{171}. (*Hint*: Generalize the methods of Section 13.5.)

d. Using advanced methods (PROC ARIMA in SAS) it can be shown that 95% prediction intervals for y^*_{169}, y^*_{170}, and y^*_{171} are:

For y^*_{169}: [6.6836, 6.7546]
For y^*_{170}: [6.6117, 6.6861]
For y^*_{171}: [6.6197, 6.6966]

Using these intervals, find 95% prediction intervals for y_{169}, y_{170}, and y_{171}.

e. Compare the prediction intervals you calculated in part (d) with the prediction intervals obtained in Exercise 13.37(d). Which model gives more precise intervals?

13.39 (Optional) Write a SAS program that will carry out the analysis of the tax form completion time data in Table 13.2 of Example 13.2.

13.40 (Optional) Write a SAS program that will carry out the analysis of the transformed telephone order model in Example 13.5. Use the data in Table 13.4.

13.41 (Optional) Write a SAS program that will carry out the Happy Valley Real Estate purchase analysis of Example 13.7. Use the data in Table 13.6.

13.42 (Optional) Write a SAS program that will carry out an analysis of the Bran-Nu Cereal demand data in Table 13.12. Use both of the models

$$y = \beta_0 x^{\beta_1} \varepsilon \qquad \text{and} \qquad y = \beta_0 \beta_1^x \varepsilon$$

13.43 (Optional) Write a SAS program that will carry out an analysis of the clothing expenditure data in Table 13.13. Use each of the models

$$\frac{y}{x} = \beta_0 \left(\frac{1}{x}\right) + \beta_1 + \eta$$

$$\frac{y}{x^2} = \beta_0 \left(\frac{1}{x^2}\right) + \beta_1 \left(\frac{1}{x}\right) + \eta$$

$$\sqrt{y} = \beta_0 + \beta_1 x + \varepsilon$$

$$\ln y = \beta_0 + \beta_1 x + \varepsilon$$

13.44 (Optional) Write a SAS program to analyze the NARHS data of Table 13.3 by using the models

$$\sqrt{y} = \beta_0 + \beta_1 x + \beta_2 x^2 + \varepsilon \qquad \text{and}$$

$$\ln y = \beta_0 + \beta_1 x + \beta_2 x^2 + \varepsilon$$

13.45 (Optional) Write a SAS program to analyze the service call data of Exercise 6.25 by using each of the models

$$\frac{y}{x} = \beta_0\left(\frac{1}{x}\right) + \beta_1 + \eta \qquad \text{and}$$

$$\frac{y}{x^2} = \beta_0\left(\frac{1}{x^2}\right) + \beta_1\left(\frac{1}{x}\right) + \eta$$

13.46 (Optional) Write a SAS program that will carry out an analysis of the certification examination data in Table 13.14 by using the procedures outlined in Exercises 13.23 and 13.24.

13.47 (Optional) Write a SAS program that will analyze the Happy Valley name recognition data in Table 13.15 by using logistic regression.

13.48 (Optional) Write a SAS program that will analyze the advertising claim data in Table 13.16 by using logistic regression.

13.49 (Optional) Write a SAS program that will analyze the two variable advertising experiment data in Table 13.17 by using multiple logistic regression.

13.50 (Optional) Write a SAS program that will analyze the microwave oven sales data in Table 13.18 by using the dummy variable regression model given in Exercise 13.31.

13.51 (Optional) Write a SAS program that will analyze the air conditioner sales data in Table 13.20 by using the dummy variable regression model given in Exercise 13.33.

13.52 (Optional) Write a SAS program that will analyze the disease data in Table 13.21 by using the growth curve model given in Exercise 13.34.

13.53 (Optional) Write a SAS program that will carry out the analysis of the hotel room occupancy data in Table 13.22 by using the procedure outlined in Exercises 13.36 through 13.38. Specifically,

a. Analyze the data using the dummy variable regression model of Exercise 13.36. Use SAS to calculate the Durbin-Watson statistic and test for positive first order autocorrelation.
b. Analyze the data using the first-order autoregressive model given in Exercise 13.37. Employ PROC ARIMA.
c. Use PROC AUTOREG to determine the autoregressive model specified in Exercise 13.38.
d. Analyze the data using the autoregressive model given in Exercise 13.38. Use PROC ARIMA to generate the prediction intervals given in Exercise 13.38(d).

CHAPTER 14

ONE-FACTOR ANALYSIS

In this chapter we begin our discussion of *designed statistical experiments.* Such an experiment is carried out to study the effects of one or more (often *qualitative*) independent variables on a dependent (*response*) variable. We begin in Sections 14.1 and 14.2 by presenting introductory concepts and the most basic type of experimental design—*the completely randomized experimental design.* Then in Section 14.3 we introduce the main topic of this chapter: *how to analyze the effect of one qualitative independent variable on a dependent variable.* This is called *one-factor analysis.* In Section 14.4 we explain the *ANOVA* (*analysis of variance*) *approach* to one-factor analysis, and in Section 14.5 we present the *regression approach* to one-factor analysis. We continue in Section 14.6 by discussing *fixed* and *random models.* We conclude this chapter with optional Sections 14.7 and 14.8. Section 14.7 presents Hartley's and Bartlett's tests for variance equality. Section 14.8 shows how to use SAS to perform one-factor analysis.

14.1

BASIC CONCEPTS OF EXPERIMENTAL DESIGN

In experimental design, the *dependent variable* is called the *response variable*, and the *independent variables* are called *factors*. We may wish to study a *single factor* or *several factors*. The different *levels* of a factor (or combination of factors) are called *treatments*.

The *purpose* of most statistical experiments is to *compare and estimate the effects of different treatments on the response variable*. That is, we wish to

1. Determine whether the effects of the various treatments are different.
2. If the treatment effects are different, estimate how different they are.

Example 14.1 Aamco Marketing is a marketing research firm. It handles the account of an electronics firm that produces VCRs and CD players. Aamco wishes to study the effects of three different advertising campaigns (A, B, and C) on VCR sales. Here, the firm wants to study a one-factor advertising campaign. There are three treatments: campaigns A, B, and C. The response variable is VCR sales. The firm's managers wish to determine whether the different advertising campaigns (treatments) have different effects on VCR sales. In addition, if the treatment effects differ, the firm wishes to estimate the size(s) of the difference(s) in the treatment effects.

In addition to the one-factor study, Aamco Marketing also wishes to study the effects of advertising campaigns A, B, and C and four different pricing strategies (M, N, O, and P) on CD player sales. Here, the firm's managers wish to study two factors—type of advertising campaign (with levels A, B, and C) and type of pricing strategy (with levels M, N, O, and P). Since Aamco wishes to study the effects of *combinations* of advertising campaign type and pricing strategy type, there are 12 treatments:

$$\begin{array}{cccc} AM & AN & AO & AP \\ BM & BN & BO & BP \\ CM & CN & CO & CP \end{array}$$

The response variable is CD player sales. Again, the firm wishes to compare and estimate the effects of the different treatments in order to find an effective advertising campaign–pricing strategy combination.

14.2

THE COMPLETELY RANDOMIZED EXPERIMENTAL DESIGN

In a designed experiment the different treatments are applied to *experimental units.* When a treatment is applied to more than one experimental unit, it is said to be *replicated.* The term *randomization* refers to the manner in which experimental units are assigned to treatments.

Suppose that we wish to assign a total of n experimental units to a total of v treatments. Let

$$n_1, n_2, \ldots, n_v$$

denote the number of experimental units assigned to treatments 1, 2, . . . , v. Then

$$n = n_1 + n_2 + \cdots + n_v$$

We can obtain what is called a *completely randomized experimental design* by assigning experimental units to treatments as follows. First, randomly select n_1 experimental units and assign them to the first treatment. Next, randomly select n_2 *different* experimental units and assign them to the second treatment. Then randomly select n_3 *different* experimental units. That is, select these units from those not assigned to either the first or the second treatment. Assign these units to the third treatment. Continue this procedure until the proper number of experimental units have been assigned to each treatment. This procedure results in n_l experimental units being assigned to treatment l for $l = 1, 2, \ldots, v$.

Once experimental units have been assigned to treatments, a value of the response variable y is observed for each experimental unit. Thus we obtain a *sample* of n_l values of the response variable

$$y_{l,1}, y_{l,2}, \ldots, y_{l,n_l}$$

corresponding to treatment l. When we employ a completely randomized design, we assume that this sample has been randomly selected from the population of all values of the response variable that could potentially be obtained by using treatment l. We also assume that the v samples of values of the response variable are independent of each other. This is usually a reasonable assumption, since the completely randomized design ensures that each different sample results from *different measurements* being taken on *different experimental units.*

Example 14.2 North American Oil Company is attempting to develop a reasonably priced unleaded gasoline that will deliver improved gasoline mileages. As part of its development process, the company would like to compare the effects of three types of unleaded gasoline (A, B, and C) on gasoline mileage. The chemical compositions of these three gasoline types differ. Since these differences in chemical composition cannot be defined to be a function of one or more *quantitative* independent variables, North American Oil defines "gasoline type" to be a *qualitative* independent variable.

For testing purposes, North American Oil will compare the effects of gasoline types A, B, and C on the gasoline mileage obtained by an automobile called the Fire-Hawk. Suppose that the company has access to 1000 Fire-Hawks that are representative of the (theoretically) infinite population of all Fire-Hawks. Using this subpopulation of 1000 Fire-Hawks, North American Oil can employ a *completely randomized experimental design* as follows. First, $n_A = 4$ Fire-Hawks will be randomly selected from the subpopulation of 1000 Fire-Hawks. These Fire-Hawks will be assigned to gasoline type A. Next, $n_B = 5$ *different* Fire-Hawks will be randomly selected from the remaining 996 Fire-Hawks in the subpopulation. These Fire-Hawks will be assigned to gasoline type B. Finally, $n_C = 3$ *different* Fire-Hawks will be randomly selected from the remaining 991 Fire-Hawks in the subpopulation. These Fire-Hawks will be assigned to gasoline type C.

Each randomly selected Fire-Hawk is test driven using the appropriate gasoline type (treatment) under normal conditions for a specified distance, and the gasoline mileage for each test drive is measured. We let

$$y_{l,k} = \text{the } k\text{th gasoline mileage obtained when using gasoline type } l$$

Here, we assume that

$$y_{l,1}, y_{l,2}, \ldots, y_{l,n_l}$$

is a sample of n_l Fire-Hawk mileages that has been randomly selected from the infinite population of all Fire-Hawk mileages that could be obtained using gasoline type l. For example,

$$y_{C,2} = \text{the second gasoline mileage obtained when using gasoline type C}$$

and

$$y_{C,1}, y_{C,2}, y_{C,3}$$

is a sample of $n_C = 3$ Fire-Hawk mileages that has been randomly selected from the infinite population of all Fire-Hawk mileages that could be obtained by using gasoline type C.

14.3 AN INTRODUCTION TO ONE-FACTOR ANALYSIS

Suppose that *we wish to study one qualitative factor* with levels $1, 2, \ldots, v$. That is, we wish to study the effects of v treatments (treatments $1, 2, \ldots, v$) on a response variable. Then, for $l = 1, 2, \ldots, v$, define *population* l to be the population of all possible values of the response variable that could potentially be observed when using treatment l. We let μ_l be the mean of this population. That is,

μ_l = the mean value of the response variable obtained by using treatment l

In addition, define σ_l to be the standard deviation of population l.

The goal of one-factor analysis is to estimate and compare the effects of the different treatments on the response variable. That is, *we wish to estimate and compare the treatment means* $\mu_1, \mu_2, \ldots, \mu_v$. To do this, we will assume that a sample has been randomly selected from each of the v populations by employing a completely randomized experimental design. For $l = 1, 2, \ldots, v$ we let n_l denote the size of the sample that has been randomly selected from population l. We write this sample as

$$y_{l,1}, y_{l,2}, \ldots, y_{l,n_l}$$

where

$y_{l,k}$ = the kth value of the response variable that is observed when using treatment l

One-factor analysis employs the following model.

The One-Factor Model

The *one-factor model* says that

$$y_{l,k} = \mu_l + \varepsilon_{l,k}$$

Here,

1. μ_l, the lth treatment mean, describes the effect of treatment l on $y_{l,k}$;
2. $\varepsilon_{l,k}$, the (l, k)th error term, describes the effects of all factors other than treatment l on $y_{l,k}$.

The obvious point estimate of the treatment mean μ_l is

$$\bar{y}_l = \frac{\sum_{k=1}^{n_l} y_{l,k}}{n_l}$$

This is simply the sample mean for sample l. The point estimate of σ_l is

$$s_l = \sqrt{\frac{\sum_{k=1}^{n_l} (y_{l,k} - \bar{y}_l)^2}{n_l - 1}}$$

This is the sample standard deviation for sample l.

Both the "analysis of variance (ANOVA) approach" and the "regression approach" to one-factor analysis allow us to test for statistically significant differences between treatment means. They also allow us to estimate differences between treatment means. The validity of both approaches depends on the *inference assumptions*. The *constant variance assumption* says that

$$\sigma_1^2 = \sigma_2^2 = \cdots = \sigma_v^2$$

This assumption is reasonable if the sample variances $s_1^2, s_2^2, \ldots, s_v^2$ are reasonably equal. There are several statistical tests that employ the sample variances to test the equality of the population variances. Two such tests are Hartley's test and Bartlett's test, which are discussed in optional Section 14.7. Since there are some drawbacks to these tests, there is controversy as to whether they should be performed. Furthermore, studies have shown that if the sizes of the v samples are equal, unequal population variances are not a serious problem. However, if the sizes of the v samples are quite different—say, if the largest sample size is at least twice the smallest—then unequal variances may cause our results to be invalid. Thus it is best to try to randomly select samples of equal sizes.

The *independence assumption* will probably hold if the data have been randomly selected at a specific time and if there is no relationship between the values in one sample and the values in another sample. In such a case we say that the samples are independent of each other and that we have performed an *independent samples experiment*. In particular, the independence assumption is reasonable when using a completely randomized experimental design. This is because each treatment is applied to a different set of experimental units. Sometimes the values in different samples result from different measurements being taken on the *same experimental units*. Then the values in different samples are related; we say that the samples are dependent upon each other. In such a situation the independence assumption would be badly violated, and the methods presented in this chapter should not be used to compare the treatment means. The appropriate methods for analyzing such data are given in Chapter 17, which discusses the *randomized block* design.

The *normality assumption* says that each of the v populations is normally distributed. This assumption is not crucial. It has been shown that the techniques of this chapter are approximately valid for "mound-shaped" distributions.

Example 14.3 Consider the North American Oil Company situation. For l = A, B, C, we let

μ_l = the mean Fire-Hawk gasoline mileage obtained when using gasoline type l

To compare μ_A, μ_B, and μ_C, the mean gasoline mileages obtained by using gasoline types A, B, and C, respectively, suppose that North American Oil employs a completely randomized experimental design. The company randomly selects samples of $n_A = 4$ Fire-Hawks driven using gasoline type A, $n_B = 5$ Fire-Hawks driven using gasoline type B, and $n_C = 3$ Fire-Hawks driven using gasoline type C. The data that is obtained is shown in Table 14.1. Note that sample means and standard deviations are also shown for each sample.

TABLE 14.1 **Three samples of Fire-Hawk mileages obtained by using gasoline types A, B, and C**

Gasoline type A	*Gasoline type B*	*Gasoline type C*
$y_{A,1} = 24.0$	$y_{B,1} = 25.3$	$y_{C,1} = 23.3$
$y_{A,2} = 25.0$	$y_{B,2} = 26.5$	$y_{C,2} = 24.0$
$y_{A,3} = 24.3$	$y_{B,3} = 26.4$	$y_{C,3} = 24.7$
$y_{A,4} = 25.5$	$y_{B,4} = 27.0$	
	$y_{B,5} = 27.6$	
$\bar{y}_A = \frac{\sum_{k=1}^{4} y_{A,k}}{4} = \frac{98.8}{4} = 24.7$	$\bar{y}_B = \frac{\sum_{k=1}^{5} y_{B,k}}{5} = \frac{132.8}{5} = 26.56$	$\bar{y}_C = \frac{\sum_{k=1}^{3} y_{C,k}}{3} = \frac{72}{3} = 24.0$
$s_A^2 = .46$	$s_B^2 = .723$	$s_C^2 = .49$
$s_A = .6782$	$s_B = .8503$	$s_C = .7$

We might conclude (somewhat arbitrarily) that the sample variances do not differ substantially. Note that the sample standard deviations, being the square roots of the sample variances, are more nearly equal. Thus although the sample sizes for this data are not equal, we might conclude that the constant variance assumption approximately holds. This says that the unequal sample sizes are not a problem. Since North American Oil has employed a completely randomized design, the independence assumption probably holds. This is because the gasoline mileages in the different samples were obtained for *different* Fire-Hawks. Since many small independent factors influence gasoline mileage, the distributions of gasoline mileages for gasoline types A, B, and C are probably mound shaped. Thus the normality assumption probably approximately holds.

The purpose of the data analysis to be carried out in Sections 14.4 and 14.5 is to (1) test to see whether there are any statistically significant differences between the treatment means μ_A, μ_B, and μ_C, (2) if such differences exist, determine which treatment means differ, and (3) estimate the magnitudes of any statistically significant differences (this will allow North American Oil to judge whether or not these differences have practical importance).

14.4 THE ANOVA APPROACH

In this section we present "sums of squares" formulas that allow us to estimate and compare v treatment means. The method we describe in this section is commonly called *one-way analysis of variance* (or *one-way ANOVA*).

TESTING FOR SIGNIFICANT DIFFERENCES BETWEEN TREATMENT MEANS

As a preliminary step in one-way ANOVA, we wish to determine whether or not there are any statistically significant differences between the treatment means μ_1, $\mu_2, \ldots, \mu_v$. To do this, we test the null hypothesis

$$H_0: \mu_1 = \mu_2 = \cdots = \mu_v$$

This hypothesis says that all the treatments have the same effect on the mean response. We test H_0 versus the alternative hypothesis

$$H_1: \text{At least two of } \mu_1, \mu_2, \ldots, \mu_v \text{ differ}$$

This alternative says that at least two treatments have different effects on the mean response. To test these hypotheses, we begin by defining "sums of squares" and "mean squares."

Sums of Squares and Mean Squares in One-Way ANOVA

We define

$$n = \sum_{l=1}^{v} n_l$$

to be the total number of experimental units employed, and

$$\bar{y} = \frac{\sum_{l=1}^{v} \sum_{k=1}^{n_l} y_{l,k}}{n}$$

to be the overall mean of all observed values of the response variable. Then we define:

1. The *treatment sum of squares* (or *explained variation*) to be

$$SS_{\text{means}} = \sum_{l=1}^{v} n_l(\bar{y}_l - \bar{y})^2$$

2. The *treatment mean square* to be

$$MS_{\text{means}} = \frac{SS_{\text{means}}}{v - 1}$$

3. The *error sum of squares* (or *unexplained variation*) to be

$$SSE = \sum_{l=1}^{v} \sum_{k=1}^{n_l} (y_{l,k} - \bar{y}_l)^2$$

4. The *error mean square* to be

$$MSE = s^2 = \frac{SSE}{n - v}$$

5. The *total sum of squares* (or *total variation*) to be

$$SS_{\text{total}} = \sum_{l=1}^{v} \sum_{k=1}^{n_l} (y_{l,k} - \bar{y})^2$$

Furthermore, it can be shown that

Total sum of squares = treatment sum of squares + error sum of squares

That is,

$$SS_{\text{total}} = SS_{\text{means}} + SSE$$

Here, the *total sum of squares*, SS_{total}, measures the total amount of variability in the observed values of the response variable. The *treatment sum of squares*, SS_{means}, measures the variability of the sample treatment means. For example, if all the sample treatment means ($\bar{y}_l$ values) were equal, then the treatment sum of squares would be equal to zero. The more the $\bar{y}_l$ values vary, the larger will be SS_{means}. In other words, the treatment sum of squares measures the amount of *between treatment variability*. The *error sum of squares, SSE*, measures the variability of the observed values of the response variable around their respective treatment means. For example, if there were no variability within each sample, the error sum of squares would be equal to zero. The more the values within the samples vary, the larger will be *SSE*. Note that an alternative way to compute the error sum of squares is to use the formula

$$SSE = \sum_{l=1}^{v} (n_l - 1) s_l^2$$

The *treatment sum of squares* and *error sum of squares* are said to *partition the total sum of squares*. That is,

$$SS_{\text{total}} = SS_{\text{means}} + SSE$$

Suppose that we wish to decide whether or not there are any statistically significant differences between the treatment means $\mu_1, \mu_2, \ldots, \mu_v$. Then it makes sense to compare the amount of between treatment variability to the amount of within treatment variability. This comparison motivates using the following F statistic and prob-value.

An F-Test for Differences Between Treatment Means

Define

$$F(\text{means}) = \frac{MS_{\text{means}}}{MSE} = \frac{SS_{\text{means}}/(v-1)}{SSE/(n-v)} = \frac{\sum_{l=1}^{v} n_l(\bar{y}_l - \bar{y})^2/(v-1)}{\sum_{l=1}^{v}\sum_{k=1}^{n_l}(y_{l,k} - \bar{y}_l)^2/(n-v)}$$

Also, define the prob-value to be the area to the right of F(means) under the curve of the F-distribution having $v - 1$ and $n - v$ degrees of freedom. Then we can reject

$$H_0: \mu_1 = \mu_2 = \cdots = \mu_v \quad \text{(all treatment means are equal)}$$

in favor of

H_1: At least two of $\mu_1, \mu_2, \ldots, \mu_v$ differ
(at least two treatment means differ)

by setting the probability of a Type I error equal to α if and only if either of the following equivalent conditions holds:

1. $F(\text{means}) > F_{[\alpha]}^{(v-1,n-v)}$
2. prob-value $< \alpha$

A large value of F(means) results when SS_{means}, which measures the between treatment variability, is large in comparison to SSE, which measures the within treatment variability. If F(means) is large enough, this implies that the null hypothesis should be rejected. The rejection point $F_{[\alpha]}^{(v-1,n-v)}$ tells us when F(means) is large enough to allow us to reject H_0 with the probability of a Type I error set equal to α. When F(means) is large, the associated prob-value is small. If this prob-value is less than α, we can reject H_0 with the probability of a Type I error set equal to α.

Example 14.4 Consider the North American Oil Company data in Table 14.1. The company wishes to determine whether or not any of gasoline types A, B, and C have different effects on mean Fire-Hawk gasoline mileage. That is, we wish to see whether or not there are any statistically significant differences between μ_A, μ_B, and μ_C. To do this, we test the null hypothesis

$$H_0: \mu_A = \mu_B = \mu_C$$

which says that gasoline types A, B, and C have the same effects on mean gasoline mileage, versus

$$H_1: \text{At least two of } \mu_{\text{A}}, \mu_{\text{B}}, \text{ and } \mu_{\text{C}} \text{ differ}$$

which says that at least two of gasoline types A, B, and C have different effects on mean gasoline mileage. To test these hypotheses, we compute the following quantities:

$$n = \sum_{l=\text{A,B,C}} n_l = n_{\text{A}} + n_{\text{B}} + n_{\text{C}} = 4 + 5 + 3 = 12$$

$$\bar{y} = \frac{\sum_{l=\text{A,B,C}} \sum_{k=1}^{n_l} y_{l,k}}{12} = \frac{24.0 + 25.0 + \cdots + 24.7}{12} = \frac{303.6}{12} = 25.3$$

$$\begin{aligned} SS_{\text{means}} &= \sum_{l=\text{A,B,C}} n_l(\bar{y}_l - \bar{y})^2 \\ &= n_{\text{A}}(\bar{y}_{\text{A}} - \bar{y})^2 + n_{\text{B}}(\bar{y}_{\text{B}} - \bar{y})^2 + n_{\text{C}}(\bar{y}_{\text{C}} - \bar{y})^2 \\ &= 4(24.7 - 25.3)^2 + 5(26.56 - 25.3)^2 + 3(24 - 25.3)^2 \\ &= 14.448 \end{aligned}$$

$$MS_{\text{means}} = \frac{SS_{\text{means}}}{v - 1} = \frac{14.448}{3 - 1} = 7.224$$

$$\begin{aligned} SSE &= \sum_{l=\text{A,B,C}} \sum_{k=1}^{n_l} (y_{l,k} - \bar{y}_l)^2 \\ &= \sum_{k=1}^{n_{\text{A}}} (y_{\text{A},k} - \bar{y}_{\text{A}})^2 + \sum_{k=1}^{n_{\text{B}}} (y_{\text{B},k} - \bar{y}_{\text{B}})^2 + \sum_{k=1}^{n_{\text{C}}} (y_{\text{C},k} - \bar{y}_{\text{C}})^2 \\ &= [(24 - 24.7)^2 + (25 - 24.7)^2 + (24.3 - 24.7)^2 + (25.5 - 24.7)^2] \\ &\quad + [(25.3 - 26.56)^2 + (26.5 - 26.56)^2 + (26.4 - 26.56)^2 \\ &\quad + (27 - 26.56)^2 + (27.6 - 26.56)^2] + [(23.3 - 24)^2 \\ &\quad + (24 - 24)^2 + (24.7 - 24)^2] \\ &= 5.252 \end{aligned}$$

$$MSE = \frac{SSE}{n - v} = \frac{5.252}{12 - 3} = .583556$$

Also note that an alternative way to calculate SSE is to compute

$$\begin{aligned} SSE &= (n_{\text{A}} - 1)s_{\text{A}}^2 + (n_{\text{B}} - 1)s_{\text{B}}^2 + (n_{\text{C}} - 1)s_{\text{C}}^2 \\ &= (4 - 1)(.46) + (5 - 1)(.723) + (3 - 1)(.49) \\ &= 5.252 \end{aligned}$$

Finally, we compute

$$F(\text{means}) = \frac{MS_{\text{means}}}{MSE} = \frac{7.224}{.583556} = 12.379$$

We find that

$$F(\text{means}) > F_{[\alpha]}^{(v-1,n-v)} = F_{[.05]}^{(3-1,12-3)} = 4.26$$

TABLE 14.2 **Analysis of variance table for testing $H_0: \mu_A = \mu_B = \mu_C$ in the North American Oil Company problem ($v = 3$ gasoline types, $n = 12$ observations)**

Source	df	Sum of squares	Mean square	F statistic	Prob-value
Model	$v - 1 = 3 - 1 = 2$	SS_{means} = explained variation $= \sum_{l=A,B,C} n_l(\bar{y}_l - \bar{y})^2 = 14.448$	$MS_{\text{means}} = \frac{SS_{\text{means}}}{v-1} = \frac{14.448}{3-1} = 7.224$	$F(\text{means}) = \frac{MS_{\text{means}}}{MSE} = \frac{7.224}{.583556} = 12.379$	.0026
Error	$n - v = 12 - 3 = 9$	SSE = unexplained variation $= \sum_{l=A,B,C} \sum_{k=1}^{n_l} (y_{l,k} - \bar{y}_l)^2 = 5.252$	$MSE = \frac{SSE}{n-v} = \frac{5.252}{12-3} = .583556$		
Total	$n - 1 = 12 - 1 = 11$	SS_{total} = total variation $= \sum_{l=A,B,C} \sum_{k=1}^{n_l} (y_{l,k} - \bar{y})^2 = 19.7$			

In addition, it can be shown that

$$\text{prob-value} = .0026$$

which is less than $\alpha = .05$. Therefore we reject

$$H_0: \mu_A = \mu_B = \mu_C$$

with $\alpha = .05$. We conclude that statistically significant differences exist between at least two of μ_A, μ_B, and μ_C. This says that at least two of gasoline types A, B, and C have different effects on mean Fire-Hawk gasoline mileage. Note that the calculation of F(means) is summarized in Table 14.2, which is called an *analysis of variance table*.

COMPARING AND ESTIMATING SPECIFIC TREATMENT MEANS

Suppose we have determined that statistically significant differences exist between some of the treatment means. It then makes sense to perform pairwise comparisons of these means. This can be done by testing the null hypothesis

$$H_0: \mu_i - \mu_j = 0 \quad \text{or} \quad H_0: \mu_i = \mu_j$$

This hypothesis says that treatments i and j have the same effects on the mean response. We test H_0 versus the alternative hypothesis

$$H_1: \mu_i - \mu_j \neq 0 \qquad \text{or} \qquad H_1: \mu_i \neq \mu_j$$

Here, H_1 says that treatments i and j have different effects on the mean response. This test is done as follows.

A t-Test for Pairwise Comparisons of Treatment Means

Define

$$t = \frac{\bar{y}_i - \bar{y}_j}{s\sqrt{(1/n_i) + (1/n_j)}}$$

Also, define the prob-value to be twice the area under the curve of the t-distribution having $n - v$ degrees of freedom to the right of $|t|$. Then we can reject

$$H_0: \mu_i - \mu_j = 0 \qquad \text{or} \qquad H_0: \mu_i = \mu_j$$

in favor of

$$H_1: \mu_i - \mu_j \neq 0 \qquad \text{or} \qquad H_1: \mu_i \neq \mu_j$$

by setting the probability of a Type I error equal to α if and only if either of the following equivalent conditions holds:

1. $|t| > t_{[\alpha/2]}^{(n-v)}$ or
2. prob-value $< \alpha$

Example 14.5 Consider the North American Oil Company problem. In Example 14.4 we concluded that at least two of μ_A, μ_B, and μ_C differ. Suppose that we wish to see whether or not μ_B and μ_C differ. To do this, we will test the null hypothesis

$$H_0: \mu_C - \mu_B = 0 \qquad \text{or} \qquad H_0: \mu_C = \mu_B$$

which says that gasoline types B and C have the same effects on mean Fire-Hawk gasoline mileage, versus the alternative hypothesis

$$H_1: \mu_C - \mu_B \neq 0 \qquad \text{or} \qquad H_1: \mu_C \neq \mu_B$$

which says that gasoline types B and C have different effects on mean Fire-Hawk gasoline mileage. To test these hypotheses, we compute

$$t = \frac{\bar{y}_C - \bar{y}_B}{s\sqrt{(1/n_C) + (1/n_B)}} = \frac{24 - 26.56}{.7639\sqrt{(1/3) + (1/5)}} = -4.59$$

We find that

$$|t| = 4.59 > t_{[\alpha/2]}^{(n-v)} = t_{[.025]}^{(12-3)} = 2.262$$

In addition, it can be shown that

$$\text{prob-value} = .0013$$

which is less than $\alpha = .05$. Therefore we reject H_0 with $\alpha = .05$. We conclude that there is a statistically significant difference between treatment means μ_C and μ_B. In a similar manner we can test for differences between μ_A and μ_B and μ_A and μ_C.

Suppose that we have determined that *statistically significant* differences exist between some of the treatment means. It then also makes sense to compute point estimates of and confidence intervals for differences between the treatment means. This allows us to assess whether or not the differences are of *practical importance*. In addition, we can compute confidence intervals for *linear combinations of treatment means.*

Estimation of Differences and Linear Combinations

1. A *point estimate of the difference* $\mu_i - \mu_j$ is $\bar{y}_i - \bar{y}_j$, and a $100(1-\alpha)\%$ *confidence interval* for $\mu_i - \mu_j$ is

$$\left[(\bar{y}_i - \bar{y}_j) \pm t_{[\alpha/2]}^{(n-v)} s \sqrt{\frac{1}{n_i} + \frac{1}{n_j}}\right]$$

2. A *point estimate of the linear combination* $\sum_{l=1}^{v} a_l \mu_l$ is

$$\sum_{l=1}^{v} a_l \bar{y}_l$$

and a $100(1-\alpha)\%$ *confidence interval* for that linear combination is

$$\left[\sum_{l=1}^{v} a_l \bar{y}_l \pm t_{[\alpha/2]}^{(n-v)} s \sqrt{\sum_{l=1}^{v} \frac{a_l^2}{n_l}}\right]$$

Example 14.6 In the North American Oil Company problem, we have found that there is a statistically significant difference between the treatment means μ_C and μ_B. A point estimate of $\mu_C - \mu_B$ is

$$\bar{y}_C - \bar{y}_B = 24 - 26.56 = -2.56$$

Furthermore, a 95% confidence interval for $\mu_C - \mu_B$ is

$$\left[(\bar{y}_C - \bar{y}_B) \pm t_{[.025]}^{(12-3)} s\sqrt{\frac{1}{n_C} + \frac{1}{n_B}}\right] = [-2.56 \pm 2.262(.7639)\sqrt{\tfrac{1}{3} + \tfrac{1}{5}}]$$
$$= [-3.822, \ -1.298]$$

We conclude that we can be 95 percent confident that the mean mileage obtained by using gasoline type B is between 1.298 and 3.822 mpg higher than the mean mileage obtained by using gasoline type C.

Similarly, a point estimate of $\mu_B - \mu_A$ is

$$\bar{y}_B - \bar{y}_A = 26.56 - 24.7 = 1.86$$

Moreover, a 95% confidence interval for $\mu_B - \mu_A$ is

$$\left[(\bar{y}_B - \bar{y}_A) \pm t_{[.025]}^{(12-3)} s\sqrt{\frac{1}{n_B} + \frac{1}{n_A}}\right] = [1.86 \pm 2.262(.7639)\sqrt{\tfrac{1}{5} + \tfrac{1}{4}}]$$
$$= [.701, \ 3.019]$$

We conclude that we can be 95 percent confident that the mean mileage obtained by using gasoline type B is between .701 and 3.019 mpg higher than the mean mileage obtained by using gasoline type A. Therefore we have evidence that gasoline type B yields a higher mean mileage than either gasoline type A or gasoline type C. To compare gasoline types A and C, we note that a point estimate of $\mu_C - \mu_A$ is

$$\bar{y}_C - \bar{y}_A = 24 - 24.7 = -.7$$

Furthermore, a 95 percent confidence interval for $\mu_C - \mu_A$ is

$$\left[(\bar{y}_C - \bar{y}_A) \pm t_{[.025]}^{(12-3)} s\sqrt{\frac{1}{n_C} + \frac{1}{n_A}}\right] = [-.7 \pm 2.262(.7639)\sqrt{\tfrac{1}{3} + \tfrac{1}{4}}]$$
$$= [-2.0197, \ .6197]$$

We conclude that we can be 95 percent confident that the mean mileage obtained by using gasoline type C is between 2.0197 mpg less than and .6197 mpg greater than the mean mileage obtained by using gasoline type A.

Gasoline type B contains a chemical—Chemical XX—that is not contained in gasoline types A or C. To assess the effect of Chemical XX on gasoline mileage, we consider

$$\mu_B - \frac{\mu_C + \mu_A}{2}$$

This is the difference between the mean mileage obtained by using gasoline type B and the mean mileage obtained by using gasoline type C one-half of the time and gasoline type A one-half of the time. We see that

$$\begin{aligned}\mu_B - (\mu_C + \mu_A)/2 &= (-\tfrac{1}{2})\mu_A + (1)\mu_B + (-\tfrac{1}{2})\mu_C \\ &= a_A\mu_A + a_B\mu_B + a_C\mu_C \\ &= \sum_{l=A,B,C} a_l\mu_l\end{aligned}$$

The point estimate of this linear combination is

$$\begin{aligned}\bar{y}_B - (\bar{y}_C + \bar{y}_A)/2 &= (-\tfrac{1}{2})\bar{y}_A + (1)\bar{y}_B + (-\tfrac{1}{2})\bar{y}_C \\ &= (-\tfrac{1}{2})(24.7) + (1)(26.56) + (-\tfrac{1}{2})(24.0) \\ &= 2.21\end{aligned}$$

Since

$$\sum_{l=1}^{v} \frac{a_l^2}{n_l} = \sum_{l=A,B,C} \frac{a_l^2}{n_l} = \frac{(-\frac{1}{2})^2}{4} + \frac{(1)^2}{5} + \frac{(-\frac{1}{2})^2}{3} = .3458$$

a 95 percent confidence interval for this linear combination is

$$\begin{aligned}[2.21 \pm t_{[.025]}^{(12-3)} s\sqrt{.3458}] &= [2.21 \pm 2.262(.7639)\sqrt{.3458}] \\ &= [2.21 \pm 1.016] = [1.194, 3.226]\end{aligned}$$

We conclude that we can be 95 percent confident that μ_B is between 1.194 mpg and 3.226 mpg greater than $(\mu_C + \mu_A)/2$. Note here that Chemical XX might be a major factor causing μ_B to be greater than $(\mu_C + \mu_A)/2$. However, this is not at all certain. The chemists at North American Oil must use the above comparison along with their knowledge of the chemical compositions of gasoline types A, B, and C, to assess the effect of Chemical XX on gasoline mileage.

We can also estimate specific treatment means and can predict individual values of the response variable.

Estimating Treatment Means and Predicting Individual Values

1. A *point estimate* of the treatment mean μ_l is $\bar{y}_l$, and a $100(1-\alpha)\%$ *confidence interval* for μ_l is

$$\left[\bar{y}_l \pm t_{[\alpha/2]}^{(n-v)}\left(\frac{s}{\sqrt{n_l}}\right)\right]$$

2. A *point prediction* of a randomly selected individual value $y_{l,0} = \mu_l + \varepsilon_{l,0}$ is $\bar{y}_l$, and a $100(1-\alpha)\%$ *prediction interval* for $y_{l,0}$ is

$$\left[\bar{y}_l \pm t_{[\alpha/2]}^{(n-v)} s\sqrt{1 + \frac{1}{n_l}}\right]$$

Example 14.7 In the North American Oil Company problem, a point estimate of μ_B, the mean gasoline mileage obtained by using gasoline type B, is $\bar{y}_B = 26.56$. Moreover, a 95% confidence interval for μ_B is

$$\left[\bar{y}_B \pm t_{[.025]}^{(12-3)}\left(\frac{s}{\sqrt{n_B}}\right)\right] = \left[26.56 \pm 2.262\left(\frac{.7639}{\sqrt{5}}\right)\right]$$
$$= [25.787, 27.333]$$

This interval says that we can be 95 percent confident that the mean mileage obtained by using gasoline type B is between 25.787 and 27.333 mpg. Furthermore, a point prediction of $y_{B,0}$, an individual mileage obtained by using gasoline type B, is $\bar{y}_B = 26.56$. A 95% prediction interval for this gasoline mileage is

$$\left[\bar{y}_B \pm t_{[.025]}^{(12-3)}s\sqrt{1 + \frac{1}{n_B}}\right] = [26.56 \pm 2.262(.7639)\sqrt{1 + \tfrac{1}{5}}]$$
$$= [24.667, 28.453]$$

This interval says that we can be 95 percent confident that the gasoline mileage obtained by an individual Fire-Hawk using gasoline type B will be between 24.667 and 28.453 mpg.

Example 14.8 Figure 14.1 presents the SAS output of an analysis of variance of the gasoline mileage data. The SAS output presents most of the quantities calculated in the previous examples. However, the output does not include the confidence intervals for $\mu_B - \mu_A$, $\mu_C - \mu_A$, $\mu_C - \mu_B$, and $\mu_B - (\mu_C + \mu_A)/2$. These intervals can be computed by using the SAS output because it includes point estimates of these linear combinations and the corresponding standard errors of these point estimates. For example, the SAS output tells us that the point estimate of $\mu_C - \mu_B$ (denoted MUC − MUB on the SAS output) is

$$\bar{y}_C - \bar{y}_B = -2.56$$

The output also tells us that the standard error of this point estimate is

$$s\sqrt{\frac{1}{n_C} + \frac{1}{n_B}} = .5579$$

It follows that a 95% confidence interval for $\mu_C - \mu_B$ is

$$\left[(\bar{y}_C - \bar{y}_B) \pm t_{[.025]}^{(12-3)}s\sqrt{\frac{1}{n_C} + \frac{1}{n_B}}\right] = [-2.56 \pm 2.262(.5579)]$$
$$= [-3.822, -1.298]$$

This interval was calculated in Example 14.6.

FIGURE 14.1 **SAS output of an analysis of variance of the gasoline mileage data in Table 14.1**

SAS

GENERAL LINEAR MODELS PROCEDURE

DEPENDENT VARIABLE: MILEAGE

SOURCE	DF	SUM OF SQUARES	MEAN SQUARE	F VALUE	PR > F	R-SQUARE	C.V.
MODEL	2	14.44800000[a]	7.22400000[b]	12.38[e]	0.0026[f]	0.733401	3.0194
ERROR	9	5.25200000[c]	0.58355556[d]		ROOT MSE		MILEAGE MEAN
CORRECTED TOTAL	11	19.70000000[g]			0.76390808		25.30000000

SOURCE	DF	TYPE I SS	F VALUE	PR > F	DF	TYPE III SS	F VALUE	PR > F
GASTYPE	2	14.44800000	12.38	0.0026	2	14.44800000	12.38	0.0026

PARAMETER	ESTIMATE	T FOR H0: PARAMETER=0	PR > |T|	STD ERROR OF ESTIMATE
MUB-MUA	1.86000000	3.63	0.0055	0.51244512
MUC-MUA	-0.70000000	-1.20	0.2609	0.58344443
MUC-MUB	-2.56000000[h]	-4.59[i]	0.0013[j]	0.55787958[k]
MUB-(MUC+MUA)/2	2.21000000	4.92	0.0008	0.44923598

OBSERVATION	OBSERVED VALUE	PREDICTED VALUE	RESIDUAL	LOWER 95% CL FOR MEAN	UPPER 95% CL FOR MEAN
1	24.00000000	24.70000000	-0.70000000	23.83595297	25.56404703
2	25.00000000	24.70000000	0.30000000	23.83595297	25.56404703
3	24.30000000	24.70000000	-0.40000000	23.83595297	25.56404703
4	25.50000000	24.70000000	0.80000000	23.83595297	25.56404703
5	25.30000000	26.56000000	-1.26000000	25.78717284[l]	27.33282716[l]
6	26.50000000	26.56000000	-0.06000000	25.78717284	27.33282716
7	26.40000000	26.56000000	-0.16000000	25.78717284	27.33282716
8	27.00000000	26.56000000	0.44000000	25.78717284	27.33282716
9	27.60000000	26.56000000	1.04000000	25.78717284	27.33282716
10	23.30000000	24.00000000	-0.70000000	23.00228443	24.99771557
11	24.00000000	24.00000000	0.00000000	23.00228443	24.99771557
12	24.70000000	24.00000000	0.70000000	23.00228443	24.99771557

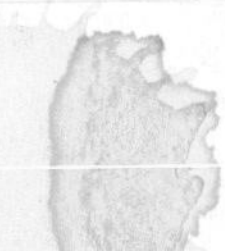

OBSERVATION	OBSERVED VALUE	PREDICTED VALUE	RESIDUAL	LOWER 95% CL INDIVIDUAL	UPPER 95% CL INDIVIDUAL
1	24.00000000	24.70000000	-0.70000000	22.76793211	26.63206789
2	25.00000000	24.70000000	0.30000000	22.76793211	26.63206789
3	24.30000000	24.70000000	-0.40000000	22.76793211	26.63206789
4	25.50000000	24.70000000	0.80000000	22.76793211	26.63206789
5	25.30000000	26.56000000	-1.26000000	24.66696781[m]	28.45303219[m]
6	26.50000000	26.56000000	-0.06000000	24.66696781	28.45303219
7	26.40000000	26.56000000	-0.16000000	24.66696781	28.45303219
8	27.00000000	26.56000000	0.44000000	24.66696781	28.45303219
9	27.60000000	26.56000000	1.04000000	24.66696781	28.45303219
10	23.30000000	24.00000000	-0.70000000	22.00456886	25.99543114
11	24.00000000	24.00000000	0.00000000	22.00456886	25.99543114
12	24.70000000	24.00000000	0.70000000	22.00456886	25.99543114

[a] SS_{means} [b] MS_{means} [c] *SSE* [d] *MSE* [e] *F(means)* [f] *prob-value for F(means)* [g] SS_{total} [h] $\bar{y}_C - \bar{y}_B$

[i] $t = \dfrac{\bar{y}_C - \bar{y}_B}{s\sqrt{(1/n_C) + (1/n_B)}}$ [j] *prob-value for t* [k] $s\sqrt{(1/n_C) + (1/n_B)}$ [l] *95% confidence interval for* μ_B

[m] *95% prediction interval for* $y_{B,0}$

Example 14.9 The Tastee Bakery Company supplies a bakery product to a large number of supermarkets in a metropolitan area. It wishes to study the effect of *shelf display height*, which has levels B (bottom), M (middle), and T (top), on monthly demand for the product in these supermarkets. Here, months are considered to be four-week periods, and demand is measured in cases of ten units each. For l = B, M, and T we define the *lth population of demands* to be the population of all possible monthly demands that could potentially be obtained at supermarkets using display height l. We let μ_l denote the mean of this population, and we call μ_l the mean demand obtained by using display height l. To compare the treatment means μ_B, μ_M, and μ_T, the bakery will employ a completely randomized experimental design. Specifically, for each treatment (l = B, M, and T) the company will randomly select a sample of $n_l = 6$ supermarkets of equal sales potential in the metropolitan area. These supermarkets will sell the bakery product for one month using shelf height l. The demand for the product for this month period will be recorded. For $k = 1, 2, \ldots, 6$ we let

$y_{l,k}$ = the monthly demand observed at the kth supermarket using display height l

Here we assume that for l = B, M, and T the sample of six demands

$$y_{l,1}, y_{l,2}, y_{l,3}, y_{l,4}, y_{l,5}, y_{l,6}$$

is randomly selected from the lth population of demands.

When the company employs this experimental design, it obtains the data in Table 14.3. The SAS output of an analysis of variance of this data is shown in Figure 14.2. To test to see whether or not there are any statistically significant differences between the treatment means μ_B, μ_M, and μ_T, we test the null hypothesis

$$H_0: \mu_B = \mu_M = \mu_T$$

We test H_0 versus the alternative hypothesis

H_1: At least two of μ_B, μ_M, and μ_T differ from each other

TABLE 14.3 **Three samples of monthly demands**

Display height		
B	***M***	***T***
$y_{B,1} = 58.2$	$y_{M,1} = 73.0$	$y_{T,1} = 52.4$
$y_{B,2} = 53.7$	$y_{M,2} = 78.1$	$y_{T,2} = 49.7$
$y_{B,3} = 55.8$	$y_{M,3} = 75.4$	$y_{T,3} = 50.9$
$y_{B,4} = 55.7$	$y_{M,4} = 76.2$	$y_{T,4} = 54.0$
$y_{B,5} = 52.5$	$y_{M,5} = 78.4$	$y_{T,5} = 52.1$
$y_{B,6} = 58.9$	$y_{M,6} = 82.1$	$y_{T,6} = 49.9$
$\bar{y}_B = 55.8$	$\bar{y}_M = 77.2$	$\bar{y}_T = 51.5$

FIGURE 14.2 **SAS output of an analysis of variance of the monthly bakery product demand data in Table 14.3**

DEPENDENT VARIABLE: DEMAND

SOURCE	DF	SUM OF SQUARES	MEAN SQUARE	F VALUE	PR > F	R-SQUARE	C.V.
MODEL	2	2273.88000000	1136.94000000	184.57	0.0001	0.960951	4.0357
ERROR	15	92.40000000	6.16000000		ROOT MSE		DEMAND MEAN
CORRECTED TOTAL	17	2366.28000000			2.48193473		61.50000000

SOURCE	DF	TYPE I SS	F VALUE	PR > F	DF	TYPE III SS	F VALUE	PR > F
DISPLAY	2	2273.88000000	184.57	0.0001	2	2273.88000000	184.57	0.0001

PARAMETER	ESTIMATE	T FOR HO: PARAMETER=0	PR > |T|	STD ERROR OF ESTIMATE
MUT-MUB	-4.30000000	-3.00	0.0090	1.43294568
MUM-MUB	21.40000000	14.93	0.0001	1.43294568
MUT-MUM	-25.70000000	-17.94	0.0001	1.43294568

OBSERVATION	OBSERVED VALUE	PREDICTED VALUE	RESIDUAL	LOWER 95% CL FOR MEAN	UPPER 95% CL FOR MEAN
1	58.20000000	55.80000000	2.40000000	53.64032731	57.95967269
2	53.70000000	55.80000000	-2.10000000	53.64032731	57.95967269
3	55.80000000	55.80000000	0.00000000	53.64032731	57.95967269
4	55.70000000	55.80000000	-0.10000000	53.64032731	57.95967269
5	52.50000000	55.80000000	-3.30000000	53.64032731	57.95967269
6	58.90000000	55.80000000	3.10000000	53.64032731	57.95967269
7	73.00000000	77.20000000	-4.20000000	75.04032731	79.35967269
8	78.10000000	77.20000000	0.90000000	75.04032731	79.35967269
9	75.40000000	77.20000000	-1.80000000	75.04032731	79.35967269
10	76.20000000	77.20000000	-1.00000000	75.04032731	79.35967269
11	78.40000000	77.20000000	1.20000000	75.04032731	79.35967269
12	82.10000000	77.20000000	4.90000000	75.04032731	79.35967269
13	52.40000000	51.50000000	0.90000000	49.34032731	53.65967269
14	49.70000000	51.50000000	-1.80000000	49.34032731	53.65967269
15	50.90000000	51.50000000	-0.60000000	49.34032731	53.65967269
16	54.00000000	51.50000000	2.50000000	49.34032731	53.65967269
17	52.10000000	51.50000000	0.60000000	49.34032731	53.65967269
18	49.90000000	51.50000000	-1.60000000	49.34032731	53.65967269

OBSERVATION	OBSERVED VALUE	PREDICTED VALUE	RESIDUAL	LOWER 95% CL INDIVIDUAL	UPPER 95% CL INDIVIDUAL
1	58.20000000	55.80000000	2.40000000	50.08604314	61.51395686
2	53.70000000	55.80000000	-2.10000000	50.08604314	61.51395686
3	55.80000000	55.80000000	0.00000000	50.08604314	61.51395686
4	55.70000000	55.80000000	-0.10000000	50.08604314	61.51395686
5	52.50000000	55.80000000	-3.30000000	50.08604314	61.51395686
6	58.90000000	55.80000000	3.10000000	50.08604314	61.51395686
7	73.00000000	77.20000000	-4.20000000	71.48604314	82.91395686
8	78.10000000	77.20000000	0.90000000	71.48604314	82.91395686
9	75.40000000	77.20000000	-1.80000000	71.48604314	82.91395686
10	76.20000000	77.20000000	-1.00000000	71.48604314	82.91395686
11	78.40000000	77.20000000	1.20000000	71.48604314	82.91395686
12	82.10000000	77.20000000	4.90000000	71.48604314	82.91395686
13	52.40000000	51.50000000	0.90000000	45.78604314	57.21395686
14	49.70000000	51.50000000	-1.80000000	45.78604314	57.21395686
15	50.90000000	51.50000000	-0.60000000	45.78604314	57.21395686
16	54.00000000	51.50000000	2.50000000	45.78604314	57.21395686
17	52.10000000	51.50000000	0.60000000	45.78604314	57.21395686
18	49.90000000	51.50000000	-1.60000000	45.78604314	57.21395686

We see from the SAS output that

$$F(\text{means}) = 184.57 > F_{[\alpha]}^{(v-1,n-v)} = F_{[.05]}^{(3-1,18-3)} = F_{[.05]}^{(2,15)} = 3.68$$

and that prob-value = .0001 < α = .05. Therefore we can reject H_0 in favor of H_1 by setting α equal to .05. We conclude that at least two of the treatment means μ_B, μ_M, and μ_T differ. This says that at least two of shelf display heights B, M, and T have different effects on mean demand.

To investigate the exact nature of the differences between μ_B, μ_M, and μ_T, we consider the t statistics corresponding to $\mu_T - \mu_B$, $\mu_M - \mu_B$, and $\mu_T - \mu_M$. These differences are denoted by MUT − MUB, MUM − MUB, and MUT − MUM on the SAS output. Each of these t statistics is greater than $t_{[\alpha/2]}^{(n-v)} = t_{[.025]}^{(18-3)} = 2.131$ (and each of the associated prob-values is less than .05). Therefore if we set $\alpha = .05$ for each t-test, we conclude that each of the differences $\mu_T - \mu_B$, $\mu_M - \mu_B$, and $\mu_T - \mu_M$ is statistically significant (that is, different from zero).

Since we have concluded that μ_T and μ_B differ, we compute a 95 percent confidence interval for $\mu_T - \mu_B$. Note that the standard error of the estimate $\bar{y}_T - \bar{y}_B$ is given on the SAS output. The 95 percent confidence interval is

$$\begin{aligned}\left[(\bar{y}_T - \bar{y}_B) \pm t_{[\alpha/2]}^{(n-v)} s\sqrt{\frac{1}{n_T} + \frac{1}{n_B}}\right] &= \left[(51.5 - 55.8) \pm t_{[.025]}^{(18-3)}(2.4819)\sqrt{\tfrac{1}{6} + \tfrac{1}{6}}\right] \\ &= [-4.3 \pm 2.131(1.43294568)] \\ &= [-4.3 \pm 3.0536] \\ &= [-7.3536, -1.2464]\end{aligned}$$

This interval says that the company can be 95 percent confident that μ_B, the mean monthly demand obtained by using a bottom display height, is between 1.2464 and 7.3536 cases greater than μ_T, the mean monthly demand obtained by using a top display height.

Also, since we have concluded that μ_M and μ_B differ, we compute a 95% confidence interval for $\mu_M - \mu_B$. This interval is

$$\begin{aligned}\left[(\bar{y}_M - \bar{y}_B) \pm t_{[\alpha/2]}^{(n-v)} s\sqrt{\frac{1}{n_M} + \frac{1}{n_B}}\right] &= \left[(77.2 - 55.8) \pm t_{[.025]}^{(18-3)}(2.4819)\sqrt{\tfrac{1}{6} + \tfrac{1}{6}}\right] \\ &= [21.4 \pm 2.131(1.43294568)] \\ &= [21.4 \pm 3.0536] \\ &= [18.3464, 24.4536]\end{aligned}$$

This interval says that the company can be 95% confident that μ_M, the mean monthly demand obtained by using a middle display height, is between 18.3464 and 24.4536 cases greater than μ_B, the mean monthly demand obtained by using a bottom display height.

Finally, since we have concluded that μ_T and μ_M differ, we compute a 95% confidence interval for $\mu_T - \mu_M$. This interval is

$$\left[(\bar{y}_T - \bar{y}_M) \pm t_{[\alpha/2]}^{(n-v)} s\sqrt{\frac{1}{n_T} + \frac{1}{n_M}}\right]$$
$$= [(51.5 - 77.2) \pm t_{[.025]}^{(18-3)}(2.4819)\sqrt{\tfrac{1}{6} + \tfrac{1}{6}}]$$
$$= [-25.7 \pm 2.131(1.43294568)]$$
$$= [-25.7 \pm 3.0536]$$
$$= [-28.7536, -22.6464]$$

This interval says that the company can be 95 percent confident that μ_M, the mean monthly demand obtained by using a middle display height, is between 22.6464 and 28.7536 cases greater than μ_T, the mean monthly demand obtained by using a top display height.

On the basis of the above results, we have strong evidence that μ_M is substantially greater than both μ_B and μ_T. Therefore suppose that the company decides to use a middle display height. Then a point estimate of μ_M is $\bar{y}_M = 77.2$, and a 95% confidence interval for μ_M is

$$\left[\bar{y}_M \pm t_{[.025]}^{(15)}\left(\frac{s}{\sqrt{n_M}}\right)\right] = \left[77.2 \pm 2.131\left(\frac{2.4819}{\sqrt{6}}\right)\right]$$
$$= [77.2 \pm 2.1592]$$
$$= [75.0408, 79.3592]$$

Note that this interval is included in the SAS output. This interval says that the company can be 95 percent confident that μ_M is between 75.0408 and 79.3592 cases. Furthermore, a point prediction of an individual monthly demand obtained when using a middle display height is $\bar{y}_M = 77.2$. A 95% prediction interval for this individual demand is

$$\left[\bar{y}_M \pm t_{[.025]}^{(15)}\left(s\sqrt{1 + \frac{1}{n_M}}\right)\right] = [77.2 \pm 2.131(2.4819)\sqrt{1 + \tfrac{1}{6}}]$$
$$= [77.2 \pm 5.7127]$$
$$= [71.4873, 82.9127]$$

Note that this interval is also included on the SAS output. This interval says that the company can be 95 percent confident that $y_{M,0}$, an individual monthly demand in a supermarket using a middle display height, will be between 71.4873 and 82.9127 cases.

MULTIPLE COMPARISONS USING SIMULTANEOUS CONFIDENCE INTERVALS

We can use the individual confidence intervals of the preceding subsection to compare all pairs (and other linear combinations) of treatment means. However, there is an objection to this procedure. While the confidence level for each individual interval may be 95%, it is not possible to determine how confident we can be that all of the intervals are (simultaneously) correct. The *Tukey, Scheffé,* and *Bonferroni formulas* of this section can be used to determine an overall confidence level.

Suppose that we are interested in studying all pairwise differences between treatment means (and not more general linear combinations of treatment means). Then the Tukey formula yields the most precise simultaneous $100(1 - \alpha)\%$ confidence intervals. The Tukey formula assumes that all sample sizes are equal: $n_1 = n_2 = \cdots = n_v$. We denote the common sample size by the symbol m.

Tukey Simultaneous Confidence Intervals for All Possible Pairs of Treatment Means

A *Tukey simultaneous* $100(1 - \alpha)\%$ *confidence interval for* $\mu_i - \mu_j$ in the set of all possible pairwise differences between treatment means is

$$\left[(\bar{y}_i - \bar{y}_j) \pm q_{[\alpha]}(v, n - v)\frac{s}{\sqrt{m}}\right]$$

Here, $q_{[\alpha]}(v, n - v)$ is obtained from the *table of percentage points of the studentized range* (see Table E.7 in Appendix E).

Example 14.10 Consider the bakery product demand data of Example 14.9. Here, $v = 3$, m $(= n_B = n_M = n_T) = 6$ (see Table 14.3), $n = 18$, and $s = 2.4819$. It follows that a Tukey simultaneous 95% confidence interval for $\mu_i - \mu_j$ in the set of all possible pairwise differences between μ_B, μ_M, and μ_T is

$$\begin{aligned}\left[(\bar{y}_i - \bar{y}_j) \pm q_{[.05]}(v, n - v)\frac{s}{\sqrt{m}}\right] &= \left[(\bar{y}_i - \bar{y}_j) \pm q_{[.05]}(3, 15)\frac{2.4819}{\sqrt{6}}\right] \\ &= \left[(\bar{y}_i - \bar{y}_j) \pm 3.67\frac{2.4819}{\sqrt{6}}\right] \\ &= [(\bar{y}_i - \bar{y}_j) \pm 3.7186]\end{aligned}$$

Specifically, Tukey simultaneous 95% confidence intervals for $\mu_T - \mu_B$, $\mu_M - \mu_B$, and $\mu_T - \mu_M$ are

$$\begin{aligned}[(\bar{y}_T - \bar{y}_B) \pm 3.7186] &= [(51.5 - 55.8) \pm 3.7186] \\ &= [-4.3 \pm 3.7186] \\ &= [-8.0186, -.5814]\end{aligned}$$

$$\begin{aligned}[(\bar{y}_M - \bar{y}_B) \pm 3.7186] &= [(77.2 - 55.8) \pm 3.7186] \\ &= [21.4 \pm 3.7186] \\ &= [17.6814, 25.1186]\end{aligned}$$

and

$$\begin{aligned}[(\bar{y}_T - \bar{y}_M) \pm 3.7186] &= [(51.5 - 77.2) \pm 3.7186] \\ &= [-25.7 \pm 3.7186] \\ &= [-29.4186, -21.9814]\end{aligned}$$

The assumption of equal sample sizes in the Tukey formula can be somewhat relaxed. When the sample sizes do not differ drastically (say, if the largest n_l is no more than twice the smallest), we can modify the Tukey formula. Here, we let

$$m = \frac{1}{\dfrac{1}{n_1} + \dfrac{1}{n_2} + \cdots + \dfrac{1}{n_v}}$$

which is the *harmonic mean* of the sample sizes. However, if the sample sizes are substantially different, we can use the Scheffé approach. This approach also allows us to compute a simultaneous confidence interval for a linear combination of means that is not a paired difference. For instance, we might compute a simultaneous interval for

$$\mu_1 - \frac{(\mu_2 + \mu_3)}{2}$$

Scheffé Simultaneous Confidence Intervals

1. We define a *contrast* to be any linear combination $\Sigma_{l=1}^{v} a_l\mu_l$ such that $\Sigma_{l=1}^{v} a_l = 0$. Suppose that we wish to find a *Scheffé simultaneous 100(1 − α)% confidence interval for a contrast in the set of all possible contrasts*. Then:
   a. The *Scheffé interval for the difference* $\mu_i - \mu_j$ (which is a contrast) is

   $$\left[(\bar{y}_i - \bar{y}_j) \pm \sqrt{(v-1)F_{[\alpha]}^{(v-1,n-v)}}\, s\sqrt{\frac{1}{n_i} + \frac{1}{n_j}}\right]$$

   b. The *Scheffé interval for the contrast* $\Sigma_{l=1}^{v} a_l\mu_l$ is

   $$\left[\sum_{l=1}^{v} a_l\bar{y}_l \pm \sqrt{(v-1)F_{[\alpha]}^{(v-1,n-v)}}\, s\sqrt{\sum_{l=1}^{v}\frac{a_l^2}{n_l}}\right]$$

2. Suppose that we wish to find a *Scheffé simultaneous 100(1 − α)% confidence interval for a linear combination in the set of all possible linear combinations* (some of which are not contrasts). Then,
   a. The *Scheffé interval for the difference* $\mu_i - \mu_j$ is

   $$\left[(\bar{y}_i - \bar{y}_j) \pm \sqrt{vF_{[\alpha]}^{(v,n-v)}}\, s\sqrt{\frac{1}{n_i} + \frac{1}{n_j}}\right]$$

b. The *Scheffé interval for the linear combination* $\Sigma_{l=1}^{v} a_l \mu_l$ is

$$\left[\sum_{l=1}^{v} a_l \bar{y}_l \pm \sqrt{v F_{[\alpha]}^{(v,n-v)}}\, s \sqrt{\sum_{l=1}^{v} \frac{a_l^2}{n_l}}\right]$$

The choice of one of the above formulas requires that we make a decision *before we observe the samples*. We must decide whether we are interested in

1. Finding simultaneous confidence intervals for *linear combinations, all of which are contrasts*, in which case we use formulas 1a and 1b.
2. Finding simultaneous confidence intervals for *linear combinations, some of which are not contrasts*, in which case we use formulas 2a and 2b.

Of course, we will not literally calculate Scheffé simultaneous confidence intervals for *all possible contrasts* (*or more general linear combinations*). However, the Scheffé simultaneous confidence interval formula applies to all possible contrasts (or more general linear combinations). This allows us to "*data snoop*." Data snooping means that we will let the data suggest which contrasts or linear combinations we will investigate further. Remember, however, we must decide whether we will study contrasts or more general linear combinations before we observe the data.

Example 14.11 Consider the North American Oil Company problem. Suppose that we had decided—before we observed the gasoline mileage data in Table 14.1—that we wished to find Scheffé simultaneous 95% confidence intervals for all contrasts in the following set of contrasts

Set I

$\mu_B - \mu_A$

$\mu_C - \mu_A$

$\mu_C - \mu_B$

$$\mu_B - \left(\frac{\mu_C + \mu_A}{2}\right)$$

Suppose that we also wish to find such intervals for other contrasts that the data might suggest. That is, we are considering all possible contrasts. To verify, for example, that $\mu_B - (\mu_C + \mu_A)/2$ is a contrast, note that

$$\begin{aligned}\mu_B - \frac{\mu_C + \mu_A}{2} &= -\frac{1}{2}\cdot\mu_A + 1\cdot\mu_B - \frac{1}{2}\cdot\mu_C \\ &= a_A\mu_A + a_B\mu_B + a_C\mu_C\end{aligned}$$

Here, $a_A = -\frac{1}{2}$, $a_B = 1$, and $a_C = -\frac{1}{2}$, which implies that

$$\begin{aligned}\sum_{l=A,B,C} a_l &= a_A + a_B + a_C \\ &= -\frac{1}{2} + 1 - \frac{1}{2} \\ &= 0\end{aligned}$$

Therefore

$$\begin{aligned}\sum_{l=A,B,C} \frac{a_l^2}{n_l} &= \frac{a_A^2}{n_A} + \frac{a_B^2}{n_B} + \frac{a_C^2}{n_C} \\ &= \frac{(-\frac{1}{2})^2}{4} + \frac{(1)^2}{5} + \frac{(-\frac{1}{2})^2}{3} \\ &= .3458\end{aligned}$$

It follows that a Scheffé simultaneous 95% confidence interval for $\mu_B - (\mu_C + \mu_A)/2$ is (using formula 1b)

$$\begin{aligned}&\left[\bar{y}_B - \frac{\bar{y}_C + \bar{y}_A}{2} \pm \sqrt{(v-1)F_{[.05]}^{(v-1,n-v)}}\, s \sqrt{\sum_{l=A,B,C} \frac{a_l^2}{n_l}}\right] \\ &= \left[26.56 - \frac{24 + 24.7}{2} \pm \sqrt{(3-1)F_{[.05]}^{(3-1,12-3)}}(.7639)\sqrt{.3458}\right] \\ &= [2.21 \pm \sqrt{2(4.26)}(.7639)\sqrt{.3458}] \\ &= [.8988, 3.5212]\end{aligned}$$

Furthermore, using formula 1a, it can be verified that the Scheffé simultaneous 95% confidence intervals for the remaining contrasts in Set I are

For $\mu_B - \mu_A$: [.3644, 3.3556]

For $\mu_C - \mu_A$: [−2.4029, 1.0029]

For $\mu_C - \mu_B$: [−4.1885, −.9315]

Next, suppose that we had decided—before we observed the gasoline mileage data in Table 14.1—that we wished to calculate Scheffé simultaneous 95% confidence intervals for all the linear combinations in Set II:

Set II

μ_A

μ_B

μ_C

$\mu_B - \mu_A$

$\mu_C - \mu_A$

$\mu_C - \mu_B$

$\mu_B - \dfrac{\mu_C + \mu_A}{2}$

In addition, suppose that we wish to find such intervals for other linear combinations that the data might suggest. Note that μ_A, μ_B, and μ_C are not contrasts. That is, these means cannot be written as $\Sigma_{l=A,B,C}\, a_l\mu_l$, where

$$\sum_{l=A,B,C} a_l = 0$$

For example,

$$\mu_B = (0)\mu_A + (1)\mu_B + (0)\mu_C$$

which implies that

$$\begin{aligned}\sum_{l=A,B,C} a_l &= 0 + 1 + 0 \\ &= 1\end{aligned}$$

Therefore we must use formulas 2a and 2b to calculate Scheffé intervals. For example, a Scheffé simultaneous 95% confidence interval for $\mu_B - (\mu_C + \mu_A)/2$ is (using formula 2b)

$$\begin{aligned}&\left[\bar{y}_B - \frac{\bar{y}_C + \bar{y}_A}{2} \pm \sqrt{vF_{[.05]}^{(v,n-v)}}\, s\sqrt{\sum_{l=A,B,C}\frac{a_l^2}{n_l}}\right] \\ &= \left[26.56 - \left(\frac{24 + 24.7}{2}\right) \pm \sqrt{3F_{[.05]}^{(3.12-3)}}(.7639)\sqrt{.3458}\right] \\ &= [2.21 \pm \sqrt{3(3.86)}(.7639)\sqrt{.3458}] \\ &= [.6814, 3.7386]\end{aligned}$$

As another example, a Scheffé simultaneous 95% confidence interval for

$$\mu_B = (0)\mu_A + (1)\mu_B + (0)\mu_C$$

is

$$\begin{aligned}\left[\sum_{l=1}^{v} a_l\bar{y}_l \pm \sqrt{vF_{[\alpha]}^{(v,n-v)}}\, s\sqrt{\sum_{l=1}^{v}\frac{a_l^2}{n_l}}\right] &= [\bar{y}_B \pm \sqrt{3F_{[.05]}^{(3,12-3)}}\, s\sqrt{.20}] \\ &= [26.56 \pm \sqrt{3(3.86)}(.7639)\sqrt{.20}] \\ &= [26.56 \pm 1.16] = [25.4, 27.72]\end{aligned}$$

Here,

$$\sum_{l=A,B,C}\frac{a_l^2}{n_l} = \frac{(0)^2}{4} + \frac{(1)^2}{5} + \frac{(0)^2}{3} = .20$$

Furthermore, by using formulas 2a and 2b it can be verified that Scheffé simultaneous 95% confidence intervals for the remaining linear combinations in Set II are

For μ_A: [23.4, 26]

For μ_C: [22.50, 25.50]

For $\mu_B - \mu_A$: [.1164, 3.6036]

For $\mu_C - \mu_A$: [−2.6853, 1.2853]

For $\mu_C - \mu_B$: [−4.4585, −.6615]

To conclude this example, note that the Scheffé 95% intervals for the *contrasts* in Set II are less precise than the Scheffé 95% intervals for the contrasts in Set I. This is because we calculated the simultaneous intervals for the contrasts in Set II by using formulas 2a and 2b. These formulas compute intervals for linear combinations in the set of all possible linear combinations (some of which are not contrasts). Thus formulas 2a and 2b employ the quantity $vF_{[\alpha]}^{(v,n-v)}$. This quantity is larger than the quantity $(v - 1)F_{[\alpha]}^{(v-1,n-v)}$, which has been employed by formulas 1a and 1b to calculate the intervals for the contrasts in Set I.

We next present the *Bonferroni* formula.

Bonferroni Simultaneous Confidence Intervals

1. A *Bonferroni simultaneous 100(1 − α)% confidence interval for* $\mu_i - \mu_j$ *in a prespecified set of g linear combinations* is

$$\left[(\bar{y}_i - \bar{y}_j) \pm t_{[\alpha/2g]}^{(n-v)} s\sqrt{\frac{1}{n_i} + \frac{1}{n_j}}\right]$$

2. A *Bonferroni simultaneous 100(1 − α)% confidence interval for* $\Sigma_{l=1}^{v} a_l\mu_l$ *in a prespecified set of g linear combinations* is

$$\left[\sum_{l=1}^{v} a_l\bar{y}_l \pm t_{[\alpha/2g]}^{(n-v)} s\sqrt{\sum_{l=1}^{v} \frac{a_l^2}{n_l}}\right]$$

The Tukey formula applies to all possible pairwise differences between treatment means. The Scheffé formula applies to all possible contrasts, or more general linear combinations, concerning treatment means. Therefore these formulas can be used for data snooping. However, the Bonferroni formula requires that we prespecify —before we observe the data—a set of g linear combinations. Thus this formula does not allow us to data snoop.

Example 14.12 Suppose that we had decided—before we observed the gasoline mileage data in Table 14.1—that we wished to calculate Bonferroni simultaneous 95% confidence intervals for all of the linear combinations in Set I:

Set I

$\mu_B - \mu_A$

$\mu_C - \mu_A$

$$\mu_C - \mu_B$$

$$\mu_B - \frac{\mu_C + \mu_A}{2}$$

Since there are $g = 4$ linear combinations in Set I, we use

$$t_{[\alpha/2g]}^{(n-v)} = t_{[.05/2(4)]}^{(12-3)} = t_{[.00625]}^{(9)}$$

Although this t-point is not in Table E.2 in Appendix E, we see that $t_{[.00625]}^{(9)}$ is between

$$t_{[.0075]}^{(9)} = 2.998 \qquad \text{and} \qquad t_{[.005]}^{(9)} = 3.250$$

To determine an approximate value for $t_{[.00625]}^{(9)}$, we could linearly interpolate between these known t-points. Alternatively, to be conservative, we could use $t_{[.005]}^{(9)} = 3.250$ in place of $t_{[.00625]}^{(9)}$. By doing the latter, for example, an approximate Bonferroni simultaneous 95% confidence interval for $\mu_B - (\mu_C + \mu_A)/2$ is (using formula 2 above)

$$\begin{aligned}
\left[\bar{y}_B - \frac{\bar{y}_C + \bar{y}_A}{2} \pm t_{[\alpha/2g]}^{(n-v)} s \sqrt{\sum_{l=A,B,C} \frac{a_l^2}{n_l}}\right]
&= \left[26.56 - \left(\frac{24 + 24.7}{2}\right) \pm t_{[.00625]}^{(9)}(.7639)\sqrt{.3458}\right] \\
&\approx [2.21 \pm 3.250(.7639)\sqrt{.3458}] \\
&= [2.21 \pm 1.4599] \\
&= [.7501, 3.6699]
\end{aligned}$$

This interval is longer than the Scheffé simultaneous 95% confidence interval for $\mu_B - (\mu_C + \mu_A)/2$ (in Set I). This Scheffé interval was calculated in Example 14.11 to be [.8988, 3.5212]. In fact, the t-point used to calculate the Bonferroni intervals for the linear combinations in Set I is

$$t_{[\alpha/2g]}^{(n-v)} = t_{[.00625]}^{(9)}$$

(which is between 2.998 and 3.250). This t-point is larger than

$$\sqrt{(v - 1)F_{[\alpha]}^{(v-1,n-v)}} = \sqrt{(3 - 1)F_{[.05]}^{(3-1,12-3)}} = 2.9189$$

which is used to calculate the Scheffé intervals for the same linear combinations. Therefore each Bonferroni interval is longer than the corresponding Scheffé interval.

As another example, consider calculating Bonferroni simultaneous 95% confidence intervals for each prespecified linear combination in Set II:

Set II

$$\mu_A$$

$$\mu_B$$

$$\mu_C$$

$\mu_B - \mu_A$

$\mu_C - \mu_A$

$\mu_C - \mu_B$

$$\mu_B - \frac{\mu_C + \mu_A}{2}$$

It can be verified that each of these Bonferroni intervals is longer than the corresponding Scheffé simultaneous 95% confidence interval. On the other hand, suppose that we have prespecified the $g = 2$ contrasts in Set III:

Set III

$\mu_C - \mu_B$

$$\mu_B - \frac{\mu_C + \mu_A}{2}$$

Then the t-point

$$t^{(n-v)}_{[\alpha/2g]} = t^{(12-3)}_{[.05/2(2)]} = t^{(9)}_{[.0125]}$$

(which is smaller than $t^{(9)}_{[.01]} = 2.821$) is used to calculate Bonferroni simultaneous 95% confidence intervals. This t-point is less than the quantity

$$\sqrt{(v-1)F^{(v-1,n-v)}_{[\alpha]}} = \sqrt{(3-1)F^{(3-1,12-3)}_{[.05]}} = 2.9189$$

that is used to calculate Scheffé simultaneous 95% confidence intervals. It follows that each Bonferroni interval would be shorter than the corresponding Scheffé interval.

We now summarize the use of Tukey, Scheffé, and Bonferroni simultaneous confidence intervals.

1. If we are interested in all pairwise comparisons of treatment means, the Tukey formula will give shorter intervals than will the Scheffé or Bonferroni formulas. If a small number of prespecified pairwise comparisons are of interest, the Bonferroni formula might give shorter intervals in some situations.
2. If we are interested in all contrasts (or more general linear combinations) of treatment means, the Scheffé formula should be used. If a small number of prespecified contrasts (or more general linear combinations) are of interest, the Bonferroni formula might give shorter intervals. This is particularly true if the number of prespecified contrasts (or more general linear combinations) is less than or equal to the number of treatments.
3. Whereas the Tukey and Scheffé formulas can be used for data snooping, the Bonferroni formula cannot.

4. It is reasonable in any given problem to use all of the formulas (Tukey, Scheffé, and Bonferroni) that apply. Then we can choose the formula that provides the shortest intervals.

We need to make several additional comments concerning simultaneous intervals. First, multiple comparisons can also be done by using hypothesis tests. Suppose, for example, that we wish to make several pairwise comparisons. We might perform several individual t-tests each with a probability of a Type I error set equal to α. Here, we say that we are setting the *comparisonwise error rate* equal to α. However, in such a case the overall probability of making at least one Type I error (called the *experimentwise error rate*) is larger than α. To control the experimentwise error rate, we can carry out hypothesis tests based on the Bonferroni, Tukey, or Scheffé methods. The rejection rules for these tests are simple modifications of the simultaneous confidence interval formulas. For example, suppose that we wish to test the null hypothesis

$$H_0: \mu_i - \mu_j = 0 \quad \text{or} \quad H_0: \mu_i = \mu_j$$

which says that the effects of treatments i and j are the same, versus the alternative hypothesis

$$H_1: \mu_i - \mu_j \neq 0 \quad \text{or} \quad H_1: \mu_i \neq \mu_j$$

which says that the effects of treatments i and j are different. Using the Bonferroni method, we would declare the difference between μ_i and μ_j to be statistically significant if and only if

$$|\bar{y}_i - \bar{y}_j| > t_{[\alpha/2g]}^{(n-v)} s \sqrt{\frac{1}{n_i} + \frac{1}{n_j}}$$

Here, g is the number of pairwise comparisons in a prespecified set. The Tukey method declares the difference between μ_i and μ_j to be statistically significant if and only if

$$|\bar{y}_i - \bar{y}_j| > q_{[\alpha]}(v, n - v) \frac{s}{\sqrt{m}}$$

Here, we are controlling the experimentwise error rate over all possible pairwise comparisons of treatment means. The Scheffé method declares the difference between μ_i and μ_j to be statistically significant if and only if

$$|\bar{y}_i - \bar{y}_j| > \sqrt{(v - 1) F_{[\alpha]}^{(v-1,n-v)}}\, s \sqrt{\frac{1}{n_i} + \frac{1}{n_j}}$$

In this case we are controlling the experimentwise error rate over all null hypotheses that set a contrast $\sum_{l=1}^{v} a_l \mu_l$ equal to zero.

14.5

THE REGRESSION APPROACH

Recall that the one-factor model is

$$y_{l,k} = \mu_l + \varepsilon_{l,k}$$

To write this model as a regression model, we write μ_l as a function of dummy variables:

$$\mu_l = \beta_1 + \beta_2 D_{l,2} + \beta_3 D_{l,3} + \cdots + \beta_v D_{l,v}$$

where

$$D_{l,2} = \begin{cases} 1 & \text{if } l = 2\text{; that is, if we use treatment 2} \\ 0 & \text{otherwise} \end{cases}$$

$$D_{l,3} = \begin{cases} 1 & \text{if } l = 3\text{; that is, if we use treatment 3} \\ 0 & \text{otherwise} \end{cases}$$

$$\vdots$$

$$D_{l,v} = \begin{cases} 1 & \text{if } l = v\text{; that is, if we use treatment } v \\ 0 & \text{otherwise} \end{cases}$$

Using the definitions of the dummy variables, we can express each of the treatment means $\mu_1, \mu_2, \ldots, \mu_v$ as linear combinations of the parameters $\beta_1, \beta_2, \ldots, \beta_v$. That is,

$$\mu_1 = \beta_1 + \beta_2(0) + \beta_3(0) + \cdots + \beta_v(0) = \beta_1$$

is the mean value of the response variable when we are using treatment 1;

$$\mu_2 = \beta_1 + \beta_2(1) + \beta_3(0) + \cdots + \beta_v(0) = \beta_1 + \beta_2$$

is the mean value of the response variable when we are using treatment 2;

$$\mu_3 = \beta_1 + \beta_2(0) + \beta_3(1) + \cdots + \beta_v(0) = \beta_1 + \beta_3$$

is the mean value of the response variable when we are using treatment 3; and so forth. Finally,

$$\mu_v = \beta_1 + \beta_2(0) + \beta_3(0) + \cdots + \beta_v(1) = \beta_1 + \beta_v$$

is the mean value of the response variable when we are using treatment v.

To summarize, $\beta_1 = \mu_1$ and $\mu_l = \beta_1 + \beta_l$ for $l = 2, 3, \ldots, v$. It follows that

$$\begin{aligned} \beta_l &= \mu_l - \beta_1 \\ &= \mu_l - \mu_1 \\ &= \text{the change in the mean response associated with changing from treatment 1 to treatment } l \end{aligned}$$

We call the regression model

$$\begin{aligned} y_{l,k} &= \mu_l + \varepsilon_{l,k} \\ &= \beta_1 + \beta_2 D_{l,2} + \beta_3 D_{l,3} + \cdots + \beta_v D_{l,v} + \varepsilon_{l,k} \end{aligned}$$

the *means model*. To compare the treatment means, we first calculate the least squares point estimates of the model parameters. It can be proven that the least squares point estimate of

$$\beta_1 = \mu_1$$

is $b_1 = \bar{y}_1$. This is the sample mean corresponding to treatment 1, which is the intuitive point estimate of μ_1. Similarly, it can be proven that the least squares point estimate of

$$\beta_l = \mu_l - \mu_1 \qquad (\text{for } l = 2, 3, \ldots, v)$$

is $b_l = \bar{y}_l - \bar{y}_1$. This is the difference between the sample mean corresponding to treatment l and the sample mean corresponding to treatment 1. Here, $\bar{y}_l - \bar{y}_1$ is the intuitive point estimate of $\mu_l - \mu_1$. Furthermore, the point estimate of μ_l, the mean response for treatment l, and the point prediction of an individual response $y_l = \mu_l + \varepsilon_{l,k}$ is

$$\hat{y}_l = b_1 + b_2 D_{l,2} + b_3 D_{l,3} + \cdots + b_v D_{l,v}$$

Specifically,

1. The point estimate of $\mu_1 = \beta_1$ and the point prediction of

$$\begin{aligned} y_{1,k} &= \mu_1 + \varepsilon_{1,k} \\ &= \beta_1 + \varepsilon_{1,k} \end{aligned}$$

is

$$\hat{y}_1 = b_1 = \bar{y}_1$$

2. For $l = 2, 3, \ldots, v$ the point estimate of

$$\mu_l = \beta_1 + \beta_l$$

and the point prediction of

$$\begin{aligned} y_{l,k} &= \mu_l + \varepsilon_{l,k} \\ &= \beta_1 + \beta_l + \varepsilon_{l,k} \end{aligned}$$

is

$$\hat{y}_l = b_1 + b_l = \bar{y}_1 + (\bar{y}_l - \bar{y}_1) = \bar{y}_l$$

Therefore $\bar{y}_l$ is the least squares point estimate of μ_l and the point prediction of $y_{l,k} = \mu_l + \varepsilon_{l,k}$.

Example 14.13 ***Part 1: Developing the Means Model***

Consider the North American Oil Company situation of Examples 14.2 and 14.3. Recall that we wish to compare μ_A, μ_B, and μ_C, the mean mileages obtained by using gasoline types A, B, and C, respectively. Considering the experimental data in Table 14.1, we let

$y_{l,k}$ = the gasoline mileage obtained by the kth Fire-Hawk that is test driven using gasoline type l

We define the means model for this problem to be

$$\begin{aligned} y_{l,k} &= \mu_l + \varepsilon_{l,k} \\ &= \beta_A + \beta_B D_{l,B} + \beta_C D_{l,C} + \varepsilon_{l,k} \end{aligned}$$

Here,

$$D_{l,B} = \begin{cases} 1 & \text{if } l = \text{B; that is, if we use gasoline type B} \\ 0 & \text{otherwise} \end{cases}$$

$$D_{l,C} = \begin{cases} 1 & \text{if } l = \text{C; that is, if we use gasoline type C} \\ 0 & \text{otherwise} \end{cases}$$

We now use the equation above for μ_l to express μ_A, μ_B, and μ_C in terms of the parameters β_A, β_B, and β_C. If l = A, then

$$\begin{aligned} \mu_A &= \beta_A + \beta_B D_{A,B} + \beta_C D_{A,C} \\ &= \beta_A + \beta_B(0) + \beta_C(0) \\ &= \beta_A \end{aligned}$$

Thus

$$\beta_A = \mu_A = \text{the mean mileage obtained by using gasoline type A}$$

Next, if l = B, then

$$\begin{aligned} \mu_B &= \beta_A + \beta_B D_{B,B} + \beta_C D_{B,C} \\ &= \beta_A + \beta_B(1) + \beta_C(0) \\ &= \beta_A + \beta_B \end{aligned}$$

This implies (since $\beta_A = \mu_A$) that

$$\mu_B = \mu_A + \beta_B$$

Thus

$\beta_B = \mu_B - \mu_A$ = the difference between μ_B, the mean mileage obtained by using gasoline type *B*, and μ_A, the mean mileage obtained by using gasoline type A

Finally, if l = C, then

$$\begin{aligned}\mu_C &= \beta_A + \beta_B D_{C,B} + \beta_C D_{C,C}\\ &= \beta_A + \beta_B(0) + \beta_C(1)\\ &= \beta_A + \beta_C\end{aligned}$$

This implies that

$$\mu_C = \mu_A + \beta_C$$

Thus

$$\beta_C = \mu_C - \mu_A = \text{the difference between } \mu_C\text{, the mean mileage obtained by using gasoline type C, and } \mu_A\text{, the mean mileage obtained by using gasoline type A}$$

Note that the equation describing μ_l uses the intercept β_A to describe the effect of the mean μ_A and uses a separate dummy variable to describe the effect of each of the other means μ_B and μ_C.

Part 2: Estimating the Regression Parameters

To calculate the least squares point estimates of β_A, β_B, and β_C, we use

$$\mathbf{y} = \begin{bmatrix} y_{A,1}\\ y_{A,2}\\ y_{A,3}\\ y_{A,4}\\ y_{B,1}\\ y_{B,2}\\ y_{B,3}\\ y_{B,4}\\ y_{B,5}\\ y_{C,1}\\ y_{C,2}\\ y_{C,3}\end{bmatrix} = \begin{bmatrix} 24.0\\ 25.0\\ 24.3\\ 25.5\\ 25.3\\ 26.5\\ 26.4\\ 27.0\\ 27.6\\ 23.3\\ 24.0\\ 24.7\end{bmatrix} \quad \text{and} \quad \mathbf{X} = \begin{array}{c} \begin{array}{ccc} 1 & D_{l,B} & D_{l,C} \end{array}\\ \begin{bmatrix} 1 & 0 & 0\\ 1 & 0 & 0\\ 1 & 0 & 0\\ 1 & 0 & 0\\ 1 & 1 & 0\\ 1 & 1 & 0\\ 1 & 1 & 0\\ 1 & 1 & 0\\ 1 & 1 & 0\\ 1 & 0 & 1\\ 1 & 0 & 1\\ 1 & 0 & 1\end{bmatrix}\end{array}$$

Using these data, we can calculate $(\mathbf{X}'\mathbf{X})^{-1}$ and $\mathbf{X}'\mathbf{y}$ to be

$$(\mathbf{X}'\mathbf{X})^{-1} = \begin{array}{c} \\ A\\ B\\ C \end{array}\begin{array}{c} \begin{array}{ccc} A & B & C \end{array}\\ \begin{bmatrix} .25 & -.25 & -.25\\ -.25 & .45 & .25\\ -.25 & .25 & .58333 \end{bmatrix}\end{array} = \begin{bmatrix} c_{AA} & & \\ & c_{BB} & \\ & & c_{CC} \end{bmatrix}$$

and

$$\mathbf{X}'\mathbf{y} = \begin{bmatrix} 303.6 \\ 132.8 \\ 72.0 \end{bmatrix}$$

It follows that

$$\begin{bmatrix} b_A \\ b_B \\ b_C \end{bmatrix} = \mathbf{b} = (\mathbf{X}'\mathbf{X})^{-1}\mathbf{X}'\mathbf{y} = \begin{bmatrix} 24.7 \\ 1.86 \\ -.7 \end{bmatrix}$$

These least squares point estimates yield the prediction equation

$$\begin{aligned} \hat{y}_l &= b_A + b_B D_{l,B} + b_C D_{l,C} \\ &= 24.7 + 1.86 D_{l,B} - .7 D_{l,C} \end{aligned}$$

Here, $\hat{y}_l$ is the point estimate of the mean mileage obtained by using gasoline type l,

$$\mu_l = \beta_A + \beta_B D_{l,B} + \beta_C D_{l,C}$$

and the point prediction of the kth mileage observed when using gasoline type l,

$$\begin{aligned} y_{l,k} &= \mu_l + \varepsilon_{l,k} \\ &= \beta_A + \beta_B D_{l,B} + \beta_C D_{l,C} + \varepsilon_{l,k} \end{aligned}$$

Specifically,

1. The point estimate of

$$\mu_A = \beta_A$$

and the point prediction of

$$\begin{aligned} y_{A,k} &= \mu_A + \varepsilon_{A,k} \\ &= \beta_A + \varepsilon_{A,k} \end{aligned}$$

is

$$\begin{aligned} \hat{y}_A &= b_A \\ &= 24.7 \end{aligned}$$

Note that $\hat{y}_A$ equals $\bar{y}_A = 24.7$ (see Table 14.1).

2. The point estimate of

$$\mu_B = \beta_A + \beta_B$$

and the point prediction of

$$\begin{aligned} y_{B,k} &= \mu_B + \varepsilon_{B,k} \\ &= \beta_A + \beta_B + \varepsilon_{B,k} \end{aligned}$$

is

$$\begin{aligned}\hat{y}_B &= b_A + b_B \\ &= 24.7 + 1.86 \\ &= 26.56\end{aligned}$$

Note that $\hat{y}_B$ equals $\bar{y}_B = 26.56$ (see Table 14.1).

3. The point estimate of

$$\mu_C = \beta_A + \beta_C$$

and the point prediction of

$$\begin{aligned}y_{C,k} &= \mu_C + \varepsilon_{C,k} \\ &= \beta_A + \beta_C + \varepsilon_{C,k}\end{aligned}$$

is

$$\begin{aligned}\hat{y}_C &= b_A + b_C \\ &= 24.7 - .7 \\ &= 24.0\end{aligned}$$

Note that $\hat{y}_C$ equals $\bar{y}_C = 24.0$ (see Table 14.1).

After calculating the least squares point estimates of the parameters in the means model, we use the model to test for differences between the treatment means. We consider testing the null hypothesis

$$H_0 : \mu_1 = \mu_2 = \mu_3 = \cdots = \mu_v$$

which says that all treatment effects are the same, versus the alternative hypothesis

$$H_1 : \text{At least two of } \mu_1, \mu_2, \mu_3, \ldots, \mu_v \text{ differ from each other}$$

which says that at least two treatment effects (means) differ. We have seen that the means model implies that

$$\mu_1 = \beta_1, \quad \mu_2 = \beta_1 + \beta_2, \quad \mu_3 = \beta_1 + \beta_3, \quad \ldots, \quad \mu_v = \beta_1 + \beta_v$$

It follows that the equality

$$\mu_1 = \mu_2 = \mu_3 = \cdots = \mu_v$$

is equivalent to the equality

$$\beta_2 = \beta_3 = \cdots = \beta_v = 0$$

Therefore H_0 is equivalent to

$$H_0 : \beta_2 = \beta_3 = \cdots = \beta_v = 0$$

We test this equivalent null hypothesis versus

$$H_1 : \text{At least one of } \beta_2, \beta_3, \ldots, \beta_v \text{ does not equal zero}$$

This alternative hypothesis is equivalent to our original H_1. As discussed in Chapter 8, this test can be carried out by using the overall F statistic for the means model

$$F(\text{model}) = \frac{MS_{\text{model}}}{MSE} = \frac{SS_{\text{model}}/(k-1)}{SSE/(n-k)}$$

Here, SS_{model} and SSE are the explained variation and unexplained variation for the means model, respectively.

To obtain computational formulas for SS_{model} and SSE, consider the point prediction of $y_{l,k}$

$$\hat{y}_l = b_1 + b_2 D_{l,2} + b_3 D_{l,3} + \cdots + b_v D_{l,v}$$

Recall that this point prediction can be proven to be equal to $\bar{y}_l$, the sample mean for treatment l. Also note that the mean of all $n = \Sigma_{l=1}^{v} n_l$ observations is

$$\bar{y} = \frac{\sum_{l=1}^{v}\sum_{k=1}^{n_l} y_{l,k}}{n}$$

It follows that

$$\begin{aligned} SS_{\text{model}} &= \text{explained variation} \\ &= \sum_{l=1}^{v}\sum_{k=1}^{n_l} (\hat{y}_l - \bar{y})^2 \\ &= \sum_{l=1}^{v} n_l (\hat{y}_l - \bar{y})^2 \\ &= \sum_{l=1}^{v} n_l (\bar{y}_l - \bar{y})^2 \end{aligned}$$

and

$$\begin{aligned} SSE &= \text{unexplained variation} \\ &= \sum_{l=1}^{v}\sum_{k=1}^{n_l} (y_{l,k} - \hat{y}_l)^2 \\ &= \sum_{l=1}^{v}\sum_{k=1}^{n_l} (y_{l,k} - \bar{y}_l)^2 \end{aligned}$$

Therefore since the means model utilizes $k = v$ parameters, we can calculate and use the overall F statistic for testing the equality of the treatment means as follows.

An F-Test for Differences Between Treatment Means

For the means model

$$\begin{aligned} y_{l,k} &= \mu_l + \varepsilon_{l,k} \\ &= \beta_1 + \beta_2 D_{l,2} + \beta_3 D_{l,3} + \cdots + \beta_v D_{l,v} + \varepsilon_{l,k} \end{aligned}$$

define

$$F(\text{model}) = \frac{MS_{\text{model}}}{MSE} = \frac{SS_{\text{model}}/(k-1)}{SSE/(n-k)} = \frac{\sum_{l=1}^{v} n_l(\bar{y}_l - \bar{y})^2/(v-1)}{\sum_{l=1}^{v}\sum_{k=1}^{n_l}(y_{l,k} - \bar{y}_l)^2/(n-v)}$$

Also, define the prob-value to be the area to the right of F(model) under the curve of the F-distribution having $v - 1$ and $n - v$ degrees of freedom. Then we can reject

$$H_0: \mu_1 = \mu_2 = \cdots = \mu_v$$

or, equivalently,

$$H_0: \beta_2 = \beta_3 = \cdots = \beta_v = 0$$

in favor of

H_1: At least two of $\mu_1, \mu_2, \ldots, \mu_v$ differ from each other

or, equivalently,

H_1: At least one of $\beta_2, \beta_3, \ldots, \beta_v$ does not equal zero

by setting the probability of a Type I error equal to α if and only if either of the following equivalent conditions holds:

1. $F(\text{model}) > F_{[\alpha]}^{(v-1,n-v)}$ 2. prob-value $< \alpha$

Note that F(model) in this section is equal to F(means) in Section 14.4. That is, the ANOVA approach and the regression approach give the same F statistics.

Example 14.14 In Example 14.13 we began to analyze the gasoline mileage data (in Table 14.1) by using regression analysis. Specifically, in Part 1 we constructed the means model

$$\begin{aligned} y_{l,k} &= \mu_l + \varepsilon_{l,k} \\ &= \beta_A + \beta_B D_{l,B} + \beta_C D_{l,C} + \varepsilon_{l,k} \end{aligned}$$

Next, in Part 2 we calculated least squares point estimates of the model parameters and least squares point estimates of the treatment means μ_A, μ_B, and μ_C. We now continue the analysis of the gasoline mileage data.

Part 3: Testing for Differences Between Gasoline Types

We now test to see whether or not there are any statistically significant differences between the effects of gasoline types A, B, and C on mean gasoline mileage. To do

FIGURE 14.3 **SAS output of a means model analysis of the gasoline mileage data in Table 14.1**

DEP VARIABLE: Y

SOURCE	DF[a]	SUM OF SQUARES	MEAN SQUARE	F VALUE	PROB>F
MODEL	2	14.448000[b]	7.224000[e]	12.379[g]	0.0026[h]
ERROR	9	5.252000[c]	0.583556[f]		
C TOTAL	11	19.700000[d]			

PARAMETER[i]	ESTIMATE[j]	T FOR HO: PARAMETER=0[k]	PROB > |T|[l]	STD ERROR OF ESTIMATE[m]
INTERCEPT	24.70000000	64.67	0.0001	0.38195404
DB	1.86000000	3.63	0.0055	0.51244512
DC	-0.70000000	-1.20	0.2609	0.58344443
MUB-MUA	1.86000000	3.63	0.0055	0.51244512
MUC-MUA	-0.70000000	-1.20	0.2609	0.58344443
MUC-MUB	-2.56000000	-4.59	0.0013	0.55787958
MUB-(MUC+MUA)/2	2.21000000	4.92	0.0008	0.44923598

[a] $k - 1 = 2$, $n - k = 9$, $n - 1 = 11$ [b] SS_{model} [c] SSE [d] *Total variation* [e] MS_{model} [f] $MSE = s^2$ [g] $F(model)$

[h] *Prob-value* [i] $\lambda'\beta$ [j] $\lambda'\mathbf{b}$ [k] t [l] *Prob-value* [m] $s\sqrt{\lambda'(\mathbf{X}'\mathbf{X})^{-1}\lambda}$

this, we test the null hypothesis

$$H_0: \mu_A = \mu_B = \mu_C \quad \text{or, equivalently,} \quad H_0: \beta_B = \beta_C = 0$$

versus the alternative hypothesis

$$H_1: \text{At least two of } \mu_A, \mu_B, \text{ and } \mu_C \text{ differ from each other}$$

or, equivalently,

$$H_1: \text{At least one of } \beta_B \text{ and } \beta_C \text{ does not equal zero}$$

The SAS output in Figure 14.3 results from using the means model (which has $k = v = 3$ parameters) to perform a regression analysis of the data in Table 14.1. Note that the SAS output gives the least squares point estimates of the parameters β_A, β_B, and β_C. These parameters are denoted INTERCEPT, DB, and DC on the SAS output. The output also gives the least squares point estimates of the differences $\mu_B - \mu_A$, $\mu_C - \mu_A$, $\mu_C - \mu_B$, and $\mu_B - (\mu_C + \mu_A)/2$. These differences are denoted MUB − MUA, MUC − MUA, MUC − MUB, and MUB − (MUC + MUA)/2 on the output. In addition, the SAS output presents the following quantities needed to test H_0 versus H_1:

$$\text{Total variation} = 19.7$$

$$SS_{model} = \text{Explained variation} = 14.448$$

$$MS_{\text{model}} = \frac{SS_{\text{model}}}{k - 1} = \frac{14.448}{3 - 1} = 7.224$$

$$SSE = \text{Unexplained variation} = 5.252$$

$$s^2 = MSE = \frac{SSE}{n - k} = \frac{5.252}{12 - 3} = .583556$$

$$F(\text{model}) = \frac{MS_{\text{model}}}{MSE} = \frac{7.224}{.583556} = 12.379$$

$$\text{prob-value} = .0026$$

To test H_0 with $\alpha = .05$, we use the rejection point

$$F_{[\alpha]}^{(v-1,n-v)} = F_{[.05]}^{(3-1,12-3)} = F_{[.05]}^{(2,9)} = 4.26$$

Since $F(\text{model}) = 12.379 > 4.26 = F_{[.05]}^{(2,9)}$ (or since prob-value $= .0026$ is less than .05), we can reject H_0 in favor of H_1 with $\alpha = .05$. We conclude that at least two of gasoline types A, B, and C have different effects on mean gasoline mileage.

Part 4: Making Pairwise Comparisons of Gasoline Types

We now investigate the exact nature of the differences between the treatment means μ_A, μ_B, and μ_C by making pairwise comparisons of the effects of gasoline types A, B, and C. To do this, we use the SAS output in Figure 14.4. Note that this SAS output differs somewhat from that in Figure 14.3 even though the same data set is being analyzed. In (optional) Section 14.8 we explain how to obtain these different SAS outputs. Also note that both SAS outputs present the t statistics corresponding to the differences $\mu_B - \mu_A$, $\mu_C - \mu_A$, $\mu_C - \mu_B$, and $\mu_B - (\mu_C + \mu_A)/2$. Both outputs also include the prob-value and standard error of the estimate $\boldsymbol{\lambda}'\mathbf{b}$ associated with each t statistic.

We first consider the difference $\mu_B - \mu_A$. The least squares point estimate of

$$\begin{aligned} \mu_B - \mu_A &= [\beta_A + \beta_B] - \beta_A \\ &= \beta_B \\ &= \boldsymbol{\lambda}'\boldsymbol{\beta}, \end{aligned}$$

where $\boldsymbol{\lambda}' = [0 \quad 1 \quad 0]$ and $\boldsymbol{\beta} = \begin{bmatrix} \beta_A \\ \beta_B \\ \beta_C \end{bmatrix}$

is

$$\boldsymbol{\lambda}'\mathbf{b} = b_B = 1.86$$

From Figure 14.4 we see that the prob-value for testing

$$H_0: \mu_B - \mu_A = 0 \quad \text{or, equivalently,} \quad H_0: \mu_B = \mu_A$$

versus

$$H_1: \mu_B - \mu_A \neq 0 \quad \text{or, equivalently,} \quad H_1: \mu_B \neq \mu_A$$

FIGURE 14.4 **SAS output of a means model analysis of the gasoline mileage data in Table 14.1**

SAS

GENERAL LINEAR MODELS PROCEDURE

DEPENDENT VARIABLE: MILEAGE

SOURCE	DF	SUM OF SQUARES	MEAN SQUARE	F VALUE	PR > F	R-SQUARE	C.V.
MODEL	2	14.44800000[c]	7.22400000[f]	12.38[a]	0.0026[b]	0.733401	3.0194
ERROR	9	5.25200000[d]	0.58355556[g]		ROOT MSE		MILEAGE MEAN
CORRECTED TOTAL	11	19.70000000[e]			0.76390808		25.30000000

SOURCE	DF	TYPE I SS	F VALUE	PR > F	DF	TYPE III SS	F VALUE	PR > F
GASTYPE	2	14.44800000	12.38[a]	0.0026[b]	2	14.44800000	12.38	0.0026

PARAMETER	ESTIMATE[h]	T FOR HO: PARAMETER=0[i]	PR > |T|[j]	STD ERROR OF ESTIMATE[k]
MUB-MUA	1.86000000	3.63	0.0055	0.51244512
MUC-MUA	-0.70000000	-1.20	0.2609	0.58344443
MUC-MUB	-2.56000000	-4.59	0.0013	0.55787958
MUB-(MUC+MUA)/2	2.21000000	4.92	0.0008	0.44923598

OBSERVATION	OBSERVED VALUE	PREDICTED VALUE	RESIDUAL	LOWER 95% CL FOR MEAN	UPPER 95% CL FOR MEAN
1	24.00000000	24.70000000	-0.70000000	23.83595297	25.56404703
2	25.00000000	24.70000000	0.30000000	23.83595297	25.56404703
3	24.30000000	24.70000000	-0.40000000	23.83595297	25.56404703
4	25.50000000	24.70000000	0.80000000	23.83595297	25.56404703
5	25.30000000	26.56000000	-1.26000000	[25.78717284	27.33282716][l]
6	26.50000000	26.56000000	-0.06000000	25.78717284	27.33282716
7	26.40000000	26.56000000	-0.16000000	25.78717284	27.33282716
8	27.00000000	26.56000000	0.44000000	25.78717284	27.33282716
9	27.60000000	26.56000000	1.04000000	25.78717284	27.33282716
10	23.30000000	24.00000000	-0.70000000	23.00228443	24.99771557
11	24.00000000	24.00000000	0.00000000	23.00228443	24.99771557
12	24.70000000	24.00000000	0.70000000	23.00228443	24.99771557

OBSERVATION	OBSERVED VALUE	PREDICTED VALUE	RESIDUAL	LOWER 95% CL INDIVIDUAL	UPPER 95% CL INDIVIDUAL
1	24.00000000	24.70000000	-0.70000000	22.76793211	26.63206789
2	25.00000000	24.70000000	0.30000000	22.76793211	26.63206789
3	24.30000000	24.70000000	-0.40000000	22.76793211	26.63206789
4	25.50000000	24.70000000	0.80000000	22.76793211	26.63206789
5	25.30000000	26.56000000	-1.26000000	[24.66696781	28.45303219][m]
6	26.50000000	26.56000000	-0.06000000	24.66696781	28.45303219
7	26.40000000	26.56000000	-0.16000000	24.66696781	28.45303219
8	27.00000000	26.56000000	0.44000000	24.66696781	28.45303219
9	27.60000000	26.56000000	1.04000000	24.66696781	28.45303219
10	23.30000000	24.00000000	-0.70000000	22.00456886	25.99543114
11	24.00000000	24.00000000	0.00000000	22.00456886	25.99543114
12	24.70000000	24.00000000	0.70000000	22.00456886	25.99543114

[a] *F(model)* [b] *prob-value for F(model)* [c] SS_{model} [d] *SSE* [e] SS_{total} [f] MS_{model} [g] *MSE* [h] $\lambda'\mathbf{b}$ [i] *t*
[j] *prob-value for t* [k] $s\sqrt{\lambda'(\mathbf{X}'\mathbf{X})^{-1}\lambda}$ [l] *95% confidence interval for* μ_B [m] *95% prediction interval for* $y_{B,0}$

is .0055. Thus we reject H_0 with $\alpha = .05$. We conclude that we have very substantial evidence that μ_B is greater than μ_A.

To assess the practical importance of the difference $\mu_B - \mu_A$, we consider a 95% individual confidence interval for this difference. This interval is computed in Table 14.4. Note that simultaneous confidence intervals—to be discussed later—are also computed in this table. Also note that the quantity $s\sqrt{\boldsymbol{\lambda}'(\mathbf{X}'\mathbf{X})^{-1}\boldsymbol{\lambda}}$, the standard error of the estimate $\boldsymbol{\lambda}'\mathbf{b}$, is obtained from the SAS output in Figure 14.4. The individual 95% confidence interval for $\mu_B - \mu_A$ is [.701, 3.019]. Thus we are 95 percent confident that μ_B is between .701 and 3.019 mpg greater than μ_A.

Next, we consider the difference $\mu_C - \mu_A$. The least squares point estimate of

$$\begin{aligned} \mu_C - \mu_A &= [\beta_A + \beta_C] - \beta_A \\ &= \beta_C \\ &= \boldsymbol{\lambda}'\boldsymbol{\beta}, \end{aligned}$$

where $\boldsymbol{\lambda}' = [0 \quad 0 \quad 1]$ and $\boldsymbol{\beta} = \begin{bmatrix} \beta_A \\ \beta_B \\ \beta_C \end{bmatrix}$

is

$$\boldsymbol{\lambda}'\mathbf{b} = b_C = -.7$$

From Figure 14.4 we see that the prob-value for testing

$$H_0: \mu_C - \mu_A = 0 \quad \text{or, equivalently,} \quad H_0: \mu_C = \mu_A$$

versus

$$H_1: \mu_C - \mu_A \neq 0 \quad \text{or, equivalently,} \quad H_1: \mu_C \neq \mu_A$$

is .2609. Thus we do not reject H_0 with $\alpha = .05$.

Moreover, the individual 95% confidence interval for $\mu_C - \mu_A$ (in Table 14.4) is [−2.0197, .6197]. This interval tells us that we can be 95 percent confident that

TABLE 14.4 **Individual and simultaneous 95% confidence intervals for differences between means in the gasoline mileage problem**

Difference $\lambda'\beta$	*Point estimate $\lambda'b$*	*Individual 95% confidence interval for $\lambda'\beta$* $[\lambda'b \pm t_{[.025]}^{(n-k)} s\sqrt{\lambda'(X'X)^{-1}\lambda}]$	*Scheffé simultaneous 95% confidence interval for $\lambda'\beta$ (in Set I)* $[\lambda'b \pm \sqrt{qF_{[\alpha]}^{(q,n-k)}} s\sqrt{\lambda'(X'X)^{-1}\lambda}]$
$\mu_B - \mu_A = \beta_B$	$b_B = 1.86$	[1.86 ± 2.262(.5124)] = [.701, 3.019]	[1.86 ± 2.9189(.5124)] = [.3644, 3.3556]
$\mu_C - \mu_A = \beta_C$	$b_C = -0.70$	[−.70 ± 2.262(.5834)] = [−2.0197, .6197]	[−.70 ± 2.9189(.5834)] = [−2.4029, 1.0029]
$\mu_C - \mu_B = \beta_C - \beta_B$	$b_C - b_B = -2.56$	−2.56 ± 2.262(.5579)] = [−3.822, −1.298]	[−2.56 ± 2.9189(.5579)] = [−4.1885, −.9315]
$\mu_B - \left(\frac{\mu_C + \mu_A}{2}\right) = \beta_B - .5\beta_C$	$b_B - .5b_C = 2.21$	[2.21 ± 2.262(.4492)] = [1.1939, 3.2261]	[2.21 ± 2.9189(.4492)] = [.8988, 3.5212]

μ_C is between 2.0197 mpg less than and .6197 mpg greater than μ_A. We conclude that there is no statistically significant difference between μ_C and μ_A.

Third, we consider the difference $\mu_C - \mu_B$. The least squares point estimate of

$$\begin{aligned} \mu_C - \mu_B &= [\beta_A + \beta_C] - [\beta_A + \beta_B] \\ &= \beta_C - \beta_B \\ &= \boldsymbol{\lambda}'\boldsymbol{\beta}, \end{aligned}$$

where $\boldsymbol{\lambda}' = [0 \quad -1 \quad 1]$ and $\boldsymbol{\beta} = \begin{bmatrix} \beta_A \\ \beta_B \\ \beta_C \end{bmatrix}$

is

$$\boldsymbol{\lambda}'\mathbf{b} = b_C - b_B = -.7 - 1.86 = -2.56$$

From Figure 14.4 we see that the prob-value for testing

$$H_0: \mu_C - \mu_B = 0 \quad \text{or, equivalently,} \quad H_0: \mu_C = \mu_B$$

versus

$$H_1: \mu_C - \mu_B \neq 0 \quad \text{or, equivalently,} \quad H_1: \mu_C \neq \mu_B$$

is .0013. Thus we reject H_0 with $\alpha = .05$. We conclude that we have very strong evidence that μ_C is less than μ_B.

To assess the practical importance of the difference $\mu_C - \mu_B$, we compute a 95% individual confidence interval for this difference to be $[-3.822, -1.298]$ (see Table 14.4). This interval says that we can be 95 percent confident that μ_C is between 3.822 and 1.298 mpg less than μ_B.

Part 5: Studying a More Complicated Difference

Gasoline type B contains a chemical—Chemical XX—that is not contained in gasoline type C or in gasoline type A. To assess the effect of Chemical XX on gasoline mileage, we consider

$$\mu_B - \frac{\mu_C + \mu_A}{2}$$

The least squares point estimate of

$$\begin{aligned} \mu_B - \frac{\mu_C + \mu_A}{2} &= [\beta_A + \beta_B] - \frac{[\beta_A + \beta_C] + \beta_A}{2} \\ &= \beta_B - .5\beta_C \\ &= \boldsymbol{\lambda}'\boldsymbol{\beta}, \end{aligned}$$

where $\boldsymbol{\lambda}' = [0 \quad 1 \quad -.5]$ and $\boldsymbol{\beta} = \begin{bmatrix} \beta_A \\ \beta_B \\ \beta_C \end{bmatrix}$

is

$$\begin{aligned} \boldsymbol{\lambda}'\mathbf{b} = b_B - .5b_C &= 1.86 - .5(-.7) \\ &= 2.21 \end{aligned}$$

From Figure 14.4 we see that the prob-value for testing

$$H_0: \mu_B - \frac{\mu_C + \mu_A}{2} = 0 \qquad \text{or, equivalently,} \qquad H_0: \mu_B = \frac{\mu_C + \mu_A}{2}$$

versus

$$H_1: \mu_B - \frac{\mu_C + \mu_A}{2} \neq 0 \qquad \text{or, equivalently,} \qquad H_1: \mu_B \neq \frac{\mu_C + \mu_A}{2}$$

is .0008. Thus we reject H_0 with $\alpha = .05$. We conclude that we have very substantial evidence that μ_B is greater than $(\mu_C + \mu_A)/2$. Furthermore, a 95% individual confidence interval for $\mu_B - (\mu_C + \mu_A)/2$ is [1.1939, 3.2261] (see Table 14.4). This interval says that we can be 95 percent confident that μ_B (the mean gasoline mileage obtained by using gasoline type B, which contains Chemical XX) is between 1.1939 and 3.2261 mpg greater than $(\mu_C + \mu_A)/2$ (the average of the mean gasoline mileages obtained by gasoline types C and A, which do not contain Chemical XX). Notice that, although these results suggest that Chemical XX might be a factor causing μ_B to be greater than $(\mu_C + \mu_A)/2$, the statistical evidence here certainly does not prove that such a causal relationship exists. Furthermore, remember that the above results apply only to Fire-Hawk automobiles driven under the test conditions employed by North American Oil.

Part 6: Computing Simultaneous Confidence Intervals

We now demonstrate how we calculated the Scheffé simultaneous 95% confidence intervals for all the linear combinations of regression parameters in Table 14.4. Call these linear combinations Set I. There are two regression parameters, β_B and β_C, from the means model involved nontrivially in at least one of the linear combinations in Set I. It follows that $q = 2$ and that

$$\begin{aligned} \sqrt{qF_{[\alpha]}^{(q,n-k)}} &= \sqrt{2F_{[.05]}^{(2,12-3)}} \\ &= \sqrt{2F_{[.05]}^{(2,9)}} \\ &= \sqrt{2(4.26)} \\ &= 2.9189 \end{aligned}$$

Thus

$$\left[\boldsymbol{\lambda}'\mathbf{b} \pm \sqrt{qF_{[\alpha]}^{(q,n-k)}}\, s\sqrt{\boldsymbol{\lambda}'(\mathbf{X}'\mathbf{X})^{-1}\boldsymbol{\lambda}}\right] = \left[\boldsymbol{\lambda}'\mathbf{b} \pm (2.9189)\, s\sqrt{\boldsymbol{\lambda}'(\mathbf{X}'\mathbf{X})^{-1}\boldsymbol{\lambda}}\right]$$

is a Scheffé simultaneous 95% confidence interval for any linear combination $\boldsymbol{\lambda}'\boldsymbol{\beta}$

TABLE 14.5 **Scheffé simultaneous 95% confidence intervals for the linear combinations in Set II**

Linear combination, $\lambda'\beta$	Point estimate, $\lambda'b$	$s\sqrt{\lambda'(X'X)^{-1}\lambda}$	Scheffé simultaneous 95% confidence interval for $\lambda'\beta$ (in Set II), $[\lambda'b \pm (3.4029)s\sqrt{\lambda'(X'X)^{-1}\lambda}]$
$\mu_A = \beta_A$	24.7	$(.7639)\sqrt{.25} = .382$	$[24.7 \pm 3.4029(.382)]$ $= [23.4, 26]$
$\mu_B = \beta_A + \beta_B$	26.56	$(.7639)\sqrt{.20} = .3416$	$[26.56 \pm 3.4029(.3416)]$ $= [25.3976, 27.7224]$
$\mu_C = \beta_A + \beta_C$	24.0	$(.7639)\sqrt{.3333} = .441$	$[24.0 \pm 3.4029(.441)]$ $= [22.4993, 25.5007]$
$\mu_B - \mu_A = \beta_B$	1.86	$(.7639)\sqrt{.4499} = .5124$	$[1.86 \pm 3.4029(.5124)]$ $= [.1164, 3.6036]$
$\mu_C - \mu_A = \beta_C$	−.7	$(.7639)\sqrt{.5833} = .5834$	$[-.7 \pm 3.4029(.5834)]$ $= [-2.6853, 1.2853]$
$\mu_C - \mu_B = \beta_C - \beta_B$	−2.56	$(.7639)\sqrt{.5334} = .5579$	$[-2.56 \pm 3.4029(.5579)]$ $= [-4.4585, -.6615]$
$\mu_B - \frac{(\mu_C + \mu_A)}{2} = \beta_B - .5\beta_C$	2.21	$(.7639)\sqrt{.3458} = .4492$	$[2.21 \pm 3.4029(.4492)]$ $= [.6814, 3.7386]$

in Set I. Note that this simultaneous interval is longer and thus less precise than the individual 95% confidence interval for $\boldsymbol{\lambda}'\boldsymbol{\beta}$

$$[\boldsymbol{\lambda}'\mathbf{b} \pm t_{[\alpha/2]}^{(n-k)} s\sqrt{\boldsymbol{\lambda}'(\mathbf{X}'\mathbf{X})^{-1}\boldsymbol{\lambda}}] = [\boldsymbol{\lambda}'\mathbf{b} \pm (2.262)s\sqrt{\boldsymbol{\lambda}'(\mathbf{X}'\mathbf{X})^{-1}\boldsymbol{\lambda}}]$$

Next, we calculate Scheffé simultaneous 95% confidence intervals for all the linear combinations of regression parameters in Table 14.5. Call these linear combinations Set II. All three regression parameters β_A, β_B, and β_C from the means model are involved nontrivially in at least one of the linear combinations in Set II. It follows that $q = 3$ and that

$$\begin{aligned} \sqrt{qF_{[\alpha]}^{(q,n-k)}} &= \sqrt{3F_{[.05]}^{(3,12-3)}} \\ &= \sqrt{3F_{[.05]}^{(3,9)}} \\ &= \sqrt{3(3.86)} \\ &= 3.4029 \end{aligned}$$

Thus

$$[\boldsymbol{\lambda}'\mathbf{b} \pm \sqrt{qF_{[\alpha]}^{(q,n-k)}}\, s\sqrt{\boldsymbol{\lambda}'(\mathbf{X}'\mathbf{X})^{-1}\boldsymbol{\lambda}}] = [\boldsymbol{\lambda}'\mathbf{b} \pm (3.4029)s\sqrt{\boldsymbol{\lambda}'(\mathbf{X}'\mathbf{X})^{-1}\boldsymbol{\lambda}}]$$

is a Scheffé simultaneous 95% confidence interval for any linear combination $\boldsymbol{\lambda}'\boldsymbol{\beta}$ in Set II. Note that $q = 3$ for Set II, while $q = 2$ for Set I. Thus this simultaneous

95% interval is longer (less precise) than

$$[\boldsymbol{\lambda}'\mathbf{b} \pm (2.9189)s\sqrt{\boldsymbol{\lambda}'(\mathbf{X}'\mathbf{X})^{-1}\boldsymbol{\lambda}}]$$

the Scheffé simultaneous 95% confidence interval for any corresponding linear combination in Set I. In Table 14.5 we calculate the Scheffé simultaneous 95% confidence intervals for the seven linear combinations in Set II.

Part 7: Using the Model to Estimate and Predict

Suppose that, on the basis of the information it has obtained in this experiment, North American Oil has concluded that μ_B is sufficiently greater than μ_A and μ_C to warrant producing gasoline type B. For this reason we find a point estimate of and a confidence interval for μ_B. These quantities would be of interest because the average gasoline mileage obtained by the fleet of all Fire-Hawks would be close to μ_B. Also note that it is reasonable to assume that purchasing a Fire-Hawk is equivalent to randomly selecting a Fire-Hawk from the population of all Fire-Hawks. It follows that a Fire-Hawk owner who plans to use gasoline type B would like to have a prediction of

$$y_{B,0} = \mu_B + \varepsilon_{B,0}$$

an individual gasoline mileage obtained by a Fire-Hawk using gasoline type B. We can express μ_B by the equation

$$\begin{aligned}\mu_B &= \beta_A + \beta_B D_{l,B} + \beta_C D_{l,C}\\ &= \beta_A + \beta_B D_{B,B} + \beta_C D_{B,C}\\ &= \beta_A + \beta_B(1) + \beta_C(0)\\ &= \beta_A + \beta_B\end{aligned}$$

It follows that we can express $y_{B,0}$ by the equation

$$\begin{aligned}y_{B,0} &= \mu_B + \varepsilon_{B,0}\\ &= \beta_A + \beta_B + \varepsilon_{B,0}\end{aligned}$$

Thus

$$\begin{aligned}\hat{y}_B &= b_A + b_B D_{l,B} + b_C D_{l,C}\\ &= b_A + b_B D_{B,B} + b_C D_{B,C}\\ &= b_A + b_B(1) + b_C(0)\\ &= b_A + b_B\\ &= 24.7 + 1.86\\ &= 26.56\end{aligned}$$

is the point estimate of μ_B and is the point prediction of $y_{B,0} = \mu_B + \varepsilon_{B,0}$. Furthermore,

$$\mathbf{x}_B' = [1 \quad 1 \quad 0]$$

is the row vector containing the numbers multiplied by b_A, b_B, and b_C in the preceding prediction equation. Thus

$$\mathbf{x}_B'(\mathbf{X}'\mathbf{X})^{-1}\mathbf{x}_B = .20$$

It follows that

$$\begin{aligned}[\hat{y}_B \pm t_{[.025]}^{(9)} s\sqrt{\mathbf{x}_B'(\mathbf{X}'\mathbf{X})^{-1}\mathbf{x}_B}] &= [26.56 \pm 2.262(.7639)\sqrt{.20}] \\ &= [26.56 \pm .7727] \\ &= [25.787, 27.333]\end{aligned}$$

is a 95% confidence interval for μ_B. In addition,

$$\begin{aligned}[\hat{y}_B \pm t_{[.025]}^{(9)} s\sqrt{1 + \mathbf{x}_B'(\mathbf{X}'\mathbf{X})^{-1}\mathbf{x}_B}] &= [26.56 \pm 2.262(.7639)\sqrt{1.20}] \\ &= [26.56 \pm 1.8931] \\ &= [24.667, 28.453]\end{aligned}$$

is a 95% prediction interval for $y_{B,0}$. Using the preceding interval for μ_B, the federal government can be 95% confident that μ_B is between 25.787 and 27.333 mpg. Using the prediction interval for $y_{B,0}$, the owner of a Fire-Hawk can be 95% confident that the individual gasoline mileage that will be obtained by his Fire-Hawk when driven using gasoline type B will be between 24.667 and 28.453 mpg. Notice that both of the above intervals are given on the SAS output of Figure 14.4.

Note that in the above problem, instead of using regression analysis, we can carry out the analysis by using the algebraic formulas of Section 14.4. These formulas give the same results as the regression techniques illustrated in Example 14.14. Furthermore, in Section 14.4 we discussed the Tukey and Bonferroni methods for computing simultaneous confidence intervals. These can be employed in addition to—or instead of—the Scheffé method when appropriate.

14.6

FIXED AND RANDOM MODELS

The methods of Sections 14.4 and 14.5 describe the situation in which the treatments (that is, factor levels) are the only treatments of interest. This is called the *fixed model* case. However, in some situations the treatments have been randomly selected from a population of treatments. In such a case we are interested in making statistical inferences about the population of treatments.

Example 14.15 Suppose that a pharmaceutical company wishes to examine the potency of a liquid medication mixed in large vats. To do this, the company randomly selects a sample of four vats from a month's production and randomly selects four separate samples from each vat. The data in Table 14.6 represents the recorded potencies. In this case

TABLE 14.6 **Liquid medication potencies from four randomly selected vats**

Vat 1	*Vat 2*	*Vat 3*	*Vat 4*
6.1	7.1	5.6	6.5
6.6	7.3	5.8	6.8
6.4	7.3	5.7	6.2
6.3	7.7	5.3	6.3
$\bar{y}_1 = 6.35$	$\bar{y}_2 = 7.35$	$\bar{y}_3 = 5.6$	$\bar{y}_4 = 6.45$

we are not interested in the potencies in only the four randomly selected vats. Rather, we are interested in the potencies in all possible vats.

Let $y_{l,k}$ denote the potency of the kth sample in the lth randomly selected vat. Then the *random model* says that

$$y_{l,k} = \mu_l + \varepsilon_{l,k}$$

Here, μ_l is the mean potency of all possible samples of liquid medication that could be randomly selected from the lth randomly selected vat. That is, μ_l is the mean potency of all of the liquid medication in the lth randomly selected vat. Moreover, since the four vats were randomly selected, μ_l is assumed to have been randomly selected from the population of all possible vat means. This population is assumed to be normally distributed with mean μ and variance σ_μ^2. Here, μ is the mean potency of all possible samples of liquid medication that could be randomly selected from all possible vats. That is, μ is the mean potency of all possible liquid medication. In addition, σ_μ^2 is the variance between all possible vat means.

We further assume that the error term $\varepsilon_{l,k}$ is randomly selected from a normally distributed population of error terms having mean zero and variance σ^2 and that the error terms $\varepsilon_{l,k}$ are independent of each other and of the randomly selected means μ_l. Under these assumptions we can test the null hypothesis

$$H_0 : \sigma_\mu^2 = 0$$

This hypothesis says that all possible vat means are equal. We test H_0 versus the alternative hypothesis

$$H_1 : \sigma_\mu^2 \neq 0$$

which says that there is some variation between the vat means. Specifically, we can reject H_0 in favor of H_1 by setting the probability of a Type I error equal to α if the F(means) statistic of Section 14.4 is greater than $F_{[\alpha]}^{(v-1, n-v)}$.

Table 14.7 tells us that we can reject

$$H_0 : \sigma_\mu^2 = 0$$

with $\alpha = .05$. Therefore we conclude that there is variation in the population of all vat means. That is, we conclude that some of the vat means differ. Furthermore,

TABLE 14.7 ANOVA table for fixed and random models

Source	df	Sum of squares	Mean square	F statistic	E(mean square) fixed model	E(mean square) random model
Model	$v-1$ $=3$	SS_{means} $=7.4$	MS_{means} $=2.4667$	$F(\text{means})$ $=45.5111$	$\sigma^2 + \frac{1}{v-1}\sum_{l=1}^{v} n_l(\mu_l-\mu)^2$	$\sigma^2 + n'\sigma_\mu^2$
Error	$n-v$ $=12$	SSE $=.65$	MSE $=.0542$		σ^2	σ^2
					$H_0: \mu_1 = \mu_2 = \cdots = \mu_v$	$H_0: \sigma_\mu^2 = 0$

Notes: 1. $n' = \frac{1}{v-1}\left[\left(\sum_{l=1}^{v} n_l\right) - \frac{\sum_{l=1}^{v} n_l^2}{\sum_{l=1}^{v} n_l}\right]$ (=m for equal sample sizes)

2. Since $F(\text{means}) = 45.5111 > F_{[\alpha]}^{(v-1,n-v)} = F_{[.05]}^{(3,12)} = 3.49$, we can reject $H_0: \sigma_\mu^2 = 0$ with $\alpha = .05$.
3. Since $E(MSE) = \sigma^2$, a point estimate of σ^2 is $MSE = .0542$.
4. Since $E(MS_{\text{means}}) = \sigma^2 + n'\sigma_\mu^2$, a point estimate of $\sigma^2 + n'\sigma_\mu^2$ is MS_{means}. Thus a point estimate of σ_μ^2 is

$$\frac{(MS_{\text{means}} - MSE)}{n'} = \frac{(MS_{\text{means}} - MSE)}{m} = \frac{2.4667 - .0542}{4} = .6031$$

5. For equal sample sizes ($n_l = m$) a point estimate of μ is

$$\bar{y} = \frac{\sum_{l=1}^{v} \bar{y}_l}{v} = \frac{6.35 + 7.35 + 5.6 + 6.45}{4} = 6.4375$$

Furthermore, a $100(1-\alpha)\% = 95\%$ confidence interval for μ is

$$\left[\bar{y} \pm t_{[\alpha/2]}^{(v-1)}\sqrt{\frac{MS_{\text{means}}}{vm}}\right] = \left[6.4375 \pm t_{[.025]}^{(4-1)}\sqrt{\frac{2.4667}{4(4)}}\right] = [6.4375 \pm 3.182(.3926)]$$
$$= [5.1881, 7.6869]$$

as illustrated in Table 14.7, we can calculate point estimates of the *variance components* σ^2 and σ_μ^2. These estimates are .0542 and .6031, respectively. Note here that the variance component σ^2 measures the "within-vat variability," while σ_μ^2 measures the "between-vat variability." In this case the between-vat variability is substantially higher than the within-vat variability. Note from Table 14.7 that we can also calculate a 95% confidence interval for μ, the mean potency of all possible liquid medication. This 95% interval is [5.1881, 7.6869]. To narrow this interval, we could randomly select more vats and more samples from each vat.

The above example illustrates that the procedure for testing

$$H_0: \mu_1 = \mu_2 = \cdots = \mu_v$$

which is appropriate when the v treatments are the only treatments of interest, is the same as the procedure for testing

$$H_0: \sigma_\mu^2 = 0$$

which is appropriate when the v treatments are randomly selected from a large population of treatments. Furthermore, each procedure is justified by the expected mean squares given in Table 14.7.

*14.7

HARTLEY'S TEST AND BARTLETT'S TEST FOR VARIANCE EQUALITY

Consider testing

$$H_0: \sigma_1^2 = \sigma_2^2 = \cdots = \sigma_v^2$$

versus

$$H_1: \text{At least two of } \sigma_1^2, \sigma_2^2, \ldots, \sigma_v^2 \text{ differ}$$

We test H_0 versus H_1 by using the sample variances $s_1^2, s_2^2, \ldots, s_v^2$. Here these sample variances are assumed to be calculated from v independent samples of sizes $n_1, n_2, \ldots, n_v$ that have been randomly selected from v normally distributed populations having variances $\sigma_1^2, \sigma_2^2, \ldots, \sigma_v^2$. Then *Hartley's test* says the following.

Hartley's Test for Variance Equality

We can reject

$$H_0: \sigma_1^2 = \sigma_2^2 = \cdots = \sigma_v^2$$

in favor of H_1 above by setting the probability of a Type I error equal to α if

$$H = \frac{\max(s_1^2, s_2^2, \ldots, s_v^2)}{\min(s_1^2, s_2^2, \ldots, s_v^2)}$$

is greater than $H_{[\alpha]}^{(v,m-1)}$ (see Table E.8 in Appendix E).

Here, all samples are assumed to be of the same size m. However, the test may be used for samples whose sizes do not differ substantially by setting m equal to the largest sample size.

Bartlett's test says the following.

Bartlett's Test for Variance Equality

We can reject

$$H_0: \sigma_1^2 = \sigma_2^2 = \cdots = \sigma_v^2$$

in favor of H_1 above by setting the probability of a Type I error equal to α if

$$B = \frac{1}{D}\left[\left(\sum_{l=1}^{v} n_l - v\right) \ln s^2 - \sum_{l=1}^{v} (n_l - 1) \ln s_l^2\right]$$

*This section is optional.

is greater than $\chi^2_{[\alpha]}(v - 1)$ (see Table E.9 in Appendix E). Here,

$$D = 1 + \frac{1}{3(v-1)}\left[\sum_{l=1}^{v} \frac{1}{(n_l - 1)} - \frac{1}{\left(\sum_{l=1}^{v} n_l - v\right)}\right]$$

and

$$s^2 = \frac{1}{\left(\sum_{l=1}^{v} n_l - v\right)} \sum_{l=1}^{v} (n_l - 1) s_l^2$$

This test should be used only if each sample size n_l is at least 5. If any n_l is less than 5, we calculate the *Box modified Bartlett statistic*

$$B' = \frac{g_2 BD}{g_1(A - BD)}$$

where

$$g_1 = v - 1$$

$$g_2 = \frac{v + 1}{(D - 1)^2}$$

$$A = \frac{g_2}{2 - D + \dfrac{2}{g_2}}$$

We reject H_0 if $B' > F_{[\alpha]}^{(g_1, g_2)}$.

One might wish to use Hartley's test or Bartlett's test to check the validity of the equal variance assumption before performing one-factor analysis. However, these tests for variance equality are more adversely affected by violations of the normality assumption than is one-factor analysis by violations of the constant variance assumption. Therefore some practitioners question whether the tests for variance equality should be performed.

*14.8 USING SAS

In Figure 14.5 we present the SAS program that yields the analysis of the North American Oil Company data that is presented in Figures 14.1 and 14.4. Note that in this program we employ a "class variable" to define the means model.

*This section is optional.

FIGURE 14.5 **SAS program to analyze the North American Oil Company data in Table 14.1 by using a class variable**

```
DATA UNLEAD;
INPUT GASTYPE $ MILEAGE @@; }⟶ Defines factor GASTYPE
                                 and response variable MILEAGE
CARDS;
A  24.0  A  25.0  A  24.3  A  25.5
B  25.3  B  26.5  B  26.4  B  27.0  B  27.6 }⟶ Data—See Table 14.1
C  23.3  C  24.0  C  24.7
 ;
PROC GLM; }⟶ Specifies General Linear Models Procedure
CLASS GASTYPE; }⟶ Defines class variable GASTYPE
MODEL MILEAGE = GASTYPE / P CLM; }⟶ Specifies means model, and CLM
                                    requests confidence intervals
ESTIMATE 'MUB-MUA' GASTYPE -1 1 ; }⟶ Estimates μB − μA
ESTIMATE 'MUC-MUA' GASTYPE -1 0 1 ; }⟶ Estimates μC − μA
ESTIMATE 'MUC-MUB' GASTYPE 0 -1 1 ; }⟶ Estimates μC − μB
ESTIMATE 'MUB-(MUC+MUA)/2' GASTYPE -.5  1  -.5 ; }⟶ Estimates μB − ((μC + μA)/2)
PROC GLM ;
CLASS GASTYPE ;
MODEL MILEAGE = GASTYPE / P CLI ; }⟶ CLI requests prediction intervals
```

Notes:

1. *The coefficients in the above ESTIMATE statements are obtained by writing the quantity to be estimated as a linear combination of the factor level means μ_A, μ_B, and μ_C with the factor levels considered in* alphabetical order. *For example, if we consider MUB − MUA (that is, $\mu_B - \mu_A$), we write this difference as*

   $$-\mu_A + \mu_B = -1(\mu_A) + 1(\mu_B) + 0(\mu_C)$$

   Here, the "trailing zero" coefficient corresponding to μ_C may be dropped to obtain

   ESTIMATE 'MUB − MUA' GASTYPE −1 1 ;

   As another example, the coefficients in the ESTIMATE statement for MUB − (MUC + MUA)/2 (that is, $\mu_B - (\mu_C + \mu_A)/2$) are obtained by writing this expression as

   $$\begin{aligned} \mu_B - (\mu_C + \mu_A)/2 &= -\tfrac{1}{2}(\mu_A) + 1(\mu_B) + (-\tfrac{1}{2})\mu_C \\ &= -.5(\mu_A) + 1(\mu_B) + (-.5)\mu_C \end{aligned}$$

   Thus we obtain

   ESTIMATE 'MUB−(MUC+MUA)/2' GASTYPE −5 1 −5 ;

2. *Expressions inside single quotes (for example, 'MUB − MUA') are labels that may be up to 16 characters in length.*
3. *Confidence intervals (CLM) and prediction intervals (CLI) may not be requested in the same MODEL statement when using PROC GLM.*

In Figure 14.6 we present the SAS program that yields the means model analysis of the North American Oil Company data that is presented in Figure 14.3. Note that in this program we employ explicitly defined dummy variables to define the means model.

FIGURE 14.6 **SAS program to obtain a means model analysis of the North American Oil Company data in Table 14.1 by using dummy variables**

```
DATA UNLEAD ;
INPUT Y DB DC ; }          → Defines response variable Y
                             and dummy variables DB and DC
CARDS ;                      for the means model
24.0 0 0                   → Dummy variable input for means model
25.0 0 0
24.3 0 0
25.5 0 0
25.3 1 0
26.5 1 0
26.4 1 0                   → Data—See Table 14.1
27.0 1 0
27.6 1 0
23.3 0 1
24.0 0 1
24.7 0 1
PROC REG ;                 } → Specifies means model
MODEL Y = DB DC / P ;      }   using PROC REG
PROC GLM ;                 } → Specifies means model using PROC GLM
MODEL Y = DB DC / P ;      }   (PROC GLM is needed for "estimation")
ESTIMATE 'MUB - MUA' DB 1 ; } → Estimates μB − μA
ESTIMATE 'MUC - MUA' DC 1 ; } → Estimates μC − μA
ESTIMATE 'MUC - MUB' DB -1 DC 1 ; } → Estimates μC − μB
ESTIMATE 'MUB-(MUC+MUA)/2' DB 1 DC -.5 ; } → Estimates μB − ((μC + μA)/2)
```

Note: The means model

$$y_{l,k} = \mu_l + \varepsilon_{l,k} = \beta_A + \beta_B D_{l,B} + \beta_C D_{l,C} + \varepsilon_{l,k}$$

implies that

$$\mu_A = \beta_A \qquad \mu_B = \beta_A + \beta_B \qquad \mu_C = \beta_A + \beta_C$$

It follows that

$$\mu_B - \mu_A = \beta_B \qquad \text{(in SAS: DB 1)}$$
$$\mu_C - \mu_A = \beta_C \qquad \text{(in SAS: DC 1)}$$
$$\mu_C - \mu_B = \beta_C - \beta_B \qquad \text{(in SAS: DB } -1 \text{ DC 1)}$$

and $\mu_B - \left(\frac{\mu_C + \mu_A}{2}\right) = \beta_B - .5\beta_C$ *(in SAS: DB 1 DC −.5)*

EXERCISES

Exercises 14.1–14.7 are based on the following situation. The Tastee Bakery Company supplies a bakery product to a large number of supermarkets in a metropolitan area. The company wishes to study the effect of shelf display height, which has levels B (bottom), M (middle), and T (top), on monthly demand for the product. Here, months are considered to be four-week periods, and demand is measured in cases of 10 units each. For l = B, M, and T we define μ_l to be the mean monthly demand at supermarkets using display height l.

To compare μ_B, μ_M, and μ_T, the company will employ a completely randomized design. Specifically, for l = B, M, and T the bakery will randomly select a sample of n_l metropolitan area supermarkets of equal sales potential. These supermarkets will sell the product for one month using display height l. For $k = 1, 2, \ldots, n_l$, we let

$y_{l,k}$ = the monthly demand observed at the kth supermarket using display height l

Suppose that when Tastee Bakery employs this experimental design, it obtains the data in Table 14.8.

TABLE 14.8 Three samples of monthly bakery product demands

Display height		
B	*M*	*T*
$y_{B,1} = 58.2$	$y_{M,1} = 73.0$	$y_{T,1} = 52.5$
$y_{B,2} = 53.7$	$y_{M,2} = 78.1$	$y_{T,2} = 49.8$
$y_{B,3} = 55.6$	$y_{M,3} = 76.2$	$y_{T,3} = 56.0$
	$y_{M,4} = 82.0$	$y_{T,4} = 51.9$
	$y_{M,5} = 78.4$	$y_{T,5} = 53.3$

14.1 Considering analyzing the bakery product data in Table 14.8 by using one-way ANOVA.

a. Set up the null and alternative hypotheses needed to test for whether or not there are any statistically significant differences between μ_B, μ_M, and μ_T.
b. Test the null hypothesis set up in part (a) with $\alpha = .05$. Can we conclude that statistically significant differences exist between μ_B, μ_M, and μ_T?

14.2 Consider analyzing the bakery product data in Table 14.8 by using one-way ANOVA.

a. Set up the null and alternative hypotheses needed to test to see whether or not μ_M and μ_B differ. Use an appropriate t statistic to test these hypotheses with $\alpha = .05$.
b. Compute a 95% (individual) confidence interval for $\mu_M - \mu_B$. Interpret the practical meaning of this interval.
c. Set up the null and alternative hypotheses needed to test for whether or not μ_T and μ_B differ. Use an appropriate t statistic to test these hypotheses with $\alpha = .05$.
d. Compute a 95% (individual) confidence interval for $\mu_T - \mu_B$. Interpret the practical meaning of this interval.
e. Set up the null and alternative hypotheses needed to test for whether or not μ_T and μ_M differ. Use an appropriate t statistic to test these hypotheses with $\alpha = .05$.
f. Compute a 95% (individual) confidence interval for $\mu_T - \mu_M$. Interpret the practical meaning of this interval.
g. Compute a 95% (individual) confidence interval for $\mu_M - (\mu_T + \mu_B)/2$. Interpret this interval.

14.3 Consider analyzing the bakery product data in Table 14.8 by using one-way ANOVA.

a. Compute a 95% (individual) confidence interval for each of μ_B, μ_M, and μ_T. Interpret these intervals.
b. Compute a 95% prediction interval for each of the individual values $y_{B,0}$, $y_{M,0}$, and $y_{T,0}$.

14.4 Figure 14.7 presents the SAS output obtained by using PROC GLM to perform a one-way ANOVA of the bakery product data in Table 14.8.

a. Identify and report the values of SS_{total}, SS_{means}, MS_{means}, SSE, and MSE.
b. Identify, report, and interpret F(means) and its associated prob-value.
c. Identify, report, and interpret the appropriate t statistics and associated prob-values for making all pairwise comparisons of μ_B, μ_M, and μ_T.
d. Use the computer output to calculate 95% (individual) confidence intervals for $\mu_M - \mu_B$, $\mu_T - \mu_B$, and $\mu_T - \mu_M$.
e. Identify, report, and interpret point estimates and 95% confidence intervals for μ_B, μ_M, and μ_T.
f. Identify, report, and interpret point predictions and 95% prediction intervals for the individual values $y_{B,0}$, $y_{M,0}$, and $y_{T,0}$.

14.5 Consider analyzing the bakery product data in Table 14.8 by using one-way ANOVA and consider the following set of linear combinations of means:

$$\mu_M - \mu_B$$

$$\mu_T - \mu_B$$

$$\mu_T - \mu_M$$

a. Calculate Tukey simultaneous 95% confidence intervals for each of the linear combinations in this set.
b. Calculate Bonferroni simultaneous 95% confidence intervals for each of the linear combinations in this set.
c. Calculate Scheffé simultaneous 95% confidence intervals for each of the linear combinations in this set.

FIGURE 14.7 SAS output of a one-way ANOVA of the bakery data in Exercise 14.1

SAS

GENERAL LINEAR MODELS PROCEDURE

DEPENDENT VARIABLE: DEMAND

SOURCE	DF	SUM OF SQUARES	MEAN SQUARE	F VALUE	PR > F
MODEL	2	1741.58441026	870.79220513	117.84	0.0001
ERROR	10	73.89866667	7.38986667		ROOT MSE
CORRECTED TOTAL	12	1815.48307692			2.71843092

SOURCE	DF	TYPE I SS	F VALUE	PR > F	DF	TYPE III SS
HEIGHT	2	1741.58441026	117.84	0.0001	2	1741.58441026

PARAMETER	ESTIMATE	T FOR HO: PARAMETER=0	PR > |T|	STD ERROR OF ESTIMATE
MUM-MUB	21.70666667	10.93	0.0001	1.98526125
MUT-MUB	-3.13333333	-1.58	0.1456	1.98526125
MUT-MUM	-24.84000000	-14.45	0.0001	1.71928667

OBSERVATION	OBSERVED VALUE	PREDICTED VALUE	RESIDUAL	LOWER 95% CL FOR MEAN	UPPER 95% CL FOR MEAN
1	58.20000000	55.83333333	2.36666667	52.33627677	59.33038990
2	53.70000000	55.83333333	-2.13333333	52.33627677	59.33038990
3	55.60000000	55.83333333	-0.23333333	52.33627677	59.33038990
4	73.00000000	77.54000000	-4.54000000	74.83119163	80.24880837
5	78.10000000	77.54000000	0.56000000	74.83119163	80.24880837
6	76.20000000	77.54000000	-1.34000000	74.83119163	80.24880837
7	82.00000000	77.54000000	4.46000000	74.83119163	80.24880837
8	78.40000000	77.54000000	0.86000000	74.83119163	80.24880837
9	52.50000000	52.70000000	-0.20000000	49.99119163	55.40880837
10	49.80000000	52.70000000	-2.90000000	49.99119163	55.40880837
11	56.00000000	52.70000000	3.30000000	49.99119163	55.40880837
12	51.90000000	52.70000000	-0.80000000	49.99119163	55.40880837
13	53.30000000	52.70000000	0.60000000	49.99119163	55.40880837

OBSERVATION	OBSERVED VALUE	PREDICTED VALUE	RESIDUAL	LOWER 95% CL INDIVIDUAL	UPPER 95% CL INDIVIDUAL
1	58.20000000	55.83333333	2.36666667	48.83922021	62.82744646
2	53.70000000	55.83333333	-2.13333333	48.83922021	62.82744646
3	55.60000000	55.83333333	-0.23333333	48.83922021	62.82744646
4	73.00000000	77.54000000	-4.54000000	70.90480169	84.17519831
5	78.10000000	77.54000000	0.56000000	70.90480169	84.17519831
6	76.20000000	77.54000000	-1.34000000	70.90480169	84.17519831
7	82.00000000	77.54000000	4.46000000	70.90480169	84.17519831
8	78.40000000	77.54000000	0.86000000	70.90480169	84.17519831
9	52.50000000	52.70000000	-0.20000000	46.06480169	59.33519831
10	49.80000000	52.70000000	-2.90000000	46.06480169	59.33519831
11	56.00000000	52.70000000	3.30000000	46.06480169	59.33519831
12	51.90000000	52.70000000	-0.80000000	46.06480169	59.33519831
13	53.30000000	52.70000000	0.60000000	46.06480169	59.33519831

14.6 Consider analyzing the bakery product data in Table 14.8 by using one-way ANOVA and consider the following set of linear combinations of means (note that each of these linear combinations is a contrast):

$$\mu_M - \mu_B$$

$$\mu_T - \mu_B$$

$$\mu_T - \mu_M$$

$$\mu_M - \frac{(\mu_T + \mu_B)}{2}$$

a. Calculate Bonferroni simultaneous 95% confidence intervals for each of the contrasts in this set.
b. Calculate Scheffé simultaneous 95% confidence intervals for each of the contrasts in this set.

14.7 Consider analyzing the bakery product data in Table 14.8 by using one-way ANOVA and consider the following set of linear combinations of means:

$$\mu_M - \mu_B \qquad \mu_M - \frac{(\mu_T + \mu_B)}{2}$$

$$\mu_T - \mu_B \qquad \mu_M$$

$$\mu_T - \mu_M$$

a. Calculate Bonferroni simultaneous 95% confidence intervals for each of the linear combinations in this set.
b. Calculate Scheffé simultaneous 95% confidence intervals for each of the linear combinations in this set.

Exercises 14.8 through 14.10 refer to the following situation. To compare the durability of four different brands of golf balls (ALPHA, BEST, CENTURY, and DIVOT), the National Golf Association randomly selects five balls of each brand and places each ball into a machine that exerts the force produced by a 250-yard drive. The number of simulated drives needed to crack or chip each ball is recorded. The results are given in Table 14.9. A one-way analysis of variance for this data is carried out by using SAS to compare the mean number of simulated drives needed to crack or chip golf balls of each brand. The PROC GLM output is given in Figure 14.8. (*Note*: MUALPHA, MUBEST, MUCENTURY, and MUDIVOT denote the appropriate means on this output.)

TABLE 14.9 **Golf ball durability test results**

Brand			
ALPHA	*BEST*	*CENTURY*	*DIVOT*
281	270	218	364
220	334	244	302
274	307	225	325
242	290	273	337
251	331	249	355

FIGURE 14.8 **SAS output of a one-way ANOVA of the golf ball durability data**

SAS

GENERAL LINEAR MODELS PROCEDURE

DEPENDENT VARIABLE: DRIVES

SOURCE	DF	SUM OF SQUARES	MEAN SQUARE	F VALUE	PR > F
MODEL	3	29860.40000000	9953.46666667	16.42	0.0001
ERROR	16	9698.40000000	606.15000000		ROOT MSE
CORRECTED TOTAL	19	39558.80000000			24.62011373

SOURCE	DF	TYPE I SS	F VALUE	PR > F	DF	TYPE III SS
BRAND	3	29860.40000000	16.42	0.0001	3	29860.40000000

PARAMETER	ESTIMATE	T FOR HO: PARAMETER=0	PR > |T|	STD ERROR OF ESTIMATE
MUDIVOT-MUALPHA	83.00000000	5.33	0.0001	15.57112713
MUDIVOT-MUBEST	30.20000000	1.94	0.0703	15.57112713
MUDIVOT-MUCENTURY	94.80000000	6.09	0.0001	15.57112713
MUCENTURY-MUBEST	-64.60000000	-4.15	0.0008	15.57112713
MUCENTURY-MUALPHA	-11.80000000	-0.76	0.4596	15.57112713
MUBEST-MUALPHA	52.80000000	3.39	0.0037	15.57112713

OBSERVATION	OBSERVED VALUE	PREDICTED VALUE	RESIDUAL	LOWER 95% CL FOR MEAN	UPPER 95% CL FOR MEAN
1	281.00000000	253.60000000	27.40000000	230.25901629	276.94098371
2	220.00000000	253.60000000	-33.60000000	230.25901629	276.94098371
3	274.00000000	253.60000000	20.40000000	230.25901629	276.94098371
4	242.00000000	253.60000000	-11.60000000	230.25901629	276.94098371
5	251.00000000	253.60000000	-2.60000000	230.25901629	276.94098371
6	270.00000000	306.40000000	-36.40000000	283.05901629	329.74098371
7	334.00000000	306.40000000	27.60000000	283.05901629	329.74098371
8	307.00000000	306.40000000	0.60000000	283.05901629	329.74098371
9	290.00000000	306.40000000	-16.40000000	283.05901629	329.74098371
10	331.00000000	306.40000000	24.60000000	283.05901629	329.74098371
11	218.00000000	241.80000000	-23.80000000	218.45901629	265.14098371
12	244.00000000	241.80000000	2.20000000	218.45901629	265.14098371
13	225.00000000	241.80000000	-16.80000000	218.45901629	265.14098371
14	273.00000000	241.80000000	31.20000000	218.45901629	265.14098371
15	249.00000000	241.80000000	7.20000000	218.45901629	265.14098371
16	364.00000000	336.60000000	27.40000000	313.25901629	359.94098371
17	302.00000000	336.60000000	-34.60000000	313.25901629	359.94098371
18	325.00000000	336.60000000	-11.60000000	313.25901629	359.94098371
19	337.00000000	336.60000000	0.40000000	313.25901629	359.94098371
20	355.00000000	336.60000000	18.40000000	313.25901629	359.94098371

OBSERVATION	OBSERVED VALUE	PREDICTED VALUE	RESIDUAL	LOWER 95% CL INDIVIDUAL	UPPER 95% CL INDIVIDUAL
1	281.00000000	253.60000000	27.40000000	196.42649983	310.77350017
2	220.00000000	253.60000000	-33.60000000	196.42649983	310.77350017
3	274.00000000	253.60000000	20.40000000	196.42649983	310.77350017
4	242.00000000	253.60000000	-11.60000000	196.42649983	310.77350017
5	251.00000000	253.60000000	-2.60000000	196.42649983	310.77350017
6	270.00000000	306.40000000	-36.40000000	249.22649983	363.57350017
7	334.00000000	306.40000000	27.60000000	249.22649983	363.57350017
8	307.00000000	306.40000000	0.60000000	249.22649983	363.57350017
9	290.00000000	306.40000000	-16.40000000	249.22649983	363.57350017
10	331.00000000	306.40000000	24.60000000	249.22649983	363.57350017
11	218.00000000	241.80000000	-23.80000000	184.62649983	298.97350017

(*continues*)

FIGURE 14.8 Continued

OBSERVATION	OBSERVED VALUE	PREDICTED VALUE	RESIDUAL	LOWER 95% CL INDIVIDUAL	UPPER 95% CL INDIVIDUAL
12	244.00000000	241.80000000	2.20000000	184.62649983	298.97350017
13	225.00000000	241.80000000	-16.80000000	184.62649983	298.97350017
14	273.00000000	241.80000000	31.20000000	184.62649983	298.97350017
15	249.00000000	241.80000000	7.20000000	184.62649983	298.97350017
16	364.00000000	336.60000000	27.40000000	279.42649983	393.77350017
17	302.00000000	336.60000000	-34.60000000	279.42649983	393.77350017
18	325.00000000	336.60000000	-11.60000000	279.42649983	393.77350017
19	337.00000000	336.60000000	0.40000000	279.42649983	393.77350017
20	355.00000000	336.60000000	18.40000000	279.42649983	393.77350017

14.8 Consider analyzing the golf ball durability data in Table 14.9 by using one-way ANOVA.

a. Set up the null and alternative hypotheses needed to test for statistically significant differences between the treatment means μ_{ALPHA}, μ_{BEST}, $\mu_{CENTURY}$, and μ_{DIVOT}.
b. Using the computer output in Figure 14.8, carry out the hypothesis test you set up in part (a). Interpret the practical meaning of the results of this test.
c. Using numbers found on the computer output in Figure 14.8, hand calculate the F(means) statistic for the golf ball durability data.

14.9 Consider analyzing the golf ball durability data in Table 14.9 by using one-way ANOVA.

a. Using the computer output in Figure 14.8, perform pairwise comparisons of the treatment means. Be sure to explain which means differ and which means do not differ. Use $\alpha = .05$ to judge statistical significance.
b. Using the computer output in Figure 14.8, compute a 95% confidence interval for each of the pairwise differences

$$\mu_{DIVOT} - \mu_{ALPHA} \qquad \mu_{CENTURY} - \mu_{BEST} \qquad \mu_{DIVOT} - \mu_{BEST}$$

$$\mu_{DIVOT} - \mu_{CENTURY} \qquad \mu_{BEST} - \mu_{ALPHA} \qquad \mu_{CENTURY} - \mu_{ALPHA}$$

c. Interpret the practical meaning of each of the confidence intervals computed in part (b).

14.10 Consider analyzing the golf ball durability data in Table 14.9 by using one-way ANOVA and consider the SAS output in Figure 14.8.

a. Using the computer output, find a point estimate of the mean number of drives required to crack or chip a DIVOT golf ball.
b. Using the computer output, find a 95% confidence interval for the mean number of drives required to crack or chip a DIVOT golf ball. Interpret this interval.
c. Using the computer output, find a point prediction of the number of drives required to crack or chip a BEST golf ball.
d. Using the computer output, find a 95% prediction interval for the number of drives required to crack or chip a BEST golf ball. Interpret this interval.

Exercises 14.11 through 14.13 refer to the following situation. A consumer preference study is conducted to examine the effects of three different bottle designs (A, B, and C)

TABLE 14.10 **Bottle design study data**

Bottle design		
A	*B*	*C*
16	33	23
18	31	27
19	37	21
17	29	28
13	34	25

on sales of a popular fabric softener. A completely randomized design is employed. Fifteen supermarkets of equal sales potential are selected, and five of these supermarkets are randomly assigned to each bottle design. The number of bottles sold in a 24-hour period at each supermarket is recorded. The data obtained is given in Table 14.10. A one-way ANOVA of this data is carried out by using SAS. The PROC GLM output is given in Figure 14.9. Note that the treatment means μ_A, μ_B, and μ_C are denoted as MUA, MUB, and MUC on the output.

14.11 Consider analyzing the bottle design study data in Table 14.10 by using one-way ANOVA.

a. Set up the null and alternative hypotheses needed to test for statistically significant differences between the treatment means μ_A, μ_B, and μ_C.
b. Using the computer output in Figure 14.9, carry out the hypothesis test you set up in part (a). Interpret the practical meaning of the result of this test.
c. Using numbers found on the computer output in Figure 14.9, hand calculate the F(means) statistic for the bottle design data.

14.12 Consider analyzing the bottle design study data in Table 14.10 by using one-way ANOVA.

a. Using the computer output in Figure 14.9, perform pairwise comparisons of the treatment means. Be sure to explain which means differ and which means do not differ. Use $\alpha = .05$ to judge statistical significance.
b. Using the computer output in Figure 14.9, compute a 95% confidence interval for each of the pairwise differences

$$\mu_C - \mu_A \qquad \mu_C - \mu_B \qquad \mu_B - \mu_A$$

c. Interpret the practical meaning of each of the confidence intervals computed in part (b).

14.13 Consider analyzing the bottle design study data in table 14.10 by using one-way ANOVA.

a. Using the computer output in Figure 14.9, find point estimates of each of the treatment means μ_A, μ_B, and μ_C.
b. Using the computer output in Figure 14.9, find 95% confidence intervals for each of the treatment means μ_A, μ_B, and μ_C. Interpret these intervals.

FIGURE 14.9 SAS output of a one-way ANOVA of the bottle design study data

SAS

GENERAL LINEAR MODELS PROCEDURE

DEPENDENT VARIABLE: SALES

SOURCE	DF	SUM OF SQUARES	MEAN SQUARE	F VALUE	PR > F	R-SQUARE	C.V.
MODEL	2	656.13333333	328.06666667	43.36	0.0001	0.878436	11.1217
ERROR	12	90.80000000	7.56666667		ROOT MSE		SALES MEAN
CORRECTED TOTAL	14	746.93333333			2.75075747		24.73333333

SOURCE	DF	TYPE I SS	F VALUE	PR > F	DF	TYPE III SS	F VALUE	PR > F
DESIGN	2	656.13333333	43.36	0.0001	2	656.13333333	43.36	0.0001

PARAMETER	ESTIMATE	T FOR HO: PARAMETER=0	PR > |T|	STD ERROR OF ESTIMATE
MUB-MUA	16.20000000	9.31	0.0001	1.73973178
MUC-MUA	8.20000000	4.71	0.0005	1.73973178
MUC-MUB	-8.00000000	-4.60	0.0006	1.73973178

OBSERVATION	OBSERVED VALUE	PREDICTED VALUE	RESIDUAL	LOWER 95% CL FOR MEAN	UPPER 95% CL FOR MEAN
1	16.00000000	16.60000000	-0.60000000	13.91967250	19.28032750
2	18.00000000	16.60000000	1.40000000	13.91967250	19.28032750
3	19.00000000	16.60000000	2.40000000	13.91967250	19.28032750
4	17.00000000	16.60000000	0.40000000	13.91967250	19.28032750
5	13.00000000	16.60000000	-3.60000000	13.91967250	19.28032750
6	33.00000000	32.80000000	0.20000000	30.11967250	35.48032750
7	31.00000000	32.80000000	-1.80000000	30.11967250	35.48032750
8	37.00000000	32.80000000	4.20000000	30.11967250	35.48032750
9	29.00000000	32.80000000	-3.80000000	30.11967250	35.48032750
10	34.00000000	32.80000000	1.20000000	30.11967250	35.48032750
11	23.00000000	24.80000000	-1.80000000	22.11967250	27.48032750
12	27.00000000	24.80000000	2.20000000	22.11967250	27.48032750
13	21.00000000	24.80000000	-3.80000000	22.11967250	27.48032750
14	28.00000000	24.80000000	3.20000000	22.11967250	27.48032750
15	25.00000000	24.80000000	0.20000000	22.11967250	27.48032750

OBSERVATION	OBSERVED VALUE	PREDICTED VALUE	RESIDUAL	LOWER 95% CL INDIVIDUAL	UPPER 95% CL INDIVIDUAL
1	16.00000000	16.60000000	-0.60000000	10.03456527	23.16543473
2	18.00000000	16.60000000	1.40000000	10.03456527	23.16543473
3	19.00000000	16.60000000	2.40000000	10.03456527	23.16543473
4	17.00000000	16.60000000	0.40000000	10.03456527	23.16543473
5	13.00000000	16.60000000	-3.60000000	10.03456527	23.16543473
6	33.00000000	32.80000000	0.20000000	26.23456527	39.36543473
7	31.00000000	32.80000000	-1.80000000	26.23456527	39.36543473
8	37.00000000	32.80000000	4.20000000	26.23456527	39.36543473
9	29.00000000	32.80000000	-3.80000000	26.23456527	39.36543473
10	34.00000000	32.80000000	1.20000000	26.23456527	39.36543473
11	23.00000000	24.80000000	-1.80000000	18.23456527	31.36543473
12	27.00000000	24.80000000	2.20000000	18.23456527	31.36543473
13	21.00000000	24.80000000	-3.80000000	18.23456527	31.36543473
14	28.00000000	24.80000000	3.20000000	18.23456527	31.36543473
15	25.00000000	24.80000000	0.20000000	18.23456527	31.36543473

c. Let $y_{l,0}$ be an individual number of bottles sold using bottle design l. Using the computer output in Figure 14.9, find point predictions of $y_{A,0}$, $y_{B,0}$, and $y_{C,0}$.
d. Using the computer output in Figure 14.9, find 95% prediction intervals for each of $y_{A,0}$, $y_{B,0}$, and $y_{C,0}$. Interpret these intervals.

14.14 Consider using dummy variable regression to analyze the bakery product data in Table 14.8. To analyze these data, we use the following model to describe $y_{l,k}$:

$$\begin{aligned} y_{l,k} &= \mu_l + \varepsilon_{l,k} \\ &= \beta_B + \beta_M D_{l,M} + \beta_T D_{l,T} + \varepsilon_{l,k} \end{aligned}$$

Here, $D_{l,M} = 1$ if a middle display height was used to obtain $y_{l,k}$ and $D_{l,M} = 0$ otherwise, and $D_{l,T} = 1$ if a top display height was used to obtain $y_{l,k}$ and $D_{l,T} = 0$ otherwise.

a. Express the following means and differences in means in terms of β_B, β_M, and β_T:

$$\mu_B \quad \mu_M \quad \mu_T \quad \mu_M - \mu_B \quad \mu_T - \mu_B \quad \mu_T - \mu_M$$

Then show that H_0: $\beta_M = \beta_T = 0$ is equivalent to H_0: $\mu_B = \mu_M = \mu_T$.

b. Specify the vector **y** and the matrix **X** used to calculate the least squares point estimates of the parameters in the preceding model.
c. When we use this model to perform a regression analysis of these data, we find that the least squares point estimates of the parameters are

$$b_B = 55.8333 \qquad b_M = 21.7067 \qquad b_T = -3.1333$$

and we also obtain the following quantities

$$SS_{\text{model}} = \text{Explained variation} = 1741.5844$$
$$SSE = \text{Unexplained variation} = 73.8987$$

Use these quantities to compute the overall F statistic for the dummy variable model.

d. Use the overall F statistic to test

$$H_0\text{: } \beta_M = \beta_T = 0 \quad \text{or} \quad \mu_B = \mu_M = \mu_T$$

versus

H_1: At least one of β_M and β_T does not equal zero

or

H_1: At least two of μ_B, μ_M, and μ_T differ from each other

with $\alpha = .05$. Can we conclude that there are statistically significant differences between the treatment means μ_B, μ_M, and μ_T? Explain.

e. The prob-value for testing H_0 versus H_1 can be calculated to be .0001. Demonstrate how this prob-value has been calculated, and use this prob-value to determine whether we can reject H_0 in favor of H_1 by setting α equal to .05 or .01.

14.15 Consider using the dummy variable regression model of Exercise 14.14 to analyze the bakery product data in Table 14.8. In Table 14.11 we give several linear combinations of parameters in the dummy variable model (each of which can be expressed in the form $\boldsymbol{\lambda}'\boldsymbol{\beta}$) along with the corresponding value of $s\sqrt{\boldsymbol{\lambda}'(\mathbf{X}'\mathbf{X})^{-1}\boldsymbol{\lambda}}$ and the prob-value for testing H_0: $\boldsymbol{\lambda}'\boldsymbol{\beta} = 0$.

a. In Exercise 14.14, each of the linear combinations in this table has been expressed in terms of β_B, β_M, and β_T. By using the least squares point estimates of β_B, β_M, and β_T given in Exercise 14.14(c), calculate a point estimate of each linear combination in the table.

TABLE 14.11 **Standard errors and prob-values for several linear combinations of parameters in the bakery product dummy variable model**

Linear combination, $\lambda'\beta$	$s\sqrt{\lambda'(X'X)^{-1}\lambda}$	*Prob-value for* H_0*:* $\lambda'\beta = 0$
$\mu_M - \mu_B$	1.9853	.0001
$\mu_T - \mu_B$	1.9853	.1456
$\mu_T - \mu_M$	1.7193	.0001
μ_M	1.2157	—

b. Each linear combination can be expressed in the form $\lambda'\beta$. Find the vector λ' for each linear combination.

c. Calculate an (individual) 95% confidence interval for each linear combination in the table.

d. Calculate the t statistic for testing H_0: $\lambda'\beta = 0$ for each of the linear combinations $\mu_M - \mu_B$, $\mu_T - \mu_B$, and $\mu_T - \mu_M$.

e. Use the t statistics of part (d) to test the appropriate null hypotheses with $\alpha = .05$. Explain the practical meanings of the results.

f. Show how each of the prob-values in the table has been calculated. What do these prob-values say about whether or not the appropriate null hypotheses should be rejected?

g. Calculate a point prediction of, and a 95% prediction interval for, $y_{M,0}$, a future (individual) demand that will be observed when using a middle display height.

14.16 Consider using the dummy variable regression model of Exercise 14.14 to analyze the bakery product data in Table 14.8.

a. Using any pertinent information in Exercises 14.14 and 14.15, calculate Scheffé simultaneous confidence intervals for the linear combinations (contrasts)

$$\mu_M - \mu_B, \quad \mu_T - \mu_B, \quad \text{and} \quad \mu_T - \mu_M$$

b. Using any pertinent information in Exercises 14.14 and 14.15, calculate Scheffé simultaneous confidence intervals for the linear combinations

$$\mu_M - \mu_B, \quad \mu_T - \mu_B, \quad \mu_T - \mu_M, \quad \text{and} \quad \mu_M$$

14.17 An oil company wishes to study the effects of four different gasoline additives on mean gasoline mileage. The company randomly selects four groups of six automobiles each and assigns a group of six automobiles to each additive type (W, X, Y, and Z). Here all 24 automobiles employed in the experiment are the same make and model. Each of the six automobiles assigned to a gasoline additive is test driven using the appropriate additive, and the gasoline mileage for the test drive is recorded. The results of the experiment are given in Table 14.12. A one-way ANOVA of this data is carried out by using SAS. The PROC GLM output is given in Figure 14.10. Note that the treatment means μ_W, μ_X, μ_Y, and μ_Z are denoted as MUW, MUX, MUY, AND MUZ on the output.

a. Identify and report the values of SS_{total}, SS_{means}, MS_{means}, SSE, and MSE.

b. Identify, report, and interpret F(means) and its associated prob-value.

TABLE 14.12 Gasoline additive test results

Gasoline additive			
W	*X*	*Y*	*Z*
31.2	27.6	35.7	34.5
32.6	28.1	34.0	36.2
30.8	27.4	35.1	35.2
31.5	28.5	33.9	35.8
32.0	27.5	36.1	34.9
30.1	28.7	34.8	35.3

FIGURE 14.10 SAS output of a one-way ANOVA of the gasoline additive test results

```
                                                  SAS
                                    GENERAL LINEAR MODELS PROCEDURE
DEPENDENT VARIABLE: MILEAGE
SOURCE           DF   SUM OF SQUARES   MEAN SQUARE   F VALUE       PR > F   R-SQUARE          C.V.
MODEL             3     213.88125000   71.29375000    127.22       0.0001   0.950205        2.3108
ERROR            20      11.20833333    0.56041667               ROOT MSE             MILEAGE MEAN
CORRECTED TOTAL  23     225.08958333                           0.74860982              32.39583333

SOURCE        DF       TYPE I SS      F VALUE    PR > F     DF     TYPE III SS     F VALUE    PR > F
ADDTYPE        3    213.88125000       127.22    0.0001      3    213.88125000      127.22    0.0001

                                        T FOR HO:         PR > |T|      STD ERROR OF
PARAMETER               ESTIMATE     PARAMETER=0                          ESTIMATE
MUZ-MUW               3.95000000            9.14          0.0001        0.43221008
MUZ-MUX               7.35000000           17.01          0.0001        0.43221008
MUZ-MUY               0.38333333            0.89          0.3857        0.43221008
MUY-MUW               3.56666667            8.25          0.0001        0.43221008
MUY-MUX               6.96666667           16.12          0.0001        0.43221008
MUX-MUW              -3.40000000           -7.87          0.0001        0.43221008
(Y+Z)/2-(X+W)/2       5.45833333           17.86          0.0001        0.30561868

OBSERVATION         OBSERVED         PREDICTED          RESIDUAL        LOWER 95% CL     UPPER 95% CL
                       VALUE             VALUE                             FOR MEAN         FOR MEAN
          1      31.20000000       31.36666667       -0.16666667        30.72916207      32.00417126
          2      32.60000000       31.36666667        1.23333333        30.72916207      32.00417126
          3      30.80000000       31.36666667       -0.56666667        30.72916207      32.00417126
          4      31.50000000       31.36666667        0.13333333        30.72916207      32.00417126
          5      32.00000000       31.36666667        0.63333333        30.72916207      32.00417126
          6      30.10000000       31.36666667       -1.26666667        30.72916207      32.00417126
          7      27.60000000       27.96666667       -0.36666667        27.32916207      28.60417126
          8      28.10000000       27.96666667        0.13333333        27.32916207      28.60417126
          9      27.40000000       27.96666667       -0.56666667        27.32916207      28.60417126
         10      28.50000000       27.96666667        0.53333333        27.32916207      28.60417126
         11      27.50000000       27.96666667        0.46666667        27.32916207      28.60417126
         12      28.70000000       27.96666667        0.73333333        27.32916207      28.60417126
         13      35.70000000       34.93333333        0.76666667        34.29582874      35.57083793
         14      34.00000000       34.93333333       -0.93333333        34.29582874      35.57083793
```

(*continues*)

FIGURE 14.10 Continued

OBSERVATION	OBSERVED VALUE	PREDICTED VALUE	RESIDUAL	LOWER 95% CL FOR MEAN	UPPER 95% CL FOR MEAN
15	35.10000000	34.93333333	0.16666667	34.29582874	35.57083793
16	33.90000000	34.93333333	-1.03333333	34.29582874	35.57083793
17	36.10000000	34.93333333	1.16666667	34.29582874	35.57083793
18	34.80000000	34.93333333	-0.13333333	34.29582874	35.57083793
19	34.50000000	35.31666667	-0.81666667	34.67916207	35.95417126
20	36.20000000	35.31666667	0.88333333	34.67916207	35.95417126
21	35.20000000	35.31666667	-0.11666667	34.67916207	35.95417126
22	35.80000000	35.31666667	0.48333333	34.67916207	35.95417126
23	34.90000000	35.31666667	-0.41666667	34.67916207	35.95417126
24	35.30000000	35.31666667	-0.01666667	34.67916207	35.95417126

OBSERVATION	OBSERVED VALUE	PREDICTED VALUE	RESIDUAL	LOWER 95% CL INDIVIDUAL	UPPER 95% CL INDIVIDUAL
1	31.20000000	31.36666667	-0.16666667	29.67998806	33.05334528
2	32.60000000	31.36666667	1.23333333	29.67998806	33.05334528
3	30.80000000	31.36666667	-0.56666667	29.67998806	33.05334528
4	31.50000000	31.36666667	0.13333333	29.67998806	33.05334528
5	32.00000000	31.36666667	0.63333333	29.67998806	33.05334528
6	30.10000000	31.36666667	-1.26666667	29.67998806	33.05334528
7	27.60000000	27.96666667	-0.36666667	26.27998806	29.65334528
8	28.10000000	27.96666667	0.13333333	26.27998806	29.65334528
9	27.40000000	27.96666667	-0.56666667	26.27998806	29.65334528
10	28.50000000	27.96666667	0.53333333	26.27998806	29.65334528
11	27.50000000	27.96666667	-0.46666667	26.27998806	29.65334528
12	28.70000000	27.96666667	0.73333333	26.27998806	29.65334528
13	35.70000000	34.93333333	0.76666667	33.24665472	36.62001194
14	34.00000000	34.93333333	-0.93333333	33.24665472	36.62001194
15	35.10000000	34.93333333	0.16666667	33.24665472	36.62001194
16	33.90000000	34.93333333	-1.03333333	33.24665472	36.62001194
17	36.10000000	34.93333333	1.16666667	33.24665472	36.62001194
18	34.80000000	34.93333333	-0.13333333	33.24665472	36.62001194
19	34.50000000	35.31666667	-0.81666667	33.62998806	37.00334528
20	36.20000000	35.31666667	0.88333333	33.62998806	37.00334528
21	35.20000000	35.31666667	-0.11666667	33.62998806	37.00334528
22	35.80000000	35.31666667	0.48333333	33.62998806	37.00334528
23	34.90000000	35.31666667	-0.41666667	33.62998806	37.00334528
24	35.30000000	35.31666667	-0.01666667	33.62998806	37.00334528

c. Identify, report, and interpret the appropriate t statistics and associated prob-values for making all pairwise comparisons of μ_W, μ_X, μ_Y, and μ_Z.
d. Identify, report, and interpret the appropriate t statistic and associated prob-value for testing the significance of $[(\mu_Y + \mu_Z)/2] - [(\mu_X + \mu_W)/2]$.
e. Identify, report, and interpret point estimates and 95% confidence intervals for μ_W, μ_X, μ_Y, and μ_Z.
f. Identify, report, and interpret point predictions and 95% prediction intervals for $y_{W,0}$, $y_{X,0}$, $y_{Y,0}$, and $y_{Z,0}$.

14.18 Consider the one-way ANOVA of the gasoline additive data in Table 14.12 and the SAS output of Figure 14.10.

a. Compute individual 95% confidence intervals for all possible pairwise differences between treatment means.
b. Compute Tukey simultaneous 95% confidence intervals for all possible pairwise differences between treatment means.

c. Compute Scheffé simultaneous 95% confidence intervals for all possible pairwise differences between treatment means.
d. Compute Bonferroni simultaneous 95% confidence intervals for the (prespecified) set of all possible pairwise differences between treatment means.
e. Which of the above intervals are the most precise?

14.19 Consider the one-way ANOVA of the gasoline additive data in Table 14.12 and the SAS output of Figure 14.10. Also consider the prespecified set of linear combinations (contrasts):

$$\mu_Z - \mu_W \quad \mu_Y - \mu_W$$

$$\mu_Z - \mu_X \quad \mu_Y - \mu_X$$

$$\mu_Z - \mu_Y \quad \mu_X - \mu_W$$

$$\frac{(\mu_Y + \mu_Z)}{2} - \frac{(\mu_X + \mu_W)}{2}$$

a. Compute an individual 95% confidence interval for the linear combination

$$\frac{(\mu_Y + \mu_Z)}{2} - \frac{(\mu_X + \mu_W)}{2}$$

b. Compute Scheffé simultaneous 95% confidence intervals for the linear combinations in the above set.
c. Compute Bonferroni simultaneous 95% confidence intervals for the linear combinations in the above set.

14.20 Consider the gasoline additive data in Table 14.12.

a. Formulate a dummy variable regression model that will appropriately analyze this data.
b. Using a computer and the dummy variable regression model, completely analyze the gasoline additive data.

14.21 Compare the results of the ANOVA analysis of the bakery product data (in Exercises 14.1–14.5) with the results of the dummy variable regression analysis of the bakery product data (in Exercises 14.14–14.16).

14.22 A study was made to compare three different display panels for use by air traffic controllers. Each display panel was tested in a simulated emergency condition. Twelve highly trained air traffic controllers took part in the study. Four controllers were randomly assigned to each display panel. The time (in seconds) needed to stabilize the emergency condition was recorded. The results of the study are given in Table 14.13. Completely analyze the data by using one-way ANOVA.

14.23 Consider the display panel study data in Table 14.13.

a. Formulate a dummy variable regression model that will appropriately analyze the data.
b. Using a computer and the dummy variable regression model, completely analyze the display panel data.

14.24 Table 14.14 presents the yields (in bushels per one-third-acre plot) for corn hybrid types W, X, and Y. Completely analyze this data by using one-way ANOVA.

TABLE 14.13 **Display panel study data (time, in seconds, required to stabilize emergency condition)**

Display panel		
A	**B**	**C**
21	24	40
27	21	36
24	18	35
26	19	32

14.25 Consider the corn yield data in Table 14.14.

a. Formulate a dummy variable regression model that will appropriately analyze the data.

b. Using a computer and the dummy variable regression model, completely analyze the corn yield data.

14.26 Modify the golf ball durability data in Table 14.9 by assuming that the four brands of golf balls have been randomly selected from the population of all brands. Then, using the random model,

a. Test H_0: $\sigma_\mu^2 = 0$ versus H_1: $\sigma_\mu^2 \neq 0$ by setting α equal to .05.
b. Find point estimates of σ^2 and σ_μ^2. Interpret.
c. Find a 95% confidence interval for μ. Interpret this interval.

14.27 Modify the corn yield data in Table 14.14 by assuming that the three corn types have been randomly selected from the population of all corn types. Then, using the random model,

a. Test H_0: $\sigma_\mu^2 = 0$ versus H_1: $\sigma_\mu^2 \neq 0$ by setting α equal to .05.
b. Find point estimates of σ^2 and σ_μ^2. Interpret.
c. Find a 95% confidence interval for μ. Interpret this interval.

TABLE 14.14 **Corn yield data (bushels per plot)**

Corn type		
W	**X**	**Y**
22.6	27.6	25.4
21.5	26.5	24.5
22.1	27.0	26.3
21.8	27.2	24.8
22.4	26.8	25.1

14.28 Consider the gasoline additive test data in Table 14.12. Employ Hartley's test for variance equality to test

$$H_0: \sigma_W^2 = \sigma_X^2 = \sigma_Y^2 = \sigma_Z^2$$

versus

$$H_1: \text{At least two of } \sigma_W^2, \sigma_X^2, \sigma_Y^2, \text{ and } \sigma_Z^2 \text{ differ}$$

14.29 Consider the gasoline additive test data in Table 14.12. Employ Bartlett's test for variance equality to test

$$H_0: \sigma_W^2 = \sigma_X^2 = \sigma_Y^2 = \sigma_Z^2$$

versus

$$H_1: \text{At least two of } \sigma_W^2, \sigma_X^2, \sigma_Y^2, \text{ and } \sigma_Z^2 \text{ differ}$$

14.30 Consider the bottle design study data in Table 14.10. Employ Hartley's test for variance equality to test

$$H_0: \sigma_A^2 = \sigma_B^2 = \sigma_C^2$$

versus

$$H_1: \text{At least two of } \sigma_A^2, \sigma_B^2, \text{ and } \sigma_C^2 \text{ differ}$$

14.31 Consider the corn yield data in Table 14.14. Employ Bartlett's test for variance equality to test

$$H_0: \sigma_W^2 = \sigma_X^2 = \sigma_Y^2$$

versus

$$H_1: \text{At least two of } \sigma_W^2, \sigma_X^2, \text{ and } \sigma_Y^2 \text{ differ}$$

14.32 Consider the display panel study data in Table 14.13. Employ the Box modified Bartlett statistic to test

$$H_0: \sigma_A^2 = \sigma_B^2 = \sigma_C^2$$

versus

$$H_1: \text{At least two of } \sigma_A^2, \sigma_B^2, \text{ and } \sigma_C^2 \text{ differ}$$

14.33 An employment agency gives an aptitude test to all clients. The results of this test are analyzed according to the type of college degree held by the clients (business, science, liberal arts, fine arts, engineering). The test results for eight randomly selected clients of each degree type are given in Table 14.15. Fully analyze this data using one-way ANOVA.

14.34 Consider the aptitude test score data in Table 14.15.

a. Formulate a dummy variable regression model that can be used to appropriately analyze the data.
b. Using a computer and the dummy variable regression model, completely analyze the aptitude test score data.

14.35 A drug company wishes to compare the effects of three different drugs (X, Y, and Z) that are being developed to reduce cholesterol levels. Each drug is administered to six patients at the recommended dosage for a six-month period. At the end of this period the reduction in cholesterol level is recorded for each patient. The results are given in Table 14.16. Fully analyze this data using one-way ANOVA.

TABLE 14.15 **Aptitude test scores**

College degree type				
Business	*Science*	*Liberal arts*	*Fine arts*	*Engineering*
49	77	50	37	88
34	65	41	26	70
45	71	65	52	95
59	85	57	39	77
74	79	62	44	81
53	91	47	50	89
51	60	59	41	62
57	72	48	35	83

14.36 Consider the cholesterol level reduction data in Table 14.16.

a. Formulate a dummy variable regression model that can be used to appropriately analyze the data.
b. Using a computer and the dummy variable regression model, completely analyze the cholesterol level reduction data.

14.37 Consider the aptitude test score data in Table 14.15. Employ Hartley's test for variance equality to test

$$H_0: \sigma^2_{\text{BUS}} = \sigma^2_{\text{SCI}} = \sigma^2_{\text{LIB}} = \sigma^2_{\text{FINE}} = \sigma^2_{\text{ENG}}$$

versus

$$H_1\text{: At least two of } \sigma^2_{\text{BUS}}, \sigma^2_{\text{SCI}}, \sigma^2_{\text{LIB}}, \sigma^2_{\text{FINE}}, \text{ and } \sigma^2_{\text{ENG}} \text{ differ}$$

14.38 Consider the cholesterol level reduction data in Table 14.16. Employ Bartlett's test for variance equality to test

$$H_0: \sigma^2_{\text{X}} = \sigma^2_{\text{Y}} = \sigma^2_{\text{Z}}$$

versus

$$H_1\text{: At least two of } \sigma^2_{\text{X}}, \sigma^2_{\text{Y}}, \text{ and } \sigma^2_{\text{Z}} \text{ differ}$$

TABLE 14.16 **Reduction of cholesterol levels using three drugs**

Drug		
X	*Y*	*Z*
22	40	15
31	35	9
19	47	14
27	41	11
25	39	21
18	33	5

14.39 (Optional) Write a SAS program that will perform a one-way analysis of variance of the bakery product data in Table 14.8. Produce the output given in Figure 14.7.

14.40 (Optional) Write a SAS program that will perform a one-way analysis of variance of the golf ball durability data in Table 14.9. Produce the output given in Figure 14.8.

14.41 (Optional) Write a SAS program that will perform a one-way analysis of variance of the bottle design study data in Table 14.10. Produce the output given in Figure 14.9.

14.42 (Optional) Write a SAS program that will perform a one-way analysis of variance of the gasoline additive data in Table 14.12. Produce the output given in Figure 14.10.

14.43 (Optional) Write a SAS program that will analyze the bakery product data in Table 14.8 by using dummy variable regression.

14.44 (Optional) Write a SAS program that will analyze the golf ball durability data in Table 14.9 by using dummy variable regression.

14.45 (Optional) Write a SAS program that will analyze the gasoline additive data in Table 14.12 by using dummy variable regression.

14.46 (Optional) Write a SAS program that will perform a one-way analysis of variance of the aptitude test score data in Table 14.15.

14.47 (Optional) Write a SAS program that will perform a one-way analysis of variance of the cholesterol level reduction data in Table 14.16.

CHAPTER 15

TWO-FACTOR ANALYSIS: I

In this chapter we begin our discussion of how to assess the effects of *two qualitative factors* on a response variable. We begin in Section 15.1 by introducing *two-factor analysis*. Then, in Section 15.2 we present the *analysis of variance approach* to analyzing *balanced data* (equal sample sizes). Section 15.3 discusses the regression approach to analyzing unbalanced data (unequal sample sizes). We conclude with optional Section 15.4, which shows how to use SAS to implement the techniques presented in this chapter.

15.1 INTRODUCTORY CONCEPTS

In *two-factor analysis* we wish to assess the effects of *two qualitative factors* on a response variable. We assume that *factor 1* has *a levels*. That is, factor 1 possesses levels 1, 2, . . . , *a*. Further, we assume that *factor 2* has *b levels*. That is, factor 2 possesses levels 1, 2, . . . , *b*. Here, a *treatment* is considered to be a *combination* of a level of factor 1 and a level of factor 2.

The goal of two-factor analysis is to estimate and compare the effects of the different treatments on the response variable. Depending on the particular situation, we may wish to test to see whether there are *statistically significant* differences (1) between the effects of the different levels of factor 1 or (2) between the effects of the different levels of factor 2, or (3) between the effects of different combinations of a level of factor 1 and a level of factor 2. If statistically significant differences exist, we wish to estimate the magnitude of the differences in order to judge whether or not the differences have any practical importance. Furthermore, we may wish to estimate the effect of a particular treatment on the mean response, and we may wish to predict individual values of the response variable.

To begin, we assume that we perform a *completely randomized experimental design*. For $i = 1, 2, \ldots, a$ and $j = 1, 2, \ldots, b$ we define the (i, j)th population to be the population of all possible values of the response variable that can potentially be observed when we employ level i of factor 1 and level j of factor 2. We define μ_{ij} to be the mean of this population. That is,

μ_{ij} = the mean value of the response variable obtained by using level i of factor 1 and level j of factor 2

For $i = 1, 2, \ldots, a$ and $j = 1, 2, \ldots, b$ we let n_{ij} be the size of the sample that has been randomly selected from the (i, j)th population. We write this sample as

$$y_{ij,1}, \quad y_{ij,2}, \quad \ldots, \quad y_{ij,n_{ij}}$$

where

$y_{ij,k}$ = the kth value of the response variable observed when using level i of factor 1 and level j of factor 2

As in one-factor analysis, two basic approaches can be used. First, algebraic "sums of squares" formulas can be employed. The "sums of squares" approach is commonly called *two-way analysis of variance* (two-way ANOVA). Two-way ANOVA requires that we carry out a balanced complete factorial experiment. A *complete factorial experiment* is carried out if we randomly select a sample corresponding to each and every treatment (combination of factor levels). When the sample sizes employed for all treatments are equal, the experiment is called a *balanced* complete factorial experiment. Thus two-way ANOVA requires that we randomly select equally sized samples corresponding to each and every factor level combination (treatment). The validity of two-way ANOVA also requires that the inference assumptions are (at least approximately) satisfied.

The second approach to two-factor analysis is to employ *dummy variable regression models*. While the regression approach also requires that the inference assumptions be (approximately) met, this approach can legitimately be employed when the experimental data is either unbalanced, incomplete, or both. When sample sizes for different factor level combinations are not equal, the data is said to be *unbalanced*. When a sample is not randomly selected for one or more treatments, the experiment is called *incomplete*.

In general, to obtain the most information from a two-factor analysis, it is best to carry out a balanced complete factorial experiment. However, sometimes data is lost or unavailable, making analysis of unbalanced or incomplete data necessary. For example, in an experiment to study the effects of several wheat types and fertilizer types on wheat yields, some experimental plots of wheat might be accidentally destroyed by application of an excessive amount of herbicide. In addition, some data is *observational* rather than *experimental*. Such data cannot be controlled by the experimenter. For example, we might wish to study the effects of several different strategic management strategies, but we cannot control the management strategies employed by the firms we observe. Such observational data is often unbalanced or incomplete. If we are to analyze unbalanced or incomplete data, we must use the "regression approach" rather than the "sums of squares" two-way ANOVA approach.

Example 15.1 The Tastee Bakery Company supplies a bakery product to a large number of supermarkets in a metropolitan area.* The company wishes to study the effects of two qualitative factors—*shelf display height* and *shelf display width*—on *monthly demand* (measured in cases of ten units each) for this product. The factor "display height" is defined to have three levels: B (Bottom), M (Middle), and T (Top). The factor "display width" is defined to have two levels: R (Regular) and W (Wide). The *treatments* in this experiment are *display height and display width combinations.* These treatments are

BR BW MR MW TR TW

Here, for example, the notation BR denotes the treatment bottom display height and regular display width.

For i = B, M, and T and j = R and W we define the *(i, j)th population of monthly demands* to be the (theoretically) infinite population of all possible monthly demands that could be observed at supermarkets that use display height i and display width j. We let μ_{ij} denote the mean of this population. That is,

μ_{ij} = the mean monthly demand obtained by using display height i and display width j

The company will carry out a *balanced complete factorial experiment.* Specifically, for i = B, M, and T and j = R and W the company will randomly select a sample of $m = 3$ metropolitan area supermarkets. Each supermarket will sell the product for one month using display height i and display width j and will record demand during the month. We let

$y_{ij,k}$ = the kth monthly demand observed when using display height i and display width j

*The context of this example is taken from a similar example in Neter, Wasserman, and Kutner (1985). The data in this example (see Table 15.1) is our own.

TABLE 15.1 **Six samples of monthly demands**

Display height, i	Display width, j: R	Display width, j: W	$\bar{y}_{i\cdot}$
B	$y_{BR,1} = 58.2$	$y_{BW,1} = 55.7$	
	$y_{BR,2} = 53.7$	$y_{BW,2} = 52.5$	
	$y_{BR,3} = 55.8$	$y_{BW,3} = 58.9$	
	$\bar{y}_{BR} = 55.9$	$\bar{y}_{BW} = 55.7$	$\bar{y}_{B\cdot} = 55.8$
M	$y_{MR,1} = 73.0$	$y_{MW,1} = 76.2$	
	$y_{MR,2} = 78.1$	$y_{MW,2} = 78.4$	
	$y_{MR,3} = 75.4$	$y_{MW,3} = 82.1$	
	$\bar{y}_{MR} = 75.5$	$\bar{y}_{MW} = 78.9$	$\bar{y}_{M\cdot} = 77.2$
T	$y_{TR,1} = 52.4$	$y_{TW,1} = 54.0$	
	$y_{TR,2} = 49.7$	$y_{TW,2} = 52.1$	
	$y_{TR,3} = 50.9$	$y_{TW,3} = 49.9$	
	$\bar{y}_{TR} = 51.0$	$\bar{y}_{TW} = 52.0$	$\bar{y}_{T\cdot} = 51.5$
$\bar{y}_{\cdot j}$	$\bar{y}_{\cdot R} = 60.8$	$\bar{y}_{\cdot W} = 62.2$	$\bar{y} = 61.5$

The six samples obtained in this experiment are presented in Table 15.1. Note that a sample has been randomly selected for each treatment and that the data is balanced.

The company wishes to determine whether or not there are statistically significant differences between the effects on mean demand of the different levels of display height and display width and/or the different combinations of display height and display width. If statistically significant differences exist, the company wishes to estimate the size of these differences in order to assess their practical importance. The company also wishes to estimate mean demand and predict individual demands when using specific display height and display width combinations. We will fully analyze the data of this example in Section 15.2.

15.2

THE ANOVA APPROACH TO THE ANALYSIS OF A BALANCED COMPLETE FACTORIAL EXPERIMENT

THE TWO-WAY ANOVA MODEL AND INTERACTION BETWEEN FACTORS

In two-way ANOVA we employ algebraic "sums of squares" formulas to carry out *two-factor analysis*. To use two-way ANOVA, we must (as stated in Section 15.1) perform a *balanced complete factorial experiment*. The validity of the analysis also depends upon the inference assumptions being (approximately) satisfied. We now state the appropriate model.

The Two-Way ANOVA Model

The *two-way ANOVA model* (sometimes called the (α, γ, θ) model) says that

$$y_{ij,k} = \mu_{ij} + \varepsilon_{ij,k}$$

where

$$\mu_{ij} = \mu + \alpha_i + \gamma_j + \theta_{ij}$$

Here,

1. $y_{ij,k}$ = the kth observed value of the response variable when we are using level i of factor 1 and level j of factor 2
2. $\varepsilon_{ij,k}$ = the kth error term when we are using level i of factor 1 and level j of factor 2
3. μ_{ij} = the mean value of the response variable when we are using level i of factor 1 and level j of factor 2
4. μ = an overall mean or intercept, which is an unknown constant
5. α_i = the effect due to level i of factor 1, which is an unknown constant
6. γ_j = the effect due to level j of factor 2, which is an unknown constant
7. θ_{ij} = the effect due to the interaction of level i of factor 1 and level j of factor 2, which is an unknown constant.

It is important to notice that this model (by employing θ_{ij}) allows for the possibility that *interaction* exists between factors. Factors 1 and 2 interact if the relationship between the mean response and the different levels of one factor depends upon the level of the other factor. For example, consider assessing the effects of two factors—advertising campaign (with levels A, B, and C) and pricing strategy (with levels M, N, and O)—on demand for a product. Suppose, for instance, that changes in mean demand associated with changing the advertising campaign type are different for the different pricing strategy types. Then interaction exists between these two factors.

To see how two-way ANOVA models such interaction, consider

$\mu_{AM} - \mu_{BM}$ = the change in mean demand associated with changing from advertising campaign B to advertising campaign A when we are using pricing strategy M

and

$\mu_{AN} - \mu_{BN}$ = the change in mean demand associated with changing from advertising campaign B to advertising campaign A when we are using pricing strategy N

The two-way ANOVA model says that

$$\begin{aligned}\mu_{AM} - \mu_{BM} &= (\mu + \alpha_A + \gamma_M + \theta_{AM}) - (\mu + \alpha_B + \gamma_M + \theta_{BM}) \\ &= (\alpha_A - \alpha_B) + (\theta_{AM} - \theta_{BM})\end{aligned}$$

The model also says that

$$\begin{aligned}\mu_{AN} - \mu_{BN} &= (\mu + \alpha_A + \gamma_N + \theta_{AN}) - (\mu + \alpha_B + \gamma_N + \theta_{BN}) \\ &= (\alpha_A - \alpha_B) + (\theta_{AN} - \theta_{BN})\end{aligned}$$

Here, $(\theta_{AM} - \theta_{BM})$ is (in general) not equal to $(\theta_{AN} - \theta_{BN})$. Thus we see that the change in mean demand associated with changing from advertising campaign B to advertising campaign A is different for the different pricing strategies M and N. That is, we can say that the two-way ANOVA model allows for the possibility that the effect of changing from advertising campaign B to advertising campaign A depends on the type of pricing strategy.

One way to detect interaction between two factors is to perform a graphical analysis. We plot the sample mean response versus the levels of one of the factors for the different levels of the other factor. This graphical method is illustrated in the following example.

Example 15.2 Consider the Tastee Bakery demand data in Table 15.1. To study possible interaction between shelf display height and shelf display width, we carry out the

FIGURE 15.1 **Graphical analysis of the Tastee Bakery demand data in Table 15.1: Little interaction**

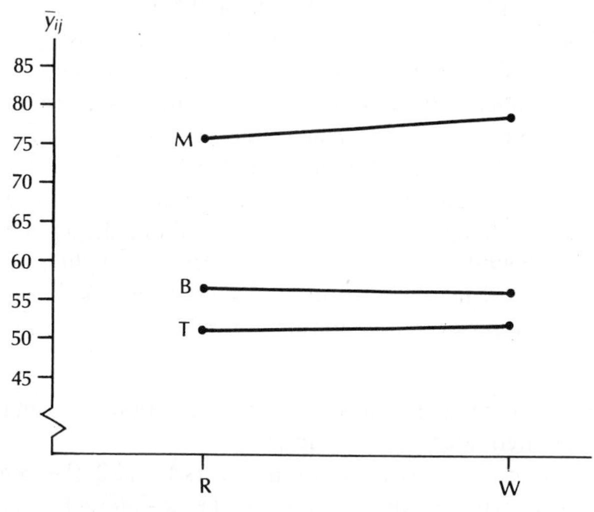

(a) Plot of the change in $\bar{y}_{ij}$ associated with changing the display width (from R to W) for each display height (B,M, and T). Note the largely parallel data patterns.

(continues)

FIGURE 15.1 **Continued**

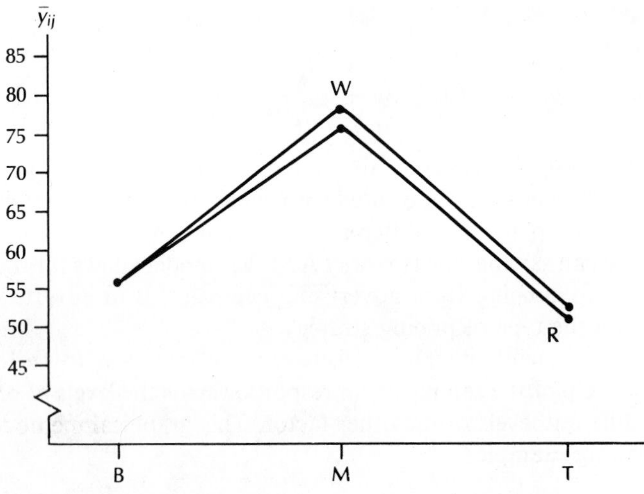

(b) Plot of the change in $\bar{y}_{ij}$ associated with changing the display height (from B to M to T) for each display width (R and W). Note the largely parallel data patterns.

graphical analysis in Figure 15.1. The largely parallel data patterns in Figure 15.1(a) indicate that the change in $\bar{y}_{ij}$ associated with changing the display width from R to W depends little on the shelf display height. The largely parallel data patterns in Figure 15.1(b) indicate that the change in $\bar{y}_{ij}$ associated with changing the display height from B to M to T depends little on the shelf display width. Together, the parallel data patterns in these figures suggest that there is little interaction between shelf display height and shelf display width. However, it is possible that a small amount of interaction exists. Although the data patterns in Figure 15.1 are largely parallel, we see that, for example, $\bar{y}_{ij}$ increases more as the shelf display width changes from R to W for a middle display height than it does for a top display height. Also, we see that $\bar{y}_{ij}$ decreases slightly as the shelf display width changes from R to W for a bottom display height. Whether this behavior is due to a small amount of interaction or to sampling variation is questionable.

An F-test can also be carried out to test for interaction between two factors. This test will be presented later in this chapter.

Finally, if no interaction exists between factors 1 and 2, the interaction effect θ_{ij} for each treatment will be equal to zero. In the no-interaction case the two-way ANOVA model can be written as

$$y_{ij,k} = \mu_{ij} + \varepsilon_{ij,k}$$

where

$$\mu_{ij} = \mu + \alpha_i + \gamma_j$$

This says that we can separate the effects of factors 1 and 2.

SUMS OF SQUARES AND MEAN SQUARES IN TWO-WAY ANOVA

Two-way analysis of variance involves *partitioning a total sum of squares*. To define the appropriate *sums of squares* and *mean squares*, we need to define the following four means:

1. $\bar{y}_{ij} = \dfrac{\sum_{k=1}^{m} y_{ij,k}}{m}$ is the mean of the sample of m values randomly selected from the (i, j)th population.
2. $\bar{y}_{i\cdot} = \dfrac{\sum_{j=1}^{b}\sum_{k=1}^{m} y_{ij,k}}{bm}$ is the mean of the bm values observed when using the ith level of factor 1.
3. $\bar{y}_{\cdot j} = \dfrac{\sum_{i=1}^{a}\sum_{k=1}^{m} y_{ij,k}}{am}$ is the mean of the am values observed when using the jth level of factor 2.
4. $\bar{y} = \dfrac{\sum_{i=1}^{a}\sum_{j=1}^{b}\sum_{k=1}^{m} y_{ij,k}}{abm}$ is the mean of the total of abm values that we have observed in the experiment.

Example 15.3 Consider the Tastee Bakery data given in Table 15.1. For this data,

$\bar{y}_{BR} = 55.9 =$ the mean of the sample of 3 demands observed when using a bottom display height and a regular display width

$\bar{y}_{B\cdot} = 55.8 =$ the mean of the six demands observed when using a bottom display height

$\bar{y}_{M\cdot} = 77.2 =$ the mean of the six demands observed when using a middle display height

$\bar{y}_{T\cdot} = 51.5 =$ the mean of the six demands observed when using a top display height

$\bar{y}_{\cdot R} = 60.8 =$ the mean of the nine demands observed when using a regular display width

$\bar{y}_{\cdot W} = 62.2 =$ the mean of the nine demands observed when using a wide display

$\bar{y} = 61.5 =$ the mean of the total of 18 demands observed in the experiment

To carry out the F-tests in two-way ANOVA, we now define the following *sums of squares* and *mean squares*.

Sums of Squares and Mean Squares in Two-Way ANOVA

Given $\bar{y}_{ij}$, $\bar{y}_{i\cdot}$, $\bar{y}_{\cdot j}$, and $\bar{y}$ as defined previously, we define

1. The *treatment sum of squares* (or *explained variation*) to be

$$SS_{\text{means}} = m \sum_{i=1}^{a} \sum_{j=1}^{b} (\bar{y}_{ij} - \bar{y})^2$$

2. The *treatment mean square* to be

$$MS_{\text{means}} = \frac{SS_{\text{means}}}{ab - 1}$$

3. The *sum of squares due to factor 1* to be

$$SS_{\text{factor 1}} = bm \sum_{i=1}^{a} (\bar{y}_{i\cdot} - \bar{y})^2$$

4. The *factor 1 mean square* to be

$$MS_{\text{factor 1}} = \frac{SS_{\text{factor 1}}}{a - 1}$$

5. The *sum of squares due to factor 2* to be

$$SS_{\text{factor 2}} = am \sum_{j=1}^{b} (\bar{y}_{\cdot j} - \bar{y})^2$$

6. The *factor 2 mean square* to be

$$MS_{\text{factor 2}} = \frac{SS_{\text{factor 2}}}{b - 1}$$

7. The *sum of squares due to interaction* to be

$$SS_{\text{interaction}} = m \sum_{i=1}^{a} \sum_{j=1}^{b} (\bar{y}_{ij} - \bar{y}_{i\cdot} - \bar{y}_{\cdot j} + \bar{y})^2$$

8. The *interaction mean square* to be

$$MS_{\text{interaction}} = \frac{SS_{\text{interaction}}}{(a - 1)(b - 1)}$$

9. The *error sum of squares* to be

$$SSE = \sum_{i=1}^{a} \sum_{j=1}^{b} \sum_{k=1}^{m} (y_{ij,k} - \bar{y}_{ij})^2$$

10. The *error mean square* to be

$$MSE = \frac{SSE}{ab(m - 1)}$$

11. The *total sum of squares* (or *total variation*) to be

$$SS_{\text{total}} = \sum_{i=1}^{a} \sum_{j=1}^{b} \sum_{k=1}^{m} (y_{ij,k} - \bar{y})^2$$

Furthermore, it can be shown that

$$SS_{\text{total}} = SS_{\text{factor 1}} + SS_{\text{factor 2}} + SS_{\text{interaction}} + SSE$$

and it can also be shown that

$$SS_{\text{means}} = SS_{\text{factor 1}} + SS_{\text{factor 2}} + SS_{\text{interaction}}$$

Here, the *total sum of squares* measures the total amount of variability in the observed values of the response variable. The *treatment sum of squares*, SS_{means}, measures the variability of the sample treatment means. For example, if all the treatment means ($\bar{y}_{ij}$ values) were equal, then the treatment sum of squares would be equal to zero. The more the $\bar{y}_{ij}$ values vary, the larger will be SS_{means}. In other words, the treatment sum of squares measures the *variability of the response between treatments*. The *sum of squares due to factor 1*, $SS_{\text{factor 1}}$, measures the amount of variability between the different $\bar{y}_{i\cdot}$ values. Recall that the $\bar{y}_{i\cdot}$ values are the mean values of the response variable at the different levels of factor 1. For example, if all of the $\bar{y}_{i\cdot}$ values were equal, then $SS_{\text{factor 1}}$ would be equal to zero. This is because $\bar{y}$ would equal $\bar{y}_{i\cdot}$ for each level i of factor 1. In other words, $SS_{\text{factor 1}}$ measures the *variability of the response between levels of factor 1*. The *sum of squares due to factor 2*, $SS_{\text{factor 2}}$, measures the amount of variability between the different $\bar{y}_{\cdot j}$ values. That is, $SS_{\text{factor 2}}$ measures the *variability of the response between levels of factor 2*. The *sum of squares due to interaction*, $SS_{\text{interaction}}$, *measures the amount of interaction between factors 1 and 2*. To see this, suppose that for $i = 1, 2, \ldots, a$ and $j = 1, 2, \ldots, b$

$$\bar{y}_{ij} - \bar{y}_{i\cdot} - \bar{y}_{\cdot j} + \bar{y} = 0$$

Then it follows that for arbitrary levels i and i' of factor 1 and arbitrary levels j and j' of factor 2,

$$\bar{y}_{i'j} - \bar{y}_{i'\cdot} - \bar{y}_{\cdot j} + \bar{y} = 0$$

$$\bar{y}_{ij'} - \bar{y}_{i\cdot} - \bar{y}_{\cdot j'} + \bar{y} = 0$$

and

$$\bar{y}_{i'j'} - \bar{y}_{i'\cdot} - \bar{y}_{\cdot j'} + \bar{y} = 0$$

This implies that

$$(\bar{y}_{ij} - \bar{y}_{i\cdot} - \bar{y}_{\cdot j} + \bar{y}) - (\bar{y}_{i'j} - \bar{y}_{i'\cdot} - \bar{y}_{\cdot j} + \bar{y}) = (\bar{y}_{ij'} - \bar{y}_{i\cdot} - \bar{y}_{\cdot j'} + \bar{y}) - (\bar{y}_{i'j'} - \bar{y}_{i'\cdot} - \bar{y}_{\cdot j'} + \bar{y})$$

which implies that

$$\bar{y}_{ij} - \bar{y}_{i'j} = \bar{y}_{ij'} - \bar{y}_{i'j'}$$

This says that the effect on the sample mean response of changing from level i' of factor 1 to level i of factor 1 is the same for level j of factor 2 as it is for level j' of factor 2. Therefore (in the sample) there is no interaction between factors 1 and 2. Generalizing, if most of the terms

$$\bar{y}_{ij} - \bar{y}_{i\cdot} - \bar{y}_{\cdot j} + \bar{y}$$

are near zero, $SS_{\text{interaction}}$ will be small. This indicates that there is little or no interaction between factors 1 and 2. On the other hand, a large value of $SS_{\text{interaction}}$ indicates that there is interaction between factors 1 and 2.

The *error sum of squares, SSE,* measures the variability of the observed values of the response variable around their respective treatment means. For example, if there were no variability within each sample, the error sum of squares would be equal to zero. The more the values within the samples vary, the larger will be the *SSE*. In other words, the *SSE* measures the amount of *within-treatment variability*.

The quantities $SS_{\text{factor 1}}$, $SS_{\text{factor 2}}$, $SS_{\text{interaction}}$, and SSE are said to *partition the total sum of squares*. That is,

$$SS_{\text{total}} = SS_{\text{factor 1}} + SS_{\text{factor 2}} + SS_{\text{interaction}} + SSE$$

Moreover, $SS_{\text{factor 1}}$, $SS_{\text{factor 2}}$, and $SS_{\text{interaction}}$ *partition the treatment sum of squares (or explained variation)*. That is,

$$SS_{\text{means}} = SS_{\text{factor 1}} + SS_{\text{factor 2}} + SS_{\text{interaction}}$$

This says that the amount of variability of the response variable explained by the model is composed of the amounts of variability of the response explained by factor 1, factor 2, and the interaction between factors 1 and 2, respectively.

Example 15.4 Consider the Tastee Bakery data given in Table 15.1. We compute the necessary sums of squares and mean squares for a two-way ANOVA of this data as follows. Note here that there are $a = 3$ levels of display height and $b = 2$ levels of display width and that Tastee Bakery has randomly selected a sample of $m = 3$ demands for each treatment.

$$SS_{\text{means}} = m \sum_{i=(\text{B,M,T})} \sum_{j=(\text{R,W})} (\bar{y}_{ij} - \bar{y})^2$$

$$\begin{aligned}
&= 3[(\bar{y}_{\text{BR}} - \bar{y})^2 + (\bar{y}_{\text{BW}} - \bar{y})^2 + (\bar{y}_{\text{MR}} - \bar{y})^2 \\
&\quad + (\bar{y}_{\text{MW}} - \bar{y})^2 + (\bar{y}_{\text{TR}} - \bar{y})^2 + (\bar{y}_{\text{TW}} - \bar{y})^2] \\
&= 3[(55.9 - 61.5)^2 + (55.7 - 61.5)^2 + (75.5 - 61.5)^2 \\
&\quad + (78.9 - 61.5)^2 + (51.0 - 61.5)^2 + (52.0 - 61.5)^2] \\
&= 2292.78
\end{aligned}$$

$$MS_{\text{means}} = \frac{SS_{\text{means}}}{ab - 1} = \frac{2292.78}{3(2) - 1} = \frac{2292.78}{5} = 458.556$$

$$\begin{aligned}
SS_{\text{factor 1}} &= bm \sum_{i=(\text{B,M,T})} (\bar{y}_{i\cdot} - \bar{y})^2 \\
&= 2 \cdot 3[(\bar{y}_{\text{B}\cdot} - \bar{y})^2 + (\bar{y}_{\text{M}\cdot} - \bar{y})^2 + (\bar{y}_{\text{T}\cdot} - \bar{y})^2] \\
&= 6[(55.8 - 61.5)^2 + (77.2 - 61.5)^2 + (51.5 - 61.5)^2] \\
&= 6[32.49 + 246.49 + 100] \\
&= 2273.88
\end{aligned}$$

$$MS_{\text{factor 1}} = \frac{SS_{\text{factor 1}}}{a - 1} = \frac{2273.88}{3 - 1} = 1136.94$$

$$\begin{aligned}
SS_{\text{factor 2}} &= am \sum_{j=(\text{R,W})} (\bar{y}_{\cdot j} - \bar{y})^2 \\
&= 3 \cdot 3[(\bar{y}_{\cdot\text{R}} - \bar{y})^2 + (\bar{y}_{\cdot\text{W}} - \bar{y})^2] \\
&= 9[(60.8 - 61.5)^2 + (62.2 - 61.5)^2] \\
&= 9[.49 + .49] \\
&= 8.82
\end{aligned}$$

$$MS_{\text{factor 2}} = \frac{SS_{\text{factor 2}}}{b - 1} = \frac{8.82}{2 - 1} = 8.82$$

$$\begin{aligned}
SS_{\text{interaction}} &= m \sum_{i=(\text{B,M,T})} \sum_{j=(\text{R,W})} (\bar{y}_{ij} - \bar{y}_{i\cdot} - \bar{y}_{\cdot j} + \bar{y})^2 \\
&= 3[(\bar{y}_{\text{BR}} - \bar{y}_{\text{B}\cdot} - \bar{y}_{\cdot\text{R}} + \bar{y})^2 + (\bar{y}_{\text{BW}} - \bar{y}_{\text{B}\cdot} - \bar{y}_{\cdot\text{W}} + \bar{y})^2 \\
&\quad + (\bar{y}_{\text{MR}} - \bar{y}_{\text{M}\cdot} - \bar{y}_{\cdot\text{R}} + \bar{y})^2 + (\bar{y}_{\text{MW}} - \bar{y}_{\text{M}\cdot} - \bar{y}_{\cdot\text{W}} + \bar{y})^2 \\
&\quad + (\bar{y}_{\text{TR}} - \bar{y}_{\text{T}\cdot} - \bar{y}_{\text{R}\cdot} + \bar{y})^2 + (\bar{y}_{\text{TW}} - \bar{y}_{\text{T}\cdot} - \bar{y}_{\cdot\text{W}} + \bar{y})^2] \\
&= 3[(55.9 - 55.8 - 60.8 + 61.5)^2 \\
&\quad + (55.7 - 55.8 - 62.2 + 61.5)^2 \\
&\quad + (75.5 - 77.2 - 60.8 + 61.5)^2 \\
&\quad + (78.9 - 77.2 - 62.2 + 61.5)^2 \\
&\quad + (51.0 - 51.5 - 60.8 + 61.5)^2 \\
&\quad + (52.0 - 51.5 - 62.2 + 61.5)^2] \\
&= 3(3.36) = 10.08
\end{aligned}$$

or

$$\begin{aligned}
SS_{\text{interaction}} &= SS_{\text{means}} - SS_{\text{factor 1}} - SS_{\text{factor 2}} \\
&= 2292.78 - 2273.88 - 8.82 \\
&= 10.08
\end{aligned}$$

$$MS_{\text{interaction}} = \frac{SS_{\text{interaction}}}{(a-1)(b-1)} = \frac{10.08}{(3-1)(2-1)} = 5.04$$

$$\begin{aligned} SSE &= \sum_{i=(\text{B,M,T})} \sum_{j=(\text{R,W})} \sum_{k=1}^{3} (y_{ij,k} - \bar{y}_{ij})^2 \\ &= [(58.2 - 55.9)^2 + (53.7 - 55.9)^2 + (55.8 - 55.9)^2] \\ &\quad + [(55.7 - 55.7)^2 + (52.5 - 55.7)^2 + (58.9 - 55.7)^2] \\ &\quad + [(73.0 - 75.5)^2 + (78.1 - 75.5)^2 + (75.4 - 75.5)^2] \\ &\quad + [(76.2 - 78.9)^2 + (78.4 - 78.9)^2 + (82.1 - 78.9)^2] \\ &\quad + [(52.4 - 51.0)^2 + (49.7 - 51.0)^2 + (50.9 - 51.0)^2] \\ &\quad + [(54.0 - 52.0)^2 + (52.1 - 52.0)^2 + (49.9 - 52.0)^2] \\ &= 73.5 \end{aligned}$$

$$MSE = \frac{SSE}{ab(m-1)} = \frac{73.5}{3(2)(3-1)} = \frac{73.5}{12} = 6.125$$

$$\begin{aligned} SS_{\text{total}} &= SS_{\text{factor 1}} + SS_{\text{factor 2}} + SS_{\text{interaction}} + SSE \\ &= 2273.88 + 8.82 + 10.08 + 73.5 \\ &= 2366.28 \end{aligned}$$

F-TESTS IN TWO-WAY ANOVA

The analysis in two-way ANOVA begins by testing the significance of factors 1 and/or 2 (and/or the interaction between factors 1 and 2). We do this by using an F statistic to test the null hypothesis

H_0: All treatment means μ_{ij} are equal

This hypothesis says that neither factor 1 nor factor 2 nor the interaction between these factors has a significant (important) effect on the mean response. We test H_0 versus

H_1: At least two of the treatment means μ_{ij} differ

Thus H_1 says that either factor 1 or factor 2 or the interaction between these factors has a significant effect on the mean response. If we reject H_0, we certainly wish to continue the analysis. If we cannot reject H_0, it is possible that neither factor 1 nor factor 2 nor the interaction between these factors is important. However, many practitioners suggest that the analysis should be continued anyway. For example, it is possible that we have erroneously failed to reject H_0 when in fact at least two treatment means differ. Further analysis might show that differences between treatment means do exist. We should realize in such a case, however, that further analysis might not reveal any significant differences between treatment means.

We now present the F-test that is used to test the above null hypothesis.

An F-Test for the Significance of Factors 1 and/or 2 (and/or the Interaction Between Factors 1 and 2)

Define

$$F(\text{means}) = \frac{MS_{\text{means}}}{MSE} = \frac{SS_{\text{means}}/(ab - 1)}{SSE/ab(m - 1)}$$

$$= \frac{m \sum_{i=1}^{a} \sum_{j=1}^{b} (\bar{y}_{ij} - \bar{y})^2/(ab - 1)}{\sum_{i=1}^{a} \sum_{j=1}^{b} \sum_{k=1}^{m} (y_{ij,k} - \bar{y}_{ij})^2/ab(m - 1)}$$

Also, define the prob-value to be the area to the right of F(means) under the curve of the F-distribution having $ab - 1$ and $ab(m - 1)$ degrees of freedom. Then we can reject

H_0: All treatment means μ_{ij} are equal

in favor of the alternative hypothesis

H_1: At least two of the treatment means μ_{ij} differ

by setting the probability of a Type I error equal to α if and only if either of the following equivalent conditions holds:

1. $F(\text{means}) > F_{[\alpha]}^{(ab-1, ab(m-1))}$
2. prob-value $< \alpha$

Example 15.5 Consider the Tastee Bakery data given in Table 15.1. We wish to test

H_0: All treatment means μ_{ij} are equal (i = B, M, and T; j = R and W)

This hypothesis says that neither shelf display height nor shelf display width nor the interaction between these factors has a significant (important) effect on mean demand. We test H_0 versus

H_1: At least two of the treatment means μ_{ij} differ

which says that either shelf display height or shelf display width or the interaction between these factors has a significant effect on mean demand. Recalling that SS_{means} and SSE were computed in Example 15.4, we test these hypotheses by computing

$$F(\text{means}) = \frac{MS_{\text{means}}}{MSE} = \frac{SS_{\text{means}}/(ab - 1)}{SSE/ab(m - 1)} = \frac{2292.78/((3)(2) - 1)}{73.5/(3)(2)(3 - 1)} = \frac{2292.78/5}{73.5/12} = 74.8663$$

Since

$$F(\text{means}) = 74.8663 > F_{[\alpha]}^{(ab-1,ab(m-1))} = F_{[.05]}^{(5,12)} = 3.11$$

and

$$\text{prob-value} = .0001 < .05$$

we reject H_0 in favor of H_1 with $\alpha = .05$. We therefore conclude that either shelf display height or shelf display width or the interaction between these factors has a significant effect on mean demand. On the basis of this conclusion, it makes sense to continue the analysis to determine which of these factors have important effects on mean demand and to study the nature of these effects.

As a second step in two-way ANOVA, we must decide whether or not interaction exists between the two factors as they affect the mean response. This can be done graphically as was illustrated in Example 15.2. It can also be done by employing the following F-test for interaction.

An F-Test for Interaction Between Factors 1 and 2

Define

$$F(\text{interaction}) = \frac{MS_{\text{interaction}}}{MSE} = \frac{SS_{\text{interaction}}/(a-1)(b-1)}{SSE/ab(m-1)}$$

$$= \frac{m \sum_{i=1}^{a} \sum_{j=1}^{b} (\bar{y}_{ij} - \bar{y}_{i\cdot} - \bar{y}_{\cdot j} + \bar{y})^2/(a-1)(b-1)}{\sum_{i=1}^{a} \sum_{j=1}^{b} \sum_{k=1}^{m} (y_{ij,k} - \bar{y}_{ij})^2/ab(m-1)}$$

Also, define the prob-value to be the area to the right of $F(\text{interaction})$ under the curve of the F-distribution having $(a-1)(b-1)$ and $ab(m-1)$ degrees of freedom. Then we can reject

H_0: No interaction exists between factors 1 and 2

in favor of the alternative hypothesis

H_1: Interaction does exist between factors 1 and 2

by setting the probability of a Type I error equal to α if and only if either of the following equivalent conditions holds:

1. $F(\text{interaction}) > F_{[\alpha]}^{((a-1)(b-1),ab(m-1))}$
2. $\text{prob-value} < \alpha$

Example 15.6 Consider the Tastee Bakery data given in Table 15.1. Suppose that we wish to test

H_0: No interaction exists between the factors shelf display height and shelf display width versus

H_1: Interaction does exist between the factors shelf display height and shelf display width

Recalling that $SS_{\text{interaction}}$ and SSE were computed in Example 15.4, we test these hypotheses by computing

$$F(\text{interaction}) = \frac{MS_{\text{interaction}}}{MSE} = \frac{SS_{\text{interaction}}/(a-1)/(b-1)}{SSE/ab(m-1)} = \frac{10.08/(3-1)(2-1)}{73.5/3(2)(3-1)} = \frac{10.08/2}{73.5/12} = .8229$$

Since

$$F(\text{interaction}) = .8229 < F_{[\alpha]}^{((a-1)(b-1),ab(m-1))} = F_{[.05]}^{(2,12)} = 3.89$$

and

$$\text{prob-value} = .4625 > .05$$

we cannot reject H_0 in favor of H_1 with $\alpha = .05$. We conclude that little or no interaction exists between shelf display height and shelf display width.

Note that this conclusion is consistent with the results of the graphical analysis in Example 15.2. We should also point out that, although we cannot reject H_0, it is possible that a small amount of interaction exists. Recall that the graphical analysis of Figure 15.1 indicates that some interaction is possible. Therefore, when we do not reject the null hypothesis of no interaction between factors 1 and 2, *we conclude that little or no interaction exists between the two factors.*

If there is no interaction between factors 1 and 2, then the relationship between the means $\mu_{1j}, \mu_{2j}, \mu_{3j}, \ldots,$ and μ_{aj} does not depend on the level j of factor 2. That is, the effects of changing the level of factor 1 (from 1 to 2 to 3 to . . . a) are the same for all levels of factor 2. Likewise, no interaction says that the relationship between the means $\mu_{i1}, \mu_{i2}, \mu_{i3}, \ldots, \mu_{ib}$ does not depend on the level i of factor 1. That is, the effects of changing the level of factor 2 (from 1 to 2 to 3 to . . . b) are the same for all levels of factor 1. This allows us to (separately) test the significance of each of factors 1 and 2. We often refer to this as *testing the significance of the main effects.*

When no interaction exists between factors 1 and 2, we can test the significance of factor 1 by testing the null hypothesis

$$H_0: \mu_{1j} = \mu_{2j} = \cdots = \mu_{aj} \quad \text{for each and every level } j \text{ of factor 2}$$

or

$$
\begin{aligned}
H_0: \mu_{11} &= \mu_{21} = \cdots = \mu_{a1} \\
\mu_{12} &= \mu_{22} = \cdots = \mu_{a2} \\
&\vdots \\
\mu_{1b} &= \mu_{2b} = \cdots = \mu_{ab}
\end{aligned}
$$

Here, H_0 says that all levels of factor 1 have the same effect on the mean response. We test H_0 versus

H_1: At least two of $\mu_{1j}, \mu_{2j}, \ldots, \mu_{aj}$ differ

which says that at least two levels of factor 1 have different effects on the mean response. Another way to express H_0 involves defining the *factor level mean for level i of factor 1*

$$\mu_{i\cdot} = \frac{\sum_{j=1}^{b} \mu_{ij}}{b} \quad \text{for} \quad i = 1, 2, \ldots, a$$

To see how factor level means can be used, note that no interaction between factors 1 and 2 implies that the relationship between the means $\mu_{1j}, \mu_{2j}, \ldots, \mu_{aj}$ for each and every level j of factor 2 is exactly the same as the relationship between the factor level means

$$\mu_{1\cdot} = \frac{\sum_{j=1}^{b} \mu_{1j}}{b}, \quad \mu_{2\cdot} = \frac{\sum_{j=1}^{b} \mu_{2j}}{b}, \quad \ldots, \quad \mu_{a\cdot} = \frac{\sum_{j=1}^{b} \mu_{aj}}{b}$$

Therefore, if there is no interaction between factors 1 and 2, testing H_0 versus H_1 above is equivalent to testing

$$H_0: \mu_{1\cdot} = \mu_{2\cdot} = \cdots = \mu_{a\cdot}$$

versus

H_1: At least two of $\mu_{1\cdot}, \mu_{2\cdot}, \ldots, \mu_{a\cdot}$ differ

Henceforth, we will test H_0 expressed in this fashion. We do this because this statement of H_0 allows us to test the significance of factor 1 when there is a small amount of interaction between factors 1 and 2.

To understand this, consider two qualitative factors—factor 1 with levels A, B, and C and factor 2 with levels D, E, and F. Suppose that there is only a small amount of interaction between factors 1 and 2 as illustrated in Figure 15.2(a). Then the relationship between μ_{AD}, μ_{BD}, and μ_{CD} differs slightly from the relationship between μ_{AE}, μ_{BE}, and μ_{CE}, and each of these relationships differs slightly from the relationship between μ_{AF}, μ_{BF}, and μ_{CF}. However, these

FIGURE 15.2 An interpretation of factor level means

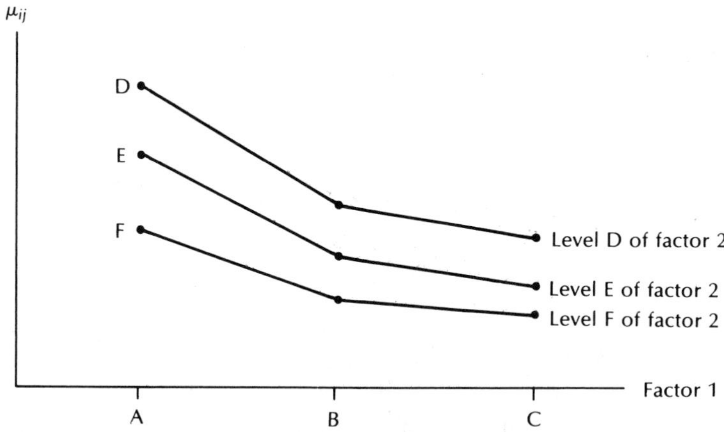

(a) Little interaction: The relationship between the factor level means $\mu_{A\bullet}$, $\mu_{B\bullet}$, and $\mu_{C\bullet}$ approximately represents the relationship between μ_{Aj}, μ_{Bj}, and μ_{Cj} for j = D, E, and F.

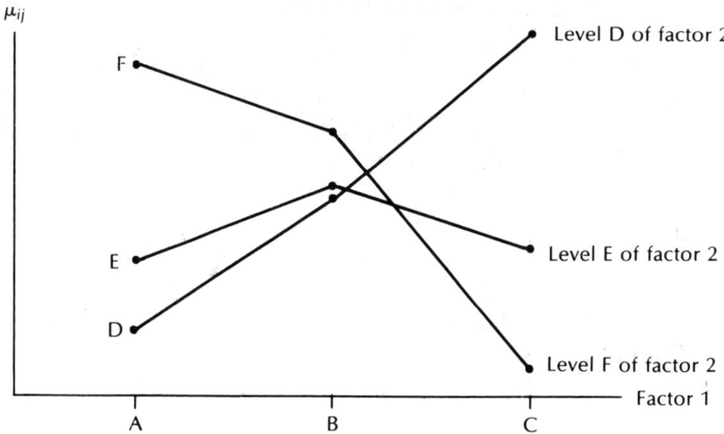

(b) Substantial interaction: The relationship between the factor level means $\mu_{A\bullet}$, $\mu_{B\bullet}$, and $\mu_{C\bullet}$ does not represent the relationship between μ_{Aj}, μ_{Bj}, and μ_{Cj} for j = D, E, and F.

relationships are similar enough that each of them is approximately represented by the relationship between the factor level means

$$\mu_{A\cdot} = \frac{\sum_{j=D,E,F} \mu_{Aj}}{3}, \quad \mu_{B\cdot} = \frac{\sum_{j=D,E,F} \mu_{Bj}}{3}, \quad \text{and} \quad \mu_{C\cdot} = \frac{\sum_{j=D,E,F} \mu_{Cj}}{3}$$

Therefore it is reasonable to test the significance of factor 1 by testing

$$H_0\colon \mu_{A\cdot} = \mu_{B\cdot} = \mu_{C\cdot}$$

versus

$$H_1\colon \text{At least two of } \mu_{A\cdot}, \mu_{B\cdot}, \text{ and } \mu_{C\cdot} \text{ differ}$$

In summary, if there is either no interaction or a small amount of interaction between factors 1 and 2, we can test the significance of factor 1 by testing

$$H_0\colon \mu_{1\cdot} = \mu_{2\cdot} = \cdots = \mu_{a\cdot}$$

versus

$$H_1\colon \text{At least two of } \mu_{1\cdot}, \mu_{2\cdot}, \ldots, \mu_{a\cdot} \text{ differ}$$

On the other hand, if there is substantial interaction between factors 1 and 2, it is not reasonable to test these hypotheses. This is because the relationship between $\mu_{1j}, \mu_{2j}, \ldots, \mu_{aj}$ depends substantially on the level j of factor 2. Therefore we cannot define a relationship between the levels of factor 1 without fixing the level of factor 2 at a specific value. It follows that the relationship between the factor level means $\mu_{1\cdot}, \mu_{2\cdot}, \ldots, \mu_{a\cdot}$ does not (even approximately) represent the relationship between $\mu_{1j}, \mu_{2j}, \ldots, \mu_{aj}$ for each and every j. This is illustrated in Figure 15.2(b). In this figure there is so much interaction between factors 1 and 2 that the relationship between μ_{AD}, μ_{BD}, and μ_{CD} differs substantially from the relationship between μ_{AE}, μ_{BE}, and μ_{CE}, and each of these relationships differs substantially from the relationship between μ_{AF}, μ_{BF}, and μ_{CF}. Thus these relationships are not even approximately represented by the relationship between the factor level means $\mu_{A\cdot}$, $\mu_{B\cdot}$, and $\mu_{C\cdot}$.

When no interaction exists between factors 1 and 2, we can test the significance of factor 2 by testing the null hypothesis

$$H_0\colon \mu_{i1} = \mu_{i2} = \cdots = \mu_{ib} \text{ for each and every level } i \text{ of factor 1}$$

This hypothesis says that all levels of factor 2 have the same effect on the mean response. We test H_0 versus

$$H_1\colon \text{At least two of } \mu_{i1}, \mu_{i2}, \ldots, \mu_{ib} \text{ differ}$$

which says that at least two levels of factor 2 have different effects on the mean response. Furthermore, we can define the factor level mean for level j of factor 2 to be (for $j = 1, 2, \ldots, b$)

$$\mu_{\cdot j} = \frac{\sum_{i=1}^{a} \mu_{ij}}{a}$$

Then, alternatively, if there is little or no interaction between factors 1 and 2, it is reasonable to test the significance of factor 2 by testing

$$H_0\colon \mu_{\cdot 1} = \mu_{\cdot 2} = \cdots = \mu_{\cdot b}$$

versus

H_1: At least two of $\mu_{\cdot 1}, \mu_{\cdot 2}, \ldots, \mu_{\cdot b}$ differ

Each of the hypothesis tests described above is carried out by using an F statistic. These F statistics are denoted $F_{\text{BCF}}(\text{factor 1})$ and $F_{\text{BCF}}(\text{factor 2})$. Here, the subscript BCF refers to the fact that the validity of these F statistics requires that we perform a balanced complete factorial experiment.

We now present the appropriate F-tests.

F-Tests for the Significance of the Main Effects (Testing the Significance of Factors 1 and 2)

1. Assume that little or no interaction exists between factors 1 and 2, and define

$$F_{\text{BCF}}(\text{factor 1}) = \frac{MS_{\text{factor 1}}}{MSE} = \frac{SS_{\text{factor 1}}/(a-1)}{SSE/ab(m-1)}$$

$$= \frac{bm \sum_{i=1}^{a} (\bar{y}_{i\cdot} - \bar{y})^2/(a-1)}{\sum_{i=1}^{a}\sum_{j=1}^{b}\sum_{k=1}^{m} (y_{ij,k} - \bar{y}_{ij})^2/ab(m-1)}$$

Also, define the prob-value to be the area to the right of $F_{\text{BCF}}(\text{factor 1})$ under the curve of the F-distribution having $a-1$ and $ab(m-1)$ degrees of freedom. Then we can reject

$H_0: \mu_{1\cdot} = \mu_{2\cdot} = \cdots = \mu_{a\cdot}$

in favor of

H_1: At least two of $\mu_{1\cdot}, \mu_{2\cdot}, \ldots, \mu_{a\cdot}$ differ

by setting the probability of a Type I error equal to α if and only if either of the following equivalent conditions holds:

a. $F_{\text{BCF}}(\text{factor 1}) > F_{[\alpha]}^{(a-1,ab(m-1))}$
b. prob-value $< \alpha$

If we reject H_0, we conclude that at least two levels of factor 1 have different effects on the mean response.

2. Assume that little or no interaction exists between factors 1 and 2, and define

$$F_{\text{BCF}}(\text{factor 2}) = \frac{MS_{\text{factor 2}}}{MSE} = \frac{SS_{\text{factor 2}}/(b-1)}{SSE/ab(m-1)}$$

$$= \frac{am \sum_{j=1}^{b} (\bar{y}_{\cdot j} - \bar{y})^2/(b-1)}{\sum_{i=1}^{a}\sum_{j=1}^{b}\sum_{k=1}^{m} (y_{ij,k} - \bar{y}_{ij})^2/ab(m-1)}$$

Also, define the prob-value to be the area to the right of F_{BCF}(factor 2) under the curve of the F-distribution having $b - 1$ and $ab(m - 1)$ degrees of freedom. Then we can reject

$$H_0: \mu_{\cdot 1} = \mu_{\cdot 2} = \cdots = \mu_{\cdot b}$$

in favor of

H_1: At least two of $\mu_{\cdot 1}, \mu_{\cdot 2}, \ldots, \mu_{\cdot b}$ differ

by setting the probability of a Type I error equal to α if and only if either of the following equivalent conditions holds:

a. $F_{BCF}(\text{factor 2}) > F_{[\alpha]}^{(b-1, ab(m-1))}$
b. prob-value $< \alpha$

If we reject H_0, we conclude that at least two levels of factor 2 have different effects on the mean response.

Example 15.7 Consider the Tastee Bakery data given in Table 15.1. In Example 15.5 we concluded that either shelf display height, shelf display width, or the interaction between these factors has a statistically significant effect on mean demand. In Example 15.6 we concluded that little or no interaction exists between shelf display height and shelf display width.

Since little or no interaction exists, it is reasonable to test the significance of shelf display height by testing the null hypothesis

$$H_0: \mu_{B\cdot} = \mu_{M\cdot} = \mu_{T\cdot}$$

This hypothesis says that shelf display heights B, M, and T have the same effect on mean demand. We test H_0 versus

H_1: At least two of $\mu_{B\cdot}$, $\mu_{M\cdot}$, and $\mu_{T\cdot}$ differ

which says that at least two of shelf display heights B, M, and T have different effects on mean demand. Recalling that $SS_{\text{factor 1}}$ and SSE were computed in Example 15.4, we test these hypotheses by computing

$$\begin{aligned} F_{BCF}(\text{factor 1}) &= \frac{MS_{\text{factor 1}}}{MSE} \\ &= \frac{SS_{\text{factor 1}}/(a-1)}{SSE/ab(m-1)} \\ &= \frac{2273.88/(3-1)}{73.5/3(2)(3-1)} \\ &= \frac{2273.88/2}{73.5/12} = 185.6229 \end{aligned}$$

Since

$$F_{BCF}(\text{factor 1}) = 185.6229 > F_{[\alpha]}^{(a-1, ab(m-1))} = F_{[.05]}^{(2,12)} = 3.89$$

and

$$\text{prob-value} = .0001 < .05$$

we reject H_0 with $\alpha = .05$. We conclude that shelf display height has a statistically significant effect on mean demand. That is, we conclude that at least two of shelf display heights B, M, and T have different effects on mean demand.

Since little or no interaction exists, it is reasonable to test the significance of shelf display width by testing the null hypothesis

$$H_0\colon \mu_{\cdot R} = \mu_{\cdot W}$$

This hypothesis says that shelf display widths R and W have the same effect on mean demand. We test H_0 versus

$$H_1\colon \mu_{\cdot R} \neq \mu_{\cdot W}$$

which says that shelf display widths R and W have different effects on mean demand.

Recalling that $SS_{\text{factor 2}}$ and SSE were computed in Example 15.4, we test these hypotheses by calculating

$$\begin{aligned} F_{\text{BCF}}(\text{factor 2}) &= \frac{MS_{\text{factor 2}}}{MSE} \\ &= \frac{SS_{\text{factor 2}}/(b-1)}{SSE/ab(m-1)} \\ &= \frac{8.82/(2-1)}{73.5/3(2)(3-1)} \\ &= \frac{8.82/1}{73.5/12} = 1.44 \end{aligned}$$

Since

$$F_{\text{BCF}}(\text{factor 2}) = 1.44 < F_{[\alpha]}^{(b-1,ab(m-1))} = F_{[.05]}^{(1,12)} = 4.75$$

and

$$\text{prob-value} = .2533 > .05$$

we cannot reject H_0 with $\alpha = .05$. We conclude that the factor shelf display width does not have a statistically significant effect on mean demand. That is, we conclude that there is little or no difference between the effects of shelf display widths R and W on mean demand.

Finally, we present in Figure 15.3 the SAS output of the two-way analysis of variance for the Tastee Bakery data in Table 15.1. Note that all of the sums of squares and F statistics discussed in this section are given on the SAS output. The output also includes several quantities that will be explained later in this section.

FIGURE 15.3 **SAS output of a two-way ANOVA for the Tastee Bakery data in Table 15.1**

SAS

GENERAL LINEAR MODELS PROCEDURE

CLASS LEVEL INFORMATION

CLASS	LEVELS	VALUES
HEIGHT	3	B M T
WIDTH	2	R W

SAS

GENERAL LINEAR MODELS PROCEDURE

DEPENDENT VARIABLE: DEMAND

SOURCE	DF	SUM OF SQUARES	MEAN SQUARE	F VALUE	PR > F	R-SQUARE	C.V.
MODEL	5	2292.78000000[a]	458.55600000[b]	74.87[i]	0.0001[i]	0.968939	4.0242
ERROR	12	73.50000000[f]	6.12500000[g]		ROOT MSE		DEMAND MEAN
CORRECTED TOTAL	17	2366.28000000[h]			2.47487373		61.50000000

SOURCE	DF	TYPE I SS	F VALUE	PR > F	DF	TYPE III SS	F VALUE	PR > F
HEIGHT	2	2273.88000000	185.62	0.0001	2	2273.88000000[c]	185.62[j]	0.0001[j]
WIDTH	1	8.82000000	1.44	0.2533	1	8.82000000[d]	1.44[k]	0.2533[k]
HEIGHT*WIDTH	2	10.08000000	0.82	0.4625	2	10.08000000[e]	0.82[l]	0.4625[l]

PARAMETER	ESTIMATE	T FOR H0: PARAMETER=0	PR > |T|	STD ERROR OF ESTIMATE
MUM-MUB	21.40000000[m]	14.98[n]	0.0001[n]	1.42886902[o]
MUT-MUB	-4.30000000	-3.01	0.0109	1.42886902[o]
MUT-MUM	-25.70000000	-17.99	0.0001	1.42886902[o]
MUW-MUR	1.40000000[p]	1.20[q]	0.2533[q]	1.16666667[r]
MUMW-MUMR	3.40000000[s]	1.68[t]	0.1183[t]	2.02072594[u]

OBSERVATION	OBSERVED VALUE	PREDICTED[v] VALUE	RESIDUAL	LOWER 95% CL[w] FOR MEAN	UPPER 95% CL[w] FOR MEAN
1	58.20000000	55.90000000	2.30000000	52.78675869	59.01324131
2	53.70000000	55.90000000	-2.20000000	52.78675869	59.01324131
3	55.80000000	55.90000000	-0.10000000	52.78675869	59.01324131
4	73.00000000	75.50000000	-2.50000000	72.38675869	78.61324131
5	78.10000000	75.50000000	2.60000000	72.38675869	78.61324131
6	75.40000000	75.50000000	-0.10000000	72.38675869	78.61324131
7	52.40000000	51.00000000	1.40000000	47.88675869	54.11324131
8	49.70000000	51.00000000	-1.30000000	47.88675869	54.11324131
9	50.90000000	51.00000000	-0.10000000	47.88675869	54.11324131
10	55.70000000	55.70000000	0.00000000	52.58675869	58.81324131
11	52.50000000	55.70000000	-3.20000000	52.58675869	58.81324131
12	58.90000000	55.70000000	3.20000000	52.58675869	58.81324131
13	76.20000000	78.90000000	-2.70000000	75.78675869	82.01324131
14	78.40000000	78.90000000	-0.50000000	75.78675869	82.01324131
15	82.10000000	78.90000000	3.20000000	75.78675869	82.01324131
16	54.00000000	52.00000000	2.00000000	48.88675869	55.11324131
17	52.10000000	52.00000000	0.10000000	48.88675869	55.11324131
18	49.90000000	52.00000000	-2.10000000	48.88675869	55.11324131

[a] SS_{means} [b] MS_{means} [c] $SS_{\text{factor 1}}$ [d] $SS_{\text{factor 2}}$ [e] $SS_{\text{interaction}}$ [f] SSE [g] MSE [h] SS_{total} [i] *F (means) and associated prob-value* [j] F_{BCF} *(factor 1) and associated prob-value* [k] F_{BCF} *(factor 2) and associated prob-value* [l] *F (interaction) and associated prob-value* [m] $\bar{y}_{M\cdot} - \bar{y}_{B\cdot}$ [n] *t related to* $\bar{y}_{M\cdot} - \bar{y}_{B\cdot}$ *and associated prob-value* [o] $\sqrt{MSE(2/bm)}$ [p] $\bar{y}_{\cdot W} - \bar{y}_{\cdot R}$ [q] *t related to* $\bar{y}_{\cdot W} - \bar{y}_{\cdot R}$ *and associated prob-value* [r] $\sqrt{MSE(2/am)}$ [s] $\bar{y}_{MW} - \bar{y}_{MR}$ [t] *t related to* $\bar{y}_{MW} - \bar{y}_{MR}$ *and associated prob-value* [u] $\sqrt{MSE(2/m)}$ [v] $\bar{y}_{ij}$ [w] *95% confidence interval for* μ_{ij}

COMPARING AND ESTIMATING SPECIFIC MEANS

We next present estimation and prediction formulas. These formulas can be used to compare and estimate specific means and to predict future individual values.

Estimation and Prediction in a Balanced Complete Factorial Experiment

1. A *point estimate of the difference* $\mu_{i\cdot} - \mu_{i'\cdot}$ is $\bar{y}_{i\cdot} - \bar{y}_{i'\cdot}$, and an *individual 100(1 − α)% confidence interval for* $\mu_{i\cdot} - \mu_{i'\cdot}$ is

$$\left[(\bar{y}_{i\cdot} - \bar{y}_{i'\cdot}) \pm t_{[\alpha/2]}^{(ab(m-1))}\sqrt{MSE\left(\frac{2}{bm}\right)}\right]$$

2. A *point estimate of the linear combination* $\Sigma_{i=1}^{a} a_i\mu_{i\cdot}$ is $\Sigma_{i=1}^{a} a_i\bar{y}_{i\cdot}$, and an *individual 100(1 − α)% confidence interval for this linear combination* is

$$\left[\sum_{i=1}^{a} a_i\bar{y}_{i\cdot} \pm t_{[\alpha/2]}^{(ab(m-1))}\sqrt{\frac{MSE}{bm}\sum_{i=1}^{a} a_i^2}\right]$$

3. For *Scheffé simultaneous intervals* in 1 and 2 above, replace $t_{[\alpha/2]}^{(ab(m-1))}$ by $\sqrt{(a-1)F_{[\alpha]}^{(a-1,ab(m-1))}}$ (which applies to all possible *contrasts* in the $\mu_{i\cdot}$ values) or by $\sqrt{aF_{[\alpha]}^{(a,ab(m-1))}}$ (which applies to all possible *linear combinations* in the $\mu_{i\cdot}$ values). For *Bonferroni simultaneous intervals* in 1 and 2 above, replace $t_{[\alpha/2]}^{(ab(m-1))}$ by $t_{[\alpha/2g]}^{(ab(m-1))}$ (which applies to *g prespecified linear combinations*). For *Tukey simultaneous intervals* in 1 above, replace $t_{[\alpha/2]}^{(ab(m-1))}$ by $q_{[\alpha]}(a, ab(m-1))$, and replace $(2/bm)$ by $(1/bm)$.
4. A *point estimate of the difference* $\mu_{\cdot j} - \mu_{\cdot j'}$ is $\bar{y}_{\cdot j} - \bar{y}_{\cdot j'}$, and an *individual 100(1 − α)% confidence interval for* $\mu_{\cdot j} - \mu_{\cdot j'}$ is

$$\left[(\bar{y}_{\cdot j} - \bar{y}_{\cdot j'}) \pm t_{[\alpha/2]}^{(ab(m-1))}\sqrt{MSE\left(\frac{2}{am}\right)}\right]$$

5. A *point estimate of the linear combination* $\Sigma_{j=1}^{b} a_j\mu_{\cdot j}$ is $\Sigma_{j=1}^{b} a_j\bar{y}_{\cdot j}$, and an *individual 100(1 − α)% confidence interval for this linear combination* is

$$\left[\sum_{j=1}^{b} a_j\bar{y}_{\cdot j} \pm t_{[\alpha/2]}^{(ab(m-1))}\sqrt{\frac{MSE}{am}\sum_{j=1}^{b} a_j^2}\right]$$

6. For *Scheffé simultaneous intervals* in 4 and 5 above, replace $t_{[\alpha/2]}^{(ab(m-1))}$ by $\sqrt{(b-1)F_{[\alpha]}^{(b-1,ab(m-1))}}$ (which applies to all possible *contrasts* in the $\mu_{\cdot j}$ values) or by $\sqrt{bF_{[\alpha]}^{(b,ab(m-1))}}$ (which applies to all possible *linear combinations* in the $\mu_{\cdot j}$ values). For *Bonferroni simultaneous intervals* in 4 and 5 above, replace $t_{[\alpha/2]}^{(ab(m-1))}$ by $t_{[\alpha/2g]}^{(ab(m-1))}$ (which applies to *g prespecified linear combinations*). For *Tukey simultaneous intervals* in 4 above, replace $t_{[\alpha/2]}^{(ab(m-1))}$ by $q_{[\alpha]}(b, ab(m-1))$, and replace $(2/am)$ by $(1/am)$.
7. A *point estimate of the difference* $\mu_{ij} - \mu_{i'j'}$ is $\bar{y}_{ij} - \bar{y}_{i'j'}$, and an *individual*

100(1 − α)% confidence interval for $\mu_{ij} - \mu_{i'j'}$ is

$$\left[(\bar{y}_{ij} - \bar{y}_{i'j'}) \pm t_{[\alpha/2]}^{(ab(m-1))} \sqrt{MSE\left(\frac{2}{m}\right)} \right]$$

8. A *point estimate of the linear combination of treatment means* $\Sigma_{i=1}^{a} \Sigma_{j=1}^{b} a_{ij}\mu_{ij}$ is $\Sigma_{i=1}^{a} \Sigma_{j=1}^{b} a_{ij}\bar{y}_{ij}$, and an *individual 100(1 − α)% confidence interval for this linear combination* is

$$\left[\sum_{i=1}^{a} \sum_{j=1}^{b} a_{ij}\bar{y}_{ij} \pm t_{[\alpha/2]}^{(ab(m-1))} \sqrt{MSE \sum_{i=1}^{a} \sum_{j=1}^{b} a_{ij}^2/m} \right]$$

9. For *Scheffé simultaneous intervals* in 7 and 8 above, replace $t_{[\alpha/2]}^{(ab(m-1))}$ by $\sqrt{(ab-1)F_{[\alpha]}^{(ab-1,ab(m-1))}}$ (which applies to all possible *contrasts* of treatment means) or by $\sqrt{abF_{[\alpha]}^{(ab,ab(m-1))}}$ (which applies to all possible *linear combinations* of treatment means). For *Bonferroni simultaneous intervals* in 7 and 8 above, replace $t_{[\alpha/2]}^{(ab(m-1))}$ by $t_{[\alpha/2g]}^{(ab(m-1))}$ (which applies to *g prespecified linear combinations*). For *Tukey simultaneous intervals* in 7 above, replace $t_{[\alpha/2]}^{(ab(m-1))}$ by $q_{[\alpha]}(ab, ab(m-1))$, and replace $(2/m)$ by $(1/m)$.

10. A *point estimate of the treatment mean* μ_{ij} is $\bar{y}_{ij}$, and an *individual 100(1 − α)% confidence interval for* μ_{ij} is

$$\left[\bar{y}_{ij} \pm t_{[\alpha/2]}^{(ab(m-1))} \sqrt{\frac{MSE}{m}} \right]$$

11. A *point prediction* of a randomly selected *individual value*

$$y_{ij,0} = \mu_{ij} + \varepsilon_{ij,0}$$

is $\bar{y}_{ij}$ and a *100(1 − α)% prediction interval for* $y_{ij,0}$ is

$$\left[\bar{y}_{ij} \pm t_{[\alpha/2]}^{(ab(m-1))} \sqrt{MSE} \sqrt{1 + \frac{1}{m}} \right]$$

Example 15.8 We have concluded in the Tastee Bakery problem that

1. There is little or no interaction between display height and display width.
2. At least two of the shelf display heights B, M, and T have different effects on mean demand.
3. There seems to be little or no difference between the effects of display widths R and W on mean demand.

To investigate the nature of the differences between the display heights, we calculate 95% Tukey simultaneous confidence intervals for all three pairwise differences $\mu_{M\cdot} - \mu_{B\cdot}$, $\mu_{T\cdot} - \mu_{B\cdot}$, and $\mu_{T\cdot} - \mu_{M\cdot}$. A 95% Tukey simultaneous confidence interval for $\mu_{M\cdot} - \mu_{B\cdot}$ is

$$\begin{aligned}
&[(\bar{y}_{M\cdot} - \bar{y}_{B\cdot}) \pm q_{[\alpha]}(a, ab(m-1))\sqrt{MSE/bm}] \\
&\quad = [(77.2 - 55.8) \pm q_{[.05]}(3, 3(2)(3-1))\sqrt{6.125/2(3)}] \\
&\quad = [21.4 \pm q_{[.05]}(3, 12)\sqrt{6.125/6}] \\
&\quad = [21.4 \pm 3.773\,(1.010363)] \\
&\quad = [21.4 \pm 3.8121] \\
&\quad = [17.5879, 25.2121]
\end{aligned}$$

See Figure 15.3 for the SAS output of $\bar{y}_{M\cdot} - \bar{y}_{B\cdot}$.

Similarly, a 95% Tukey simultaneous confidence interval for $\mu_{T\cdot} - \mu_{B\cdot}$ is

$$\begin{aligned}
[(\bar{y}_{T\cdot} - \bar{y}_{B\cdot}) \pm 3.8121] &= [(51.5 - 55.8) \pm 3.8121] \\
&= [-4.3 \pm 3.8121] \\
&= [-8.1121, -0.4879]
\end{aligned}$$

and a 95% Tukey simultaneous confidence interval for $\mu_{T\cdot} - \mu_{M\cdot}$ is

$$\begin{aligned}
[(\bar{y}_{T\cdot} - \bar{y}_{M\cdot}) \pm 3.8121] &= [(51.5 - 77.2) \pm 3.8121] \\
&= [-25.7 \pm 3.8121] \\
&= [-29.5121, -21.8879]
\end{aligned}$$

We conclude that we can be 95% confident that

1. Changing from using a bottom shelf display height to using a middle shelf display height will increase mean demand by between 17.5879 and 25.2121 cases.
2. Changing from using a bottom shelf display height to using a top shelf display height will decrease mean demand by between .4879 and 8.1121 cases.
3. Changing from using a middle shelf display height to using a top shelf display height will decrease mean demand by between 21.8879 and 29.5121 cases.

We therefore have strong evidence suggesting that using a middle shelf display height is substantially more effective than using either a top or a bottom shelf display height. In addition, we have strong evidence that using a bottom shelf display height is somewhat more effective than using a top shelf display height.

In spite of the conclusion that $\mu_{\cdot W}$ and $\mu_{\cdot R}$ do not differ significantly, it might be interesting to compare μ_{MW} with μ_{MR}. A point estimate of

$\mu_{MW} - \mu_{MR}$ = the change in mean demand associated with changing from a regular display width to a wide display while using a middle display height

is $\bar{y}_{MW} - \bar{y}_{MR} = 78.9 - 75.5 = 3.4$. Notice that this estimate is given on the SAS output—the difference $\mu_{MW} - \mu_{MR}$ is denoted "MUMW – MUMR." In addition, an 80% confidence interval for $\mu_{MW} - \mu_{MR}$ is

$$\begin{aligned}
&[(\bar{y}_{MW} - \bar{y}_{MR}) \pm t_{[\alpha/2]}^{(ab(m-1))} \sqrt{MSE(2/m)}] \\
&\quad = [(78.9 - 75.5) \pm t_{[.10]}^{(12)} \sqrt{6.125(2/3)}] \\
&\quad = [3.4 \pm 1.356(2.02072594)] \\
&\quad = [3.4 \pm 2.7401] \\
&\quad = [.6599, 6.1401]
\end{aligned}$$

This interval says that we can be 80% confident that μ_{MW} is between .6599 cases (about 7 units) and 6.1401 cases (about 61 units) greater than μ_{MR}. This provides a reasonable amount of evidence suggesting that we can maximize mean monthly demand by using a middle display height and a wide display.

Next, we find that a point estimate of, and a 95% confidence interval for, the mean μ_{MW} are

$$\bar{y}_{MW} = 78.9$$

and

$$\begin{aligned}
\left[\bar{y}_{MW} \pm t_{[.025]}^{(12)} \sqrt{\frac{MSE}{m}}\right] &= \left[78.9 \pm 2.179 \sqrt{\frac{6.125}{3}}\right] \\
&= [78.9 \pm 3.1135] \\
&= [75.7865, 82.0135]
\end{aligned}$$

This interval says that we can be 95% confident that μ_{MW} is between 75.7865 and 82.0135 cases.

Finally, we find that a point prediction of, and a 95% prediction interval for, an individual demand

$$y_{MW,0} = \mu_{MW} + \varepsilon_{MW,0}$$

that will be observed when using a middle display height and a wide display are

$$\bar{y}_{MW} = 78.9$$

and

$$\begin{aligned}
\left[\bar{y}_{MW} \pm t_{[.025]}^{(12)} \sqrt{MSE} \sqrt{1 + \frac{1}{m}}\right] &= \left[78.9 \pm 2.179(2.4749) \sqrt{1 + \frac{1}{3}}\right] \\
&= [78.9 \pm 6.2271] \\
&= [72.6729, 85.1271]
\end{aligned}$$

This interval says that we can be 95% confident that the individual demand $y_{MW,0}$ will be at least 72.6729 cases and no more than 85.1271 cases.

15.3

THE REGRESSION APPROACH TO THE ANALYSIS OF A COMPLETE FACTORIAL EXPERIMENT

The general approach to analyzing an *unbalanced complete factorial experiment* ($n_{ij} \geq 1$ for each and every combination of i and j) is similar to analyzing a balanced complete factorial experiment. However, the calculation and interpretation of the F statistics are best accomplished by writing the two-factor—or (α, γ, θ)—model

$$\begin{aligned} y_{ij,k} &= \mu_{ij} + \varepsilon_{ij,k} \\ &= \mu + \alpha_i + \gamma_j + \theta_{ij} + \varepsilon_{ij,k} \end{aligned}$$

as a dummy variable regression model. To do this, we let

$$\begin{aligned} \alpha_i &= \alpha_2 D_{i,2} + \alpha_3 D_{i,3} + \cdots + \alpha_a D_{i,a} \\ \gamma_j &= \gamma_2 D_{j,2} + \gamma_3 D_{j,3} + \cdots + \gamma_b D_{j,b} \\ \theta_{ij} &= \theta_{22} D_{i,2} D_{j,2} + \theta_{23} D_{i,2} D_{j,3} + \cdots + \theta_{2b} D_{i,2} D_{j,b} \\ &\quad + \theta_{32} D_{i,3} D_{j,2} + \theta_{33} D_{i,3} D_{j,3} + \cdots + \theta_{3b} D_{i,3} D_{j,b} \\ &\quad \vdots \\ &\quad + \theta_{a2} D_{i,a} D_{j,2} + \theta_{a3} D_{i,a} D_{j,3} + \cdots + \theta_{ab} D_{i,a} D_{j,b} \end{aligned}$$

Here, rather than defining the above dummy variables to be the usual 0, 1 dummy variables, we define

$$D_{i,2} = \begin{cases} 1 & \text{if } i = 2\text{; that is, if we use level 2 of factor 1} \\ -1 & \text{if } i = 1\text{; that is, if we use level 1 of factor 1} \\ 0 & \text{otherwise} \end{cases}$$

$$D_{i,3} = \begin{cases} 1 & \text{if } i = 3\text{; that is, if we use level 3 of factor 1} \\ -1 & \text{if } i = 1\text{; that is, if we use level 1 of factor 1} \\ 0 & \text{otherwise} \end{cases}$$

$$\vdots$$

$$D_{i,a} = \begin{cases} 1 & \text{if } i = a\text{; that is, if we use level } a \text{ of factor 1} \\ -1 & \text{if } i = 1\text{; that is, if we use level 1 of factor 1} \\ 0 & \text{otherwise} \end{cases}$$

$$D_{j,2} = \begin{cases} 1 & \text{if } j = 2\text{; that is, if we use level 2 of factor 2} \\ -1 & \text{if } j = 1\text{; that is, if we use level 1 of factor 2} \\ 0 & \text{otherwise} \end{cases}$$

$$D_{j,3} = \begin{cases} 1 & \text{if } j = 3\text{; that is, if we use level 3 of factor 2} \\ -1 & \text{if } j = 1\text{; that is, if we use level 1 of factor 2} \\ 0 & \text{otherwise} \end{cases}$$

$$\vdots$$

$$D_{j,b} = \begin{cases} 1 & \text{if } j = b\text{; that is, if we use level } b \text{ of factor 2} \\ -1 & \text{if } j = 1\text{; that is, if we use level 1 of factor 2} \\ 0 & \text{otherwise} \end{cases}$$

Example 15.9 Suppose that in another product demand experiment the Tastee Bakery Company observes the data in Table 15.2. Here, the company has performed an unbalanced complete factorial experiment. Thus it is appropriate to analyze this data by using the (α, γ, θ) model

$$\begin{aligned} y_{ij,k} &= \mu_{ij} + \varepsilon_{ij,k} \\ &= \mu + \alpha_i + \gamma_j + \theta_{ij} + \varepsilon_{ij,k} \\ &= \mu + \alpha_M D_{i,M} + \alpha_T D_{i,T} + \gamma_W D_{j,W} \\ &\quad + \theta_{MW} D_{i,M} D_{j,W} + \theta_{TW} D_{i,T} D_{j,W} + \varepsilon_{ij,k} \end{aligned}$$

Here, $D_{i,M}$, $D_{i,T}$, and $D_{j,W}$ are dummy variables defined as follows:

$$D_{i,M} = \begin{cases} 1 & \text{if } i = \text{M; that is, if we use a middle display height} \\ -1 & \text{if } i = \text{B; that is, if we use a bottom display height} \\ 0 & \text{otherwise} \end{cases}$$

$$D_{i,T} = \begin{cases} 1 & \text{if } i = \text{T; that is, if we use a top display height} \\ -1 & \text{if } i = \text{B; that is, if we use a bottom display height} \\ 0 & \text{otherwise} \end{cases}$$

$$D_{j,W} = \begin{cases} 1 & \text{if } j = \text{W; that is, if we use a wide display} \\ -1 & \text{if } j = \text{R; that is, if we use a regular display width} \end{cases}$$

It follows that we use the following **y** vector and **X** matrix to calculate the least squares point estimates of the model parameters.

$$\mathbf{y} = \begin{bmatrix} y_{BR,1} \\ y_{BR,2} \\ y_{BW,1} \\ y_{MR,1} \\ y_{MR,2} \\ y_{MW,1} \\ y_{MW,2} \\ y_{MW,3} \\ y_{TR,1} \\ y_{TR,2} \\ y_{TW,1} \\ y_{TW,2} \\ y_{TW,3} \end{bmatrix} = \begin{bmatrix} 58.2 \\ 53.7 \\ 55.6 \\ 73.0 \\ 78.1 \\ 76.2 \\ 82.0 \\ 78.4 \\ 52.5 \\ 49.8 \\ 56.0 \\ 51.9 \\ 53.3 \end{bmatrix}$$

and

$$\begin{array}{cccccc} 1 & D_{i,\mathrm{M}} & D_{i,\mathrm{T}} & D_{j,\mathrm{W}} & (D_{i,\mathrm{M}} D_{j,\mathrm{W}}) & (D_{i,\mathrm{T}} D_{j,\mathrm{W}}) \end{array}$$

$$\mathbf{X} = \begin{bmatrix} 1 & -1 & -1 & -1 & 1 & 1 \\ 1 & -1 & -1 & -1 & 1 & 1 \\ 1 & -1 & -1 & 1 & -1 & -1 \\ 1 & 1 & 0 & -1 & -1 & 0 \\ 1 & 1 & 0 & -1 & -1 & 0 \\ 1 & 1 & 0 & 1 & 1 & 0 \\ 1 & 1 & 0 & 1 & 1 & 0 \\ 1 & 1 & 0 & 1 & 1 & 0 \\ 1 & 0 & 1 & -1 & 0 & -1 \\ 1 & 0 & 1 & -1 & 0 & -1 \\ 1 & 0 & 1 & 1 & 0 & 1 \\ 1 & 0 & 1 & 1 & 0 & 1 \\ 1 & 0 & 1 & 1 & 0 & 1 \end{bmatrix}$$

TABLE 15.2 **Six samples of monthly demands**

Display height, i	Display width, j: R	Display width, j: W
B	$y_{BR,1} = 58.2$ $y_{BR,2} = 53.7$	$y_{BW,1} = 55.6$
M	$y_{MR,1} = 73.0$ $y_{MR,2} = 78.1$	$y_{MW,1} = 76.2$ $y_{MW,2} = 82.0$ $y_{MW,3} = 78.4$
T	$y_{TR,1} = 52.5$ $y_{TR,2} = 49.8$	$y_{TW,1} = 56.0$ $y_{TW,2} = 51.9$ $y_{TW,3} = 53.3$

Using **y** and **X**, we obtain the prediction equation

$$\begin{aligned}\hat{y}_{ij} &= \hat{\mu} + \hat{\alpha}_i + \hat{\gamma}_j + \hat{\theta}_{ij} \\ &= \hat{\mu} + \hat{\alpha}_M D_{i,M} + \hat{\alpha}_T D_{i,T} + \hat{\gamma}_W D_{j,W} \\ &\quad + \hat{\theta}_{MW} D_{i,M} D_{j,W} + \hat{\theta}_{TW} D_{i,T} D_{j,W} \\ &= 61.8083 + 15.4002 D_{i,M} - 9.3668 D_{i,T} + .925 D_{j,W} \\ &\quad + .7335 D_{i,M} D_{j,W} + .3665 D_{i,T} D_{j,W}\end{aligned}$$

For example, a point estimate of the mean monthly demand that would be obtained by using a medium display height and a regular display width

$$\begin{aligned}\mu_{MR} &= \mu + \alpha_M D_{M,M} + \alpha_T D_{M,T} + \gamma_W D_{R,W} + \theta_{MW} D_{M,M} D_{R,W} \\ &\quad + \theta_{TW} D_{M,T} D_{R,W} \\ &= \mu + \alpha_M(1) + \alpha_T(0) + \gamma_W(-1) + \theta_{MW}(1)(-1) + \theta_{TW}(0)(-1) \\ &= \mu + \alpha_M - \gamma_W - \theta_{MW}\end{aligned}$$

and a point prediction of a corresponding individual monthly demand

$$\begin{aligned}y_{MR,k} &= \mu_{MR} + \varepsilon_{MR,k} \\ &= \mu + \alpha_M - \gamma_W - \theta_{MW} + \varepsilon_{MR,k}\end{aligned}$$

is

$$\begin{aligned}\hat{y}_{MR} &= \hat{\mu} + \hat{\alpha}_M - \hat{\gamma}_W - \hat{\theta}_{MW} \\ &= 61.8083 + 15.4002 - .925 - .7335 \\ &= 75.55\end{aligned}$$

Note that $\hat{y}_{MR}$ equals the sample mean

$$\begin{aligned}\bar{y}_{MR} &= \frac{y_{MR,1} + y_{MR,2}}{2} \\ &= \frac{73.0 + 78.1}{2} \\ &= 75.55\end{aligned}$$

In general, it can be proven that $\hat{y}_{ij}$, the point estimate of μ_{ij} and point prediction of $y_{ij,k}$, is the sample mean $\bar{y}_{ij}$ *when using the* (α, γ, θ) *model.* It follows that the unexplained variation for the (α, γ, θ) model can be calculated to be

$$SSE_{(\alpha,\gamma,\theta)} = \sum_{i=\text{B,M,T}} \sum_{j=\text{R,W}} \sum_{k=1}^{n_{ij}} (y_{ij,k} - \bar{y}_{ij})^2 = 52.6083$$

In general, it can be proven that using the previously defined dummy variables implies that the parameters of the (α, γ, θ) model satisfy the following Σ-*conditions*:

$$\sum_{i=1}^{a} \alpha_i = 0 \qquad \sum_{j=1}^{b} \gamma_j = 0$$

$$\sum_{i=1}^{a} \theta_{ij} = 0 \qquad \text{for } j = 1, 2, \ldots, b$$

$$\sum_{j=1}^{b} \theta_{ij} = 0 \qquad \text{for } i = 1, 2, \ldots, a$$

For example, if we set i equal to 1 in the equation

$$\alpha_i = \alpha_2 D_{i,2} + \alpha_3 D_{i,3} + \cdots + \alpha_a D_{i,a}$$

we obtain

$$\begin{aligned} \alpha_1 &= \alpha_2 D_{1,2} + \alpha_3 D_{1,3} + \cdots + \alpha_a D_{1,a} \\ &= \alpha_2(-1) + \alpha_3(-1) + \cdots + \alpha_a(-1) \end{aligned}$$

This implies that

$$\sum_{i=1}^{a} \alpha_i = 0$$

Verifications of the other conditions are obtained similarly.

We now express the meanings of the parameters in the (α, γ, θ) model in terms of the treatment means μ_{ij} (also called the *cell means*). We first note that the mean of all the treatment means is

$$\begin{aligned} \mu_{..} &= \frac{\sum_{i=1}^{a} \sum_{j=1}^{b} \mu_{ij}}{ab} \\ &= \frac{\sum_{i=1}^{a} \sum_{j=1}^{b} (\mu + \alpha_i + \gamma_j + \theta_{ij})}{ab} \\ &= \frac{ab\mu + b \sum_{i=1}^{a} \alpha_i + a \sum_{j=1}^{b} \gamma_j + \sum_{i=1}^{a} \left[\sum_{j=1}^{b} \theta_{ij} \right]}{ab} \\ &= \mu \end{aligned}$$

Next, the factor level mean for level i of factor 1 (originally discussed in Section 15.2) is

$$\begin{aligned}
\mu_{i\cdot} &= \frac{\sum_{j=1}^{b} \mu_{ij}}{b} \\
&= \frac{\sum_{j=1}^{b} (\mu + \alpha_i + \gamma_j + \theta_{ij})}{b} \\
&= \frac{b\mu + b\alpha_i + \sum_{j=1}^{b} \gamma_j + \sum_{j=1}^{b} \theta_{ij}}{b} \\
&= \mu + \alpha_i
\end{aligned}$$

This implies that

$$\begin{aligned}
\alpha_i &= \mu_{i\cdot} - \mu \\
&= \mu_{i\cdot} - \mu_{\cdot\cdot}
\end{aligned}$$

Similarly, the factor level mean for level j of factor 2 is

$$\begin{aligned}
\mu_{\cdot j} &= \frac{\sum_{i=1}^{a} \mu_{ij}}{a} \\
&= \frac{\sum_{i=1}^{a} (\mu + \alpha_i + \gamma_j + \theta_{ij})}{a} \\
&= \frac{a\mu + \sum_{i=1}^{a} \alpha_i + a\gamma_j + \sum_{i=1}^{a} \theta_{ij}}{a} \\
&= \mu + \gamma_j
\end{aligned}$$

This implies that

$$\begin{aligned}
\gamma_j &= \mu_{\cdot j} - \mu \\
&= \mu_{\cdot j} - \mu_{\cdot\cdot}
\end{aligned}$$

Finally, the equation

$$\mu_{ij} = \mu + \alpha_i + \gamma_j + \theta_{ij}$$

implies that

$$\begin{aligned}
\theta_{ij} &= \mu_{ij} - \alpha_i - \gamma_j - \mu \\
&= \mu_{ij} - (\mu_{i\cdot} - \mu_{\cdot\cdot}) - (\mu_{\cdot j} - \mu_{\cdot\cdot}) - \mu_{\cdot\cdot} \\
&= \mu_{ij} - \mu_{i\cdot} - \mu_{\cdot j} + \mu_{\cdot\cdot}
\end{aligned}$$

To summarize, for $i = 1, 2, \ldots, a$ and $j = 1, 2, \ldots, b$,

$$\begin{aligned}\mu &= \mu_{..}\\ \alpha_i &= \mu_{i.} - \mu_{..}\\ \gamma_j &= \mu_{.j} - \mu_{..}\\ \theta_{ij} &= \mu_{ij} - \mu_{i.} - \mu_{.j} + \mu_{..}\end{aligned}$$

These relationships and the previous discussion allow us to use the (α, γ, θ) model to make the following four hypothesis tests:

1. The equation

$$\mu_{ij} = \mu + \alpha_i + \gamma_j + \theta_{ij}$$

implies that all of the treatment means μ_{ij} are equal to each other if and only if all of the parameters α_i, γ_j, and θ_{ij} are equal to zero. Therefore we can test

H_0: All treatment means μ_{ij} are equal

versus

H_1: At least two of the treatment means μ_{ij} differ

by using the (α, γ, θ) model to test

H_0: All of the parameters α_i, γ_j, and θ_{ij} equal zero

versus

H_1: At least one of the parameters α_i, γ_j, and θ_{ij} does not equal zero

Specifically, let $SS_{(\alpha,\gamma,\theta)}$ and $SSE_{(\alpha,\gamma,\theta)}$ denote the explained variation and the unexplained variation, respectively, for the (α, γ, θ) model. Then we can reject H_0 in favor of H_1 by setting the probability of a Type I error equal to α if and only if

$$F((\alpha, \gamma, \theta) \text{ model}) = \frac{SS_{(\alpha,\gamma,\theta)}/(ab - 1)}{SSE_{(\alpha,\gamma,\theta)}/(n - ab)}$$

is greater than $F_{[\alpha]}^{(ab-1, n-ab)}$. Here,

$$n = \sum_{i=1}^{a} \sum_{j=1}^{b} n_{ij}$$

is the total number of observations in the two-factor experiment.

2. The equation

$$\theta_{ij} = \mu_{ij} - \mu_{i.} - \mu_{.j} + \mu_{..}$$

implies that for arbitrary levels i and i' of factor 1 and arbitrary levels j

and j' of factor 2

$$\theta_{i'j} = \mu_{i'j} - \mu_{i'\cdot} - \mu_{\cdot j} + \mu_{\cdot\cdot}$$

$$\theta_{ij'} = \mu_{ij'} - \mu_{i\cdot} - \mu_{\cdot j'} + \mu_{\cdot\cdot}$$

and

$$\theta_{i'j'} = \mu_{i'j'} - \mu_{i'\cdot} - \mu_{\cdot j'} + \mu_{\cdot\cdot}$$

It follows that

H_0: All parameters θ_{ij} equal zero

is equivalent to

H_0: for arbitrary levels i and i' of factor 1 and arbitrary levels j and j' of factor 2

$$\theta_{ij} = \theta_{i'j} = \theta_{ij'} = \theta_{i'j'} = 0$$

or

$$\theta_{ij} - \theta_{i'j} = \theta_{ij'} - \theta_{i'j'}$$

or

$$(\mu_{ij} - \mu_{i\cdot} - \mu_{\cdot j} + \mu_{\cdot\cdot}) - (\mu_{i'j} - \mu_{i'\cdot} - \mu_{\cdot j} + \mu_{\cdot\cdot}) = (\mu_{ij'} - \mu_{i\cdot} - \mu_{\cdot j'} + \mu_{\cdot\cdot}) - (\mu_{i'j'} - \mu_{i'\cdot} - \mu_{\cdot j'} + \mu_{\cdot\cdot})$$

or

$$\mu_{ij} - \mu_{i'j} = \mu_{ij'} - \mu_{i'j'}$$

The last equality in H_0 says that the effect on the mean response of changing from level i' of factor 1 to level i of factor 1 is the same for level j of factor 2 as it is for level j' of factor 2. This says that there is no interaction between factors 1 and 2. To summarize, we can test for interaction between the two qualitative factors by using the (α, γ, θ) model to test

H_0: All parameters θ_{ij} equal zero (no interaction exists)

versus

H_1: At least one θ_{ij} does not equal zero (interaction does exist)

Specifically, we can reject H_0 in favor of H_1 by setting the probability of a Type I error equal to α if and only if

$$F(\theta \mid \alpha, \gamma) = \frac{\{SSE_{(\alpha,\gamma)} - SSE_{(\alpha,\gamma,\theta)}\}/(a-1)(b-1)}{SSE_{(\alpha,\gamma,\theta)}/(n-ab)}$$

is greater than $F_{[\alpha]}^{((a-1)(b-1), n-ab)}$. Here, $SSE_{(\alpha,\gamma)}$ is the unexplained variation for the (α, γ) *model*

$$y_{ij,k} = \mu + \alpha_i + \gamma_j + \varepsilon_{ij,k}$$

3. The condition $\Sigma_{i=1}^{a} \alpha_i = 0$ implies that

$$\alpha_1 = -\alpha_2 - \alpha_3 - \cdots - \alpha_a$$

It follows that if each of $\alpha_2, \alpha_3, \ldots, \alpha_a$ equals zero, then α_1 also equals zero. Therefore if each of $\alpha_2, \alpha_3, \ldots, \alpha_a$ equals zero, each factor level mean $\mu_{i\cdot}$ equals $\mu_{\cdot\cdot}$ because for $i = 1, 2, \ldots, a$

$$\alpha_i = \mu_{i\cdot} - \mu_{\cdot\cdot}$$

This says that

$$\mu_{1\cdot} = \mu_{2\cdot} = \cdots = \mu_{a\cdot}$$

To summarize, we can test

$$H_0 : \mu_{1\cdot} = \mu_{2\cdot} = \cdots = \mu_{a\cdot}$$

versus

H_1 : At least two of $\mu_{1\cdot}, \mu_{2\cdot}, \ldots, \mu_{a\cdot}$ differ

by using the (α, γ, θ) model to test

$$H_0 : \alpha_2 = \alpha_3 = \cdots = \alpha_a = 0$$

versus

H_1 : At least one of $\alpha_2, \alpha_3, \ldots, \alpha_a$ does not equal zero

Specifically, we can reject H_0 in favor of H_1 by setting the probability of a Type I error equal to α if and only if

$$F(\alpha \mid \gamma, \theta) = \frac{\{SSE_{(\gamma,\theta)} - SSE_{(\alpha,\gamma,\theta)}\}/(a-1)}{SSE_{(\alpha,\gamma,\theta)}/(n-ab)}$$

is greater than $F_{[\alpha]}^{(a-1,n-ab)}$. Here, $SSE_{(\gamma,\theta)}$ is the unexplained variation for the (γ, θ) *model*

$$y_{ij,k} = \mu + \gamma_j + \theta_{ij} + \varepsilon_{ij,k}$$

If we reject H_0, we conclude that at least two levels of factor 1 have different effects on the mean response.

4. The condition $\Sigma_{j=1}^{b} \gamma_j = 0$ implies that

$$\gamma_1 = -\gamma_2 - \gamma_3 - \cdots - \gamma_b$$

It follows that if each of $\gamma_2, \gamma_3, \ldots, \gamma_b$ equals zero, then γ_1 also equals zero. Therefore if each of $\gamma_2, \gamma_3, \ldots, \gamma_b$ equals zero, each factor level mean $\mu_{\cdot j}$ equals $\mu_{\cdot\cdot}$ because for $j = 1, 2, \ldots, b$

$$\gamma_j = \mu_{\cdot j} - \mu_{\cdot\cdot}$$

This says that

$$\mu_{\cdot 1} = \mu_{\cdot 2} = \cdots = \mu_{\cdot b}$$

To summarize, we can test

$$H_0: \mu_{\cdot 1} = \mu_{\cdot 2} = \cdots = \mu_{\cdot b}$$

versus

H_1: At least two of $\mu_{\cdot 1}, \mu_{\cdot 2}, \ldots, \mu_{\cdot b}$ differ

by using the (α, γ, θ) model to test

$$H_0: \gamma_2 = \gamma_3 = \cdots = \gamma_b = 0$$

versus

H_1: At least one of $\gamma_2, \gamma_3, \ldots, \gamma_b$ does not equal zero

Specifically, we can reject H_0 in favor of H_1 by setting the probability of a Type I error equal to α if and only if

$$F(\gamma \mid \alpha, \theta) = \frac{\{SSE_{(\alpha,\theta)} - SSE_{(\alpha,\gamma,\theta)}\}/(b - 1)}{SSE_{(\alpha,\gamma,\theta)}/(n - ab)}$$

is greater than $F_{[\alpha]}^{(b-1, n-ab)}$. Here, $SSE_{(\alpha,\theta)}$ is the unexplained variation for the (α,θ) *model*

$$y_{ij,k} = \mu + \alpha_i + \theta_{ij} + \varepsilon_{ij,k}$$

If we reject H_0, we conclude that at least two levels of factor 2 have different effects on the mean response.

Finally, note that for a *balanced complete factorial experiment*, the statistics $F((\alpha, \gamma, \theta)$ model), $F(\theta \mid \alpha, \gamma)$, $F(\alpha \mid \gamma, \theta)$, and $F(\gamma \mid \alpha, \theta)$ of this section become the statistics F(means), F(interaction), F_{BCF}(factor 1), and F_{BCF}(factor 2), respectively, of Section 15.2. That is, the (α, γ, θ) model is the theoretical model underlying the ANOVA approach of Section 15.2. However, the ANOVA approach requires balanced data and cannot be used to analyze an *unbalanced complete factorial experiment*. One reason for this is that the partition of the treatment sum of squares

$$SS_{\text{means}} = SS_{\text{factor 1}} + SS_{\text{factor 2}} + SS_{\text{interaction}}$$

does not hold for unbalanced complete data.

Example 15.10 Figure 15.4 presents the SAS output resulting from using the (α, γ, θ) model to analyze the data in Table 15.2. From this output we see that:

1. Since the prob-value related to $F((\alpha, \gamma, \theta)$ model) equals .0001, we have overwhelming evidence that at least two of the treatment means μ_{ij} differ.

FIGURE 15.4 SAS output of an (α, γ, θ) model analysis of the data in Table 15.2

SAS

GENERAL LINEAR MODELS PROCEDURE

DEPENDENT VARIABLE: DEMAND

SOURCE	DF	SUM OF SQUARES	MEAN SQUARE	F VALUE	PR > F	R-SQUARE	C.V.
MODEL	5	1762.87474359	352.57494872	46.91[a]	0.0001[b]	0.971022	4.3531
ERROR	7	52.60833333	7.51547619		ROOT MSE		DEMAND MEAN
CORRECTED TOTAL	12	1815.48307692			2.74143688		62.97692308

SOURCE	DF	TYPE I SS	F VALUE	PR > F	DF	TYPE III SS	F VALUE	PR > F
HEIGHT	2	1741.58441026	115.87	0.0001	2	1643.05762319	109.31[c]	0.0001[d]
WIDTH	1	15.28592754	2.03	0.1969	1	9.72710526	1.29[e]	0.2927[f]
HEIGHT*WIDTH	2	6.00440580	0.40	0.6850	2	6.00440580	0.40[g]	0.6850[h]

PARAMETER	ESTIMATE	T FOR HO: PARAMETER=0	PR > |T|	STD ERROR OF ESTIMATE
MUM-MUB	21.43333333	10.24	0.0001	2.09380701
MUT-MUB	-3.33333333	-1.59	0.1554	2.09380701
MUT-MUM	-24.76666667[i]	-14.00[j]	0.0001[k]	1.76958990[l]
MUW-MUR	1.85000000	1.14	0.2927	1.62614090
MUMW-MUMR	3.31666667[m]	1.33[n]	0.2267[o]	2.50257804[p]

OBSERVATION	OBSERVED VALUE	PREDICTED VALUE	RESIDUAL	LOWER 95% CL FOR MEAN	UPPER 95% CL FOR MEAN	LOWER 95% CL INDIVIDUAL	UPPER 95% CL INDIVIDUAL
1	58.20000000	55.95000000	2.25000000	51.36616602	60.53383398	48.01056665	63.88943335
2	53.70000000	55.95000000	-2.25000000	51.36616602	60.53383398	48.01056665	63.88943335
3	55.60000000	55.60000000	0.00000000	49.11747982	62.08252018	46.43233204	64.76766796
4	73.00000000	75.55000000	-2.55000000	70.96616602	80.13383398	67.61056665	83.48943335
5	78.10000000	75.55000000	2.55000000	70.96616602	80.13383398	67.61056665	83.48943335
6	76.20000000	78.86666667[q]	-2.66666667	75.12398189[r]	82.60935144[r]	71.38129712[s]	86.35203621[s]
7	78.40000000	78.86666667	-0.46666667	75.12398189	82.60935144	71.38129712	86.35203621
8	82.00000000	78.86666667	3.13333333	75.12398189	82.60935144	71.38129712	86.35203621
9	52.50000000	51.15000000	1.35000000	46.56616602	55.73383398	43.21056665	59.08943335
10	49.80000000	51.15000000	-1.35000000	46.56616602	55.73383398	43.21056665	59.08943335
11	56.00000000	53.73333333	2.26666667	49.99064856	57.47601811	46.24796379	61.21870288
12	51.90000000	53.73333333	-1.83333333	49.99064856	57.47601811	46.24796379	61.21870288
13	53.30000000	53.73333333	-0.43333333	49.99064856	57.47601811	46.24796379	61.21870288

[a] *$F((\alpha, \gamma, \theta)$ model)* [b] *Prob-value for $F((\alpha, \gamma, \theta)$ model)* [c] *$F(\alpha \mid \gamma, \theta)$* [d] *Prob-value for $F(\alpha \mid \gamma, \theta)$* [e] *$F(\gamma \mid \alpha, \theta)$* [f] *Prob-value for $F(\gamma \mid \alpha, \theta)$* [g] *$F(\theta \mid \alpha, \gamma)$*
[h] *Prob-value for $F(\theta \mid \alpha, \gamma)$* [i] *Point estimate of $\mu_{T\cdot} - \mu_{M\cdot} = \hat{\alpha}_T - \hat{\alpha}_M = \boldsymbol{\lambda}'\mathbf{b}$* [j] *t related to estimate of $\mu_{T\cdot} - \mu_{M\cdot}$* [k] *Prob-value for t* [l] $s_{(\alpha,\gamma,\theta)}\sqrt{\boldsymbol{\lambda}'(\mathbf{X}'\mathbf{X})^{-1}\boldsymbol{\lambda}}$
[m] *Point estimate of $\mu_{MW} - \mu_{MR} = 2\hat{\gamma}_W + 2\hat{\theta}_{MW} = \boldsymbol{\lambda}'\mathbf{b}$* [n] *t related to estimate of $\mu_{MW} - \mu_{MR}$* [o] *Prob-value for t* [p] $s_{(\alpha,\gamma,\theta)}\sqrt{\boldsymbol{\lambda}'(\mathbf{X}'\mathbf{X})^{-1}\boldsymbol{\lambda}}$
[q] $\hat{y}_{MW} = \hat{\mu} + \hat{\alpha}_M + \hat{\gamma}_W + \hat{\theta}_{MW}$ = *point estimate of μ_{MW}*
= point prediction of $y_{MW,0}$
[r] *95% confidence interval for μ_{MW}* [s] *95% prediction interval for $y_{MW,0}$*

2. Since the prob-value related to $F(\theta \mid \alpha, \gamma)$ equals .6850, we have little or no evidence to suggest that there is interaction between display height and display width.
3. Since the prob-value related to $F(\alpha \mid \gamma, \theta)$ equals .0001, we have overwhelming evidence that at least two of the display heights have different effects on mean demand.
4. Since the prob-value related to $F(\gamma \mid \alpha, \theta)$ equals .2927, there is not much evidence to suggest that a regular display width and a wide display have different effects on mean demand.

Since 3 above indicates that we can reject

$$H_0: \mu_{B\cdot} = \mu_{M\cdot} = \mu_{T\cdot}$$

we conclude that statistically significant differences exist between at least two of the means $\mu_{B\cdot}$, $\mu_{M\cdot}$, and $\mu_{T\cdot}$. To judge the practical importance of these differences, we can study pairwise differences between means. For instance, the equation

$$\mu_{ij} = \mu + \alpha_M D_{i,M} + \alpha_T D_{i,T} + \gamma_W D_{j,W} + \theta_{MW} D_{i,M} D_{j,W} + \theta_{TW} D_{i,T} D_{j,W}$$

implies that

$$\begin{aligned} \mu_{T\cdot} &= \frac{\mu_{TR} + \mu_{TW}}{2} \\ &= \frac{(\mu + \alpha_T - \gamma_W - \theta_{TW}) + (\mu + \alpha_T + \gamma_W + \theta_{TW})}{2} \\ &= \mu + \alpha_T \end{aligned}$$

and

$$\begin{aligned} \mu_{M\cdot} &= \frac{\mu_{MR} + \mu_{MW}}{2} \\ &= \frac{(\mu + \alpha_M - \gamma_W - \theta_{MW}) + (\mu + \alpha_M + \gamma_W + \theta_{MW})}{2} \\ &= \mu + \alpha_M \end{aligned}$$

It follows that

$$\begin{aligned} \mu_{T\cdot} - \mu_{M\cdot} &= (\mu + \alpha_T) - (\mu + \alpha_M) \\ &= \alpha_T - \alpha_M \\ &= 0\mu - 1\alpha_M + 1\alpha_T + 0\gamma_W + 0\theta_{MW} + 0\theta_{TW} \\ &= [0 \quad -1 \quad 1 \quad 0 \quad 0 \quad 0] \begin{bmatrix} \mu \\ \alpha_M \\ \alpha_T \\ \gamma_W \\ \theta_{MW} \\ \theta_{TW} \end{bmatrix} \\ &= \boldsymbol{\lambda}'\boldsymbol{\beta} \end{aligned}$$

where $\boldsymbol{\lambda}' = [0 \quad -1 \quad 1 \quad 0 \quad 0 \quad 0]$. Therefore noting that $\mu_{T\cdot} - \mu_{M\cdot}$ is denoted "MUT − MUM" on the SAS output in Figure 15.4, it follows that a point estimate of $\mu_{T\cdot} - \mu_{M\cdot}$ is

$$\boldsymbol{\lambda}'\mathbf{b} = \hat{\alpha}_T - \hat{\alpha}_M = -24.7667$$

Here, $\hat{\alpha}_T$ and $\hat{\alpha}_M$ are the least squares point estimates of α_T and α_M. Furthermore, note that Table 15.2 contains $n = 13$ observations and that the (α, γ, θ) model utilizes $k = 6$ parameters. It follows that a 95% confidence interval for $\mu_{T\cdot} - \mu_{M\cdot}$ is

$$\begin{aligned}[\boldsymbol{\lambda}'\mathbf{b} \pm t_{[.025]}^{(13-6)} s_{(\alpha,\gamma,\theta)} \sqrt{\boldsymbol{\lambda}'(\mathbf{X}'\mathbf{X})^{-1}\boldsymbol{\lambda}}] &= [-24.7667 \pm 2.365(1.7696)] \\ &= [-28.9518, -20.5816]\end{aligned}$$

Here, the standard error of the estimate $s_{(\alpha,\gamma,\theta)} \sqrt{\boldsymbol{\lambda}'(\mathbf{X}'\mathbf{X})^{-1}\boldsymbol{\lambda}}$ is obtained from the SAS output (see Figure 15.4). This interval says that we can be 95 percent confident that changing from using a middle display height to using a top display height will decrease mean demand by between 20.5816 and 28.9518 cases.

As another example, the output labeled "MUMW − MUMR" in Figure 15.4 tells us that a point estimate of

$$\begin{aligned}\mu_{MW} - \mu_{MR} &= (\mu + \alpha_M + \gamma_W + \theta_{MW}) - (\mu + \alpha_M - \gamma_W - \theta_{MW}) \\ &= 2\gamma_W + 2\theta_{MW}\end{aligned}$$

is

$$2\hat{\gamma}_W + 2\hat{\theta}_{MW} = 3.3167$$

Other estimation and prediction results are labeled in Figure 15.4.

Not all of the F statistics given in this section can be (easily) expressed by using algebraic (instead of regression) formulas. However, estimation and prediction results for an unbalanced complete factorial experiment can be expressed algebraically.

Estimation and Prediction in a Balanced or Unbalanced Complete Factorial Experiment

Let

$$\bar{y}_{ij} = \frac{\sum_{k=1}^{n_{ij}} y_{ij,k}}{n_{ij}}, \qquad \bar{y}_{i\cdot} = \frac{\sum_{j=1}^{b} \bar{y}_{ij}}{b}, \qquad \bar{y}_{\cdot j} = \frac{\sum_{i=1}^{a} \bar{y}_{ij}}{a}$$

$$MSE = \frac{\sum_{i=1}^{a} \sum_{j=1}^{b} \sum_{k=1}^{n_{ij}} (y_{ij,k} - \bar{y}_{ij})^2}{n - ab}$$

1. A *point estimate of the difference* $\mu_{i\cdot} - \mu_{i'\cdot}$ is $\bar{y}_{i\cdot} - \bar{y}_{i'\cdot}$, and an *individual 100(1 − α)% confidence interval for* $\mu_{i\cdot} - \mu_{i'\cdot}$ is

$$\left[(\bar{y}_{i\cdot} - \bar{y}_{i'\cdot}) \pm t_{[\alpha/2]}^{(n-ab)} \sqrt{\frac{MSE}{b^2} \sum_{j=1}^{b} \left(\frac{1}{n_{ij}} + \frac{1}{n_{i'j}}\right)}\right]$$

2. A *point estimate of the linear combination* $\Sigma_{i=1}^{a} a_i \mu_{i\cdot}$ is $\Sigma_{i=1}^{a} a_i \bar{y}_{i\cdot}$, and an *individual 100(1 − α)% confidence interval for this linear combination* is

$$\left[\sum_{i=1}^{a} a_i \bar{y}_{i\cdot} \pm t_{[\alpha/2]}^{(n-ab)} \sqrt{\frac{MSE}{b^2} \sum_{i=1}^{a} a_i^2 \sum_{j=1}^{b} \frac{1}{n_{ij}}}\right]$$

3. For *Scheffé simultaneous intervals* in 1 and 2 above, replace $t_{[\alpha/2]}^{(n-ab)}$ by $\sqrt{(a-1)F_{[\alpha]}^{(a-1,n-ab)}}$ (which applies to all possible *contrasts* in the $\mu_{i\cdot}$ values) or by $\sqrt{aF_{[\alpha]}^{(a,n-ab)}}$ (which applies to all possible *linear combinations* in the $\mu_{i\cdot}$ values). For *Bonferroni simultaneous intervals* in 1 and 2 above, replace $t_{[\alpha/2]}^{(n-ab)}$ by $t_{[\alpha/2g]}^{(n-ab)}$ (which applies to *g prespecified linear combinations*).
4. A *point estimate of the difference* $\mu_{\cdot j} - \mu_{\cdot j'}$ is $\bar{y}_{\cdot j} - \bar{y}_{\cdot j'}$, and an *individual 100(1 − α)% confidence interval for* $\mu_{\cdot j} - \mu_{\cdot j'}$ is

$$\left[(\bar{y}_{\cdot j} - \bar{y}_{\cdot j'}) \pm t_{[\alpha/2]}^{(n-ab)} \sqrt{\frac{MSE}{a^2} \sum_{i=1}^{a} \left(\frac{1}{n_{ij}} + \frac{1}{n_{ij'}}\right)}\right]$$

5. A *point estimate of the linear combination* $\Sigma_{j=1}^{b} a_j \mu_{\cdot j}$ is $\Sigma_{j=1}^{b} a_j \bar{y}_{\cdot j}$, and an *individual 100(1 − α)% confidence interval for this linear combination* is

$$\left[\sum_{j=1}^{b} a_j \bar{y}_{\cdot j} \pm t_{[\alpha/2]}^{(n-ab)} \sqrt{\frac{MSE}{a^2} \sum_{j=1}^{b} a_j^2 \sum_{i=1}^{a} \frac{1}{n_{ij}}}\right]$$

6. For *Scheffé simultaneous intervals* in 4 and 5 above, replace $t_{[\alpha/2]}^{(n-ab)}$ by $\sqrt{(b-1)F_{[\alpha]}^{(b-1,n-ab)}}$ (which applies to all possible *contrasts* in the $\mu_{\cdot j}$ values) or by $\sqrt{bF_{[\alpha]}^{(b,n-ab)}}$ (which applies to all possible *linear combinations* in the $\mu_{\cdot j}$ values). For *Bonferroni simultaneous intervals* in 4 and 5 above, replace $t_{[\alpha/2]}^{(n-ab)}$ by $t_{[\alpha/2g]}^{(n-ab)}$ (which applies to *g prespecified linear combinations*).
7. A *point estimate of the difference* $\mu_{ij} - \mu_{i'j'}$ is $\bar{y}_{ij} - \bar{y}_{i'j'}$, and an *individual 100(1 − α)% confidence interval for* $\mu_{ij} - \mu_{i'j'}$ is

$$\left[(\bar{y}_{ij} - \bar{y}_{i'j'}) \pm t_{[\alpha/2]}^{(n-ab)} \sqrt{MSE\left(\frac{1}{n_{ij}} + \frac{1}{n_{i'j'}}\right)}\right]$$

8. A *point estimate of the linear combination of treatment means* $\Sigma_{i=1}^{a} \Sigma_{j=1}^{b} a_{ij}\mu_{ij}$ is $\Sigma_{i=1}^{a} \Sigma_{j=1}^{b} a_{ij}\bar{y}_{ij}$, and an *individual 100(1 − α)% confidence interval for this linear combination* is

$$\left[\sum_{i=1}^{a} \sum_{j=1}^{b} a_{ij}\bar{y}_{ij} \pm t_{[\alpha/2]}^{(n-ab)} \sqrt{MSE \sum_{i=1}^{a} \sum_{j=1}^{b} a_{ij}^2/n_{ij}}\right]$$

9. For *Scheffé simultaneous intervals* in 7 and 8 above, replace $t_{[\alpha/2]}^{(n-ab)}$ by $\sqrt{(ab-1)F_{[\alpha]}^{(ab-1,n-ab)}}$ (which applies to all possible *contrasts* of treatment means) or $\sqrt{abF_{[\alpha]}^{(ab,n-ab)}}$ (which applies to all possible *linear combinations* of treatment means). For *Bonferroni simultaneous intervals* in 7 and 8 above, replace $t_{[\alpha/2]}^{(n-ab)}$ by $t_{[\alpha/2g]}^{(n-ab)}$ (which applies to *g prespecified linear combinations*).
10. A *point estimate of the treatment mean* μ_{ij} is $\bar{y}_{ij}$, and an *individual 100(1 − α)% confidence interval for* μ_{ij} is

$$\left[\bar{y}_{ij} \pm t_{[\alpha/2]}^{(n-ab)}\sqrt{\frac{MSE}{n_{ij}}}\right]$$

11. A *point prediction* of a randomly selected *individual value*

$$y_{ij,0} = \mu_{ij} + \varepsilon_{ij,0}$$

is $\bar{y}_{ij}$, and a *100(1 − α)% prediction interval for* $y_{ij,0}$ is

$$\left[\bar{y}_{ij} \pm t_{[\alpha/2]}^{(n-ab)}\sqrt{MSE}\sqrt{1+\frac{1}{n_{ij}}}\right]$$

Example 15.11 For the data in Table 15.2, $n = 13$ and $ab = 3 \cdot 2 = 6$. It follows that a 95% confidence interval for $\mu_{T\cdot} - \mu_{M\cdot}$ is

$$\left[(\bar{y}_{T\cdot} - \bar{y}_{M\cdot}) \pm t_{[.025]}^{(13-6)}\sqrt{\frac{MSE}{b^2}\sum_{j=1}^{b}\left(\frac{1}{n_{Tj}}+\frac{1}{n_{Mj}}\right)}\right]$$
$$= [-24.7667 \pm 2.365(1.7696)]$$
$$= [-28.9518, -20.5816]$$

This is the same interval that we calculated using regression in Example 15.10.

*15.4 USING SAS

In Figure 15.5 we present the SAS program that yields the Tastee Bakery demand data analysis in Figure 15.3. Note that the coefficients in the ESTIMATE statements are obtained by considering the appropriate *factor levels* in *alphabetical order*.

For example, the alphabetically ordered levels of display height are

B M T

Thus the coefficients in the ESTIMATE statement for MUM − MUB (that is, $\mu_{M\cdot} - \mu_{B\cdot}$) are

−1 1 0

*This section is optional.

FIGURE 15.5 **SAS program to perform the two-way ANOVA of the bakery demand data given in Figure 15.3**

```
DATA BAKERY;
INPUT HEIGHT $ WIDTH $ DEMAND @@ ; } ——→ Defines factors HEIGHT and WIDTH and
                                          the response variable DEMAND
CARDS;
B R 58.2 B R 53.7 B R 55.8 ⎫
B W 55.7 B W 52.5 B W 58.9 ⎪
M R 73.0 M R 78.1 M R 75.4 ⎬ ——→ Data—See Table 15.1
M W 76.2 M W 78.4 M W 82.1 ⎪
T R 52.4 T R 49.7 T R 50.9 ⎪
T W 54.0 T W 52.1 T W 49.9 ⎭
;
PROC GLM; } ——————→ Specifies General Linear Models Procedure
CLASS HEIGHT WIDTH; } ——→ Defines class variables HEIGHT and WIDTH
MODEL DEMAND = HEIGHT WIDTH HEIGHT*WIDTH/P CLM; }—┐
                                    └——→ Specifies model y_{ij,k} = μ_{ij} + ε_{ij,k}
                                         where μ_{ij} = μ + α_i + γ_j + θ_{ij}
                                         HEIGHT*WIDTH defines interaction term θ_{ij}

ESTIMATE 'MUM-MUB' HEIGHT -1 1 0; } ——→ Estimates μ_{M.} − μ_{B.}
ESTIMATE 'MUT-MUB' HEIGHT -1 0 1; } ——→ Estimates μ_{T.} − μ_{B.}
ESTIMATE 'MUT-MUM' HEIGHT 0 -1 1; } ——→ Estimates μ_{T.} − μ_{M.}
ESTIMATE 'MUW-MUR' WIDTH -1 1; } ————→ Estimates μ_{.W} − μ_{.R}
ESTIMATE 'MUMW-MUMR' HEIGHT 0 0 0 WIDTH -1 1 HEIGHT*WIDTH 0 0 -1 1 0 0; }┐
                                                  └——→ Estimates μ_{MW} − μ_{MR}
PROC GLM;
CLASS HEIGHT WIDTH;
MODEL DEMAND = HEIGHT WIDTH HEIGHT*WIDTH/ P CLI; } ——→ CLI requests prediction
                                                       intervals
```

As another example, the alphabetically ordered levels of display width are

R W

Thus the coefficients in the ESTIMATE statement for MUW − MUR (that is, $\mu_{.W} - \mu_{.R}$) are

−1 1

To define the coefficients in the ESTIMATE statement for MUMW − MUMR (that is, $\mu_{MW} - \mu_{MR}$), note that

$$\begin{aligned}
\mu_{MW} - \mu_{MR} &= (\mu + \alpha_M + \gamma_W + \theta_{MW}) - (\mu + \alpha_M + \gamma_R + \theta_{MR}) \\
&= (\gamma_W - \gamma_R) + (\theta_{MW} - \theta_{MR}) \\
&= 0(\alpha_B) + 0(\alpha_M) + 0(\alpha_T) + (-1)(\gamma_R) + (1)(\gamma_W) \\
&\quad + 0(\theta_{BR}) + 0(\theta_{BW}) + (-1)(\theta_{MR}) + (1)(\theta_{MW}) + 0(\theta_{TR}) \\
&\quad + 0(\theta_{TW})
\end{aligned}$$

Therefore the coefficients corresponding to the display heights

B M T

are

0 0 0

The coefficients corresponding to the display widths

R W

are

-1 1

The coefficients corresponding to the alphabetically ordered factor level combinations (which represent interaction)

BR BW MR MW TR TW

are

0 0 -1 1 0 0

Note that "trailing zero coefficients" could be omitted if so desired.

Also, note that this type of SAS program would be appropriate if one were analyzing an *unbalanced complete* factorial experiment. For example, if we replace the data in the above program by the unbalanced data in Table 15.2, then we obtain the SAS output in Figure 15.4.

EXERCISES

15.1 Consider Table 15.3. Here, A and B are two fertilizer types; M, N, O, and P are four wheat types; and the $y_{ij,k}$ values are wheat yields in bushels per plot (one third of an

TABLE 15.3 **Results of a two-factor wheat yield experiment**

Fertilizer type, i	*Wheat type, j*			
	M	*N*	*O*	*P*
A	$y_{AM,1} = 19.4$ $y_{AM,2} = 20.6$ $y_{AM,3} = 20.0$	$y_{AN,1} = 25.0$ $y_{AN,2} = 24.0$ $y_{AN,3} = 24.5$	$y_{AO,1} = 24.8$ $y_{AO,2} = 26.0$ $y_{AO,3} = 25.4$	$y_{AP,1} = 23.1$ $y_{AP,2} = 24.3$ $y_{AP,3} = 23.7$
B	$y_{BM,1} = 22.6$ $y_{BM,2} = 21.6$ $y_{BM,3} = 22.1$	$y_{BN,1} = 25.6$ $y_{BN,2} = 26.8$ $y_{BN,3} = 26.2$	$y_{BO,1} = 27.6$ $y_{BO,2} = 26.4$ $y_{BO,3} = 27.0$	$y_{BP,1} = 25.4$ $y_{BP,2} = 24.5$ $y_{BP,3} = 26.3$

acre) corresponding to the different combinations of fertilizer type and wheat type. Also, assume that this data was obtained by using a completely randomized experimental design. Use the ANOVA approach of Section 15.2 to test each of the following null hypotheses by setting α equal to .05.

H_0: All treatment means μ_{ij} are equal

H_0: There is no interaction between fertilizer type and wheat type

H_0: $\mu_{A\cdot} = \mu_{B\cdot}$

H_0: $\mu_{\cdot M} = \mu_{\cdot N} = \mu_{\cdot O} = \mu_{\cdot P}$

Also, perform a graphical analysis to check for interaction.

15.2 Consider Exercise 15.1.

a. Calculate (individual) 95% confidence intervals for the following differences:

$\mu_{\cdot N} - \mu_{\cdot M}$ $\quad$ $\mu_{\cdot O} - \mu_{\cdot N}$

$\mu_{\cdot O} - \mu_{\cdot M}$ $\quad$ $\mu_{\cdot P} - \mu_{\cdot N}$

$\mu_{\cdot P} - \mu_{\cdot M}$ $\quad$ $\mu_{\cdot P} - \mu_{\cdot O}$

b. Calculate a 95% confidence interval for $\mu_{B\cdot} - \mu_{A\cdot}$.
c. Calculate a 95% confidence interval for $\mu_{BO} - \mu_{AN}$.
d. Calculate a 95% confidence interval for μ_{BO}.
e. Calculate a 95% prediction interval for $y_{BO,0} = \mu_{BO} + \varepsilon_{BO,0}$.
f. Calculate Tukey and Bonferroni 95% simultaneous confidence intervals for the differences in part (a).
g. Calculate a 95% confidence interval for

$$\frac{\mu_{\cdot N} + \mu_{\cdot O}}{2} - \left[\frac{\mu_{\cdot M} + \mu_{\cdot P}}{2}\right]$$

Also, calculate a Scheffé 95% simultaneous confidence interval for this contrast in the set of all possible contrasts in the $\mu_{\cdot j}$ values.

h. Calculate a Scheffé 95% simultaneous confidence interval for μ_{BO} in the set of all possible linear combinations of treatment means.

15.3 A study was made to compare three display panels for use by air traffic controllers. Each display panel was tested for four different simulated emergency conditions. Twenty-four highly trained air traffic controllers were used in the study. Two controllers were randomly assigned to each display panel–emergency condititition combination. The time (in seconds) required to stabilize the emergency condition was recorded. The data in Table 15.4 was observed. Figure 15.6 presents the SAS output obtained by using PROC GLM to perform a two-way ANOVA of this data. Use the output to test each of the following null hypotheses by setting α equal to .05:

H_0: All treatment means μ_{ij} are equal

H_0: There is no interaction between display panel and emergency condition

H_0: $\mu_{A\cdot} = \mu_{B\cdot} = \mu_{C\cdot}$

H_0: $\mu_{\cdot 1} = \mu_{\cdot 2} = \mu_{\cdot 3} = \mu_{\cdot 4}$

Also, perform a graphical analysis to check for interaction.

TABLE 15.4 Results of a two-factor display panel experiment

Display panel	Emergency condition 1	2	3	4
A	17	25	31	14
	14	24	34	13
B	15	22	28	9
	12	19	31	10
C	21	29	32	15
	24	28	37	19

FIGURE 15.6 SAS output of a two-way ANOVA of the display panel data

```
                                         SAS

                          GENERAL LINEAR MODELS PROCEDURE
DEPENDENT VARIABLE: TIME

SOURCE            DF  SUM OF SQUARES    MEAN SQUARE   F VALUE       PR > F  R-SQUARE            C.V.
MODEL             11   1482.45833333   134.76893939     32.67       0.0001  0.967688          9.3201
ERROR             12     49.50000000     4.12500000               ROOT MSE                 TIME MEAN
CORRECTED TOTAL   23   1531.95833333                            2.03100960               21.79166667

SOURCE             DF        TYPE I SS   F VALUE        PR > F   DF      TYPE III SS   F VALUE    PR > F
PANEL               2    218.58333333      26.49        0.0001    2     218.58333333     26.49    0.0001
CONDI               3   1247.45833333     100.80        0.0001    3    1247.45833333    100.80    0.0001
PANEL*CONDI         6     16.41666667       0.66        0.6809    6      16.41666667      0.66    0.6809

                                       T FOR HO:   PR > |T|  STD ERROR OF
PARAMETER               ESTIMATE    PARAMETER=0               ESTIMATE

MUB-MUA              -3.25000000          -3.20    0.0076    1.01550480
MUC-MUA               4.12500000           4.06    0.0016    1.01550480
MUC-MUB               7.37500000           7.26    0.0001    1.01550480
MU3-MU1              15.00000000          12.79    0.0001    1.17260394

OBSERVATION          OBSERVED        PREDICTED          RESIDUAL       LOWER 95% CL       UPPER 95% CL
                        VALUE            VALUE                             FOR MEAN           FOR MEAN

          1       17.00000000      15.50000000        1.50000000        12.37091369        18.62908631
          2       14.00000000      15.50000000       -1.50000000        12.37091369        18.62908631
          3       25.00000000      24.50000000        0.50000000        21.37091369        27.62908631
          4       24.00000000      24.50000000       -0.50000000        21.37091369        27.62908631
          5       31.00000000      32.50000000       -1.50000000        29.37091369        35.62908631
          6       34.00000000      32.50000000        1.50000000        29.37091369        35.62908631
          7       14.00000000      13.50000000        0.50000000        10.37091369        16.62908631
          8       13.00000000      13.50000000       -0.50000000        10.37091369        16.62908631
          9       15.00000000      13.50000000        1.50000000        10.37091369        16.62908631
         10       12.00000000      13.50000000       -1.50000000        10.37091369        16.62908631
         11       22.00000000      20.50000000        1.50000000        17.37091369        23.62908631
         12       19.00000000      20.50000000       -1.50000000        17.37091369        23.62908631
         13       28.00000000      29.50000000       -1.50000000        26.37091369        32.62908631
```

(continues)

FIGURE 15.6 Continued

OBSERVATION	OBSERVED VALUE	PREDICTED VALUE	RESIDUAL	LOWER 95% CL FOR MEAN	UPPER 95% CL FOR MEAN
14	31.00000000	29.50000000	1.50000000	26.37091369	32.62908631
15	9.00000000	9.50000000	-0.50000000	6.37091369	12.62908631
16	10.00000000	9.50000000	0.50000000	6.37091369	12.62908631
17	21.00000000	22.50000000	-1.50000000	19.37091369	25.62908631
18	24.00000000	22.50000000	1.50000000	19.37091369	25.62908631
19	29.00000000	28.50000000	0.50000000	25.37091369	31.62908631
20	28.00000000	28.50000000	-0.50000000	25.37091369	31.62908631
21	32.00000000	34.50000000	-2.50000000	31.37091369	37.62908631
22	37.00000000	34.50000000	2.50000000	31.37091369	37.62908631
23	15.00000000	17.00000000	-2.00000000	13.87091369	20.12908631
24	19.00000000	17.00000000	2.00000000	13.87091369	20.12908631

OBSERVATION	OBSERVED VALUE	PREDICTED VALUE	RESIDUAL	LOWER 95% CL INDIVIDUAL	UPPER 95% CL INDIVIDUAL
1	17.00000000	15.50000000	1.50000000	10.08026353	20.91973647
2	14.00000000	15.50000000	-1.50000000	10.08026353	20.91973647
3	25.00000000	24.50000000	0.50000000	19.08026353	29.91973647
4	24.00000000	24.50000000	-0.50000000	19.08026353	29.91973647
5	31.00000000	32.50000000	-1.50000000	27.08026353	37.91973647
6	34.00000000	32.50000000	1.50000000	27.08026353	37.91973647
7	14.00000000	13.50000000	0.50000000	8.08026353	18.91973647
8	13.00000000	13.50000000	-0.50000000	8.08026353	18.91973647
9	15.00000000	13.50000000	1.50000000	8.08026353	18.91973647
10	12.00000000	13.50000000	-1.50000000	8.08026353	18.91973647
11	22.00000000	20.50000000	1.50000000	15.08026353	25.91973647
12	19.00000000	20.50000000	-1.50000000	15.08026353	25.91973647
13	28.00000000	29.50000000	-1.50000000	24.08026353	34.91973647
14	31.00000000	29.50000000	1.50000000	24.08026353	34.91973647
15	9.00000000	9.50000000	-0.50000000	4.08026353	14.91973647
16	10.00000000	9.50000000	0.50000000	4.08026353	14.91973647
17	21.00000000	22.50000000	-1.50000000	17.08026353	27.91973647
18	24.00000000	22.50000000	1.50000000	17.08026353	27.91973647
19	29.00000000	28.50000000	0.50000000	23.08026353	33.91973647
20	28.00000000	28.50000000	-0.50000000	23.08026353	33.91973647
21	32.00000000	34.50000000	-2.50000000	29.08026353	39.91973647
22	37.00000000	34.50000000	2.50000000	29.08026353	39.91973647
23	15.00000000	17.00000000	-2.00000000	11.58026353	22.41973647
24	19.00000000	17.00000000	2.00000000	11.58026353	22.41973647

15.4 Consider Exercise 15.3. Use the SAS output in Figure 15.6 to do the following.

a. Calculate (individual) 95% confidence intervals for the following differences:

$\mu_{B\cdot} - \mu_{A\cdot}$

$\mu_{C\cdot} - \mu_{A\cdot}$

$\mu_{C\cdot} - \mu_{B\cdot}$

b. Calculate a 95% confidence interval for $\mu_{\cdot 3} - \mu_{\cdot 1}$.

c. Find 95% confidence intervals for μ_{B1}, μ_{B2}, μ_{B3}, and μ_{B4}.

d. Find 95% prediction intervals for $y_{B1,0} = \mu_{B1} + \varepsilon_{B1,0}$, $y_{B2,0} = \mu_{B2} + \varepsilon_{B2,0}$, $y_{B3,0} = \mu_{B3} + \varepsilon_{B3,0}$, and $y_{B4,0} = \mu_{B4} + \varepsilon_{B4,0}$.

e. Calculate Tukey and Bonferroni 95% simultaneous confidence intervals for the differences in part (a).

f. Calculate Scheffé 95% simultaneous confidence intervals for $\mu_{B\cdot} - \mu_{A\cdot}$, $\mu_{C\cdot} - \mu_{A\cdot}$, $\mu_{C\cdot} - \mu_{B\cdot}$, μ_{B1}, μ_{B2}, μ_{B3}, and μ_{B4}.

TABLE 15.5 Results of a two-factor gasoline mileage experiment

Premium gasoline type, *i*	Gasoline additive type, *j*			
	Q	R	S	T
D	$y_{DQ,1} = 17.4$ $y_{DQ,2} = 18.6$	$y_{DR,1} = 23.0$ $y_{DR,2} = 22.0$	$y_{DS,1} = 23.5$ $y_{DS,2} = 22.3$	$y_{DT,1} = 20.8$ $y_{DT,2} = 19.7$
E	$y_{EQ,1} = 23.3$ $y_{EQ,2} = 24.5$	$y_{ER,1} = 25.6$ $y_{ER,2} = 24.4$	$y_{ES,1} = 23.4$ $y_{ES,2} = 23.1$	$y_{ET,1} = 19.6$ $y_{ET,2} = 20.6$
F	$y_{FQ,1} = 23.0$ $y_{FQ,2} = 23.5$	$y_{FR,1} = 24.7$ $y_{FR,2} = 23.3$	$y_{FS,1} = 23.0$ $y_{FS,2} = 22.0$	$y_{FT,1} = 18.6$ $y_{FT,2} = 19.8$

15.5 Consider Table 15.5. Here, D, E, and F are three premium gasoline types; Q, R, S, and T are four gasoline additive types; and the $y_{ij,k}$ values are gasoline mileages in miles per gallon corresponding to the different combinations of premium gasoline type and gasoline additive type. Also, this data was obtained by using a completely randomized design. Use two-way ANOVA to analyze the data. Show by hypothesis testing and graphical analysis that interaction exists between premium gasoline type and gasoline additive type. Find (individual) 95% confidence intervals for μ_{ER}, $\mu_{ER} - \mu_{DQ}$, and $\mu_{ER} - \mu_{FS}$.

15.6 A record company has performed an experiment to assess the effects of time of day and positioning of advertisements on telemarketing response. The data in Table 15.6 represents the number of calls placed to the 800-number following a sample broadcast of the advertisement. Fully analyze this data using two-way ANOVA.

TABLE 15.6 Results of a two-factor telemarketing response experiment

	Position of advertisement			
Time of day	*On the hour*	*On the half-hour*	*Early in program*	*Late in program*
10:00 morning: Rerun of The Honeymooners	42 37 41	36 41 38	62 68 64	51 47 48
*4:00 afternoon: Rerun of M*A*S*H*	62 60 58	57 60 55	88 85 81	67 60 66
9:00 evening: First run of Cheers	100 96 103	97 96 101	127 120 126	105 101 107

TABLE 15.7 Results of an experiment to study the effects of two factors on hourly defective box rates

Machine type, *i*	Cardboard grade, *j*: *Y*	*Z*
A	9	8
	10	11
	11	12
B	3	4
	5	5
	7	5

FIGURE 15.7 SAS output of a two-way ANOVA of the defective box data

```
                                        SAS

                          GENERAL LINEAR MODELS PROCEDURE
DEPENDENT VARIABLE: DEFECTS

SOURCE             DF   SUM OF SQUARES    MEAN SQUARE   F VALUE        PR > F   R-SQUARE           C.V.
MODEL               3      90.91666667    30.30555556     11.02        0.0033   0.805166        21.8679
ERROR               8      22.00000000     2.75000000                 ROOT MSE             DEFECTS MEAN
CORRECTED TOTAL    11     112.91666667                              1.65831240               7.58333333

SOURCE             DF        TYPE I SS    F VALUE         PR > F   DF     TYPE III SS   F VALUE   PR > F
MACHTYPE            1      90.75000000      33.00         0.0004    1     90.75000000     33.00   0.0004
CBGRADE             1       0.08333333       0.03         0.8661    1      0.08333333      0.03   0.8661
MACHTYPE*GRADE      1       0.08333333       0.03         0.8661    1      0.08333333      0.03   0.8661

                                      T FOR HO:   PR > |T|  STD ERROR OF
PARAMETER              ESTIMATE    PARAMETER=0                ESTIMATE

MUB-MUA             -5.50000000          -5.74    0.0004     0.95742711
MUZ-MUY             -0.16666667          -0.17    0.8661     0.95742711
MUBZ-MUAY           -5.66666667          -4.19    0.0031     1.35400640

OBSERVATION         OBSERVED        PREDICTED              RESIDUAL      LOWER 95% CL       UPPER 95% CL
                       VALUE            VALUE                              FOR MEAN           FOR MEAN

          1       9.00000000      10.33333333           -1.33333333        8.12548116        12.54118550
          2      10.00000000      10.33333333           -0.33333333        8.12548116        12.54118550
          3      12.00000000      10.33333333            1.66666667        8.12548116        12.54118550
          4       8.00000000      10.33333333           -2.33333333        8.12548116        12.54118550
          5      11.00000000      10.33333333            0.66666667        8.12548116        12.54118550
          6      12.00000000      10.33333333            1.66666667        8.12548116        12.54118550
          7       3.00000000       5.00000000           -2.00000000        2.79214783         7.20785217
          8       5.00000000       5.00000000            0.00000000        2.79214783         7.20785217
          9       7.00000000       5.00000000            2.00000000        2.79214783         7.20785217
         10       4.00000000       4.66666667           -0.66666667        2.45881450         6.87451884
         11       5.00000000       4.66666667            0.33333333        2.45881450         6.87451884
         12       5.00000000       4.66666667            0.33333333        2.45881450         6.87451884
```

15.7 An experiment is conducted to study the effect of machine type (A or B) and cardboard grade (Y or Z) on the number of defective cardboard boxes produced in an hour of production. The data in Table 15.7 is obtained by using a completely randomized design. Figure 15.7 presents the SAS output obtained by using PROC GLM to perform a two-way ANOVA of this data. Use the output to test each of the following null hypotheses by setting α equal to .05.

H_0: All treatment means μ_{ij} are equal

H_0: There is no interaction between machine type and cardboard grade

H_0: $\mu_{A\cdot} = \mu_{B\cdot}$

H_0: $\mu_{\cdot Y} = \mu_{\cdot Z}$

Also, find (individual) 95% confidence intervals for $\mu_{B\cdot} - \mu_{A\cdot}$, $\mu_{\cdot Z} - \mu_{\cdot Y}$, $\mu_{BZ} - \mu_{AY}$, and μ_{BZ}.

15.8 An experiment is conducted to study the effect of a high- or low-pressure sales approach (H or L) and the effect of sales pitch (1 or 2) on the weekly sales of a product. The data in Table 15.8 is obtained by using a completely randomized design. Figure 15.8 presents the SAS output obtained by using PROC GLM to perform a two-way ANOVA of this data. Use the output to test each of the following null hypotheses by setting α equal to .05.

H_0: All treatment means μ_{ij} are equal

H_0: There is no interaction between sales pressure and sales pitch

H_0: $\mu_{H\cdot} = \mu_{L\cdot}$

H_0: $\mu_{\cdot 1} = \mu_{\cdot 2}$

Also, find (individual) 95% confidence intervals for $\mu_{H\cdot} - \mu_{L\cdot}$, $\mu_{\cdot 1} - \mu_{\cdot 2}$, $\mu_{H1} - \mu_{L2}$, and μ_{H1}.

TABLE 15.8 Results of an experiment to study the effects of sales pressure and sales pitch on weekly sales of a product

Sales pressure, *i*	Sales pitch, *j*: 1	2
H	32	32
	29	30
	30	28
L	28	25
	25	24
	23	23

FIGURE 15.8 **SAS output of a two-way ANOVA of the sales approach data**

SAS

GENERAL LINEAR MODELS PROCEDURE

DEPENDENT VARIABLE: SALES

SOURCE	DF	SUM OF SQUARES	MEAN SQUARE	F VALUE	PR > F	R-SQUARE	C.V.
MODEL	3	93.58333333	31.19444444	9.13	0.0058	0.773949	6.7420
ERROR	8	27.33333333	3.41666667		ROOT MSE		SALES MEAN
CORRECTED TOTAL	11	120.91666667			1.84842275		27.41666667

SOURCE	DF	TYPE I SS	F VALUE	PR > F	DF	TYPE III SS	F VALUE	PR > F
PRESSURE	1	90.75000000	26.56	0.0009	1	90.75000000	26.56	0.0009
PITCH	1	2.08333333	0.61	0.4574	1	2.08333333	0.61	0.4574
PRESSURE*PITCH	1	0.75000000	0.22	0.6519	1	0.75000000	0.22	0.6519

PARAMETER	ESTIMATE	T FOR H0: PARAMETER=0	PR > |T|	STD ERROR OF ESTIMATE
MUH-MUL	5.50000000	5.15	0.0009	1.06718737
MU1-MU2	0.83333333	0.78	0.4574	1.06718737
MUH1-MUL2	6.33333333	4.20	0.0030	1.50923086

OBSERVATION	OBSERVED VALUE	PREDICTED VALUE	RESIDUAL	LOWER 95% CL FOR MEAN	UPPER 95% CL FOR MEAN
1	32.00000000	30.33333333	1.66666667	27.87237110	32.79429557
2	32.00000000	30.00000000	2.00000000	27.53903776	32.46096224
3	28.00000000	25.33333333	2.66666667	22.87237110	27.79429557
4	25.00000000	24.00000000	1.00000000	21.53903776	26.46096224
5	29.00000000	30.33333333	-1.33333333	27.87237110	32.79429557
6	30.00000000	30.00000000	0.00000000	27.53903776	32.46096224
7	25.00000000	25.33333333	-0.33333333	22.87237110	27.79429557
8	24.00000000	24.00000000	0.00000000	21.53903776	26.46096224
9	30.00000000	30.33333333	-0.33333333	27.87237110	32.79429557
10	28.00000000	30.00000000	-2.00000000	27.53903776	32.46096224
11	23.00000000	25.33333333	-2.33333333	22.87237110	27.79429557
12	23.00000000	24.00000000	-1.00000000	21.53903776	26.46096224

15.9 A dummy variable regression model describing the display panel data in Exercise 15.3 is the (α, γ, θ) model

$$
\begin{aligned}
y_{ij,k} &= \mu_{ij} + \varepsilon_{ij,k} \\
&= \mu + \alpha_i + \gamma_j + \theta_{ij} + \varepsilon_{ij,k} \\
&= \mu + \alpha_B D_{i,B} + \alpha_C D_{i,C} \\
&\quad + \gamma_2 D_{j,2} + \gamma_3 D_{j,3} + \gamma_4 D_{j,4} \\
&\quad + \theta_{B2} D_{i,B} D_{j,2} + \theta_{B3} D_{i,B} D_{j,3} + \theta_{B4} D_{i,B} D_{j,4} \\
&\quad + \theta_{C2} D_{i,C} D_{j,2} + \theta_{C3} D_{i,C} D_{j,3} + \theta_{C4} D_{i,C} D_{j,4} \\
&\quad + \varepsilon_{ij,k}
\end{aligned}
$$

Here, for example,

$$
D_{i,B} = \begin{cases} 1 & \text{if } i = \text{B; that is, if we use display panel B} \\ -1 & \text{if } i = \text{A; that is, if we use display panel A} \\ 0 & \text{otherwise} \end{cases}
$$

$$D_{j,3} = \begin{cases} 1 & \text{if } j = 3\text{; that is, if a display panel is tested using emergency condition 3} \\ -1 & \text{if } j = 1\text{; that is, if a display panel is tested using emergency condition 1} \\ 0 & \text{otherwise} \end{cases}$$

a. Define the other dummy variables.
b. Express the following parameters in terms of the parameters of the (α, γ, θ) regression model.

μ_{A1} μ_{A2} μ_{A3} μ_{A4}

μ_{B1} μ_{B2} μ_{B3} μ_{B4}

μ_{C1} μ_{C2} μ_{C3} μ_{C4}

$$\mu_{A\cdot} = \frac{\sum_{j=1}^{4} \mu_{Aj}}{4} \qquad \mu_{B\cdot} = \frac{\sum_{j=1}^{4} \mu_{Bj}}{4} \qquad \mu_{C\cdot} = \frac{\sum_{j=1}^{4} \mu_{Cj}}{4}$$

$$\mu_{\cdot 1} = \frac{\sum_{i=A,B,C} \mu_{i1}}{3} \qquad \mu_{\cdot 2} = \frac{\sum_{i=A,B,C} \mu_{i2}}{3}$$

$$\mu_{\cdot 3} = \frac{\sum_{i=A,B,C} \mu_{i3}}{3} \qquad \mu_{\cdot 4} = \frac{\sum_{i=A,B,C} \mu_{i4}}{3}$$

c. Figure 15.6 (see Exercise 15.3) tells us that

$F((\alpha, \gamma, \theta) \text{ model}) = 32.67$

$F(\theta \mid \alpha, \gamma) = .66$

$F(\alpha \mid \gamma, \theta) = 26.49$

$F(\gamma \mid \alpha, \theta) = 100.80$

Discuss how each of these F statistics is calculated by using regression analysis. Be sure to specify exactly which models are being used to calculate the F statistics.

d. Figure 15.6 gives the point estimate $\boldsymbol{\lambda}'\mathbf{b}$ and the standard error of the estimate $s\sqrt{\boldsymbol{\lambda}'(\mathbf{X}'\mathbf{X})^{-1}\boldsymbol{\lambda}}$ for each of the parameters

$\mu_{B\cdot} - \mu_{A\cdot}$ $\mu_{C\cdot} - \mu_{A\cdot}$ $\mu_{C\cdot} - \mu_{B\cdot}$ $\mu_{\cdot 3} - \mu_{\cdot 1}$

Identify $\boldsymbol{\lambda}'$ for each parameter by expressing the parameter in terms of the parameters of the (α, γ, θ) model. Also, explain how $\boldsymbol{\lambda}'\mathbf{b}$ is calculated for each parameter.

e. Figure 15.6 presents a 95% confidence interval for μ_{B3}

$$[\hat{y}_{B3} \pm t_{[\alpha/2]}^{(n-k)} s\sqrt{\mathbf{x}'_{B3}(\mathbf{X}'\mathbf{X})^{-1}\mathbf{x}_{B3}}]$$

and also presents a 95% prediction interval for $y_{B3,0} = \mu_{B3} + \varepsilon_{B3,0}$

$$[\hat{y}_{B3} \pm t_{[\alpha/2]}^{(n-k)} s\sqrt{1 + \mathbf{x}'_{B3}(\mathbf{X}'\mathbf{X})^{-1}\mathbf{x}_{B3}}]$$

Explain how $\hat{y}_{B3}$ is calculated by using the (α, γ, θ) model, and identify $\mathbf{x}'_{B3}$.

15.10 (Optional) Write a SAS program using a CLASS statement to obtain the output in Figure 15.6.

15.11 (Optional) Write a SAS program that uses the dummy variable regression model of Exercise 15.9 to obtain the same results as are given in Figure 15.6. In the program, actually define and use the appropriate values of the dummy variables.

15.12 Answer the questions in Exercises 15.1 and 15.2 by using the unbalanced data in Table 15.9.

TABLE 15.9 **Results of a two-factor wheat yield experiment in which the data is unbalanced**

Fertilizer type, *i*	Wheat type, *j*			
	M	*N*	*O*	*P*
A	$y_{AM,1} = 19.4$ $y_{AM,2} = 20.6$ $y_{AM,3} = 20.0$	$y_{AN,1} = 25.0$ $y_{AN,2} = 24.0$	$y_{AO,1} = 24.8$ $y_{AO,2} = 26.0$ $y_{AO,3} = 25.4$	$y_{AP,1} = 23.1$ $y_{AP,2} = 24.3$
B	$y_{BM,1} = 22.3$	$y_{BN,1} = 25.6$ $y_{BN,2} = 26.8$ $y_{BN,3} = 26.2$	$y_{BO,1} = 27.6$ $y_{BO,2} = 26.4$ $y_{BO,3} = 27.0$	$y_{BP,1} = 25.4$ $y_{BP,2} = 24.5$ $y_{BP,3} = 26.3$

15.13 Answer the questions in Exercises 15.3 and 15.4 by using the unbalanced data in Table 15.10.

TABLE 15.10 **Results of a two-factor display panel experiment in which the data is unbalanced**

Display panel, *i*	Emergency condition, *j*			
	1	2	3	4
A	17 14	25	31 34	14
B	15 12	22 19	28 31	9 10
C	21 24	29	32 37	15 19

15.14 Answer the questions in Exercise 15.5 by using the unbalanced data in Table 15.11.

15.15 (Optional) Write a SAS program to analyze the wheat yield data in Table 15.3. Use a CLASS statement.

15.16 (Optional) Write a SAS program to analyze the gasoline mileage data in Table 15.5. Use dummy variables.

TABLE 15.11 Results of a two-factor gasoline mileage experiment in which the data is unbalanced

Premium gasoline type, *i*	Gasoline additive type, *j*			
	Q	*R*	*S*	*T*
D	$y_{DQ,1} = 17.4$ $y_{DQ,2} = 18.6$	$y_{DR,1} = 23.0$	$y_{DS,1} = 23.5$ $y_{DS,2} = 22.3$	$y_{DT,1} = 20.8$ $y_{DT,2} = 19.7$
E	$y_{EQ,1} = 23.3$ $y_{EQ,2} = 24.5$	$y_{ER,1} = 25.6$ $y_{ER,2} = 24.4$	$y_{ES,1} = 23.4$ $y_{ES,2} = 23.1$	$y_{ET,1} = 19.6$ $y_{ET,2} = 20.6$
F	$y_{FQ,1} = 23.0$ $y_{FQ,2} = 23.5$	$y_{FR,1} = 24.7$ $y_{FR,2} = 23.3$	$y_{FS,1} = 23.0$	$y_{FT,1} = 18.6$ $y_{FT,2} = 19.8$

15.17 (Optional) Write a SAS program to analyze the telemarketing data in Table 15.6. Use a CLASS statement.

15.18 (Optional) Write a SAS program to analyze the cardboard data in Table 15.7. Use dummy variables.

15.19 (Optional) Write a SAS program to analyze the unbalanced wheat yield data in Table 15.9.

15.20 (Optional) Write a SAS program to analyze the unbalanced display panel data in Table 15.10.

CHAPTER 16

TWO-FACTOR ANALYSIS: II

In this chapter we discuss several important topics in two-factor analysis. We begin in Section 16.1 by presenting the regression approach to analyzing an *incomplete factorial experiment*. We will see that this approach can be used to analyze any type of factorial experiment. However, as will be discussed, it differs from the approach discussed in Section 15.3 for analyzing a complete factorial experiment. In Section 16.2 we discuss *fixed*, *random*, and *mixed* models, and in Section 16.3 we discuss the difference between *crossed* and *nested* factors. We also show how to analyze nested factors using both ANOVA and regression. Note that Sections 16.2 and 16.3 can be understood without reading Section 16.1. We conclude this chapter with optional Section 16.4, which shows how to use SAS to implement the techniques we have presented.

16.1

THE REGRESSION APPROACH TO THE ANALYSIS OF AN INCOMPLETE FACTORIAL EXPERIMENT

If we are analyzing an *incomplete factorial experiment* ($n_{ij} = 0$ for at least one combination of i and j), then we cannot use the (α, γ, θ) model to carry out the entire analysis. The reason for this will be discussed later. To explain the analysis of incomplete data, we consider two dummy variable regression models. The first, the *(α, γ) model*, assumes that there is *no interaction* between factors 1 and 2. The second, the *two-factor means model*, assumes that there is *interaction* between factors 1 and 2. In this section we begin by defining these models and demonstrating their use. Then we discuss how to determine which model should be used in a given situation.

THE (α, γ) MODEL

If there is no interaction between factors 1 and 2, it is reasonable to describe $y_{ij,k}$ by using the following model.

The (α, γ) Model

The (α, γ) model says that

$$\begin{aligned} y_{ij,k} &= \mu_{ij} + \varepsilon_{ij,k} \\ &= \mu + \alpha_i + \gamma_j + \varepsilon_{ij,k} \end{aligned}$$

Intuitively, μ is an overall mean, α_i is the effect of level i of factor 1 on $y_{ij,k}$, and γ_j is the effect of level j of factor 2 on $y_{ij,k}$. To give precise meaning to the parameters μ, α_i, and γ_j, we let

$$\alpha_i = \alpha_2 D_{i,2} + \alpha_3 D_{i,3} + \cdots + \alpha_a D_{i,a}$$

and

$$\gamma_j = \gamma_2 D_{j,2} + \gamma_3 D_{j,3} + \cdots + \gamma_b D_{j,b}$$

Here, $D_{i,2}, D_{i,3}, \ldots, D_{i,a}, D_{j,2}, D_{j,3}, \ldots, D_{j,b}$ are dummy variables. We define

$$D_{i,2} = \begin{cases} 1 & \text{if } i = 2\text{; that is, if we use level 2 of factor 1} \\ 0 & \text{otherwise} \end{cases}$$

$$D_{i,3} = \begin{cases} 1 & \text{if } i = 3\text{; that is, if we use level 3 of factor 1} \\ 0 & \text{otherwise} \end{cases}$$

and similarly for $D_{i,4}, \ldots, D_{i,a}$. We define

$$D_{j,2} = \begin{cases} 1 & \text{if } j = 2\text{; that is, if we use level 2 of factor 2} \\ 0 & \text{otherwise} \end{cases}$$

$$D_{j,3} = \begin{cases} 1 & \text{if } j = 3\text{; that is, if we use level 3 of factor 2} \\ 0 & \text{otherwise} \end{cases}$$

and similarly for $D_{j,4}, \ldots, D_{j,b}$.

To summarize, we may write the (α, γ) model as follows:

$$\begin{aligned} y_{ij,k} &= \mu_{ij} + \varepsilon_{ij,k} \\ &= \mu + \alpha_i + \gamma_j + \varepsilon_{ij,k} \\ &= \mu + \alpha_2 D_{i,2} + \alpha_3 D_{i,3} + \cdots + \alpha_a D_{i,a} \\ &\quad + \gamma_2 D_{j,2} + \gamma_3 D_{j,3} + \cdots + \gamma_b D_{j,b} + \varepsilon_{ij,k} \end{aligned}$$

Before considering an example, we note that we could have defined the above dummy variables by using the 1, −1, 0 coding of Section 15.3. This coding is, in fact, necessary to define the parameters of the (α, γ, θ) model. However, as will be demonstrated in the following example, we can define the parameters of the (α, γ) model (which does not use an interaction term θ_{ij}) by using the simpler 1, 0 coding.

Example 16.1 ***Part 1: The Problem and Data***

Suppose that North American Oil Company is attempting to develop a regular gasoline that will deliver improved gasoline mileages. As part of its development process, the company would like to study the effects of two qualitative factors on the gasoline mileage obtained by an automobile called the Fire-Hawk. These factors are regular gasoline type (which has levels A, B, and C) and gasoline additive type (which has levels M, N, O, and P). To carry out the study, the company test drove three Fire-Hawks using each treatment. However, upon completion of the experiment the company found that several Fire-Hawks had not been driven under the proper test conditions. Rather than running more tests, the company decided (because of limited time) to analyze the data that remained after the data for the improperly tested Fire-Hawks was dropped from the data set. The remaining data is given in Table 16.1. Here,

$y_{ij,k}$ = the kth gasoline mileage obtained when using regular gasoline type i and additive type j

In addition to the data, Table 16.1 gives the sample mean $\bar{y}_{ij}$ for each treatment. Thus we can check for interaction between gasoline type and additive type by performing the graphical analysis of Figure 16.1. The parallel data patterns in Figure 16.1(a) indicate that the change in $\bar{y}_{ij}$ associated with changing the additive

TABLE 16.1 **Data for the regular gasoline mileage experiment**

Regular gasoline type, i	*Gasoline additive type, j*			
	M	*N*	*O*	*P*
A	$n_{AM} = 3$ $y_{AM,1} = 19.4$ $y_{AM,2} = 20.6$ $y_{AM,3} = 20.0$ $\bar{y}_{AM} = 20.0$	$n_{AN} = 2$ $y_{AN,1} = 25.0$ $y_{AN,2} = 24.0$ $\bar{y}_{AN} = 24.5$	$n_{AO} = 2$ $y_{AO,1} = 24.3$ $y_{AO,2} = 25.5$ $\bar{y}_{AO} = 24.9$	$n_{AP} = 0$
B	$n_{BM} = 2$ $y_{BM,1} = 22.6$ $y_{BM,2} = 21.6$ $\bar{y}_{BM} = 22.1$	$n_{BN} = 2$ $y_{BN,1} = 25.3$ $y_{BN,2} = 26.5$ $\bar{y}_{BN} = 25.9$	$n_{BO} = 3$ $y_{BO,1} = 27.6$ $y_{BO,2} = 26.4$ $y_{BO,3} = 27.0$ $\bar{y}_{BO} = 27.0$	$n_{BP} = 1$ $y_{BP,1} = 25.4$ $\bar{y}_{BP} = 25.4$
C	$n_{CM} = 2$ $y_{CM,1} = 18.6$ $y_{CM,2} = 19.8$ $\bar{y}_{CM} = 19.2$	$n_{CN} = 0$	$n_{CO} = 3$ $y_{CO,1} = 24.0$ $y_{CO,2} = 24.7$ $y_{CO,3} = 23.3$ $\bar{y}_{CO} = 24.0$	$n_{CP} = 2$ $y_{CP,1} = 22.0$ $y_{CP,2} = 23.0$ $\bar{y}_{CP} = 22.5$

type depends little on the gasoline type. The parallel data patterns in Figure 16.1(b) indicate that the change in $\bar{y}_{ij}$ associated with changing the gasoline type depends little on the additive type. Together, the parallel data patterns in these figures suggest that there is little interaction between gasoline type and additive type.

Part 2: Developing the (α, γ) Model

Since there is little or no interaction between gasoline type and additive type, it is reasonable to describe $y_{ij,k}$ by the (α, γ) *model*

$$\begin{aligned} y_{ij,k} &= \mu_{ij} + \varepsilon_{ij,k} \\ &= \mu + \alpha_i + \gamma_j + \varepsilon_{ij,k} \\ &= \mu + \alpha_B D_{i,B} + \alpha_C D_{i,C} + \gamma_N D_{j,N} + \gamma_O D_{j,O} + \gamma_P D_{j,P} + \varepsilon_{ij,k} \end{aligned}$$

Here, μ_{ij} denotes the mean gasoline mileage obtained by using gasoline type i and additive type j. Furthermore, for example,

$$D_{i,C} = \begin{cases} 1 & \text{if } i = \text{C; that is, if we are using gasoline type C} \\ 0 & \text{otherwise} \end{cases}$$

$$D_{j,N} = \begin{cases} 1 & \text{if } j = \text{N; that is, if we are using additive type N} \\ 0 & \text{otherwise} \end{cases}$$

FIGURE 16.1 **Graphical analysis of the regular gasoline mileage data in Table 16.1: Little interaction**

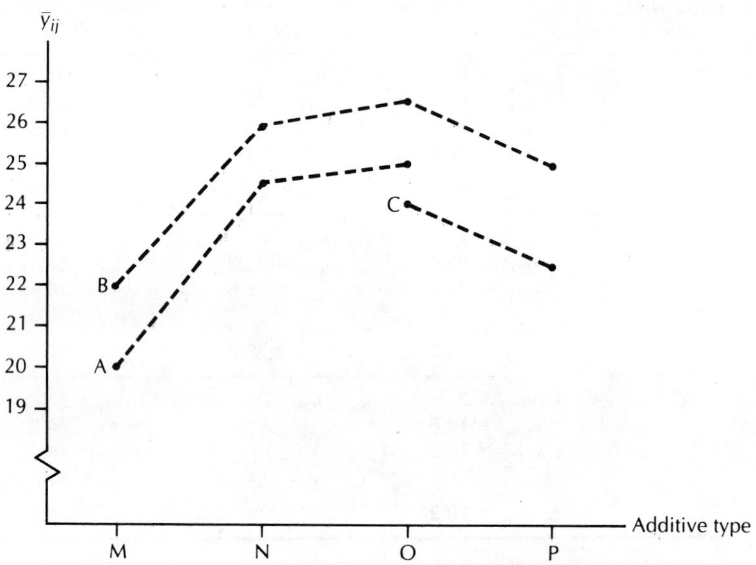

(a) Plot of the change in $\bar{y}_{ij}$ associated with changing the additive type (from M to N to O to P) for each gasoline type (A, B, and C). Note the parallel data patterns.

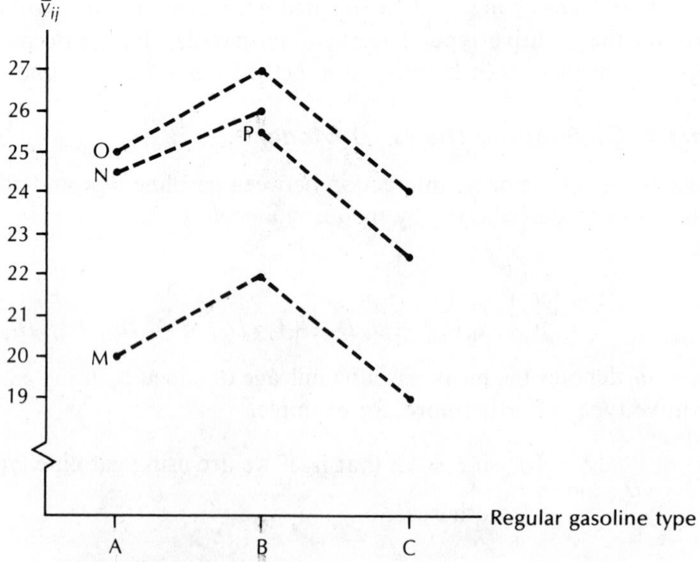

(b) Plot of the change in $\bar{y}_{ij}$ associated with changing the gasoline type (from A to B to C) for each additive type (M, N, O, and P). Note the parallel data patterns.

The other dummy variables are defined similarly. Considering the above model, note that

$$\alpha_i = \alpha_B D_{i,B} + \alpha_C D_{i,C}$$

This implies that

$$\begin{aligned} \alpha_A &= \alpha_B D_{A,B} + \alpha_C D_{A,C} \\ &= \alpha_B(0) + \alpha_C(0) \\ &= 0 \end{aligned}$$

Also note that

$$\gamma_j = \gamma_N D_{j,N} + \gamma_O D_{j,O} + \gamma_P D_{j,P}$$

This implies that

$$\begin{aligned} \gamma_M &= \gamma_N D_{M,N} + \gamma_O D_{M,O} + \gamma_P D_{M,P} \\ &= \gamma_N(0) + \gamma_O(0) + \gamma_P(0) \\ &= 0 \end{aligned}$$

Therefore for i = A, B, and C and j = M, N, O, and P,

$$\begin{aligned} \mu_{ij} - \mu_{Aj} &= [\mu + \alpha_i + \gamma_j] - [\mu + \alpha_A + \gamma_j] \\ &= \alpha_i \qquad (\text{since } \alpha_A = 0) \end{aligned}$$

In other words, $\alpha_i = \mu_{ij} - \mu_{Aj}$ equals the change in mean gasoline mileage associated with changing from gasoline type A to gasoline type i while the additive type remains the same. Similarly, for i = A, B, and C and j = M, N, O, and P,

$$\begin{aligned} \mu_{ij} - \mu_{iM} &= [\mu + \alpha_i + \gamma_j] - [\mu + \alpha_i + \gamma_M] \\ &= \gamma_j \qquad (\text{since } \gamma_M = 0) \end{aligned}$$

In other words, $\gamma_j = \mu_{ij} - \mu_{iM}$ equals the change in mean gasoline mileage associated with changing from additive type M to additive type j while the gasoline type remains the same.

Note that since the change $(\mu_{ij} - \mu_{Aj})$ equals α_i, this change is independent of j, the level of the additive type. Moreover, since the change $(\mu_{ij} - \mu_{iM})$ equals γ_j, this change is independent of i, the level of the gasoline type. It follows that the equation

$$\mu_{ij} = \mu + \alpha_i + \gamma_j$$

assumes that no interaction exists between gasoline type and additive type.

Part 3: Estimating the Regression Parameters

To use the (α, γ) model to carry out a regression analysis of the data in Table 16.1, we use the following column vector **y** and matrix **X**:

$$\mathbf{y} = \begin{bmatrix} y_{AM,1} \\ y_{AM,2} \\ y_{AM,3} \\ y_{AN,1} \\ y_{AN,2} \\ y_{AO,1} \\ y_{AO,2} \\ y_{BM,1} \\ y_{BM,2} \\ y_{BN,1} \\ y_{BN,2} \\ y_{BO,1} \\ y_{BO,2} \\ y_{BO,3} \\ y_{BP,1} \\ y_{CM,1} \\ y_{CM,2} \\ y_{CO,1} \\ y_{CO,2} \\ y_{CO,3} \\ y_{CP,1} \\ y_{CP,2} \end{bmatrix} = \begin{bmatrix} 19.4 \\ 20.6 \\ 20.0 \\ 25.0 \\ 24.0 \\ 24.3 \\ 25.5 \\ 22.6 \\ 21.6 \\ 25.3 \\ 26.5 \\ 27.6 \\ 26.4 \\ 27.0 \\ 25.4 \\ 18.6 \\ 19.8 \\ 24.0 \\ 24.7 \\ 23.3 \\ 22.0 \\ 23.0 \end{bmatrix} \quad \text{and} \quad \mathbf{X} = \begin{array}{c} \begin{array}{cccccc} 1 & D_{i,B} & D_{i,C} & D_{j,N} & D_{j,O} & D_{j,P} \end{array} \\ \begin{bmatrix} 1 & 0 & 0 & 0 & 0 & 0 \\ 1 & 0 & 0 & 0 & 0 & 0 \\ 1 & 0 & 0 & 0 & 0 & 0 \\ 1 & 0 & 0 & 1 & 0 & 0 \\ 1 & 0 & 0 & 1 & 0 & 0 \\ 1 & 0 & 0 & 0 & 1 & 0 \\ 1 & 0 & 0 & 0 & 1 & 0 \\ 1 & 1 & 0 & 0 & 0 & 0 \\ 1 & 1 & 0 & 0 & 0 & 0 \\ 1 & 1 & 0 & 1 & 0 & 0 \\ 1 & 1 & 0 & 1 & 0 & 0 \\ 1 & 1 & 0 & 0 & 1 & 0 \\ 1 & 1 & 0 & 0 & 1 & 0 \\ 1 & 1 & 0 & 0 & 1 & 0 \\ 1 & 1 & 0 & 0 & 0 & 1 \\ 1 & 0 & 1 & 0 & 0 & 0 \\ 1 & 0 & 1 & 0 & 0 & 0 \\ 1 & 0 & 1 & 0 & 1 & 0 \\ 1 & 0 & 1 & 0 & 1 & 0 \\ 1 & 0 & 1 & 0 & 1 & 0 \\ 1 & 0 & 1 & 0 & 0 & 1 \\ 1 & 0 & 1 & 0 & 0 & 1 \end{bmatrix} \end{array}$$

Figure 16.2 gives the SAS output of the least squares point estimates of the parameters in the (α, γ) model (along with the least squares point estimates of several differences between means that will be analyzed in the next example).

For example, $\hat{\gamma}_N = 4.1452$ is the least squares point estimate of γ_N. Since

$$\gamma_N = \mu_{iN} - \mu_{iM}$$

FIGURE 16.2 **SAS output of the least squares point estimates for an (α, γ) model analysis of the regular gasoline mileage data in Table 16.1**

PARAMETER[a]	ESTIMATE[b]	T FOR HO: PARAMETER=0[c]	PROB > |T|[d]	STD ERROR OF ESTIMATE[e]
INTERCEPT	20.10290282	67.28	0.0001	0.29879024
DB	1.90382836	5.66	0.0001	0.33614296
DC	-0.96398822	-2.56	0.0208	0.37610087
DN	4.14518300	10.15	0.0001	0.40828051
DO	4.89465713	14.69	0.0001	0.33310546
DP	3.37181321	7.27	0.0001	0.46365176
MUB-MUA	1.90382836	5.66	0.0001	0.33614296
MUC-MUA	-0.96398822	-2.56	0.0208	0.37610087
MUC-MUB	-2.86781658	-8.28	0.0001	0.34632013
MUB-(MUC+MUA)/2	2.38582247	8.38	0.0001	0.25478394
MUN-MUM	4.14518300	10.15	0.0001	0.40828051
MUO-MUM	4.89465713	14.69	0.0001	0.33310546
MUP-MUM	3.37181321	7.27	0.0001	0.46365176
MUO-MUN	0.74947413	1.84	0.0848	0.40786538
MUP-MUN	-0.77336979	-1.45	0.1674	0.53474573
MUP-MUO	-1.52284392	-3.45	0.0033	0.44176097

[a] $\boldsymbol{\lambda}'\boldsymbol{\beta}$ [b] $\boldsymbol{\lambda}'\mathbf{b}$ [c] t [d] *Prob-value* [e] $s_{(\alpha,\gamma)}\sqrt{\boldsymbol{\lambda}'(\mathbf{X}'\mathbf{X})^{-1}\boldsymbol{\lambda}}$

we estimate that the change in mean gasoline mileage associated with changing from additive type M to additive type N while the gasoline type remains the same is 4.1452 mpg. Similar interpretations can be made for the other least squares point estimates.

The least squares point estimates of the (α, γ) model parameters yield the prediction equation

$$\begin{aligned}\hat{y}_{ij} &= \hat{\mu} + \hat{\alpha}_i + \hat{\gamma}_j \\ &= \hat{\mu} + \hat{\alpha}_B D_{i,B} + \hat{\alpha}_C D_{i,C} + \hat{\gamma}_N D_{j,N} + \hat{\gamma}_O D_{j,O} + \hat{\gamma}_P D_{j,P} \\ &= 20.1029 + 1.9038 D_{i,B} - .9640 D_{i,C} + 4.1452 D_{j,N} \\ &\quad + 4.8947 D_{j,O} + 3.3718 D_{j,P}\end{aligned}$$

For example, a point prediction of a gasoline mileage obtained by using gasoline type B and additive type O

$$\begin{aligned}y_{BO,k} &= \mu_{BO} + \varepsilon_{BO,k} \\ &= \mu + \alpha_B D_{B,B} + \alpha_C D_{B,C} + \gamma_N D_{O,N} + \gamma_O D_{O,O} + \gamma_P D_{O,P} + \varepsilon_{BO,k} \\ &= \mu + \alpha_B(1) + \alpha_C(0) + \gamma_N(0) + \gamma_O(1) + \gamma_P(0) + \varepsilon_{BO,k} \\ &= \mu + \alpha_B + \gamma_O + \varepsilon_{BO,k}\end{aligned}$$

is

$$\begin{aligned}\hat{y}_{BO} &= \hat{\mu} + \hat{\alpha}_B + \hat{\gamma}_O \\ &= 20.1029 + 1.9038 + 4.8947 \\ &= 26.9014\end{aligned}$$

By using such predictions, the unexplained variation is calculated to be

$$
\begin{aligned}
SSE_{(\alpha,\gamma)} &= \sum_{i=\mathrm{A,B,C}} \sum_{j=\mathrm{M,N,O,P}} \sum_{k=1}^{n_{ij}} (y_{ij,k} - \hat{y}_{ij})^2 \\
&= 6.4428
\end{aligned}
$$

Part 4: Testing Main Effects by Using Partial F-Tests

To compare the effects of the gasoline types and the additive types, we first perform two partial F-tests. To perform the first test, consider the following complete (α, γ) model:

$$
\begin{aligned}
y_{ij,k} &= \mu_{ij} + \varepsilon_{ij,k} \\
&= \mu + \alpha_i + \gamma_j + \varepsilon_{ij,k} \\
&= \mu + \alpha_{\mathrm{B}} D_{i,\mathrm{B}} + \alpha_{\mathrm{C}} D_{i,\mathrm{C}} + \gamma_{\mathrm{N}} D_{j,\mathrm{N}} + \gamma_{\mathrm{O}} D_{j,\mathrm{O}} + \gamma_{\mathrm{P}} D_{j,\mathrm{P}} + \varepsilon_{ij,k}
\end{aligned}
$$

This model has $k = 6$ parameters and an unexplained variation equal to $SSE_{\mathrm{C}} = SSE_{(\alpha,\gamma)} = 6.4428$. Since for j = M, N, O, and P,

$$
\begin{aligned}
\alpha_{\mathrm{B}} &= \mu_{\mathrm{B}j} - \mu_{\mathrm{A}j} \\
\alpha_{\mathrm{C}} &= \mu_{\mathrm{C}j} - \mu_{\mathrm{A}j}
\end{aligned}
$$

the null hypothesis

$$H_0\colon \alpha_{\mathrm{B}} = \alpha_{\mathrm{C}} = 0$$

is equivalent to

$$H_0\colon \mu_{\mathrm{A}j} = \mu_{\mathrm{B}j} = \mu_{\mathrm{C}j}$$

This hypothesis says that gasoline types A, B, and C have the same effects on mean gasoline mileage. The alternative hypothesis,

H_1: At least one of α_{B} or α_{C} does not equal zero

or, equivalently,

H_1: At least two of $\mu_{\mathrm{A}j}$, $\mu_{\mathrm{B}j}$, and $\mu_{\mathrm{C}j}$ differ from each other

says that at least two of gasoline types A, B, and C have different effects on mean gasoline mileage. To test H_0 versus H_1, first note that $p - g = 2$, since two parameters (α_{B} and α_{C}) are set equal to zero in the statement of H_0. Also note that under the assumption that H_0 is true, the complete model becomes the reduced γ *model*:

$$
\begin{aligned}
y_{ij,k} &= \mu + \gamma_j + \varepsilon_{ij,k} \\
&= \mu + \gamma_{\mathrm{N}} D_{j,\mathrm{N}} + \gamma_{\mathrm{O}} D_{j,\mathrm{O}} + \gamma_{\mathrm{P}} D_{j,\mathrm{P}} + \varepsilon_{ij,k}
\end{aligned}
$$

For this model the unexplained variation is $SSE_R = SSE_\gamma = 36.8210$. Thus we use the following partial F statistic and prob-value:

$$\begin{aligned} F(\alpha|\gamma) &= F(D_{i,\mathrm{B}}, D_{i,\mathrm{C}} | D_{j,\mathrm{N}}, D_{j,\mathrm{O}}, D_{j,\mathrm{P}}) \\ &= \frac{\{SSE_\mathrm{R} - SSE_\mathrm{C}\}/(p-g)}{SSE_\mathrm{C}/(n-k)} \\ &= \frac{\{SSE_\gamma - SSE_{(\alpha,\gamma)}\}/(p-g)}{SSE_{(\alpha,\gamma)}/(n-k)} \\ &= \frac{\{36.8210 - 6.4428\}/2}{6.4428/(22-6)} \\ &= \frac{30.3782/2}{6.4428/16} \\ &= \frac{15.1891}{.4027} \\ &= 37.7182 \end{aligned}$$

$$\text{prob-value} = .0001$$

Since $F(\alpha|\gamma) = 37.7182 > 3.63 = F_{[.05]}^{(2,16)}$ and prob-value $= .0001$ is less than .05, we can reject H_0 in favor of H_1 by setting α equal to .05. We conclude that at least two of gasoline types A, B, and C have different effects on mean gasoline mileage. Note that Figure 16.3 presents the SAS output of $F(\alpha|\gamma)$ and its related prob-value.

To perform the second partial F-test, again consider the complete model to be the (α, γ) model. Since for i = A, B, and C,

$$\gamma_\mathrm{N} = \mu_{i\mathrm{N}} - \mu_{i\mathrm{M}}$$

$$\gamma_\mathrm{O} = \mu_{i\mathrm{O}} - \mu_{i\mathrm{M}}$$

$$\gamma_\mathrm{P} = \mu_{i\mathrm{P}} - \mu_{i\mathrm{M}}$$

the null hypothesis

$$H_0: \gamma_\mathrm{N} = \gamma_\mathrm{O} = \gamma_\mathrm{P} = 0$$

is equivalent to

$$H_0: \mu_{i\mathrm{M}} = \mu_{i\mathrm{N}} = \mu_{i\mathrm{O}} = \mu_{i\mathrm{P}}$$

This hypothesis says that additive types M, N, O, and P have the same effects on mean gasoline mileage. The alternative hypothesis,

H_1: At least one of γ_N, γ_O, or γ_P does not equal zero

or

H_1: At least two of $\mu_{i\mathrm{M}}$, $\mu_{i\mathrm{N}}$, $\mu_{i\mathrm{O}}$, and $\mu_{i\mathrm{P}}$ differ from each other

says that at least two of additive types M, N, O, and P have different effects on mean gasoline mileage. To test H_0 versus H_1, first note that $p - g = 3$, since three parameters (γ_N, γ_O, and γ_P) are set equal to zero in the statement of H_0. Also note

FIGURE 16.3 An (α, γ) model analysis of the regular gasoline mileage data in Table 16.1

(a) SAS output of the (α, γ) model analysis

SAS

GENERAL LINEAR MODELS PROCEDURE

DEPENDENT VARIABLE: MILEAGE

SOURCE	DF	SUM OF SQUARES	MEAN SQUARE	F VALUE	PR > F	R-SQUARE	C.V.
MODEL	5	137.82997591	27.56599516	68.46	0.0001	0.955343	2.7024
ERROR	16	6.44275137[j]	0.40267196		ROOT MSE		MILEAGE MEAN
CORRECTED TOTAL	21	144.27272727			0.63456439[k]		23.48181818

SOURCE	DF	TYPE I SS	F VALUE	PR > F	DF	TYPE III SS	F VALUE	PR > F
GASTYPE	2	42.38415584	52.63	0.0001	2	30.37820101	37.72[a]	0.0001[b]
ADDTYPE	3	95.44582006	79.01	0.0001	3	95.44582006	79.01[c]	0.0001[d]

PARAMETER[e]	ESTIMATE[f]	T FOR HO: PARAMETER=0[g]	PR > |T|[h]	STD ERROR OF ESTIMATE[i]
MUB-MUA	1.90382836	5.66	0.0001	0.33614296
MUC-MUA	-0.96398822	-2.56	0.0208	0.37610087
MUC-MUB	-2.86781658	-8.28	0.0001	0.34632013
MUB-(MUC+MUA)/2	2.38582247	8.38	0.0001	0.28478394
MUN-MUM	4.14518300	10.15	0.0001	0.40828051
MUO-MUM	4.89465713	14.69	0.0001	0.33310546
MUP-MUM	3.37181321	7.27	0.0001	0.46365176
MUO-MUN	0.74947413	1.84	0.0848	0.40786538
MUP-MUN	-0.77336979	-1.45	0.1674	0.53474573
MUP-MUO	-1.52284392	-3.45	0.0033	0.44176097

OBSERVATION	OBSERVED VALUE	PREDICTED VALUE	RESIDUAL	LOWER 95% CL FOR MEAN	UPPER 95% CL FOR MEAN
1	19.40000000	20.10290282	-0.70290282	19.46949924	20.73630640
2	20.60000000	20.10290282	0.49709718	19.46949924	20.73630640
3	20.00000000	20.10290282	-0.10290282	19.46949924	20.73630640
4	25.00000000	24.24808582	0.75191418	23.48694058	25.00923106
5	24.00000000	24.24808582	-0.24808582	23.48694058	25.00923106
6	24.30000000	24.99755995	-0.69755995	24.31053731	25.68458259
7	25.50000000	24.99755995	0.50244005	24.31053731	25.68458259
8	22.60000000	22.00673117	0.59326883	21.33881184	22.67465051
9	21.60000000	22.00673117	-0.40673117	21.33881184	22.67465051
10	25.30000000	26.15191418	-0.85191418	25.39076894	26.91305942
11	26.50000000	26.15191418	0.34808582	25.39076894	26.91305942
12	27.60000000	26.90138830[l]	0.69861170	[26.29124811	27.51152850][m]
13	26.40000000	26.90138830	-0.50138830	26.29124811	27.51152850
14	27.00000000	26.90138830	0.09861170	26.29124811	27.51152850
15	25.40000000	25.37854438	0.02145562	24.46053099	26.29655778
16	18.60000000	19.13891460	-0.53891460	18.43115111	19.84667808
17	19.80000000	19.13891460	0.66108540	18.43115111	19.84667808
18	24.00000000	24.03357173	-0.03357173	23.39776888	24.66937457
19	24.70000000	24.03357173	0.66642827	23.39776888	24.66937457
20	23.30000000	24.03357173	-0.73357173	23.39776888	24.66937457
21	22.00000000	22.51072781	-0.51072781	21.69642809	23.32502753
22	23.00000000	22.51072781	0.48927219	21.69642809	23.32502753

[a] $F(\alpha|\gamma)$ [b] *Prob-value related to* $F(\alpha|\gamma)$ [c] $F(\gamma|\alpha)$ [d] *Prob-value related to* $F(\gamma|\alpha)$ [e] $\boldsymbol{\lambda}'\boldsymbol{\beta}$ [f] $\boldsymbol{\lambda}'\mathbf{b}$ [g] t
[h] *Prob-value related to* t [i] $s_{(\alpha,\gamma)}\sqrt{\boldsymbol{\lambda}'(\mathbf{X}'\mathbf{X})^{-1}\boldsymbol{\lambda}}$ [j] $SSE_{(\alpha,\gamma)}$ [k] $s_{(\alpha,\gamma)}$ [l] $\hat{y}_{BO}$ [m] *95% confidence interval for* μ_{BO}
[n] *95% prediction interval for* $y_{BO,0}$

OBSERVATION	OBSERVED VALUE	PREDICTED VALUE	RESIDUAL	LOWER 95% CL INDIVIDUAL	UPPER 95% CL INDIVIDUAL
1	19.40000000	20.10290282	-0.70290282	18.61603038	21.58977526
2	20.60000000	20.10290282	0.49709718	18.61603038	21.58977526
3	20.00000000	20.10290282	-0.10290282	18.61603038	21.58977526
4	25.00000000	24.24808582	0.75191418	22.70246886	25.79370279
5	24.00000000	24.24808582	-0.24808582	22.70246886	25.79370279
6	24.30000000	24.99755995	-0.69755995	23.48706674	26.50805316
7	25.50000000	24.99755995	0.50244005	23.48706674	26.50805316
8	22.60000000	22.00673117	0.59326883	20.50483043	23.50863192
9	21.60000000	22.00673117	-0.40673117	20.50483043	23.50863192
10	25.30000000	26.15191418	-0.85191418	24.60629721	27.69753114
11	26.50000000	26.15191418	0.34808582	24.60629721	27.69753114
12	27.60000000	26.90138830′	0.69861170	[25.42427606	28.37850055]″
13	26.40000000	26.90138830	-0.50138830	25.42427606	28.37850055
14	27.00000000	26.90138830	0.09861170	25.42427606	28.37850055
15	25.40000000	25.37854438	0.02145562	23.74994420	27.00714457
16	18.60000000	19.13891460	-0.53891460	17.61887552	20.65895368
17	19.80000000	19.13891460	0.66108540	17.61887552	20.65895368
18	24.00000000	24.03357173	-0.03357173	22.54567562	25.52146783
19	24.70000000	24.03357173	0.66642827	22.54567562	25.52146783
20	23.30000000	24.03357173	-0.73357173	22.54567562	25.52146783
21	22.00000000	22.51072781	-0.51072781	20.93825402	24.08320160
22	23.00000000	22.51072781	0.48927219	20.93825402	24.08320160

(b) Individual and Scheffé simultaneous confidence intervals for parameters of interest

Parameter $\lambda'\beta$	*Individual 95% confidence interval for* $\lambda'\beta$ $[\lambda'\mathbf{b} \pm 2.12 s_{(\alpha,\gamma)}\sqrt{\lambda'(\mathbf{X}'\mathbf{X})^{-1}\lambda}]$	*Scheffé simultaneous 95% confidence interval for* $\lambda'\beta$ $[\lambda'\mathbf{b} \pm 3.7749 s_{(\alpha,\gamma)}\sqrt{\lambda'(\mathbf{X}'\mathbf{X})^{-1}\lambda}]$
$\mu_{Bj} - \mu_{Aj} = \alpha_B$	[1.1913, 2.6163]	[.6351, 3.1725]
$\mu_{Cj} - \mu_{Aj} = \alpha_C$	[−1.7613, −.1667]	[−2.3837, .4557]
$\mu_{iN} - \mu_{iM} = \gamma_N$	[3.2796, 5.0108]	[2.6039, 5.6865]
$\mu_{iO} - \mu_{iM} = \gamma_O$	[4.1894, 5.6008]	[3.6372, 6.152]
$\mu_{iP} - \mu_{iM} = \gamma_P$	[2.389, 4.3546]	[1.6218, 5.1218]
$\mu_{Cj} - \mu_{Bj} = \alpha_C - \alpha_B$	[−3.6019, −2.1337]	[−4.175, −1.5606]
$\mu_{Bj} - (\mu_{Cj} + \mu_{Aj})/2 = \alpha_B - .5\alpha_C$	[1.7820, 2.9896]	[1.3107, 3.4609]
$\mu_{iO} - \mu_{iN} = \gamma_O - \gamma_N$	[−.1152, 1.6142]	[−.7903, 2.2393]
$\mu_{iP} - \mu_{iN} = \gamma_P - \gamma_N$	[−1.907, .3602]	[−2.7918, 1.245]
$\mu_{iP} - \mu_{iO} = \gamma_P - \gamma_O$	[−2.4594, −.5862]	[−3.1906, .145]

that under the assumption that H_0 is true, the complete model becomes the reduced *α model*:

$$\begin{aligned} y_{ij,k} &= \mu + \alpha_i + \varepsilon_{ij,k} \\ &= \mu + \alpha_B D_{i,B} + \alpha_C D_{i,C} + \varepsilon_{ij,k} \end{aligned}$$

For this model the unexplained variation is $SSE_R = SSE_\alpha = 101.8886$. Thus we use the partial F statistic and prob-value:

$$\begin{aligned} F(\gamma|\alpha) &= F(D_{j,N}, D_{j,O}, D_{j,P} | D_{i,B}, D_{i,C}) \\ &= \frac{\{SSE_R - SSE_C\}/(p-g)}{SSE_C/(n-k)} \\ &= \frac{\{SSE_\alpha - SSE_{(\alpha,\gamma)}\}/(p-g)}{SSE_{(\alpha,\gamma)}/(n-k)} \end{aligned}$$

$$= \frac{\{101.8886 - 6.4428\}/3}{6.4428/(22 - 6)}$$

$$= \frac{95.4458/3}{6.4428/16}$$

$$= \frac{31.8153}{.4027}$$

$$= 79.005$$

$$\text{prob-value} = .0001$$

Since the prob-value is less than .05, we can reject H_0 by setting α equal to .05. We conclude that at least two of additive types M, N, O, and P have different effects on mean gasoline mileage. Note that Figure 16.3 presents the SAS output of $F(\gamma|\alpha)$ and its related prob-value.

Part 5: Testing and Estimating Differences Using Linear Combinations of Regression Parameters

Below we present several linear combinations of parameters that will allow us to investigate the exact nature of the differences between the effects of the gasoline types and the additive types:

$$\mu_{Bj} - \mu_{Aj} = \alpha_B$$

$$\mu_{Cj} - \mu_{Aj} = \alpha_C$$

$$\mu_{iN} - \mu_{iM} = \gamma_N$$

$$\mu_{iO} - \mu_{iM} = \gamma_O$$

$$\mu_{iP} - \mu_{iM} = \gamma_P$$

$$\mu_{Cj} - \mu_{Bj} = \alpha_C - \alpha_B$$

$$\mu_{Bj} - \frac{\mu_{Cj} + \mu_{Aj}}{2} = \alpha_B - .5\alpha_C$$

$$\mu_{iO} - \mu_{iN} = \gamma_O - \gamma_N$$

$$\mu_{iP} - \mu_{iN} = \gamma_P - \gamma_N$$

$$\mu_{iP} - \mu_{iO} = \gamma_P - \gamma_O$$

Note that the above relationships can be easily established by using the equation

$$\mu_{ij} = \mu + \alpha_i + \gamma_j$$

For example,

$$\begin{aligned}\mu_{Cj} - \mu_{Bj} &= [\mu + \alpha_C + \gamma_j] - [\mu + \alpha_B + \gamma_j] \\ &= \alpha_C - \alpha_B \\ &= \text{the change in mean gasoline mileage associated with changing from gasoline type B to gasoline type C while the additive type remains the same}\end{aligned}$$

$$\begin{aligned}\mu_{Bj} - \frac{\mu_{Cj} + \mu_{Aj}}{2} &= [\mu + \alpha_B + \gamma_j] - \frac{[\mu + \alpha_C + \gamma_j] + [\mu + \gamma_j]}{2} \\ &= \alpha_B - .5\alpha_C \qquad (\text{recall that } \alpha_A = 0) \\ &= \text{the change in mean gasoline mileage associated with changing from using gasoline type C half the time and gasoline type A half the time to using only gasoline type B while the additive type remains the same}\end{aligned}$$

Similarly,

$$\begin{aligned}\mu_{iO} - \mu_{iN} &= [\mu + \alpha_i + \gamma_O] - [\mu + \alpha_i + \gamma_N] \\ &= \gamma_O - \gamma_N \\ &= \text{the change in mean gasoline mileage associated with changing from additive type N to additive type O while the gasoline type remains the same}\end{aligned}$$

Figure 16.3 presents the SAS output of the least squares point estimates of each of the above linear combinations. Recall that we have previously computed $\hat{\mu}$, $\hat{\alpha}_B$, $\hat{\alpha}_C$, $\hat{\gamma}_N$, $\hat{\gamma}_O$, and $\hat{\gamma}_P$. Here, for example, the linear combination

$$\mu_{Bj} - \mu_{Aj} = \alpha_B$$

is denoted "MUB − MUA" on the SAS output. Besides the point estimate $\boldsymbol{\lambda}'\mathbf{b}$, the SAS output includes the t statistic and related prob-value for testing H_0: $\boldsymbol{\lambda}'\boldsymbol{\beta} = 0$ versus H_1: $\boldsymbol{\lambda}'\boldsymbol{\beta} \neq 0$ for each of the linear combinations. The figure also presents individual and Scheffé simultaneous 95% confidence intervals for each of the linear combinations. These intervals are not included in the SAS output—they have been hand calculated.

The unexplained variation for the (α, γ) model is $SSE_{(\alpha,\gamma)} = 6.4428$, and the standard error for this model is $s_{(\alpha,\gamma)} = .6346$. Thus the individual 95% confidence intervals in Figure 16.3 are computed by using the formula

$$\begin{aligned}[\boldsymbol{\lambda}'\mathbf{b} \pm t_{[.025]}^{(n-k)} s_{(\alpha,\gamma)} \sqrt{\boldsymbol{\lambda}'(\mathbf{X}'\mathbf{X})^{-1}\boldsymbol{\lambda}}] &= [\boldsymbol{\lambda}'\mathbf{b} \pm t_{[.025]}^{(22-6)} s_{(\alpha,\gamma)} \sqrt{\boldsymbol{\lambda}'(\mathbf{X}'\mathbf{X})^{-1}\boldsymbol{\lambda}}] \\ &= [\boldsymbol{\lambda}'\mathbf{b} \pm (2.12)\, s_{(\alpha,\gamma)} \sqrt{\boldsymbol{\lambda}'(\mathbf{X}'\mathbf{X})^{-1}\boldsymbol{\lambda}}]\end{aligned}$$

For instance,

$$\begin{aligned}
\mu_{iO} - \mu_{iN} &= \gamma_O - \gamma_N \\
&= 0 \cdot \mu + 0 \cdot \alpha_B + 0 \cdot \alpha_C + (-1) \cdot \gamma_N + 1 \cdot \gamma_O + 0 \cdot \gamma_P \\
&= [0 \quad 0 \quad 0 \quad -1 \quad 1 \quad 0] \begin{bmatrix} \mu \\ \alpha_B \\ \alpha_C \\ \gamma_N \\ \gamma_O \\ \gamma_P \end{bmatrix} \\
&= \boldsymbol{\lambda}'\boldsymbol{\beta}
\end{aligned}$$

where $\boldsymbol{\lambda}' = [0 \quad 0 \quad 0 \quad -1 \quad 1 \quad 0]$. Also note that there are $q = 5$ parameters —$\alpha_B, \alpha_C, \gamma_N, \gamma_O$, and γ_P—from the (α, γ) model involved nontrivially in at least one of the linear combinations above. It follows that

$$\begin{aligned}
\sqrt{qF_{[\alpha]}^{(q,n-k)}} &= \sqrt{5F_{[.05]}^{(5,22-6)}} \\
&= \sqrt{5F_{[.05]}^{(5,16)}} \\
&= \sqrt{5(2.85)} \\
&= 3.7749
\end{aligned}$$

Thus the Scheffé simultaneous 95% confidence intervals in Figure 16.3 are computed by using the formula

$$\left[\boldsymbol{\lambda}'\mathbf{b} \pm \sqrt{qF_{[\alpha]}^{(q,n-k)}}\, s_{(\alpha,\gamma)}\sqrt{\boldsymbol{\lambda}'(\mathbf{X}'\mathbf{X})^{-1}\boldsymbol{\lambda}}\right] = \left[\boldsymbol{\lambda}'\mathbf{b} \pm (3.7749)\, s_{(\alpha,\gamma)}\sqrt{\boldsymbol{\lambda}'(\mathbf{X}'\mathbf{X})^{-1}\boldsymbol{\lambda}}\right]$$

The results in Figure 16.3 provide overwhelming evidence that μ_{Bj} is greater than μ_{Aj} and μ_{Cj}. For example, the small prob-value related to

$$H_0: \alpha_B = \mu_{Bj} - \mu_{Aj} = 0 \qquad \text{(prob-value} = .0001)$$

and the individual 95% confidence interval for $\mu_{Bj} - \mu_{Aj}$, [1.1913, 2.6163], lead us to conclude that μ_{Bj} is greater than μ_{Aj}. Similarly, the small prob-value related to

$$H_0: \alpha_C - \alpha_B = \mu_{Cj} - \mu_{Bj} = 0 \qquad \text{(prob-value} = .0001)$$

and the individual 95% confidence interval for $\mu_{Cj} - \mu_{Bj}$, [−3.6019, −2.1337], lead us to conclude that μ_{Bj} is greater than μ_{Cj}. Similarly, the small prob-value related to

$$H_0: \gamma_O = \mu_{iO} - \mu_{iM} = 0 \qquad \text{(prob-value} = .0001)$$

and the individual 95% confidence interval for $\mu_{iO} - \mu_{iM}$, [4.1894, 5.6008], provide overwhelming evidence that μ_{iO} is greater than μ_{iM}. The small prob-value related to

$$H_0: \gamma_P - \gamma_O = \mu_{iP} - \mu_{iO} = 0 \qquad \text{(prob-value} = .0033)$$

and the individual 95% confidence interval for $\mu_{iP} - \mu_{iO}$, $[-2.4594, -.5862]$, provide strong evidence that μ_{iO} is greater than μ_{iP}.

To compare μ_{iO} and μ_{iN}, first note that $\hat{\gamma}_O - \hat{\gamma}_N = .7495$ is the least squares point estimate of $\gamma_O - \gamma_N = \mu_{iO} - \mu_{iN}$. Thus we estimate that the effect of changing from additive type N to additive type O (while the gasoline type remains the same) is to increase mean gasoline mileage by .7495 mpg. However, the prob-value related to

$$H_0: \gamma_O - \gamma_N = \mu_{iO} - \mu_{iN} = 0 \qquad \text{(prob-value } = .0848)$$

does not allow us to reject H_0 with $\alpha = .05$. In spite of this, the individual 95% confidence interval for $\mu_{iO} - \mu_{iN}$, $[-.1152, 1.6142]$, is "mostly positive." Thus we have some evidence that μ_{iO} is greater than μ_{iN}. On the basis of these comparisons, and supposing that North American Oil has theoretical reasons for believing that additive type O is the best additive, the company will produce gasoline type B and will use additive type O.

Next, suppose that North American Oil has been producing gasoline type A using additive type N. Note that

$$\begin{aligned} \mu_{BO} - \mu_{AN} &= (\mu + \alpha_B + \gamma_O) - (\mu + \gamma_N) \\ &= \alpha_B + (\gamma_O - \gamma_N) \\ &= 0 \cdot \mu + 1 \cdot \alpha_B + 0 \cdot \alpha_C + (-1) \cdot \gamma_N + 1 \cdot \gamma_O + 0 \cdot \gamma_P \\ &= [0 \quad 1 \quad 0 \quad -1 \quad 1 \quad 0] \begin{bmatrix} \mu \\ \alpha_B \\ \alpha_C \\ \gamma_N \\ \gamma_O \\ \gamma_P \end{bmatrix} \\ &= \boldsymbol{\lambda}'\boldsymbol{\beta} \end{aligned}$$

where $\boldsymbol{\lambda}' = [0 \quad 1 \quad 0 \quad -1 \quad 1 \quad 0]$. It follows that a point estimate of, and an individual 95% confidence interval for $\mu_{BO} - \mu_{AN}$ are

$$\hat{\alpha}_B + (\hat{\gamma}_O - \hat{\gamma}_N) = 1.9038 + .7495 = 2.6533$$

and

$$\begin{aligned} [\boldsymbol{\lambda}'\mathbf{b} \pm t_{[.025]}^{(22-6)} s_{(\alpha,\gamma)} \sqrt{\boldsymbol{\lambda}'(\mathbf{X}'\mathbf{X})^{-1}\boldsymbol{\lambda}}] &= [2.6533 \pm 2.12(.6346)\sqrt{.6386}] \\ &= [2.6533 \pm 1.0751] \\ &= [1.5782, 3.7284] \end{aligned}$$

This interval says that we can be 95% confident that μ_{BO}, the mean gasoline mileage obtained by using gasoline type B and additive type O, is between 1.5782 mpg and 3.7284 mpg greater than μ_{AN}, the mean gasoline mileage obtained by using gasoline type A and additive type N.

Part 6: Using the Model to Estimate and Predict

Finally, we can describe an individual gasoline mileage obtained when using gasoline type B and additive type O by the equation

$$\begin{aligned} y_{BO,0} &= \mu_{BO} + \varepsilon_{BO,0} \\ &= \mu + \alpha_B D_{B,B} + \alpha_C D_{B,C} + \gamma_N D_{O,N} + \gamma_O D_{O,O} + \gamma_P D_{O,P} + \varepsilon_{BO,0} \\ &= \mu + \alpha_B(1) + \alpha_C(0) + \gamma_N(0) + \gamma_O(1) + \gamma_P(0) + \varepsilon_{BO,0} \\ &= \mu + \alpha_B + \gamma_O + \varepsilon_{BO,0} \end{aligned}$$

It follows that

$$\begin{aligned} \hat{y}_{BO} &= \hat{\mu} + \hat{\alpha}_B + \hat{\gamma}_O \\ &= 20.1029 + 1.9038 + 4.8947 \\ &= 26.9014 \end{aligned}$$

is the point estimate of μ_{BO} and is the point prediction of $y_{BO,0}$. Furthermore, the equation describing $y_{BO,0}$ implies that

$$\mathbf{x}'_{BO} = [1 \quad 1 \quad 0 \quad 0 \quad 1 \quad 0]$$

Therefore a 95% confidence interval for μ_{BO} is

$$\begin{aligned} [\hat{y}_{BO} \pm t_{[.025]}^{(22-6)} s_{(\alpha,\gamma)} \sqrt{\mathbf{x}'_{BO}(\mathbf{X}'\mathbf{X})^{-1}\mathbf{x}_{BO}}] &= [26.9014 \pm 2.12(.6346)\sqrt{.2057}] \\ &= [26.9014 \pm .6102] \\ &= [26.2912, 27.5116] \end{aligned}$$

This interval says that we can be 95 percent confident that μ_{BO} is between 26.2912 mpg and 27.5116 mpg. Also, a 95% prediction interval for $y_{BO,0}$ is

$$\begin{aligned} [\hat{y}_{BO} \pm t_{[.025]}^{(22-6)} s_{(\alpha,\gamma)} \sqrt{1 + \mathbf{x}'_{BO}(\mathbf{X}'\mathbf{X})^{-1}\mathbf{x}_{BO}}] &= [26.9014 \pm 2.12(.6346)\sqrt{1.2057}] \\ &= [26.9014 \pm 1.4773] \\ &= [25.4241, 28.3787] \end{aligned}$$

This interval says that the owner of a Fire-Hawk can be 95 percent confident that $y_{BO,0}$ will be between 25.4241 mpg and 28.3787 mpg. Note that both of the above intervals are given in the SAS output of Figure 16.3.

Before continuing, we note that it can in general be proven that we can obtain least squares point estimates of the parameters in the (α, γ) *model* if the two-factor data set under consideration is *connected.* To understand this term, note that the cell corresponding to level i of factor 1 and level j of factor 2 is said to be *filled* if $n_{ij} > 0$. A two-factor data set is said to be *connected* if the filled cells (as they are positioned in the data set) can be joined by a continuous line. This line must consist only of vertical and horizontal line segments and must have direction changes only in filled cells. Figure 16.4 illustrates that the gasoline mileage data in Table 16.1 is connected.

FIGURE 16.4 An illustration showing that the regular gasoline mileage data in Table 16.1 is connected

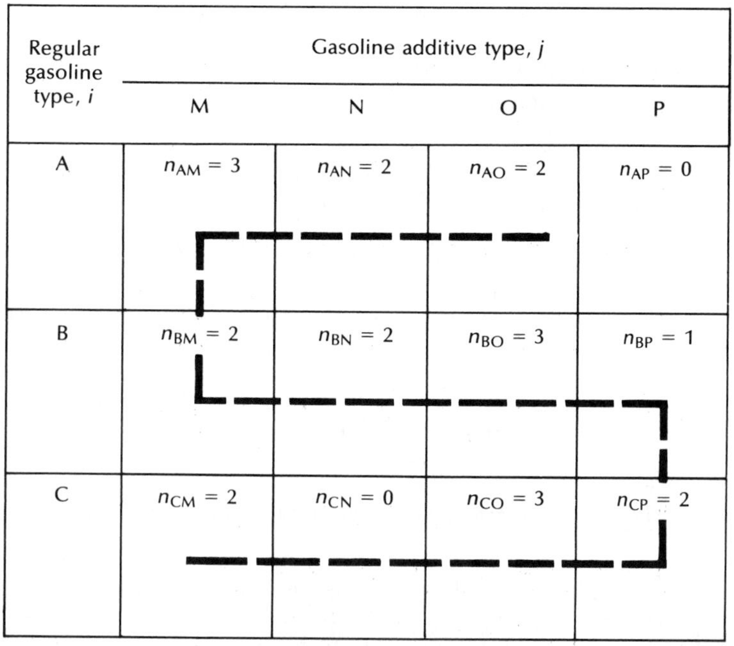

Regular gasoline type, i	Gasoline additive type, j			
	M	N	O	P
A	$n_{AM} = 3$	$n_{AN} = 2$	$n_{AO} = 2$	$n_{AP} = 0$
B	$n_{BM} = 2$	$n_{BN} = 2$	$n_{BO} = 3$	$n_{BP} = 1$
C	$n_{CM} = 2$	$n_{CN} = 0$	$n_{CO} = 3$	$n_{CP} = 2$

THE TWO-FACTOR MEANS MODEL

If there is interaction between factors 1 and 2, it is reasonable to describe $y_{ij,k}$ by the *two-factor means model.*

The Two-Factor Means Model

The *two-factor means model* says that

$$y_{ij,k} = \mu_{ij} + \varepsilon_{ij,k}$$

where

$$\begin{aligned}\mu_{ij} = {} & \beta_{11} + \beta_{12}D_{ij,12} + \beta_{13}D_{ij,13} + \cdots + \beta_{1b}D_{ij,1b} \\ & + \beta_{21}D_{ij,21} + \beta_{22}D_{ij,22} + \beta_{23}D_{ij,23} + \cdots + \beta_{2b}D_{ij,2b} \\ & \vdots \\ & + \beta_{a1}D_{ij,a1} + \beta_{a2}D_{ij,a2} + \beta_{a3}D_{ij,a3} + \cdots + \beta_{ab}D_{ij,ab}\end{aligned}$$

Here, for example,

$$D_{ij,12} = \begin{cases} 1 & \text{if } i = 1 \text{ and } j = 2\text{; that is, if we are using level 1} \\ & \text{of factor 1 and level 2 of factor 2} \\ 0 & \text{otherwise} \end{cases}$$

and the other dummy variables are defined similarly. Using the definitions of the dummy variables, we find that

$$\mu_{11} = \beta_{11}$$

We also find that for any $\mu_{ij} \neq \mu_{11}$

$$\mu_{ij} = \beta_{11} + \beta_{ij}$$

which implies that

$$\begin{aligned} \beta_{ij} &= \mu_{ij} - \beta_{11} \\ &= \mu_{ij} - \mu_{11} \end{aligned}$$

= the change in the mean response associated with changing from using level 1 of factor 1 and level 1 of factor 2 to using level i of factor 1 and level j of factor 2

To see that the means model implies that there is interaction between factors 1 and 2, note that

1. The relationship between the levels of factor 1—that is, between

$$\begin{aligned} \mu_{1j} &= \beta_{11} + \beta_{1j} \\ \mu_{2j} &= \beta_{11} + \beta_{2j} \\ &\vdots \\ \mu_{aj} &= \beta_{11} + \beta_{aj} \end{aligned}$$

depends upon the level j of factor 2. This implies that if we compare μ_{1j}, $\mu_{2j}, \ldots, \mu_{aj}$ by first fixing j at one value and then fixing j at another value, the two comparisons of $\mu_{1j}, \mu_{2j}, \ldots, \mu_{aj}$ might yield different results.

2. The relationship between the levels of factor 2—that is, between

$$\begin{aligned} \mu_{i1} &= \beta_{11} + \beta_{i1} \\ \mu_{i2} &= \beta_{11} + \beta_{i2} \\ &\vdots \\ \mu_{ib} &= \beta_{11} + \beta_{ib} \end{aligned}$$

depends upon the level i of factor 1. This implies that if we compare μ_{i1}, $\mu_{i2}, \ldots, \mu_{ib}$ by first fixing i at one value and then fixing i at another value, the two comparisons of $\mu_{i1}, \mu_{i2}, \ldots, \mu_{ib}$ might yield different results.

Therefore we cannot separate the analyses of factors 1 and 2. Here, one reasonable approach is to simply compare combinations of the levels of factors 1 and 2.

Example 16.2

Part 1: The Problem and Data

To illustrate the two-factor means model, suppose that North American Oil is also attempting to develop a high-mileage premium gasoline. The company wishes to study the effects of two qualitative factors on the gasoline mileage obtained by an automobile called the GT-500. These factors are premium gasoline type (which has levels D, E, and F) and gasoline additive type (which has levels Q, R, S, and T). Suppose that when a mileage experiment was performed, some GT-500s were not test driven under the proper test conditions. After dropping the unusable data from the data set, the remaining data is given in Table 16.2. Here,

$y_{ij,k}$ = the kth gasoline mileage obtained when using premium gasoline type i and additive type j

Also note that Table 16.2 gives the sample mean $\bar{y}_{ij}$ for each treatment. To check for interaction between premium gasoline type and additive type, we perform the graphical analysis in Figure 16.5.

Since the data patterns in Figure 16.5(a) are not parallel, the change in $\bar{y}_{ij}$ associated with changing the additive type depends substantially on the gasoline

TABLE 16.2 **Data for the premium gasoline mileage experiment**

Premium gasoline type, i	*Gasoline additive type, j*			
	Q	*R*	*S*	*T*
D	$n_{DQ} = 3$ $y_{DQ,1} = 18.0$ $y_{DQ,2} = 18.6$ $y_{DQ,3} = 17.4$ $\bar{y}_{DQ} = 18.0$	$n_{DR} = 2$ $y_{DR,1} = 23.0$ $y_{DR,2} = 22.0$ $\bar{y}_{DR} = 22.5$	$n_{DS} = 2$ $y_{DS,1} = 23.5$ $y_{DS,2} = 22.3$ $\bar{y}_{DS} = 22.9$	$n_{DT} = 0$
E	$n_{EQ} = 2$ $y_{EQ,1} = 23.3$ $y_{EQ,2} = 24.5$ $\bar{y}_{EQ} = 23.9$	$n_{ER} = 3$ $y_{ER,1} = 25.6$ $y_{ER,2} = 24.4$ $y_{ER,3} = 25.0$ $\bar{y}_{ER} = 25.0$	$n_{ES} = 1$ $y_{ES,1} = 23.4$ $\bar{y}_{ES} = 23.4$	$n_{ET} = 2$ $y_{ET,1} = 19.6$ $y_{ET,2} = 20.6$ $\bar{y}_{ET} = 20.1$
F	$n_{FQ} = 0$	$n_{FR} = 3$ $y_{FR,1} = 24.0$ $y_{FR,2} = 23.3$ $y_{FR,3} = 24.7$ $\bar{y}_{FR} = 24.0$	$n_{FS} = 2$ $y_{FS,1} = 23.0$ $y_{FS,2} = 22.0$ $\bar{y}_{FS} = 22.5$	$n_{FT} = 2$ $y_{FT,1} = 18.6$ $y_{FT,2} = 19.8$ $\bar{y}_{FT} = 19.2$

type. For example, when changing from additive type R to additive type S, $\bar{y}_{ij}$ increases when using premium gasoline type D but decreases when using premium gasoline type E or F. Since the data patterns in Figure 16.5(b) are not parallel, the change in $\bar{y}_{ij}$ associated with changing the gasoline type depends substantially on the additive type. For example, when changing from premium gasoline type D to premium gasoline type E, $\bar{y}_{ij}$ increases much more when using additive type Q than it increases when using additive type S. Together, the nonparallel data patterns in these figures suggest that there is substantial interaction between premium gasoline type and additive type.

Part 2: The Two-Factor Means Model

Since there seems to be substantial interaction between premium gasoline type and additive type, it is reasonable to describe $y_{ij,k}$ by the two-factor means model

$$y_{ij,k} = \mu_{ij} + \varepsilon_{ij,k}$$

where

$$\begin{aligned}\mu_{ij} = {} & \beta_{\mathrm{DQ}} + \beta_{\mathrm{DR}} D_{ij,\mathrm{DR}} + \beta_{\mathrm{DS}} D_{ij,\mathrm{DS}} + \beta_{\mathrm{EQ}} D_{ij,\mathrm{EQ}} + \beta_{\mathrm{ER}} D_{ij,\mathrm{ER}} + \beta_{\mathrm{ES}} D_{ij,\mathrm{ES}} \\ & + \beta_{\mathrm{ET}} D_{ij,\mathrm{ET}} + \beta_{\mathrm{FR}} D_{ij,\mathrm{FR}} + \beta_{\mathrm{FS}} D_{ij,\mathrm{FS}} + \beta_{\mathrm{FT}} D_{ij,\mathrm{FT}}\end{aligned}$$

FIGURE 16.5 **Graphical analysis of the premium gasoline mileage data in Table 16.2: Substantial interaction**

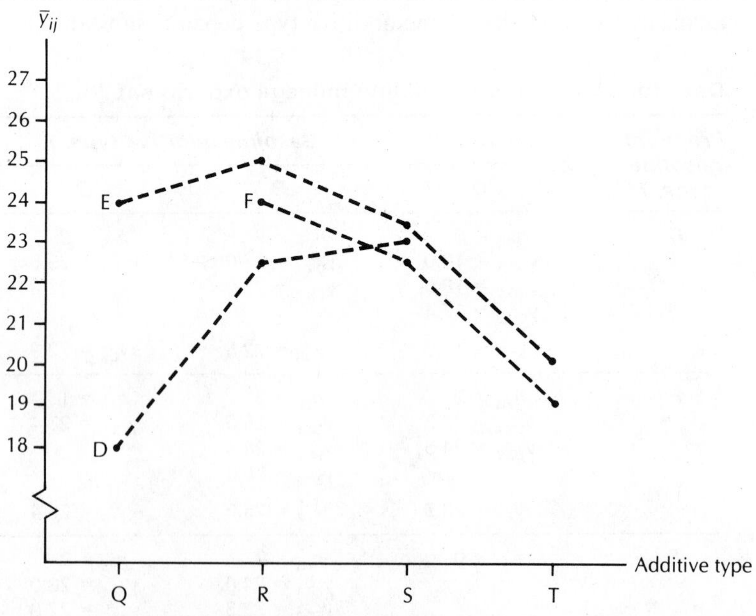

(a) Plot of the change in $\bar{y}_{ij}$ associated with changing the additive type (from Q to R to S to T) for each premium gasoline type (D, E, and F). Note the nonparallel data patterns.

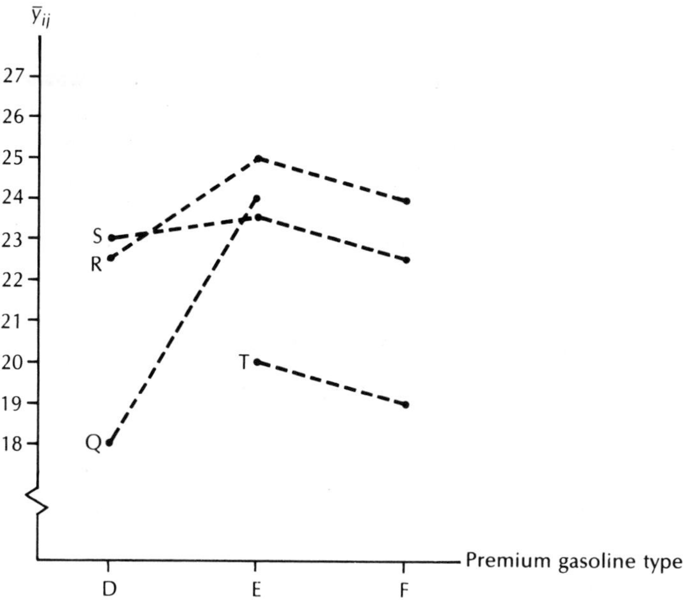

(b) Plot of the change in $\bar{y}_{ij}$ associated with changing the premium gasoline type (from D to E to F) for each additive type (Q, R, S, and T). Note the nonparallel data patterns.

Here, for example,

$$D_{ij,\text{ER}} = \begin{cases} 1 & \text{if } i = \text{E and } j = \text{R; that is, if we are using gasoline} \\ & \text{type E and additive type R} \\ 0 & \text{otherwise} \end{cases}$$

The other dummy variables are defined similarly. Using the definitions of the dummy variables, we find that

$$\mu_{\text{DQ}} = \beta_{\text{DQ}}$$

Also, for any ij not equal to DQ,

$$\mu_{ij} = \beta_{\text{DQ}} + \beta_{ij}$$

This implies that for any ij not equal to DQ,

$$\begin{aligned} \beta_{ij} &= \mu_{ij} - \beta_{\text{DQ}} \\ &= \mu_{ij} - \mu_{\text{DQ}} \end{aligned}$$

= the change in mean gasoline mileage associated with changing from using gasoline type D and additive type Q to using gasoline type i and additive type j

Note that the equation describing μ_{ij} uses the intercept β_{DQ} to describe the effect of μ_{DQ}. This equation also uses a separate dummy variable to describe the effect of each of the treatment means for filled cells ($n_{ij} \neq 0$). However, the equation does not use a dummy variable to describe the effect of any treatment mean for an empty cell ($n_{ij} = 0$). This is because there would be no $y_{ij,k}$ values that we would model by using such a dummy variable. Since this model uses a separate dummy variable to describe each μ_{ij}, we call it the *two-factor means model.* We let SS_{means} and SSE_{means} denote the explained variation and the unexplained variation, respectively, for this model.

Part 3: Estimating the Regression Parameters

To use the means model to carry out a regression analysis of the data in Table 16.2, we use the following column vector $\mathbf{y}$ and matrix $\mathbf{X}$:

$$\begin{matrix} 1 & D_{ij,DR} & D_{ij,DS} & D_{ij,EQ} & D_{ij,ER} & D_{ij,ES} & D_{ij,ET} & D_{ij,FR} & D_{ij,FS} & D_{ij,FT} \end{matrix}$$

$$\mathbf{y} = \begin{bmatrix} y_{DQ,1} \\ y_{DQ,2} \\ y_{DQ,3} \\ y_{DR,1} \\ y_{DR,2} \\ y_{DS,1} \\ y_{DS,2} \\ y_{EQ,1} \\ y_{EQ,2} \\ y_{ER,1} \\ y_{ER,2} \\ y_{ER,3} \\ y_{ES,1} \\ y_{ET,1} \\ y_{ET,2} \\ y_{FR,1} \\ y_{FR,2} \\ y_{FR,3} \\ y_{FS,1} \\ y_{FS,2} \\ y_{FT,1} \\ y_{FT,2} \end{bmatrix} = \begin{bmatrix} 18.0 \\ 18.6 \\ 17.4 \\ 23.0 \\ 22.0 \\ 23.5 \\ 22.3 \\ 23.3 \\ 24.5 \\ 25.6 \\ 24.4 \\ 25.0 \\ 23.4 \\ 19.6 \\ 20.6 \\ 24.0 \\ 23.3 \\ 24.7 \\ 23.0 \\ 22.0 \\ 18.6 \\ 19.8 \end{bmatrix} \quad \mathbf{X} = \begin{bmatrix} 1 & 0 & 0 & 0 & 0 & 0 & 0 & 0 & 0 & 0 \\ 1 & 0 & 0 & 0 & 0 & 0 & 0 & 0 & 0 & 0 \\ 1 & 0 & 0 & 0 & 0 & 0 & 0 & 0 & 0 & 0 \\ 1 & 1 & 0 & 0 & 0 & 0 & 0 & 0 & 0 & 0 \\ 1 & 1 & 0 & 0 & 0 & 0 & 0 & 0 & 0 & 0 \\ 1 & 0 & 1 & 0 & 0 & 0 & 0 & 0 & 0 & 0 \\ 1 & 0 & 1 & 0 & 0 & 0 & 0 & 0 & 0 & 0 \\ 1 & 0 & 0 & 1 & 0 & 0 & 0 & 0 & 0 & 0 \\ 1 & 0 & 0 & 1 & 0 & 0 & 0 & 0 & 0 & 0 \\ 1 & 0 & 0 & 0 & 1 & 0 & 0 & 0 & 0 & 0 \\ 1 & 0 & 0 & 0 & 1 & 0 & 0 & 0 & 0 & 0 \\ 1 & 0 & 0 & 0 & 1 & 0 & 0 & 0 & 0 & 0 \\ 1 & 0 & 0 & 0 & 0 & 1 & 0 & 0 & 0 & 0 \\ 1 & 0 & 0 & 0 & 0 & 0 & 1 & 0 & 0 & 0 \\ 1 & 0 & 0 & 0 & 0 & 0 & 1 & 0 & 0 & 0 \\ 1 & 0 & 0 & 0 & 0 & 0 & 0 & 1 & 0 & 0 \\ 1 & 0 & 0 & 0 & 0 & 0 & 0 & 1 & 0 & 0 \\ 1 & 0 & 0 & 0 & 0 & 0 & 0 & 1 & 0 & 0 \\ 1 & 0 & 0 & 0 & 0 & 0 & 0 & 0 & 1 & 0 \\ 1 & 0 & 0 & 0 & 0 & 0 & 0 & 0 & 1 & 0 \\ 1 & 0 & 0 & 0 & 0 & 0 & 0 & 0 & 0 & 1 \\ 1 & 0 & 0 & 0 & 0 & 0 & 0 & 0 & 0 & 1 \end{bmatrix}$$

Using **y** and **X**, we obtain the prediction equation

$$\begin{aligned}\hat{y}_{ij} &= b_{DQ} + b_{DR}D_{ij,DR} + b_{DS}D_{ij,DS} + b_{EQ}D_{ij,EQ} + b_{ER}D_{ij,ER} \\ &\quad + b_{ES}D_{ij,ES} + b_{ET}D_{ij,ET} + b_{FR}D_{ij,FR} + b_{FS}D_{ij,FS} + b_{FT}D_{ij,FT} \\ &= 18.0 + 4.5D_{ij,DR} + 4.9D_{ij,DS} + 5.9D_{ij,EQ} + 7.0D_{ij,ER} + 5.4D_{ij,ES} \\ &\quad + 2.1D_{ij,ET} + 6.0D_{ij,FR} + 4.5D_{ij,FS} + 1.2D_{ij,FT}\end{aligned}$$

The SAS output for this problem is given in Figure 16.6.

FIGURE 16.6 **SAS output of a two-factor means model analysis of the premium gasoline mileage data in Table 16.2**

SAS

GENERAL LINEAR MODELS PROCEDURE

DEPENDENT VARIABLE: MILEAGE

SOURCE	DF	SUM OF SQUARES	MEAN SQUARE	F VALUE	PR > F	R-SQUARE	C.V.
MODEL	9	121.39272727[c]	13.48808081[d]	26.62[a]	0.0001[b]	0.952304	3.2182
ERROR	12	6.08000000[e]	0.50666667[f]		ROOT MSE		MILEAGE MEAN
CORRECTED TOTAL	21	127.47272727			0.71180522[g]		22.11818182

SOURCE	DF	TYPE I SS	F VALUE	PR > F	DF	TYPE III SS	F VALUE	PR > F
GASTYPE*ADDTYPE	9	121.39272727	26.62	0.0001	9	121.39272727	26.62	0.0001

PARAMETER	ESTIMATE[h]	T FOR HO: PARAMETER=0[i]	PR > |T|[j]	STD ERROR OF ESTIMATE[k]
MUER-MUDS	2.10000000	3.23	0.0072	0.64978629

OBSERVATION	OBSERVED VALUE	PREDICTED VALUE	RESIDUAL	LOWER 95% CL FOR MEAN	UPPER 95% CL FOR MEAN
1	18.00000000	18.00000000	0.00000000	17.10459213	18.89540787
2	18.60000000	18.00000000	0.60000000	17.10459213	18.89540787
3	17.40000000	18.00000000	-0.60000000	17.10459213	18.89540787
4	23.00000000	22.50000000	0.50000000	21.40335381	23.59664619
5	22.00000000	22.50000000	-0.50000000	21.40335381	23.59664619
6	23.50000000	22.90000000	0.60000000	21.80335381	23.99664619
7	22.30000000	22.90000000	-0.60000000	21.80335381	23.99664619
8	23.30000000	23.90000000	-0.60000000	22.80335381	24.99664619
9	24.50000000	23.90000000	0.60000000	22.80335381	24.99664619
10	25.60000000	25.00000000[l]	0.60000000	[24.10459213	25.89540787][m]
11	24.40000000	25.00000000	-0.60000000	24.10459213	25.89540787
12	25.00000000	25.00000000	0.00000000	24.10459213	25.89540787
13	23.40000000	23.40000000	0.00000000	21.84910808	24.95089192
14	19.60000000	20.10000000	-0.50000000	19.00335381	21.19664619
15	20.60000000	20.10000000	0.50000000	19.00335381	21.19664619
16	24.00000000	24.00000000	0.00000000	23.10459213	24.89540787
17	23.30000000	24.00000000	-0.70000000	23.10459213	24.89540787
18	24.70000000	24.00000000	0.70000000	23.10459213	24.89540787
19	23.00000000	22.50000000	0.50000000	21.40335381	23.59664619
20	22.00000000	22.50000000	-0.50000000	21.40335381	23.59664619
21	18.60000000	19.20000000	-0.60000000	18.10335381	20.29664619
22	19.80000000	19.20000000	0.60000000	18.10335381	20.29664619

[a] *F(means model)* [b] *Prob-value for F(means model)* [c] SS_{means} [d] MS_{means} [e] SSE_{means} [f] s^2_{means} [g] s_{means}
[h] $\lambda' b$ [i] t [j] *Prob-value related to t* [k] $s_{means}\sqrt{\lambda'(X'X)^{-1}\lambda}$ [l] $\hat{y}_{ER}$ [m] *95% confidence interval for* μ_{ER}
[n] *95% prediction interval for* $y_{ER,0}$

(continues)

FIGURE 16.6 **Continued**

OBSERVATION	OBSERVED VALUE	PREDICTED VALUE	RESIDUAL	LOWER 95% CL INDIVIDUAL	UPPER 95% CL INDIVIDUAL
1	18.00000000	18.00000000	0.00000000	16.20918345	19.79081655
2	18.60000000	18.00000000	0.60000000	16.20918345	19.79081655
3	17.40000000	18.00000000	-0.60000000	16.20918345	19.79081655
4	23.00000000	22.50000000	0.50000000	20.60055221	24.39944779
5	22.00000000	22.50000000	-0.50000000	20.60055221	24.39944779
6	23.50000000	22.90000000	0.60000000	21.00055221	24.79944779
7	22.30000000	22.90000000	-0.60000000	21.00055221	24.79944779
8	23.30000000	23.90000000	-0.60000000	22.00055221	25.79944779
9	24.50000000	23.90000000	0.60000000	22.00055221	25.79944779
10	25.60000000	25.00000000[l]	0.60000000	[23.20918345	26.79081655][n]
11	24.40000000	25.00000000	-0.60000000	23.20918345	26.79081655
12	25.00000000	25.00000000	0.00000000	23.20918345	26.79081655
13	23.40000000	23.40000000	0.00000000	21.20670662	25.59329338
14	19.60000000	20.10000000	-0.50000000	18.20055221	21.99944779
15	20.60000000	20.10000000	0.50000000	18.20055221	21.99944779
16	24.00000000	24.00000000	0.00000000	22.20918345	25.79081655
17	23.30000000	24.00000000	-0.70000000	22.20918345	25.79081655
18	24.70000000	24.00000000	0.70000000	22.20918345	25.79081655
19	23.00000000	22.50000000	0.50000000	20.60055221	24.39944779
20	22.00000000	22.50000000	-0.50000000	20.60055221	24.39944779
21	18.60000000	19.20000000	-0.60000000	17.30055221	21.09944779
22	19.80000000	19.20000000	0.60000000	17.30055221	21.09944779

Part 4: Comparing Means by Using the Overall F-Test

The means model implies that $\beta_{DQ} = \mu_{DQ}$ and that for any ij not equal to DQ, $\beta_{ij} = \mu_{ij} - \mu_{DQ}$. It follows that

$$H_0\text{: All } \beta_{ij} \text{ (except } \beta_{DQ}\text{) equal zero}$$

is equivalent to

$$H_0\text{: All } \{\mu_{ij} - \mu_{DQ}\} \text{ equal zero}$$

which is equivalent to

$$H_0\text{: All treatment means } \mu_{ij} \text{ are equal to each other}$$

We test H_0 versus

$$H_1\text{: At least two of the treatment means } \mu_{ij} \text{ differ from each other}$$

Since the prob-value for F(means model) equals .0001 (see Figure 16.6), we can reject H_0 with $\alpha = .05$. We conclude that there are statistically significant differences between at least two of the treatment means.

Part 5: Estimating a Difference Between Treatment Means Using a Linear Combination of Regression Parameters

We note that $\bar{y}_{ER} = 25$ is the largest sample mean in Table 16.2. Therefore it is reasonable to believe that gasoline type E and additive type R maximize mean gasoline mileage. Supposing that North American Oil has been producing gasoline type D with additive type S, the company wishes to estimate $\mu_{ER} - \mu_{DS}$. The means model implies that

$$\begin{aligned}
\mu_{ER} - \mu_{DS} &= (\beta_{DQ} + \beta_{ER}) - (\beta_{DQ} + \beta_{DS}) \\
&= \beta_{ER} - \beta_{DS} \\
&= 0 \cdot \beta_{DQ} + 0 \cdot \beta_{DR} + (-1) \cdot \beta_{DS} + 0 \cdot \beta_{EQ} + 1 \cdot \beta_{ER} \\
&\quad + 0 \cdot \beta_{ES} + 0 \cdot \beta_{ET} + 0 \cdot \beta_{FR} + 0 \cdot \beta_{FS} + 0 \cdot \beta_{FT} \\
&= [0 \quad 0 \quad -1 \quad 0 \quad 1 \quad 0 \quad 0 \quad 0 \quad 0 \quad 0] \begin{bmatrix} \beta_{DQ} \\ \beta_{DR} \\ \beta_{DS} \\ \beta_{EQ} \\ \beta_{ER} \\ \beta_{ES} \\ \beta_{ET} \\ \beta_{FR} \\ \beta_{FS} \\ \beta_{FT} \end{bmatrix} \\
&= \boldsymbol{\lambda}'\boldsymbol{\beta}
\end{aligned}$$

where $\boldsymbol{\lambda}' = [0 \quad 0 \quad -1 \quad 0 \quad 1 \quad 0 \quad 0 \quad 0 \quad 0 \quad 0]$. It follows that a point estimate of

$$\mu_{ER} - \mu_{DS} = \beta_{ER} - \beta_{DS}$$

is

$$b_{ER} - b_{DS} = 7 - 4.9 = 2.1$$

Note that $\mu_{ER} - \mu_{DS}$ is denoted MUER − MUDS on the SAS output in Figure 16.6. We see that the prob-value for testing

$$H_0\colon \mu_{ER} - \mu_{DS} = 0$$

versus

$$H_1\colon \mu_{ER} - \mu_{DS} \neq 0$$

is .0072. Thus we reject H_0 with $\alpha = .05$. We conclude that we have very strong evidence that μ_{ER} is greater than μ_{DS}. To assess the practical importance of $\mu_{ER} - \mu_{DS}$, we calculate the 95% confidence interval for this difference. This interval is

$$\begin{aligned}
[\boldsymbol{\lambda}'\mathbf{b} \pm t_{[.025]}^{(22-10)} s_{\text{means}} \sqrt{\boldsymbol{\lambda}'(\mathbf{X}'\mathbf{X})^{-1}\boldsymbol{\lambda}}] &= [2.1 \pm 2.179(.7118)\sqrt{.83333}] \\
&= [2.1 \pm 1.4159] \\
&= [.6841, 3.5159]
\end{aligned}$$

The interval says that we can be 95 percent confident that μ_{ER}, the mean gasoline mileage obtained by using gasoline type E and additive type R, is between .6841 mpg and 3.5159 mpg greater than μ_{DS}, the mean gasoline mileage obtained by using gasoline type D and additive type S.

Part 6: Using the Means Model to Estimate and Predict

Finally, we can describe an individual GT-500 gasoline mileage obtained by using gasoline type E and additive type R by the equation

$$\begin{aligned} y_{ER,0} &= \mu_{ER} + \varepsilon_{ER,0} \\ &= \beta_{DQ} + \beta_{DR} D_{ij,DR} + \beta_{DS} D_{ij,DS} + \beta_{EQ} D_{ij,EQ} + \beta_{ER} D_{ij,ER} \\ &\quad + \beta_{ES} D_{ij,ES} + \beta_{ET} D_{ij,ET} + \beta_{FR} D_{ij,FR} + \beta_{FS} D_{ij,FS} + \beta_{FT} D_{ij,FT} + \varepsilon_{ER,0} \\ &= \beta_{DQ} + \beta_{DR}(0) + \beta_{DS}(0) + \beta_{EQ}(0) + \beta_{ER}(1) + \beta_{ES}(0) \\ &\quad + \beta_{ET}(0) + \beta_{FR}(0) + \beta_{FS}(0) + \beta_{FT}(0) + \varepsilon_{ER,0} \\ &= \beta_{DQ} + \beta_{ER} + \varepsilon_{ER,0} \end{aligned}$$

Therefore

$$\begin{aligned} \hat{y}_{ER} &= b_{DQ} + b_{ER} \\ &= 18.0 + 7.0 \\ &= 25.0 \end{aligned}$$

is the point estimate of μ_{ER} and the point prediction of $y_{ER,0}$. Moreover, the equation describing $y_{ER,0}$ implies that

$$\mathbf{x}'_{ER} = [1 \quad 0 \quad 0 \quad 0 \quad 1 \quad 0 \quad 0 \quad 0 \quad 0 \quad 0]$$

Therefore a 95% confidence interval for μ_{ER} is

$$\begin{aligned} [\hat{y}_{ER} \pm t_{[.025]}^{(22-10)} s_{\text{means}} \sqrt{\mathbf{x}'_{ER}(\mathbf{X}'\mathbf{X})^{-1}\mathbf{x}_{ER}}] &= [25.0 \pm 2.179(.7118)\sqrt{\tfrac{1}{3}}] \\ &= [25.0 \pm .8955] \\ &= [24.1045, 25.8955] \end{aligned}$$

This interval says that we can be 95 percent confident that μ_{ER}, the mean gasoline mileage obtained by using gasoline type E and additive type R, is between 24.1045 mpg and 25.8955 mpg. Also, a 95% prediction interval for $y_{ER,0}$ is

$$\begin{aligned} [\hat{y}_{ER} \pm t_{[.025]}^{(22-10)} s_{\text{means}} \sqrt{1 + \mathbf{x}'_{ER}(\mathbf{X}'\mathbf{X})^{-1}\mathbf{x}_{ER}}] &= \left[25.0 \pm 2.179(.7118)\sqrt{1 + \frac{1}{3}}\right] \\ &= [25.0 \pm 1.7910] \\ &= [23.209, 26.791] \end{aligned}$$

This interval says that we can be 95 percent confident that $y_{ER,0}$, an individual GT-500 gasoline mileage obtained by using gasoline type E and additive type R, will be between 23.209 mpg and 26.791 mpg. Note that both of the above intervals are given in the SAS output in Figure 16.6.

In the preceding example,

$$\begin{aligned} \hat{y}_{ER} &= b_{DQ} + b_{ER} \\ &= 25.0 \end{aligned}$$

is the least squares point estimate of $\mu_{ER} = \beta_{DQ} + \beta_{ER}$ obtained by using the means model. This estimate equals $\bar{y}_{ER} = 25.0$, the mean of the sample of $n_{ER} = 3$ GT-500 mileages obtained by using gasoline type E and additive type R (see Table 16.2). This is because it can be proved that, in general, the *least squares point estimate of the treatment mean* μ_{ij} *is the sample mean* $\bar{y}_{ij}$ *when we use the means model to analyze a factorial experiment.* In contrast, recall that in Example 16.1,

$$\begin{aligned}\hat{y}_{BO} &= \hat{\mu} + \hat{\alpha}_B + \hat{\gamma}_O \\ &= 26.9014\end{aligned}$$

is the least squares point estimate of $\mu_{BO} = \mu + \alpha_B + \gamma_O$. This estimate does not equal $\bar{y}_{BO} = 27.0$, the mean of the sample of $n_{BO} = 3$ mileages obtained by using gasoline type B and additive type O (see Table 16.1). This difference illustrates that the *least squares point estimate of* μ_{ij} *usually is not* $\bar{y}_{ij}$ *when we use the* (α, γ) *model to analyze a factorial experiment.*

We can perform an hypothesis test to investigate the presence of interaction by using an F statistic that employs SSE_{means} and $SSE_{(\alpha,\gamma)}$. This statistic compares the SSE for the two-factor means model, which assumes that there is interaction between factors 1 and 2, with the SSE for the (α, γ) model, which assumes no interaction. We now present this test.

Testing for Interaction Between Factors 1 and 2

Assume that the inference assumptions are satisfied and consider testing

H_0: No interaction exists between factors 1 and 2

versus

H_1: Interaction does exist between factors 1 and 2

Recall that SSE_{means} and $SSE_{(\alpha,\gamma)}$ denote the unexplained variations for, respectively, the two-factor means model and the (α, γ) model. Define k_{means} and $k_{(\alpha,\gamma)}$ to be the number of parameters in the two-factor means model and the (α,γ) model, respectively. Then we define the F(interaction) statistic to be

$$\begin{aligned}F(\text{interaction}) &= \frac{MS_{\text{interaction}}}{MSE_{\text{means}}} = \frac{SS_{\text{interaction}}/(k_{\text{means}} - k_{(\alpha,\gamma)})}{SSE_{\text{means}}/(n - k_{\text{means}})} \\ &= \frac{(SSE_{(\alpha,\gamma)} - SSE_{\text{means}})/(k_{\text{means}} - k_{(\alpha,\gamma)})}{SSE_{\text{means}}/(n - k_{\text{means}})}\end{aligned}$$

Also, define the prob-value to be the area to the right of F(interaction) under the curve of the F-distribution having $(k_{\text{means}} - k_{(\alpha,\gamma)})$ and $(n - k_{\text{means}})$ degrees of freedom. Then we can reject H_0 in favor of H_1 by setting the probability of a Type I error equal to α if and only if either of the following equivalent conditions holds:

$$F(\text{interaction}) > F_{[\alpha]}^{(k_{\text{means}} - k_{(\alpha,\gamma)}, n - k_{\text{means}})}$$

or

$$\text{prob-value} < \alpha$$

The first condition, which says that we should reject H_0 in favor of H_1 if F(interaction) is large, is intuitively reasonable. This is because a large value of F(interaction) would be caused by a large value of

$$SS_{\text{interaction}} = SSE_{(\alpha,\gamma)} - SSE_{\text{means}}$$

Such a value of $SS_{\text{interaction}}$ would be obtained if the two-factor means model, which models interaction, yields an unexplained variation (SSE_{means}) much smaller than the unexplained variation ($SSE_{(\alpha,\gamma)}$) yielded by the (α, γ) model, which does not model interaction. However, although this test is reasonable, it does not provide an exact test for interaction for certain types of incomplete factorial experiments (see Searle (1971)). In spite of this, it is common practice to use this test for incomplete factorial experiments.

Example 16.3 Consider Table 16.2. For i = D, E, and F and j = Q, R, S, and T the two-factor means model describing $y_{ij,k}$ is

$$\begin{aligned} y_{ij,k} &= \mu_{ij} + \varepsilon_{ij,k} \\ &= \beta_{\text{DQ}} + \beta_{\text{DR}} D_{ij,\text{DR}} + \beta_{\text{DS}} D_{ij,\text{DS}} \\ &\quad + \beta_{\text{EQ}} D_{ij,\text{EQ}} + \beta_{\text{ER}} D_{ij,\text{ER}} + \beta_{\text{ES}} D_{ij,\text{ES}} + \beta_{\text{ET}} D_{ij,\text{ET}} \\ &\quad + \beta_{\text{FR}} D_{ij,\text{FR}} + \beta_{\text{FS}} D_{ij,\text{FS}} + \beta_{\text{FT}} D_{ij,\text{FT}} + \varepsilon_{ij,k} \end{aligned}$$

The (α, γ) model describing $y_{ij,k}$ is

$$\begin{aligned} y_{ij,k} &= \mu_{ij} + \varepsilon_{ij,k} \\ &= \mu + \alpha_i + \gamma_j + \varepsilon_{ij,k} \\ &= \mu + \alpha_{\text{E}} D_{i,\text{E}} + \alpha_{\text{F}} D_{i,\text{F}} + \gamma_{\text{R}} D_{j,\text{R}} + \gamma_{\text{S}} D_{j,\text{S}} + \gamma_{\text{T}} D_{j,\text{T}} + \varepsilon_{ij,k} \end{aligned}$$

When we use these models to perform regression analyses of the data in Table 16.2, we obtain

$$SSE_{\text{means}} = 6.08 \quad \text{and} \quad SSE_{(\alpha,\gamma)} = 22.4096$$

To test

H_0: No interaction exists between premium gasoline type and additive type

versus

H_1: Interaction does exist between premium gasoline type and additive type

we use

$$\begin{aligned} F(\text{interaction}) &= \frac{MS_{\text{interaction}}}{MSE_{\text{means}}} \\ &= \frac{SS_{\text{interaction}}/(k_{\text{means}} - k_{(\alpha,\gamma)})}{SSE_{\text{means}}/(n - k_{\text{means}})} \\ &= \frac{\{SSE_{(\alpha,\gamma)} - SSE_{\text{means}}\}/(k_{\text{means}} - k_{(\alpha,\gamma)})}{SSE_{\text{means}}/(n - k_{\text{means}})} \end{aligned}$$

$$= \frac{\{22.4096 - 6.08\}/(10 - 6)}{6.08/(22 - 10)}$$
$$= \frac{16.3296/4}{6.08/12}$$
$$= \frac{4.0824}{.5067}$$
$$= 8.0568$$

and

$$\text{prob-value} = .0021$$

Since this prob-value is less than .05, we reject H_0 with $\alpha = .05$. This test and the graphical analysis of Figure 16.5 indicate that substantial interaction exists between premium gasoline type and additive type. Note, however, that Figure 16.5 also indicates that if we eliminate all sample information concerning gasoline type D, then little interaction remains. Thus it would probably be appropriate to use the (α, γ) model to analyze the data in Table 16.2 for gasoline types E and F and additive types Q, R, S, and T. However, if we wish to include gasoline type D in our study, we should use the two-factor means model.

If we have performed a *connected* incomplete factorial experiment, it is possible to obtain least squares point estimates of the parameters in the (α, γ, θ) model

$$\begin{aligned} y_{ij,k} &= \mu_{ij} + \varepsilon_{ij,k} \\ &= \mu + \alpha_i + \gamma_j + \theta_{ij} + \varepsilon_{ij,k} \end{aligned}$$

In addition, we can (approximately) test for interaction between factors 1 and 2 by testing

H_0: All parameters θ_{ij} equal zero (no interaction exists)

versus

H_1: At least one θ_{ij} does not equal zero (interaction does exist)

Specifically, we can reject H_0 in favor of H_1 by setting the probability of a Type I error equal to α if and only if

$$F(\theta|\alpha, \gamma) = \frac{\{SSE_{(\alpha,\gamma)} - SSE_{(\alpha,\gamma,\theta)}\}/(f - a - b + 1)}{SSE_{(\alpha,\gamma,\theta)}/(n - f)}$$

is greater than $F_{[\alpha]}^{(f-a-b+1,n-f)}$. Here, f is the total number of filled cells, and n is the total number of observations in the two-factor data set.

It can be shown that using the previously defined (see Section 15.3) 1, −1, 0 dummy variables to define α_i, γ_j, and θ_{ij} in the (α, γ, θ) model is appropriate for some connected incomplete factorial experiments. However, it is necessary to use other types of dummy variables for other connected incomplete factorial experiments. This all depends on the exact nature of the data. We will not discuss this

FIGURE 16.7 **SAS output that results from using the (α, γ, θ) model and the data in Table 16.2 to test for interaction between premium gasoline type and additive type**

```
                                        SAS

                           GENERAL LINEAR MODELS PROCEDURE
DEPENDENT VARIABLE: MILEAGE
SOURCE            DF   SUM OF SQUARES   MEAN SQUARE   F VALUE      PR > F   R-SQUARE          C.V.
MODEL              9     121.39272727   13.48808081     26.62      0.0001   0.952304        3.2182
ERROR             12       6.08000000    0.50666667               ROOT MSE             MILEAGE MEAN
CORRECTED TOTAL   21     127.47272727                           0.71180522              22.11818182

SOURCE            DF      TYPE I SS    F VALUE   PR > F    DF    TYPE III SS    F VALUE     PR > F
GASTYPE            2    25.58415584      25.25   0.0001     2    25.67470209      25.34     0.0001
ADDTYPE            3    79.47898655      52.29   0.0001     3    71.26827409      46.89     0.0001
GASTYPE*ADDTYPE    4    16.32958488       8.06   0.0021     4    16.32958488       8.06[a]   0.0021[b]
```

[a] $F(\theta|\alpha, \gamma)$ [b] *Prob-value for* $F(\theta|\alpha, \gamma)$

point further, but in Figure 16.7 we present the SAS output resulting from using the (α, γ, θ) model and the data in Table 16.2 to test for interaction between premium gasoline type and additive type. Note that $F(\theta|\alpha, \gamma)$ and its associated prob-value in Figure 16.7 are precisely the same as F(interaction) and its associated prob-value in Example 16.3.

A SUMMARY OF THE ANALYSIS OF INCOMPLETE DATA

We begin the analysis of incomplete data by using the F(means model) statistic (illustrated in Example 16.2) to test the null hypothesis that all treatment means μ_{ij} are equal. If we reject this null hypothesis, we certainly wish to continue the analysis. If we cannot reject the null hypothesis, it is possible that neither factor 1 nor factor 2 nor the interaction between these factors is important. However, many practitioners suggest that the analysis should be continued anyway. If we continue the analysis, the next step is to check for interaction between factors 1 and 2. We can use theoretical knowledge of the problem under study, graphical analysis of the observed data, or an F-test. Here we employ the F(interaction) statistic (illustrated in Example 16.3) or, equivalently, the $F(\theta|\alpha, \gamma)$ statistic (see Figure 16.7).

If we conclude that interaction exists, we should continue the analysis by using the two-factor means model (as discussed in Example 16.2). If we conclude that little or no interaction exists, then it might be appropriate to use the (α, γ) model (as discussed in Example 16.1). However, some statisticians feel that the (α, γ) model should be used only if there are strong theoretical reasons to believe that no interaction exists between factors 1 and 2 in the problem under study. These statisticians regard graphical analysis and hypothesis testing only as verifications of this theoretical knowledge. Furthermore, they believe that it is always possible that "a little interaction" exists. Therefore in the absence of theoretical knowledge against interaction they use the two-factor means model regardless of the results obtained in checking for interaction. The debate as to whether to use the (α, γ) model when "checks for interaction" (as opposed to theoretical knowledge) tell us that little or no interaction exists is an interesting one. In fact, statisticians who argue for the use of the (α, γ) model feel that when checks indicate that little or no interaction exists in any type of factorial experiment, then it does not make sense to test for main effects by using any model that assumes that there is interaction. Therefore these statisticians would argue against the common practice of using the (α, γ, θ) model to test for main effects in a complete factorial experiment when checks suggest that little or no interaction exists. They would instead use the (α, γ) model. Their logic certainly has some merit (and is more consistent with the model-building approach employed in regression analysis). However, we do not wish to take sides because both points of view have merit.

To conclude this section, recall that we can use the (α, γ, θ) model to (approximately) test for *interaction* in a *connected incomplete* factorial experiment. However, it can be difficult or impossible to use this model to test for *main effects* in a connected (or any other) *incomplete* factorial experiment. This depends on which cells are empty. The reasons for this are based on the fact that not all of the $\mu_{i\cdot}$ values and $\mu_{\cdot j}$ values can be estimated for an incomplete factorial experiment. Therefore when little or no interaction exists in an incomplete factorial experiment, we must choose between the (α, γ) model and the two-factor means model. Clearly, the two-factor means model is the most similar to the (α, γ, θ) model because both of these models assume that there is interaction between factors 1 and 2.

16.2 FIXED, RANDOM, AND MIXED MODELS

Consider performing a balanced complete factorial experiment. The methods of Section 15.2 are appropriate when (1) the levels $1, 2, \ldots, a$ of factor 1 are the only levels of factor 1 that we wish to study and (2) the levels $1, 2, \ldots, b$ of factor 2 are the only levels of factor 2 that we wish to study. This is the *fixed model* case. The ANOVA table summarizing much of the methodology of Section 15.2 that is used to analyze this case is given in Table 16.3.

If we randomly select the levels $1, 2, \ldots, a$ of factor 1 from a large population of levels of factor 1 and randomly select the levels $1, 2, \ldots, b$ of factor 2 from a large population of levels of factor 2, this is the *random model* case. For example, we might randomly select $a = 4$ machine operators from the population of all operators who work in a factory and might randomly select $b = 5$ machines of a certain type from the population of all such machines in the factory. Suppose that we wish to employ these randomly selected operators and machines to make statistical inferences about the effects of the many machine operators and machines in the factory on the number of items produced in a day. We would analyze the

TABLE 16.3 ANOVA table for the fixed two-factor model

Source	Degrees of freedom	Sum of squares	Mean square	F statistic	E(mean square)
Factor 1	$a-1$	$SS_{\text{factor 1}}$	$MS_{\text{factor 1}}$	$F_{BCF}(\text{factor 1}) = \dfrac{MS_{\text{factor 1}}}{MSE}$	$\sigma^2 + \dfrac{bm}{a-1}\sum_{i=1}^{a}(\mu_{i\cdot} - \mu)^2$
Factor 2	$b-1$	$SS_{\text{factor 2}}$	$MS_{\text{factor 2}}$	$F_{BCF}(\text{factor 2}) = \dfrac{MS_{\text{factor 2}}}{MSE}$	$\sigma^2 + \dfrac{am}{b-1}\sum_{j=1}^{b}(\mu_{\cdot j} - \mu)^2$
Interaction	$(a-1)(b-1)$	$SS_{\text{interaction}}$	$MS_{\text{interaction}}$	$F(\text{interaction}) = \dfrac{MS_{\text{interaction}}}{MSE}$	$\sigma^2 + \dfrac{m}{(a-1)(b-1)} \times \sum_{i=1}^{a}\sum_{j=1}^{b}(\mu_{ij} - \mu_{i\cdot} - \mu_{\cdot j} + \mu)^2$
Error	$ab(m-1)$	SSE	MSE		σ^2

Notes:

1. $\mu_{i\cdot} = \dfrac{\sum_{j=1}^{b}\mu_{ij}}{b}$, $\mu_{\cdot j} = \dfrac{\sum_{i=1}^{a}\mu_{ij}}{a}$, $\mu = \dfrac{\sum_{i=1}^{a}\sum_{j=1}^{b}\mu_{ij}}{ab}$
2. If $F(\text{interaction}) > F_{[\alpha]}^{((a-1)(b-1), ab(m-1))}$, reject H_0: No interaction exists between factor 1 and factor 2
3. If $F_{BCF}(\text{factor 1}) > F_{[\alpha]}^{(a-1, ab(m-1))}$, reject H_0: $\mu_{1\cdot} = \mu_{2\cdot} = \cdots = \mu_{a\cdot}$
4. If $F_{BCF}(\text{factor 2}) > F_{[\alpha]}^{(b-1, ab(m-1))}$, reject H_0: $\mu_{\cdot 1} = \mu_{\cdot 2} = \cdots = \mu_{\cdot b}$
5. A $100(1-\alpha)\%$ confidence interval for $\mu_{i\cdot} - \mu_{i'\cdot}$ is $\left[\bar{y}_{i\cdot} - \bar{y}_{i'\cdot} \pm t_{[\alpha/2]}^{(ab(m-1))}\sqrt{MSE\left(\dfrac{2}{bm}\right)}\right]$

data by using the random model

$$y_{ij,k} = \mu + \alpha_i + \gamma_j + \theta_{ij} + \varepsilon_{ij,k}$$

This model says that $y_{ij,k}$, the number of items produced in day k by the ith randomly selected machine operator using the jth randomly selected machine is the sum of

1. a constant μ.
2. The main effect α_i of operator i on output, where α_i is assumed to have been randomly selected from a population of operator effects that is normally distributed with mean zero and variance σ_α^2.
3. The main effect γ_j of machine j on output, where γ_j is assumed to have been randomly selected from a population of machine effects that is normally distributed with mean zero and variance σ_γ^2.
4. The interaction effect θ_{ij}, which is assumed to have been randomly selected from a population of interaction effects that is normally distributed with mean zero and variance $\sigma_{\alpha\gamma}^2$.
5. A random error component $\varepsilon_{ij,k}$, which is assumed to have been randomly selected from a population of random error components that is normally distributed with mean zero and variance σ^2.

If we further assume that all of the α_i, γ_j, θ_{ij}, and $\varepsilon_{ij,k}$ values are statistically independent, then we should analyze the data as shown in Table 16.4. Here, $MS_{\text{factor 1}}$, $MS_{\text{factor 2}}$, $MS_{\text{interaction}}$, and MSE are as defined in Section 15.2. Then,

1. By testing $H_0: \sigma^2_{\alpha\gamma} = 0$, we can test for interaction between factors 1 and 2.
2. By testing $H_0: \sigma^2_{\alpha} = 0$, we can test the significance of factor 1.
3. By testing $H_0: \sigma^2_{\gamma} = 0$, we can test the significance of factor 2.

Note that the main effects tests of 2 and 3 above should be carried out whether or not we reject $H_0: \sigma^2_{\alpha\gamma} = 0$ because all tests concern variances. Furthermore, by finding point estimates of the *variance components* $\sigma^2_{\alpha\gamma}$, σ^2_{α}, and σ^2_{γ} we can determine the main sources of variation in the $y_{ij,k}$ values (see Table 16.4).

If the levels 1, 2, . . . , a of factor 1 are the only levels of factor 1 we wish to study and if we randomly select the levels 1, 2, . . . , b of factor 2 from a large population

TABLE 16.4 **ANOVA table for the random two-factor model**

Source	*Degrees of freedom*	*Sum of squares*	*Mean square*	*F statistic*	*E(mean square)*
Factor 1	$a - 1$	$SS_{\text{factor 1}}$	$MS_{\text{factor 1}}$	$F_{\text{RBCF}}(\text{factor 1}) = \dfrac{MS_{\text{factor 1}}}{MS_{\text{interaction}}}$	$\sigma^2 + mb\sigma^2_{\alpha} + m\sigma^2_{\alpha\gamma}$
Factor 2	$b - 1$	$SS_{\text{factor 2}}$	$MS_{\text{factor 2}}$	$F_{\text{RBCF}}(\text{factor 2}) = \dfrac{MS_{\text{factor 2}}}{MS_{\text{interaction}}}$	$\sigma^2 + ma\sigma^2_{\gamma} + m\sigma^2_{\alpha\gamma}$
Interaction	$(a - 1)(b - 1)$	$SS_{\text{interaction}}$	$MS_{\text{interaction}}$	$F_{\text{RBCF}}(\text{interaction}) = \dfrac{MS_{\text{interaction}}}{MSE}$	$\sigma^2 + m\sigma^2_{\alpha\gamma}$
Error	$ab(m - 1)$	SSE	MSE		σ^2

Notes:

1. If $F_{\text{RBCF}}(\text{interaction}) > F_{[\alpha]}^{((a-1)(b-1),ab(m-1))}$, reject $H_0: \sigma^2_{\alpha\gamma} = 0$ (no interaction).
2. If $F_{\text{RBCF}}(\text{factor 1}) > F_{[\alpha]}^{(a-1,(a-1)(b-1))}$, reject $H_0: \sigma^2_{\alpha} = 0$.
3. If $F_{\text{RBCF}}(\text{factor 2}) > F_{[\alpha]}^{(b-1,(a-1)(b-1))}$, reject $H_0: \sigma^2_{\gamma} = 0$.
4. Since $E(MSE) = \sigma^2$, a point estimate of σ^2 is MSE.
5. Since $E(MS_{\text{interaction}}) = \sigma^2 + m\sigma^2_{\alpha\gamma}$, a point estimate of $\sigma^2_{\alpha\gamma}$ is

$$\frac{MS_{\text{interaction}} - MSE}{m}$$

6. Since $E(MS_{\text{factor 1}}) = \sigma^2 + mb\sigma^2_{\alpha} + m\sigma^2_{\alpha\gamma}$, a point estimate of σ^2_{α} is

$$\frac{MS_{\text{factor 1}} - MSE - m\left[\dfrac{MS_{\text{interaction}} - MSE}{m}\right]}{mb} = \frac{MS_{\text{factor 1}} - MS_{\text{interaction}}}{mb}$$

7. Since $E(MS_{\text{factor 2}}) = \sigma^2 + ma\sigma^2_{\gamma} + m\sigma^2_{\alpha\gamma}$, a point estimate of σ^2_{γ} is

$$\frac{MS_{\text{factor 2}} - MSE - m\left[\dfrac{MS_{\text{interaction}} - MSE}{m}\right]}{ma} = \frac{MS_{\text{factor 2}} - MS_{\text{interaction}}}{ma}$$

of levels of factor 2, this is a *mixed model* case. For instance, suppose that we wish to study the effects of $a = 3$ teaching methods and $b = 4$ randomly selected instructors on student understanding of the meaning of confidence intervals in basic statistics. If we are interested in making inferences about only the $a = 3$ teaching methods and about all possible instructors, we would analyze the data by using the mixed model

$$y_{ij,k} = \mu + \alpha_i + \gamma_j + \theta_{ij} + \varepsilon_{ij,k}$$

This model says that $y_{ij,k}$, the degree of understanding of the kth class taught by the ith method and the jth instructor, is the sum of

1. A constant μ.
2. The fixed main effect α_i of the ith teaching method.
3. The random main effect γ_j of the jth instructor, where γ_j is assumed to have been randomly selected from a population of instructor effects that is normally distributed with mean zero and variance σ_γ^2.
4. The interaction effect θ_{ij}, which is assumed to have been randomly selected from a population of interaction effects that is normally distributed with mean 0 and variance $((a - 1)/a)\,\sigma_{\alpha\gamma}^2$.
5. A random error component $\varepsilon_{ij,k}$, which is assumed to have been randomly selected from a population of random error components that is normally distributed with mean zero and variance σ^2.

If we further assume that all of the γ_j, θ_{ij}, and $\varepsilon_{ij,k}$ values are statistically independent, then we should analyze the data as shown in Table 16.5. Here, $MS_{\text{factor 1}}$, $MS_{\text{factor 2}}$, $MS_{\text{interaction}}$, and MSE are as defined in Section 15.2.

Comparing Tables 16.3 and 16.5, we note that

1. When factor 2 is *fixed* (Table 16.3), we test for differences between the levels of fixed factor 1 by testing

   $$H_0\colon \mu_{1\cdot} = \mu_{2\cdot} = \cdots = \mu_{a\cdot}$$

   using the F statistic

   $$F_{\text{BCF}}(\text{factor 1}) = \frac{MS_{\text{factor 1}}}{MSE}$$

   Further, we calculate a $100(1 - \alpha)\%$ confidence interval for $\mu_{i\cdot} - \mu_{i'\cdot}$ by using the formula

   $$\left[(\bar{y}_{i\cdot} - \bar{y}_{i'\cdot}) \pm t_{[\alpha/2]}^{(ab(m-1))} \sqrt{MSE\left(\frac{2}{bm}\right)}\right]$$

2. When factor 2 is *random* (Table 16.5), we test for differences between the levels of fixed factor 1 by testing

   $$H_0\colon \mu_{1\cdot} = \mu_{2\cdot} = \cdots = \mu_{a\cdot}$$

using the F statistic

$$F_{\text{MBCF}}(\text{factor 1}) = \frac{MS_{\text{factor 1}}}{MS_{\text{interaction}}}$$

Further, we calculate a $100(1-\alpha)\%$ confidence interval for $\mu_{i\cdot} - \mu_{i'\cdot}$ by using the formula

$$\left[(\bar{y}_{i\cdot} - \bar{y}_{i'\cdot}) \pm t_{[\alpha/2]}^{((a-1)(b-1))}\sqrt{MS_{\text{interaction}}\left(\frac{2}{bm}\right)}\right]$$

Also note that when factor 2 is *random*, we test the significance of factor 2 by testing the null hypothesis

$$H_0: \sigma_\gamma^2 = 0$$

which says that there is no variation in the population of factor 2 effects.

TABLE 16.5 ANOVA table for the mixed two-factor model (factor 1 fixed, factor 2 random)

Source	Degrees of freedom	Sum of squares	Mean square	F statistic	E(mean square)
Factor 1	$a-1$	$SS_{\text{factor 1}}$	$MS_{\text{factor 1}}$	$F_{\text{MBCF}}(\text{factor 1}) = \frac{MS_{\text{factor 1}}}{MS_{\text{interaction}}}$	$\sigma^2 + mb\frac{\sum_{i=1}^{a}\alpha_i^2}{a-1} + m\sigma_{\alpha\gamma}^2$
Factor 2	$b-1$	$SS_{\text{factor 2}}$	$MS_{\text{factor 2}}$	$F_{\text{MBCF}}(\text{factor 2}) = \frac{MS_{\text{factor 2}}}{MSE}$	$\sigma^2 + ma\sigma_\gamma^2$
Interaction	$(a-1)(b-1)$	$SS_{\text{interaction}}$	$MS_{\text{interaction}}$	$F_{\text{MBCF}}(\text{interaction}) = \frac{MS_{\text{interaction}}}{MSE}$	$\sigma^2 + m\sigma_{\alpha\gamma}^2$
Error	$ab(m-1)$	SSE	MSE		σ^2

Notes:

1. If $F_{\text{MBCF}}(\text{interaction}) > F_{[\alpha]}^{((a-1)(b-1),ab(m-1))}$, reject $H_0: \sigma_{\alpha\gamma}^2 = 0$ (no interaction)
2. If $F_{\text{MBCF}}(\text{factor 1}) > F_{[\alpha]}^{(a-1,(a-1)(b-1))}$, reject $H_0: \mu_{1\cdot} = \mu_{2\cdot} = \cdots = \mu_{a\cdot}$.
3. If $F_{\text{MBCF}}(\text{factor 2}) > F_{[\alpha]}^{(b-1,ab(m-1))}$, reject $H_0: \sigma_\gamma^2 = 0$
4. Since $E(MSE) = \sigma^2$, a point estimate of σ^2 is MSE
5. Since $E(MS_{\text{interaction}}) = \sigma^2 + m\sigma_{\alpha\gamma}^2$, a point estimate of $\sigma_{\alpha\gamma}^2$ is $\frac{MS_{\text{interaction}} - MSE}{m}$
6. Since $E(MS_{\text{factor 1}}) = \sigma^2 + ma\sigma_\gamma^2$, a point estimate of σ_γ^2 is $\frac{MS_{\text{factor 1}} - MSE}{ma}$
7. A $100(1-\alpha)\%$ confidence interval for $\mu_{i\cdot} - \mu_{i'\cdot}$ is $\left[(\bar{y}_{i\cdot} - \bar{y}_{i'\cdot}) \pm t_{[\alpha/2]}^{((a-1)(b-1))}\sqrt{MS_{\text{interaction}}\left(\frac{2}{bm}\right)}\right]$

16.3

NESTED FACTORS

Suppose that the plant supervisor of a small electronics firm wishes to compare three methods for assembling an electronic device. The plant has a total of six different workstations. To carry out the study, the supervisor randomly assigns $b = 2$ workstations to each of the $a = 3$ assembly methods. Here, the workstations assigned to any one method are different from those assigned to any other method. At each workstation, $m = 5$ randomly selected production workers (chosen from all of the workers who normally work at the workstation) assemble the device for one hour using the appropriate assembly method. Note that all workers have been thoroughly trained in the assembly method they are to use. The data in Table 16.6 is observed. Here, for $i = 1, 2, 3, j = 1, 2$, and $k = 1, 2, 3, 4, 5$,

$y_{ij,k}$ = the number of devices produced in an hour by the kth production worker at workstation j using assembly method i

Although it might be tempting to use the methods of Section 15.2 to analyze this data, these methods are appropriate for analyzing *factorial* experiments. In factorial experiments, each level of factor 1 appears with every level of factor 2. Thus factors 1 and 2 are said to be *crossed.* On the other hand, in Table 16.6, each level of factor 2 (that is, each specific workstation) is assigned to only one level of factor 1 (assembly method). That is, each of the total of six different workstations utilized is used to test one and only one assembly method. Therefore factor 2 (workstation) is said to be *nested* within factor 1 (assembly method).

The Two-Factor Nested Model

The two-factor nested model says that

$$
\begin{aligned}
y_{ij,k} &= \mu_{ij} + \varepsilon_{ij,k} \\
&= \mu + \alpha_i + \gamma_{j(i)} + \varepsilon_{ij,k}
\end{aligned}
$$

Here,

1. $y_{ij,k}$ = the kth value of the response variable when level j of factor 2 is nested within level i of factor 1
2. $\varepsilon_{ij,k}$ = the error term corresponding to $y_{ij,k}$
3. μ_{ij} = the mean value of the response variable when level j of factor 2 is nested within level i of factor 1
4. μ = an overall mean
5. α_i = the effect due to level i of factor 1
6. $\gamma_{j(i)}$ = the effect of level j of factor 2 nested within level i of factor 1

TABLE 16.6 **Assembly method data**

	Assembly method 1		2		3	
Workstation	*1*	*2*	*1*	*2*	*1*	*2*
	$y_{11,1} = 16$	$y_{12,1} = 14$	$y_{21,1} = 21$	$y_{22,1} = 24$	$y_{31,1} = 25$	$y_{32,1} = 31$
	$y_{11,2} = 7$	$y_{12,2} = 24$	$y_{21,2} = 25$	$y_{22,2} = 28$	$y_{31,2} = 35$	$y_{32,2} = 31$
	$y_{11,3} = 7$	$y_{12,3} = 13$	$y_{21,3} = 16$	$y_{22,3} = 27$	$y_{31,3} = 33$	$y_{32,3} = 38$
	$y_{11,4} = 13$	$y_{12,4} = 17$	$y_{21,4} = 18$	$y_{22,4} = 25$	$y_{31,4} = 31$	$y_{32,4} = 36$
	$y_{11,5} = 16$	$y_{12,5} = 21$	$y_{21,5} = 16$	$y_{22,5} = 21$	$y_{31,5} = 28$	$y_{32,5} = 35$

THE ANOVA APPROACH TO THE ANALYSIS OF A BALANCED TWO-FACTOR NESTED DESIGN

Let

$$\bar{y} = \frac{\sum_{i=1}^{a}\sum_{j=1}^{b}\sum_{k=1}^{m} y_{ij,k}}{abm} \qquad \bar{y}_{i\cdot} = \frac{\sum_{j=1}^{b}\sum_{k=1}^{m} y_{ij,k}}{bm} \qquad \bar{y}_{ij} = \frac{\sum_{k=1}^{m} y_{ij,k}}{m}$$

and note that $\bar{y}_{i\cdot}$ is the point estimate of the factor level mean

$$\mu_{i\cdot} = \frac{\sum_{j=1}^{b} \mu_{ij}}{b}$$

Then the ANOVA approach for analyzing a two-factor nested experiment is summarized in Table 16.7. Here, we assume that both factors are fixed.

Example 16.4 The SAS output resulting from using the ANOVA approach to analyze the assembly method data in Table 16.6 is given in Figure 16.8. Since the prob-value for $F_{BN}(1)$ is .0001, we have overwhelming evidence that we should reject

$$H_0\colon \mu_{1\cdot} = \mu_{2\cdot} = \mu_{3\cdot}$$

in favor of

$$H_1\colon \text{At least two of } \mu_{1\cdot}, \mu_{2\cdot}, \text{ and } \mu_{3\cdot} \text{ differ}$$

We conclude that at least two of the assembly methods differ. From Figure 16.8 we see that the prob-value for testing

$$H_0\colon \mu_{3\cdot} - \mu_{1\cdot} = 0$$

is .0001 and that the point estimate of $\mu_{3\cdot} - \mu_{1\cdot}$ is 17.5. Thus we have overwhelming evidence that $\mu_{3\cdot}$ is greater than $\mu_{1\cdot}$. Furthermore, the prob-value for testing

$$H_0\colon \mu_{3\cdot} - \mu_{2\cdot} = 0$$

is .0001, and the point estimate of $\mu_{3\cdot} - \mu_{2\cdot}$ is 10.2. Thus we also have overwhelming

TABLE 16.7 ANOVA table for the fixed two-factor nested model

Source	Degrees of freedom	Sum of squares	Mean square	F statistic	E(mean square)
Factor 1	$a-1$	$SS_1 = bm\sum_{i=1}^{a}(\bar{y}_{i\cdot} - \bar{y})^2$	$MS_1 = \frac{SS_1}{a-1}$	$F_{BN}(1) = \frac{MS_1}{MSE}$	$\sigma^2 + bm\frac{\sum_{i=1}^{a}\alpha_i^2}{a-1}$
Factor 2 (factor 1)	$a(b-1)$	$SS_{2(1)} = m\sum_{i=1}^{a}\sum_{j=1}^{b}(\bar{y}_{ij} - \bar{y}_{i\cdot})^2$	$MS_{2(1)} = \frac{SS_{2(1)}}{a(b-1)}$	$F_{BN}(2(1)) = \frac{MS_{2(1)}}{MSE}$	$\sigma^2 + \frac{m}{a(b-1)}\sum_{i=1}^{a}\sum_{j=1}^{b}\gamma_{j(i)}^2$
Error	$ab(m-1)$	$SSE = \sum_{i=1}^{a}\sum_{j=1}^{b}\sum_{k=1}^{m}(y_{ij,k} - \bar{y}_{ij})^2$	$MSE = \frac{SSE}{ab(m-1)}$		σ^2

Notes:

1. If $F_{BN}(1) > F_{[\alpha]}^{(a-1,ab(m-1))}$, reject $H_0: \mu_{1\cdot} = \mu_{2\cdot} = \cdots = \mu_{a\cdot}$ (all levels of factor 1 have the same effect on the mean response).
2. If $F_{BN}(2(1)) > F_{[\alpha]}^{(a(b-1),ab(m-1))}$, reject $H_0: \mu_{11} = \mu_{12} = \cdots = \mu_{1b}; \mu_{21} = \mu_{22} = \cdots = \mu_{2b}; \cdots; \mu_{a1} = \mu_{a2} = \cdots = \mu_{ab}$ (within each level of factor 1, all levels of factor 2 have the same effect on the mean response).
3. A point estimate of $\mu_{i\cdot} - \mu_{i'\cdot}$ is $\bar{y}_{i\cdot} - \bar{y}_{i'\cdot}$, and an individual $100(1-\alpha)\%$ confidence interval for $\mu_{i\cdot} - \mu_{i'\cdot}$ is $\left[(\bar{y}_{i\cdot} - \bar{y}_{i'\cdot}) \pm t_{[\alpha/2]}^{(ab(m-1))}\sqrt{MSE\left(\frac{2}{bm}\right)}\right]$. For Tukey simultaneous intervals, replace $t_{[\alpha/2]}^{(ab(m-1))}$ by $\frac{1}{\sqrt{2}}q_{[\alpha]}(a, ab(m-1))$.
4. A point estimate of $\mu_{ij} - \mu_{ij'}$ is $\bar{y}_{ij} - \bar{y}_{ij'}$, and an individual $100(1-\alpha)\%$ confidence interval for $\mu_{ij} - \mu_{ij'}$ is $\left[(\bar{y}_{ij} - \bar{y}_{ij'}) \pm t_{[\alpha/2]}^{(ab(m-1))}\sqrt{MSE\left(\frac{2}{m}\right)}\right]$. For Tukey simultaneous intervals, replace $t_{[\alpha/2]}^{(ab(m-1))}$ by $\frac{1}{\sqrt{2}}q_{[\alpha]}(ab, ab(m-1))$. For Bonferroni simultaneous intervals, replace $t_{[\alpha/2]}^{(ab(m-1))}$ by $t_{[\alpha/2g]}^{(ab(m-1))}$ (which applies to g prespecified linear combinations).
5. A point estimate of μ_{ij} is $\bar{y}_{ij}$, and a $100(1-\alpha)\%$ confidence interval for μ_{ij} is $\left[\bar{y}_{ij} \pm t_{[\alpha/2]}^{(ab(m-1))}\sqrt{\frac{MSE}{m}}\right]$.

evidence that $\mu_{3\cdot}$ is greater than $\mu_{2\cdot}$. We conclude that assembly method 3 yields the highest mean hourly output.

Furthermore, for example, an individual 95% confidence interval for $\mu_{3\cdot} - \mu_{2\cdot}$ is

$$
\begin{aligned}
\left[(\bar{y}_{3\cdot} - \bar{y}_{2\cdot}) \pm t_{[\alpha/2]}^{(ab(m-1))}\sqrt{MSE\left(\frac{2}{bm}\right)}\right] &= [10.2 \pm t_{[.025]}^{(3(2)(5-1))}(1.7330)] \\
&= [10.2 \pm 2.064(1.7330)] \\
&= [10.2 \pm 3.58] \\
&= [6.62, 13.78]
\end{aligned}
$$

This interval says that we are 95 percent confident that the mean hourly output obtained by using assembly method 3 is between 6.62 and 13.78 electronic devices greater than the mean hourly output obtained by using assembly method 2.

Since the prob-value for $F_{BN}(2(1))$ is .0105, we have strong evidence that we should reject

$$H_0: \mu_{11} = \mu_{12};\ \mu_{21} = \mu_{22};\ \mu_{31} = \mu_{32}$$

FIGURE 16.8 **SAS output of a nested model ANOVA of the assembly method data in Table 16.6**

SAS

GENERAL LINEAR MODELS PROCEDURE

DEPENDENT VARIABLE: UNITS

SOURCE	DF	SUM OF SQUARES	MEAN SQUARE	F VALUE	PR > F	R-SQUARE	C.V.
MODEL	5	1755.46666667	351.09333333	23.38	0.0001	0.829668	16.7997
ERROR	24	360.40000000[a]	15.01666667[b]		ROOT MSE		UNITS MEAN
CORRECTED TOTAL	29	2115.86666667			3.87513441		23.06666667

SOURCE	DF	TYPE I SS	F VALUE	PR > F	DF	TYPE III SS	F VALUE	PR > F
METHOD	2	1545.26666667[c]	51.45[d]	0.0001[e]	2	1545.26666667	51.45	0.0001
STATION(METHOD)	3	210.20000000[f]	4.67[g]	0.0105[h]	3	210.20000000	4.67	0.0105

PARAMETER	ESTIMATE[i]	T FOR HO: PARAMETER=0[j]	PR > |T|[k]	STD ERROR OF ESTIMATE	
MU2-MU1	7.30000000	4.21	0.0003	1.73301279	} [l]
MU3-MU1	17.50000000	10.10	0.0001	1.73301279	
MU3-MU2	10.20000000	5.89	0.0001	1.73301279	
MU12-MU11	6.00000000	2.45	0.0220	2.45085019	} [m]
MU22-MU21	5.80000000	2.37	0.0264	2.45085019	
MU32-MU31	3.80000000	1.55	0.1341	2.45085019	

OBSERVATION	OBSERVED VALUE	PREDICTED VALUE	RESIDUAL	LOWER 95% CL FOR MEAN	UPPER 95% CL FOR MEAN
1	16.00000000	11.80000000	4.20000000	8.22326620	15.37673380
2	7.00000000	11.80000000	-4.80000000	8.22326620	15.37673380
3	7.00000000	11.80000000	-4.80000000	8.22326620	15.37673380
4	13.00000000	11.80000000	1.20000000	8.22326620	15.37673380
5	16.00000000	11.80000000	4.20000000	8.22326620	15.37673380
6	14.00000000	17.80000000	-3.80000000	14.22326620	21.37673380
7	24.00000000	17.80000000	6.20000000	14.22326620	21.37673380
8	13.00000000	17.80000000	-4.80000000	14.22326620	21.37673380
9	17.00000000	17.80000000	-0.80000000	14.22326620	21.37673380
10	21.00000000	17.80000000	3.20000000	14.22326620	21.37673380
11	21.00000000	19.20000000	1.80000000	15.62326620	22.77673380
12	25.00000000	19.20000000	5.80000000	15.62326620	22.77673380
13	16.00000000	19.20000000	-3.20000000	15.62326620	22.77673380
14	18.00000000	19.20000000	-1.20000000	15.62326620	22.77673380
15	16.00000000	19.20000000	-3.20000000	15.62326620	22.77673380
16	24.00000000	25.00000000	-1.00000000	21.42326620	28.57673380
17	28.00000000	25.00000000	3.00000000	21.42326620	28.57673380
18	27.00000000	25.00000000	2.00000000	21.42326620	28.57673380
19	25.00000000	25.00000000	0.00000000	21.42326620	28.57673380
20	21.00000000	25.00000000	-4.00000000	21.42326620	28.57673380
21	25.00000000	30.40000000	-5.40000000	26.82326620	33.97673380
22	35.00000000	30.40000000	4.60000000	26.82326620	33.97673380
23	33.00000000	30.40000000	2.60000000	26.82326620	33.97673380
24	31.00000000	30.40000000	0.60000000	26.82326620	33.97673380
25	28.00000000	30.40000000	-2.40000000	26.82326620	33.97673380
26	31.00000000	34.20000000[n]	-3.20000000	[30.62326620	37.77673380][o]
27	31.00000000	34.20000000	-3.20000000	30.62326620	37.77673380
28	38.00000000	34.20000000	3.80000000	30.62326620	37.77673380
29	36.00000000	34.20000000	1.80000000	30.62326620	37.77673380
30	35.00000000	34.20000000	0.80000000	30.62326620	37.77673380

[a] *SSE* [b] *MSE* [c] SS_1 [d] $F_{BN}(1)$ [e] *Prob-value for* $F_{BN}(1)$ [f] $SS_{2(1)}$ [g] $F_{BN}(2(1))$ [h] *Prob-value for* $F_{BN}(2(1))$

[i] *Point estimates* $\bar{y}_{2\cdot} - \bar{y}_{1\cdot}$, $\bar{y}_{3\cdot} - \bar{y}_{1\cdot}$, $\bar{y}_{3\cdot} - \bar{y}_{2\cdot}$, $\bar{y}_{12} - \bar{y}_{11}$, $\bar{y}_{22} - \bar{y}_{21}$, $\bar{y}_{32} - \bar{y}_{31}$

[j,k] *t statistics and prob-values for testing* $\begin{cases} H_0: \mu_{2\cdot} - \mu_{1\cdot} = 0,\ H_0: \mu_{3\cdot} - \mu_{1\cdot} = 0,\ H_0: \mu_{3\cdot} - \mu_{2\cdot} = 0, \\ H_0: \mu_{12} - \mu_{11} = 0,\ H_0: \mu_{22} - \mu_{21} = 0,\ H_0: \mu_{32} - \mu_{31} = 0 \end{cases}$ [l] $\sqrt{MSE\left(\frac{2}{bm}\right)}$

[m] $\sqrt{MSE\left(\frac{2}{m}\right)}$ [n] *Point estimate* $\bar{y}_{32}$ *of* μ_{32} [o] *95% confidence interval for* μ_{32}

To see which workstations differ, note from Figure 16.8 that:

1. The prob-value for testing

   $$H_0: \mu_{11} - \mu_{12} = 0$$

   is .0220. This indicates that the workstations at which assembly method 1 was tested differ with respect to mean hourly output.
2. The prob-value for testing

   $$H_0: \mu_{21} - \mu_{22} = 0$$

   is .0264. This indicates that the workstations at which assembly method 2 was tested differ with respect to mean hourly output.
3. The prob-value for testing

   $$H_0: \mu_{31} - \mu_{32} = 0$$

   is .1341. Therefore there is not much evidence that the workstations at which assembly method 3 was tested differ with respect to mean hourly output. Also, note that an individual 95% confidence interval for $\mu_{32} - \mu_{31}$ is

$$\begin{aligned}\left[(\bar{y}_{32} - \bar{y}_{31}) \pm t_{[\alpha/2]}^{(ab(m-1))}\sqrt{MSE\left(\frac{2}{bm}\right)}\right] &= [3.8 \pm t_{[.025]}^{(3(2)(5-1))}(2.4509)] \\ &= [3.8 \pm 2.064(2.4509)] \\ &= [3.8 \pm 5.06] \\ &= [-1.26, 8.86]\end{aligned}$$

This interval says that we are 95 percent confident that the mean hourly output when we use assembly method 3 at the second workstation at which assembly method 3 was tested is between 1.26 devices less and 8.86 devices more than the mean hourly output when we use assembly method 3 at the first workstation at which method 3 was tested. Again, we conclude that there is no statistically significant difference between these means.

It also follows from Figure 16.8 that a 95% confidence interval for μ_{32} is

$$\begin{aligned}\left[\bar{y}_{32} \pm t_{[\alpha/2]}^{(ab(m-1))}\sqrt{\frac{MSE}{m}}\right] &= \left[34.2 \pm t_{[.025]}^{(24)}\sqrt{\frac{15.0167}{5}}\right] \\ &= [34.2 \pm 2.064(1.7330)] \\ &= [30.62, 37.78]\end{aligned}$$

This interval says that we are 95 percent confident that the mean hourly output when using assembly method 3 at the second workstation at which assembly method 3 was tested is between 30.62 and 37.78 electronic devices.

To conclude this example, recall that the plant supervisor randomly selected $m = 5$ production workers from each workstation. Thus we do not know whether the differences between workstations are due to the physical nature of the workstations or are due to differences between production workers who normally work at the workstations. As an alternative strategy, the plant supervisor could put all production workers into a common pool and randomly assign five to each workstation for the testing. Then differences between workstations would probably be due to the physical nature of the workstations. However, in this case, differences in the familiarity of the production workers with the workstations could confound our ability to compare assembly methods and workstations.

In the above example, suppose that we had randomly selected two workstations for testing each assembly method from a great many workstations in the plant. Then the factor workstation would be a *random* factor. In this case we would analyze the data in Table 16.6 by using the *mixed two-factor nested model*

$$y_{ij,k} = \mu + \alpha_i + \gamma_{j(i)} + \varepsilon_{ij,k}$$

In this model, μ, α_i, and $\varepsilon_{ij,k}$ are as previously defined. However, $\gamma_{j(i)}$ is the random effect of workstation j nested within assembly method i. Here, $\gamma_{j(i)}$ is assumed to have been randomly selected from a population of workstation effects that is normally distributed with mean zero and variance σ_γ^2. The ANOVA approach to analyzing the mixed two-factor nested model is as summarized in Table 16.8.

If the assembly methods had been randomly selected from a population of many assembly methods and the workstations had been randomly selected from a

TABLE 16.8 ANOVA table for the mixed two-factor nested model (factor 1 fixed, factor 2 random)

Source	Degrees of freedom	Sum of squares	Mean square	F statistic	E(mean square)
Factor 1	$a-1$	$SS_1 = bm\sum_{i=1}^{a}(\bar{y}_{i\cdot}-\bar{y})^2$	$MS_1 = \dfrac{SS_1}{a-1}$	$F_{MBN}(1) = \dfrac{MS_1}{MS_{2(1)}}$	$\sigma^2 + \dfrac{bm}{a-1}\sum_{i=1}^{a}\alpha_i^2 + m\sigma_\gamma^2$
Factor 2 (factor 1)	$a(b-1)$	$SS_{2(1)} = m\sum_{i=1}^{a}\sum_{j=1}^{b}(\bar{y}_{ij}-\bar{y}_{i\cdot})^2$	$MS_{2(1)} = \dfrac{SS_{2(1)}}{a(b-1)}$	$F_{MBN}(2(1)) = \dfrac{MS_{2(1)}}{MSE}$	$\sigma^2 + m\sigma_\gamma^2$
Error	$ab(m-1)$	$SSE = \sum_{i=1}^{a}\sum_{j=1}^{b}\sum_{k=1}^{m}(y_{ij,k}-\bar{y}_{ij})^2$	$MSE = \dfrac{SSE}{ab(m-1)}$		σ^2

Notes:

1. If $F_{MBN}(1) > F_{[\alpha]}^{(a-1,a(b-1))}$, reject H_0: $\mu_{1\cdot} = \mu_{2\cdot} = \cdots = \mu_{a\cdot}$ (all levels of factor 1 have the same effect on the mean response).
2. If $F_{MBN}(2(1)) > F_{[\alpha]}^{(a(b-1),ab(m-1))}$, reject H_0: $\sigma_\gamma^2 = 0$ (within each level of factor 1 there is no variation in the effects of all possible levels of factor 2 on the mean response).
3. A point estimate of $\mu_{i\cdot} - \mu_{i'\cdot}$ is $\bar{y}_{i\cdot} - \bar{y}_{i'\cdot}$, and an individual $100(1-\alpha)$% confidence interval for $\mu_{i\cdot} - \mu_{i'\cdot}$ is $\left[(\bar{y}_{i\cdot} - \bar{y}_{i'\cdot}) \pm t_{[\alpha/2]}^{(a(b-1))}\sqrt{MS_{2(1)}\left(\dfrac{2}{bm}\right)}\right]$. For Tukey simultaneous intervals, replace $t_{[\alpha/2]}^{(a(b-1))}$ by $\dfrac{1}{\sqrt{2}}q_{[\alpha]}(a, a(b-1))$.
4. Since $E(MSE) = \sigma^2$, a point estimate of σ^2 is MSE.
5. Since $E(MS_{2(1)}) = \sigma^2 + m\sigma_\gamma^2$, a point estimate of σ_γ^2 is $(MS_{2(1)} - MSE)/m$.

population of many workstations, both factors would be *random*. Then, we would analyze the data in Table 16.6 by using the *random two-factor nested* model

$$y_{ij,k} = \mu + \alpha_i + \gamma_{j(i)} + \varepsilon_{ij,k}$$

Here, μ and $\varepsilon_{ij,k}$ are as previously defined, $\gamma_{j(i)}$ is as defined for the mixed model, and α_i is the random effect of assembly method i. Note that α_i is assumed to have been randomly selected from a population of assembly method effects that is normally distributed with mean zero and variance σ_α^2. The ANOVA approach to analyzing the random two-factor nested model is as summarized in Table 16.9. It should be noted that this model has great application in studying the variability of production processes. For example, suppose that a pelletized chemical product is produced in a batch process. After production, each batch is blended in a large mixing chamber. Then randomly selected batches might be random factor 1 and randomly selected locations within the mixing chamber might be random factor 2.

TABLE 16.9 **ANOVA table for the random two-factor nested model**

Source	Degrees of freedom	Sum of squares	Mean square	F statistic	E(mean square)
Factor 1	$a-1$	$SS_1 = bm\sum_{i=1}^{a}(\bar{y}_{i\cdot} - \bar{y})^2$	$MS_1 = \frac{SS_1}{a-1}$	$F_{RBN}(1) = \frac{MS_1}{MS_{2(1)}}$	$\sigma^2 + bm\sigma_\alpha^2 + m\sigma_\gamma^2$
Factor 2 (factor 1)	$a(b-1)$	$SS_{2(1)} = m\sum_{i=1}^{a}\sum_{j=1}^{b}(\bar{y}_{ij} - \bar{y}_{i\cdot})^2$	$MS_{2(1)} = \frac{SS_{2(1)}}{a(b-1)}$	$F_{RBN}(2(1)) = \frac{MS_{2(1)}}{MSE}$	$\sigma^2 + m\sigma_\gamma^2$
Error	$ab(m-1)$	$SSE = \sum_{i=1}^{a}\sum_{j=1}^{b}\sum_{k=1}^{m}(y_{ij,k} - \bar{y}_{ij})^2$	$MSE = \frac{SSE}{ab(m-1)}$		σ^2

Notes:

1. If $F_{RBN}(1) > F_{[\alpha]}^{(a-1,a(b-1))}$, reject H_0: $\sigma_\alpha^2 = 0$ (there is no variation in the effects of all possible levels of factor 1 on the mean response).
2. If $F_{RBN}(2(1)) > F_{[\alpha]}^{(a(b-1),ab(m-1))}$, reject H_0: $\sigma_\gamma^2 = 0$ (within each possible level of factor 1 there is no variation in the effects of all possible levels of factor 2 on the mean response).
3. Since $E(MSE) = \sigma^2$, a point estimate of σ^2 is MSE.
4. Since $E(MS_{2(1)}) = \sigma^2 + m\sigma_\gamma^2$, a point estimate of σ_γ^2 is

   $$\frac{MS_{2(1)} - MSE}{m}$$

5. Since $E(MS_1) = \sigma^2 + bm\sigma_\alpha^2 + m\sigma_\gamma^2$, a point estimate of σ_α^2 is

   $$\frac{MS_1 - MSE - m\left[\frac{MS_{2(1)} - MSE}{m}\right]}{bm} = \frac{MS_1 - MS_{2(1)}}{bm}$$

6. A point estimate of μ (the mean response over all possible combinations of levels of factors 1 and 2) is $\bar{y}$, and a $100(1-\alpha)\%$ confidence interval for μ is

   $$\left[\bar{y} \pm t_{[\alpha/2]}^{(a-1)}\sqrt{\frac{MS_1}{abm}}\right]$$

Samples might then be randomly selected from locations. The objective of the study might be to examine the sources of variability in the viscosity of the product when melted at a given temperature. Here, the between-batch variability measures the consistency of the batch process, and the within-chamber variability measures the effectiveness of the blending operation.

THE REGRESSION APPROACH

Consider an assembly method experiment yielding the data in Table 16.10. In this situation the number of levels of factor 2 (workstation) nested within the levels of factor 1 (assembly method) varies for different levels of factor 1. In addition, the number of $y_{ij,k}$ values observed for different combinations of levels of factors 1 and 2 varies. In such a case we must use a dummy variable regression model called the $(\alpha, \gamma(\alpha))$ model.

Example 16.5 The $(\alpha, \gamma(\alpha))$ model describing the data in Table 16.10 is

$$\begin{aligned} y_{ij,k} &= \mu_{ij} + \varepsilon_{ij,k} \\ &= \mu + \alpha_i + \gamma_{j(i)} + \varepsilon_{ij,k} \\ &= \mu + (\alpha_2 D_{i,2} + \alpha_3 D_{i,3}) + (\gamma_{2(1)} D_{ij,12}) \\ &\quad + (\gamma_{2(2)} D_{ij,22}) + (\gamma_{2(3)} D_{ij,32} + \gamma_{3(3)} D_{ij,33}) \\ &\quad + \varepsilon_{ij,k} \end{aligned}$$

Here,

$$D_{i,2} = \begin{cases} 1 & \text{if the observation is from assembly method 2 } (i = 2) \\ -1 & \text{if the observation is from assembly method 1 } (i = 1) \\ 0 & \text{otherwise} \end{cases}$$

$$D_{i,3} = \begin{cases} 1 & \text{if the observation is from assembly method 3 } (i = 3) \\ -1 & \text{if the observation is from assembly method 1 } (i = 1) \\ 0 & \text{otherwise} \end{cases}$$

TABLE 16.10 Assembly method data

	Assembly method 1		Assembly method 2		Assembly method 3		
Workstation	1	2	1	2	1	2	3
	$y_{11,1} = 16$	$y_{12,1} = 14$	$y_{21,1} = 25$	$y_{22,1} = 24$	$y_{31,1} = 25$	$y_{32,1} = 31$	$y_{33,1} = 32$
	$y_{11,2} = 7$	$y_{12,2} = 24$	$y_{21,2} = 20$	$y_{22,2} = 28$	$y_{31,2} = 35$	$y_{32,2} = 31$	$y_{33,2} = 30$
	$y_{11,3} = 7$	$y_{12,3} = 13$	$y_{21,3} = 16$	$y_{22,3} = 27$	$y_{31,3} = 33$	$y_{32,3} = 38$	$y_{33,3} = 39$
	$y_{11,4} = 13$	$y_{12,4} = 17$		$y_{22,4} = 25$	$y_{31,4} = 31$	$y_{32,4} = 36$	$y_{33,4} = 36$
		$y_{12,5} = 21$		$y_{22,5} = 21$	$y_{31,5} = 28$	$y_{32,5} = 35$	

$$D_{ij,12} = \begin{cases} 1 & \text{if the observation is from workstation 2 using assembly method 1 } (i = 1, j = 2) \\ -1 & \text{if the observation is from workstation 1 using assembly method 1 } (i = 1, j = 1) \\ 0 & \text{otherwise} \end{cases}$$

$$D_{ij,22} = \begin{cases} 1 & \text{if the observation is from workstation 2 using assembly method 2 } (i = 2, j = 2) \\ -1 & \text{if the observation is from workstation 1 using assembly method 2 } (i = 2, j = 1) \\ 0 & \text{otherwise} \end{cases}$$

$$D_{ij,32} = \begin{cases} 1 & \text{if the observation is from workstation 2 using assembly method 3 } (i = 3, j = 2) \\ -1 & \text{if the observation is from workstation 1 using assembly method 3 } (i = 3, j = 1) \\ 0 & \text{otherwise} \end{cases}$$

$$D_{ij,33} = \begin{cases} 1 & \text{if the observation is from workstation 3 using assembly method 3 } (i = 3, j = 3) \\ -1 & \text{if the observation is from workstation 1 using assembly method 3 } (i = 3, j = 1) \\ 0 & \text{otherwise} \end{cases}$$

Using the equation,

$$\begin{aligned} \mu_{ij} &= \mu + (\alpha_2 D_{i,2} + \alpha_3 D_{i,3}) + (\gamma_{2(1)} D_{ij,12}) \\ &\quad + (\gamma_{2(2)} D_{ij,22}) + (\gamma_{2(3)} D_{ij,32} + \gamma_{3(3)} D_{ij,33}) \end{aligned}$$

and the previous definitions of the dummy variables, we find that

$$\left.\begin{aligned} \mu_{11} &= \mu - \alpha_2 - \alpha_3 - \gamma_{2(1)} \\ \mu_{12} &= \mu - \alpha_2 - \alpha_3 + \gamma_{2(1)} \end{aligned}\right\}$$

Note that $\mu_{11} = \mu_{12}$ implies that $\mu - \alpha_2 - \alpha_3 - \gamma_{2(1)} = \mu - \alpha_2 - \alpha_3 + \gamma_{2(1)}$, which implies that $\gamma_{2(1)} = 0$.

$$\left.\begin{aligned} \mu_{21} &= \mu + \alpha_2 - \gamma_{2(2)} \\ \mu_{22} &= \mu + \alpha_2 + \gamma_{2(2)} \end{aligned}\right\}$$

Note that $\mu_{21} = \mu_{22}$ implies that $\mu + \alpha_2 - \gamma_{2(2)} = \mu + \alpha_2 + \gamma_{2(2)}$, which implies that $\gamma_{2(2)} = 0$.

$$\left.\begin{aligned} \mu_{31} &= \mu + \alpha_3 - \gamma_{2(3)} - \gamma_{3(3)} \\ \mu_{32} &= \mu + \alpha_3 + \gamma_{2(3)} \\ \mu_{33} &= \mu + \alpha_3 + \gamma_{3(3)} \end{aligned}\right\}$$

Note that $\mu_{32} = \mu_{33}$ implies that $\mu + \alpha_3 + \gamma_{2(3)} = \mu + \alpha_3 + \gamma_{3(3)}$, which implies that $\gamma_{2(3)} = \gamma_{3(3)}$. Then $\mu_{31} = \mu_{32} = \mu_{33}$ implies that $\mu + \alpha_3 - \gamma_{2(3)} - \gamma_{2(3)} = \mu + \alpha_3 + \gamma_{2(3)}$, which implies that $\gamma_{2(3)} = 0 = \gamma_{3(3)}$.

Hence we can test

$$H_0: \mu_{11} = \mu_{12};\ \mu_{21} = \mu_{22};\ \mu_{31} = \mu_{32} = \mu_{33}$$

by testing

$$H_0: \gamma_{2(1)} = \gamma_{2(2)} = \gamma_{2(3)} = \gamma_{3(3)} = 0$$

Specifically, if

$$F(\gamma(\alpha)|\alpha) = \frac{\{SSE_\alpha - SSE_{(\alpha,\gamma(\alpha))}\}/4}{SSE_{(\alpha,\gamma(\alpha))}/(31 - 7)}$$

is greater than $F_{[\alpha]}^{(4,24)}$, we can reject H_0 by setting the probability of a Type I error equal to α. Here, $SSE_{(\alpha,\gamma(\alpha))}$ is the unexplained variation for the $(\alpha, \gamma(\alpha))$ model, and SSE_α is the unexplained variation for the α *model*

$$y_{ij,k} = \mu + \alpha_2 D_{i,2} + \alpha_3 D_{i,3} + \varepsilon_{ij,k}$$

Also note that we are using $n = 31$ observations and that the $(\alpha, \gamma(\alpha))$ model has seven parameters. The SAS output of the analysis of the data in Table 16.10 is given in Figure 16.9. This output indicates that the prob-value for $F(\gamma(\alpha)|\alpha)$ is .0298. Therefore we reject H_0 with $\alpha = .05$ and conclude that workstations nested within assembly methods have different effects on mean hourly output.

To see which workstations differ, we consider the pairwise differences

$$\mu_{12} - \mu_{11}$$

$$\mu_{22} - \mu_{21}$$

$$\mu_{32} - \mu_{31}$$

$$\mu_{33} - \mu_{31}$$

$$\mu_{33} - \mu_{32}$$

Examining the t statistics and prob-values for testing the significance of these differences, we conclude that the most significant difference is $\mu_{12} - \mu_{11}$ ($t = 2.68$ and prob-value $= .0131$). A point estimate of

$$\begin{aligned}\mu_{12} - \mu_{11} &= (\mu - \alpha_2 - \alpha_3 + \gamma_{2(1)}) - (\mu - \alpha_2 - \alpha_3 - \gamma_{2(1)}) \\ &= 2\gamma_{2(1)}\end{aligned}$$

is $2\hat{\gamma}_{2(1)} = 7.05$. Here, $\hat{\gamma}_{2(1)}$ is the least squares point estimate of $\gamma_{2(1)}$. Furthermore, an individual 95% confidence interval for

$$\begin{aligned}\mu_{12} - \mu_{11} &= 2\gamma_{2(1)} \\ &= 0\mu + 0\alpha_2 + 0\alpha_3 + 2\gamma_{2(1)} + 0\gamma_{2(2)} + 0\gamma_{2(3)} + 0\gamma_{3(3)} \\ &= \boldsymbol{\lambda}'\boldsymbol{\beta} \quad \text{where } \boldsymbol{\lambda}' = [0\ \ 0\ \ 0\ \ 2\ \ 0\ \ 0\ \ 0]\end{aligned}$$

FIGURE 16.9 **SAS output of an (α, γ(α)) model analysis of the assembly method data in Table 16.10**

SAS

GENERAL LINEAR MODELS PROCEDURE

DEPENDENT VARIABLE: UNITS

SOURCE	DF	SUM OF SQUARES	MEAN SQUARE	F VALUE	PR > F	R-SQUARE	C.V.
MODEL	6	2047.74301075	341.29050179	22.20	0.0001	0.847327	15.6232
ERROR	24	368.96666667	15.37361111		ROOT MSE		UNITS MEAN
CORRECTED TOTAL	30	2416.70967742			3.92091968		25.09677419

SOURCE	DF	TYPE I SS	F VALUE	PR > F	DF	TYPE III SS	F VALUE	PR > F
METHOD	2	1849.49539171	60.15	0.0001	2	1938.14358068	63.03[a]	0.0001[b]
STATION(METHOD)	4	198.24761905	3.22	0.0298	4	198.24761905	3.22[c]	0.0298[d]

PARAMETER	ESTIMATE	T FOR HO: PARAMETER=0	PR > |T|	STD ERROR OF ESTIMATE
MU2-MU1	8.39166667	4.12	0.0002	1.94405403
MU3-MU1	18.67500000	11.08	0.0001	1.68518474
MU3-MU2	10.28333333[e]	5.78	0.0001	1.77767578[f]
MU12-MU11	7.05000000[g]	2.68	0.0131	2.63023288[h]
MU22-MU21	4.66666667	1.63	0.1162	2.86343487
MU32-MU31	3.80000000	1.53	0.1385	2.47980734
MU33-MU31	3.85000000	1.46	0.1562	2.63023288
MU33-MU32	0.05000000	0.02	0.9850	2.63023288

OBSERVATION	OBSERVED VALUE	PREDICTED VALUE	RESIDUAL	LOWER 95% CL FOR MEAN	UPPER 95% CL FOR MEAN
1	16.00000000	10.75000000	5.25000000	6.70384234	14.79615766
2	7.00000000	10.75000000	-3.75000000	6.70384234	14.79615766
3	7.00000000	10.75000000	-3.75000000	6.70384234	14.79615766
4	13.00000000	10.75000000	2.25000000	6.70384234	14.79615766
5	14.00000000	17.80000000	-3.80000000	14.18100657	21.41899343
6	24.00000000	17.80000000	6.20000000	14.18100657	21.41899343
7	13.00000000	17.80000000	-4.80000000	14.18100657	21.41899343
8	17.00000000	17.80000000	-0.80000000	14.18100657	21.41899343
9	21.00000000	17.80000000	3.20000000	14.18100657	21.41899343
10	25.00000000	20.33333333	4.66666667	15.66123291	25.00543376
11	16.00000000	20.33333333	-4.33333333	15.66123291	25.00543376
12	20.00000000	20.33333333	-0.33333333	15.66123291	25.00543376
13	24.00000000	25.00000000	-1.00000000	21.38100657	28.61899343
14	28.00000000	25.00000000	3.00000000	21.38100657	28.61899343
15	27.00000000	25.00000000	2.00000000	21.38100657	28.61899343
16	25.00000000	25.00000000	0.00000000	21.38100657	28.61899343
17	21.00000000	25.00000000	-4.00000000	21.38100657	28.61899343
18	25.00000000	30.40000000	-5.40000000	26.78100657	34.01899343
19	35.00000000	30.40000000	4.60000000	26.78100657	34.01899343
20	33.00000000	30.40000000	2.60000000	26.78100657	34.01899343
21	31.00000000	30.40000000	0.60000000	26.78100657	34.01899343
22	28.00000000	30.40000000	-2.40000000	26.78100657	34.01899343
23	31.00000000	34.20000000	-3.20000000	30.58100657	37.81899343
24	31.00000000	34.20000000	-3.20000000	30.58100657	37.81899343
25	38.00000000	34.20000000	3.80000000	30.58100657	37.81899343
26	36.00000000	34.20000000	1.80000000	30.58100657	37.81899343
27	35.00000000	34.20000000	0.80000000	30.58100657	37.81899343
28	32.00000000	34.25000000	-2.25000000	30.20384234	38.29615766
29	30.00000000	34.25000000	-4.25000000	30.20384234	38.29615766
30	39.00000000	34.25000000	4.75000000	30.20384234	38.29615766
31	36.00000000	34.25000000[i]	1.75000000	[30.20384234	38.29615766][j]

[a] $F(\alpha|\gamma(\alpha))$ [b] *Prob-value for* $F(\alpha|\gamma(\alpha))$ [c] $F(\gamma(\alpha)|\alpha)$ [d] *Prob-value for* $F(\gamma(\alpha)|\alpha)$
[e] $\hat{\alpha}_3 - \hat{\alpha}_2$ = *point estimate of* $\{\mu_{3.} - \mu_{2.} = \alpha_3 - \alpha_2\}$ [f] $s_{(\alpha,\gamma(\alpha))}\sqrt{\boldsymbol{\lambda}'(\mathbf{X}'\mathbf{X})^{-1}\boldsymbol{\lambda}}$
[g] $2\hat{\gamma}_{2(1)}$ = *point estimate of* $\{\mu_{12} - \mu_{11} = 2\gamma_{2(1)}\}$ [h] $s_{(\alpha,\gamma(\alpha))}\sqrt{\boldsymbol{\lambda}'(\mathbf{X}'\mathbf{X})^{-1}\boldsymbol{\lambda}}$ [i] $\hat{y}_{33}$ [j] *95% confidence interval for* μ_{33}

is

$$\begin{aligned}[2\hat{\gamma}_{2(1)} &\pm t_{[\alpha/2]}^{(n-k)} s_{(\alpha,\gamma(\alpha))}\sqrt{\boldsymbol{\lambda}'(\mathbf{X}'\mathbf{X})^{-1}\boldsymbol{\lambda}}] \\ &= [7.05 \pm t_{[.025]}^{(31-7)}(2.6302)] \\ &= [7.05 \pm 2.064(2.6302)] \\ &= [7.05 \pm 5.43] \\ &= [1.62, 12.48]\end{aligned}$$

We next note that

$$\begin{aligned}\mu_{1\cdot} &= \frac{\mu_{11} + \mu_{12}}{2} = \frac{(\mu - \alpha_2 - \alpha_3 - \gamma_{2(1)}) + (\mu - \alpha_2 - \alpha_3 + \gamma_{2(1)})}{2} \\ &= \mu - \alpha_2 - \alpha_3\end{aligned}$$

$$\begin{aligned}\mu_{2\cdot} &= \frac{\mu_{21} + \mu_{22}}{2} = \frac{(\mu + \alpha_2 - \gamma_{2(2)}) + (\mu + \alpha_2 + \gamma_{2(2)})}{2} \\ &= \mu + \alpha_2\end{aligned}$$

$$\begin{aligned}\mu_{3\cdot} &= \frac{\mu_{31} + \mu_{32} + \mu_{33}}{3} \\ &= \frac{(\mu + \alpha_3 - \gamma_{2(3)} - \gamma_{3(3)}) + (\mu + \alpha_3 + \gamma_{2(3)}) + (\mu + \alpha_3 + \gamma_{3(3)})}{3} \\ &= \mu + \alpha_3\end{aligned}$$

Setting $\mu_{2\cdot} = \mu_{3\cdot}$ implies that $\alpha_2 = \alpha_3$. Setting $\mu_{1\cdot} = \mu_{2\cdot}$ implies that $\mu - \alpha_2 - \alpha_3 = \mu + \alpha_2$. Thus setting $\mu_{1\cdot} = \mu_{2\cdot} = \mu_{3\cdot}$ implies that $\mu - 2\alpha_2 = \mu + \alpha_2$, which implies that $\alpha_2 = 0$. Hence we can test

$$H_0\colon \mu_{1\cdot} = \mu_{2\cdot} = \mu_{3\cdot}$$

by testing

$$H_0\colon \alpha_2 = \alpha_3 = 0$$

Specifically, if

$$F(\alpha|\gamma(\alpha)) = \frac{\{SSE_{\gamma(\alpha)} - SSE_{(\alpha,\gamma(\alpha))}\}/2}{SSE_{(\alpha,\gamma(\alpha))}/(31 - 7)}$$

is greater than $F_{[\alpha]}^{(2,24)}$, we can reject H_0 by setting the probability of a Type I error equal to α. Here, $SSE_{\gamma(\alpha)}$ is the unexplained variation for the $\gamma(\alpha)$ *model*

$$y_{ij,k} = \mu + \gamma_{2(1)}D_{ij,12} + \gamma_{2(2)}D_{ij,22} + \gamma_{2(3)}D_{ij,32} + \gamma_{3(3)}D_{ij,33} + \varepsilon_{ij,k}$$

Since the SAS output in Figure 16.9 indicates that the prob-value for $F(\alpha|\gamma(\alpha))$ is .0001, we have overwhelming evidence that we should reject H_0. We conclude that assembly methods have different effects on mean hourly output.

To see which assembly methods differ, we consider the pairwise differences

$$\mu_{2\cdot} - \mu_{1\cdot}$$

$$\mu_{3\cdot} - \mu_{1\cdot}$$

$$\mu_{3\cdot} - \mu_{2\cdot}$$

Examining the t statistics and prob-values for testing the significance of these differences (given in Figure 16.9), we conclude that all three of these differences are statistically significant. We can also estimate these pairwise differences. For example, a point estimate of

$$\mu_{3\cdot} - \mu_{2\cdot} = (\mu + \alpha_3) - (\mu + \alpha_2) = \alpha_3 - \alpha_2$$

is $\hat{\alpha}_3 - \hat{\alpha}_2 = 10.2833$. Here, $\hat{\alpha}_3$ and $\hat{\alpha}_2$ are the least squares point estimates of α_3 and α_2. Furthermore, an individual 95% confidence interval for

$$\begin{aligned} \mu_{3\cdot} - \mu_{2\cdot} &= \alpha_3 - \alpha_2 \\ &= 0\mu + (-1)\alpha_2 + 1\alpha_3 + 0\gamma_{2(1)} + 0\gamma_{2(2)} + 0\gamma_{2(3)} + 0\gamma_{3(3)} \\ &= \boldsymbol{\lambda}'\boldsymbol{\beta} \quad \text{where } \boldsymbol{\lambda}' = [0 \quad -1 \quad 1 \quad 0 \quad 0 \quad 0 \quad 0] \end{aligned}$$

is

$$\begin{aligned} [(\hat{\alpha}_3 - \hat{\alpha}_2) \pm t_{[\alpha/2]}^{(n-k)} s_{(\alpha,\gamma(\alpha))}\sqrt{\boldsymbol{\lambda}'(\mathbf{X}'\mathbf{X})^{-1}\boldsymbol{\lambda}}] &= [10.2833 \pm t_{[.025]}^{(31-7)}(1.7777)] \\ &= [10.2833 \pm 2.064(1.7777)] \\ &= [10.2833 \pm 3.6692] \\ &= [6.6141, 13.9525] \end{aligned}$$

Examining Figure 16.9, we conclude that assembly method 3 yields the largest mean hourly output.

Finally, we note from Figure 16.9 that a point estimate of μ_{33} is

$$\begin{aligned} \hat{y}_{33} &= \hat{\mu} + \hat{\alpha}_2(0) + \hat{\alpha}_3(1) + \hat{\gamma}_{2(1)}(0) + \hat{\gamma}_{2(2)}(0) + \hat{\gamma}_{2(3)}(0) + \hat{\gamma}_{3(3)}(1) \\ &= 34.25 \end{aligned}$$

Moreover, a 95% confidence interval for μ_{33} is

$$[\hat{y}_{33} \pm t_{[.025]}^{(31-7)} s\sqrt{\mathbf{x}_0'(\mathbf{X}'\mathbf{X})^{-1}\mathbf{x}_0}] = [30.2038, 38.2962]$$

where $\mathbf{x}_0' = [1 \quad 0 \quad 1 \quad 0 \quad 0 \quad 0 \quad 1]$.

*16.4 USING SAS

In Figure 16.10 we present the SAS program that yields the North American Oil Company gasoline mileage analysis in Figure 16.3.

In Figure 16.11 we present the SAS program that yields the North American Oil Company premium gasoline mileage analysis in Figures 16.6 and 16.7.

In Figure 16.12 we show how to use SAS to analyze the North American Oil Company regular gasoline mileage data by explicitly coding dummy variables. Here, we enter the data in the order that it is specified in the **y**-vector in Example 16.1. Note that we show how to employ both the (α, γ) model and the means model.

Finally, in Figure 16.13 we present the SAS program that yields the assembly method data analysis in Figure 16.8.

*This section is optional.

FIGURE 16.10 **SAS program that produces the (α, γ) model analysis of the regular gasoline mileage data shown in Figure 16.3**

```
DATA GASONE ;
INPUT GASTYPE $ ADDTYPE $ MILEAGE @ @ ; } ——→ Defines factors GASTYPE
                                               and ADDTYPE and the
CARDS ;                                        response variable MILEAGE
A  M  19.4  A  M  20.6  A  M  20.0  )
A  N  25.0  A  N  24.0              |
A  O  24.3  A  O  25.5              |
B  M  22.6  B  M  21.6              |
B  N  25.3  B  N  26.5              |
B  O  27.6  B  O  26.4  B  O  27.0  } ——→ Data—see Table 16.1
B  P  25.4                          |
C  M  18.6  C  M  19.8              |
C  O  24.0  C  O  24.7  C  O  23.3  |
C  P  22.0  C  P  23.0              )
;
PROC GLM ; } ——→ Specifies General Linear Models Procedure
CLASS GASTYPE ADDTYPE ; } ——→ Defines class variables GASTYPE and ADDTYPE
MODEL MILEAGE = GASTYPE ADDTYPE / P CLM ; } ——→ Specifies model
ESTIMATE 'MUB-MUA' GASTYPE -1 1 ; } ——→ Estimates
ESTIMATE 'MUC-MUA' GASTYPE -1 0 1 ; } ——→ Estimates
ESTIMATE 'MUC-MUB' GASTYPE 0 -1 1 ; } ——→ Estimates
ESTIMATE 'MUB-(MUC+MUA)/2' GASTYPE -.5 1 -.5 ; } ——→ Estimates
ESTIMATE 'MUN-MUM' ADDTYPE -1 1 ; } ——→ Estimates
ESTIMATE 'MUO-MUM' ADDTYPE -1 0 1 ; } ——→ Estimates
ESTIMATE 'MUP-MUM' ADDTYPE -1 0 0 1 ; } ——→ Estimates
ESTIMATE 'MUO-MUN' ADDTYPE 0 -1 1 ; } ——→ Estimates
ESTIMATE 'MUP-MUN' ADDTYPE 0 -1 0 1 ; } ——→ Estimates
ESTIMATE 'MUP-MUO' ADDTYPE 0 0 -1 1 ; } ——→ Estimates
PROC GLM ;
CLASS GASTYPE ADDTYPE ;
MODEL MILEAGE = GASTYPE ADDTYPE / P CLI ; } ——→ CLI requests prediction intervals
```

Annotations (in order): Specifies model $y_{ij,k} = \mu_{ij} + \varepsilon_{ij,k}$ where $\mu_{ij} = \mu + \alpha_i + \gamma_j$; Estimates $\mu_{Bj} - \mu_{Aj}$; Estimates $\mu_{Cj} - \mu_{Aj}$; Estimates $\mu_{Cj} - \mu_{Bj}$; Estimates $\mu_{Bj} - \left(\frac{\mu_{Cj} + \mu_{Aj}}{2}\right)$; Estimates $\mu_{iN} - \mu_{iM}$; Estimates $\mu_{iO} - \mu_{iM}$; Estimates $\mu_{iP} - \mu_{iM}$; Estimates $\mu_{iO} - \mu_{iN}$; Estimates $\mu_{iP} - \mu_{iN}$; Estimates $\mu_{iP} - \mu_{iO}$

Note that the coefficients in the above ESTIMATE statements are obtained by considering the alphabetically ordered factor levels of GASTYPE (A, B, C) and the alphabetically ordered factor levels of ADDTYPE (M, N, O, P). "Trailing zero coefficients" have been dropped.

FIGURE 16.11 **SAS program that produces the two-factor means model analysis of the premium gasoline mileage data (in Figure 16.6) and the test for interaction between premium gasoline type and additive type (in Figure 16.7)**

```
DATA GASTWO ;
INPUT GASTYPE $ ADDTYPE $ MILEAGE @ @ ; }  ——→ Defines factors GASTYPE
                                               and ADDTYPE and response
                                               variable MILEAGE
CARDS ;
D  Q  18.0  D  Q  18.6  D  Q  17.4
D  R  23.0  D  R  22.0
D  S  23.5  D  S  22.3
E  Q  23.3  E  Q  24.5
E  R  25.6  E  R  24.4  E  R  25.0
E  S  23.4                                 ——→ Data—see Table 16.2
E  T  19.6  E  T  20.6
F  R  24.0  F  R  23.3  F  R  24.7
F  S  23.0  F  S  22.0
F  T  18.6  F  T  19.8
;

PROC GLM ;                                         Produces output in
CLASS GASTYPE ADDTYPE ;                            Figure 16.6 including
MODEL MILEAGE = GASTYPE*ADDTYPE / P CLM ;          F(means model)
ESTIMATE 'MUER-MUDS' GASTYPE*ADDTYPE 0 0 -1 0 1 ;  and related prob-value
                                        └→ Estimates μER − μDS

PROC GLM ;                                                   Produces output in
CLASS GASTYPE ADDTYPE ;                                      Figure 16.7 including
MODEL MILEAGE = GASTYPE ADDTYPE GASTYPE*ADDTYPE ;            F(interaction) and
                                                             related prob-value
```

Notes:

1. *The coefficients in the above ESTIMATE statement for MUER-MUDS (that is,* $\mu_{ER} - \mu_{DS}$*) follow from the alphabetically ordered factor level combinations of GASTYPE and ADDTYPE:*

DQ	DR	DS	EQ	ER	ES	ET	FR	FS	FT
0	0	−1	0	1	0	0	0	0	0

   In the above listing we omit DT and FQ because no data has been obtained for these combinations.

2. *MODEL MILEAGE = GASTYPE ADDTYPE GASTYPE*ADDTYPE;* *allows us to test for interaction by using the* (α, γ, θ) *model.*

FIGURE 16.12 **SAS program that produces an (α, γ) model analysis and a two-factor means model analysis of the regular gasoline mileage data in Table 16.1 by using coded dummy variables**

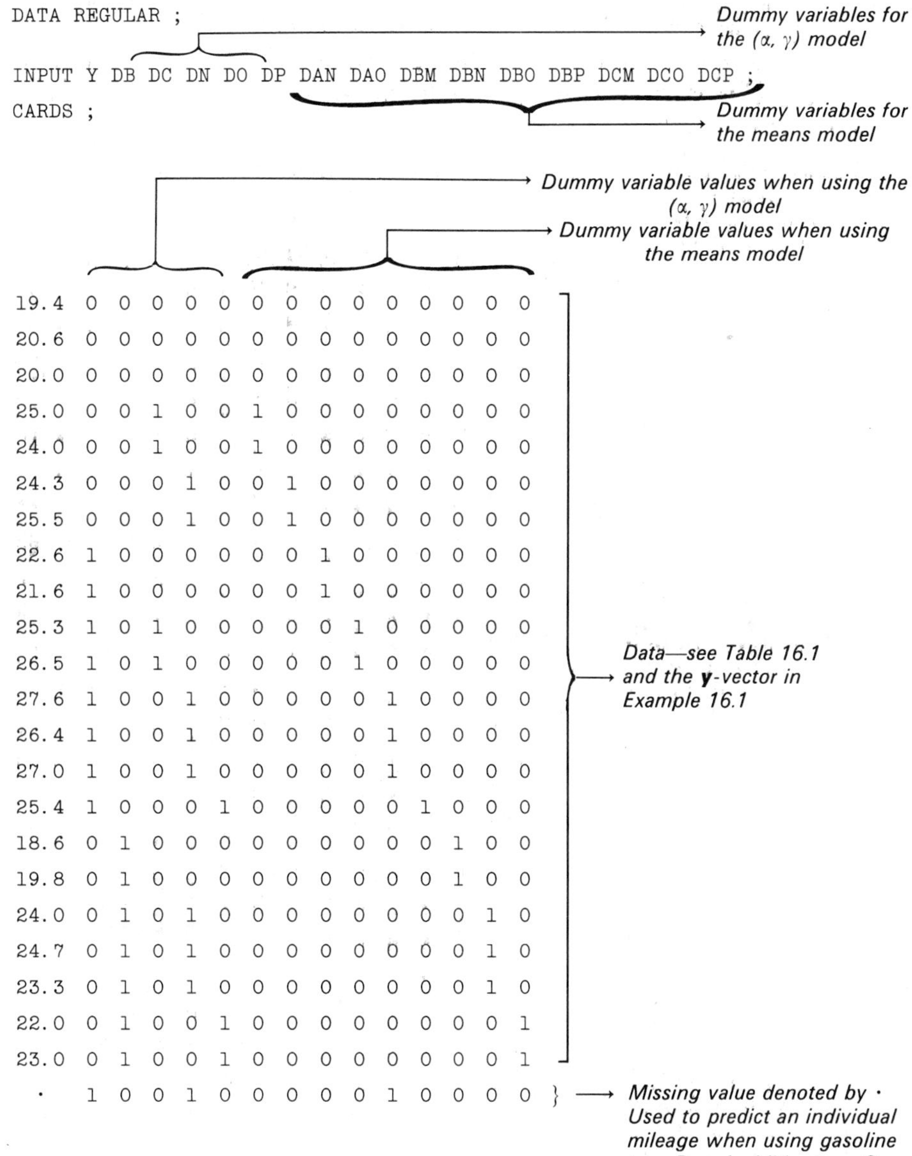

```
DATA REGULAR ;
INPUT Y DB DC DN DO DP DAN DAO DBM DBN DBO DBP DCM DCO DCP ;
CARDS ;
19.4 0 0 0 0 0 0 0 0 0 0 0 0 0 0
20.6 0 0 0 0 0 0 0 0 0 0 0 0 0 0
20.0 0 0 0 0 0 0 0 0 0 0 0 0 0 0
25.0 0 0 1 0 0 1 0 0 0 0 0 0 0 0
24.0 0 0 1 0 0 1 0 0 0 0 0 0 0 0
24.3 0 0 0 1 0 0 1 0 0 0 0 0 0 0
25.5 0 0 0 1 0 0 1 0 0 0 0 0 0 0
22.6 1 0 0 0 0 0 0 1 0 0 0 0 0 0
21.6 1 0 0 0 0 0 0 1 0 0 0 0 0 0
25.3 1 0 1 0 0 0 0 0 1 0 0 0 0 0
26.5 1 0 1 0 0 0 0 0 1 0 0 0 0 0
27.6 1 0 0 1 0 0 0 0 0 1 0 0 0 0
26.4 1 0 0 1 0 0 0 0 0 1 0 0 0 0
27.0 1 0 0 1 0 0 0 0 0 1 0 0 0 0
25.4 1 0 0 0 1 0 0 0 0 0 1 0 0 0
18.6 0 1 0 0 0 0 0 0 0 0 0 1 0 0
19.8 0 1 0 0 0 0 0 0 0 0 0 1 0 0
24.0 0 1 0 1 0 0 0 0 0 0 0 0 1 0
24.7 0 1 0 1 0 0 0 0 0 0 0 0 1 0
23.3 0 1 0 1 0 0 0 0 0 0 0 0 1 0
22.0 0 1 0 0 1 0 0 0 0 0 0 0 0 1
23.0 0 1 0 0 1 0 0 0 0 0 0 0 0 1
  .  1 0 0 1 0 0 0 0 0 1 0 0 0 0
```


(*continues*)

FIGURE 16.12 Continued

```
PROC REG ; } ——→ Specifies Regression Procedure

MODEL Y = DB DC DN DO DP / P CLM CLI ; } ——→ Specifies the (α, γ) model and
                                             calculates the least squares point
                                             estimates in Figure 16.2

T1: TEST DB = 0, DC = 0 ;            } ——→ Carries out the partial F-tests
T2: TEST DN = 0, DO = 0, DP = 0 ;    }     in Example 16.1

PROC GLM ; } ——→ Specifies the General Linear Models Procedure

MODEL Y = DB DC DN DO DP / P CLM ; } ——→ Specifies the (α, γ) model and
                                         calculates the least squares point
                                         estimates given in Figure 16.2

ESTIMATE 'MUB-MUA' DB 1 ; } ——→ Estimates μBj − μAj
ESTIMATE 'MUC-MUA' DC 1 ; } ——→ Estimates μCj − μAj
ESTIMATE 'MUC-MUB' DB -1 DC 1 ; } ——→ Estimates μCj − μBj

ESTIMATE 'MUB-(MUC+MUA)/2' DB 1 DC -.5 ; } ——→ Estimates μBj − (μCj + μAj)/2

ESTIMATE 'MUN-MUM' DN 1 ; } ——→ Estimates μiN − μiM
ESTIMATE 'MUO-MUM' DO 1 ; } ——→ Estimates μiO − μiM
ESTIMATE 'MUP-MUM' DP 1 ; } ——→ Estimates μiP − μiM
ESTIMATE 'MUO-MUN' DN -1 DO 1 ; } ——→ Estimates μiO − μiN
ESTIMATE 'MUP-MUN' DN -1 DP 1 ; } ——→ Estimates μiP − μiN
ESTIMATE 'MUP-MUO' DO -1 DP 1 ; } ——→ Estimates μiP − μiO
ESTIMATE 'MUBO-MUAN' DB 1 DN -1 DO 1 ; } ——→ Estimates μBO − μAN
                                             using the (α, γ) model

PROC GLM ; } ——→ Specifies the general linear models procedure
                 (Multiple model statements cannot be used in PROC GLM)

MODEL Y = DAN DAO DBM DBN DBO DBP DCM DCO DCP / P CLM ; } ——→ Specifies the
                                                              Means Model

ESTIMATE 'MUBO-MUAN' DBO 1 DAN -1 ; } ——→ Estimates μBO − μAN
                                         using the means model
```

FIGURE 16.13 **SAS program that produces the nested model ANOVA of the assembly method data in Figure 16.8**

```
DATA ASSEMBLY ;
INPUT METHOD $ STATION $ UNITS @ @ ; } ——→ Defines factors METHOD and
                                           STATION and response variable UNITS
CARDS ;
1 1 16
1 1  7
1 1  7
1 1 13
1 1 16
1 2 14
1 2 24
1 2 13
1 2 17
1 2 21
2 1 21
2 1 25
2 1 16
2 1 18
2 1 16
2 2 24      } ——→ Data—see Table 16.6
2 2 28
2 2 27
2 2 25
2 2 21
3 1 25
3 1 35
3 1 33
3 1 31
3 1 28
3 2 31
3 2 31
3 2 38
3 2 36
3 2 35
PROC GLM ; } ——→ Specifies the General Linear Models Procedure
```

(*continues*)

FIGURE 16.13 Continued

```
CLASS METHOD STATION ; }
```
→ *Defines class variables METHOD and STATION*

```
MODEL UNITS = METHOD STATION(METHOD) / P CLM ; }
```
→ *Specifies the model* $y_{ij,k} = \mu_{ij} + \varepsilon_{ij,k}$ *where* $\mu_{ij} = \alpha_i + \gamma_{j(i)}$

```
ESTIMATE 'MU2-MU1' METHOD -1 1 ; }
```
→ *Estimates* $\mu_{2\cdot} - \mu_{1\cdot}$

```
ESTIMATE 'MU3-MU1' METHOD -1 0 1 ; }
```
→ *Estimates* $\mu_{3\cdot} - \mu_{1\cdot}$

```
ESTIMATE 'MU3-MU2' METHOD 0 -1 1 ; }
```
→ *Estimates* $\mu_{3\cdot} - \mu_{2\cdot}$

```
ESTIMATE 'MU12-MU11' METHOD 0 0 0 STATION(METHOD) -1 1 ; }
```
→ *Estimates* $\mu_{12} - \mu_{11}$

```
ESTIMATE 'MU22-MU21' METHOD 0 0 0 STATION(METHOD) 0 0 -1 1 ; }
```
→ *Estimates* $\mu_{22} - \mu_{21}$

```
ESTIMATE 'MU32-MU31' METHOD 0 0 0 STATION(METHOD) 0 0 0 0 -1 1 ; }
```
→ *Estimates* $\mu_{32} - \mu_{31}$

The ordered levels of assembly method are

1 2 3

Thus the coefficients in the ESTIMATE statement for MU3 − MU2 (that is, $\mu_{3\cdot} - \mu_{2\cdot}$*) are*

0 −1 1

As another example, consider defining the coefficients in the estimate statement for MU22 − MU21 (that is, $\mu_{22} - \mu_{21}$*).*

The ordered levels of station within method are

1(1) 2(1) 1(2) 2(2) 1(3) 2(3)

Note that

$$\begin{aligned}\mu_{22} - \mu_{21} &= (\mu + \alpha_2 + \gamma_{2(2)}) - (\mu + \alpha_2 + \gamma_{1(2)}) \\ &= \gamma_{2(2)} - \gamma_{1(2)} \\ &= 0(\alpha_1) + 0(\alpha_2) + 0(\alpha_3) + 0(\gamma_{1(1)}) + 0(\gamma_{2(1)}) + (-1)\gamma_{1(2)} \\ &\quad + 1(\gamma_{2(2)}) + 0(\gamma_{1(3)}) + 0(\gamma_{2(3)})\end{aligned}$$

Hence the coefficients for METHODS are

0 0 0

and the coefficients for STATION(METHOD) are

0 0 −1 1

Here, we have dropped the "trailing zero coefficients."

EXERCISES

Note: In Section 16.1 we showed how to use the (α, γ) model and the two-factor means model to analyze an incomplete factorial experiment. We also stated that these models can be used to analyze a (balanced or unbalanced) complete factorial experiment. This will be illustrated in Exercises 16.1 through 16.10.

16.1 The Tastee Bakery wishes to study the effect of two qualitative factors on monthly demand for a bakery product (measured in cases of 10 units each). These factors are shelf display height, which has levels B (Bottom), M (Middle), and T (Top); and shelf display width, which has levels R (Regular) and W (Wide). For i = B, M, and T and j = R and W we define *the (i, j)th population of demands* to be the infinite population of all possible monthly demands for the product that could be obtained at supermarkets that use display height i and display width j. Furthermore, we let μ_{ij} denote the mean of this population. To compare the treatment means, Tastee Bakery will, for i = B, M, and T, and j = R and W, randomly select a sample of n_{ij} metropolitan area supermarkets. These supermarkets will sell the product for one month using display height i and display width j. Consider the kth randomly selected supermarket (where $k = 1, 2, \ldots, n_{ij}$) that uses display height i and display width j. Let

$$y_{ij,k} = \text{the monthly demand that occurs in this supermarket}$$

Suppose that when Tastee Bakery employs this completely randomized design, it obtains the six samples summarized in Table 16.11.

a. Calculate the means of the six samples. Using these sample means, perform a graphical analysis that indicates whether or not interaction exists between display height and display width.

b. Since the graphical analysis performed in part (a) implies that little or no interaction exists between display height and display width, we analyze these data by considering the (α, γ) model:

$$\begin{aligned} y_{ij,k} &= \mu_{ij} + \varepsilon_{ij,k} \\ &= \mu + \alpha_i + \gamma_j + \varepsilon_{ij,k} \\ &= \mu + \alpha_M D_{i,M} + \alpha_T D_{i,T} + \gamma_W D_{j,W} + \varepsilon_{ij,k} \end{aligned}$$

TABLE 16.11 **Six samples of monthly demands**

Display height, *i*	Display width, *j*: R	W
B	$y_{BR,1} = 58.2$ $y_{BR,2} = 53.7$	$y_{BW,1} = 55.6$
M	$y_{MR,1} = 73.0$ $y_{MR,2} = 78.1$	$y_{MW,1} = 76.2$ $y_{MW,2} = 82.0$ $y_{MW,3} = 78.4$
T	$y_{TR,1} = 52.5$ $y_{TR,2} = 49.8$	$y_{TW,1} = 56.0$ $y_{TW,2} = 51.9$ $y_{TW,3} = 53.3$

Here, $D_{i,\text{M}} = 1$ if $i = \text{M}$, and $D_{i,\text{M}} = 0$ otherwise; $D_{i,\text{T}} = 1$ if $i = \text{T}$, and $D_{i,\text{T}} = 0$ otherwise; and $D_{j,\text{W}} = 1$ if $j = \text{W}$, and $D_{j,\text{W}} = 0$ otherwise. Using the fact that the (α, γ) model implies that

$$\begin{aligned} \mu_{ij} &= \mu + \alpha_i + \gamma_j \\ \alpha_i &= \alpha_{\text{M}} D_{i,\text{M}} + \alpha_{\text{T}} D_{i,\text{T}} \\ \gamma_j &= \gamma_{\text{W}} D_{j,\text{W}} \end{aligned}$$

show that $\alpha_{\text{B}} = 0$ and $\gamma_{\text{R}} = 0$. Also, express the following means and differences in means in terms of μ, α_{M}, α_{T}, and γ_{W}:

$$\mu_{\text{MW}} \quad \mu_{\text{B}j} \quad \mu_{\text{M}j} \quad \mu_{\text{T}j} \quad \mu_{\text{M}j} - \mu_{\text{B}j} \quad \mu_{\text{T}j} - \mu_{\text{B}j}$$

$$\mu_{\text{T}j} - \mu_{\text{M}j} \quad \mu_{i\text{R}} \quad \mu_{i\text{W}} \quad \mu_{i\text{W}} - \mu_{i\text{R}}$$

Then show that $H_0: \alpha_{\text{M}} = \alpha_{\text{T}} = 0$ is equivalent to $H_0: \mu_{\text{B}j} = \mu_{\text{M}j} = \mu_{\text{T}j}$, and show that $H_0: \gamma_{\text{W}} = 0$ is equivalent to $H_0: \mu_{i\text{R}} = \mu_{i\text{W}}$.

c. Specify the vector $\mathbf{y}$ and the matrix $\mathbf{X}$ used to calculate the least squares point estimates of the parameters in the (α, γ) model.

d. When we use the (α, γ) model to perform a regression analysis of this data, we find that the least squares point estimates of the parameters in this model are

$$\hat{\mu} = 55.0891 \quad \hat{\alpha}_{\text{M}} = 21.1113 \quad \hat{\alpha}_{\text{T}} = -3.7287 \quad \hat{\gamma}_{\text{W}} = 2.2326$$

We also find that the unexplained variation is $SSE_{(\alpha,\gamma)} = 58.6127$. Furthermore, when we use the α model

$$\begin{aligned} y_{ij,k} &= \mu_{ij} + \varepsilon_{ij,k} \\ &= \mu + \alpha_i + \varepsilon_{ij,k} \\ &= \mu + \alpha_{\text{M}} D_{i,\text{M}} + \alpha_{\text{T}} D_{i,\text{T}} + \varepsilon_{ij,k} \end{aligned}$$

and the γ model

$$\begin{aligned} y_{ij,k} &= \mu_{ij} + \varepsilon_{ij,k} \\ &= \mu + \gamma_j + \varepsilon_{ij,k} \\ &= \mu + \gamma_{\text{W}} D_{i,\text{W}} + \varepsilon_{ij,k} \end{aligned}$$

to perform regression analyses of this data, we obtain unexplained variations of $SSE_\alpha = 73.8987$ and $SSE_\gamma = 1766.6426$, respectively. Using these unexplained variations, calculate the $F(\alpha|\gamma)$ statistic used to test

$$H_0: \alpha_{\text{M}} = \alpha_{\text{T}} = 0 \quad \text{or} \quad \mu_{\text{B}j} = \mu_{\text{M}j} = \mu_{\text{T}j}$$

versus

H_1: At least one of α_{M} or α_{T} does not equal zero

or

H_1: At least two of $\mu_{\text{B}j}$, $\mu_{\text{M}j}$, and $\mu_{\text{T}j}$ differ from each other

Also, calculate the $F(\gamma|\alpha)$ statistic used to test

$$H_0: \gamma_{\text{W}} = 0 \quad \text{or} \quad \mu_{i\text{R}} = \mu_{i\text{W}}$$

versus

$$H_1: \gamma_{\text{W}} \neq 0 \quad \text{or} \quad \mu_{i\text{R}} \neq \mu_{i\text{W}}$$

Then use these partial F statistics and the appropriate rejection points to determine whether we can reject these null hypotheses by setting α equal to .05. The prob-values related to these null hypotheses can be calculated to be .0001 and .1599, respectively.

TABLE 16.12 **Standard errors and prob-values for several linear combinations in the bakery product demand study**

Parameter number	*Parameter* $= \lambda'\beta$	$s\sqrt{\lambda'(X'X)^{-1}\lambda}$	*Prob-value for* $H_0\colon \lambda'\beta = 0$
1	$\mu_{Mj} - \mu_{Bj}$	1.9038	.0001
2	$\mu_{Tj} - \mu_{Bj}$	1.9038	.0818
3	$\mu_{Tj} - \mu_{Mj}$	1.614	.0001
4	$\mu_{iW} - \mu_{iR}$	1.4573	.1599
5	μ_{MW}	1.2815	—

Demonstrate how these prob-values have been calculated, and use these prob-values to determine whether we can reject the above null hypotheses by setting α equal to .05 or .01.

e. Consider Table 16.12. In part (b) the reader expressed each of these parameters in terms of μ, α_M, α_T, and γ_W. By using the least squares point estimates of μ, α_M, α_T, and γ_W given in part (d), calculate a point estimate of each parameter in the table and find a 95% confidence interval for each parameter by using the formula

$$[\boldsymbol{\lambda}'\mathbf{b} \pm t_{[\alpha/2]}^{(n-k)} s\sqrt{\boldsymbol{\lambda}'(\mathbf{X}'\mathbf{X})^{-1}\boldsymbol{\lambda}}]$$

Also, find $\boldsymbol{\lambda}'$ for each parameter.

f. Calculate the t statistic for testing $H_0\colon \boldsymbol{\lambda}'\boldsymbol{\beta} = 0$ for each of parameters 1, 2, 3, and 4 in Table 16.12. Also, show how the prob-values in the table have been calculated, and discuss what these prob-values say about the validity of the appropriate null hypotheses.

g. Using any pertinent information in this problem, calculate Scheffé simultaneous 95% confidence intervals for parameters 1, 2, 3, and 4 in Table 16.12, calculate Scheffé simultaneous 95% confidence intervals for parameters 1, 2, 3, 4, and 5 in this table and calculate Scheffé simultaneous 95% confidence intervals for parameters 1, 2, and 3 in Table 16.12.

h. Calculate a point prediction of, and a 95% prediction interval for, $y_{MW,0}$, a future (individual) demand that will be observed when using a middle display height and a wide display.

i. If we remove the subscript j from the data in this exercise, the data become the data in Exercise 14.1. Compare the individual and Scheffé simultaneous confidence intervals for $\mu_M - \mu_B$, $\mu_T - \mu_B$, and $\mu_T - \mu_M$ calculated in Exercise 14.15 with the corresponding intervals for $\mu_{Mj} - \mu_{Bj}$, $\mu_{Tj} - \mu_{Bj}$, and $\mu_{Tj} - \mu_{Mj}$ calculated in this exercise. Does the model of Exercise 14.15 or the model of this exercise seem best for calculating these intervals? Explain.

16.2 Figure 16.14 presents the SAS output obtained by using PROC GLM and the (α, γ) model to analyze the demand data in Exercise 16.1. Show that the values of $F(\alpha|\gamma)$ and $F(\gamma|\alpha)$ on the output are the same as those calculated in part (d) of Exercise 16.1. Show that the point estimate $\boldsymbol{\lambda}'\mathbf{b}$ and standard error of the estimate $s\sqrt{\boldsymbol{\lambda}'(\mathbf{X}'\mathbf{X})^{-1}\boldsymbol{\lambda}}$ for each of the parameters

$$\mu_{Mj} - \mu_{Bj} \qquad \mu_{Tj} - \mu_{Bj} \qquad \mu_{Tj} - \mu_{Mj} \qquad \mu_{iW} - \mu_{iR}$$

FIGURE 16.14 SAS output of an (α, γ) model analysis of the bakery product demand data in Table 16.11

SAS

GENERAL LINEAR MODELS PROCEDURE

DEPENDENT VARIABLE: DEMAND

SOURCE	DF	SUM OF SQUARES	MEAN SQUARE	F VALUE	PR > F	R-SQUARE	C.V.
MODEL	3	1756.87033779	585.62344593	89.92	0.0001	0.967715	4.0522
ERROR	9	58.61273913	6.51252657		ROOT MSE		DEMAND MEAN
CORRECTED TOTAL	12	1815.48307692			2.55196524		62.97692308

SOURCE	DF	TYPE I SS	F VALUE	PR > F	DF	TYPE III SS	F VALUE	PR > F
HEIGHT	2	1741.58441026	133.71	0.0001	2	1708.02987992	131.13	0.0001
WIDTH	1	15.28592754	2.35	0.1599	1	15.28592754	2.35	0.1599

PARAMETER	ESTIMATE	T FOR HO: PARAMETER=0	PR > |T|	STD ERROR OF ESTIMATE
MUM-MUB	21.11130435	11.09	0.0001	1.90377588
MUT-MUB	-3.72869565	-1.96	0.0818	1.90377588
MUT-MUM	-24.84000000	-15.39	0.0001	1.61400453
MUW-MUR	2.23260870	1.53	0.1599	1.45727484

OBSERVATION	OBSERVED VALUE	PREDICTED VALUE	RESIDUAL	LOWER 95% CL FOR MEAN	UPPER 95% CL FOR MEAN
1	58.20000000	55.08913043	3.11086957	51.57961814	58.59864273
2	53.70000000	55.08913043	-1.38913043	51.57961814	58.59864273
3	55.60000000	57.32173913	-1.72173913	53.32934280	61.31413546
4	73.00000000	76.20043478	-3.20043478	72.94807385	79.45279571
5	78.10000000	76.20043478	1.89956522	72.94807385	79.45279571
6	76.20000000	78.43304348	-2.23304348	75.53402287	81.33206408
7	82.00000000	78.43304348	3.56695652	75.53402287	81.33206408
8	78.40000000	78.43304348	-0.03304348	75.53402287	81.33206408
9	52.50000000	51.36043478	1.13956522	48.10807385	54.61279571
10	49.80000000	51.36043478	-1.56043478	48.10807385	54.61279571
11	56.00000000	53.59304348	2.40695652	50.69402287	56.49206408
12	51.90000000	53.59304348	-1.69304348	50.69402287	56.49206408
13	53.30000000	53.59304348	-0.29304348	50.69402287	56.49206408

OBSERVATION	OBSERVED VALUE	PREDICTED VALUE	RESIDUAL	LOWER 95% CL INDIVIDUAL	UPPER 95% CL INDIVIDUAL
1	58.20000000	55.08913043	3.11086957	48.33308468	61.84517619
2	53.70000000	55.08913043	-1.38913043	48.33308468	61.84517619
3	55.60000000	57.32173913	-1.72173913	50.30271454	64.34076372
4	73.00000000	76.20043478	-3.20043478	69.57432511	82.82654445
5	78.10000000	76.20043478	1.89956522	69.57432511	82.82654445
6	76.20000000	78.43304348	-2.23304348	71.97302814	84.89305882
7	82.00000000	78.43304348	3.56695652	71.97302814	84.89305882
8	78.40000000	78.43304348	-0.03304348	71.97302814	84.89305882
9	52.50000000	51.36043478	1.13956522	44.73432511	57.98654445
10	49.80000000	51.36043478	-1.56043478	44.73432511	57.98654445
11	56.00000000	53.59304348	2.40695652	47.13302814	60.05305882
12	51.90000000	53.59304348	-1.69304348	47.13302814	60.05305882
13	53.30000000	53.59304348	-0.29304348	47.13302814	60.05305882

on the output are the same as those calculated in part (e) of Exercise 16.1. Show that the point estimate of, and 95% confidence interval for, μ_{MW} on the output are the same as those calculated in part (e) of Exercise 16.1. Show that the point prediction of, and 95% prediction interval for, $y_{MW,0}$ on the output are the same as those calculated in part (h) of Exercise 16.1.

16.3 (Optional) Write a SAS program using a CLASS statement to obtain the output in Figure 16.14.

16.4 (Optional) Write a SAS program that uses the dummy variable regression model of Exercise 16.1 to obtain the same results given in Figure 16.14. In the program, actually define and use the appropriate values of the dummy variables.

16.5 In this exercise, consider analyzing the data in Table 16.11 by using the two-factor means model:

$$\begin{aligned} y_{ij,k} &= \mu_{ij} + \varepsilon_{ij,k} \\ &= \beta_{BR} + \beta_{BW} D_{ij,BW} + \beta_{MR} D_{ij,MR} + \beta_{MW} D_{ij,MW} + \beta_{TR} D_{ij,TR} + \beta_{TW} D_{ij,TW} + \varepsilon_{ij,k} \end{aligned}$$

Here, for example, $D_{ij,MW} = 1$ if $i = $ M and $j = $ W, and $D_{ij,MW} = 0$ otherwise.

a. Express the following means and differences in means in terms of β_{BR}, β_{BW}, β_{MR}, β_{MW}, β_{TR}, and β_{TW}:

$$\mu_{BR} \quad \mu_{BW} \quad \mu_{MR} \quad \mu_{MW} \quad \mu_{TR} \quad \mu_{TW} \quad \mu_{BW} - \mu_{BR} \quad \mu_{MR} - \mu_{BR}$$

$$\mu_{MW} - \mu_{BR} \quad \mu_{TR} - \mu_{BR} \quad \mu_{TW} - \mu_{BR}$$

Then show that

$$H_0: \beta_{BW} = \beta_{MR} = \beta_{MW} = \beta_{TR} = \beta_{TW} = 0$$

is equivalent to

$$H_0: \mu_{BR} = \mu_{BW} = \mu_{MR} = \mu_{MW} = \mu_{TR} = \mu_{TW}$$

b. Specify the vector **y** and the matrix **X** used to calculate the least squares point estimates of the parameters in the means model.

c. When we use the means model to perform a regression analysis of the data in Table 16.11, we find that the least squares point estimates of the parameters in this model are

$$b_{BR} = 55.95 \qquad b_{BW} = -.35 \qquad b_{MR} = 19.6 \qquad b_{MW} = 22.9167$$

$$b_{TR} = -4.8 \qquad b_{TW} = -2.2167$$

We also obtain the following quantities:

$$SS_{\text{means}} = \text{explained variation} = 1762.8747$$

$$SSE_{\text{means}} = \text{unexplained variation} = 52.6083$$

By using these quantities to calculate the F(means model) statistic and by using the appropriate rejection point, determine whether we can, by setting α equal to .05, reject

$$H_0: \beta_{BW} = \beta_{MR} = \beta_{MW} = \beta_{TR} = \beta_{TW} = 0$$

or

$$H_0: \mu_{BR} = \mu_{BW} = \mu_{MR} = \mu_{MW} = \mu_{TR} = \mu_{TW}$$

in favor of

H_1: At least one of β_{BW}, β_{MR}, β_{MW}, β_{TR}, and β_{TW} does not equal zero

or

H_1: At least two of μ_{BR}, μ_{BW}, μ_{MR}, μ_{MW}, μ_{TR}, and μ_{TW} differ from each other

The prob-value for testing H_0 versus H_1 can be calculated to be .0001. Demonstrate how this prob-value has been calculated, and use this prob-value to determine whether we can reject H_0 in favor of H_1 by setting α equal to .05 or .01.

d. Consider the following table.

Parameter number	Parameter $= \lambda'\beta$	$s\sqrt{\lambda'(X'X)^{-1}\lambda}$
1	$\mu_{MW} - \mu_{BR}$	2.5026
2	μ_{MW}	1.5827

In part (a) the reader expressed each of these parameters in terms of β_{BR}, β_{BW}, β_{MR}, β_{MW}, β_{TR}, and β_{TW}. By using the least squares point estimates of these parameters given in part (c), calculate a point estimate of each parameter in the table, and find a 95% confidence interval for each parameter by using the formula

$$[\boldsymbol{\lambda}'\mathbf{b} \pm t_{[\alpha/2]}^{(n-k)} s\sqrt{\boldsymbol{\lambda}'(\mathbf{X}'\mathbf{X})^{-1}\boldsymbol{\lambda}}]$$

Also, find $\boldsymbol{\lambda}'$ for each parameter.

e. Calculate a point prediction of, and a 95% prediction interval for, $y_{MW,0}$, a future (individual) product demand that will be observed when using a middle display height and a wide display.

f. By using any needed information from this exercise and from Exercise 16.1, calculate the F(interaction) statistic. Then, by using this statistic and the appropriate rejection point, determine whether we can, by setting α equal to .05, reject

H_0: No interaction exists between display height and display width

in favor of

H_1: Interaction does exist between display height and display width

Next, compare the length of the 95% confidence interval for μ_{MW} calculated by using the means model in this exercise with the length of the 95% confidence interval for μ_{MW} calculated by using the (α, γ) model in Table 16.11. Which model seems to best describe the data in Table 16.11—the (α, γ) model or the means model? Explain.

16.6 Figure 16.15 presents the SAS output obtained by using PROC GLM and the two-factor means model to analyze the demand data in Table 16.11. Identify on the output the quantities referred to in Exercise 16.5. Figure 16.16 presents the SAS output obtained by using PROC GLM and the (α, γ, θ) model to calculate the $F(\theta|\alpha, \gamma)$ statistic. Identify this statistic and note that it equals the F(interaction) statistic calculated in part (f) of Exercise 16.5.

16.7 (Optional) Write a SAS program using a CLASS statement to obtain the outputs in Figures 16.15 and 16.16.

16.8 (Optional) Write a SAS program that uses the dummy variable regression model of

FIGURE 16.15 SAS output of a two-factor means model analysis of the bakery product demand data in Table 16.11

```
                                          SAS
                          GENERAL LINEAR MODELS PROCEDURE
DEPENDENT VARIABLE: DEMAND

SOURCE           DF  SUM OF SQUARES    MEAN SQUARE   F VALUE      PR > F   R-SQUARE          C.V.
MODEL             5   1762.87474359   352.57494872     46.91      0.0001   0.971022        4.3531
ERROR             7     52.60833333     7.51547619               ROOT MSE               DEMAND MEAN
CORRECTED TOTAL  12   1815.48307692                            2.74143688               62.97692308

SOURCE         DF      TYPE I SS     F VALUE    PR > F    DF    TYPE III SS     F VALUE    PR > F
HEIGHT*WIDTH    5   1762.87474359      46.91    0.0001     5   1762.87474359      46.91    0.0001

                              T FOR HO:       PR > |T|     STD ERROR OF
PARAMETER        ESTIMATE    PARAMETER=0                     ESTIMATE
MUMW-MUBR     22.91666667           9.16        0.0001       2.50257804

OBSERVATION      OBSERVED        PREDICTED        RESIDUAL      LOWER 95% CL     UPPER 95% CL
                   VALUE           VALUE                          FOR MEAN         FOR MEAN

      1        58.20000000     55.95000000      2.25000000     51.36616602      60.53383398
      2        53.70000000     55.95000000     -2.25000000     51.36616602      60.53383398
      3        55.60000000     55.60000000      0.00000000     49.11747982      62.08252018
      4        73.00000000     75.55000000     -2.55000000     70.96616602      80.13383398
      5        78.10000000     75.55000000      2.55000000     70.96616602      80.13383398
      6        76.20000000     78.86666667     -2.66666667     75.12398189      82.60935144
      7        82.00000000     78.86666667      3.13333333     75.12398189      82.60935144
      8        78.40000000     78.86666667     -0.46666667     75.12398189      82.60935144
      9        52.50000000     51.15000000      1.35000000     46.56616602      55.73383398
     10        49.80000000     51.15000000     -1.35000000     46.56616602      55.73383398
     11        56.00000000     53.73333333      2.26666667     49.99064856      57.47601811
     12        51.90000000     53.73333333     -1.83333333     49.99064856      57.47601811
     13        53.30000000     53.73333333     -0.43333333     49.99064856      57.47601811

OBSERVATION      OBSERVED        PREDICTED        RESIDUAL      LOWER 95% CL     UPPER 95% CL
                   VALUE           VALUE                         INDIVIDUAL       INDIVIDUAL

      1        58.20000000     55.95000000      2.25000000     48.01056665      63.88943335
      2        53.70000000     55.95000000     -2.25000000     48.01056665      63.88943335
      3        55.60000000     55.60000000      0.00000000     46.43233204      64.76766796
      4        73.00000000     75.55000000     -2.55000000     67.61056665      83.48943335
      5        78.10000000     75.55000000      2.55000000     67.61056665      83.48943335
      6        76.20000000     78.86666667     -2.66666667     71.38129712      86.35203621
      7        82.00000000     78.86666667      3.13333333     71.38129712      86.35203621
      8        78.40000000     78.86666667     -0.46666667     71.38129712      86.35203621
      9        52.50000000     51.15000000      1.35000000     43.21056665      59.08943335
     10        49.80000000     51.15000000     -1.35000000     43.21056665      59.08943335
     11        56.00000000     53.73333333      2.26666667     46.24796379      61.21870288
     12        51.90000000     53.73333333     -1.83333333     46.24796379      61.21870288
     13        53.30000000     53.73333333     -0.43333333     46.24796379      61.21870288
```

Exercise 16.5 to obtain the same results as those given in Figure 16.15. In the program, actually define and use the appropriate values of the dummy variables.

16.9 Note that the data in Table 16.11 is the same data that is given in Table 15.2 and analyzed in Section 15.3 by the (α, γ, θ) model. Compare $F(\alpha|\gamma)$ and $F(\gamma|\alpha)$ in Figure 16.14 with $F(\alpha|\gamma, \theta)$ and $F(\gamma|\alpha, \theta)$ in Figure 15.4. Also compare the 95% confidence interval for $\mu_{Tj} - \mu_{Mj}$ calculated in Exercise 16.1 with the 95% confidence interval for

FIGURE 16.16 SAS output that results from using the (α, γ, θ) model and the data in Table 16.11 to test for interaction between display height and display width

SAS

GENERAL LINEAR MODELS PROCEDURE

DEPENDENT VARIABLE: DEMAND

SOURCE	DF	SUM OF SQUARES	MEAN SQUARE	F VALUE	PR > F	R-SQUARE	C.V.
MODEL	5	1762.87474359	352.57494872	46.91	0.0001	0.971022	4.3531
ERROR	7	52.60833333	7.51547619		ROOT MSE		DEMAND MEAN
CORRECTED TOTAL	12	1815.48307692			2.74143688		62.97692308

SOURCE	DF	TYPE I SS	F VALUE	PR > F	DF	TYPE III SS	F VALUE	PR > F
HEIGHT	2	1741.58441026	115.87	0.0001	2	1643.05762319	109.31	0.0001
WIDTH	1	15.28592754	2.03	0.1969	1	9.72710526	1.29	0.2927
HEIGHT*WIDTH	2	6.00440580	0.40	0.6850	2	6.00440580	0.40	0.6850

FIGURE 16.17 SAS output of an (α, γ) model analysis of the bakery product demand data in Table 15.1

SAS

GENERAL LINEAR MODELS PROCEDURE

DEPENDENT VARIABLE: DEMAND

SOURCE	DF	SUM OF SQUARES	MEAN SQUARE	F VALUE	PR > F	R-SQUARE	C.V.
MODEL	3	2282.70000000	760.90000000	127.45	0.0001	0.964679	3.9729
ERROR	14	83.58000000	5.97000000		ROOT MSE		DEMAND MEAN
CORRECTED TOTAL	17	2366.28000000			2.44335834		61.50000000

SOURCE	DF	TYPE I SS	F VALUE	PR > F	DF	TYPE III SS	F VALUE	PR > F
HEIGHT	2	2273.88000000	190.44	0.0001	2	2273.88000000	190.44	0.0001
WIDTH	1	8.82000000	1.48	0.2443	1	8.82000000	1.48	0.2443

PARAMETER	ESTIMATE	T FOR HO: PARAMETER=0	PR > |T|	STD ERROR OF ESTIMATE
MUM-MUB	21.40000000	15.17	0.0001	1.41067360
MUT-MUB	-4.30000000	-3.05	0.0087	1.41067360
MUT-MUM	-25.70000000	-18.22	0.0001	1.41067360
MUW-MUR	1.40000000	1.22	0.2443	1.15181017
MUMW-MUMR	1.40000000	1.22	0.2443	1.15181017

OBSERVATION	OBSERVED VALUE	PREDICTED VALUE	RESIDUAL	LOWER 95% CL FOR MEAN	UPPER 95% CL FOR MEAN
1	58.20000000	55.10000000	3.10000000	52.62961870	57.57038130
2	53.70000000	55.10000000	-1.40000000	52.62961870	57.57038130
3	55.80000000	55.10000000	0.70000000	52.62961870	57.57038130
4	73.00000000	76.50000000	-3.50000000	74.02961870	78.97038130
5	78.10000000	76.50000000	1.60000000	74.02961870	78.97038130
6	75.40000000	76.50000000	-1.10000000	74.02961870	78.97038130
7	52.40000000	50.80000000	1.60000000	48.32961870	53.27038130
8	49.70000000	50.80000000	-1.10000000	48.32961870	53.27038130
9	50.90000000	50.80000000	0.10000000	48.32961870	53.27038130
10	55.70000000	56.50000000	-0.80000000	54.02961870	58.97038130
11	52.50000000	56.50000000	-4.00000000	54.02961870	58.97038130
12	58.90000000	56.50000000	2.40000000	54.02961870	58.97038130
13	76.20000000	77.90000000	-1.70000000	75.42961870	80.37038130
14	78.40000000	77.90000000	0.50000000	75.42961870	80.37038130
15	82.10000000	77.90000000	4.20000000	75.42961870	80.37038130
16	54.00000000	52.20000000	1.80000000	49.72961870	54.67038130
17	52.10000000	52.20000000	-0.10000000	49.72961870	54.67038130
18	49.90000000	52.20000000	-2.30000000	49.72961870	54.67038130

$\mu_{T\cdot} - \mu_{M\cdot}$ calculated in Example 15.10. Which interval is shorter? Then compare the 95% confidence intervals for μ_{MW} in Figure 15.10 and in Figures 16.14 and 16.15. Note that the (α, γ) model provides a shorter interval than do the (α, γ, θ) model and the two-factor means model, which yield the same interval. Which model seems to best describe the data in Table 16.11?

16.10 Consider the bakery demand data in Table 15.1. Note that this data is balanced (that is, consists of samples of equal sizes). Compare the (α, γ) model analysis of this data in Figure 16.17 with the (α, γ, θ) model analysis of this data in Figure 15.3.

16.11 Use the (α, γ) model and the two-factor means model to analyze the incomplete bakery product demand data in Table 16.13.

TABLE 16.13 **Results of a two-factor product demand experiment in which the data is incomplete**

Display height, *i*	Display width, *j*: *R*	*W*
B	58.2 53.7	
M	73.0 78.1	76.2 82.0 78.4
T	52.5 49.8	56.0 51.9 53.3

16.12 Use the (α, γ) model and the two-factor means model to analyze the incomplete display panel data in Table 16.14. Recall from Exercise 15.3 that the response variable is the time in seconds to stabilize the emergency condition.

TABLE 16.14 **Results of a two-factor display panel experiment in which the data is incomplete**

Display panel, *i*	Emergency condition, *j*: *1*	*2*	*3*	*4*
A	17 14	25 24	31 34	
B	15 12		28 31	9 10
C	21 24	29	32 37	15 19

16.13 Use the (α, γ) model and the two-factor means model to analyze the incomplete wheat yield data in Table 16.15.

TABLE 16.15 **Results of a two-factor wheat yield experiment in which the data is incomplete**

Fertilizer type, *i*	Wheat type, *j*			
	M	*N*	*O*	*P*
A	$y_{AM,1} = 19.4$ $y_{AM,2} = 20.6$ $y_{AM,3} = 20.0$	$y_{AN,1} = 25.0$ $y_{AN,2} = 24.0$	$y_{AO,1} = 24.8$ $y_{AO,2} = 26.0$ $y_{AO,3} = 25.4$	$y_{AP,1} = 23.1$ $y_{AP,2} = 24.3$
B		$y_{BN,1} = 25.6$ $y_{BN,2} = 26.8$ $y_{BN,3} = 26.2$	$y_{BO,1} = 27.6$ $y_{BO,2} = 26.4$ $y_{BO,3} = 27.0$	$y_{BP,1} = 25.4$ $y_{BP,2} = 24.5$ $y_{BP,3} = 26.3$

16.14 Use the (α, γ) model and the two-factor means model to analyze the incomplete gasoline mileage data in Table 16.16.

TABLE 16.16 **Results of a two-factor gasoline mileage experiment in which the data is incomplete**

Premium gasoline type, *i*	Gasoline additive type, *j*			
	Q	*R*	*S*	*T*
D	$y_{DQ,1} = 27.4$ $y_{DQ,2} = 28.6$	$y_{DR,1} = 32.0$	$y_{DS,1} = 33.5$ $y_{DS,2} = 32.3$	$y_{DT,1} = 30.8$ $y_{DT,2} = 29.7$
E	$y_{EQ,1} = 33.3$ $y_{EQ,2} = 34.5$	$y_{ER,1} = 35.6$ $y_{ER,2} = 34.4$	$y_{ES,1} = 33.4$ $y_{ES,2} = 33.1$	$y_{ET,1} = 29.6$ $y_{ET,2} = 30.6$
F		$y_{FR,1} = 34.7$ $y_{FR,2} = 33.3$	$y_{FS,1} = 33.0$	$y_{FT,1} = 28.6$ $y_{FT,2} = 29.8$

16.15 Recall the display panel data from Table 15.4 (reproduced in Table 16.17).

TABLE 16.17 **Display panel data (originally given in Table 15.4)**

Display panel	Emergency condition			
	1	2	3	4
A	17 14	25 24	31 34	14 13
B	15 12	22 19	28 31	9 10
C	21 24	29 28	32 37	15 19

a. Assume that both display panel and emergency condition are random factors. Analyze the data by using the techniques of Table 16.4.
b. Assume that display panel is a fixed factor and that emergency condition is a random factor. Analyze the data by using the technique of Table 16.5.

Hint: Some of the quantities needed in the analyses are given in Figure 15.6.

FIGURE 16.18 **SAS output of a nested model ANOVA of the assembly method data in Table 16.18**

SAS

GENERAL LINEAR MODELS PROCEDURE

DEPENDENT VARIABLE: UNITS

SOURCE	DF	SUM OF SQUARES	MEAN SQUARE	F VALUE	PR > F	R-SQUARE	C.V.
MODEL	5	1646.20833333	329.24166667	21.73	0.0001	0.857866	16.7127
ERROR	18	272.75000000	15.15277778		ROOT MSE		UNITS MEAN
CORRECTED TOTAL	23	1918.95833333			3.89265690		23.29166667

SOURCE	DF	TYPE I SS	F VALUE	PR > F	DF	TYPE III SS	F VALUE	PR > F
METHOD	2	1464.08333333	48.31	0.0001	2	1464.08333333	48.31	0.0001
PLACE(METHOD)	3	182.12500000	4.01	0.0239	3	182.12500000	4.01	0.0239

PARAMETER	ESTIMATE	T FOR HO: PARAMETER=0	PR > |T|	STD ERROR OF ESTIMATE
MU2-MU1	9.12500000	4.69	0.0002	1.94632845
MU3-MU1	19.12500000	9.83	0.0001	1.94632845
MU3-MU2	10.00000000	5.14	0.0001	1.94632845
MU12-MU11	6.25000000	2.27	0.0357	2.75252409
MU22-MU21	6.00000000	2.18	0.0428	2.75252409
MU32-MU31	4.00000000	1.45	0.1634	2.75252409

OBSERVATION	OBSERVED VALUE	PREDICTED VALUE	RESIDUAL	LOWER 95% CL FOR MEAN	UPPER 95% CL FOR MEAN
1	16.00000000	10.75000000	5.25000000	6.66094268	14.83905732
2	7.00000000	10.75000000	-3.75000000	6.66094268	14.83905732
3	7.00000000	10.75000000	-3.75000000	6.66094268	14.83905732
4	13.00000000	10.75000000	2.25000000	6.66094268	14.83905732
5	14.00000000	17.00000000	-3.00000000	12.91094268	21.08905732
6	24.00000000	17.00000000	7.00000000	12.91094268	21.08905732
7	13.00000000	17.00000000	-4.00000000	12.91094268	21.08905732
8	17.00000000	17.00000000	0.00000000	12.91094268	21.08905732
9	21.00000000	20.00000000	1.00000000	15.91094268	24.08905732
10	25.00000000	20.00000000	5.00000000	15.91094268	24.08905732
11	16.00000000	20.00000000	-4.00000000	15.91094268	24.08905732
12	18.00000000	20.00000000	-2.00000000	15.91094268	24.08905732
13	24.00000000	26.00000000	-2.00000000	21.91094268	30.08905732
14	28.00000000	26.00000000	2.00000000	21.91094268	30.08905732
15	27.00000000	26.00000000	1.00000000	21.91094268	30.08905732
16	25.00000000	26.00000000	-1.00000000	21.91094268	30.08905732
17	25.00000000	31.00000000	-6.00000000	26.91094268	35.08905732
18	35.00000000	31.00000000	4.00000000	26.91094268	35.08905732
19	33.00000000	31.00000000	2.00000000	26.91094268	35.08905732
20	31.00000000	31.00000000	0.00000000	26.91094268	35.08905732
21	31.00000000	35.00000000	-4.00000000	30.91094268	39.08905732
22	38.00000000	35.00000000	3.00000000	30.91094268	39.08905732
23	36.00000000	35.00000000	1.00000000	30.91094268	39.08905732
24	35.00000000	35.00000000	0.00000000	30.91094268	39.08905732

16.16 Suppose that the fifth observation is invalid for each assembly method and workstation in Table 16.6 except for assembly method 3 and workstation 2. For this assembly method and workstation the first observation is invalid. This implies that we might analyze the data in Table 16.18. Figure 16.18 presents the SAS output obtained by using PROC GLM and the techniques of Table 16.7 to analyze this data. Fully interpret the output and calculate 95% confidence intervals for the differences in means referred to in the output.

TABLE 16.18 **Results of a two-factor nested assembly method experiment**

	Assembly method					
	1		2		3	
Workstation	*1*	*2*	*1*	*2*	*1*	*2*
	16	14	21	24	25	31
	7	24	25	28	35	38
	7	13	16	27	33	36
	13	17	18	25	31	35

16.17 Using the methodology summarized in Table 16.8, fully analyze the data in Table 16.18 by assuming that "assembly method" is fixed and "workstation" is random.

16.18 Using the methodology summarized in Table 16.9, fully analyze the data in Table 16.18 by assuming that "assembly method" is random and "workstation" is random.

16.19 Figure 16.19 presents the SAS output obtained by using PROC GLM and the $(\alpha, \gamma(\alpha))$ model to analyze the unbalanced, nested assembly method data in Table 16.19.

a. Fully interpret the output and calculate 95% confidence intervals for the differences in means referred to in the output.

b. Specify the $(\alpha, \gamma(\alpha))$ dummy variable regression model that can be used to obtain the output in Figure 16.19. Also, explain how this dummy variable model would be used to obtain these results.

TABLE 16.19 **Results of a two-factor assembly method experiment in which the data is nested and unbalanced**

	Assembly method							
	1		2			3		
Workstation	*1*	*2*	*1*	*2*	*3*	*1*	*2*	*3*
	16	14	25	24	21	25	31	32
	7	24	16	28	29	35	31	30
	7	13	20	27	30	33	38	39
	13	17		25	26	31	36	36
		21		21		28	35	

FIGURE 16.19 SAS output of an $(\alpha, \gamma(\alpha))$ model analysis of the assembly method data in Table 16.19

SAS

GENERAL LINEAR MODELS PROCEDURE

DEPENDENT VARIABLE: UNITS

SOURCE	DF	SUM OF SQUARES	MEAN SQUARE	F VALUE	PR > F	R-SQUARE	C.V.
MODEL	7	2054.71904762	293.53129252	18.96	0.0001	0.830967	15.5777
ERROR	27	417.96666667	15.48024691		ROOT MSE		UNITS MEAN
CORRECTED TOTAL	34	2472.68571429			3.93449449		25.25714286

SOURCE	DF	TYPE I SS	F VALUE	PR > F	DF	TYPE III SS	F VALUE	PR > F
METHOD	2	1828.30476190	59.05	0.0001	2	1910.16300921	61.70	0.0001
PLACE(METHOD)	5	226.41428571	2.93	0.0309	5	226.41428571	2.93	0.0309

PARAMETER	ESTIMATE	T FOR HO: PARAMETER=0	PR > |T|	STD ERROR OF ESTIMATE
MU2-MU1	9.66944444	5.50	0.0001	1.75752172
MU3-MU1	18.67500000	11.04	0.0001	1.69101910
MU3-MU2	9.00555556	5.74	0.0001	1.57015052
MU12-MU11	7.05000000	2.67	0.0127	2.63933914
MU22-MU21	4.66666667	1.62	0.1160	2.87334851
MU23-MU21	6.16666667	2.05	0.0500	3.00501981
MU23-MU22	1.50000000	0.57	0.5745	2.63933914
MU32-MU31	3.80000000	1.53	0.1384	2.48839281
MU33-MU31	3.85000000	1.46	0.1562	2.63933914
MU33-MU32	0.05000000	0.02	0.9850	2.63933914

OBSERVATION	OBSERVED VALUE	PREDICTED VALUE	RESIDUAL	LOWER 95% CL FOR MEAN	UPPER 95% CL FOR MEAN
1	16.00000000	10.75000000	5.25000000	6.71357343	14.78642657
2	7.00000000	10.75000000	-3.75000000	6.71357343	14.78642657
3	7.00000000	10.75000000	-3.75000000	6.71357343	14.78642657
4	13.00000000	10.75000000	2.25000000	6.71357343	14.78642657
5	14.00000000	17.80000000	-3.80000000	14.18971032	21.41028968
6	24.00000000	17.80000000	6.20000000	14.18971032	21.41028968
7	13.00000000	17.80000000	-4.80000000	14.18971032	21.41028968
8	17.00000000	17.80000000	-0.80000000	14.18971032	21.41028968
9	21.00000000	17.80000000	3.20000000	14.18971032	21.41028968
10	25.00000000	20.33333333	4.66666667	15.67246940	24.99419726
11	16.00000000	20.33333333	-4.33333333	15.67246940	24.99419726
12	20.00000000	20.33333333	-0.33333333	15.67246940	24.99419726
13	24.00000000	25.00000000	-3.00000000	21.38971032	28.61028968
14	28.00000000	25.00000000	3.00000000	21.38971032	28.61028968
15	27.00000000	25.00000000	2.00000000	21.38971032	28.61028968
16	25.00000000	25.00000000	0.00000000	21.38971032	28.61028968
17	21.00000000	25.00000000	-4.00000000	21.38971032	28.61028968
18	21.00000000	26.50000000	-5.50000000	22.46357343	30.53642657
19	29.00000000	26.50000000	2.50000000	22.46357343	30.53642657
20	30.00000000	26.50000000	3.50000000	22.46357343	30.53642657
21	26.00000000	26.50000000	-0.50000000	22.46357343	30.53642657
22	25.00000000	30.40000000	-5.40000000	26.78971032	34.01028968
23	35.00000000	30.40000000	4.60000000	26.78971032	34.01028968
24	33.00000000	30.40000000	2.60000000	26.78971032	34.01028968
25	31.00000000	30.40000000	0.60000000	26.78971032	34.01028968
26	28.00000000	30.40000000	-2.40000000	26.78971032	34.01028968
27	31.00000000	34.20000000	-3.20000000	30.58971032	37.81028968
28	31.00000000	34.20000000	-3.20000000	30.58971032	37.81028968
29	38.00000000	34.20000000	3.80000000	30.58971032	37.81028968
30	36.00000000	34.20000000	1.80000000	30.58971032	37.81028968
31	35.00000000	34.20000000	0.80000000	30.58971032	37.81028968
32	32.00000000	34.25000000	-2.25000000	30.21357343	38.28642657
33	30.00000000	34.25000000	-4.25000000	30.21357343	38.28642657
34	39.00000000	34.25000000	4.75000000	30.21357343	38.28642657
35	36.00000000	34.25000000	1.75000000	30.21357343	38.28642657

16.20 (Optional) Write a SAS program to fully analyze the display panel data in Table 16.14 using both the (α, γ) model and the two-factor means model.

16.21 (Optional) Write a SAS program to fully analyze the wheat yield data in Table 16.15 using both the (α, γ) model and the two-factor means model.

16.22 (Optional) Write a SAS program to fully analyze the gasoline mileage data in Table 16.16 using both the (α, γ) model and the two-factor means model.

16.23 (Optional) Write a SAS program to analyze the nested assembly method data in Table 16.18.

16.24 (Optional) Write a SAS program to analyze the unbalanced, nested assembly method data in Table 16.19.

CHAPTER 17

THE RANDOMIZED BLOCK AND LATIN SQUARE DESIGNS

In this chapter we discuss the *randomized block design* and *Latin square design*. These experimental designs can be more effective than the completely randomized design when (possible) differences between the experimental units may be concealing any true differences between the treatments. We begin in Section 17.1 by introducing the randomized block design and by presenting the *randomized block model*. Section 17.2 explains the ANOVA approach to analyzing this design, while Section 17.3 presents the regression approach to the randomized block analysis. In Section 17.4 we discuss the Latin square design. We conclude this chapter with optional Section 17.5, which demonstrates how to use SAS to implement these techniques.

17.1 INTRODUCTORY CONCEPTS

As the reader might suspect, not all experiments employ a completely randomized design. To see why this is true, consider the following example.

Example 17.1 The Universal Paper Company manufactures cardboard boxes. The company wishes to perform an experiment to investigate the effects of four production methods (methods 1, 2, 3, and 4) on the number of defective boxes produced in an hour of production. To perform the experiment, the company could utilize a completely randomized design. To do this, for $l = 1, 2, 3$, and 4 the company would randomly select three machine operators (three is chosen arbitrarily) from the pool of all machine operators that it employs, train each operator thoroughly to use production method l, have each operator produce boxes for one hour by using production method l, and record the number of defective boxes produced. The three operators using any one production method would be *different* from those using any other production method. That is, the completely randomized design would utilize a total of 12 machine operators. However, the abilities of the machine operators could differ substantially. These differences would tend to conceal any real differences between the production methods. To overcome this disadvantage, the company will employ a *randomized block experimental design.* This involves randomly selecting three machine operators from the pool of all machine operators and training each operator thoroughly to use all four production methods. Then each of the three operators will produce boxes for one hour using each of the four production methods. The order in which each operator uses the four methods should be random. We record the number of defective boxes produced by each operator using each method. The advantage of the randomized block design is that the defective rates obtained by using the four methods result from employing the *same* operators (a total of three machine operators). Thus any true differences in the effectiveness of the methods would not be concealed by differences in the abilities of the operators.

Suppose that when the company employs the randomized block design, it observes the data in Table 17.1. Here, for $l = 1, 2, 3$, and 4, and $h = 1, 2$, and 3,

y_{lh} = the number of defective boxes obtained when machine operator h produces boxes for one hour by using production method l

TABLE 17.1 **Numbers of defective cardboard boxes obtained by production methods 1, 2, 3, and 4 and machine operators 1, 2, and 3**

	Machine operator, h			
Production method, l	*1*	*2*	*3*	$\bar{y}_{l\cdot}$
1	$y_{11} = 9$	$y_{12} = 10$	$y_{13} = 12$	$\bar{y}_{1\cdot} = 10.3333$
2	$y_{21} = 8$	$y_{22} = 11$	$y_{23} = 12$	$\bar{y}_{2\cdot} = 10.3333$
3	$y_{31} = 3$	$y_{32} = 5$	$y_{33} = 7$	$\bar{y}_{3\cdot} = 5.0$
4	$y_{41} = 4$	$y_{42} = 5$	$y_{43} = 5$	$\bar{y}_{4\cdot} = 4.6667$
$\bar{y}_{\cdot h}$	$\bar{y}_{\cdot 1} = 6.0$	$\bar{y}_{\cdot 2} = 7.75$	$\bar{y}_{\cdot 3} = 9.0$	$\bar{y} = 7.5833$

This value is assumed to have been randomly selected from the (theoretically) infinite population of all possible hourly defective box rates when machine operator h uses production method l. We denote the mean of this population by the symbol μ_{lh}. That is, μ_{lh} is the mean number of defective *boxes produced per hour by machine operator h using production method l.*

In general, a *randomized block design* compares v treatments (for example, production methods) by using d blocks (for example, machine operators). Each block is used exactly once to measure the effect of each and every treatment. The advantage of the randomized block design over the completely randomized design is that we are comparing the treatments by using the same experimental units. Thus any true differences in the treatments will not be concealed by differences in the experimental units.

In some experiments a block consists of *homogeneous* (similar) experimental units. For example, to compare the effects of five wheat types (the treatments) on wheat yield, an experimenter might choose four different plots of soil (the blocks) on which to make comparisons. Then each of the five wheat types would be randomly assigned to a subplot within each of the four different plots of soil. If the experimenter carefully selected the four different plots of soil so that the five subplots within each plot were of roughly the same soil fertility, any true differences in the wheat types would not be concealed by different soil fertility conditions.

Suppose that when we employ a randomized block design, we observe the data in Table 17.2. Here, for $l = 1, 2, \ldots, v$, and $h = 1, 2, \ldots, d$,

y_{lh} = the value of the response variable observed when block h used treatment l

This value is assumed to have been randomly selected from the (theoretically) infinite population of all possible values of the response variable that could be observed when block h uses treatment l. We denote the mean of this population by the symbol μ_{lh}, and we consider the following model.

TABLE 17.2 **Data resulting from a randomized block design**

	Block				
Treatment	***1***	***2***	$\cdots$	***d***	$\bar{y}_{l\cdot}$
1	y_{11}	y_{12}	$\cdots$	y_{1d}	$\bar{y}_{1\cdot}$
2	y_{21}	y_{22}	$\cdots$	y_{2d}	$\bar{y}_{2\cdot}$
$\vdots$	$\vdots$	$\vdots$		$\vdots$	$\vdots$
v	y_{v1}	y_{v2}	$\cdots$	y_{vd}	$\bar{y}_{v\cdot}$
$\bar{y}_{\cdot h}$	$\bar{y}_{\cdot 1}$	$\bar{y}_{\cdot 2}$	$\cdots$	$\bar{y}_{\cdot d}$	$\bar{y}$

The Randomized Block Model

The *randomized block model* says that

$$y_{lh} = \mu_{lh} + \varepsilon_{lh}$$

where

$$\mu_{lh} = \mu + \tau_l + \delta_h$$

Here,

1. y_{lh} = the value of the response variable when block h uses treatment l
2. ε_{lh} = the error term when block h uses treatment l
3. μ_{lh} = the mean value of the response variable when block h uses treatment l
4. μ = an overall mean or intercept, which is an unknown constant
5. τ_l = the effect due to treatment l, which is an unknown constant
6. δ_h = the effect due to block h, which is an unknown constant

It is important to notice that this model assumes that no interaction exists between treatments and blocks. This says that the relationship between the effects of the treatments on the mean value of the response variable is the same for each and every block. Likewise, the relationship between the effects of the blocks on the mean value of the response variable is the same for each and every treatment. For example, the randomized block model says that

$$\begin{aligned} \mu_{lh} - \mu_{l'h} &= (\mu + \tau_l + \delta_h) - (\mu + \tau_{l'} + \delta_h) \\ &= \tau_l - \tau_{l'} \end{aligned}$$

does not depend on block h (no interaction). It follows that we may say that

$\mu_{lh} - \mu_{l'h}$ = the change in the mean value of the response variable associated with *any particular block* changing from using treatment l' to using treatment l

Similarly, the randomized block model says that

$$\begin{aligned} \mu_{lh} - \mu_{lh'} &= (\mu + \tau_l + \delta_h) - (\mu + \tau_l + \delta_{h'}) \\ &= \delta_h - \delta_{h'} \end{aligned}$$

does not depend on treatment l. Therefore we may write

$\mu_{lh} - \mu_{lh'}$ = the change in the mean value of the response variable associated with changing from block h' using *any particular treatment* to block h using the same treatment

FIGURE 17.1 **Graphical analysis of the defective box data in Table 17.1: Little interaction**

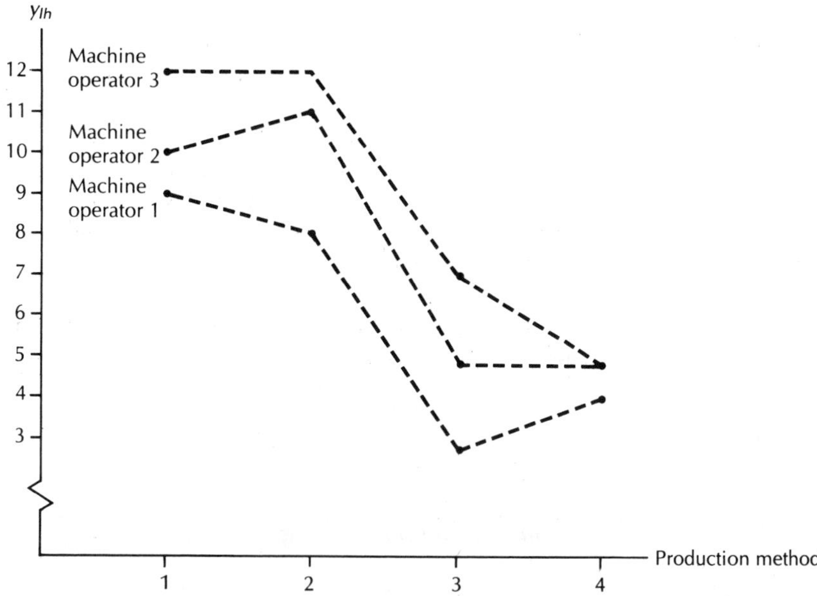

For each machine operator a plot is made of the change in y_{lh} associated with changing the production method. Little interaction means this change in the y_{lh} values depends little on the machine operator.

To check for interaction between treatments and blocks, we can use graphical analysis or a statistical test called *Tukey's test for additivity*—see Miller and Wichern (1977). To illustrate the appropriate graphical analysis, consider the data in Table 17.1 and Figure 17.1. The largely parallel data patterns in the figure indicate that the change in the number of defective boxes associated with changing the production method depends little on the machine operator. This indicates that there is little or no interaction between production method and machine operator.

17.2

THE ANOVA APPROACH TO RANDOMIZED BLOCKS

The ANOVA approach to analyzing a randomized block design involves partitioning a total sum of squares. We define the appropriate sums of squares and mean squares as following.

Sums of Squares and Mean Squares in a Randomized Block Design

Let

$$\bar{y}_{l\cdot} = \frac{\sum_{h=1}^{d} y_{lh}}{d} = \text{the mean of the } d \text{ values of the response variable observed when using treatment } l$$

$$\bar{y}_{\cdot h} = \frac{\sum_{l=1}^{v} y_{lh}}{v} = \text{the mean of the } v \text{ values of the response variable observed when using block } h$$

$$\bar{y} = \frac{\sum_{l=1}^{v}\sum_{h=1}^{d} y_{lh}}{dv} = \text{the mean of the total of the } dv \text{ values of the response variable that we have observed in the experiment}$$

Then we define

1. The *treatment sum of squares* to be

$$SS_{\text{treatments}} = d\sum_{l=1}^{v} (\bar{y}_{l\cdot} - \bar{y})^2$$

2. The *treatment mean square* to be

$$MS_{\text{treatments}} = \frac{SS_{\text{treatments}}}{v-1}$$

3. The *block sum of squares* to be

$$SS_{\text{blocks}} = v\sum_{h=1}^{d} (\bar{y}_{\cdot h} - \bar{y})^2$$

4. The *block mean square* to be

$$MS_{\text{blocks}} = \frac{SS_{\text{blocks}}}{d-1}$$

5. The *error sum of squares* to be

$$SSE = \sum_{l=1}^{v}\sum_{h=1}^{d} (y_{lh} - \bar{y}_{l\cdot} - \bar{y}_{\cdot h} + \bar{y})^2$$

6. The *error mean square* to be

$$MSE = \frac{SSE}{(v-1)(d-1)}$$

7. The *total sum of squares* to be

$$SS_{\text{total}} = \sum_{l=1}^{v}\sum_{h=1}^{d} (y_{lh} - \bar{y})^2$$

It can be shown that

$$SS_{\text{total}} = SS_{\text{treatments}} + SS_{\text{blocks}} + SSE$$

Generally speaking, SSE is calculated by the formula

$$SSE = SS_{\text{total}} - SS_{\text{treatments}} - SS_{\text{blocks}}$$

Note that the treatment sum of squares measures the variability of the sample treatment means (the $\bar{y}_{l\cdot}$ values) and that the block sum of squares measures the variability of the sample block means (the $\bar{y}_{\cdot h}$ values). We now summarize how to test for differences in the effects of the treatments and how to test for differences in the effects of the blocks.

F-Tests in a Randomized Block Design

1. *Testing the significance of the treatments.* Define

$$F(\text{treatments}) = \frac{MS_{\text{treatments}}}{MSE} = \frac{SS_{\text{treatments}}/(v - 1)}{SSE/(v - 1)(d - 1)}$$

Also, define the prob-value to be the area to the right of F(treatments) under the curve of the F-distribution having $v - 1$ and $(v - 1)(d - 1)$ degrees of freedom. To test for significant differences between treatment effects, we test

$$H_0\colon \mu_{1h} = \mu_{2h} = \cdots = \mu_{vh}$$

which says that all treatments have the same effects on the mean response, versus

H_1: At least two of $\mu_{1h}, \mu_{2h}, \ldots, \mu_{vh}$ differ

which says that at least two treatments have different effects on the mean response. We can reject H_0 by setting the probability of a Type I error equal to α if and only if either of the following equivalent conditions holds:

a. $F(\text{treatments}) > F_{[\alpha]}^{(v-1,(v-1)(d-1))}$
b. prob-value $< \alpha$

If we reject H_0, we conclude that there are statistically significant differences between treatment effects.

2. *Testing the significance of the blocks.* Define

$$F(\text{blocks}) = \frac{MS_{\text{blocks}}}{MSE} = \frac{SS_{\text{blocks}}/(d - 1)}{SSE/(v - 1)(d - 1)}$$

Also, define the prob-value to be the area to the right of F(blocks) under the curve of the F-distribution having $d - 1$ and $(v - 1)(d - 1)$

degrees of freedom. To test block significance, we test

H_0: $\mu_{l1} = \mu_{l2} = \cdots = \mu_{ld}$

which says that all blocks have the same effect on the mean response, versus

H_1: At least two of $\mu_{l1}, \mu_{l2}, \ldots, \mu_{ld}$ differ

which says that at least two blocks have different effects on the mean response. We can reject H_0 by setting the probability of a Type I error equal to α if and only if either of the following equivalent conditions holds:

a. $F(\text{blocks}) > F_{[\alpha]}^{(d-1,(v-1)(d-1))}$
b. prob-value $< \alpha$

If we reject H_0, we conclude that there are statistically significant differences between blocks.

Before presenting an example, we note that in the defective box experiment the machine operators (the blocks) have been randomly selected from the pool of all machine operators. In general, when the blocks in a randomized block design have been randomly selected from a population of many possible blocks, the factor "block" is a *random* factor. We discussed random factors in Section 16.2. However, when there is no interaction between treatments and blocks, then assuming that the treatments are *fixed*, it follows that the above F-test for treatment effects (and confidence intervals for differences between treatment effects, to be discussed later) are the same whether or not the blocks are fixed or random. In general, assuming that there is no interaction between treatments and blocks, then the above F-tests are valid whether the treatments and/or blocks are fixed or random. Of course, as discussed in Section 16.2, the F-test for a random factor (be it treatments or blocks) tests the null hypothesis that there is no variation between the effects on the response variable of all possible levels of the factor. Also, note that the same comments apply to analyzing a *Latin square* model (see Section 17.4). This model also assumes that there is no interaction between treatments and blocks.

Example 17.2 The quantities $\bar{y}$, $\bar{y}_{l\cdot}$ ($l = 1, 2, 3, 4$), and $\bar{y}_{\cdot h}$ ($h = 1, 2, 3$) are calculated for the defective cardboard box data in Table 17.1. It follows that F(treatments), F(blocks), and the related prob-values for this data are calculated and utilized as follows.

$$SS_{\text{treatments}} = d \sum_{l=1}^{v} (\bar{y}_{l\cdot} - \bar{y})^2$$

$$= 3 \sum_{l=1}^{4} (\bar{y}_{l\cdot} - \bar{y})^2$$

$$\begin{aligned}
&= 3[(\bar{y}_{1\cdot} - \bar{y})^2 + (\bar{y}_{2\cdot} - \bar{y})^2 + (\bar{y}_{3\cdot} - \bar{y})^2 + (\bar{y}_{4\cdot} - \bar{y})^2] \\
&= 3[(10.3333 - 7.5833)^2 + (10.3333 - 7.5833)^2 \\
&\quad + (5.0 - 7.5833)^2 + (4.6667 - 7.5833)^2] \\
&= 90.9167
\end{aligned}$$

$$MS_{\text{treatments}} = \frac{SS_{\text{treatments}}}{v - 1} = \frac{90.9167}{4 - 1} = 30.3056$$

$$\begin{aligned}
SS_{\text{blocks}} &= v \sum_{h=1}^{d} (\bar{y}_{\cdot h} - \bar{y})^2 \\
&= 4[(\bar{y}_{\cdot 1} - \bar{y})^2 + (\bar{y}_{\cdot 2} - \bar{y})^2 + (\bar{y}_{\cdot 3} - \bar{y})^2] \\
&= 4[(6.0 - 7.5833)^2 + (7.75 - 7.5833)^2 \\
&\quad + (9.0 - 7.5833)^2] \\
&= 18.1667
\end{aligned}$$

$$MS_{\text{blocks}} = \frac{SS_{\text{blocks}}}{d - 1} = \frac{18.1667}{3 - 1} = 9.08335$$

$$\begin{aligned}
SS_{\text{total}} &= \sum_{l=1}^{v} \sum_{h=1}^{d} (y_{lh} - \bar{y})^2 \\
&= \sum_{l=1}^{4} \sum_{h=1}^{3} (y_{lh} - 7.5833)^2 \\
&= 112.9167
\end{aligned}$$

$$\begin{aligned}
SSE &= SS_{\text{total}} - SS_{\text{treatments}} - SS_{\text{blocks}} \\
&= 112.9167 - 90.9167 - 18.1667 \\
&= 3.8333
\end{aligned}$$

$$MSE = \frac{SSE}{(v - 1)(d - 1)} = \frac{3.8333}{(4 - 1)(3 - 1)} = .6389$$

$$F(\text{treatments}) = \frac{MS_{\text{treatments}}}{MSE} = \frac{30.3056}{.6389} = 47.4348$$

$$F(\text{blocks}) = \frac{MS_{\text{blocks}}}{MSE} = \frac{9.08335}{.6389} = 14.2172$$

We find that

$$F(\text{treatments}) = 47.4348 > F_{[\alpha]}^{(v-1,(v-1)(d-1))} = F_{[.05]}^{(3,6)} = 4.76$$

and

$$\text{prob-value} = .0001 < .05$$

Thus we reject

$$H_0\colon \mu_{1h} = \mu_{2h} = \mu_{3h} = \mu_{4h}$$

in favor of

$$H_1\colon \text{At least two of } \mu_{1h}, \mu_{2h}, \mu_{3h}, \text{ and } \mu_{4h} \text{ differ}$$

with $\alpha = .05$. We conclude that at least two production methods have different effects on the mean number of defective boxes produced per hour.

In addition,

$$F(\text{blocks}) = 14.2172 > F_{[\alpha]}^{(d-1,(v-1)(d-1))} = F_{[.05]}^{(2,6)} = 5.14$$

and

$$\text{prob-value} = .0053 < .05$$

FIGURE 17.2 **SAS output of a randomized block ANOVA (and a regression analysis) of the defective cardboard box data in Table 17.1**

SAS

GENERAL LINEAR MODELS PROCEDURE

DEPENDENT VARIABLE: PRDN

SOURCE	DF	SUM OF SQUARES	MEAN SQUARE	F VALUE	PR > F	R-SQUARE	C.V.
MODEL	5	109.08333333	21.81666667	34.15	0.0002	0.966052	10.5403
ERROR	6	3.83333333	0.63888889		ROOT MSE		PRDN MEAN
CORRECTED TOTAL	11	112.91666667			0.79930525		7.58333333

SOURCE	DF	TYPE I SS	F VALUE	PR > F	DF	TYPE III SS	F VALUE	PR > F
BLOCK	2	18.16666667	14.22[a]	0.0053[b]	2	18.16666667	14.22	0.0053
METHOD	3	90.91666667	47.43[c]	0.0001[d]	3	90.91666667	47.43	0.0001

PARAMETER	ESTIMATE	T FOR HO: PARAMETER=0	PR > |T|	STD ERROR OF ESTIMATE
MU4-MU1	-5.66666667[e]	-8.68	0.0001	0.65263001
MU4-MU2	-5.66666667	-8.68	0.0001	0.65263001
MU4-MU3	-0.33333333	-0.51	0.6278	0.65263001
MU3-MU2	-5.33333333	-8.17	0.0002	0.65263001
MU3-MU1	-5.33333333	-8.17	0.0002	0.65263001
MU2-MU1	2.2204460E-16	0.00	1.0000	0.65263001
MUTHREE-MUTWO	1.25000000	2.21	0.0690	0.56519417
MUTHREE-MUONE	3.00000000	5.31	0.0018	0.56519417
MUTWO-MUONE	1.75000000[f]	3.10	0.0212	0.56519417

OBSERVATION	OBSERVED VALUE	PREDICTED VALUE	RESIDUAL	LOWER 95% CL FOR MEAN	UPPER 95% CL FOR MEAN
1	9.00000000	8.75000000	0.25000000	7.36701931	10.13298069
2	8.00000000	8.75000000	-0.75000000	7.36701931	10.13298069
3	3.00000000	3.41666667	-0.41666667	2.03368598	4.79964735
4	4.00000000	3.08333333	0.91666667	1.70035265	4.46631402
5	10.00000000	10.50000000	-0.50000000	9.11701931	11.88298069
6	11.00000000	10.50000000	0.50000000	9.11701931	11.88298069
7	5.00000000	5.16666667	-0.16666667	3.78368598	6.54964735
8	5.00000000	4.83333333[g]	0.16666667	[3.45035265	6.21631402][h]
9	12.00000000	11.75000000	0.25000000	10.36701931	13.13298069
10	12.00000000	11.75000000	0.25000000	10.36701931	13.13298069
11	7.00000000	6.41666667	0.58333333	5.03368598	7.79964735
12	5.00000000	6.08333333	-1.08333333	4.70035265	7.46631402

[a]*F (blocks)* [b]*Prob-value for F (blocks)* [c]*F (treatments)* [d]*Prob-value for F (treatments)* [e]$\bar{y}_{4\cdot} - \bar{y}_{1\cdot}$ [f]$\bar{y}_{\cdot 2} - \bar{y}_{\cdot 1}$
[g]*Point estimate of* μ_{42} and point prediction of $y_{42,0} = \mu_{42} + \varepsilon_{42,0}$ [h]*95% confidence interval for* μ_{42}

Thus we reject

$$H_0: \mu_{l1} = \mu_{l2} = \mu_{l3}$$

in favor of

$$H_1: \text{At least two of } \mu_{l1}, \mu_{l2}, \text{ and } \mu_{l3} \text{ differ}$$

with $\alpha = .05$. We conclude that at least two machine operators have different effects on the mean number of defective boxes produced per hour.

Figure 17.2 presents the SAS output of a randomized block ANOVA of the defective cardboard box data.

We next present estimation formulas for a randomized block design.

Estimation in a Randomized Block Design

Let

$$s = \sqrt{MSE} = \sqrt{\frac{SSE}{(v-1)(d-1)}}$$

Then,

1. A *point estimate of* $\mu_{lh} - \mu_{l'h}$ is $\bar{y}_{l\cdot} - \bar{y}_{l'\cdot}$, and a *100(1 − α)% confidence interval* for $\mu_{lh} - \mu_{l'h}$ is

$$\left[(\bar{y}_{l\cdot} - \bar{y}_{l'\cdot}) \pm t_{[\alpha/2]}^{((v-1)(d-1))}\, s\sqrt{\frac{2}{d}}\right]$$

2. A *point estimate of* $\mu_{lh} - \mu_{lh'}$ is $\bar{y}_{\cdot h} - \bar{y}_{\cdot h'}$, and a *100(1 − α)% confidence interval* for $\mu_{lh} - \mu_{lh'}$ is

$$\left[(\bar{y}_{\cdot h} - \bar{y}_{\cdot h'}) \pm t_{[\alpha/2]}^{((v-1)(d-1))}\, s\sqrt{\frac{2}{v}}\right]$$

3. A *Tukey simultaneous 100(1 − α)% confidence interval for* $\mu_{lh} - \mu_{l'h}$ in the set of all possible paired differences between $\mu_{1h}, \mu_{2h}, \ldots, \mu_{vh}$ is

$$\left[(\bar{y}_{l\cdot} - \bar{y}_{l'\cdot}) \pm q_{[\alpha]}\,(v, (v-1)(d-1))\frac{s}{\sqrt{d}}\right]$$

where $q_{[\alpha]}(v, (v-1)(d-1))$ is obtained from the table of percentage points of the studentized range (see Appendix E).

4. A *Tukey simultaneous 100(1 − α)% confidence interval for* $\mu_{lh} - \mu_{lh'}$ in the set of all possible paired differences between $\mu_{l1}, \mu_{l2}, \ldots, \mu_{ld}$ is

$$\left[(\bar{y}_{\cdot h} - \bar{y}_{\cdot h'}) \pm q_{[\alpha]}(d, (v-1)(d-1))\frac{s}{\sqrt{v}}\right]$$

where $q_{[\alpha]}(d, (v-1)(d-1))$ is obtained from the table of percentage points of the studentized range (see Appendix E).

5. A *point estimate of the contrast* $\Sigma_{l=1}^{v} a_l \mu_{lh}$ is $\Sigma_{l=1}^{v} a_l \bar{y}_{l\cdot}$, and a $100(1-\alpha)\%$ *confidence interval for this contrast* is

$$\left[\sum_{l=1}^{v} a_l \bar{y}_{l\cdot} \pm t_{[\alpha/2]}^{((v-1)(d-1))} s\sqrt{\sum_{l=1}^{v} a_l^2/d}\right]$$

Note that $\Sigma_{l=1}^{v} a_l \mu_{lh}$ is a constant if $\Sigma_{l=1}^{v} a_l = 0$.

6. For *Scheffé simultaneous intervals* in 1 and 5 above, replace $t_{[\alpha/2]}^{(v-1)(d-1)}$ by $\sqrt{(v-1)F_{[\alpha]}^{(v-1,(v-1)(d-1))}}$ (which applies to all possible contrasts in $\mu_{1h}, \mu_{2h}, \ldots, \mu_{vh}$). For *Bonferroni simultaneous intervals* in 1 and 5 above, replace $t_{[\alpha/2]}^{((v-1)(d-1))}$ by $t_{[\alpha/2g]}^{((v-1)(d-1))}$ (which applies to g prespecified linear combinations).

7. A *point estimate of the contrast* $\Sigma_{h=1}^{d} a_h \mu_{lh}$ is $\Sigma_{h=1}^{d} a_h \bar{y}_{\cdot h}$, and a $100(1-\alpha)\%$ *confidence interval for this contrast* is

$$\left[\sum_{h=1}^{d} a_h \bar{y}_{\cdot h} \pm t_{[\alpha/2]}^{((v-1)(d-1))} s\sqrt{\sum_{h=1}^{d} a_h^2/v}\right]$$

Note that $\Sigma_{h=1}^{d} a_h \mu_{lh}$ is a constant if $\Sigma_{h=1}^{d} a_h = 0$.

8. For *Scheffé simultaneous intervals* in 2 and 7 above, replace $t_{[\alpha/2]}^{((v-1)(d-1))}$ by $\sqrt{(d-1)F_{[\alpha]}^{(d-1,(v-1)(d-1))}}$ (which applies to all possible contrasts in $\mu_{l1}, \mu_{l2}, \ldots, \mu_{ld}$). For *Bonferroni simultaneous intervals* in 2 and 7 above, replace $t_{[\alpha/2]}^{((v-1)(d-1))}$ by $t_{[\alpha/2g]}^{((v-1)(d-1))}$ (which applies to g prespecified linear combinations).

We do not give a point estimate of or a confidence interval for a specific treatment mean μ_{lh}, nor do we give a point prediction of or a prediction interval for an individual value $y_{lh} = \mu_{lh} + \varepsilon_{lh}$. This is because, since the randomized block model assumes that there is no interaction between treatments and blocks, regression analysis is needed to obtain these quantities (see Section 17.3). However, SAS can be used to obtain these quantities (as will be demonstrated in the following example).

Example 17.3 We now study the exact nature of the differences between production methods and machine operators in the defective cardboard box example. First, note that

$$\begin{aligned} s &= \sqrt{MSE} \\ &= \sqrt{.6389} \\ &= .7993 \end{aligned}$$

Since we have concluded that little interaction exists between production method and machine operator, we estimate (for example) $\mu_{4\cdot} - \mu_{1\cdot}$. A point estimate of

$\mu_{4\cdot} - \mu_{1\cdot}$ is

$$\bar{y}_{4\cdot} - \bar{y}_{1\cdot} = 4.6667 - 10.3333 = -5.6666$$

(see Table 17.1) and a 95% confidence interval for this difference is

$$\begin{aligned}\left[(\bar{y}_{4\cdot} - \bar{y}_{1\cdot}) \pm t_{[.025]}^{(v-1)(d-1)}\, s\sqrt{\frac{2}{d}}\right] &= \left[-5.6666 \pm t_{[.025]}^{(6)}(.7993)\sqrt{\frac{2}{3}}\right] \\ &= \left[-5.6666 \pm 2.447(.7993)\sqrt{\frac{2}{3}}\right] \\ &= [-5.6666 \pm 1.5970] \\ &= [-7.2636, -4.0696]\end{aligned}$$

This interval says that we can be 95 percent confident that the effect of a particular machine operator changing from production method 1 to production method 4 is to decrease the mean number of defective boxes produced per hour by between 7.2636 boxes and 4.0696 boxes.

Next, consider estimating (for example) $\mu_{\cdot 2} - \mu_{\cdot 1}$. A point estimate of $\mu_{\cdot 2} - \mu_{\cdot 1}$ is

$$\bar{y}_{\cdot 2} - \bar{y}_{\cdot 1} = 7.75 - 6 = 1.75$$

(see Table 17.1) and a 95% confidence interval for this difference is

$$\begin{aligned}\left[(\bar{y}_{\cdot 2} - \bar{y}_{\cdot 1}) \pm t_{[.025]}^{(v-1)(d-1)}\, s\sqrt{\frac{2}{v}}\right] &= \left[1.75 \pm t_{[.025]}^{(6)}(.7993)\sqrt{\frac{2}{4}}\right] \\ &= \left[1.75 \pm 2.447(.7993)\sqrt{\frac{2}{4}}\right] \\ &= [1.75 \pm 1.3830] \\ &= [.367, 3.133]\end{aligned}$$

This interval says that we can be 95 percent confident that the effect of changing from machine operator 1 using a particular production method to machine operator 2 using the same production method is to increase the mean number of defective boxes produced per hour by between .367 boxes and 3.133 boxes.

Confidence intervals for other differences can be computed similarly. Furthermore, note that in the SAS program employed to obtain the output of Figure 17.2 (see Section 17.5) y_{42} is "observation 9." It follows from Figure 17.2 that a point estimate of μ_{42} is 4.8333 and that a 95% confidence interval for μ_{42} is [3.4504, 6.2163]. Hence we are 95 percent confident that μ_{42}, the mean number of defective boxes that would be produced per hour by machine operator 2 using production method 4, is between 3.4504 and 6.2163 boxes. It also follows that 4.8333 is a point prediction of $y_{42} = \mu_{42} + \varepsilon_{42}$, the number of defective boxes that will be produced in a (randomly selected) hour by machine operator 2 using production method 4. Although a 95% prediction interval for y_{42} is not given in Figure 17.2, this interval can be easily obtained by using SAS (see Section 17.5).

17.3

THE REGRESSION APPROACH TO RANDOMIZED BLOCKS

We can implement the regression approach to analyzing the randomized block design by using the usual 0, 1 dummy variables to define τ_l and δ_h in the randomized block model. Therefore we are led to consider the regression model

$$\begin{aligned} y_{lh} &= \mu_{lh} + \varepsilon_{lh} \\ &= \mu + \tau_l + \delta_h + \varepsilon_{lh} \end{aligned}$$

where

$$\tau_l = \tau_2 D_{l,2} + \tau_3 D_{l,3} + \cdots + \tau_v D_{l,v}$$

and

$$\delta_h = \delta_2 D_{h,2} + \delta_3 D_{h,3} + \cdots + \delta_d D_{h,d}$$

Here, for example,

$$D_{l,3} = \begin{cases} 1 & \text{if } l = 3\text{; that is, if we use treatment 3} \\ 0 & \text{otherwise} \end{cases}$$

$$D_{h,2} = \begin{cases} 1 & \text{if } h = 2\text{; that is, if we use block 2} \\ 0 & \text{otherwise} \end{cases}$$

and the other dummy variables are defined similarly. The analysis using this model is similar to the analysis of two factor data using the (α, γ) model. This model was discussed in Section 16.1. Calling the model defined above the *(τ, δ) model,* we can analyze a randomized block experiment as follows.

1. *Testing for Block Importance.* We first wish to see whether the blocks (experimental units) have a statistically significant effect on the mean response. The (τ, δ) model says that

   $$\mu_{lh} = \mu + \tau_l + \delta_h$$

   This implies (using the definitions of the dummy variables) that

   $$\begin{aligned} \mu_{l1} &= \mu + \tau_l \\ \mu_{l2} &= \mu + \tau_l + \delta_2 \quad \text{and} \quad \delta_2 = \mu_{l2} - (\mu + \tau_l) = \mu_{l2} - \mu_{l1} \\ \mu_{l3} &= \mu + \tau_l + \delta_3 \quad \text{and} \quad \delta_3 = \mu_{l3} - \mu_{l1} \\ &\vdots \\ \mu_{ld} &= \mu + \tau_l + \delta_d \quad \text{and} \quad \delta_d = \mu_{ld} - \mu_{l1} \end{aligned}$$

Thus the null hypothesis

$$H_0\colon \mu_{l1} = \mu_{l2} = \cdots = \mu_{ld}$$

which says that all d blocks have the same effect on the mean response, is equivalent to

$$H_0\colon \delta_2 = \delta_3 = \cdots = \delta_d = 0$$

We test H_0 versus

H_1: At least one of $\delta_2, \delta_3, \ldots, \delta_d$ does not equal zero

or

H_1: At least two of $\mu_{l1}, \mu_{l2}, \ldots, \mu_{ld}$ differ

The test is carried out by using the partial F statistic

$$F(\delta \mid \tau) = F(D_{h,2}, D_{h,3}, \ldots, D_{h,d} \mid D_{l,2}, D_{l,3}, \ldots, D_{l,v})$$

Here, we define the complete and reduced models to be

Complete model: $y_{lh} = \mu + \tau_l + \delta_h + \varepsilon_{lh}$

Reduced model: $y_{lh} = \mu + \tau_l + \varepsilon_{lh}$

2. *Testing for Differences Between Treatments.* Next, we wish to see whether the treatments have a statistically significant effect on the mean response. Again, the (τ, δ) model says that

$$\mu_{lh} = \mu + \tau_l + \delta_h$$

Therefore using the definitions of the dummy variables, we find that

$$\mu_{1h} = \mu + \delta_h$$

$$\mu_{2h} = \mu + \tau_2 + \delta_h \quad \text{and} \quad \tau_2 = \mu_{2h} - (\mu + \delta_h) = \mu_{2h} - \mu_{1h}$$

$$\mu_{3h} = \mu + \tau_3 + \delta_h \quad \text{and} \quad \tau_3 = \mu_{3h} - \mu_{1h}$$

$$\vdots$$

$$\mu_{vh} = \mu + \tau_v + \delta_h \quad \text{and} \quad \tau_v = \mu_{vh} - \mu_{1h}$$

Thus the null hypothesis

$$H_0\colon \mu_{1h} = \mu_{2h} = \cdots = \mu_{vh}$$

which says that all v treatments have the same effect on the mean response, is equivalent to

$$H_0\colon \tau_2 = \tau_3 = \cdots = \tau_v = 0$$

We test H_0 versus

H_1: At least one of $\tau_2, \tau_3, \ldots, \tau_v$ does not equal zero

or

H_1: At least two of $\mu_{1h}, \mu_{2h}, \ldots, \mu_{vh}$ differ

The test is carried out by using the partial F statistic

$$F(\tau \mid \delta) = F(D_{l,2}, D_{l,3}, \ldots, D_{l,v} \mid D_{h,2}, D_{h,3}, \ldots, D_{h,d})$$

Here, we define the complete and reduced models to be

Complete model: $y_{lh} = \mu + \tau_l + \delta_h + \varepsilon_{lh}$

Reduced model: $y_{lh} = \mu + \delta_h + \varepsilon_{lh}$

Example 17.4 Again consider the defective cardboard box data given in Table 17.1. For $l = 1, 2, 3, 4$, and $h = 1, 2, 3$ we let

y_{lh} = the number of defective boxes produced per hour when machine operator h employs production method l

Then the (τ, δ) model describing this data is

$$\begin{aligned} y_{lh} &= \mu_{lh} + \varepsilon_{lh} \\ &= \mu + \tau_l + \delta_h + \varepsilon_{lh} \\ &= \mu + (\tau_2 D_{l,2} + \tau_3 D_{l,3} + \tau_4 D_{l,4}) + (\delta_2 D_{h,2} + \delta_3 D_{h,3}) + \varepsilon_{lh} \end{aligned}$$

Here,

$$D_{l,2} = \begin{cases} 1 & \text{if } l = 2\text{; that is, if the machine operator uses production method 2} \\ 0 & \text{otherwise} \end{cases}$$

$$D_{h,2} = \begin{cases} 1 & \text{if } h = 2\text{; that is, if machine operator 2 produces boxes} \\ 0 & \text{otherwise} \end{cases}$$

and the other dummy variables are defined similarly.

To test for block importance, we test

$$H_0: \delta_2 = \delta_3 = 0$$

or

$$H_0: \mu_{l1} = \mu_{l2} = \mu_{l3}$$

which says that all three machine operators have the same effects on the mean response. We test H_0 versus

H_1: At least one of δ_2 and δ_3 does not equal zero

or

H_1: At least two of μ_{l1}, μ_{l2}, and μ_{l3} differ which says that at least two machine operators have different effects on the mean response. To do this, we define:

1. The complete model to be

$$y_{lh} = \mu + \tau_l + \delta_h + \varepsilon_{lh}$$

For this model, $SSE_C = 3.8333$ (see Figure 17.2, which gives the SAS output for the regression analysis of the data).

2. The reduced model to be

$$\begin{aligned} y_{lh} &= \mu + \tau_l + \varepsilon_{lh} \\ &= \mu + \tau_2 D_{l,2} + \tau_3 D_{l,3} + \tau_4 D_{l,4} + \varepsilon_{lh} \end{aligned}$$

It can be shown that for this model, $SSE_R = 22.0$.

Thus

$$\begin{aligned} F(\delta \mid \tau) &= F(D_{h,2}, D_{h,3} \mid D_{l,2}, D_{l,3}, D_{l,4}) \\ &= \frac{(SSE_R - SSE_C)/(p - g)}{SSE_C/(n - k_{(\tau,\delta)})} \\ &= \frac{(22.0 - 3.8333)/2}{3.8333/(12 - 6)} = 14.22 \end{aligned}$$

and

prob-value = .0053 (see Figure 17.2)

Since this prob-value is less than .05, we reject H_0 with $\alpha = .05$. We conclude that at least two machine operators have different effects on the mean number of defective boxes produced per hour. That is, we conclude that "blocking is important."

To test for differences between production methods, we test

$$H_0\colon \tau_2 = \tau_3 = \tau_4 = 0$$

or

$$H_0\colon \mu_{1h} = \mu_{2h} = \mu_{3h} = \mu_{4h}$$

which says that all four production methods have the same effects on the mean response. We test H_0 versus

H_1: At least one of τ_2, τ_3, and τ_4 does not equal zero

or

H_1: At least two of μ_{1h}, μ_{2h}, μ_{3h}, and μ_{4h} differ

which says that at least two of the production methods have different effects on the mean response. To do this, we define a new reduced model to be

$$\begin{aligned} y_{lh} &= \mu + \delta_h + \varepsilon_{lh} \\ &= \mu + \delta_2 D_{h,2} + \delta_3 D_{h,3} + \varepsilon_{lh} \end{aligned}$$

It can be shown that for this model $SSE_R = 94.75$. Thus,

$$\begin{aligned} F(\tau \mid \delta) &= F(D_{l,2}, D_{l,3}, D_{l,4} \mid D_{h,2}, D_{h,3}) \\ &= \frac{(SSE_R - SSE_C)/(p - g)}{SSE_C/(n - k_{(\tau,\delta)})} \\ &= \frac{(94.75 - 3.8333)/3}{3.8333/(12 - 6)} = 47.43 \end{aligned}$$

and

$$\text{prob-value} = .0001 \qquad \text{(see Figure 17.2).}$$

Since this prob-value is less than .05, we reject H_0 with $\alpha = .05$. We conclude that at least two production methods have different effects on the mean number of defective boxes produced per hour.

Since we have concluded that there are statistically significant differences between at least two of the machine operators and between at least two of the production methods, we might wish to make pairwise comparisons of machine operators and production methods. For example,

$$\begin{aligned} \mu_{l3} - \mu_{l1} &= (\mu + \tau_l + \delta_3) - (\mu + \tau_l) \\ &= \delta_3 = [0 \ \ 0 \ \ 0 \ \ 0 \ \ 0 \ \ 1] \begin{bmatrix} \mu \\ \tau_2 \\ \tau_3 \\ \tau_4 \\ \delta_2 \\ \delta_3 \end{bmatrix} = \boldsymbol{\lambda}'\boldsymbol{\beta} \end{aligned}$$

where

$$\boldsymbol{\lambda}' = [0 \ \ 0 \ \ 0 \ \ 0 \ \ 0 \ \ 1]$$

Therefore referring to Figure 17.2, a point estimate of $\mu_{l3} - \mu_{l1}$ is

$$\boldsymbol{\lambda}'\mathbf{b} = \hat{\delta}_3 = 3$$

and a 95% confidence interval for $\mu_{l3} - \mu_{l1}$ is

$$\begin{aligned} [\boldsymbol{\lambda}'\mathbf{b} \pm t_{[\alpha/2]}^{(n-k_{(\tau,\delta)})} s_{(\tau,\delta)} \sqrt{\boldsymbol{\lambda}'(\mathbf{X}'\mathbf{X})^{-1}\boldsymbol{\lambda}}] &= [3 \pm t_{[.025]}^{(12-6)} (0.56519417)] \\ &= [3 \pm 2.447(0.56519417)] \\ &= [3 \pm 1.383] \\ &= [1.617, 4.383] \end{aligned}$$

This interval says that we can be 95 percent confident that changing from machine operator 1 to machine operator 3 while the production method remains the same will increase the mean number of defective boxes produced per hour by between

1.617 and 4.383 boxes. As another example,

$$\mu_{3h} - \mu_{2h} = (\mu + \tau_3 + \delta_h) - (\mu + \tau_2 + \delta_h)$$

$$= \tau_3 - \tau_2 = [0 \quad -1 \quad 1 \quad 0 \quad 0 \quad 0]\begin{bmatrix} \mu \\ \tau_2 \\ \tau_3 \\ \tau_4 \\ \delta_2 \\ \delta_3 \end{bmatrix} = \boldsymbol{\lambda}'\boldsymbol{\beta}$$

where

$$\boldsymbol{\lambda}' = [0 \quad -1 \quad 1 \quad 0 \quad 0 \quad 0]$$

Therefore, again referring to Figure 17.2, a point estimate of $\mu_{3h} - \mu_{2h}$ is

$$\boldsymbol{\lambda}'\mathbf{b} = \hat{\tau}_3 - \hat{\tau}_2 = -5.333$$

and a 95% confidence interval for $\mu_{3h} - \mu_{2h}$ is

$$\begin{aligned}\left[\boldsymbol{\lambda}'\mathbf{b} \pm t_{[\alpha/2]}^{(n-k_{(\tau,\delta)})} s_{(\tau,\delta)} \sqrt{\boldsymbol{\lambda}'(\mathbf{X}'\mathbf{X})^{-1}\boldsymbol{\lambda}}\right] &= [-5.333 \pm t_{[.025]}^{(12-6)} (.65263001)] \\ &= [-5.333 \pm 2.447\,(.65263001)] \\ &= [-5.333 \pm 1.597] \\ &= [-6.930, \ -3.736]\end{aligned}$$

This interval says that we can be 95 percent confident that changing from production method 2 to production method 3 while the machine operator remains the same will reduce the mean number of defective boxes produced per hour by between 3.736 and 6.930 boxes.

Finally, suppose that we wish to estimate

$$\mu_{33} = \mu + \tau_3 + \delta_3$$

A point estimate of μ_{33} is

$$\hat{y}_{33} = \hat{\mu} + \hat{\tau}_3 + \hat{\delta}_3 = 7$$

where $\hat{\mu}$, $\hat{\tau}_3$, and $\hat{\delta}_3$ are the least squares point estimates of the parameters μ, τ_3, and δ_3, respectively. Note that while $\hat{\mu}$, $\hat{\tau}_3$, and $\hat{\delta}_3$ are not given on the SAS output of Figure 17.2, the point estimate $\hat{y}_{33}$ is given on the output—see observation 11. In addition, a 95% confidence interval for μ_{33} is

$$\left[\hat{y}_{33} \pm t_{[\alpha/2]}^{(n-k_{(\tau,\delta)})} s_{(\tau,\delta)} \sqrt{\mathbf{x}_{33}'(\mathbf{X}'\mathbf{X})^{-1}\mathbf{x}_{33}}\right]$$

Here,

$$\mathbf{x}_{33}' = [1 \quad 0 \quad 1 \quad 0 \quad 0 \quad 1]$$

since $\mu_{33} = \mu + \tau_3 + \delta_3$. We omit the detailed calculations, noting that the SAS output of Figure 17.2 shows that this interval is

$$[5.0337, 7.7996] \qquad \text{(see observation 11)}$$

This interval says that we can be 95% confident that μ_{33}, the mean number of defective boxes produced per hour when machine operator 3 employs production method 3, is between 5.0337 and 7.7996 boxes.

We now extend the previous example and consider using a "model-building approach" to assessing the effects of two factors in a randomized block design.

Example 17.5 ***Part 1: The Problem and Data***

Consider the defective box data. Suppose that each of the production methods consists of a machine type and cardboard grade combination. The Universal Paper Company wishes to investigate the effects of machine type, which has levels A and B, and of cardboard grade, which has levels Y and Z, on the mean number of defective boxes produced per hour. Here, production methods 1, 2, 3, and 4 are defined to be the four combinations AY, AZ, BY, and BZ, respectively. Suppose that the observed data is given in Table 17.3. Here, for i = A and B, j = Y and Z, and h = 1, 2, and 3,

y_{ijh} = the number of defective boxes obtained when machine operator h produces boxes for one hour by using machine type i and cardboard grade j

This observation is assumed to have been randomly selected from the (theoretically) infinite population of all possible hourly defective box rates that could be observed when machine operator h uses machine type i and cardboard grade j. We denote the mean of this population by the symbol μ_{ijh}. That is, μ_{ijh} is the *mean number of defective boxes produced per hour by machine operator h using machine type i and cardboard grade j.*

TABLE 17.3 **The number of defective cardboard boxes obtained by machine types A and B, cardboard grades Y and Z, and machine operators 1, 2, and 3**

Production method, ij	Machine operator, h: 1	2	3	$\bar{y}_{ij\cdot}$
AY	$y_{AY1} = 9$	$y_{AY2} = 10$	$y_{AY3} = 12$	$\bar{y}_{AY\cdot} = 10.3333$
AZ	$y_{AZ1} = 8$	$y_{AZ2} = 11$	$y_{AZ3} = 12$	$\bar{y}_{AZ\cdot} = 10.3333$
BY	$y_{BY1} = 3$	$y_{BY2} = 5$	$y_{BY3} = 7$	$\bar{y}_{BY\cdot} = 5.0$
BZ	$y_{BZ1} = 4$	$y_{BZ2} = 5$	$y_{BZ3} = 5$	$\bar{y}_{BZ\cdot} = 4.6667$
$\bar{y}_{\cdot\cdot h}$	$\bar{y}_{\cdot\cdot 1} = 6.0$	$\bar{y}_{\cdot\cdot 2} = 7.75$	$\bar{y}_{\cdot\cdot 3} = 9.0$	$\bar{y} = 7.5833$

Part 2: Developing a Regression Model

Examining Figure 17.1, recall that graphical analysis of the data in Table 17.3 indicates that there is little interaction between production method and machine operator. Therefore one possible model describing y_{ijh} is

$$y_{ijh} = \mu_{ijh} + \varepsilon_{ijh}$$

where

$$\begin{aligned} \mu_{ijh} &= \mu + \tau_{ij} + \delta_h \\ &= \mu + \tau_{AZ} D_{ij,AZ} + \tau_{BY} D_{ij,BY} + \tau_{BZ} D_{ij,BZ} + \delta_2 D_{h,2} + \delta_3 D_{h,3} \end{aligned}$$

Here, for example,

$$D_{ij,BY} = \begin{cases} 1 & \text{if } i = \text{B and } j = \text{Y; that is, if the machine operator uses machine type B and cardboard grade Y} \\ 0 & \text{otherwise} \end{cases}$$

$$D_{h,3} = \begin{cases} 1 & \text{if } h = 3\text{; that is, if machine operator 3 produces boxes for one hour} \\ 0 & \text{otherwise} \end{cases}$$

FIGURE 17.3 **Graphical analysis of the defective box data in Table 17.3: Little interaction**

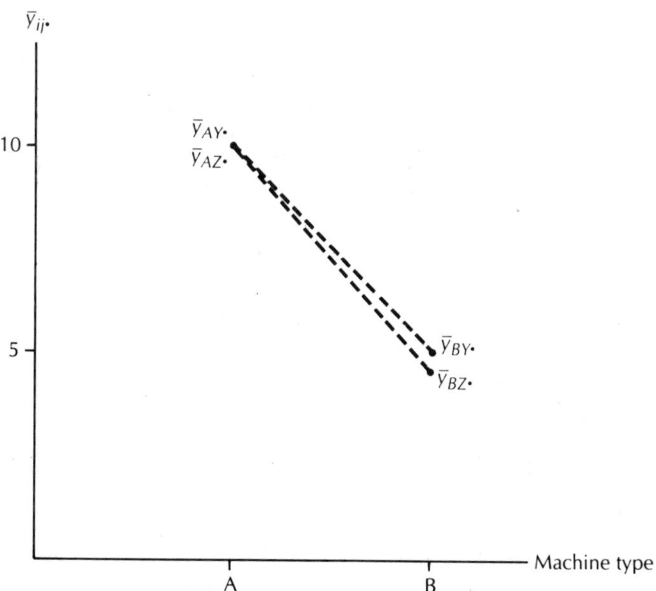

For each cardboard grade a plot is made of the change in the $\bar{y}_{ij\cdot}$ values associated with changing the machine type. Little interaction means this change in the $\bar{y}_{ij\cdot}$ values depends little on the cardboard grade.

and the other dummy variables are defined similarly. We refer to this model as the (τ, δ) *model*. This model, however, is not the only model that we might use to describe the data in Table 17.3. To see this, note that the (τ, δ) model correctly assumes that no interaction exists between production method and machine operator. However, the equation describing μ_{ijh} in this model uses a separate dummy variable to describe the effect of each of the combinations of machine type and cardboard grade. Therefore the (τ, δ) model is the appropriate model to use if interaction exists between machine type and cardboard grade. One way to check for such interaction is to carry out the graphical analysis of Figure 17.3. Since this figure indicates that little or no interaction exists between machine type and cardboard grade, we should express μ_{ijh} by the equation

$$\begin{aligned} \mu_{ijh} &= \mu + \alpha_i + \gamma_j + \delta_h \\ &= \mu + \alpha_B D_{i,B} + \gamma_Z D_{j,Z} + \delta_2 D_{h,2} + \delta_3 D_{h,3} \end{aligned}$$

Here, for example,

$$D_{i,B} = \begin{cases} 1 & \text{if } i = \text{B; that is, if the machine operator uses machine type B} \\ 0 & \text{otherwise} \end{cases}$$

$$D_{j,Z} = \begin{cases} 1 & \text{if } j = \text{Z; that is, if the machine operator uses cardboard grade Z} \\ 0 & \text{otherwise} \end{cases}$$

$$D_{h,3} = \begin{cases} 1 & \text{if } h = 3\text{; that is, if machine operator 3 produces boxes for one hour} \\ 0 & \text{otherwise} \end{cases}$$

and the other dummy variable is defined similarly. We call this model the (α, γ, δ) *model*. Note from the equation describing μ_{ijh} that

$$\alpha_i = \alpha_B D_{i,B}$$

which implies that

$$\begin{aligned} \alpha_A &= \alpha_B(0) \\ &= 0 \end{aligned}$$

Also note that

$$\gamma_j = \gamma_Z D_{j,Z}$$

which implies that

$$\begin{aligned} \gamma_Y &= \gamma_Z(0) \\ &= 0 \end{aligned}$$

In addition,

$$\delta_h = \delta_2 D_{h,2} + \delta_3 D_{h,3}$$

which implies that

$$\begin{aligned}\delta_1 &= \delta_2(0) + \delta_3(0)\\ &= 0\end{aligned}$$

Therefore the (α, γ, δ) model implies that for i = A and B, j = Y and Z, and h = 1, 2, and 3,

$$\begin{aligned}\mu_{\mathrm{B}jh} - \mu_{\mathrm{A}jh} &= [\mu + \alpha_{\mathrm{B}} + \gamma_j + \delta_h] - [\mu + \alpha_{\mathrm{A}} + \gamma_j + \delta_h]\\ &= \alpha_{\mathrm{B}} \quad (\text{since } \alpha_{\mathrm{A}} = 0)\end{aligned}$$

= the change in the mean number of defective boxes produced per hour associated with a particular machine operator changing from machine type A to machine type B while the cardboard grade remains the same

$$\begin{aligned}\mu_{i\mathrm{Z}h} - \mu_{i\mathrm{Y}h} &= [\mu + \alpha_i + \gamma_{\mathrm{Z}} + \delta_h] - [\mu + \alpha_i + \gamma_{\mathrm{Y}} + \delta_h]\\ &= \gamma_{\mathrm{Z}} \quad (\text{since } \gamma_{\mathrm{Y}} = 0)\end{aligned}$$

= the change in the mean number of defective boxes produced per hour associated with a particular machine operator changing from cardboard grade Y to cardboard grade Z while the machine type remains the same

Also, for h = 2 and 3,

$$\begin{aligned}\mu_{ijh} - \mu_{ij1} &= [\mu + \alpha_i + \gamma_j + \delta_h] - [\mu + \alpha_i + \gamma_j + \delta_1]\\ &= \delta_h \quad (\text{since } \delta_1 = 0)\end{aligned}$$

= the change in the mean number of defective boxes produced per hour associated with changing from machine operator 1 using a particular machine type and cardboard grade combination to machine operator h using the same machine type and cardboard grade combination

Note that the change $(\mu_{\mathrm{B}jh} - \mu_{\mathrm{A}jh})$ is equal to α_{B} and thus is independent of j and of h. Likewise, the change $(\mu_{i\mathrm{Z}h} - \mu_{i\mathrm{Y}h})$ is equal to γ_{Z} and thus is independent of i and h. Also, the change $(\mu_{ijh} - \mu_{ij1})$ is equal to δ_h and thus is independent of i and j. It follows that the (α, γ, δ) model assumes that no interactions exist between machine type, cardboard grade, and machine operator.

Part 3: Estimating the Regression Parameters

To use the (α, γ, δ) model

$$\begin{aligned}y_{ijh} &= \mu_{ijh} + \varepsilon_{ijh}\\ &= \mu + \alpha_i + \gamma_j + \delta_h + \varepsilon_{ijh}\\ &= \mu + \alpha_{\mathrm{B}} D_{i,\mathrm{B}} + \gamma_{\mathrm{Z}} D_{j,\mathrm{Z}} + \delta_2 D_{h,2} + \delta_3 D_{h,3} + \varepsilon_{ijh}\end{aligned}$$

to carry out a regression analysis of the defective box data in Table 17.3, we use

$$\mathbf{y} = \begin{bmatrix} y_{AY1} \\ y_{AY2} \\ y_{AY3} \\ y_{AZ1} \\ y_{AZ2} \\ y_{AZ3} \\ y_{BY1} \\ y_{BY2} \\ y_{BY3} \\ y_{BZ1} \\ y_{BZ2} \\ y_{BZ3} \end{bmatrix} = \begin{bmatrix} 9 \\ 10 \\ 12 \\ 8 \\ 11 \\ 12 \\ 3 \\ 5 \\ 7 \\ 4 \\ 5 \\ 5 \end{bmatrix} \qquad \mathbf{X} = \begin{array}{c} \begin{matrix} 1 & D_{i,B} & D_{j,Z} & D_{h,2} & D_{h,3} \end{matrix} \\ \begin{bmatrix} 1 & 0 & 0 & 0 & 0 \\ 1 & 0 & 0 & 1 & 0 \\ 1 & 0 & 0 & 0 & 1 \\ 1 & 0 & 1 & 0 & 0 \\ 1 & 0 & 1 & 1 & 0 \\ 1 & 0 & 1 & 0 & 1 \\ 1 & 1 & 0 & 0 & 0 \\ 1 & 1 & 0 & 1 & 0 \\ 1 & 1 & 0 & 0 & 1 \\ 1 & 1 & 1 & 0 & 0 \\ 1 & 1 & 1 & 1 & 0 \\ 1 & 1 & 1 & 0 & 1 \end{bmatrix} \end{array}$$

We obtain the prediction equation

$$\begin{aligned} \hat{y}_{ijh} &= \hat{\mu} + \hat{\alpha}_i + \hat{\gamma}_j + \hat{\delta}_h \\ &= \hat{\mu} + \hat{\alpha}_B D_{i,B} + \hat{\gamma}_Z D_{j,Z} + \hat{\delta}_2 D_{h,2} + \hat{\delta}_3 D_{h,3} \\ &= 8.8333 - 5.5D_{i,B} - .1667D_{j,Z} + 1.75D_{h,2} + 3.0D_{h,3} \end{aligned}$$

In addition, the unexplained variation is calculated to be

$$\begin{aligned} SSE_{(\alpha,\gamma,\delta)} &= \sum_{i=A,B} \sum_{j=Y,Z} \sum_{h=1}^{3} (y_{ijh} - \hat{y}_{ijh})^2 \\ &= 3.9167 \end{aligned}$$

Part 4: Testing for Interaction Between Machine Type and Cardboard Grade

The unexplained variation for the (τ, δ) model, which is the appropriate model to use if interaction exists between machine type and cardboard grade, can be calculated to be $SSE_{(\tau,\delta)} = 3.8333$. We wish to use this quantity to test

H_0: No interaction exists between machine type and cardboard grade

versus

H_1: Interaction does exist between machine type and cardboard grade

We can test H_0 by intuitively modifying the test for interaction discussed in Section 16.1 (see the box titled "Testing for Interaction Between Two Qualitative

Factors"). To do this, we let

$$k_{(\tau,\delta)} = 6 = \text{the number of parameters in the } (\tau, \delta) \text{ model}$$

and

$$k_{(\alpha,\gamma,\delta)} = 5 = \text{the number of parameters in the } (\alpha, \gamma, \delta) \text{ model}$$

We now define the following F(interaction) statistic and prob-value:

$$\begin{aligned} F(\text{interaction}) &= \frac{MS_{\text{interaction}}}{MSE_{(\tau,\delta)}} \\ &= \frac{SS_{\text{interaction}}/(k_{(\tau,\delta)} - k_{(\alpha,\gamma,\delta)})}{SSE_{(\tau,\delta)}/(n - k_{(\tau,\delta)})} \\ &= \frac{(SSE_{(\alpha,\gamma,\delta)} - SSE_{(\tau,\delta)})/(k_{(\tau,\delta)} - k_{(\alpha,\gamma,\delta)})}{SSE_{(\tau,\delta)}/(n - k_{(\tau,\delta)})} \\ &= \frac{(3.9167 - 3.8333)/(6 - 5)}{3.8333/(12 - 6)} \\ &= \frac{.0834/1}{3.8333/6} \\ &= \frac{.0834}{.6389} \\ &= .13 \end{aligned}$$

and

prob-value = the area under the curve of the F-distribution having 1 and 6 degrees of freedom to the right of .13
= .7304

Since this prob-value is greater than .05, we cannot reject H_0 with $\alpha = .05$. Thus we have little evidence of interaction between machine type and cardboard grade.

Part 5: Testing for Block Importance

Since we have concluded that little or no interaction exists between machine type and cardboard grade, we further analyze the data in Table 17.3 by using the (α, γ, δ) model. As was shown previously, this model implies that for $h = 1, 2,$ and 3,

$$\delta_h = \mu_{ijh} - \mu_{ij1}$$

and thus that

$$\delta_2 = \mu_{ij2} - \mu_{ij1}$$

$$\delta_3 = \mu_{ij3} - \mu_{ij1}$$

Therefore we will test

$$H_0: \delta_2 = \delta_3 = 0$$

which is equivalent to

$$H_0: \mu_{ij1} = \mu_{ij2} = \mu_{ij3}$$

This null hypothesis says that the three machine operators have the same effects on the mean number of defective boxes produced per hour. We test H_0 versus

$$H_1\text{: At least one of } \delta_2 \text{ and } \delta_3 \text{ does not equal zero}$$

FIGURE 17.4 **SAS output of an (α, γ, δ) model analysis of the cardboard box data in Table 17.3**

SAS

GENERAL LINEAR MODELS PROCEDURE

DEPENDENT VARIABLE: DEFECTS

SOURCE	DF	SUM OF SQUARES	MEAN SQUARE	F VALUE	PR > F	R-SQUARE	C.V.
MODEL	4	109.00000000	27.25000000	48.70	0.0001	0.965314	9.8639
ERROR	7	3.91666667[g]	0.55952381		ROOT MSE		DEFECTS MEAN
CORRECTED TOTAL	11	112.91666667			0.74801324[h]		7.58333333

SOURCE	DF	TYPE I SS	F VALUE	PR > F	DF	TYPE III SS	F VALUE	PR > F
MACHTYPE	1	90.75000000	162.19	0.0001	1	90.75000000	162.19[a]	0.0001[d]
CBGRADE	1	0.08333333	0.15	0.7110	1	0.08333333	0.15[b]	0.7110[e]
OPERATOR	2	18.16666667	16.23	0.0023	2	18.16666667	16.23[c]	0.0023[f]

PARAMETER	ESTIMATE	T FOR HO: PARAMETER=0	PR > |T|	STD ERROR OF ESTIMATE
MUBJH-MUAJH	-5.50000000[i]	-12.74	0.0001	0.43186565
MUIZH-MUIYH	-0.16666667	-0.39	0.7110	0.43186565
MUIJ3-MUIJ1	3.00000000	5.67	0.0008	0.52892524
MUIJ2-MUIJ1	1.75000000	3.31	0.0130	0.52892524
MUIJ3-MUIJ2	1.25000000	2.36	0.0501	0.52892524

OBSERVATION	OBSERVED VALUE	PREDICTED VALUE	RESIDUAL	LOWER 95% CL FOR MEAN	UPPER 95% CL FOR MEAN
1	9.00000000	8.83333333	0.16666667	7.69158839	9.97507828
2	10.00000000	10.58333333	-0.58333333	9.44158839	11.72507828
3	12.00000000	11.83333333	0.16666667	10.69158839	12.97507828
4	8.00000000	8.66666667	-0.66666667	7.52492172	9.80841161
5	11.00000000	10.41666667	0.58333333	9.27492172	11.55841161
6	12.00000000	11.66666667	0.33333333	10.52492172	12.80841161
7	3.00000000	3.33333333	-0.33333333	2.19158839	4.47507828
8	5.00000000	5.08333333	-0.08333333	3.94158839	6.22507828
9	7.00000000	6.33333333	0.66666667	5.19158839	7.47507828
10	4.00000000	3.16666667	0.83333333	2.02492172	4.30841161
11	5.00000000	4.91666667[j]	0.08333333	[3.77492172	6.05841161][k]
12	5.00000000	6.16666667	-1.66666667	5.02492172	7.20841161

*OBSERVATION WAS NOT USED IN THIS ANALYSIS

[a]$F(\alpha|\gamma, \delta)$ [b]$F(\gamma|\alpha, \delta)$ [c]$F(\delta|\alpha, \gamma)$ [d]*Prob-value related to* $F(\alpha|\gamma, \delta)$ [e]*Prob-value related to* $F(\gamma|\alpha, \delta)$
[f]*Prob-value related to* $F(\delta|\alpha, \gamma)$ [g]$SSE_{(\alpha,\gamma,\delta)}$ [h]$s_{(\alpha,\gamma,\delta)}$ [i]*Point estimate of* $\mu_{Bjh} - \mu_{Ajh}$ [j]$\hat{y}_{BZ2}$
[k]*95% confidence interval for* μ_{BZ2}

or

H_1: At least two of μ_{ij1}, μ_{ij2}, and μ_{ij3} differ from each other

which says that at least two of the machine operators have different effects on the mean number of defective boxes produced per hour. We consider the (α, γ, δ) model to be the complete model (for which $SSE_C = SSE_{(\alpha,\gamma,\delta)} = 3.9167$), and we note that H_0 assumes $p - g = 2$ parameters (δ_2 and δ_3) to be zero. Thus if H_0 is true, the complete model becomes the reduced model

$$y_{ijh} = \mu + \alpha_B D_{i,B} + \gamma_Z D_{j,Z} + \varepsilon_{ijh}$$

It can be shown that for this model, $SSE_R = 22.0834$. We can test H_0 versus H_1 by using the following partial F statistic

$$\begin{aligned} F(\delta \mid \alpha, \gamma) = F(D_{h,2}, D_{h,3} \mid D_{i,B}, D_{j,Z}) &= \frac{SS_{\text{drop}}/(p - g)}{SSE_C/(n - k_{(\alpha,\gamma,\delta)})} \\ &= \frac{\{SSE_R - SSE_C\}/(p - g)}{SSE_C/(n - k_{(\alpha,\gamma,\delta)})} \\ &= \frac{\{22.0834 - 3.9167\}/2}{3.9167/(12 - 5)} \\ &= \frac{18.1667/2}{3.9167/7} \\ &= \frac{9.0833}{.5595} \\ &= 16.2347 \end{aligned}$$

As shown in Figure 17.4, which presents the SAS output that results from using the (α, γ, δ) model to analyze the data in Table 17.3, the prob-value related to $F(\delta \mid \alpha, \gamma)$ is .0023. Since this prob-value is less than .05, we reject H_0 with $\alpha = .05$. Thus there is very substantial evidence that at least two of the machine operators have different effects on the mean number of defective boxes produced per hour. This indicates that blocking is important.

Part 6: Studying Differences Between Means

To test for differences between machine types, recall that

$$\mu_{Bjh} - \mu_{Ajh} = \alpha_B$$

Thus consider testing

$$H_0: \alpha_B = 0 \quad \text{or} \quad H_0: \mu_{Bjh} = \mu_{Ajh}$$

versus

$$H_1: \alpha_B \neq 0 \quad \text{or} \quad H_1: \mu_{Bjh} \neq \mu_{Ajh}$$

We consider the (α, γ, δ) model to be the complete model, and we note that H_0 assumes $p - g = 1$ parameter (α_B) to be zero. Thus if H_0 is true, the complete model becomes the reduced model

$$y_{ijh} = \mu + \gamma_Z D_{j,Z} + \delta_2 D_{h,2} + \delta_3 D_{h,3} + \varepsilon_{ijh}$$

For this model it can be shown that $SSE_R = 94.6707$. It follows that we can test H_0 versus H_1 by using the partial F statistic

$$\begin{aligned} F(\alpha \mid \gamma, \delta) &= F(D_{i,B} \mid D_{j,Z}, D_{h,2}, D_{h,3}) \\ &= \frac{(SSE_R - SSE_C)/(p-g)}{SSE_C/(n - k_{(\alpha,\gamma,\delta)})} \\ &= \frac{(94.6707 - 3.9167)/1}{3.9167/(12-5)} \\ &= 162.197 \end{aligned}$$

As is shown in Figure 17.4, the prob-value related to $F(\alpha \mid \gamma, \delta)$ is .0001. Since this prob-value is less than .05, we reject H_0 with $\alpha = .05$. We conclude that there is overwhelming evidence that machine types A and B have different effects on the mean number of defective boxes produced per hour.

To test for differences between cardboard grades, recall that

$$\mu_{iZh} - \mu_{iYh} = \gamma_Z$$

Thus we test

$$H_0: \gamma_Z = 0 \quad \text{or} \quad H_0: \mu_{iZh} = \mu_{iYh}$$

versus

$$H_1: \gamma_Z \neq 0 \quad \text{or} \quad H_1: \mu_{iZh} \neq \mu_{iYh}$$

We consider the (α, γ, δ) model to be the complete model, and we note that H_0 assumes $p - g = 1$ parameter (γ_Z) to be zero. Thus if H_0 is true, the complete model becomes the reduced model

$$y_{ijh} = \mu + \alpha_B D_{i,B} + \delta_2 D_{h,2} + \delta_3 D_{h,3} + \varepsilon_{ijh}$$

For this model it can be shown that $SSE_R = 4.00007$. It follows that we can test H_0 versus H_1 by using the partial F statistic

$$\begin{aligned} F(\gamma \mid \alpha, \delta) &= F(D_{j,Z} \mid D_{i,B}, D_{h,2}, D_{h,3}) \\ &= \frac{(SSE_R - SSE_C)/(p-g)}{SSE_C/(n - k_{(\alpha,\gamma,\delta)})} \\ &= \frac{(4.00007 - 3.9167)/1}{3.9167/(12-5)} \\ &= 0.1490 \end{aligned}$$

As is shown in Figure 17.4, the prob-value related to $F(\gamma \mid \alpha, \delta)$ is .7110. Since this prob-value is greater than .05, we cannot reject H_0 with $\alpha = .05$. We conclude that there is little evidence to suggest that cardboard grades Y and Z have different effects on the mean number of defective boxes produced per hour.

Since we have found that there is a statistically significant difference between the effects of machine types A and B, we will attempt to assess the practical importance of this difference by computing a 95% confidence interval for $\mu_{Bjh} - \mu_{Ajh}$. Since $SSE_{(\alpha,\gamma,\delta)} = 3.9167$, it follows that $s_{(\alpha,\gamma,\delta)} = .7480$. Furthermore, the diagonal element of $(\mathbf{X}'\mathbf{X})^{-1}$ corresponding to α_B in this model can be shown to be $c_{BB} = \frac{1}{3}$. A point estimate of $\mu_{Bjh} - \mu_{Ajh} = \alpha_B$ is $\hat{\alpha}_B = -5.5$ (see Figure 17.4). Moreover, a 95% confidence interval for $\mu_{Bjh} - \mu_{Ajh}$ is

$$\begin{aligned}
[\hat{\alpha}_B \pm t_{[.025]}^{(n-k_{(\alpha,\gamma,\delta)})} s_{(\alpha,\gamma,\delta)} \sqrt{c_{BB}}] &= \left[-5.5 \pm t_{[.025]}^{(7)} (.748) \sqrt{\frac{1}{3}}\right] \\
&= \left[-5.5 \pm 2.365(.748) \sqrt{\frac{1}{3}}\right] \\
&= [-5.5 \pm 1.0214] \\
&= [-6.5214, -4.4786]
\end{aligned}$$

This interval says that we can be 95 percent confident that the effect of a particular machine operator changing from machine type A to machine type B while the cardboard grade remains the same is to decrease the mean number of defective boxes produced per hour by between 4.4786 and 6.5214 boxes.

Part 7: Using the Model to Estimate and Predict

We have concluded that machine type B produces a smaller mean defective box rate than does machine type A. We have also concluded that changing cardboard grades has little effect on the mean defective box rate. Suppose that since cardboard grade Z is known to be less expensive and more durable, Universal Paper decides to produce boxes by using machine type B and cardboard grade Z. Consider, for instance, y_{BZ2}, the number of defective boxes that will be produced in an hour by machine operator 2 using machine type B and cardboard grade Z. Here, y_{BZ2} can be described by the equation

$$\begin{aligned}
y_{BZ2} &= \mu_{BZ2} + \varepsilon_{BZ2} \\
&= \mu + \alpha_B D_{B,B} + \gamma_Z D_{Z,Z} + \delta_2 D_{2,2} + \delta_3 D_{2,3} + \varepsilon_{BZ2} \\
&= \mu + \alpha_B(1) + \gamma_Z(1) + \delta_2(1) + \delta_3(0) + \varepsilon_{BZ2} \\
&= \mu + \alpha_B + \gamma_Z + \delta_2 + \varepsilon_{BZ2}
\end{aligned}$$

It follows that

$$\begin{aligned}
\hat{y}_{BZ2} &= \hat{\mu} + \hat{\alpha}_B + \hat{\gamma}_Z + \hat{\delta}_2 \\
&= 8.8333 - 5.5 - .1667 + 1.75 \\
&= 4.9166
\end{aligned}$$

is the point estimate of μ_{BZ2} and is the point prediction of y_{BZ2}. Furthermore, the equation describing y_{BZ2} implies that

$$\mathbf{x}'_{BZ2} = [1 \quad 1 \quad 1 \quad 1 \quad 0]$$

Therefore a 95% confidence interval for μ_{BZ2} is

$$\begin{aligned}[\hat{y}_{BZ2} \pm t^{(7)}_{[.025]} s_{(\alpha,\gamma,\delta)} \sqrt{\mathbf{x}'_{BZ2}(\mathbf{X}'\mathbf{X})^{-1}\mathbf{x}_{BZ2}}] &= [4.9166 \pm 2.365(.748)\sqrt{.4166}] \\ &= [4.9166 \pm 1.1418] \\ &= [3.7748, 6.0584]\end{aligned}$$

This interval says that we can be 95 percent confident that μ_{BZ2}, the mean number of defective boxes produced per hour by machine operator 2 using machine type B and cardboard grade Z, is between 3.7748 and 6.0584 boxes.

Moreover, a 95% prediction interval for y_{BZ2} is

$$\begin{aligned}[\hat{y}_{BZ2} \pm t^{(7)}_{[.025]} s_{(\alpha,\gamma,\delta)} \sqrt{1 + \mathbf{x}'_{BZ2}(\mathbf{X}'\mathbf{X})^{-1}\mathbf{x}_{BZ2}}] &= [4.9166 \pm 2.365(.748)\sqrt{1.4166}] \\ &= [4.9166 \pm 2.1055] \\ &= [2.8111, 7.0221]\end{aligned}$$

This interval says that we can be 95 percent confident that y_{BZ2} will be between 2.8111 and 7.0221 boxes.

Part 8: An Alternative Analysis Using the $(\alpha, \gamma, \theta, \delta)$ Model

We saw in Section 15.3 that it is customary to analyze a complete factorial experiment by using the (α, γ, θ) model. Therefore since the data in Table 17.3 is a balanced complete factorial experiment in a randomized block design, it might be reasonable to analyze this data by the $(\alpha, \gamma, \theta, \delta)$ model. This model is

$$\begin{aligned}y_{ijh} &= \mu_{ijh} + \varepsilon_{ijh} \\ &= \mu + \alpha_i + \gamma_j + \theta_{ij} + \delta_h + \varepsilon_{ijh} \\ &= \mu + \alpha_B D_{i,B} + \gamma_Z D_{j,Z} \\ &\quad + \theta_{BZ} D_{i,B} D_{j,Z} + \delta_2 D_{h,2} + \delta_3 D_{h,3} \\ &\quad + \varepsilon_{ijh}\end{aligned}$$

Here:

$$D_{i,B} = \begin{cases} 1 & \text{if } i = \text{B; that is, if the machine operator uses machine type B} \\ -1 & \text{if } i = \text{A; that is, if the machine operator uses machine type A} \end{cases}$$

$$D_{j,Z} = \begin{cases} 1 & \text{if } j = \text{Z; that is, if the machine operator uses cardboard grade Z} \\ -1 & \text{if } j = \text{Y; that is, if the machine operator uses cardboard grade Y} \end{cases}$$

$$D_{h,2} = \begin{cases} 1 & \text{if } h = 2\text{; that is, if machine operator 2 produces boxes for one hour} \\ 0 & \text{otherwise} \end{cases}$$

$$D_{h,3} = \begin{cases} 1 & \text{if } h = 3\text{; that is, if machine operator 3 produces boxes for one hour} \\ 0 & \text{otherwise} \end{cases}$$

The reduced model for testing $H_0: \theta_{BZ} = 0$ is the (α, γ, δ) model

$$y_{ijh} = \mu + \alpha_B D_{i,B} + \gamma_Z D_{j,Z} + \delta_2 D_{h,2} + \delta_3 D_{h,3} + \varepsilon_{ijh}$$

FIGURE 17.5 **SAS output of an $(\alpha, \gamma, \theta, \delta)$ model analysis of the cardboard box data in Table 17.3**

SAS

GENERAL LINEAR MODELS PROCEDURE

DEPENDENT VARIABLE: DEFECTS

SOURCE	DF	SUM OF SQUARES	MEAN SQUARE	F VALUE	PR > F	R-SQUARE	C.V.
MODEL	5	109.08333333	21.81666667	34.15	0.0002	0.966052	10.5403
ERROR	6	3.83333333	0.63888889		ROOT MSE		DEFECTS MEAN
CORRECTED TOTAL	11	112.9166667			0.79930525		7.58333333

SOURCE	DF	TYPE I SS	F VALUE	PR > F	DF	TYPE III SS	F VALUE	PR > F
MACHTYPE	1	90.75000000	142.04	0.0001	1	90.75000000	142.04[a]	0.0001
CBGRADE	1	0.08333333	0.13	0.7304	1	0.08333333	0.13[b]	0.7304
MACHTYPE*CBGRADE	1	0.08333333	0.13	0.7304	1	0.08333333	0.13[c]	0.7304
OPERATOR	2	18.16666667	14.22	0.0053	2	18.16666667	14.22[d]	0.0053

PARAMETER	ESTIMATE	T FOR H0: PARAMETER=0	PR > |T|	STD ERROR OF ESTIMATE
MUBH-MUAH	-5.50000000[e]	-11.92	0.0001	0.46147910
MUZH-MUYH	-0.16666667	-0.36	0.7304	0.46147910
MUIJ3-MUIJ1	3.00000000	5.31	0.0018	0.56519417
MUIJ2-MUIJ1	1.75000000	3.10	0.0212	0.56519417
MUIJ3-MUIJ2	1.25000000	2.21	0.0690	0.56519417
MUBZH-MUAYH	-5.66666667	-8.68	0.0001	0.65263001

OBSERVATION	OBSERVED VALUE	PREDICTED VALUE	RESIDUAL	LOWER 95% CL FOR MEAN	UPPER 95% CL FOR MEAN
1	9.00000000	8.75000000	0.25000000	7.36701870	10.13298130
2	10.00000000	10.50000000	-0.50000000	9.11701870	11.88298130
3	12.00000000	11.75000000	0.25000000	10.36701870	13.13298130
4	8.00000000	8.75000000	-0.75000000	7.36701870	10.13298130
5	11.00000000	10.50000000	0.50000000	9.11701870	11.88298130
6	12.00000000	11.75000000	0.25000000	10.36701870	13.13298130
7	3.00000000	3.41666667	-0.41666667	2.03368537	4.79964797
8	5.00000000	5.16666667	-0.16666667	3.78368537	6.54964797
9	7.00000000	6.41666667	0.58333333	5.03368537	7.79964797
10	4.00000000	3.08333333	0.91666667	1.70035203	4.46631463
11	5.00000000	4.83333333	0.16666667	[3.45035203	6.21631463][f]
12	5.00000000	6.08333333	-1.08333333	4.70035203	7.46631463

[a] $F(\alpha \mid \gamma, \theta, \delta)$ [b] $F(\gamma \mid \alpha, \theta, \delta)$ [c] $F(\theta \mid \alpha, \gamma, \delta)$ [d] $F(\delta \mid \alpha, \gamma, \theta)$ [e] *Point estimate of* $\mu_{B \cdot h} - \mu_{A \cdot h}$ [f] *95% confidence interval for* μ_{BZ2}

Figure 17.5 tells us that

$$F(\theta \mid \alpha, \gamma, \delta) = \frac{(SSE_{(\alpha,\gamma,\delta)} - SSE_{(\alpha,\gamma,\theta,\delta)})/1}{SSE_{(\alpha,\gamma,\theta,\delta)}/(12 - 6)}$$

equals .13 and that the associated prob-value equals .7304. Therefore we have little evidence of interaction between machine type and cardboard grade.

The reduced model for testing $H_0: \delta_2 = \delta_3 = 0$ is the (α, γ, θ) model

$$y_{ijh} = \mu + \alpha_B D_{i,B} + \gamma_Z D_{j,Z} + \theta_{BZ} D_{i,B} D_{j,Z} + \varepsilon_{ijh}$$

Figure 17.5 tells us that

$$F(\delta \mid \alpha, \gamma, \theta) = \frac{(SSE_{(\alpha,\gamma,\theta)} - SSE_{(\alpha,\gamma,\theta,\delta)})/2}{SSE_{(\alpha,\gamma,\theta,\delta)}/(12 - 6)}$$

equals 14.22 and that the associated prob-value equals .0053. Thus there is very substantial evidence that at least two of the machine operators have different effects on the mean number of defective boxes produced per hour.

The reduced model for testing $H_0: \alpha_B = 0$ is the (γ, θ, δ) model

$$y_{ijh} = \mu + \gamma_Z D_{j,Z} + \theta_{BZ} D_{i,B} D_{j,Z} + \delta_2 D_{h,2} + \delta_3 D_{h,3} + \varepsilon_{ijh}$$

Figure 17.5 tells us that

$$F(\alpha \mid \gamma, \theta, \delta) = \frac{(SSE_{(\gamma,\theta,\delta)} - SSE_{(\alpha,\gamma,\theta,\delta)})/1}{SSE_{(\alpha,\gamma,\theta,\delta)}/(12 - 6)}$$

equals 142.04 and that the associated prob-value equals .0001. We conclude that there is overwhelming evidence that machine types A and B have different effects on the mean number of defective boxes produced per hour.

The reduced model for testing $H_0: \gamma_Z = 0$ is the (α, θ, δ) model

$$y_{ijh} = \mu + \alpha_B D_{i,B} + \theta_{BZ} D_{i,B} D_{j,Z} + \delta_2 D_{h,2} + \delta_3 D_{h,3} + \varepsilon_{ijh}$$

Figure 17.5 tells us that

$$F(\gamma \mid \alpha, \theta, \delta) = \frac{(SSE_{(\alpha,\theta,\delta)} - SSE_{(\alpha,\gamma,\theta,\delta)})/1}{SSE_{(\alpha,\gamma,\theta,\delta)}/(12 - 6)}$$

equals .13 and that the associated prob-value equals .7304. We conclude that there is little evidence to suggest that cardboard grades Y and Z have different effects on the mean number of defective boxes produced per hour.

Next, note that the $(\alpha, \gamma, \theta, \delta)$ model implies that

$$\mu_{AYh} = \mu + \alpha_B(-1) + \gamma_Z(-1) + \theta_{BZ}(-1)(-1) + \delta_2 D_{h,2} + \delta_3 D_{h,3}$$

$$\mu_{AZh} = \mu + \alpha_B(-1) + \gamma_Z(1) + \theta_{BZ}(-1)(1) + \delta_2 D_{h,2} + \delta_3 D_{h,3}$$

$$\mu_{BYh} = \mu + \alpha_B(1) + \gamma_Z(-1) + \theta_{BZ}(1)(-1) + \delta_2 D_{h,2} + \delta_3 D_{h,3}$$

$$\mu_{BZh} = \mu + \alpha_B(1) + \gamma_Z(1) + \theta_{BZ}(1)(1) + \delta_2 D_{h,2} + \delta_3 D_{h,3}$$

Therefore the factor level means $\mu_{A\cdot h}$ and $\mu_{B\cdot h}$ are given by

$$\mu_{A\cdot h} = \frac{\mu_{AYh} + \mu_{AZh}}{2} = \frac{2\mu - 2\alpha_B + 2\delta_2 D_{h,2} + 2\delta_3 D_{h,3}}{2}$$
$$= \mu - \alpha_B + \delta_2 D_{h,2} + \delta_3 D_{h,3}$$

and

$$\mu_{B\cdot h} = \frac{\mu_{BYh} + \mu_{BZh}}{2} = \frac{2\mu + 2\alpha_B + 2\delta_2 D_{h,2} + 2\delta_3 D_{h,3}}{2}$$
$$= \mu + \alpha_B + \delta_2 D_{h,2} + \delta_3 D_{h,3}$$

Figure 17.5 tells us that a point estimate of

$$\mu_{B\cdot h} - \mu_{A\cdot h} = 2\alpha_B$$

is -5.5 and that the standard error of the point estimate, $s\sqrt{\boldsymbol{\lambda}'(\mathbf{X}'\mathbf{X})^{-1}\boldsymbol{\lambda}}$, is .4615. Therefore a 95% confidence interval for $\mu_{B\cdot h} - \mu_{A\cdot h}$ is

$$[\boldsymbol{\lambda}'\mathbf{b} \pm t_{[.025]}^{(12-6)} s\sqrt{\boldsymbol{\lambda}'(\mathbf{X}'\mathbf{X})^{-1}\boldsymbol{\lambda}}] = [-5.5 \pm 2.447(.4615)]$$
$$= [-6.6293, \ -4.3707]$$

This 95% interval is longer than the 95% interval, $[-6.5214, -4.4786]$, for $\mu_{Bjh} - \mu_{Ajh}$ calculated in Part 6 by using the (α, γ, δ) model.

Figure 17.5 also provides information that can be used to calculate confidence intervals for the differences $\mu_{\cdot Zh} - \mu_{\cdot Yh}$, $\mu_{ij3} - \mu_{ij1}$, $\mu_{ij2} - \mu_{ij1}$, $\mu_{ij3} - \mu_{ij2}$, and $\mu_{BZh} - \mu_{AYh}$. Also, the figure tells us that a 95% confidence interval for μ_{BZ2} is [3.4504, 6.2163]. This 95% interval is longer than the 95% interval, [3.7748, 6.0584], for μ_{BZ2} calculated in Part 7 by using the (α, γ, δ) model. Since the F-test for interaction indicates that there is little interaction between machine type and cardboard grade, it seems reasonable that the (α, γ, δ) model would give shorter intervals than the $(\alpha, \gamma, \theta, \delta)$ model. However, it is common practice to use the $(\alpha, \gamma, \theta, \delta)$ model to analyze a complete factorial experiment in a randomized block experiment, no matter what the result of the F-test for interaction.

17.4 LATIN SQUARES

In this section we briefly discuss the *Latin square design*. This design can be used to compare d treatments in the presence of two extraneous sources of variability, which the Latin square design blocks off into d rows and d columns. The d treatments are then randomly assigned to the rows and columns so that each treatment appears in every row and every column. One way to do this is discussed later in this section.

For example, suppose that North American Oil Company wishes to compare the gasoline mileage obtained by $d = 4$ different gasolines (G1, G2, G3, and G4).

TABLE 17.4 **The gasoline mileage data for a Latin square design**

	Car model			
Driver	***1***	***2***	***3***	***4***
1	G4 17.6	G2 37.8	G3 15.3	G1 31.0
2	G2 20.4	G3 28.7	G1 21.3	G4 24.7
3	G3 12.7	G1 33.0	G4 19.0	G2 34.4
4	G1 16.8	G4 36.1	G2 23.8	G3 23.7

Because there can be substantial variability due to test drivers and car models, these factors are used as rows and columns in a Latin square design. When the gasolines are randomly assigned to the drivers (rows) and car models (columns), the gasoline mileages in Table 17.4 are observed.

We now present the appropriate model for a Latin square design.

The Latin Square Model

The *Latin square model* says that

$$\begin{aligned} y_{ijk} &= \mu_{ijk} + \varepsilon_{ijk} \\ &= \mu + \alpha_i + \gamma_j + \delta_k + \varepsilon_{ijk} \end{aligned}$$

where

$$\mu_{ijk} = \mu + \alpha_i + \gamma_j + \delta_k$$

Here,

1. y_{ijk} = the value of the response variable when treatment i is used in row j and column k
2. ε_{ijk} = the error term when treatment i is used in row j and column k
3. μ_{ijk} = the mean value of the response variable when treatment i is used in row j and column k
4. μ = an overall mean
5. α_i = the effect due to treatment i
6. γ_j = the effect due to row j
7. δ_k = the effect due to column k

The Latin square model assumes that there are no interactions between treatments, rows, and columns. This is because the equation

$$\mu_{ijk} = \mu + \alpha_i + \gamma_j + \delta_k$$

implies that

1. $\mu_{ijk} - \mu_{i'jk} = (\mu + \alpha_i + \gamma_j + \delta_k) - (\mu + \alpha_{i'} + \delta_j + \delta_k) = \alpha_i - \alpha_{i'}$
2. $\mu_{ijk} - \mu_{ij'k} = (\mu + \alpha_i + \gamma_j + \delta_k) - (\mu + \alpha_i + \gamma_{j'} + \delta_k) = \gamma_j - \gamma_{j'}$
3. $\mu_{ijk} - \mu_{ijk'} = (\mu + \alpha_i + \gamma_j + \delta_k) - (\mu + \alpha_i + \gamma_j + \delta_{k'}) = \delta_k - \delta_{k'}$

THE ANOVA APPROACH

Let

$\bar{y}_{i\cdot\cdot}$ = the average of the sample values obtained by using treatment i

$\bar{y}_{\cdot j\cdot}$ = the average of the sample values in row j

$\bar{y}_{\cdot\cdot k}$ = the average of the sample values in column k

$\bar{y}$ = the average of all of the sample values

Then the ANOVA approach for analyzing a Latin square design is as summarized in Table 17.5.

TABLE 17.5 **ANOVA table for analyzing a Latin square design**

Source	Degrees of freedom	Sum of squares	Mean square	F statistic
Treatments	$d-1$	$SS_{\text{treatments}} = d\sum_{i=1}^{d}(\bar{y}_{i\cdot\cdot} - \bar{y})^2$	$MS_{\text{treatments}} = \frac{SS_{\text{treatments}}}{d-1}$	$F(\text{treatments}) = \frac{MS_{\text{treatments}}}{MSE}$
Rows	$d-1$	$SS_{\text{rows}} = d\sum_{j=1}^{d}(\bar{y}_{\cdot j\cdot} - \bar{y})^2$	$MS_{\text{rows}} = \frac{SS_{\text{rows}}}{d-1}$	$F(\text{rows}) = \frac{MS_{\text{rows}}}{MSE}$
Columns	$d-1$	$SS_{\text{columns}} = d\sum_{k=1}^{d}(\bar{y}_{\cdot\cdot k} - \bar{y})^2$	$MS_{\text{columns}} = \frac{SS_{\text{columns}}}{d-1}$	$F(\text{columns}) = \frac{MS_{\text{columns}}}{MSE}$
Error	$d^2 - 3d + 2$	$SSE = \sum_{j=1}^{d}\sum_{k=1}^{d}(y_{ijk} - \bar{y})^2 - SS_{\text{treatments}} - SS_{\text{rows}} - SS_{\text{columns}}$	$MSE = \frac{SSE}{d^2 - 3d + 2}$	

Notes:

1. If $F(\text{treatments}) > F_{[\alpha]}^{(d-1, d^2-3d+2)}$, reject H_0: $\mu_{1jk} = \mu_{2jk} = \cdots = \mu_{djk}$ (all treatments have the same effect on the mean response).
2. If $F(\text{rows}) > F_{[\alpha]}^{(d-1, d^2-3d+2)}$, reject H_0: $\mu_{i1k} = \mu_{i2k} = \cdots = \mu_{idk}$ (all rows have the same effect on the mean response).
3. If $F(\text{columns}) > F_{[\alpha]}^{(d-1, d^2-3d+2)}$, reject H_0: $\mu_{ij1} = \mu_{ij2} = \cdots = \mu_{ijd}$ (all columns have the same effect on the mean response).
4. A point estimate of $\mu_{ijk} - \mu_{i'jk}$ is $\bar{y}_{i\cdot\cdot} - \bar{y}_{i'\cdot\cdot}$, and an individual $100(1-\alpha)\%$ confidence interval for $\mu_{ijk} - \mu_{i'jk}$ is $\left[(\bar{y}_{i\cdot\cdot} - \bar{y}_{i'\cdot\cdot}) \pm t_{[\alpha/2]}^{(d^2-3d+2)}\sqrt{MSE\left(\frac{2}{d}\right)}\right]$. For Tukey simultaneous intervals, replace $t_{[\alpha/2]}^{(d^2-3d+2)}$ by $\frac{1}{\sqrt{2}}q_{[\alpha]}(d, d^2-3d+2)$.

Example 17.6 The SAS output from using the ANOVA approach to analyze the gasoline mileage data in Table 17.4 is given in Figure 17.6. Since the prob-value related to F(treatments) is .0045, we conclude that at least two gasolines have different effects on mean mileage. Furthermore, for example, an individual 95% confidence interval

FIGURE 17.6 **SAS output of a Latin square analysis of the gasoline mileage data in Table 17.4**

```
                                        SAS

                          GENERAL LINEAR MODELS PROCEDURE

DEPENDENT VARIABLE: MILES

SOURCE           DF   SUM OF SQUARES    MEAN SQUARE    F VALUE       PR > F   R-SQUARE          C.V.

MODEL             9     905.07562500   100.56395833      24.51       0.0005   0.973520        8.1781
ERROR             6      24.61875000a    4.10312500b                ROOT MSE                MILES MEAN
CORRECTED TOTAL  15     929.69437500                              2.02561719               24.76875000

SOURCE           DF        TYPE I SS   F VALUE    PR > F         DF       TYPE III SS   F VALUE   PR > F

GASTYPE           3   165.21687500c    13.42d    0.0045e          3      165.21687500     13.42   0.0045
DRIVER            3     6.11187500f     0.50g    0.6980h          3        6.11187500      0.50   0.6980
CAR               3   733.74687500i    59.61j    0.0001k          3      733.74687500     59.61   0.0001

                                    T FOR HO:    PR > |T|    STD ERROR OF
PARAMETER           ESTIMATE    PARAMETER=0                   ESTIMATE

MUGAS2-MUGAS1     3.57500000            2.50      0.0468      1.43232765
MUGAS2-MUGAS3     9.00000000            6.28      0.0008      1.43232765
MUGAS2-MUGAS4     4.75000000l           3.32      0.0161      1.43232765m

OBSERVATION        OBSERVED        PREDICTED           RESIDUAL      LOWER 95% CL     UPPER 95% CL
                      VALUE            VALUE                           FOR MEAN          FOR MEAN

          1     17.60000000      17.11250000         0.48750000       13.19403456      21.03096544
          2     37.80000000      38.88750000        -1.08750000       34.96903456      42.80596544
          3     15.30000000      15.83750000        -0.53750000       11.91903456      19.75596544
          4     31.00000000      29.86250000         1.13750000       25.94403456      33.78096544
          5     20.40000000      20.21250000         0.18750000       16.29403456      24.13096544
          6     28.70000000      28.23750000         0.46250000       24.31903456      32.15596544
          7     21.30000000      19.61250000         1.68750000       15.69403456      23.53096544
          8     24.70000000      27.03750000        -2.33750000       23.11903456      30.95596544
          9     12.70000000      12.21250000         0.48750000        8.29403456      16.13096544
         10     33.00000000      34.66250000        -1.66250000       30.74403456      38.58096544
         11     19.00000000      19.43750000        -0.43750000       15.51903456      23.35596544
         12     34.40000000      32.78750000         1.61250000       28.86903456      36.70596544
         13     16.80000000      17.96250000        -1.16250000       14.04403456      21.88096544
         14     36.10000000      33.81250000         2.28750000       29.89403456      37.73096544
         15     23.80000000      24.51250000        -0.71250000       20.59403456      28.43096544
         16     23.70000000      24.11250000        -0.41250000       20.19403456      28.03096544
```

```
SUM OF RESIDUALS                              0.00000000
SUM OF SQUARED RESIDUALS                     24.61875000
SUM OF SQUARED RESIDUALS - ERROR SS          -0.00000000
PRESS STATISTIC                             175.06666667
FIRST ORDER AUTOCORRELATION                  -0.45570576
DURBIN-WATSON D                               2.89484641
```

[a]*SSE* [b]*MSE* [c]$SS_{\text{treatments}}$ [d]*F (treatments)* [e]*Prob-value for F (treatments)* [f]SS_{rows} [g]*F (rows)*

[h]*Prob-value for F (rows)* [i]SS_{columns} [j]*F (columns)* [k]*Prob-value for F (columns)* [l]$\bar{y}_{2\cdot\cdot} - \bar{y}_{4\cdot\cdot}$ [m]$\sqrt{MSE\left(\frac{2}{d}\right)}$

for $\mu_{2jk} - \mu_{4jk}$ is

$$\begin{aligned}
&\left[(\bar{y}_{2\cdot\cdot} - \bar{y}_{4\cdot\cdot}) \pm t_{[\alpha/2]}^{(d^2-3d+2)} \sqrt{MSE\left(\frac{2}{d}\right)}\right] \\
&= [4.75 \pm t_{[.025]}^{(16-12+2)} (1.4323)] \\
&= [4.75 \pm 2.447\ (1.4323)] \\
&= [1.25, 8.25]
\end{aligned}$$

This interval says that we are 95 percent confident that the mean mileage obtained by gasoline G2 is between 1.25 mpg and 8.25 mpg greater than the mean mileage obtained by gasoline G4. Since the prob-value related to F(rows) is .6980, there is little evidence to suggest that test driver effects differ. Since the prob-value related to F(columns) is .0001, we have overwhelming evidence that at least two car models have different effects on mean gasoline mileage.

THE REGRESSION APPROACH

A dummy variable regression model can also be used to analyze a Latin square design. This model is

$$\begin{aligned}
y_{ijk} &= \mu_{ijk} + \varepsilon_{ijk} \\
&= \mu + \alpha_i + \gamma_j + \delta_k + \varepsilon_{ijk}
\end{aligned}$$

where

$$\alpha_i = \alpha_2 D_{i,2} + \alpha_3 D_{i,3} + \cdots + \alpha_d D_{i,d}$$

$$\gamma_j = \gamma_2 D_{j,2} + \gamma_3 D_{j,3} + \cdots + \gamma_d D_{j,d}$$

and

$$\delta_k = \delta_2 D_{k,2} + \delta_3 D_{k,3} + \cdots + \delta_d D_{k,d}$$

Here, for example,

$$D_{i,3} = \begin{cases} 1 & \text{if } i = 3\text{; that is, if the observation results from using treatment 3} \\ 0 & \text{otherwise} \end{cases}$$

$$D_{j,2} = \begin{cases} 1 & \text{if } j = 2\text{; that is, if the observation is in row 2} \\ 0 & \text{otherwise} \end{cases}$$

$$D_{k,3} = \begin{cases} 1 & \text{if } k = 3\text{; that is, if the observation is in column 3} \\ 0 & \text{otherwise} \end{cases}$$

The other dummy variables are defined similarly. Using this model we can, for example, test

$$H_0: \mu_{1jk} = \mu_{2jk} = \cdots = \mu_{djk}$$

by testing

$$H_0: \alpha_2 = \alpha_3 = \cdots = \alpha_d = 0$$

TABLE 17.6 **Some standard Latin squares**

3 × 3

A	B	C
B	C	A
C	A	B

4 × 4

1

A	B	C	D
B	A	D	C
C	D	B	A
D	C	A	B

2

A	B	C	D
B	C	D	A
C	D	A	B
D	A	B	C

3

A	B	C	D
B	D	A	C
C	A	D	B
D	C	B	A

4

A	B	C	D
B	A	D	C
C	D	A	B
D	C	B	A

5 × 5

A	B	C	D	E
B	A	E	C	D
C	D	A	E	B
D	E	B	A	C
E	C	D	B	A

6 × 6

A	B	C	D	E	F
B	F	D	C	A	E
C	D	E	F	B	A
D	A	F	E	C	B
E	C	A	B	F	D
F	E	B	A	D	C

7 × 7

A	B	C	D	E	F	G
B	C	D	E	F	G	A
C	D	E	F	G	A	B
D	E	F	G	A	B	C
E	F	G	A	B	C	D
F	G	A	B	C	D	E
G	A	B	C	D	E	F

8 × 8

A	B	C	D	E	F	G	H
B	C	D	E	F	G	H	A
C	D	E	F	G	H	A	B
D	E	F	G	H	A	B	C
E	F	G	H	A	B	C	D
F	G	H	A	B	C	D	E
G	H	A	B	C	D	E	F
H	A	B	C	D	E	F	G

9 × 9

A	B	C	D	E	F	G	H	I
B	C	D	E	F	G	H	I	A
C	D	E	F	G	H	I	A	B
D	E	F	G	H	I	A	B	C
E	F	G	H	I	A	B	C	D
F	G	H	I	A	B	C	D	E
G	H	I	A	B	C	D	E	F
H	I	A	B	C	D	E	F	G
I	A	B	C	D	E	F	G	H

ONE WAY TO RANDOMLY ASSIGN THE d TREATMENTS TO THE d ROWS AND d COLUMNS

Table 17.6 presents *standard $d \times d$ Latin square designs* for $d = 3, 4, 5, 6, 7, 8$, and 9. One way to randomly assign the d treatments to the rows and columns is to start with a standard $d \times d$ Latin square. For $d = 4$ we randomly select a standard Latin square. Likewise, for $d \geqslant 5$ there is more than one standard Latin square. However, for simplicity's sake, it is sufficient to consider only one standard Latin square when $d \geqslant 5$. Then we *randomly permute* the rows and columns of the standard Latin square.

For example, to assign $d = 4$ treatments (A, B, C, and D) to rows and columns, suppose that

1. We randomly select from Table 17.6 the standard Latin square

   A B C D

   B D A C

   C A D B

   D C B A

2. We choose a random permutation of the numbers 1, 2, 3, 4 to rearrange the rows. If the random permutation is 3, 1, 4, 2, then we rearrange the rows as follows:

   C A D B

   A B C D

   D C B A

   B D A C

3. We choose a random permutation of the numbers 1, 2, 3, 4 to rearrange the columns. If the random permutation is 4, 2, 1, 3, then we rearrange the columns as follows

   B A C D

   D B A C

   A C D B

   C D B A

   This is the final assignment of the $d = 4$ treatments to the rows and columns.

*17.5 USING SAS

The SAS program in Figure 17.7 produces the randomized block ANOVA of the defective cardboard box data that is shown in Figure 17.2.

The program in Figure 17.8 produces the analysis of the defective cardboard box data (as given in Table 17.3) that is presented in Figures 17.4 and 17.5.

FIGURE 17.7 **SAS program to produce the randomized block ANOVA of the cardboard box data in Figure 17.2**

```
DATA BOXES;
INPUT BLOCK $ METHOD $ PRDN @@; } ——→ Defines variables BLOCK, METHOD, PRDN
CARDS;
ONE    1  9
ONE    2  8
ONE    3  3
ONE    4  4
TWO    1 10
TWO    2 11
TWO    3  5
TWO    4  5
THREE  1 12
THREE  2 12
THREE  3  7
THREE  4  5
;
```
(data lines above) ——→ *Data — See Table 17.1*

```
PROC GLM; } ——→ Specifies General Linear Models Procedure
CLASS BLOCK METHOD; } ——→ Defines class variables BLOCK and METHOD
MODEL PRDN=BLOCK METHOD/P CLM; } ——→ Specifies randomized block — that is, (τ, δ) — model; CLM requests confidence intervals
ESTIMATE 'MU4-MU1' METHOD -1  0  0 1; } ——→ Estimates μ4h − μ1h
ESTIMATE 'MU4-MU2' METHOD  0 -1  0 1; } ——→ Estimates μ4h − μ2h
ESTIMATE 'MU4-MU3' METHOD  0  0 -1 1; } ——→ Estimates μ4h − μ3h
ESTIMATE 'MU3-MU2' METHOD  0 -1  1 0; } ——→ Estimates μ3h − μ2h
ESTIMATE 'MU3-MU1' METHOD -1  0  1; } ——→ Estimates μ3h − μ1h
ESTIMATE 'MU2-MU1' METHOD -1  1; } ——→ Estimates μ2h − μ1h
ESTIMATE 'MUTHREE-MUTWO' BLOCK  0 1 -1; } ——→ Estimates μl3 − μl2
ESTIMATE 'MUTHREE-MUONE' BLOCK -1 1  0; } ——→ Estimates μl3 − μl1
ESTIMATE 'MUTWO-MUONE' BLOCK   -1 0  1; } ——→ Estimates μl2 − μl1
PROC GLM;
CLASS BLOCK METHOD;
MODEL PRDN=BLOCK METHOD/P CLI; } ——→ CLI requests prediction intervals
```

*This section is optional.

FIGURE 17.8 **SAS program to produce (1) the (α, γ, δ) model analysis of the cardboard box data given in Figure 17.4 and (2) the ($\alpha, \gamma, \theta, \delta$) model analysis of the cardboard box data given in Figure 17.5**

```
DATA CARDBD;
INPUT DEFECTS MACHTYPE $ CBGRADE $ OPERATOR $ @@; }
```
→ *Defines response variable DEFECTS, factors MACHTYPE and CBGRADE, and block OPERATOR*

```
CARDS;
 9  A  Y  1
10  A  Y  2
12  A  Y  3
 8  A  Z  1
11  A  Z  2
12  A  Z  3
 3  B  Y  1
 5  B  Y  2
 7  B  Y  3
 4  B  Z  1
 5  B  Z  2
 5  B  Z  3
;
```
} *Data — See Table 17.3*

```
PROC GLM; }
CLASS MACHTYPE CBGRADE OPERATOR; }
MODEL DEFECTS=MACHTYPE CBGRADE OPERATOR/P CLM; }
ESTIMATE 'MUBJH-MUAJH' MACHTYPE -1  1; }
ESTIMATE 'MUIZH-MUIYH' CBGRADE  -1  1; }
ESTIMATE 'MUIJ3-MUIJ1' OPERATOR -1  0 1; }
ESTIMATE 'MUIJ2-MUIJ1' OPERATOR -1  1 0; }
ESTIMATE 'MUIJ3-MUIJ2' OPERATOR  0 -1 1; }
```
- PROC GLM → *Specifies General Linear Models Procedure*
- CLASS → *Defines class variables MACHTYPE, CBGRADE, and OPERATOR*
- MODEL → *Specifies* (α, γ, δ) model
- → *Estimates* $\mu_{Bjh} - \mu_{Ajh}$
- → *Estimates* $\mu_{iZh} - \mu_{iYh}$
- → *Estimates* $\mu_{ij3} - \mu_{ij1}$
- → *Estimates* $\mu_{ij2} - \mu_{ij1}$
- → *Estimates* $\mu_{ij3} - \mu_{ij2}$

```
PROC GLM;
CLASS MACHTYPE CBGRADE OPERATOR;
MODEL DEFECTS=MACHTYPE CBGRADE OPERATOR/P CLI;
PROC GLM;
CLASS MACHTYPE CBGRADE OPERATOR;
MODEL DEFECTS=MACHTYPE CBGRADE MACHTYPE*CBGRADE OPERATOR/P CLM; }
```
→ *Specifies* ($\alpha, \gamma, \theta, \delta$) *model*

```
ESTIMATE 'MUBH-MUAH' MACHTYPE -1  1; }
ESTIMATE 'MUZH-MUYH' CBGRADE  -1  1; }
ESTIMATE 'MUIJ3-MUIJ1' OPERATOR -1  0 1;
ESTIMATE 'MUIJ2-MUIJ1' OPERATOR -1  1 0;
ESTIMATE 'MUIJ3-MUIJ2' OPERATOR  0 -1 1;
ESTIMATE 'MUBZH-MUAYH' MACHTYPE -1  1 CBGRADE -1  1
                  MACHTYPE*CBGRADE -1  0  0  1 OPERATOR 0  0  0  0; }
```
- 'MUBH-MUAH' → *Estimates* $\mu_{B\cdot h} - \mu_{A\cdot h}$
- 'MUZH-MUYH' → *Estimates* $\mu_{\cdot Zh} - \mu_{\cdot Yh}$

→ *Estimates* $\mu_{BZh} - \mu_{AYh} = (\mu + \alpha_B + \gamma_Z + \theta_{BZ} + \delta_h) - (\mu + \alpha_A + \gamma_Y + \theta_{AY} + \delta_h)$
$= \alpha_B - \alpha_A + \gamma_Z - \gamma_Y + \theta_{BZ} - \theta_{AY}$

The program in Figure 17.9 produces the SAS output in Figure 17.6 of a Latin square analysis of the gasoline mileage data in Table 17.4.

FIGURE 17.9 **SAS program to produce the Latin square analysis of the gasoline mileage data given in Figure 17.6**

```
DATA GASOLINE;
INPUT GASTYPE $ DRIVER $ CAR $ MILES @@; } Defines variables GASTYPE, DRIVER,
                                           and CAR and response variable MILES
CARDS;
4 1 1 17.6 \
2 1 2 37.8  |
3 1 3 15.3  |
1 1 4 31.0  |
2 2 1 20.4  |
3 2 2 28.7  |
1 2 3 21.3  |
4 2 4 24.7  |
3 3 1 12.7   > Data — See Table 17.4
1 3 2 33.0  |
4 3 3 19.0  |
2 3 4 34.4  |
1 4 1 16.8  |
4 4 2 36.1  |
2 4 3 23.8  |
3 4 4 23.7 /
PROC GLM; } ——————→ Specifies General Linear Models Procedure
CLASS GASTYPE DRIVER CAR; } ——→ Defines class variables GASTYPE, DRIVER, and CAR
MODEL MILES=GASTYPE DRIVER CAR/P CLM; } ——→ Specifies (α, γ, δ) model
ESTIMATE 'MUGAS2-MUGAS1' GASTYPE -1 1  0  0; } ——→ Estimates μ2jk − μ1jk
ESTIMATE 'MUGAS2-MUGAS3' GASTYPE  0 1 -1  0; } ——→ Estimates μ2jk − μ3jk
ESTIMATE 'MUGAS2-MUGAS4' GASTYPE  0 1  0 -1; } ——→ Estimates μ2jk − μ4jk
```

EXERCISES

17.1 A consumer preference study involving three different bottle designs (A, B, and C) for the jumbo size of a new liquid laundry detergent was carried out using a randomized block experimental design, with supermarkets as blocks. Specifically, four supermarkets were supplied with all three bottle designs, which were priced the same. The data shown in Table 17.7 represent the number of bottles sold in a 24-hour period.

Let

μ_{lh} = the mean number of bottles sold per 24-hour period when selling bottle design l in supermarket h

a. Set up the appropriate null and alternative hypotheses needed to test for differences between bottle design effects.

b. Set up the appropriate null and alternative hypotheses needed to test for differences between supermarket effects.

c. Perform a graphical analysis to determine whether or not interaction exists between the bottle designs and supermarkets. What do you conclude about the existence of interaction?

TABLE 17.7 Results of a bottle design experiment employing a randomized block design

	Supermarket, h			
Bottle design, l	**1**	**2**	**3**	**4**
A	16	14	1	6
B	33	30	19	23
C	23	21	8	12

17.2 Consider the consumer preference data given in Table 17.7. Use the ANOVA approach to carry out the following analysis.

a. Test for statistically significant differences between bottle design effects with $\alpha = .05$.
b. Test for statistically significant differences between supermarket effects with $\alpha = .05$.
c. Compute an (individual) 95% confidence interval for each of the following differences:

(i) $\mu_{Bh} - \mu_{Ah}$ (ii) $\mu_{Ch} - \mu_{Ah}$ (iii) $\mu_{Ch} - \mu_{Bh}$

Interpret each of these intervals.

d. Compute an (individual) 95% confidence interval for each of the following differences:

(i) $\mu_{l1} - \mu_{l4}$ (ii) $\mu_{l1} - \mu_{l3}$
(iii) $\mu_{l1} - \mu_{l2}$ (iv) $\mu_{l2} - \mu_{l4}$
(v) $\mu_{l2} - \mu_{l3}$ (vi) $\mu_{l3} - \mu_{l4}$

Interpret each of these intervals.

e. Compute an (individual) 95% confidence interval for

$$\mu_{Bh} - \frac{(\mu_{Ah} + \mu_{Ch})}{2}$$

Interpret the meaning of this interval.

f. Calculate a Scheffé 95% simultaneous confidence interval for the contrast

$$\mu_{Bh} - \frac{(\mu_{Ah} + \mu_{Ch})}{2}$$

in the set of all possible contrasts in μ_{Ah}, μ_{Bh}, and μ_{Ch}.

g. Calculate Tukey 95% simultaneous confidence intervals for the differences

$\mu_{Bh} - \mu_{Ah}$, $\mu_{Ch} - \mu_{Ah}$, and $\mu_{Ch} - \mu_{Bh}$

h. Calculate Bonferroni 95% simultaneous confidence intervals for the differences

$\mu_{Bh} - \mu_{Ah}$, $\mu_{Ch} - \mu_{Ah}$, and $\mu_{Ch} - \mu_{Bh}$

17.3 Consider the consumer preference data given in Table 17.7.

a. Define appropriate dummy variables and an appropriate dummy variable regression model to analyze the data given in Table 17.7.
b. Using the regression model of part (a), define an appropriate partial F statistic to test for statistically significant differences between bottle design effects.

c. Using the regression approach and SAS, test for statistically significant differences between bottle design effects with $\alpha = .05$.
d. Using the regression model of part (a), define an appropriate partial F statistic to test for statistically significant differences between supermarket effects.
e. Using the regression approach and SAS, test for statistically significant differences between supermarket effects with $\alpha = .05$.
f. Using the regression approach and SAS, compute the confidence intervals specified in parts (c) through (h) of Exercise 17.2.
g. Using the regression approach and SAS, find a point estimate of μ_{B1} and a 95% confidence interval for μ_{B1}. Interpret the 95% confidence interval.
h. Using the regression approach and SAS, find a point prediction of $y_{B1} = \mu_{B1} + \varepsilon_{B1}$ and a 95% prediction interval for y_{B1}. Interpret the 95% prediction interval.

17.4 A marketing organization wishes to study the effects of four sales methods on weekly sales of a product. The organization employs a randomized block design in which three salesmen use each of the sales methods. The results obtained are given in Table 17.8.

TABLE 17.8 **Results of a sales method experiment employing a randomized block design**

	Salesman, h		
Sales method, l	A	B	C
1	32	29	30
2	32	30	28
3	28	25	23
4	25	24	23

Let

μ_{lh} = the mean weekly sales of the product when salesmen h employs sales method l

a. Carry out a graphical analysis that shows that little interaction exists between sales method and the blocks (salesmen).
b. Set up the appropriate null and alternative hypotheses needed to test for differences between blocks (salesmen).
c. Set up the appropriate null and alternative hypotheses needed to test for differences between sales method effects.

17.5 Figure 17.10 presents the SAS output obtained by using PROC GLM to perform a randomized block ANOVA of the data in Table 17.8. Using the output, answer the following questions.

a. Test for statistically significant differences between sales method effects with $\alpha = .05$.
b. Test for block importance with $\alpha = .05$. That is, test for statistically significant differences between blocks (salesmen) with $\alpha = .05$.
c. Compute an (individual) 95% confidence interval for each of the following differences:

(i) $\mu_{4h} - \mu_{1h}$ (ii) $\mu_{4h} - \mu_{2h}$

FIGURE 17.10 SAS output of a randomized block ANOVA of the sales method data in Table 17.8

```
                                    SAS

                      GENERAL LINEAR MODELS PROCEDURE

DEPENDENT VARIABLE: SALES

SOURCE           DF  SUM OF SQUARES    MEAN SQUARE   F VALUE       PR > F  R-SQUARE          C.V.
MODEL             5    115.75000000    23.15000000     26.88       0.0005  0.957271        3.3847
ERROR             6      5.16666667     0.86111111                 ROOT MSE            SALES MEAN
CORRECTED TOTAL  11    120.91666667                              0.92796073           27.41666667

SOURCE           DF       TYPE I SS   F VALUE      PR > F   DF     TYPE III SS   F VALUE   PR > F
BLOCK             2     22.16666667     12.87      0.0068    2     22.16666667     12.87   0.0068
METHOD            3     93.58333333     36.23      0.0003    3     93.58333333     36.23   0.0003

                                  T FOR HO:    PR > |T|   STD ERROR OF
PARAMETER             ESTIMATE  PARAMETER=0                  ESTIMATE

MU4-MU1            -6.33333333        -8.36     0.0002     0.75767676
MU4-MU2            -6.00000000        -7.92     0.0002     0.75767676
MU4-MU3            -1.33333333        -1.76     0.1289     0.75767676
MU3-MU2            -4.66666667        -6.16     0.0008     0.75767676
MU3-MU1            -5.00000000        -6.60     0.0006     0.75767676
MU2-MU1            -0.33333333        -0.44     0.6754     0.75767676
MUC-MUB             1.00000000         1.52     0.1783     0.65616732
MUB-MUA            -2.25000000        -3.43     0.0140     0.65616732
MUC-MUA            -3.25000000        -4.95     0.0026     0.65616732

OBSERVATION        OBSERVED        PREDICTED          RESIDUAL       LOWER 95% CL      UPPER 95% CL
                      VALUE            VALUE                             FOR MEAN          FOR MEAN

          1     32.00000000      32.16666667       -0.16666667        30.56108191       33.77225142
          2     32.00000000      31.83333333        0.16666667        30.22774858       33.43891809
          3     28.00000000      27.16666667        0.83333333        25.56108191       28.77225142
          4     25.00000000      25.83333333       -0.83333333        24.22774858       26.52225142
          5     29.00000000      29.91666667       -0.91666667        28.31108191       31.52225142
          6     30.00000000      29.58333333        0.41666667        27.97774858       31.18891809
          7     25.00000000      24.91666667        0.08333333        23.31108191       26.52225142
          8     24.00000000      23.58333333        0.41666667        21.97774858       25.18891809
          9     30.00000000      28.91666667        1.08333333        27.31108191       30.52225142
         10     28.00000000      28.58333333       -0.58333333        26.97774858       30.18891809
         11     23.00000000      23.91666667       -0.91666667        22.31108191       25.52225142
         12     23.00000000      22.58333333        0.41666667        20.97774858       24.18891809

OBSERVATION        OBSERVED        PREDICTED          RESIDUAL       LOWER 95% CL      UPPER 95% CL
                      VALUE            VALUE                           INDIVIDUAL        INDIVIDUAL

          1     32.00000000      32.16666667       -0.16666667        29.38571229       34.94762104
          2     32.00000000      31.83333333        0.16666667        29.05237896       34.61428771
          3     28.00000000      27.16666667        0.83333333        24.38571229       29.94762104
          4     25.00000000      25.83333333       -0.83333333        23.05237896       28.61428771
          5     29.00000000      29.91666667       -0.91666667        27.13571229       32.69762104
          6     30.00000000      29.58333333        0.41666667        26.80237896       32.36428771
          7     25.00000000      24.91666667        0.08333333        22.13571229       27.69762104
          8     24.00000000      23.58333333        0.41666667        20.80237896       26.36428771
          9     30.00000000      28.91666667        1.08333333        26.13571229       31.69762104
         10     28.00000000      28.58333333       -0.58333333        25.80237896       31.36428771
         11     23.00000000      23.91666667       -0.91666667        21.13571229       26.69762104
         12     23.00000000      22.58333333        0.41666667        19.80237896       25.36428771
```

(iii) $\mu_{4h} - \mu_{3h}$ (iv) $\mu_{3h} - \mu_{1h}$
(iv) $\mu_{3h} - \mu_{2h}$ (vi) $\mu_{2h} - \mu_{1h}$

Interpret these intervals.

d. Compute an (individual) 95% confidence interval for each of the following differences:

(i) $\mu_{lA} - \mu_{lB}$ (ii) $\mu_{lA} - \mu_{lC}$ (iii) $\mu_{lB} - \mu_{lC}$

Interpret these intervals.

e. Calculate Tukey 95% simultaneous confidence intervals for the differences

$$\mu_{lA} - \mu_{lB}, \quad \mu_{lA} - \mu_{lC}, \quad \text{and} \quad \mu_{lB} - \mu_{lC}$$

f. Calculate Bonferroni 95% simultaneous confidence intervals for the differences

$$\mu_{lA} - \mu_{lB}, \quad \mu_{lA} - \mu_{lC}, \quad \text{and} \quad \mu_{lB} - \mu_{lC}$$

g. Find a point estimate of μ_{A1}.

h. Find a 95% confidence interval for μ_{A1} and a 95% prediction interval for $y_{A1} = \mu_{A1} + \varepsilon_{A1}$.

17.6 Consider Exercises 17.4 and 17.5.

a. Compute an (individual) 95% confidence interval for

$$\frac{(\mu_{1h} + \mu_{2h})}{2} - \frac{(\mu_{3h} + \mu_{4h})}{2}$$

Interpret the meaning of this interval.

b. Calculate a Scheffé 95% simultaneous confidence interval for

$$\frac{(\mu_{1h} + \mu_{2h})}{2} - \frac{(\mu_{3h} + \mu_{4h})}{2}$$

in the set of all possible contrasts in μ_{1h}, μ_{2h}, μ_{3h}, and μ_{4h}.

17.7 (Optional) Write a SAS program using a CLASS statement to obtain the output in Figure 17.10.

17.8 (Optional) Write a SAS program that uses a dummy variable regression model to obtain the same results given in Figure 17.10. In the program actually define and use the appropriate values of the dummy variables.

17.9 To compare three brands of typewriters (A, B, and C), four typists were randomly selected. Each typist used all three typewriters (in a random order) to type the same material for ten minutes, and the numbers of words typed per minute were recorded.

TABLE 17.9 **Results of a typewriter experiment employing a randomized block design**

	Typewriter brand		
Typist	***A***	***B***	***C***
1	77	67	63
2	71	62	59
3	74	63	59
4	67	57	54

The data are given in Table 17.9.

a. Use the ANOVA approach to carry out a complete analysis of this data.
b. Use the regression approach to carry out a complete analysis of this data.

17.10 In the typewriter experiment of Exercise 17.9, suppose that typist 3 became ill and was not able to use typewriter B. Then the data in Table 17.10 were obtained. Use regression analysis to carry out a complete analysis of this data.

TABLE 17.10 **Results of a typewriter experiment employing a randomized block design with a missing observation**

	Typewriter brand		
Typist	***A***	***B***	***C***
1	77	67	63
2	71	62	59
3	74	–	59
4	67	57	54

17.11 Consider the sales method study data given in Table 17.8 and suppose that a sales method consists of a combination of a "high or low pressure" approach (H or L) and a "sales pitch" (1 or 2). Sales methods 1, 2, 3 and 4 consist of the following combinations.

Method 1: High pressure (H) and sales pitch 1

Method 2: High pressure (H) and sales pitch 2

Method 3: Low pressure (L) and sales pitch 1

Method 4: Low pressure (L) and sales pitch 2

Let

μ_{ijh} = the mean weekly sales of the product when salesman h employs pressure level i and sales pitch j

a. Figure 17.11 presents the SAS output obtained by using PROC GLM to analyze the data in Table 17.8 by using the (α, γ, δ) model

$$\begin{aligned} y_{ijh} &= \mu_{ijh} + \varepsilon_{ijh} \\ &= \mu + \alpha_i + \gamma_j + \delta_h + \varepsilon_{ijh} \\ &= \mu + \alpha_L D_{i,L} + \gamma_2 D_{j,2} + \delta_B D_{h,B} + \delta_C D_{h,C} + \varepsilon_{ijh} \end{aligned}$$

Define the dummy variables. Fully interpret the output.

b. Use the output to compute 95% confidence intervals for $\mu_{Hjh} - \mu_{Ljh}$, $\mu_{i1h} - \mu_{i2h}$, $\mu_{ijC} - \mu_{ijB}$, $\mu_{ijB} - \mu_{ijA}$, $\mu_{ijC} - \mu_{ijA}$, and μ_{H1A}.

c. Use Figures 17.11 and 17.12 to test for interaction between pressure level (H or L) and sales pitch (1 or 2).

17.12 (Optional) Write a SAS program using a CLASS statement to obtain the output in Figure 17.11.

FIGURE 17.11 SAS output of an (α, γ, δ) model analysis of the sales method data in Table 17.8 (as described in Exercise 17.11)

```
                                   SAS

                     GENERAL LINEAR MODELS PROCEDURE

DEPENDENT VARIABLE: SALES

SOURCE            DF   SUM OF SQUARES    MEAN SQUARE   F VALUE      PR > F   R-SQUARE           C.V.
MODEL              4     115.00000000    28.75000000     34.01      0.0001   0.951068         3.3533
ERROR              7       5.91666667     0.84523810               ROOT MSE                SALES MEAN
CORRECTED TOTAL   11     120.91666667                            0.91936831               27.41666667

SOURCE            DF      TYPE I SS   F VALUE        PR > F   DF     TYPE III SS   F VALUE    PR > F
PRESSURE           1    90.75000000    107.37        0.0001    1     90.75000000    107.37    0.0001
PITCH              1     2.08333333      2.46        0.1604    1      2.08333333      2.46    0.1604
BLOCK              2    22.16666667     13.11        0.0043    2     22.16666667     13.11    0.0043

                                    T FOR HO:    PR > |T|    STD ERROR OF
PARAMETER             ESTIMATE   PARAMETER=0                   ESTIMATE

MUH-MUL             5.50000000         10.36      0.0001      0.53079754
MU1-MU2             0.83333333          1.57      0.1604      0.53079754
MUC-MUB            -1.00000000         -1.54      0.1679      0.65009157
MUB-MUA            -2.25000000         -3.46      0.0105      0.65009157
MUC-MUA            -3.25000000         -5.00      0.0016      0.65009157

OBSERVATION        OBSERVED         PREDICTED          RESIDUAL        LOWER 95% CL       UPPER 95% CL
                      VALUE             VALUE                              FOR MEAN           FOR MEAN

         1      32.00000000       32.41666667       -0.41666667         31.01336986        33.81996347
         2      32.00000000       31.58333333        0.41666667         30.18003653        32.98663014
         3      28.00000000       26.91666667        1.08333333         25.51336986        28.31996347
         4      25.00000000       26.08333333       -1.08333333         24.68003653        27.48663014
         5      29.00000000       30.16666667       -1.16666667         28.76336986        31.56996347
         6      30.00000000       29.33333333        0.66666667         27.93003653        30.73663014
         7      25.00000000       24.66666667        0.33333333         23.26336986        26.06996347
         8      24.00000000       23.83333333        0.16666667         22.43003653        25.23663014
         9      30.00000000       29.16666667        0.83333333         27.76336986        30.56996347
        10      28.00000000       28.33333333       -0.33333333         26.93003653        29.73663014
        11      23.00000000       23.66666667       -0.66666667         22.26336986        25.06996347
        12      23.00000000       22.83333333        0.16666667         21.43003653        24.23663014
```

17.13 (Optional) Write a SAS program using the actual values of the dummy variables to obtain the same results in Figure 17.11.

17.14 Figure 17.12 presents the SAS output obtained by using PROC GLM to analyze the data in Table 17.8 by using the (α, γ, θ, δ) model

$$\begin{aligned} y_{ijh} &= \mu_{ijh} + \varepsilon_{ijh} \\ &= \mu + \alpha_i + \gamma_j + \theta_{ij} + \delta_h + \varepsilon_{ijh} \\ &= \mu + \alpha_L D_{i,L} + \gamma_2 D_{j,2} + \theta_{L2} D_{i,L} D_{j,2} + \delta_B D_{h,B} + \delta_C D_{h,C} + \varepsilon_{ijh} \end{aligned}$$

a. Define the dummy variables. Fully interpret the output.

b. Use the output to compute 95% confidence intervals for $\mu_{H\cdot h} - \mu_{L\cdot h}$, $\mu_{\cdot 1h} - \mu_{\cdot 2h}$, $\mu_{ijC} - \mu_{ijB}$, $\mu_{ijB} - \mu_{ijA}$, $\mu_{ijC} - \mu_{ijA}$, $\mu_{H1h} - \mu_{L2h}$, and μ_{H1A}.

FIGURE 17.12 **SAS output of an $(\alpha, \gamma, \theta, \delta)$ model analysis of the sales method data in Table 17.8 (as described in Exercise 17.11)**

```
                                         SAS

                            GENERAL LINEAR MODELS PROCEDURE

DEPENDENT VARIABLE:  SALES

SOURCE              DF    SUM OF SQUARES      MEAN SQUARE    F VALUE        PR > F   R-SQUARE            C.V.

MODEL                5      115.75000000      23.15000000      26.88        0.0005   0.957271          3.3847

ERROR                6        5.16666667       0.86111111                  ROOT MSE                  SALES MEAN

CORRECTED TOTAL     11      120.91666667                                0.92796073                 27.41666667

SOURCE              DF         TYPE I SS    F VALUE         PR > F   DF      TYPE III SS    F VALUE    PR > F

PRESSURE             1       90.75000000     105.39         0.0001    1      90.75000000     105.39    0.0001
PITCH                1        2.08333333       2.42         0.1708    1       2.08333333       2.42    0.1708
PRESSURE*PITCH       1        0.75000000       0.87         0.3867    1       0.75000000       0.87    0.3867
BLOCK                2       22.16666667      12.87         0.0068    2      22.16666667      12.87    0.0068

                                      T FOR HO:     PR > |T|      STD ERROR OF
PARAMETER               ESTIMATE   PARAMETER=0                       ESTIMATE

MUH-MUL               5.50000000         10.27      0.0001         0.53575838
MU1-MU2               0.83333333          1.56      0.1708         0.53575838
MUC-MUB              -1.00000000         -1.52      0.1783         0.65616732
MUB-MUA              -2.25000000         -3.43      0.0140         0.65616732
MUC-MUA              -3.25000000         -4.95      0.0026         0.65616732
MUH1-MUL2             6.33333333          8.36      0.0002         0.75767676

OBSERVATION           OBSERVED          PREDICTED            RESIDUAL         LOWER 95% CL        UPPER 95% CL
                         VALUE              VALUE                                FOR MEAN            FOR MEAN

          1        32.00000000        32.16666667         -0.16666667          30.56108191         33.77225142
          2        32.00000000        31.83333333          0.16666667          30.22774858         33.43891809
          3        28.00000000        27.16666667          0.83333333          25.56108191         28.77225142
          4        25.00000000        25.83333333         -0.83333333          24.22774858         27.43891809
          5        29.00000000        29.91666667         -0.91666667          28.31108191         31.52225142
          6        30.00000000        29.58333333          0.41666667          27.97774858         31.18891809
          7        25.00000000        24.91666667          0.08333333          23.31108191         26.52225142
          8        24.00000000        23.58333333          0.41666667          21.97774858         25.18891809
          9        30.00000000        28.91666667          1.08333333          27.31108191         30.52225142
         10        28.00000000        28.58333333         -0.58333333          26.97774858         30.18891809
         11        23.00000000        23.91666667         -0.91666667          22.31108191         25.52225142
         12        23.00000000        22.58333333          0.41666667          20.97774858         24.18891809
```

17.15 (Optional) Write a SAS program using a CLASS statement to obtain the output in Figure 17.12.

17.16 (Optional) Write a SAS program using the actual values of the dummy variables to obtain the same results in Figure 17.12.

17.17 By comparing Figures 17.11 and 17.12, does the (α, γ, δ) model or the $(\alpha, \gamma, \theta, \delta)$ model better describe the data in Table 17.8? Justify your answer.

17.18 Suppose that the gasoline mileages in Table 17.4 corresponding to gasoline G2 have been incorrectly measured. The correct mileages are circled and given in Table 17.11 with the other mileages in Table 17.4.

TABLE 17.11 Results of a gasoline mileage experiment employing a Latin square design

	Car model			
Driver	***1***	***2***	***3***	***4***
1	G4 17.6	(G2 37.6)	G3 15.3	G1 31.0
2	(G2 20.3)	G3 28.7	G1 21.3	G4 24.7
3	G3 12.7	G1 33.0	G4 19.0	(G2 34.2)
4	G1 16.8	G4 36.1	(G2 23.5)	G3 23.7

FIGURE 17.13 SAS output of a Latin square analysis of the gasoline mileage data in Table 17.11

```
                                        SAS

                          GENERAL LINEAR MODELS PROCEDURE

DEPENDENT VARIABLE: MILES

SOURCE            DF   SUM OF SQUARES    MEAN SQUARE   F VALUE        PR > F   R-SQUARE           C.V.

MODEL              9     897.41562500    99.71284722     24.12        0.0005   0.973099         8.2262

ERROR              6      24.80875000     4.13479167                ROOT MSE                MILES MEAN

CORRECTED TOTAL   15     922.22437500                             2.03341871               24.71875000

SOURCE            DF        TYPE I SS    F VALUE       PR > F   DF      TYPE III SS   F VALUE    PR > F

GAS                3     158.40687500      12.77       0.0051    3     158.40687500     12.77    0.0051
DRIVER             3       5.85187500       0.47       0.7131    3       5.85187500      0.47    0.7131
CAR                3     733.15687500      59.10       0.0001    3     733.15687500     59.10    0.0001

                                   T FOR HO:     PR > |T|    STD ERROR OF
PARAMETER              ESTIMATE   PARAMETER=0                  ESTIMATE

MUGAS2-MUGAS1        3.37500000          2.35      0.0573      1.43784416
MUGAS2-MUGAS3        8.80000000          6.12      0.0009      1.43784416
MUGAS2-MUGAS4        4.55000000          3.16      0.0195      1.43784416

OBSERVATION          OBSERVED        PREDICTED              RESIDUAL       LOWER 95% CL        UPPER 95% CL
                        VALUE            VALUE                                 FOR MEAN            FOR MEAN

         1        17.60000000      17.13750000            0.46250000        13.20394287         21.07105713
         2        37.60000000      38.68750000           -1.08750000        34.75394287         42.62105713
         3        15.30000000      15.81250000           -0.51250000        11.87894287         19.74605713
         4        31.00000000      29.86250000            1.13750000        25.92894287         33.79605713
         5        20.30000000      20.06250000            0.23750000        16.12894287         23.99605713
         6        28.70000000      28.26250000            0.43750000        24.32894287         32.19605713
         7        21.30000000      19.61250000            1.68750000        15.67894287         23.54605713
         8        24.70000000      27.06250000           -2.36250000        23.12894287         30.99605713
         9        12.70000000      12.23750000            0.46250000         8.30394287         16.17105713
        10        33.00000000      34.66250000           -1.66250000        30.72894287         38.59605713
        11        19.00000000      19.41250000           -0.41250000        15.47894287         23.34605713
        12        34.20000000      32.58750000            1.61250000        28.65394287         36.52105713
        13        16.80000000      17.96250000           -1.16250000        14.02894287         21.89605713
        14        36.10000000      33.78750000            2.31250000        29.85394287         37.72105713
        15        23.50000000      24.26250000           -0.76250000        20.32894287         28.19605713
        16        23.70000000      24.08750000           -0.38750000        20.15394287         28.02105713
```

a. Figure 17.13 presents the SAS output obtained by using PROC GLM and the techniques of Table 17.5 to analyze this data. Fully interpret the output.
b. Calculate 95% confidence intervals for the differences in means referred to in the output.

17.19 Figure 17.14 presents the SAS output obtained by using PROC GLM and the appropriate dummy variable regression model to analyze the Latin square gasoline mileage data in Table 17.12. Note that this data has a missing observation.

a. Fully interpret the output and calculate 95% confidence intervals for the differences in means referred to in the output.
b. Specify the dummy variable regression model that can be used to obtain the results in Figure 17.14.

FIGURE 17.14 **SAS output of a Latin square analysis of the gasoline mileage data in Table 17.12**

```
                                          SAS

                           GENERAL LINEAR MODELS PROCEDURE

DEPENDENT VARIABLE: MILES

SOURCE            DF   SUM OF SQUARES     MEAN SQUARE    F VALUE        PR > F   R-SQUARE          C.V.

MODEL              9     889.16100000     98.79566667      20.54        0.0020   0.973666        8.9490

ERROR              5      24.04833333      4.80966667                  ROOT MSE                MILES MEAN

CORRECTED TOTAL   14     913.20933333                               2.19309523                24.50666667

SOURCE            DF       TYPE I SS    F VALUE         PR > F   DF      TYPE III SS   F VALUE    PR > F

GAS                3    247.34516667      17.14         0.0046    3     134.33388889      9.31    0.0173
DRIVER             3     71.12527778       4.93         0.0592    3       6.18055556      0.43    0.7417
CAR                3    570.69055556      39.55         0.0007    3     570.69055556     39.55    0.0007

                                    T FOR HO:    PR > |T|   STD ERROR OF
PARAMETER               ESTIMATE   PARAMETER=0                ESTIMATE

MUGAS2-MUGAS1         3.57500000          2.31    0.0693     1.55075251
MUGAS2-MUGAS3         9.30833333          5.20    0.0035     1.79065475
MUGAS2-MUGAS4         4.75000000          3.06    0.0280     1.55075251

OBSERVATION            OBSERVED          PREDICTED             RESIDUAL        LOWER 95% CL        UPPER 95% CL
                          VALUE              VALUE                                FOR MEAN            FOR MEAN

          1         17.60000000        17.26666667           0.33333333         12.66370880         21.86962453
          2         37.80000000        38.73333333          -0.93333333         34.13037547         43.33629120
          3         15.30000000        15.68333333          -0.38333333         11.08037547         20.28629120
          4         31.00000000        30.01666667           0.98333333         25.41370880         34.61962453
          5         20.40000000        20.05833333           0.34166667         15.45537547         24.66129120
          6         21.30000000        19.45833333           1.84166667         14.85537547         24.06129120
          7         24.70000000        26.88333333          -2.18333333         22.28037547         31.48629120
          8         12.70000000        12.05833333           0.64166667          7.45537547         16.66129120
          9         33.00000000        34.50833333          -1.50833333         29.90537547         39.11129120
         10         19.00000000        19.59166667          -0.59166667         14.98870880         24.19462453
         11         34.40000000        32.94166667           1.45833333         28.33870880         37.54462453
         12         16.80000000        18.11666667          -1.31666667         13.51370880         22.71962453
         13         36.10000000        33.65833333           2.44166667         29.05537547         38.26129120
         14         23.80000000        24.66666667          -0.86666667         20.06370880         29.26962453
         15         23.70000000        23.95833333          -0.25833333         19.35537547         28.56129120
```

TABLE 17.12 **Results of a gasoline mileage experiment employing a Latin square design with a missing observation**

	Car model			
Driver	***1***	***2***	***3***	***4***
1	G4 17.6	G2 37.8	G3 15.3	G1 31.0
2	G2 20.4		G1 21.3	G4 24.7
3	G3 12.7	G1 33.0	G4 19.0	G2 34.4
4	G1 16.8	G4 36.1	G2 23.8	G3 23.7

17.20 (Optional) Write a SAS program to obtain the output in Figure 17.13.

17.21 (Optional) Write a SAS program to analyze the data in Table 17.12. The program should produce the output in Figure 17.14. Hint: Use the dummy variable regression model you specified in Exercise 17.19b.

17.22 (Optional) Write a SAS program using a CLASS statement to analyze the randomized block typewriter experiment data in Table 17.9.

17.23 (Optional) Write a SAS program that uses a dummy variable regression model to analyze the randomized block typewriter experiment data in Table 17.9. Actually define and use the appropriate values of the dummy variables.

17.24 (Optional) Write a SAS program using a CLASS statement to analyze the randomized block bottle design data in Table 17.7.

17.25 (Optional) Write a SAS program that uses a dummy variable regression model to analyze the randomized block typewriter experiment data in Table 17.10.

APPENDIX A

DERIVATIONS OF THE MEAN AND VARIANCE OF $\bar{y}$

In this appendix we derive the formulas for $\mu_{\bar{y}}$ and $\sigma^2_{\bar{y}}$ given in Chapter 3, using four properties of means and variances to do the necessary proofs. Before we randomly select the sample

$$\{y_1, y_2, \ldots, y_n\}$$

from a single population of values with mean μ and variance σ^2, we note that for $i = 1, 2, \ldots, n$ the ith sample value, y_i, can potentially be any of the values in the population. Thus y_i is a *random variable*, which we define to be a variable whose value is determined by the outcome of an experiment. Moreover, it can be proven (and is intuitive) that:

1. the mean (or expected value) of y_i, denoted μ_{y_i}, is μ, the mean of the population from which y_i is randomly selected
2. the variance of y_i, denoted $\sigma^2_{y_i}$, is σ^2, the variance of the population from which y_i is randomly selected

That is, for $i = 1, 2, \ldots, n$,

$$\mu_{y_i} = \mu \qquad \text{(or equivalently, } \mu_{y_1} = \mu_{y_2} = \cdots = \mu_{y_n} = \mu\text{)}$$

$$\sigma^2_{y_i} = \sigma^2 \qquad \text{(or equivalently, } \sigma^2_{y_1} = \sigma^2_{y_2} = \cdots = \sigma^2_{y_n} = \sigma^2\text{)}$$

The following are some important properties of the means and variances of random variables. For proofs of these properties, see Wonnacott and Wonnacott (1977).

Assume that y is a random variable and K is a constant. Then:

1. $\mu_{Ky} = K\mu_y$.
2. $\sigma^2_{Ky} = K^2\sigma^2_y$.

Assume that $y_1, y_2, \ldots, y_n$ are random variables. Then:

3. $\mu_{(y_1+y_2+\cdots+y_n)} = \mu_{y_1} + \mu_{y_2} + \cdots + \mu_{y_n}$.
4. If $y_1, y_2, \ldots, y_n$ are statistically independent, then

$$\sigma^2_{(y_1+y_2+\cdots+y_n)} = \sigma^2_{y_1} + \sigma^2_{y_2} + \cdots + \sigma^2_{y_n}$$

We now use these properties to prove that if we randomly select the sample

$$\{y_1, y_2, \ldots, y_n\}$$

from a single infinite population of values with mean μ and variance σ^2, and if

$$\bar{y} = \frac{\sum_{i=1}^{n} y_i}{n}$$

then $\mu_{\bar{y}} = \mu$ and $\sigma^2_{\bar{y}} = \sigma^2/n$. This implies that $\sigma_{\bar{y}} = \sigma/\sqrt{n}$.

The first proof is as follows:

$$\begin{aligned}
\mu_{\bar{y}} &= \mu_{\left(\sum_{i=1}^{n} y_i/n\right)} \\
&= \frac{1}{n}\mu_{\left(\sum_{i=1}^{n} y_i\right)} && \text{(see Property 1)} \\
&= \frac{1}{n}(\mu_{y_1} + \mu_{y_2} + \cdots + \mu_{y_n}) && \text{(see Property 3)} \\
&= \frac{1}{n}(\mu + \mu + \cdots + \mu) \\
&= \frac{n\mu}{n} \\
&= \mu
\end{aligned}$$

The second proof is as follows:

$$\begin{aligned}
\sigma^2_{\bar{y}} &= \sigma^2_{\left(\sum_{i=1}^{n} y_i/n\right)} \\
&= \left(\frac{1}{n}\right)^2 \sigma^2_{\left(\sum_{i=1}^{n} y_i\right)} \qquad \text{(see Property 2)} \\
&= \frac{1}{n^2}(\sigma^2_{y_1} + \sigma^2_{y_2} + \cdots + \sigma^2_{y_n}) \qquad \text{(see Property 4)} \\
&= \frac{1}{n^2}(\sigma^2 + \sigma^2 + \cdots + \sigma^2) \\
&= \frac{n\sigma^2}{n^2} \\
&= \frac{\sigma^2}{n}
\end{aligned}$$

Here, $y_1, y_2, \ldots, y_n$ are statistically independent, since we assume that the population of values from which $y_1, y_2, \ldots, y_n$ are randomly selected is infinite.

As an exercise, let $\bar{y}_1$ be the mean of a sample of n_1 observations randomly selected from a population having mean μ_1 and variance σ_1^2, and let $\bar{y}_2$ be the mean of a sample of n_2 observations randomly selected from a population having mean μ_2 and variance σ_2^2. We know that

$$\mu_{\bar{y}_1} = \mu_1, \qquad \sigma^2_{\bar{y}_1} = \frac{\sigma_1^2}{n_1}$$

and

$$\mu_{\bar{y}_2} = \mu_2, \qquad \sigma^2_{\bar{y}_2} = \frac{\sigma_2^2}{n_2}$$

Assuming that the two samples, and thus $\bar{y}_1$ and $\bar{y}_2$, are statistically independent of each other, prove that

$$\mu_{(\bar{y}_1 - \bar{y}_2)} = \mu_1 - \mu_2$$

and

$$\sigma^2_{(\bar{y}_1 - \bar{y}_2)} = \frac{\sigma_1^2}{n_1} + \frac{\sigma_2^2}{n_2}$$

A P P E N D I X 

DERIVATION OF THE LEAST SQUARES POINT ESTIMATES

In this appendix we show that the least squares point estimates $b_0, b_1, b_2, \ldots, b_p$ are calculated by using the matrix algebra equation

$$\begin{bmatrix} b_0 \\ b_1 \\ b_2 \\ \cdot \\ \cdot \\ \cdot \\ b_p \end{bmatrix} = \mathbf{b} = (\mathbf{X}'\mathbf{X})^{-1}\mathbf{X}'\mathbf{y}$$

We do this by differentiating

$$\begin{aligned} SSE &= \sum_{i=1}^{n} (y_i - \hat{y}_i)^2 \\ &= \sum_{i=1}^{n} (y_i - (b_0 + b_1 x_{i1} + b_2 x_{i2} + \cdots + b_p x_{ip}))^2 \end{aligned}$$

with respect to $b_0, b_1, b_2, \ldots,$ and b_p as follows:

$$\frac{\partial SSE}{\partial b_0} = -2 \sum_{i=1}^{n} \left(y_i - b_0 - \sum_{j=1}^{p} b_j x_{ij} \right)$$

$$\frac{\partial SSE}{\partial b_k} = -2\sum_{i=1}^{n}\left(y_i - b_0 - \sum_{j=1}^{p} b_j x_{ij}\right)x_{ik} \qquad \text{for } k = 1, 2, \ldots, p$$

Setting these partial derivatives equal to zero, we obtain

$$\sum_{i=1}^{n}\left(y_i - b_0 - \sum_{j=1}^{p} b_j x_{ij}\right) = 0$$

$$\sum_{i=1}^{n}\left(y_i - b_0 - \sum_{j=1}^{p} b_j x_{ij}\right)x_{ik} = 0 \qquad \text{for } k = 1, 2, \ldots, p$$

These equations imply that

$$nb_0 + \sum_{j=1}^{p} b_j \sum_{i=1}^{n} x_{ij} = \sum_{i=1}^{n} y_i$$

$$b_0 \sum_{i=1}^{n} x_{ik} + \sum_{j=1}^{p} b_j \sum_{i=1}^{n} x_{ij}x_{ik} = \sum_{i=1}^{n} x_{ik}y_i \qquad \text{for } k = 1, 2, \ldots, p$$

These $p + 1$ equations are called the *normal equations* and can usually be solved simultaneously for $b_0, b_1, b_2, \ldots$, and b_p. To do this easily, consider the **X** matrix and **y** vector

$$\mathbf{X} = \begin{bmatrix} 1 & x_{11} & \cdots & x_{1p} \\ 1 & x_{21} & \cdots & x_{2p} \\ \vdots & \vdots & & \vdots \\ 1 & x_{n1} & \cdots & x_{np} \end{bmatrix} \quad \text{and} \quad \mathbf{y} = \begin{bmatrix} y_1 \\ y_2 \\ \vdots \\ y_n \end{bmatrix}$$

We see that

$$\mathbf{X'X} = \begin{bmatrix} 1 & 1 & \cdots & 1 \\ x_{11} & x_{21} & \cdots & x_{n1} \\ \vdots & \vdots & & \vdots \\ x_{1p} & x_{2p} & \cdots & x_{np} \end{bmatrix} \begin{bmatrix} 1 & x_{11} & \cdots & x_{1p} \\ 1 & x_{21} & \cdots & x_{2p} \\ \vdots & \vdots & & \vdots \\ 1 & x_{n1} & \cdots & x_{np} \end{bmatrix}$$

$$= \begin{bmatrix} n & \sum_{i=1}^{n} x_{i1} & \sum_{i=1}^{n} x_{i2} & \cdots & \sum_{i=1}^{n} x_{ip} \\ \sum_{i=1}^{n} x_{i1} & \sum_{i=1}^{n} x_{i1}x_{i1} & \sum_{i=1}^{n} x_{i2}x_{i1} & \cdots & \sum_{i=1}^{n} x_{ip}x_{i1} \\ \vdots & \vdots & \vdots & & \vdots \\ \sum_{i=1}^{n} x_{ip} & \sum_{i=1}^{n} x_{i1}x_{ip} & \sum_{i=1}^{n} x_{i2}x_{ip} & \cdots & \sum_{i=1}^{n} x_{ip}x_{ip} \end{bmatrix}$$

$$\mathbf{X}'\mathbf{y} = \begin{bmatrix} 1 & 1 & \cdots & 1 \\ x_{11} & x_{21} & \cdots & x_{n1} \\ \vdots & \vdots & & \vdots \\ x_{1p} & x_{2p} & \cdots & x_{np} \end{bmatrix} \begin{bmatrix} y_1 \\ y_2 \\ \vdots \\ y_n \end{bmatrix} = \begin{bmatrix} \sum_{i=1}^{n} y_i \\ \sum_{i=1}^{n} x_{i1} y_i \\ \vdots \\ \sum_{i=1}^{n} x_{ip} y_i \end{bmatrix}$$

Thus letting

$$\mathbf{b} = \begin{bmatrix} b_0 \\ b_1 \\ \vdots \\ b_p \end{bmatrix}$$

we see that the normal equations can be written as

$$\begin{bmatrix} n & \sum_{i=1}^{n} x_{i1} & \sum_{i=1}^{n} x_{i2} & \cdots & \sum_{i=1}^{n} x_{ip} \\ \sum_{i=1}^{n} x_{i1} & \sum_{i=1}^{n} x_{i1}x_{i1} & \sum_{i=1}^{n} x_{i2}x_{i1} & \cdots & \sum_{i=1}^{n} x_{ip}x_{i1} \\ \vdots & \vdots & \vdots & & \vdots \\ \sum_{i=1}^{n} x_{ip} & \sum_{i=1}^{n} x_{i1}x_{ip} & \sum_{i=1}^{n} x_{i2}x_{ip} & \cdots & \sum_{i=1}^{n} x_{ip}x_{ip} \end{bmatrix} \begin{bmatrix} b_0 \\ b_1 \\ \vdots \\ b_p \end{bmatrix} = \begin{bmatrix} \sum_{i=1}^{n} y_i \\ \sum_{i=1}^{n} x_{i1} y_i \\ \vdots \\ \sum_{i=1}^{n} x_{ip} y_i \end{bmatrix}$$

or

$$(\mathbf{X}'\mathbf{X})\mathbf{b} = \mathbf{X}'\mathbf{y}$$

Assuming that the columns of the matrix **X** are linearly independent, which implies that the inverse of the matrix **X′X** exists, then **b**, the vector of least squares point estimates, is given by

$$\mathbf{b} = (\mathbf{X}'\mathbf{X})^{-1}\mathbf{X}'\mathbf{y}$$

Again, using calculus it can be shown that the values of $b_0, b_1, b_2, \ldots, b_p$ given by the preceding matrix algebra equation do in fact minimize

$$SSE = \sum_{i=1}^{n} (y_i - \hat{y}_i)^2$$

(rather than, for example, maximizing SSE).

APPENDIX C

DERIVATION OF THE COMPUTATIONAL FORMULA FOR *SSE*

In this appendix we prove that

$$SSE = \sum_{i=1}^{n} (y_i - \hat{y}_i)^2 = \sum_{i=1}^{n} y_i^2 - \mathbf{b}'\mathbf{X}'\mathbf{y}$$

First recall that

$$\mathbf{b} = \begin{bmatrix} b_0 \\ b_1 \\ \cdot \\ \cdot \\ \cdot \\ b_p \end{bmatrix}, \quad \mathbf{y} = \begin{bmatrix} y_1 \\ y_2 \\ \cdot \\ \cdot \\ \cdot \\ y_n \end{bmatrix}, \quad \text{and} \quad \mathbf{X} = \begin{bmatrix} 1 & x_{11} & \cdots & x_{1p} \\ 1 & x_{21} & \cdots & x_{2p} \\ \cdot & \cdot & & \cdot \\ \cdot & \cdot & & \cdot \\ \cdot & \cdot & & \cdot \\ 1 & x_{n1} & \cdots & x_{np} \end{bmatrix}$$

Then we let

$$\mathbf{Xb} = \begin{bmatrix} 1 & x_{11} & \cdots & x_{1p} \\ 1 & x_{21} & \cdots & x_{2p} \\ \vdots & \vdots & & \vdots \\ 1 & x_{n1} & \cdots & x_{np} \end{bmatrix} \begin{bmatrix} b_0 \\ b_1 \\ \vdots \\ b_p \end{bmatrix} = \begin{bmatrix} b_0 + b_1 x_{11} + \cdots + b_p x_{1p} \\ b_0 + b_1 x_{21} + \cdots + b_p x_{2p} \\ \vdots \\ b_0 + b_1 x_{n1} + \cdots + b_p x_{np} \end{bmatrix}$$

$$= \begin{bmatrix} \hat{y}_1 \\ \hat{y}_2 \\ \vdots \\ \hat{y}_n \end{bmatrix} = \hat{\mathbf{y}}$$

Since

$$\mathbf{y} - \hat{\mathbf{y}} = \begin{bmatrix} y_1 \\ y_2 \\ \vdots \\ y_n \end{bmatrix} - \begin{bmatrix} \hat{y}_1 \\ \hat{y}_2 \\ \vdots \\ \hat{y}_n \end{bmatrix} = \begin{bmatrix} y_1 - \hat{y}_1 \\ y_2 - \hat{y}_2 \\ \vdots \\ y_n - \hat{y}_n \end{bmatrix}$$

and

$$(\mathbf{y} - \hat{\mathbf{y}})' = [(y_1 - \hat{y}_1) \quad (y_2 - \hat{y}_2) \quad \cdots \quad (y_n - \hat{y}_n)]$$

it follows that

$$(\mathbf{y} - \hat{\mathbf{y}})'(\mathbf{y} - \hat{\mathbf{y}}) = [(y_1 - \hat{y}_1) \quad (y_2 - \hat{y}_2) \quad \cdots \quad (y_n - \hat{y}_n)] \begin{bmatrix} y_1 - \hat{y}_1 \\ y_2 - \hat{y}_2 \\ \vdots \\ y_n - \hat{y}_n \end{bmatrix}$$

$$= (y_1 - \hat{y}_1)^2 + (y_2 - \hat{y}_2)^2 + \cdots + (y_n - \hat{y}_n)^2$$

$$= \sum_{i=1}^{n} (y_i - \hat{y}_i)^2$$

Thus since $\hat{\mathbf{y}} = \mathbf{Xb}$, we see that

$$SSE = \sum_{i=1}^{n} (y_i - \hat{y}_i)^2 = (\mathbf{y} - \hat{\mathbf{y}})'(\mathbf{y} - \hat{\mathbf{y}})$$

$$= (\mathbf{y} - \mathbf{Xb})'(\mathbf{y} - \mathbf{Xb})$$

Now,

$$SSE = (\mathbf{y} - \mathbf{Xb})'(\mathbf{y} - \mathbf{Xb}) = (\mathbf{y}' - \mathbf{b}'\mathbf{X}')(\mathbf{y} - \mathbf{Xb})$$

$$= \mathbf{y}'\mathbf{y} - \mathbf{b}'\mathbf{X}'\mathbf{y} - \mathbf{y}'\mathbf{Xb} + \mathbf{b}'\mathbf{X}'\mathbf{Xb}$$

Using the fact that $\mathbf{b} = (\mathbf{X}'\mathbf{X})^{-1}\mathbf{X}'\mathbf{y}$, we see that

$$\mathbf{b}'\mathbf{X}'\mathbf{X}\mathbf{b} = \mathbf{b}'\mathbf{X}'\mathbf{X}(\mathbf{X}'\mathbf{X})^{-1}\mathbf{X}'\mathbf{y} = \mathbf{b}'\mathbf{X}'\mathbf{y}$$

Therefore

$$\begin{aligned} SSE &= \mathbf{y}'\mathbf{y} - \mathbf{b}'\mathbf{X}'\mathbf{y} - \mathbf{y}'\mathbf{X}\mathbf{b} + \mathbf{b}'\mathbf{X}'\mathbf{y} \\ &= \mathbf{y}'\mathbf{y} - \mathbf{b}'\mathbf{X}'\mathbf{y} \end{aligned}$$

Here, we have used the fact that

$$\mathbf{y}'\mathbf{X}\mathbf{b} = \mathbf{b}'\mathbf{X}'\mathbf{y}$$

This follows because $\mathbf{b}'\mathbf{X}'\mathbf{y}$ is a number, $(\mathbf{b}'\mathbf{X}'\mathbf{y})' = \mathbf{y}'\mathbf{X}\mathbf{b}$, and the transpose of a number is the number itself.

Finally, noting that

$$\begin{aligned} \mathbf{y}'\mathbf{y} &= [y_1 \quad y_2 \quad \cdots \quad y_n] \begin{bmatrix} y_1 \\ y_2 \\ \cdot \\ \cdot \\ \cdot \\ y_n \end{bmatrix} \\ &= y_1^2 + y_2^2 + \cdots + y_n^2 = \sum_{i=1}^{n} y_i^2 \end{aligned}$$

we have

$$SSE = \mathbf{y}'\mathbf{y} - \mathbf{b}'\mathbf{X}'\mathbf{y} = \sum_{i=1}^{n} y_i^2 - \mathbf{b}'\mathbf{X}'\mathbf{y}$$

APPENDIX 

DERIVATIONS OF THE MEANS AND VARIANCES OF b_j, $\hat{y}_0$, AND $y_0 - \hat{y}_0$

In this appendix we derive the formulas for μ_{b_j}, $\sigma^2_{b_j}$, $\mu_{\hat{y}_0}$, $\sigma^2_{\hat{y}_0}$, $\mu_{(y_0-\hat{y}_0)}$, and $\sigma^2_{(y_0-\hat{y}_0)}$ given in Chapter 8. To do the necessary proofs, which we number (i) through (vi) as we proceed, we use properties of means and variances, which we number (1) through (6) as we proceed. Furthermore, as we present each proof, we indicate which of the six properties implies each equality in the proof by placing the number of the property above the equality.

We now consider regression analysis. Before the sample

$$\{y_1, y_2, \ldots, y_n\}$$

is randomly selected from the n populations of potential values of the dependent variable, it follows that for $i = 1, 2, \ldots, n$ the ith observed value of the dependent variable

$$\begin{aligned} y_i &= \mu_i + \varepsilon_i \\ &= \beta_0 + \beta_1 x_{i1} + \cdots + \beta_p x_{ip} + \varepsilon_i \end{aligned}$$

can be any of the potential values of the dependent variable in the ith population. This population has mean μ_i and (in accordance with inference assumption 1) has variance σ^2. Thus we define the mean of y_i (denoted by μ_{y_i}) and the variance of y_i

(denoted by $\sigma^2_{y_i}$) to be the mean and variance, respectively, of the ith population. That is, for $i = 1, 2, \ldots, n$,

$$\mu_{y_i} = \mu_i = \beta_0 + \beta_1 x_{i1} + \cdots + \beta_p x_{ip}$$

$$\sigma^2_{y_i} = \sigma^2 \qquad \text{(or equivalently, } \sigma^2_{y_1} = \sigma^2_{y_2} = \cdots = \sigma^2_{y_n} = \sigma^2\text{)}$$

Moreover, for $i = 1, 2, \ldots, n$, y_i is a *random variable*, which we define to be a variable whose value is determined by the outcome of an experiment.

In general, if $y_1, y_2, \ldots, y_n$ are random variables, and if

$$\mathbf{y} = \begin{bmatrix} y_1 \\ y_2 \\ \vdots \\ y_n \end{bmatrix}$$

then we define the mean of $\mathbf{y}$ (denoted by $\mu_{\mathbf{y}}$) and the variance of $\mathbf{y}$ (denoted by $\sigma^2_{\mathbf{y}}$) as follows:

$$\mu_{\mathbf{y}} = \begin{bmatrix} \mu_{y_1} \\ \mu_{y_2} \\ \vdots \\ \mu_{y_n} \end{bmatrix}$$

$$\sigma^2_{\mathbf{y}} = \begin{bmatrix} \sigma^2_{y_1} & \sigma^2_{y_1 y_2} & \cdots & \sigma^2_{y_1 y_n} \\ \sigma^2_{y_2 y_1} & \sigma^2_{y_2} & \cdots & \sigma^2_{y_2 y_n} \\ \vdots & \vdots & & \vdots \\ \sigma^2_{y_n y_1} & \sigma^2_{y_n y_2} & \cdots & \sigma^2_{y_n} \end{bmatrix}$$

We call $\sigma^2_{y_i y_j}$ (for $i = 1, 2, \ldots, n$, and $j = 1, 2, \ldots, n$, and $i \neq j$) the *covariance between y_i and y_j*. For a mathematical definition of $\sigma^2_{y_i y_j}$, see Wonnacott and Wonnacott (1977). For now, suffice it to say that the covariance $\sigma^2_{y_i y_j}$ is a measure of the linear relationship between the random variables y_i and y_j. If y_i and y_j are statistically independent, then $\sigma^2_{y_i y_j}$ equals zero.

Now we have seen that in regression analysis, for $i = 1, 2, \ldots, n$,

$$\mu_{y_i} = \mu_i = \beta_0 + \beta_1 x_{i1} + \cdots + \beta_p x_{ip}$$

$$\sigma^2_{y_i} = \sigma^2 \qquad \text{(or equivalently, } \sigma^2_{y_1} = \sigma^2_{y_2} = \cdots = \sigma^2_{y_n} = \sigma^2\text{)}$$

Thus in regression analysis,

$$\boldsymbol{\mu}_{\mathbf{y}} = \begin{bmatrix} \mu_{y_1} \\ \mu_{y_2} \\ \vdots \\ \mu_{y_n} \end{bmatrix} = \begin{bmatrix} \beta_0 + \beta_1 x_{11} + \cdots + \beta_p x_{1p} \\ \beta_0 + \beta_1 x_{21} + \cdots + \beta_p x_{2p} \\ \vdots \\ \beta_0 + \beta_1 x_{n1} + \cdots + \beta_p x_{np} \end{bmatrix}$$

$$= \begin{bmatrix} 1 & x_{11} & \cdots & x_{1p} \\ 1 & x_{21} & \cdots & x_{2p} \\ \vdots & \vdots & & \vdots \\ 1 & x_{n1} & \cdots & x_{np} \end{bmatrix} \begin{bmatrix} \beta_0 \\ \beta_1 \\ \vdots \\ \beta_p \end{bmatrix}$$

$$= \mathbf{X}\boldsymbol{\beta}$$

Here, $\boldsymbol{\beta}$ is the column vector containing the parameters $\beta_0, \beta_1, \ldots, \beta_p$. Moreover, inference assumption 2 says that $y_1, y_2, \ldots, y_n$, and y_0 are statistically independent. This implies that for $i = 1, 2, \ldots, n$, and $j = 1, 2, \ldots, n$, and $i \neq j$, $\sigma^2_{y_i y_j}$ equals zero. It follows that

$$\boldsymbol{\sigma}^2_{\mathbf{y}} = \begin{bmatrix} \sigma^2_{y_1} & \sigma^2_{y_1 y_2} & \cdots & \sigma^2_{y_1 y_n} \\ \sigma^2_{y_2 y_1} & \sigma^2_{y_2} & \cdots & \sigma^2_{y_2 y_n} \\ \vdots & \vdots & & \vdots \\ \sigma^2_{y_n y_1} & \sigma^2_{y_n y_2} & \cdots & \sigma^2_{y_n} \end{bmatrix}$$

$$= \begin{bmatrix} \sigma^2 & 0 & \cdots & 0 \\ 0 & \sigma^2 & \cdots & 0 \\ \vdots & \vdots & & \vdots \\ 0 & 0 & \cdots & \sigma^2 \end{bmatrix}$$

$$= \sigma^2 \begin{bmatrix} 1 & 0 & \cdots & 0 \\ 0 & 1 & \cdots & 0 \\ \vdots & \vdots & & \vdots \\ 0 & 0 & \cdots & 1 \end{bmatrix}$$

$$= \sigma^2 \mathbf{I}$$

Recall that $\mathbf{I}$ is the n dimensional identity matrix. To summarize, in regression analysis

$$\boldsymbol{\mu}_{\mathbf{y}} = \mathbf{X}\boldsymbol{\beta}$$

$$\boldsymbol{\sigma}^2_{\mathbf{y}} = \sigma^2 \mathbf{I}$$

Two important properties concerning $\mu_{\mathbf{y}}$ and $\sigma^2_{\mathbf{y}}$ follow. For proofs of these properties, see Searle (1971).

Assume that **y** is a column vector containing the random variables $y_1, y_2, \ldots, y_n$, and assume that **A** is a matrix of constants such that the matrix product **Ay** exists (note that **Ay** is a column vector). Then:

(1) $\mu_{\mathbf{Ay}} = \mathbf{A}\mu_{\mathbf{y}}$
(2) $\sigma^2_{\mathbf{Ay}} = \mathbf{A}\sigma^2_{\mathbf{y}}\mathbf{A}'$

We first use these properties to prove that if b_j is the least squares point estimate of the parameter β_j in the linear regression model, then:

(i) $\mu_{b_j} = \beta_j$
(ii) $\sigma^2_{b_j} = \sigma^2 c_{jj}$

Here, c_{jj} is the diagonal element of $(\mathbf{X}'\mathbf{X})^{-1}$ corresponding to β_j. The proofs of these results are as follows.

(i) Recall that $\mathbf{b} = (\mathbf{X}'\mathbf{X})^{-1}\mathbf{X}'\mathbf{y}$ is a column vector containing the least squares point estimates $b_0, b_1, \ldots, b_p$. Also, recall that we have previously shown that $\boldsymbol{\mu}_{\mathbf{y}} = \mathbf{X}\boldsymbol{\beta}$. It follows that

$$\begin{bmatrix} \mu_{b_0} \\ \mu_{b_1} \\ \cdot \\ \cdot \\ \cdot \\ \mu_{b_p} \end{bmatrix} = \mu_{\mathbf{b}} = \mu_{(\mathbf{X}'\mathbf{X})^{-1}\mathbf{X}'\mathbf{y}}$$

$$\overset{(1)}{=} (\mathbf{X}'\mathbf{X})^{-1}\mathbf{X}'\mu_{\mathbf{y}}$$
$$= (\mathbf{X}'\mathbf{X})^{-1}\mathbf{X}'[\mathbf{X}\boldsymbol{\beta}]$$
$$= (\mathbf{X}'\mathbf{X})^{-1}(\mathbf{X}'\mathbf{X})\boldsymbol{\beta}$$
$$= \mathbf{I}\boldsymbol{\beta}$$
$$= \boldsymbol{\beta}$$
$$= \begin{bmatrix} \beta_0 \\ \beta_1 \\ \cdot \\ \cdot \\ \cdot \\ \beta_p \end{bmatrix}$$

Thus we have shown that for $j = 0, 1, \ldots, p$,

$$\mu_{b_j} = \beta_j$$

In terms of matrices we have shown that

$$\boldsymbol{\mu}_{\mathbf{b}} = \boldsymbol{\beta}$$

(ii) To prove that $\sigma^2_{b_j} = \sigma^2 c_{jj}$, we first need to perform some matrix manipulations. To do this, we note that it can be proved that if **A** and **B** are matrices such that **A** has an inverse and the matrix product **AB** exists, then

$$(\mathbf{A}')' = \mathbf{A}$$

$$(\mathbf{AB})' = \mathbf{B}'\mathbf{A}'$$

$$(\mathbf{A}^{-1})' = (\mathbf{A}')^{-1}$$

Thus

$$\begin{aligned}[(\mathbf{X}'\mathbf{X})^{-1}\mathbf{X}']' &= (\mathbf{X}')'[(\mathbf{X}'\mathbf{X})^{-1}]' \\ &= \mathbf{X}[(\mathbf{X}'\mathbf{X})']^{-1} \\ &= \mathbf{X}[\mathbf{X}'(\mathbf{X}')']^{-1} \\ &= \mathbf{X}(\mathbf{X}'\mathbf{X})^{-1}\end{aligned}$$

We next recall that we have shown that inference assumptions 1 and 2 imply that $\sigma^2_{\mathbf{y}} = \sigma^2\mathbf{I}$. Hence, if $\sigma^2_{b_ib_j}$ denotes the covariance between b_i and b_j, it follows that

$$\begin{aligned}\begin{bmatrix} \sigma^2_{b_0} & \sigma^2_{b_0b_1} & \cdots & \sigma^2_{b_0b_p} \\ \sigma^2_{b_1b_0} & \sigma^2_{b_1} & \cdots & \sigma^2_{b_1b_p} \\ \vdots & \vdots & & \vdots \\ \sigma^2_{b_pb_0} & \sigma^2_{b_pb_1} & \cdots & \sigma^2_{b_p} \end{bmatrix} &= \sigma^2_{\mathbf{b}} \\ &= \sigma^2_{(\mathbf{X}'\mathbf{X})^{-1}\mathbf{X}'\mathbf{y}} \\ &= [(\mathbf{X}'\mathbf{X})^{-1}\mathbf{X}']\sigma^2_{\mathbf{y}}[(\mathbf{X}'\mathbf{X})^{-1}\mathbf{X}']' \\ &= [(\mathbf{X}'\mathbf{X})^{-1}\mathbf{X}']\sigma^2\mathbf{I}[\mathbf{X}(\mathbf{X}'\mathbf{X})^{-1}] \\ &= \sigma^2[(\mathbf{X}'\mathbf{X})^{-1}\mathbf{X}'\mathbf{I}\mathbf{X}(\mathbf{X}'\mathbf{X})^{-1}] \\ &= \sigma^2[(\mathbf{X}'\mathbf{X})^{-1}\mathbf{X}'\mathbf{X}(\mathbf{X}'\mathbf{X})^{-1}] \\ &= \sigma^2[(\mathbf{X}'\mathbf{X})^{-1}\mathbf{I}] \\ &= \sigma^2(\mathbf{X}'\mathbf{X})^{-1} \\ &= \sigma^2\begin{bmatrix} c_{00} & c_{01} & \cdots & c_{0p} \\ c_{10} & c_{11} & \cdots & c_{1p} \\ \vdots & \vdots & & \vdots \\ c_{p0} & c_{p1} & \cdots & c_{pp} \end{bmatrix}\end{aligned}$$

Thus we have shown that for $j = 0, 1, \ldots, p$,

$$\sigma^2_{b_j} = \sigma^2 c_{jj}$$

Moreover, we have shown that for $i = 0, 1, \ldots, p$, and $j = 0, 1, \ldots, p$, and $i \neq j$,

$$\sigma^2_{b_i b_j} = \sigma^2 c_{ij}$$

In terms of matrices we have shown that

$$\sigma^2_{\mathbf{b}} = \sigma^2(\mathbf{X}'\mathbf{X})^{-1}$$

Next, recall that

$$\hat{y}_0 = b_0 + b_1 x_{01} + \cdots + b_p x_{0p}$$

is the point estimate of the mean value of the dependent variable

$$\mu_0 = \beta_0 + \beta_1 x_{01} + \cdots + \beta_p x_{0p}$$

and is the point prediction of the individual value of the dependent variable

$$\begin{aligned} y_0 &= \mu_0 + \varepsilon_0 \\ &= \beta_0 + \beta_1 x_{01} + \cdots + \beta_p x_{0p} + \varepsilon_0 \end{aligned}$$

We now show that if

$$\mathbf{x}_0' = [1 \quad x_{01} \quad \cdots \quad x_{0p}]$$

is a row vector containing the specified values of the independent variables, then:

(iii) $\mu_{\hat{y}_0} = \mu_0$
(iv) $\sigma^2_{\hat{y}_0} = \sigma^2 \mathbf{x}_0'(\mathbf{X}'\mathbf{X})^{-1}\mathbf{x}_0$
(v) $\mu_{(y_0 - \hat{y}_0)} = 0$
(vi) $\sigma^2_{(y_0 - \hat{y}_0)} = \sigma^2(1 + \mathbf{x}_0'(\mathbf{X}'\mathbf{X})^{-1}\mathbf{x}_0)$

To prove these results, we note that

$$\begin{aligned} \hat{y}_0 &= b_0 + b_1 x_{01} + \cdots + b_p x_{0p} \\ &= [1 \quad x_{01} \quad \cdots \quad x_{0p}] \begin{bmatrix} b_0 \\ b_1 \\ \vdots \\ b_p \end{bmatrix} \\ &= \mathbf{x}_0'\mathbf{b} \end{aligned}$$

Thus the proofs for (iii) and (iv) are as follows.

(iii) Since we have shown that $\mu_{\mathbf{b}} = \boldsymbol{\beta}$, we have

$$\begin{aligned} \mu_{\hat{y}_0} &= \mu_{\mathbf{x}_0'\mathbf{b}} \\ &\overset{(1)}{=} \mathbf{x}_0'\mu_{\mathbf{b}} \\ &= \mathbf{x}_0'\boldsymbol{\beta} \end{aligned}$$

$$
\begin{aligned}
&= [1 \quad x_{01} \quad \cdots \quad x_{0p}] \begin{bmatrix} \beta_0 \\ \beta_1 \\ \cdot \\ \cdot \\ \cdot \\ \beta_p \end{bmatrix} \\
&= \beta_0 + \beta_1 x_{01} + \cdots + \beta_p x_{0p} \\
&= \mu_0
\end{aligned}
$$

(iv) Since we have shown that $\sigma^2_{\mathbf{b}} = \sigma^2(\mathbf{X}'\mathbf{X})^{-1}$, we have

$$
\begin{aligned}
\sigma^2_{\hat{y}_0} &= \sigma^2_{\mathbf{x}_0'\mathbf{b}} \\
&\overset{(2)}{=} \mathbf{x}_0'\sigma^2_{\mathbf{b}}(\mathbf{x}_0')' \\
&= \mathbf{x}_0'[\sigma^2(\mathbf{X}'\mathbf{X})^{-1}]\mathbf{x}_0 \\
&= \sigma^2\mathbf{x}_0'(\mathbf{X}'\mathbf{X})^{-1}\mathbf{x}_0
\end{aligned}
$$

To prove (v) and (vi), we first summarize some important properties of the means and variances of random variables. For proofs of these properties, see Wonnacott and Wonnacott (1977).

Assume that y is a random variable and K is a constant. Then:

(3) $\mu_{Ky} = K\mu_y$

(4) $\sigma^2_{Ky} = K^2\sigma^2_y$

Assume that $y_1, y_2, \ldots, y_n$ are random variables. Then:

(5) $\mu_{(y_1+y_2+\cdots+y_n)} = \mu_{y_1} + \mu_{y_2} + \cdots + \mu_{y_n}$

(6) If $y_1, y_2, \ldots, y_n$ are statistically independent,

$$\sigma^2_{(y_1+y_2+\cdots+y_n)} = \sigma^2_{y_1} + \sigma^2_{y_2} + \cdots + \sigma^2_{y_n}$$

To prove (v) and (vi), we note that before we observe

$$y_0 = \mu_0 + \varepsilon_0$$

y_0 can be any of the values in the population of potential values of the dependent variable. This population has mean μ_0 and (in accordance with inference assumption 1) variance σ^2. Thus we define the mean of y_0 (denoted by μ_{y_0}) and the variance of y_0 (denoted by $\sigma^2_{y_0}$) to be the mean and variance, respectively, of the population of potential values of the dependent variable. That is,

$$
\begin{aligned}
\mu_{y_0} &= \mu_0 \\
\sigma^2_{y_0} &= \sigma^2
\end{aligned}
$$

Thus:

(v) $$\mu_{(y_0-\hat{y}_0)} \overset{(5)}{=} \mu_{y_0} + \mu_{(-\hat{y}_0)} \overset{(3)}{=} \mu_{y_0} + (-1)\mu_{\hat{y}_0} = \mu_0 - \mu_0 = 0$$

(vi) Since

$$\hat{y}_0 = b_0 + b_1x_{01} + \cdots + b_px_{0p}$$

is a function of the least squares point estimates $b_0, b_1, \ldots, b_p$, which are in turn functions of $y_1, y_2, \ldots, y_n$, it follows that $\hat{y}_0$ is a function of $y_1, y_2, \ldots, y_n$. Since inference assumption 2 says that y_0 is statistically independent of $y_1, y_2, \ldots, y_n$, it follows that y_0 and $\hat{y}_0$ are statistically independent. Hence,

$$\sigma^2_{(y_0-\hat{y}_0)} \overset{(6)}{=} \sigma^2_{y_0} + \sigma^2_{(-\hat{y}_0)} \overset{(4)}{=} \sigma^2_{y_0} + (-1)^2\sigma^2_{\hat{y}_0} = \sigma^2 + \sigma^2\mathbf{x}_0'(\mathbf{X}'\mathbf{X})^{-1}\mathbf{x}_0 = \sigma^2(1 + \mathbf{x}_0'(\mathbf{X}'\mathbf{X})^{-1}\mathbf{x}_0)$$

APPENDIX 

STATISTICAL TABLES

TABLE E.1 Normal curve areas

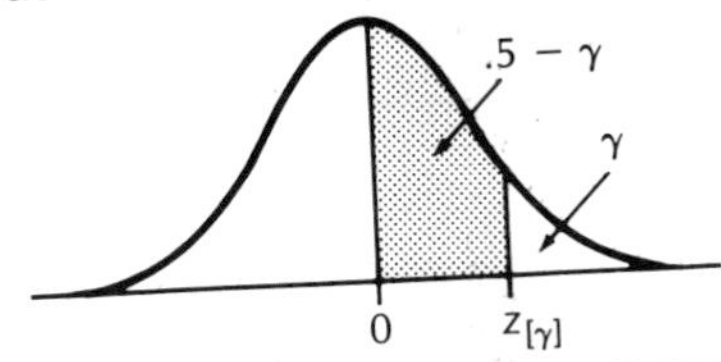

$z_{[\gamma]}$	.00	.01	.02	.03	.04	.05	.06	.07	.08	.09
0.0	.0000	.0040	.0080	.0120	.0160	.0199	.0239	.0279	.0319	.0359
0.1	.0398	.0438	.0478	.0517	.0557	.0596	.0636	.0675	.0714	.0753
0.2	.0793	.0832	.0871	.0910	.0948	.0987	.1026	.1064	.1103	.1141
0.3	.1179	.1217	.1255	.1293	.1331	.1368	.1406	.1443	.1480	.1517
0.4	.1554	.1591	.1628	.1664	.1700	.1736	.1772	.1808	.1844	.1879
0.5	.1915	.1950	.1985	.2019	.2054	.2088	.2123	.2157	.2190	.2224
0.6	.2257	.2291	.2324	.2357	.2389	.2422	.2454	.2486	.2517	.2549
0.7	.2580	.2611	.2642	.2673	.2704	.2734	.2764	.2794	.2823	.2852
0.8	.2881	.2910	.2939	.2967	.2995	.3023	.3051	.3078	.3106	.3133
0.9	.3159	.3186	.3212	.3238	.3264	.3289	.3315	.3340	.3365	.3389
1.0	.3413	.3438	.3461	.3485	.3508	.3531	.3554	.3577	.3599	.3621
1.1	.3643	.3665	.3686	.3708	.3729	.3749	.3770	.3790	.3810	.3830
1.2	.3849	.3869	.3888	.3907	.3925	.3944	.3962	.3980	.3997	.4015
1.3	.4032	.4049	.4066	.4082	.4099	.4115	.4131	.4147	.4162	.4177
1.4	.4192	.4207	.4222	.4236	.4251	.4265	.4279	.4292	.4306	.4319
1.5	.4332	.4345	.4357	.4370	.4382	.4394	.4406	.4418	.4429	.4441
1.6	.4452	.4463	.4474	.4484	.4495	.4505	.4515	.4525	.4535	.4545
1.7	.4554	.4564	.4573	.4582	.4591	.4599	.4608	.4616	.4625	.4633
1.8	.4641	.4649	.4656	.4664	.4671	.4678	.4686	.4693	.4699	.4706
1.9	.4713	.4719	.4726	.4732	.4738	.4744	.4750	.4756	.4761	.4767
2.0	.4772	.4778	.4783	.4788	.4793	.4798	.4803	.4808	.4812	.4817
2.1	.4821	.4826	.4830	.4834	.4838	.4842	.4846	.4850	.4854	.4857
2.2	.4861	.4864	.4868	.4871	.4875	.4878	.4881	.4884	.4887	.4890
2.3	.4893	.4896	.4898	.4901	.4904	.4906	.4909	.4911	.4913	.4916
2.4	.4918	.4920	.4922	.4925	.4927	.4929	.4931	.4932	.4934	.4936
2.5	.4938	.4940	.4941	.4943	.4945	.4946	.4948	.4949	.4951	.4952
2.6	.4953	.4955	.4956	.4957	.4959	.4960	.4961	.4962	.4963	.4964
2.7	.4965	.4966	.4967	.4968	.4969	.4970	.4971	.4972	.4973	.4974
2.8	.4974	.4975	.4976	.4977	.4977	.4978	.4979	.4979	.4980	.4981
2.9	.4981	.4982	.4982	.4983	.4984	.4984	.4985	.4985	.4986	.4986
3.0	.4987	.4987	.4987	.4988	.4988	.4989	.4989	.4989	.4990	.4990

Source: A. Hald, *Statistical Tables and Formulas* (New York: Wiley, 1952), abridged from Table 1. Reproduced by permission of the publisher.

TABLE E.2 Critical values of t

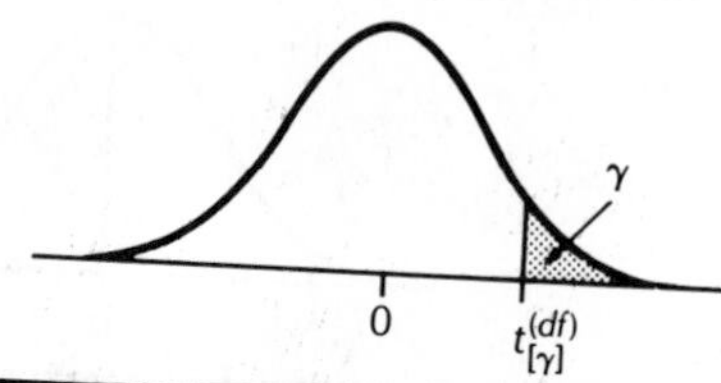

df	$t^{(df)}_{[.40]}$	$t^{(df)}_{[.30]}$	$t^{(df)}_{[.20]}$	$t^{(df)}_{[.15]}$	$t^{(df)}_{[.10]}$	$t^{(df)}_{[.05]}$	$t^{(df)}_{[.025]}$
1	0.325	0.727	1.376	1.963	3.078	6.314	12.706
2	0.289	0.617	1.061	1.386	1.886	2.920	4.303
3	0.277	0.584	0.978	1.250	1.638	2.353	3.182
4	0.271	0.569	0.941	1.190	1.533	2.132	2.776
5	0.267	0.559	0.920	1.156	1.476	2.015	2.571
6	0.265	0.553	0.906	1.134	1.440	1.943	2.447
7	0.263	0.549	0.896	1.119	1.415	1.895	2.365
8	0.262	0.546	0.889	1.108	1.397	1.860	2.306
9	0.261	0.543	0.883	1.100	1.383	1.833	2.262
10	0.260	0.542	0.879	1.093	1.372	1.812	2.228
11	0.260	0.540	0.876	1.088	1.363	1.796	2.201
12	0.259	0.539	0.873	1.083	1.356	1.782	2.179
13	0.259	0.537	0.870	1.079	1.350	1.771	2.160
14	0.258	0.537	0.868	1.076	1.345	1.761	2.145
15	0.258	0.536	0.866	1.074	1.341	1.753	2.131
16	0.258	0.535	0.865	1.071	1.337	1.746	2.120
17	0.257	0.534	0.863	1.069	1.333	1.740	2.110
18	0.257	0.534	0.862	1.067	1.330	1.734	2.101
19	0.257	0.533	0.861	1.066	1.328	1.729	2.093
20	0.257	0.533	0.860	1.064	1.325	1.725	2.086
21	0.257	0.532	0.859	1.063	1.323	1.721	2.080
22	0.256	0.532	0.858	1.061	1.321	1.717	2.074
23	0.256	0.532	0.858	1.060	1.319	1.714	2.069
24	0.256	0.531	0.857	1.059	1.318	1.711	2.064
25	0.256	0.531	0.856	1.058	1.316	1.708	2.060
26	0.256	0.531	0.856	1.058	1.315	1.706	2.056
27	0.256	0.531	0.855	1.057	1.314	1.703	2.052
28	0.256	0.530	0.855	1.056	1.313	1.701	2.048
29	0.256	0.530	0.854	1.055	1.311	1.699	2.045
30	0.256	0.530	0.854	1.055	1.310	1.697	2.042
40	0.255	0.529	0.851	1.050	1.303	1.684	2.021
60	0.254	0.527	0.848	1.045	1.296	1.671	2.000
120	0.254	0.526	0.845	1.041	1.289	1.658	1.980
∞	0.253	0.524	0.842	1.036	1.282	1.645	1.960

df	$t^{(df)}_{[.02]}$	$t^{(df)}_{[.015]}$	$t^{(df)}_{[.01]}$	$t^{(df)}_{[.0075]}$	$t^{(df)}_{[.005]}$	$t^{(df)}_{[.0025]}$	$t^{(df)}_{[.0005]}$
1	15.895	21.205	31.821	42.434	63.657	127.322	636.590
2	4.849	5.643	6.965	8.073	9.925	14.089	31.598
3	3.482	3.896	4.541	5.047	5.841	7.453	12.924
4	2.999	3.298	3.747	4.088	4.604	5.598	8.610
5	2.757	3.003	3.365	3.634	4.032	4.773	6.869
6	2.612	2.829	3.143	3.372	3.707	4.317	5.959
7	2.517	2.715	2.998	3.203	3.499	4.029	5.408
8	2.449	2.634	2.896	3.085	3.355	3.833	5.041
9	2.398	2.574	2.821	2.998	3.250	3.690	4.781
10	2.359	2.527	2.764	2.932	3.169	3.581	4.587
11	2.328	2.491	2.718	2.879	3.106	3.497	4.437
12	2.303	2.461	2.681	2.836	3.055	3.428	4.318
13	2.282	2.436	2.650	2.801	3.012	3.372	4.221
14	2.264	2.415	2.624	2.771	2.977	3.326	4.140
15	2.249	2.397	2.602	2.746	2.947	3.286	4.073
16	2.235	2.382	2.583	2.724	2.921	3.252	4.015
17	2.224	2.368	2.567	2.706	2.898	3.222	3.965
18	2.214	2.356	2.552	2.689	2.878	3.197	3.922
19	2.205	2.346	2.539	2.674	2.861	3.174	3.883
20	2.197	2.336	2.528	2.661	2.845	3.153	3.849
21	2.189	2.328	2.518	2.649	2.831	3.135	3.819
22	2.183	2.320	2.508	2.639	2.819	3.119	3.792
23	2.177	2.313	2.500	2.629	2.807	3.104	3.768
24	2.172	2.307	2.492	2.620	2.797	3.091	3.745
25	2.167	2.301	2.485	2.612	2.787	3.078	3.725
26	2.162	2.296	2.479	2.605	2.779	3.067	3.707
27	2.158	2.291	2.473	2.598	2.771	3.057	3.690
28	2.154	2.286	2.467	2.592	2.763	3.047	3.674
29	2.150	2.282	2.462	2.586	2.756	3.038	3.659
30	2.147	2.278	2.457	2.581	2.750	3.030	3.646
40	2.123	2.250	2.423	2.542	2.704	2.971	3.551
60	2.099	2.223	2.390	2.504	2.660	2.915	3.460
120	2.076	2.196	2.358	2.468	2.617	2.860	3.373
∞	2.054	2.170	2.326	2.432	2.576	2.807	3.291

Source: Reprinted from Neter, J., W. Wasserman, and M. H. Kutner. *Applied Linear Statistical Models*. 2nd ed. Homewood, Ill.: Richard Irwin, 1985.

TABLE E.3 Percentage points of the *F*-distribution ($\gamma = .05$)

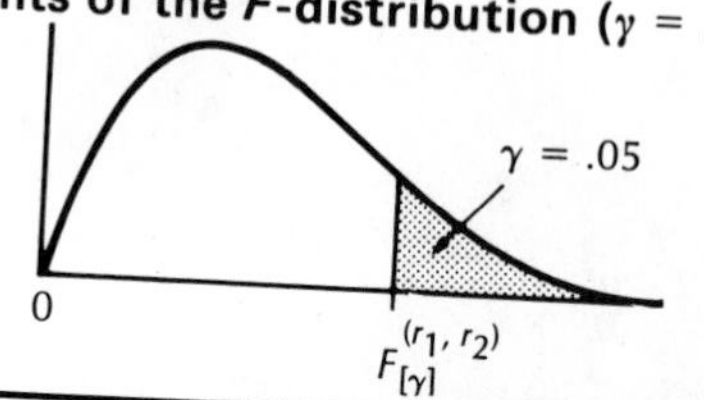

Denominator degrees of freedom, r_2	Numerator degrees of freedom, r_1								
	1	**2**	**3**	**4**	**5**	**6**	**7**	**8**	**9**
1	161.4	199.5	215.7	224.6	230.2	234.0	236.8	238.9	240.5
2	18.51	19.00	19.16	19.25	19.30	19.33	19.35	19.37	19.38
3	10.13	9.55	9.28	9.12	9.01	8.94	8.89	8.85	8.81
4	7.71	6.94	6.59	6.39	6.26	6.16	6.09	6.04	6.00
5	6.61	5.79	5.41	5.19	5.05	4.95	4.88	4.82	4.77
6	5.99	5.14	4.76	4.53	4.39	4.28	4.21	4.15	4.10
7	5.59	4.74	4.35	4.12	3.97	3.87	3.79	3.73	3.68
8	5.32	4.46	4.07	3.84	3.69	3.58	3.50	3.44	3.39
9	5.12	4.26	3.86	3.63	3.48	3.37	3.29	3.23	3.18
10	4.96	4.10	3.71	3.48	3.33	3.22	3.14	3.07	3.02
11	4.84	3.98	3.59	3.36	3.20	3.09	3.01	2.95	2.90
12	4.75	3.89	3.49	3.26	3.11	3.00	2.91	2.85	2.80
13	4.67	3.81	3.41	3.18	3.03	2.92	2.83	2.77	2.71
14	4.60	3.74	3.34	3.11	2.96	2.85	2.76	2.70	2.65
15	4.54	3.68	3.29	3.06	2.90	2.79	2.71	2.64	2.59
16	4.49	3.63	3.24	3.01	2.85	2.74	2.66	2.59	2.54
17	4.45	3.59	3.20	2.96	2.81	2.70	2.61	2.55	2.49
18	4.41	3.55	3.16	2.93	2.77	2.66	2.58	2.51	2.46
19	4.38	3.52	3.13	2.90	2.74	2.63	2.54	2.48	2.42
20	4.35	3.49	3.10	2.87	2.71	2.60	2.51	2.45	2.39
21	4.32	3.47	3.07	2.84	2.68	2.57	2.49	2.42	2.37
22	4.30	3.44	3.05	2.82	2.66	2.55	2.46	2.40	2.34
23	4.28	3.42	3.03	2.80	2.64	2.53	2.44	2.37	2.32
24	4.26	3.40	3.01	2.78	2.62	2.51	2.42	2.36	2.30
25	4.24	3.39	2.99	2.76	2.60	2.49	2.40	2.34	2.28
26	4.23	3.37	2.98	2.74	2.59	2.47	2.39	2.32	2.27
27	4.21	3.35	2.96	2.73	2.57	2.46	2.37	2.31	2.25
28	4.20	3.34	2.95	2.71	2.56	2.45	2.36	2.29	2.24
29	4.18	3.33	2.93	2.70	2.55	2.43	2.35	2.28	2.22
30	4.17	3.32	2.92	2.69	2.53	2.42	2.33	2.27	2.21
40	4.08	3.23	2.84	2.61	2.45	2.34	2.25	2.18	2.12
60	4.00	3.15	2.76	2.53	2.37	2.25	2.17	2.10	2.04
120	3.92	3.07	2.68	2.45	2.29	2.17	2.09	2.02	1.96
∞	3.84	3.00	2.60	2.37	2.21	2.10	2.01	1.94	1.88

Denominator degrees of freedom, r_2	Numerator degrees of freedom, r_1									
	10	12	15	20	24	30	40	60	120	∞
1	241.9	243.9	245.9	248.0	249.1	250.1	251.1	252.2	253.3	254.3
2	19.40	19.41	19.43	19.45	19.45	19.46	19.47	19.48	19.49	19.50
3	8.79	8.74	8.70	8.66	8.64	8.62	8.59	8.57	8.55	8.53
4	5.96	5.91	5.86	5.80	5.77	5.75	5.72	5.69	5.66	5.63
5	4.74	4.68	4.62	4.56	4.53	4.50	4.46	4.43	4.40	4.36
6	4.06	4.00	3.94	3.87	3.84	3.81	3.77	3.74	3.70	3.67
7	3.64	3.57	3.51	3.44	3.41	3.38	3.34	3.30	3.27	3.23
8	3.35	3.28	3.22	3.15	3.12	3.08	3.04	3.01	2.97	2.93
9	3.14	3.07	3.01	2.94	2.90	2.86	2.83	2.79	2.75	2.71
10	2.98	2.91	2.85	2.77	2.74	2.70	2.66	2.62	2.58	2.54
11	2.85	2.79	2.72	2.65	2.61	2.57	2.53	2.49	2.45	2.40
12	2.75	2.69	2.62	2.54	2.51	2.47	2.43	2.38	2.34	2.30
13	2.67	2.60	2.53	2.46	2.42	2.38	2.34	2.30	2.25	2.21
14	2.60	2.53	2.46	2.39	2.35	2.31	2.27	2.22	2.18	2.13
15	2.54	2.48	2.40	2.33	2.29	2.25	2.20	2.16	2.11	2.07
16	2.49	2.42	2.35	2.28	2.24	2.19	2.15	2.11	2.06	2.01
17	2.45	2.38	2.31	2.23	2.19	2.15	2.10	2.06	2.01	1.96
18	2.41	2.34	2.27	2.19	2.15	2.11	2.06	2.02	1.97	1.92
19	2.38	2.31	2.23	2.16	2.11	2.07	2.03	1.98	1.93	1.88
20	2.35	2.28	2.20	2.12	2.08	2.04	1.99	1.95	1.90	1.84
21	2.32	2.25	2.18	2.10	2.05	2.01	1.96	1.92	1.87	1.81
22	2.30	2.23	2.15	2.07	2.03	1.98	1.94	1.89	1.84	1.78
23	2.27	2.20	2.13	2.05	2.01	1.96	1.91	1.86	1.81	1.76
24	2.25	2.18	2.11	2.03	1.98	1.94	1.89	1.84	1.79	1.73
25	2.24	2.16	2.09	2.01	1.96	1.92	1.87	1.82	1.77	1.71
26	2.22	2.15	2.07	1.99	1.95	1.90	1.85	1.80	1.75	1.69
27	2.20	2.13	2.06	1.97	1.93	1.88	1.84	1.79	1.73	1.67
28	2.19	2.12	2.04	1.96	1.91	1.87	1.82	1.77	1.71	1.65
29	2.18	2.10	2.03	1.94	1.90	1.85	1.81	1.75	1.70	1.64
30	2.16	2.09	2.01	1.93	1.89	1.84	1.79	1.74	1.68	1.62
40	2.08	2.00	1.92	1.84	1.79	1.74	1.69	1.64	1.58	1.51
60	1.99	1.92	1.84	1.75	1.70	1.65	1.59	1.53	1.47	1.39
120	1.91	1.83	1.75	1.66	1.61	1.55	1.50	1.43	1.35	1.25
∞	1.83	1.75	1.67	1.57	1.52	1.46	1.39	1.32	1.22	1.00

Source: From "Tables of Percentage Points of the Inverted Beta (*F*)-Distribution," by Maxine Merrington and Catherine M. Thompson, *Biometrika* 33 (1943), 73–88. Reproduced by permission of the *Biometrika* Trustees.

TABLE E.4 **Percentage points of the F-distribution ($\gamma = .01$)**

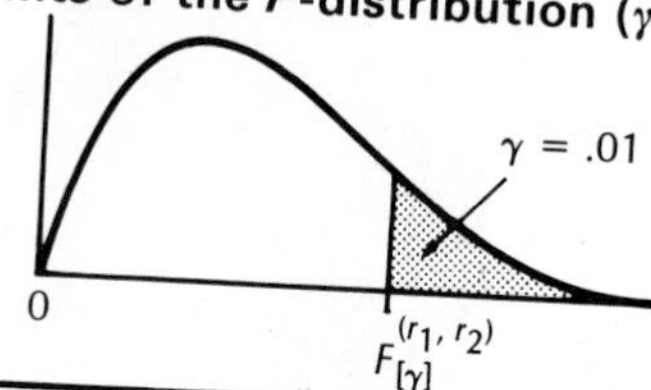

Denominator degrees of freedom, r_2	Numerator degrees of freedom, r_1								
	1	2	3	4	5	6	7	8	9
1	4052	4999.5	5403	5625	5764	5859	5928	5982	6022
2	98.50	99.00	99.17	99.25	99.30	99.33	99.36	99.37	99.39
3	34.12	30.82	29.46	28.71	28.24	27.91	27.67	27.49	27.35
4	21.20	18.00	16.69	15.98	15.52	15.21	14.98	14.80	14.66
5	16.26	13.27	12.06	11.39	10.97	10.67	10.46	10.29	10.16
6	13.75	10.92	9.78	9.15	8.75	8.47	8.26	8.10	7.98
7	12.25	9.55	8.45	7.85	7.46	7.19	6.99	6.84	6.72
8	11.26	8.65	7.59	7.01	6.63	6.37	6.18	6.03	5.91
9	10.56	8.02	6.99	6.42	6.06	5.80	5.61	5.47	5.35
10	10.04	7.56	6.55	5.99	5.64	5.39	5.20	5.06	4.94
11	9.65	7.21	6.22	5.67	5.32	5.07	4.89	4.74	4.63
12	9.33	6.93	5.95	5.41	5.06	4.82	4.64	4.50	4.39
13	9.07	6.70	5.74	5.21	4.86	4.62	4.44	4.30	4.19
14	8.86	6.51	5.56	5.04	4.69	4.46	4.28	4.14	4.03
15	8.68	6.36	5.42	4.89	4.56	4.32	4.14	4.00	3.89
16	8.53	6.23	5.29	4.77	4.44	4.20	4.03	3.89	3.78
17	8.40	6.11	5.18	4.67	4.34	4.10	3.93	3.79	3.68
18	8.29	6.01	5.09	4.58	4.25	4.01	3.84	3.71	3.60
19	8.18	5.93	5.01	4.50	4.17	3.94	3.77	3.63	3.52
20	8.10	5.85	4.94	4.43	4.10	3.87	3.70	3.56	3.46
21	8.02	5.78	4.87	4.37	4.04	3.81	3.64	3.51	3.40
22	7.95	5.72	4.82	4.31	3.99	3.76	3.59	3.45	3.35
23	7.88	5.66	4.76	4.26	3.94	3.71	3.54	3.41	3.30
24	7.82	5.61	4.72	4.22	3.90	3.67	3.50	3.36	3.26
25	7.77	5.57	4.68	4.18	3.85	3.63	3.46	3.32	3.22
26	7.72	5.53	4.64	4.14	3.82	3.59	3.42	3.29	3.18
27	7.68	5.49	4.60	4.11	3.78	3.56	3.39	3.26	3.15
28	7.64	5.45	4.57	4.07	3.75	3.53	3.36	3.23	3.12
29	7.60	5.42	4.54	4.04	3.73	3.50	3.33	3.20	3.09
30	7.56	5.39	4.51	4.02	3.70	3.47	3.30	3.17	3.07
40	7.31	5.18	4.31	3.83	3.51	3.29	3.12	2.99	2.89
60	7.08	4.98	4.13	3.65	3.34	3.12	2.95	2.82	2.72
120	6.85	4.79	3.95	3.48	3.17	2.96	2.79	2.66	2.50
∞	6.63	4.61	3.78	3.32	3.02	2.80	2.64	2.51	2.41

Denominator degrees of freedom, r_2	Numerator degrees of freedom, r_1									
	10	12	15	20	24	30	40	60	120	∞
1	6056	6106	6157	6209	6235	6261	6287	6313	6339	6366
2	99.40	99.42	99.43	99.45	99.46	99.47	99.47	99.48	99.49	99.50
3	27.23	27.05	26.87	26.69	26.60	26.50	26.41	26.32	26.22	26.13
4	14.55	14.37	14.20	14.02	13.93	13.84	13.75	13.65	13.56	13.46
5	10.05	9.89	9.72	9.55	9.47	9.38	9.29	9.20	9.11	9.02
6	7.87	7.72	7.56	7.40	7.31	7.23	7.14	7.06	6.97	6.88
7	6.62	6.47	6.31	6.16	6.07	5.99	5.91	5.82	5.74	5.65
8	5.81	5.67	5.52	5.36	5.28	5.20	5.12	5.03	4.95	4.86
9	5.26	5.11	4.96	4.81	4.73	4.65	4.57	4.48	4.40	4.31
10	4.85	4.71	4.56	4.41	4.33	4.25	4.17	4.08	4.00	3.91
11	4.54	4.40	4.25	4.10	4.02	3.94	3.86	3.78	3.69	3.60
12	4.30	4.16	4.01	3.86	3.78	3.70	3.62	3.54	3.45	3.36
13	4.10	3.96	3.82	3.66	3.59	3.51	3.43	3.34	3.25	3.17
14	3.94	3.80	3.66	3.51	3.43	3.35	3.27	3.18	3.09	3.00
15	3.80	3.67	3.52	3.37	3.29	3.21	3.13	3.05	2.96	2.87
16	3.69	3.55	3.41	3.26	3.18	3.10	3.02	2.93	2.84	2.75
17	3.59	3.46	3.31	3.16	3.08	3.00	2.92	2.83	2.75	2.65
18	3.51	3.37	3.23	3.08	3.00	2.92	2.84	2.75	2.66	2.57
19	3.43	3.30	3.15	3.00	2.92	2.84	2.76	2.67	2.58	2.49
20	3.37	3.23	3.09	2.94	2.86	2.78	2.69	2.61	2.52	2.42
21	3.31	3.17	3.03	2.88	2.80	2.72	2.64	2.55	2.46	2.36
22	3.26	3.12	2.98	2.83	2.75	2.67	2.58	2.50	2.40	2.31
23	3.21	3.07	2.93	2.78	2.70	2.62	2.54	2.45	2.35	2.26
24	3.17	3.03	2.89	2.74	2.66	2.58	2.49	2.40	2.31	2.21
25	3.13	2.99	2.85	2.70	2.62	2.54	2.45	2.36	2.27	2.17
26	3.09	2.96	2.81	2.66	2.58	2.50	2.42	2.33	2.23	2.13
27	3.06	2.93	2.78	2.63	2.55	2.47	2.38	2.29	2.20	2.10
28	3.03	2.90	2.75	2.60	2.52	2.44	2.35	2.26	2.17	2.06
29	3.00	2.87	2.73	2.57	2.49	2.41	2.33	2.23	2.14	2.03
30	2.98	2.84	2.70	2.55	2.47	2.39	2.30	2.21	2.11	2.01
40	2.80	2.66	2.52	2.37	2.29	2.20	2.11	2.02	1.92	1.80
60	2.63	2.50	2.35	2.20	2.12	2.03	1.94	1.84	1.73	1.60
120	2.47	2.34	2.19	2.03	1.95	1.86	1.76	1.66	1.53	1.38
∞	2.32	2.18	2.04	1.88	1.79	1.70	1.59	1.47	1.32	1.00

Source: From "Tables of Percentage Points of the Inverted Beta (F)-Distribution," by Maxine Merrington and Catherine M. Thompson, *Biometrika* 33 (1943), 73–88. Reproduced by permission of the *Biometrika* Trustees.

TABLE E.5 Critical values for the Durbin–Watson d statistic ($\alpha = .05$)

n	$k-1=1$		$k-1=2$		$k-1=3$		$k-1=4$		$k-1=5$	
	$d_{L,.05}$	$d_{U,.05}$	$d_{L,.05}$	$d_{U,.05}$	$d_{L,.05}$	$d_{U,.05}$	$d_{L,.05}$	$d_{U,.05}$	$d_{L,.05}$	$d_{U,.05}$
15	1.08	1.36	0.95	1.54	0.82	1.75	0.69	1.97	0.56	2.21
16	1.10	1.37	0.98	1.54	0.86	1.73	0.74	1.93	0.62	2.15
17	1.13	1.38	1.02	1.54	0.90	1.71	0.78	1.90	0.67	2.10
18	1.16	1.39	1.05	1.53	0.93	1.69	0.82	1.87	0.71	2.06
19	1.18	1.40	1.08	1.53	0.97	1.68	0.86	1.85	0.75	2.02
20	1.20	1.41	1.10	1.54	1.00	1.68	0.90	1.83	0.79	1.99
21	1.22	1.42	1.13	1.54	1.03	1.67	0.93	1.81	0.83	1.96
22	1.24	1.43	1.15	1.54	1.05	1.66	0.96	1.80	0.86	1.94
23	1.26	1.44	1.17	1.54	1.08	1.66	0.99	1.79	0.90	1.92
24	1.27	1.45	1.19	1.55	1.10	1.66	1.01	1.78	0.93	1.90
25	1.29	1.45	1.21	1.55	1.12	1.66	1.04	1.77	0.95	1.89
26	1.30	1.46	1.22	1.55	1.14	1.65	1.06	1.76	0.98	1.88
27	1.32	1.47	1.24	1.56	1.16	1.65	1.08	1.76	1.01	1.86
28	1.33	1.48	1.26	1.56	1.18	1.65	1.10	1.75	1.03	1.85
29	1.34	1.48	1.27	1.56	1.20	1.65	1.12	1.74	1.05	1.84
30	1.35	1.49	1.28	1.57	1.21	1.65	1.14	1.74	1.07	1.83
31	1.36	1.50	1.30	1.57	1.23	1.65	1.16	1.74	1.09	1.83
32	1.37	1.50	1.31	1.57	1.24	1.65	1.18	1.73	1.11	1.82
33	1.38	1.51	1.32	1.58	1.26	1.65	1.19	1.73	1.13	1.81
34	1.39	1.51	1.33	1.58	1.27	1.65	1.21	1.73	1.15	1.81
35	1.40	1.52	1.34	1.58	1.28	1.65	1.22	1.73	1.16	1.80
36	1.41	1.52	1.35	1.59	1.29	1.65	1.24	1.73	1.18	1.80
37	1.42	1.53	1.36	1.59	1.31	1.66	1.25	1.72	1.19	1.80
38	1.43	1.54	1.37	1.59	1.32	1.66	1.26	1.72	1.21	1.79
39	1.43	1.54	1.38	1.60	1.33	1.66	1.27	1.72	1.22	1.79
40	1.44	1.54	1.39	1.60	1.34	1.66	1.29	1.72	1.23	1.79
45	1.48	1.57	1.43	1.62	1.38	1.67	1.34	1.72	1.29	1.78
50	1.50	1.59	1.46	1.63	1.42	1.67	1.38	1.72	1.34	1.77
55	1.53	1.60	1.49	1.64	1.45	1.68	1.41	1.72	1.38	1.77
60	1.55	1.62	1.51	1.65	1.48	1.69	1.44	1.73	1.41	1.77
65	1.57	1.63	1.54	1.66	1.50	1.70	1.47	1.73	1.44	1.77
70	1.58	1.64	1.55	1.67	1.52	1.70	1.49	1.74	1.46	1.77
75	1.60	1.65	1.57	1.68	1.54	1.71	1.51	1.74	1.49	1.77
80	1.61	1.66	1.59	1.69	1.56	1.72	1.53	1.74	1.51	1.77
85	1.62	1.67	1.60	1.70	1.57	1.72	1.55	1.75	1.52	1.77
90	1.63	1.68	1.61	1.70	1.59	1.73	1.57	1.75	1.54	1.78
95	1.64	1.69	1.62	1.71	1.60	1.73	1.58	1.75	1.56	1.78
100	1.65	1.69	1.63	1.72	1.61	1.74	1.59	1.76	1.57	1.78

Source: From J. Durbin and G. S. Watson, "Testing for Serial Correlation in Least Squares Regression, II," *Biometrika* 30 (1951), 159–178. Reproduced by permission of the *Biometrika* Trustees.

TABLE E.6

Critical values for the Durbin–Watson d statistic ($\alpha = .01$)

n	$k-1=1$		$k-1=2$		$k-1=3$		$k-1=4$		$k-1=5$	
	$d_{L,.01}$	$d_{U,.01}$	$d_{L,.01}$	$d_{U,.01}$	$d_{L,.01}$	$d_{U,.01}$	$d_{L,.01}$	$d_{U,.01}$	$d_{L,.01}$	$d_{U,.01}$
15	0.81	1.07	0.70	1.25	0.59	1.46	0.49	1.70	0.39	1.96
16	0.84	1.09	0.74	1.25	0.63	1.44	0.53	1.66	0.44	1.90
17	0.87	1.10	0.77	1.25	0.67	1.43	0.57	1.63	0.48	1.85
18	0.90	1.12	0.80	1.26	0.71	1.42	0.61	1.60	0.52	1.80
19	0.93	1.13	0.83	1.26	0.74	1.41	0.65	1.58	0.56	1.77
20	0.95	1.15	0.86	1.27	0.77	1.41	0.68	1.57	0.60	1.74
21	0.97	1.16	0.89	1.27	0.80	1.41	0.72	1.55	0.63	1.71
22	1.00	1.17	0.91	1.28	0.83	1.40	0.75	1.54	0.66	1.69
23	1.02	1.19	0.94	1.29	0.86	1.40	0.77	1.53	0.70	1.67
24	1.04	1.20	0.96	1.30	0.88	1.41	0.80	1.53	0.72	1.66
25	1.05	1.21	0.98	1.30	0.90	1.41	0.83	1.52	0.75	1.65
26	1.07	1.22	1.00	1.31	0.93	1.41	0.85	1.52	0.78	1.64
27	1.09	1.23	1.02	1.32	0.95	1.41	0.88	1.51	0.81	1.63
28	1.10	1.24	1.04	1.32	0.97	1.41	0.90	1.51	0.83	1.62
29	1.12	1.25	1.05	1.33	0.99	1.42	0.92	1.51	0.85	1.61
30	1.13	1.26	1.07	1.34	1.01	1.42	0.94	1.51	0.88	1.61
31	1.15	1.27	1.08	1.34	1.02	1.42	0.96	1.51	0.90	1.60
32	1.16	1.28	1.10	1.35	1.04	1.43	0.98	1.51	0.92	1.60
33	1.17	1.29	1.11	1.36	1.05	1.43	1.00	1.51	0.94	1.59
34	1.18	1.30	1.13	1.36	1.07	1.43	1.01	1.51	0.95	1.59
35	1.19	1.31	1.14	1.37	1.08	1.44	1.03	1.51	0.97	1.59
36	1.21	1.32	1.15	1.38	1.10	1.44	1.04	1.51	0.99	1.59
37	1.22	1.32	1.16	1.38	1.11	1.45	1.06	1.51	1.00	1.59
38	1.23	1.33	1.18	1.39	1.12	1.45	1.07	1.52	1.02	1.58
39	1.24	1.34	1.19	1.39	1.14	1.45	1.09	1.52	1.03	1.58
40	1.25	1.34	1.20	1.40	1.15	1.46	1.10	1.52	1.05	1.58
45	1.29	1.38	1.24	1.42	1.20	1.48	1.16	1.53	1.11	1.58
50	1.32	1.40	1.28	1.45	1.24	1.49	1.20	1.54	1.16	1.59
55	1.36	1.43	1.32	1.47	1.28	1.51	1.25	1.55	1.21	1.59
60	1.38	1.45	1.35	1.48	1.32	1.52	1.28	1.56	1.25	1.60
65	1.41	1.47	1.38	1.50	1.35	1.53	1.31	1.57	1.28	1.61
70	1.43	1.49	1.40	1.52	1.37	1.55	1.34	1.58	1.31	1.61
75	1.45	1.50	1.42	1.53	1.39	1.56	1.37	1.59	1.34	1.62
80	1.47	1.52	1.44	1.54	1.42	1.57	1.39	1.60	1.36	1.62
85	1.48	1.53	1.46	1.55	1.43	1.58	1.41	1.60	1.39	1.63
90	1.50	1.54	1.47	1.56	1.45	1.59	1.43	1.61	1.41	1.64
95	1.51	1.55	1.49	1.57	1.47	1.60	1.45	1.62	1.42	1.64
100	1.52	1.56	1.50	1.58	1.48	1.60	1.46	1.63	1.44	1.65

Source: From J. Durbin and G. S. Watson, "Testing for Serial Correlation in Least Squares Regression, II," *Biometrika* 30 (1951), 159–178. Reproduced by permission of the *Biometrika* Trustees.

TABLE E.7 Percentiles of the studentized range distribution.

Entry is $q_{[.10]}(r, v)$

v	r = 2	3	4	5	6	7	8	9	10	11	12	13	14	15	16	17	18	19	20
1	8.93	13.4	16.4	18.5	20.2	21.5	22.6	23.6	24.5	25.2	25.9	26.5	27.1	27.6	28.1	28.5	29.0	29.3	29.7
2	4.13	5.73	6.77	7.54	8.14	8.63	9.05	9.41	9.72	10.0	10.3	10.5	10.7	10.9	11.1	11.2	11.4	11.5	11.7
3	3.33	4.47	5.20	5.74	6.16	6.51	6.81	7.06	7.29	7.49	7.67	7.83	7.98	8.12	8.25	8.37	8.48	8.58	8.68
4	3.01	3.98	4.59	5.03	5.39	5.68	5.93	6.14	6.33	6.49	6.65	6.78	6.91	7.02	7.13	7.23	7.33	7.41	7.50
5	2.85	3.72	4.26	4.66	4.98	5.24	5.46	5.65	5.82	5.97	6.10	6.22	6.34	6.44	6.54	6.63	6.71	6.79	6.86
6	2.75	3.56	4.07	4.44	4.73	4.97	5.17	5.34	5.50	5.64	5.76	5.87	5.98	6.07	6.16	6.25	6.32	6.40	6.47
7	2.68	3.45	3.93	4.28	4.55	4.78	4.97	5.14	5.28	5.41	5.53	5.64	5.74	5.83	5.91	5.99	6.06	6.13	6.19
8	2.63	3.37	3.83	4.17	4.43	4.65	4.83	4.99	5.13	5.25	5.36	5.46	5.56	5.64	5.72	5.80	5.87	5.93	6.00
9	2.59	3.32	3.76	4.08	4.34	4.54	4.72	4.87	5.01	5.13	5.23	5.33	5.42	5.51	5.58	5.66	5.72	5.79	5.85
10	2.56	3.27	3.70	4.02	4.26	4.47	4.64	4.78	4.91	5.03	5.13	5.23	5.32	5.40	5.47	5.54	5.61	5.67	5.73
11	2.54	3.23	3.66	3.96	4.20	4.40	4.57	4.71	4.84	4.95	5.05	5.15	5.23	5.31	5.38	5.45	5.51	5.57	5.63
12	2.52	3.20	3.62	3.92	4.16	4.35	4.51	4.65	4.78	4.89	4.99	5.08	5.16	5.24	5.31	5.37	5.44	5.49	5.55
13	2.50	3.18	3.59	3.88	4.12	4.30	4.46	4.60	4.72	4.83	4.93	5.02	5.10	5.18	5.25	5.31	5.37	5.43	5.48
14	2.49	3.16	3.56	3.85	4.08	4.27	4.42	4.56	4.68	4.79	4.88	4.97	5.05	5.12	5.19	5.26	5.32	5.37	5.43
15	2.48	3.14	3.54	3.83	4.05	4.23	4.39	4.52	4.64	4.75	4.84	4.93	5.01	5.08	5.15	5.21	5.27	5.32	5.38
16	2.47	3.12	3.52	3.80	4.03	4.21	4.36	4.49	4.61	4.71	4.81	4.89	4.97	5.04	5.11	5.17	5.23	5.28	5.33
17	2.46	3.11	3.50	3.78	4.00	4.18	4.33	4.46	4.58	4.68	4.77	4.86	4.93	5.01	5.07	5.13	5.19	5.24	5.30
18	2.45	3.10	3.49	3.77	3.98	4.16	4.31	4.44	4.55	4.65	4.75	4.83	4.90	4.98	5.04	5.10	5.16	5.21	5.26
19	2.45	3.09	3.47	3.75	3.97	4.14	4.29	4.42	4.53	4.63	4.72	4.80	4.88	4.95	5.01	5.07	5.13	5.18	5.23
20	2.44	3.08	3.46	3.74	3.95	4.12	4.27	4.40	4.51	4.61	4.70	4.78	4.85	4.92	4.99	5.05	5.10	5.16	5.20
24	2.42	3.05	3.42	3.69	3.90	4.07	4.21	4.34	4.44	4.54	4.63	4.71	4.78	4.85	4.91	4.97	5.02	5.07	5.12
30	2.40	3.02	3.39	3.65	3.85	4.02	4.16	4.28	4.38	4.47	4.56	4.64	4.71	4.77	4.83	4.89	4.94	4.99	5.03
40	2.38	2.99	3.35	3.60	3.80	3.96	4.10	4.21	4.32	4.41	4.49	4.56	4.63	4.69	4.75	4.81	4.86	4.90	4.95
60	2.36	2.96	3.31	3.56	3.75	3.91	4.04	4.16	4.25	4.34	4.42	4.49	4.56	4.62	4.67	4.73	4.78	4.82	4.86
120	2.34	2.93	3.28	3.52	3.71	3.86	3.99	4.10	4.19	4.28	4.35	4.42	4.48	4.54	4.60	4.65	4.69	4.74	4.78
∞	2.33	2.90	3.24	3.48	3.66	3.81	3.93	4.04	4.13	4.21	4.28	4.35	4.41	4.47	4.52	4.57	4.61	4.65	4.69

Entry is $q_{[.05]}(r, v)$

v	r = 2	3	4	5	6	7	8	9	10	11	12	13	14	15	16	17	18	19	20
1	18.0	27.0	32.8	37.1	40.4	43.1	45.4	47.4	49.1	50.6	52.0	53.2	54.3	55.4	56.3	57.2	58.0	58.8	59.6
2	6.08	8.33	9.80	10.9	11.7	12.4	13.0	13.5	14.0	14.4	14.7	15.1	15.4	15.7	15.9	16.1	16.4	16.6	16.8
3	4.50	5.91	6.82	7.50	8.04	8.48	8.85	9.18	9.46	9.72	9.95	10.2	10.3	10.5	10.7	10.8	11.0	11.1	11.2
4	3.93	5.04	5.76	6.29	6.71	7.05	7.35	7.60	7.83	8.03	8.21	8.37	8.52	8.66	8.79	8.91	9.03	9.13	9.23
5	3.64	4.60	5.22	5.67	6.03	6.33	6.58	6.80	6.99	7.17	7.32	7.47	7.60	7.72	7.83	7.93	8.03	8.12	8.21
6	3.46	4.34	4.90	5.30	5.63	5.90	6.12	6.32	6.49	6.65	6.79	6.92	7.03	7.14	7.24	7.34	7.43	7.51	7.59
7	3.34	4.16	4.68	5.06	5.36	5.61	5.82	6.00	6.16	6.30	6.43	6.55	6.66	6.76	6.85	6.94	7.02	7.10	7.17
8	3.26	4.04	4.53	4.89	5.17	5.40	5.60	5.77	5.92	6.05	6.18	6.29	6.39	6.48	6.57	6.65	6.73	6.80	6.87
9	3.20	3.95	4.41	4.76	5.02	5.24	5.43	5.59	5.74	5.87	5.98	6.09	6.19	6.28	6.36	6.44	6.51	6.58	6.64
10	3.15	3.88	4.33	4.65	4.91	5.12	5.30	5.46	5.60	5.72	5.83	5.93	6.03	6.11	6.19	6.27	6.34	6.40	6.47
11	3.11	3.82	4.26	4.57	4.82	5.03	5.20	5.35	5.49	5.61	5.71	5.81	5.90	5.98	6.06	6.13	6.20	6.27	6.33
12	3.08	3.77	4.20	4.51	4.75	4.95	5.12	5.27	5.39	5.51	5.61	5.71	5.80	5.88	5.95	6.02	6.09	6.15	6.21
13	3.06	3.73	4.15	4.45	4.69	4.88	5.05	5.19	5.32	5.43	5.53	5.63	5.71	5.79	5.86	5.93	5.99	6.05	6.11
14	3.03	3.70	4.11	4.41	4.64	4.83	4.99	5.13	5.25	5.36	5.46	5.55	5.64	5.71	5.79	5.85	5.91	5.97	6.03
15	3.01	3.67	4.08	4.37	4.59	4.78	4.94	5.08	5.20	5.31	5.40	5.49	5.57	5.65	5.72	5.78	5.85	5.90	5.96
16	3.00	3.65	4.05	4.33	4.56	4.74	4.90	5.03	5.15	5.26	5.35	5.44	5.52	5.59	5.66	5.73	5.79	5.84	5.90
17	2.98	3.63	4.02	4.30	4.52	4.70	4.86	4.99	5.11	5.21	5.31	5.39	5.47	5.54	5.61	5.67	5.73	5.79	5.84
18	2.97	3.61	4.00	4.28	4.49	4.67	4.82	4.96	5.07	5.17	5.27	5.35	5.43	5.50	5.57	5.63	5.69	5.74	5.79
19	2.96	3.59	3.98	4.25	4.47	4.65	4.79	4.92	5.04	5.14	5.23	5.31	5.39	5.46	5.53	5.59	5.65	5.70	5.75
20	2.95	3.58	3.96	4.23	4.45	4.62	4.77	4.90	5.01	5.11	5.20	5.28	5.36	5.43	5.49	5.55	5.61	5.66	5.71
24	2.92	3.53	3.90	4.17	4.37	4.54	4.68	4.81	4.92	5.01	5.10	5.18	5.25	5.32	5.38	5.44	5.49	5.55	5.59
30	2.89	3.49	3.85	4.10	4.30	4.46	4.60	4.72	4.82	4.92	5.00	5.08	5.15	5.21	5.27	5.33	5.38	5.43	5.47
40	2.86	3.44	3.79	4.04	4.23	4.39	4.52	4.63	4.73	4.82	4.90	4.98	5.04	5.11	5.16	5.22	5.27	5.31	5.36
60	2.83	3.40	3.74	3.98	4.16	4.31	4.44	4.55	4.65	4.73	4.81	4.88	4.94	5.00	5.06	5.11	5.15	5.20	5.24
120	2.80	3.36	3.68	3.92	4.10	4.24	4.36	4.47	4.56	4.64	4.71	4.78	4.84	4.90	4.95	5.00	5.04	5.09	5.13
∞	2.77	3.31	3.63	3.86	4.03	4.17	4.29	4.39	4.47	4.55	4.62	4.68	4.74	4.80	4.85	4.89	4.93	4.97	5.01

Entry is $q_{[.01]}(r, v)$.

v	r = 2	3	4	5	6	7	8	9	10	11	12	13	14	15	16	17	18	19	20
1	90.0	135	164	186	202	216	227	237	246	253	260	266	272	277	282	286	290	294	298
2	14.0	19.0	22.3	24.7	26.6	28.2	29.5	30.7	31.7	32.6	33.4	34.1	34.8	35.4	36.0	36.5	37.0	37.5	37.9
3	8.26	10.6	12.2	13.3	14.2	15.0	15.6	16.2	16.7	17.1	17.5	17.9	18.2	18.5	18.8	19.1	19.3	19.5	19.8
4	6.51	8.12	9.17	9.96	10.6	11.1	11.5	11.9	12.3	12.6	12.8	13.1	13.3	13.5	13.7	13.9	14.1	14.2	14.4
5	5.70	6.97	7.80	8.42	8.91	9.32	9.67	9.97	10.2	10.5	10.7	10.9	11.1	11.2	11.4	11.6	11.7	11.8	11.9
6	5.24	6.33	7.03	7.56	7.97	8.32	8.61	8.87	9.10	9.30	9.49	9.65	9.81	9.95	10.1	10.2	10.3	10.4	10.5
7	4.95	5.92	6.54	7.01	7.37	7.68	7.94	8.17	8.37	8.55	8.71	8.86	9.00	9.12	9.24	9.35	9.46	9.55	9.65
8	4.74	5.63	6.20	6.63	6.96	7.24	7.47	7.68	7.87	8.03	8.18	8.31	8.44	8.55	8.66	8.76	8.85	8.94	9.03
9	4.60	5.43	5.96	6.35	6.66	6.91	7.13	7.32	7.49	7.65	7.78	7.91	8.03	8.13	8.23	8.32	8.41	8.49	8.57
10	4.48	5.27	5.77	6.14	6.43	6.67	6.87	7.05	7.21	7.36	7.48	7.60	7.71	7.81	7.91	7.99	8.07	8.15	8.22
11	4.39	5.14	5.62	5.97	6.25	6.48	6.67	6.84	6.99	7.13	7.25	7.36	7.46	7.56	7.65	7.73	7.81	7.88	7.95
12	4.32	5.04	5.50	5.84	6.10	6.32	6.51	6.67	6.81	6.94	7.06	7.17	7.26	7.36	7.44	7.52	7.59	7.66	7.73
13	4.26	4.96	5.40	5.73	5.98	6.19	6.37	6.53	6.67	6.79	6.90	7.01	7.10	7.19	7.27	7.34	7.42	7.48	7.55
14	4.21	4.89	5.32	5.63	5.88	6.08	6.26	6.41	6.54	6.66	6.77	6.87	6.96	7.05	7.12	7.20	7.27	7.33	7.39
15	4.17	4.83	5.25	5.56	5.80	5.99	6.16	6.31	6.44	6.55	6.66	6.76	6.84	6.93	7.00	7.07	7.14	7.20	7.26
16	4.13	4.78	5.19	5.49	5.72	5.92	6.08	6.22	6.35	6.46	6.56	6.66	6.74	6.82	6.90	6.97	7.03	7.09	7.15
17	4.10	4.74	5.14	5.43	5.66	5.85	6.01	6.15	6.27	6.38	6.48	6.57	6.66	6.73	6.80	6.87	6.94	7.00	7.05
18	4.07	4.70	5.09	5.38	5.60	5.79	5.94	6.08	6.20	6.31	6.41	6.50	6.58	6.65	6.72	6.79	6.85	6.91	6.96
19	4.05	4.67	5.05	5.33	5.55	5.73	5.89	6.02	6.14	6.25	6.34	6.43	6.51	6.58	6.65	6.72	6.78	6.84	6.89
20	4.02	4.64	5.02	5.29	5.51	5.69	5.84	5.97	6.09	6.19	6.29	6.37	6.45	6.52	6.59	6.65	6.71	6.76	6.82
24	3.96	4.54	4.91	5.17	5.37	5.54	5.69	5.81	5.92	6.02	6.11	6.19	6.26	6.33	6.39	6.45	6.51	6.56	6.61
30	3.89	4.45	4.80	5.05	5.24	5.40	5.54	5.65	5.76	5.85	5.93	6.01	6.08	6.14	6.20	6.26	6.31	6.36	6.41
40	3.82	4.37	4.70	4.93	5.11	5.27	5.39	5.50	5.60	5.69	5.77	5.84	5.90	5.96	6.02	6.07	6.12	6.17	6.21
60	3.76	4.28	4.60	4.82	4.99	5.13	5.25	5.36	5.45	5.53	5.60	5.67	5.73	5.79	5.84	5.89	5.93	5.98	6.02
120	3.70	4.20	4.50	4.71	4.87	5.01	5.12	5.21	5.30	5.38	5.44	5.51	5.56	5.61	5.66	5.71	5.75	5.79	5.83
∞	3.64	4.12	4.40	4.60	4.76	4.88	4.99	5.08	5.16	5.23	5.29	5.35	5.40	5.45	5.49	5.54	5.57	5.61	5.65

Source: Reproduced with permission from Henry Scheffé, *The Analysis of Variance* (New York: John Wiley & Sons, 1959) pp. 414–416.

TABLE E.8 Values of $H_{[\alpha]}^{(v,m-1)}$

Upper 5% points

$m-1$ \ v	2	3	4	5	6	7	8	9	10	11	12
2	39.0	87.5	142	202	266	333	403	475	550	626	704
3	15.4	27.8	39.2	50.7	62.0	72.9	83.5	93.9	104	114	124
4	9.60	15.5	20.6	25.2	29.5	33.6	37.5	41.1	44.6	48.0	51.4
5	7.15	10.8	13.7	16.3	18.7	20.8	22.9	24.7	26.5	28.2	29.9
6	5.82	8.38	10.4	12.1	13.7	15.0	16.3	17.5	18.6	19.7	20.7
7	4.99	6.94	8.44	9.70	10.8	11.8	12.7	13.5	14.3	15.1	15.8
8	4.43	6.00	7.18	8.12	9.03	9.78	10.5	11.1	11.7	12.2	12.7
9	4.03	5.34	6.31	7.11	7.80	8.41	8.95	9.45	9.91	10.3	10.7
10	3.72	4.85	5.67	6.34	6.92	7.42	7.87	8.28	8.66	9.01	9.34
12	3.28	4.16	4.79	5.30	5.72	6.09	6.42	6.72	7.00	7.25	7.48
15	2.86	3.54	4.01	4.37	4.68	4.95	5.19	5.40	5.59	5.77	5.93
20	2.46	2.95	3.29	3.54	3.76	3.94	4.10	4.24	4.37	4.49	4.59
30	2.07	2.40	2.61	2.78	2.91	3.02	3.12	3.21	3.29	3.36	3.39
60	1.67	1.85	1.96	2.04	2.11	2.17	2.22	2.26	2.30	2.33	2.36
∞	1.00	1.00	1.00	1.00	1.00	1.00	1.00	1.00	1.00	1.00	1.00

Upper 1% points

$m-1$ \ v	2	3	4	5	6	7	8	9	10	11	12
2	199	448	729	1036	1362	1705	2063	2432	2813	3204	3605
3	47.5	85	120	151	184	21(6)	24(9)	28(1)	31(0)	33(7)	36(1)
4	23.2	37	49	59	69	79	89	97	106	113	120
5	14.9	22	28	33	38	42	46	50	54	57	60
6	11.1	15.5	19.1	22	25	27	30	32	34	36	37
7	8.89	12.1	14.5	16.5	18.4	20	22	23	24	26	27
8	7.50	9.9	11.7	13.2	14.5	15.8	16.6	17.9	18.9	19.8	21
9	6.54	8.5	9.9	11.1	12.1	13.1	13.9	14.7	15.3	16.0	16.6
10	5.85	7.4	8.6	9.6	10.4	11.1	11.8	12.4	12.9	13.4	13.9
12	4.91	6.1	6.9	7.6	8.2	8.7	9.1	9.5	9.9	10.2	10.6
15	4.07	4.9	5.5	6.0	6.4	6.7	7.1	7.3	7.5	7.8	8.0
20	3.32	3.8	4.3	4.6	4.9	5.1	5.3	5.5	5.6	5.8	5.9
30	2.63	3.0	3.3	3.4	3.6	3.7	3.8	3.9	4.0	4.1	4.2
60	1.96	2.2	2.3	2.4	2.4	2.5	2.5	2.6	2.6	2.7	2.7
∞	1.00	1.0	1.0	1.0	1.0	1.0	1.0	1.0	1.0	1.0	1.0

Source: From *Biometrika Tables for Statisticians*, Vol. I, edited by E. S. Pearson and H. O. Hartley (New York: Cambridge University Press, 1966) Table, p. 202. Reproduced by permission of the *Biometrika Trustees*.

TABLE E.9 Critical values of chi-square

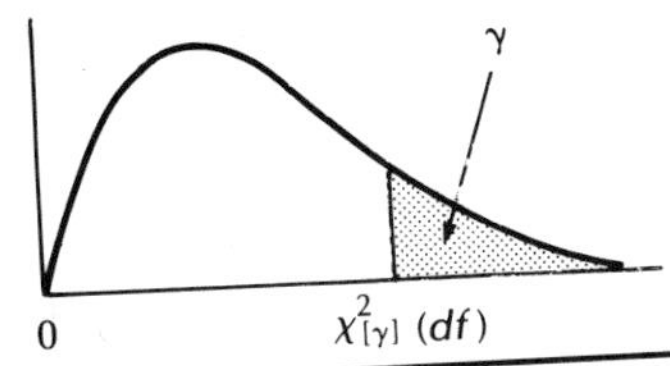

df	$\chi^2_{[.995]}(df)$	$\chi^2_{[.99]}(df)$	$\chi^2_{[.975]}(df)$	$\chi^2_{[.95]}(df)$	$\chi^2_{[.90]}(df)$
1	0.0000393	0.0001571	0.0009821	0.0039321	0.0157908
2	0.0100251	0.0201007	0.0506356	0.102587	0.210720
3	0.0717212	0.114832	0.215795	0.341846	0.584375
4	0.206990	0.297110	0.484419	0.710721	0.063623
5	0.411740	0.554300	0.831211	1.145476	1.61031
6	0.675727	0.872085	1.237347	1.63539	2.20413
7	0.989265	1.239043	1.68987	2.16735	2.83311
8	1.344419	1.646482	2.17973	2.73264	3.48954
9	1.734926	2.087912	2.70039	3.32511	4.16816
10	2.15585	2.55821	3.24697	3.94030	4.86518
11	2.60321	3.05347	3.81575	4.57481	5.57779
12	3.07382	3.57056	4.40379	5.22603	6.30380
13	3.56503	4.10691	5.00874	5.89186	7.04150
14	4.07468	4.66043	5.62872	6.57063	7.78953
15	4.60094	5.22935	6.26214	7.26094	8.54675
16	5.14224	5.81221	6.90766	7.96164	9.31223
17	5.69724	6.40776	7.56418	8.67176	10.0852
18	6.26481	7.01491	8.23075	9.39046	10.8649
19	6.84398	7.63273	8.90655	10.1170	11.6509
20	7.43386	8.26040	9.59083	10.8508	12.4426
21	8.03366	8.89720	10.28293	11.5913	13.2396
22	8.64272	9.54249	10.9823	12.3380	14.0415
23	9.26042	10.19567	11.6885	13.0905	14.8479
24	9.88623	10.8564	12.4011	13.8484	15.6587
25	10.5197	11.5240	13.1197	14.6114	16.4734
26	11.1603	12.1981	13.8439	15.3791	17.2919
27	11.8076	12.8786	14.5733	16.1513	18.1138
28	12.4613	13.5648	15.3079	16.9279	18.9392
29	13.1211	14.2565	16.0471	17.7083	19.7677
30	13.7867	14.9535	16.7908	18.4926	20.5992
40	20.7065	22.1643	24.4331	26.5093	29.0505
50	27.9907	29.7067	32.3574	34.7642	37.6886
60	35.5346	37.4848	40.4817	43.1879	46.4589
70	43.2752	45.4418	48.7576	51.7393	55.3290
80	51.1720	53.5400	57.1532	60.3915	64.2778
90	59.1963	61.7541	65.6466	69.1260	73.2912
100	67.3276	70.0648	74.2219	77.9295	82.3581

(continues)

TABLE E.9 Continued

$\chi^2_{[.10]}(df)$	$\chi^2_{[.05]}(df)$	$\chi^2_{[.025]}(df)$	$\chi^2_{[.01]}(df)$	$\chi^2_{[.005]}(df)$	df
2.70554	3.84146	5.02389	6.63490	7.87944	1
4.60517	5.99147	7.37776	9.21034	10.5966	2
6.25139	7.81473	9.34840	11.3449	12.8381	3
7.77944	9.48773	11.1433	13.2767	14.8602	4
9.23635	11.0705	12.8325	15.0863	16.7496	5
10.6446	12.5916	14.4494	16.8119	18.5476	6
12.0170	14.0671	16.0128	18.4753	20.2777	7
13.3616	15.5073	17.5346	20.0902	21.9550	8
14.6837	16.9190	19.0228	21.6660	23.5893	9
15.9871	18.3070	20.4831	23.2093	25.1882	10
17.2750	19.6751	21.9200	24.7250	26.7569	11
18.5494	21.0261	23.3367	26.2170	28.2995	12
19.8119	22.3621	24.7356	27.6883	29.8194	13
21.0642	23.6848	26.1190	29.1413	31.3193	14
22.3072	24.9958	27.4884	30.5779	32.8013	15
23.5418	26.2962	28.8454	31.9999	34.2672	16
24.7690	27.5871	30.1910	33.4087	35.7185	17
25.9894	28.8693	31.5264	34.8053	37.1564	18
27.2036	30.1435	32.8523	36.1908	38.5822	19
28.4120	31.4104	34.1696	37.5662	39.9968	20
29.6151	32.6705	35.4789	38.9321	41.4010	21
30.8133	33.9244	36.7807	40.2894	42.7956	22
32.0069	35.1725	38.0757	41.6384	44.1813	23
33.1963	36.4151	39.3641	42.9798	45.5585	24
34.3816	37.6525	40.6465	44.3141	46.9278	25
35.5631	38.8852	41.9232	45.6417	48.2899	26
36.7412	40.1133	43.1944	46.9630	49.6449	27
37.9159	41.3372	44.4607	48.2782	50.9933	28
39.0875	42.5569	45.7222	49.5879	52.3356	29
40.2560	43.7729	46.9792	50.8922	53.6720	30
51.8050	55.7585	59.3417	63.6907	66.7659	40
63.1671	67.5048	71.4202	76.1539	79.4900	50
74.3970	79.0819	83.2976	88.3794	91.9517	60
85.5271	90.5312	95.0231	100.425	104.215	70
96.5782	101.879	106.629	112.329	116.321	80
107.565	113.145	118.136	124.116	128.299	90
118.498	124.342	129.561	135.807	140.169	100

Source: From "Tables of the Percentage Points of the χ^2-Distribution," by Catherine M. Thompson, *Biometrika* *32* (1941), 188–189. Reproduced by permission of the *Biometrika* Trustees.

APPENDIX 

DESCRIPTIONS OF DATA BASES

DATA BASE 1: MANUFACTURING DATA

This data base contains manufacturing data from the *1982 United States Census of Manufactures* (Subject Series). Twelve variables are recorded for each of 150 Standard Metropolitan Statistical Areas (SMSAs). The variables are as follows:

REGION = the region of the United States in which the SMSA is located (East North Central, Mid Atlantic, etc.)
NMANES = total number of manufacturing establishments
NMAN20 = number of manufacturing establishments employing 20 or more people
NEMP = number of employees (thousands)
PAY = total payroll (millions of dollars)
NPROD = number of production workers (thousands)
NPRODH = number of production worker hours (millions)
PRODPAY = production wages (millions of dollars)
VALMAN = value added by manufacture (millions of dollars)

COSTMAT = cost of materials (millions of dollars)
VALSHIP = value of shipments (millions of dollars)
NCAPEX = new capital expenditures (millions of dollars)

DATA BASE 2: OHIO LOCAL GOVERNMENT AND PAYROLL DATA

This data base contains October 1982 Ohio county government and payroll data (*1982 Census of Governments*, Vol. 3, *Government Employment. Employment, Payrolls, and Average Earnings in Industrial County Government by States*, October 1982). Twenty-five variables are recorded for each of the 88 counties in Ohio. The variables are as follows:

POP = 1980 population
EMP = number of employees
FEMP = number of full-time employees
FEVEMP = number of full-time equivalent employees
EDEMP = number of employees in elementary, secondary, and higher education
INSTEMP = number of instructional employees
PWEMP = number of public welfare employees
HOSPEMP = number of hospital employees
HEALEMP = number of health employees
TRANSEMP = number of transportation employees
POLEMP = number of employees in police protection
FIREMP = number of employees in fire protection
RECEMP = number of employees in parks and recreation
COMEMP = number of employees in housing and community development
SEWEMP = number of employees in sewage
SANEMP = number of employees in sanitation other than sewage
FINEMP = number of employees in financial administration
JLEMP = number of judicial and legal employees
GOVEMP = number of other government administration employees
UTILEMP = number of employees in utilities
WATEMP = number of employees in water supply
OTHEMP = number of other government employees
OCTPAY = October payroll (thousands of dollars)
AVPAYIN = average October earnings per instructional employee
AVPAYOTH = average October earnings per other employee

DATA BASE 3: POPULATION DATA

This data base contains population characteristics from the *1980 United States Census of Population* (Characteristics of Population). Thirty variables are recorded for each of 150 Standard Metropolitan Statistical Areas (SMSAs). The variables are as follows:

REGION = the region of the United States in which the SMSA is located (East North Central, Mid Atlantic, etc.)
AREA = land area (square miles)
POPTOT = total population
POPSQMI = total population per square mile
URBPOP = urban population
RURPOP = rural population
PCTPOPCH = percentage change in population 1970–1980
PCTUND18 = percentage under 18 years old in central city of SMSA
PCT18T64 = percentage from 18 to 64 years old in central city of SMSA
PCTOV65 = percent 65 years old and older in central city of SMSA
PCTBLK = percent black in central city of SMSA
PCTSPAN = percent of Spanish origin in central city of SMSA
MEDAGE = median age in central city of SMSA
PCTMARML = percent of married males in central city of SMSA
PCTMARFM = percent of married females in central city of SMSA
16T19NSC = civilian persons 16 to 19 years old, percent not enrolled in school, not high school graduates
18T24SC = persons 18 to 24 years old, percent enrolled in school
OV25HSG = persons 25 years old and older, percent high school graduates
OV25CG = persons 25 years old and older, percent completed 4 or more years of college
79MFIN = 1979 median family income
79PIN = 1979 per capita income
PCTFPOV = percent of 1979 families below poverty level
PCTPPOV = percent of persons with poverty status
PCTMLAB = percent of 16 years old and older males in labor force
PCTFLAB = percent of 16 years old and older females in labor force
PCTUNEM = civilian labor force, percent unemployed

PCTMAN = percent of 16 years old and older employed persons in manufacturing industries
PCTWOR = percent of workers who work outside area of residence
PCTWCAR = percent of workers in carpools
PCTWPUB = percent of workers using public transportation

REFERENCES

Anderson, T.W. *The Statistical Analysis of Time Series*. New York: Wiley, 1971.

Anscombe, F.J., and J.W. Tukey. "The Examination and Analysis of Residuals." *Technometrics* 5 (1963): 141–160.

Belsley, D. A., E. Kuh, and R.E. Welsch. *Regression Diagnostics: Identifying Influential Data and Sources of Collinearity*. New York: Wiley, 1980.

Bowerman, B.L., and R.T. O'Connell. *Time Series and Forecasting: An Applied Approach*. Boston: Duxbury Press, 1979.

———. *Time Series Forecasting: Unified Concepts and Computer Implementation*, 2d ed. Boston: Duxbury Press, 1987.

Box, G.E.P., and D.R. Cox. "An Analysis of Transformations." *Journal of Royal Statistical Society B* 26 (1964): 211–243.

——— and G.M. Jenkins. *Time Series Analysis: Forecasting and Control*. 2d ed. San Francisco: Holden-Day, 1977.

Cochran, G.W., and G.M. Cox. *Experimental Designs*. 2d ed. New York: Wiley, 1957.

Cramer and Applebaum. "Orthogonal and Nonorthogonal Anova." Speech at 1977 Meeting of the Joint Statistical Societies, Chicago, Ill.

Davis, O.L. *The Design and Analysis of Industrial Experiments*. New York: Hafner, 1956.

Draper, N., and H. Smith. *Applied Regression Analysis*. 2d ed. New York: Wiley, 1981.

Durbin, J., and G.S. Watson. "Testing for Serial Correlation in Least Squares Regression I." *Biometrika 37* (1950): 409–428.

———. "Testing for Serial Correlation in Least Squares Regression, II." *Biometrika* 38 (1951): 159–179.

Fuller, W.A. *Introduction to Statistical Time Series*. New York: Wiley, 1976.

Graybill, F.A. *Theory and Application of the Linear Model*. Boston: Duxbury Press, 1976.

Kennedy, W.J., Jr., and J.E. Gentle. *Statistical Computing*. New York: Dekker, 1980.

Kleinbaum, D., and L. Kupper. *Applied Regression Analysis and Other Multivariable Methods*. Boston: Duxbury Press, 1978.

McClave, J.T., and P.G. Benson. *Statistics for Business and Economics*. 4th ed. San Francisco: Dellen Publishing Company, 1988.

Mendenhall, W. *Introduction to Linear Models and the Design and Analysis of Experiments*. Belmont, Mass.: Wadsworth, 1968.

——— and J. Reinmuth. *Statistics for Management and Economics*. 4th ed. Boston: Duxbury Press, 1982.

——— and T. Sincich. *A Second Course in Business Statistics: Regression Analysis*. 3rd ed. San Francisco: Dellen Publishing Company, 1989.

———, J. Reinmuth, R. Beaver, and D. Duhan. *Statistics for Management and Economics*. 5th ed. Boston: Duxbury Press, 1986.

Miller, R.B., and D.W. Wichern. *Intermediate Business Statistics: Analysis of Variance, Regression, and Time Series*. New York: Holt, Rinehart, and Winston, 1977.

Myers, R. *Classical and Modern Regression with Applications*. Boston: Duxbury Press, 1986.

Nelson, C.R. *Applied Time Series Analysis for Managerial Forecasting*. San Francisco: Holden-Day, 1973.

Neter, J., W. Wasserman, and M.H. Kutner. *Applied Linear Statistical Models*. 2d ed. Homewood, Ill.: Richard D. Irwin, 1985.

———. *Applied Linear Regression Models*. 3rd ed. Homewood, Ill.: Richard D. Irwin, 1989.

Neter, J., W. Wasserman, and G.A. Whitmore. *Applied Statistics*. 2d ed. Boston: Allyn and Bacon, 1982.

Ott, Lyman. *An Introduction to Statistical Methods and Data Analysis*. 2d ed. Boston: Duxbury Press, 1984.

———. *An Introduction to Statistical Methods and Data Analysis*. 3d ed. Boston: Duxbury Press, 1987.

Pfaffenberger, R.C., and J.H. Patterson. *Statistical Methods for Business and Economics*. 3d ed. Homewood, Ill.: Richard D. Irwin, 1987.

SAS User's Guide, 1982 Edition. Cary, North Carolina: SAS Institute, 1982.

Scheffé, H. *The Analysis of Variance*. New York: Wiley, 1959.

Searle, S.R. *Linear Models*. New York: Wiley, 1971.

Winer, B.J. *Statistical Principles in Experimental Design*. New York: McGraw-Hill, 1962.

Wonnacott, T.H., and R.J. Wonnacott. *Introductory Statistics for Business and Economics*. 2d ed. New York: Wiley, 1977.

———. *Regression: A Second Course in Statistics*. New York: Wiley, 1981.

Younger, M.S. *A First Course in Linear Regression*. 2d ed. Boston: Duxbury Press, 1985.

INDEX